MAP OF THE WORLD
MAR DEL NORT
THE WESTERNE OCEAN
UNKNOWNE

Nonindigenous Fishes Introduced into Inland Waters of the United States

Funding for the publication of this book was provided by

U.S. Geological Survey

U.S. Fish and Wildlife Service, Office of Fisheries Assistance

The Regina Bauer Frankenberg Foundation for Animal Welfare through their support of The Nature Conservancy's National Rivers Program

Dedication

This book is dedicated to Dr. Walter R. Courtenay, Jr. and Dr. James A. McCann in recognition of their contributions to our knowledge and understanding of nonindigenous fishes.

Nonindigenous Fishes Introduced into Inland Waters of the United States

Pam L. Fuller
Leo G. Nico
James D. Williams

with special thanks to
Amy Benson and Charles Boydstun
for graphics and technical support

U.S. Geological Survey
Biological Resources Division
Florida Caribbean Science Center
Gainesville, Florida

Bethesda, Maryland
1999

Suggested citation format follows.

Fuller, P. L., L. G. Nico, and J. D. Williams. 1999. Nonindigenous fishes introduced into inland waters of the United States. American Fisheries Society, Special Publication 27, Bethesda, Maryland.

Cover design and map graphic by Pam Fuller, USGS. Fish illustrations by Doug Weaver, USGS.

Library of Congress Catalog Card Number: 98-88571
ISBN: 1-888569-14-X

Printed in the United States of America.

American Fisheries Society
5410 Grosvenor Lane, Suite 110
Bethesda, Maryland 20814-2199, USA

Table of Contents

Acknowledgments

The compilation of this book was the result of contributions and cooperation of many individuals and institutions. Special thanks must go to Carter R. Gilbert for critically reading and commenting on a late version of the manuscript, and to Brooks Burr and Walter Courtenay, Jr. for reviewing and providing advice on large portions of an earlier draft. We also thank our colleagues William Smith-Vaniz and Robert H. Robins for reading selected sections of the manuscript and for their subsequent comments, discussions, and technical advice. Other experts graciously provided comments and corrections on particular families or portions of the document. For their review work, we thank Robert Behnke, Noel Burkhead, Glenn Clemmer, Anthony Echelle, David Etnier, Carl H. Ferraris, Jr., Robert Jenkins, Gary Meffe, Larry Page, Wayne Starnes, Stephen J. Walsh, and Stanley Weitzman. We retain full responsibility for any shortcomings or errors in the work.

Many other researchers, ichthyologists, and museum curators provided important information, shared their expertise on taxonomic and systematic issues, assisted in the identification of specimens, answered various queries, loaned material, or supplied critical records and literature including their own published or unpublished data. For their contribution, we are especially indebted to Jonathan Armbruster, Bruce Bauer, Gerald T. Bergman, Richard Bryant, George Burgess, Noel Burkhead, Brooks Burr, Robert Cashner, Ted Cavender, John Chick, Wilson J. E. M. Costa, Walter Courtenay, Jr., Kevin Cummings, George Dalrymple, Robert Daniels, Neil Douglas, William Eschmeyer, David Etnier, Carter Gilbert, Karsten Hartel, Dannie Hensley, Jeffrey E. Hill, Clark Hubbs, Howard Jelks, Robert Jenkins, Thomas Keevin, Hiram Li, William Loftus, Daniel Logan, Richard Mayden, Joseph Nelson, Larry Page, William Pflieger, Joseph M. Quattro, C. Richard Robins, Steve Ross, Wayne Starnes, Arnold Suzumoto, Jamie E. Thomerson, Joel Trexler, Kathryn Vaughan, Ronald Ward, Andrew White, and Jeffrey Williams.

Many and various representatives of local, state, and federal agencies, as well as many students, private citizens and industry employees, answered requests for information on nonindigenous fishes found in their states and regions. For their assistance, we gratefully acknowledge Kevin Aitkin, Chuck Bever, Tom Busiahn, Allen Carter, Almo J. Cordone, Kenneth M. Cox, John Crutchfield, William Devick, Alan Dextrase, Jack Erickson, Kevin Fleming, Douglas H. Fletcher, Jim Francis, Dave B. Fuller, David B. Halliwell, Robert Hanten, Robert Hartman, Michael Hatch, Jim Heinrich, Donald C. Hicks, Don Horak, Rodney Horner, William D. Horton, Robert G. Howells, Cecil Jennings, Richard M. Jennings, Doug Jensen, Howard E. Johnson, Robert A. Jones, David Jude, Steven W. Kelsch, Sandra Keppner, Frederick W. Kircheis, Carey Knight, William J. Lapin, William C. Latta, Tom Lauer, Joseph Logan, John Lyons, Ellen Marsden, Catherine C. Martin, R. Trent Martin, Franklin T. McBride, Carol McIvor, William A. Molini, Scott Morrison, Mary Moser, Peter W. Pfeiffer, Jimmie Pigg, Wayne Pollock, Maynard S. Raasch, Michael Ray, Fitz Rhode, Karl J. Scheidegger, Bruce Schmidt, Konrad Schmitd, Dave Seiler, Paul L. Shafland, Mark Shaw, Clark N. Shiffer, Al Smith, Robert H. Soldwedel, Ron Southwick, Michael Stone, Michael Thomas, Gary A. Tilyou, Thomas N. Todd, Jay Troxel, Dennis Unkenholz, Michael Vaughn, W. Waldo Wakefield, Peter Walker, Patrick Walsh, Rocky Ward, Charley Wheeler, Robert White, Paul J. Wingate, Bill Woodward, Michael Yamamoto, Larry Zadina, Lawrence Zuckerman, and to others that remain unnamed. We are additionally very grateful to those individuals who sent us voucher specimens.

We are particularly grateful to Amy J. Benson for help in preparing many of the graphics. In addition, she was instrumental in writing the text portion of species' accounts in early drafts of the families Percidae and Salmonidae. Douglas Weaver enthusiastically gave us aid by preparing the fish illustrations seen on the front cover. Charles Boydstun and Bruce Bush were critical in providing computer network support.

PLF wishes to than her father, Donald B. Fuller, for instilling a love of nature in her as a small child by taking her on "nature walks" in the back yard. She also thanks her father; mother, Francis L. Fuller; and husband, Noel F. Ocampo for encouragement through the years. She hopes her brother, Dave B. Fuller, will find a career in biology as rewarding as she has.

A special thanks goes to Linda S. Nico for her great patience, support, and encouragement over the six or so years during which time her husband spent numerous weekends and late evenings at the lab researching and writing the book. LGN also acknowledges and thanks his wife and two step-daughters, Jessica and Lindsey, for the many meals prepared for him and delivered to the lab. Their attention and kindness will not be forgotten. In addition, LGN wishes to give recognition to his parents, Teresa A. Nico and the late Frank T. Nico. In many ways, this book is a tribute to them.

Dawn Jennings, with the support of James McCann, was instrumental in creating the prototype database that formed part of the framework of this book. In those instances when the private library of Jim Williams failed us, we relied on the aid of Joyce Mann, Lora McKenzie, and Vi Catrow (USGS Leetown Science Library) in securing many references, including rare published and unpublished printed materials. Sherry Bostick, Amy Hester, Abby Renegar, and Leslie Straub contributed by entering records in the database or by providing assistance in manuscript preparation. During the shaping of this book, Kristin King and Lisa Spinella were attentive and competent in their editorial work. Jason Evert helped with the glossary.

We were immensely fortunate to have had the meticulous services of the editors and staff of the American Fisheries Society. Susan Braxton and Janet Harry did the final editing. Beth Staehle, Robert Kendall, Robert Rand, and Eric Wurzbacher coordinated production of this book. For all of their efforts we express our appreciation and gratitude.

Introduction

Throughout history, humans have transported and released plants and animals from one ecosystem to another for various reasons. Plants or animals moved into areas outside their historic or natural geographic range are considered nonindigenous species. This includes species transferred into the United States from foreign countries as well as those species moved from one region or watershed to another within the United States. In the past, nonindigenous species have been referred to as "exotics," "transplants," "nonnatives," or "introduced species" (Shafland and Lewis 1984); in foreign countries, they are often called "alien" species. Within the United States alone, humans have intentionally or unintentionally introduced over 4,500 foreign species (U.S. Congress, Office of Technology Assessment (OTA) 1993). Although many terrestrial introductions are viewed as providing economic and social benefits to humans, all but a few intentional aquatic introductions have proven to be mixed blessings (U.S. Congress, OTA 1993; Steirer 1992; Courtenay and Williams 1992). This is not surprising, because considerable effort has been made to document possible beneficial aspects, whereas comparably little has been done to assess harmful impacts, especially over the entire lifespan of the introduced species. No unintentional aquatic introductions have been considered beneficial (Steirer 1992). In general, the environmental consequences of these introductions have been harmful; in some cases, they have been catastrophic (Taylor et al. 1984; U.S. Congress, OTA 1993).

Both intentional and unintentional introductions have enabled nonindigenous fishes to become temporary, and often permanent, residents in nearly every major aquatic system in the United States. Complete eradication or exclusion is often neither economically plausible nor socially justifiable (U.S. Congress, OTA 1993). Whether temporary or permanent, beneficial or harmful, nonindigenous fishes are, and will continue to be, components of our nation's aquatic systems. Their interactions within the aquatic community, however, must be monitored and analyzed to ensure that proactive, effective management actions are taken before crisis situations arise.

This report provides species accounts for all known nonindigenous fishes in inland, open waters of the United States on record at the U.S. Geological Survey, Biological Resources Division's laboratory in Gainesville, Florida (USGS/BRD-G). The report considers freshwater species not native to any part of the United States, as well as species native to one area of the United States but introduced through human activities to other open-water localities within the country. Accounts also are included for marine species introduced to inland waters and for many cultured hybrid fish that have been stocked in or have escaped into U.S. open waters.

The Database

In 1978, the U.S. Fish and Wildlife Service's National Fisheries Research Center in Gainesville, Florida, began monitoring the status and distribution of nonindigenous fish species in U.S. waters. Originally, this information was maintained as paper files, and it was not until 1989 that these records were computerized. Williams and Jennings (1991) gave a description of the early database; Nico and Fuller (1999) provided an updated account. In 1993, the laboratory was transferred along with many other facilities to the newly created National Biological Survey (later renamed the National Biological Service). In October 1996, the National Biological Service was moved into the U.S. Geological Survey as the Biological Resources Division.

To facilitate information transfer about nonindigenous species, we continue to computerize all records of nonindigenous fishes taken in open waters of the United States. Information on each introduction has been obtained by searching published and unpublished reports, by contacting state and federal agencies, fisheries biologists, ichthyologists, and museum curators, and by searching laboratory and museum collections, as well as computer databases. Additional data have been obtained as a result of our field inventory and monitoring efforts.

The database currently contains more than 16,000 records. Each record includes fields for information on taxonomy, locality, method of collection, disposition of specimens, and status of the introduction, where known, as well as fields for other pertinent data. Recently, the database has been integrated with geographical information system (GIS) technology to facilitate improved monitoring and mapping of nonindigenous species distributions. On-line access to the dataset is available on the Internet at http://nas.er.usgs.gov.

There is an obvious need for synthesis, organization, and verification of records of nonindigenous fishes. A past source of information has been Lee et al. (1980 et seq.); however, compilers of the distribution atlas did not always consider the distributional status of all populations, consequently, some accounts in that tome did not distinguish native from introduced fish populations. Present information on distribution is scattered and much of it is relatively inaccessible. For instance, some natural history museum records have never been reported, while other records have been overlooked in files and unpublished documents. Because of their subtropical climates, Florida, Hawaii, and the southwestern United States support a greater number and diversity of nonindigenous species than northern locations. Nevertheless, other regions of the United States have been successfully colonized by many nonindigenous fishes from more temperate climates.

At present, our database contains many records from various sources that list the names of different nonindigenous fishes from particular regions or states. However, many of these sources do not provide documentation concerning dates, specific locality information, or references to voucher material. Records in this book are based on a variety of sources. The majority are from published accounts. Records taken from unpublished reports are accompanied by appropriate citations. Species occurrences not reported in the literature but represented by museum specimens are indicated by means of museum vouchers. In most cases, personal communications are given if they represent the most recent or only documentation for a species' occurrence at a particular locality.

Status of Nonindigenous Fishes

We have more than 16,000 records that document 536 unique fish taxa (i.e., fish species, hybrids, and certain unidentified forms) introduced outside their native ranges within U.S. waters. This diverse group of 75 families of fishes includes species from every continent except Antarctica (Table 1). Of the 536 taxa, **316 (61%)** are native to the United States but occur, or have been found, outside their native ranges, and **185 (35%)** are from other countries; at least 75 of these foreign forms are known to have become established in the United States. The remainder are hybrids.

Our total of 536 is considerably higher than the 127 nonindigenous fish species (70 U.S. and 57 non-U.S. species) reported in the United States in 1992 by the Office of Technology Assessment (OTA) (U.S. Congress, OTA 1993). Courtenay and Williams (1992) reported 99 exotic (non-U.S.) nonindigenous fish species in waters of the contiguous United States in 1992, of which 46 were established as sustaining populations. The disparity between our results and these numbers is influenced primarily by our intent to include all occurrences of nonindigenous fishes within the United States since about 1800, regardless of current status (i.e., established, nonestablished, or unknown).

Sport and associated forage fishes are the most widely distributed nonindigenous fishes. These include salmonids (salmon and trout), ictalurids (catfishes), centrarchids (basses and sunfishes), percids (walleye and sauger), and cyprinids (minnows). The three most widely distributed nonindigenous fishes are common carp *Cyprinus carpio*, goldfish *Carassius auratus*, and rainbow trout *Oncorhynchus mykiss*. The first two species have been reported or collected from every state except Alaska (Table 2). Goldfish introductions have resulted from the release of bait and aquarium fish and from forage fish stocking for sport fishes. Rainbow trout have been stocked as game fish. Widespread distribution of common carp is due primarily to the stocking program implemented by the U.S. Fish Commission in the late 1800s and early 1900s, and to the later use of juvenile carp as bait.

Summary of Reported Occurrences of Nonindigenous Fishes

All 50 states have recorded nonindigenous fishes in their open waters (Figure 1). California (162), Florida (127), Texas (105), Colorado (106), and Nevada (93) are the top five states in terms of number of different nonindigenous fish taxa. Of the total 536 taxa, 381 (71%) are recorded as occurring or having been found within the 11 states crossing or below the 35th parallel (i.e., Hawaii, California, Arizona, New Mexico, Texas, Oklahoma, Arkansas, Louisiana, Mississippi, Alabama, Georgia, and Florida). In fact, 317 different taxa (59%) have been recorded from just four states; Hawaii, California, Texas, and Florida.

Figure 2 shows the number of nonindigenous fish taxa by major drainage basin. The drainages depicted are based on the U.S. Geological Survey's (USGS) Hydrologic Unit Code (HUC) system. The South Atlantic–Gulf basin has more introduced fishes (210) by far than any other region — mainly due to the presence of tropical or ornamental fish farms and the warm climate of the region, which allows more introduced fishes to survive. The California basin (151) and the Mid Atlantic basin (118) are next in terms of the number of introduced fish taxa. The Alaska (9) and the Souris–Red–Rainey basins (27) have the fewest nonindigenous fishes, probably a

Table 1. List of families of nonindigenous fishes recorded for inland waters of the United States and the numbers of taxa, origins, and states of occurrence.

Family	Number of Taxa[a]	Origin[b]	States of Occurrence[c]
Acipenseridae	3	NA	AL, AZ, CA, ID, OR, WI
Adrianichthyidae	1	AS	CA, HI, NY
Albulidae	1	NA, SA	CA
Amblyopsidae	1	NA	IL
Amiidae	1	NA	CT, IA, IL, KS, KY, MA, MD, MO, NC, NJ, NY, PA, VA, WI, WV
Anabantidae	2	AF, AS	FL
Anguillidae	4	AU, EU, NA	AZ, CA, CO, IL, IN, NE, NV, OH, OR, PA, UT, WI
Anostomidae	1	SA	FL, HI
Aphredoderidae	1	NA	WI
Aplocheilidae	2	AF, AS	HI
Atherinidae	5	NA	AR, CA, IL, MO, NE, NM, OK, TX, WV
Auchenipteridae	1	SA	AZ
Belonidae	2	AS, NA	AL, HI, TN
Belontiidae	8	AS	CT, FL, HI, PA
Callichthyidae	4	SA	CO, FL, HI, NY
Carangidae	2	NA, SA	CA
Catostomidae	27	NA	AL, AZ, CA, CO, CT, GA, IL, MA, MD, MT, NC, NE, NM, NV, OH, OR, SC, TN, TX, UT, VA, WI, WV, WY
Centrarchidae	25	NA	AL, AR, AZ, CA, CO, CT, DE, FL, GA, HI, IA, ID, IL, IN, KS, KY, LA, MA, MD, ME, MI, MN, MO, MS, MT, NC, ND, NE, NH, NJ, NM, NV, NY, OH, OK, OR, PA, RI, SC, SD, TN, TX, UT, VA, VT, WA, WI, WV, WY
Centropomidae	3	AF	TX
Chanidae	1	AS	CA
Channidae	2	AS	HI, MA, ME, RI
Characidae	14	SA	AL, AR, AZ, CA, CO, CT, FL, GA, HI, IA, ID, IL, KY, LA, MA, MI, MN, MO, MS, MT, NC, NE, NH, NM, NY, OH, OK, OR, PA, SC, TX, UT, VA, WA
Cichlidae	44	AF, NA, SA	AL, AR, AZ, CA, CO, CT, FL, GA, HI, ID, IL, KS, KY, LA, MA, MS, MT, NC, NJ, NV, NY, OK, PA, RI, SC, SD, TX, UT, VA, WY
Clariidae	2	AS	CA, CT, FL, GA, HI, MA, NV
Clupeidae	7	NA, SP	AK, AL, AR, AZ, CA, CO, DE, FL, GA, HI, IA, ID, IL, IN, KS, KY, LA, MA, MD, ME, MI, MN, MO, MS, NC, NE, NH, NM, NV, NY, OH, OK, OR, PA, SC, TN, TX, UT,VA, VT, WA, WI, WV, WY
Cobitidae	3	AS	CA, FL, HI, ID, IL, MI, OR, TN
Cottidae	4	NA	AZ, CA, CO, ID, NC, NV, UT, WY
Cyprinidae	130	NA	AL, AR, AZ, CA, CO, CT, DE, FL, GA, HI, IA, ID, IL, IN, KS, KY, LA, MA, MD, ME, MI, MN, MO, MS, MT, NC, ND, NE, NH, NJ, NM, NV, NY, OH, OK, OR, PA, RI, SC, SD, TN, TX, UT, VA, VT, WA, WI, WV, WY
Cyprinodontidae	8	NA	AZ, CA, FL, NV, SC, TX

Family	Number of Taxa[a]	Origin[b]	States of Occurrence[c]
Doradidae	4	SA	FL
Embiotocidae	1	NA	CA
Engraulidae	3	NA, SA	CA
Erythrinidae	1	SA	FL
Esocidae	7	EU, NA	AK, AL, AR, AZ, CA, CO, CT, DE, FL, GA, IA, ID, IL, IN, KS, KY, MA, MD, ME, MI, MN, MO, MT, NC, ND, NE, NH, NJ, NM, NV, NY, OH, OK, OR, PA, RI, SC, SD, TN, TX, UT, VA, VT, WA, WI, WV, WY
Fundulidae	13	NA	AZ, CA, CO, FL, GA, HI, KY, MO, MT, NC, NE, NH, NJ, NM, NV, OH, OK, OR, PA, SC, SD, TX, UT, WA, WV, WY
Gadidae	1	NA	CT, NJ, OH
Gasterosteidae	4	NA	AL, CA, CO, CT, IL, KY, MA, ME, MI, MN, NJ, NM, OH, OK, PA, TN, UT, WI
Gerreidae	2	NA, SA	CA
Gobiidae	9	AS, AU, EU, NA, SP	AZ, CA, IL, IN, MI, MN, OH, PA, TX, WI
Goodeidae	4	NA	NV
Haemulidae	1	NA	CA, NM
Helostomatidae	1	AS	FL
Hiodontidae	2	NA	MD, NE
Ictaluridae	15	NA	AL, AR, AZ, CA, CO, CT, DE, FL, GA, HI, IA, ID, IL, IN, KS, KY, LA, MA, MD, MI, MN, MO, MS, MT, NC, NE, NH, NJ, NM, NV, NY, OH, OK, OR, PA, RI, SC, TN, TX, UT, VA, VT, WA, WI, WV, WY
Kuhliidae	1	SP	HI
Kyphosidae	1	NA, SA	CA
Labridae	2	NA	CA, NE
Lepisosteidae	4	NA	CA, MA, OR
Loricariidae	8	SA	AZ, CO, FL, HI, LA, NV, PA, SC, TX
Megalopidae	1	NA	TX
Melanotaeniidae	1	AU	IL
Moronidae	8	NA	AL, AR, AZ, CA, CO, DE, FL, GA, HI, IA, IL, IN, KS, KY, LA, MA, MD, MI, MN, MO, MS, MT, NC, ND, NE, NH, NJ, NM, NV, NY, OH, OK, OR, PA, SC, SD, TN, TX, UT, VA, WA, WI, WV
Mugilidae	2	NA, SA	CA
Nemastistiidae	1	NA, SA	CA
Notopteridae	1	AF	FL
Osmeridae	4	AS, NA	AR, CA, CO, CT, GA, HI, IA, ID, IL, IN, KS, KY, LA, MA, MD, MI, MN, MO, MT, NC, ND, NE, NH, NY, OH, PA, SD, TN, UT, VA, VT, WI
Osphronemidae	1	AS	HI, WA
Osteoglossidae	1	SA	CA, NV
Percidae	25	AS, EU, NA	AL, AR, AZ, CA, CO, CT, DE, FL, GA, IA, ID, IL, IN, KS, KY, LA, MA, MD, ME, MI, MN, MO, MS, MT, NC, ND, NE, NH, NJ, NM, NV, NY, OH, OK, OR, PA, SC, SD, TN, TX, UT, VA, VT, WA, WI, WV, WY
Percopsidae	1	NA	CT, MA, UT

Family	Number of Taxa[a]	Origin[b]	States of Occurrence[c]
Petromyzontidae	3	NA	CT, IL, IN, MI, MN, NY, OH, PA, WI
Pimelodidae	3	SA	FL, TX
Paralichthyidae	5	NA, SA	CA, TX
Pleuronectidae	3	EU, NA	CA, MI, OH
Poeciliidae	24	NA, SA	AK, AL, AZ, CA, CO, CT, FL, HI, IA, ID, IL, IN, KS, KY, MA, MI, MN, MO, MS, MT, NC, NE, NJ, NM, NV, NY, OH, OK, OR, PA, TN, TX, UT, VA, WA, WI, WY
Polyodontidae	1	NA	FL, GA
Rivulidae	4	SA	CA
Salmonidae	30	EU, NA	AK, AL, AR, AZ, CA, CO, CT, DE, FL, GA, HI, IA, ID, IL, IN, KS, KY, LA, MA, MD, ME, MI, MN, MO, MS, MT, NC, ND, NE, NH, NJ, NM, NV, NY, OH, OK, OR, PA, RI, SC, SD, TN, TX, UT, VA, VT, WA, WI, WV, WY
Sciaenidae	16	NA, SA	CA, CO, IL, NM, PA, TX, WI, WY
Scombridae	1	NA	CA
Serranidae	1	NA	CA
Stromateidae	1	NA	CA
Synbranchidae	1	AS	GA, HI
Umbridae	3	NA	AK, CT, MA, MT, OK, TX, WA

([a]) Total number of taxa (i.e., species, hybrids, and unidentified forms) per family.

([b]) Region of native range. AF = Africa, AS = Asia, AU = Australia, EU = Europe, NA = North America, SA = South America, SP = South Pacific.

([c]) States of nonindigenous fish occurrence (established, not established, or reproductive status unknown) in open waters between about 1800 and 1998. State abbreviations follow U.S. Postal Service usage (below).

AL	Alabama	LA	Louisiana	OH	Ohio
AK	Alaska	ME	Maine	OK	Oklahoma
AZ	Arizona	MD	Maryland	OR	Oregon
AR	Arkansas	MA	Massachusetts	PA	Pennsylvania
CA	California	MI	Michigan	RI	Rhode Island
CO	Colorado	MN	Minnesota	SC	South Carolina
CT	Connecticut	MS	Mississippi	SD	South Dakota
DE	Delaware	MO	Missouri	TN	Tennessee
FL	Florida	MT	Montana	TX	Texas
GA	Georgia	NE	Nebraska	UT	Utah
HI	Hawaii	NV	Nevada	VT	Vermont
ID	Idaho	NH	New Hampshire	VA	Virginia
IL	Illinois	NJ	New Jersey	WA	Washington
IN	Indiana	NM	New Mexico	WV	West Virginia
IA	Iowa	NY	New York	WI	Wisconsin
KS	Kansas	NC	North Carolina	WY	Wyoming
KY	Kentucky	ND	North Dakota		

Table 2. Nonindigenous fishes (foreign and native transplants) recorded as introduced into 10 or more states during the period *ca.* 1800 to 1998 (includes established and not established introductions).

Common name	Scientific name	Number of states
common carp	*Cyprinus carpio*	49
goldfish	*Carassius auratus*	49
rainbow trout	*Oncorhynchus mykiss*	48
brown trout	*Salmo trutta*	47
grass carp	*Ctenopharyngodon idella*	45
largemouth bass	*Micropterus salmoides*	43
walleye	*Stizostedion vitreum*	43
smallmouth bass	*Micropterus dolomieu*	41
Atlantic salmon	*Salmo salar*	40
white crappie	*Pomoxis annularis*	38
tench	*Tinca tinca*	38
striped bass	*Morone saxatilis*	37
northern pike	*Esox lucius*	36
mosquitofish	*Gambusia affinis/holbrooki**	35
bluegill	*Lepomis macrochirus*	35
fathead minnow	*Pimephales promelas*	35
brook trout	*Salvelinus fontinalis*	35
American shad	*Alosa sapidissima*	34
coho salmon	*Oncorhynchus kisutch*	34
black crappie	*Pomoxis nigromaculatus*	33
muskellunge	*Esox masquinongy*	32
white bass	*Morone chrysops*	32
yellow perch	*Perca flavescens*	32
green sunfish	*Lepomis cyanellus*	31
lake trout	*Salvelinus namaycush*	31
chinook salmon	*Oncorhynchus tshawytscha*	30
rock bass	*Ambloplites rupestris*	29
channel catfish	*Ictalurus punctatus*	29
rainbow smelt	*Osmerus mordax*	29
redear sunfish	*Lepomis microlophus*	27
threadfin shad	*Dorosoma petenense*	26
pumpkinseed	*Lepomis gibbosus*	25
sockeye salmon/kokanee	*Oncorhynchus nerka*	25
Arctic grayling	*Thymallus arcticus*	24
golden shiner	*Notemigonus crysoleucas*	23
tiger muskellunge	*Esox lucius* × *E. masquinongy*	21
rudd	*Scardinius erythrophthalmus*	21
brown bullhead	*Ameiurus nebulosus*	20
chain pickerel	*Esox niger*	20
striped bass hybrid	*Morone chrysops* × *M. saxatilis*	20
cutthroat trout	*Oncorhynchus clarki*	20
warmouth	*Chaenobryttus gulosus*	19
black bullhead	*Ameiurus melas*	18

Common name	Scientific name	Number of states
alewife	*Alosa pseudoharengus*	18
lake whitefish	*Coregonus clupeaformis*	18
flathead catfish	*Pylodictis olivaris*	18
white catfish	*Ameiurus catus*	17
bighead carp	*Hypophthalmichthys nobilis*	17
blue catfish	*Ictalurus furcatus*	17
spotted bass	*Micropterus punctulatus*	17
pirapatinga	*Colossoma brachypomum*	15
guppy	*Poecilia reticulata*	15
yellow bullhead	*Ameiurus natalis*	14
unidentified pacus	*Colossoma* or *Piaractus* spp.	14
unidentified piranhas	*Pygocentrus* or *Serrasalmus* spp.	14
bowfin	*Amia calva*	13
gizzard shad	*Dorosoma cepedianum*	13
redbreast sunfish	*Lepomis auritus*	13
white perch	*Morone americana*	13
spottail shiner	*Notropis hudsonius*	13
Mozambique tilapia	*Oreochromis mossambicus*	13
sauger	*Stizostedion canadense*	13
American eel	*Anguilla rostrata*	12
silver carp	*Hypophthalmichthys molitrix*	12
golden trout	*Oncorhynchus aguabonita*	12
blue tilapia	*Oreochromis aureus*	12
Sacramento perch	*Archoplites interruptus*	11
Oscar	*Astronotus ocellatus*	11
cutbow trout	*Oncorhynchus clarki* × *O. mykiss*	11
splake	*Salvelinus fontinalis* × *S. namaycush*	11
redbelly tilapia	*Tilapia zillii*	11
green swordtail	*Xiphophorus hellerii*	11
brook stickleback	*Culaea inconstans*	10
green sunfish hybrid	*Lepomis cyanellus* × *L. macrochirus*	10
margined madtom	*Noturus insignis*	10
pink salmon	*Oncorhynchus gorbuscha*	10
red piranha	*Pygocentrus nattereri*	10

* Because it often was not possible to distinguish species; both species were combined.

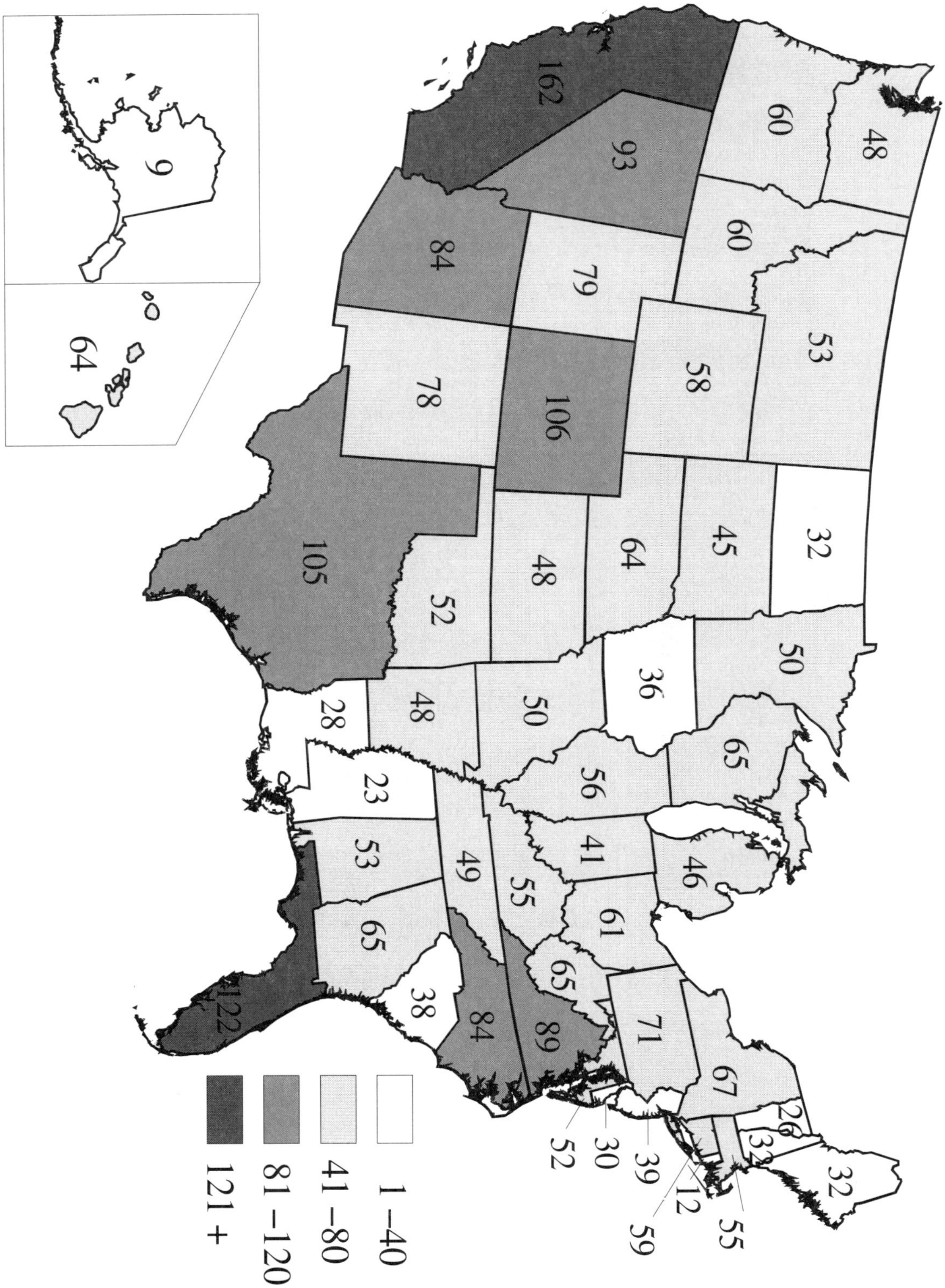

Figure 1. Number of nonindigenous fish taxa (foreign and native transplants) by state (includes established and not established introductions).

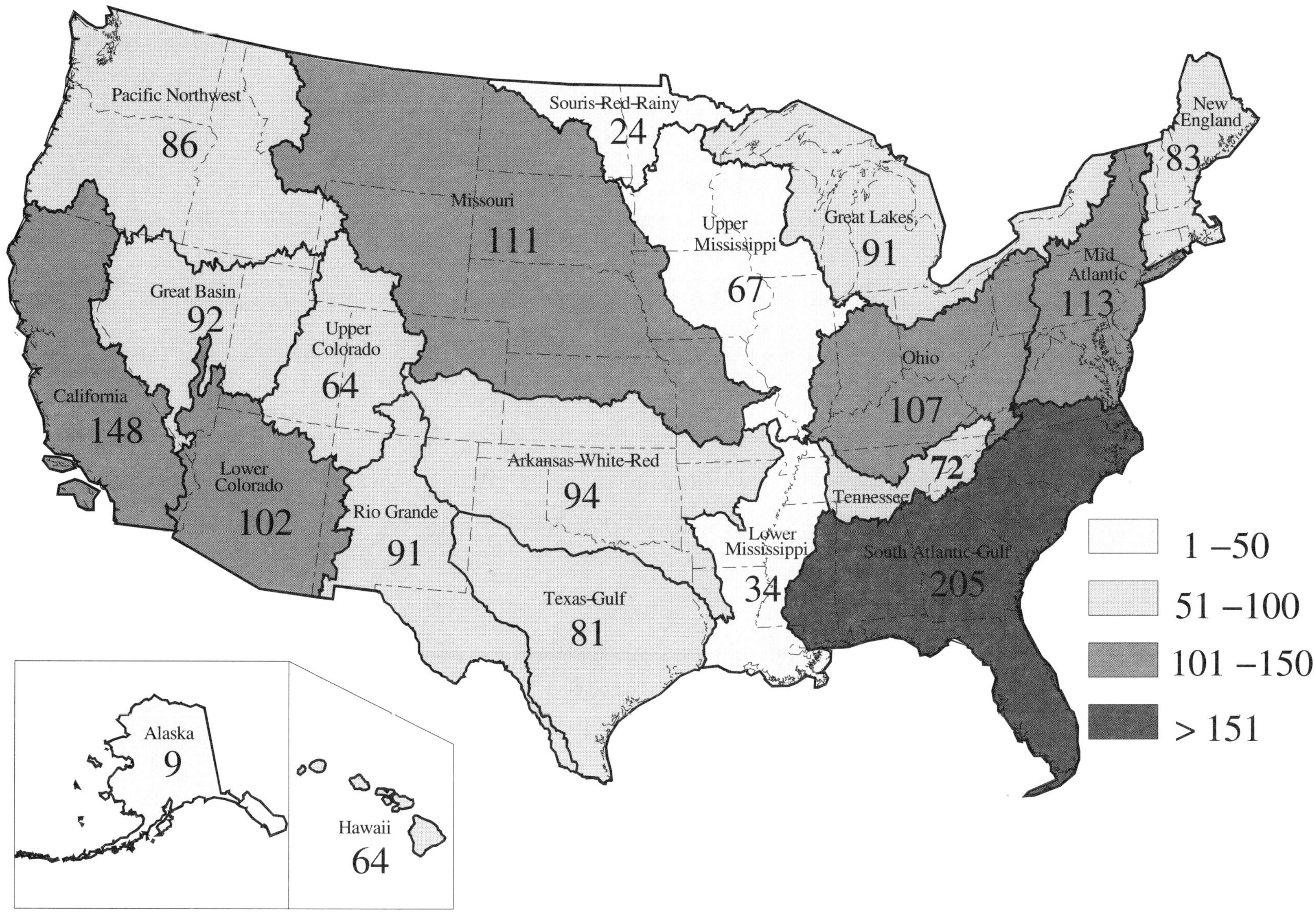

Figure 2. Approximate number of nonindigenous fish taxa (foreign and native transplants) by region (USGS 2-digit HUC [hydrologic unit code]) (includes established and not established introductions).

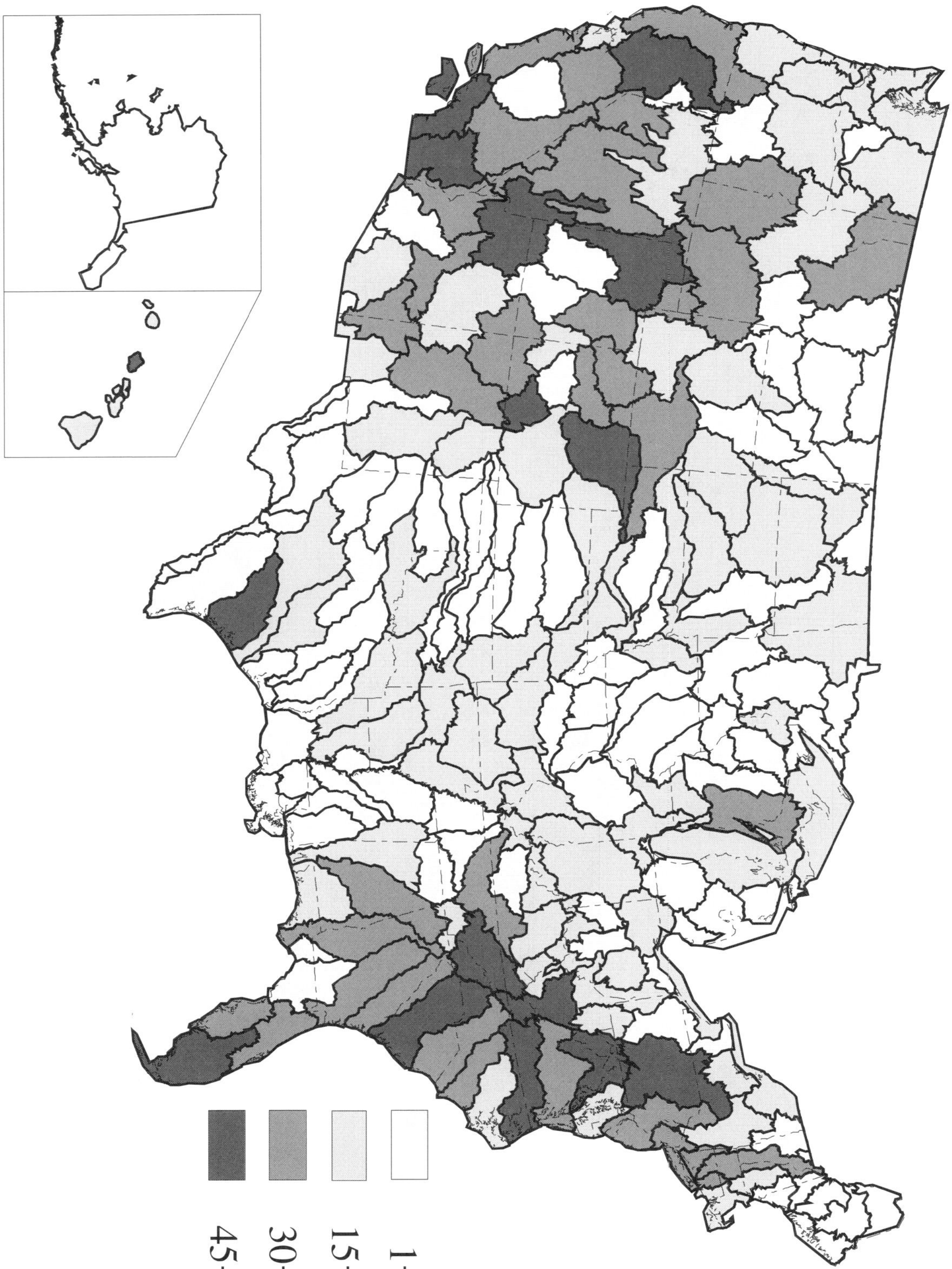

Figure 3. Approximate number of nonindigenous fish taxa (foreign and transplants) by sub-region (USGS 4-digit HUC [hydrologic unit code]) (includes established and not established introductions).

result of their cold climates, low human population densities, and remote locations. Somewhat surprisingly, relatively few introductions have been recorded for the upper and lower Mississippi basins. Similarly, there have been few nonindigenous fishes (**63**) recorded from the Great Lakes basin compared to other basins in the country.

Figure 3 shows the number of nonindigenous fish taxa by drainage at the USGS 4-digit HUC level. Drainages with the most nonindigenous fishes (darkest shading) are Oahu (69), the Kanawha (62), the upper Tennessee (60), South Florida (54), the Chowan–Roanoke (52), the Lower Colorado–Lake Mead (51), the Southern California coastal (50), the Rio Grande headwaters (47), the Sacramento (47), the Edisto–Santee (46), and the Central Texas coastal (45). These are conservative figures because some taxa, recorded only at the state level or from some region of a state, could not be assigned to a drainage at this level. Drainages on the east coast have a consistently greater number of nonindigenous fishes than drainages in other areas of the country. There are few nonindigenous fishes recorded from the central part of the United States.

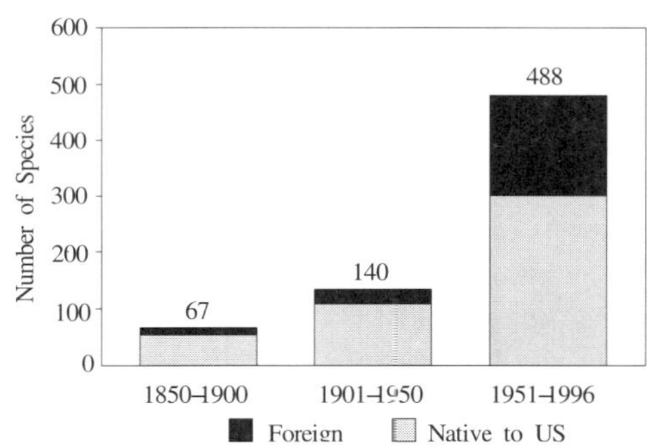

Figure 4. Rates of introduction of nonindigenous fish taxa into inland waters of the United States.

During the past 45 years, there has been a dramatic increase in the numbers of fishes introduced (Figure 4), many of them due to the release of aquarium or ornamental fishes and escapes from fish farms. Air transport has increased the number of ornamental fish species entering the United States, greatly contributing to the increase in the number and diversity of foreign fish species introduced into U.S. waters. In recent decades, improved shipping techniques, as well as the ease and availability of transportation, have also contributed to the transplantation of other fish species.

Nonindigenous foreign fishes have come from all areas of the world (Figure 5). Most species come from South America (largely cichlids), followed by Asia (many cyprinids), Africa (mostly cichlids), and Central America

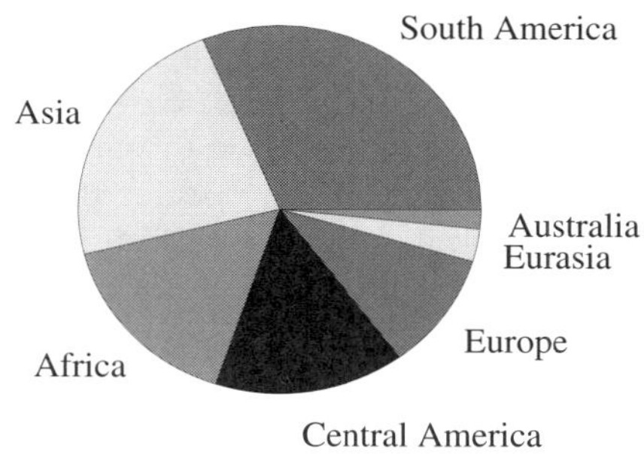

Figure 5. Geographic origin of foreign nonindigenous fishes recorded from U.S. waters (includes established and not established species).

(cichlids and poeciliids). The majority of these fishes are associated with the aquarium trade.

For purposes of analysis, eight categories of introduction methods were chosen (Figure 6). Percent of nonindigenous fishes introduced by each of these methods are as follows: stocked for sport (44.1%), aquarium release (26%), bait release (16.4%), stocked for conservation purposes (3.5%), method unknown (3.7%), miscellaneous (2.5%), stocked for biocontrol (2.1%), and release with ballast water (1.7%).

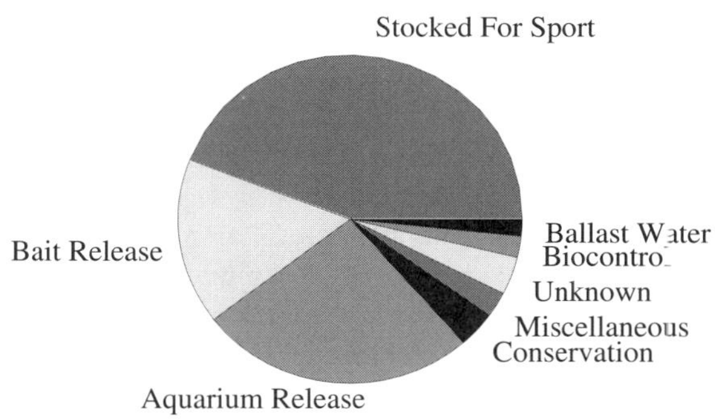

Figure 6. Proportion of nonindigenous fishes introduced through various methods or pathways. For those species introduced by multiple pathways, the most common pathway is represented. Graph does not distinguish between successful versus unsuccessful pathways.

Intentional stocking for sport is the most common reason for introductions of fishes. This category also includes stocking of forage fishes for sport fishes and accidental introduction of species by way of stock contamination. Most sportfish related stockings have been authorized introductions carried out by state and federal agencies. In addition, most sport-related stockings involve species that are native to a portion of the United States that are subsequently transplanted into nonnative drainages.

The category "aquarium release" is primarily composed of species that are released by aquarium hobbyists, although it also includes species that may have escaped from tropical or ornamental fish farms. These two methods are combined because in some cases it is not possible to distinguish between the two. "Bait release" introductions occur when anglers release live baitfish, mostly minnows. Ludwig and Leach (1996) surveyed a number of bait shops for bait contaminants that were not native to the surrounding area and discussed the likelihood of bait bucket transfers based on their findings.

Stocking for conservation or preservation purposes, as defined here, includes the introduction of endangered or threatened species into new habitats outside of their native ranges. Any resulting introduced populations of this sort are often called refuge populations. In such cases, stocks of some rare species are typically taken from natural populations or bred in captivity and released into natural habitats, such as fishless springs, or artificial reservoirs or ponds, some created specifically for the imperiled species (Williams 1991). Recovery plans commonly have proposed such transfers to increase the number of populations and increase the chance of survival and the American Fisheries Society has provided guidelines to increase the success of such stockings (Williams et al. 1988). Stocking for conservation purposes has most commonly involved desert or spring species, such as certain cyprinodontids, goodeids, and cyprinids inhabiting the southwest (e.g., Minckley and Deacon 1991). In our present work, we do not provide information on transplant reintroductions back into native ranges, except anecdotally, even though the practice of restocking rare species into their native habitat has been common (e.g., Minckley and Deacon 1991).

A few species have been stocked as biological control agents, usually to control aquatic vegetation or aquatic insects, such as mosquitoes. Ballast water is the most common method of introduction in major shipping ports. Water taken up in foreign ports and held in a ship's ballast tank is discharged when an empty ship is loaded with cargo in a destination port. The category "miscellaneous" includes other methods that do not fall under the above categories, such as introductions resulting from the construction of canals and aqueducts, releases from research facilities, and aquaculture escapes not related to the aquarium trade. If a species were introduced into U.S. water bodies for more than one reason (e.g., both by way of bait bucket release and also stocked as forage), the most prevalent category of introduction was assigned to that species.

Figure 7 shows a breakdown by origins of introductions in each state. States with the majority of foreign introductions include Florida (89), Hawaii (51), California (41), and Texas (36). Alaska is the only state with no introduced foreign species. Western states, in general, have a greater proportion of species transplanted from another state, many of these sport fishes from the eastern United States. North Carolina has an especially large number of nonindigenous fishes recorded, a large proportion are native fishes, mostly minnows, introduced across drainages. However, there is much speculation and disagreement among researchers concerning the true natural range of a fair number of these species. As a consequence, the total number of nonindigenous fishes recorded for North Carolina may be artificially high.

Terminology

Definitions of terms used in this report are given in Appendix B. We use the terms *nonindigenous* and *indigenous* to describe a species' geographic range. These terms are always dependent on the scale of reference and geographic units used. The scale may be based on hydrologic units (e.g., watershed, drainage, basin), geopolitical boundaries (e.g., county, state, country), or some other geographical unit (e.g., continent, hemisphere). For example, the flathead catfish *Pylodictis olivaris* is indigenous to the central and south central United States and northeastern Mexico. Each of the following statements is true, depending on scale of reference: the flathead catfish is indigenous to northwestern Georgia; it is indigenous to the Mississippi River basin; and it is indigenous to North America. Flathead catfish have been released by humans into rivers of southeastern Georgia, and these introductions represent nonindigenous populations (i.e., the species is not indigenous to southeastern Georgia). We generally avoid using the terms *exotic* and *transplanted*, partially because of the misleading connotation that introduction of a species into an adjacent watershed poses a less serious threat than establishment of a foreign species in the same system.

Information on reproductive status, population size changes, dispersal rates, and range expansion of introduced fishes is especially important. In this report, we generally rely on three terms to describe a nonindigenous species' reproductive status: established, locally established, and reported (see Glossary for definitions).

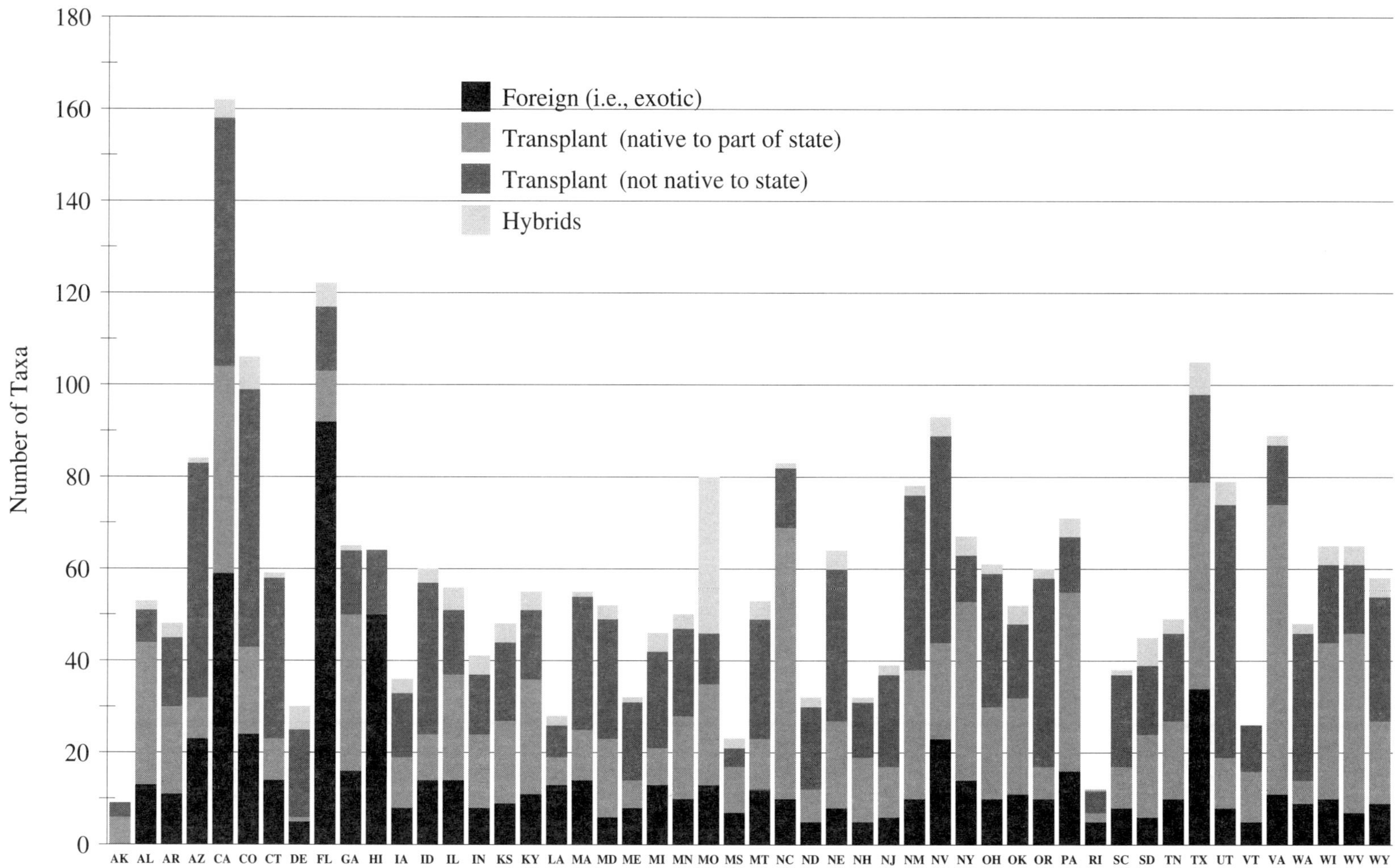

Figure 7. Origin of nonindigenous fish taxa by state (includes established and not established species). Legend definition: "foreign" (species from another country), "transplant (native to part of state)" (species native to part of the state but introduced to a nonnative area(s) of the state, source of introduction may be from either within or outside the state), "transplant (not native to state)" (species native to the United States but not to that state where introduced), and "hybrid" (stocked as hybrid).

Arrangement and Format of Species Accounts

Most of the information in this report is presented in individual species accounts. Species are grouped by order and family, and the order of presentation generally follows tradition, with groups considered more primitive treated first. Family names follow Nelson (1994), except where noted. A short account of each family precedes the accounts of the fish belonging to that family. Each species account begins with the generally accepted scientific name, its author, and the year of description. It was sometimes difficult or impossible to assign a scientific name because of nonspecific identification or later taxonomic revision. In these cases, designation was made at the lowest taxonomic level possible (e.g., genus or subfamily) based on available data. Scientific names used follow the American Fisheries Society checklist (Robins et al. 1991a), Robins et al. (1991b), and published updates (e.g., Kendall 1997), unless otherwise discussed, or an alternate spelling is used in accordance with the International Commission on Zoological Nomenclature. We have also attempted to correct the spelling of scientific names and dates of description based on information provided in Kendall (1997) and the species catalogues of Eschmeyer (1998) and Gilbert (1998).

Common Names: This section gives the common or vernacular names of a species. Some are names that are in common usage, but more local names are also included in some cases. The preferred common name, as provided by the Committee on Names of Fishes (i.e., Robins et al. 1991a, 1991b), is given first. Other vernacular names are given in parentheses. A few of the foreign species treated in our work are not included in Robins et al. (1991a, 1991b). As such, all common names listed for these species appear in parentheses.

Identification: This section provides references considered useful in accurately identifying a fish specimen. We recommend Page and Burr (1991) for most freshwater fishes native to the portion of North America north of Mexico. In addition, most states have guides for identification of fishes found within their political boundaries; however, older publications do not include recent name changes and may not include all species found in a particular area. Where relevant, we mention similar species that may confuse proper identification. For many fishes introduced into the United States from other regions of the world, there is a greater likelihood of misidentification. In cases of known identification problems, we have either referred the reader to references that will assist in correct identification or provided updated information on a species' taxonomic status. Unfortunately, because many foreign fish groups are poorly known taxonomically, assigning names to fishes from these groups remains problematic. Mention is also made if the reported identification of a fish is highly questionable and cannot be verified.

Native Range: This section summarizes the known native geographic distribution of each species. For most native fishes and some established foreign fishes, this information is based on Lee et al. (1980 et seq.) and modifications to that work as given by Page and Burr (1991). If the species is not generally considered a freshwater fish, we identify its habitat as being either marine or brackish. If a species' native distribution includes parts of the United States, the states within its native range are indicated on the species map. For many fishes originating in the Neotropics, Africa, or Asia, native ranges are not well known and range statements are relatively general.

Size: In this section, we provide maximum size for adults, where known. Our primary reference source was information compiled by Page and Burr (1991). All lengths given are total length (TL) unless otherwise indicated. In some cases, length may be given as standard length (SL) or fork length (FL). Metric units used include millimeters (mm) or centimeters (cm) for small-sized species, and meters (m) for larger species.

Nonindigenous Occurrences: This section summarizes the nonnative geographic distribution of each species and supplements information provided in species maps. In general, we provide information on the places, states, and counties or drainages where introductions have occurred or where the species has spread following introduction. The level of detail varies greatly depending on the quality of information sources and the number of locations. More generalized information is provided for species that have been widely introduced and firmly established.

Means of Introduction: This section provides dates and primary reasons, if known, for initial and subsequent introductions. Primary reasons for past introductions usually have been based on the species' value (real or intended), typically related to one or more of the following uses: as sport fish, commercial fish, forage fish, baitfish, insect control agents, aquatic weed control agents, and aquarium releases. In addition, some introductions have occurred for purposes of species preservation or experimentation (list from Moyle 1976b). Introductions are also sometimes categorized as authorized, unauthorized, or accidental, but precise designation of this type was not always possible based on available information. Courtenay (1993) categorized and reviewed the reasons for fish introductions: new food resources; sport purposes (includes sport fish, forage fish, and bait releases); biological control; aquarium trade and hobbyist releases; aquaculture releases; species preservation or restoration; canal and aqueduct construction; and ship ballast releases. Appendix B gives definitions of terms used in species accounts relating to means of introduction.

Status: In this section, we summarize information about the reproductive status and persistence of each nonindigenous species. We also provide known data on relative abundance and on whether the fish is actively expanding its range. Although some introduced fish species formerly had locally established breeding populations in open waters of the United States, their populations are now considered extirpated, either through intentional eradication or by natural causes. Unfortunately, for most species there are little or no data indicating whether populations are stable, increasing, declining, or extirpated.

Impact of Introduction: This section documents evidence of impacts resulting from each species' introduction. For the most part, this section only deals with negative effects resulting from introduction. Intended beneficial uses of a species generally are covered in the section on the means of introduction. Due to a paucity of data in most cases, it usually was not possible to adequately summarize the impacts of a species as detrimental or beneficial, or to determine whether impacts were major or minor. In many cases, we characterize the effects of an introduction as unknown simply because there have been no follow-up studies. Our use of the phrase "impact unknown" does not imply that actual or potential impacts are insignificant, but that we are unaware of any investigations to determine negative impacts. Possible, but undocumented, impacts are discussed in either this section or in the remarks section.

Remarks: This section provides additional comments and important references. For species that are collected sporadically, and for many that are frequently misidentified, we indicate the states where a fish has been reported, followed by a reference listing either a voucher specimen or a photograph of the fish in question. In these cases, we reference one or a few (even if there are many) catalogued museum specimens by state. Museum and institutional abbreviations follow Leviton et al. (1985). We also have attempted to point out those cases where a species is reported from a state, but the identification is questionable and there are no known voucher specimens or photographs to indicate the uncertainty of a particular record. Except for selected cases, museum voucher specimens listed in species accounts were not examined by us. Museum catalogue numbers are supplied simply to provide a tracking mechanism for other researchers and also to indicate the existence of museum material for those wishing to verify specimen identification.

Hybrids

Hybridization is not uncommon and can occur between closely related species within virtually any fish family (Hubbs 1955; Schwartz 1981; Jenkins and Burkhead 1994). Among nonindigenous fishes, there is evidence of both artificial and natural hybrids occurring in open waters. Our records on introduced hybrids are obviously incomplete. Even though there are many unpublished reports documenting introductions of hybrids, the identity of stocked hybrids is not always certain. The possibility of back-crosses further complicates the issue. Nevertheless, hybridization and introgression between introduced and native fishes, some of which are endangered or threatened, is an issue of increasing concern. Hybridization between introduced and native species is known to result in loss of genetic purity and reduced reproductive output in native fish populations. The consequences are especially serious when it involves rare species or subspecies (e.g., Echelle and Connor 1989; Behnke 1992). Introgression (incorporation of genes from one population into another) is also a concern because it may lead to the breakdown of locally adapted gene complexes and loss of genetic diversity (e.g., Hindar et al. 1991). As such, we felt it important to provide some of the more common or unusual cases involving hybrid introductions, both artificial and natural.

Hybrids artificially produced in laboratories or fish hatcheries have commonly been stocked in open waters (e.g., *Morone chrysops* × *M. saxatilis*). Although somewhat less common, natural hybrids involving introduced fishes are also known. In such cases, an introduced pure species hybridizes in the wild with a closely related species, either with a native fish or with another introduced species. For instance, hybridization between introduced brook trout *Salvelinus fontinalis* and native bull trout *S. confluentus* is quite common (Leary et al. 1995). Similarly, certain introduced foreign fishes, for instance some of the tilapias, are also known to produce hybrids in nature (e.g., *Tilapia zillii* and *T. mariae*), as reported by Taylor et al. (1986). We include accounts for those hybrids which are stocked as such. Information for hybrids resulting from the introduction of one or both parental species is included under "Impacts" of the introduced parental species.

Scientists typically refer to hybrids by the scientific names of both parental species. A few of these hybrids even have accepted common names (e.g., *Esox lucius* × *E. masquinongy* is the tiger muskellunge). For artificial hybrids, authors often give the male parent species name first. However, parental sexes are not always known and it has become more common to list parental species alphabetically (Robins et al. 1991a). We follow Robins et al. (1991a) in listing all hybrid combinations in alphabetical order.

Maps

For each nonindigenous species we provide a map of the continental United States, with insets of Hawaii and Alaska, showing the native (if applicable) and nonnative occurrence by state. Maps were created at the state level because of the often imprecise locality data available for many records. In instances in which a fish is native to one section of a state but is introduced to another, that state is indicated as both native and nonnative. Shading of a state does not imply that a species is distributed statewide, but that it has been reported from at least one locality.

Species Erroneously Reported as Introduced

Species erroneously reported as introduced were addressed either in the **Remarks** section, if it had been introduced elsewhere, or in the section at the end of the species accounts entitled "Species Erroneously Reported as Introduced" if the species had no valid introductions. Erroneous reports were based on species misidentifications or a change in the perceived introduction status due to more recent information.

Reporting Problems and Requests for Information

Positive identification of a nonindigenous fish species is an essential first step in determining its status and predicting potential negative effects. Positive identification is also critical for the successful prevention, containment, and control of any nonindigenous nuisance fish species. Unfortunately, there has been considerable confusion surrounding the identification of some introduced fish species, especially many that are of foreign origin. Examples of particularly problematic groups include the tilapias, foreign poeciliids, some members of the genera *Coregonus* and *Hypostomus*, and the pacus and piranhas, among others. In cases involving hybrids and back-crosses there is an added dimension of confusion.

Uncertainty of identification is sometimes unavoidable, reflecting the fact that the state of our knowledge of the systematics and taxonomy of certain fish groups is poor and these groups are in need of additional scientific study. Nevertheless, incorrect or questionable identifications have made their way into unpublished and published accounts of introduced fishes. If voucher species were taken and are available, questions about the identification of a particular nonindigenous species can be resolved. However, in many cases involving reports of nonindigenous fishes, voucher material is unavailable because the fish specimens were never preserved or were later lost. In these cases, we may never know with certainty the true identity of the fish species involved (Nico and Fuller 1999).

Our database includes reports based on personal communications, many of which lack supporting documentation such as preserved specimens or photographs; such identifications should be accepted with caution. For species that are difficult to identify, voucher specimens or photographs (at a minimum) are essential. Unfortunately, once an erroneous name enters the literature, that record continues to be cited. Ludlow and Walsh (1991) discussed some of the problems encountered in researching the occurrence of a nonindigenous species while dealing with early reports that did not include sufficient voucher materials. As our work with museums progresses, we expect to find many more specimens representing nonindigenous species that have been misidentified or whose names need to be updated.

It is crucial that reports of nonindigenous fish introductions be based on reliable identifications. Please contact us if you have preserved or frozen specimens and collection data (precise locality and date). All future reports to us should include as much information as possible. If voucher specimens are unavailable, please indicate the reference used for identification, whether it is a published key or simply a comparison of photographs or information with those from aquarium literature.

Because the status of nonindigenous fish populations can change significantly over time, it is especially meaningful to monitor their abundance and distribution.

Although we provide summary information on the status in the species accounts, it should be realized that this information may be out of date. Often the most recent information on a species' status is based solely on field work carried out many years ago. In such instances, there is a critical need to relocate sites where nonindigenous species were previously reported and determine if the population is still present, and if present, whether or not it has expanded into new areas.

Species Accounts

ORDER PETROMYZONTIFORMES

Petromyzontidae - Lampreys

The lampreys comprise 6 genera and 41 extant species (Nelson 1994). These eel-like, jawless fishes occur in cool zones of the world including temperate marine and freshwaters. Most species are found in the Northern Hemisphere; of these, 19 are native to North America north of Mexico (Page and Burr 1991). Four species are known from southern South America and southeastern Australia, although these are usually considered to represent two additional families, the Geotriidae and the Mordaciidae (Berra 1981). Some species are parasitic as adults, and they use their suction disk mouths and rasping teeth to attach to other fish, thereby feeding on the body fluids. Free-living species use their suction disk mouths to hold onto rocks and move stones during spawning. All species require freshwater to breed; the resulting young are called ammocoetes. Ammocoetes filter feed and take three to eight years to change into adults.

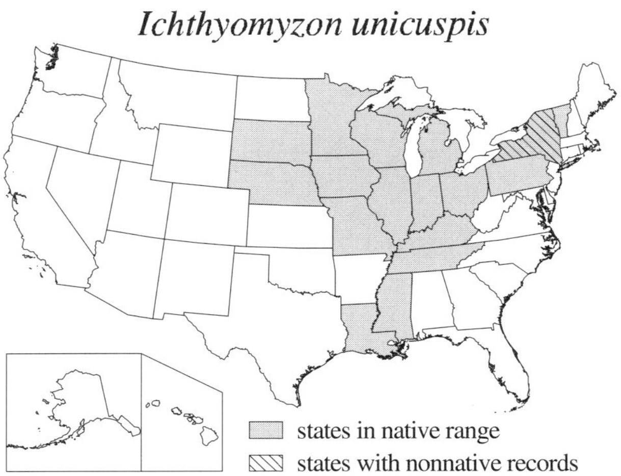

Ichthyomyzon unicuspis Hubbs and Trautman 1937

Common Name: silver lamprey.

Identification: Trautman (1981); Smith (1985); Page and Burr (1991); Etnier and Starnes (1993); Pflieger (1997). Maximum size: 39 cm.

Native Range: St. Lawrence–Great Lakes basin from Quebec to southwestern Ontario and south through upper Mississippi ans Ohio River basins to central Tennessee. Nelson River, Manitoba; Missouri River, Nebraska; and Mississippi River, Mississippi (Page and Burr 1991). Not reported from Arkansas (Robison and Buchanan 1988).

Nonindigenous Occurrences: Introduced into the Hudson River basin, **New York**, where it was first collected in 1974 (Mills et al. 1997). It has been collected from the Hudson River at Stillwater, Albany, and Nyack (Smith 1985).

Means of Introduction: Canal connection. Probably gained access from Lake Champlain to the Hudson River via the Champlain-Hudson Canal (Smith 1985; Mills et al. 1997).

Status: Reported in New York.

Impact of Introduction: Unknown, however, this species is parasitic and feeds on trout, whitefish, smelt, pike, white sucker, black buffalo, brown bullhead, carp, rock bass, walleye, paddlefish, sturgeons, and gars (Smith 1985).

Remarks: None.

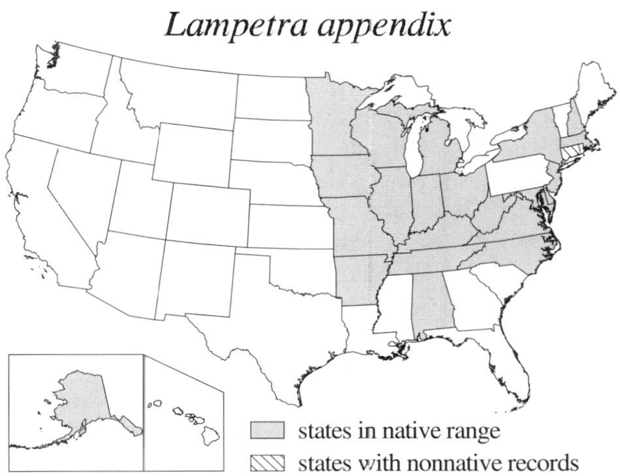

Lampetra appendix (DeKay 1842)

Common Name: American brook lamprey.

Identification: Becker (1983); Page and Burr (1991); Jenkins and Burkhead (1994); Pflieger (1997). Two, possibly three, subspecies; *L. a. appendix* on Atlantic Slope and *L. a. wilderi* throughout the St. Lawrence-Great Lakes and Mississippi basins and in Roanoke River, Virginia. *L. a. alaskense* in Northern Territories and Alaska may be a third subspecies (Page and Burr 1991). Maximum size: 15–22 cm, depending on subspecies.

Native Range: Fragmented distribution includes Atlantic, Great Lakes, and Mississippi River basins from St. Lawrence River, Quebec, west to Minnesota, and south to Roanoke River drainage, Virginia, Tennessee River basin, northern Alabama, and St. Francis and White River systems, Missouri and Arkansas; Martin River (Arctic slope), Northern Territories to Ugashik and Chatanika rivers (Pacific slope), Alaska (Page and Burr 1991).

Nonindigenous Occurrences: This species is known from a small stream in **Connecticut**, Kettle Brook, a tributary of the Connecticut River near the Windsor Locks in Hartford County (Whitworth 1996).

Means of Introduction: A hatchery existed along Kettle Brook during the early 1900s. Whitworth (1996) suggested that this lamprey may have been inadvertently introduced to the hatchery (and subsequently to the brook) through stock contamination of fish brought from out of state.

Status: Established in Connecticut; the population has existed since about the mid-1950s (Whitworth 1996).

Impact of Introduction: Unknown.

Remarks: Schmidt (1986) listed this nonparasitic lamprey as native to the Connecticut River drainage.

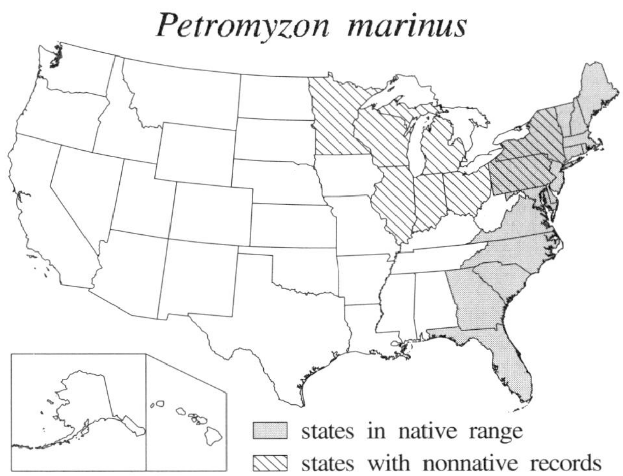

Petromyzon marinus Linnaeus 1758

Common Names: sea lamprey.

Identification: Becker (1983); Page and Burr (1991); Jenkins and Burkhead (1994); Vladykov and Kott (1980) provided a key to the ammocoetes of lampreys found in the Great Lakes region. Maximum size: 120 cm anadromous; 64 cm landlocked.

Native Range: Generally marine but ascends freshwater rivers to spawn. Atlantic Coast from Labrador to Gulf of Mexico, Florida; landlocked in Great Lakes and several New York lakes. Also along Atlantic coast of Europe and Mediterranean Sea (Page and Burr 1991).

Nonindigenous Occurrences: Lampreys were first discovered in Lake Ontario in 1835, Lake Erie in 1921, Lake Michigan in 1936, Lake Huron in 1937, and Lake Superior in 1946 (Applegate 1950; Lawrie 1970; Smith 1979; Smith and Tibbles 1980; Smith 1985). Its distribution in all five Great Lakes and selected tributaries includes the states **Illinois** (Smith 1979; Emery 1985); **Indiana** (Gerking 1955; Emery 1985); **Michigan** (Applegate 1950; Smith 1979); **Minnesota** (Eddy and Underhill 1974; Phillips et al. 1982; Emery 1985); **New York** (Smith 1985); **Ohio** (Trautman 1981; Emery 1985); **Pennsylvania** (Emery 1985); and **Wisconsin** (Becker 1983; Emery 1985).

Means of Introduction: First reported from Lake Ontario in the 1830s, the lamprey may have entered that lake from its native habitat in the Atlantic drainage by migrating through the Erie Canal after the canal was opened between 1819 and 1825 (Emery 1985), or by hitching rides on boats passing through the Erie and St. Lawrence canal systems (Morman et al. 1980). Previously blocked from Lake Erie and all the upper Great Lakes by Niagara Falls, the sea lamprey apparently gained access to Lake Erie from Lake Ontario via the Welland Canal. It has often been noted that the Welland Canal was originally opened in 1829, but that the sea lamprey was not reported from Lake Erie until 1921, nearly a century later (Applegate 1950). Several reasons have been put forth to explain the 92-year delay. For instance, Morman et al. (1980) indicated that the original Welland Canal was relatively small and may not have allowed sea lampreys easy negotiation; the chance of dispersal was increased with construction of the second Welland Canal, built between 1914 and 1932 for handling larger ships. Similar to its possible passage through the Erie Canal, the sea lamprey may have migrated through the Welland Canal by navigating the locks or by attaching to the hulls of upstream bound ships (Courtenay 1993). Although the available literature does not mention the possibility, lamprey also may have been transported through canals and open lakes by host fish. In the Great Lakes, there are reports of sea lampreys attaching to fish that swim long distances (Scott and Crossman 1973). Morman et al. (1980) provided a map of the Great Lakes basin showing the major waterways of possible importance in the spread of this species. Reproduction was first documented in the Lake Erie drainage in 1932 (Smith 1985). Reproducing populations were found in all of these upper lakes by 1947 (Applegate 1950; Trautman 1981). It also is conceivable that lamprey spread was facilitated because of its use as bait during the 1800s and early 1900s. For instance, several early authors reported that larval lampreys were commonly sold and used as bait in parts of New York (Adams and Hankinson 1928).

Status: Established throughout the Great Lakes region. This species was common in lakes Michigan and Huron by the 1930s and in eastern Lake Superior by the 1940s; it never became abundant in Lake Ontario (Applegate 1950; Emery 1985) or Lake Erie (Smith 1985).

Impact of Introduction: Attack and parasitic feeding on other fishes by adult lampreys often results in death of the prey, either directly from the loss of fluids and tissues or indirectly from secondary infection of the wound (Phillips et al. 1982). Of the fish that survived attacks by lampreys, 85% of various species had been attacked up to five times (Scott and Crossman 1973). The species' introduction to the Great Lakes and its later abundance, combined with water pollution and overfishing, resulted in the decline of several large native species, including several ciscoes *Coregonus* spp., lake trout *Salvelinus namaycush*, and walleye *Stizostedion vitreum*, among others. Consequently, there was a collapse in the commercial fisheries during the 1940s and 1950s in many parts of the Great Lakes, particularly in lakes Huron and Michigan, and in eastern Lake Superior (e.g., Lawrie 1970; Scott and Crossman 1973; Christie 1974; Lee et al. 1980 et seq.; Smith and Tibbles 1980; Becker 1983; Emery 1985; Courtenay 1993). Lake trout catch in Lake Huron fell from 3.4 million pounds in 1937 to virtual failure in 1947. In Lake Michigan, U.S. catch fell from 5.5 million pounds in 1946 to 402 pounds in 1953. In Lake Superior, catch dropped from an average of 4.5 million pounds to 368 thousand pounds in 1961 (Scott and Crossman 1973).

In freshwater, sea lampreys are also known to attack white sucker *Catostomus commersoni*, longnose sucker *Catostomus catostomus*, redhorse *Moxostoma* spp., yellow perch *Perca flavescens*, rainbow trout *Oncorhynchus mykiss*, burbot *Lota lota*, channel catfish *Ictalurus punctatus*, northern pike *Esox lucius*, and common carp *Cyprinus carpio* (Scott and Crossman 1973). Sea lamprey predation, in combination with other factors (i.e., overfishing and hybridization with more common cisco species), led to the extinction of three endemics in the Great Lakes; the longjaw cisco *Coregonus alpenae*, the deepwater cisco *C. johannae*, and the blackfin cisco *C. nigripinnis* (Miller et al. 1989). During the late 1940s, the alewife *Alosa pseudoharengus* invaded the Great Lakes from the Atlantic Ocean. Because the sea lamprey had greatly reduced the population of large predators, alewife populations exploded and were followed by tremendous die-offs, resulting in additional changes to fish species composition in the lakes (Smith and Tibbles 1980). Sea lampreys also took a toll on the introduced salmon in the Great Lakes, much to the dismay of anglers and state fish agencies. Although the number of sea lamprey in the Great Lakes has been reduced, they still kill substantial numbers of lake trout in some areas and thus are impeding the rebuilding of established populations (Schneider et al. 1996, and references therein).

Remarks: Controversy exists as to whether the sea lamprey is native to Lake Ontario. Several believe it is native (e.g., Lawrie 1970; Smith 1985), suggesting that sea lampreys found in Lake Ontario and its tributaries, the Finger Lakes, and Lake Champlain represent relict populations from the last Pleistocene glaciation. Those contending that it is not native believe that this species, unknown in Lake Ontario prior to the 1830s, had most likely entered the inland lake from Atlantic coastal drainages via the artificially created Erie Canal (e.g., Emery 1985). Early methods to control this species included mechanical weirs and electrical barriers (Scott and Crossman 1973; Smith and Tibbles 1980). Beginning in the late 1950s, sea lampreys began to be successfully controlled by use of the lampricide 3-trifluoromethyl-4-nitrophenol (TFM), a chemical agent that kills larval lampreys in their stream habitats (Smith and Tibbles 1980). The lampricide has reduced the population by over 90% of the 1961 peak (Scott and Crossman 1973). As a result, commercial fisheries reportedly have shown some recovery (Smith 1985; Page and Laird 1993) and the sea lamprey's impact on native fishes has been reduced (Page and Laird 1993). However, continued use of TFM is apparently required to keep sea lamprey populations under control (Scott and Crossman 1973; Becker 1983). TFM is sometimes harmful to other fish (e.g., walleye) (Becker 1983), as well as to the larvae of nonparasitic lamprey species. Results of an international symposium on the sea lamprey were published in the Canadian Journal of Fisheries and Aquatic Sciences in 1980. The demise of lake trout led to development of the splake, a hybrid between lake trout and brook trout. It was hoped that the hybrid would better avoid lampreys and mature faster, hence spawn at least once before becoming parasitized (Scott and Crossman 1973). As of 1991, it was estimated that the U.S. and Canada were spending $8 million per year on lamprey control and another $12 million per year on lake trout restoration (Newman 1991).

ORDER ACIPENSERIFORMES

Acipenseridae - Sturgeons

The family Acipenseridae contains 4 genera with 24 species (Nelson 1994). Sturgeons are distributed in the Northern Hemisphere, with most species in central and eastern Europe and a few species in Asia and North America (Berra 1981; Page and Burr 1991). These primitive fishes are characterized by rows of bony scutes on the body, prominent snouts, inferior mouths with barbels, heterocercal tails, and largely cartilaginous skeletons. Sturgeons are among the largest of living fishes, with adult lengths ranging from 3 to 30 feet. Many species are commercially taken for their flesh and ripe eggs, the latter sold as caviar. Unfortunately, overharvesting and habitat modification, coupled with a naturally slow rate of reproduction, have resulted in significant population declines of several species. Although some species spend most of their lives in the ocean, all breed in freshwater.

Acipenser fulvescens

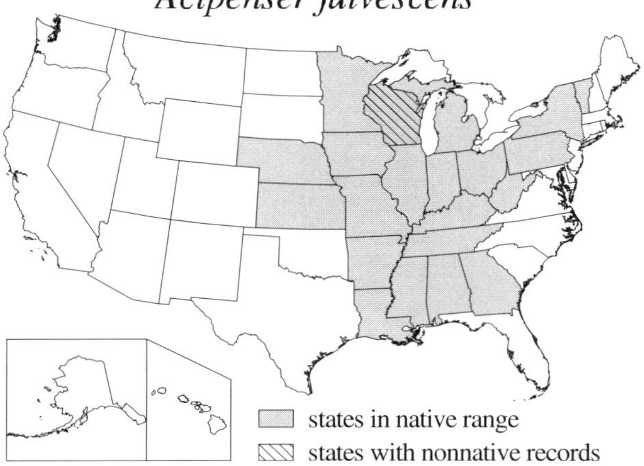

Acipenser fulvescens Rafinesque 1817

Common Name: lake sturgeon.

Identification: Trautman (1981); Becker (1983); Page and Burr (1991); Etnier and Starnes (1993). Maximum size: 274 cm.

Native Range: St. Lawrence-Great Lakes, Hudson Bay, and Mississippi River basins from Quebec to Alberta and south to Alabama and Louisiana; Coosa River system (Mobile Bay drainage), Alabama and Georgia (Page and Burr 1991).

Nonindigenous Occurrences: Lake sturgeon have been introduced into several lakes in **Wisconsin** where they were not originally found, including Big Cedar Lake in Washington County, Madison lakes in Dane County, Chain of Lakes in Waupaca County, and Pear Lake in Washburn County, among others (Becker 1983).

Means of Introduction: Intentionally stocked.

Status: Unknown.

Impact of Introduction: Unknown.

Remarks: In 1901, 20,000 lake sturgeon were stocked in the Missisquoi River at Swanton, Vermont (Ravenel 1902). The stocking locality is just upriver from Lake Champlain, its native range. Lake sturgeon are rare throughout most of their range (Becker 1983).

Acipenser transmontanus

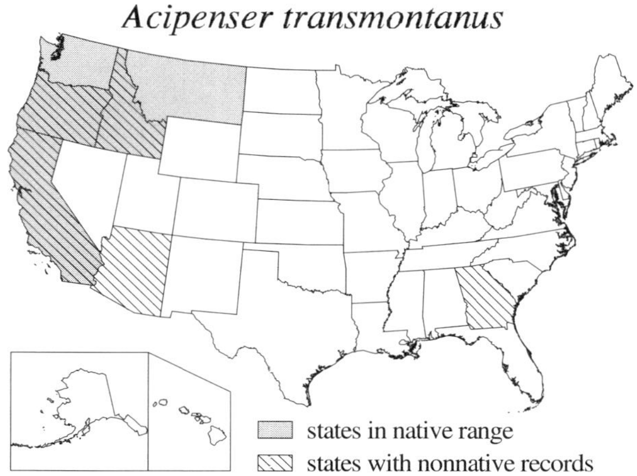

Acipenser transmontanus Richardson 1836

Common Names: white sturgeon.

Identification: Moyle (1976a); Wydoski and Whitney (1979); Page and Burr (1991). Maximum size: 6.1 m and nearly one ton.

Native Range: Pacific Coast from Alaska Bay, Alaska, to central California and extreme northwestern Montana (Page and Burr 1991).

Nonindigenous Occurrences: White sturgeon may have been released in the Coosa River in **Georgia** (J. D. Williams, personal communication) and stocked in Lake Havasu on the Colorado River (**Arizona** and **California**) in 1967 and 1968 (Minckley 1973; Lee et al. 1980 et seq.; Swift et al. 1993; Rinne 1995). Several collections of single fish were taken downstream at Imperial Dam in 1967 and 1969 (Swift et al. 1993). Another was taken at the mouth of the Bill Williams River, **Arizona** in 1976 (Swift et al. 1993). More than 200 white sturgeon were stocked in Irvine Lake and the Santa Ana River lakes of **California** in 1987; several were later captured (Swift et al. 1993). The species was stocked in the Snake River above the Shoshone Falls, **Idaho** (Idaho Fish and Game 1990), where it is native below the falls, but not above (Idaho Fish and Game 1990). It also was introduced to upper Klamath Lake, **Oregon** (Bond 1994; H. Li, personal communication).

Means of Introduction: Intentionally stocked in the Colorado River, on the Arizona–California border, in Oregon, and presumably in Idaho. The possibility of introduction for the Coosa River collection is not clear. We have some reports that several hundred white sturgeon were intentionally released from a fish farm on Cohulla Creek in the Conasauga system, in the Coosa River drainage in Georgia (J. D. Williams, personal communication; B. Freeman, personal communicaton in Walters 1997). However, G. Beisser (personal communication) says they were confiscated and does not believe any were intentionally released. Game and Fish biologists searched the area and no sturgeon were found downstream of the fish farm that had them. Walters (1997) reported a second fish that was captured from the Coosa but does not say if it was in Alabama or Georgia.

Status: Status of Lake Havasu population is not known; there is no evidence of reproduction (Minckley 1973; Lee et al. 1980 et seq.; Swift et al. 1993; Rinne 1995). The last specimen from the Colorado River was caught in 1976 (Swift et al. 1993). Reported from California, Georgia, Idaho, and Oregon.

Impact of Introduction: Unknown.

Remarks: A single subadult individual taken in 1992 in the Coosa River, Talladega County, Alabama, (Anonymous 1992b, 1992c; D. Catchins, personal communication) was previously identified as this species. It was later identified as a lake sturgeon *A. fulvescens*, a species native to the Coosa River. The specimen is deposited at the Florida Museum of Natural History (UF 93520).

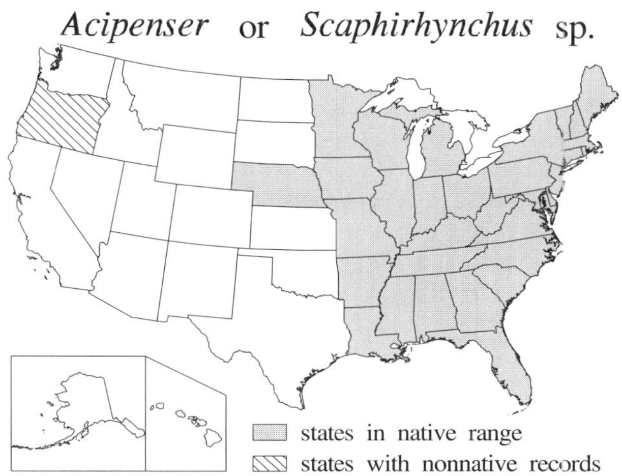

Acipenser or *Scaphirhynchus* sp.

Common Name: unidentified sturgeon (eastern species).

Identification: Vladykov and Greeley (1963); Pflieger (1975); Becker (1983); Jenkins and Burkhead (1994); see Page and Burr (1991) for distinguishing characteristics of these genera. Maximum size: Eastern sturgeon species range from 86 cm to 4.3 m.

Native Range: Eastern North America.

Nonindigenous Occurrence: Two individuals of an unidentified species of sturgeon were released into Guild's Lake on the Willamette River, Portland, **Oregon**, in November 1905 (Lampman 1946).

Means of Introduction: The two fish were released at the close of a fair exhibiting freshwater fishes from the eastern United States (Lampman 1946).

Status: Reported in Oregon.

Impact of Introduction: Unknown.

Remarks: None.

Polyodontidae - Paddlefishes

The family Polyodontidae, commonly called the paddlefishes, is a very primitive family that contains two species. One species is native to North America; the other is native to China's Yangzte River (Page and Burr 1991; Nelson 1994). These fish are characterized by the long, flat snout. Jenkins and Burkhead (1994) produce a vivid image, describing them as "piscine Pinocchios". The snout is not to stir up the bottom or to dislodge food from vegetation, as previously thought (Rosen and Hales 1981). But rather it is covered with taste buds, which are thought to aid the fish in finding its planktonic food (Robison and Buchanan 1988; Jenkins and Burkhead 1994). The North American paddlefish is planktivorous, whereas the Chinese paddlefish is predaceous (Jenkins and Burkhead 1994). The species in North America has been commercially harvested for its roe, a source of caviar (Robison and Buchanan 1988). Paddlefish mainly inhabit freshwater, but are rarely found in brackish water (Nelson 1994).

Polyodon spathula

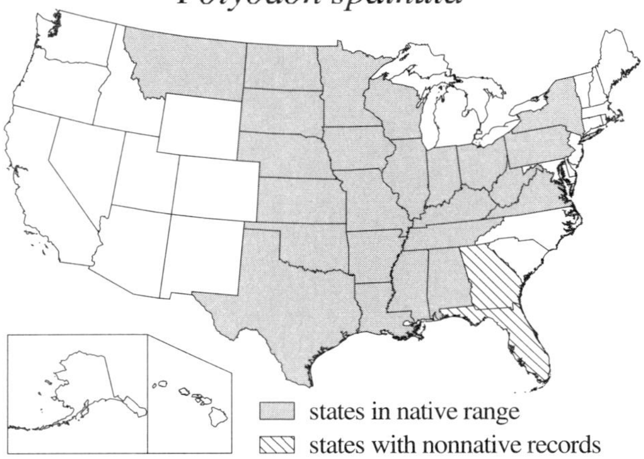

Polyodon spathula (Walbaum 1792)

Common Name: paddlefish.

Identification: Robison and Buchanan (1988); Page and Burr (1991); Etnier and Starnes (1993); Jenkins and Burkhead (1994); Mettee et al. (1996). Maximum size: 221 cm.

Native Range: Mississippi River basin from southwestern New York to central Montana and south to Louisiana; Gulf Slope drainages from Mobile Bay, Alabama, to Galveston Bay, Texas. Formerly in the Lake Erie drainage in Ohio (and possibly lakes Huron and Michigan). Extirpated from Great Lakes basin, Galveston Bay and Sabine River drainages in Texas, and Calcasieu drainage in Louisiana (Page and Burr 1991).

Nonindigenous Occurrences: Introduced into the Flint River below Newton, **Georgia**. The fish have dispersed downstream to Lake Seminole and the Apalachicola River, **Florida** (R. Ober and F. Paruka, personal communication).

Means of Introduction: Approximately 1,200 fish, 10-15" in length, escaped an aquaculture facility along the Flint River in Georgia, during Tropical Storm Alberto in early July 1994 (Ober and Paruka, personal communication).

Status: Four or five fish have been collected from the stretch from Newton to the Apalachicola River. One individual was collected in the spring of 1997 and measured 30" (Paruka, personal communication). Two were collected below Jim Woodruff Dam on the Apalachicola River in 1997 and weighed 10–11 lbs (Ober, personal communication).

Impact of Introduction: Unknown.

Remarks: Voucher specimen: The U.S. Army Corps of Engineers office at Lake Seminole has one of the paddlefish mounted on their wall (Ober, personal communication).

ORDER SEMIONTIFORMES

Lepisosteidae - Gars

The primitive family Lepisosteidae is composed of two genera with seven species (Nelson 1994). Although fossils have been found in many parts of the world, including South America, Africa, and western Asia (C. R. Gilbert, personal communication), living gars are restricted to eastern North America, Central America, and Cuba (Berra 1981; Nelson 1994). Five species are native to the United States (Page and Burr 1991). Gars are medium- to large-sized predators, cylindrical in shape, and with needle-sharp teeth. The largest is the alligator gar (*Atractosteus spatula*), reaching about 3 m and 135 kg. All gars are found in freshwater, but can enter brackish or even marine waters.

Atractosteus spatula

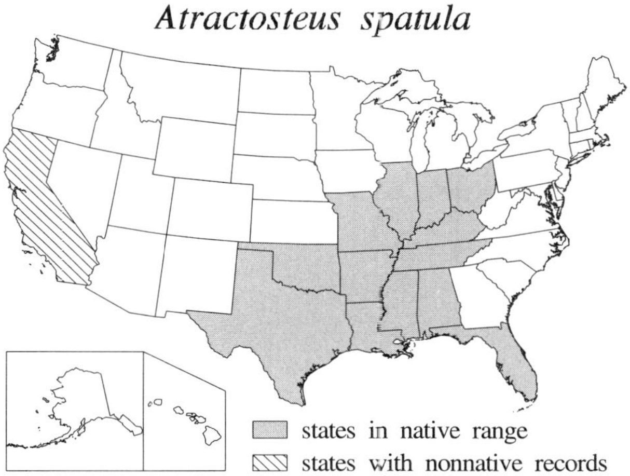

Atractosteus spatula (Lacepède 1803)

Common Name: alligator gar.

Identification: Suttkus (1963); Page and Burr (1991); Etnier and Starnes (1993). Synonym *Lepisosteus spatula*. Maximum size: 3 m.

Native Range: Mississippi River basin from southwestern Ohio and southern Illinois south to Gulf of Mexico; Gulf Coastal Plain from Econfina River, Florida, to Veracruz, Mexico (Page and Burr 1991).

Reported Nonindigenous Occurrence: A single alligator gar measuring 145.5 cm TL and weighing 18.6 kg was reported from Clifton Court Forebay, a water storage reservoir located at the head of the California Aqueduct, near Byron in Contra Costa County, **California**, on 17 September 1991 (Raquel 1992).

Means of Introduction: The introduction of this individual was attributed to release by an aquarium hobbyist (Raquel 1992).

Status: Reported in California.

Impact of Introduction: Unknown.

Remarks: The California specimen was recorded in water with a temperature of 25°C and a salinity of approximately 4.5 ppt (Raquel 1992).

Lepisosteus cf. oculatus

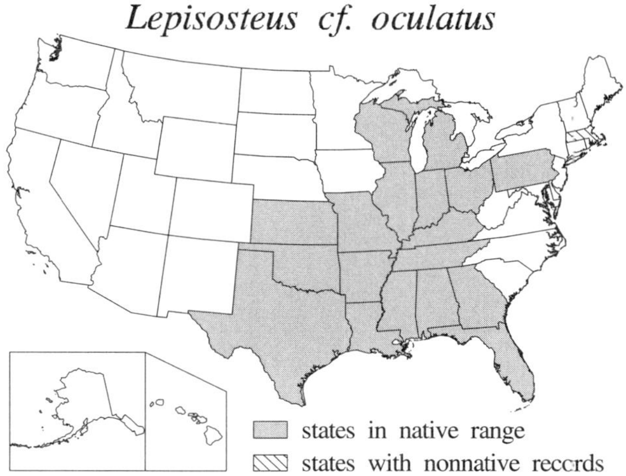

Lepisosteus cf. oculatus (Winchell 1864)

Common Names: spotted gar.

Identification: Suttkus (1963); Becker (1983); Page and Burr (1991); Etnier and Starnes (1993). Maximum size: 112 cm.

Native Range: Lake Erie and southern Lake Michigan drainages south through Mississippi River basin to Gulf Coast; Gulf Slope drainages from lower Apalachicola River, Florida, to Nueces River, Texas (Page and Burr 1991).

Nonindigenous Occurrence: Gar were reported from two sites within the Connecticut River drainage in **Massachusetts**, but specimens were not collected and specific identity could not be confirmed (Hartel 1992; Cardoza et al. 1993; Hartel et al. 1996). The gars were observed in a pond on the Amherst campus, University of Massachusetts in Hampshire County (repeated sightings of a single individual) and in Stanley Park, Northampton, Hampshire County (four to five individuals measuring 400–520 mm TL) (Hartel 1992; Hartel et al. 1996).

Means of Introduction: Unknown.

Status: Reported from Massachusetts.

Impact of Introduction: Unknown.

Remarks: None.

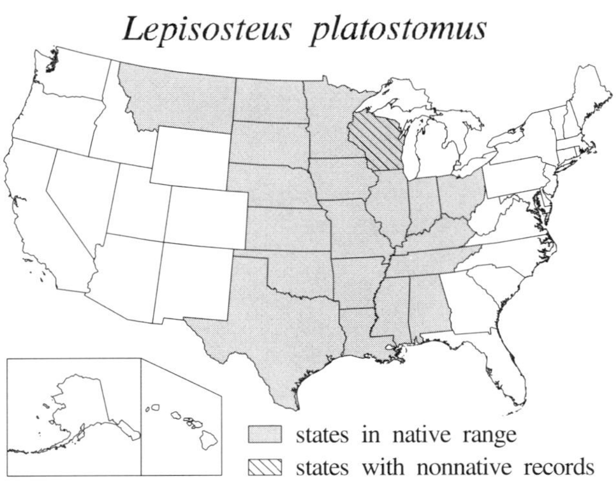

Lepisosteus platostomus Rafinesque 1820

Common Name: shortnose gar.

Identification: Suttkus (1963); Becker (1983); Page and Burr (1991); Etnier and Starnes (1993). Maximum size: 83 cm.

Native Range: Mississippi River basin from southcentral Ohio, northern Indiana and Wisconsin to Montana and south to northern Alabama and Louisiana (Page and Burr 1991).

Nonindigenous Occurrences: This species was first collected in Lake Winnebago, Great Lakes basin, **Wisconsin** in 1962 (Priegel 1963). It is now also found it the upper and lower Fox River, the lower Wolf River, and in lower Green Bay (Becker 1983).

Means of Introduction: This species likely reached Lake Winnebago via the Wisconsin-Fox Canal, a canal connection from the Mississippi River basin to the Fox River in the Great Lakes basin (Priegel 1963).

Status: Established in Wisconsin (Becker 1983; Page and Burr 1991).

Impact of Introduction: Unknown.

Remarks: None.

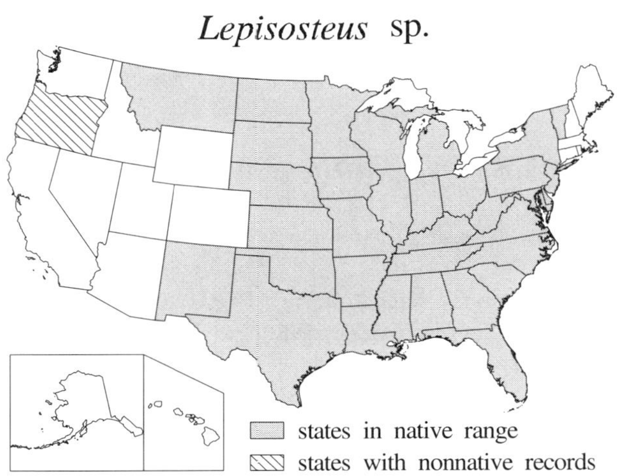

Lepisosteus sp.

Common Name: gar.

Identification: See Suttkus (1963) or Page and Burr (1991) for descriptions of species belonging to this genus. Maximum size: Unknown.

Native Range: Unknown; however, no *Lepisosteus* species are native to the West Coast. See Page and Burr (1991) for range descriptions of species within this genus.

Nonindigenous Occurrences: One unidentified gar was found dead at Crane Prairie Reservoir, Deschutes County, **Oregon**, in 1991 (Li, personal communication).

Means of Introduction: Unknown.

Status: Reported from Oregon.

Impact of Introduction: Unknown.

Remarks: None.

ORDER AMIIFORMES

Amiidae - Bowfin

This primitive family includes a single living species, *Amia calva* (Nelson 1994). Although fossils have been found in marine and freshwater deposits in many parts of the world, *Amia calva* is restricted to the eastern half of North America (Berra 1981; Page and Burr 1991; Nelson 1994). Bowfin can tolerate water with low levels of dissolved oxygen by gulping air and using the swimbladder as a primitive lung; they can survive intermittent drought periods by aestivation (Lee et al. 1980 et seq.; Jenkins and Burkhead 1994). This species is a medium-sized predator with a long, nearly cylindrical body. It is known only from freshwater.

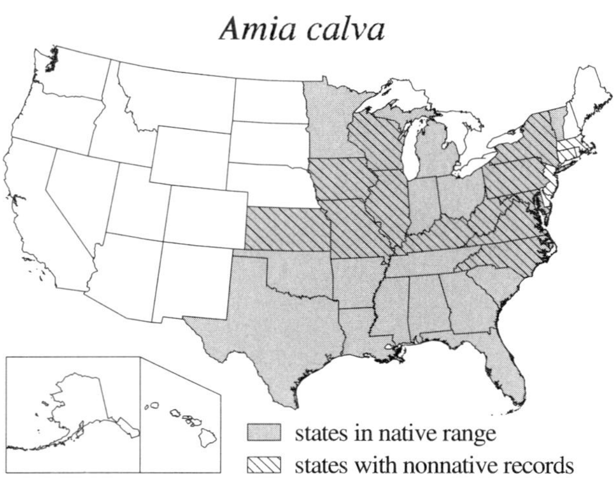

Amia calva Linnaeus 1766

Common Name: bowfin.

Identification: Becker (1983); Page and Burr (1991); Jenkins and Burkhead (1994). Maximum size: 109 cm.

Native Range: Eastern North America. St. Lawrence-Great Lakes (including Georgian Bay and lakes Nipissing and Simcoe, Ontario) and Mississippi River Basin drainages from Quebec to northern Minnesota and south to the Gulf of Mexico; Atlantic and Gulf Coastal Plain from the Susquehanna River drainage in southeastern Pennsylvania to the Colorado River in Texas (Page and Burr 1991).

Nonindigenous Occurrences: Bowfin have been introduced into the Connecticut River and the Housatonic drainage, **Connecticut** (Behnke and Wetzel 1960; Whitworth et al. 1968; Lee et al. 1980 et seq.; Smith 1985; Schmidt 1986; Whitworth 1996); a number of localities in **Illinois** (Lee et al. 1980 et seq.); the Little Sioux drainage, **Iowa** (Lee et al. 1980 et seq.); Hughes Lake in Douglas County, **Kansas** (Cross and Collins 1995) and perhaps the mouth of Independence Creek in Atchison County (Pflieger 1971, 1975; Cross and Collins 1995); ponds in Scott County, and possibly the Kentucky River, Licking River, and Tygarts Creek, **Kentucky** (Burr and Warren 1986); ponds in western and northern **Maryland** and the Gunpowder River, Maryland (Pearson and Ward 1972); the Connecticut, Housatonic, and Merrimack drainages, **Massachusetts** (Schmidt 1986; Hartel 1992; Hartel et al. 1996); private lakes and the upper Missouri River near the Kansas border in **Missouri** (Pflieger 1971, 1975; Cross et al. 1986); northern **New Jersey** (Stiles 1978; Lee et al. 1980 et seq.); southern **New York** (Lee et al. 1980 et seq.; Smith 1985; Schmidt 1986); the Tennessee (Lee et al. 1980 et seq.; Menhinick 1991; Jenkins and Burkhead 1994), Catawba, and upper Yadkin drainages (Lee et al. 1980 et seq.) in **North Carolina**; Black Moshannon Lake in Centre County, Glendale Lake in Cambria County, the Susquehanna River (Denoncourt et al. 1975a), and Somerset Lake in the Youghiogheny drainage (Hendricks et al. 1979) in **Pennsylvania**; the New and upper Roanoke drainages, **Virginia** (Lee et al. 1980 et seq.; Hocutt et al. 1986; Jenkins and Burkhead 1994); the South Branch Potomac, **West Virginia** (Jenkins and Burkhead 1994); and the Lake Michigan drainage including the Fox and Wolf rivers and Green Bay, **Wisconsin** (Becker 1983). Cross et al. (1986) listed this species as introduced into the Chariton-Nishnabotna and the Kansas drainages; however, these drainages occur in more than one state and they did not indicate states of occurrence.

Means of Introduction: Past bowfin introductions have been attributed mainly to intentional stockings in ponds, lakes, and rivers (Jenkins and Burkhead 1994). Jenkins and Burkhead (1994) provided detailed background information on bowfin introductions into Virginia and several other eastern states. They reported that bowfin stocking was apparently popular 20–40 years ago. The Pennsylvania Fish Commission stocked bowfin in two lakes mentioned above (Denoncourt et al. 1975a). Bowfin were introduced to the Gunpowder River in Maryland in 1971 when stocked ponds overflowed (Pearson and Ward 1972). They were also intentionally stocked in ponds in northern and western Maryland in the 1960s (Pearson and Ward 1972). Many of the original introductions into Connecticut were illegal stockings and were eradicated (Whitworth 1996). The Kansas report is of a single fish stocked accidentally into a fee-fishing pond with other fishes taken from Minnesota or northwestern Iowa (Cross and Collins 1995). In Wisconsin, bowfin may have been stocked as part of the fish rescue operations from the Mississippi River backwaters in the 1930s, and/or it may have gained access to the Lake Michigan drainage via the Wisconsin-Fox Canal (Becker 1983). Bowfin have been intentionally stocked in private lakes in Missouri and records from the Missouri River in this state may be the result of escapes from these lakes (Pflieger 1971, 1975).

Status: This species has been introduced into or reported in Connecticut, Illinois, Iowa, Kansas, Kentucky, Maryland, Massachusetts, New Jersey, New York, North Carolina, Pennsylvania, Virginia, West Virginia, and Wisconsin; it is probably established in most areas of the states into which it has been introduced. Not established in Kansas; the report is of a single fish.

Impact of Introduction: Unknown.

Remarks: Burr and Warren (1986) reported that Kentucky records of this species from east of the Green River represent waif and/or probable introductions. Pflieger (1978) reported that Missouri River records resulted from bowfin escapes from privately stocked lakes; however, Burgess and Gilbert (in Lee et al. 1980 et seq.) viewed these records as representing natural occurrences. Although bowfin are voracious, piscivorous predators, they will consume virtually any animal small enough to eat (Lee et al. 1980 et seq.). Consequently, introduced bowfin pose a potential threat to native fishes and their prey.

ORDER OSTEOGLOSSIFORMES

Osteoglossidae - Bonytongues

The family Osteoglossidae, whose members are known as bonytongues, is composed of four genera with seven species (Nelson 1994). These tropical fishes are found in South America, Africa, Southeast Asia, and northern Australia (Berra 1981; Nelson 1994). Most species are carnivorous or omnivorous and grow to about 1 m, although *Arapaima gigas* from South America grows to over 2 m. All represent locally important food fishes. The family is restricted to freshwater.

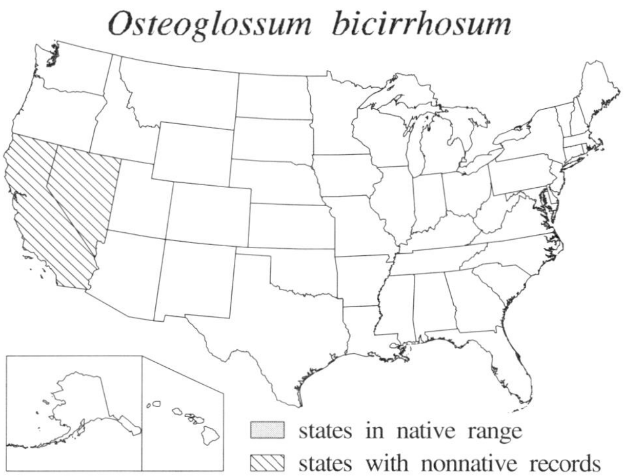

Osteoglossum bicirrhosum Vandelli 1829

Common Names: arawana (silver aruana, green arowana, aruana, arowhana).

Identification: Distinguishing characteristics and illustrations were given by Kanazawa (1966) and Goulding (1980). Photographs appeared in Kanazawa (1966) and Axelrod et al. (1985). Maximum size: 100 cm.

Native Range: Tropical America; northern South America including the Amazon basin and the Guianas (Robins et al. 1991b).

Nonindigenous Occurrences: Two specimens have been recorded from **California**. The first was a single fish taken from Lake Berryessa, Napa County, by an angler ca. 1972 (Shapovalov et al. 1981; Courtenay et al. 1991); a second specimen was netted from Lake Merced, San Francisco County, by anglers in August 1994 (Dill and Cordone 1997). The species was unsuccessfully introduced at Forest Spring in Ash Meadows, Nye County, **Nevada**, during the early 1960s (Soltz and Naiman 1978; Deacon and Williams 1984).

Means of Introduction: Probably aquarium releases in California (Dill and Cordone 1997); introduced by an illegal ornamental fish farming operation in Nevada (Soltz and Naiman 1978).

Status: Reported in California and Nevada.

Impact of Introduction: Unknown.

Remarks: Devick (1991b) indicated that there was a report of an *Osteoglossum* species from Hawaii but it was not collected. Juveniles and small adults are found in the ornamental fish trade. This species is a mouthbrooder, with males performing the brooding role. Although the arawana will take fish and other vertebrates, it feeds primarily on insects and other arthropods (Goulding 1980). There are no known voucher specimens.

Hiodontidae - Mooneyes

The family Hiodontidae contains one genus with two living species, the mooneye and the goldeye (Nelson 1994). Mooneyes are found in North America, primarily in the Mackenzie, Saskatchewan, Mississippi, and St. Lawrence river basins (Nelson 1994). These fishes are characterized by a strongly compressed body with an untoothed keel along the belly, a lateral line and large eyes. They superficially resemble the herrings in that they have adipose eyelids, as well as a few other herring-like features. Mooneyes are medium-sized fishes, attaining a size of about one-half meter. They are active, fast-moving fishes that typically feed near the surface of the water. Diets are somewhat varied, consisting mainly of invertebrates (e.g, aquatic insects) and small fishes. Both species occur only in freshwater.

Hiodon alosoides

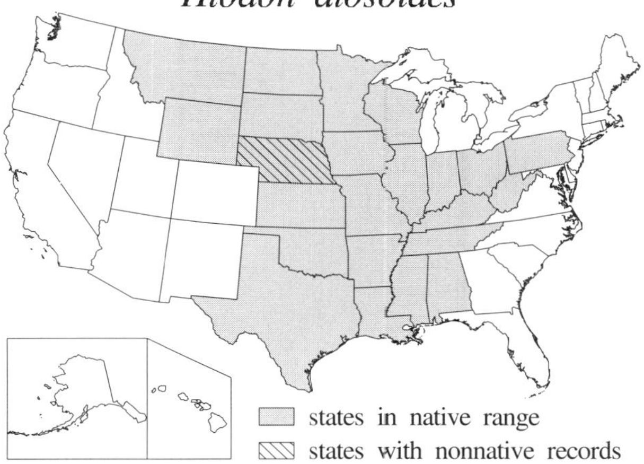

Hiodon alosoides (Rafinesque 1819)

Common Name: goldeye.

Identification: Becker (1983); Page and Burr (1991); Etnier and Starnes (1993); Pflieger (1997). Maximum size: 51 cm.

Native Range: Tributaries of James Bay, Quebec and Ontario; the Arctic, Missouri, Mississippi, and Ohio basins from the Mackenzie River drainage, Northern Territories, to western Pennsylvania and south to Louisiana (Page and Burr 1991).

Nonindigenous Occurrences: This species was stocked in the upper White River in Dawes County, **Nebraska**, in 1896; and in Lake Minatare, Scotts Bluff County, Nebraska, (North Platte drainage) in 1952 (Jones 1963). It is native to the Platte drainage farther downstream.

Means of Introduction: Intentionally stocked, probably as forage.

Status: Presumably the stockings failed.

Impact of Introduction: Unknown.

Remarks: This species is mainly considered a forage fish; however, smoked goldeye is considered a delicacy (Scott and Crossman 1973; Becker 1983).

Hiodon tergisus

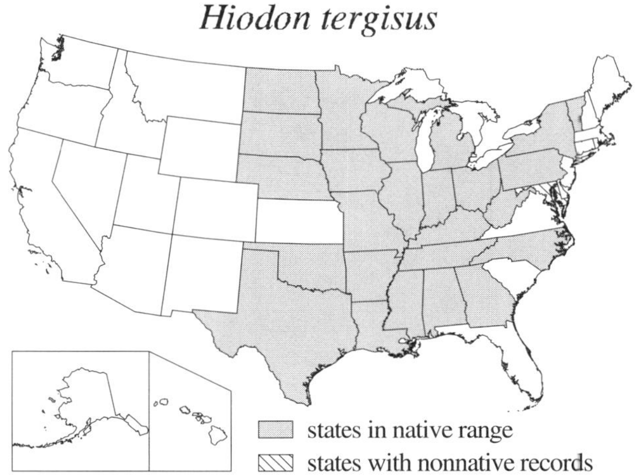

Hiodon tergisus Lesueur 1818

Common Name: mooneye.

Identification: Smith (1985); Page and Burr (1991); Etnier and Starnes (1993); Mettee et al. (1996); Pflieger (1997). Maximum size: 47 cm.

Native Range: St. Lawrence–Great Lakes (except Superior), Mississippi River, and Hudson Bay basins from Quebec to Alberta and south to Gulf; Gulf Slope drainages from Mobile Bay, Alabama, to Lake Pontchartrain, Louisiana (Page and Burr 1991).

Nonindigenous Occurrences: Introduced into a canal in Washington County, **Maryland** (Uhler and Lugger 1876).

Means of Introduction: Unknown. Possibly intentionally stocked.

Status: Extirpated in Maryland.

Impact of Introduction: Unknown.

Remarks: None.

Notopteridae - Featherfin knifefishes

The osteoglossiform family Notopteridae, whose members are also known as featherbacks or featherfin knifefishes, comprises eight species in four genera from tropical Africa and Southeast Asia (Roberts 1992b; Kottelat et al. 1993; Nelson 1994). These "Old World knifefishes" have a highly specialized body form and fins that include an elongate anal fin joined to the caudal fin, a reduced dorsal fin, and ventral scutes (Roberts 1992b). The body is not bent during swimming and the long anal fin provides propulsion (Berra 1981). Notopterids are able to breath atmospheric air; they are largely nocturnal and prey on small fishes and invertebrates (Berra 1981; Kottelat et al. 1993). They inhabit both fresh and brackish waters.

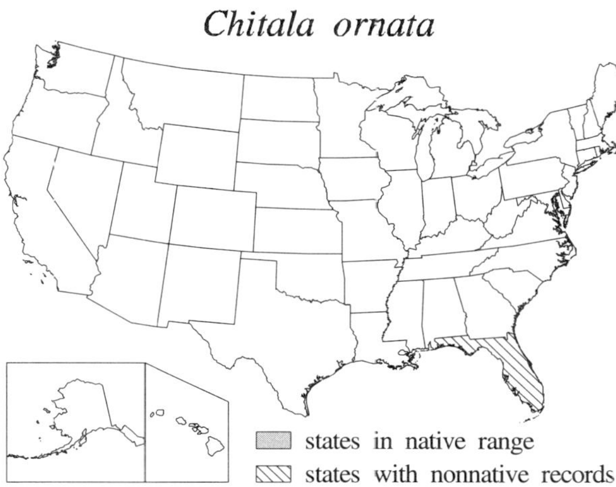

Chitala ornata (Gray 1831)

Common Names: clown knifefish (featherback).

Identification: The commonly used synonym for this species is *Notopterus chitala*. See Roberts (1992b) for recent revision, identification key, diagnostic and distinguishing characteristics, and photographs. He also discussed the different color varieties produced by artificial breeding. Maximum size: 100 cm.

Native Range: Tropical Asia. The Mekong and Chao Phraya basins of Indochina and Thailand (Roberts 1992b).

Nonindigenous Occurrences: A single fish (56 cm TL, 1.6 kg) was taken in early 1994 by an angler from a small lake south of Winter Garden in Lake County, **Florida** (Hawkins 1994; W. Icenhour, personal communication). A small population of clown knifefish was recently discovered in Lake Osborne to Lake Ida system, Palm Beach County, Florida; the species was first recorded at the site in 1994 (W. Courtenay, personal communication).

Means of Introduction: Probable aquarium release.

Status: Reported from Florida, possibly established in a lake system in southeastern part of state.

Impact of Introduction: Unknown.

Remarks: This is an important food fish in Thailand (Berra 1981) and a popular aquarium fish in the United States. Voucher specimens: The Florida Lake County record is supported by a photograph in Hawkins (1994); the collector's father related that the specimen was made into a trophy mount (W. Icenhour, personal communication).

Voucher specimens taken from Lake Osborne are apparently being held by the Florida Game and Fresh Water Fish Commission.

ORDER ELOPIFORMES

Megalopidae - Tarpons

The family Megalopidae is composed of two extant species, both in the genus *Megalops* (Nelson 1994). Some authors group tarpons with ladyfish in the family Elopidae. The two species are found along ocean coastlines in tropical and subtropical regions of the Indo-Pacific and western Atlantic, off tropical West Africa, and sometimes off southern Europe (Berra 1981; Nelson 1994). They are medium- to large-sized carnivorous fishes. Although mainly marine, tarpons frequently enter freshwater.

Megalops atlanticus

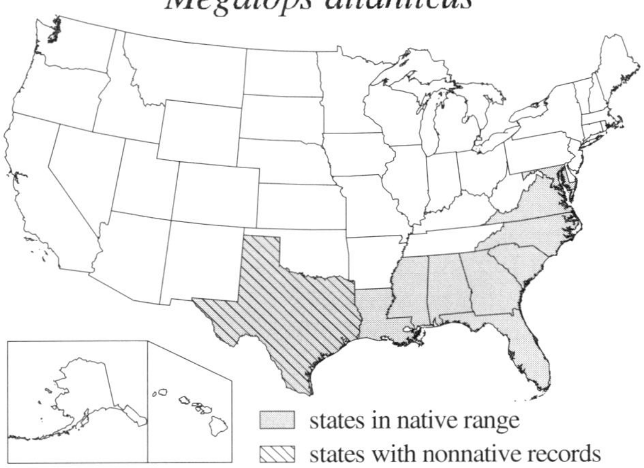

Megalops atlanticus Valenciennes 1847

Common Names: tarpon.

Identification: Hoese and Moore (1977); Manooch (1984); Robins and Ray (1986); Murdy et al. (1997). Maximum size: 2.4 m.

Native Range: Marine and estuarine. Virginia (occasionally Nova Scotia), Bermuda, and Gulf of Mexico to Brazil; also eastern Atlantic; occasional near Pacific terminus of Panama Canal (Robins and Ray 1986). Murdy et al. (1997) reported this species occasionally gets as far north as Chesapeake Bay, Maryland, as a result of seasonal movement via the Gulf Stream.

Nonindigenous Occurrences: The **Texas** Parks and Wildlife Department stocked 34 young tarpon from a drainage ditch in Calhoun County into Victor Braunig Reservoir, Bexar County, during 1984 and 1985. One fish was recovered by an angler in 1986, and another found dead in 1989. The latter fish died either from cold or from the effects of explosives being used to dissipate dying blue tilapia from a power plant intake. The tarpon had several blue tilapia in its stomach and one in its mouth (Howells and Garrett 1992).

Means of Introduction: These fish were part of a group of stranded tarpons used for cold tolerance studies. Tarpon were released into the Texas reservoir after the experiments ended (Howells and Garrett 1992).

Status: Some of the released tarpon may still be surviving in reservoirs, but it is unlikely that they are capable of spawning in the freshwater impoundment (Howells and Garrett 1992).

Impact of Introduction: Except for some predation on introduced blue tilapia *Oreochromis aureus*, effects of the introduction are unknown.

Remarks: The tarpon is a prized sport fish.

ORDER ALBULIFORMES

Albulidae - Bonefishes

The family Albulidae, also known as bonefishes, are medium-sized, silvery fishes with conical, cartilaginous snouts, and inferior mouths (Eschmeyer et al. 1983; Robins and Ray 1986). The family includes about 4 species (Eschmeyer et al. 1983; Nelson 1994), although Nelson (1994) pointed out that *A. vulpes* may actually represent several species. Bonefishes are primarily marine, but may occur in brackish or freshwater and are found in tropical seas worldwide (Nelson 1994).

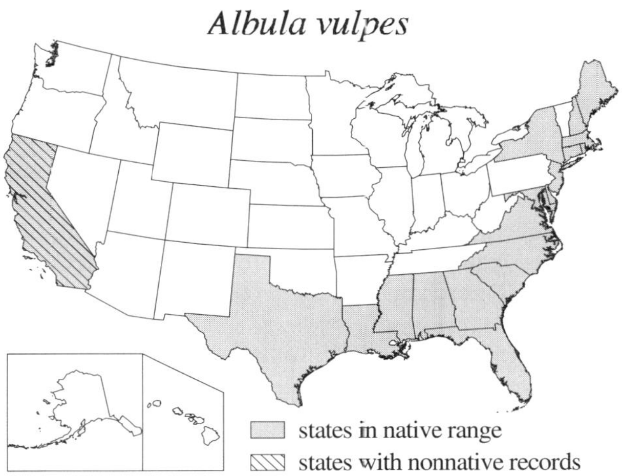

Albula vulpes (Linnaeus 1758)

Common Name: bonefish.

Identification: Miller and Lea (1972); Eschmeyer et al. (1983). Maximum size: 46 cm in eastern Pacific; 91 cm elsewhere in range (Eschmeyer et al. 1983). May actually be several different species (Nelson 1994). Maximum size: 46 cm in eastern Pacific; over 91 cm elsewhere (Eschmeyer et al. 1983).

Native Range: Marine; worldwide in tropical waters of the Atlantic and Pacific (Robins and Ray 1986). In the Pacific, San Francisco to Peru, uncommon north of Baja (Eschmeyer et al. 1983). In the western Atlantic, New Brunswick and Bermuda to southern Brazil; rare north of Florida and the Bahamas (Robins and Ray 1986).

Nonindigenous Occurrences: Introduced into Salton Sea, **California**, in 1950 and 1951 (Walker et al. 1961; Dill and Cordone 1997).

Means of Introduction: Intentionally stocked as a sport fish. This species was stocked twice; 29 fish were stocked in 1950, and 48 fish in 1951. Stock was obtained from San Felipe, Mexico (Walker et al. 1961).

Status: The introduction failed to produce a viable population and the species is extirpated in the Salton Sea.

Impact of Introduction: Unknown.

Remarks: None.

ORDER ANGUILLIFORMES

Anguillidae - Freshwater eels

This family of eels is composed of a single genus, *Anguilla*, with 15 species (Nelson 1994). The family has a wide distribution in tropical and temperate areas, including every continent except Antarctica (Berra 1981) and every sea except the eastern Pacific and the southern Atlantic Ocean (Nelson 1994). A single species, *A. rostrata*, is native to the United States (Page and Burr 1991). Diets range from omnivory to carnivory. Adult anguillids live in freshwater or in estuaries, but migrate to selected ocean areas to spawn. The specialized larvae, called leptocephali, move back to coastal areas and undergo morphological changes. They then enter freshwater as young eels, or elvers (Nelson 1994).

Anguilla anguilla

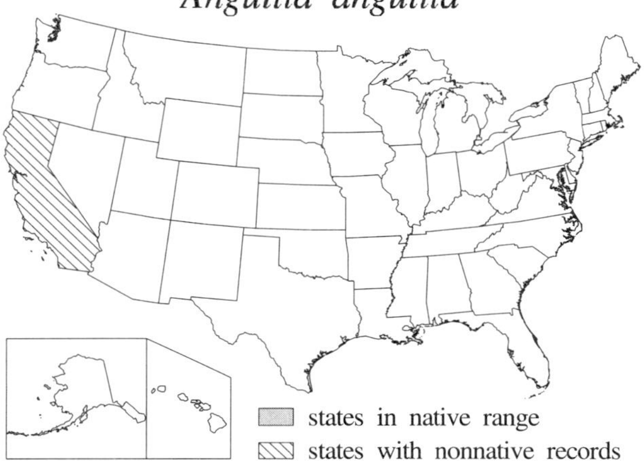

Anguilla anguilla (Linnaeus 1758)

Common Name: European eel.

Identification: McCosker (1989) presents a key to five *Anguilla* species most likely to be imported to the United States. Also see Maitland (1977) and Wheeler (1978). Maximum size: 200 cm.

Native Range: Marine, estuarine, and freshwater. Eastern Atlantic Ocean throughout Europe and North Africa (Sterba 1967).

Nonindigenous Occurrences: One specimen was collected from the Sacramento-Joaquin Delta, Contra Costa County, **California** in January 1969 (Skinner 1971). Another was collected from San Pablo Bay, in or before 1977 (McCosker 1989; Williamson and Tabeta 1991).

Means of Introduction: Skinner (1971) suggested that some California anguillids may have arrived in ballast water of foreign ships. However, McCosker (1989) discounted ballast water as a vector. In general, eels caught in California waters are believed to have been imported by Japanese or Chinese restaurants, or by fish farms, and the eels either escaped or were released (McCosker 1989; Williamson and Tabeta 1991).

Status: Reported from California.

Impact of Introduction: Unknown.

Remarks: There is still some uncertainty associated with the identification of the two California *A. anguilla* specimens. Because each had 114 vertebrae, Williamson and Tabeta (1991) indicated that these fish cannot be reliably distinguished from *A. japonica*. Anguillids are a popular food fish, particularly among Asian Americans. Introduced anguillids may prey on trout and young salmon, and also may compete with salmonids and other native fishes for food (McCosker 1989). Because eels do not reproduce in captivity, their culture requires removal of live eels from the wild. Consequently, there is a high probability for introduction of new diseases, pathogens, and parasites (McCosker 1989). Eels can also travel overland during rainy periods, demonstrating that they are not easily contained (McCosker 1989). Because of potential impacts, California has made the importation, sale, or culture of live anguillids illegal. According to Williamson and Tabeta (1991), the Pacific Ocean near California does not seem to have an area suitable for *Anguilla* spawning, and, therefore, eels introduced into the western United States probably would not reproduce successfully.

Voucher specimens: California (CAS 27136, 39012).

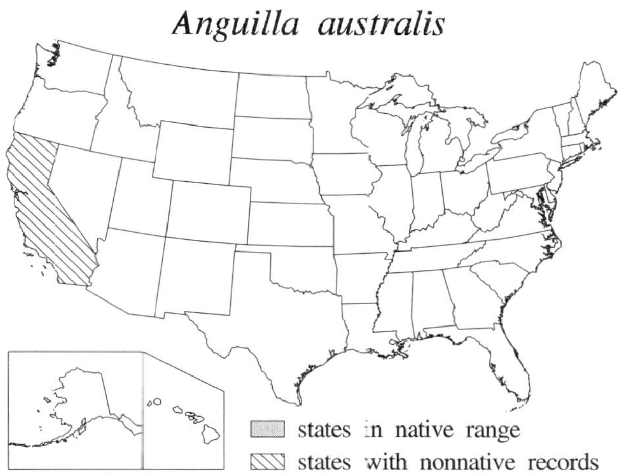

Anguilla australis Richardson 1841

Common Name: shortfin eel.

Identification: McCosker (1989) presented a key to several *Anguilla* including this species. Distinguishing characteristics and photographs appeared in Merrick and Schmida (1984). Maximum size: 110 cm and 6.8 kg.

Native Range: Marine, estuarine, and freshwater. In Australia, restricted to the Tasmanian, Southeast Coast, and southernmost Northeast Coast drainages; also extends to Lord Howe and Norfolk islands, New Zealand, and to the Chatham Islands (Merrick and Schmida 1984).

Nonindigenous Occurrence: Two specimens were collected from Puddingstone Reservoir, Los Angeles County, **California**, in June and July 1983. A seven - to eight-year-old eel was collected from Legg Lake, Los Angeles County, in 1985 (McCosker 1989; Williamson and Tabeta 1991; Swift et al. 1993).

Means of Introduction: Unknown. Possible escapes or releases from fish farms or markets (McCosker 1989). Many live eels were imported during the period when these fish were discovered in open waters (Swift et al. 1993).

Status: Reported from California.

Impact of Introduction: Unknown.

Remarks: *Anguilla australis* greater than 50 cm in length are highly piscivorous and may significantly prey on native fishes (Swift et al. 1993). Also see remarks for *A. anguilla*.

Voucher specimens: LACM 44140, 43527-1, LACM unnumbered.

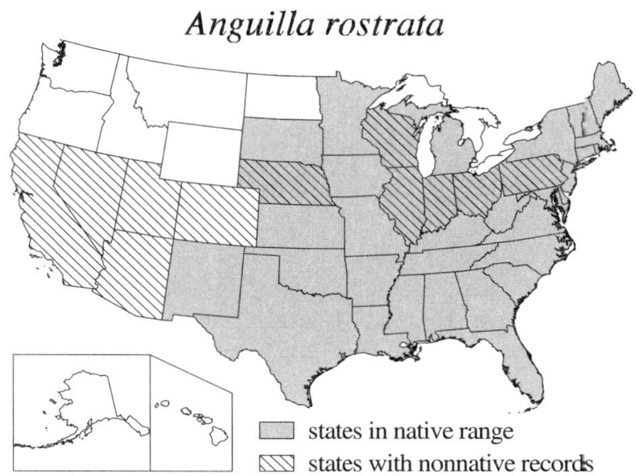

Anguilla rostrata (Lesueur 1817)

Common Name: American eel.

Identification: Becker (1983); Page and Burr (1991); Etnier and Starnes (1993); McCosker (1989) presented a key to several *Anguilla*, including this species. Maximum size: 152 cm.

Native Range: Marine, estuarine, and freshwater. Spawns in the Atlantic Ocean in the Sargasso Sea and ascends streams in North and South America. Atlantic, Great Lakes, Mississippi, and Gulf basins from Newfoundland to South Dakota and south to South America (Page and Burr 1991).

Nonindigenous Occurrences: This species has been recorded from Lake Mead on the Colorado River, on the **Arizona** border (Minckley 1973). It was stocked on several occasions in Sacramento and San Francisco bays, **California**, in the late 1800s, apparently with no evidence of survival (Smith 1896; Shebley 1917; Shapovalov et al. 1981; McCosker 1989). A total of 13 specimens were collected in the San Francisco area during the period 1978 to 1984 (McCosker 1989; Williamson and Tabeta 1991). Specimens have been reported from the San Luis Valley, Conejos County, **Colorado**, where they escaped from an aquaculture facility (Zuckerman and Behnke 1986). Eels were stocked in the Calumet River south of Chicago, **Illinois**, in 1873 (Milner 1874b; Goode 1884). It was also reported to be established in several localities in **Indiana** in 1945 (Gerking

1945). Eels have been collected in Lake Michigan in Illinois and Indiana (Lee et al. 1980 et seq.). The species was accidentally introduced into the Elkhorn River near Omaha, **Nebraska**, in 1873, when a railroad car transporting East Coast fish to the West Coast lost its cargo to the river after a railroad bridge collapsed; all told, an estimated fifteen hundred eels from Martha's Vineyard and 40,000 from the Hudson River were released in the Elkhorn (Smith 1896). There are no subsequent records. Eels have been collected from Lake Mead on the Colorado River, **Nevada** (Minckley 1973). In the late 1800s, **Ohio** received transplants of hundreds of thousands of eels (Trautman 1981). Introduced into Lake Erie in Ohio and **Pennsylvania** (Scott and Crossman 1973; Busch et al. 1977; Trautman 1980). Eels were stocked in Utah Lake and Salt Lake in **Utah** in the late 1800s, but soon disappeared (Popov and Low 1953; Sigler and Miller 1963). They were also stocked in the Fox River at Appleton, **Wisconsin**, in 1873 (Milner 1874b; Goode 1884).

Means of Introduction: Intentionally stocked in most locations; accidentally stocked in the Elkhorn River, Nebraska; escaped from an aquaculture facility in Colorado. Since 1874, there have been many shipments of *A. rostrata* from the eastern United States to California to stock rivers and for fish farming purposes (Williamson and Tabeta 1991). Eels caught in California waters during the 1970s and 1980s are believed to have been imported by Japanese or Chinese restaurants or by fish farms, and the eels either escaped or were released (McCosker 1989; Williamson and Tabeta 1991). In Texas and South Carolina eels were intentionally introduced for aquaculture and then escaped the aquaculture facility. Goode (1884) stated that captains of vessels on the Great Lakes commonly carried a tub of eels for food while on Lake Ontario. These eels were frequently thrown overboard when the crew tired of them or procured better fishes as the vessels sailed through the Great Lakes. Several thousand eels were stocked into rivers in Illinois draining into Lake Michigan and into a river near Eaton, Michigan (Goode 1884). They are also thought to have gained access to the upper Great Lakes through the Welland Canal (Scott and Crossman 1973). The first record from above Lake Ontario was from Lake Erie in 1844; the canal opened in 1829 (Trautman 1980).

Status: Possibly established in Indiana and Ohio; reported from Arizona, California, Colorado, and Nevada; extirpated from Nebraska and Utah. Present but uncommon in the upper Great Lakes (Becker 1983). Scott and Crossman (1973) listed them as uncommon in the lakes Erie, Huron, and Superior, with no mention of Lake Michigan. These individuals are ones that have migrated to the upper Great Lakes through the Welland Canal.

Impact of Introduction: Eels in Texas and South Carolina have been found to be infected with the Asian eel nematode *Anguicolla crassus* (M. Ray, personal communication). This parasite infests the swimbladder, causing retarded growth and maturation, and is fatal to a high percentage of eels. It has also been found to infect fin fish, with unknown consequences (R. Ward, personal communication). The nematode was originally known only in Japanese eels *Anguilla japonica*. As a result of transporting eels around the continent, it spread into European eel *A. anguilla*. It has now been found in our native American eel (Ray, personal communication). Undoubtedly it arrived in shipments of foreign eels and spread to the native eel. Impacts in other states are unknown.

Remarks: An adult eel was captured from the Sacramento–San Joaquin Delta, near Stockton, San Joaquin County, California, in 1964, but there is confusion surrounding its identify. The voucher specimen (SIO 64-219) is a skin only and was reported to be *A. rostrata* by Skinner (1971); however, the same specimen was reported as being either *A. rostrata* or *A. anguilla* by McCosker (1989), and as *Anguilla* sp. by Williamson and Tabeta (1991). Also see remarks under *A. anguilla*. Although not outside their native range, eels tentatively identified as this species have been used extensively for aquaculture in several areas, including Texas and South Carolina. The exact origin, and therefore positive identity of eels cultured in Texas has been somewhat uncertain. The Texas fish are believed to be native eels from the southeast coast, but there is a possibility that the fish actually may represent one of the foreign species, *A. japonica* or *A. anguilla*. Many of these eels escaped a southern Texas aquaculture facility into nearby Arroyo Colorado. In addition, before it was realized the eels from the Texas facility were infected with a parasite, *Anguillicola crassus*, live specimens were shipped to South Carolina (Ray, personal communication). Possibly escapees from South Carolina facilities, infected *A. rostrata* were captured later off the coast of South Carolina (Ray, personal communication).

At least three American eels have been collected from Lake Superior on the Canadian side . These were likely ballast water introductions (Mandrak and Crossman 1992).

Voucher specimens: California (CAS 41042-45, 62246, 44056, 53928, 62206), Colorado (MSB, uncataloged (LDZ-50, 51, 86)), Texas (Texas Parks and Wildlife, Palacios).

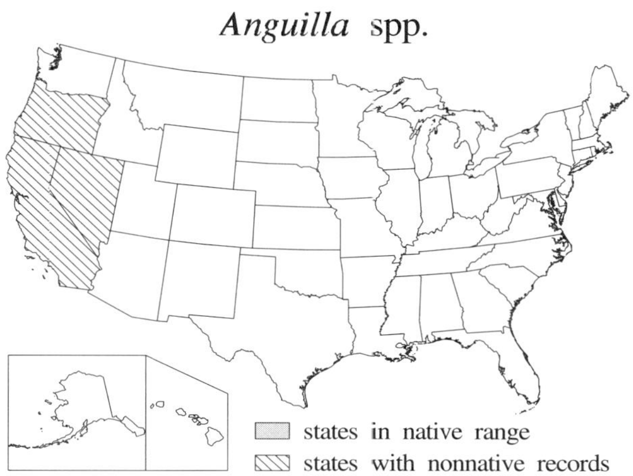

Anguilla spp.

Common Name: anguillid eels.

Identification: Genus is currently recognized as containing 15 species. McCosker (1989) provided a key to the five *Anguilla* species most likely to have been imported. Maximum size: To over 200 cm.

Native Range: Marine, estuarine, and freshwater. Most temperate and tropical regions of the world (Berra 1981; Nelson 1994).

Nonindigenous Occurrences: Williamson and Tabeta (1991) listed an eel taken from the San Joaquin River, **California**, in 1964 as *Anguilla* sp. Several unidentified eels have been caught from Lake Mead, Clark County, **Nevada** (Minckley 1973; Deacon and Williams 1984). In their review of West Coast anguillid records, Williamson and Tabeta (1991) listed three unidentified *Anguilla* specimens taken from the Portland, **Oregon** area, during the period, 1981 to 1983. These records included two specimens from the Willamette River and one from the Lake River in the Willamette River drainage.

Means of Introduction: These eels were probably imported by Japanese or Chinese restaurants, or by fish farms, and the eels either escaped or were released (Williamson and Tabeta 1991).

Status: Reported from California, Nevada, and Oregon.

Impact of Introduction: Unknown.

Remarks: There is some confusion concerning the correct identity of the eel taken from the San Joaquin River, California. The voucher specimen (SIO 64-219) is a skin only; it was reported as *A. rostrata* by Skinner (1971) and as being either *A. rostrata* or *A. anguilla* by McCosker (1989) (see species account for *A. rostrata*). One of the three Oregon specimens was reported as being kept alive in an aquarium (as of 1990); the other two specimens were not preserved (Williamson and Tabeta 1991).

ORDER CLUPEIFORMES

Engraulidae - Anchovies

The family Engraulidae, also known as anchovies, includes 16 genera with 139 species. Anchovies are found in the Atlantic, Indian, and Pacific oceans, and occasionally in freshwater (Nelson 1994). Anchovies are small-sized fish (most less than 20 cm) that are generally planktivorous, although a few of the large species are piscivorous (Nelson 1994). Whitehead et al. (1988) provided a synopsis of the family.

Anchoa mundeoloides

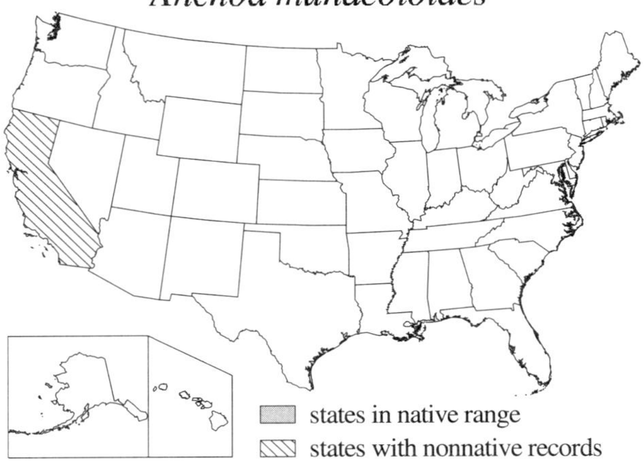

Anchoa mundeoloides (Breder 1928)

Common Name: (Northern Gulf anchovy, anchovy).

Identification: Whitehead et al. (1988); Fischer et al. (1995a). Maximum size: 6.4 cm.

Native Range: Marine; Northern Gulf of California (Whitehead et al. 1988; Fischer et al. 1995a).

Nonindigenous Occurrences: Introduced into Salton Sea, **California**, in 1948 and 1951 (Walker et al. 1961; Dill and Cordone 1997).

Means of Introduction: Intentionally stocked as forage. Forty-three fish were stocked in 1948 and six more were stocked in 1951. Stock was obtained from Guaymas, Mexico, for the first stocking, and San Felipe, Mexico for the second (Walker et al. 1961).

Status: The introduction failed to produce a viable population and the species is extirpated in the Salton Sea.

Impact of Introduction: Unknown.

Remarks: None.

Cetengraulis mysticetus

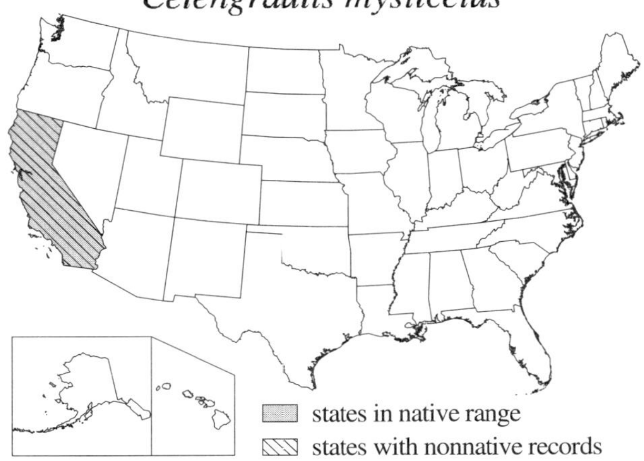

Cetengraulis mysticetus (Günther 1867)

Common Name: anchoveta (Pacific anchoveta).

Identification: Miller and Lea (1972); Eschmeyer et al. (1983); Whitehead et al. (1988); Fischer et al. (1995a). Maximum size: 18 cm.

Native Range: Marine; Southern California to Peru; rare north of southern Baja (Eschmeyer et al. 1983; Whitehead et al. 1988; Fischer et al. 1995a).

Nonindigenous Occurrences: Introduced into Salton Sea, **California**, six times between 1948 and 1955 (Walker et al. 1961; Dill and Cordone 1997).

Means of Introduction: Several thousand fish were intentionally stocked as forage.

Status: The introduction failed to produce a viable population and the species is extirpated in the Salton Sea.

Impact of Introduction: Unknown.

Remarks: This species' occurrence off California may be due to introductions by tuna bait boats (Miller and Lea (1972).

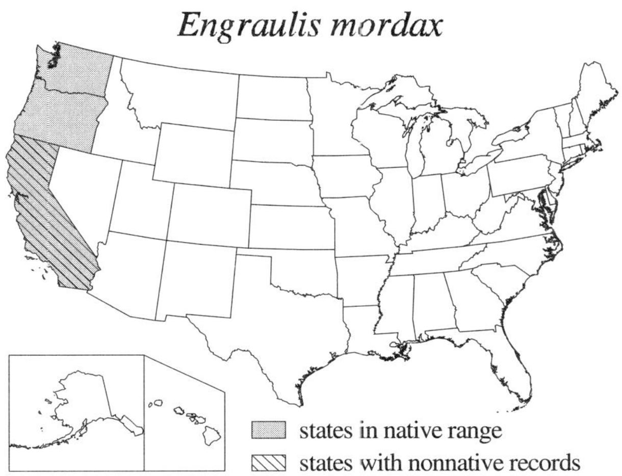

Engraulis mordax Girard 1854

Common Name: northern anchovy (California anchovy).

Identification: Miller and Lea (1972); Eschmeyer et al. (1983); Whitehead et al. (1988); Fischer et al. (1995a). Maximum size: 18 cm.

Native Range: Marine; Queen Charlotte Island, British Columbia, to tip of Baja and Gulf of California (Ramirez-Hernández and Gonzales Pages 1976; Eschmeyer et al. 1983; Fischer et al. 1995a).

Nonindigenous Occurrences: Introduced into Salton Sea, **California**, in 1953 (Walker et al. 1961; Dill and Cordone 1997).

Means of Introduction: Six thousand fish were intentionally stocked as forage. Stock was obtained from Los Angeles Harbor (Walker et al. 1961).

Status: The introduction failed to produce a viable population and the species is extirpated in the Salton Sea.

Impact of Introduction: Unknown.

Remarks: None.

Clupeidae - Herrings

This large family is composed of about 56 genera and 181 species and includes the shads, sprats, sardines, pilchards, and menhadens (Nelson 1994). Although mostly tropical, clupeids are distributed worldwide (Berra 1981; Nelson 1994). Eight species are native to the United States (Page and Burr 1991). Clupeids are generally small fish (usually less than 25 cm), have compressed bodies with a keel, and feed on plankton. Although most species are marine, some are anadromous, and about 50 species are freshwater forms that occasionally enter brackish waters (Nelson 1994). Whitehead (1985) provided a synopsis of the family.

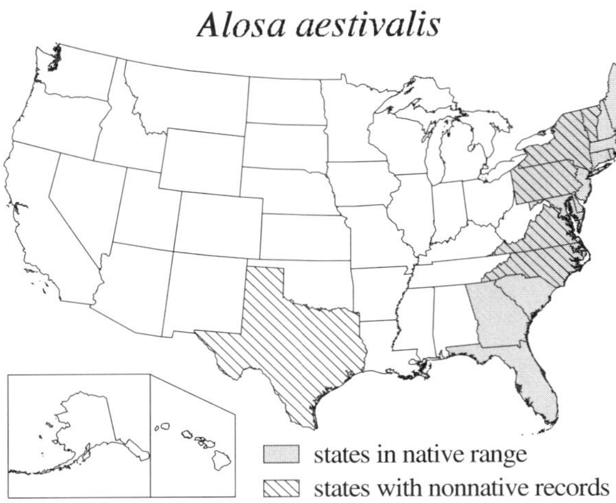

Alosa aestivalis (Mitchill 1814)

Common Name: blueback herring.

Identification: Smith (1985); Whitehead (1985); Page and Burr (1991); Jenkins and Burkhead (1994). Maximum size: 40 cm.

Native Range: Atlantic Coast from Cape Breton, Nova Scotia, to the St. Johns River, Florida. Ascends coastal rivers during spawning season (Page and Burr 1991).

Nonindigenous Occurrences: Blueback herring have been collected from Oneida Lake, the Oswego River in Minetto, Lake Champlain, and Lake Ontario, **New York** (D. MacNeill and R. Owens, personal communication). In **North Carolina**, blueback herring were introduced into the Savannah, Broad, and Yadkin drainages, and into nonnative areas of the Cape Fear and Roanoke drainages (Menhinick 1991). Stock obtained from the Cooper River, South Carolina was released in **Texas** by the Texas Parks and Wildlife Department in Lake Theo, Briscoe County, and at an unidentified research site in north Texas in 1982 (Guest 1983; Howells 1992a). Blueback herring have been collected from Lake Champlain, **Vermont** (S. Good, personal communication). Blueback herring have been stocked in several inland reservoirs in **Virginia**, including Smith Mountain Lake, Occoquan Reservoir, Kerr Reservoir, and lakes Anna, Brittle, and Chesdin (Jenkins and Burkhead 1994).

Means of Introduction: In areas other than New York, these fish were intentionally stocked for forage. In New York these fish are expanding their range using canals. Blueback herrings were first recorded in the Mohawk River in 1978 by NYDEC personnel. They were reported from Lake Champlain on the New York side in the late 1970s, and from the Vermont side in 1997. Juveniles were apparently present in Oneida Lake by 1981 or 1982. Adults were first documented in 1994 by Cornell researchers based at Shackleton Point. Several thousand immature fish were also documented in 1994 at a power plant in Minetto on the Oswego River. Two immature fish were caught in Lake Ontario near Oswego in October 1995 by U.S. Geological Survey personnel conducting fish surveys (Owens, personal communication).

Status: Established in Texas, New York, North Carolina, Vermont, and Virginia.

Impact of Introduction: Unknown.

Remarks: None.

Alosa chrysochloris

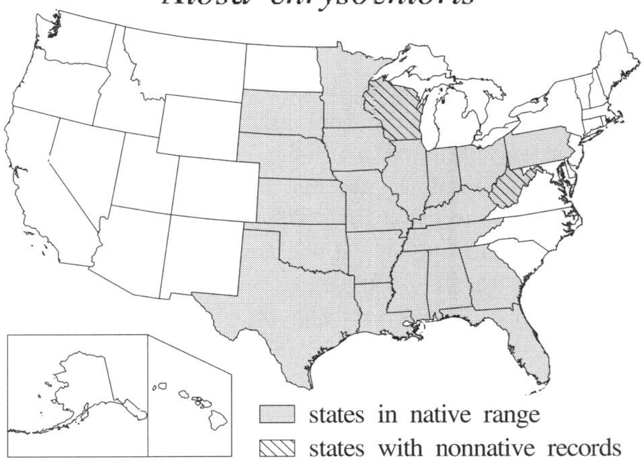

Alosa chrysochloris (Rafinesque 1820)

Common Name: skipjack herring.

Identification: Becker (1983); Whitehead (1985); Page and Burr (1991); Etnier and Starnes (1993). Maximum size: 53 cm.

Native Range: Red River drainage (Hudson Bay basin) and Mississippi River basin from central Minnesota south to the Gulf of Mexico, and from southwestern Pennsylvania west to eastern South Dakota, Nebraska, Kansas, Oklahoma, and Texas; Gulf Slope drainages from the Apalachicola River, Florida, to the Colorado River, Texas (Page and Burr 1991).

Nonindigenous Occurrences: Skipjack herring were collected above the falls of the Kanawha drainage in **West Virginia** prior to 1993 (Stauffer et al. 1995). Skipjack herring were recently collected in Lake Michigan in **Wisconsin** (Fago 1993). The first collection was a single fish taken in Green Bay north of Dyckeysville, Kewaunee County on 2 August 1989. A second fish was caught in Lake Michigan just east of Kenosha, Kenosha County in January of 1991. A third was caught east of Bailey's Harbor near the outlet of Moonlight Bay in Door County (Fago 1993).

Means of Introduction: The skipjack herring was reported to have entered lake Michigan and Lake Erie via canals (Goode 1884). The herring likely gained access to Lake Michigan via the Chicago Shipping Canal (Fago 1993). Unknown in West Virginia.

Status: Reported from Wisconsin and West Virginia.
Impact of Introduction: Unknown.

Remarks: The report of skipjack herring from Lake Erie was rejected by Trautman (1981).

Voucher specimens: Wisconsin (UWM 9962, Wisconsin Department of Natural Resources, Monona, catalogue numbers 18070, 18071).

Alosa pseudoharengus

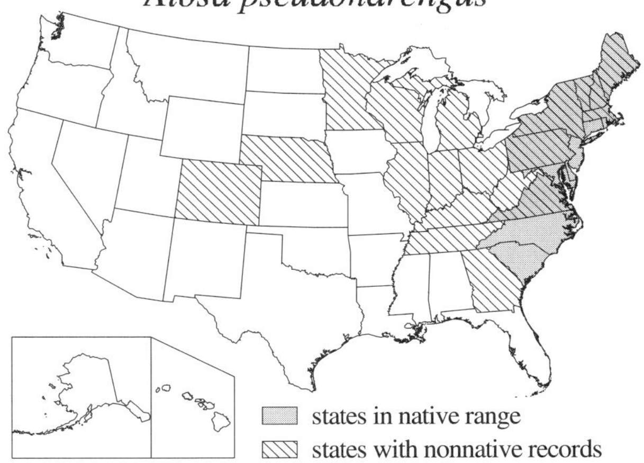

Alosa pseudoharengus (Wilson 1811)

Common Name: alewife.

Identification: Whitehead (1985); Page and Burr (1991); Etnier and Starnes (1993); Jenkins and Burkhead (1994). Maximum size: To 38 cm, but inland populations usually less than 25 cm.

Native Range: Atlantic Coast from Red Bay, Labrador, to South Carolina; many landlocked populations (Page and Burr 1991).

Nonindigenous Occurrences: Alewife were introduced into **Colorado** (Minckley 1973); **Georgia** (Dahlberg and Scott 1971a); Lake Michigan, **Illinois** (Miller 1957; Smith 1979; Phillips et al. 1982; Emery 1985); Lake Michigan, **Indiana** (Miller 1957; Phillips et al. 1982; Emery 1985); **Kentucky** (Burr and Page 1986; Burr and Warren 1986); **Maine** (Smith 1985); **Massachusetts** (Hartel 1992); the Great Lakes, **Michigan** (Miller 1957; Eddy and Underhill 1974; Phillips et al. 1982; Emery 1985; Smith 1985); Lake Superior, **Minnesota** (Miller 1957; Eddy and Underhill 1974; Phillips et al. 1982; Emery 1985); **Nebraska** (Morris et al. 1974; Bouc 1987); **New Hampshire** (Smith 1985); Lake Erie (Miller 1957; Eddy and Underhill 1974; Emery 1985; Smith 1985), and possibly Lake Ontario (Smith 1970; Smith

1985), **New York**; Lake Erie, **Ohio** (Miller 1957; Emery 1985); Lake Erie (Miller 1957; Eddy and Underhill 1974; Emery 1985), Youghiogheny River (Hendricks et al. 1979), and Colyer Lake in Centre County (Denoncourt et al. 1975a), **Pennsylvania**; Dale Hollow and Watauga Reservoirs, **Tennessee** (Etnier and Starnes 1993; W. Pollock, personal communication); Lake St. Catherine, **Vermont** (Hauser 1998); several reservoirs in **Virginia** (Hocutt et al. 1986; Jenkins and Burkhead 1994); Bluestone Reservoir, New drainage, **West Virginia** (Hocutt et al. 1986; Stauffer et al. 1995; Jenkins and Burkhead 1994); and lakes Michigan and Superior, Kangaroo Lake, Pigeon River, Pigeon Lake, East Twin River, Sheboygan River, Sauk Creek, and Milwaukee River, **Wisconsin** (Miller 1957; Phillips et al. 1982; Becker 1983; Emery 1985).

Means of Introduction: There is apparently disagreement concerning the native status of alewife in Lake Ontario. Miller (1957) and Smith (1970) point out the first record from Lake Ontario was in 1873. Smith (1970) is of the opinion that it was introduced into the lake. Although Smith (1970) brings up the possibility that alewife were introduced into Lake Ontario with American shad stockings in the 1880s, he discounts this possibility in favor of the hypothesis that they reached the lake via the Erie Canal from the Hudson River. He contends that alewife were only able to invade the lake after the decline of predators such as lake trout and Atlantic salmon in the 1860s. Other authors believe, this species was probably native to Lake Ontario (Lee et al. 1980 et seq.) and spread through the Great Lakes via the Welland Canal (Lee et al. 1980 et seq.). The species was first reported from Lake Erie in 1931, Lake Huron in 1933, Lake Michigan in 1949, and Lake Superior in 1954. The alewife was intentionally stocked in inland waters. The population in the New River, West Virginia, resulted from stockings in Claytor Lake, New River, Virginia (Jenkins and Burkhead 1994). The recently discovered population in Lake St. Catherine, Vermont, is likely a result of an illegal stocking (Good, personal communication).

Status: Established in many states. Introduction to the Youghioghenay River was unsuccessful (Hendricks et al. 1979).

Impact of Introduction: Disappearance of native planktivorous salmonids, such as whitefish, in the Great Lakes has been attributed in part to the introduction of alewives, which reduced zooplankton populations (Crowder and Binkowski 1983; Todd 1986b; Page and Laird 1993). Crowder (1984) speculated that a cisco native to Lake Michigan, the bloater *Coregonus hoyi*, evolved fewer and shorter gill rakers, and shifted to benthic habitat and diet as a result of competition with alewifes. Smith (1970) attributed the extermination of the lake herring and decline of chub species in the Great Lakes to the alewife. Smith also talks about the various interrelated changes that took place in each of the Great Lakes as alewife abundance increased. Christie (1972), on the other hand, argues that the alewife was not responsible for these changes. The alewife is the dominant fish in Lake Michigan. It accounts for 70–90% of the fish weight (Becker 1983). Alewives periodically undergo large-scale die-offs and litter the beaches of the Great Lakes with rotting fish, posing both a nuisance and a health hazard (Becker 1983). Pacific salmonids have been introduced, in part, to control the alewife populations.

Remarks: Although there is a report of two small alewives taken from the Colorado River, Texas (Bean 1882), we believe this record is in error. Bean (1882) reported that the specimens were sent to Professor Baird at the National Museum. However, a query of the museum's holdings did not return these specimens. We believe the fish are more likely either misidentified *A. chrysochloris* or *A. sapidissima*. *Alosa sapidissima* were stocked in the Colorado River in 1874 (Bean 1882).

Voucher specimens: Michigan (UMMZ 157215, 160969, 167872, 171308, 170945), Wisconsin (UMMZ 162861, 167945).

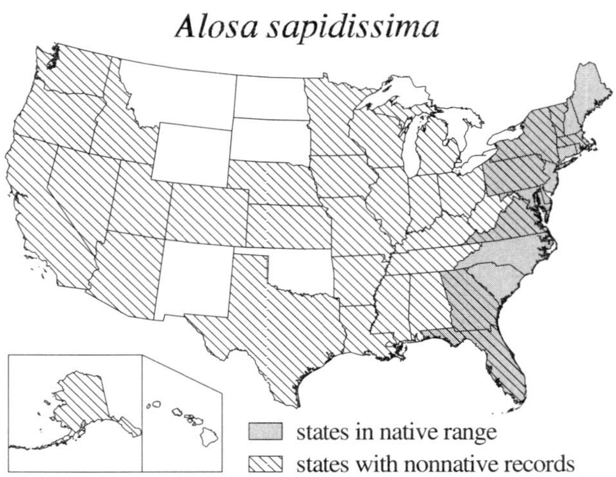

Alosa sapidissima (Wilson 1811)

Common Name: American shad.

Identification: Smith (1985); Whitehead (1985); Page and Burr (1991); Jenkins and Burkhead (1994). Maximum size: 75 cm.

Native Range: Atlantic Coast from the Sand Hill River, Labrador, to the St. Johns River, Florida; ascends coastal rivers to spawn (Page and Burr 1991).

Nonindigenous Occurrences: American shad were first transported to the west coast when the U.S. Fish Commission relocated the species from the Hudson River in New York to the Sacramento River in California in 1871. By 1891, the species had expanded its range along the coast as far north as **Alaska** (Smith 1896; Skinner 1962; Moyle 1976a; Wydoski and Whitney 1979; Lee et al. 1980 et seq.). Shad were stocked in the Alabama and Cahaba rivers, **Alabama** (Baird 1878; Bowers 1904). During the late 1800s, American shad were stocked extensively throughout the western states. Close to three million fish were planted in the Colorado River in **Arizona** between 1884 and 1886 (Smith 1896), but failed to establish a reproducing population. They were also stocked in the White River, **Arkansas**, in the 1870s (Baird 1878). American shad now occur in **California** as a result of dispersal from initial stockings in the late 1800s (Baird 1874a, 1878; Shebley 1917; Moyle 1976a; Dill and Cordone 1997). They were first stocked in **Colorado** in 1872 in the Platte River near Denver (Baird 1878; Wiltzius 1985). In the 1870s, shad were stocked in several drainages in **Florida** including the Aucilla, Chattahoochee, New River in Ft. Lauderdale, Ochlockonee, Suwannee, and St. Lucie (Bowers 1901, 1904). During the 1870s and 1880s, they were also placed in the Flint and Chattahoochee rivers in **Georgia** (Smiley 1885a; Bowers 1901, 1905). They were stocked in the Bear River, **Idaho**, several times between 1871 and 1892 with no success (Smith 1896; Popov and Low 1953; Linder 1963; Sigler and Miller 1963; Simpson and Wallace 1978). The Snake River in Idaho now has an established population that is extending its range upstream to Lewiston as impoundments are completed (Simpson and Wallace 1978; Idaho Fish and Game 1990). No date was found to indicate when American shad were introduced to this river. Two hundred shad from the Hudson River were stocked in Lake Michigan at Chicago, **Illinois**, in 1871 (a joint venture between the New York and California Fish Commissions) (Smith 1896). Many thousands more were stocked in the following few years (Milner 1874a; Baird 1878). Shad were stocked in the Wabash River, **Indiana** (Milner 1874a). Shad were found in the Ohio River between **Indiana** and **Kentucky** in the late 1800s (Baird 1878; Clay 1975). American shad were stocked in **Kansas** in the 1880s but did not survive (Miller and Robison 1973). Hartel (1992) reported that shad had been transplanted to nonnative areas of **Massachusetts**. Shad were stocked in **Michigan** tributaries to lakes Michigan, Erie, and Huron in the 1870s (Milner 1874a; Baird 1878). The Pearl and Yazoo rivers in **Mississippi** were stocked in the mid 1870s (Baird 1878). They were stocked in the Missouri, Black, Chariton, Meramec, and Kaw rivers in **Missouri** in 1872–1884 (Baird 1878; Pflieger 1971). American shad were stocked in 1872 into the Mississippi River at St. Paul, **Minnesota** (Baird 1878; Becker 1983). This species was part of an 1873 shipment of East Coast fishes sent to the West Coast for stocking. The shipment was lost in the Elkhorn River near Omaha, **Nebraska**, as the result of a railroad bridge collapse (Smith 1896; Morris et al. 1974). Shad were stocked unsuccessfully in the Colorado River, **Nevada**, in 1884–1886 and in 1946 (Miller and Alcorn 1946; Courtenay 1970; Deacon and Williams 1984). Christie (1972) cites Dymond et al. (1929) in stating the shad were introduced into Lake Ontario, **New York**, in the 1800s. Smith (1970) reports that 80,000 fry were stocked in Lake Ontario from 1870–1872. Smith (1985) only says that early records from Lake Ontario may represent stocked fish. The last shad was collected in 1931 (Smith 1985). They were stocked in the Allegheny River, New York, in 1872 (Baird 1878). In 1871, 200 were stocked in Lake Erie, **Ohio** (by the New York and California Fish Commissions). The population did not persist (Smith 1896). Many thousands of shad were stocked in Lake Erie and Ohio River tributaries in the 1870s (Milner 1874a; Baird 1878; Jordan 1882). Between 1870 and 1900, hundreds of thousands of American shad were stocked in Ohio with no success (Jordan 1882; Trautman 1981). Shad from Maryland were planted in the Willamette River, **Oregon**, in 1886. The Columbia River also was stocked several times in the late 1800s (Smith 1896, Linder 1963; Wydoski and Whitney 1979). Shad also were stocked, unsuccessfully, in the Youghiogheny River, **Pennsylvania** (Baird 1878; Hendricks et al. 1979). The Cumberland, Holston, Eastanalbee, Tennessee, and Forked Deer rivers in **Tennessee** were stocked in the 1870s (Baird 1878). **Texas** was stocked repeatedly in the late 1800s without success (Baird 1878; Bean 1882; Baughman 1950; Howells 1992a). The New York and California Fish Commissions formed an alliance through which stock was obtained from the Hudson River and planted in the Bear River, Jordan River, Weber River, Bear Lake and Utah Lake, **Utah**, several times between 1871 and 1892. This effort was unsuccessful (Baird 1878; Smith 1896; Popov and Low 1953; Linder 1963; Sigler and Miller 1963; Simpson and Wallace 1978). Stocked in the Lake Champlain system, **Vermont**, in the late 1800s (Milner 1874a; Baird 1878). Shad have been stocked outside of their native range in **Virginia** in the New (Ohio basin) drainage (Jenkins and Burkhead 1994). American shad are found in **Washington** as a result of spread from the introductions into California and stocking in the Columbia River (Smith 1896; Linder 1963; Moyle 1976a; Wydoski and Whitney 1979). Shad were stocked into the Greenbrier River in **West Virginia** in the late 1800s, but now are extirpated (Baird 1878; Clay 1975; Stauffer et al. 1995). Shad were introduced twice in the late 1800s into **Wisconsin**. They were introduced in 1872 into the Mississippi River at St. Paul, Minnesota, and again in 1873, into the Fox River at Appleton (Milner 1874a; Baird 1878; Becker 1983).

Means of Introduction: This species was stocked intentionally in California starting in 1871 (Dill and Cordone 1997), and then spread to Oregon, Washington, and Alaska. It was intentionally stocked in many other areas as well, for forage, food, sport, and commercial fishing. It was accidentally stocked in Nebraska (Smith 1896).

Status: Established in coastal states, including; Alaska, California, Idaho (Snake River), Oregon, Virginia, and Washington. Stocked but extirpated in Arizona, Arkansas, Idaho (Bear Lake and Bear River), Iowa, Kansas, Illinois, Louisiana, Massachusetts, Minnesota, Mississippi, Nebraska, Nevada, Ohio, Pennsylvania, Tennessee, Texas, Utah, West Virginia, and Wisconsin. Large spawning runs observed in Oregon and Washington as early as 1876 (Smith 1896; Moyle 1976a; Wydoski and Whitney 1979).

Impact of Introduction: Unknown.

Remarks: Intensive stocking took place in the 1800s in eastern coastal states where populations had declined from overharvesting (Ferguson 1876; Baird 1878; Morse 1905). One such example is the Delaware River, Pennsylvania, which was stocked to help the overfished population recover (Morse 1905). Dill and Cordone (1997) provided a detailed account of shad introductions in California.

Dorosoma cepedianum

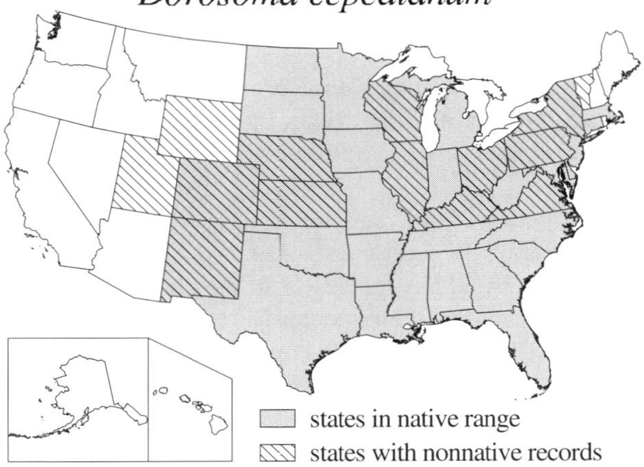

Dorosoma cepedianum (Lesueur 1818)

Common Names: gizzard shad (hickory shad).

Identification: Becker (1983); Whitehead (1985); Page and Burr (1991); Etnier and Starnes (1993). Maximum size: 52 cm.

Native Range: St. Lawrence–Great Lakes(?), Mississippi, Atlantic, and Gulf Slope drainages from Quebec to central North Dakota and New Mexico, and south to central Florida and Mexico (Page and Burr 1991).

Nonindigenous Occurrences: Gizzard shad were introduced into **Colorado** in the Arkansas, South Platte, and Republican drainages (Beckman 1952; Woodling 1985). The species is native to **Illinois** but has increased its range and abundance within the state due to construction of reservoirs (Smith 1979). In **Kansas**, gizzard shad have been stocked in Fall River Reservoir, in sand pits along Prairie Dog Creek, and in Kirwin, Webster, and Cedar Bluff reservoirs (Cross 1967). They were stocked in Martin's Fork Lake and in other small impoundments above Cumberland Falls in **Kentucky** (Minckley and Krumholz 1960; Burr and Warren 1986), and in one or more unspecified areas (presumably the Platte River) of **Nebraska** (Bouc 1987). They have become established in the upper Rio Grande River in **New Mexico** (J. Wilbur, personal communication). It is unclear whether they are native to the Great Lakes or gained access through canals and rivers in **New York** such as the Mohawk River, Oswego River, and Barge Canal (Smith 1985). Miller (1957) reviewed the evidence for native versus introduced status. Gizzard shad gained access to Lake Erie, **Ohio**, via the Ohio Canal (Jordan 1882). Gizzard shad were accidentally stocked in Conowingo

Pond, Lancaster County, **Pennsylvania** (Denoncourt et al. 1975a). They also were stocked in **Utah** (B. Schmidt, personal communication). They have expanded into Lake Champlain and the Connecticut River, **Vermont** (K. Cox, personal communication). They were stocked in Lake Anna, Smith Mountain Lake, and Kerr, Philpot, and Leesville reservoirs, **Virginia** (Jenkins and Burkhead 1994). Gizzard shad reached Lake Michigan, **Wisconsin**, either through the Chicago River Canal or the Fox-Wisconsin Canal at Portage (Becker 1983). The species has expanded its range into **Wyoming** from introductions into Nebraska (Baxter and Simon 1970), and from intentional stockings in the state east of the Continental Divide (Hubert 1994).

Means of Introduction: These fish were stocked intentionally for forage. The Wyoming populations also spread from introductions into Nebraska (Baxter and Simon 1970). In Pennsylvania, gizzard shad were stocked accidentally with American shad (Denoncourt et al. 1975a). In Vermont, they have expanded their range through the Connecticut River assisted by fishways that were constructed for American shad *Alosa sapidissima* and Atlantic salmon *Salmo salar* restoration (Cox, personal communication). They likely gained access to Lake Champlain through the Hudson Barge Canal that links the lake to the Hudson River (Cox, personal communication). They gained access to Lake Michigan through either the Chicago River Canal or the Fox-Wisconsin Canal (Becker 1983), and to Lake Erie through the Ohio Canal (Jordan 1882).

Status: Established in many states.

Impact of Introduction: Competition for food between gizzard shad and other fish species may occur (Burns 1966b; Moyle 1976a). Jenkins (1954) found that gizzard shad directly compete with centrarchids resulting in decreased growth and size of the centrarchids. Gizzard shad show tremendous invasion potential. After only two plantings totally 1020 fish in Lake Havasu, the species spread through the Colorado River from Davis Dam southward to the Mexican border, the Salton Sea, and associated irrigation ditches within only 18 months (Burns 1966b).

Remarks: The gizzard shad has expanded its range naturally since the 1600s to include **Massachusetts** (O'Leary and Smith 1987; Hartel 1992). Jordan (1882) stated that the gizzard shad was not found in Lake Erie prior to completion of the Ohio Canal. Cold weather limits this species' northern range (Becker 1983). Propst and Carlson (1986) believe the gizzard shad may be native to the South Platte drainage in Colorado.

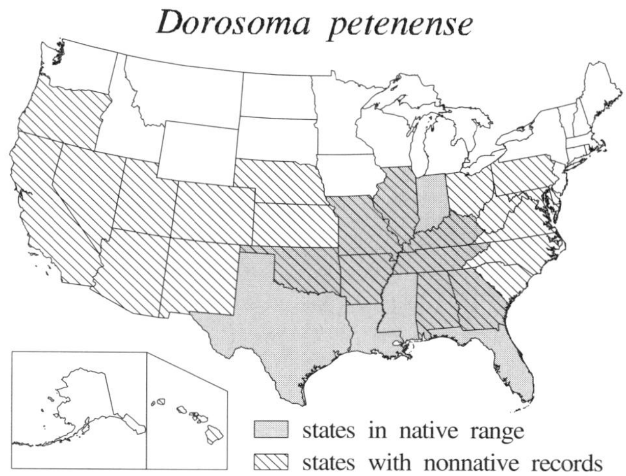

Dorosoma petenense (Günther 1867)

Common Name: threadfin shad.

Identification: Etnier and Starnes (1993); Moyle (1976a); Whitehead (1985); Page and Burr (1991). Maximum size: 23 cm.

Native Range: The Ohio River, Indiana and Illinois, and the Mississippi River, southern Illinois, south through the Mississippi River basin to the Gulf; Atlantic Slope drainages of Florida; Gulf drainages from south central Florida (?) to northern Guatemala (Page and Burr 1991).

Nonindigenous Occurrences: Threadfin shad have been stocked in the Cahaba, Coosa, and Tallapoosa rivers, and in the Tennessee drainage, **Alabama** (Lee et al. 1980 et seq.; Swift et al. 1986; Boschung 1992) and perhaps in Big Creek Lake, a municipal reservoir of Mobile (Mettee et al. 1996); the upper White River, **Arkansas** (Pflieger 1975; Lee et al. 1980 et seq.); many **Arizona** lakes and rivers (Haskell 1959; Minckley 1973; Tyus et al. 1982); throughout **California** (La Rivers 1962; Moyle 1976a, 1976b; Dill and Cordone 1997); Valmont Reservoir, the Platte drainage, and the San Luis Valley of **Colorado** (Zuckerman and Behnke 1986; Woodling 1985; Walker 1993); Coursey Pond, **Delaware** (Raasch and Altemus 1991); Gulf drainages in the panhandle, and peninsular **Florida** (Finucane 1965; Gilbert and Williams, unpublished manuscript); the Savannah, Ogeechee, Altamaha, and Apalachicola drainages and coastal areas in **Georgia** (Miller and Jorgenson 1969; Dahlberg and Scott 1971a, 1971b; Swift et al. 1986); **Hawaii** on several of the major islands (Brock 1960; Burns 1966b; Maciolek 1984); the Ohio, Wabash, and Mississippi rivers in **Illinois**, and southern Illinois reservoirs (Smith 1979; Burr 1991); **Kansas** (where stocking failed) (Cross 1967; Cross et al. 1986); **Kentucky** (Minckley and Krumholz 1960; Clay 1975; Burr and Page 1986; Burr and Warren 1986); the Potomac River in **Maryland** (Lee et al. 1980 et seq.); the upper White River, lower Missouri drainage, Montrose Lake in Henry County, and other lakes in of **Missouri** (Pflieger 1975; Lee et al. 1980 et seq.; Cross et al. 1986); **Nebraska** (Bouc 1987); in the San Juan, Canadian, Pecos, and Rio Grande drainages of **New Mexico** (Sublette et al. 1990); **Nevada** (La Rivers 1962; Minckley 1973; Moyle 1976a; Deacon and Williams 1984); many drainages in **North Carolina** (Hocutt et al. 1986; Menhinick 1991; Jenkins and Burkhead 1994); **Ohio** (Moyle 1976a); the Canadian drainage, Fort Gibson and several other lakes in **Oklahoma** (Miller and Robison 1973; Cross et al. 1986); **Oregon** as far north as Yaquina Bay, where they have even spread from stockings in California (Lee et al. 1980 et seq.); **Pennsylvania** (Cooper 1983); coastal areas of **South Carolina** (Lee et al. 1980 et seq.; Hocutt et al. 1986); the Ohio River, Tennessee Valley Authority (TVA) impoundments, Dale Hollow, and Center Hill reservoirs in **Tennessee** (Clay 1975); the Colorado River in **Utah** (Tyus et al. 1982); the Potomac River and other areas in **Virginia** (Moyle 1976a; Lee et al. 1980 et seq.; Hocutt et al. 1986; Jenkins and Burkhead 1994); and the Potomac River and possibly the lower Kanawha and Ohio drainages of **West Virginia** (Lee et al. 1980 et seq.; Hocutt et al. 1986; Stauffer et al. 1995).

Means of Introduction: It is unknown whether the populations on the east coast of Georgia resulted from past stocking or from dispersal through estuarine waters (Miller and Jorgenson 1969; Dahlberg and Scott 1971b). Populations in other locations were intentionally stocked as forage.

Status: Canadian River population possibly extirpated (Sublette et al. 1990). Populations in several West Virginia lakes extirpated by cold weather (Stauffer et al. 1995). Probably not established in the Platte drainage of Colorado (Walker 1993). Introduced and abundant in reservoirs of the Colorado River (Deacon and Williams 1984). Established in other areas.

Impact of Introduction: Concern exists regarding possible impacts on other fish species with planktonic larvae, such as minnows and suckers, and on young centrarchids. Dill and Cordone (1997) stated that threadfin compete with young centrarchids for food and that they have destroyed kokanee fishing in some areas.

Remarks: Stock introduced into the Colorado River was from the Tennessee River (Minckley 1973). Dill and Cordone (1997) gave the history of the introduction of this species into California. Although Lee et al. (1980 et seq.) listed the species as native to Florida, Gilbert (personal communication) believes there is a considerable amount of circumstantial evidence to indicate that this species is not native east of the Mississippi River, but was introduced as a forage fish beginning in the early part of this century. Gilbert cites the fact that there are no published records of the species east of the Mississippi River prior to the 1940s. Expansion of the range of this species during the past half century likely resulted from a combination of natural range extension and human introduction (Gilbert, personal communication).

Opisthonema libertate

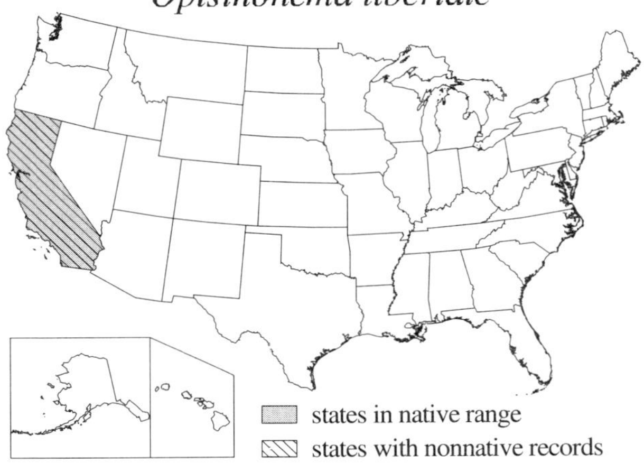

Opisthonema libertate (Günther 1867)

Common Name: deepbody thread herring (Pacific thread herring).

Identification: Eschmeyer et al. (1983); Whitehead (1985); Fischer et al. (1995a). Maximum size: 26 cm.

Native Range: Marine; Southern Baja southward to Peru. Occasionally seen off southern California (Berry and Barrett 1963; Eschmeyer et al. 1983; Whitehead 1985; Fischer et al. 1995a).

Nonindigenous Occurrences: Introduced into Salton Sea, **California**, in 1953 (Walker et al. 1961; Dill and Cordone 1997).

Means of Introduction: Intentionally stocked. Twenty-six fish were stocked in 1953. Stock was obtained from San Felipe, Mexico (Walker et al. 1961).

Status: The introduction failed to produced a viable population and the species is extirpated in the Salton Sea.

Impact of Introduction: Unknown.

Remarks: None.

ORDER GONORYNCHIFORMES

Chanidae - Milkfishes

The milkfish, *Chanos chanos*, is the only species of this tropical Indo-Pacific family. It occurs near continental shelves and around oceanic islands and has a long history in southeast Asia as an important food fish and as a fish used in aquaculture (Bagarinao 1994; Nelson 1994). Milkfish grow to over 1 m and adults feed primarily on algae (Nelson 1994). Although mostly oceanic, this species is known to enter coastal bays, estuaries, and sometimes lakes and rivers (Burgess and Axelrod 1974; Berra 1981; Bagarinao 1994).

Chanos chanos

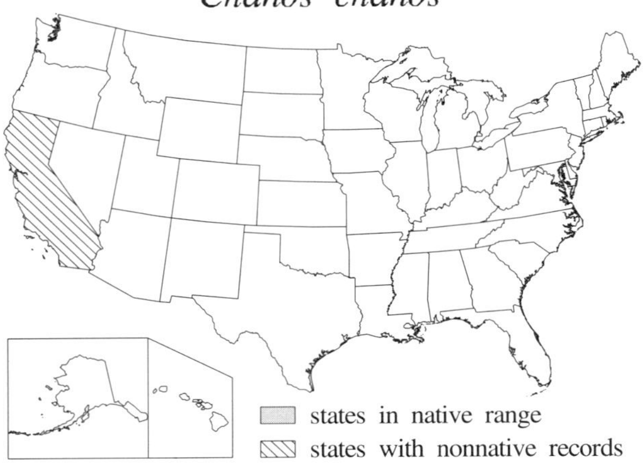

Chanos chanos (Forsskål 1775)

Common Names: milkfish (bangdang, bangos, giant herring, salmon-herring, awa).

Identification: A few distinguishing characteristics and a discussion of similar species were given by Eschmeyer et al. (1983). A photograph and distinguishing characteristics were given in Burgess and Axelrod (1974). This fish's taxonomy and morphology have been summarized recently by Bagarinao (1994). The species resembles a toothless herring with a deeply forked tail fin. Maximum size: 150 cm.

Native Range: Tropical marine. Widespread in Pacific and Indian oceans from the eastern coast of Africa to the tropical western coast of the Americas (Berra 1981; Bagarinao 1994).

Nonindigenous Occurrences: This species was stocked by the state into a small stream at Bridgeport, Solano County, **California**, in 1877; however, there were no records of survival (Smith 1896; Shebley 1917; Shapovalov et al. 1981).

Means of Introduction: The Solano County, California, introduction of 1877 consisted of about 100 fish from Hawaii (in exchange for salmon and trout eggs) that were stocked by the California Fish and Game Commission (Shebley 1917; Duffy and Bernard 1985).

Status: An attempt by the state to stock these fish in inland waters of California in 1877 apparently failed. Other records are from coastal marine habitats; furthermore, these reports most likely represent natural strays and not introductions.

Impact of Introduction: Unknown.

Remarks: Hubbs et al. (1979) listed *Chanos chanos* as having been introduced but not established in California. However, Berra (1981) and Bagarinao (1994) included the California coast in the species' native range. Reports of milkfishes from San Pedro and San Diego bays in 1929 and 1979 and between 1982 and 1983 are likely the result of natural strays from Mexican populations (Duffy and Bernard 1985; Bagarinao 1994). There is no mention of *Chanos chanos* in inland waters of California by Moyle (1976a) or by Swift et al. (1993).

Voucher specimens: California (SIO 82-22, LACM 42887-1).

ORDER CYPRINIFORMES

Cyprinidae - Minnows

The ostariophysan family Cyprinidae represents the largest family of freshwater fishes, comprising about 210 genera and about 2,010 species (Nelson 1994). It is widely distributed in Eurasia, North America, and Africa, with the largest number of species present in China and Southeast Asia (Berra 1981; Nelson 1994). About 251 described species are found in North America north of Mexico (Gilbert 1998). During recent years there has been considerable instability and some confusion in the nomenclature of North American minnows, particularly at the generic level. These shifts in status and composition of genus groups (and other taxa) are based on new research or new interpretations of old data that bear on the distinctiveness or relationships of taxa (Snelson 1991; Mayden 1992). Cyprinids are morphologically, ecologically, and behaviorally diverse. Although most are small fishes, a few species grow to well over 1 m. Most cyprinids are omnivorous, some are carnivorous, and others herbivorous. Many are important commercially as food fishes or in the aquarium trade. Some small cyprinids are considered important prey of sport fish and have been widely introduced through stocking or by way of bait bucket releases. Several large herbivorous species have been brought to the United States from Asia for the purpose of controlling aquatic plants. Cyprinids are freshwater fishes very rarely found in brackish water.

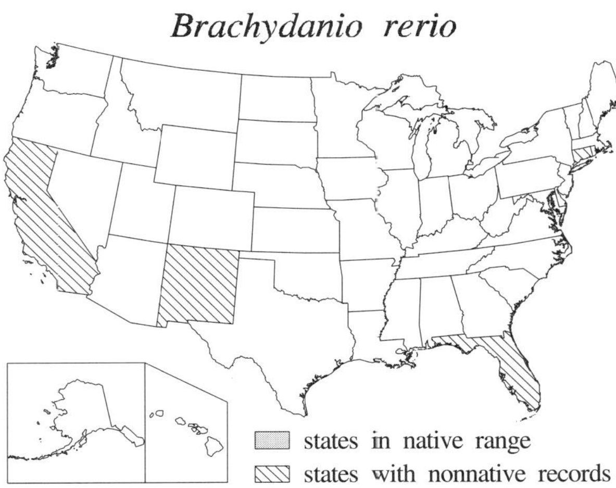

Brachydanio rerio (Hamilton 1822)

Common Names: zebra danio (zebrafish, striped danio).

Identification: Another commonly used name is *Danio rerio*. Keys and distinguishing characteristics were given in Sublette et al. (1990) and Talwar and Jhingran (1992). Color photographs of live fish appeared in Axelrod et al. (1985). Maximum size: 4.5 cm.

Native Range: Tropical Asia. Pakistan, India, Bangladesh, and Nepal (Talwar and Jhingran 1992).

Nonindigenous Occurrences: This species was reported from the Westminster flood control channel near a fish farm in Westminster, Orange County, **California**, in 1968 (St. Amant and Hoover 1969; Courtenay et al. 1984, 1991). Specimens ranging from 2-4 cm were captured in the Thames River drainage in **Connecticut** in 1985 (Whitworth 1996). It was recorded from Lake Worth Drainage District canal L-15 adjacent to fish farm in Palm Beach County, **Florida**, in the early 1970s (Courtenay and Robins 1973; Courtenay et al. 1974). Specimens also were taken from two sites adjacent to fish farms in Hillsborough County, including a ditch in Gibsonton, and from a site in Adamsville (Courtenay and Hensley 1979a; museum specimen). The species is locally established in McCauley Spring in Sandoval County, **New Mexico** (Sublette et al. 1990; M. Hatch, personal communication).

Means of Introduction: California (St. Amant and Hoover 1969) and Florida occurrences probably due to probable release or escape from local fish farms. Source of New Mexico population not known; possible aquarium release. Probably aquarium release in Connecticut (Whitworth 1996).

Status: Locally established in New Mexico; reported from California, Connecticut, and Florida.

Impact of Introduction: Unknown.

Remarks: This species is a popular aquarium fish and has been widely used in experimental and developmental research.

Voucher specimens: Florida (UF uncatalogued).

Campostoma anomalum

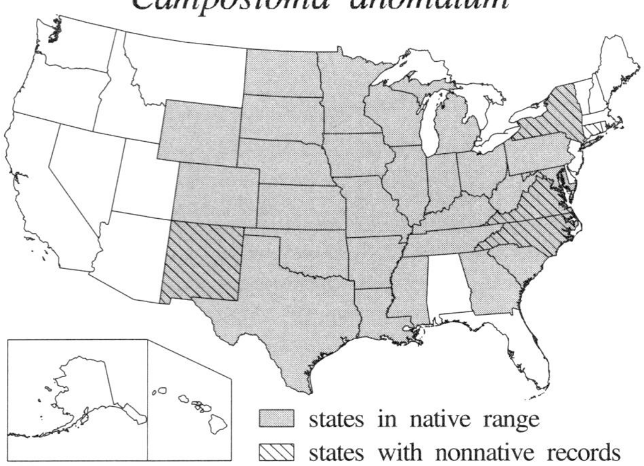

Campostoma anomalum (Rafinesque 1820)

Common Name: central stoneroller.

Identification: Becker (1983); Page and Burr (1991); Etnier and Starnes (1993). Three subspecies are recognized. Maximum size: 22 cm.

Native Range: Widespread across most of eastern and central United States in Atlantic, Great Lakes, Mississippi River, and Hudson Bay (Red River) basins from New York west to North Dakota and Wyoming, and south to South Carolina and Texas; present in Thames River system (Great Lakes basin), Ontario; found in Gulf Slope drainages from Galveston Bay, Texas, to Rio Grande, Mexico; isolated population in southwestern Mississippi and eastern Louisiana. Generally absent on Piedmont and Coastal Plain (Page and Burr 1991).

Nonindigenous Occurrences: This species has been recorded from Byram River (Bronx) watershed of **Connecticut** (Whitworth et al. 1968; Schmidt 1986; Whitworth 1996; Jones, personal communication); the Rio San Jose drainage near Bluewater Lake, Taylor Creek in the Gila River drainage, and the Rio Grande near Albuquerque and Truth or Consequences, **New Mexico** (Sublette et al. 1990); Adirondacks in Franklin County, **New York** (Smith 1985); the Yadkin River region (Pee Dee River drainage) in Wilkes County, **North Carolina** (Hocutt et al. 1986; Menhinick 1991; Jenkins and Burkhead 1994); and Paul's Creek, Carroll County, in the upper Pee Dee drainage, **Virginia** (Hocutt et al. 1986; Jenkins and Burkhead 1994).

Means of Introduction: Unknown. Probable bait bucket releases.

Status: Established in New Mexico near Albuquerque (Sublette et al. 1990). Reported from New York (Smith 1985), and North Carolina (Menhinick 1991), and Virginia (Jenkins and Burkhead 1994). Populations in the Gila drainage and near Truth or Consequences, New Mexico, apparently extirpated (Sublette et al. 1990). The Connecticut population found in the 1960s consisted of several age classes, but intensive sampling in 1987 and 1988 revealed that the species did not survive (Whitworth 1996).

Impact of Introduction: Unknown.

Remarks: According to Jenkins and Burkhead (1994), the two records of this species from the Pee Dee drainage of Virginia and North Carolina, although possibly native, are thought to be the result two separate introductions. Menhinick (1991) also concluded that the species was probably introduced into the Pee Dee drainage of North Carolina.

Voucher specimens: North Carolina (DU JRB 60-20), Virginia (RC-REJ 970).

Campostoma oligolepis

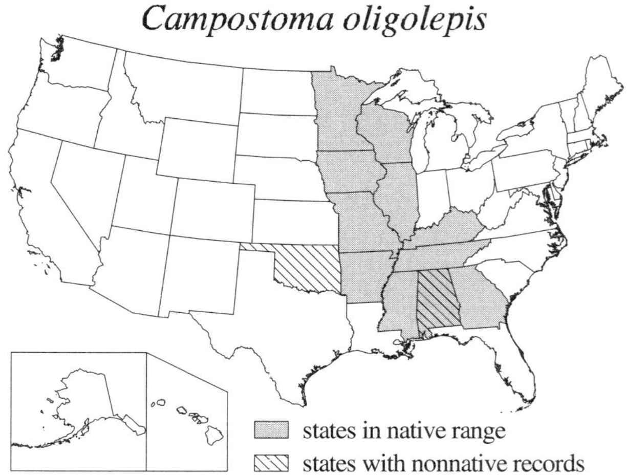

Campostoma oligolepis Hubbs and Green 1935

Common Name: largescale stoneroller.

Identification: Becker (1983); Robison and Buchanan (1988); Page and Burr (1991); Etnier and Starnes (1993); Pflieger (1997). This species requires additional taxonomic study. Maximum size: 22 cm.

Native Range: Upper Mississippi River and Lake Michigan drainages of Wisconsin, eastern Minnesota, eastern Iowa, and northern Illinois; Ozarkian streams of central and southern Missouri, and northern Arkansas; Mobile Bay drainage, Georgia, Alabama, and eastern Mississippi; parts of Green, Cumberland, and Tennessee River drainages of Kentucky, Tennessee, Georgia, and Alabama (Page and Burr 1991).

Nonindigenous Occurrences: This species was taken from the Escambia River drainage of **Alabama** (Gilbert 1992), and we suspect its presence to be the result of an introduction. It also was (circa 1979) discovered in the Illinois River in **Oklahoma** (middle Arkansas drainage) and may be introduced there (Lee et al. 1980 et seq.; Cross et al. 1986).

Means of Introduction: Unknown.

Status: Reported from Alabama and Oklahoma.

Impact of Introduction: Largely unknown. Burr and Cashner (1983) noted that stonerollers are popular bait minnows and stated that hybrids between *Campostoma oligolepis* and *C. pauciradii* found in some sites are likely the result of the introduction of one or the other species into the natural range of the other by man.

Remarks: Gilbert (1992) suggested that several different minnow species recently found in the Escambia River drainage, including *Luxilus chrysocephalus*, *Nocomis leptocephalus*, and *Notropis baileyi*, may have been introduced there. As support for introduction, he indicated that these minnows were absent in all early samples taken from that drainage. In his opinion (Gilbert, personal communication), the same case can be made for *Campostoma oligolepis*. Unlike the other three cyprinids, *Campostoma oligolepis* is not yet known from the Florida portion of the Escambia River drainage. In their summary table on Alabama fishes, Mettee et al. (1996) listed it as native to the Conecuh (Escambia) River drainage. Concerning its native range, Etnier and Starnes (1993) questioned the identification of lower Tennessee River *Campostoma* as *oligolepis*. Consequently, this species may not be part of the Tennessee fauna. Menhinick (1991) reported no records from North Carolina but said that it may be present in the Little Tennessee system in the state.

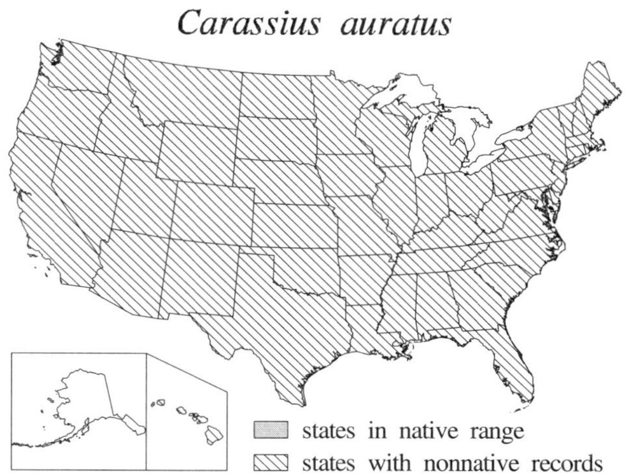

Carassius auratus (Linnaeus 1758)

Common Name: goldfish.

Identification: Distinguishing characteristics were provided in Wheeler (1978), Raicu et al. (1981), Trautman (1981) and Page and Burr (1991). Identification keys that include this species and photographs or illustrations are provided in most state and regional fish books (e.g., Hubbs and Lagler 1958; Becker 1983; Etnier and Starnes 1993). There has been considerable confusion concerning the taxonomic status of this species. Many authors have recognized two subspecies in its native range: *C. a. auratus* (the goldfish, Chinese goldfish, or Asian goldfish) from Asia, and *C. a. gibelio* (the Prussian carp, gibele carp, or European goldfish) from eastern Europe (Raicu et al. 1981). Others have concluded that the goldfish is a subspecies of the crucian carp *Carassius carassius* (i.e., *C. c. auratus*). More recently, it has been reported that *C. auratus* is a tetraploid derivative of *C. carassius* (references in Jenkins and Burkhead 1994). There are many mutant goldfish varieties and these exhibit a broad range of body forms and colors. Howells (1992b) reported that some exotic fish experts believe that "goldfish" typically observed in U.S. waters are actually crucian carp × goldfish hybrids. In addition, goldfish commonly hybridize with common carp *Cyprinus carpio*, giving rise to individuals that are intermediate in morphology between the two parent species. Goldfish have been widely and repeatedly stocked in the United States from many points of origin, including both Asia and Europe. As such, U.S. populations represent a complex of morphologically and taxonomically diverse forms. Maximum size: 41 cm.

Native Range: Eastern Asia, including China and perhaps adjacent regions; also possibly parts of Europe if *C. auratus gibelio* is a valid subspecies and not just a feral introduction (Raicu et al. 1981).

Nonindigenous Occurrences: This species has been recorded from virtually every state. Documented cases were available for **Alabama** (Smith-Vaniz 1968; Boschung 1992; Mettee et al. 1996); **Arizona** (Minckley 1973); **Arkansas** (Buchanan 1973); **California** (Moyle 1976a; Smith 1982; Swift et al. 1993; Dill and Cordone 1997); **Colorado** (Ellis 1974; Wiltzius 1981; Woodling 1985);); **Connecticut** (Webster 1942; Whitworth et al. 1968; Schmidt 1986; Whitworth 1996); **Delaware** (Lee et al. 1976; Raasch and Altemus 1991); **Florida** (Courtenay et al. 1974; Shafland 1976); **Georgia** (Ramsey 1965; Dahlberg and Scott 1971a, 1971b; Walters 1997); **Hawaii** (Cobb 1902; Jordan and Evermann 1902, 1905; Brock 1960; Maciolek 1984; Devick 1991b); **Idaho** (Linder 1963; Simpson and Wallace 1978; Idaho Fish and Game 1990); **Illinois** (O'Donnell 1935; Smith 1979; Burr and Page 1986; Burr et al. 1996; Laird and Page 1996); **Indiana** (Blatchley 1938; Gerking 1945; Nelson and Gerking 1968; Simon et al. 1992); **Iowa** (Burr and Page 1986; Harlan et al. 1987); **Kansas** (Dyche 1914; Cross 1967); **Kentucky** (Clay 1975; Burr and Page 1986; Burr and Warren 1986; Conner and Suttkus 1986); **Louisiana** (Sweeney 1902; Douglas and Davis 1967; Douglas 1974); **Maine** (Kendall 1914a); **Maryland** (Uhler and Lugger 1876; Truitt et al. 1929; Schwartz 1963; Lee et al. 1976; Jenkins and Burkhead 1994); **Massachusetts** (Smith 1833; Hartel 1992; Hartel et al. 1996); **Michigan** (Hubbs and Cooper 1936; Hubbs and Lagler 1958); **Minnesota** (McDonald 1893; Eddy and Underhill 1974; Phillips et al. 1982); **Mississippi** (Cook 1959; Ross and Brenneman 1991); **Missouri** (McDonald 1893; Ravenel 1896, 1898; Pflieger 1971, 1975, 1997); **Montana** (Brown 1971; Holton 1990); **Nebraska** (Morris et al. 1974); **Nevada** (Miller and Alcorn 1946; La Rivers 1962; Hubbs et al. 1974; U.S. Fish and Wildlife Service 1983a; Deacon and Williams 1984); **New Hampshire** (Scarola 1973); **New Jersey** (Nelson 1890; Ravenel 1898; Fowler 1906, 1952; Stiles 1978); **New Mexico** (Koster 1957; Propst et al. 1987; Sublette et al. 1990); **New York** (Bean 1903; Smith 1985; Schmidt 1986); **North Carolina** (Schmidt 1980; Menhinick 1991); **North Dakota** (Owen et al. 1981; Courtenay et al. 1984); **Ohio** (Jordan 1882; Trautman 1981; Burr and Page 1986); **Oklahoma** (Hall 1956; Miller and Robison 1973); **Oregon** (Schultz 1929; Bond 1994); **Pennsylvania** (Bean 1903; Hendricks et al. 1979; Cooper 1983; Pearson and Krumholz 1984); **Rhode Island** (Lapin, personal communication; museum specimen); **South Carolina** (Loyacano 1975; Courtenay and Hensley 1979a; Lee et al. 1980 et seq.); **South Dakota** (Churchill and Over 1933; Bailey and Allum 1962); **Tennessee** (Kuhne 1939; Starnes and Etnier 1986; Ryon and Loar 1988; Etnier and Starnes 1993); **Texas** (Kuehne 1955; Conner and Suttkus 1986; Edwards and Contreras-Balderas 1991; Hubbs et al. 1991; Howells 1992a); **Utah** (Popov and Low 1953; Sigler and Miller 1963); **Vermont** (Countryman 1975; Courtenay and Hensley 1979a); **Virginia** (Ravenel 1896; Jenkins and Burkhead 1994); **Washington** (Chapman 1942; Wydoski and Whitney 1979); **West Virginia** (Pearson and Krumholz 1984; Stauffer et al. 1995); **Wisconsin** (Johnson and Becker 1980; Becker 1983; Burr and Page 1986; Fago 1992); and **Wyoming** (Baxter and Simon 1970; Stone 1995).

Means of Introduction: DeKay (1842) reported that goldfish first were brought into this country shortly after the early part of the seventeenth century. Citing that work, Courtenay and Stauffer (1990) reported that the first recorded goldfish releases in the United States occurred during the late 1600s, and they suggested that these earliest introductions resulted from intentional releases by settlers wanting to add it to the North American fish fauna, as opposed to goldfish escaping from ponds. DeKay (1842) reported, likely in reference to the early 1800s, that goldfish in the United States were considered an ornamental species rather than a food fish, even though the fish freely reproduced in ponds in New York and adjacent states. During the late 1800s the U.S. Fish Commission raised the species and was responsible for distributing it to many states, mainly to meet the demand for fish for aquaria, fountains, and ornamental lakes (McDonald 1887, 1893). Johnson and Becker (1980) stated that some goldfish were introduced to Wisconsin through a fish exchange program with the Nebraska Fish Commission in the early 1900s. Brock (1960) stated that this species was established in Hawaii before 1900 and that there had doubtless been many reintroductions. Jordan and Evermann (1905) indicated that these fish were introduced into Hawaii from China, but stated that there is no record as to the date. They did state that shipments of goldfish were being made to San Francisco (California) as early as 1867. More recent introductions of *C. auratus* in the United States were the result of escapes from hatcheries and ponds, escapes and releases of baitfish, and aquarium releases (Knapp 1953;

Courtenay and Hensley 1979a; Courtenay et al. 1984; Pflieger 1997).

Status: Established or reported in all states except Alaska. Even though the species is one of the most widely distributed foreign fishes in North America, much of its established range is restricted to only portions of certain drainages (Courtenay and Stauffer 1990). Some areas may represent repeated escapes or releases rather than established populations (Lee et al. 1980 et seq.). Courtenay et al. (1984) reviewed the literature and concluded that the species had been taken in the wild in all states except Alaska. They denoted the species as being established in 16 of the lower 48 states, including California, Delaware, Georgia, Idaho, Illinois, Indiana, Iowa, Kentucky, Maryland, Massachusetts, Michigan, Nebraska, Nevada, New Hampshire, New York, Ohio, Oklahoma, Pennsylvania, Rhode Island, South Carolina, South Dakota, Tennessee, Texas, Virginia, Washington, and Wisconsin. Courtenay et al. (1984) listed it as possibly established, status uncertain, in 17 additional states: Alabama, Arizona, Arkansas, Colorado, Connecticut, Kansas, Louisiana, Minnesota, Mississippi, Missouri, Montana, New Jersey, New Mexico, North Carolina, North Dakota, Oregon, and Wyoming. In recent years, the existence of breeding populations has been confirmed for some of the above names states, for example, Connecticut (Whitworth 1996), Missouri (Pflieger 1997), and Wyoming (Stone 1995). Bond (1994) noted that it is present locally in warm, still waters of Oregon. The species is also known from all the major islands of Hawaii and Brock (1960) stated that it was established in that state before 1900. Devick (1991b) listed it as definitely established in Hawaii. This species has been recorded from Florida, Maine, Utah, Vermont, and West Virginia, but there is as yet no available data indicating the presence of self-sustaining populations. Countryman (1975) listed it as "naturalized" in Vermont, which suggests presence of reproducing populations. In a recent paper, Courtenay (1993) listed goldfish as established in most states except Alaska and Florida.

Impact of Introduction: Little is known about the ecology of the goldfish in North America. In most states it is not considered a "pest species", this term generally referring to species whose populations become very large and highly visible to the public. Nevertheless, this species has the potential to produce large populations in U.S. waters. In the United States most of the large established populations are recorded from the vicinity of western Lake Erie (references in Courtenay and Hensley 1979a; Trautman 1981) and in parts of southern and central California (Moyle 1976a). In Nevada, its establishment in Manse Spring was believed to be a major reason for the initial decline during the early 1960s of populations of a subspecies of the now endangered Pahrump poolfish *Empetrichthys latos latos* (Deacon et al. 1964; Lee et al. 1980 et seq.), possibly competing with adult *E. l. latos* and preying on their young. According to Moyle (1976a), goldfish probably compete with native fishes for food and space; and similar to common carp, large populations of this species can greatly disturb sportfish habitats. The Sacramento sucker *Catostomus occidentalis*, is one native species that suffers in the presence of goldfish (Moyle 1976a). However, Laird and Page (1996) concluded that goldfish in Illinois appear unable to compete with native fishes and reported that the species becomes established only in severely disturbed areas. Based on a series of artificial pond experiments, and observations of a feral population, Richardson et al. (1995) found that goldfish are benthic herbivores whose behavior often results in visible increases in turbidity and decreases in aquatic vegetation.

Remarks: The goldfish is thought to be the first foreign fish species introduced to North America (DeKay 1842; Courtenay et al. 1984). In the United States, large numbers are cultured as bait and as forage for sport fishes, as well as for the production of young fish that are then sold in the aquarium trade, mostly as live food (i.e., feeder fish) for carnivorous ornamental fishes. Although brightly-colored forms popular in the aquarium trade are occasionally taken in North American waters, the surviving progeny of these colorful individuals typically are the natural, cryptic olive-green coloration, partly as a result of selective predation by other animals. In nature, goldfish often hybridize with common carp *Cyprinus carpio* producing reproductively fertile offspring; cross fertilization and back-crossing is common in some areas, for instance polluted habitats, and in such places hybrid offspring may outnumber the parent species (Trautman 1981; Page and Burr 1991). Goldfish grow rapidly and thus are limited in their usefulness as a forage fish. The use of goldfish as baitfish is prohibited in some states. Detailed background information on the occurrence of this species in the United States was provided by Becker (1983). Knapp (1953) stated that this species is used as a hardy bait fish in many areas or crossed with carp for same purpose. He reported that a population of these hybrids was found in Buffalo Lake, Randall County, in the Texas panhandle. In their summary table, Bailey and Smith (1981) indicated that *Carassius auratus* is widely distributed in the Great Lakes basin.

Voucher specimens: Alabama (TU 16398, 51965, 52008, 52022, many others), Arkansas (TU 7071, 44838, 46903), California (USNM 4485, 38016), Georgia (UGAMNH, USNM 110111), Hawaii (BPBM 1803, 3623, specimens discarded in1969), Illinois (INHS 710,

726, many others including hybrids with carp), Maryland (USNM 85073, 85217, 85795, 238723, 271219, 271221), Massachusetts (USNM 020091, 77787), Michigan (TNHC 671), Nevada (TU 94343), New York (USNM 020271, TU 36678), Ohio (USNM 28416, TU 6566), Pennsylvania (USNM 335461), Rhode Island (USNM 21658), South Carolina (USNM 271220), Texas (TCWC 0455.01, 1045.01, 1030.01, TNHC 6969, many others), Virginia (USNM 37789, 85694, 283639), West Virginia (USNM 64464).

Carassius carassius

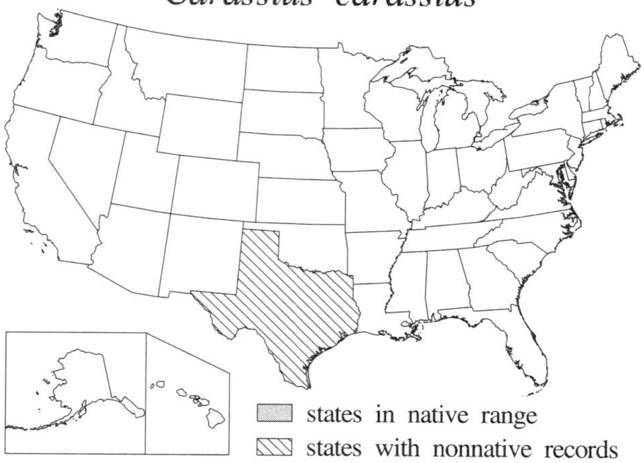

Carassius carassius (Linnaeus 1758)

Common Names: crucian carp (English carp, golden carp, gibele).

Identification: Distinguishing characteristics were provided by Berg (1949), Muus and Dahlstrom (1978), Wheeler (1978), Raicu et al. (1981), and Howells (1992b). Maximum size: To about 50 cm.

Native Range: Europe and Siberia (Raicu et al. 1981).

Nonindigenous Occurrences: The crucian carp, or a Crucian carp × goldfish hybrid, may have been introduced into **Texas** waters (Howells 1992b).

Means of Introduction: Howells (1992b) implied that this species (or a hybrid) was introduced into Texas waters, allegedly the result of being propagated and used as a baitfish.

Status: Unconfirmed report of this species, or a hybrid with goldfish, in Texas (Howells 1992b).

Impact of Introduction: Unknown.

Remarks: According to Howells (1991b), many of the goldfish *Carassius auratus* found in U.S. waters may actually be crucian carp × goldfish hybrids. Because of this species' similarity to goldfish, and because of possible hybridization, characters may overlap and positive identification may be difficult. Similar to the goldfish *C. auratus*, *C. carassius* is also known to hybridize with the common carp *Cyprinus carpio* (Berg 1949; Muus and Dahlstrom 1978; Wheeler 1978). Although prohibited in some states, crucian carp are a legal baitfish in Texas (Howells 1992b). Eddy and Underhill (1974) reported that both the goldfish and the crucian carp had been introduced into the United States, but they provided no additional details concerning the latter species. Welcomme (1988) reported that *C. carassius* was established in Chicago in the 1900s but later died out; however, he did not provide documentation for that record and we have found no additional information to support it. There is some confusion in the literature surrounding the use of the names crucian carp and Prussian carp. Lever (1996) listed Prussian Carp as an alternative or local vernacular name sometimes used for the crucian carp; however, Berg (1949) and most others use the name Prussian carp for *Carassius auratus gibelio*. In the 1800s Baird witnessed fish taken out of the Hudson River, New York; Baird later wrote that these fish appeared to be "hybrids between goldfish and the Prussian carp" (Redding 1884). In that instance it is not certain as to which species Baird is referring to in using the term Prussian carp. Cole (1905) quoted from one of Baird's reports, in which Prussian carp is treated as synonymous with *Cyprinus carassius* (=*Carassius carassius*?). The species is known for its remarkable hardiness, able to survive cold conditions and polluted waters (Muus and Dahlstrom 1978).

No known voucher specimens.

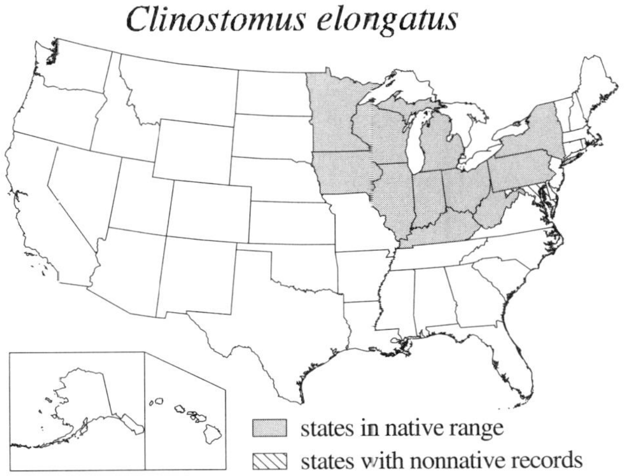

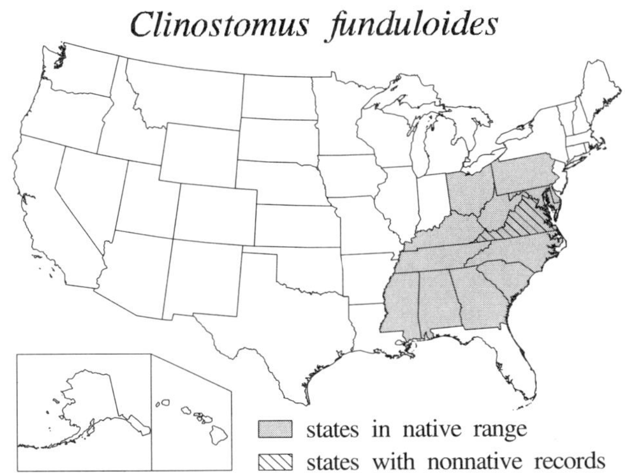

Clinostomus elongatus (Kirtland 1838)

Common Name: redside dace.

Identification: Page and Burr (1991). Maximum size: 12 cm.

Native Range: Upper Susquehanna drainage, New York and Pennsylvania; Great Lakes (except Lake Superior) and Mississippi River basins from New York and southern Ontario to Minnesota and south to West Virginia, northern Kentucky, Indiana, Wisconsin, and Iowa (Page and Burr 1991).

Nonindigenous Occurrences: Hendricks et al. (1979) reported the collection of a single specimen of the redside dace *Clinostomus elongatus* in Bear Creek, a tributary of the Youghiogheny River (Monongahela River drainage) in **Maryland**. Although these authors noted that the species is native to the drainage, they stated that the Maryland specimen may represent a bait bucket introduction. One of several reasons given was that the site, Bear Creek, is a popular fishing stream stocked with trout. Lee et al. (1980 et seq.) apparently considered the Maryland site to represent part of the native range of *C. elongatus*.

Means of Introduction: Probable bait release.

Status: Reported in Maryland.

Remarks: The collection, made in 1977, is the first record of this species for the state of Maryland (Hendricks et al. 1979).

Clinostomus funduloides Girard 1856

Common Name: rosyside dace.

Identification: Page and Burr (1991); Etnier and Starnes (1993); Jenkins and Burkhead (1994). Gilbert (1998) recognized at least two subspecies (*C. f. funduloides* and *C. f. estor*). Maximum size: 11 cm.

Native Range: Atlantic Slope (mostly above the Fall Line) from lower Delaware River drainage, Pennsylvania, to Savannah River drainage, Georgia; Ohio River basin, West Virginia, Virginia, southern Ohio, Kentucky, Tennessee, North Carolina, Alabama, and Mississippi. Absent in Ohio River basin between central Ohio and Cumberland River (including most of Kentucky) (Page and Burr 1991).

Nonindigenous Occurrences: The subspecies *C. f. funduloides* is known from the upper Tennessee drainage in **Virginia**, possibly the result of a past introduction (Lee et al. 1980 et seq.).

Means of Introduction: Unknown.

Status: Established, possibly introduced, in extreme western part of Virginia.

Impact of Introduction: Unknown.

Remarks: Although Lee et al. (1980 et seq.) suggested that this species was possibly introduced into the upper Tennessee drainage of Virginia, Jenkins and Burkhead (1994) concluded that *Clinostomus funduloides* was native to that area. Gilbert (personal communication) stated that the species is clearly native to the middle and lower Tennessee River drainage, and he believes there is a good chance that the fish is native to the upper reaches as well. There is also some uncertainty concerning the native versus introduced distribution of this species in other areas. For example, the distribution map of Lee et al. (1980 et seq.) seems to indicate the occurrence of this species in the Delaware drainage (Pennsylvania), an occurrence that seems to be based on a single record. Consequently, Gilbert (personal communication) is suspicious of its natural occurrence there, particularly considering that the fish faunas of the Delaware and Susquehanna drainages are substantially different.

Couesius plumbeus

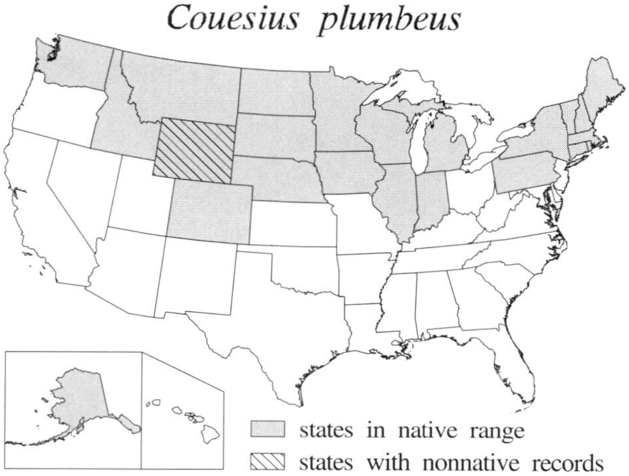

Couesius plumbeus (Agassiz 1850)

Common Names: lake chub.

Identification: Scott and Crossman (1973); Becker (1983); Page and Burr (1991). Gilbert (1998) recognized three distinct subspecies: *C. p. plumbeus*, *C. p. greeni*, and one formerly known as *C. p. dissimilis* (a name no longer available). Maximum size: 23 cm.

Native Range: Throughout much of Canada and northern United States; south to Delaware River, New York; southern end of Lake Michigan, Illinois; Platte River system, Colorado; and Columbia River drainage, Washington. Relict populations in Mississippi River basin, Iowa (Page and Burr 1991).

Nonindigenous Occurrences: This species is known from the Big and Little Sandy creeks, Green River drainage (Colorado River basin), and Yellowstone Lake, **Wyoming** (Baxter and Simon 1970; C. Wheeler, personal communication).

Means of Introduction: Probably introduced by bait bucket release (Baxter and Simon 1970).

Status: Established in the Green River drainage and in Yellowstone Lake, Wyoming (Baxter and Simon 1970).

Impact of Introduction: Unknown.

Remarks: Wydoski and Whitney (1979) stated the species is not native to Idaho and that it has been introduced into the panhandle region, however, Simpson and Wallace (1978), Lee et al. (1980 et seq.), and Page and Burr (1991) depicted the species' range to include northern Idaho and made no mention of it being introduced there. We believe it is probably native to the Kootenai drainage in northern Idaho. According to C. Gilbert (personal communication), its occurrence in the upper Colorado basin may have involved an introduction; however, he also noted that the species may have entered the Colorado basin by way of a localized stream capture involving a tributary of the Colorado and the upper Platte River. Gilbert (1998) discussed the distribution of this species in his account of *Ceratichthys squamilentus*.

Voucher specimens: Wyoming (currently in C. Wheeler's possession but will be placed in the UWL museum).

Ctenopharyngodon idella

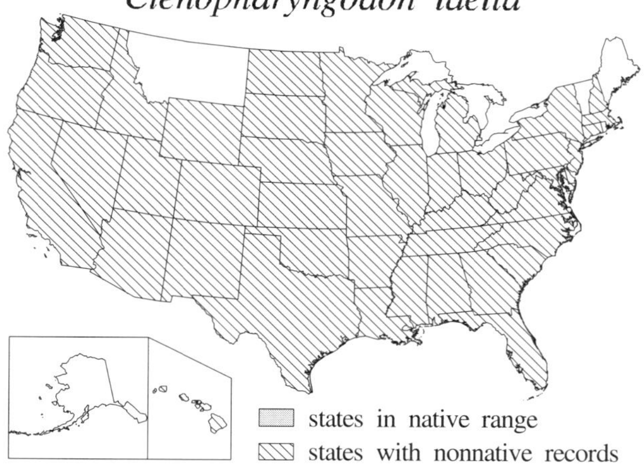

Ctenopharyngodon idella (Valenciennes 1844)

Common Names: grass carp (white amur).

Identification: Distinguishing characteristics were given in Berg (1949), Shireman and Smith (1983), and Page and Burr (1991). Keys that include this species and photographs or illustrations were provided in most of the more recently published state and regional fish books (e.g., Robison and Buchanan 1988; Etnier and Starnes 1993; Jenkins and Burkhead 1994; Pflieger 1997). A few authors present scientific name as *Ctenopharyngodon idellus*. Maximum size: 125 cm.

Native Range: Eastern Asia from the Amur River of eastern Russia and China south to West River of southern China (Lee et al. 1980 et seq.; Shireman and Smith 1983).

Nonindigenous Occurrences: Grass carp have been recorded from **Alabama** (Guillory and Gasaway 1978; Boschung 1992; Kirk et al. 1994; Mettee et al. 1996); **Arizona** (Minckley 1973; Guillory and Gasaway 1978; Courtenay et al. 1984, 1991); **Arkansas** (Buchanan 1973; Guillory and Gasaway 1978; Zimpfer et al. 1987); **California** (Guillory and Gasaway 1978; Courtenay et al. 1984, 1991; Dill and Cordone 1997); **Colorado** (Guillory and Gasaway 1978; Courtenay et al. 1984, 1991; Woodling 1985); **Connecticut** (Whitworth 1996); **Delaware** (Courtenay et al. 1984, 1991; Raasch and Altemus 1991; Rohde et al. 1994); **Florida** (Guillory and Gasaway 1978; Courtenay and Stauffer 1984; Florida Game and Freshwater Fish Commission 1989, 1994; Shafland 1995b); **Georgia** (Guillory and Gasaway 1978; Courtenay et al. 1984; Walters 1997); **Hawaii** (Maciolek 1984); **Idaho** (Courtenay et al. 1984, 1991; Idaho Fish and Game 1990); **Illinois** (Pflieger 1975; Anonymous 1977; Guillory and Gasaway 1978; Smith 1979; Phillips et al. 1982; Burr and Page 1986; Burr et al. 1996; Laird and Page 1996); **Indiana** (Anonymous 1977; Guillory and Gasaway 1978; Simon et al. 1992); **Iowa** (Guillory and Gasaway 1978; Burr and Page 1986; Harlan et al. 1987; Courtenay et al. 1991); **Kansas** (Guillory and Gasaway 1978; Courtenay and Williams 1992; Cross and Collins 1995); **Kentucky** (Conner et al. 1980; Courtenay et al. 1984, 1991; Burr and Page 1986; Burr and Warren 1986); **Louisiana** (Guillory and Gasaway 1978; Conner and Suttkus 1986; Zimpfer et al. 1987; Carp Task Force 1989); **Maryland** (Guillory and Gasaway 1978; Courtenay et al. 1984, 1991; Rohde et al. 1994); **Massachusetts** (Courtenay et al. 1984, 1991; Hartel 1992; Hartel et al. 1996); **Michigan** (Guillory and Gasaway 1978; Lee et al. 1980 et seq.; Courtenay et al. 1984; Emery 1985); **Minnesota** (Phillips et al. 1982; Courtenay et al. 1984, 1991); **Mississippi** (Guillory and Gasaway 1978; Courtenay et al. 1991; Courtenay 1993); **Missouri** (Pflieger 1975, 1978, 1997; Guillory and Gasaway 1978; Brown and Coon 1991); **Nebraska** (Guillory and Gasaway 1978; Courtenay et al. 1984, 1991); **Nevada** (Courtenay et al. 1984, 1991; Deacon and Williams 1984); **New Hampshire** (Guillory and Gasaway 1978; Lee et al. 1980 et seq.; Schmidt 1986); **New Jersey** (Guillory and Gasaway 1978; D. Mitchell and Soldwedel, personal communication); **New Mexico** (Guillory and Gasaway 1978; Courtenay et al. 1984, 1991; Cowley and Sublette 1987; Sublette et al. 1990); **New York** (Guillory and Gasaway 1978; Courtenay et al. 1984, 1991; Smith 1985; Schmidt 1986); **North Carolina** (Guillory and Gasaway 1978; Courtenay et al. 1984, 1991; Menhinick 1991; Rohde et al. 1994); **North Dakota** (Lee et al. 1980 et seq.; Owen et al. 1981; Power and Ryckman 1998); **Ohio** (Guillory and Gasaway 1978; Courtenay et al. 1984, 1991); **Oklahoma** (Guillory and Gasaway 1978; Courtenay et al. 1984, Cashner and Matthews 1988; Pigg et al. 1992); **Oregon** (Lee et al. 1980 et seq.; Pauley et al. 1994); **Pennsylvania** (C. N. Shiffer, personal communication); **South Carolina** (Guillory and Gasaway 1978; Courtenay et al. 1984, 1991; Foltz and Kirk 1994; Rohde et al. 1994); **South Dakota** (Guillory and Gasaway 1978; Lee et al. 1980 et seq.; Owen et al. 1981); **Tennessee** (Guillory and Gasaway 1978; Ryon and Loar 1988; Etnier and Starnes 1993); **Texas** (Guillory and Gasaway 1978; Conner and Suttkus 1986; Trimm et al. 1989; Howells 1992a); **Utah** (Courtenay et al. 1984, 1991; Sigler and Sigler 1996); **Virginia** (Guillory and Gasaway 1978; Courtenay et al. 1984, 1991; Jenkins and Burkhead 1994; Rohde et al. 1994); **Washington** (Pauley et al. 1994; Fletcher, personal communication); **West Virginia** (Guillory and Gasaway 1978; Courtenay et al. 1991); **Wisconsin** (Guillory and Gasaway 1978; Becker 1983; Emery 1985; Burr and Page 1986; Mulvey 1990; Fago 1992); and **Wyoming** (Courtenay et al. 1984, 1991; Stone 1995).

Means of Introduction: Both authorized and unauthorized stockings of grass carp have taken place for biological control of vegetation. This species was first imported to the United States in 1963 to aquaculture facilities in Auburn, Alabama, and Stuttgart, Arkansas. The Auburn stock came from Taiwan, and the Arkansas stock was imported from Malaysia (Courtenay et al. 1984). The first release of this species into open waters took place at Stuttgart, Arkansas, when fish escaped the Fish Farming Experimental Station (Courtenay et al. 1984). However, many of the early stockings in Arkansas were in lakes or reservoirs open to stream systems, and by the early 1970s there were many reports of grass carp captured in the Missouri and Mississippi rivers (Pflieger 1975, 1997). During the past few decades, the species has spread rapidly as a result of widely scattered research projects, stockings by federal, state, and local government agencies, legal and illegal interstate transport and release by individuals and private groups, escapes from farm ponds and aquaculture facilities; and natural dispersal from introduction sites (e.g., Pflieger 1975; Lee et al. 1980 et seq.; Dill and Cordone 1997). Some of the agencies that have stocked grass carp in the past include the Arkansas Game and Fish Commission, the Tennessee Valley Authority, the U.S. Fish and Wildlife Service, the Delaware Division of Fish and Wildlife, the Florida Game and Fresh Water Fish Commission, the Iowa Conservation Commission, the New Mexico Department of Fish and Game, and the Texas Parks and Wildlife Department. The species also has been stocked by private individuals and organizations. In some cases, grass carp have escaped from stocked waterbodies and appeared in nearby waterbodies. Stocking of grass carp as a biological control against nuisance aquatic plants in ponds and lakes continues. For instance, Pflieger (1997) stated that thousands of grass carp are reared and sold by fish farmers in Missouri and Arkansas.

Status: Grass carp have been recorded from 45 states; there are no reports of introductions in Alaska, Maine, Montana, Rhode Island, and Vermont. It is known to have established populations in a number of states in the Mississippi River basin. Breeding populations have been recorded for the Mississippi River in Kentucky (Conner et al. 1980; Burr and Warren 1986), the Illinois and upper Mississippi rivers of Illinois and Missouri (Raibley et al. 1995), the lower Missouri River in Missouri (Raibley et al. 1995), the Mississippi River or its tributaries in the states of Arkansas (Conner et al. 1980), Louisiana (Conner et al. 1980; Zimpfer et al. 1987), Tennessee (Etnier and Starnes 1993), and presumably Mississippi (Courtenay et al. 1991). It is also established in the Ohio River in Illinois (Burr, personal communication); it was listed as established in Minnesota (Courtenay et al. 1991, but see Courtenay 1993), and in the Trinity River of Texas (Waldrip 1992; Webb et al. 1994; Elder and Murphy 1997). Courtenay (1993) listed grass carp as established in eight states, Arkansas, Kentucky, Illinois, Louisiana, Missouri, Mississippi, Tennessee, and Texas; an additional one, Minnesota, was included in an earlier listing of states with established populations (Courtenay et al. 1991). Stone (1995) listed this species as being established in Wyoming; however, Stone (personal communication) clarified his earlier report by stating that, as of early 1997, there is no evidence of natural reproduction in that state. Similar to a few other authors, he used the term 'established' to indicate that grass carp populations have persisted for many years, presumably because of their long life span and because of long-term maintenance of wild populations through continued stockings. Pearson and Krumholz (1984) mentioned several records from the Ohio River, including river mile 963 on the Illinois-Kentucky border and from the Falls of the Ohio, at Louisville, along the Kentucky-Ohio border. They also stated that the species had been stocked in many private ponds and lakes in the Ohio River basin. Sigler and Sigler (1996) stated that this species is no longer found in Utah, but they provide no details. Harvest of grass carp by commercial fishermen in the Missouri and Mississippi rivers of Missouri has exhibited a general climb. In 1996, the most recent available data, there was a record reported harvest, about 44,000 pounds, 8 percent of the total commercial fish harvest (J. W. Robinson, personal communication).

Impact of Introduction: Various authors (e.g., Shireman and Smith 1983; Chilton and Muoneke 1992; Bain 1993) have reviewed the literature on grass carp; most also discuss actual and potential impacts caused by the species' introduction. Shireman and Smith (1983) concluded that the effects of grass carp introduction on a water body are complex and apparently depend on the stocking rate, macrophyte abundance, and community structure of the ecosystem. They indicated that numerous contradictory results are reported in the literature concerning grass carp interaction with other species. Negative effects involving grass carp reported in the literature and summarized by these authors included interspecific competition for food with invertebrates (e.g., crayfish) and other fishes, significant changes in the composition of macrophyte, phytoplankton, and invertebrate communities, interference with the reproduction of other fishes, decreases in refugia for other fishes, and so on. In their overview, Chilton and Muoneke (1992) reported that grass carp seem to affect other animal species by modifying preferred habitat, an indirect effect. However, they also indicated that grass carp may directly influence other animals through either predation or competition when plant food is scarce. In his review, Bain (1993) stated that grass carp have significantly altered the food web and trophic structure of aquatic systems by inducing changes in plant, invertebrate, and fish communities. He indicated that effects are largely secondary consequences of decreases in the density and composition of aquatic plant communities. Organisms requiring limnetic habitats and food webs based on phytoplankton tend to benefit from the presence of grass carp. On the other hand, Bain reported that declines have occurred in the diversity and density of organisms that require structured littoral habitats and food chains based on plant detritus, macrophytes, and attached algae. Removal of vegetation can have negative effects on native fish, such as elimination of food sources, shelter, and spawning substrates (Taylor et al. 1984). Hubert (1994) cited a study that found vegetation removal by grass carp lead to better growth of rainbow trout due to increases in phytoplankton and zooplankton production, but it also lead to higher predation on rainbow trout by cormorants *Pha-*

lacrocorax auritus due to lack of cover, and changes in diet, densities, and growth of native fishes. Although grass carp are often used to control selected aquatic weeds, these fish sometimes feed on preferred rather than on target plant species (Taylor et al. 1984). Increases in phytoplankton populations is a secondary effect of grass carp presence. A single grass carp can digest only about half of the approximately 45 kg of plant material that it consumes each day. The remaining material is expelled into the water, enriching it and promoting algal blooms (Rose 1972). These blooms can reduce water clarity and decrease oxygen levels (Bain 1993). In addition to the above, grass carp may carry several parasites and diseases known to be transmissible or potentially transmissible to native fishes. For instance, it is believed that grass carp imported from China were the source of introduction of the Asian tapeworm *Bothriocephalus opsarichthydis* (Hoffman and Schubert 1984; Ganzhorn et al. 1992). As such, the species may have been responsible indirectly for the infection of the endangered woundfin *Plagopterus argentissimus* (by way of the red shiner *Cyprinella lutrensis*) (Moyle 1993).

Remarks: Triploids are considered to be sterile and incapable of reproduction. However, some researchers have questioned the sterility of triploids. Howells (1992b) referred to a study in which milt from triploids was used to successfully fertilize normal diploid eggs, but he found no published information on the viability of eggs from triploid females. Available information (e.g., Chilton and Muoneke 1992; Exotic Species Workgroup 1994; W. Shelton, personal communication), indicate that triploid grass carp can produce some viable gametes, but the proportion of such gametes is extremely low. Because of this, the reproductive potential of triploids is considered to be very low (Chilton and Muoneke 1992). In general with triploid fish, the triploid females never fully develop ovaries, but triploid males can sometimes have the "false appearance of being fully developed." Nevertheless, techniques used to induce triploidy are not always totally effective and every individual needs to be genetically checked. Triploid grass carp are indistinguishable in external morphology from normal (fertile) diploids. Triploidy can be confirmed by blood or tissue analysis, commonly by the use of a Coulter counter with a channelyzer. Although diploids are banned in Florida, three adults, including a gravid female (UF 85587), were captured in the Suwannee River, Florida (Burkhead and Williams 1991). Similarly, Webb et al. (1994) reported on the capture of young grass carp in Texas whose ploidy and age indicated some were escapees from illegal stockings. Many states no longer permit import of diploid grass carp, although triploids are allowed. On the other hand, Texas now bans grass carp for fear they will remove too much vegetation and thus destroy fish and wildlife habitat. As of 1994, only the states of Alaska, Oregon, Montana, North Dakota, Minnesota, Wisconsin, Michigan, Massachusetts, Vermont, Maine, Maryland, and Rhode Island prohibit grass carp, diploid and triploid, in their state. The states of Hawaii, Iowa, Kansas, Missouri, Oklahoma, Arkansas, Mississippi, and Alabama have no restrictions, meaning that both diploids and triploids can be used and there is no permit required. The remaining states have some type of restriction on the use of grass carp. Typical restrictions include one or more of the following: use of only verified triploids, use only in public waters, or use requires a permit (Wattendorf and Phillippy 1996). The numbers of grass carp legally stocked in a region can be fairly high. For instance, in January 1992, the Texas Parks and Wildlife Commission approved the use of certified triploid grass carp for vegetation control in state waters. In March 1994, Durocher (1994) reported that the relatively new program already had issued nearly 2,900 permits and allowed the stocking of more than 78,000 grass carp. Foltz and Kirk (1994) indicated that 300,000 triploid grass carp were stocked into Lake Marion, South Carolina, between 1989 and 1991. Since the late 1970s more than 2,000 water bodies in Florida have been stocked with grass carp (Haller 1994). Lake Istokpoga, Florida, was stocked with 125,000 triploid grass carp during a two-year period in the early 1990s (Thomas 1994). Guillory and Gasaway (1978) gave dot map detailing the known distribution of grass carp at that time. Pflieger (1978) described the invasion of Missouri streams. Dill and Cordone (1997) detailed the history of grass carp introductions in California. As part of their account, they reported the removal of about 1,500 diploid grass carp from golf course ponds that had been illegally stocked during the 1980s.

Voucher specimens: Alabama (UF 17284), Florida (UF 48686, 74774, 85587, 86438, 91910, 92224, 228599), Illinois (SIUC 19962, INHS 30406), Louisiana (TU 125915, 125981, 129532), New York (NYSM 35119).

Cyprinella analostana

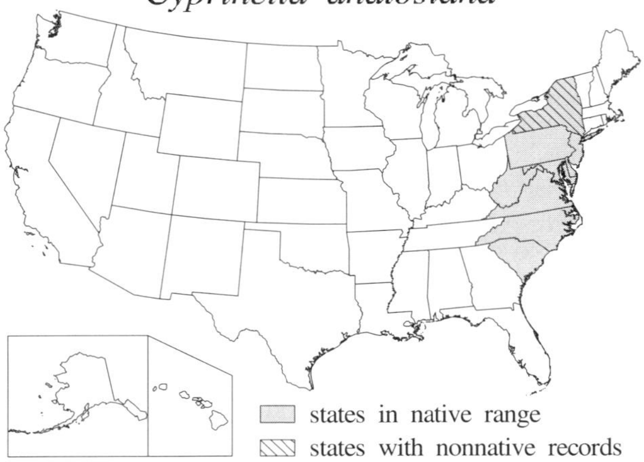

Cyprinella analostana Girard 1859

Common Name: satinfin shiner.

Identification: Page and Burr (1991); Jenkins and Burkhead (1994); commonly used name is *Notropis analostanus*. Maximum size: 11 cm.

Native Range: Atlantic Slope from Hudson River drainage, New York, to Pee Dee River drainage, North Carolina; Lake Ontario drainage, New York (Page and Burr 1991).

Nonindigenous Occurrences: Three isolated populations of this species in **New York** may be the result of past introductions; the three sites include the Black River (a tributary of Lake Ontario in northern New York), the southern end of Cayuga Lake, and a Lake Ontario tributary west of Rochester (Gibbs 1963; Smith 1985).

Means of Introduction: If the above records represent introductions, the species may have gained access to the sites by way of bait bucket release or by way of a canal; however, the populations may be natural, simply remnants of former natural distributions (Smith 1985). Gibbs (1963) suggested that the presence of this species in the Black River, New York, may have been the result of dispersal through the Black River Canal; however, Smith (1985) indicated that there are few records of the satinfin in the Mohawk basin other than Choharie Creek.

Status: Established, possibly introduced, in several localities in New York (Smith 1985).

Impact of Introduction: Unknown.

Remarks: This species is a known bait fish (Cooper 1983). Gibbs (1963) considered the record of this species from a Lake Ontario tributary to be a relict population and not the result of introduction. Gilbert (1993) also concluded that *Cyprinella analostana* could well be native to Lake Ontario and tributaries. In contrast to situation to the south, where only one-way transfers are considered possible, there is abundant evidence that two-way post-glacial transfers occurred in areas such as the upper Susquehanna River and glacial overflow streams in the Finger Lakes region of New York (see Gilbert 1980). On the other hand, the many canals constructed in New York during the early 1800's (plus of course the omnipresent bait bucket) make it very possible that species with basically Atlantic Slope affinities and restricted distributions in Lake Ontario and its tributaries (e.g., *Cyprinella analostana, Notropis procne, Notropis bifrenatus*) did not reach there naturally. However, Gilbert (personal communication) agrees that Atlantic Slope species with broad distributions in the St. Lawrence basin (e.g., *Semotilus corporalis, Exoglossum maxillingua, Etheostoma olmstedi*) clearly did reach there by natural post-glacial dispersal.

Cyprinella galactura

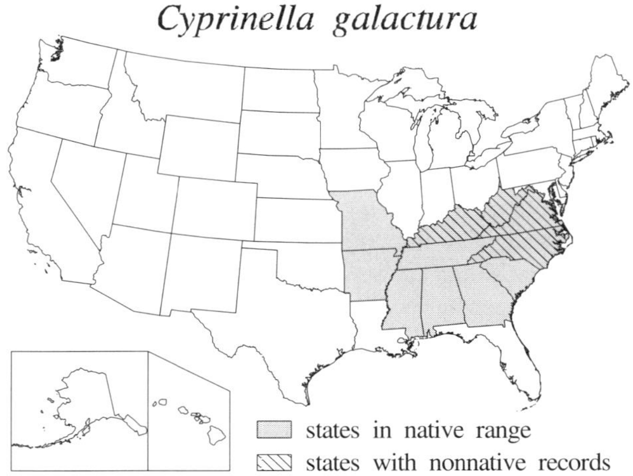

Cyprinella galactura (Cope 1868)

Common Name: whitetail shiner.

Identification: Page and Burr (1991); Etnier and Starnes (1993); a commonly used name is *Notropis galacturus*. Maximum size: 15 cm.

Native Range: Disjunct range east and west of Former Mississippi Embayment. Cumberland and Tennessee River drainages, Virginia, North Carolina, Kentucky,

Tennessee, Georgia, Alabama, and Mississippi; upper Savannah and Santee drainages (Atlantic Slope), North Carolina, South Carolina, and Georgia; and upper New River drainage, Virginia. St. Francis and White River drainages, Missouri and Arkansas (Page and Burr 1991).

Nonindigenous Occurrences: Possibly the result of past introduction, this shiner has been recorded in **Kentucky** from several localities in the Big Sandy system, Ohio River drainage, in Pike County (Warren 1981; Burr and Warren 1986; Jenkins and Burkhead 1994). In **North Carolina**, its presence in the Broad system probably represents an introduction, and its presence in the Catawba and Savannah drainages, although regarded as native, is possibly attributable to introductions (Menhinick 1991). The species occurs in the Big Sandy system (Ohio River drainage) and middle New drainages, **Virginia,** likely the result of introductions (Lee et al. 1980 et seq.; Jenkins and Burkhead 1994), and the middle New drainage, **West Virginia** (Jenkins and Burkhead 1994; Stauffer et al. 1995).

Means of Introduction: Populations of this species, interpreted as the result of human introductions, were most likely the result of bait bucket releases. Jenkins and Burkhead (1994) believed that the West Virginia population is the result of dispersal downstream from the introduced population in the New River in Virginia.

Status: Established, possibly introduced, in parts of Kentucky, North Carolina, Virginia, and West Virginia.

Impact of Introduction: Unknown.

Remarks: Jenkins and Burkhead (1994) provided a series of arguments to support their position that this species was introduced into both the Big Sandy and New drainages. Both Lee et al. (1980 et seq.) and Jenkins and Burkhead (1994) concluded that its presence in the New River drainage was likely the result of human introduction because the species was not present in early collections. Previously, a few researchers had believed that the Big Sandy and New drainages were part of the species' natural distribution, possibly relict populations or the result of recent stream capture events (references in Jenkins and Burkhead 1994). Similarly, Tsai and Raney (1974) stated that *C. galactura* may have entered the Savannah River drainage of Atlantic Slope via stream capture.

Voucher specimens: Virginia (USNM 104157).

Cyprinella lutrensis

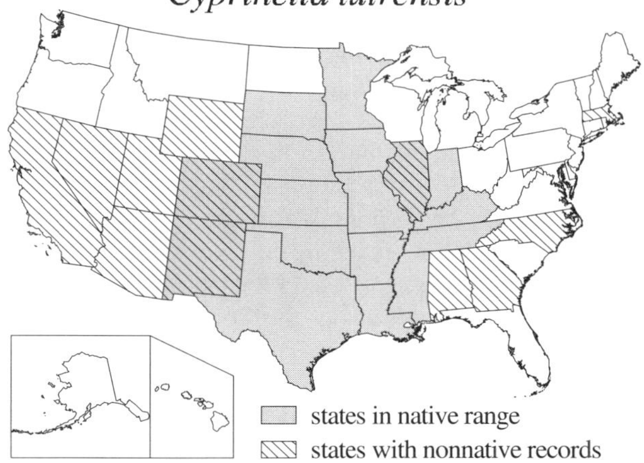

Cyprinella lutrensis (Baird and Girard 1853)

Common Name: red shiner.

Identification: Robison and Buchanan (1988); Page and Burr (1991); Pflieger (1997); commonly used name is *Notropis lutrensis*. Maximum size: 9 cm.

Native Range: Mississippi River basin from southern Wisconsin and eastern Indiana to South Dakota and Wyoming and south to Louisiana; Gulf drainages west of Mississippi River to Rio Grande, Texas, New Mexico, and Colorado. Absent in Ozark and Ouachita uplands (Page and Burr 1991).

Nonindigenous Occurrences: This species is known from the upper Tombigbee River, the Coosa River drainage, and reservoirs of the Chattahoochee River, **Alabama** (Boschung 1987, 1992; Mettee et al. 1996); the Colorado River and its major tributaries in **Arizona** including the Gila and Virgin drainages (Minckley and Deacon 1968; Minckley 1973; Tyus et al. 1982; Greger and Deacon 1988; Stolzenburg 1992); the Colorado River, the San Joaquin River drainage, Salton Sea drainages, and the Los Angeles basin and certain surroundings areas in **California** (Hubbs 1954; Moyle 1976a, 1976b; Jennings and Saiki 1990; Swift et al. 1993; Dill and Cordone 1997); the Colorado River and some of its tributaries in western **Colorado** (Everhart and Seaman 1971; Tyus et al. 1982; Woodling 1985); the Ocmulgee, Coosa, and Chattahoochee river drainages in **Georgia** (Couch et al. 1995; Devivo and Freeman 1995); in northeastern **Illinois** including lagoons of Lake Michigan in Chicago (Hubbs and Lagler 1958) and Channel Lake, Fox River drainage, in Lake County (Smith 1979); Dickey Brook in New Salem, **Massachusetts** (Hartel et al. 1996); portions of the Colorado River basin in **Nevada** including the lower Virgin River (Branson 1968; Cross 1975; Deacon and Williams 1984; Greger and Deacon 1988), and the Moapa River (Deacon and Bradley 1972; Cross 1976); portions of the Colorado River basin in **New Mexico** including the San Juan, Gila, and San Francisco drainages (Tyus et al. 1982; Sublette et al. 1990; Stolzenburg 1992); portions of Atlantic Coastal drainages in **North Carolina** including the Yadkin and Roanoke river drainages (Moore et al. 1976; Hocutt et al. 1986; Menhinick 1991); portions of the Colorado River basin in **Utah** including the Green, White, and Virgin river drainages (Holden and Stalnaker 1975b; Tyus et al. 1982; Deacon 1988; B. Schmidt, personal communication); and from the Green River drainage, Colorado River basin, **Wyoming** (Tyus et al. 1982).

Means of Introduction: The origin of most introduced red shiner populations can be attributed to bait bucket releases; however, initial introduction is often followed by the species' rapid multiplication, dispersal, and aggressive colonization (e.g., Hubbs and Lagler 1958; Minckley and Deacon 1968; Minckley 1973). In some areas dispersal of introduced populations has been aided by the presence of irrigation ditches and canals (e.g., Jennings and Saiki 1990). Koehn (1965) mentioned that the species has been introduced as a forage fish. According to Dill and Cordone (1997), it was introduced into northern California as forage, not as a bait minnow as Kimsey and Fisk (1964) had suggested. The introduction into the Yadkin drainage, North Carolina, was possibly the result of an aquarium release (Moore et al. 1976). Hubbs (1954) reported this species as established in the lower Colorado River basin by 1953. He attributed the source of the introduction to escapes from the Arizona Fish Farms in Ehrenburg, Arizona. There apparently has been more than one subspecies introduced into the southwestern United States. Hubbs (1954) also noted that red shiners found in the lower Colorado River basin were intergrades between the subspecies *N. l. lutrensis* and *N. l. suavis*. In contrast, Minckley (1973) reported that the Arizona specimens he examined more closely resembled the typical subspecies, *C. l. lutrensis*. Gilbert (1998) also referred it to the typical subspecies (*C. l. lutrensis*).

Status: Established in areas outside their native range in Alabama, Arizona, California, Colorado, Georgia, Illinois, Nevada, New Mexico, North Carolina, Utah, and Wyoming. In contrast to Hubbs and Lagler's statement (1958), Becker (1983) found that there was no evidence to substantiate the presence of this species in lagoons of Lake Michigan at Chicago.

Impact of Introduction: The red shiner is very aggressive and where introduced may dilute the gene pools of native *Cyprinella* via hybridization (Mayden 1989). The red shiner is probably hybridizing with the blacktail shiner *C. venusta* in Alabama (Mettee et al. 1996). The red shiner has also affected the distribution and abundance of native fishes. For example, populations in the Moapa and Virgin rivers, Nevada, have been implicated in the decline of the native fish of this region, including spikedace *Meda fulgida*, woundfin *Plagopterus argentissimus*, and Virgin River chub *Gila seminuda* (Moyle 1976a; Deacon 1988; U.S. Fish and Wildlife Service 1990a, 1995). Members of this species may compete with and affect adversely young Colorado pikeminnow *Ptychocheilus lucius*, an endangered species (Karp and Tyus 1990). The introduced redside shiner *Richardsonius balteatus* declined when the red shiner became common in the Green River near the boundary of Dinosaur National Monument, Utah, in 1971 (Holden and Stalnaker 1975b). In degraded streams in Georgia, introduced red shiners have become one of the most abundant species (Devivo and Freeman 1995). The introduction of red shiners into Utah was probably the means by which the Asian tapeworm entered the Virgin River; subsequent tapeworm infestation of woundfin *Plagopterus argentissimus*, an endangered species, may be primarily responsible for the woundfin's decline during the 1980s (Deacon 1988). Dill and Cordone (1997) called the red shiner the second greatest threat to the welfare of indigenous southwestern fishes, after the mosquitofish.

Remarks: The red shiner is a widespread and commonly used bait fish; it is also in the aquarium trade (Becker 1983; Etnier and Starnes 1993). It has been marketed in a pet shop under the name "rainbow dace" (Moore et al. 1976). Several attempts have been made to eradicate the red shiner from a portion of the Virgin River as part of the recovery plan for woundfin and Virgin River chubs. It was successfully eliminated from the river between Washington Fields Diversion and Johnson Diversion, but have re-invaded below Johnson Diversion (U.S. Fish and Wildlife Service 1995). Tyus et al. (1982) gave a distribution map of the this species in the upper Colorado basin. Swift et al. (1993) and Dill and Cordone (1997) detailed the history of this species in California.

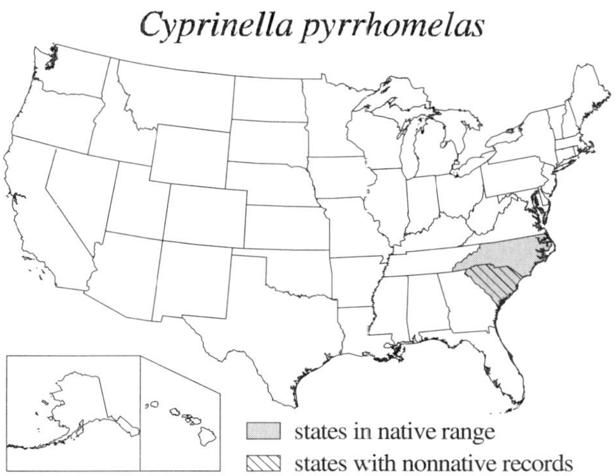

Cyprinella pyrrhomelas (Cope 1870)

Common Name: fieryblack shiner.

Identification: Menhinick (1991); Page and Burr (1991); another name is *Notropis pyrrhomelas*. Maximum size: 11 cm.

Native Range: Pee Dee and Santee river drainages, North Carolina and South Carolina (Page and Burr 1991).

Nonindigenous Occurrences: Known from (and probably introduced to) the Chattooga River (Savannah River drainage), Oconee County, **South Carolina** (Johnston et al. 1995).

Means of Introduction: Most likely the result of a bait bucket release (Johnston et al. 1995).

Status: Reported, probably established, in Savannah River drainage, South Carolina.

Impact of Introduction: Unknown.

Remarks: Johnston et al. (1995) collected five specimens in the Chattooga River as part of their 1990 fish survey of the Southern Appalachians. They stated that *Cyprinella pyrrhomelas* was previously unknown from this drainage (apparently meaning the entire Savannah River drainage). The Chattooga River forms the Georgia/South Carolina state border at the site where this species was taken.

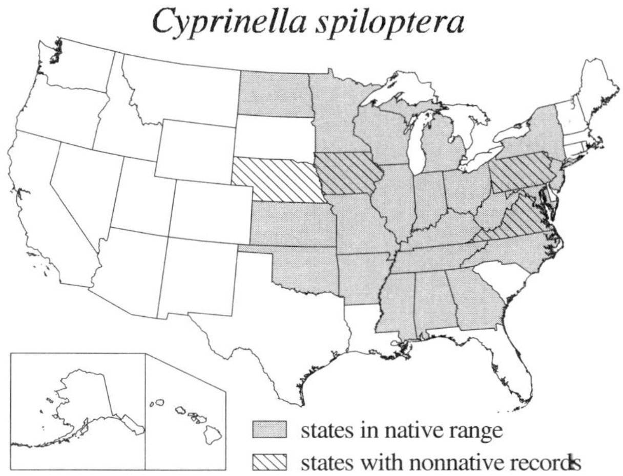

Cyprinella spiloptera (Cope 1868)

Common Name: spotfin shiner.

Identification: Becker (1983); Page and Burr (1991); Etnier and Starnes (1993); commonly used name is *Notropis spilopterus*. Maximum size: 12 cm.

Native Range: Atlantic Slope from St. Lawrence drainage, Quebec, to Potomac River drainage, Virginia; Great Lakes (except Lake Superior), Hudson Bay (Red River), and Mississippi River basins from Ontario and New York to southeastern North Dakota and south to Alabama and eastern Oklahoma; isolated populations in Ozarks (Page and Burr 1991).

Nonindigenous Occurrences: Spotfin shiners have been stocked in **Nebraska** (Bouc 1987). The species is also known from the Missouri River on the Nebraska-Iowa border, and from several localities within **Iowa** including a tributary of the Missouri (possibly the Keg River drainage) in the southwestern part of the state, and the upper Chariton River drainage (possibly Rathbun Reservoir) (Lee et al. 1980 et seq.; Cross et al. 1986). In **Virginia**, a 1973 record of this species from the Sherando Lake, Shenandoah drainage, Augusta County is regarded as an introduction; its presence in Smith Mountain Lake, Roanoke drainage, is also apparently the result of an introduction (Hocutt et al. 1986; Jenkins and Burkhead 1994).

Means of Introduction: Spotfin shiners have been stocked intentionally as a forage fish in Nebraska (Bouc 1987). It most likely entered Smith Mountain Lake, Virginia, from the nearby New drainage, possibly transferred with white bass from Claytor Lake (Jenkins and

Burkhead 1994). The method of introduction into other areas has not been reported although introduction via bait bucket releases is a possibility in some areas.

Status: Established in parts of Nebraska (Bouc 1987; Lee et al. 1980 et seq.). Established and abundant in Smith Mountain Lake, Virginia; reported from Sherando Lake, Virginia (Jenkins and Burkhead 1994). Reported, probably established, in Iowa outside its native range (Lee et al. 1980 et seq.).

Impact of Introduction: Unknown.

Remarks: This species has received limited use as a baitfish (Scott and Crossman 1973; Becker 1983; Harlan et al. 1987). This species may have spread via the Erie Canal (Gibbs 1963; Smith 1985).

Cyprinella venusta

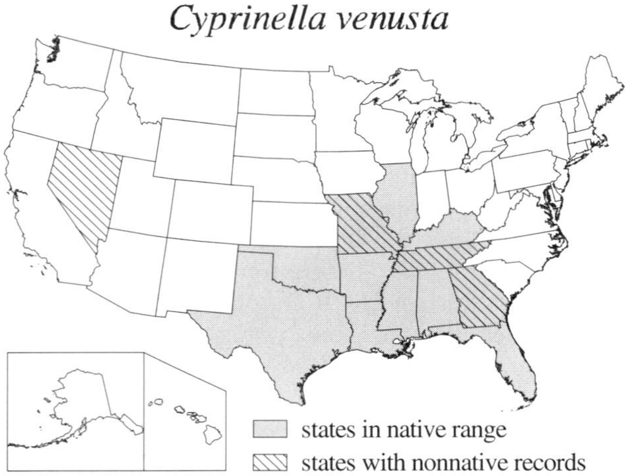

Cyprinella venusta Girard 1856

Common Name: blacktail shiner.

Identification: Smith (1979); Page and Burr (1991); Etnier and Starnes (1993); Pflieger (1997); another name commonly used is *Notropis venustus*. During recent years *Cyprinella venusta* has been recognized as a species complex, possibly representing as many as five different species (Gilbert, personal communication; also see Kristmundsdottir and Gold 1996). Recently, Gilbert (1998) formally elevated a few of the forms to species. Maximum size: 19 cm.

Native Range: Gulf drainages from Suwannee River, Georgia and Florida, to Rio Grande, Texas; Mississippi River basin (mostly on Former Mississippi Embayment) from southern Illinois to Louisiana and west in Red River drainage to western Oklahoma (Page and Burr 1991).

Nonindigenous Occurrences: This species has been found in the upper Ocmulgee River (Atlantic Slope) in **Georgia**, possibly the result of an introduction (Bart et al. 1994). An introduced population occurs in Stockton Lake in the Sac River system in **Missouri** (Pflieger 1997). Four specimens were taken in **Nevada** from the lower Virgin River, Colorado River basin, near Bunkerville, Clark County, in 1964 (Branson 1968; Minckley 1973; Deacon and Williams 1984). It has recently invaded Kentucky Lake on the lower Tennessee River, **Tennessee** (Etnier and Starnes 1993).

Means of Introduction: This species gained access to Kentucky Lake, Tennessee, via a canal connection, presumably the Tennessee-Tombigbee waterway (Etnier and Starnes 1993). Although not reported from Stockton Lake, Missouri, until 1986, Pflieger (1997) suspected that it was inadvertently introduced into the lake as a stock contaminant along with an intentional introduction of the inland silverside *Menidia beryllina* from Lake Texoma, Oklahoma, in 1970. Branson (1968) concluded that blacktail shiners found in the Utah portion of the Virgin River probably migrated upstream from Lake Mead, but he did not speculate on the possible means of introduction into that lake. It is not certain when and how this species gained access to the Ocmulgee River drainage of Georgia. Bart et al. (1994) concluded that the species came from the nearby Flint River, but they were unable to determine if the species' presence in the Ocmulgee was the result of some natural event or of a bait bucket release.

Status: Established in Tennessee (Etnier and Starnes 1993) and Missouri (Pflieger 1997); introduced, possibly established, in Georgia (Bart et al. 1994); reported from Nevada (Branson 1968; Deacon and Williams 1984).

Impact of Introduction: In the Ocmulgee of Georgia, there is the potential for the blacktail shiner to negatively affect closely related species native to the system through hybridization or competitive displacement (Bart et al. 1994). According to Bart et al. (1994), three native congeners are present in the Ocmulgee, the Ocmulgee shiner *Cyprinella callisema*, the bannerfin shiner *C. leedsi*, and the Altamaha shiner *C. xaenura*; the latter is listed as a state endangered species.

Remarks: The Ocmulgee River collections in Georgia represent the first record of this species from the Atlantic Slope (Bart et al. 1994). The introduced population in Kentucky Reservoir, Tennessee, represent the subspecies *C. v. stigmatura* (Etnier and Starnes 1993); specimens taken in Nevada were the subspecies *C. v. venusta* (Branson 1968). If the source of the Ocmulgee fish was the Flint River, those fish may be currently treated as representing the subspecies *C. v. eurystoma* (Gilbert, personal communication). Recently, Gilbert (1998) elevated two of the above named forms to species level (i.e., *Cyprinella stigmaturus* and *C. eurystomus*).

Voucher specimens: Georgia (Georgia Department of Natural Resources, Fort Valley; UGAMNH); Nevada (Nevada Southern University, Las Vegas 1756).

Status: Established in Oklahoma.

Impact of Introduction: Unknown.

Remarks: Smith-Vaniz (1968) reported that a few collections were known from the upper Black Warrior River system (Mobile Bay drainage) of Alabama, and attributed its occurrence to introductions by fishermen. However, Barclay and Howell (1973) argued that stream capture rather than bait bucket introductions may have been the source of invasion. Page and Burr (1991) apparently considered the Black Warrior River system to be part of the species' native range.

Cyprinella whipplei

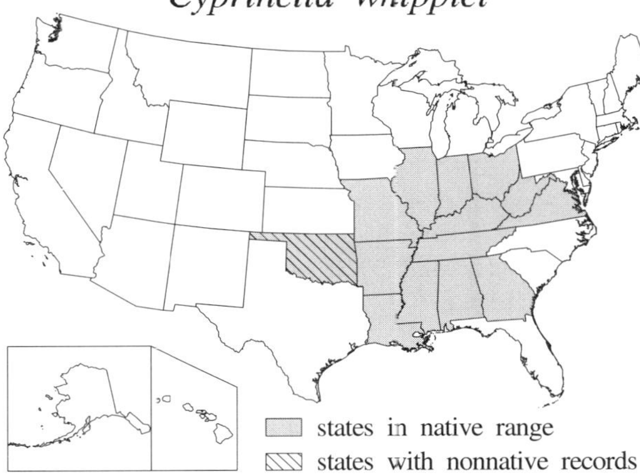

Cyprinella whipplei Girard 1856

Common Name: steelcolor shiner.

Identification: Trautman (1981); Page and Burr (1991); Etnier and Starnes (1993); commonly used name *Notropis whipplei*. Maximum size: 16 cm.

Native Range: Mississippi River basin, from Ohio and West Virginia to Illinois, Missouri, and eastern Oklahoma, and south to northern Alabama and northern Louisiana; Black Warrior River system (Mobile Bay drainage), Alabama. Mostly absent from Coastal Plain (Page and Burr 1991).

Nonindigenous Occurrences: There are sporadic records of this species from western parts of **Oklahoma** that likely represent introductions (Miller and Robison 1973).

Means of Introduction: Likely bait bucket release (Miller and Robison 1973).

Cyprinus carpio

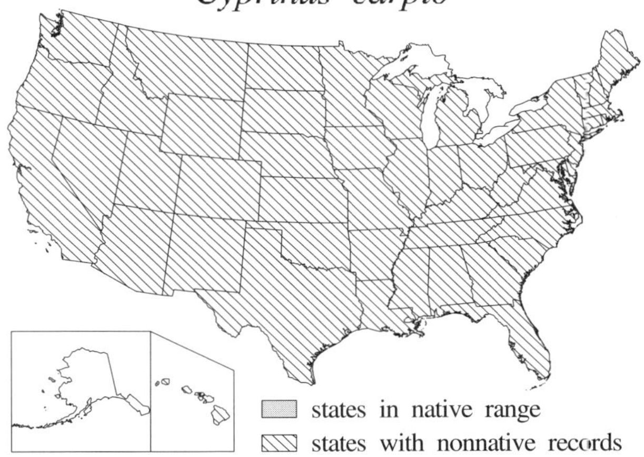

Cyprinus carpio Linnaeus 1758

Common Names: common carp (German carp, European carp, mirror carp, leather carp, koi).

Identification: Wheeler (1978); Becker (1983); Page and Burr (1991); Etnier and Starnes (1993); Jenkins and Burkhead (1994); Balon (1995). In Eurasia there are two poorly defined subspecies *C. c. carpio* and *C. c. haematopterus*; unfortunately, feral common carp, descendants of earlier escapees or introductions, have greatly confused the picture (Balon 1995). Several genetic strains, some bred in aquaculture or used as ornamentals (e.g., leather carp, mirror carp, Israeli carp, koi), are recognized by some as separate varieties (Robison and Buchanan 1988; Balon 1995). Maximum size: 122 cm.

Native Range: Eurasia (Page and Burr 1991; Balon 1995).

Nonindigenous Occurrences: Records of introductions are available for the following states: **Alabama** (Smiley 1886; Baird 1887; McDonald 1887, 1893; Spencer et al. 1964; Smith-Vaniz 1968; Dahlberg and Scott 1971a; Swift et al. 1977; Mettee et al. 1987, 1996; Boschung 1992); **Arizona** (Rule 1885; Taggart 1885; Evermann and Rutter 1895; Gilbert and Scofield 1898; Minckley 1973); **Arkansas** (Baird 1887; McDonald 1887, 1893; Courtenay 1970; Lee et al. 1980 et seq.; Robison and Buchanan 1988); **California** (Smiley 1886; Smith 1896; Shebley 1917; Lampman 1946; La Rivers 1962; Moyle 1976a; Moyle and Daniels 1982; Smith 1982); **Colorado** (Smiley 1886; Baird 1887; McDonald 1887; Ellis 1974; Wiltzius 1981; Woodling 1985); **Connecticut** (Smiley 1886; Ravenel 1896; Webster 1942; Whitworth 1996); **Delaware** (Smiley 1886; Baird 1887; McDonald 1887; Lee et al. 1976; Raasch and Altemus 1991); **Florida** (Baird 1887; McDonald 1887; Anonymous 1892; Courtenay et al. 1974; Swift et al. 1977; Shafland 1995b); **Georgia** (Smiley 1886; Baird 1887; McDonald 1887, 1893; Anonymous 1892; Worth 1895; Ravenel 1896, 1898; Hildebrand 1923; Dahlberg and Scott 1971a, 1971b; Walters 1997); **Hawaii** (Cobb 1902; Jordan and Evermann 1902, 1905; Brock 1960; Maciolek 1984; Devick 1991b); **Idaho** (Smith 1896; Lampman 1946; Linder 1963; Simpson and Wallace 1978; Wydoski and Whitney 1979; Idaho Fish and Game 1990; Sigler and Sigler 1996); **Illinois** (Smiley 1886; Baird 1887; McDonald 1887; Hay 1894; Sweeney 1902; Smith 1979; Emery 1985; Laird and Page 1996); **Indiana** (Smiley 1886; Anonymous 1892; Hay 1894; Sweeney 1902; Blatchley 1938; Gerking 1945; Nelson and Gerking 1968; Emery 1985; Burr and Page 1986); **Iowa** (Cleary 1956; Bailey and Allum 1962; Burr and Page 1986; Harlan et al. 1987); **Kansas** (Smiley 1886; Anonymous 1892; Ravenel 1896; Dyche 1914; Breukelman 1946; Call 1961; Cross 1967; Cross and Collins 1995); **Kentucky** (McDonald 1893; Worth 1895; Ravenel 1896; Clay 1975; Burr and Page 1986; Burr and Warren 1986), **Louisiana** (Baird 1887; McDonald 1887; Anonymous 1892; Douglas and Davis 1967; Douglas 1974); **Maine** (Everhart 1976); **Maryland** (Ferguson 1876; Smiley 1886; McDonald 1887, 1893; Anonymous 1892; Ravenel 1898; Truitt et al. 1929; Schwartz 1963; Lee et al. 1976, 1980 et seq.); **Massachusetts** (Baird 1887; McDonald 1887; Lee et al. 1980 et seq.; Hartel 1992; Hartel et al. 1996); **Michigan** (Smiley 1886; McDonald 1893; Hubbs and Cooper 1936; Hubbs and Lagler 1958; Emery 1985); **Minnesota** (Baird 1887; McDonald 1887, 1893; Moore and Bream 1965; Eddy and Underhill 1974; Phillips et al. 1982; Emery 1985; Burr and Page 1986); **Mississippi** (Smiley 1886; Baird 1887; McDonald 1887, 1893; Cook 1959; Ross and Brenneman 1991); **Missouri** (Smiley 1886; McDonald 1893; Ravenel 1896, 1898; Pflieger 1971, 1975, 1997; Burr and Page 1986); **Montana** (Brown 1971; Courtenay 1985; Holton 1990); **Nebraska** (Bailey and Allum 1962; Morris et al. 1974); **Nevada** (Smith 1896; Miller and Alcorn 1946; Lampman 1946; La Rivers 1962; Deacon and Williams 1984; Scoppettone et al. 1998); **New Hampshire** (Scarola 1973); **New Jersey** (Smiley 1886; Nelson 1890; Ravenel 1898; Fowler 1906, 1952; Stiles 1978); **New Mexico** (Baird 1887; McDonald 1887; Koster 1957; Sublette et al. 1990; Platania 1991); **New York** (Smiley 1886; McDonald 1893; Bean 1903; Werner 1980; Emery 1985; Smith 1985); **North Carolina** (Anonymous 1892; Worth 1895; Ravenel 1896; Cahoon 1953; Menhinick 1991); **North Dakota** (Owen et al. 1981; Power and Ryckman 1998); **Ohio** (Jordan 1882; Smiley 1886; Baird 1887; McDonald 1887, 1893; Anonymous 1892; Trautman 1981; Emery 1985; Burr and Page 1986); **Oklahoma** (McDonald 1893; Bean 1896; Hall 1956; Miller and Robison 1973); **Oregon** (Smith 1896; Lampman 1946; Wydoski and Whitney 1979; Bonds 1994); **Pennsylvania** (Smiley 1886; McDonald 1893; Hendricks et al. 1979; Cooper 1983; Pearson and Krumholz 1984); **Rhode Island** (Lapin, personal communication); **South Carolina** (Baird 1887; McDonald 1887, 1893; Ravenel 1898; Hildebrand 1923; Dahlberg and Scott 1971a, 1971b; Loyacano 1975); **South Dakota** (Anonymous 1892; Worth 1895; Cleary 1956; Shields 1958a, 1958b; Moyle and Clothier 1959; Underhill 1959; Bailey and Allum 1962); **Tennessee** (Baird 1887; McDonald 1887; Anonymous 1892; Bean 1896; Ravenel 1896; Kuhne 1939; Ryon and Loar 1988; Etnier and Starnes 1993); **Texas** (Smiley 1886; Baird 1887; McDonald 1887, 1893; Bean 1896; Baughman 1950; Conner and Suttkus 1986; Howells 1992a); **Utah** (Tanner 1936; Sigler and Miller 1963; Vanicek et al. 1970; Sigler and Sigler 1996); **Vermont** (Countryman 1975; Lee et al. 1980 et seq.); **Virginia** (Smiley 1886; Baird 1887, 1889; McDonald 1887, 1893; Anonymous 1892; Worth 1895; Bean 1896; Ravenel 1896, 1898; Jenkins and Burkhead 1994); **Washington** (Smith 1896; Chapman 1942; Lampman 1946; Wydoski and Whitney 1979; Beecher and Fernau 1983); **West Virginia** (Clay 1962; Stauffer et al. 1995); **Wisconsin** (Johnson and Becker 1980; Becker 1983; Emery 1985; Burr and Page 1986; Fago 1992); and **Wyoming** (Baxter and Simon 1970; Stone 1995).

Means of Introduction: There is some question as to when and where common carp were first introduced into the United States. DeKay (1842) reported that the species was first brought into the United States from France by Henry Robinson of Orange County, New York in 1831 and 1832. In a letter to DeKay, Robinson detailed that he kept the fish in ponds and for several years released one to two dozen carp during the spring in the Hudson River near his residence, thereby creating a commercial fishery for the species. S. F. Baird of the U.S. Fish Commission examined fish taken from the Hudson River, as well as area fish then being sold on the New York markets, and reported that they were goldfish or goldfish hybrids and not true common carp (Redding 1884; Cole 1905). Whitworth (1996) cited early literature indicating common carp had been introduced into Connecticut as early as the 1840s; however, we question the positive identity of the species. Smith (1896) reported that common carp first appeared in the United States in 1872 when J. A. Poppe of Sonoma, California, imported five specimens from Germany and propagated them in private ponds for commercial purposes, mainly distributing them to applicants as a food fish (Smith 1896; Lampman 1946). In 1877, the U.S. Fish Commission imported common carp from Germany and for the next two decades the agency began stocking and distributing the species as food fish throughout much of the United States and its territories (Smiley 1886; Smith 1896; Cole 1905). State fish commissions also were commonly involved in distributing the species (e.g., Johnson and Becker 1980). Records from the early 1880s indicate that common carp stocked in farm ponds frequently escaped into open waters as a result of dam breaks or flood events (Smiley 1886). By 1885, the U.S. Fish Commission was actively stocking lakes and rivers throughout the country, often the fish were released from railroad tank cars at bridge crossing directly into streams (e.g., McDonald 1887). As a result of subsequent population growth and dispersal, common carp spread even further. More recently introductions of common carp have resulted because of the use of juvenile carp as bait fish (e.g., Swift et al. 1977). Various unusual genetic strains of common carp have been introduced into open waters the United States. In addition to the normal scaled carp, the U.S. Fish Commission distributed both mirror carp and leather carp varieties in the late 1800s (Smiley 1886; Cole 1905). Colorful varieties of common carp (i.e., nishikigoi or koi) are kept as pets in garden ponds and some have been introduced to ponds and public water bodies (Balon 1995). However, only a small percentage of common carp records in U.S. open waters are based on koi. Another cultured variety occasionally found in open waters is the Israeli carp (Robison and Buchanan 1988).

Status: Recorded from all states except Alaska; believed to be established in all states listed above except Maine.

Impact of Introduction: The common carp is regarded as a pest fish because of its widespread abundance and because of its tendency to destroy vegetation and increase water turbidity by dislodging plants and rooting around in the substrate, causing a deterioration of habitat for species requiring vegetation and clean water (Cole 1905; Cahoon 1953; Bellrichard 1996; Laird and Page 1996). Available literature indicates common carp may destroy aquatic macrophytes directly by uprooting or consuming the plants, or indirectly by increasing turbidity and thereby reducing light for photosynthesis. Bellrichard (1996) found that alterations in macrophyte biomass are due more to direct effects of common carp as opposed to direct effect In their review of the literature, Richardson et al. (1995) concluded that common carp has had noted adverse effects on biological systems including destruction of vegetated breedings habitat used by both fish and birds, and an increase in turbidity. It stirs up the bottom during feeding, resulting in increased siltation and turbidity (Lee et al. 1980 et seq.). This feeding behavior also destroys rooted aquatic plants that provide habitat for native fish species and food for waterfowl (Dentler 1993). There is also evidence that common carp prey on the eggs of other fish species (Moyle 1976a; Taylor et al. 1984; Miller and Beckman 1996). For this reason, it may be responsible for the decline of the razorback sucker *Xyrauchen texanus* in the Colorado River basin (Taylor et al. 1984). In another case, Miller and Beckman (1996) documented white sturgeon *Acipenser transmontanus* eggs in the stomachs of common carp in the Columbia River. In California, carp have been implicated in the decrease in water clarity in Clear Lake, Lake County, and in the gradual disappearance of native fishes (Moyle 1976a). McCarraher and Gregory (1970) wrote that in 1894 there was documentation that Sacramento perch *Archoplites interruptus* were becoming more scare because carp were destroying their spawning grounds. The Sacramento sucker *Catostomus occidentalis*, is one native species that suffers in the presence of goldfish (Moyle 1976a). Similarly, Laird and Page (1996) stated that common carp may compete with ecologically similar species such as carpsuckers and buffalos. Because this species has been present in many areas since the first surveys, its impacts on many of the native fishes are difficult to determine. Once established in a water body, common carp are difficult and expensive to eliminate (e.g., Cahoon 1953).

Remarks: Balon (1995) reviewed the origin and history of domestication of common carp in Europe and elsewhere. Several agents of the U.S. Fish Commission documented the early years of common carp propagation and stocking in the United States (e.g., Smiley 1886; Smith 1896; Cole 1905). Although this species was popular in the early 1870s as a food fish, common carp fell into wide disfavor soon after and is now considered a nuisance fish because of its abundance and detrimental effects on aquatic habitats. Trautman (1981) found common carp most abundant in streams enriched with sewage or with substantial runoff from agricultural land, but he reported it to be rare or absent in clear, cold waters, and streams of high gradient. Pflieger (1997) reported that the total weight and value of common carp taken by commercial fishermen in Missouri exceeded that of any other fish. Hartel et al. (1996) noted that more than 20,000 common carp were killed by a bacterial disease over a short period of time in the Merrimack River in the late 1970s. Because common carp have a higher salinity tolerance than most freshwater fishes, Swift et al. (1977) hypothesized that it may be spreading from one coastal stream to another through fresh or nearly fresh coastal waters in the Gulf area during periods of heavy rainfall and run-off, periods when salinities are greatly reduced. In their summary table, Bailey and Smith (1981) indicated that *Cyprinus carpio* is widely distributed in the Great Lakes basin.

Danio malabaricus

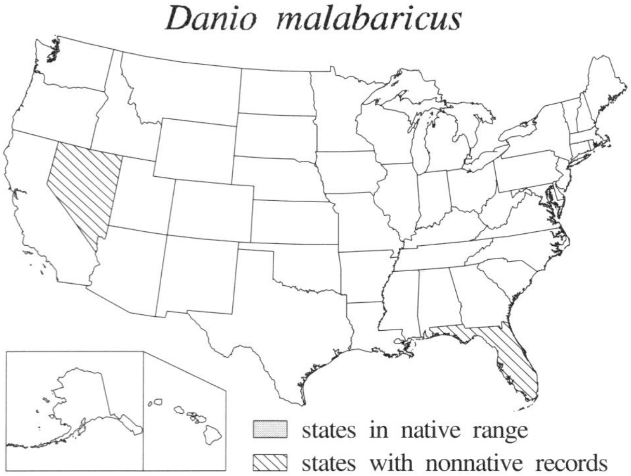

Danio malabaricus (Jerdon 1849)

Common Names: Malabar danio (giant danio).

Identification: Recent systematic review and distinguishing characteristics were provided by Jayaram (1991). A key, distinguishing characteristics, and figure were given in Talwar and Jhingran (1992). Color photographs of live fish appeared in Axelrod et al. (1985). Maximum size: 12 cm.

Native Range: Tropical Asia, India and Sri Lanka (Talwar and Jhingran 1992).

Nonindigenous Occurrences: This species was collected near a fish farm in a canal in Ruskin, Hillsborough County, **Florida**, in or before 1979 (Courtenay and Hensley 1979a). Two specimens were taken in the same county from a ditch adjacent to the Tampa Bypass Canal in November 1993 (museum specimen). It was also discovered in a small roadside borrow pit south of Miami (Shafland 1976). A single specimen was captured in Rogers Spring, Clark County, **Nevada**, on 15 November 1983 (Courtenay, personal communication); that record is apparently the basis for other reports from this site and state (e.g., Courtenay et al. 1984, 1991).

Means of Introduction: The Hillsborough County, Florida, introductions were probably due to releases or escapes from local fish farms. The Nevada introduction likely represents the result of an aquarium release.

Status: Reported from Florida and Nevada.

Impact of Introduction: Unknown.

Remarks: The Malabar danio has long been popular in the aquarium trade and is widely available.

Voucher specimens: Florida (UF 98915); the Nevada specimen was reported as deposited at FAU (Courtenay, personal communication 1983), but that fish is now unavailable, possibly destroyed.

Eremichthys acros

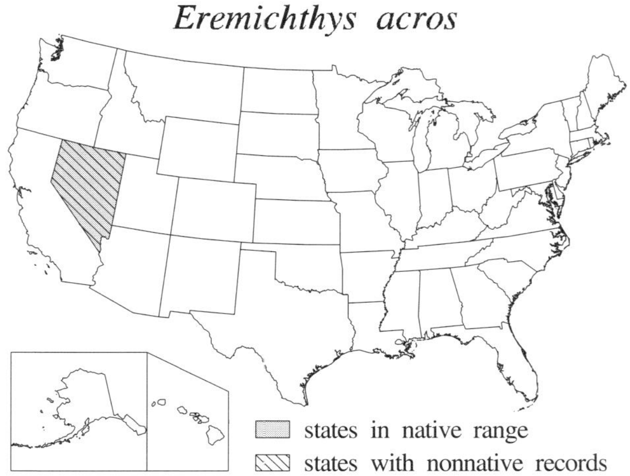

Eremichthys acros Hubbs and Miller 1948

Common Name: desert dace.

Identification: La Rivers (1962); Page and Burr (1991). Maximum size: 7.7 cm.

Native Range: Endemic to warm springs and outflows in Soldier Meadow, Humboldt County, Nevada (Page and Burr 1991).

Nonindigenous Occurrences: Specimens of this species were transplanted, in 1984, into several springs in Soldier Meadow, near its native habitat in Humboldt County, **Nevada** (Clemmer, personal communication).

Means of Introduction: Intentionally introduced for conservation purposes (Hendrickson and Brooks 1991).

Status: Established in Nevada. Populations currently stable, and springs devoid of predaceous exotics (Clemmer, personal communication).
Impact of Introductions: Unknown.

Remarks: Introduction sites were on public lands. Listed as a federally threatened species in 1986 (U.S. Fish and Wildlife Service 1993a). In their list of native southwestern fishes transplanted for conservation purposes, Hendrickson and Brooks (1991) indicated that this species was stocked in one locality with a single successful stocking. Locality information was not provided.

Exoglossum laurae

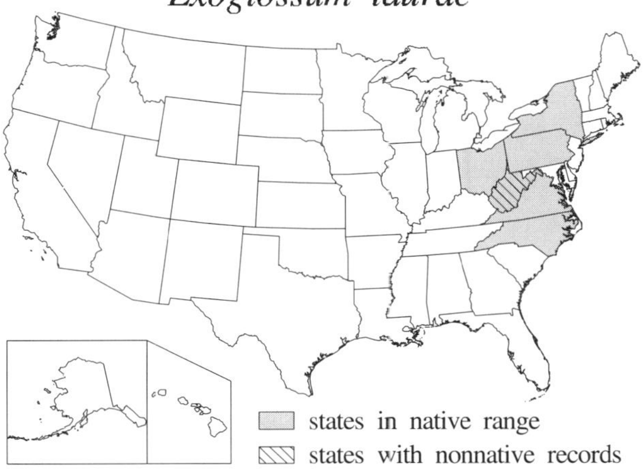

Exoglossum laurae (Hubbs 1931)

Common Name: tonguetied minnow.

Identification: Smith (1985); Page and Burr (1991); Jenkins and Burkhead (1994). Maximum size: 16 cm.

Native Range: Three disjunct areas of upper Ohio River basin; upper Allegheny River drainage, New York and Pennsylvania; upper New River drainage, West Virginia, Virginia, and North Carolina; and Great Miami and Little Miami River systems, Ohio; also in upper Genesee River (Lake Ontario drainage), New York and Pennsylvania (Page and Burr 1991).

Nonindigenous Occurrences: Although the tonguetied minnow is generally regarded as native to the Monongahela system, Ohio River drainage, of **West Virginia**, Stauffer et al. (1995) indicated its presence in that water body may have been the result of a past introduction.

Means of Introduction: Unknown; possible bait bucket release.

Status: Based on information given in Stauffer et al. (1995), this species is apparently established, possibly introduced, in Monongahela drainage, West Virginia.

Impact of Introduction: Unknown.

Remarks: In their discussion of the species, Stauffer et al. (1995) reported that there is one record of *E. laurae* from the Tygart Valley River, a Monongahela River tributary, which they indicated was the only record of the species in West Virginia outside of the New and lower Kanawha drainages. However, only in their table on page 14 of their book did Stauffer et al. (1995) hint that the species may have been introduced to the Monongahela drainage.

Exoglossum maxillingua

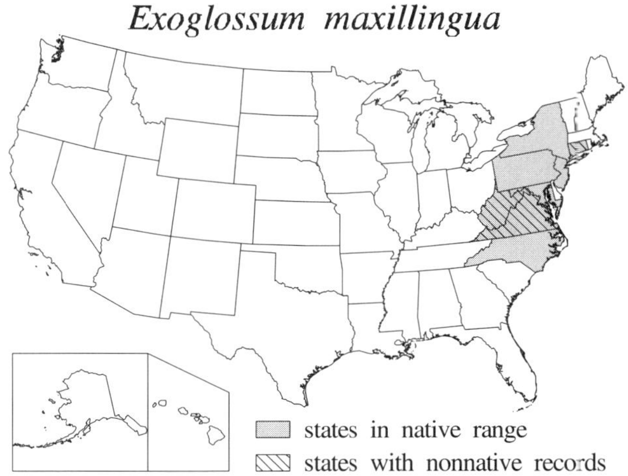

Exoglossum maxillingua (Lesueur 1817)

Common Name: cutlips minnow.

Identification: Smith (1985); Page and Burr (1991); Jenkins and Burkhead (1994). Maximum size: 16 cm.

Native Range: Atlantic Slope from St. Lawrence River drainage, Quebec, to upper Roanoke River, North Carolina (absent in Connecticut River and other rivers in Northeast); Lake Ontario drainage, Ontario and New York (Page and Burr 1991).

Nonindigenous Occurrences: Probably the result of an introduction, a single specimen was taken from the Farmington River, **Connecticut**, in 1988 (Whitworth 1996). This minnow is present in the New River system and Ohio River drainage, of **Virginia** and **West Virginia**, possibly the result of introduction (Page and Burr 1991; Jenkins and Burkhead 1994; Stauffer et al. 1995).

Means of Introduction: Probable bait bucket release. The first records of this species (and hybrids with *C. laurae*) in the New drainage are from 1956 from the Walker Creek system, a heavily fished area (Jenkins and Burkhead 1994).

Status: Established in Virginia and West Virginia. Reported in Connecticut.

Impact of Introduction: Cutlips minnows in the New drainage are known to hybridize with the native tonguetied minnow *E. laurae* (Jenkins and Burkhead 1994).

Remarks: In their treatment of this species in the New-Kanawha drainage, Jenkins and Burkhead (1994) listed it as introduced possibly established, whereas Stauffer et al. (1995) listed the species as regarded as native, but possibly introduced. Jenkins and Burkhead (1994), citing Smith (1905), reported that the occurrence of *E. maxillingua* in the Passaic drainage, New Jersey, may be due to dispersal via a canal.

Gila atraria

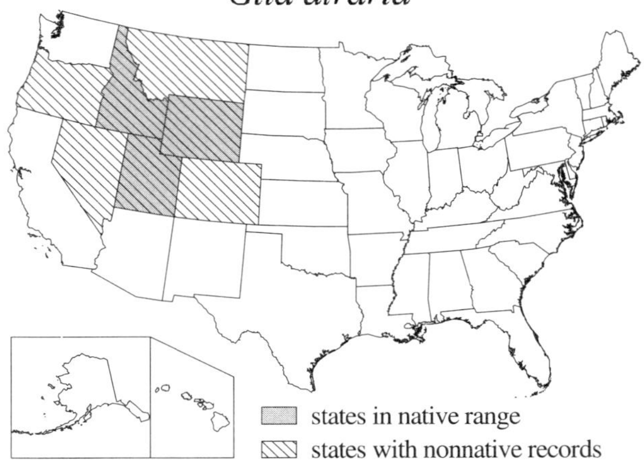

Gila atraria (Girard 1856)

Common Name: Utah chub.

Identification: Simpson and Wallace (1978); Sigler and Sigler (1987); Page and Burr (1991); Bond (1994). Maximum size: 56 cm.

Native Range: Upper Snake River system, Wyoming and Idaho, and Lake Bonneville basin (including Great Salt Lake drainage and Sevier River system), southeastern Idaho and Utah (Page and Burr 1991).

Nonindigenous Occurrences: This species is known from the Green and Yampa rivers in **Colorado** (Tyus et al. 1982). In **Idaho**, it has been recorded from the Snake River near Boise (circa 1970s) (Whitney and Wydoski 1979; Lee et al. 1980 et seq.), and Island Park Reservoir and several other small reservoirs in the upper part of Henrys Fork in the Snake River drainage (Simpson and Wallace 1978). The species was introduced into Hebgen Lake, **Montana**, where it became established and eventually spread to most of the Madison River and as far downstream as Canyon River Reservoir on the Missouri River (Brown 1971; Cross et al. 1986; Holton 1990). It has been introduced into several sites in **Nevada** including Duck Lake (Lincoln County), Shoshone Springs (White Pine County), and both Murphy Spring and Comins Lake in Steptoe Valley (White Pine County) (Miller and Alcorn 1946; La Rivers 1962; Hubbs et al. 1974; Sigler and Sigler 1987; Miller et al. 1991). In **Oregon**, it was introduced to the Owyhee River system (Snake River drainage) (Bond 1994). In **Utah**, it is known from Panguitch Lake in the Sevier drainage, Fish Lake, and reservoirs and streams in the Colorado River drainage (Sigler and Miller 1963; Vanicek et al. 1970;

Holden and Stalnaker 1975b; Lee et al. 1980 et seq.; Tyus et al. 1982; Sigler and Sigler 1987). It has also been recorded from the Green River in **Wyoming** (Baxter and Simon 1970; Lee et al. 1980 et seq.; Tyus et al. 1982; Hubert 1994).

Means of Introduction: Many introductions have been the result of bait bucket releases (Holden and Stalnaker 1975b; Simpson and Wallace 1978; Sigler and Sigler 1987). Hubbs et al. (1974) found evidence that the species may have been introduced to certain sites in the Great Basin by early Mormon settlers. These researchers also speculated that Native Americans may have brought Utah chubs into Shoshone Spring. The species may have been introduced to Murphy Spring, in Steptoe Valley, as forage for sportfish, although its establishment may have resulted from bait bucket releases (Hubbs et al. 1974). In some areas the species has become widespread because of natural dispersal from original points of introduction (e.g., Madison and Missouri rivers in Montana). Holden (1991) stated that it first appeared in Flaming Gorge Reservoir on the Wyoming-Utah border in 1964.

Status: Established in Colorado, Idaho, Nevada, Montana, Utah, Wyoming, and apparently Oregon. It is considered rare or incidental in the Green and Yampa rivers in Colorado, and in the Dolores, Green, and Price rivers in Utah. It is abundant in Flaming Gorge Reservoir in Utah and Wyoming (Tyus et al. 1982).

Impact of Introduction: Introduced populations often reach great abundance and become serious competitors with sport fish, especially trout (Sigler and Miller 1963). For instance, this species has been found to depress growth of kokanee salmon *Oncorhynchus nerka* through competition for food (Teuscher and Luecke 1996). Hubbs et al. (1974) also noted that it has a tendency toward population explosion and habitat dominance in artificial impoundments. Utah chub became a major management concern in Flaming Gorge Reservoir by the late 1960s, in part, because it appeared to compete with trout for planktonic foods and because the species established growth records of its own (Holden 1991). Introduced Utah chub, along with other introduced species, may have replaced the relict dace *Relictus solitarius* at Murphy Spring, Nevada (Hubbs et al. 1974). Predation by, and hybridization with, the Utah chub are considered some of the most serious hazards to the least chub *Iotichthys phlegethontis* in Utah (Sigler and Sigler 1987), a species proposed for federal listing as an endangered species (U.S. Fish and Wildlife Service 1997). Attempts at eradication have been largely unsuccessful and costly (Sigler and Sigler 1987).

Remarks: Hubbs et al. (1974) detailed the introduction history of Utah chub in parts of the Great Basin of western North America. Tyus et al. (1982) gave a distribution map of this species in the upper Colorado basin. There is some uncertainty concerning the exact natural range of this species. For instance, Hubbs et al. (1974) stated that it might be theorized that an ancient outlet into the Bonneville system (where Utah chub is native), or some other stream connection, could have led to the spread of this species into Spring Valley; however, they noted that there is no definitive evidence of such a discharge, and no other indication of the occurrence of Bonneville fishes in Spring Valley. Most authors (e.g., Lee et al. 1980 et seq.; Sigler and Sigler 1987, 1996) have concluded that the Utah chub is native to the Snake River above Shoshone Falls but not below. Simpson and Wallace (1978) noted that the native range of Utah chub in Henrys Fork of the Snake River was restricted to below Mesa Falls; however, the fish was later introduced to Island park Reservoir and several small reservoirs in the area. Many western states have outlawed the use of live fish as bait to prevent the spread of this and other bait fish (Sigler and Sigler 1987).

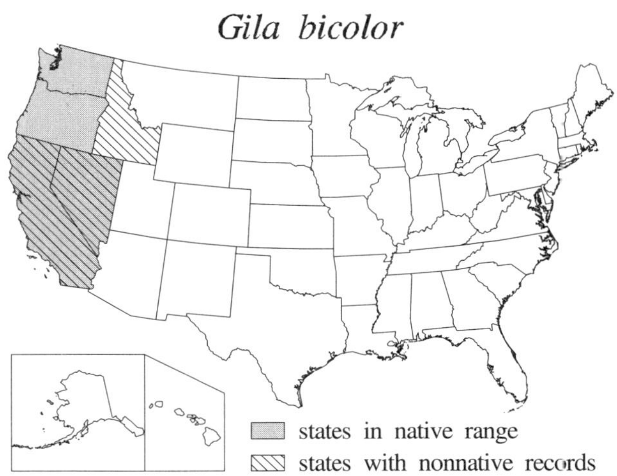

Gila bicolor (Girard 1856)

Common Name: tui chub.

Identification: Moyle (1976a); Sigler and Sigler (1987); Page and Burr (1991). Gilbert (1998) listed 12 valid subspecies. Maximum size: 45 cm.

Native Range: Columbia River drainage, Washington, Oregon, and Idaho, south in Klamath and upper Pit River (Sacramento River drainage), and interior drainages of

Nevada and California to Mohave River, southern California (Page and Burr 1991).

Nonindigenous Occurrences: The tui chub has been introduced into several areas of the Pit and Fall River drainages in **California**, as well as into the Surprise Valley and Central Valley drainages (Moyle 1976a; Moyle and Daniels 1982). A small number of individuals (then treated as *Gila mohavensis*) were introduced to San Felipe Creek in Sentenac Canyon, a western tributary of the Salton Sea, San Diego County, in 1936 and again in 1940 (Miller 1968). In **Idaho** it has been collected from Lucy Peak Reservoir on the Boise River (Simpson and Wallace 1978) and from the Snake River below Shoshone Falls (Idaho Fish and Game 1990). Some records were reported at the subspecies level: *G. b. mohavensis*, the Mohave tui chub, was unsuccessfully stocked at Paradise Spa south of Las Vegas, Clark County, **Nevada** in June 1967 (Hoover and St. Amant 1983; Deacon and Williams 1984; U.S. Fish and Wildlife Service 1984b). It was introduced into at least 12 different sites in **California**; a number of these habitats were ponds or springs associated with developments (e.g., Busch Gardens) (Hoover and St. Amant 1983; U.S. Fish and Wildlife Service 1984b). *Gila b. obesa*, the Lahontan tui chub, was introduced by bait bucket and is now established in Crowley Lake (Mono County), **California** (Moyle 1976a). It also has been collected from the headwaters of the Sacramento River and the Railroad Valley in **Nevada** (La Rivers 1962). *Gila b. snyderi*, the Owens tui chub, was stocked as a precautionary measure in a native fish sanctuary in Inyo County, **California**, where a population is now established (Moyle 1976a). *Gila b.* ssp., an undescribed subspecies referred to as the Hot Creek Valley tui chub, was introduced into Stone Cabin (=Willow Creek) Valley, **Nevada**, in the 1940s (Hubbs et al. 1974; Williams and Williams 1981; Clemmer, personal communication).

Means of Introduction: Bait bucket releases and intentional stockings for the purpose of restoring or preserving a known rare species. The stocking of San Felipe Creek in California was part of a series of experiments to test the effects of changed environment on meristic and morphometric characters (Miller 1968). Fish introduced into Stone Cabin Valley had been brought in from Twin Springs, in Hot Creek Valley (Hubbs et al. 1974); the purpose of that introduction was not made clear.

Status: Established in California, Nevada; reported from Idaho. Populations of *G. b. obesa* in Railroad Valley, Nevada now extirpated; eradicated by Nevada Department of Wildlife (Clemmer, personal communication). In California, the San Felipe Creek population survived for two decades before perishing (Miller 1968). Of the 13 introductions in California and Nevada reported by Hoover and St. Amant (1983), only three were deemed a success (U.S. Fish and Wildlife Service 1984b).

Impact of Introduction: In Crowley Lake, California, introduced Lahontan tui chubs (*G. b. obesa*) have hybridized with endangered Owens tui chubs (*G. b. snyderi*) (Moyle 1976a, 1976b). Tui chubs often overpopulate lakes and reduce or eliminate trout, and compete with trout and largemouth bass for food (Burns 1966e).

Remarks: The Mohave tui chub *G. b. mohavensis* was nearly extirpated through hybridization with the introduced arroyo chub *G. orcutti* (Moyle 1976b). Both the Mohave tui chub and the Owens tui chub are federally listed endangered subspecies (U.S. Fish and Wildlife Service 1993a). Prior to the elimination of the U.S. Fish and Wildlife Service candidate species list, sixteen subspecies of *Gila bicolor* were under review for Federal listing as endangered or threatened taxa (U.S. Fish and Wildlife Service 1994a). Only the Cowhead Lake tui chub *G. b. vaccaceps* is on the current candidate list (U.S. Fish and Wildlife Service 1997). Some native populations of the Owens tui chub were eliminated by introduced trout and largemouth bass (Swift et al. 1993).

Gila coerulea

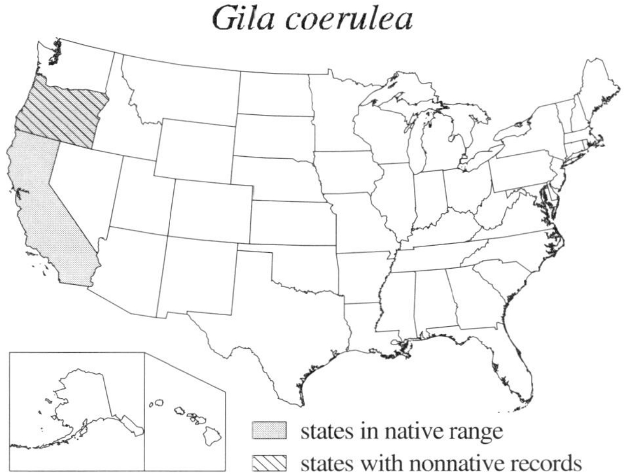

Gila coerulea Girard 1856

Common Name: blue chub.

Identification: Moyle (1976a); Page and Burr (1991); Bond (1994). Maximum size: 41 cm.

Native Range: Klamath and Lost River systems of Oregon and California (Lee et al. 1980 et seq.; Page and Burr 1991).

Nonindigenous Occurrences: This species was introduced into some Cascade lakes in **Oregon** (Bonds 1994); specific sites included Paulina Lake and East Lake (C. Bond, personal communication; D. Logan, personal communication). It also may have been introduced into Diamond Lake, Oregon (P. M. Harris, personal communication).

Means of Introduction: Unknown. According to Logan (personal communication), Oregon introductions took place during the 1920s.

Status: Introductions into Paulina and East lakes in Oregon were unsuccessful (Logan, personal communication). Status of Diamond Lake introduction unknown.

Impact of Introduction: Unknown.

Remarks: Lee et al. (1980 et seq.) noted that it had been introduced outside its native range, but nonindigenous sites apparently were not mapped. McPhail and Lindsey (1986) listed it as introduced to the Middle Columbia drainage, but did not provide details. Paulina and East lakes are calderas and both are within the same crater, a remnant of Mt. Newberry. According to Logan (personal communication), neither lake had naturally occurring fish, though a few salmonid species have been successfully introduced. East Lake does not have an outlet, but Paulina Lake does have an outlet (Paulina Creek) which is in the Deschutes-Columbia River drainage.

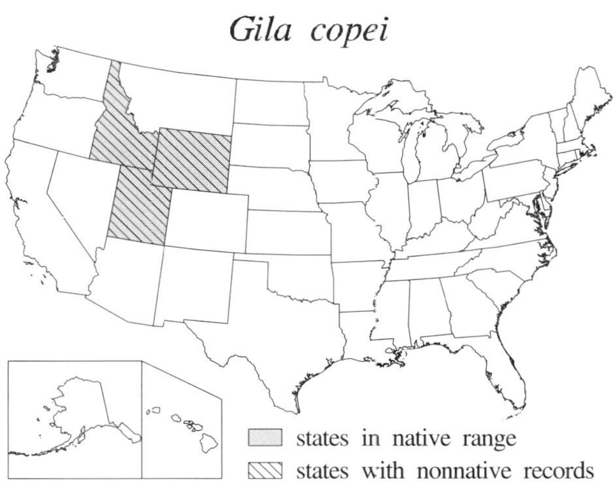

Gila copei (Jordan and Gilbert 1881)

Common Name: leatherside chub.

Identification: Simpson and Wallace (1978); Sigler and Sigler (1987); Page and Burr (1991). Maximum size: 15 cm.

Native Range: Upper Snake River system, Wyoming and Idaho, south to Sevier River system, southern Utah (Page and Burr 1991).

Nonindigenous Occurrences: The leatherside chub was introduced into Strawberry Reservoir, **Utah**, circa 1948 and into the Price River (Colorado drainage) prior to 1963 (Sigler and Miller 1963; Tyus et al. 1982). It also has been introduced into the Dirty Devil River (a Colorado River tributary) in southeastern Utah (specimens as NCSM) (W. Starnes, personal communication), and the Fremont River, a tributary of the Dirty Devil (Tyus et al. 1982). It may have been introduced into the Snake River drainage and the Bonneville basin, **Idaho** (Wydoski and Whiney 1979; see remarks); and the upper part of the Snake River in **Wyoming**, where it first was collected in 1934 (Sigler and Miller 1963).

Means of Introduction: Probably bait bucket introductions.

Status: Established in Utah; locally established in Wyoming. Presumably established in Idaho.

Impact of Introduction: Unknown.

Remarks: Although Simpson and Wallace (1978) listed this species as native to the upper South Fork of the Snake River in Idaho, we believe it is possible that it was introduced because of the fact that it was not collected from the upper Snake until the 1930s. In their list of fishes found in Idaho, Wydoski and Whitney (1979, Appendix 7) seemed to suggest that this species was introduced into the Snake River drainage and Bonneville basin in Idaho. However, Sigler and Sigler (1987) listed it as native to the eastern Bonneville basin which includes Idaho.

Gila orcutti

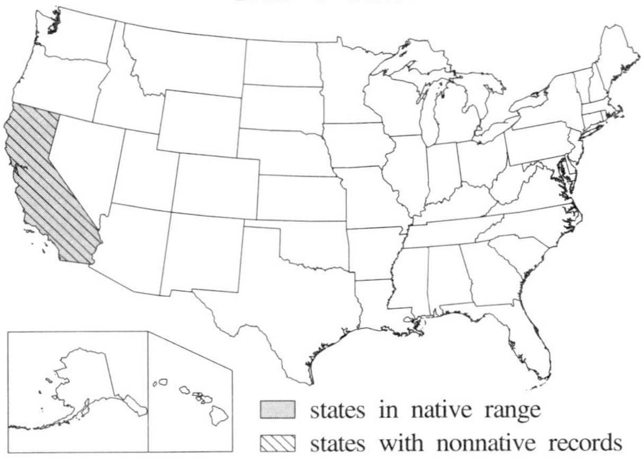

Gila orcutti (Eigenmann and Eigenmann 1890)

Common Name: arroyo chub.

Identification: Moyle (1976a); Page and Burr (1991). Maximum size: 40 cm.

Native Range: Malibu Creek, Santa Clara, San Luis Rey, and Santa Margarita River drainages, southern California (Page and Burr 1991).

Nonindigenous Occurrences: The arroyo chub was introduced into various sites in **California** including the Santa Maria–St. Inez (Santa Ynez), Mojave, and Cuyama river drainages (Miller 1968; Moyle 1976a; Sigler and Sigler 1987; Page and Burr 1991). The species has also been recorded from the Santa Clara River system where it was first collected in 1934 (Miller 1968; Bell 1978). A small number of individuals were planted in the lower part of Sentenac Canyon, in San Felipe Creek, a western tributary of the Salton Sea, San Diego County, in 1950 (Miller 1968).

Means of Introduction: Bait bucket release, stock contamination with trout and mosquitofish in the 1930s and 1940s (Swift et al. 1993). Although originally thought to be native to the Mohave River, it was later discovered that the species' presence was the result of bait introductions about 1930 by trout anglers (Miller et al. 1991). The stocking of San Felipe Creek was part of a series of experiments to test the effects of changed environment on meristic and morphometric characters (Miller 1968).

Status: Established in parts of California outside its native range. The San Felipe population did not survive (Miller 1968).

Impact of Introduction: Introduced *G. orcutti* hybridized with endangered Mohave tui chub *G. bicolor mohavensis*; since its introduction in the 1930s the continued mass hybridization is believed to have replaced all, or most, pure stock of the tui chub in the Mohave River drainage (Hubbs and Miller 1943; Moyle 1976a, 1976b; Miller et al. 1991; Swift et al. 1993).

Remarks: Species of special concern in California. There is some disagreement in the literature concerning the extent of this species' native distribution. Miller (1968) and Bell (1978) concluded that its occurrence in the Santa Clara River system probably represented an introduction. Lee et al. (1980 et seq.) indicated that *G. orcutti* was possibly introduced into Malibu Creek and that it was widely introduced north to Santa Inez River and Mojave River of Death Valley drainage. In contrast, Moyle (1976a) included the upper Santa Clara River system as part of this species' native distribution. Page and Burr (1991) also included the Santa Clara drainage in its native range. Hubbs and Miller (1943) provided a detailed description of *G. bicolor mohavensis* × *G. orcutti* hybrids. They also found evidence of backcrossing.

Gila pandora

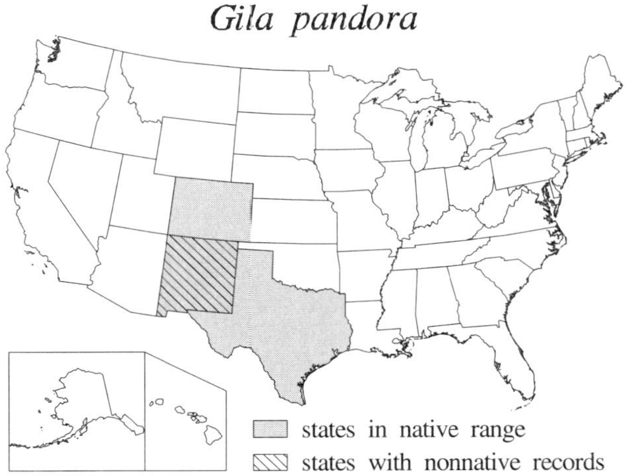

Gila pandora (Cope 1872)

Common Name: Rio Grande chub.

Identification: Woodling (1985); Sublette et al. (1990); Page and Burr (1991). Maximum size: 18 cm.

Native Range: Upper Rio Grande and Pecos River systems, Colorado and New Mexico; isolated population in Davis Mountains (Pecos River system), Texas (Page and Burr 1991).

Nonindigenous Occurrences: The Rio Grande chub was introduced into the headwaters of the Canadian River in **New Mexico** (Cross et al. 1986; Page and Burr 1991).

Means of Introduction: Unknown.

Status: Reported from New Mexico from an area thought to be outside its native range.

Impact of Introduction: Unknown.

Remarks: Sublette et al. (1990) stated that the *Gila pandora* is "possibly native to the Canadian drainage although it may be introduced there." They also noted that the species hybridizes with longnose dace *Rhinichthys cataractae*.

Gila robusta

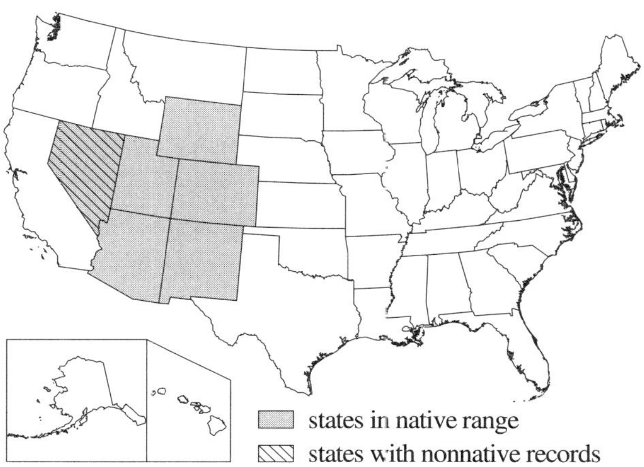

Gila robusta Baird and Girard 1853

Common Name: roundtail chub

Identification: La Rivers (1962); Minckley (1973); Sublette et al. (1990); Page and Burr (1991); Sigler and Sigler (1996). Composed of four subspecies: *G. r. robusta*, *G. r. grahami*, *G. r. jordani*, and *G. r. minacae* (Gilbert 1998). Maximum size: 43 cm.

Native Range: The Colorado River basin in Wyoming, Colorado, Utah, Nevada, New Mexico, and Arizona.

Nonindigenous Occurrences: The White River form of the Virgin River chub (identified as *G. r. jordani*) was introduced into spring-fed refugium ponds constructed at Shoshone, White Pine County, **Nevada**, in September 1972 (Hubbs et al. 1974).

Means of Introduction: Intentionally stocked for conservation purposes as part of an effort to insure the survival of an imperiled species and subspecies of desert fishes (Hubbs et al. 1974). According to Hubbs et al. (1974) 20 young specimens were transplanted, these taken from the White River, Pahranagat Valley, between the outlets of Crystal and Ash springs.

Status: Hubbs et al. (1974) noted that some fish were still living in the pond nearly a month after introduction. However, the species supposedly did not persist after 1979 (see Remarks section).

Impact of Introduction: Unknown.

Remarks: This fish is protected as federally endangered (as *G. r. jordani*, Page and Burr 1991; U.S. Fish and Wildlife Service 1994a). Information in our files presents an apparent conflict concerning the identity of the *Gila* introduced into Shoshone Ponds. It was suggested that a Moapa population of the Virgin River chub *Gila seminuda* may have been introduced into Shoshone Ponds in 1972 (the stocked population reportedly did not persist after 1979) (Clemmer, personal communication). Until additional documentation is made available, we are following the identification made by Hubbs et al. (1994) and we are assuming only one introduction (of a single *Gila* form) took place. Until recently, authors treated the Virgin River chub as a subspecies of the roundtail chub *Gila robusta* (i.e., *G. r. seminuda*). The form *seminuda*, which is considered to be of hybrid origin (*Gila robusta* × *Gila elegans*), was recently elevated to a full species by DeMarais et al. (1992); this was confirmed by Dowling and DeMarais (1993).

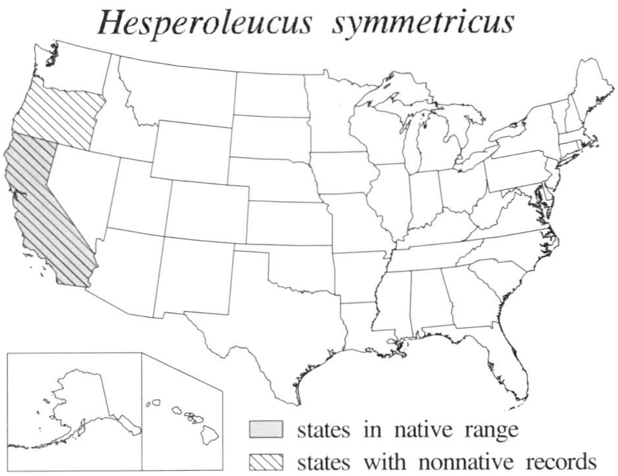

Hesperoleucus symmetricus (Baird and Girard 1854)

Common Name: California roach.

Identification: Moyle (1976a); McGinnis (1984); Page and Burr (1991); Bond (1994). Placed in the genus *Lavinia* by some researchers. Maximum size: 11 cm.

Native Range: Sacramento–San Joaquin and Pajaro-Salinas River drainages, California (Page and Burr 1991).

Nonindigenous Occurrences: This species is known from the Cuyama River, San Luis Obispo and Santa Barbara counties, **California** (Moyle 1976a). It may also have been introduced into the Eel River of California (Fite 1973, cited by Moyle 1976a). In **Oregon**, it is known from the Warner Valley (Great Basin) (Sigler and Sigler 1987) and, probably in reference to the same record or site, from the Summer Lake Basin (Bond 1994).

Means of Introduction: Populations in the Cuyama River, California are presumably the result of bait-bucket introduction (Moyle 1976a). Unknown in Oregon.

Status: Established in at least one river system in California outside its native range (Moyle 1976a). Reported from Oregon (Sigler and Sigler 1987).

Impact of Introduction: Unknown.

Remarks: Sigler and Sigler (1987), apparently citing Moyle (1976a), stated that California roach may have been introduced into the Cuyama River of California, although earlier natural stream capture cannot be ruled out. We found no mention of possible stream capture in that earlier reference. The California roach can survive where most other fishes cannot because of their small size and tolerance of low oxygen levels and high temperatures. Nevertheless, it has declined or disappeared from parts of its native range (the upper San Joaquin, Fresno, and Chowchilla river systems in central California) as a result of a combination of factors, one being predation by the introduced green sunfish *Lepomis cyanellus* (Moyle 1976b).

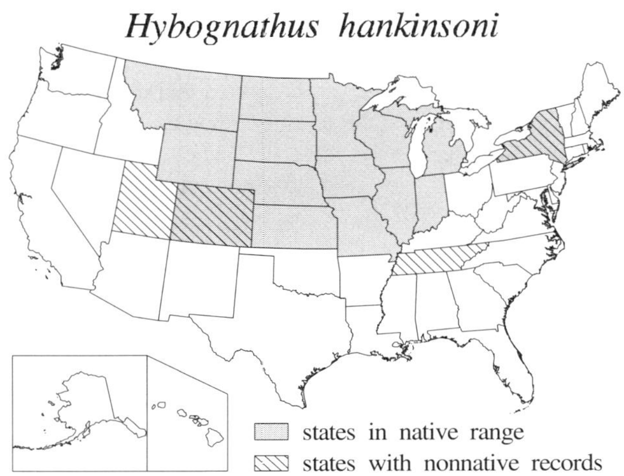

Hybognathus hankinsoni Hubbs 1929

Common Name: brassy minnow.

Identification: Becker (1983); Holton (1990); Page and Burr (1991); Pflieger (1997). Maximum size: 9.7 cm.

Native Range: Upper St. Lawrence River and Lake Champlain drainages, Quebec and New York, across Great Lakes, Hudson Bay, and Missouri–upper Mississippi River basins of southern Canada and northern United States south to Kansas; Fraser River system (Pacific Slope), Alberta and British Columbia (Page and Burr 1991).

Nonindigenous Occurrences: The brassy minnow has been recorded from the Colorado River, **Colorado** (Tyus et al. 1982; Walker 1993), the Susquehanna River in **New York** (Smith 1985), the Powell River in **Tennessee** (Etnier and Starnes 1993), and from an unspecified location in **Utah** (B. Schmidt, personal communication).

Means of Introduction: Its presence in the Susquehanna River drainage, New York, may have been by way of dispersal through the Chenango Canal (Smith 1985). According to Etnier and Starnes (1993), the Tennessee record was based on a single specimen taken in 1973; they considered it a bait bucket introduction. Records in Colorado also may be the result of bait bucket releases. There is no information for means of its introduction into Utah.

Status: Reported from Colorado, New York, Tennessee, and Utah.

Impact of Introduction: Unknown.

Remarks: Pflieger (1997) noted that small populations of this species found in two tributaries of the upper Mississippi River in Missouri may be the result of recent bait bucket releases.

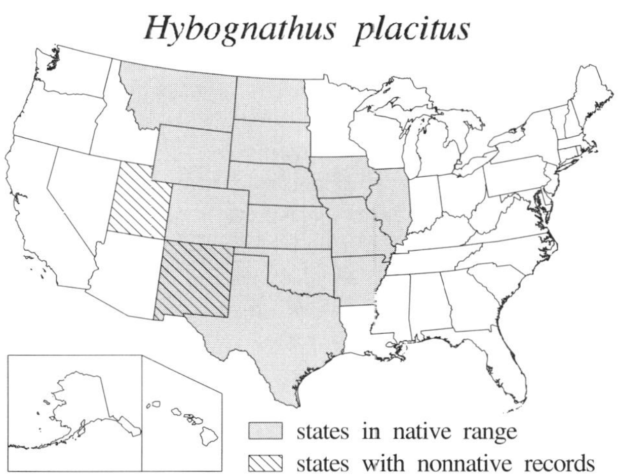

Hybognathus placitus Girard 1856

Common Name: plains minnow.

Identification: Woodling (1985); Robison and Buchanan (1988); Sublette et al. (1990); Page and Burr (1991). Maximum size: 13 cm.

Native Range: Missouri, Arkansas, Red, Brazos, and Colorado River drainages, from Montana and North Dakota south to New Mexico and Texas; Mississippi River from mouth of Missouri River to mouth of Ohio River (Page and Burr 1991).

Nonindigenous Occurrences: This species is known from the Pecos River of **New Mexico** (Bestgen et al. 1989; Sublette et al. 1990; B. Burr, personal communication). It has also been found in the San Juan River drainage, Colorado River basin, in **Utah** (Tyus et al. 1982).

Means of Introduction: Bait bucket release. It was apparently introduced in 1968 into the Pecos River system of New Mexico from the Canadian drainage (Sublette et al. 1990).

Status: Established and very abundant in the Pecos River in New Mexico. Bestgen et al. (1989) stated that the species is now the most abundant fish in the Pecos River and accounted for more than 80% of the catch at some sites. Tyus et al. (1982) listed it as an incidental record in the Colorado River basin, a category that was defined as occurring in very low numbers and known from only a few point collections.

Impact of Introduction: The disappearance of the native Rio Grande silvery minnow, *Hybognathus amarus*, has been attributed to wholesale displacement by introduced *H. placitus*. *Hybognathus amarus* was last taken in the Pecos River in 1968 (Bestgen et al. 1989).

Remarks: None.

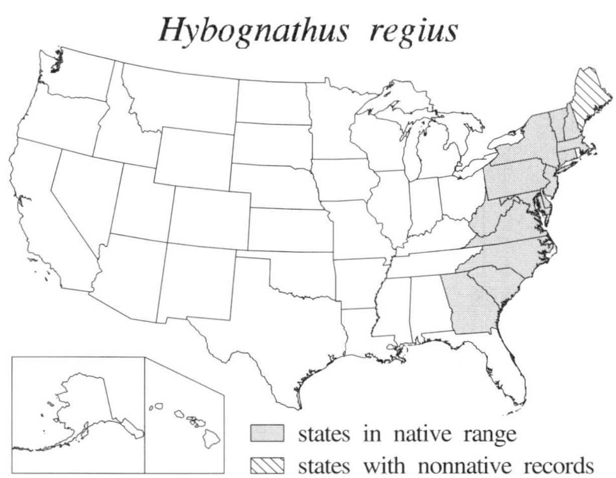

Hybognathus regius Girard 1856

Common Name: eastern silvery minnow.

Identification: Smith (1985); Page and Burr (1991); Jenkins and Burkhead (1994). Maximum size: 12 cm.

Native Range: Atlantic Slope from St. Lawrence River drainage, Quebec, to Altamaha River drainage, Georgia; Lake Ontario drainage, Ontario and New York (Page and Burr 1991).

Nonindigenous Occurrences: This species is known from Fish Brook in the Kennebec River drainage in Fairfield, Kennebec County, **Maine** (Kircheis 1994).

Means of Introduction: First observed in Maine in 1982, probably bait bucket release of illegally imported baitfish (Kircheis 1994).

Status: Established in Maine. The species was first observed in the Kennebec River drainage in 1982; first confirmed reproduction also was 1982 (Kircheis 1994).

Impact of Introduction: Unknown.

Remarks: None.

Hybopsis amblops

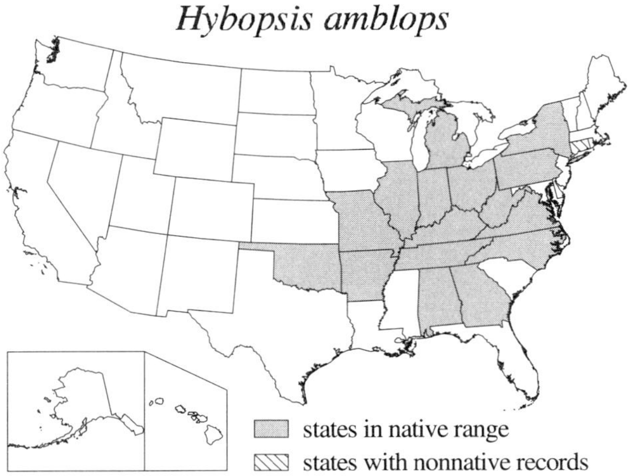

Hybopsis amblops (Rafinesque 1820)

Common Name: bigeye chub.

Identification: Robison and Buchanan (1988); Page and Burr (1991); Jenkins and Burkhead (1994). Mayden (1989) and Mayden et al. (1992) use *Hybopsis*, other authorities (e.g., Coburn and Cavender 1992; Gilbert 1998) have argued that this and related species should be in the genus *Notropis* (i.e., *Notropis amblops*). It is presently uncertain which of the two generic names will ultimately be accepted as valid. Maximum size: 9 cm.

Native Range: Lake Erie drainage, New York, Pennsylvania, Ohio, and Michigan; Ohio River basin from New York to eastern Illinois, and south to Tennessee River drainage, Georgia and Alabama; Ozarks of southern Missouri, northern Arkansas, and northeastern Oklahoma (absent in Missouri River drainage) (Page and Burr 1991).

Nonindigenous Occurrences: A single specimen was taken from Mopus Brook, a tributary of the lower Hudson River, Fairfield County, **Connecticut**, in 1987 (Whitworth 1996).

Means of Introduction: Fish was probably introduced as a result of a bait bucket release.

Status: Reported in Connecticut; more recent sampling in state has failed to produce any additional specimens (Whitworth 1996).

Impact of Introduction: Unknown.

Remarks: Swift et al. (1977) reported two collections, of what was tentatively referred to as *Hybopsis amblops*, from the Ochlockonee drainage in Florida in 1967 (voucher specimens, UF 65225). These fish are now considered to represent a new species, one closely allied to *Hybopsis winchelli* (=*Notropis winchelli*), whose native range is thought to extend from the Perdido River in Alabama and Florida to the Apalachicola basin of Alabama, Florida, and Georgia (Clemmer 1970; Gilbert 1998). The Ochlockonee specimens of this undescribed species are most likely the result of introduction. Support for this conclusion, as stated by Swift et al. (1977), is based on the "recent vintage of these collections and the fact that extensive previous collecting by many investigators did not disclose the species..."

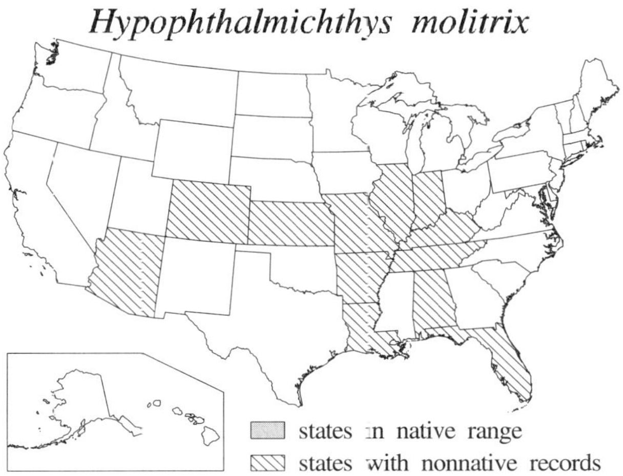

Hypophthalmichthys molitrix (Valenciennes 1844)

Common Name: silver carp.

Identification: Distinguishing characteristics were given in Berg (1949). Keys that include this species and photographs or illustrations are provided in several of the more recently published state and regional fish books (e.g., Robison and Buchanan 1988; Etnier and Starnes 1993; Pflieger 1997). Maximum size: about 1 m and 27 kg.

Native Range: Several major Pacific drainages in eastern Asia from the Amur River of far eastern Russia south through much of eastern half of China to Pearl River, possibly including northern Vietnam (Berg 1949; Li and Fang 1990).

Nonindigenous Occurrences: This species has been recorded from the Black Warrior and Tallapoosa river drainages of the Mobile Basin, including Yates Reservoir, in **Alabama** (Mettee et al. 1996; J. Hornsby and M. Pierson, personal communication); from **Arizona** (Courtenay et al. 1991; Rinne 1995); and from the Arkansas and White river systems and the Mississippi River mainstem in **Arkansas** (Freeze and Henderson 1982; Carter and Beadles 1983; Courtenay et al. 1984; Robison and Buchanan 1988). It has been stocked in water treatment ponds on the East Slope of **Colorado** (D. Horak, personal communication). A single specimen was taken in August 1994 from St. Andrews Bay at the Deer Point Lake spillway, Bay County, **Florida** (Middlemas 1994). It has also been collected or reported from several water bodies in, or bordering, **Illinois**, including the Mississippi and Ohio rivers and several of their tributaries, the Muddy River, Horseshoe Lake and vicinity in the Cache River Drainage (Burr 1991; Burr et al. 1996; Laird and Page 1996) and the Embarras River below Lake Charleston (K. Cummings, personal communication). There are also records of this species from the southeastern part of **Indiana** (presumably the Ohio River) (Courtenay et al. 1991; Simon et al. 1992); **Kansas** (Courtenay et al. 1991); from the Ohio River in **Kentucky** (Pearson and Krumholz 1984; Burr and Warren 1986); from the lower Mississippi River and many tributary sites in **Louisiana** including the Atchafalaya, Red, Boeuf, Old, Ouachita, and Little river drainages, LaFourche Canal, Miller Lake, and Loggy Bayou (Freeze and Henderson 1982; Carp Task Force 1989; Douglas et al. 1996; F. Bryan and J. Hughes Little, personal communication); from the Mississippi and Missouri river mainstems in **Missouri** (Courtenay et al. 1991; Robinson 1995; Pflieger 1997); and from a Mississippi River outflow in **Tennessee** (C. Saylor, personal communication).

Means of Introduction: This species was imported and stocked for phytoplankton control in eutrophic water bodies and also apparently as a food fish. It was first brought into the United States in 1973 when a private fish farmer imported silver carp into Arkansas (Freeze and Henderson 1982). By the mid 1970s the silver carp was being raised at six state, federal, and private facilities, and by the late 1970s it had been stocked in several municipal sewage lagoons (Robison and Buchanan 1988). By 1980 the species was discovered in natural waters, probably a result of escapes from fish hatcheries and other types of aquaculture facilities (Freeze and Henderson 1982). The occurrence of silver carp in the Ouachita River of the Red River system in Louisiana was likely the result of an escape from an aquaculture facility upstream in Arkansas (Freeze and Henderson 1982). The Florida introduction was probably a result of stock contamination, a silver carp having been inadvertently released with a stock of grass carp being used for aquatic plant control (Middlemas 1994). In a similar case, the species was apparently introduced accidentally to an Arizona lake as part of an intentional, albeit illegal, stock of diploid grass carp (W. Silvey, personal communication). Pearson and Krumholz (1984) suggested that individuals taken from the Ohio River may have come from plantings in local ponds or entered the Ohio River from populations originally introduced in Arkansas.

Status: Records are available for 12 states. It is apparently established in Louisiana (Douglas et al. 1996) and is possibly established in Illinois; silver carp have been reported in Alabama, Arizona, Arkansas, Colorado, Florida, Indiana, Kansas, Kentucky, Missouri, and Tennessee. Douglas et al. (1996) collected more than 1600 larvae of this genus from a backwater outlet of the Black River in Louisiana in 1994. Burr et al. (1996) found young-of-the-year in a ditch near Horseshoe Lake and reported this as the first evidence of successful spawning of silver carp in Illinois waters and the United States. They felt that the species would be 'established' in the state within the next ten years. Based on the occurrence of juvenile fish in Illinois waters, Pflieger (1997) felt that successful spawning of silver carp in Missouri seems inevitable. In the early 1980s commercial fishermen in Arkansas had caught 166 silver carp from seven different sites; however, during an intensive 1980-1981 survey to determine the distribution and status of bighead and silver carp in state open waters, Arkansas Game and Fish Commission personnel were unsuccessful in procuring any additional specimens (Freeze and Henderson 1982). Although Arkansas state personnel did not find young-of-the-year fish, several specimens taken by the commercial fishermen were sexually mature and exhibited secondary sexual characteristics (Freeze and Henderson 1982). Nevertheless, Robison and Buchanan (1988) reported that there was still no evidence of natural reproduction in Arkansas waters. Rinne (1995) listed silver carp as introduced to Arizona in 1972 and denoted it as established. Apparently in reference to the same record, William Silvey of the Arizona Game and Fish Department recently informed us that the only silver carp documented in Arizona open waters was a population inhabiting an urban lake in Chandler during the early 1970s. That population, along with a large population of diploid grass carp, was exterminated in 1975 or 1976 by personnel from the Arizona Game and Fish Department and Arizona State University (Silvey, personal communication). Pearson and Krumholz (1984) documented records from the Ohio River, but they did not include it as one of the species that exist in well-established, reproducing populations. Etnier and Starnes (1993) provided information on silver carp, but by publication they were unaware of any records of the species in the state of Tennessee.

Impact of Introduction: Pflieger (1997) considered the impact of this species difficult to predict because of its place in the food web. In numbers, the silver carp has the potential to cause enormous damage to native species because it feeds on plankton required by larval fish and native mussels (Laird and Page 1996). This species would also be a potential competitor with adults of some native fishes, for instance, gizzard shad, that also rely on plankton for food (Pflieger 1997).

Remarks: Jenkins and Burkhead (1994) reported on the use of silver carp in a wastewater treatment pond in the upper James River drainage of Virginia. However, there is as yet no record of the species in Virginia open waters. One of the so-called Chinese carps, the silver carp is a filter-feeder capable of taking large amounts of phytoplankton. Its diet also includes zooplankton, bacteria, and detritus (Leventer 1987). This species has been intensively cultured in many parts of the world, often raised in combination with other fishes.

Voucher specimens: Florida (UF 98162); Illinois (SIUC 17716, 23043, 23046, 24415; INHS 88425); Louisiana (NLU 65811, 66858, 66859).

Hypophthalmichthys nobilis

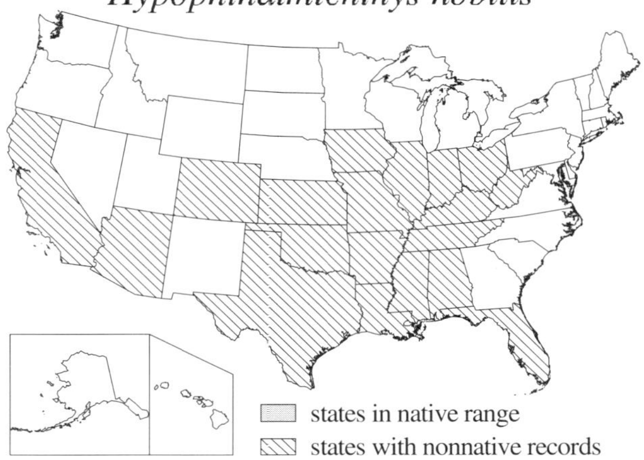

Hypophthalmichthys nobilis (Richardson 1845)

Common Name: bighead carp.

Identification: Distinguishing characteristics were given in Berg (1949) and Jennings (1988). Distinguishing characteristics, along with keys that include this species and photographs or illustrations also were included in a few of the more recently published state fish books (e.g., Robison and Buchanan 1988; Etnier and Starnes 1993; Pflieger 1997). A commonly used name is *Aristichthys nobilis*. Maximum size: 40 kg and 71.6 cm (Jennings 1988).

Native Range: Southern and central China (Li and Fang 1990; Robins et al. 1991b).

Nonindigenous Occurrences: This species has been recorded from the Black Warrior and Tallapoosa river drainages of the Mobile Basin, including Yates Reservoir, in **Alabama** (Mettee et al. 1996; Hornsby and Pierson, personal communication). This species has been recorded from **Arizona** (Silvey, personal communication). In **Arkansas**, it has been taken from the Arkansas River just upstream of Pine Bluff and stocked in several municipal sewage lagoons in the state (Robison and Buchanan 1988; Courtenay et al. 1991). In **California**, 26 specimens were taken from one of three ponds in a small drainage adjacent to Brannin Creek, Tehama County, in the Sacramento River basin, in 1992 (Dill and Cordone 1997). It also has been stocked in water treatment ponds on the East Slope of **Colorado** (Horak, personal communication). This species has been recorded from **Florida** (Shafland 1995b) including Lake Okeechobee (D. Fox, personal communication); from several water bodies in, or bordering, **Illinois**, including the Mississippi and Ohio rivers and several of their tributaries, the Cache, Big Muddy, and Kaskaskia river drainages, and Horseshoe Lake in Madison County (Burr 1991; Burr et al. 1996; Tucker et al. 1996), the Illinois River (Cummings, personal communication), and Chain Lake, Schuyler County, Illinois (Jennings 1988); the Ohio River at mile marker 919 on the **Indiana-Kentucky** border (Freeze and Henderson 1982; Jennings 1988); the Chariton River below Rathbun Lake, and the Des Moines River, in **Iowa** (J. Bruce, personal communication); the Kansas, Delaware, and Wakarusa rivers in **Kansas** (Cross and Collins 1995); several water bodies in **Louisiana**, including the Atchafalaya River and Turkey Creek (Carp Task Force 1989) and the Red-Ouachita River (Douglas et al. 1996); from several water bodies in, or bordering, **Missouri**, including the Mississippi River mainstem, the Missouri Osage, and Salt rivers, among others (Bennett 1988; Robinson 1995; Tucker et al. 1996; Pflieger 1997); various water bodies in **Mississippi**, including the Yazoo and Pacagoula river drainages in **Mississippi** (Douglas et al. 1996) and along the main channel of the Mississippi River of the state below a wing dike (L. G. Nico, personal communication); Lake Erie, **Ohio**, off Cedar Point (T. Cavender, personal communication); several sites in the Grand River drainage of **Oklahoma**, including the Neosho River in Ottawa County, Grand River in Mayes County, upper Grand Lake, and Lake Hudson (Pigg et al. 1993, 1997); the Mississippi River mainstem, the Hatchie River, and the Tennessee River near Lake Barkley, **Tennessee** (Etnier and Starnes 1993; Anonymous 1995a); Victor Braunig Reservoir, **Texas** (Howells 1992a); and the Ohio River at Moundsville, **West Virginia** (M. Hoeft, personal communication).

Means of Introduction: Bighead carp were first imported into the United States in 1972 by a private fish farmer in Arkansas who wanted to use them in combination with other phytophagous fishes to improve water quality and increase fish production in culture ponds. In 1974 the Arkansas Game and Fish Commission and Auburn University, Alabama, obtained stock to assess their potential benefits and impacts (Jennings 1988). The species first began to appear in open waters, the Ohio and Mississippi rivers, in the early 1980s, likely as a result of escapes from aquaculture facilities (Jennings 1988). In April 1994, several thousand bighead carp, along with a few black carp *Mylopharyngodon piceus*, escaped into the Osage River, Missouri, when high water flooded hatchery ponds at an aquaculture facility near Lake of the Ozarks (Anonymous 1994b). Fish that escaped into the Missouri River have increased and spread, since 1990, into the lower Kansas River of Kansas, and elsewhere (Cross and Collins 1995). The species may have dispersed into Oklahoma waters from fish illegally brought into southeast Kansas by a commercial fish farmer in 1988 (Pigg et al. 1993). According to Pigg et al. (1997), collections in the Grand River of Oklahoma since 1991 indicate a gradual downstream dispersal. The species was illegally stocked along with grass carp in one or a few ponds in California; these were brought into the state by a commercial aquaculturist. The live fish were reportedly transported in a concealed compartment under a load of black bass in the fall of 1989 from a fish grower in Oklahoma or Arkansas (Dill and Cordone 1997).

Status: This species has been recorded from within, or along the borders of, at least 18 states. There is evidence of reproducing populations in the middle and lower Mississippi and Missouri rivers and the species is apparently firmly established in the states of Illinois and Missouri (Burr et al. 1996; Pflieger 1997). Pflieger (1997) received first evidence of natural reproduction, capture of young bighead carp, in Missouri in 1989. Burr and Warren (1993) reported on the taking of a postlarval fish in southern Illinois in 1992. Subsequently, Burr et al. (1996) noted that bighead carp appeared to be using the lower reaches of the Big Muddy, Cache, and Kaskaskia rivers in Illinois as spawning areas. Tucker et al. (1996) also found young-of-the-year in their 1992 and 1994 collections in the Mississippi River of Illinois and Missouri. Douglas et al. (1996) collected more than 1600 larvae of this genus from a backwater outlet of the Black River in Louisiana in 1994. The first open water record of this species in Arkansas is based on two specimens taken from the Arkansas River in 1986; however, as of the late 1980s there has been no evidence of natural reproduction in that state (Robison and Buchanan 1988). According to Dill and Cordone (1997), there is evidence that the California ponds containing Chinese carp have spilled since 1989, opening the door for bighead carp and grass carp to gain access to the Sacramento River. The West Virginia record involved a single fish taken in 1997 (Hoeft, personal communication). Harvest of bighead carp by commercial fishermen in Missouri has been somewhat erratic. In 1993, the species accounted for 0.6 percent (3,348 pounds) of the reported commercial fish harvest, a decline from the previous year (Robinson 1995).

Impact of Introduction: The impact of this species in the United States is not adequately known. Because bighead carp are planktivorous and attain a large size, Laird and Page (1996) suggested these carp have the potential to deplete zooplankton populations. As Laird and Page pointed out, a decline in the availability of plankton can lead to reductions in populations of native species that rely on plankton for food, including all larval fishes, some adult fishes, and native mussels. Adult fishes most at risk from such competition in the Mississippi and Missouri rivers are paddlefish *Polyodon spathula*, bigmouth buffalo *Ictiobus cyprinellus*, and gizzard shad *Dorosoma petenense* (Burr et al. 1996; Pflieger 1997; Whitmore 1997).

Remarks: Similar to the closely-related silver carp *Hypophthalmichthys molitrix*, the bighead carp is a filter feeder that prefers large river habitats. One of the so-called Chinese carps, it has been used in many parts of the world as a food fish and sometimes introduced in combination with silver carp into sewage lagoons and aquaculture ponds (Jennings 1988). In the United States bighead carp are frequently stocked into catfish culture ponds. According to Stickney (1996), studies have not confirmed that bighead carp actually do improve water quality in culture ponds.

Voucher specimens: Illinois (INHS 62722, 62723, 59111, 62721, 27896; SIUC 19280, 19282, 19289, 20308, 21715, 22240, 23040, 23042, 23045, 23919, 24003, 24403, 24414); Kansas (KU 21956); Louisiana (NLU 74346, 74349, 65810, 72488, 72473); Mississippi (photograph); Ohio (OSUM 77818); Tennessee (photograph).

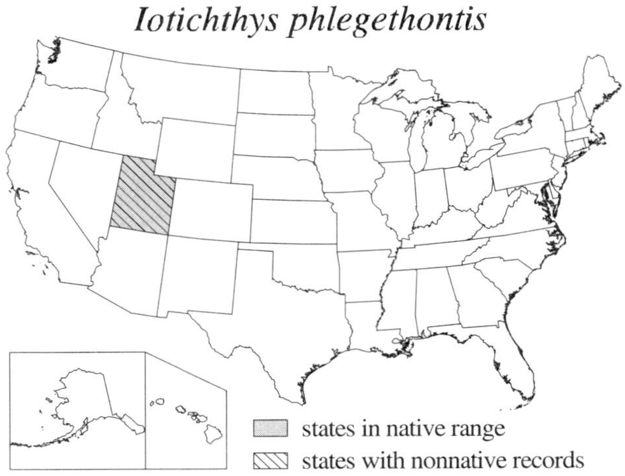

Iotichthys phlegethontis (Cope 1874)

Common Name: least chub.

Identification: Sigler and Miller (1963); Sigler and Sigler (1987; 1996); Page and Burr (1991). Maximum size: 6.4 cm.

Native Range: Streams near Salt Lake City and freshwater ponds and swamps around Great Salt Lake, Utah Lake, Beaver River, Parowan Creek, Clear Creek, and Provo River (Sigler and Miller 1963) in the Bonneville basin, northern Utah (Page and Burr 1991).

Nonindigenous Occurrences: This species was stocked in a pond west of Salt Lake City, **Utah**, in 1979, in a waterbody where it was not originally found (Baugh 1980; Sigler and Sigler 1987). Apparently it was stocked again, probably in other locations, in the late 1980s (Sigler and Sigler 1987, 1996).

Means of Introduction: Intentionally stocked for conservation purposes to create another population of a rare species.

Status: Status of stockings conducted in the late 1980s has not been evaluated (Sigler and Sigler 1996).

Impact of Introduction: Unknown.

Remarks: The least chub is under review for listing as a federally endangered species (U.S. Fish and Wildlife Service 1997).

Lavinia exilicauda

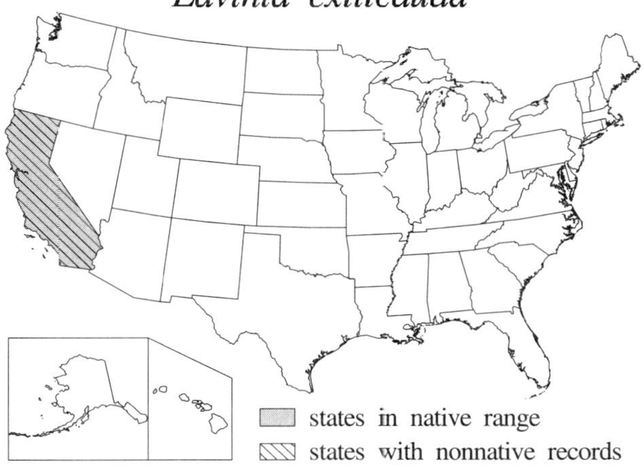

Lavinia exilicauda Baird and Girard 1854

Common Name: Hitch.

Identification: Moyle (1976a); Boschung et al. 1983; McGinnis (1984); Page and Burr (1991). Maximum size: 36 cm.
Native Range: Clear Lake, Russian River, Sacramento-San Joaquin, San Francisco, and Monterey Bay drainages, California (Page and Burr 1991).

Nonindigenous Occurrences: Reported from several sites in southern **California** including Pyramid Lake, Lake Silverwood, Ramer Lake, and a tributary of the Santa Clara River (Kimsey and Fisk 1960; Swift et al. 1993).

Means of Introduction: Some records may represent regular immigrants from central California (Swift et al. 1993). Some records may be the result of bait bucket releases. The species was reportedly sold as bait in the lower Colorado River (Miller 1952). Kimsey and Fisk (1960) suggested that it was one of several species in California widely transported as bait. They reported that it was collected in 1954 from Ramer Lake, Imperial County, about 320 road miles south of its regular range.

Status: Established in parts of southern California (Swift et al. 1993).

Impact of Introduction: Unknown. Possibly in reference to introduced populations, Moyle (1976a) stated that the species is largely regarded as a nuisance by fisheries managers.

Remarks: McGinnis (1984) stated that this species was apparently introduced into the Russian River of California. Other authors (e.g., Page and Burr 1991) apparently consider it to be native to that drainage.

Voucher specimens: California (LACM 45699-1, 45701-1).

Lepidomeda albivallis

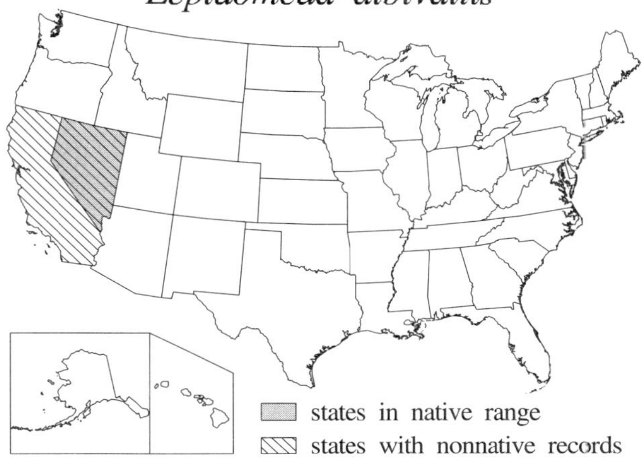

Lepidomeda albivallis Miller and Hubbs 1960

Common Name: White River spinedace.

Identification: La Rivers (1962); Sigler and Sigler (1987); Page and Burr (1991). Maximum size: 15 cm.

Native Range: Historically in the White River system, Nevada. Most populations are extirpated, and the species now exists only at Flag Springs in Nye County, Nevada (Page and Burr 1991; U.S. Fish and Wildlife Service 1994e).

Nonindigenous Occurrences: This species is known from the lower Colorado River, where four adult fish were collected from "Murphy's Windmill Camp," **California**, in February 1951 (Miller 1952; La Rivers 1962; U.S. Fish and Wildlife Service 1994e). It was also introduced into Railroad Valley, **Nevada** (La Rivers 1962; U.S. Fish and Wildlife Service 1994e).

Means of Introduction: Reportedly used as bait in 1951 in lower Colorado River (Miller 1952; La Rivers 1962). Transplanted from White River below Adams-McGill Reservoir to Railroad Valley in 1957 by Nevada Department of Wildlife (La Rivers 1962; U.S. Fish and Wildlife Service 1994e).

Status: Extirpated from sites where introduced in California and Nevada.

Impact of Introduction: Unknown.

Remarks: Listed as federally endangered species (U.S. Fish and Wildlife Service 1994e). Species has been adversely affected by introduced fishes including brown trout *Salmo trutta*, brook trout *Salvelinus fontinalis*, rainbow trout *Oncorhynchus mykiss*, cutthroat trout *O. clarki*, largemouth bass *Micropterus salmoides*, guppies *Poecilia reticulata*, mosquitofish *Gambusia affinis*, and goldfish *Carassius auratus* (U.S. Fish and Wildlife Service 1994e).

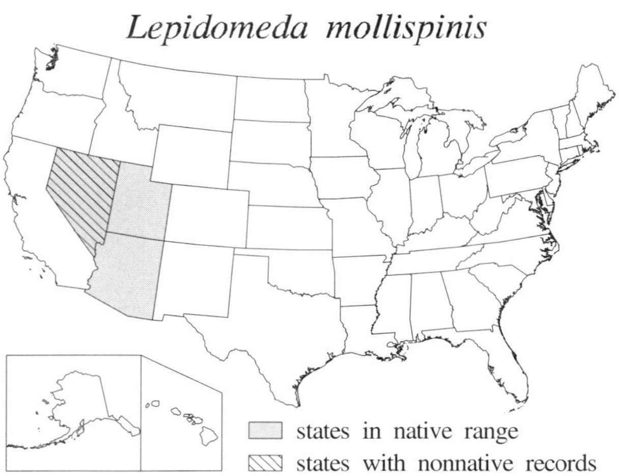

Lepidomeda mollispinis Miller and Hubbs 1960

Common Names: Virgin spinedace (Big Spring spinedace, Virgin River spinedace, Colorado River spinedace).

Identification: La Rivers (1962); Page and Burr (1991). Two subspecies: *L. m. pratensis*, Big Spring spinedace, and *L. m. mollispinis*, Virgin River spinedace (U.S. Fish and Wildlife Service 1993d). Maximum size: 15 cm.

Native Range: *L. m. pratensis*: Meadow Valley Wash and Panaca Spring, Lincoln County, Nevada (U.S. Fish and Wildlife Service 1993d). *L. m. mollispinis*: Virgin River system, Utah, Nevada, and Arizona (Page and Burr 1991).

Nonindigenous Occurrences: The subspecies *Lepidomeda m. pratensis* was transplanted to instream pools 1.5 mi above the waterfalls of the Meadow Valley Wash, Lincoln County, and to a pond at the Bureau of Land Management's Shoshone Pond Resource Area, White Pine County, **Nevada** (U.S. Fish and Wildlife Service 1993d). The subspecies *L. m. mollispinis* has been collected twice from Lake Mead, Nevada, once in December 1948 (two specimens) and once in February 1951 (three specimens) (La Rivers 1962).

Means of Introduction: *Lepidomeda m. pratensis* was transplanted for the purpose of preserving a threatened species. Larval fish from the plunge pool below the falls of Meadow Valley Wash were transplanted above the falls in 1980. Seventy-five fish were moved to Shoshone Pond Resource Area in July 1987 as an emergency protective measure after largemouth bass were discovered in Meadow Valley Wash. Six individuals were recovered in August 1989, none in July 1991 (U.S. Fish and Wildlife Service 1993d; Bureau of Land Management, Ely, records). The occurrence of *L. m. mollispinis* in Lake Mead, Nevada, was the result of bait bucket introductions of fish taken from the Santa Clara River in Utah (La Rivers 1962).

Status: In Nevada, *Lepidomeda m. pratensis* is established above the falls of Meadow Valley Wash; the population introduced to Shoshone Pond Resource Area is considered extirpated (U.S. Fish and Wildlife Service 1993d; Bureau of Land Management, Ely, records). The subspecies *L. m. mollispinis* was reported from Lake Mead, Nevada (La Rivers 1962); there is no evidence that it ever became established at that site.

Impact of Introduction: Unknown.

Remarks: Hendrickson and Brooks (1991) included the subspecies *L. m. pratensis* in their list of native southwestern fishes transplanted for conservation purposes. These researchers indicated that the fish was stocked once, successfully, but they did not provide details. The Big Spring spinedace is listed as a federally threatened subspecies (U.S. Fish and Wildlife Service 1993a). Prior to the elimination of the U.S. Fish and Wildlife Service candidate species list, the Virgin River spinedace was under review for Federal listing as an endangered or threatened species (U.S. Fish and Wildlife Service 1994a). It is not known whether this fish occurred above the waterfalls of Meadow Valley Wash (Nevada) prior to the transplant (U.S. Fish and Wildlife Service 1993d).

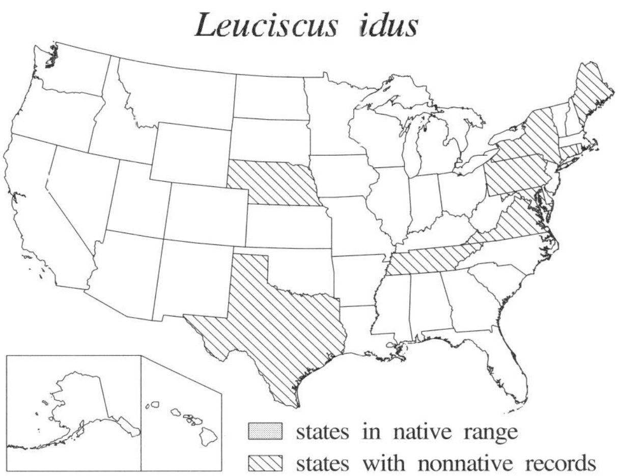

Leuciscus idus (Linnaeus 1758)

Common Names: ide (orfe, silver orfe, golden orfe, golden ide).

Identification: Distinguishing characteristics and photographs or illustrations appeared in Berg (1949), Muus and Dahlstrom (1978), Wheeler (1978), Phillips and Rix (1985), Ladiges and Vogt (1986), and Smith (1995). Also refer to Page and Burr (1991). Another name used in some of the earlier literature is *Idus idus*. Maximum size: 102 cm.

Native Range: Native from northern Europe through Siberia (Berg 1949; Robins et al. 1991b).

Nonindigenous Occurrences: This species was introduced to a private pond in East Lyme Township, New London County, **Connecticut** (Whitworth et al. 1968; Whitworth 1996). It was discovered in a private pond in Holden, Penobscot County, **Maine** in late March 1983 (Courtenay et al. 1984; Page and Burr 1991). In **Maryland**, it was introduced into the Potomac River near Baltimore and Washington, D.C., and possibly into the Monocacy River (McDonald 1893; Schwartz 1963; Lee et al. 1976, 1980 et seq.). This species was unsuccessfully stocked in **Nebraska** in 1884 (Morris et al. 1974). It was recorded from **New York** (Bean 1903); the earlier introduction was in Long Island, and it was collected in the Chenango River, a tributary of the Susquehanna River, between Hamilton and Norwich, New York, in the early 1950s (Courtenay et al. 1984) and also Cortland County ca. 1954 (Courtenay, personal communication). It has been reported from several ponds in **Pennsylvania** including sites in Delaware and York counties; there also were unconfirmed records from other private waters of that state (Lee et al. 1980 et seq.; Cooper 1983; Courtenay et al. 1984). This species was introduced to a pond off the Clinch River, **Tennessee** (Saylor, personal communication). It was reportedly stocked in **Texas** during the late 1800s (Howells 1992a). The species has been recorded from the Potomac River and possibly other areas of **Virginia** (Jenkins and Burkhead 1994; also see Smith and Bean 1899).

Means of Introduction: The ide was first imported in 1877 by the U.S. Fish Commission (Baird 1879). It has been stocked intentionally by the U.S. Fish Commission (historically), and by a state agency (recently), and through escapes from commercial and government ponds (Lee et al. 1980 et seq.). In 1889, an estimated 20 ide, along with other foreign cyprinids, escaped into the Potomac River from fish ponds in Washington, D.C., during a flood event (McDonald 1893). Similarly, Schwartz (1963) stated that it may have escaped from commercial ponds at Thurmont, Maryland into the Monocacy River; but he did not provide a date nor other details. Ide were also consigned to applicants in Virginia from 1892 to 1894 (Jenkins and Burkhead 1994). According to Courtenay et al. (1984, 1986), the U.S. Fish Commission gave no specific reason for importing and distributing this species, although they assumed that the intended use was both as an ornamental and food fish. Bean (1903) indicated that this species was introduced into American ponds for ornamental purposes. Schwartz (1963) stated that it may have escaped from commercial ponds at Thurmont, Maryland into the Monocacy River; but he did not provide a date or other details. Arkansas fish farmers have been raising and exporting ide for bait (Fletcher and Hallock 1992). In the early 1980s this species was reportedly being used as a bait fish in Tennessee (Courtenay et al. 1984, 1986). Because this species is sometimes misidentified as goldfish, Courtenay et al. (1984) argued that there was an increased probability of its spread and possible establishment.

Status: The ide has been recorded from nine states, but the documentation of its true status in the United States is poor and often contradictory. Courtenay and Stauffer (1990), Courtenay et al. (1991) and Courtenay (1993) listed it as established in Maine; Page and Burr (1991) stated that the species is locally established in a pond in Maine. However, in some early accounts the only known Maine population was thought to have been eradicated (e.g., Courtenay et al. 1984, 1986). In fact, Courtenay et al. (1986) stated that the Maine Department of Inland Fisheries and Wildlife eradicated the population in May 1983. According to Whitworth et al. (1968) one small pond in Connecticut has maintained a population since 1962 or 1963; Schmidt (1986) also listed *L.*

idus as introduced to lowland lacustrine habitat in the Connecticut River drainage. Subsequently, Courtenay et al. (1984, 1986) and Whitworth (1996) stated that the population in Connecticut had been eradicated. It was formerly established in one or a few water bodies in Pennsylvania and New York; however, according to Courtenay et al. (1984, 1986), those populations are no longer extant. The fish that escaped from Washington, D.C. ponds in 1889 apparently persisted for some time. That escape also apparently resulted in the species being listed as occurring in several states along Potomac River, including Virginia and Maryland. Although the literature seems to suggest that the ide population in the Potomac was self sustaining, it is not known when ide actually disappeared completely from the Potomac drainage. For instance, Schwartz (1963) apparently considered the ide to be established in the Potomac River; and Hocutt et al. (1986) listed it as introduced to the Potomac River. However, Musick (1972) doubted that this species still survived there, and Jenkins and Burkhead (1994) also concluded that there was no evidence indicating the ide persists in the Potomac drainage. Stockings of this species during the late 1800s were unsuccessful in both Nebraska (Morris et al. 1974) and Texas (Howells 1992a). There is only a single record documenting its occurrence in open waters of Tennessee (Saylor, personal communication). Howells (1992a) indicated that the U.S. Bureau of Fisheries (i.e., the U.S. Fish Commission) distributed the species to private individuals for introduction, but the actual introduction sites are largely unrecorded. The species never became established in Texas and has not been reported since their original introduction (Howells 1992a).

Impact of Introduction: Unknown. Seeley (1962) recommended against the introduction of the ide into California. Based on his review of the literature, he indicated the species had the potential of becoming established in state rivers, lakes, and reservoirs, and also of entering brackish and estuarine waters. Because of its tolerance to a wide range of conditions, Seeley believed that it had the potential of becoming more of a problem than either goldfish or common carp.

Remarks: In its native range, the ide is migratory and found in both freshwater and brackish water habitats (Muus and Dahlstrom 1978). The fish typically prey on larval and adult insects, snails, and other invertebrates; larger individuals also take small fish (Phillips and Rix 1985). The golden orfe or golden ide is a domestic form with portions of the body and fins pinkish-gold or red-orange. This colorful variety has received some degree of attention as an ornamental pond fish since its first introduction into this country near the turn of the century (e.g., Bean 1896, 1903). This specie still is occasionally kept in garden ponds, sometimes in combination with goldfish and koi, or with other orfe varieties, such as the blue orfe (Smith 1995). Dill and Cordone (1997) stated that this fish is not known to have been introduced into wild waters of California; however, they also indicated that the domesticated form, the golden orfe, has been present in garden pools and commercial aquaculture facilities in that state for a number of years.

Voucher specimens: New York (CU 28332). Based on information in Smith and Bean (1899), one or more specimens taken from the Washington, D.C. area were preserved and are likely deposited in the U.S. National Museum.

Luxilus albeolus

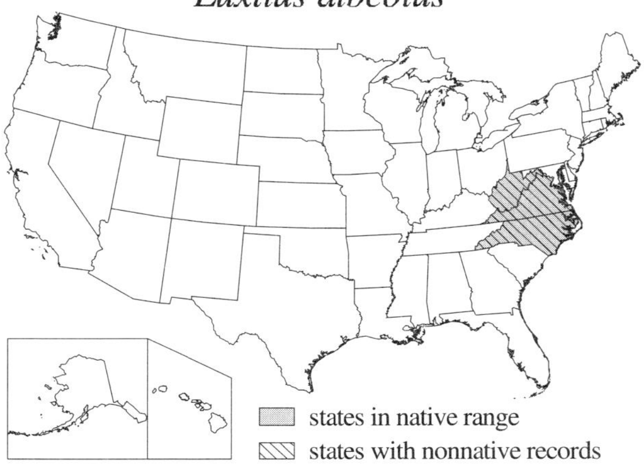

Luxilus albeolus (Jordan 1889)

Common Name: white shiner.

Identification: Gilbert (1964); Page and Burr (1991); Jenkins and Burkhead (1994); commonly used name is *Notropis albeolus*. Maximum size: 13 cm.

Native Range: Atlantic Slope from Chowan River system, Virginia, to the Cape Fear River drainage, North Carolina; upper New River drainage (Ohio River basin), West Virginia, Virginia, and North Carolina (Page and Burr 1991; but see Lee et al. 1980 et seq.).

Nonindigenous Occurrences: The white shiner may have been introduced into the New drainage of **North Carolina**, **Virginia**, and **West Virginia** (Lee et al. 1980 et seq.; Menhinick 1991; Gilbert, personal communication).

Means of Introduction: Unknown; probable bait bucket release.

Status: Established in North Carolina.

Impact of Introduction: Unknown.

Remarks: The species is widely distributed in the upper New River drainage (Ohio River basin), including sites in North Carolina, Virginia, and West Virginia (Lee et al. 1980 et seq.). Jenkins and Burkhead (1994) considered the white shiner to be native to the New drainage in Virginia. Starnes and R. E. Jenkins (personal communications) believe the same should be true in North Carolina. These researchers have concluded that the species distribution may be better explained as being locally restricted. The native range given by Page and Burr (1991) is in agreement with the conclusions of these investigators. In contrast, Gilbert (personal communication) contends that *albeolus* was introduced into the New system very early, prior to the days when any scientific collections were made (also see Lee et al. 1980 et seq.). Menhinick (1991) listed this species as considered native in the New drainage, but possibly introduced.

Luxilus cerasinus

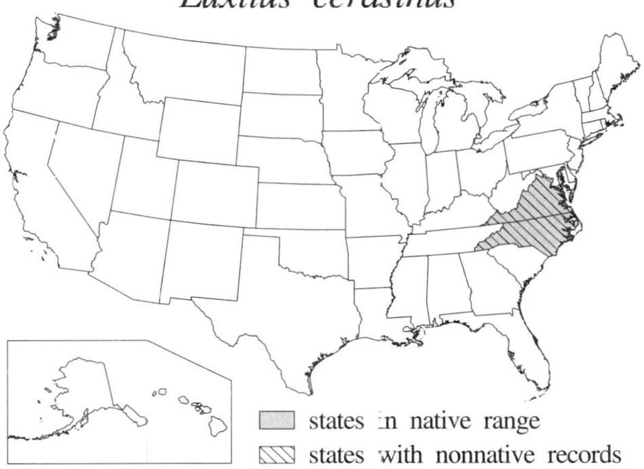

Luxilus cerasinus (Cope 1868)

Common Name: crescent shiner.

Identification: Gilbert (1964); Page and Burr (1991); Jenkins and Burkhead (1994); another commonly used name is *Notropis cerasinus*. Maximum size: 11 cm.

Native Range: Upper and middle Roanoke (and adjacent tributaries of Meherrin River–Chowan River system), and New River drainage, Virginia and North Carolina (Page and Burr 1991).

Nonindigenous Occurrences: This species is known from the Cape Fear drainage in Rockingham and Guilford counties and the Neuse drainage in Person County, **North Carolina** (Hocutt et al. 1986; Menhinick 1991). It has also been found in the headwaters of the Appomattox River drainage, and in several creeks of the middle and upper James drainages, **Virginia** (Abbott et al. 1977; Hocutt et al. 1986; Jenkins and Burkhead 1994).

Means of Introduction: Probable bait bucket release in Virginia, possible bait bucket release (or a natural stream capture) in North Carolina. The earliest record for Virginia was from 1965 in the Jennings Creek system of the middle James River drainage (Jenkins and Burkhead 1994). Abbott et al. (1977) stated that the presence of this species in Holiday Creek (Appomattox drainage) probably resulted from a bait bucket release into Holiday Lake, a recreational impoundment.

Status: Established in Virginia (in lower Barbours Creek, James drainage). Reported from North Carolina and from other areas of Virginia.

Impact of Introduction: Unknown.

Remarks: The earliest record from the Cape Fear drainage is from 1959 from the upper Haw system in Virginia; the earliest North Carolina Cape Fear records are from 1968 (Jenkins and Burkhead 1994). In their summary table of Virginia fishes, Jenkins and Burkhead (1994) listed this species as "regarded as introduced, possibly native" for the James River drainage (Atlantic Slope) and the New River drainage (Ohio River basin). Contrary to the opinions of several other authors (i.e., Page and Burr 1991; Jenkins and Burkhead 1994), Gilbert (1980) argued that *cerasinus* is not native to the New River system of Virginia. He noted that the species was first reported from that system about 40 years ago. Since that time this shiner has exhibited the rapid dispersal and expanded range so typical of other fishes introduced into the New system (Gilbert, personal communication).

Luxilus chrysocephalus

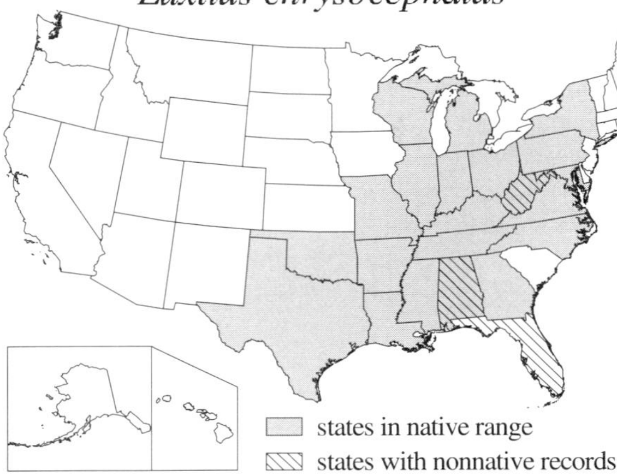

Luxilus chrysocephalus Rafinesque 1820

Common Name: striped shiner.

Identification: Gilbert (1964); Becker (1983); Page and Burr (1991); Etnier and Starnes (1993); Jenkins and Burkhead (1994); Pflieger (1997); another commonly used name is *Notropis chrysocephalus*. There are two distinct subspecies, *L. c. chrysocephalus* and *L. c. isolepis* (Gilbert 1998). Maximum size: 18 cm.

Native Range: Great Lakes and Mississippi River basins, from western New York and southeastern Wisconsin south to Alabama, Louisiana, and eastern Texas; Gulf Coast drainage from Mobile Bay, Georgia and Alabama, to Lake Pontchartrain, Louisiana (Page and Burr 1991).

Nonindigenous Occurrences: This is known from, and was probably introduced into, the Escambia River drainage of **Alabama** and **Florida** (Gilbert 1992; Gilbert, personal communication). It also is known from, and probably introduced into, the Potomac drainage and the Kanawha River drainage in **West Virginia** above the falls (Hocutt et al. 1986; Stauffer et al. 1995).

Means of Introduction: Unknown; possible bait bucket release.

Status: Established in West Virginia.

Impact of Introduction: Unknown.

Remarks: There is some uncertainty concerning the native versus nonnative geographic distribution of this species. In their summary table on the distribution and status of West Virginia fishes, Stauffer et al. (1995) considered this species to be native to the Kanawha River drainage below the falls. Gilbert (personal communication) concluded that the population of *chrysocephalus isolepis* in the Escambia River drainage of Alabama and Florida to be introduced. The same case has been made for other minnow species recently found in the Escambia, including *Nocomis leptocephalus* and *Notropis baileyi*. Gilbert (1992) noted that these species were absent in all early samples taken from that drainage. In their summary table on Alabama fishes, Mettee et al. (1996) listed this species as native to the Conecuh (Escambia) River drainage. Although not considered a true introduction in our treatment, Pflieger (1997) noted that the species had disappeared from certain streams in Missouri but it became re-established, possibly the result of an undocumented reintroduction.

Luxilus coccogenis

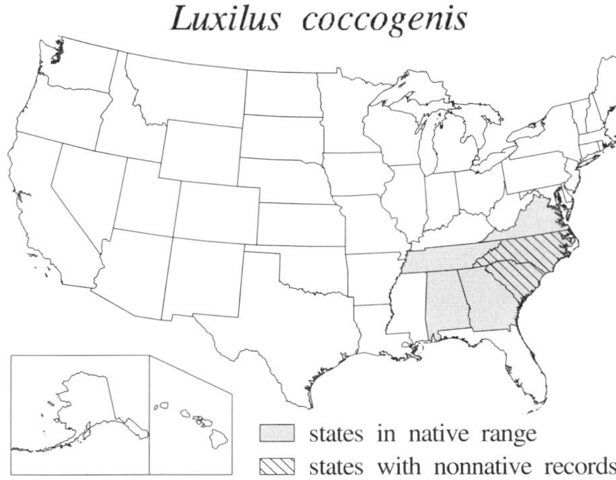

Luxilus coccogenis (Cope 1868)

Common Name: warpaint shiner.

Identification: Gilbert (1964); Page and Burr (1991); Etnier and Starnes (1993); Jenkins and Burkhead (1994); another commonly used name is *Notropis coccogenis*. Maximum size: 14 cm.

Native Range: Upper Tennessee River drainage, western Virginia, western North Carolina, Tennessee, northern Georgia, and northern Alabama, and adjacent tributaries of the Savannah River in North and South Carolina (Page and Burr 1991). Also found in the Catawba, Broad, and Savannah drainages in North Carolina (Menhinick 1991).

Nonindigenous Occurrences: The occurrence of this species in the River New drainage in **North Carolina** is probably the result of an introduction (Menhinick 1991). It has been reported only five times from the upper New drainage in **Virginia**: in 1941, 1949, 1970, 1976, and 1986 (Hocutt et al. 1986; Jenkins and Burkhead 1994). Its presence in the Santee drainage in **South Carolina** is probably the result of an introduction (Hocutt et al. 1986; Page and Burr 1991).

Means of Introduction: Unknown; probable bait bucket release.

Status: Established in Santee drainage, South Carolina; reported, possibly established in the New River drainage, North Carolina; reported from the New River drainage in Virginia.

Impact of Introduction: Unknown.

Remarks: Stauffer et al. (1995) listed this species as "present; probably introduced" in the Kanawha River drainage above the falls (i.e., the New River drainage), but they indicated that the shiner has never been recorded in West Virginia.

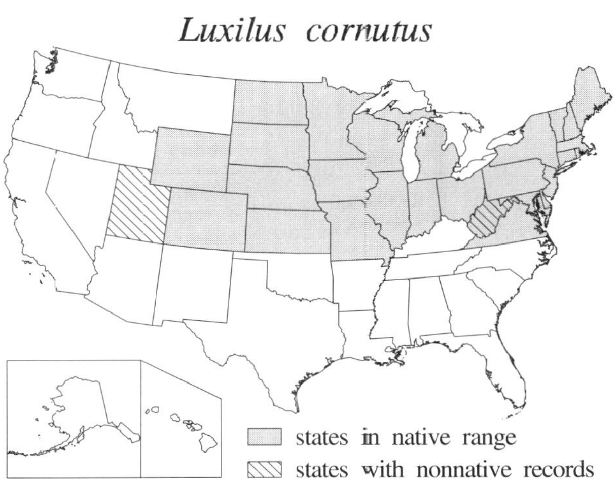

Luxilus cornutus (Mitchill 1817)

Common Name: common shiner.

Identification: Gilbert (1964); Scott and Crossman 1973; Becker (1983); Page and Burr (1991); Jenkins and Burkhead (1994); Pflieger (1997); another commonly used name is *Notropis cornutus*. Maximum size: 18 cm.

Native Range: Atlantic, Great Lakes, Hudson Bay, and Mississippi River basins, from Nova Scotia to southeastern Saskatchewan, and south to James River drainage, Virginia, northern Ohio, central Missouri, and Wyoming (Page and Burr 1991).

Nonindigenous Occurrences: The common shiner was introduced into Willard Bay Reservoir, **Utah**, in 1982 (Sigler and Sigler 1987). This shiner is regarded as native but may have been introduced in **West Virginia** into the main channel of the Ohio River, the Big Sandy, the Guyandotte, and the Kanawha below the falls (Stauffer et al. 1995).

Means of Introduction: Probable stocking for forage in Utah. Unknown in West Virginia; probable bait bucket release.

Status: Extirpated in Utah (Sigler and Sigler 1996). Established in West Virginia.

Impact of Introduction: Unknown.

Remarks: This species is known to hybridize with several other native minnows (Trautman 1981). Scott and Crossman (1973) noted that the common shiner is a very commonly used bait fish.

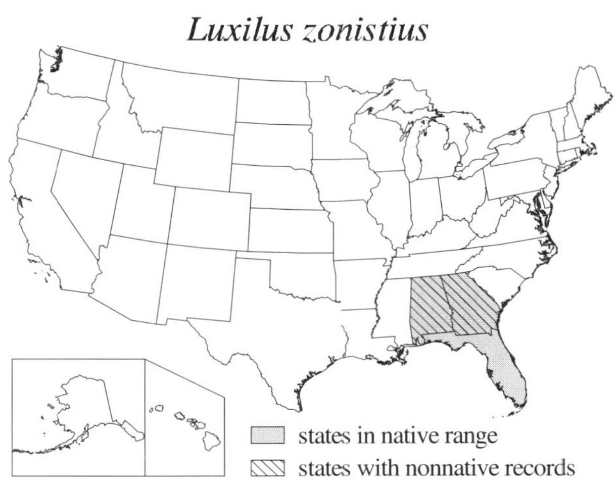

Luxilus zonistius Jordan 1880

Common Name: bandfin shiner.

Identification: Gilbert (1964); Page and Burr (1991); Mettee et al. (1996); commonly used name is *Notropis zonistius*. Maximum size: 10 cm.

Native Range: Apalachicola River drainage, Georgia, Alabama, Florida; adjacent tributaries of Savannah River, Georgia, Altamaha River, Georgia (Page and Burr 1991).

Nonindigenous Occurrences: The bandfin shiner is known from the Coosa and Tallapoosa river drainages (Mobile basin) in **Alabama** and **Georgia** (Williams 1967; Dahlberg and Scott 1971a, 1971b; Swift et al. 1986); the Tallulah River system, Rathbun County, Georgia; the Savannah drainage (CU 25170) (Gilbert 1964; Swift et al. 1986); headwaters of the Oconee system (Altamaha drainage), Georgia (Dahlberg and Scott 1971a, 1971b; Swift et al. 1986); and Town Creek (Hiwassee-Tennessee River drainage), Georgia (Johnston et al. 1995).

Means of Introduction: Unknown, possible bait bucket release.

Status: Established in Alabama and Georgia. A single collection consisting of ten specimens were taken from the Savannah River drainage in 1953.

Impact of Introduction: Unknown.

Remarks: There is some uncertainty concerning the native versus nonindigenous distribution of this species.
According to Williams (1967), the occurrence of this species in the Tallapoosa River system may have been the result of either a bait bucket release or stream capture. Smith-Vaniz (1968) indicated that its presence in the upper Tallapoosa River system is probably the result of stream capture. Gilbert (1964) and Swift et al. (1986) regarded the single record from the Savannah system as probably the result of stream capture, although these researchers did not rule out the possibility of introduction as an explanation for it presence. In their summary table on Alabama fishes, Mettee et al. (1996) listed this species as native to the Coosa and Tallapoosa river drainages, and also to the Chattahoochee River drainage. Johnston et al. (1995) discovered it in the Hiwassee River during a 1990 fish survey and considered it to be a probable introduction.

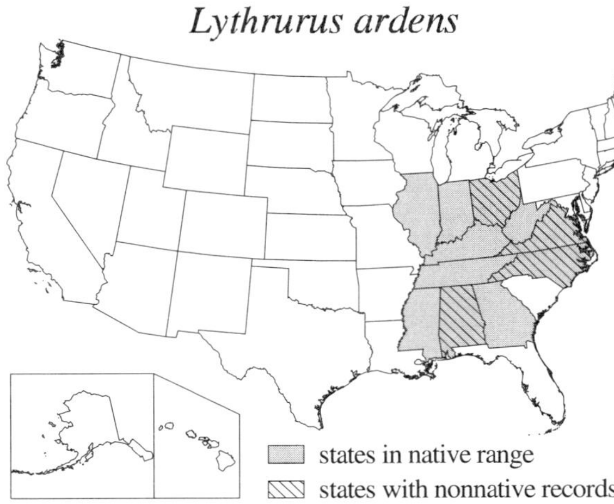

Lythrurus ardens (Cope 1868)

Common Name: rosefin shiner.

Identification: Page and Burr (1991); Etnier and Starnes (1993); Jenkins and Burkhead (1994); another commonly used name is *Notropis ardens*. Snelson (1990) recognized two subspecies, *L. a. ardens* and *L. a. fasciolaris*. Dimmick et al. (1996) recognized *L. ardens* as a species complex; as part of a reevaluation, they elevated *L. a. fasciolaris* to species status, and resurrected *L. matutinus* from synonymy. Mayden et al. (1992) and Gilbert (1998) treated all three species as valid. Maximum size: 8.5 cm.

Native Range: Atlantic Slope from York River drainage (where probably introduced), Virginia, to Roanoke River drainage, North Carolina; Ohio River basin from Scioto River drainage, Ohio, southwest to extreme southeastern Illinois (now extirpated) and south to Tennessee River drainage, Alabama; New River drainage, Virginia and West Virginia; upper Black Warrior River system (Gulf basin), Alabama (Snelson 1990; Page and Burr 1991; but see Remarks).

Nonindigenous Occurrences: Swift et al. (1986) listed this species as possibly introduced into the Black Warrior system in **Alabama**. There is a single record of several specimens collected in the Cape Fear drainage in Rockingham County, **North Carolina**, in 1963 (Hocutt et al. 1986; Snelson 1990; also see Menhinick 1991). Its presence in the Muskingum River drainage in **Ohio** may also be the result of past introduction (Snelson 1990). It is also known from the North and South Anna river systems (York River drainage), **Virginia** (Hocutt et al. 1986; Snelson 1990; Jenkins and Burkhead 1994; Dimmick et al. 1996). Hocutt et al. (1986) listed this species as probably introduced into the Potomac drainage.

Means of Introduction: Unknown; probable bait bucket release. In Virginia, it was first discovered in the South Anna system (Deep Creek) of the York River drainage in 1968, and probably later spread into the North Anna system where it was first discovered in 1979; the species is presumed to have come from an adjacent portion of the James River drainage (Snelson 1990; Jenkins and Burkhead 1994). In North Carolina, the Cape Fear introduction is presumed to have been from the adjacent Neuse drainage (Snelson 1990).

Status: Unknown in North Carolina. Established in Alabama and Virginia. Reported in Ohio.

Impact of Introduction: Unknown.

Remarks: In their taxonomic revision, Dimmick et al. (1996) recognized three distinct species in the *L. ardens* complex and they restricted the range of each of the species (these changes are obviously not reflected in the native range account given above). Based on the distribution map given by Dimmick et al. (1996) for the three species, we can assume that records from the Muskingum River drainage, Ohio, represent *Lythrurus fasciolaris*, those from the York drainage, Virginia, are the true *L. ardens*, and those found in the Cape Fear drainage likely represent *L. matutinus*. However, among the *L. ardens* complex, Dimmick et al. (1996) apparently only recognized populations found in the York River drainage of Virginia as introduced. Snelson (1990) provided details on the distribution of native and introduced populations. However, Snelson was uncertain of the origin of the subspecies *fasciolaris* (now *L. fasciolaris*) population occurring in the Black Warrior River system of the Tombigbee River drainage, Alabama; he indicated that it may be native or simply the result of recent introduction. Similarly, Swift et al. (1986) listed this species as possibly introduced into the Black Warrior system in Alabama; however, Mettee et al. (1996) listed it as native to that river system. Stauffer et al. (1978) and Hocutt et al. (1986) listed this species as probably introduced into the Potomac drainage. However, Stauffer et al. (1978) indicated that it could feasibly be native to that drainage.

Lythrurus atrapiculus

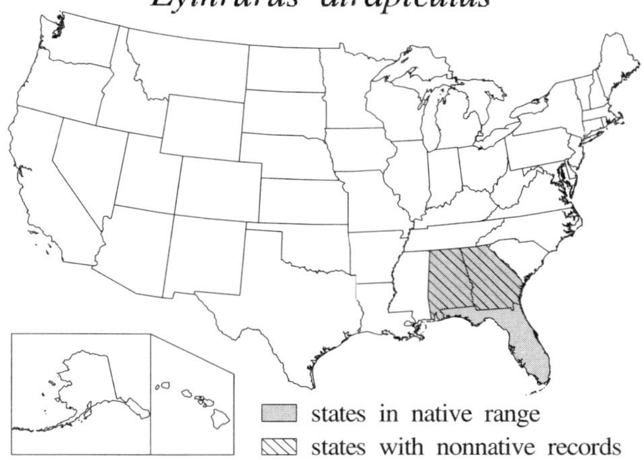

Lythrurus atrapiculus (Snelson 1972)

Common Name: blacktip shiner.

Identification: Snelson (1972); Page and Burr (1991); Mettee et al. (1996); a commonly used name is *Notropis atrapiculus*. Maximum size: 6.5 cm.

Native Range: Apalachicola, Choctawhatchee, Yellow, and Escambia drainages in western Georgia, southeastern Alabama, and Florida panhandle (Page and Burr 1991).

Nonindigenous Occurrences: This species is known from Old Town Creek (Tallapoosa River system; Mobile basin), Bullock County, **Alabama** (Lee et al. 1980 et seq.; Swift et al. 1986; Page and Burr 1991), and the upper Ocmulgee River system (Atlantic Slope) in **Georgia**, where it is very localized (Bart et al. 1994).

Means of Introduction: The introduction of the blacktip shiner into the Mobile basin of Alabama is the result of artificial diversion of the Conecuh River (Lee et al. 1980 et seq.). All information on the appearance of this species in the Ocmulgee system of Georgia came from Bart et al. (1994). According to those investigators, the species may have been introduced to the upper Ocmulgee system either by stream capture from the Flint River, or by bait bucket release. In any case, because the species is restricted to the river reach above High Falls dam, it was likely introduced after 1904, the year of dam construction. The first record of this species in the Ocmulgee River system (and in the Atlantic Slope) is based on the collection of sixteen specimens taken during the period 1988-1990 from the headwaters of the Towaliga River system, a small western tributary of the Ocmulgee River (Bart et al. 1994).

Status: Established in Georgia. A single collection was recorded in Alabama.

Impact of Introduction: Unknown.

Remarks: In their summary table on Alabama fishes, Mettee et al. (1996) listed this species as native to the Tallapoosa River system.

Margariscus margarita

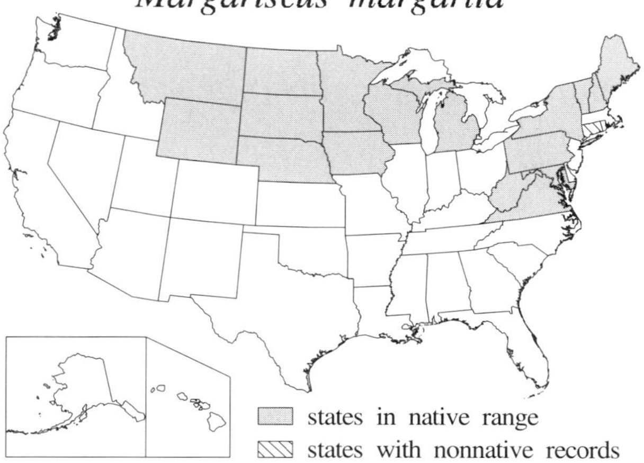

Margariscus margarita (Cope 1868)

Common Name: pearl dace.

Identification: Scott and Crossman (1973), Becker (1983), Smith (1985), Page and Burr (1991); another name is *Semotilus margarita*. Two subspecies are currently recognized, *M. m. margarita* in the Allegheny region, and *M. m. nachtriebi* in the rest of the range (Page and Burr 1991). Maximum size: 16 cm.

Native Range: Atlantic, Hudson Bay, Great Lakes, and Mississippi River basins in southern Canada and northern United States from Atlantic Coast to southern Northern Territories, eastern British Columbia, and Montana; south to Virginia, Michigan, Wisconsin, and Iowa. Isolated population in upper Missouri River basin, South Dakota, Nebraska, and Wyoming (Page and Burr 1991).

Nonindigenous Occurrences: This species was reported from the Byram River system in the Lower Hudson drainage, southwestern **Connecticut** (Whitworth et al. 1968; Whitworth 1996).

Means of Introduction: Unknown. Possible bait bucket release.

Status: Reported from Connecticut. The Byram River drainage was sampled in 1987 and 1988 but no additional specimens were found. Consequently, Whitworth (1996) concluded that the species was no longer present.

Impact of Introduction: Unknown.

Remarks: In some areas this species is considered important as a bait minnow (Scott and Crossman 1973).

Meda fulgida

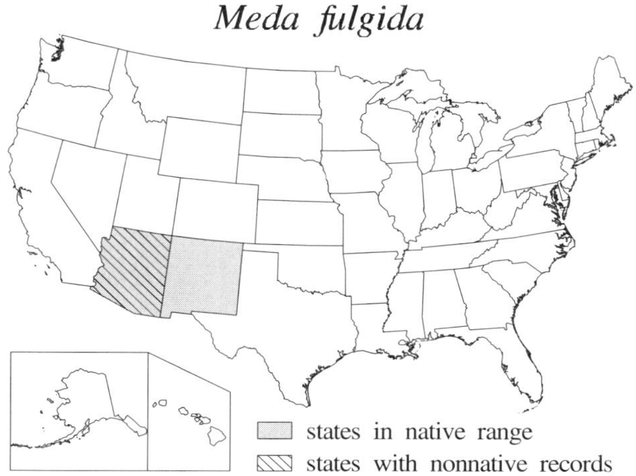

Meda fulgida Girard 1856

Common Name: spikedace.

Identification: Minckley (1973); Sublette et al. (1990); Page and Burr (1991). Maximum size: 9.1 cm.

Native Range: Gila River system, Arizona and New Mexico (Page and Burr 1991).

Nonindigenous Occurrences: This species was introduced into Sonoita Creek in Santa Cruz County, **Arizona** (U.S. Fish and Wildlife Service 1990a).

Means of Introduction: Intentionally stocked for the purpose of restoring or preserving an imperiled species. Stock obtained from Arivaipa Creek was placed in Sonoita Creek in 1968 (U.S. Fish and Wildlife Service 1990a).

Status: The Senoita Creek population is considered extirpated (U.S. Fish and Wildlife Service 1990a).

Impact of Introduction: Unknown.

Remarks: The spikedace is listed as a federally threatened species (U.S. Fish and Wildlife Service 1993a). It has been extirpated throughout much of its native range apparently as a result of competition or displacement by the introduced red shiner *Cyprinella lutrensis* (Minckley and Deacon 1968).

Moapa coriacea

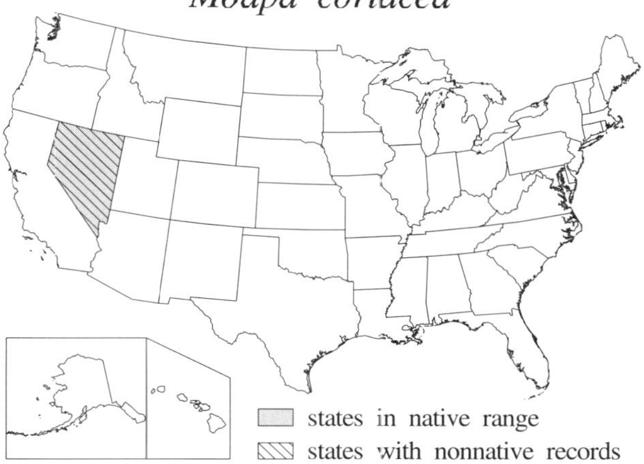

Moapa coriacea Hubbs and Miller 1948

Common Name: Moapa dace.

Identification: La Rivers (1962); Page and Burr (1991). Maximum size: 9 cm.

Native Range: Endemic to headwaters of Moapa (Muddy) River, Clark County, Nevada (U.S. Fish and Wildlife Service 1983a; Page and Burr 1991).

Nonindigenous Occurrences: Twenty Moapa dace were transplanted into Shoshone Ponds, near Ely, in White Pine County, **Nevada**, in September 1972 (Hubbs et al. 1974; U.S. Fish and Wildlife Service 1983a). The species was also transplanted into Hot Creek Valley in Nye County (Clemmer, personal communication).

Means of Introduction: This species was intentionally stocked for the purpose of preserving an endangered species.

Status: The fish at Shoshone Ponds survived at least one month (Hubbs et al. 1974); however, the population later died out for unknown reasons (U.S. Fish and Wildlife Service 1983a). The population at Hot Creek Valley also failed (Clemmer, personal communication).

Impact of Introduction: Unknown.

Remarks: Listed as federally endangered species in 1967 (U.S. Fish and Wildlife Service 1983a; Page and Burr 1991). The Shoshone Ponds facility was specially constructed by the Bureau of Land Management for endangered fish conservation. Species' decline coincided with introduction of shortfin molly, *Poecilia mexicana*, which occurred circa 1963, and with habitat alteration (U.S. Fish and Wildlife Service 1983a).

Morulius chrysophekadion

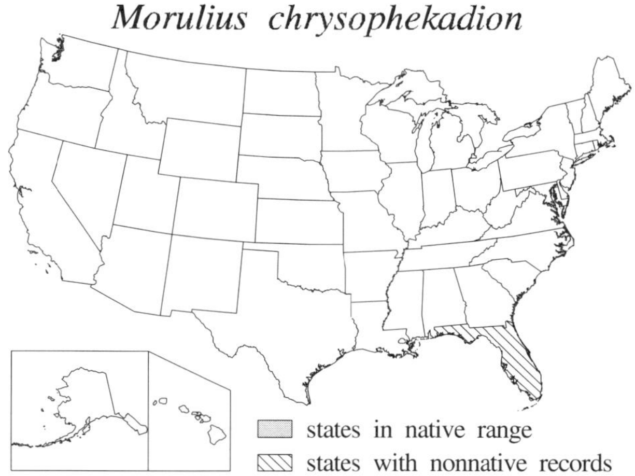

Morulius chrysophekadion (Bleeker 1850)

Common Names: black sharkminnow, black shark.

Identification: Identification key and references were given by Kottelat (1985). Distinguishing characteristics were also given by Sterba (1973). Color photographs appeared in Axelrod et al. (1985). Another commonly used name is *Labeo chrysophekadion* (see Roberts 1989). Maximum size: 60 cm.

Native Range: Tropical Asia. Kampuchea, Java, Borneo, Sumatra, Malaysia, Thailand, Laos, and Vietnam (Kottelat 1985).

Nonindigenous Occurrences: This species was reported from an unspecified locality in **Florida** (Courtenay et al. 1984, 1991; Courtenay and Stauffer 1990), probably between 1979 and 1984.

Means of Introduction: Probable aquarium release or escape from fish farm.

Status: Reported in Florida.

Impact of Introduction: Unknown.

Remarks: The black sharkminnow is a popular aquarium fish.

There are no known voucher specimens.

Mylopharyngodon piceus

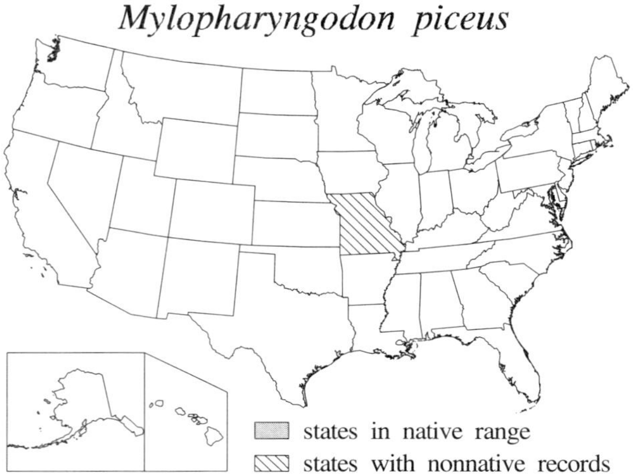

Mylopharyngodon piceus (Richardson 1846)

Common Names: black carp (snail carp, black amur, Chinese roach).

Identification: A few distinguishing characteristics were provided by Berg (1949), Masuda et al. (1984), and Howells (1992b). A color photograph appeared in Masuda et al. (1984). This species superficially resembles the grass carp *Ctenopharyngodon idella*. Maximum size: 131 cm and 36 kg (possibly larger).

Native Range: Most major Pacific drainages of eastern Asia from the Pearl River (Zhu Jiang) basin in China north to the Amur River (Heilong Jiang) and its major tributaries of China and far eastern Russia; possibly native to the Honghe or Red River of northern Vietnam (Nico and Williams 1996).

Nonindigenous Occurrences: Approximately 30 black carp escaped from a fish farm in **Missouri** into the Osage River, Missouri River basin, in April 1994 (Anonymous 1994b; W. Pflieger, personal communication).

Means of Introduction: This species was first brought into the United States in the early 1970s as a "contaminant" in imported grass carp stocks. These fish came from Asia and were sent to a private fish farm in Arkansas (Nico and Williams 1996). The second introduction of black carp into this country occurred in the early 1980s. During this period it was imported as a a food fish and as a biological control agent to combat the spread of yellow grub (*Clinostomum margaritum*) in aquaculture ponds (Nico and Williams 1996). The first and only known record of an introduction of black carp into open waters occurred in Missouri in 1994 when thirty or more black carp along with several thousand bighead carp escaped into the Osage River, Missouri River drainage, when high water flooded hatchery ponds at an aquaculture facility near Lake of the Ozarks (Anonymous 1994b; Nico and Williams 1996).

Status: Reported in Missouri. As of early 1998, there have been no reports from Missouri of the escaped black carp being recaptured.

Impact of Introduction: There is high potential that the black carp would negatively impact native aquatic communities by feeding on, and reducing, populations of native mussels and snails, many of which are considered endangered or threatened (Nico and Williams 1996).

Remarks: The black carp that escaped in Missouri may have been triploid and thus considered sterile (Anonymous 1994b). However, it also was rumored that these fish may have been brood stock. The black carp is a bottom-dwelling molluscivore that has been used by U.S. fish farmers to prey on and control disease-carrying snails in their farm ponds; more recently, this species has been proposed as a biological control for the introduced zebra mussel *Dreissena polymorpha*. Although the subject has been debated (e.g., French 1993; Rubinshtein 1994; Ricciardi 1994), to date, there is no experimental evidence that indicates black carp would be effective in controlling zebra mussels. Because black carp do not have jaw teeth and their mouths are relatively small, it is unlikely that these fish are capable of breaking apart zebra mussel rafts (Nico and Williams 1996). Black carp are superficially very similar in appearance to grass carp, *Ctenopharyngodon idella*, specifically in terms of body size and shape, position and size of fins, and position and size of the eyes. Juveniles, in particular, are difficult to distinguish from grass carp young. As such, Nico and Williams (1996) expressed concern that if black carp become more common in U.S. aquaculture, there will be an increased risk that the species be misidentified and unintentionally introduced as "grass carp" to some areas. Available information indicates that the black carp is, or has been in the recent past, maintained in research or production facilities in six states including Arkansas, Louisiana, Mississippi, Missouri, North Carolina, Oklahoma, and Texas (Nico and Williams 1996).

There are no known voucher specimens.

Nocomis biguttatus

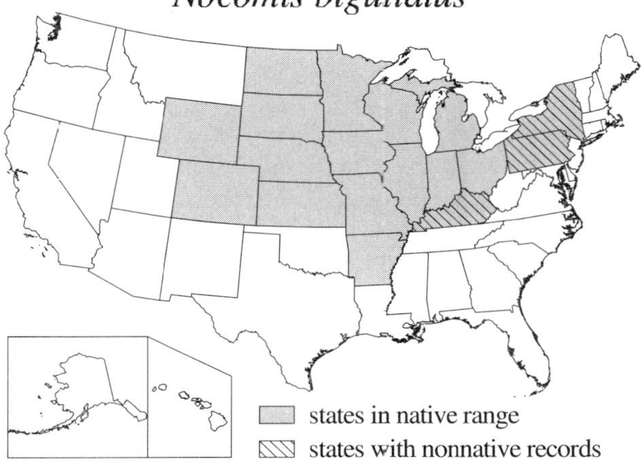

Nocomis biguttatus (Kirtland 1841)

Common Name: hornyhead chub.

Identification: Becker (1983); Robison and Buchanan (1988); Page and Burr (1991). Maximum size: 26 cm.

Native Range: Mohawk River system, New York, west through Great Lakes and Mississippi River basin to Red River drainage (Hudson Bay basin), Manitoba and North Dakota, south to Ohio River and lower Kentucky River system, Kentucky, but rare in unglaciated areas; south into Ozarks. Isolated populations in Platte and Cheyenne River systems, New Brunswick, Wyoming, and Colorado (Page and Burr 1991).

Nonindigenous Occurrences: This species is known from, and may have been introduced into, the Kentucky River system, Ohio River drainage, **Kentucky** (Lachner and Jenkins 1971; Burr and Page 1986; Burr and Warren 1986), and the Susquehanna and Mohawk river drainages in **New York** (Smith 1985).

Means of Introduction: Lachner and Jenkins (1971) stated that collections from Elkhorn Creek of the lower Kentucky River system may represent introduction via bait fishermen or transport through stocking programs and farm pond culture. Smith (1985) suggested that its presence in the Susquehanna and Mohawk rivers of New York may be the result of dispersal through artificial canal systems (Chenango Canal). There are no historical records of *Nocomis biguttatus* from the Susquehanna drainage (Lee et al. 1980 et seq.). Consequently, Gilbert (personal communication) agrees that the species has been introduced there. However, since it has only appeared in that drainage very recently, Gilbert concluded that its presence is more likely the result of a bait bucket transfer rather than migration through the canal system.

Status: Established in Kentucky and New York in areas outside its native range.

Impact of Introduction: Unknown.

Remarks: There is uncertainty concerning the native versus nonnative geographic distribution of this species. According to Gilbert (personal communication), it would be virtually impossible to determine whether or not this species was introduced or originally native in the Mohawk system of New York. In their summary table of Kentucky fishes, Burr and Warren (1986) listed this species as "regarded as native, but possibly introduced" in the lower Kentucky River system of Kentucky. In their zoogeographic analysis, Burr and Page (1986) listed this species as "regarded as native, but possibly introduced" in the Kentucky River system. Page and Burr (1991) included the lower Kentucky River system as part of the range of this species. These authors did not comment ion the possibility of introduction.

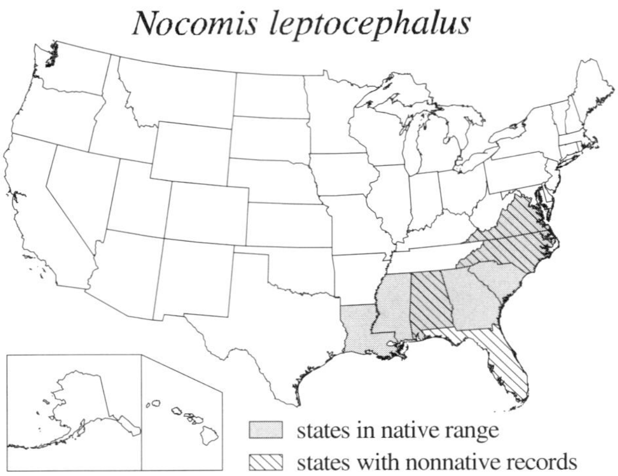

Nocomis leptocephalus (Girard 1856)

Common Name: bluehead chub.

Identification: Page and Burr (1991); Jenkins and Burkhead (1994). Maximum size: 26 cm.

Native Range: Atlantic and Gulf Slope drainages from Shenandoah River, Virginia, to Pearl River, Mississippi; mostly above Fall Line except in Alabama and Mississippi. Lower tributaries of Mississippi River, Mississippi and Louisiana; upper New River drainage, West Virginia, Virginia, and North Carolina; Bear Creek (Tennessee River drainage), Alabama and Mississippi (Page and Burr 1991).

Nonindigenous Occurrences: This species is known from, and probably introduced into, the Escambia River drainage of **Alabama** and **Florida** (Gilbert 1992) and the Tennessee River drainage of **North Carolina** (Menhinick 1991; Page and Burr 1991). In **Virginia**, it has been introduced into the Rappahannock, South Branch Potomac (Straight Creek), and possibly into the South Fork Shenandoah drainages (Jenkins and Burkhead 1994).

Means of Introduction: All or most records are likely the result of bait bucket releases. In Virginia, the bluehead chub was probably introduced into the upper Rapidan around 1947 and then spread downstream to the Rappahannock. Straight Creek is a favored fishing area and was likely populated by a bait bucket release. Jenkins and Burkhead (1994) considered the Shenandoah population to be probably native; however, others have considered it to be introduced (Jenkins et al. 1972, and others). The first record for the Potomac drainage is from 1956, from the South Fork Shenandoah. The records from Straight Creek are from 1975. The first record from the Rappahannock drainage is from 1947, from the Upper Rapidan. The chub then spread to the lower part of the drainage by 1972, and to the upper part by 1974. It reached the extreme upper tributaries of the Rappahannock in 1982 (Jenkins and Burkhead 1994). Based on collections during the past several decades (Gilbert 1992), the species may have been introduced into the Escambia River system in Alabama and subsequently spread downstream into Florida.

Status: Established in Alabama, Florida, North Carolina, and Virginia.

Impact of Introduction: Unknown.

Remarks: There is some uncertainty concerning the native versus nonnative geographic distributions of this species. Hocutt et al. (1986) listed it as 'regarded as introduced, but possibly native" for the Potomac. Menhinick (1991) listed it as "probably introduced" into the Tennessee River system of North Carolina. Burr (1991) stated that it was introduced into the Little Tennessee River system of North Carolina. Swift et al. (1986) listed it as "native freshwater species, possibly introduced" for the Escambia, a river in Alabama and Florida. Gilbert (1992) noted that *Nocomis leptocephalus bellicus* and several other minnow species (i.e., *Luxilus chrysocephalus isolepis* and *Notropis baileyi*) were not found in earlier collections taken from the Escambia River. Consequently, Gilbert presumed that these species had been introduced into that river system. In his unpublished checklist of Florida freshwater fishes, Gilbert indicated that this species was introduced and successfully established in the Escambia River drainage. In their summary table on Alabama fishes, Mettee et al. (1996) listed this species as native to the Conecuh (Escambia) River system.

Nocomis micropogon

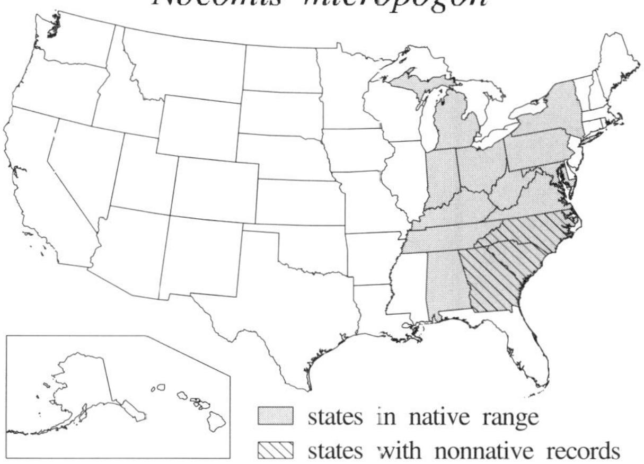

Nocomis micropogon (Cope 1865)

Common Names: river chub.

Identification: Page and Burr (1991); Etnier and Starnes (1993); Jenkins and Burkhead (1994). Maximum size: 32 cm.

Native Range: Atlantic drainages from Susquehanna River, New York, to James River, Virginia; Great Lakes basin, New York to Michigan; Ohio River basin, New York to eastern Illinois and south to northern Georgia and Alabama, but absent in southwestern Indiana, western two-thirds of Kentucky, and most of western Tennessee (Page and Burr 1991).

Nonindigenous Occurrences: This species is known from the Coosa River drainage (Mobile Basin) and Savannah River drainage of **Georgia** (Dahlberg and Scott 1971a; Page and Burr 1991); the Catawba River system (Santee River drainage) of **North Carolina** (Menhinick 1991); and the Savannah River drainage in **South Carolina** (Page and Burr 1991).

Means of Introduction: Unknown; probable bait bucket releases.

Status: Established in Georgia, North Carolina, and South Carolina.

Impact of Introduction: Unknown.

Remarks: The literature is somewhat contradictory concerning the native versus nonnative distribution of this species. It has been suggested that this species is one of several involved that provide evidence for "multiple piracies" of the upper Tennessee by headwaters of the Santee and Savannah river drainages (Starnes and Etnier 1986). Likewise, Tsai and Raney (1974) stated that *N. micropogon* may have entered the Savannah River drainage of the Atlantic Slope via stream capture. Swift et al. (1986) listed the species as native to the Savannah and possibly introduced to the Tallapoosa drainage. However, Starnes and Etnier (1986) interpreted the presence of this species in restricted portions of the upper Mobile Basin as being the result of an accidental introduction. Hocutt et al. (1986) listed it as introduced to the Santee River drainage.

Nocomis raneyi

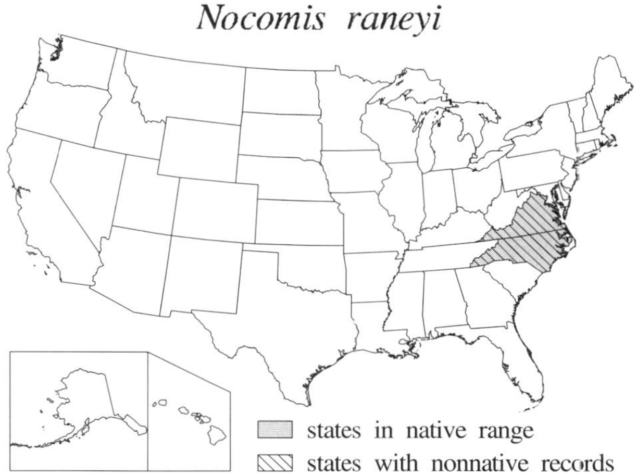

Nocomis raneyi Lachner and Jenkins 1971

Common Name: bull chub.

Identification: Page and Burr (1991); Jenkins and Burkhead (1994). Maximum size: 32 cm.

Native Range: Atlantic Slope drainages from James River, Virginia, to Neuse River, North Carolina (Page and Burr 1991).

Nonindigenous Occurrences: This species is known from, and was probably introduced into, the Cape Fear drainage in Orange County, **North Carolina** (Menhinick 1991). It may have been introduced into the James drainage, **Virginia**, where it has undergone a recent range extension (Jenkins and Burkhead 1994).

Means of Introduction: Unknown; probable bait bucket release.

Status: Established in North Carolina and Virginia.

Impact of Introduction: The spread of bull chub in the James drainage may be linked to the apparent disappearance of the closely-related river chub *N. micropogon* from much of the drainage during 1970s and 1980s (Jenkins and Burkhead 1994).

Remarks: Jenkins and Burkhead (1994) discussed the native versus nonnative distribution of this species in Virginia. Jenkins et al. (1972), Hocutt et al. (1986), and Jenkins and Burkhead (1994) listed *N. raneyi* as "regarded as native, but possibly introduced" for the James River drainage. *Nocomis raneyi* and *N. micropogon* are known to hybridize (Jenkins and Burkhead 1994).

Notemigonus crysoleucas

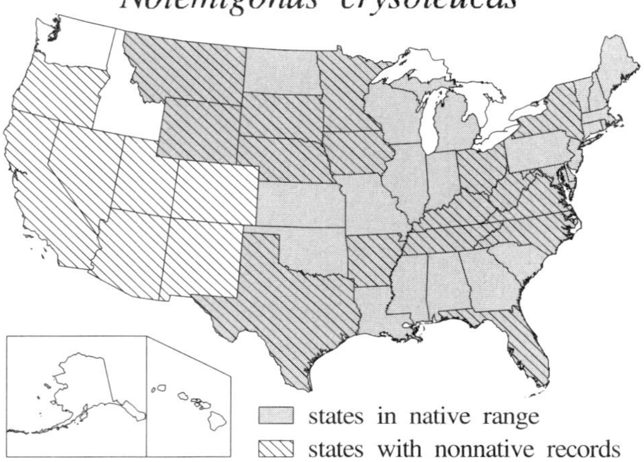

Notemigonus crysoleucas (Mitchill 1814)

Common Name: golden shiner.

Identification: Moyle (1976a); Becker (1983); Robison and Buchanan (1988); Page and Burr (1991); Jenkins and Burkhead (1994). Maximum size: 30 cm.

Native Range: Atlantic and Gulf Slope drainages from Nova Scotia to southern Texas; Great Lakes, Hudson Bay (Red River), and Mississippi River basins west to Alberta, Montana, Wyoming, and western Oklahoma (Page and Burr 1991).

Nonindigenous Occurrences: This species has been introduced into nonnative areas of the following states: **Arizona** (La Rivers 1962; Minckley 1973); **Arkansas** (Robison and Buchanan 1988); **California** (Hubbs 1919; McKechnie 1966a; Moyle 1976a; Moyle and Daniels 1982; Dill and Cordone 1997); **Colorado** (Wiltzius 1981; Woodling 1985; Propst and Carlson 1986); **Florida** (Swift et al. 1977); **Iowa** (Harlan et al. 1987); **Kentucky** (Burr and Warren 1986); **Minnesota** (Eddy and Underhill 1974); **Montana** (Brown 1971; Cross et al. 1986; Holton 1990); **North Carolina** (Menhinick 1991); **Nebraska** (Bouc 1987); **New Mexico** (Sublette et al. 1990); **Nevada** (Miller 1952; La Rivers 1962; Minckley and McNatt 1974; Deacon and Williams 1984); **New York** (possibly not native to Adirondacks) (George 1981; Smith 1985); **Ohio** (Trautman 1981); **Oregon** (Bond 1994); **South Dakota** (Bailey and Allum 1962; Cross et al. 1986); **Tennessee** (Kuhne 1939); **Texas** (Howells and Prentice 1991); **Utah** (Minckley 1973; Sigler and Sigler 1987, 1996); **Virginia** (Hocutt et al. 1986; Jenkins and Burkhead 1994); **West Virginia** (Hocutt et al. 1986; Stauffer et al. 1995); and **Wyoming** (Baxter and Simon 1970; Cross et al. 1986; Stone 1995).

Means of Introduction: The golden shiner is widely used as bait and as an ornamental and therefore has been transplanted into many areas, including parks in the United States (e.g., Lee et al. 1980 et seq.; LoVullo and Stauffer 1993). According to Swift et al. (1993), the golden shiner arrived in Sand Diego County, California, in 1891 and quickly became widespread. Dill and Cordone (1997) stated this species was planted in parts of California as forage by the California Fish Commission during the late 1890s. It is the most commonly used bait fish in many regions, for instance California (Dill and Cordone 1997) and the Midwest (Burr, personal communication). Jenkins and Burkhead (1994) and Pflieger (1997) also noted that many populations probably became established through bait bucket introductions. The golden shiner has been stocked in the Great Basin as a forage fish (Sigler and Sigler 1987). According to Jenkins and Burkhead (1994), it has been sparingly introduced to the New River drainage (earliest records: North Carolina 1949; Virginia 1971; West Virginia 1970). These researchers also noted that the earliest record of this species in the Virginia portion of the upper Tennessee drainage was 1966. That record was from a site near a hatchery that used this species as forage.

Status: Supposedly established in all states listed above.

Impact of Introduction: According to Moyle (1976a), the introduction of this species into California lakes typically results in decreased growth and reproduction of trout because of competition with golden shiners for food and space. He also noted that golden shiners may compete with young sunfish and bass for food. Eradication programs were reported as common in California (Sigler and Sigler 1996).

Remarks: This species is similar in appearance to the rudd *Scardinius erythrophthalmus*, a European species that has been introduced into the United States. The two readily hybridize in laboratory conditions (Burkhead and Williams 1991). Dill and Cordone (1997) detailed the introduction history of this species in California. They speculated that some of the "minnows" used as food for fish being transported by the U.S. Fish Commission in the late 1800s may have been golden shiners. There is some debate concerning the native versus nonnative distribution of this species in the eastern United States. For example, Etnier and Starnes (1993) stated that this species was probably rare or absent from east and middle Tennessee before reservoir construction, but later became established in larger waters of these areas as a result of bait bucket introductions or expansion of "once-scattered populations." Jenkins and Burkhead (1994) noted that the species is probably more prevalent in Virginia than their distribution maps show because their data included few samples from farm ponds, habitats where the species has often been introduced. They also stated that most records from the Blue Ridge and Valley and Ridge in Virginia probably are of introductions. Sigler and Sigler (1996) stated that introduced golden shiners in Utah Lake, Utah, are barely maintaining population levels. In their analysis of fish zoogeography, Hocutt et al. (1986) listed this species as "introduced but possibly native" for an Appalachian drainage, the Kanawha River above the falls.

Notropis amoenus

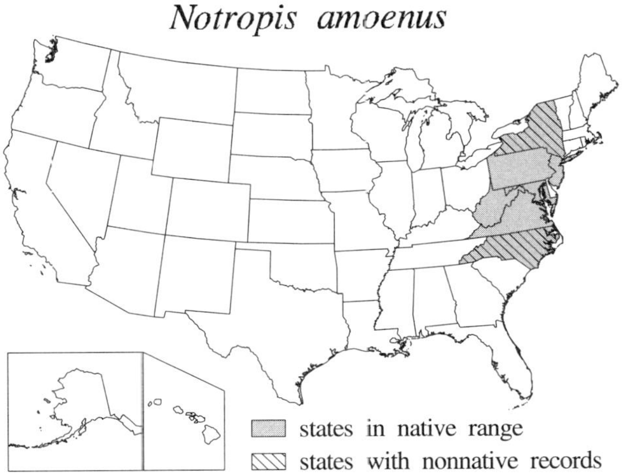

Notropis amoenus (Abbott 1874)

Common Name: comely shiner.

Identification: Smith (1985); Page and Burr (1991); Jenkins and Burkhead (1994). Maximum size: 11 cm.

Native Range: Atlantic Slope drainages from Hudson River, New York, to Cape Fear River, North Carolina (Page and Burr 1991).

Nonindigenous Occurrences: The comely shiner has been collected once from Seneca Lake (Lake Ontario drainage), **New York** (Snelson 1968; Smith 1985; Page and Burr 1991); it is also known from the Shawangunk Kill in New York (Snelson 1968; Smith 1985). It has been recorded from the Yadkin drainage of **North Carolina** (Menhinick 1991; Page and Burr 1991).

Means of Introduction: The occurrence of this species in Seneca Lake, New York, may have resulted from spread via the Chemung Canal (Snelson 1968; Smith 1985). Smith (1985) stated that the presence of this species in the Shawangunk Kill of New Yark was probably the result of dispersal through the Delaware-Hudson Canal; however, we could find no mention of the Shawangunk Kill, or the possibility of its introduction into that site, in Snelson (1968, 1971). Method of introduction in the Yadkin drainage of North Carolina is unknown; possible bait bucket release.

Status: Reported from New York and North Carolina.

Impact of Introduction: Unknown.

Remarks: There is some uncertainty concerning the native versus nonnative distribution of this species. For example, the comely shiner may be native to the Yadkin in North Carolina. Its native range extends into the Pee Dee River in South Carolina and the main rivers in the Pee Dee, but these were not adequately sampled during the statewide survey in the 1960s (Jenkins, personal communication).

Notropis ariommus

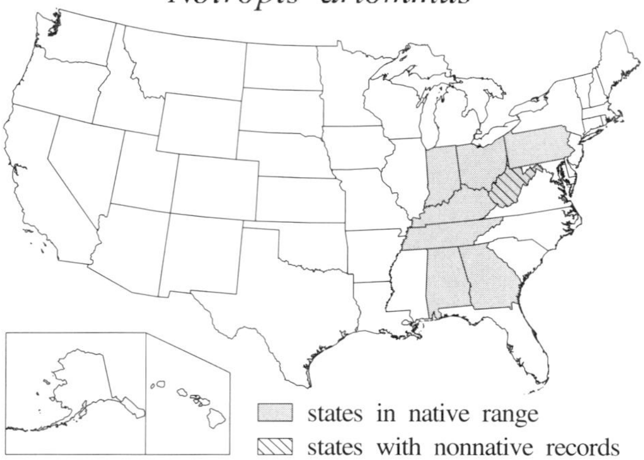

Notropis atherinoides

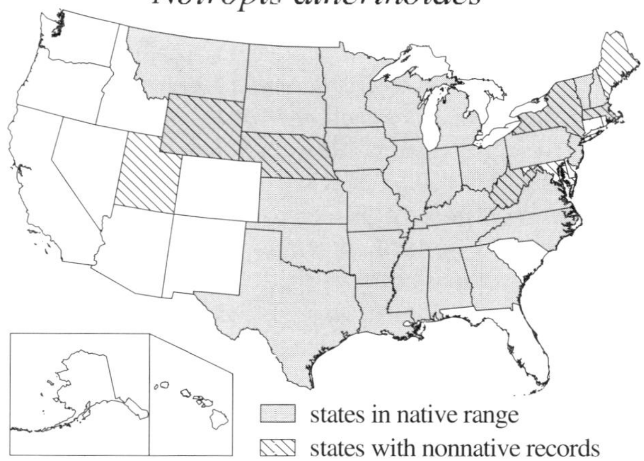

Notropis ariommus (Cope 1867)

Common Name: popeye shiner.

Identification: Trautman (1981); Page and Burr (1991); Etnier and Starnes (1993). Maximum size: 9.5 cm.

Native Range: Ohio River basin from Pennsylvania to Indiana and south to Tennessee River drainage, Georgia and Alabama. One record from Maumee River (Lake Erie drainage), Ohio; population now extirpated (Page and Burr 1991).

Nonindigenous Occurrences: This species is known from the Kanawha drainage above the falls in **West Virginia**, possibly the result of an introduction (Stauffer et al. 1995).

Means of Introduction: Unknown; possible bait bucket release.

Status: Established in West Virginia.

Impact of Introduction: Unknown.

Remarks: In their summary table of West Virginia fishes, Stauffer et al. (1995) regarded this species as "native but possibly introduced" for the Kanawha above falls. Hocutt et al. (1986) did not record it as present from above the falls. Both Hocutt et al. (1986) and Stauffer et al. (1995) regarded it as native below the falls.

Notropis atherinoides Rafinesque 1818

Common Name: emerald shiner.

Identification: Becker (1983); Robison and Buchanan (1988); Page and Burr (1991); Jenkins and Burkhead (1994); Pflieger (1997). Maximum size: 13 cm.

Native Range: St. Lawrence drainage, Quebec; Hudson River drainage, New York to Mackenzie River drainage (Arctic basin), Northwest Territories, and south through Great Lakes and Mississippi River basins to Gulf; Gulf Slope drainages from Mobile Bay, Alabama, to Galveston Bay, Texas (Page and Burr 1991).

Nonindigenous Occurrences: This species is known from several sites in central and southern **Maine** (Kircheis 1994); Jennings Randolph Reservoir in **Maryland** (J. Christmas, personal communication), and several large impoundments in **Massachusetts** (Hartel 1992; Hartel et al. 1996). Although it is native to most of eastern **Nebraska**, the species has been stocked in western parts of the state, such as Lake McConaughy (Bouc 1987). It is known from Lebanon and Erieville reservoirs (Susquehanna River drainage), **New York** (Snelson 1968; Smith 1985; Hocutt et al. 1986), and from Willard Bay Reservoir, **Utah** (Sigler and Sigler 1987, 1996). In **West Virginia**, it is considered native but may have been introduced into the Kanawha drainage above the falls (Stauffer et al. 1995). It has been introduced into **Wyoming** (Stone 1995) including many reservoirs east of the Continental Divide in that state (Hubert 1994).

Means of Introduction: This species was stocked as forage in Jennings Randolph Reservoir, Maryland

(Christmas, personal communicaton), Lake McConaughy, Nebraska (Bouc 1987), and in certain Wyoming reservoirs (Hubert 1994). Its presence in two reservoirs in New York may be due either to bait bucket releases or to movement through the Old Chenango Canal (Snelson 1968; Smith 1985). Snelson (1968) also hypothesized that the widespread occurrence of this species in the Mohawk-Hudson system of New York may have been a result of dispersal through the Erie Canal system. If records from West Virginia represent introductions, the method of entry was likely by way of bait bucket releases. The species probably was brought to Maine illegally as a baitfish; however, information on the location and exact date of its first occurrence in the state has been lost (Kircheis 1994). It was introduced into Willard Bay Reservoir in Utah in 1983 (Sigler and Sigler 1987).

Status: Introduced populations considered established in Maine, Maryland, Massachusetts, Nebraska, New York, Utah, West Virginia, and Wyoming.

Impact of Introduction: Unknown.

Remarks: According to Hartel et al. (1996), the Massachusetts specimens have never been examined by specialists. However, this species is sold in bait stores in that state (Hartel 1992; Hartel et al. 1996). Schmidt (1986) listed this species as introduced into the Housatonic and Connecticut drainages, probably in the Massachusetts portion of these waters. Hocutt et al. (1986) regarded it as "introduced, but possibly native" for the Susquehanna River drainage. In their summary table of West Virginia fishes, Stauffer et al. (1995) regarded this species as "native but possibly introduced" for the Kanawha above the falls. Pflieger (1997) noted that the progressive and dramatic increase in abundance of this sight-feeding minnow in the Missouri River (part of its native range) was the result of decreases in turbidity and other factors relating to the construction of upstream reservoirs.

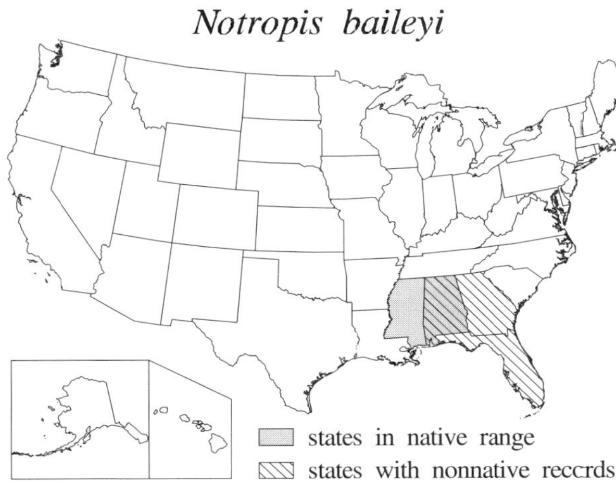

Notropis baileyi Suttkus and Raney 1955

Common Name: rough shiner.

Identification: Page and Burr (1991); Mettee et al. (1996). Maximum size: 9 cm.

Native Range: Mobile Bay and Pascagoula River drainages, mostly below the Fall Line, Alabama and Mississippi; Bear Creek system (Tennessee River drainage), Alabama and Mississippi (Page and Burr 1991).

Nonindigenous Occurrences: The rough shiner is known from the Chattahoochee River drainage in **Alabama** and **Georgia** (Dahlberg and Scott 1971a, 1971b; Yerger 1977; Swift et al. 1986; Gilbert 1992). It also has been recorded from the Escambia River drainage in **Florida** and Georgia (Lee et al. 1980 et seq.; Swift et al. 1986; Page and Burr 1991; Gilbert 1992).

Means of Introduction: Unknown; probable bait bucket release. The source of introduced fish found in for the Chattahoochee River was possibly the Tennessee or Coosa drainage in Georgia.

Status: Established in Alabama, Florida, and Georgia.

Impact of Introduction: Unknown.

Remarks: There is some uncertainty concerning the nonindigenous versus native distributions of this species. Lee et al. (1980 et seq.) provided a spot-distribution map and indicated that the shiner was possibly introduced into the Escambia River of Florida and Georgia. Gilbert (1992) stated that the presence of this species (as well as

Luxilus chrysocephalus and *Nocomis leptocephalus*) in the Escambia drainage is suspected to be the result of an introduction. As support, Gilbert noted that *N. baileyi* was only first discovered in the Escambia drainage after 1959. Recent collections seem to indicate the species has become more abundant and widespread. In their summary table on Alabama fishes, Mettee et al. (1996) listed this species as native to the Conecuh (Escambia) and Chattahoochee river drainages.

Notropis bairdi

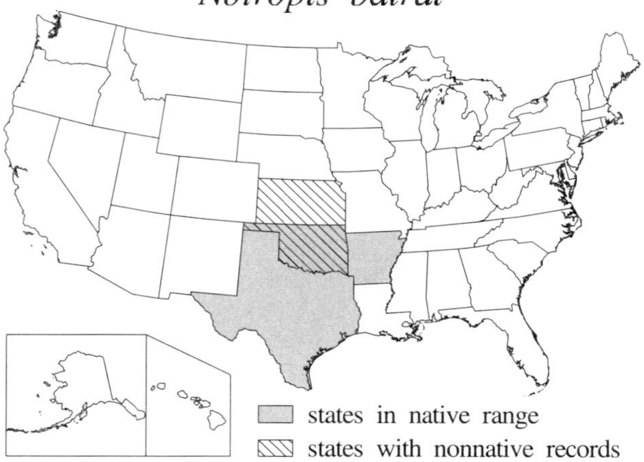

Notropis bairdi Hubbs and Ortenburger 1929

Common Name: Red River shiner.

Identification: Robison and Buchanan (1988); Page and Burr (1991). Maximum size: 8 cm.

Native Range: Red River drainage from southwestern Arkansas to western Oklahoma and northwestern Texas (Page and Burr 1991).

Nonindigenous Occurrences: This species is known from the Cimarron River of the Arkansas River drainage in **Kansas** and **Oklahoma** (Cross et al. 1983, 1986; Page and Burr 1991).

Means of Introduction: Unknown; probable bait bucket release. It was first collected from the Cimarron in Oklahoma in 1976 (Cross et al. 1983).

Status: Established in Kansas and Oklahoma.

Impact of Introduction: *Notropis bairdi* has become common in the Cimarron River and is apparently displacing the native Arkansas River shiner *Notropis girardi* which is proposed for listing as an endangered species (Cross et al. 1983; U.S. Fish and Wildlife Service 1997).

Remarks: *Notropis girardi* is being replaced by *Notropis bairdi* in its native range, but *N. girardi* is doing very well in the Pecos River drainage, where introduced (Gilbert, personal communication).

Notropis bifrenatus

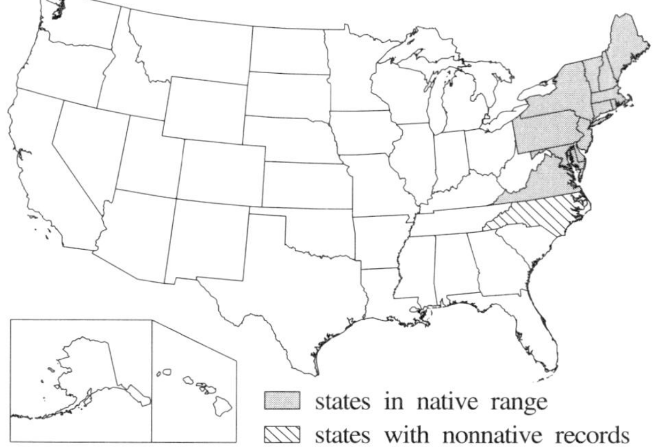

Notropis bifrenatus (Cope 1867)

Common Name: bridle shiner.

Identification: Smith (1985); Page and Burr (1991); Jenkins and Burkhead (1994). Maximum size: 6.5 cm.

Native Range: St. Lawrence–Lake Ontario drainage, southern Quebec, eastern Ontario, and New York; Atlantic Slope drainages from southern Maine to Santee River, South Carolina (Page and Burr 1991; Jenkins and Burkhead 1994).

Nonindigenous Occurrences: A single specimen collected from Tucker Creek, Craven County, in the Neuse drainage in 1964 (Jenkins and Zorach 1970) in **North Carolina,** may have been introduced (Menhinick 1991).

Means of Introduction: Unknown.

Status: Reported from North Carolina.

Impact of Introduction: Unknown.

Remarks: Rather than an introduction, the Neuse drainage record may represent a native occurrence. According to Jenkins (personal communication), the unusual distribution may be explained as a species that was once more widespread and is now widely extirpated. This species has since been identified even farther south, in the Santee River, South Carolina, from collections taken in 1954 (Jenkins and Burkhead 1994), adding credence to the argument that it may once have had a wider distribution and now exists in refugia.

Jenkins and Burkhead (1994) noted that the bridle shiner is very delicate and is an unlikely bait bucket release candidate.

Voucher specimen: North Carolina (CU 52932).

Notropis blennius

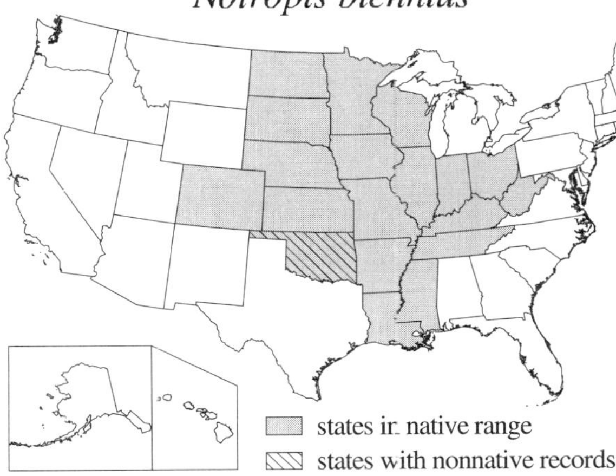

Notropis blennius (Girard 1856)

Common Name: river shiner.

Identification: Scott and Crossman (1973); Trautman (1981); Becker (1983); Page and Burr (1991); Etnier and Starnes (1993); Pflieger (1997). Maximum size: 13 cm.

Native Range: Hudson Bay basin from Manitoba to Alberta (absent in eastern Saskatchewan) in Canada, and south through Red River, Minnesota and North Dakota; Mississippi River basin from Wisconsin and Minnesota south to Gulf; in the United States occurs as far east as West Virginia and as far west as eastern Colorado (Page and Burr 1991).

Nonindigenous Occurrences: A specimen was taken from Bricken Spring (Red River drainage), 24 miles south of Ada, **Oklahoma** in April 1947 (Hubbs and Bonham 1951).

Means of Introduction: One of the collectors, G. A. Moore, thought that the Bricken Spring specimen represented a bait introduction, based on the fact that it was taken on the premises of Thomas Bricken, a bait dealer, who obtained his fish from the Arkansas and Red river drainages (Hubbs and Bonham 1951).

Status: Reported from a part of Oklahoma.

Impact of Introduction: Unknown.

Remarks: Hubbs and Bonham (1951) noted that the Bricken Spring specimen to be typical of *Notropis blennius blennius*, one two subspecies recognized by some authors. Gilbert (1998) apparently did not recognize subspecies. In commenting on the belief that the Bricken Spring specimen represented a bait introduction, Suttkus and Clemmer (1968) remarked "perhaps this is true."

Notropis boops

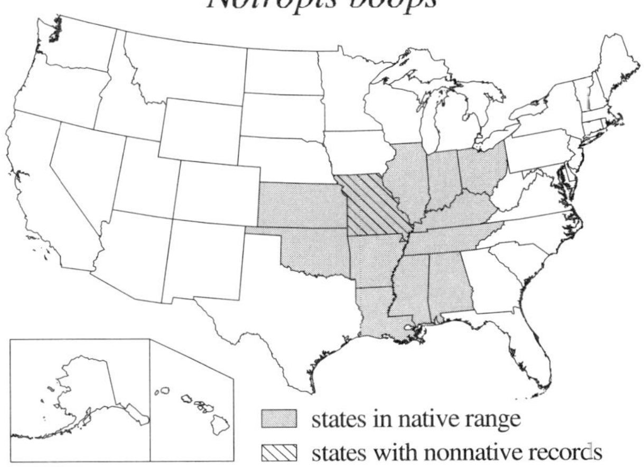

Notropis boops Gilbert 1884

Common Name: bigeye shiner.

Identification: Trautman (1981); Page and Burr (1991); Etnier and Starnes (1993); Pflieger (1997). Maximum size: 9 cm.

Native Range: Lake Erie drainage, northwest Ohio; Mississippi River basin from central Ohio to eastern

Kansas and south to northern Alabama, northern Louisiana, and southern Oklahoma (Page and Burr 1991).

Nonindigenous Occurrences: This species is known from various tributaries of the lower Missouri River in **Missouri** including the Gasconade and Loutre rivers systems (Pflieger 1997).

Means of Introduction: Probable bait bucket introduction into the Gasconade River during the 1950s and subsequent dispersal (Pflieger 1997).

Status: Established in parts of Missouri outside its native range.

Impact of Introduction: Unknown.

Remarks: No records exist for this species in tributaries of the Missouri River before the 1950s. During the last 30 years it has become abundant and widespread in the Gasconade River drainage and other tributaries of the lower Missouri (Pflieger 1997).

Notropis buccatus

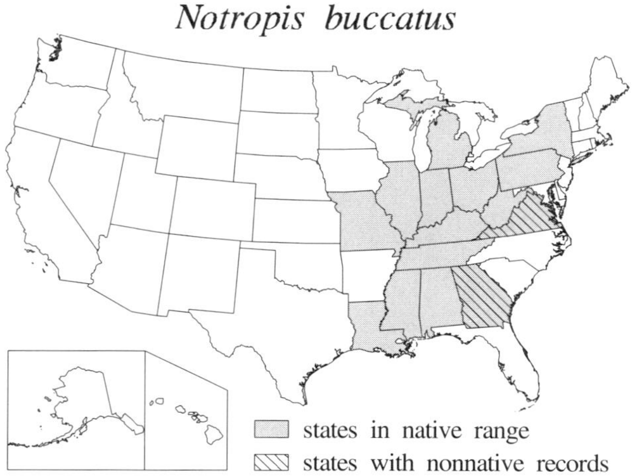

Notropis buccatus (Cope 1865)

Common Name: silverjaw minnow.

Identification: Smith (1985); Page and Burr (1991); Etnier and Starnes (1993); Pflieger (1997); another name is *Ericymba buccata*. Maximum size: 9.8 cm.

Native Range: Disjunct distribution: Atlantic, lower Great Lakes, and Mississippi River drainages from New York, Pennsylvania, and Maryland to eastern Missouri; Gulf drainages from Apalachicola River, Georgia and Florida, to Mississippi River, Mississippi and Louisiana. Absent from Tennessee River drainage and Cumberland River drainage below Cumberland Falls (Lee et al. 1980 et seq.; Page and Burr 1991).

Nonindigenous Occurrences: This species is known from the upper Ocmulgee River drainage (Atlantic Slope) in **Georgia** (Bart et al. 1994). It was taken from the uppermost Clinch River and a tributary in **Virginia** in 1972 (Jenkins and Burkhead 1994).

Means of Introduction: Unknown; possible bait bucket introductions. According to Bart et al. (1994), there are two plausible explanations for this species' present distribution within the Ocmulgee River system. The original entry may have been below Lake Jackson after 1971, with subsequent dispersal to other areas (and an independent introduction into Tussahaw Creek above Lake Jackson). Or, the entry could have occurred above Lake Jackson after the late 1970s, with subsequent dispersal to areas below the lake. The Clinch River, Virginia records probably represent one or more introductions form the Big Sandy (Jenkins and Burkhead 1994)

Status: Reported from Georgia and Virginia.

Impact of Introduction: Unknown.

Remarks: The absence of this species in earlier fish collections lend support to the idea that it was a recent introduction to the Ocmulgee River system of Georgia (Bart et al. 1994). In addition, Bart et al. (1994) considered it unlikely that such a distinctive species would have been overlooked by earlier investigators. This species may have entered the Susquehanna as a bait bucket release (Denoncourt et al. 1975b). Bean and Weed (1911) stated that this species was probably introduced into the upper portion of the Potomac River by the U.S. Bureau of Fisheries. According to Jenkins and Burkhead (1994), introduced status in the Potomac would indicate that this species also was introduced into the Rappahannock, for which the first record is 1962. However, Hocutt et al. (1986) listed it as native to the Rappahannock and Susquehanna drainages. In his recent work on Missouri fishes, Pflieger (1997) stated that isolated records of this species from the St. Francis drainage in Madison County and a tributary of Logan Creek in Reynolds County are fairly recent and possibly the result of bait bucket releases.

Notropis buccula

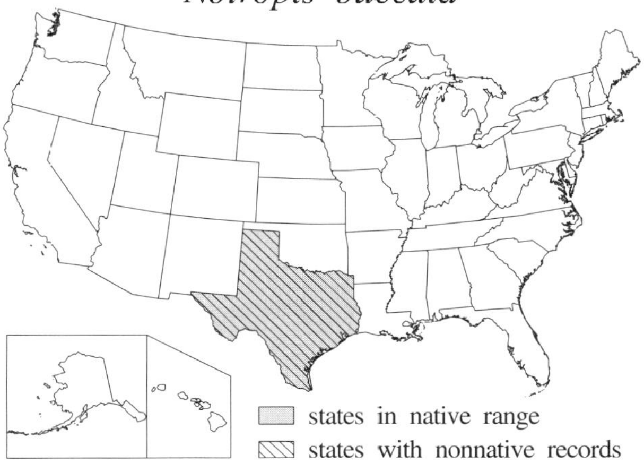

Notropis buchanani

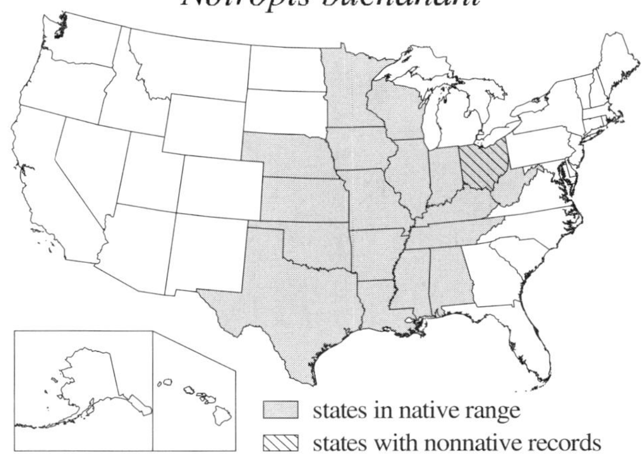

Notropis buccula Cross 1953

Common Name: smalleye shiner.

Identification: Hubbs et al. (1991); Page and Burr (1991). Maximum size: 7.1 cm.

Native Range: Brazos River drainage, Texas (Page and Burr 1991).

Nonindigenous Occurrences: This species is known from the Colorado River drainage in **Texas** (Lee et al. 1980 et seq.; Conner and Suttkus 1986; Hubbs et al. 1991).

Means of Introduction: Unknown; possible bait bucket release.

Status: Reported from Texas.

Impact of Introduction: Unknown.

Remarks: Lee et al. (1980 et seq.) stated that this species was apparently introduced into the Colorado River and their map denotes a single location. Hubbs et al. (1991) stated that the smalleye shiner is presumed to have been introduced into the Colorado River near Austin. These authors added that the fish had not been collected for more than 20 years and may be extinct.

Notropis buchanani Meek 1896

Common Name: ghost shiner.

Identification: Becker (1983); Robison and Buchanan (1988); Page and Burr (1991); Etnier and Starnes (1993); Pflieger (1997). Maximum size 6.4 cm.

Native Range: Mississippi River basin from Pennsylvania, Ohio, and West Virginia to eastern Kansas, and western Oklahoma, and from Minnesota and Wisconsin south to northern Alabama and Louisiana; Gulf Slope drainages from Calcasieu River, Louisiana, to Rio Grande, Texas and Mexico (Lee et al. 1980 et seq.; Page and Burr 1991).

Nonindigenous Occurrences: A single specimen was taken from Sandusky Harbor, Lake Erie, **Ohio**, in June 1997 (S. Keppner, personal communication).

Means of Introduction: Probable bait bucket release.

Status: Reported from Ohio.

Impact of Introduction: Unknown.

Remarks: The Lake Erie species was identified by T. Cavender. According to Mills et al. (1993), a population of the ghost shiner *Notropis buchanani* was found in the Canadian portion of the Great Lakes drainage in 1979 in backwaters of the Thames River (a tributary of Lake St. Clair), Kent County, Ontario. That introduction was thought to have been the result of a bait bucket release. In an apparent error, Page and Burr (1991) listed New Brunswick as part of this species' range.

Voucher specimen: OSUM.

Notropis chiliticus

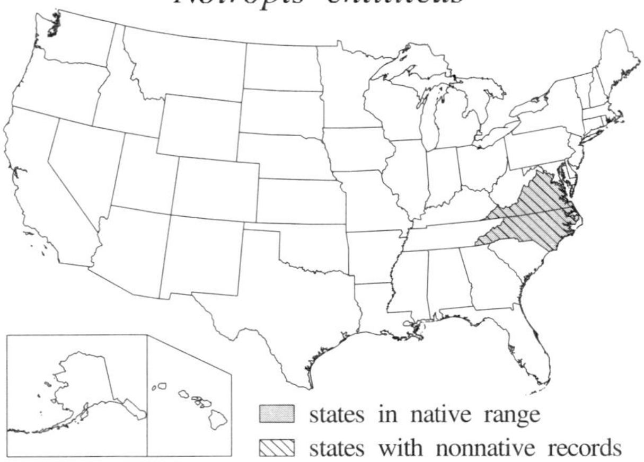

Notropis chiliticus (Cope 1870)

Common Name: redlip shiner.

Identification: Page and Burr (1991); Jenkins and Burkhead (1994). Maximum size: 7.2 cm.

Native Range: Dan River (Roanoke River drainage) and Pee Dee River drainages, Virginia and North Carolina (Page and Burr 1991).

Nonindigenous Occurrences: This species is known from several drainages of the Atlantic Slope and Ohio River basin. These include the Catawba, Pee Dee (Lumber River system) (Menhinick 1991), Cape Fear (Deep River system) (Jenkins and Burkhead 1994), and New (including Little River system) drainages (Menhinick 1991; Page and Burr 1991; Jenkins and Burkhead 1994) in **North Carolina**, and upper Big Chestnut Creek in the New drainage in **Virginia** (Hocutt et al. 1986; Page and Burr 1991; Jenkins and Burkhead 1994).

Means of Introduction: Possible bait bucket releases. Richardson and Carnes (1964, in Jenkins and Burkhead 1994) speculated that it probably was introduced into the Little River system by bait fishermen. It was first discovered in the Little River system in 1963 (Richardson and Carnes 1964); in Big Chestnut Creek, Virginia, in 1976; and in the Deep River system, North Carolina, in 1985 (Jenkins and Burkhead 1994).

Status: Established in North Carolina and Virginia. The Big Chestnut Creek population is "flourishing" (Jenkins and Burkhead 1994).

Impact of Introduction: Unknown.

Remarks: Lee et al. (1980 et seq.) and Page and Burr (1991) stated that it was recently introduced in New River drainage, North Carolina and Virginia. Hocutt et al. (1986) regarded it as introduced (but possibly native) in the Kanahwa above falls. In his summary table on North Carolina fishes, Menhinick (1991) listed this species as probably introduced into the New, the Catawba, and the Lumber and Waccamaw drainages. In their summary table on Virginia fishes, Jenkins and Burkhead (1994) regarded this species as introduced (but possibly native) in the New drainage and native (but possibly introduced) in the Roa drainage. Lee et al. (1980 et seq.) and Hocutt et al. (1986) considered it native to the Dan system (Roanoke River drainage). Jenkins and Burkhead (1994) agreed with that conclusion, but they also speculated that it may have been introduced there. These latter authors noted that its disjunctive distribution in the middle Dan and throughout the Smith, and its apparent absence in the lower Dan, may have resulted from separate introductions in upper portions of the system. Nevertheless, Jenkins and Burkhead concluded that the Dan system was scantily sampled prior to the 1950s and that the range of this species in the Dan can be deemed relictive (native). Jenkins and Burkhead (1994) noted that the upper New drainage has two tributary populations that are considered to be introduced, perhaps separately.

Notropis chrosomus

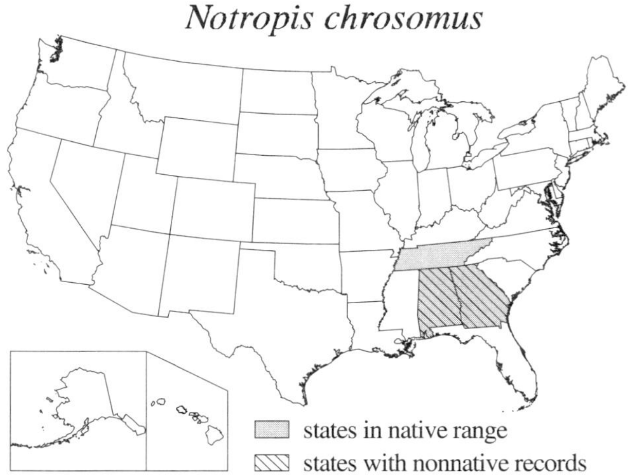

Notropis chrosomus (Jordan 1877)

Common Name: rainbow shiner.

Identification: Page and Burr (1991); Etnier and Starnes (1993); Mettee et al. (1996). Maximum size: 8.1 cm.

Native Range: Mobile Bay basin, including the Coosa, Cahaba and Alabama river drainages, and possibly the upper Black Warrior River system (Lee et al. 1980 et seq.; Page and Burr 1991).

Nonindigenous Occurrences: This species is known from, and was possibly introduced into, the Black Warrior River system of **Alabama** (Lee et al. 1980 et seq.; Swift et al. 1986; Etnier and Starnes 1993). It is also known from, and may have been introduced into, tributaries of the Tennessee River, including Town Creek in northern **Alabama** (Etnier and Starnes 1993) and the South Chickamauga Creek system in north **Georgia** (Starnes and Etnier 1986).

Means of Introduction: Unknown; possible bait bucket release (Starnes and Etnier 1986).

Status: Established in Alabama; possibly established in Georgia

Impact of Introduction: Unknown.

Remarks: Lee et al. (1980) listed this species as native to the Mobile Bay basin but noted that it was possibly introduced into the Black Warrior. Swift et al. (1986) listed it as "native, possibly introduced" to the Black Warrier system. Page and Burr (1991) apparently considered it native to the Black Warrior drainage. In their summary table on Alabama fishes, Mettee et al. (1996) listed it as native to the Black Warrior River system. Starnes and Etnier (1986) considered the presence of this species, and also *Fundulus stellifer,* to represent "recent accidental (bait bucket?) introductions" into the South Chickamauga Creek system (Tennessee River drainage) of Georgia. These two species occur together in small streams and could easily be taken by individuals seining for bait and subsequently transferred. In a footnote to their table summary of Tennessee fishes, Etnier and Starnes (1993) stated that this species is present in the Tennessee River drainage in northern Alabama (native?) and Georgia (introduced?). The Tennessee River drainage site in Alabama is mentioned as being Town Creek near Fort Payne. Within the state of Tennessee, Etnier and Starnes found it to be confined to tributaries to the Conasauga River (Mobile basin) Mettee et al. (1996) noted that isolated populations exist in two sites in the Tennessee River drainage in Alabama, the Town and Short creek systems. They do not discuss the possibility the possibility of introduction. In our opinion, the records from the Tennessee River drainage in Alabama and Georgia represent introductions. The Town Creek site (northern Alabama) is considerable distance from the drainage divide where a stream capture would be expected to occur. If it was in the Tennessee drainage via stream captures, then the species would likely be much more widespread. That region has been fairly well sampled, so if it were more widely distributed then it would most likely been collected. Reports of this species as possibly introduced into the Black Warrior system are questionable. We base this on the fact that *N. chrosomus* is typically found in streams with limestone outcrops in the watershed. Based on the distribution map given by Mettee et al. (1996), it is found along the southeast edge of the Black Warrior system. These are streams that have their headwaters in limestone formations. It likely entered the Black Warrior system via stream capture a long time ago or, if introduced, was transplanted by Native Americans. In their report on Georgia fishes, Dahlberg and Scott (1971a) made no mention of the occurrence of this species outside the Coosa River drainage.

Notropis dorsalis

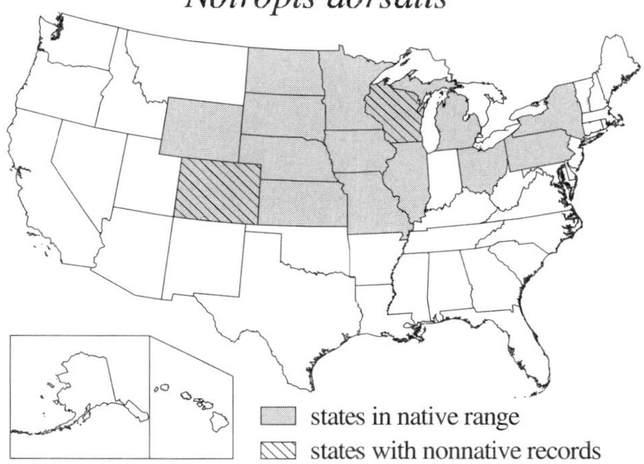

Notropis dorsalis (Agassiz 1854)

Common Name: bigmouth shiner.

Identification: Becker (1983); Smith (1985); Page and Burr (1991); Etnier and Starnes (1993). Three subspecies have been recognized *N. d. piptolepis*, *N. d. keimi*, and *N. d. dorsalis* (Page and Burr 1991; but see Gilbert 1998). Maximum size: 8 cm.

Native Range: Great Lakes, Hudson Bay (Red River), and Mississippi River basins from northern Michigan to southern Manitoba, and from eastern Illinois to the Platte River system, eastern Wyoming and northern Colorado; disjunct populations in western New York and Pennsylvania, northern West Virginia, northern Ohio, and western Michigan. *Notropis d. piptolepis* is native to the Platte River system in Wyoming and Colorado; *N. d. keimi* is native to Lake Ontario and Allegheny River drainages in New York and Pennsylvania; and *N. d. dorsalis* is throughout the rest of the range (Page and Burr 1991).

Nonindigenous Occurrences: This species is known from the upper Arkansas River drainage, **Colorado** (Miller 1983). It also is known from several areas in the Lake Michigan drainage of **Wisconsin** including the upper Wolf River watershed in Langlade County, and the Root and Pike river systems in Racine and Kenosha counties (and possibly elsewhere in state) (Becker 1983).

Means of Introduction: Means of introduction into the Arkansas River is not known with certainty, may have been natural or the result of an introduction. Miller (1983) hypothesized that the population found in Monument Creek (Arkansas River drainage) in Colorado may have resulted by way of a bait-bucket transfer from the range of the western subspecies. Becker (1983) speculated that the introduction into the upper Wolf River of Wisconsin was likely due to a bait release since it is unlikely that this minnow could move over the dam at Shawano. He also concluded that recent collections in the Root and Pike river systems are more likely to have resulted from bait bucket releases than by fish crossing the low divide between the headwaters of the Root and Des Plaines rivers. Becker also made reference to the possible passage of this species into the Fox-Wolf system through the connection at Portage. The author was apparently making reference to the Wisconsin-Fox Canal which connects the Wisconsin River to the Lake Michigan drainage, although he apparently did not rule out an earlier, natural connection, that may have existed.

Status: Presumably established in the Arkansas River drainage of Colorado (see Woodling 1985). Established in the upper Wolf watershed of Wisconsin; reported from other localities in Wisconsin outside its native range (Becker 1983).

Impact of Introduction: Unknown.

Remarks: Miller (1983) discussed the presence of this species in the Arkansas River drainage of Colorado. He noted that two subspecies of *Notropis dorsalis* are recognized, one with scales on the nape occurring in the central and eastern parts of the species range and one without scales on the nape from the western part. The specimens from Mounument Creek (Arkansas River drainage), Colorado, are referable to the subspecies without scales on the nape. According to Miller (1983), such a finding is counter-intuitive, because if *N. dorsalis* gained access to the Arkansas River naturally via a past water connection, it should be the other subspecies. As such, Miller suggested the species was introduced to the Arkansas drainage. Nevertheless, Miller stated that the presence of this species in the Arkansas River drainage remains problematical. Except for Miller (1983), the literature is unclear and somewhat contradictory concerning the status of this species in the Arkansas River drainage and in Colorado. Cross et al. (1986) listed it as introduced to the upper Arkansas drainage but did not specify the state. Presumably, their listing of it as introduced is based on the Colorado record of Miller (1983). According to Cross et al. (1986), the upper Arkansas River drainage includes portions of Oklahoma, Kansas, Colorado, and very small section of New Mexico. We are unaware of any records of this species for portions of the Arkansas River drainage outside of Colorado (see for example, Cross 1967; Lee et al. 1980 et seq.; Cross and Collins 1995). Walker (1993) noted that the bigmouth shiner was present in the Arkansas drainage in Colorado, but listed it as native to the drainage. Woodling (1985) provided a map of the species' distribution in Colorado and noted that a population had been found by Miller (see Miller 1983) in Fountain Creek, a tributary of the Arkansas River (El Paso County). However, Woodling did not specify if the Fountain Creek population was native or introduced. Becker (1983) stated that the occurrence of this fish in the Fox River (Green Lake County) in the Lake Michigan drainage of Wisconsin may have resulted from a crossover at Portage, through either glacial or modern connections. He noted that this species has appeared at a number of sites in the Fox and Wolf river watershed, many of which Becker attributed to passage of fish through the connection at Portage. This species is commonly used as a bait fish in some areas (Beckman 1974; Becker 1983), although Woodling (1985) noted that it is not as hardy as other minnows.

Notropis girardi

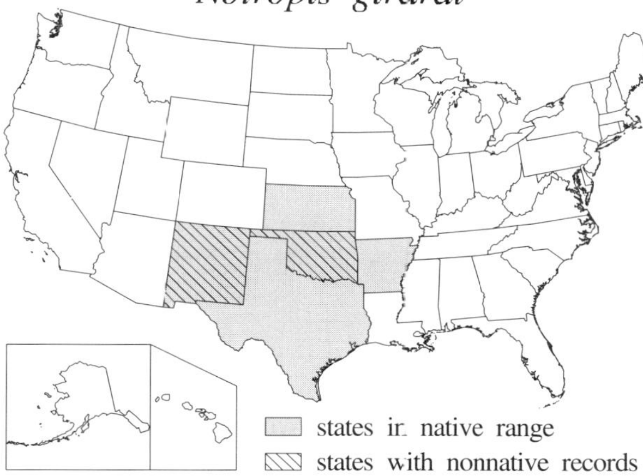

Notropis girardi Hubbs and Ortenburger 1929

Common Name: Arkansas River shiner.

Identification: Robison and Buchanan (1988); Sublette et al. (1990); Page and Burr (1991). Maximum size: 8 cm.

Native Range: Arkansas River drainage from western Arkansas to western Kansas, western Oklahoma, Texas panhandle, and northeastern New Mexico (Page and Burr 1991).

Nonindigenous Occurrences: This species has been known from the Pecos River in **New Mexico** (Bestgen et al. 1989) since 1978 and has become established from Fort Sumner Reservoir to Carlsbad (Sublette et al. 1990). It is listed as introduced into the upper Red River drainage (Cross et al. 1986); that conclusion was apparently based on a single specimen taken from the Washita River system in Garvin County, **Oklahoma** in 1968 (Cross 1970; Lee et al. 1980 et seq.).

Means of Introduction: According to Bestgen et al. (1989), introduction of *Notropis girardi* into the Pecos River probably resulted from release of baitfish by anglers below Sumner Dam in 1978. Its use as a bait fish may have been inadvertent, for it was speculated that the species may have been included in a bait shipment of plains minnow, *Hybognathus placitus*. The initial stock is unknown, but it may have come from the nearby Canadian River drainage. From the presumed point of introduction, the species dispersed downstream at least 260 km by the late 1981 and was common in collections made in the river in 1986–1987. However, Bestgen et al. (1989) admitted that dispersal may have been aided by multiple introductions throughout the Pecos River. Cross (1970) presumed that the species gained access to the Red River in Oklahoma by way of introduction by man. However, he could not rule out natural transfer from the South Canadian River (Arkansas drainage) to the adjacent Washita drainage.

Status: Established in New Mexico; reported from Oklahoma outside its native range.

Impact of Introduction: The population in the Pecos River may be posing a problem for the survival of the endemic bluntnose shiner *Notropis simus pecosensis* (Gilbert, personal communication).

Remarks: Bestgen et al. (1989) discussed the dispersal of this species in the Pecos River. The Pecos River serves as a reservoir for *Notropis girardi* since this species has been displaced throughout much of its native range by the Red River shiner *Notropis bairdi* (Gilbert, personal communication). Lee et al. (1980 et seq.) noted that the specimen taken from Washita Creek, Oklahoma, presumably represented an introduction. In their table summarizing the distribution of fishes in the western Mississippi Basin, Cross et al. (1986) regarded it as introduced to the upper Red River drainage. *Notropis girardi* is proposed for listing as an endangered species in its native range (U.S. Fish and Wildlife Service 1997).

Notropis harperi

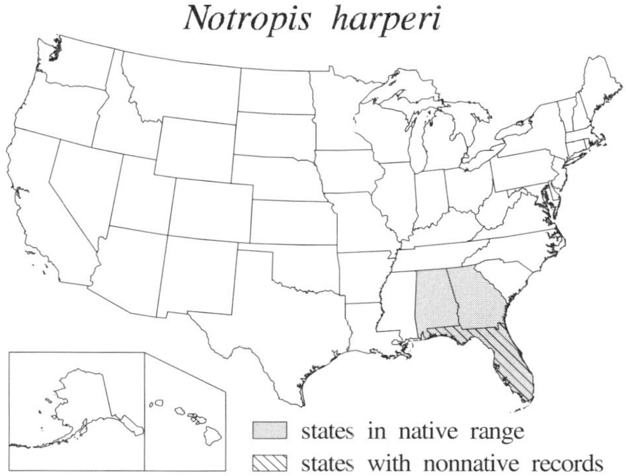

Notropis harperi Fowler 1941

Common Name: redeye chub.

Identification: Page and Burr (1991); Mettee et al. (1996). Maximum size: 6 cm.

Native Range: Below Fall Line in Atlantic and Gulf Slope drainages from Altamaha River, Georgia, to Escambia River, Alabama; south in Florida to St. Johns and Withlacoochee River drainages (Page and Burr 1991).

Nonindigenous Occurrences: This species was taken from the west end of Lake Jackson in the Ochlockonee drainage, **Florida** in 1954 (Swift et al. 1977).

Means of Introduction: According to Swift et al. (1977), the Lake Jackson record is probably a bait fish release. Lake Jackson is located near the city of Tallahassee. These researchers noted that at least one Tallahassee bait dealer regularly collects minnows (mostly *Notropis petersoni* and *Notropis harperi*) from the Wakulla River (a spring-fed river just south of Tallahassee).

Status: Reported from Florida outside its native range.

Impact of Introduction: Unknown.

Remarks: Swift et al. (1986) listed it as introduced to the Ochlockonee drainage.

Notropis hudsonius

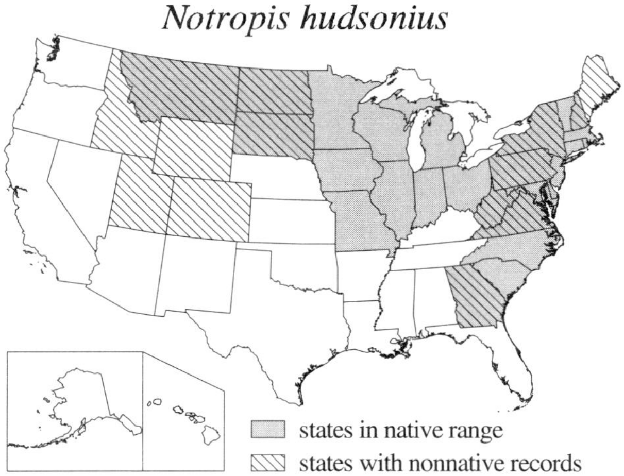

Notropis hudsonius (Clinton 1824)

Common Name: spottail shiner.

Identification: Becker (1983); Page and Burr (1991); Jenkins and Burkhead (1994); Pflieger (1997); Gilbert (1998). Two distinct morphological forms are known (Gilbert 1998). Maximum size: 15 cm.

Native Range: Atlantic and Gulf Slope drainages from St. Lawrence River, Quebec to Altamaha and upper Chattahoochee River, Georgia; Hudson Bay, Great Lakes, and Mississippi River basins from Ontario to Mackenzie River drainage (Arctic basin), Northwest Territories and Alberta, and south to northern Ohio, southern Illinois, and northeastern Montana (Page and Burr 1991).

Nonindigenous Occurrences: This species is known from Quincy and Lon Hagler reservoirs in the South Platte drainage, **Colorado** (Woodling 1985; Walker 1993); the upper Flint River drainage in **Georgia** (S. Ruessler, personal communication; museum specimen); the Snake River above and below the falls, and the Bear River in **Idaho** (Idaho Fish and Game 1990); lower Kennebec River, **Maine** (Kircheis 1994); several sites in the Missouri River drainage in **Montana** (Holton 1990); the Connecticut and Merrimack rivers in **New Hampshire** (where it may have been introduced) (Scarola 1973); the Allegheny Reservoir on the **New York–Pennsylvania** border (Ichthyological Associates 1976; Cooper 1983); Lakes Oahe and Sakakawea on the Missouri River, and the mouth of Heart River, **North Dakota** (Owen et al. 1981; Cross et al. 1986); Lake Oahe, **South Dakota** (Owen et al. 1981); Willard Bay and Utah Lake, **Utah** (Sigler and Sigler 1987; 1996); the New River drainage of **Virginia** (Hocutt et al. 1986; Jenkins and Burkhead 1994) and **West Virginia** (Hambrick et al. 1973; Hocutt et al. 1986; Jenkins and Burkhead 1994; Stauffer et al. 1995); and **Wyoming** (Stone 1995) including several reservoirs east of the Continental Divide in that state (Hubert 1994).

Means of Introduction: Commonly stocked as a forage fish. According to Cooper (1983), the recent successful introduction of this species (and white bass *Morone chrysops*) into the Allegheny Reservoir in Warren County, Pennsylvania, possibly was derived from attempted introductions of the emerald shiner *Notropis atherinoides* into this reservoir from Lake Erie stocks. Jenkins and Burkhead (1994) summarized the literature dealing with its spread in the New River drainage of Virginia and West Virginia. They speculated that it was introduced to the New drainage of Virginia, probably in Claytor Lake, either as bait, purposely for forage, or inadvertently along with sportfish stockings. After its discovery in Claytor Lake in the late 1940s, the species spread widely in the New River below Claytor Lake. The first West Virginia record is dated 1972 from the lower East River near the Virginia border. From there it spread into other parts of the New River drainage deeper in West Virginia. In Maine, the species was probably imported illegally as a baitfish; it was first observed in the state 1979 in the Cathance River, Kennebec River drainage (Kircheis 1994). It was stocked as a forage fish in Willard Bay, Utah, in 1982, and in Utah Lake in 1983 (Sigler and Sigler 1987). Specimens were taken from several locations in the upper Flint drainage in late 1997 (Ruessler, personal communication), these likely represent the result of a bait bucket introduction rather than from natural dispersal.

Status: Established, or presumably established, in Georgia, Idaho, Maine, Montana, New Hampshire, New York, North Dakota, Pennsylvania, South Dakota, Utah, Virginia, West Virginia, and Wyoming; reported from Colorado.

Impact of Introduction: Unknown.

Remarks: Although this species has been recorded from the headwaters of the Chattahoochee River in Georgia (Lee et al. 1980 et seq.; Mettee et al. 1996), there are no records of this fish from the lower reaches of that river (Mettee et al. 1996). Consequently, it is very unlikely that *Notropis hudsonius* recently discovered in the upper Flint drainage were the result of individuals dispersing naturally downstream into Lake Seminole and up the Flint River. Jenkins et al. (1972) listed it as "probably or possibly introduced" to the Kanawha River drainage above the falls.

Voucher specimens: Georgia (UF uncatalogued).

Notropis hypsilepis

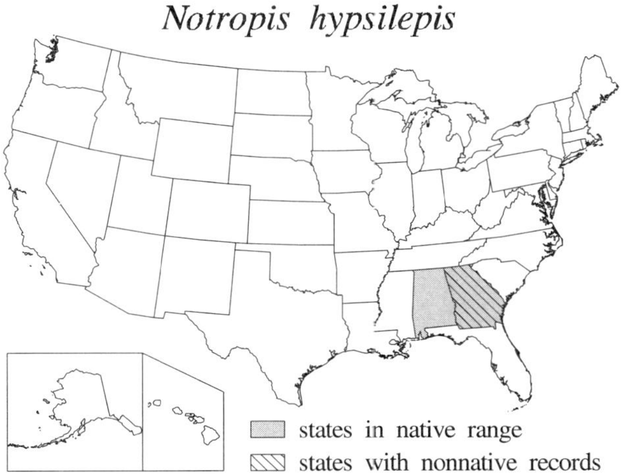

Notropis hypsilepis Suttkus and Raney 1955

Common Name: highscale shiner.

Identification: Page and Burr (1991); Mettee et al. (1996). Maximum size: 6.4 cm.

Native Range: Apalachicola River drainage, Georgia and eastern Alabama; possibly native to the upper Savannah River drainage, northeastern Georgia (Lee et al. 1980 et seq.; Page and Burr 1991).

Nonindigenous Occurrences: This species is known from the upper Savannah drainage, **Georgia** (Lee et al. 1980 et seq.; Dahlberg and Scott 1971b).

Means of Introduction: Unknown; possible bait bucket release.

Status: Established in Georgia.

Impact of Introduction: Unknown.

Remarks: It is uncertain whether this species is native to, or was introduced into, the upper Savannah drainage (Lee et al. 1980 et seq.; Dahlberg and Scott 1971b). Page and Burr (1991) apparently considered it native to the upper Savannah. According to Gilbert (personal communication), *Notropis hypsilepis* is not a common species and occurs in a very specific type of habitat (i.e., over shifting-sand substrate), thus, it would not be a species that would frequently enter bait buckets. Gilbert (personal communication) visited the site in the upper Savannah drainage (Timpson Creek) where the fish had been collected and noted that the site was not the species' typical habitat. Consequently, Gilbert favored the idea of introduction, although he cannot rule out the possibility of stream capture. The Timpson Creek site is very close to headwaters of the Chattahoochee River, and not particularly far from localities in the fish has been found (see Lee et al. 1980 et seq.).

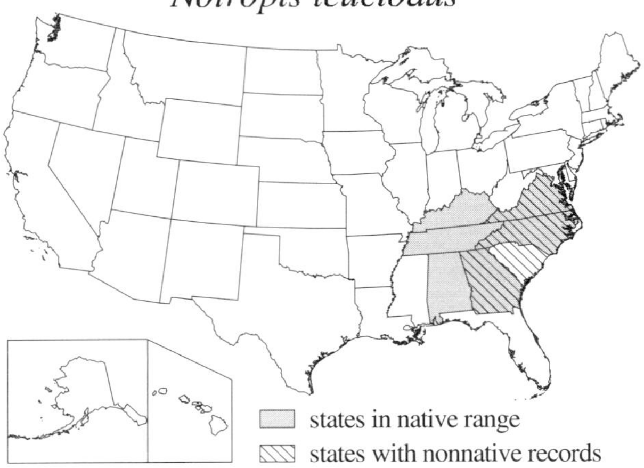

Notropis leuciodus (Cope 1868)

Common Name: Tennessee shiner.

Identification: Page and Burr (1991); Etnier and Starnes (1993); Jenkins and Burkhead (1994). Maximum size: 8.2 cm.

Native Range: Green, Cumberland, and Tennessee River drainages, Virginia, North Carolina, Kentucky, Tennessee, Georgia, and Alabama. In Green River drainage, restricted to extreme upper Green and upper Barren systems. Also in extreme upper Savannah River drainage, North Carolina and Georgia, and New River drainage, Virginia and North Carolina (latter may be result of bait bucket introduction) (Page and Burr 1991; see below).

Nonindigenous Occurrences: This species is known from, and possibly introduced into, the Chestatee River (Chattahoochee River system) of **Georgia** (Johnston et al. 1995); the New and the Catawba drainages of **North Carolina** (Lee et al. 1980 et seq.; Menhinick 1991; Page and Burr 1991; Jenkins and Burkhead 1994); and the New River drainage of **Virginia** (Lee et al. 1980 et seq.; Page and Burr 1991; but see Jenkins and Burkhead 1994).

Means of Introduction: Unknown; probable bait bucket releases.

Status: Apparently established in the Chattahoochee River system of Georgia and the New River drainage of North Carolina (and possibly the New River drainage Virginia).

Impact of Introduction: Unknown.

Remarks: There is some uncertainty concerning the native versus nonnative distribution of this species. In his summary table of North Carolina fishes, Menhinick (1991) regarded it as native but possibly introduced to the Savannah River drainage. If the presence of this species in the upper Savannah of North Carolina is the result of past introduction, then its occurrence in the upper Savannah of South Carolina possibly is also due to introduction. However, most authors believe it to be native to the Savannah. For example, Tsai and Raney (1974) stated that *N. leuciodus* may have entered the Savannah River drainage of the Atlantic Slope via stream capture. Similarly, Lee et al. (1980 et seq.) considered that its presence in the upper Savannah of Georgia and North Carolina to be the result of natural stream capture from the Tennessee River drainage. Page and Burr (1991) and Jenkins and Burkhead (1994) also considered it native to the Savannah. There is more consensus concerning its possible introduction into the New River drainage. For example, Menhinick (1991) listed this species as probably introduced to the New and Catawba drainages. In their summary table of Virginia fishes, Jenkins and Burkhead (1994) regarded it as introduced, but possibly native, to the New. They noted that it occurred in the New River drainage of North Carolina, and perhaps Virginia also. However, it should be emphasized that these researchers indicated there had been as yet no records of the fish being taken from the New drainage within Virginia borders. The species is known from the Little River system (New River drainage) in North Carolina, very near to the Virginia border, and this may explain why some authors have listed it as occurring in Virginia. For example, Page and Burr (1991) stated that the species occurred in the New River drainage of North Carolina and Virginia, and noted that its presence in that drainage may be the result of a bait bucket release. Lee et al. (1980 et seq.) also stated that the occurrence of this species in the New-Kanawha River drainage was believed to be due to human introduction. Hocutt et al. (1986) also regarded it as introduced (but possibly native) for the Kanawha (New) River drainage above the falls. However, they stated that the historical absence of this and other minnow species in earlier collections may have been due to inadequate sampling.

Notropis longirostris

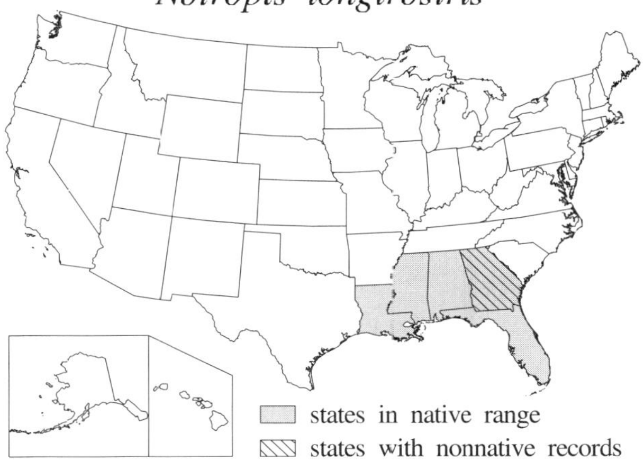

Notropis longirostris (Hay 1881)

Common Name: longnose shiner.

Identification: Page and Burr (1991); Mettee et al. (1996). Maximum size: 6.5 cm.

Native Range: Gulf Slope drainage from Apalachicola River, Georgia and Florida, to Mississippi River, Louisiana (except Mobile Bay); north in Mississippi River basin to Big Black River, Mississippi, and lower Ouachita River drainage, Louisiana (Page and Burr 1991).

Nonindigenous Occurrences: This species was recently discovered in Tussahaw Creek above Lake Jackson in the Ocmulgee River system (Atlantic Slope), **Georgia** (Bart et al. 1994).

Means of Introduction: Unknown; possible bait bucket introduction. Bart et al. (1994) suggested that Tussahaw Creek may be the point of entry of this species into the Ocmulgee River system and that Lake Jackson is currently preventing its dispersal to other parts of the water system.

Status: Reported from Georgia.

Impact of Introduction: Unknown.

Remarks: Tussahaw Creek is the only known record of this species in the Ocmulgee River system. That record is based on six specimens taken by Georgia Department of Natural Resources' personnel (Bart et al. 1994). There is a questionable previous report of this species from the Altamaha drainage (which includes the Ocmulgee) by Ramsey (1965), but his report was not based on museum specimens (Bart et al. 1994). Dahlberg and Scott (1971a) also included this species, perhaps on the basis of Ramsey's report, but with no details. This species was not reported in surveys conducted in this area during the late 1970s (Bart et al. 1994). The native population east of the Mobile Bay basin are genetically distinct from those found elsewhere and probably represent an undescribed species. In terms of external morphology, however, the eastern Gulf slope population seems to be totally indistinguishable from specimens taken from other areas (Gilbert 1998).

Notropis ludibundus

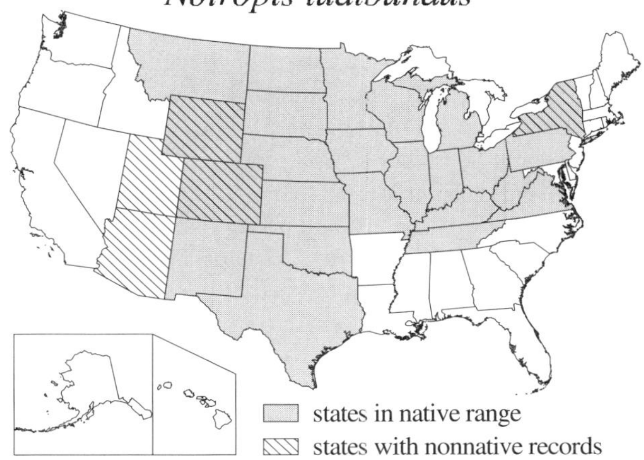

Notropis ludibundus (Girard 1856)

Common Name: sand shiner.

Identification: Becker (1983); Sublette et al. (1990); Page and Burr (1991); Jenkins and Burkhead (1994); Pflieger (1997); a commonly used synonym is *Notropis stramineus* (Mayden and Gilbert 1989). Maximum size: 8.1 cm.

Native Range: Saint Lawrence–Great Lakes, Hudson Bay (Red River), and Mississippi River basins from St. Lawrence River, southern Quebec, to eastern Saskatchewan, and south to Tennessee and Texas; west to eastern Montana, Wyoming, Colorado, and New Mexico; Gulf Slope from Trinity River to Rio Grande, Texas, New Mexico, and Mexico (Page and Burr 1991).

Nonindigenous Occurrences: This species is known from East Clear and Chevalon creeks in **Arizona** (Miller and Lowe 1967; Minckley 1973). It is also known from the Colorado, Dolores (Holden and Stalnaker 1975b; Tyus et al. 1982; Woodling 1985), and Yampa (Tyus et al. 1982) rivers in **Colorado**; the Shawangunk Kill, **New York** (Smith 1985; Schmidt 1986); the Colorado, Green, and Dolores rivers, **Utah** (Holden and Stalnaker 1975b; Tyus et al. 1982); and the Bighorn drainage, **Wyoming** (Baxter and Simon 1970; Gorges 1994).

Means of Introduction: In Arizona, this species was reportedly stocked by accident along with shipments of largemouth bass from Dexter, New Mexico, in 1935, into East Clear Creek and perhaps Chevalon Creek (Minckley 1973). In his list of introduced fishes of Arizona and the lower Colorado River, Rinne (1995) reported it as an accidental introduction and gave the year as 1938. Although uncertain, it may have entered the Shawangunk Kill of New York via the Delaware-Hudson Canal system (Smith 1985). This species may have been introduced into other western states as a bait fish. It was introduced to Willard Bay Reservoir, Utah in 1982 (Sigler and Sigler 1987). The sand shiner was first collected in the upper Colorado basin (Dolores River) in 1971 (Holden and Stalnaker 1975a, 1975b). Based on its numbers and distribution, Holden and Stalnaker (1975b) suggested that this species had been present in the Colorado basin for several years, perhaps introduced as a baitfish, as it is common in the South Platte River system of eastern Colorado.

Status: Established in Arizona, Colorado, New York, Utah, and Wyoming.

Impact of Introduction: Largely unknown. This species was reported to be "a problem" in the Yampa River in Colorado (Whitmore 1997).

Remarks: Most references use the name *Notropis stramineus* for this species. According to Minckley (1973), it has been collected only once in Arizona, in 1955 in Navajo County. However, the species was observed in the mouth of Chevalon Creek, Arizona, in 1938 (Miller and Lowe 1967). Baxter and Simon (1970) stated that this species is known from Ocean Lake in the Bighorn drainage of Wyoming, where it may have been introduced. Tyus et al. (1982) gave a distribution map of this species in the upper Colorado basin. Schmidt (1986) listed this species as native but possibly introduced to the Hudson drainage. This species is commonly used as bait and also as a bioassay animal (Baxter and Simon 1970; Sigler and Sigler 1996).

Notropis lutipinnis

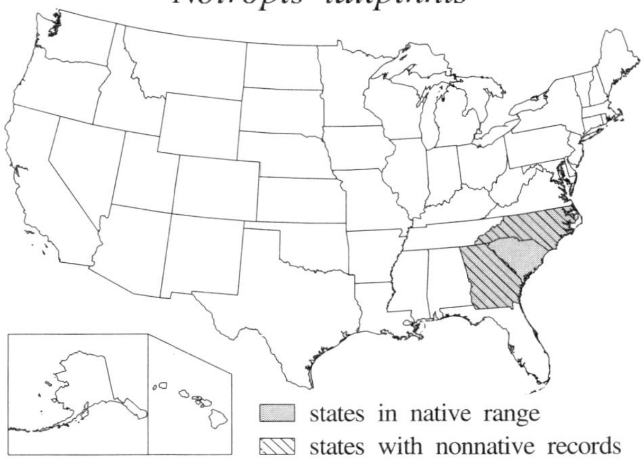

Notropis lutipinnis (Jordan and Brayton 1878)

Common Name: yellowfin shiner.

Identification: Menhinick (1991); Page and Burr (1991). According to Wood and Mayden (1992), *N. lutipinnis* is composed of several different forms. Maximum size: 7.5 cm.

Native Range: Atlantic and Gulf Slopes from Santee River, North Carolina, to Altamaha River, upper Chattahoochee River, and upper Coosa River systems, Georgia (Page and Burr 1991).

Nonindigenous Occurrences: This shiner is known from (and probably introduced into) the Tennessee River drainage in **North Carolina** (Menhinick 1991; Johnston et al. 1995). In **Georgia**, it has been recorded from (and possibly introduced into) the extreme headwaters of the Chattahoochee River and the Coosa River drainage (Lee et al. 1980 et seq.; Swift et al. 1986) and the Little Tennessee River system (Tennessee River drainage) (Johnston et al. 1995).

Means of Introduction: Unknown; possible bait bucket releases.

Status: Established in Georgia and North Carolina. Johnston et al. (1995) noted that the species appears to be expanding its range and abundance in the Tennessee River drainage.

Impact of Introduction: Unknown.

Remarks: There is some uncertainty in the literature concerning the native versus introduced range of this species. Concerning the Tennessee drainage occurrence, Menhinick (1991) listed it as "probably introduced" into the North Carolina portion of the drainage. Lee et al. (1980 et seq.) did not depict or mention a record for the Tennessee drainage. Similarly, Etnier and Starnes (1993) did not list it as occurring in the Tennessee drainage of Tennessee. The reason for its occurrence in parts of Georgia is also uncertain. Lee et al. (1980) noted that it is known from extreme headwaters of the Chattahoochee River system (Apalachicola River drainage) and Coosa River drainage, and added that it was possibly native in both drainages. However, they did not discuss the probability of it being introduced to these water bodies. Johnston et al. (1995) found it in two sites in the Little Tennessee River, Georgia and North Carolina, during 1990 surveys. Swift et al. (1986) noted that *Notropis lutipinnis* and *N. baileyi* are rare and perhaps introduced in the Apalachicola drainage. In their summary table on fishes of the southeastern United States, these authors listed *N. lutipinnis* as "native, possibly introduced" for both the Chattahoochee and Coosa River systems. Bryant et al. (1979) suggested that this species gained access to the Mobile River basin through stream capture, or less likely, by way of an introduction by man.

Notropis nubilus

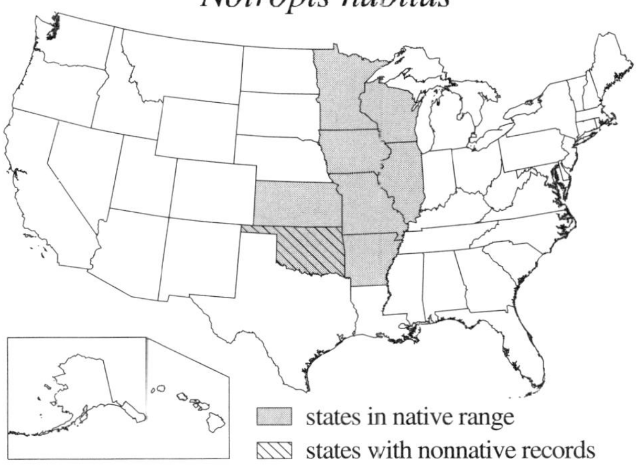

Notropis nubilus (Forbes 1878)

Common Name: Ozark minnow.

Identification: Becker (1983); Robison and Buchanan (1988); Page and Burr (1991); Pflieger (1997). Another name is *Dionda nubila*. Maximum size: 9.3 cm.

Native Range: Disjunct populations in upper Red River system, northern Wisconsin; Mississippi River tributaries in southeastern Minnesota, eastern Iowa, southern Wisconsin, and northern Illinois; Ozark Mountain drainages in southern Missouri, southeastern Kansas, northern Arkansas, and northeastern Oklahoma; and Boggy Creek system, Oklahoma (Page and Burr 1991).

Nonindigenous Occurrences: This species has been reported from (and possibly introduced into) the Blue River (Red River drainage) in **Oklahoma** (Miller and Robison 1973).

Means of Introduction: Possible bait bucket release (Miller and Robison 1973).

Status: Reported from Oklahoma.

Impact of Introduction: Unknown.

Remarks: There is some confusion in the literature concerning the presence of this species in southern Oklahoma and whether it is native or introduced to this region. Miller and Robison (1973) stated that it is "possibly present in the Blue River and elsewhere as a result of bait bucket introductions." The Blue River is a tributary of the Red River and is near a pair of other Red River tributaries, Clear Boggy and Muddy Boggy creeks. Page and Burr (1991) apparently considered it native to the Boggy Creek system of Oklahoma, but added that the species is possibly extinct in that water system. Presumably their "Boggy Creek system" represents Clear Boggy Creek, Muddy Boggy Creek, or both. Lee et al. (1980 et seq.) did not discuss nor map records of this species from Blue River, Boggy Creek, nor any other part of southern Oklahoma. Similarly, Cross et al. (1986) did not list it as occurring in the Red River drainage. Fossil remains have been found in southwestern Kansas, providing evidence of a more western distribution during glaciation (Becker 1983; Cross et al. 1986).

Notropis oxyrhynchus

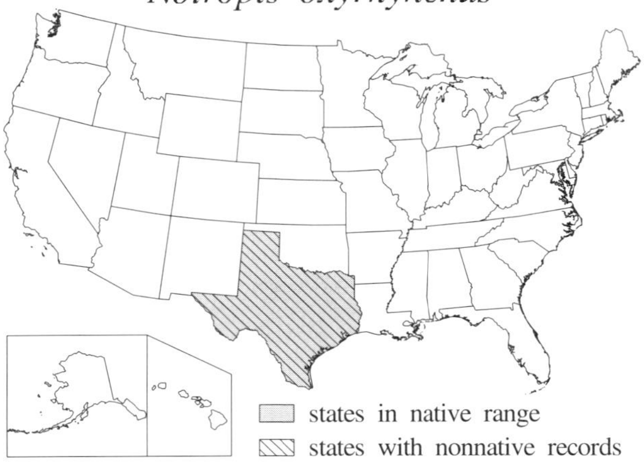

Notropis oxyrhynchus Hubbs and Bonham 1951

Common Name: sharpnose shiner.

Identification: Hubbs et al. (1991); Page and Burr (1991). Maximum size: 6.5 cm.

Native Range: Brazos River and lower portions of large tributaries, Texas (Page and Burr 1991).

Nonindigenous Occurrences: This species has been recorded from (and apparently introduced into) the Colorado River drainage of **Texas** near Austin (Yurgens 1954; Lee et al. 1980 et seq.; Conner and Suttkus 1986).

Means of Introduction: Unknown. Yurgens (1954) speculated that this species and several other recently discovered minnows were likely introduced as a bait release into the Colorado River drainage. He based his conclusion on the fact that bait dealers had been transporting Brazos River minnows in large numbers to fishing camps on Lake Travis near Austin.

Status: Apparently established in Texas outside its native range.

Impact of Introduction: Unknown.

Remarks: Yurgens (1954) based his report on the capture of a single specimen taken from Lake Travis. Although he argued that the Lake Travis record was likely the result of an introduction, Yurgens did not rule out the possibility that *N. oxyrhynchus* is native to the upper part of the Colorado River. Hubbs et al. (1991) stated that a presumed introduced population exists in the Colorado River above Buchanan Reservoir. *Notropis oxyrhynchus* is also known from the upper Red River (Miller 1955). Conner and Suttkus (1986) noted that its occurrences in both the Colorado River and Red River are almost certainly the result of introductions by man. In contrast, Cross et al. (1986) listed it as native to the upper Red River and attributed its presence to past capture by the Red River of drainage that formerly entered the Brazos.

Notropis ozarcanus

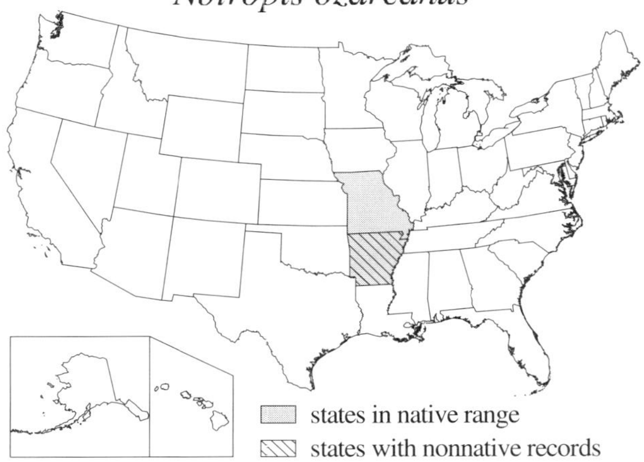

Notropis ozarcanus Meek 1891

Common Name: Ozark shiner.

Identification: Robison and Buchanan (1988); Page and Burr (1991); Pflieger (1997). Maximum size: 7.5 cm.

Native Range: Above Fall Line in White and Black River systems, Missouri and Arkansas. Formerly in upper St. Francis River drainage, Missouri, but now extirpated (Page and Burr 1991).

Nonindigenous Occurrences: This species is known from Osage Creek, a tributary to the Illinois River (Arkansas River drainage), **Arkansas** (Burr et al. 1979).

Means of Introduction: Unknown; possible bait bucket release.

Status: Reported from middle Arkansas River drainage.

Impact of Introduction: Unknown.

Remarks: In their summary table on fishes of western Mississippi River drainages, Cross et al. (1986) listed this species as introduced to the Middle Arkansas River drainage. That record is apparently based on Burr et al. (1979) who reported *Notropis ozarcanus* and *Campostoma oligolepis* from the Illinois River system of the Arkansas drainage. Both were previously known from all Ozark drainages except the Arkansas. Because of the absence of both species from earlier records, Burr et al. (1979) suggested that these fishes were bait introductions, although these researchers could not rule out the possibility of both being native to the Illinois River system. According to Cross et al. (1986), these records have been considered introductions, but the same records may also be evidence of past stream capture. These authors did not note the locality or state and they provided no additional information concerning the introduction. In their drainage map, Cross et al. (1986) showed the middle Arkansas River as including portions of northwestern Arkansas, southwestern Missouri, northeastern Oklahoma, and southeastern Kansas. Robison and Buchanan (1988), also in reference to Burr et al. (1979) stated that the species has been discovered occurring disjunctly in the Illinois River (Arkansas River system). However, Robison and Buchanan did not discuss the possibility of it being introduced to that system. According to Burr (personal communication), there are no voucher specimens to verify this record, and it may be based on a misidentification.

Notropis potteri

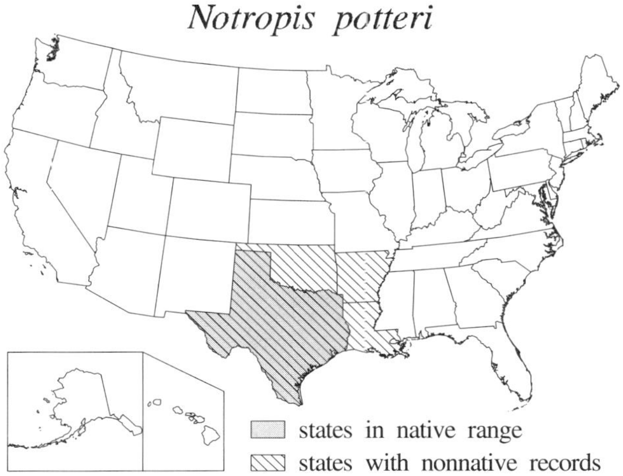

Notropis potteri Hubbs and Bonham 1951

Common Name: chub shiner.

Identification: Hubbs et al. (1991); Page and Burr (1991). Maximum size: 11 cm.

Native Range: Brazos, San Jacinto, Trinity, and Colorado rivers, Texas; possibly includes the Red River drainage in Oklahoma, Texas, Arkansas, and Louisiana (Lee et al. 1980 et seq.; Page and Burr 1991; Gilbert, personal communication).

Nonindigenous Occurrences: The chub shiner may have been introduced into the Red River drainage in **Oklahoma** and **Texas** (Hubbs and Bonham 1951; Hall 1956) and spread downriver through **Arkansas**, to the Mississippi River at the mouth of the Red River in **Louisiana** (Gilbert, personal communication), where Page and Burr (1991) list it as occurring. This species is also known from (and possibly introduced into) the Colorado River drainage of Texas (Yurgens 1954; Conner and Suttkus 1986) and the Galveston Bay drainage of Texas (Conner and Suttkus 1986).

Means of Introduction: The first record of this species in the Red River drainage is that of Hubbs and Bonham (1951) who reported it as existing in and about Lake Texoma, an artificial impoundment on the Texas-Oklahoma border. These researchers interpreted its occurrence in the Lake Texoma area to be the result of the establishment of escaped bait minnows. Hall (1956) also considered a bait bucket release as the possible mode of entry into the Red River drainage based on the fact that *Notropis potteri* was being used as a bait fish. If its occurrence in the Red River drainage near Lake Texoma is the result of an introduction, then records of this species in downstream areas of that drainage may represent subsequent spread of introduced populations and their offspring (Gilbert, personal communication). Yurgens (1954) reported *N. potteri* from Lake Travis (Colorado River drainage) in Texas and speculated that this species and several other recently discovered minnows were likely introduced as a bait release into the Colorado River drainage. He based his conclusion on recent captures and on the fact that bait dealers had been transporting Brazos River minnows in large numbers to fishing camps on Lake Travis near Austin.

Status: Established in Arkansas, Louisiana, Oklahoma, and Texas. In their treatment of Arkansas fishes, Robison and Buchanan (1988) noted that it was confined to the main channel of the Red River where it is locally abundant.

Impact of Introduction: Unknown. Suttkus and Clemmer (1968) believed that *Notropis potteri* is native to the Red River system. However, they speculated that, if man introduced this species to the Red, "it has swamped out and completely, or nearly so, replaced [*Notropis*] *blennius* in the shallow marginal areas of the lower Red River."

Remarks: There is uncertainty concerning the native versus nonindigenous distribution of this species. Hubbs and Bonham (1951) considered it introduced to the Red River drainage. In contrast, Suttkus and Clemmer (1968) believed that *N. potteri* gained access into the Red River system through natural means as the result of tributary captures or stream connections. Nevertheless, these researchers discussed the possibility of its introduction to the Red system. Cross et al. (1986) also listed it as native to the Red River drainage. However, Gilbert (personal communication) informed us that it is not certain if this species is native to, or introduced into the Red River drainage. If the Red River population is not the result of introduction, then it may have entered Red River by stream capture from the Brazos, or the species may have been native but overlooked in the Red River (Gilbert, personal communication). Hall (1956) also did not rule out the possibility of this species occurring naturally in the Red River drainage. Conner and Suttkus (1986) listed it as "native but possibly introduced" to the Colorado and Galveston Bay drainages of Texas. Other authors considered it native to the Colorado River drainage of Texas (e.g., Lee et al. 1980 et seq.; Page and Burr 1991).

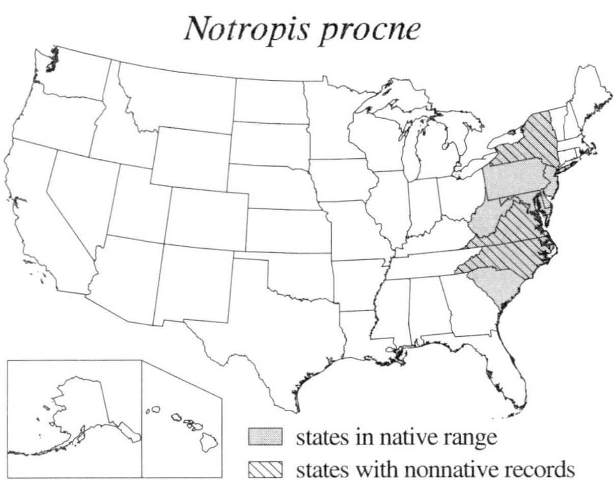

Notropis procne (Cope 1865)

Common Name: swallowtail shiner.

Identification: Smith (1985); Page and Burr (1991); Jenkins and Burkhead (1994). Two subspecies are recognized, *N. p. procne* and *N. p. longiceps* (Gilbert 1998; but see Jenkins and Burkhead 1994).

Native Range: Atlantic drainages, above and below Fall Line, from Delaware and Susquehanna rivers, New York, to Santee River, South Carolina (Page and Burr 1991); also Lake Ontario drainage, New York (Page and Burr 1991; but see Smith 1985). Maximum size: 7.2 cm.

Nonindigenous Occurrences: This species is known from (and possibly introduced into) Catherine Creek, an inlet to Seneca Lake, in **New York** (Smith 1985). In **North Carolina** this species is known from (and probably introduced into) the Yadkin drainage (Menhinick 1991) and Linville River (Santee River drainage) (Johnston et al. 1995). It is also known from (and probably introduced into) the New River drainage (Ohio River basin), **Virginia** (Stauffer et al. 1975; Page and Burr 1991; Jenkins and Burkhead 1994).

Means of Introduction: Unknown; possible bait bucket releases in North Carolina and Virginia. Although, Jenkins and Burkhead (1994) speculated that it had been introduced into the New River drainage of Virginia during the 1960s, they noted that the first valid record of this species from that drainage was from 1971. The first record of this species in the Linville River of North Carolina was 1990 (Johnston et al. 1995). Johnston et al. (1995) discovered this species in the Linville River, North Carolina, in 1990. They stated that the new record perhaps indicates the "spread of this species or introduction by fishermen." Smith (1985) speculated that the occurrence of the subspecies *Notropis procne procne* in Catherine Creek, the inlet to Seneca Lake in New York, may have resulted from dispersal via an artificial canal; however, Smith also hypothesized that it may have entered the area as a result of natural events during postglacial times. In their summary table on fishes of the Great Lakes basin, Bailey and Smith (1981) indicated that *Notropis procne* had colonized tributaries of Lake Ontario recently via canal or by natural dispersal following introduction.

Status: Established in North Carolina, Virginia, and West Virginia.

Impact of Introduction: Unknown.

Remarks: There is some uncertainty concerning the native versus nonindigenous distribution of this species. Jenkins and Burkhead (1994) discussed its distribution in Virginia and speculated on its introduction. They concluded that the New drainage population was probably introduced, based on the fact that the first valid record was from 1971 and the confinement of the shiner to the reach bounded by Bluestone Reservoir and Claytor Dam. Stauffer et al. (1975) reported a single collection from the lower East River, a tributary of the New River below Claytor Dam. That site was from the Virginia portion of the East River, near the West Virginia border. In their summary table of Virginia fishes, Jenkins and Burkhead (1994) listed it as introduced (but possibly native) to the New River drainage. Hocutt et al. (1986) listed it as "introduced, possibly native" to the Kanawha River above the falls. Similarly, Stauffer et al. (1995) listed *Notropis procne longiceps* as present and probably introduced into the Kanawha River drainage above the falls; however, they noted that it had not been found yet in the West Virginia portion of that drainage. Gilbert (personal communication) believes it possible that *Notropis procne* is native to the Yadkin system. Menhinick (1991) listed it as probably introduced to the Yadkin River drainage of North Carolina.

Notropis rubellus

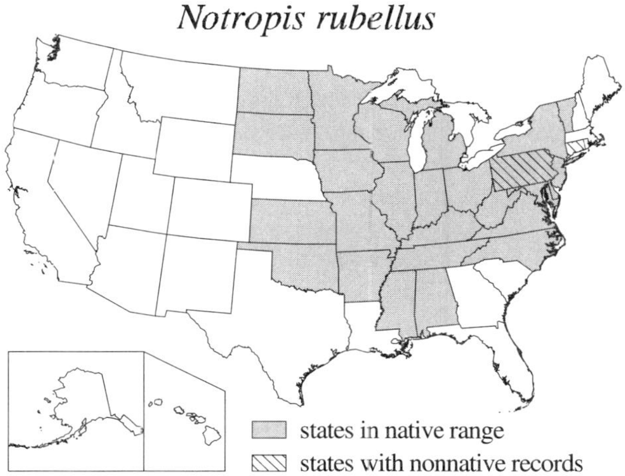

Notropis rubellus (Agassiz 1850)

Common Name: rosyface shiner.

Identification: Becker (1983), Page and Burr (1991); Etnier and Starnes (1993), Jenkins and Burkhead (1994); Pflieger (1997). *Notropis rubellus* represents a species complex (Humphries and Cashner 1994; Gilbert 1998). Maximum size: 9 cm.

Native Range: Atlantic Slope from St. Lawrence River drainage, Quebec, to James River drainage, Virginia; Great Lakes, Hudson Bay) Red River, and Mississippi River basins from Quebec to Manitoba, and south to North Carolina, Alabama, Arkansas, and Oklahoma (Page and Burr 1991).

Nonindigenous Occurrences: This species was reported from the Connecticut River in **Connecticut** prior to 1971 (Schmidt 1986; Whitworth 1996). It is also known from (and probably introduced into) the Delaware drainage, **Pennsylvania** (Jenkins and Burkhead 1994; Gilbert, personal communication).

Means of Introduction: Unknown. Possible bait bucket releases.

Status: Reported in Connecticut; no recent collections known (Whitworth 1996). Reported from Delaware drainage.

Impact of Introduction: Unknown.

Remarks: According to Jenkins and Burkhead (1994), the population reported by Cooper (1983) in the Delaware drainage, Pennsylvania, may be introduced. Gilbert (personal communication) also agrees that it probably was introduced, for reasons discussed in Gilbert (1980). Schmidt (1986) did not list this species as occurring in the Delaware drainage and makes no mention of a possible introduction there.

Notropis rubricroceus

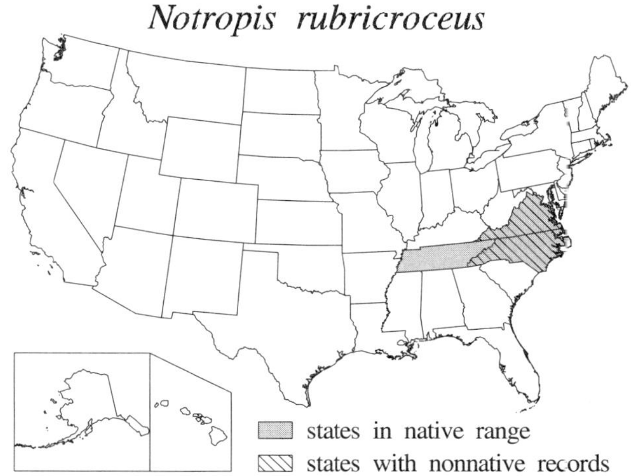

Notropis rubricroceus (Cope 1868)

Common Name: saffron shiner.

Identification: Page and Burr (1991); Etnier and Starnes (1993); Jenkins and Burkhead (1994). Maximum size: 8.4 cm.

Native Range: Mountainous area drained by upper Tennessee River drainage, Virginia, North Carolina, and Tennessee; headwaters of Santee and Savannah rivers, North Carolina (Page and Burr 1991).

Nonindigenous Occurrences: The saffron shiner was probably introduced into the upper New, Broad, and Catawba drainages, and possibly into the headwaters of the Savannah drainage in **North Carolina** (Menhinick 1991; Page and Burr 1991; Jenkins and Burkhead 1994), as well as into the upper New drainage of **Virginia** (Hocutt et al. 1986; Page and Burr 1991; Jenkins and Burkhead 1994).

Means of Introduction: Probable bait bucket release or stock contamination (with bass or trout) from hatchery in upper Holston system (Jenkins and Burkhead 1994). First found in upper New drainage, Virginia, in 1963. Became widespread soon afterwards (Jenkins and Burkhead 1994).

Status: Established in North Carolina and Virginia.

Impact of Introduction: Unknown.

Remarks: There is uncertainty concerning the native versus nonindigenous distribution of this species. In his summary table on North Carolina fishes, Menhinick (1991) listed this species as "probably introduced" to the New, Broad, and Catawba river drainages; he listed it as "native and probably introduced" to the Tennessee River drainage (presumably meaning that he regarded it as native to one part of that drainage and introduced into another part); and he listed it as "native but possibly introduced" to the Savannah drainage. In their summary table on Virginia fishes, Jenkins and Burkhead (1994) listed it as "introduced, but possibly native" to the New drainage. Hocutt et al. (1986) also listed it has "introduced, but possibly native" in the Kanawha above the falls; however, they commented that its historical absence in collections may have been due to inadequate sampling. Tsai and Raney (1974) stated that *N. rubricroceus* may have entered the Savannah River drainage of the Atlantic Slope via stream capture. Lee et al. (1980 et seq.) and Swift et al. (1986) noted that this species was probably native to the headwaters of the Savannah and Santee river drainages. Similarly, Starnes and Etnier (1986) indicated that this species was present naturally in the Savannah and Santee drainages because of stream capture.

Voucher specimens: Virginia (AMNH 69305, 69312).

Notropis shumardi

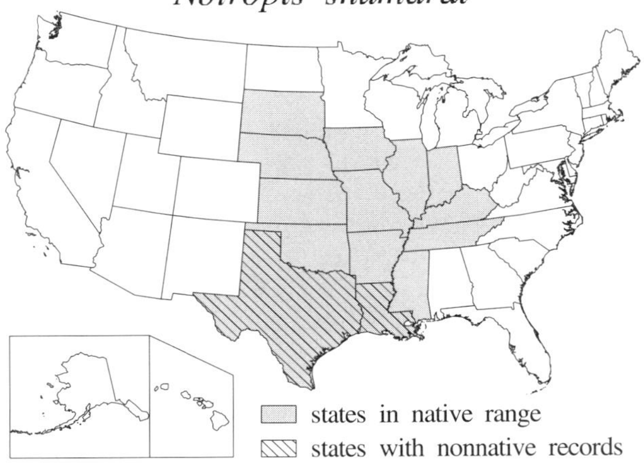

Notropis shumardi (Girard 1856)

Common Name: silverband shiner.

Identification: Page and Burr (1991); Etnier and Starnes (1993); Pflieger (1997). Gilbert and Bailey (1962) synonymized *N. brazosensis*, the Brazos River shiner, with *N. shumardi*. However, it is possible that the two are distinct (Gilbert 1998). Maximum size: 10 cm.

Native Range: Large rivers of Missouri-Mississippi basin (mainly Missouri, Mississippi, Illinois, Kaskaskia, lower Ohio, Arkansas, and Red rivers), from South Dakota and central Illinois south to Gulf; Gulf Coast drainages, Texas, from Sabine Lake (apparently introduced) to Colorado River (Page and Burr 1991). The *Notropis brazosensis* form apparently is restricted to the Brazos River, Texas (Hall 1956).

Nonindigenous Occurrences: This species is known from (and apparently introduced into) Sabine Lake, on the **Louisiana-Texas** border (Conner and Suttkus 1986; Page and Burr 1991). The *brazosensis* form is known from (and possibly introduced into) the Red River drainage in **Oklahoma** (Hall 1956). The Brazos River form is also known from (and possibly introduced into) the Colorado River drainage of **Texas** (Yurgens 1954).

Means of Introduction: Unknown; possible bait bucket releases. The first record of this species in the Red River drainage is that of D. G. Miller (1955) who reported its occurrence near the mouth of the Blue River in Bryan

County, Oklahoma. He pointed out the close proximity of tributaries of the Brazos and Red rivers and the possibility of the movement of this and other minnows across the divide during torrential rains. However, he admitted that such a conclusion in no way contradicts the possibility of a bait bucket release. Hubbs and Bonham (1951) regarded the *N. brazosensis* form as being largely endemic to the Brazos River drainage of Texas (also see *Notropis potteri* account. Consequently, Hall (1956) considered a bait bucket release as the possible mode of entry into the Red River drainage. He also noted that the *N. brazosensis* form was being used as a bait fish. However, Hall did not rule out the possibility of natural distribution. Yurgens (1954) reported the *N. brazosensis* form from Lake Travis (Colorado River drainage) in Texas and speculated that this fish and several other recently discovered minnows were likely introduced as a bait release into the Colorado River drainage. He based his conclusion on the fact that bait dealers had been transporting Brazos River minnows in large numbers to fishing camps on Lake Travis near Austin. In discussing various alternatives to explain its presence in the Red River, Gilbert and Bailey (1962) noted that the Red River may have been "repopulated" from the Texas coastal streams by transfer by man. They reported that there was no positive evidence for the presence of *Notropis shumardi* in the middle part of the Red River prior to 1953, but added that this is not compelling since the area in question had previously been little collected.

Status: Reported in Louisiana and Texas. Established in Oklahoma.

Impact of Introduction: Unknown.

Remarks: There is uncertainty concerning the native versus nonnative distribution of this species. The *N. brazosensis* form is apparently in the Red River below Lake Texoma. It may be native to that area, or could be introduced (Gilbert, personal communication). In their book on Oklahoma fishes, Miller and Robison (1973) made no mention the possible introduction of *N. shumardi* into the Red River. Cross et al. (1986) listed it as native to the lower Red River drainage. Conner and Suttkus (1986) listed it as introduced to the Sabine Lake drainage. Contrary to Yurgens (1954), both Gilbert and Bailey (1962) and Conner and Suttkus (1986) regarded this species as native to the Colorado River drainage of Texas.

Notropis spectrunculus

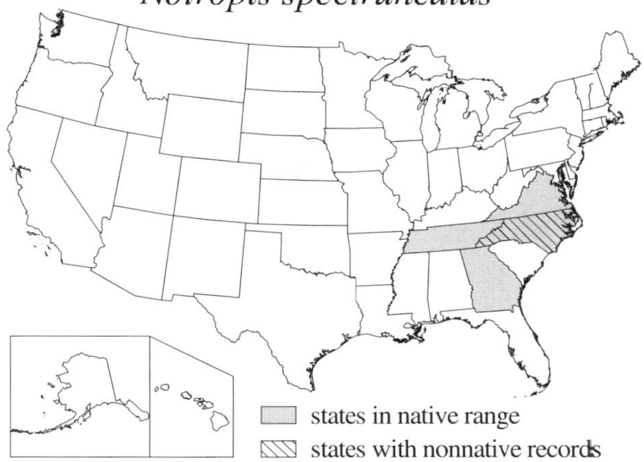

Notropis spectrunculus (Cope 1868)

Common Name: mirror shiner.

Identification: Page and Burr (1991); Etnier and Starnes (1993); Jenkins and Burkhead (1994). Maximum size: 7.5 cm.

Native Range: Upper Tennessee River drainage, Virginia, North Carolina, Tennessee, and Georgia (Page and Burr 1991).

Nonindigenous Occurrences: This species is known from (and may have been introduced into) several Atlantic Slope drainages, including the Broad, Catawba, and Savannah river drainages in **North Carolina** (Swift et al. 1986; Menhinick 1991).

Means of Introduction: Unknown, possible bait bucket release (but see Remarks section).

Status: Established in North and South Carolina.

Impact of Introduction: Unknown.

Remarks: The designation of this species as introduced into the above named Atlantic Slope drainages is considered highly tenuous. In his summary table on North Carolina fishes, Menhinick (1991) listed this species as "native but possibly introduced" to the Savannah, Broad, and Catawba river drainages. Swift et al. (1986) regarded it as native (possibly introduced) to the Savannah drainage. Other authors did not specifically discuss introduction as an alternative explanation to stream capture. For instance, Tsai and Raney (1974) stated that *N. spectrunculus* may have entered the Savannah River drainage stream capture. Similarly, Lee et al. (1980 et seq.) noted that it occurred in the upper Savannah and Catawba-Santee river drainages of North Carolina, and presumed that it reached these waters through stream capture. Hocutt et al. (1986) regarded it as native to the Santee River drainage. Starnes and Etnier (1986) stated that *N. spectrunculus* is an upland shiner, a virtual Tennessee endemic, that has recently gained access via stream capture to various drainages from the Savannah River northward.

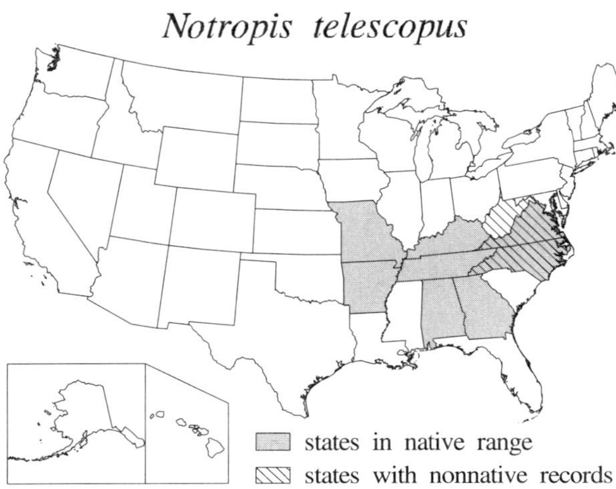

Notropis telescopus (Cope 1868)

Common Name: telescope shiner.

Identification: Page and Burr (1991); Etnier and Starnes (1993); Jenkins and Burkhead (1994). It was previously considered a subspecies of the popeye shiner *Notropis ariommus*. Maximum size: 9.4 cm.

Native Range: Disjunct; Cumberland and Tennessee River drainages, Virginia, North Carolina, Kentucky, Tennessee, Georgia, and Alabama; Little, St. Francis, and White river systems, Missouri and Arkansas (Page and Burr 1991).

Nonindigenous Occurrences: The telescope shiner is known from and may have been introduced into the Catawba (Santee) and probably into the Yadkin drainages in **North Carolina** (Hocutt et al. 1986; Menhinick 1991). It is known from, and has probably been introduced into, the New River drainage (Ohio River basin) and the upper James River drainage (Atlantic Slope) in **Virginia** (Lee et al. 1980 et seq.; Hocutt et al. 1986; Jenkins and Burkhead 1994; Stauffer et al. 1995), and into the Big Sandy drainage, Upper James drainage, New River, and Kanawha River (above and below the falls) in **West Virginia** (Hocutt et al. 1986; Jenkins and Burkhead 1994; Stauffer et al. 1995).

Means of Introduction: Unknown; probable bait bucket release. According to Jenkins and Burkhead (1994), the species was first taken in the New River drainage from Big Walker Creek, Virginia, in 1958. They noted that the pattern of these and subsequent records indicates that *Notropis telescopus* was established in the Virginia portion of the New drainage in the mid- or late 1950s. The likely source of the introduction was given as the Tennessee River drainage, but Jenkins and Burkhead were uncertain in which stream the New drainage population was founded. They believed it to be in the middle third of the Virginia section below Claytor Dam and that the species subsequently dispersed into upstream and downstream areas. It was first taken in West Virginia from East River and Greenbrier system (both New drainage) in 1972 (Jenkins and Burkhead 1994). According to Jenkins and Burkhead (1994), its appearance in the upper James was likely the result of a single introduction, probably occurring before closure in 1980 of Gathright Dam on the Jackson River.

Status: Established in North Carolina, Virginia, and West Virginia.

Impact of Introduction: Unknown.

Remarks: Lee et al. (1980 et seq.) stated that it was apparently introduced to the New (upper Kanawha) drainage of Virginia and West Virginia. Hocutt et al. (1986) listed it as introduced to the James River drainage and introduced (but possibly native) to the Santee and to the Kanawha above and below the falls. Starnes and Etnier (1986) noted that its presence in the New and Santee (Catawba) drainages was the result of recent introductions. In his summary table on North Carolina fishes, Menhinick (1991) listed it as probably introduced to the Yadkin River drainage and regarded it as native (possibly introduced) to the Catawba River drainage. Jenkins and Burkhead (1994) regarded it as introduced (but possibly native) to the James and New river drainages. Stauffer et al. (1995) listed it as introduced (possibly native) to the Kanawha above and below the falls; they regarded it as introduced to the James River drainage, but noted that it was not known from that portion of the drainage in West Virginia. Hocutt et al. (1986) considered it as probably introduced to the New,; however, they noted that the absence of this and other minnows in historical collections may have been due to inadequate sampling. Jenkins and Burkhead (1994) discussed the distribution of this species in Virginia and surrounding areas and detailed its probable introduction and spread in the New River drainage. Concerning the New population, these authors stated that the data now available from many sources supports its introduced status and, furthermore, suggest rapid dispersal. Because of its recent introduction, Jenkins and Burkhead (1994) predicted that the species appeared poised to explode in the upper James drainage.

Notropis texanus

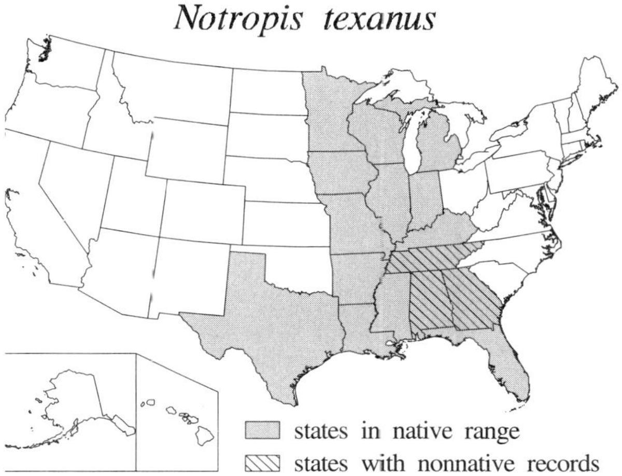

Notropis texanus (Girard 1856)

Common Name: weed shiner.

Identification: Becker (1983); Page and Burr (1991); Etnier and Starnes (1993); Pflieger (1997). Maximum size: 8.6 cm.

Native Range: Lowlands in Great Lakes, Hudson Bay (Red River), and Mississippi River basins from Michigan, Wisconsin, and Minnesota south to Gulf; Gulf Slope drainages from Suwannee River, Georgia and Florida, to Nueces River, Texas (Page and Burr 1991).

Nonindigenous Occurrences: This species is known from Second Creek, Lauderdale County, **Alabama**, near Pickwick Lake in the Tennessee River drainage (Mettee et al. 1996); the upper Ocmulgee River system (Altamaha River drainage, Atlantic Slope) in **Georgia** (Bart et al. 1994), and Robinson Creek (Pickwick Lake-Tennessee River drainage) in Hardin County, **Tennessee** (Etnier and Starnes 1993).

Means of Introduction: According to Bart et al. (1994), this species may have entered the Ocmulgee River by stream capture from the Flint River, or as an introduction via bait bucket release. These authors noted that Ocmulgee specimens exhibit the same unique features (deep bodies and higher numbers of body circumferential scales) as do Flint River specimens. Georgia Department of Natural Resources personnel found the species in the Ocmulgee system during 1988-1990 fish survey work. Bart et al. (1994) speculated that it appeared in the Ocmulgee system after construction of the High Falls dam because the species is apparently restricted to the area above the reservoir. The dam was completed in the early 1900s. *Notropis texanus* apparently invaded the Tennessee drainage from the Mobile basin by way of a canal connection, the Tennessee-Tombigbee Waterway (Etnier and Starnes 1993).

Status: Established in Alabama, Georgia, and Tennessee.

Impact of Introduction: Unknown. According to Etnier and Starnes (1993), range expansion of *Notropis texanus* in Tennessee would bring it into potential contact with the very similar bigeye shiner *N. boops.*

Remarks: Samples from the Ocmulgee system, Georgia, represent the first record of this species from the Atlantic Slope (Bart et al. 1994). Mettee et al. (1996), in reference to Etnier and Starnes (1993), mentioned the possible introduction of this species into the Tennessee River drainage via the Tennessee-Tombigbee Waterway. However, in their summary table on Alabama fishes, Mettee et al. (1996) listed it as native to the Tennessee drainage.

Notropis volucellus

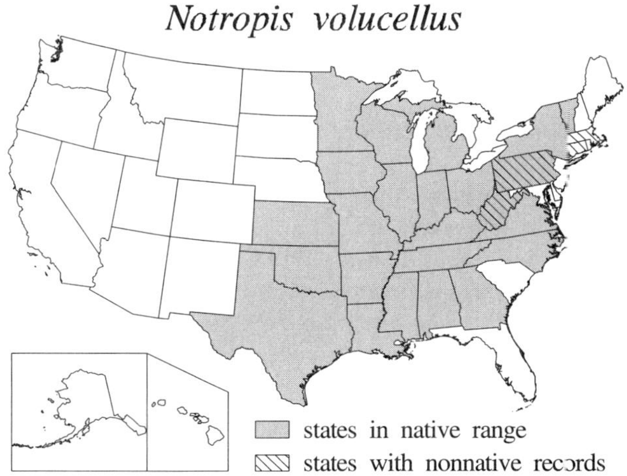

Notropis volucellus (Cope 1865)

Common Name: mimic shiner.

Identification: Scott and Crossman (1973); Becker (1983); Page and Burr (1991); Etnier and Starnes (1993); Jenkins and Burkhead (1994); Pflieger (1997). *Notropis volucellus* probably represents a complex of several cryptic species (Etnier and Starnes 1993; Gilbert 1998). Maximum size: 7.6 cm.

Native Range: St. Lawrence–Great Lakes, Hudson Bay, and Mississippi River basins from Quebec and Vermont to west-central Manitoba, and south to Gulf; Atlantic Slope drainages from James River, Virginia, to Neuse River, North Carolina; Gulf Slope drainages from Mobile Bay, Georgia and Alabama, to Nueces River, Texas (Page and Burr 1991).

Nonindigenous Occurrences: This shiner is known from the Connecticut and Housatonic rivers, **Connecticut** (Schmidt 1986; Page and Burr 1991); the Connecticut River and one site on the Squanacook River at Townsend Harbor, **Massachusetts** (Schmidt 1986; Hartel 1992; Cardoza et al. 1993; Hartel et al. 1996); the Susquehanna River, **Pennsylvania** (Malick 1978; Cooper 1983; Hocutt et al. 1986), and the Potomac River, **West Virginia** (Stauffer et al. 1995).

Means of Introduction: Unknown; possible bait bucket releases. Probably introduced accidentally into Massachusetts with other bait fish (Hartel 1992; Cardoza et al. 1993).

Status: Established in Massachusetts (Hartel 1992; Hartel et al. 1996), West Virginia (Stauffer et al. 1995), and Susquehanna drainage, Pennsylvania (Cooper 1983). Page and Burr (1991) indicated that the species was introduced into Connecticut (see Remarks section).

Impact of Introduction: Unknown.

Remarks: Whitworth (1996) made no mention of this species in his recent book on Connecticut fishes. The distribution map of Lee et al. (1980 et seq.) shows a nonindigenous record centered on the Connecticut River in southern Massachusetts near the Connecticut state border. Schmidt (1986) listed it as introduced to the Housatonic and Connecticut river drainages. Hocutt et al. (1986) regarded it as introduced (but possibly native) to the Susquehanna River drainage and as native (but possibly introduced) to the James River drainage. Page and Burr (1991) and Etnier and Starnes (1993) reported it as introduced to the Connecticut and Housatonic rivers; their conclusions were apparently based, at least in part, on the map of Lee et al. (1980 et seq.). The species has been taken from several sites in the Massachusetts portion of the Connecticut River drainage, the first dating to the 1940s (Hartel 1992; K. Hartel, personal communication). Although it is possible that the fish subsequently spread south into the Connecticut portion of the Connecticut River, we have not found any primary literature or additional data that would verify its occurrence in that state. Hartel (personal communication) is also unaware of any Connecticut records. In contrast to other researchers, Jenkins and Burkhead (1994) believed the species to be native to the Susquehanna River system of Pennsylvania. Propst and Carlson (1986) stated that reports of *N. volucellus* in Colorado are based on misidentifications. Based on a search of various databases, Walters (1997) noted that this species was reported as taken from the Conasauga River system of Georgia. Among several possible explanations (e.g., a misidentification), he speculated that the record may be based on a bait bucket introduction of the Tennessee drainage form of the mimic shiner. Scott and Crossman (1973) concluded that its value as a bait minnow, at least in northern regions, is unknown since the species is seldom recognized either by bait dealers or investigating biologists. Becker (1983) believed its use as a bait minnow is probably low; however, he did report that mimic shiners from a Wisconsin lake were used in huge quantities as bait and also mentioned an instance of its use as a food supply for fishes at one hatchery. Jenkins and Burkhead (1994) stated that this small shiner is quite frail and therefore not likely to be successfully introduced.

Notropis xaenocephalus

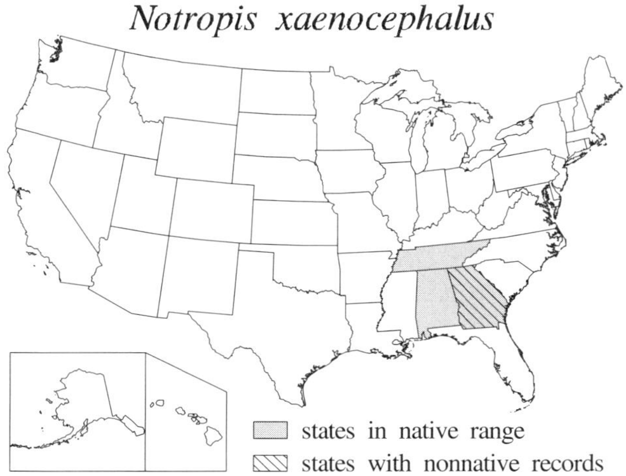

Notropis xaenocephalus (Jordan 1877)

Common Name: Coosa shiner.

Identification: Page and Burr (1991); Mettee et al. (1996). Maximum size: 7.9 cm.

Native Range: Above Fall Line in Coosa and Tallapoosa river drainages (Mobile Bay basin), southeastern Tennessee, northwestern Georgia, and eastern Alabama (Page and Burr 1991).

Nonindigenous Occurrences: This species is known from, and probably was introduced into, the Chestatee River (Chattahoochee River drainage) in north **Georgia** (Lee et al. 1980 et seq.).

Means of Introduction: Unknown; possible bait bucket release.

Status: Reported from Georgia.

Impact of Introduction: Unknown.

Remarks: Lee et al. (1980 et seq.) depicted the Chestatee River record on their dot distribution map, but these authors did not provide any additional details. Based on unpublished files, originally gathered to generate the Atlas account and map, the site is identified as Chestatee River near Tate bridge, Lumpkin County, Georgia. That record is apparently based on a collection deposited at the University of Georgia (Dahlberg and Scott 1971a). In their list of freshwater fishes known from Georgia, Dahlberg and Scott (1971a) recorded *Notropis xaenocephalus* from both the Alabama and Chattahoochee river drainages. However, Dahlberg and Scott (1971a, 1971b) apparently did not consider it to be introduced to the Chattahoochee. In their summary table on fishes found in the southeastern United States, Swift et al. (1986) listed this species as native to the Coosa and Cahaba river drainages, but they did not make note of its occurrence in the Chattahoochee drainage.

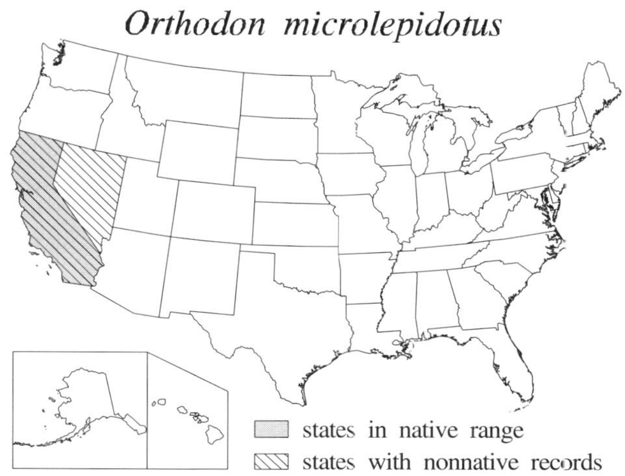

Orthodon microlepidotus (Ayres 1854)

Common Name: Sacramento blackfish.

Identification: Moyle (1976a); Page and Burr (1991). Maximum size: 55 cm.

Native Range: Sacramento–San Joaquin, Pajaro, and Salinas River drainages; Clear Lake, California (Page and Burr 1991).

Nonindigenous Occurrences: In **California**, this species is known from the Los Angeles basin, reservoirs on the Carmel River, and the Russian River (Sacramento-San Joaquin basin) (Burns 1966a; Moyle 1976a; Lee et al. 1980 et seq.). It has been recorded from a number of sites in the Los Angeles County area of southern California, including Tweedy Lake, Munz and Elizabeth lakes, Lake Hughes, Echo Park Lake, and the lower Santa Ana River below Prado Reservoir (Swift et al. 1993). In **Nevada**, it is known from a pond in the Truckee Meadows area near Reno (La Rivers and Trelease 1952; La Rivers 1962; Deacon and Williams 1984), and from the Carson River near Lahontan Reservoir, Lyon and Churchill counties (La Rivers 1962; Deacon and Williams 1984; Sigler and Sigler 1987).

Means of Introduction: The record from Truckee Valley pond is based on a single specimen and several hybrids taken in the late 1940s (La Rivers and Trelease 1952); it was first discovered in the Lahontan Reservoir area in 1962 (downstream of the dam) (La Rivers 1962). R. R. Miller tentatively suggested that *Orthodon* may have gained access to the Truckee Meadow pond through fish rescue material sent up from the Sacramento Valley (La Rivers and Trelease 1952; La Rivers

1962). The means introduction into other areas apparently is unknown (possibly unauthorized stockings by private individuals). According to Swift et al. (1993) this species was in Tweedy Lake, southern California in 1918 and subsequently spread into other Los Angeles County lakes.

Status: This species was eradicated from Lake Hughes, and (apparently) other Los Angeles County lakes in the early 1940s; it has been established in the lower Santa Ana River, California, since about 1986 (Swift et al. 1993). According to information given by Moyle (1976a), the species apparently is established in the Russian River; however the status of the Carmel River population is questionable. La Rivers (1962) stated that all cyprinids in the Truckee Meadow pond, Nevada, were "winterkilled" during 1948–1949 (the pond was subsequently drained). Even though it was suggested that the species may exist in drainage ditches east of Sparks, Nevada (La Rivers and Trelease 1952; La Rivers 1962); we are unaware of any reports verifying its occurrence in that area. Based on collections made in 1962, La Rivers (1962) concluded that the population was established in the area of the Lahontan Reservoir. Deacon and Williams (1984) also noted that it was established in Lahontan Reservoir.

Impact of Introduction: According to Burns (1966a), Sacramento blackfish compete with young game fish for food and space. Specimens taken from the Truckee Meadow pond included hybrids, presumably between *Orthodon* and "*Siphateles bicolor obesus*" (La Rivers and Trelease 1952; La Rivers 1962) (the latter form now recognized as *Gila bicolor obesa*).

Remarks: There is some uncertainty concerning the native versus nonindigenous status of this species in the Russian River of California. Moyle (1976a) stated that *Orthodon microlepidotus* was "apparently introduced into the Russian River." Based on their distribution map, Lee et al. (1980 et seq.) apparently considered it to be native to the Russian River. Minckley et al. (1986) listed it as native to the Sacramento River drainage; they made no mention of introductions. In its native range, hybrids have been reported with *Lavinia exilicauda* (Swift et al. 1993). A commercial fishery exists in California for this species, centered around the Chinese-American food trade (Moyle 1976a; McGinnis 1984).

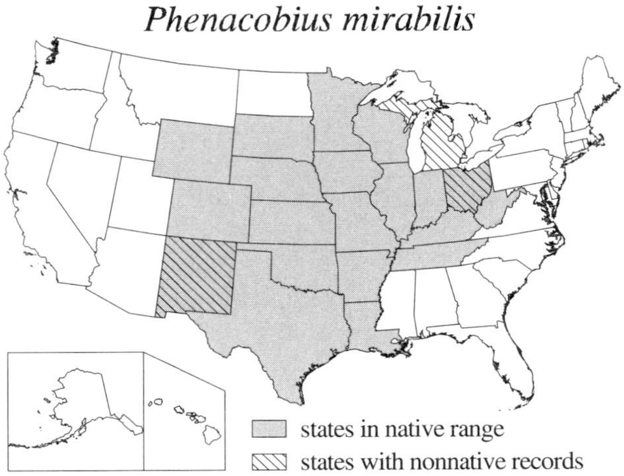

Phenacobius mirabilis (Girard 1856)

Common Name: suckermouth minnow.

Identification: Becker (1983); Hubbs et al. (1991); Page and Burr (1991); Etnier and Starnes (1993); Pflieger (1997). Maximum size: 12 cm.

Native Range: Mississippi River basin from Ohio and West Virginia to Wyoming, Colorado, and New Mexico, and from southeastern Minnesota to northern Alabama and southern Oklahoma; western Lake Erie drainage, Ohio; isolated populations in Gulf Coast drainages (Sabine Lake, Louisiana and Texas, Galveston Bay, Texas, Colorado River, Texas, and upper Pecos River, New Mexico) (Page and Burr 1991).

Nonindigenous Occurrences: This species has been recorded from (and possibly introduced into) Lake Erie tributaries of **Michigan** (Bailey and Smith 1992); the Pecos River (Pecos drainage) and the Dry Cimarron River (Canadian drainage) in **New Mexico** (Sublette et al. 1990); and Buckeye Lake in the Muskingum River drainage (Ohio River basin) of **Ohio** (Trautman 1981).

Means of Introduction: Unknown; possibly the result of bait bucket releases.

Status: Reported from Michigan and New Mexico. Apparently established in Buckeye Lake, Ohio (Trautman 1981).

Impact of Introduction: Unknown.

Remarks: It is not certain if the records of this species from Michigan and parts of Ohio represent introductions or were simply the result of natural dispersal. *Phenacobius mirabilis* is native to a large part of Ohio; however, Trautman (1981) stated that the population in Buckeye Lake, Ohio, "possibly had been introduced inadvertently from the bait buckets of fishermen." The suckermouth minnow did not occur in Michigan, or Ohio and West Virginia for that matter, before the early 1900s; natural eastward dispersal of the species has been rapid, and has occurred as a result of increased water turbidity and siltation following conversion of the prairies to farming. These changes have apparently benefited the suckermouth minnow as well as a few other native fishes. In the list of Michigan fishes given Bailey and Smith (1992), *Phenacobius mirabilis* was denoted as one of several fishes "established through the direct or indirect intervention of humans." However, these authors provided no additional details. We interpreted the words of Bailey and Smith (1992) to mean that the species was introduced to Michigan. Nevertheless, the authors may have simply meant that *P. mirabilus* spread naturally into Michigan as a result of human farming practises and other human-induced changes to the aquatic environment. In their summary table on fishes of the Great Lakes basin, Bailey and Smith (1981) indicated that *Phenacobius mirabilis* had colonized tributaries of Lake Erie recently via canal or by natural dispersal following introduction. Hubbs and Lagler (1958) and Underhill (1986) made no mention of the possible introduction of this species into the northern United States. Hocutt et al. (1986), apparently based on Cavender and Ciola (1981), stated that it was one of several species that invaded the Muskingum River drainage since 1930. However, in their summary table on fishes of the Central Appalachians and Central Atlantic Coastal Plain, Hocutt et al. (1986) listed *Phenacobius mirabilis* as native to the Muskingum drainage.

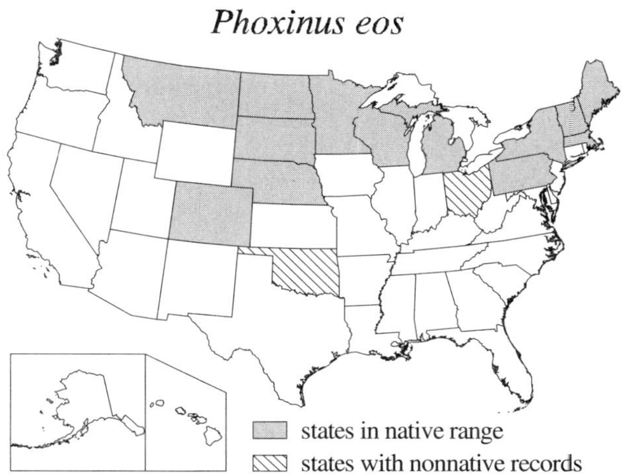

Phoxinus eos (Cope 1862)

Common Name: northern redbelly dace.

Identification: Becker (1983); Smith (1985); Holton (1990); Page and Burr (1991). Another name is *Chrosomus eos*. Maximum size: 8 cm.

Native Range: Atlantic, Great Lakes, Hudson Bay, upper Mississippi, Missouri, and Peace-Mackenzie River drainages, from Nova Scotia west to Northwest Territories and British Columbia, south to northern Pennsylvania, Wisconsin, Nebraska, and Colorado (Page and Burr 1991).

Nonindigenous Occurrences: This species was stocked unsuccessfully in **Ohio** (Trautman 1981). It was reported from **Oklahoma** (Moore and Riggs 1963).

Means of Introduction: In a summary table on fishes liberated in Ohio waters, Trautman (1981), using the name *Chrosomus eos*, noted that 7,000 northern redbelly dace from Michigan were unsuccessfully introduced into Ohio waters in October 1935. Reason for introduction into Ohio and Oklahoma apparently not known. It may have been introduced into Ohio as a forage fish and into Oklahoma, possibly, as a bait bucket release (but see Remarks section).

Status: Extirpated in Ohio. Reported from Oklahoma.

Impact of Introduction: Unknown.

Remarks: Except for the information given above, Trautman (1981) provided no details concerning the Ohio introduction. In their check list of Oklahoma fishes, Moore and Riggs (1963) denoted "*Chrosomus eos*" as an exotic fish recently introduced into Oklahoma waters, but not known to have become established. They did not provide any additional information. Based on their distribution map for *P. eos*, Cross et al. (1986) apparently considered records of this species in the upper Red River drainage of Oklahoma (and elsewhere) to represent post-glacial relict populations. However, in their summary table on the distribution of fishes in selected western Mississippi River drainages, Cross et al. (1986) listed *P. eos* as "uncertain" for the upper Arkansas River drainage. Unfortunately, it is not clear as to how the authors are using the term "uncertain" (possibly in reference to their regarding *P. eos* as native versus nonindigenous in that drainage). Despite their small size, this species, and the closely related southern redbelly dace *Phoxinus erythrogaster*, are used as bait fish. Both have the reputation of being hardy in the bait bucket or minnow pail (Scott and Crossman 1973; Becker 1983; Pflieger 1997). Becker (1983) noted that *P. erythrogaster* had begun appearing in parts of Wisconsin where it had not previously been known. However, Becker did not mention the possibility of introduction.

Phoxinus oreas

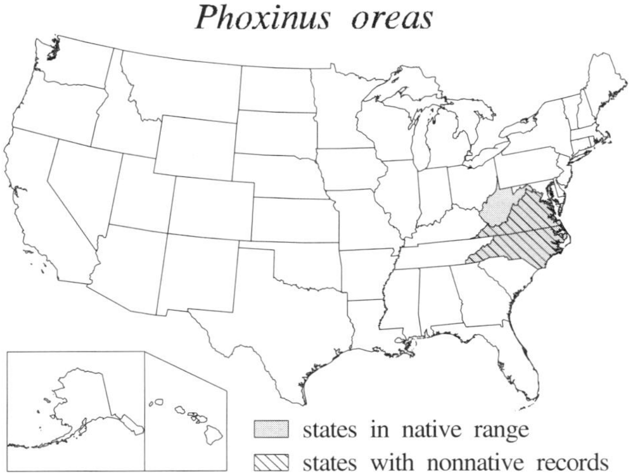

Phoxinus oreas (Cope 1868)

Common Name: mountain redbelly dace.

Identification: Page and Burr (1991); Jenkins and Burkhead (1994). Maximum size: 7.2 cm.

Native Range: Mountain and Piedmont regions of Atlantic Slope from Shenandoah River (Potomac River drainage), Virginia, to Neuse River drainage, North Carolina, upper New River drainage, West Virginia, Virginia, and North Carolina (Page and Burr 1991).

Nonindigenous Occurrences: This species is known from (and possibly introduced into) the Yadkin River system (Pee Dee River drainage) of **North Carolina** (Bortone 1972; Lee et al. 1980 et seq.; Menhinick 1991; Jenkins and Burkhead 1994). In **Virginia**, it is known from (and possibly introduced into) several Atlantic slope drainages including the Potomac River drainage (i.e., Straight Creek of the South Branch system and South Fork of the Shenandoah River system); the Rappahannock River drainage (sites in the Rapidan River system and upper Rappahannock proper); and the Pee Dee River drainage (Hocutt et al. 1986; Jenkins and Burkhead 1994). It is also known from (and possibly introduced into) the upper Holston River system (Tennessee-Ohio River basin) of Virginia (Page and Burr 1991; Jenkins and Burkhead 1994).

Means of Introduction: Probable bait bucket releases, multiple introductions to a drainage seem possible (Jenkins and Burkhead (1994). According to Jenkins and Burkhead (1994), the species was first recorded in the South Fork of the Shenandoah in 1956, Straight Creek in 1975, the Rapidan in 1951, the upper Rappahannock in 1974, the Pee Dee in North Carolina in 1970, the Pee Dee in Virginia in 1983–1984, and the Holston in 1966 (possibly as early as 1955). The population in the Rapidan likely came from the nearby James River. Jenkins and Burkhead noted that the availability of this species to bait seiners and its spectacular coloration increase the probability of this fish being moved about in bait buckets.

Status: Reported from, possibly established in, the Pee Dee River drainage of North Carolina. Established in Virginia in all drainages mentioned except, possibly, the Holston.

Impact of Introduction: Unknown.

Remarks: There is some uncertainty concerning the native versus nonindigenous distribution of this species. Citing Ross and Carico (1963), Starnes and Etnier (1986) stated that Holston River records of *Phoxinus oreas oreas* (and that of *Clinostomus funduloides funduloides*) were likely the result of accidental introduction, since each taken was based on a single collection of a very small number of individuals. Hocutt et al. (1986) listed it as introduced to the Pee Dee River drainage and as introduced (but possibly native) to the Rappahannock and Potomac river drainages. In his summary table on North Carolina fishes, Menhinick (1991) listed it as "probably introduced" to the Yadkin River drainage. In their summary table on Virginia fishes, Jenkins and Burkhead (1994) listed this species as "regarded as introduced, but possibly native" for three Atlantic Slope drainages (Potomac, Rappahannock, and Pee Dee) and one Ohio River basin drainage (Holston); they listed it as "regarded as native, but possibly introduced" to another Atlantic Slope drainage, the York. Jenkins and Burkhead (1994) summarized the distribution of this species in detail and discussed at length its introduction, or possible introduction, into various drainages.

Pimephales notatus

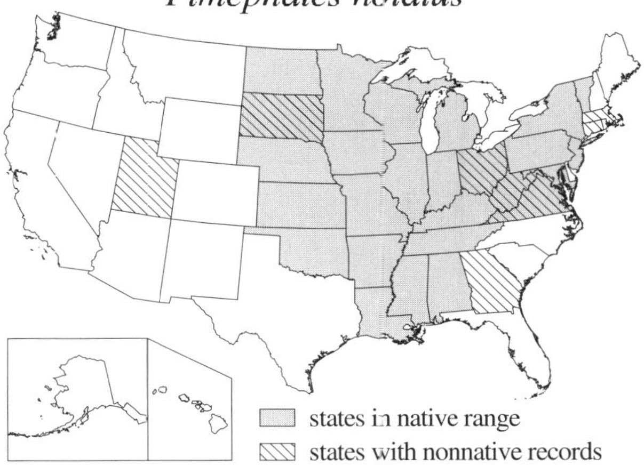

*Pimephales notatus (*Rafinesque 1820)

Common Name: bluntnose minnow.

Identification: Becker (1983); Page and Burr (1991); Etnier and Starnes (1993); Jenkins and Burkhead (1994). Maximum size: 11 cm.

Native Range: Great Lakes, Hudson Bay (Red River), and Mississippi River basins from southern Quebec to southern Manitoba, and south to Louisiana; Atlantic Slope from St. Lawrence River, Quebec, to Roanoke River, Virginia (absent from most of New England); Gulf Slope from Mobile Bay drainage, Alabama, to Mississippi River (Page and Burr 1991).

Nonindigenous Occurrences: This minnow is known from the Housatonic River drainage in **Connecticut** (Whitworth 1996); the Conasauga River system (Mobile-Coosa Basin) in **Georgia** (Walters 1997); the Housatonic River drainage and Quabbin Reservoir (Connecticut River drainage) in **Massachusetts** (Hartel 1992; Hartel et al. 1996). It has been propagated and distributed throughout **Ohio** (Trautman 1981). It also is known from Flat Creek Lake near Shadehill Reservoir on the Grand River (Missouri River drainage) in **South Dakota** (Bailey and Allum 1962); Willard Bay Reservoir (Great Basin) in **Utah** (Sigler and Sigler 1987); several river drainages in **Virginia** including the Potomac, James, Rappahannock, York, and Roanoke (Hocutt et al. 1986; Jenkins and Burkhead 1994); and the Potomac River drainage in **West Virginia** (Stauffer et al. 1995).

Means of Introduction: This species has been transplanted commonly by bait bucket release and sometimes stocked as a forage fish or introduced by way of stock contamination. As indicated by Whitworth (1996), *Pimephales notatus* was introduced to a small stream in the Housatonic River drainage and that population was probably the source for other populations that have become established in other sites in the Housatonic drainage. Walters (1997) noted that the occurrence of this species in the middle section of the Conasauga River, Georgia, during the 1980s was probably the result of bait bucket introduction. Probably a bait bucket release in Massachusetts, it was first found in the Housatonic in 1979 and first noticed in Quabbin Reservoir in the early 1980s (Hartel 1992; Hartel et al. 1996). Trautman (1981) reported that, in addition to being propagated by bait dealers, *Pimephales notatus* was propagated and stocked throughout Ohio after 1925 by the state conservation department. He added that millions were seined annually from Ohio streams and sold and used as bait. Sigler and Sigler (1987) noted that it was introduced to Williard Bay Reservoir, Utah in 1982, but they did not provide additional details. Jenkins and Burkhead (1994) provided details on introductions, and possible introductions, for drainages associated with the state of Virginia. The presence of *Pimephales notatus* in the Potomac River drainage probably resulted from stockings, naturalization, and rapid spread. The first known record for the Potomac River drainage was from a Shenandoah River tributary in 1897. It was common to abundant in the lower Potomac by the early 1900s. Jenkins and Burkhead noted that it may have been used as forage in bass and sunfish culture ponds operated in Washington, D.C. by the former U.S. Fish Commission, and the species may have been disseminated with centrarchids. The first known record for the Rappahannock drainage was from 1972. In the James River, the fish was collected above the Blue Ridge in 1928; it may have been introduced with shipments of centrarchids and subsequently dispersed downstream. The first known record from the Roanoke came in 1947, the first from the Dan system (Roanoke drainage) in 1951. Jenkins and Burkhead (1994) interpreted the distribution pattern in the

Roanoke proper and Dan systems of the Roanoke drainage to be the result of at least two introductions.

Status: Established in nonnative areas in Connecticut, Georgia, Massachusetts, Ohio, South Dakota, Virginia, and West Virginia; reported from Utah.

Impact of Introduction: Largely unknown. Based on 1989 shorefish samples, Hartel et al. (1996) found it to be the most common minnow in Quabbin Reservoir, Massachusetts. Possibly in reference to native populations, Trautman (1981) noted that the species hybridizes with *Pimephales promelas* and *P. vigilax*.

Remarks: Page and Burr (1991) commented that *Pimephales notatus* is probably the most common freshwater fish in eastern North America. There is some uncertainty concerning the native versus nonindigenous distribution of this species. For example, Hocutt et al. (1986) listed it as introduced (but possibly native) to several Atlantic Slope river drainages including the Roanoke, James, York, and Rappahannock. In their summary table on Virginia fishes, Jenkins and Burkhead (1994) listed this species as introduced to the York River drainage. In addition, these researchers listed it as introduced (but possibly native) to the Potomac, Rappahannock, James, and Roanoke river drainages; and as native (but possibly introduced) to several Ohio River basin drainages including the New, Holston, Clinch-Powell, and Big Sandy. In their summary table on West Virginia fishes, Stauffer et al. (1995) listed it as introduced (but possibly native) to the Potomac and James river drainages; however, they did not find this species in the James River drainage within West Virginia. In their analysis of southeastern fishes, Swift et al. (1986) listed it as native (but possibly introduced) to the Escambia River drainage. Although considered native to much of the lower Missouri River drainage, Cross et al. (1986) listed this species as introduced to the White-Little Missouri River system. Most likely, their listing was based on a record from Flat Creek Lake record (northwestern South Dakota) given by Bailey and Allum (1962). Bailey and Allum noted that *Pimephales notatus* was presumably introduced to the South Dakota lake. Lee et al. (1980 et seq.) and Page and Burr (1991) apparently included this site in their range maps for this species; however, these authors did not distinguish the northwestern South Dakota area (nor any other area for that matter) as an introduction. Based on recent records, Walters (1997) listed *Pimephales notatus* as introduced to the Conasauga River system in Georgia. Such a conclusion is supported by the absence of any earlier records from the Coosa-Conasauga River drainage) (see distribution map in Mettee et al. 1996). Bailey and Allum (1962) stated that this species is common in Iowa in the Big Sioux, Floyd, and Little Sioux drainages; they noted that elsewhere in the Missouri River drainage it is rare and perhaps occurs only as a result of introductions form the Des Moines drainage. However, it is unclear what they mean by the term "introduction," for Bailey and Allum added that it seems evident that the species entered the middle Missouri drainage across the divide from the upper Mississippi. It is unclear if the many stockings and releases documented within Ohio represent true introductions or simply stocking of sites where the species was already present. Trautman (1981) noted that the species was apparently as widely distributed throughout Ohio waters before 1900 as it had been in recent years. In addition, at a broad geographic scale, Ohio is well within the boundaries of its native range. Nevertheless, it is likely that this minnow was introduced to sites within parts of one or more Ohio river systems where it was previously absent. Jenkins and Burkhead (1994) detailed the distribution of this species in Virginia (and surrounding areas) and discussed at length its native versus possible nonindigenous status in various Atlantic Slope drainages and certain upper Tennessee River drainages. They stated that the hypothesis that *Pimephales notatus* became established by introduction in certain Atlantic Slope drainages is supported by "its high esteem and availability as bait and often by the recency of first capture in drainages." A single specimen was discovered in a shipment of channel catfish *Ictalurus punctatus* sent to California from Arkansas in 1968, but the species has not been documented from California waters (Swift et al. 1993; Dill and Cordone 1997).

Pimephales promelas

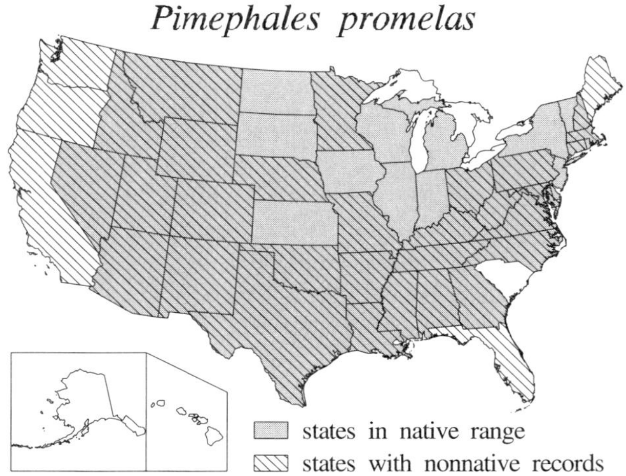

Pimephales promelas Rafinesque 1820

Common Name: fathead minnow.

Identification: Moyle (1976a); Becker (1983); Page and Burr (1991); Etnier and Starnes (1993); Pflieger (1997). Maximum size: 10 cm.

Native Range: Over much of North America from Quebec to Northern Territories, and south to Alabama, Texas, and New Mexico (Page and Burr 1991).

Nonindigenous Occurrences: This species is known from the Mobile Bay basin and the Chattahoochee and Tennessee river drainages in **Alabama** (Smith-Vaniz 1968; Lee et al. 1980 et seq.; Boschung 1992; Mettee et al. 1996); the Colorado River basin in **Arizona** (Minckley 1973; Lee et al. 1980 et seq.); various drainages in **Arkansas** including the Red, White, Ouachita, and St. Francis, and other Mississippi River tributaries (Lee et al. 1980 et seq.; Cross et al. 1986); various sites in **California** including the lower Colorado River in southern part of state and the Klamath, Sacramento, San Joaquin, and Pajaro drainages in central and northern part of state (Shapovalov et al. 1959; Moyle 1976a; Bell 1978; Lee et al. 1980 et seq.; Smith 1982; Swift et al. 1993; Dill and Cordone 1997); the upper Colorado River basin in **Colorado** (Vanicek et al. 1970; Tyus et al. 1982; Woodling 1985; Walker 1993); all of the major and most of the coastal drainages in **Connecticut** (Whitworth et al. 1968; Lee et al. 1980 et seq.; Whitworth 1996); Mill Creek in **Delaware** (Raasch and Altemus 1991; C. Martin, personal communication); Stiles Pond and Jackson Lake (Ochlocknee drainage) in **Florida** (Swift et al. 1977); Atlantic Slope and upper Mobile Bay basin drainages in **Georgia** (Miller and Jorgenson 1969; Dahlberg and Scott 1971a, 1971b; Walters 1997); various Pacific Northwest drainages in **Idaho** (Linder 1963; Simpson and Wallace 1978; Idaho Fish and Game 1990); various sites (most unspecified) in **Kentucky** (Branson and Batch 1983; Burr and Warren 1986); the Ouachita River drainage and other lower Mississippi River tributaries in **Louisiana** (Douglas 1974; Lee et al. 1980 et seq.); the Penobscot and (possibly) Kennebec river drainages in **Maine** (Schmidt 1986; museum specimen); **Maryland** including the Potomac and upper Chesapeake drainages (Schwartz 1963; Lee et al. 1976) and the Youghiogheny River system (Monongahela River drainage) (Hendricks et al. 1979); several water bodies in **Massachusetts** including the Housatonic River drainage, the Concord system, the Connecticut River drainage, and a pond in Amherst (Hartel 1992; Hartel et al. 1996); unspecified sites in the Twin Cities area of **Minnesota** (Eddy and Underhill 1974; Phillips et al. 1982; Becker 1983); several river drainages in **Mississippi** including the Pearl, Pascagoula, and Yazoo (Lee et al. 1980 et seq.; Ross and Brenneman 1991); various sites in the Ozarks of **Missouri** (Pflieger 1975, 1997); portions of the upper Missouri River drainage and a site in the Flathead River system (Columbia River drainage) in **Montana** (Lee et al. 1980 et seq.); unspecified sites in **Nebraska** (Bouc 1987); several river drainages in **North Carolina** including the New, Broad, Catawba, Roanoke, and Tennessee (Hocutt et al. 1986; Menhinick 1991); the Androscoggin River system (Kennebec River drainage) in **New Hampshire** (Scarola 1973; Schmidt 1986); several tributaries of the Colorado River basin in **New Mexico** (Koster 1957; Lee et al. 1980 et seq.; Tyus et al. 1982; Sublette et al. 1990); the lower Colorado River and Moapa River in **Nevada** (Lee et al. 1980 et seq.; Deacon and Williams 1984); various sites in **Ohio** including waters east of the Allegheny Front Escarpment and possibly elsewhere (Trautman 1981); various water bodies in southeastern **Oklahoma** (Miller and Robison 1973; Lee et al. 1980 et seq.); **Oregon** including the Upper Klamath Lake basin and a site in the Snake River drainage (U.S. Fish and Wildlife Service 1993e; Bond 1994; DeLong 1997; Simon and Markle 1997); several river drainages in **Pennsylvania** including the Monongahela (Ohio River basin) and the Delaware and Susquehanna (Hendricks et al. 1979; Cooper 1983; Hocutt et al. 1986); scattered areas in **Tennessee** (Ryon and Loar 1988; Etnier and Starnes 1993); portions of river drainages in **Texas** including the lower Rio Grande, the lower Brazos, the Sabine, the San Antonio Bay, the Nueces, and a tributary of the Red River (Lee et al. 1980 et seq.; Conner and Suttkus 1986); the upper Colorado River basin and Utah Lake (Great Basin) in **Utah** (Sigler and Miller 1963; Vanicek et al. 1970; Lee et al. 1980 et seq.; Tyus et al. 1982; Sigler and Sigler 1987, 1996); various drainages in **Virginia** including three in the Atlantic Slope, the Potomac, York, and Roanoke, and as many as four major drainages in the Ohio River basin, the New, Holston, Clinch-Powell, and Big Sandy (Hocutt et al. 1986; Jenkins and Burkhead 1994); an unspecified site in **Washington** (Fletcher, personal communication); Atlantic Slope and Ohio River basin drainages in **West Virginia** (Hocutt et al. 1986; Stauffer et al. 1995); and Colorado River basin and tributaries of the Missouri River drainage in **Wyoming** (Lee et al. 1980 et seq.; Tyus et al. 1982).

Means of Introduction: The popularity of this species as a baitfish and forage fish has led to widespread introductions. For example, Smith-Vaniz (1968) noted that this species is commonly used as a bait minnow and as a forage species in many farm ponds throughout Alabama; he concluded that records from that state's streams and impoundments were probably the result of introductions by anglers or escapes from nearby ponds. First recorded in Arizona in 1952, Minckley (1973) noted that the morphology of introduced populations in that state indicates a southwestern origin, most likely New Mexico or Texas. According to Shapovalov et al. (1959), the first record of *Pimephales promelas* in California is from a bait tank near the Colorado River in 1950. Shortly thereafter private interests, and also the Department of Fish and Game, imported the species for propagation. The resulting fish were distributed to a number of California waters to serve as forage for sport fishes (Shapovalov et al. 1959). Moyle (1976a) noted that as the bait industry grew, the species spread throughout California. According to Bell, the species may have spread into the Santa Clara River system, California, in imported water. Native to east slope rivers of Colorado, this species apparently was introduced to western parts of the state via bait bucket releases or as fish stock contaminants, or both (Woodling 1985). *Pimephales promelas* was found in Mill Creek, Delaware, in 1988, where it was probably introduced as a bait bucket release (Raasch and Altemus 1991). Swift et al. (1977) noted that specimens taken from two sites in Leon County, Florida, including Stiles Pond in 1958 and Jackson Lake in 1968, "are unquestionably the result of bait fish introductions." In Idaho, this species was introduced to two sites as a forage fish for bass, including ponds on the Perrine Trout Farm in Twin Falls County in about 1945 and Quayle Lake in Jefferson County in 1955 (Simpson and Wallace 1978). Burr and Warren (1986) noted that many Kentucky records are probably the result of deliberate introduction, but they did not provide details. Branson and Batch (1983) attributed the occurrence of *Pimephales promelas* in Meadow Creek, Kentucky, to a probable bait bucket release. Douglas (1974) noted that *Pimephales promelas* has been transported from northern states and used as a forage fish in state and private fish hatcheries in Louisiana. He also reported the discovery of specimens in waters adjacent to some hatcheries and suspected that increased numbers entering Louisiana as a commercial bait import will, perhaps, result in wider distribution. Hendricks et al. (1979) remarked that the presence of this species in the Youghiogheny River drainage in Maryland may be the result of bait bucket introduction. Similarly, Hartel et al. (1996) attributed Massachusetts records to probable bait bucket releases. Eddy and Underhill (1974) noted that this species was important in the Minnesota bait industry and sold to local dealers and to out-of-state customers as far away as Florida and New Mexico. It also has been introduced to various water bodies in the Twin Cities area of Minnesota as a mosquito control agent (Eddy and Underhill 1974; Phillips et al. 1982). Ross and Brenneman (1991) considered its occurrence in parts of Mississippi to be due to introduction, most apparently the result of bait fish release; however, they noted that *Pimephales promelas* in the Bogue Chitto system was "unfortunately released in a restoration effort by the Mississippi Department of Wildlife, Fisheries and Parks following a fish kill." Pflieger (1997) stated that most records from the Missouri Ozarks are probably the result of bait bucket releases or individuals that have escaped from minnow hatcheries. Although native to much of Nebraska, this species also has been introduced as food for gamefish in that state (Bouc 1987). Scarola (1973) stated that the single New Hampshire record, from a tributary of the Androscoggin River, may well be traced to the importation of this species as a bait fish. In their book on New Mexico fishes, Sublette et al. (1990) stated that it was introduced (apparently as a bait release) into the San Juan and Gila drainages by at least the 1950s, and into the Zuni and San Francisco drainages by the 1960s. Trautman (1981) remarked that the species was introduced to parts of Ohio outside its native range as a result of plantings by the conservation department and escapes from minnow buckets of anglers. Hendricks et al. (1979) stated that the presence of this species in the Youghiogheny drainage of Pennsylvania may be the result of bait bucket introductions. Ryon and Loar (1988) noted that a reproducing population in the White Oak Creek area (Clinch River system) of Tennessee resulted from escaped fish that had been used as food in various experiments and as test species for toxicological evaluations (also see Simon and Markle 1997). The presence of this species in one or more other areas in Tennessee may have been the result of bait bucket release (Etnier and Starnes 1993). Hubbs et al. (1991) presumed that the presence of this minnow throughout much of Texas is a result of bait releases. Sigler and Sigler (1987) stated that it was introduced in Utah Lake, Utah, as forage in 1969. Jenkins and Burkhead (1994) stated that the earliest record of capture in Virginia is 1950 and that Virginia populations likely are the result of bait bucket release or hatchery escapes. For example, these researchers noted that the York drainage record in Virginia may represent an escape from the Stevensville Hatchery and that the probable source of the South Fork Holston records is ponds at Buller Hatchery. According to Pearson and Krumholz (1984), reports of this species from the upper 300 miles of the Ohio River are possibly the result of bait bucket releases since the fathead is imported and sold extensively throughout the Ohio River basin. Becker (1983) reported the stocking of *Pimephales promelas* in sewage treatment ponds in Wisconsin and presumably elsewhere; however, he did not make clear reference to introductions into open waters of that state.

Status: Established in most states where introduced. Pflieger (1997) noted that few self-sustaining populations occur in natural waters of the Missouri Ozarks.

Jenkins and Burkhead (1994) remarked that it is unknown how many, if any, Virginia populations are stable. They commented that the species probably exists in numerous reservoirs and farm ponds. Swift et al. (1977) remarked that the establishment of this species in Florida is uncertain (Swift et al. 1977). Gilbert (personal communication) informed us that the species has never become established in Florida, even though it is sold in bait shops throughout that state.

Impact of Introduction: Largely unknown. This species has been implicated as an adverse threat to young of the Colorado pikeminnow *Ptychocheilus lucius*, an endangered species (Karp and Tyus 1990). There is also concern that this species, as well as other introduced fishes, may compete with or prey on young suckers and thereby contribute to the decline of imperiled species including the Lost River sucker *Deltistes luxatus* and shortnose sucker *Chasmistes brevirostris* in Oregon and northern California (U.S. Fish and Wildlife Service 1993e). Introduced fishes, including *Pimephales promelas*, are likely at least partially responsible for the decline of the Chiracahua leopard frog *Rana chiricahuensis* in southeastern Arizona (Rosen et al. 1995).

Remarks: The popularity of *Pimephales promelas* as a bait fish and the ease with which it is propagated have led to widespread introductions both within and outside its native range. However, because the species has been so widely introduced, its natural range is somewhat obscure. Consequently, the literature is somewhat confusing and, in part, contradictory concerning the native versus nonnative distribution. For instance, Boschung (1992) noted that it apparently is not native to Alabama, and Etnier and Starnes (1993) indicated that its status as native in Tennessee also is uncertain. In addition, some early reports of this species actually pertain to *Pimephales notatus* (Jenkins and Burkhead 1994). Schmidt (1986) listed this species as introduced to several river drainages in the northern Appalachians including the Delaware, Hudson, Connecticut, Kennebec, and St. John; however, he stated that the "status of *P. promelas* as native or introduced in the Hudson River is problematic." Based on their dot distribution map, Lee et al. (1980 et seq.) considered records from the Hudson River drainage in New York and a record from the Androscoggin River system (Kennebec drainage) in northern New Hampshire to be part of its native distribution. Hocutt et al. (1986) listed *Pimephales promelas* as introduced to nine river drainages in the Central Appalachians and Central Atlantic coastal plain including the Pee Dee, Roanoke, York, Potomac, Susquehanna, Muskingum, Allegheny, Monongahela, and Little Kanawha; they listed it as introduced (but possibly native) to the Kanawha above and below the falls. In their summary table of West Virginia fishes, Stauffer et al. (1995) listed it as introduced to several river drainages including the Potomac, Monongahela, Little Kanawha, and the Kanawha above and below the falls. Jenkins and Burkhead (1994) noted that it was native to the Tennessee and Big Sandy drainages, but they considered all populations in Virginia as well as those in extralimital parts of the New drainage to have been introduced. In their summary table of Virginia fishes, Jenkin and Burkhead listed *Pimephales promelas* as introduced to the Potomac, York, Roanoke, and York river drainages, and introduced (but possibly native) to the Holston, Clinch-Powell, and Big Sandy river drainages. In his summary table on fishes introduced to North Carolina, Menhinick (1991) listed this species as introduced to the New, Broad, Catawba, and Roanoke river drainages, and probably introduced to the Tennessee River drainage. Etnier and Starnes (1993) indicated that substantial reproducing populations are not known from natural habitats in Tennessee and "suggests the fathead may not be indigenous to the region." Although currently widespread in Tennessee, Etnier and Starnes also remarked that the species was not reported or collected during early fish surveys and reports. Walters (1997) listed this species as introduced to the Conasauga River system in Georgia. Ross and Brenneman (1991) considered it to be introduced to two systems (Leaf River and Bogue Chitto systems) that are part of the Gulf of Mexico Basin in Mississippi and likely introduced to the Yazoo River drainage (Mississippi River basin).

On a broad geographic scale, Minnesota and Nebraska are well within the boundaries of the native range of *Pimephales promelas*. Nevertheless, it is likely that this minnow was introduced to sites within parts of one or more sites in these state where it was previously absent. Cross et al. (1986) listed it as introduced to several river drainages in the western Mississippi River basin including Ouachita, the lower Arkansas, the White, and the St. Francis-Little. In their summary of fishes of the western Gulf Slope, Conner and Suttkus (1986) listed *Pimephales promelas* as introduced to several drainages including the Nueces, San Antonio Bay, and Sabine Lake; they listed it as native to the Colorado, Brazos, and Galveston Bay drainages. Lee et al. (1980) indicated that the species was native to eastern Montana and introduced to many sites in the central part of the state; however, Holton (1990) listed it as native to Montana and did not discuss the possibility of introduction. Tyus et al. (1982) gave a map showing the distribution of this species in the upper Colorado basin. Lee et al. (1980 et seq.) provided a dot distribution map distinguishing between what was considered native records versus those thought to represent introduced records.

Pimephales tenellus

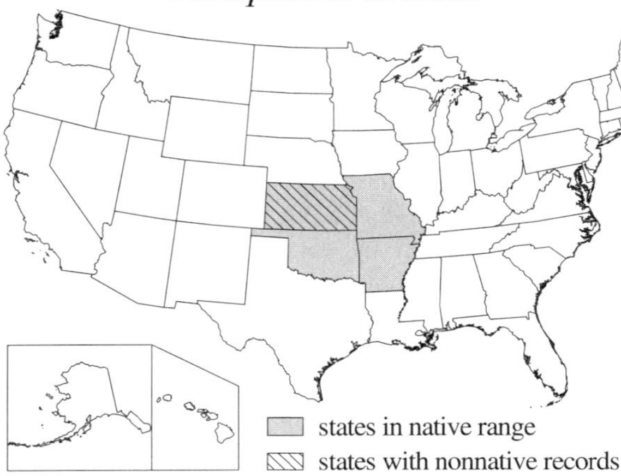

Pimephales tenellus (Girard 1856)

Common Name: slim minnow.

Identification: Robison and Buchanan (1988); Page and Burr (1991); Pflieger (1997). A name used in some of the earlier literature is *Ceratichthys tenellus* (e.g., Hubbs and Black 1947). Two subspecies are recognized, the western slim minnow *Pimephales tenellus tenellus* and the eastern slim minnow *Pimephales tenellus parviceps* (Page and Burr 1991; Pflieger 1997; Gilbert 1998). Maximum size: 7 cm.

Native Range: Southern Missouri, eastern Kansas, Arkansas, and northeastern Oklahoma (Page and Burr 1991). Restricted to the Ozarkian areas of the above named states, in Red and Arkansas river drainages and independent tributaries of Mississippi River (Lee et al. 1980 et seq.).

Nonindigenous Occurrences: This species was recorded from a pond in the Osage River system (Missouri River drainage) in **Kansas** (Hubbs and Black 1947).

Means of Introduction: Unknown, possible bait bucket release or stock contaminant.

Status: Reported from Kansas.

Impact of Introduction: Unknown.

Remarks: Hubbs and Black (1947) noted the taking of a single specimen of *"Ceratichthys perspicuus"* from a pond in the Osage River system in Kansas. In addition, they reported the finding of at least one specimen of "*Ceratichthys tenellus*" from the same station and stated that it also doubtless represented an introduction from the Neosho River system. In their account for "*Ceratichthys perspicuus*," Hubbs and Black (1947) identified the station as Jones Pond, Coffee County. That water body was apparently an artificial impoundment, probably a farm pond. In reference to the single record of Hubbs and Black (1947), Cross (1967) stated that he had never obtained *Pimephales tenellus* from the Osage drainage.

Pimephales vigilax

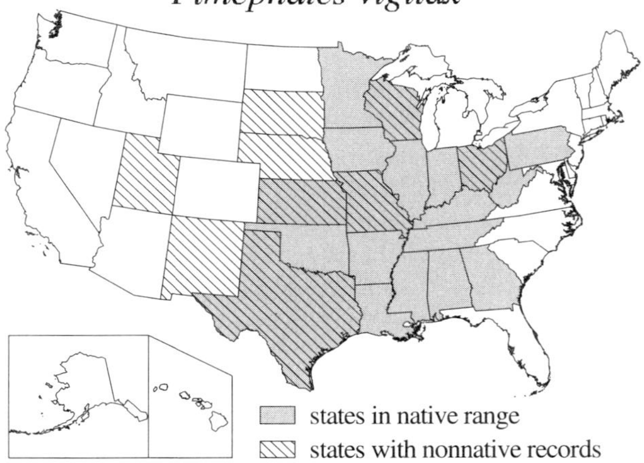

Pimephales vigilax (Baird and Girard 1853)

Common Name: bullhead minnow.

Identification: Becker (1983); Hubbs et al. (1991); Page and Burr (1991); Etnier and Starnes (1993); Pflieger (1997). Hubbs and Black (1947) revised group and recognized two closely related forms as separate species *Ceratichthys vigilax* and *Ceratichthys perspicuus*. The two are now considered a single species under the name *Pimephales vigilax* (Lee et al. 1980 et seq.); however, some authors (e.g., Trautman 1981; Becker 1983) have treated them as subspecies (i.e., *Pimephales vigilax vigilax* and *Pimephales vigilax perspicuus).* Maximum size: 8.9 cm.

Native Range: Mississippi River basin from western Pennsylvania to Minnesota, and south to Gulf; Gulf Slope drainages from Mobile Bay, Georgia and Alabama, to Rio Grande, Texas, New Mexico, and Mexico. Populations in Rio Grande of New Mexico and western Texas may have been introduced (Page and Burr 1991).

Nonindigenous Occurrences: This species is known from the Missouri River drainage in **Kansas** including the Osage River system (Hubbs and Black 1947; Cross 1967) and the Kansas River system (Cross and Haslouer 1984; Cross and Collins 1995); the Missouri River drainage of **Missouri**, including the Chariton River (Chariton County) and scour holes connected to the Missouri River (Howard and Saline counties) (Pflieger 1997); the Missouri River drainage in **Nebraska** (Cross et al. 1986; see map in Lee et al. 1980 et seq.); the Rio Grande in **New Mexico** (Lee et al. 1980 et seq.; Propst et al. 1987; Sublette et al. 1990); Lake St. Marys (Mercer County) (Lake Erie drainage) in **Ohio** (Hubbs and Black 1947; Hubbs and Lagler 1958; Trautman 1981); the James drainage in **South Dakota** (Cross et al. 1986; see map in Lee et al. 1980 et seq.); the upper Red River and Canadian River drainages and the Rio Grande in **Texas** (Lee et al. 1980 et seq.; Hubbs et al. 1991). It has been introduced to two closed drainages within the Great Basin in **Utah** including the Sevier River drainage (Sigler and Miller 1963) and Lake Utah (Sigler and Sigler 1987). It has been recorded from two sites in the Lake Michigan drainage in **Wisconsin** including the upper Fox River (Marquette County) and, possibly, the Menomonee River (a tributary of the Milwaukee River) (Johnson and Becker 1980; Becker 1983).

Means of Introduction: Hubbs and Black (1947) noted that the one known locality record for this species in the Osage River system of Kansas, was based on material taken from Jones Pond, Coffee County, and "no doubt represents an introduction from the Neosho River system." This species apparently was introduced into the Kansas River system of eastern Kansas before 1976 (Cross and Haslouer 1984). It was first recorded in the Missouri River in Missouri in 1994, apparently the result of dispersal of populations introduced into the Kansas River system of Kansas (Pflieger 1997). It is uncertain if the records from the Missouri River drainage in Nebraska and South Dakota (and also Iowa) are valid, and, if valid, if they represent introductions (see discussion in Remarks section). Hubbs and Black (1947) stated that the single record (under the name *Ceratichthys perspicuus*) from Lake St. Marys (Great Lakes drainage) in Ohio should be considered valid. These researchers also mentioned that M. B. Trautman suspected it of being a bait introduction. Maintaining that same position years later, Trautman (1981) stated that the Lake St. Marys record "probably was an escape from the minnow bucket of a fisherman." He also remarked that by 1945 it was of common occurrence in the tanks of commercial bait dealers in Ohio. In their summary table on fishes of the Great Lakes basin, Bailey and Smith (1981) indicated that *Pimephales vigilax perspicuus* had colonized tributaries of Lake Erie recently via canal or by natural dispersal following introduction. Sublette et al. (1990) stated that it was introduced to the Rio Grande drainage of New Mexico during and since the 1970s. The possible method of introduction was not discussed. Similarly, Hubbs et al. (1991) did not provide explanation for its apparent introduction into the Rio Grande and the Red and Canadian river drainages in Texas. Presumably these were by bait-bucket release or stock contaminants. According to Sigler and Miller (1963), *Pimephales vigilax vigilax* was accidentally introduced into the Sevier River drainage in Utah in 1953 along with several thousand channel catfish received from San Angelo, Texas. Sigler and Sigler (1987) indicated that it was introduced to Utah Lake, Utah, in 1969, but these authors did not provide additional details. The first record of this species in the Fox River, Wisconsin, is based on nine specimens taken in July 1960. Johnson and Becker (1980) and Becker (1983) suspected that the fish entered that drainage from the Mississippi River basin via the old Fox-Wisconsin Canal at Portage, also known as the Portage Canal. The Menomonee River record is based on a report from the early 1900s. Greene (1935, cited by Becker 1983) suggested that the record may have been based on a bait bucket release, or a recent immigrant via highwater connections between the Rock and Milwaukee river systems.

Status: Established in Kansas, New Mexico and Texas; possibly established in western Missouri; reported from Nebraska, South Dakota, Utah, and Wisconsin. Sigler and Miller (1963) were uncertain if the fish became established in the Servier River drainage of Utah. In reference to the Utah Lake introduction, Sigler and Sigler (1987) noted that it may be reproducing in small inlet streams but its presence had not been reported recently. In their Appendix on Utah fishes, Sigler and Sigler (1996) noted that the species is no longer found in Utah.

Impact of Introduction: Unknown.

Remarks: The literature is unclear and somewhat contradictory concerning the native versus nonindigenous distribution of this species. Lee et al. (1980 et seq.) stated that the Upper Rio Grande and northern Texas (Red River drainage) populations were presumably introduced. The dot distribution maps given by Hubbs and Black (1947; as *Ceratichthys perspicuus*) and Lee et al. (1980 et seq.) also show records from the Missouri River and tributaries (e.g., lower James River) within the states of Nebraska, South Dakota, and Iowa. However, these researchers apparently considered the Missouri River records to represent part of its native distribution. In contrast, Cross et al. (1986) stated that *Pimephales vigi-*

lax "did not occur naturally in the Missouri River system." In their summary table, Cross et al. listed it as introduced to the Kansas River and Sioux-James River systems and as native to the Red River and elsewhere. Page and Burr (1991) stated that populations in the Rio Grande of New Mexico and western Texas may be introduced. They did not mention the possibility of its introduction into other areas. Furthermore, in their range map for this species, Page and Burr (1991) did not show it as occurring in the Missouri River drainage. Similar to Lee et al. (1980 et seq.), Hubbs et al. (1991) stated that "populations apparently introduced into the upper Rio Grande Basin and upper Red and Canadian basins in Texas." Smith and Miller (1986) recognized it as native to the lower Rio Grande and Pecos river drainages. They made no mention of its possible introduction into the upper Rio Grande. Cross and Collins (1995) noted that the bullhead minnow has become common in the Kansas River in Kansas as far west as Saline County. Records from the Missouri River system in South Dakota (and possibly adjacent states) may be erroneous. For example, without providing details, Hubbs and Black (1947) had doubts about records from South Dakota. Apparently in reference to all, or many, of these same South Dakota records, Bailey and Allum (1962) made note of *Pimephales notatus* specimens that had been misidentified as "*Ceratichthys vigilax*" in the earlier literature. Unfortunately, they made no further mention of "*vigilax*" in their work. (Bailey and Allum also stated that *P. notatus* was one of a group of species that had become established in the upper Missouri through introduction; however, they also speculated on natural events to explain observed fish distributions.) Trautman (1981) found *Pimephales vigilax* in bait pails of anglers at Buckeye Lake, Ohio, and later captured a specimen in the lake. However, he added "Despite the recent increase in abundance and unintentional introductions into many sections by bait fishermen there appears to have been no notable recent extension of its range in Ohio." Trautman did note that the specimen taken in 1927 from Lake St. Mary constituted the only known Lake Erie drainage record for Ohio. According to Becker (1983) noted that no voucher specimens are known to substantiate the Menomonee River record (Wisconsin). In apparent conflict with Page and Burr (1991), Underhill (1986) listed this species as native to tributaries of Lake Michigan. Hubbs and Black (1947) noted the taking of a single specimen of "*Ceratichthys perspicuus*" from a pond in the Osage River system in Kansas. In addition, they reported the finding of at least one specimen of "*Ceratichthys tenellus*" (now recognized as the slim minnow *Pimephales tenellus*) from the same station and stated that it also doubtless represented an introduction from the Neosho River system.

Plagopterus argentissius

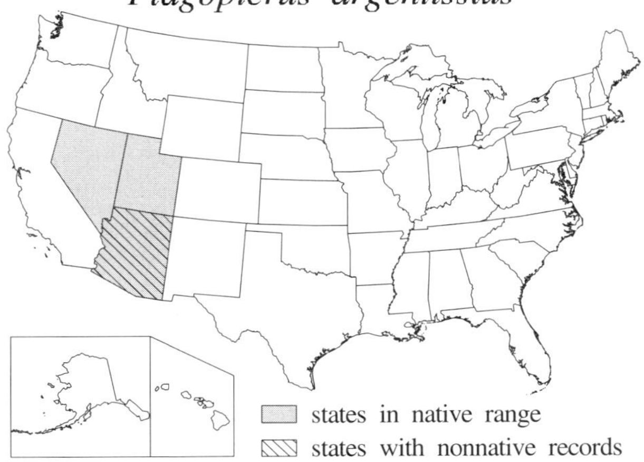

Plagopterus argentissimus Cope 1874

Common Name: woundfin.

Identification: La Rivers (1962); Minckley (1973); Page and Burr (1991). Maximum size: 9 cm.

Native Range: Historically, in drainages of the lower Colorado River basin, including the Virgin and Gila river in Utah, Nevada, and Arizona. Now found only in the Virgin River, Arizona, Nevada, and Utah (U.S. Fish and Wildlife Service 1984a; Page and Burr 1991).

Nonindigenous Occurrences: The woundfin was introduced into four localities in the lower Colorado River basin at the periphery of its native range in **Arizona**: the Hassayampa River, a Gila River tributary; the Salt River, also a Gila River tributary; Sycamore Creek, an Agua Fria tributary; and the Paria River, a Colorado tributary along the **Arizona-Utah** border (Minckley 1973; U.S. Fish and Wildlife Service 1984a, 1994c; Hendrickson and Brooks 1991).

Means of Introduction: Stocked intentionally for conservation purposes, the goal of which was to establish self-sustaining populations of this endangered species in its historic range so as to reduce the chance of woundfin extinction (Ono et al. 1983). Stocking was carried out by the Arizona Game and Fish Department. The species was stocked in the Hassayampa River in February 1972 (500 fish), in the Salt River in March 1972 (350 fish), in Sycamore Creek in the spring of spring 1972 (a "few" fish), and in the Paria River several times between 1969 and 1972 (a total of 650 fish) (U.S. Fish and Wildlife Service 1984a, 1994c).

Status: All introduced populations failed to survive (Hendrickson and Brooks 1991). No woundfin have been seen since their release in the Salt River. The population in Hassayampa River reproduced in the summer of 1972 but was destroyed in a flood in September of that year. The Sycamore Creek population survived the 1972–1973 flooding. Two gravid individuals were collected in late August 1973, but none have been seen since. No woundfin were found in the Paria River in surveys conducted in 1974 and 1975 (U.S. Fish and Wildlife Service 1984a, 1994c).

Impact of Introduction: Unknown.

Remarks: Listed as a federally endangered species in 1970 (U.S. Fish and Wildlife Service 1984a, 1993a). Recent accounts indicated that wild populations of woundfin are now found only in to the Virgin River where they are threatened by invasion of introduced red shiner *Cyprinella lutrensis*, Asian tapeworm, and extensive water development (Deacon 1988).

Platygobio gracilis

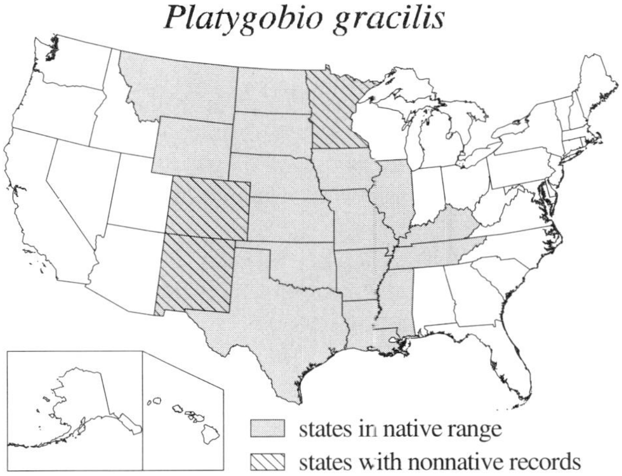

Platygobio gracilis (Richardson 1836)

Common Name: flathead chub.

Identification: Woodling (1985); Hubbs et al. (1991); Page and Burr (1991); Etnier and Starnes (1993); commonly used name is *Hybopsis gracilis*. Maximum size: 32 cm.

Native Range: Mackenzie, Saskatchewan, and Lake Winnipeg drainages in Yukon, Northwest Territories, Manitoba, Saskatchewan, Alberta, and British Columbia; Missouri-Mississippi River basin from southern Alberta and Montana to Louisiana; upper Rio Grande (including Pecos) drainage, New Mexico. Restricted to Mississippi River proper in Missouri, Illinois, and south; localized in Arkansas River drainage in Oklahoma, Texas, and New Mexico (Page and Burr 1991).

Nonindigenous Occurrences: This species was stocked in the Russell Lakes area of the San Luis Valley, **Colorado** in the 1940s (Zuckerman and Behnke 1986). It is known from the Red River drainage in northern **Minnesota** (Underhill et al. 1997). The species is known from the Gila River drainage of **New Mexico** (Koster 1957); it was recorded as having been collected in the state outside its natural range from Taylor Creek upstream from Wall Lake in the Gila drainage during the 1950s (Sublette et al. 1990).

Means of Introduction: Introduced in the Gila River drainage of New Mexico via bait-bucket release (Koster 1957; Sublette et al. 1990). This species was privately stocked as prey for a trout hatchery in the Russell Lakes area of Colorado; the fish came from the Arkansas River drainage (Zuckerman and Behnke 1986). Probably bait-bucket release in Minnesota.

Status: Recorded as having been established, at least temporarily, in part of the Gila drainage of New Mexico (Koster 1957); in contrast, Sublette et al. (1990) reported that the species did not became established there. In Colorado, the species apparently persisted for 15 years in the San Luis Closed Basin, but it is no longer considered extant in that region (Zuckerman and Behnke 1986). Recorded as occurring in Minnesota.

Impact of Introduction: Unknown.

Remarks: The only New Mexico site where this species was mentioned as being introduced is Taylor Creek (Sublette et al. 1990). Koster (1957) did not mention the specific locality where this species occurred within the Gila River drainage of New Mexico. The flathead chub is used as a bait fish in selected areas of the country (Koster 1957; Scott and Crossman 1973). Crossman and McAllister (1986) listed this species as possibly present in the U.S. section of the Red River drainage but gave no indication it was introduced there. Lee et al. (1980 et seq.) did not depict the species in that region.

Pogonichthys macrolepidotus

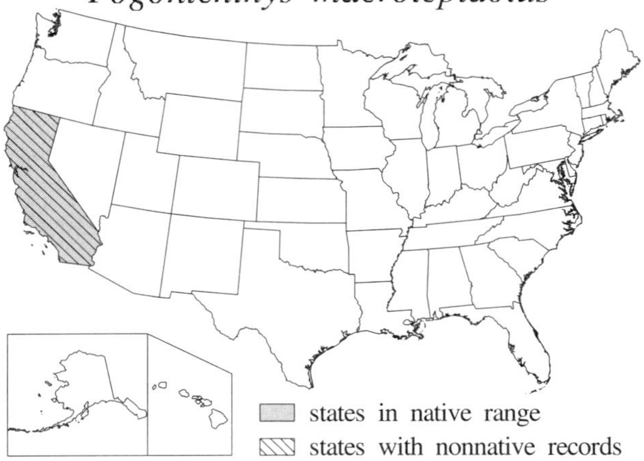

Pogonichthys macrolepidotus (Ayres 1854)

Common Name: splittail.

Identification: Moyle (1976a); McGinnis (1984); Page and Burr (1991). Maximum size: 44 cm.

Native Range: Historically throughout Sacramento–San Joaquin River drainage, California; native populations now restricted to San Francisco Bay Delta and lower Sacramento River (Page and Burr 1991).

Nonindigenous Occurrences: A single fish was reported as taken from Lake Silverwood (Mojave River drainage), San Bernadino County, in southern **California**, in 1977 (Swift et al. 1993).

Means of Introduction: Possibly came in with aqueduct water (Swift et al. 1993).

Status: Reported from California outside of its native range (Swift et al. 1993).

Impact of Introduction: Unknown.

Remarks: The Lake Silverwood specimen was not retained (Swift et al. 1993). Lee et al. (1980 et seq.) provided a dot distribution map, but the Lake Silverwood record is not shown. However, a single dot mapped is shown in west-central California (possibly representing a record from San Luis Reservoir, Merced County) that is outside the boundaries of what the authors identify as the "former range." However, it is not clear if the dot represents an introduction or a printing artifact. Lee et al. (1980 et seq.) gave no indication in text that the dot represents an introduced population.

Ptychocheilus grandis

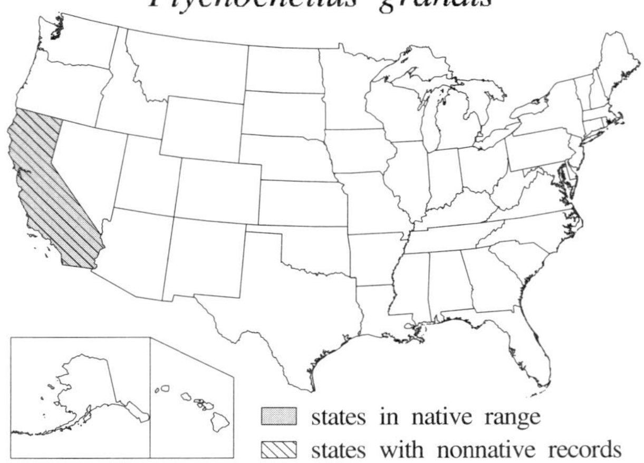

Ptychocheilus grandis (Ayres 1854)

Common Name: Sacramento pikeminnow (Sacramento squawfish).

Identification: Moyle (1976a); Page and Burr (1991). The American Fisheries Society Names Committee has changed the common name of all squawfish to pikeminnow (Nelson et al. 1998). Maximum size: 36 cm.

Native Range: Clear Lake, Russian, Sacramento–San Joaquin, Pajaro-Salinas, and upper Pit River drainages, California (Page and Burr 1991).

Nonindigenous Occurrences: This species is known from several sites in **California**, including Lake Arrowhead, Pyramid Lake, and tributaries of Morro Bay in the southern part of the state, and from the Eel River drainage in the north (Brown and Moyle 1991; Swift et al. 1993).

Means of Introduction: Illegally introduced into the Eel River in 1979 or 1980; the species spread rapidly within the drainage after introduction (Brown and Moyle 1991). Fish found in Pyramid Lake probably arrived through the aqueduct with central California water (Swift et al. 1993).

Status: Established in California outside of its native range (Brown and Moyle 1991; Swift et al. 1993).

Impact of Introduction: Field observations indicated that the introduction of Sacramento pikeminnow into the Eel River, California, resulted in changes in habitat and microhabitat use by resident fishes (natives and introduced) but no loss of species (Brown and Moyle 1991). In its native range, this piscivorous cyprinid is known to compete with trout and also prey upon young salmonids (Burns 1966d; but see Brown and Moyle 1981).

Remarks: Voucher specimens: California (SIO 66–75).

Ptychocheilus oregonensis

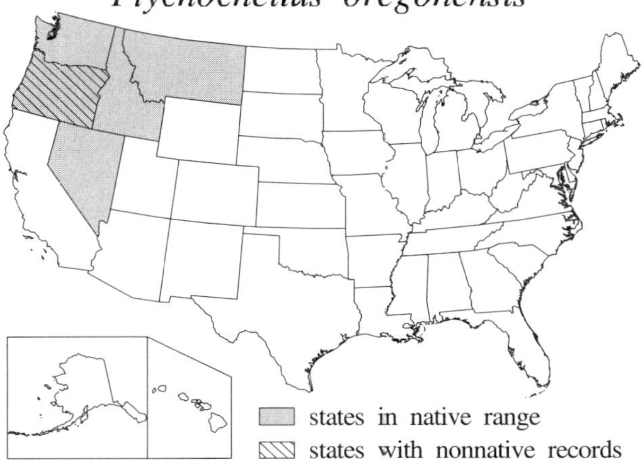

Ptychocheilus oregonensis (Richardson 1836)

Common Name: northern pikeminnow (northern squawfish).

Identification: Scott and Crossman (1973); Sigler and Sigler (1987); Page and Burr (1991). The American Fisheries Society Names Committee has changed the common name of all squawfish to pikeminnow (Nelson et al. 1998). Maximum size: 63 cm.

Native Range: Pacific drainages from Nass River, British Columbia, to Columbia River, Nevada; Peace River system (Arctic basin), British Columbia and Alberta (Page and Burr 1991); also Harney River basin, Oregon (Page and Burr 1991; but see Sigler and Sigler 1987).

Nonindigenous Occurrences: This species is known from the Harney basin of the Great Basin in **Oregon** (Sigler and Sigler 1987).

Means of Introduction: Unknown.

Status: Unknown; apparently still present in Oregon ca. 1990 (Page and Burr 1991).

Impact of Introduction: Unknown. This species is considered a nuisance by some because it preys on game species and may negatively impact hatchery-raised salmon (Wydoski and Whitney 1979; Sigler and Sigler 1987; but see Brown and Moyle 1981).

Remarks: There is some uncertainty concerning the designation of this species (i.e., native or introduced) relative to the Harney basin. Sigler and Sigler (1987) stated that this species had been introduced into the Harney system, but they did not provide any additional details. In contrast, several authors (Bond 1973; Lee et al. 1980 et seq.; Page and Burr 1991) presumably included the Harney River basin as part of its native range, since they made no mention of a possible introduction. Because this species is known to compete with salmon and trout for food, and to prey on salmonid young, it has come into conflict with humans. Past attempts at controlling pikeminnow populations (in their native range) have included the use of chemicals, dynamite, and the lowering of lake levels (Scott and Crossman 1973), as well as electrical barriers and traps (Brown and Moyle 1981). Brown and Moyle (1981) concluded that pikeminnow do not appear to be significant predators of salmon and trout except under highly localized, seasonal or unusual circumstances. They also found little evidence that they compete strongly with salmonids. Hybrids have been reported with the chiselmouth *Acrocheilus alutaceus* (Sigler and Sigler 1987).

Ptychocheilus umpquae

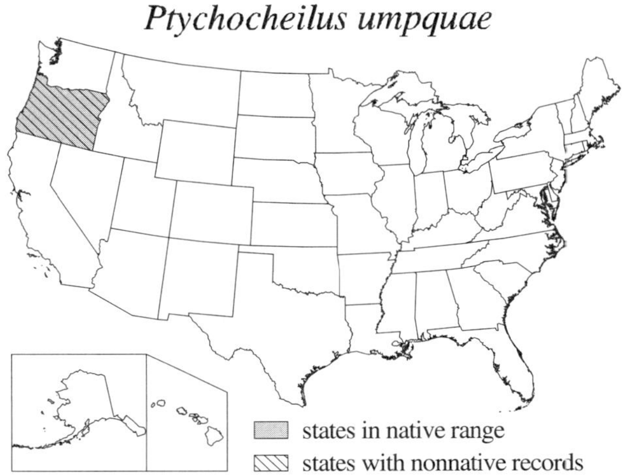

Ptychocheilus umpaquae (Snyder 1908)

Common Name: Umpqua pikeminnow (Umpqua squawfish).

Identification: Page and Burr (1991). The American Fisheries Society Names Committee has changed the common name of all squawfish to pikeminnow (Nelson et al. 1998). Maximum size: 44 cm.

Native Range: Umpqua and Diuslaw River drainages, Oregon (Page and Burr 1991).

Nonindigenous Occurrences: Introduced into the Rogue River, **Oregon** (Bond 1994).

Means of Introduction:

Status:

Remarks: None.

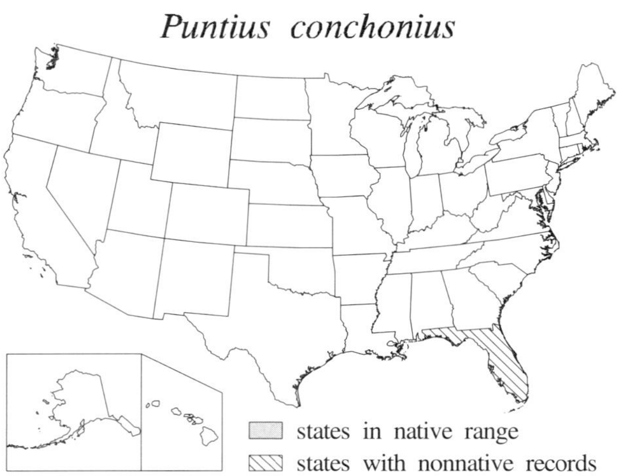

Puntius conchonius (Hamilton 1822)

Common Names: rosy barb (red barb).

Identification: Another commonly used name is *Barbus conchonius*. Distinguishing characteristics and key were given in Talwar and Jhingran (1992). Color photographs appeared in Axelrod et al. (1985). Maximum size: 14 cm.

Native Range: Asia. Afghanistan, Pakistan, India, Nepal, and Bangladesh (Talwar and Jhingran 1992).

Nonindigenous Occurrences: A few specimens were taken from a Lake Worth Drainage District canal L-15, adjacent to a fish farm in Palm Beach County, **Florida**, in the late 1960s or early 1970s (Courtenay et al. 1974; Courtenay and Hensley 1979a).

Means of Introduction: Probable escape or release from fish farm.

Status: Reported from Florida.

Impact of Introduction: Unknown.

Remarks: This species is an important ornamental fish (Talwar and Jhingran 1992).

There are no known voucher specimens.

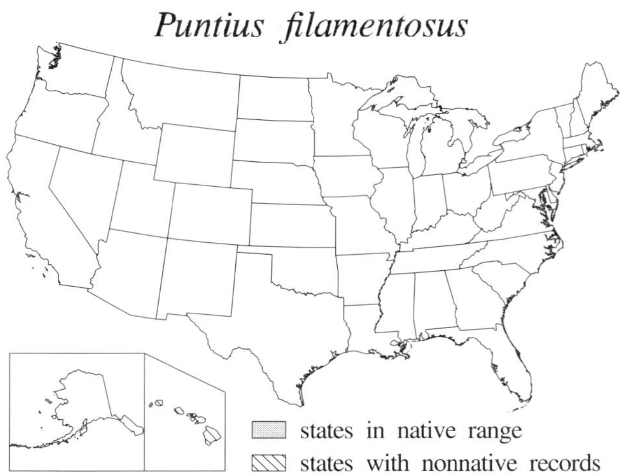

Puntius filamentosus (Valenciennes 1844)

Common Names: blackspot barb (Indian tiger barb, mahecola, filament barb, longfin barb, featherfin barb).

Identification: Another name for this species is *Barbus filamentosus*. Distinguishing characteristics, a key, and a figure were given in Talwar and Jhingran (1992). Color photographs appeared in Axelrod et al. (1985) and in Petrovicky (1988). Maximum size: 15 cm.

Native Range: Asia. India, Sri Lanka, Burma, and possibly Thailand (Talwar and Jhingran 1992).

Nonindigenous Occurrences: This species has been established in a reservoir and in ponds on Oahu, **Hawaii**, since 1984 (Devick 1991a, 1991b).

Means of Introduction: Accidental introduction, aquarium escape (Devick 1991b).

Status: Established in Hawaii.

Impact of Introduction: Unknown.

Remarks: This is a popular ornamental species. There are no known voucher specimens.

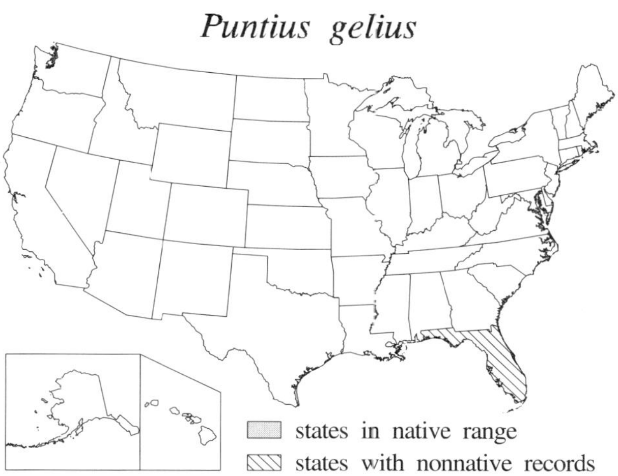

Puntius gelius (Hamilton 1822)

Common Names: dwarf barb (golden barb, golden dwarf barb).

Identification: A commonly used name is *Barbus gelius*. Distinguishing characteristics and key were given in Talwar and Jhingran (1992). Photographs appeared in Sterba (1973). Maximum size: 4 cm.

Native Range: Asia. India, Pakistan, and Bangladesh (Talwar and Jhingran 1992).

Nonindigenous Occurrences: A few specimens were taken from Lake Worth Drainage District canal L-15, adjacent to a fish farm in Palm Beach County, **Florida**, in 1969 and 1970 (Courtenay et al. 1974; Courtenay and Hensley 1979a).

Means of Introduction: Probable escape from fish farm.

Status: Reported from Florida.

Impact of Introduction: Unknown.

Remarks: This species is in the ornamental fish trade. There are no known voucher specimens.

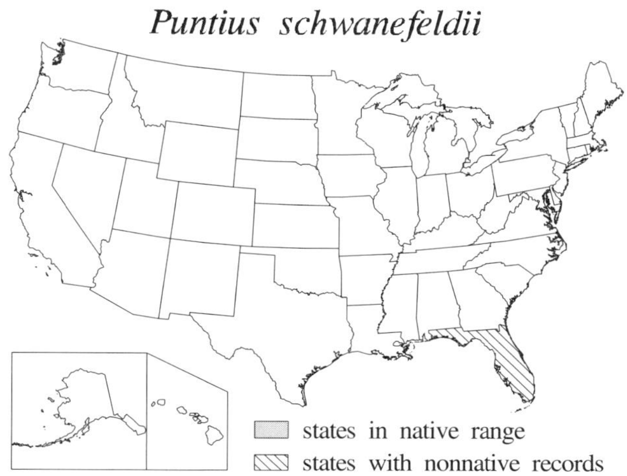

Puntius schwanefeldii (Bleeker 1853)

Common Names: tinfoil barb (goldfoil barb, Schwanefeld's barb).

Identification: Distinguishing characteristics were given by Roberts (1989) and by Kottelat et al. (1993, as *Barbodes schwanenfeldii*). It is included in the keys of Smith (1945) and Kottelat (1985). Color photographs appeared in Axelrod et al. (1985) and in Kottelat et al. (1993). Other modifications of name used include *Barbodes schwanefeldi*, *Barbus schwanenfeldii*, *Barbus schwanefeldi*, *Puntius schwanenfeldi*, and *P. schwanefeldi*. Maximum size: 35 cm.

Native Range: Tropical Asia. Thailand, Malay Peninsula, Sumatra, and Borneo (Roberts 1989), also Kampuchea, Laos, and Vietnam (Kottelat 1985).

Nonindigenous Occurrences: A single specimen was taken from Lake Worth Drainage District canal L-15, adjacent to a fish farm in Palm Beach County, **Florida**, during the period 1968 to 1972 (Courtenay et al. 1974; Courtenay and Hensley 1979a).

Means of Introduction: Probable escape from fish farm.

Status: Reported from Florida.

Impact of Introduction: Unknown.

Remarks: This species is a popular ornamental fish. In published lists of introduced species taken in Florida, this species is identified as the tinfoil barb (*Barbus* sp.) (e.g., Courtenay and Robins 1973). The exact date of collection is uncertain; dates given in the literature are in the periods 1968 to 1970 (Courtenay 1979b) and 1970 to 1972 (Courtenay and Robins 1973).

There are no known voucher specimens.

Puntius semifasciolatus

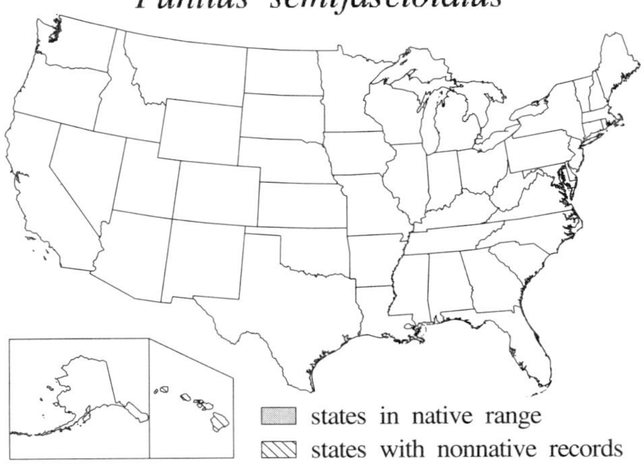

Puntius semifasciolatus (Günther 1868)

Common Names: green barb (half-striped barb, China barb, gold barb, halfbanded barb, Schubert's barb).

Identification: A related species, *Puntius sachsi*, is very similar in appearance (Petrovicky 1988). Color photographs appeared in Axelrod et al. (1985); color variants were pictured in Petrovicky (1988). Other modifications of name used include *Barbus semifasciolatus*, *Capoeta semifasciolata*, and *C. semifasciolatus*. Maximum size: 10 cm.

Native Range: Asia. Southeastern China from Hong Kong to the island of Hainan (Petrovicky 1988).

Nonindigenous Occurrences: This species has been established in Oahu, **Hawaii**, since 1940 (Maciolek 1984; Devick 1991a, 1991b). Brock (1960) indicated it was released into Nuuanu Reservoir 3 on Oahu.

Means of Introduction: According to Devick (1991b), this species was introduced from Asia for food and decoration.

Status: Listed as established in Hawaii by Devick (1991b). Brock (1960) also considered it established, but speculated that the species could be exterminated should Nuuanu Reservoir ever be drained.

Impact of Introduction: Unknown.

Remarks: This species is in the ornamental fish trade. It is listed as *Barbus semifasciolatus* in Hawaiian literature on introduced fishes.

There are no known voucher specimens.

Puntius tetrazona

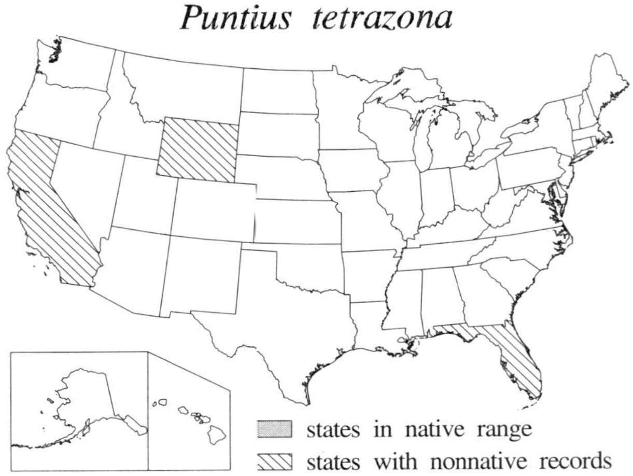

Puntius tetrazona (Bleeker 1855)

Common Names: tiger barb (Sumatra barb, partbelt barb).

Identification: Distinguishing characteristics were given by Kottelat et al. (1993). Other names used include *Barbus tetrazona* and *Capoeta tetrazona*. The color pattern is nearly identical to that of *P. anchisporus*, but *P. tetrazona* differs in having an incomplete lateral line and a more elongate body (Roberts 1989). Color photographs appeared in Axelrod et al. (1985) and in Kottelat et al. (1993); illustrations were provided in Petrovicky (1988), including examples of several of the color variants that exist in the aquarium trade. Maximum size: 7 cm.

Native Range: Tropical Asia. Sumatra (Roberts 1989).

Nonindigenous Occurrences: Two specimens, a sexually mature pair, were taken from a small stream flowing from Warm Springs Sanctuary in Owens Valley, Inyo County, **California** in July 1973 (Naiman and Pister 1974; Dill and Cordone 1997). A single large specimen was collected at Perrine Wayside Park in Perrine, Dade County, **Florida**, prior to 1979 (Courtenay and Hensley 1979a). Possibly in reference to the same record, the species was reported from a small roadside borrow pit south of Miami (Shafland 1976). This species was collected from Kelly Warm Springs, **Wyoming**, in 1990 (M. Stone, personal communication).

Means of Introduction: Possible fish farm escape or release in Florida; probable aquarium releases elsewhere. Dill and Cordone (1997) concluded that the California fish were presumably introduced by an aquarist or fish dealer wishing to use the spring as a brood pond.

Status: Reported from California, Florida, and Wyoming. Shapovalov et al. (1981) stated that no additional specimens had been taken from the California site since 1961, despite repeated collecting efforts; Hubbs et al. (1979) did not consider this fish established in that state.

Impact of Introduction: Unknown.

Remarks: This species is a popular ornamental fish. Tiger barbs are very aggressive fish, and there was concern that the fish found in California could adversely affect one of only two known populations of Owens pupfish *Cyprinodon radiosus* (Naiman and Pister 1974). The two tiger barbs collected from Owens Valley were a male and a female, both in breeding condition.

Voucher specimens: California (ASU 6239).

Relictus solitarius

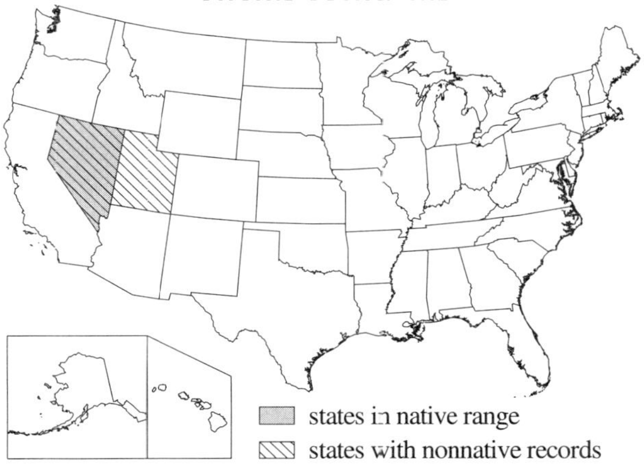

Relictus solitarius Hubbs and Miller 1972

Common Name: relict dace (Steptoe dace).

Identification: Sigler and Sigler (1987); Page and Burr (1991). Maximum size: 13 cm.

Native Range: Lakes Franklin, Gale, Waring, Steptoe and Spring basins in eastern Nevada (Page and Burr 1991).

Nonindigenous Occurrences: Forty-two relict dace were stocked into a warm-water pond at the Shoshone Ponds Natural Area southeast of Ely in White Pine County, **Nevada**, in 1977 (Clemmer, personal communication). This species may also have been introduced into Spring Valley, Nevada, presumably at Stone House (Hubbs et al. 1974); apparently in reference to this same record, Lee et al. (1980 et seq.) suggested that the Lake Spring population may also be introduced. There is also evidence that the relict dace was introduced to a spring in the Bonneville system of **Utah** (Hubbs et al. 1974; Lee et al. 1980 et seq.).

Means of Introduction: The Shoshone Ponds population was intentionally introduced for conservation purposes with stock transplanted from ponds in Steptoe Valley, Nevada. The plant created an additional population of a rare species with a restricted range. The introduction of this species into Spring Valley, Nevada, may have been for mosquito control, or possibly as a curiosity (Hubbs et al. 1974). In 1942, a rancher testified that he had taken this minnow from Butte Valley to Lake Mead, where the fish was used as bait (Hubbs et al. 1974); however, we are unaware of any records of this species in Lake Mead.

Status: Locally established. The relict dace population at Shoshone Ponds had increased to an estimated 502 individuals by August 1989 (Clemmer, personal communication). The status of this species in other sites where introduced is unknown.

Impact of Introduction: Unknown.

Remarks: Prior to elimination of the U.S. Fish and Wildlife Service candidate species list, the relict dace was under review for Federal listing as an endangered or threatened species (U.S. Fish and Wildlife Service 1994a).

Rhinichthys atratulus

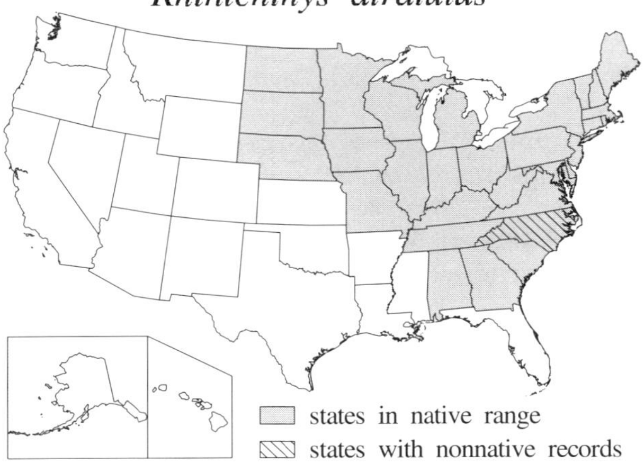

Rhinichthys cataractae

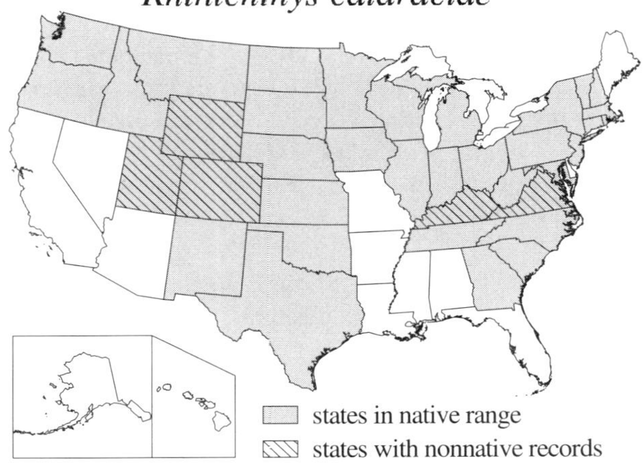

Rhinichthys atratulus (Hermann 1804)

Common Name: blacknose dace.

Identification: Becker (1983); Page and Burr (1991); Etnier and Starnes (1993); Jenkins and Burkhead (1994). Maximum size: 10 cm.

Native Range: Atlantic, Great Lakes, Hudson Bay, Mississippi, and upper Mobile Bay drainages from Nova Scotia to Manitoba and south to northern Georgia and northern Alabama (Page and Burr 1991).

Nonindigenous Occurrences: This species is present in the Yadkin drainage, **North Carolina**, probably as the result of introduction (Menhinick 1991).

Means of Introduction: Unknown, possible bait bucket release.

Status: Established, probably as a result of an introduction, in the Yadkin drainage, North Carolina (Menhinick 1991).

Impact of Introduction: Unknown.

Remarks: This species may be native to the Yadkin and have a localized distribution limited by physiography (Jenkins and Starnes, personal communication).

Rhinichthys cataractae (Valenciennes 1842)

Common Name: longnose dace.

Identification: Wydoski and Whitney (1979); Smith (1985); Hubbs et al. (1991); Page and Burr (1991); Jenkins and Burkhead (1994). In his type catalogue, Gilbert (1998) recognized three subspecies: *R. c. cataractae, R. c. dulcis*, and *R. c. smithi*; he noted that considerable work still needs to be done on the taxonomy of the species. Maximum size: 16 cm.

Native Range: Generally distributed above 40° N from coast to coast; occurs as far north as Arctic Circle in Mackenzie River drainage; south in Appalachian Mountains to northern Georgia and in Rocky Mountains south into Rio Grande drainage of Texas and northern Mexico (Page and Burr 1991).

Nonindigenous Occurrences: The longnose dace has been introduced into the Colorado River in **Colorado** (Walker 1993) and into Russell Fork (Pike County) in the Big Sandy system, Ohio River drainage, **Kentucky** (Jenkins and Burkhead 1994). It has been introduced into the upper Roanoke drainage in **Virginia** (Jenkins and Burkhead 1994). The species has been recorded from several sites in the Green River drainage of the upper Colorado River basin. These sites include Strawberry Reservoir, **Utah** (Sigler and Miller 1963; Tyus et al. 1982); and the upper Green River drainage in **Wyoming** including Hams Fork Creek (Baxter and Simon 1970), Lake Powell system (Tyus et al. 1982), and Flaming Gorge Reservoir system (Tyus et al. 1982) (the latter reservoir situated on the Utah-Wyoming border). It was also recorded from the San Juan River drainage of the

Colorado River basin (Tyus et al. 1982), but specific locality data and state were not given.

Means of Introduction: Evidently a bait bucket release in Strawberry Reservoir in 1951 (Sigler and Miller 1963). Suspected bait bucket introduction in Wyoming (Baxter and Simon 1970) and in Kentucky (Jenkins and Burkhead 1994); possible bait bucket releases in other sites.

Status: Established in Colorado where it is abundant and widespread in the main-stem Colorado River and in transition zone riffles (Walker 1993); Tyus et al. (1982) listed it as rare in the upper Colorado River basin. An eradication program was carried out in Strawberry Reservoir, Utah (Sigler and Miller 1963), but the results of those efforts were unclear. Established an spreading in the upper Roanoke, Virginia (Jenkins and Burkhead 1994). Reported from Kentucky (Jenkins and Burkhead 1994).

Impact of Introduction: Longnose dace hybridize with native speckled dace *R. osculus* in areas where longnose dace have been introduced (Sigler and Miller 1963). This hybrid has been recorded from Utah in the Provo River in 1942 and from the Strawberry Reservoir in Wasatch County in 1951 (Sigler and Miller 1963). It has also been recorded from areas where both species are native, such as, the Bear River Wyoming, Little Wood River, Idaho and Ross Fork at Fort Hall, Idaho (Sigler and Miller 1963).

Remarks: This species is sometimes used as a baitfish (Baxter and Simon 1970; Scott and Crossman 1973). Tyus et al. (1982) gave a distribution map of the this species in the upper Colorado basin.

Rhinichthys chrysogaster

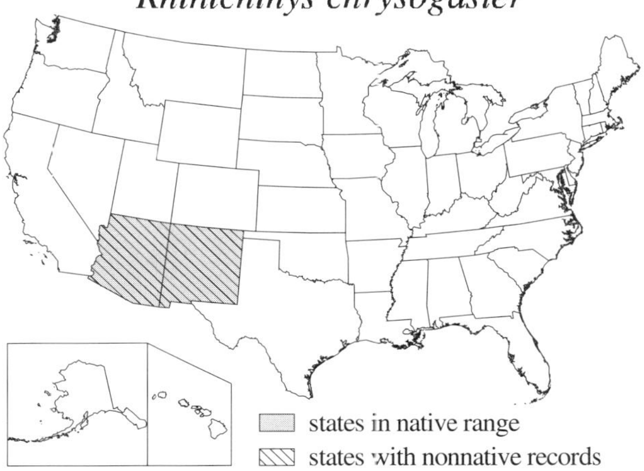

Rhinichthys chrysogaster Girard 1856

Common Name: longfin dace.

Identification: Minckley (1973); Sublette et al. (1990); Page and Burr (1991). Another name is *Agosia chrysogaster*. Maximum size: 10 cm.

Native Range: Lower Colorado River drainage, New Mexico and Arizona, and south through southern Arizona and Pacific drainages of western Mexico (Page and Burr 1991).

Nonindigenous Occurrences: This species is known from the Virgin River in northwestern **Arizona** (Minckley 1973; Lee et al. 1980 et seq.) and from several areas in **New Mexico** including the Rio Grande Basin, Rio Hondo (Pecos drainage), the Mimbres River, Largo Creek (Little Colorado River drainage), and the Zuni River (Koster 1957; Minckley 1973; Lee et al. 1980 et seq.; Propst et al. 1987; Sublette et al. 1990).

Means of Introduction: Unknown; probable bait bucket releases. Introduced as early as the 1950s into parts of the Rio Grande basin of New Mexico. The Rio Hondo, Mimbres, and Largo Creek introductions all occurred in the 1960s (Sublette et al. 1990).

Status: The Largo Creek population was extirpated by 1990, but the New Mexico populations in Rio Hondo and the Mimbres remain established (Sublette et al. 1990). It is locally established in a portion of the Rio Grande basin (Sublette et al. 1990). Reported but not known to be established in the Zuni River, New Mexico, and in the Virgin River, Arizona (Lee et al. 1980 et seq.). The status of the Virgin River population is unknown.

Impact of Introduction: Unknown.

Remarks: Koster (1957) reported that the species is used as a baitfish.

Rhinichthys osculus

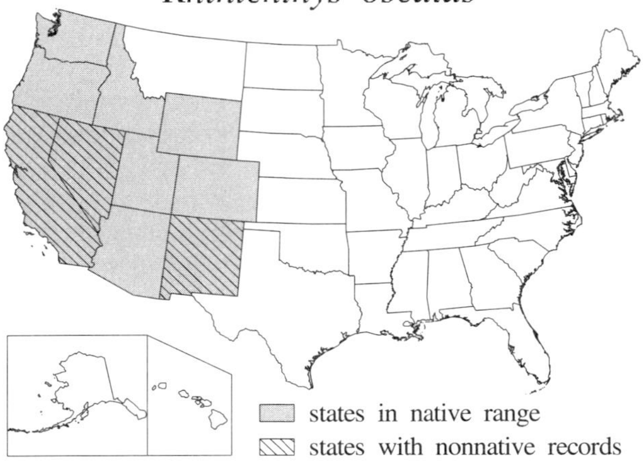

Rhinichthys osculus (Girard 1856)

Common Name: speckled dace (western dace, spring dace, dusky dace, Pacific dace).

Identification: Moyle (1976a); Wydoski and Whitney (1979); Sublette et al. (1990); Page and Burr (1991); Bond (1994). It is considered a species complex; some subspecies are protected as endangered (Page and Burr 1991). Gilbert (1998) recognized 15 subspecies. One subspecies is *R. o. nubilis* from Washington and closely adjacent areas, sometimes reported as *Rhinichthys nubilus* (Miller 1952). Maximum size: 11 cm.

Native Range: Western drainages (Pacific and endorheic) from Columbia River, British Columbia, to Colorado River, Arizona and New Mexico, and south into Sonora, Mexico (Page and Burr 1991).

Nonindigenous Occurrences: This species was recorded from the Santa Clara River basin just north of Los Angeles, and from tributaries of the Cuyama River in southern San Luis Obispo County, **California** (Miller 1968). The presence of this species in San Luis Obispo Creek in San Luis Obispo County, California, also may have been the result of an introduction (Moyle 1976a; but see Bell 1978). The subspecies, *R. o. robustus*, was recorded from Lake Webber, Sierra County, California, a high-altitude water body in the Truckee-Lahontan drainage system (La Rivers 1962). Also in California, small numbers of this species were introduced into Willow Creek, east of Independence, Inyo County, in 1939; into River Springs, Mono County, in 1940; and to the Old (Eagle) Borax Works on the west side of Death Valley in 1940 (Miller 1968). The species is known from the Mimbres River in southwestern **New Mexico** (Sublette et al. 1990). The Lahontan speckled dace, *R. o. robustus*, was introduced into Ruby Marsh and Ruby Marsh complex in northeastern **Nevada** twice during the early 1950s (La Rivers 1962; Hubbs et al. 1974). The subspecies *R. o. klamathensis* has been introduced to the Rogue River drainage in **Oregon** (Bond 1994).

Means of Introduction: Several transplants into parts of California during 1939 and 1940 were considered experiments to test the effects of changed environment on meristic and morphometric characters (Miller 1968). This species was intentionally stocked by Nevada Fish and Game officials in Ruby Marsh-Ruby Lake complex as forage for introduced largemouth bass *Micropterus salmoides*; the first introduction, in 1950, involved *R. osculus robustus* from a headwater of the Humboldt River, and the second, in 1951, from nearby Diamond Valley, Eureka County (La Rivers 1962; Hubbs et al. 1974). Although La Rivers (1962) recognized that *R. osculus robustus* was native to the Truckee River system, he concluded that the high-altitude lake had no native fishes and that small fishes present in the lake had likely arrived as a result of being stocked along with trout. Introductions of this species in other areas were probably the result of bait bucket releases (e.g., Miller 1952; La Rivers 1962). For instance, Miller (1946) indicated that the presence of *Rhinichthys osculus* in the Santa Clara River system (California) was possibly the result of its introduction as bait by trout fishermen.

Status: Established in San Luis Obispo Creek and Webber Lake, California (Moyle 1976a), and in Ruby Marsh, Nevada (La Rivers 1962). Reported from New Mexico (Sublette et al. 1990). Fishes introduced into Willow Creek and the Old Borax Works of California failed to survive; however, the species was common at River Springs, California, in 1967, more than 25 years after its initial introduction (Miller 1968). Although Miller (1968) reported on its occurrence in the Santa Clara River system, California, Bell (1978) did not collect the species there.

Impact of Introduction: Unknown in introduced areas. Speckled dace are known to hybridize with least chubs *Iotichthys phlegethontis* (Sigler and Sigler 1987), a species under review for federal listing, and therefore present a threat to this rare species.

Remarks: Widely used as a baitfish in certain parts of the western United States (Miller 1952; La Rivers 1962; Baxter and Simon 1970). Contrary to Moyle (1976a), Miller (1968) and Bell (1978) believed that *R. osculus* is native to San Luis Obispo Creek. They based that conclusion on Jordan (1894), who included this species under the name *Agosia nubila* in his early list of fishes found in San Luis Creek.

Rhodeus sericeus

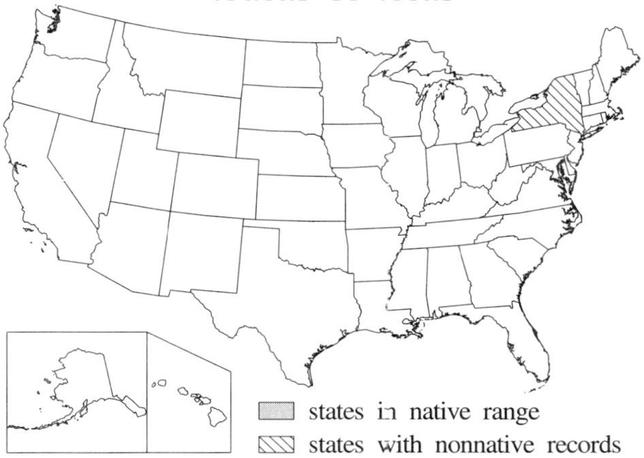

Rhodeus sericeus (Pallas 1776)

Common Names: bitterling (bitter carp, European bitterling).

Identification: Distinguishing characteristics were provided by Berg (1949), Moore (1968), and Page and Burr (1991). Keys were included in Berg (1949) and Moore (1968). Illustrations or photographs appeared in Berg (1949), Maitland (1977), and Axelrod et al. (1985). Another name used for this species is *Rhodeus amarus*. Berg (1949) apparently recognized two species: *R. s. amarus* and *R. s. sericeus*. Maximum size: 11 cm.

Native Range: Europe from the Seine and other rivers of France eastward to Asia Minor, and northern China (there is a very wide geographical gap in the northern part of the Asian continent separating the ranges of the two subspecies) (Berg 1949; Lee et al. 1980 et seq.).

Nonindigenous Occurrences: The first records of this species (as *Rhodeus amarus*) were from the Sawmill River, a tributary of the Hudson River, at Tarrytown, Westchester County, **New York** in the early 1920s (Dence 1925; Myers 1925; Bade 1926). Although Breder (1933) claimed it disappeared from this locality shortly after 1925, additional specimens were taken in subsequent years (Greeley 1937), with the last collection made in 1951 (Schmidt et al. 1981). Two specimens were taken from the Bronx River at Bronxville, Westchester County, New York, in 1933 (Greeley 1937); subsequent collections indicated the species was established in a localized reach of the river (Lee et al. 1980 et seq.; Schmidt et al. 1981; Schmidt and Samaritan 1984; Smith 1985).

Means of Introduction: Probable aquarium release (Myers 1925; Bade 1926; Lee et al. 1980 et. seq.; Schmidt et al. 1981).

Status: Established in the Bronx River, New York, since at least the early 1930s (Greeley 1937; Schmidt et al. 1981; Smith 1985). Previously established and abundant in Sawmill River, New York; that population is assumed to be extirpated (Schmidt et al. 1981).

Impact of Introduction: Unknown.

Remarks: *Rhodeus* requires freshwater bivalves as spawning sites; eggs are deposited, fertilized, and hatch in live mussels. Laboratory evidence has shown that this European fish will use certain U.S. native mussels (*Anadonta cataracta* and *Unio complanatus*) (Breder 1933). The reported recent decline in population of bitterling in the Bronx River apparently has resulted from a declining freshwater mussel population brought about by water pollution (R. Schmidt to Courtenay, personal communication).

Voucher specimens: New York (AMNH 39116, 39117, 42444; CU 5112, 23611, 26823, NYSM 11744, 11746).

Richardsonius balteatus

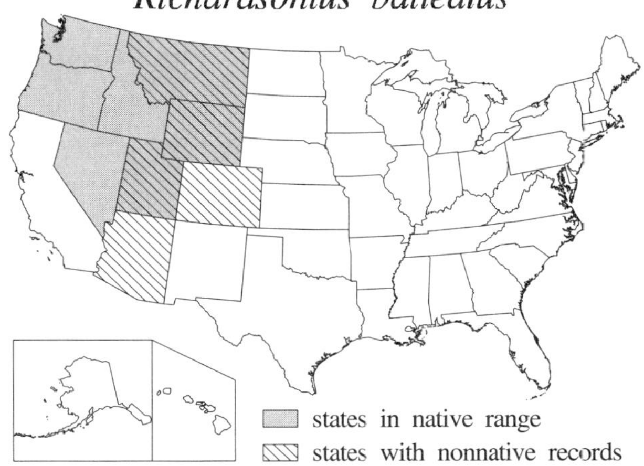

Richardsonius balteatus (Richardson 1836)

Common Names: redside shiner (Bonneville redside shiner, subspecies *R. b. hydrophlox*).

Identification: Wydoski and Whitney (1979); Sigler and Sigler (1987); Page and Burr (1991). Maximum size: 18 cm.

Native Range: Pacific Slope drainage from Nass River, British Columbia, to Rogue, Klamath, and Columbia River drainages, Oregon, Idaho, Nevada, and Wyoming; Bonneville basin, southern Idaho, western Wyoming, and Utah; Peace River system (Arctic basin), Alberta and British Columbia (Page and Burr 1991).

Nonindigenous Occurrences: The redside shiner has been recorded from **Arizona** in the Virgin River drainage, Colorado River basin (Minckley 1973; Lee et al. 1980 et seq.; Haynes et al. 1982; Page and Burr 1991). In **Colorado** it is known from several drainages in the Colorado River basin, including the Yampa, White, and Green rivers, and the upper Colorado main stem (Vanicek et al. 1970; Holden and Stalnaker 1975b; Haynes et al. 1982; Tyus et al. 1982; Woodling 1985). In **Montana** the species was introduced into the upper Missouri River drainage (Brown 1971; Lee et al. 1980 et seq.; Cross et al. 1986; Page and Burr 1991). In **Utah** it has been recorded from several sites in the Colorado River basin including the Virgin River (Haynes et al. 1982) and the Green River drainage (Simon 1946; Sigler and Miller 1963; Vanicek et al. 1970; Haynes et al. 1982; Tyus et al. 1982). In **Wyoming** it is known from the upper Green River, Colorado River basin (Simon 1946; Sigler and Miller 1963; Haynes et al. 1982; Tyus et al. 1982).

Means of Introduction: Presumably through bait bucket release (Minckley 1973) and probably by way of subsequent natural dispersal.

Status: This species has been present in the Colorado River basin since the 1930s (Simon 1946; Sigler and Miller 1963), and has continued to expand its range (Haynes et al. 1982; Tyus et al. 1982). It is established in Arizona, Colorado, Montana, Utah, and Wyoming.

Impact of Introduction: Largely unknown. It has been suggested that nonindigenous fishes, including *R. balteatus*, have contributed to the decline of native Colorado River species such as the Colorado pikeminnow *Ptychocheilus lucius* and the humpback chub *Gila cypha* (Haynes et al. 1982). These shiners are known to feed on the eggs of other species and possibly compete with the young of other fish for food and space (Woodling 1985). Sigler and Miller (1963) noted that it preys on the young of sport fishes. The introduced redside shiner appears to be replacing native Virgin River spinedace *Lepidomeda m. mollispinis* in the Virgin River (Minckley 1973). In areas where it is introduced, redside shiners can hybridize with speckled dace *Rhinichthys osculus* (Sigler and Miller 1963). This hybrid has been recorded from Utah in the Price River, Utah County in 1953; and Sheep Creek, Uintah County in 1960 (Sigler and Miller 1963). It has also been reported from Washington and the Provo River in Utah, where both species are native (Sigler and Miller 1963). Note: the subspecies *R. b. hydrophlox* is not native to Washington but the typical subspecies is. In British Columbia, introduced redside shiners adversely affected populations of a subspecies of introduced rainbow trout, the Kamloops trout *Oncorhynchus mykiss kamloops*; in addition to having a negative influence on trout growth and diet, the shiner was found to prey on trout fry (Larkin and Smith 1954).

Remarks: Introductions into Arizona, Utah, and Wyoming were recorded as the subspecies *R. b. hydrophlox* (Sigler and Miller 1963; Baxter and Simon 1970; Minckley 1973). This species is a popular bait fish (Sigler and Miller 1963; Baxter and Simon 1970; Minckley 1973). Tyus et al. (1982) gave a distribution map of the this species in the upper Colorado basin.

Richardsonius egregius

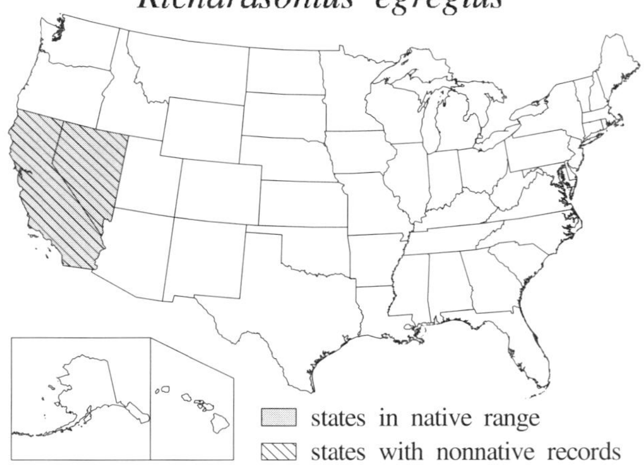

Richardsonius egregius (Girard 1858)

Common Name: Lahontan redside (Lahontan redside shiner).

Identification: La Rivers (1962); Moyle (1976a); Page and Burr (1991). It should be noted that its congener,

Richardsonius balteatus, is called the redside shiner. Maximum size: 17 cm.

Native Range: Lahontan and other interior basins in northern and western Nevada, and northern California, including Humboldt, Walker, Carson, Truckee, Susan, Quinn, and Reese River systems; Walker, Tahoe, and Pyramid lakes (Page and Burr 1991).

Nonindigenous Occurrences: The Lahontan redside has been recorded from various locations in **California** outside its native range, in the Sacramento River drainage (Page and Burr 1991), including Miller Lake on Mill Creek–Rubicon River system (Kimsey 1950) and Bucks Lake and Warner Creek, both in the North Fork of the Feather River system (Moyle 1976a). The species was apparently stocked along with trout into Lake Webber, Sierra County, in the Lahontan drainage system, California; its presence in Eagle Lake in Lassen County, California, another site within the Lahontan drainage system, may have also been the result of an introduction (La Rivers 1962). In **Nevada**, it was introduced into Railroad Valley Pond, Nye County, in 1956 (La Rivers 1962) and into Ruby Lake in 1950 (Hubbs et al. 1974); the species was recorded from Summit Lake in Humboldt County (Deacon and Williams 1984).

Means of Introduction: The species was stocked in Nye County, Nevada, in 1956 and 1957, and into Weber Lake, California (La Rivers 1962); the specific reasons for these introductions were not given. This species and *Rhinichthys osculus robustus* were introduced to Ruby Lake, Nevada, from a headwater of the Humboldt River, presumably to serve as forage for introduced largemouth bass (Hubbs et al. 1974). Most other records may have been the result of bait bucket releases (Kimsey 1950; Moyle 1976a) although Kimsey (1950) also argued that its occurrence in Miller Lake, California, a site near the Lahontan system, was possibly the result of a natural stream capture event.

Status: Established in California (Moyle 1976a); reported, possibly established, in Nevada (Deacon and Williams 1984). The Ruby Lake population apparently did not survive (Hubbs et al. 1974).

Impact of Introduction: Unknown.

Remarks: Although Wydoski and Whitney (1979) reported the Lahontan redside as introduced into the Bonneville basin, Idaho, we found no substantiating evidence to support this. Simpson and Wallace (1978), Lee et al. (1980 et seq.), Sigler and Sigler (1987), and Page and Burr (1991) make no mention of such an introduction. Because of several other errors in the same table, we chose not to include this record.

Scardinius erythrophthalmus

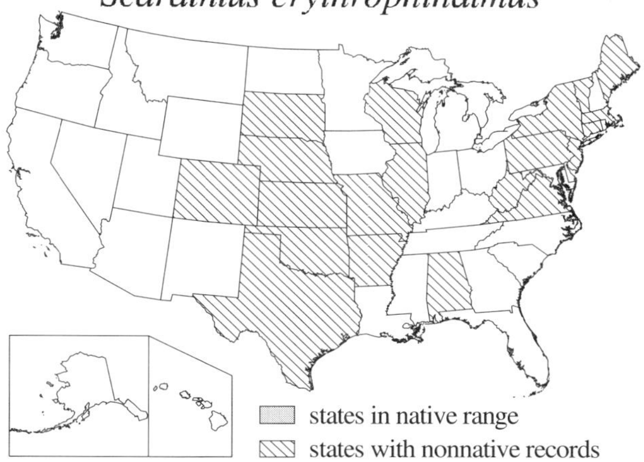

Scardinius erythrophthalmus (Linnaeus 1758)

Common Name: rudd.

Identification: Berg (1949); Maitland (1977); Muus and Dahlstrom (1978); Wheeler (1978); Smith (1985); Burkhead and Williams (1991); Page and Burr (1991); Crossman et al. (1992); Pflieger (1997). Burkhead and Williams (1991) provided diagnostic characters for the recognition of rudd × golden shiner crosses.
Maximum size: 48 cm.

Native Range: Western Europe to the Caspian and Aral sea basins (Berg 1949; Robins et al. 1991b).

Nonindigenous Occurrences: This species was reported from First Creek in Lauderdale County, **Alabama**, ca. 1987 (J. D. Williams, personal communication). It also has been recorded from Lonoke County, the White River drainage (single fish), and Horseshoe Lake in Crittenden County **Arkansas** (Robison and Buchanan 1993; J. D. Williams, personal communication). In **Colorado**, rudd have been taken from lakes in the Arkansas and in the South Platte drainages (Walker 1993). There is a report of this species from **Connecticut** (R. A. Jones, personal communication; but see Whitworth 1996). It is known from several sites in **Illinois** including the Fox, Illinois, and Kaskaskia drainages (Burr et al. 1996; Laird and Page 1996). The species is known from at least four counties in **Kansas** (Cross and Collins 1995); specific drainages mentioned for that state included Wilson Reservoir, a tributary of the Neosho River, and a farm pond near of tributary of Deep Creek, Kansas River drainage (Kansas Department of Wildlife and Parks 1990; C. Bever, personal communication). In **Massachusetts**, the species was discovered in the Charles River in Cambridge in the spring of 1991; there is also a record from Benton Lake, Otis (Hartel 1992; Hartel et al. 1996). Rudd have been known from Lake Cobbosseecontee near Augusta, Kennebec County, **Maine**, since 1973 (Courtenay et al. 1984; Burkhead and Williams 1991; Kircheis 1994). It has been recorded from several water bodies in **Missouri** including Mark Twain Reservoir, Watkins Mill Lake in Clay County, and Longview Lake and Lake Jacomo in Jackson County (Pflieger 1997; R. Mykre, personal communication). Rudd were reported from Lake Ogallala (= Lake C. W. McConaughy), Platte River drainage, near Ogallala in **Nebraska** in 1993 or 1994 (M. Peyton, personal communication). In addition to Lake Ogallala, the rudd recently has been recorded from other reservoirs in the western two-thirds of the state including Calamus, Sherman, and Southerland, and possibly others (L. Zadina, personal communication). This species was recorded from a lake in Hudson County Park in Jersey City, **New Jersey** (Myers 1925; Courtenay et al. 1984). It has been recorded from numerous locations in **New York**. These include an early record from Central Park Lake, New York City (Myers 1925); several collections, the first during the 1930s, from two sites in the Roeliff-Jansen Kill portion of the Hudson River drainage including Copake Lake and the outlet of Robinson Pond north of Copake Lake in Columbia County (Greeley 1937 cited in Courtenay et al. 1984; Smith 1985; museum specimens); and a collection from Cascadilla Creek near Ithaca in the Great Lakes basin in the early 1950s (Courtenay et al. 1984, 1986). More recent records in the New York area include several collections from Lake Ontario and the St. Lawrence River, and a population found in Oneida Lake, New York, in the Lake Ontario drainage (Klindt 1990; Mills et al. 1993), as well as collections from the Hudson River in Columbia County, Ringneck Marsh in Genesee County, and Schroon Lake in Essex County (museum specimens). This species has been recorded from at least eleven places in **Oklahoma** since 1989, including several sites in Lake Texoma, various segments of the Arkansas River, and the Salt Fork River, Lake Eufaula, South Canadian River, and Lake Watonga (Pigg and Pham 1990; Pigg et al. 1992). A single adult rudd was taken from Lake Winola, west of Scranton, **Pennsylvania** in 1991 (C. N. Shiffer, personal communication). Since about 1996, rudd have been recorded from at least eight water bodies in **South Dakota**, including Lake Vermillion, Lake Madison, Lake Francis Case, Pactola Reservoir in the Black Hills, Sheridan Lake, Newell Lake, Interstate Lake near Brookings, and Lake Alice in Deuel County (Stone, personal communication to J. Rendall; D. Unkenholz and J. Erickson, personal communication). The species has been recorded from at least five reservoirs in **Texas**, including Lake Texoma (Red River drainage), Grayson County, in 1989; Victor Braunig Reservoir (Guadalupe River drainage), Bexar County, in 1989; Calaveras Reservoir (Guadalupe River drainage), Bexar County, in 1989; Lake Whitney (Brazos River drainage), Hill County, in 1989; and Canyon Lake, Comal County in 1990 (Muoneke 1990; Howells et al. 1991b; Whiteside and Berkhouse 1992). This species is known from at least two water bodies in **Vermont**. Specimens taken from Lake Champlain in 1991 were documented by personnel of the Vermont Department of Fish and Wildlife. The species was also captured in Burr Pond in Rutland County, Vermont, in 1992. It was taken from at least four sites in **Virginia** including Lake Burke (Potomac River drainage), Fairfax County, in 1989; Gardy's Mill Pond (Potomac River drainage), Northumberland County, in 1989; and Lake Anna (York River drainage), Spotsylvania County, in 1989 (R. Southwick, personal communication); and Lake Whitehurst (= Little Creek Reservoir) northwest Virginia Beach in early 1993 (museum specimen). A single specimen was taken from the New River system, Fayette County, **West Virginia**, in 1991 (Burkhead, personal communication). The species was introduced to Oconomowoc Lake in Waukesha County, **Wisconsin**, ca. 1916 where it was present for many years (Cahn 1925; Becker 1983; Courtenay et al. 1984; Burkhead and Williams 1991). There are reports of single specimens taken from two or three different Wisconsin sites after the period when rudd were sold as bait (i.e., since about the mid-1980s) (J. Lyons, personal communication). The only specific information on file in our office involving a recent record in Wisconsin is that of a collection of rudd taken from a lagoon at the southern end of Lake Winnebago, Fond du Lac County, in 1988 (J. D. Williams, personal communication).

Means of Introduction: Bait bucket release seems to be the primary mechanism by which rudd have gained access into open waters. However, the history of its introduction is complex and only partly known. The species apparently entered the United States during two widely separated periods of introduction. It was initially brought to this country either in the late 1800s or early 1900s. A second period of introduction presumably began in the late 1960s or early 1970s. According to Courtenay and Williams (1992) and Gilbert (1998), rudd first appeared in the United States during the late 1800s. Main support for their conclusion is based on a footnote of Hubbs (1921). In his footnote, reference is made to a brief description by Bean (1897) of a variety of golden shiner *Notemigonus crysoleucas* from Central Park Lake, New York City, which Hubbs concluded to be, in reality, introduced *Scardinius erythrophthalmus*. Hubbs' opinion was based on the "permanent vermilion color of the pectorals, ventrals and anal" described by Bean for specimens apparently held in the New York Aquarium. Later, Bean (1903), repeating the same description as in the 1897 paper, elevated the race to a subspecies (i.e., *Abramis chrysoleucas roseus*). The principal character on which the new form was based was the blood red pigmentation of the fins (hence the name *roseus*). In his investigation on the history of rudd in the United States, Burkhead (personal communication) was unable to locate types for the subspecies, but he did find a single *N. crysoleucas*, dated 1894, from Central Park Lake which he examined and concluded as typical in every respect. According to Burkhead, the available information does not support or refute Hubbs' comments. Furthermore, since Hubbs' remarks presumably are based solely on Bean's color description, Burkhead concluded that it seems prudent to discard Hubbs' inference. The first verifiable U.S. record dates to 1916. In that year, some 300 specimens obtained from the New York aquarium by B. O. Webster, the Wisconsin superintendent of fisheries, were transplanted to Lake Oconomowoc, Wisconsin (Cahn 1927; Greene 1935). Myers (1925) reported on a population in Central Park Lake in New York City. The Central Park population, now extirpated, may have originated from Copake Lake, Roeliff-Jansen Kill system, which also supported a population in 1916 and possibly earlier (M. N. Feinberg, personal communication to Burkhead). The origin of the very first New York fish is not known; however, because rudd is popular in Europe as a food and game species, it may have been introduced for both purposes. The Central Park population also may have been the source for the transplant to a New Jersey park just across the Hudson (Myers 1925; Greeley 1937). The early 1950s capture of rudd by C. R. Robins in Cascadilla Creek, Tompkins County, New York, appears to be the only and last known record from the creek (Courtenay et al. 1984; Smith 1985). The origin of this population is not documented, but the relatively close Roeliff-Jansen Kill population is a logical possibility. The earliest of the more recent records is a 1973 capture of rudd from Cobbosseecontee Lake in Maine. Neither year nor origin of the Maine introduction is known. At about the same time, a bait dealer in Suffolk, Virginia, began rearing rudd for bait. The source of the Suffolk, Virginia fish farmer's stock is uncertain. The interest in bait culture of rudd dramatically intensified in the early 1980s. The central Arkansas region of Lonoke and Prairie counties, an area known for its active fish farming industry, apparently became the largest producer of rudd in the United States. It appears that the greatest dispersal of rudd has been through interstate traffic rather than direct European import. In fact, much of its recent culture and spread can be attributed to its popularity as a bait among striped bass *Morone saxatilis* anglers. As a result, rudd have been widely introduced through a combination of bait bucket releases, escapes from aquaculture facilities and farm ponds, and, presumably, by dispersal from various points of introduction (e.g., Burkhead and Williams 1991). Although many rudd introductions are considered accidental, it is likely that rudd also have been intentionally released into public waters during the past few decades.

Status: This species has been recorded as introduced to 20 states. In a numbers of other states it has been used as a bait fish, but there is as yet no record of it being found in open waters. Available data indicate established populations still survive in Maine and New York, and, more recently, evidence indicates it is established in Massachusetts, Nebraska, and South Dakota. The species had breeding populations in New Jersey, Wisconsin, and Kansas, but it is apparently no longer extant in these states. It has been established in Lake Cobboseecontee, Maine, since 1973 (Courtenay et al. 1984; Burkhead and Williams 1991; Kircheis 1994). It has been established in the Roeliff-Jansen Kill drainage, New York, since the first half of the 1900s (Smith 1985). Probably in reference to that region, Schmidt (1986) listed rudd as present in the Hudson River drainage in the northern Appalachian region. Other New York populations are considered extirpated (Courtenay et al. 1984; Smith 1985); however, Mills et al. (1993) stated that an established population was discovered in Oneida Lake, New York, in the Lake Ontario drainage, in 1990. It may also be established in the St. Lawrence River and Lake Ontario, but that conclusion awaits confirmation. Hartel et al. (1996) stated that the presence of both young and adults over several years indicated that it is reproducing in the lower Charles River in Massachusetts. According to Zadina (personal communication), rudd are considered to be established in one or more lakes in Nebraska as of 1998. Presumably this includes Lake Ogallala. There also is evidence of rudd reproduction in parts of western South Dakota; during the summer of 1997 several year classes were discovered in Pactola Reservoir, Sheraton Lake, and Newall Lake (Unkenholz, personal communication). The species was established and later became extirpated in New Jersey and Wisconsin (Becker 1983; Courtenay et al. 1984; Courtenay and Williams 1992). There has been at least one established population in Kansas. In Spring 1996, a large breeding population was discovered in a 0.6-acre farm pond in the Deep

Creek-Kansas River drainage near Manhattan, Raleigh County. However, that population was exterminated by Kansas Department of Wildlife and Parks personnel in Summer of 1996; as of early 1998, there has been no record of rudd in sites downstream of the pond (Bever, personal communication). Although they mapped its distribution in Kansas, Cross and Collins (1995) did not provide information on reproductive status. In general, the past literature on rudd distribution and status is fragmentary and somewhat contradictory. Courtenay et al. (1986, 1991), Courtenay and Stauffer (1990), and Courtenay and Williams (1992) considered it established in Maine and New York. Similarly, Page and Burr (1991) stated that it was established in Maine and in the lower Hudson River drainage in New York. In a relatively recent work, Courtenay (1993) indicated that the rudd was established in Maine, New York, Kansas, and presumably Nebraska (postal abbreviation given as NB). Jenkins and Burkhead (1994), citing unpublished information, stated that rudd had become established in Indiana, New York, and Maine. However, we have found no evidence that the rudd was ever introduced to Indiana waters. It is likely that inclusion of Indiana was an error (Burkhead, personal communication). Muoneke (1990) incorrectly cited unpublished information in stating that rudd apparently had become established in Alabama. In many states it has been recorded but not known to be reproducing. For instance, Jenkins and Burkhead (1994) stated that it is unknown whether escapees have established a population in Virginia. Although it has been reported from several reservoirs in Missouri, Pflieger (1997) stated there is no evidence yet that rudd is established in the state. Although we have a relatively recent report of this species from Connecticut (Jones, personal communication; Whitworth (1996) stated that the species has not been found in that state.

Impact of Introduction: Largely unknown. In a laboratory setting, Burkhead and Williams (1991) demonstrated that rudd readily hybridize with native golden shiner, *Notemigonus crysoleucas*, a primary forage species of many native game fishes. As such, the probability exists that rudd introduced to open waters will hybridize with golden shiner, with unknown consequences to wild populations of the native species. First generation hybrids offspring should show heterosis (or hybrid vigor), but the "genetic pollution" in subsequent generations could prove detrimental due to a variety of factors, for instance spawning behavior and recruitment success, and general loss of fitness (Burkhead and Williams 1991; Courtenay and Williams 1992). Nevertheless, the interactions of the two species in nature are not known (Burkhead, personal communication). Cadwallader (1977) reviewed the potential impacts of the rudd in waters of the North Island of New Zealand. He concluded, in part, that rudd can be expected to compete for invertebrate food sources with native fishes. In addition, being omnivorous, the rudd can shift its diet to plants, unlike most native fishes. Because rudd are fairly hardy, Cadwallader also indicated that the fish will fare better than many native fishes in waters that are eutrophic or polluted.

Remarks: Courtenay et al. (1986) described the distribution of rudd in the eastern United States. They listed it among the species with declining populations. However, Courtenay et al. (1991) and Burkhead and Williams (1991) reviewed rudd introductions and documented the more recent and rapid spread of the species as a result of its wide use as a bait fish for white bass. Many states now outlaw the use of rudd as a live bait. As a result, its rapid spread appears to have slowed.

Although there is no evidence that rudd have been introduced to California, Dill and Cordone (1997) expressed concern that rudd may find its way into the state as a contaminant in golden shiner shipments imported as bait from Arkansas. This species has been cultured in Arkansas and Virginia (and possibly elsewhere) as baitfish and distributed to bait stores in at least 16 states (Courtenay and Williams 1992). Two specimens were taken from the Canadian side of Lake Erie at Crystal Beach (near Port Abino) in June 1997 (A. Dextrase, personal communication). That record is the first report of rudd from Lake Erie.

Voucher specimens: Massachusetts (MCZ 95616, MCZ 96072, MCZ 99432), New York (AMNH 58445, NYSM 25122, 35122, 40871, 41752, 41850), Vermont (NYSM 42040).

Semotilus atromaculatus

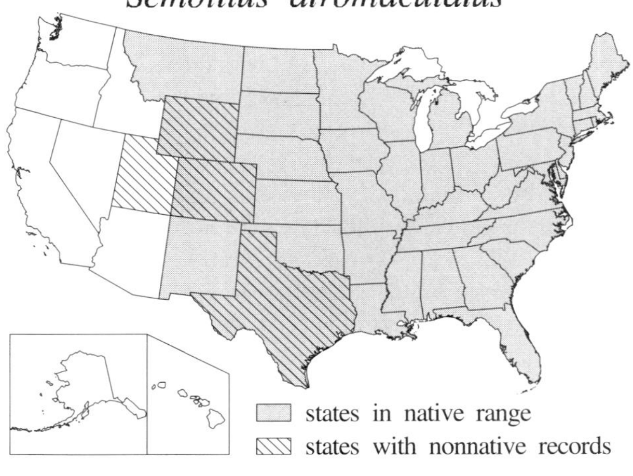

Semotilus atromaculatus (Mitchill 1818)

Common Name: creek chub.

Identification: Becker (1983); Page and Burr (1991); Etnier and Starnes (1993); Jenkins and Burkhead (1994). Maximum size: 30 cm.

Native Range: Most of eastern United States and southeastern Canada in Atlantic, Great Lakes, Hudson Bay, Mississippi, and Gulf basins as far west as Manitoba, eastern Montana, Wyoming, Colorado, eastern Oklahoma, and eastern Texas, but absent from southern Georgia and peninsular Florida; isolated population in upper Pecos and Canadian River systems, New Mexico (Page and Burr 1991).

Nonindigenous Occurrences: This species has been recorded from west slope drainages of **Colorado** in the lower Yampa and Green rivers systems, Colorado River basin (Vanicek et al. 1970; Holden and Stalnaker 1975b; Tyus et al. 1982; Woodling 1985; Walker 1993). It was introduced into the Colorado River drainage in **Texas** (Conner and Suttkus 1986; Page and Burr 1991). The species is known from **Utah** in the Green River (Vanicek et al. 1970; Tyus et al. 1982). It was reported from **Wyoming** in the Little Snake River drainage, in Willow Creek of the Green River drainage (Baxter and Simon 1970), and in Muddy Creek in the White-Yampa drainage (C. Wheeler, personal communication).

Means of Introduction: Most introductions probably via bait bucket release. Woodling (1985) stated that the Colorado West Slope populations are the result of accidental introduction by man. Hubbs (1957) stated that western records in Texas, and a report from Lake Texoma "have undoubtedly resulted from bait release."

Status: Introduced populations are established in Colorado and Wyoming; probably established in Utah. The species has been present in the Snake River, Green River drainage, since at least the 1940s, but never in abundance (Holden and Stalnaker 1975). Status unknown in Texas.

Impact of Introduction: According to Baxter and Simon (1970), in some situations populations of this species may compete with trout, but the authors did not give details. Beckman (1974) reported that creek chub prey on trout. Magnan and Fitzgerald (1982) provided evidence that introduced creek chub are a potential competitor with brook charr *Salvelinus fontinalis*. In their study of a Quebec stream, they found that charr changed their diet from benthic invertebrates to zooplankton in the presence of chub. Subsequent laboratory experiments suggested that higher relative abundance of creek chub was one of several important factors in the observed niche shift of brook charr in nature (Magnan and Fitzgerald 1984).

Remarks: This species is a commonly used baitfish (Baxter and Simon 1970; Scott and Crossman 1973; Beckman 1974). Tyus et al. (1982) gave a distribution map of the this species in the upper Colorado basin.

Semotilus corporalis

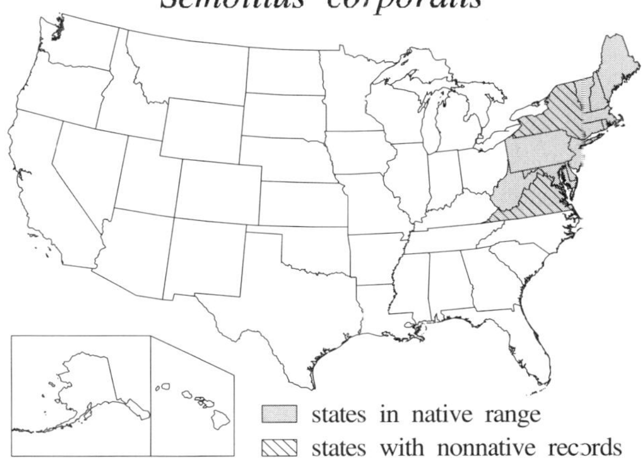

Semotilus corporalis (Mitchill 1817)

Common Name: fallfish.

Identification: Smith (1985); Page and Burr (1991); Jenkins and Burkhead (1994). Maximum size: 51 cm.

Native Range: Atlantic Slope from New Brunswick to James River drainage, Virginia; Hudson Bay, Lake Ontario, and St. Lawrence drainages, Quebec, Ontario, and New York (Page and Burr 1991).

Nonindigenous Occurrences: This species was recorded as introduced outside its native range in **New York**, including Tonawanda Creek near Buffalo and possibly the Adirondack Mountains (Smith 1985). It was discovered in the upper Falling River in the Roanoke drainage, Campbell County, **Virginia** in 1987 (Jenkins and Burkhead 1994).

Means of Introduction: Some believe that the fallfish may not have been present in the Adirondacks of New

York before canals and associated waterways were constructed (Smith 1985). As such, Smith (1985) indicated its presence is possibly the result of introduction, perhaps by way of the Erie Canal system or by bait bucket dumping. Smith did not speculate or provide details on its method of introduction into Tonawanda Creek, New York; that population may also have originated through bait bucket release. Jenkins and Burkhead (1994) strongly suspected that the source of the Falling River population in Virginia started with fish introduced from the adjacent James drainage. Presumably these introductions were a result of bait bucket release as well. Illegally introduced into California (Moyle 1976a).

Status: Established, possibly introduced, in the Adirondack Mountains, New York (Smith 1985); established in Falling River, Virginia (Jenkins and Burkhead 1994); reported from Tonawanda Creek, New York (Smith 1985).

Impact of Introduction: Unknown.

Remarks: This species is known to hybridize with *Luxilus cornutus* (Scott and Crossman 1973).

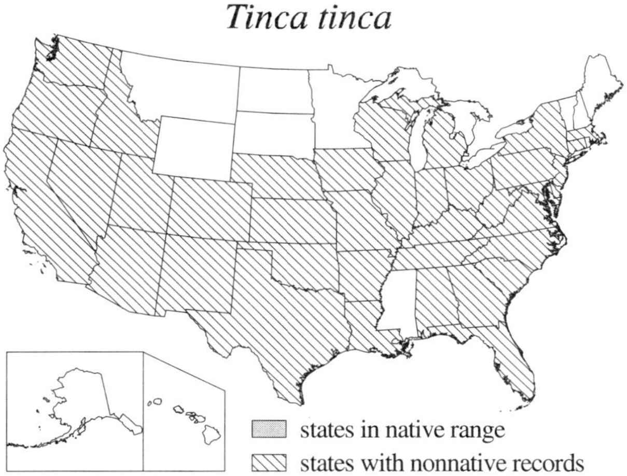

Tinca tinca (Linnaeus 1758)

Common Names: tench (green tench).

Identification: Distinguishing characteristics were given in Berg (1949), Scott and Crossman (1973), Muus and Dahlstrom (1978), Wheeler (1978), and Page and Burr (1991). Identification keys that include this species and photographs or illustrations were provided in a few published state fish books (e.g., Whitworth et al. 1968; Moyle 1976a; Wydoski and Whitney 1979; Woodling 1985; Bond 1994). A name used in some of the early literature for this species is *Tinca vulgaris*. Maximum size: 84 cm.

Native Range: Most of Europe, including the British Isles, and parts of western Asia (Berg 1949).

Nonindigenous Occurrences: States in which this species has been stocked or reported include **Alabama** (Baughman 1947); **Arizona** (Baughman 1947; Courtenay et al. 1991); **Arkansas** (Baughman 1947); **California** (Shapovalov 1944; Kimsey and Fisk 1964; Skinner 1972; Moyle 1976a; Wydoski and Whitney 1979; Courtenay et al. 1984, 1991; Dill and Cordone 1997); **Colorado** (Baughman 1947; Beckman 1952; Everhart and Seaman 1971; Ellis 1974; Wydoski and Whitney 1979; Woodling 1985; Zuckerman and Behnke 1986); **Connecticut** (Webster 1942; Connecticut State Board of Fisheries and Game 1959; Whitworth et al. 1968; Courtenay et al. 1986); **Delaware** (Ravenel 1898; Baughman 1947; Lee et al. 1976; Raasch and Altemus 1991); **Florida** (Baughman 1947); **Georgia** (Bean 1896; Ravenel 1898; Baughman 1947); **Idaho** (Smith 1896; Ravenel 1898; Lampman 1946; Baughman 1947; Linder 1963; Simpson and Wallace 1978; Wydoski and Whitney 1979; Idaho Fish and Game 1990, 1996); **Illinois** (Ravenel 1898; Baughman 1947); **Indiana** (Baughman 1947); **Iowa** (Bean 1896; Ravenel 1898; Baughman 1947); **Kansas** (Ravenel 1898; Baughman 1947); **Kentucky** (Baughman 1947); **Louisiana** (Baughman 1947); **Maryland** (Ferguson 1876; Anonymous 1892; Bean 1884, 1903; McDonald 1893; Ravenel 1898; Truitt et al. 1929; Baughman 1947; Schwartz 1963, 1964; Courtenay el al. 1991); **Massachusetts** (Baughman 1947); **Michigan** (Baughman 1947); **Missouri** (McDonald 1893; Evermann and Kendall 1895; Worth 1895; Bean 1896; Ravenel 1896, 1898; Baughman 1947; Pflieger 1971); **Nebraska** (Baughman 1947; Morris et al. 1974); **Nevada** (Smith 1896; Miller and Alcorn 1946; Baughman 1947; Deacon and Williams 1984); **New Jersey** (Ravenel 1896); **New Mexico** (Ravenel 1896; Baughman 1947; Koster 1957; Sublette et al. 1990); **New York** (Bean 1903; Baughman 1947; Courtenay et al. 1986; Schmidt 1986); **North Carolina** (Ravenel 1896; Baughman 1947); **Ohio** (Ravenel 1898; Baughman 1947; Trautman 1981); **Oklahoma** (McDonald 1893; Bean 1896; Baughman 1947; Hall 1956); **Oregon** (Smith 1896; Ravenel 1898; Chapman 1942; Lampman 1946; Baughman 1947; Wydoski and Whitney 1979; Courtenay et al. 1986); **Pennsylvania** (Ravenel 1898; Baughman 1947; Hendricks et al. 1979); **South Carolina** (Ravenel 1896, 1898; Baughman 1947); **Tennessee** (Ravenel 1898; Baughman 1947); **Texas** (Worth 1895; Bean 1896; Ravenel 1896, 1898; Baughman 1947;

Hubbs 1954; Howells 1992a); **Utah** (Tanner 1936; Baughman 1947); **Virginia** (Baird 1887; Ravenel 1896, 1898; Baughman 1947; Hocutt et al. 1986; Courtenay et al. 1991; Jenkins and Burkhead 1994); **Washington** (Smith 1896; Ravenel 1898; Chapman 1942; Lampman 1946; Baughman 1947; Gray and Dauble 1977; Wydoski and Whitney 1979; McPhail and Lindsey 1986; Killgore et al. 1991); **West Virginia** (Baughman 1947); and **Wisconsin** (Baughman 1947; Becker 1983).

Means of Introduction: This species was imported into North America from Germany by the U.S. Fish Commission in 1877 apparently for use as a food and sport fish (Baird 1879). The Commission apparently spent several years learning to culture tench, for it was not until well into the 1880s that the agency started to seriously distribute the species in the United States. According to Baughman (1947), the Commission planted more than 138,000 tench across North America during the period 1886 to 1896. By the end of that period, the Commission had provided tench to at least 36 different states. Shortly thereafter, the agency discontinued working with tench and turned over their hatchery ponds to the rearing of bass (Baughman 1947). The U.S. Fish Commission stocked tench into lakes and ponds in the Pacific states, including Idaho, Oregon, and Washington, in 1895 (Smith 1896). Additional introductions occurred in Washington when tench exhibited at the 1909 Worlds Fair, held in Seattle, were dumped into a large pond on the University of Washington campus. Some of these fish later were transferred to Lake Washington; the population eventually spread to Lake Union (Wydoski and Whitney 1979). Although most tench introductions were the result of intentional stockings, some introductions were the result of escape from holding facilities. In 1889, many of the fish, including an estimated 25 tench, held in federal ponds in Washington, D.C., escaped into the Potomac River during a flood (McDonald 1893). However, tench had been introduced to the Potomac River prior to that date. For instance, Smiley (1889) recorded the taking of tench from that river during March 1887. Evermann and Kendall (1895) reported the escape of tench from the Neosho fish hatchery into Spring Branch near Neosho in southwestern Missouri. Baughman (1947) discussed the escape of tench into the Olentangy River of Ohio after the banks of an artificial lake collapsed in 1898. Schwartz (1964) stated that tench had escaped from commercial ponds into creeks in Maryland. Tench were first brought to California in 1922. At that time, specimens obtained in Italy were illegally released into a private reservoir near Half Moon Bay, San Mateo County; the species was later spread to other California waters by ranchers (Shapovalov 1944; Dill and Cordone 1997).

Status: The tench has been documented for 38 states. Baughman (1947) indicated that this species was established in California, Colorado, Idaho, Washington, and apparently Oregon; he also quoted information of previously breeding populations in Maryland. In a more recent work, Page and Burr (1991) considered it established in California, Colorado, Connecticut, and Washington, and possibly Delaware, Maryland, and New York. Courtenay et al. (1991) believed it to be established in California, Colorado, Connecticut, Idaho, and Washington. Dill and Cordone (1997) found evidence that the tench is still established in California, but described its distribution as very limited. Although the data are more than a decade old, Woodling (1985) confirmed that the species continued to survive in several Colorado waters. Zuckerman and Benhke (1986) also noted that the species persisted in Colorado, but they indicated populations in the San Luis Valley had declined dramatically in recent years. Whitworth (1996) reported that no specimens had been obtained in Connecticut for over fifteen years. There is recent literature indicating the species is still established in Washington and, probably, Idaho. For instance, in their analysis of Northwest drainages, McPhail and Lindsey (1986) listed tench as present in the upper, middle, and lower Columbia River system and in the Chehalis river system. Although Idaho was not listed by Page and Burr (1991), the continued presence or established status of tench in that state has been documented by several authors (i.e., Simpson and Wallace 1978; Courtenay et al. 1991, Idaho Fish and Game 1996). However, it likely has been extirpated in Delaware (contrary to Page and Burr 1991), where surveys since 1950 have failed to find the species (M. Raasch, personal communication). Koster (1957) stated that this species occurred in the middle Rio Grande Valley of New Mexico, and Bond (1973) described its distribution in Oregon as the Columbia River and probably the lower Willamette River. Lee et al. (1980 et seq.) also recorded tench as established in Oregon and New Mexico. However, in more recent works, Sublette et al. (1990) considered it extirpated from New Mexico, and Courtenay et al. (1986) believed that tench had been extirpated from both Oregon and New Mexico. In his revised treatment of Oregon fishes, Bond (1994) noted tench as introduced to the Columbia River and stated that the species was once present in lower Willamette River. The Willamette is a tributary of the Columbia River. Although it seems to be the case that tench are no longer present in Oregon, Bond (1973, 1994) did not definitively state that the Willamette River population was the only Columbia River site with tench known in that state. Its present status in both Maryland and New York is unclear. According to Baughman (1947) it was temporarily established in Maryland. Schwartz (1964) concluded that wild populations no longer exist in Maryland. In fact, he indicated that the last wild specimens known from the state were taken from the C & O Canal at Buzzard's Hole, Maryland, in 1911; however, Schwartz did note that some tench may exist in Catoctin and Huntington creeks, Frederick County, as a result of escape from commercial ponds where the fish were reared for sale to the "dime-store goldfish trade." Apparently Schwartz did not consider tench in the two creeks to be reproducing. However, Lee et al. (1984) noted that it may occur in isolated areas of Piedmont in Maryland. Rohde et al. (1994) did not mention tench in their recent work on fishes of Delaware, Maryland, and Virginia. Similarly, Smith (1985) did not mention tench in his treatment of New York fishes. Nevertheless, Schmidt (1986), in his analysis of fish distribution in 14 Atlantic coastal drainages, listed tench as introduced to lowland lacustrine habitats in the Delaware, Long Island, and Housatonic drainages. Shortly after it introduction, Smith and Bean (1899) indicated that the species was becoming common in the Potomac River. Much more recently, Hocutt et al. (1986) listed it as introduced to the Potomac River. However, Jenkins and Burkhead (1994) concluded that there was no evidence that the species persists in the Potomac drainage. Although the tench has maintained breeding populations in a few places, the frequent absence of this species in most fish samples suggests that the tench is no longer present in many areas where it had previously been introduced and, in some cases, was temporarily established (e.g., Hall 1956; Morris et al. 1974; Hendricks et al. 1979; Courtenay et al. 1986; Menhinick 1991). Ravenel (1896) reported that the U.S Fish Commission stocked 1,600 tench in the Musconetcong River in New Jersey. That is the only record of the species in that state. During the late 1800s, the U.S. Fish Commission distributed tench to many states. In many cases the Commission documented that tench were released into specific rivers or lakes. However, for a number of states the Commission simply noted that tench were distributed to "various applicants" without indicating whether or not these fish were ever released into open waters (Baughman 1947). Later data indicated that tench were found or collected in open waters of a few of these states (e.g., Arizona), but for ten states (i.e., Alabama, Arkansas, Florida, Kentucky, Louisiana, Massachusetts, Michigan, West Virginia, Wisconsin, and possibly Utah) the presence of this species in open waters has not been adequately confirmed. Nevertheless, it is conceivable that some of the tench delivered by the Commission to various applicants eventually found their way to open waters as a result of floods, dam breaks, or because of intentional releases. Such losses were frequent with common carp, another cyprinid widely distributed by the U.S. Fish Commission during the late 1800s (Smiley 1886). Tanner (1936) listed tench as one of the species introduced into Utah streams, but he provided no details.

Impact of Introduction: For the most part, unknown. In the 1940s this species was reported to be a nuisance because of high abundance in certain parts of Maryland and Idaho (Baughman 1947). The diet consists mainly of aquatic insect larvae and molluscs (Scott and Crossman 1973). Moyle (1976a) considered it a potential competitor for food with sport fishes and native cyprinids. In their discussion of tench introduced to Africa, de Moor and Bruton (1988) noted that the species is known to stir up bottom sediments, possibly affecting water quality, but not to the extent of common carp *Cyprinus carpio*.

Remarks: Baughman (1947) reviewed the history of tench introductions in North America. He also presented evidence suggesting the presence of centrarchids somehow prevented more widespread establishment of tench. Zuckerman and Behnke (1986) noticed that the decline of tench in Colorado coincided with the spread and establishment of the common carp. These authors also noted the occurrence of tench at two sites in Colorado at elevations greater than 2,850 meters. Shapovalov (1944) and Dill and Cordone (1997) reviewed the history of tench in California. In addition to the normal- or wild-colored tench, the U.S. Fish Commission distributed an orange-yellow or reddish variety, the golden tench, to various applicants in the United State during the late 1800s (Bean 1896). That genetic strain apparently was only distributed as an ornamental. There is no evidence that this ornamental variety was introduced to open waters. The golden tench is still used as an ornamental fish in European ponds (Scott and Crossman 1973; Muus and Dahlstrom 1978).

Voucher specimens: Colorado (USNM 055569), Maryland (USNM 31003, 37850, 30336, 27234), Missouri (USNM 073666), New Mexico (UMMZ 118218), Virginia (USNM 30336, 31003, 37850).

HYBRIDS

Ctenopharyngodon idella ×
Hypophthalmichthys nobilis

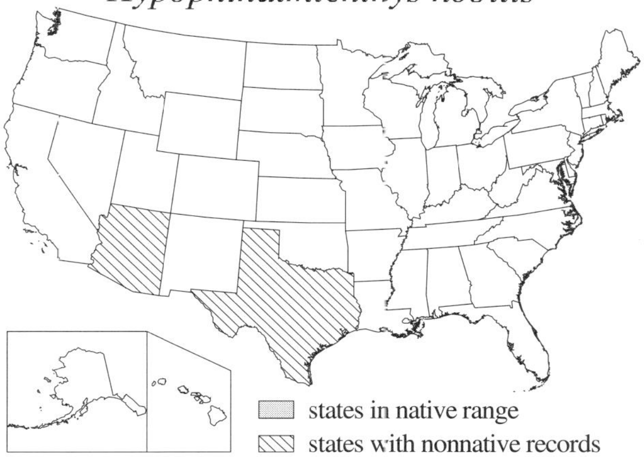

Ctenopharyngodon idella × *Hypophthalmichthys nobilis*

Common Name: grass carp × bighead carp.

Identification: Similar to pure grass carp but with many features intermediate between the two parent species. Howells (1992b) provided distinguishing features. Maximum size: Probably similar to grass carp, 75–125 cm.

Native Range: None; artificial hybrid.

Nonindigenous Occurrences: This hybrid has been reported from the Central Arizona Project (CAP) and interconnected Salt River Project (SRP) canals in **Arizona** (U.S. Fish and Wildlife Service 1994f). It was also sold and stocked in several locations in **Texas** in 1989 (Howells 1992a).

Means of Introduction: Intentionally stocked by the Texas Parks and Wildlife Department in 1989 (Howells 1992a).

Status: Reported in Texas.

Impact of Introduction: Unknown.

Remarks: This hybrid is believed, but not proven, to be sterile. As of 1990 it is illegal to possess this hybrid in Texas (Howells 1992a). From 1980 to 1984 the Florida Game and Fresh Water Fish Commission implemented a permit system for limited use of triploid hybrids in private waters (Thomas 1994). That information suggests that a few hybrids may have been introduced into open waters of Florida.

Hypophthalmichthys molitrix
× *H. nobilis*

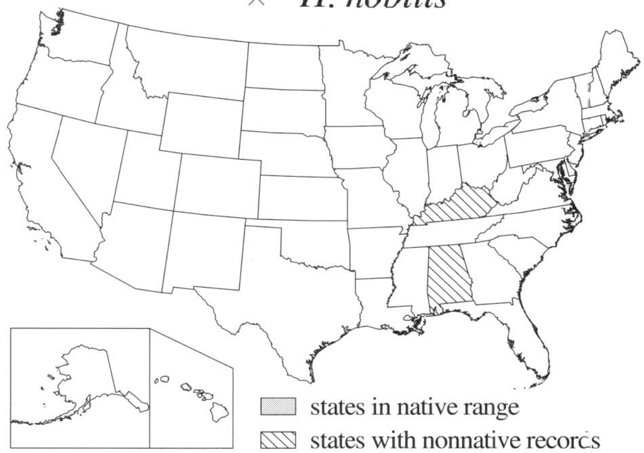

Hypophthalmichthys molitrix × *H. nobilis*

Common Name: silver carp × bighead carp.

Identification: Hybrid. Burr and Warren (1986). Maximum size: Probably similar to parent species: 71.6–??

Native Range: None; artificial hybrid.

Nonindigenous Occurrences: This hybrid has been collected in Yates Reservoir, **Alabama** (Hornsby and Pierson, personal communication). A single specimen was taken by a commercial fisherman from the Jonathan Creek arm of Kentucky Lake, **Kentucky** in 1985 (Burr and Warren 1986).

Means of Introduction: Intentionally stocked for food and to improve water quality by reducing phytoplankton.

Status: Reported from Alabama and Kentucky.

Impact of Introduction: Unknown.

Remarks: These two species are capable of hybridizing in the wild. Characteristics of hybrids are intermediate between the parents, but most closely resemble those of the bighead carp (Jennings 1988).

Catostomidae - Suckers

The family Catostomidae, whose members are known as suckers, contains 13 genera with 68 (Nelson 1994) to 70 (Gilbert 1998) species. Most species (63 or more) are distributed in North America north of Mexico, although two are found in Northeast Asia and China (Berra 1981; Page and Burr 1991; Nelson 1994). Suckers are small to large fishes with elongate to deep-bodied shapes, depending on species. Suckers are bottom dwellers and use their large, thick lips to feed on invertebrates. All but one species has a protrusible mouth. Because of their large size and abundance, suckers make up a large proportion of the biomass in many North American streams and lakes (Page and Burr 1991). The family is restricted to freshwater.

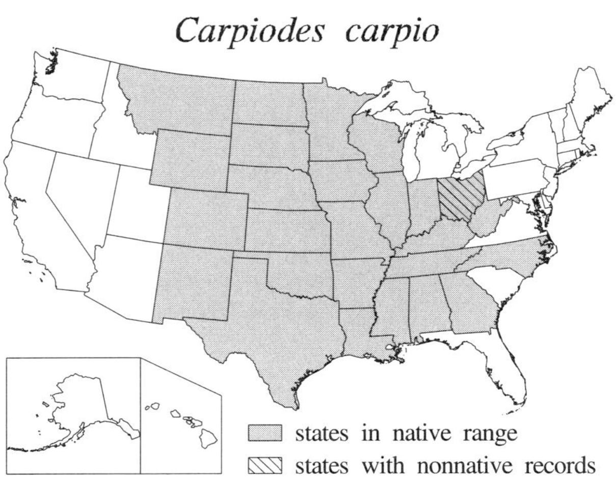

Carpiodes carpio (Rafinesque 1820)

Common Name: river carpsucker.

Identification: Becker (1983); Hubbs et al. (1991); Page and Burr (1991); Etnier and Starnes (1993). Maximum size: 64 cm.

Native Range: Mississippi River basin from Pennsylvania to Montana and south to Louisiana; Gulf Slope drainage from Calcasieu River, Louisiana, to Rio Grande, Texas and New Mexico. Also in Mexico (Page and Burr 1991).

Nonindigenous Occurrences: River carpsuckers were introduced into western Lake Erie and the lower Maumee River, **Ohio** (Lee et al. 1980 et seq.; Trautman 1981; Page and Burr 1991).

Means of Introduction: It is likely that individuals of this species were included in shipments of buffalofishes *Ictiobus* spp., which were intentionally stocked in Ohio in western Lake Erie between 1920 and 1930 (Lee et al. 1980 et seq.; Trautman 1981).

Status: Established in Ohio.

Impact of Introduction: Unknown.

Remarks: Possibly hybridizes with highfin carpsucker *Carpiodes velifer* and quillback *C. cyprinus* (Trautman 1981).

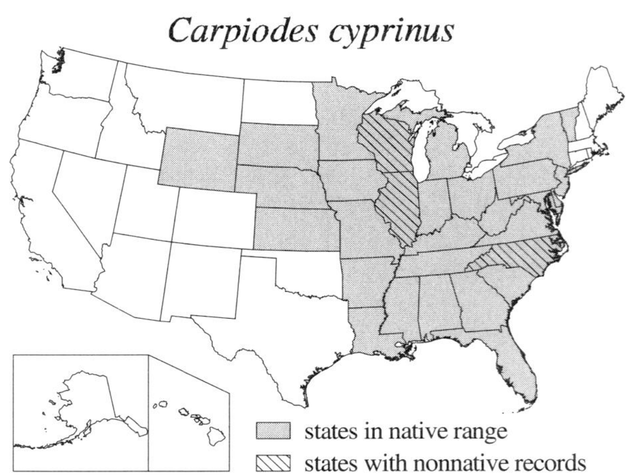

Carpiodes cyprinus (Lesueur 1817)

Common Name: quillback.

Identification: Becker (1983); Page and Burr (1991); Etnier and Starnes (1993); Jenkins and Burkhead (1994). Taxonomy of the genus Carpiodes on the lower Atlantic Slope is uncertain. Some of the forms resembling *C. cyprinus* and *C. velifer* may represent undescribed species (Gilbert, personal communication). Maximum size: 66 cm.

Native Range: Great Lakes and St. Lawrence River, Hudson Bay, and Mississippi River basins from Quebec to Alberta and south to Louisiana; Atlantic Slope drainages from Delaware River, New York, to Altamaha River, South Carolina (except apparently absent from Rappahannock and York drainages in Virginia, and Tar and Neuse drainages in North Carolina). Gulf Slope drainages from Apalachicola River, Florida and Georgia, to Pearl River, Louisiana (Page and Burr 1991).

Nonindigenous Occurrences: A quillback was collected in Jackson Park Harbor, Lake Michigan, at Chicago, **Illinois**, on 21 June 1989 (Savitz et al. 1989b). Although it is regarded as native, the quillback may have been introduced into the Roanoke, Broad, Catawba, Yadkin, and Neuse drainages of **North Carolina** (Menhinick 1991). Another introduction in Lake Michigan was at the extreme southwestern end of Green Bay in **Wisconsin** (Savitz et al. 1989b).

Means of Introduction: Unknown.

Status: Established in North Carolina. Reported from Illinois and Wisconsin.

Impact of Introduction: Unknown.

Remarks: Voucher specimen: Illinois (INHS 64805).

Carpiodes velifer

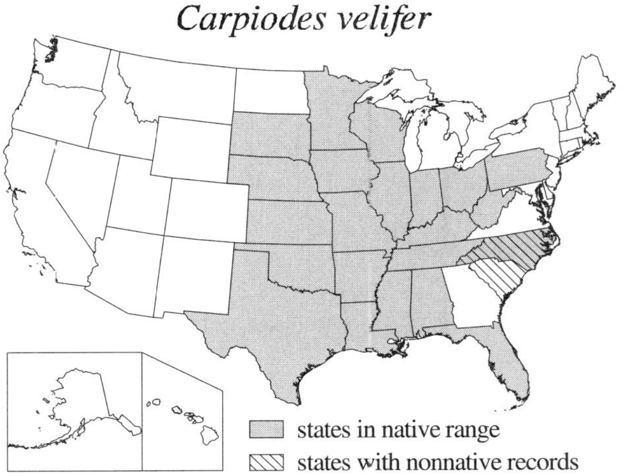

Carpiodes velifer (Rafinesque 1820)

Common Name: highfin carpsucker.

Identification: Becker (1983); Page and Burr (1991); Etnier and Starnes (1993). Taxonomy of the genus Carpiodes on the lower Atlantic Slope is uncertain. Some of the forms resembling *C. cyprinus* and *C. velifer* may represent undescribed species (Gilbert, personal communication). Maximum size: 50 cm.

Native Range: Lake Michigan drainage and Mississippi River basin from Pennsylvania to Minnesota and southeastern South Dakota, and south to Louisiana; Atlantic Slope in Cape Fear and Santee River drainages, North Carolina (Page and Burr 1991); Gulf Slope drainages from Texas east to the Choctawhatchee River, Alabama and Florida.

Nonindigenous Occurrences: The highfin carpsucker was probably introduced into the lower Cape Fear, Catawba, and Yadkin drainages of **North Carolina** (Menhinick 1991), and into the Pee Dee and Catawba drainages of **South Carolina** (Rohde et al. 1994; F. Rohde, personal communication).

Means of Introduction: Unknown.

Status: Established in North Carolina and South Carolina.

Impact of Introduction: Unknown.

Remarks: None.

Catostomus ardens

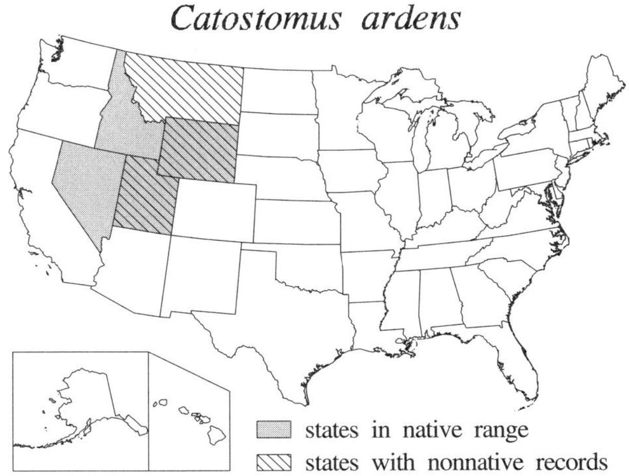

Catostomus ardens Jordan and Gilbert 1881

Common Name: Utah sucker.

Identification: Page and Burr (1991); Sigler and Sigler (1996). Maximum size: 65 cm.

Native Range: Lake Bonneville basin in Utah, Idaho, Wyoming and Nevada and the Snake River above Shoshone Falls (Sigler and Sigler 1987, 1996).

Nonindigenous Occurrences: A single specimen was collected from lower Red Rock Lake in Beaverhead County, Beaverhead drainage, **Montana** in 1951 (museum specimen). Also introduced to the Green, Strawberry, Duchesne, and Fremont rivers in the upper Colorado River basin in **Utah** (Tyus et al. 1982; Sigler and Sigler 1987, 1996), and to Fontanelle Reservoir on the Green River in **Wyoming** in or before the early 1980s (Tyus et al. 1982).

Means of Introduction: Probably introduced to the Colorado River drainage via bait bucket (Sigler and Sigler 1996). Means of introduction in Montana unknown.

Status: Established in the Colorado basin, Utah and Wyoming. Single specimen collected from Montana in 1951.

Impact of Introduction: Unknown.

Remarks: Voucher specimens: Montana (UMMZ 173861). Montana specimen identified by R.R. Miller and R.M. Bailey.

Catostomus catostomus

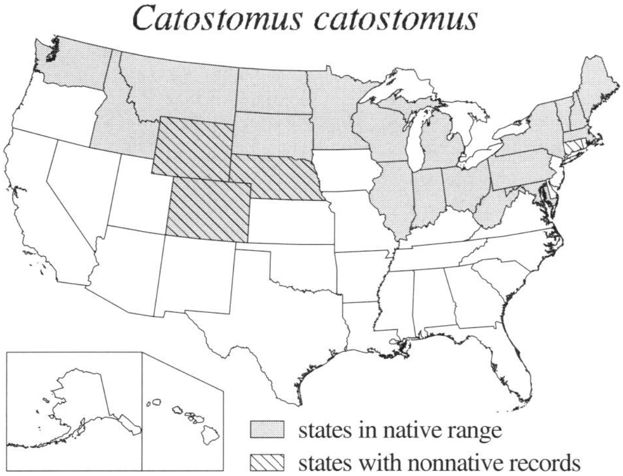

Catostomus catostomus (Forster 1773)

Common Name: longnose sucker.

Identification: Scott and Crossman (1973); Smith (1979); Smith (1985); Page and Burr (1991). Maximum size: 64 cm; some populations are dwarfed (Lee et al. 1980 et seq.).

Native Range: Atlantic, Arctic, and Pacific basins throughout most of Canada and Alaska; Atlantic Slope south to Delaware River drainage, New York; Great Lakes basin; upper Monongahela River drainage, Maryland and West Virginia; Missouri River drainage south to northeastern and central Colorado. Also in Arctic basin of eastern Siberia (Page and Burr 1991).

Nonindigenous Occurrences: The longnose sucker was introduced into the Rio Grande and Colorado drainages of **Colorado** in the 1860s and 1870s (Woodling 1985), the Housatonic drainage of **Connecticut** (Whitworth 1996), and into Big Sandy Creek in the Upper Green River drainage and Yellowstone Lake in **Wyoming** (Brown and Graham 1954; Baxter and Simon 1970; Lee et al. 1980 et seq.; Hubert 1994).

Means of Introduction: Bait bucket release and contamination of trout stock (Brown and Graham 1954; Woodling 1985). Beckman (1952) reported a recent intentional introduction of this species into the Colorado River drainage in Colorado. No method of introduction given for Connecticut. First found in that state in 1992 (Whitworth 1996).

Status: Established in Colorado and Wyoming. Collected in Connecticut.

Impact of Introduction: Unknown.

Remarks: Tyus et al. (1982) gave a distribution map of the this species in the upper Colorado basin.

Catostomus commersoni

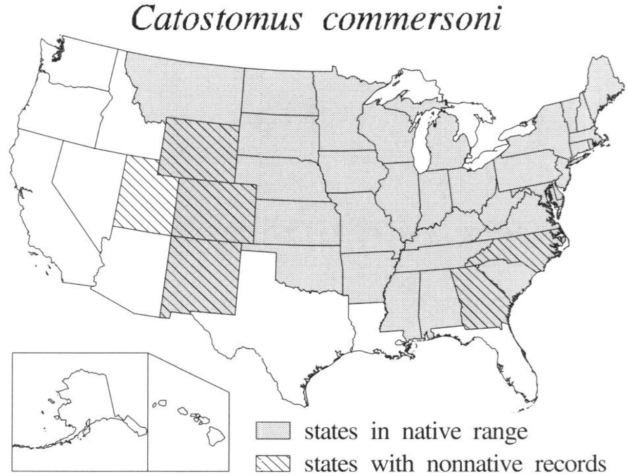

Catostomus commersoni (Lacepède 1803)

Common Name: white sucker.

Identification: Becker (1983); Page and Burr (1991); Etnier and Starnes (1993); Jenkins and Burkhead (1994). Maximum size: 64 cm.

Native Range: Atlantic, Arctic, Great Lakes, and Mississippi River basins from continental Newfoundland (i.e. Labrador) to Mackenzie River, Northwest Territories; south to the Tennessee River drainage, northern Alabama, and Arkansas River drainage, New Mexico; south on Atlantic Slope to Santee River drainage, South Carolina; upper Rio Grande drainage, New Mexico; Skeena and Fraser River drainages (Pacific Slope), British Columbia (Page and Burr 1991).

Nonindigenous Occurrences: The white sucker was introduced into the Colorado drainage in **Colorado** in the 1860s and 1870s (Vanicek et al. 1970; Tyus et al. 1982; Woodling 1985). It also now occurs in the Chattahoochee River, **Georgia** (Couch et al. 1995), Rio Grande (Sublette et al. 1990; Platania 1991), San Juan, Gila, and Rio Chama of **New Mexico** (Tyus et al. 1982; Sublette et al. 1990), **Utah**, in the Green and Colorado rivers (Vanicek et al. 1970; Tyus et al. 1982), and in the Green River drainage, **Wyoming** (Baxter and Simon 1970; Tyus et al. 1982; Page and Burr 1991; Hubert 1994). It was probably introduced in **North Carolina** into the Neuse and Cape Fear drainages (Menhinick 1991).

Means of Introduction: Bait bucket release and trout stock contamination in Colorado. Probably as bait bucket release in Utah, and Wyoming. Unknown means in Georgia, North Carolina, and New Mexico. Woodling (1985) reported the first introductions in the Colorado River took place in the 1860s and 1870s. Holden and Stalnaker (1975) reported later introductions: one in 1926, and another in 1938. They also assumed a population in the Yampa River was from a separate introduction.

Status: Established in all locations.

Impact of Introduction: In Colorado, the white sucker is replacing native suckers in the Rio Grande and Colorado river drainages (Walker 1993).

Remarks: Tyus et al. (1982) gave a distribution map of the this species in the upper Colorado basin.

Catostomus fumeiventris

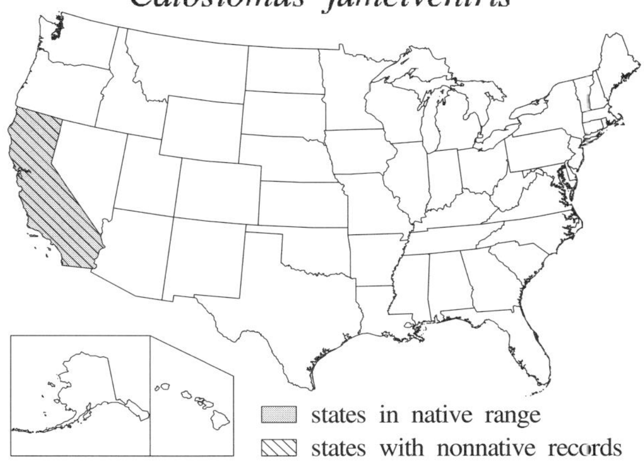

Catostomus fumeiventris Miller 1973

Common Name: Owens sucker.

Identification: Miller (1973); Moyle (1976a); Page and Burr (1991). Maximum size: 50 cm.

Native Range: Endemic to the Owens River, California (Page and Burr 1991).

Nonindigenous Occurrences: It has been introduced to June Lake in Mono Lake basin and the Santa Clara River system, **California** (Miller 1973; Moyle 1976a; Bell 1978; Swift et al. 1993). It was also stocked in a native fish sanctuary in the Owens Valley.

Means of Introduction: Apparently introduced to June Lake and to the Santa Clara River system as a result of a release of Owens River water from the Los Angeles Aqueduct (Moyle 1976a; Bell 1978). Introduced into the Owens River sanctuary as a conservation precautionary measure (Moyle 1976a).

Status: Established in parts of California outside its native range, including populations in June Lake and those in the Owens Valley sanctuary (Moyle 1976a). The status of the introduced Santa Clara River population is uncertain (Moyle 1976a; Bell 1978).

Impact of Introduction: Possible hybridization with the Santa Ana sucker *Catostomus santaanae* in the Santa Clara River system (Moyle 1976a; Bell 1978).

Remarks: None.

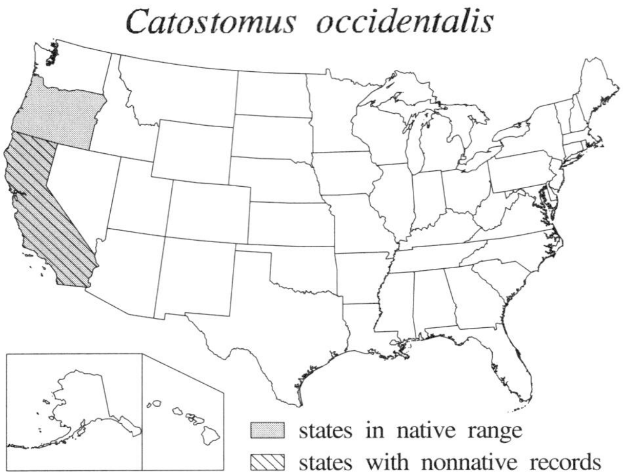

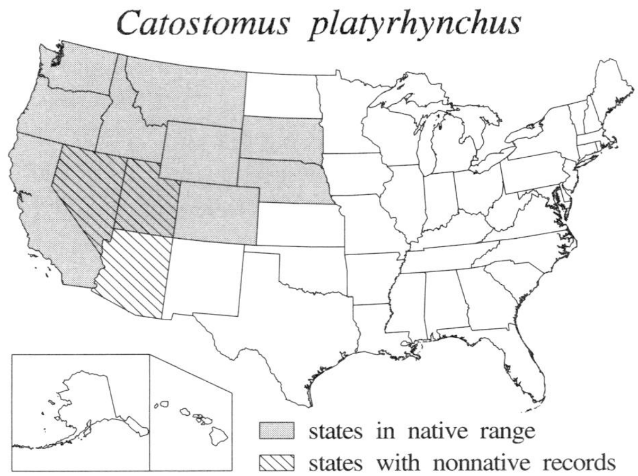

Catostomus occidentalis Ayres 1854

Common Name: Sacramento sucker.

Identification: Moyle (1976a); Page and Burr (1991). Maximum size: 60 cm.

Native Range: Pacific Slope from Mad River, northern California, to Salinas River, central California; throughout the Sacramento–San Joaquin drainage, southern Oregon and California; south in endorheic basins of central California to Kern River (Page and Burr 1991).

Nonindigenous Occurrences: It has been introduced into the upper Klamath River in **California** (Moyle 1976a).

Means of Introduction: Unknown.

Status: Reported from California.

Impact of Introduction: These suckers may compete with trout for food and space, and their feeding activities may disturb trout nests (Burns 1966c).

Remarks: None.

Catostomus platyrhynchus (Cope 1874)

Common Names: mountain sucker (northern mountain sucker, plains mountain sucker).

Identification: Moyle (1976a); Wydoski and Whitney (1979); Sigler and Sigler (1987); Page and Burr (1991). Maximum size: 25 cm.

Native Range: Western Canada and United States from Saskatchewan River system (Hudson Bay basin), Saskatchewan and Alberta, and Fraser River drainage, British Columbia, south through upper Missouri and Colorado River drainages to southern Colorado and Utah, and through Columbia River drainage, Oregon; Lahontan basin, Oregon, Nevada, and California; upper Sacramento River system, northern California (Page and Burr 1991).

Nonindigenous Occurrences: This species was documented as introduced in the Colorado River in **Arizona** and **Nevada** in the 1940s (La Rivers 1962; Rinne 1995). Its occurrence in Spring Valley, Nevada, was also attributed to an introduction (Hubbs et al. 1974). The species was introduced into the Strawberry (1951), Duchesne, and Price river systems, **Utah** (Sigler and Miller 1963).

Means of Introduction: Introduced via bait bucket into Scofield Reservoir on the Price River, Utah, and established there by 1953. Accidentally stocked in Nevada. Hubbs et al. (1974) concluded that the sucker was brought into Spring Valley as bait or forage.

Status: Established in all locations.

Impact of Introduction: Establishment of mountain suckers in the Price River has led to hybridization with the bluehead sucker *Catostomus discobolus* (formerly *Pantosteus delphinus*) (Sigler and Miller 1963).

Remarks: None.

Catostomus plebeius

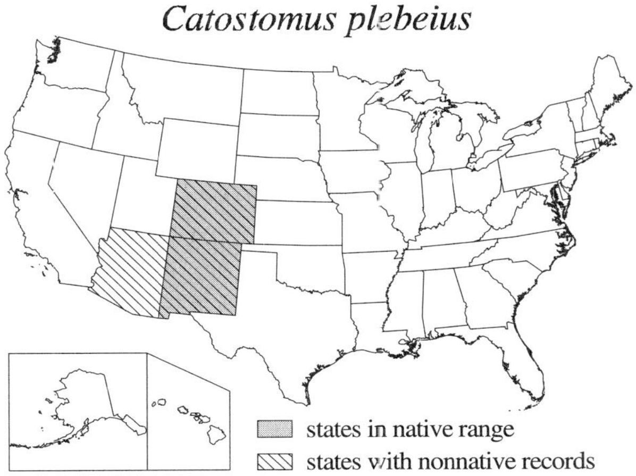

Catostomus plebeius Baird and Girard 1854

Common Name: Rio Grande sucker.

Identification: Sublette et al. (1990); Page and Burr (1991). Maximum size: 20 cm.

Native Range: Upper Rio Grande drainage in southern Colorado and New Mexico, south through endorheic basins and headwaters of Pacific Slope drainages to Durango and Zacatecas, Mexico (Page and Burr 1991).

Nonindigenous Occurrences: The Rio Grande sucker has been introduced into the San Francisco drainage in both **Arizona** and **New Mexico** (Minckley 1973; Lee et al. 1980 et seq.). It has also been introduced into the Rio Hondo along its headwater tributary streams in the Pecos drainage (Sublette et al. 1990), and Sapello Creek, a tributary of the Gila River (Buth and Crabtree 1985; Sublette et al. 1990) in New Mexico. The Rio Grande sucker may have been introduced into the Platte drainage in **Colorado** (Propst and Carlson 1986).

Means of Introduction: Introduced via bait bucket into the Rio Hondo and its tributaries. It is uncertain whether the Sapello Creek population originated naturally from stream capture via the Mimbres River or was introduced. Accidental introduction into the San Francisco River in the 1940s (Rinne 1995).

Status: Established in all locations in Arizona and New Mexico. Apparently extirpated in the Platte drainage, Colorado.

Impact of Introduction: Unknown.

Remarks: Smith (1966) believed that an earlier record of *Catostomus platyrhnchus* reported from the Platte River drainage was actually *Catostomus plebeius*. According to Propst and Carlson (1986), *C. plebeius* is not native to the Platte River drainage and, therefore, they stated that the reported specimen probably represents an introduction.

Catostomus santaanae

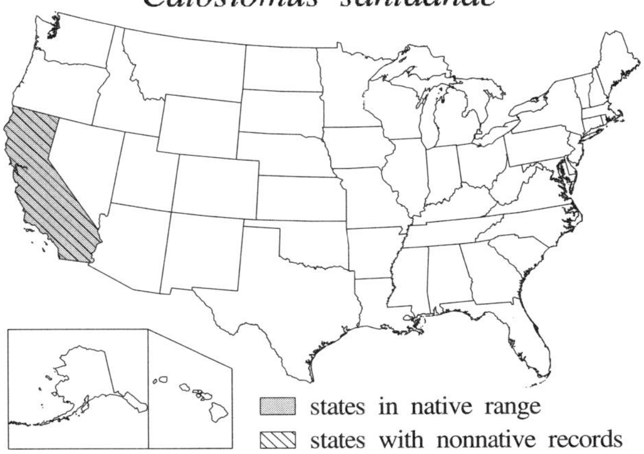

Catostomus santaanae (Snyder 1908)

Common Name: Santa Ana sucker.

Identification: Moyle (1976a); Page and Burr (1991). Maximum size: 25 cm.

Native Range: Los Angeles, San Gabriel, and Santa Ana River drainages, southern California (Page and Burr 1991).

Nonindigenous Occurrences: This species was introduced in **California** into the Santa Clara River in the Los Angeles basin by 1934 (Miller 1968; Bell 1978; Swift et al. 1993) and the St. Maria–St. Inez drainage some time before 1976 (Moyle 1976a). Two individuals of this sucker were introduced into River Springs in Adobe Valley, Mono County, California, in 1940 (Miller 1968; Swift et al. 1993).

Means of Introduction: The two suckers introduced into River Springs, California, were accidentally introduced from the San Gabriel River during the transplant of speckled dace *Rhinichthys osculus* (Miller 1968). The means of introduction in the other instances are unknown.

Status: Established in parts of California outside of its native range. Later visits to River Springs indicated the species did not survive at that site (Miller 1968).

Impact of Introduction: Unknown.

Remarks: The Santa Ana sucker is a candidate for Federal listing as an endangered or threatened species (U.S. Fish and Wildlife Service 1997).

Catostomus sp. (undescribed)

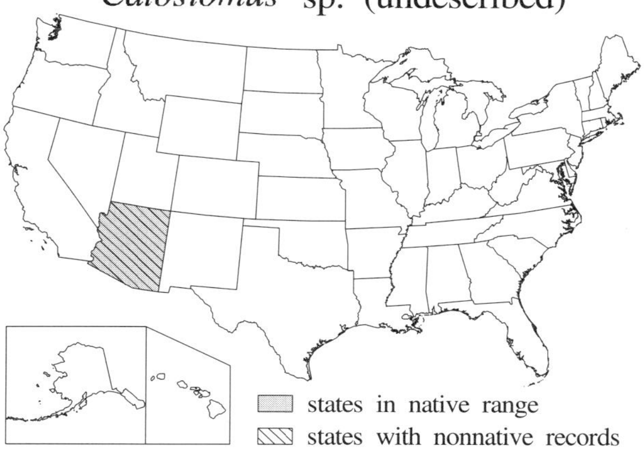

Catostomus sp. (undescribed)

Common Name: Little Colorado River sucker.

Identification: Minckley (1973); Page and Burr (1991). Maximum size: 50 cm.

Native Range: Upper Little Colorado River system, Arizona (Page and Burr 1991).

Nonindigenous Occurrences: This sucker has been introduced into the Salt River in **Arizona** (Minckley 1973; Page and Burr 1991).

Means of Introduction: Unknown.

Status: Established in Arizona.

Impact of Introduction: Unknown.

Remarks: None.

Catostomus tahoensis

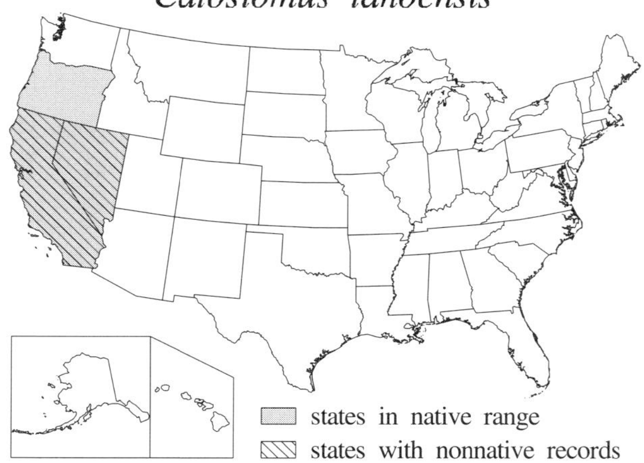

Catostomus tahoensis Gill and Jordan 1878

Common Name: Tahoe sucker.

Identification: Moyle (1976a); Sigler and Sigler (1987); Page and Burr (1991). Maximum size: 61 cm.

Native Range: Lahontan basin, southeastern Oregon, Nevada, and northeastern California (Page and Burr 1991).

Nonindigenous Occurrences: The Tahoe sucker was collected in 1950 from the headwaters of the Sacramento drainage in **Nevada** (La Rivers 1962) and in the upper reaches of the Feather and Rubicon rivers in **California** (Moyle 1976a). It may have been introduced into the Central Valley drainage, California (Moyle 1976a).

Means of Introduction: The source of these fish is either transfer by anglers or natural stream capture.

Status: Reported from the Sacramento drainage (Nevada) and the Feather and Rubicon rivers (California) in 1950, but more recent records are lacking. Established in the Central Valley, California.

Impact of Introduction: Unknown.

Remarks: None.

Erimyzon sucetta

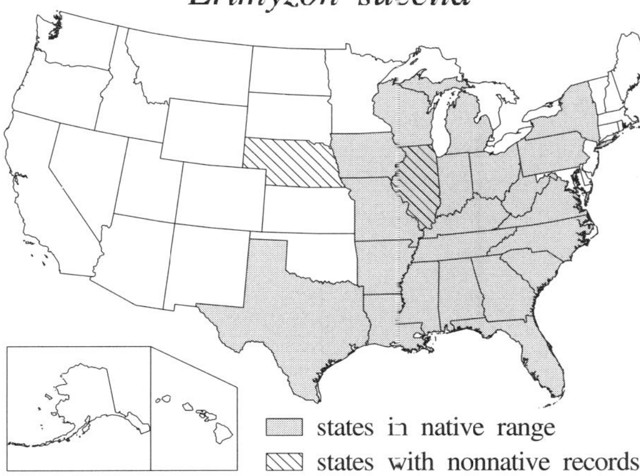

Erimyzon sucetta (Lacepède 1803)

Common Name: lake chubsucker.

Identification: Becker (1983); Smith (1985); Page and Burr (1991); Etnier and Starnes (1993). Maximum size: 41 cm.

Native Range: Great Lakes and Mississippi River basin lowlands from southern Ontario to Wisconsin and south to Gulf of Mexico; Atlantic Slope from southern Virginia to southern Florida; Gulf Slope drainage from Charlotte Harbor, Florida, to Guadalupe River, Texas (Page and Burr 1991).

Nonindigenous Occurrences: Lake chubsuckers have been introduced in four locations in southern **Illinois**. Specimens were collected from a stripmine pond in Perry County in the 1950s (Gunning and Lewis 1956), from LaRue–Pine Hills Swamp, Union County, in 1952, from Oakwood Bottoms, Jackson County, in 1978, and from Yellowbank Slough, Gallatin County in 1987 (Burr et al. 1988). This species was also stocked in **Nebraska** from 1979 to 1981 (Bouc 1987).

Means of Introduction: Intentionally stocked as a forage fish in Nebraska. Unknown means in Illinois.

Status: Established in Illinois and Nebraska. The population in Gallatin County, Illinois, was known to be reproducing in 1988 (Burr et al. 1988).

Impact of Introduction: Unknown.

Remarks: Earlier records of this species in southern Illinois are based on creek chubsuckers *Erimyzon oblongus* (Burr et al. 1988).

Voucher specimens: Illinois (SIUC 8267, 8268, 15263).

Hypentelium etowanum

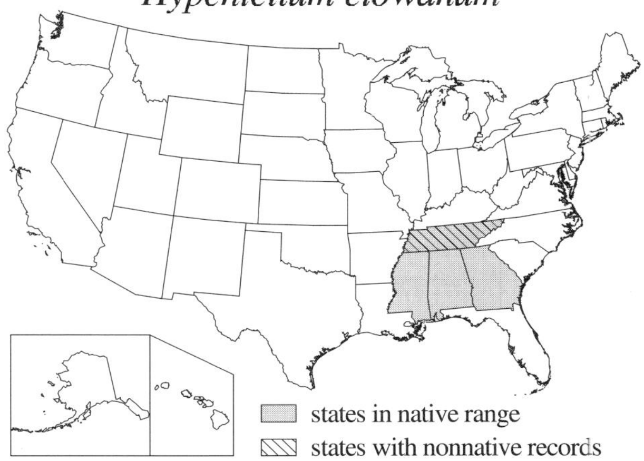

Hypentelium etowanum (Jordan 1877)

Common Name: Alabama hog sucker.

Identification: Page and Burr (1991); Etnier and Starnes (1993); Mettee et al. (1996). Maximum size: 23 cm.

Native Range: Chattahoochee River and Mobile Bay drainages in Georgia, Alabama, Mississippi, and southeastern Tennessee (Page and Burr 1991).

Nonindigenous Occurrences: Introduced into Baker Creek, Polk County, in the Ocoee system of the Tennessee drainage in **Tennessee** (Starnes and Etnier 1986; Page and Burr 1991; Etnier and Starnes 1993).

Means of Introduction: "Inadvertent" (Etnier and Starnes 1993).

Status: Established in Tennessee.

Impact of Introduction: Unknown.

Remarks: None.

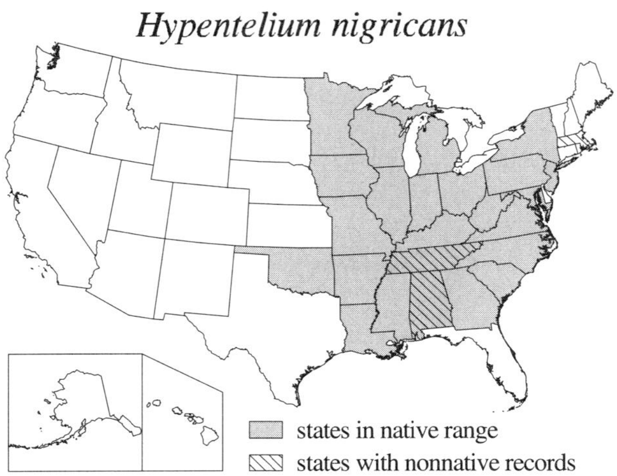

Hypentelium nigricans (Lesueur 1817)

Common Name: northern hog sucker.

Identification: Becker (1983); Page and Burr (1991); Etnier and Starnes (1993); Jenkins and Burkhead (1994). Maximum size: 61 cm.

Native Range: Great Lakes, Hudson Bay (Red River), and Mississippi River basins from New York and southern Ontario to Minnesota, and south to northern Alabama, southern Arkansas, and eastern Louisiana; Atlantic Slope drainages from Mohawk-Hudson River, New York, to Altamaha River, northern Georgia; Gulf Slope drainages from Pascagoula River, Mississippi, to Comite River, Louisiana (Page and Burr 1991).

Nonindigenous Occurrences: Swift et al. (1986) listed this species as introduced into the Black Warrior drainage in **Alabama**. One individual was collected from Hadley in a tributary of the Connecticut River, **Massachusetts**, in 1953 (Hartel 1992; Hartel et al. 1996). Swift et al. (1986) and Walters (1997) reported this species as introduced into the upper Coosa drainage (Conasauga system) in **Tennessee**.

Means of Introduction: Probably introduced via bait bucket in Massachusetts (Hartel 1992; Hartel et al. 1996).

Status: Reported from Alabama, Georgia, Massachusetts, and Tennessee.

Impact of Introduction: Unknown.

Remarks: Etnier and Starnes (1993) and Mettee et al. (1996) mapped the locations of the collections in Tennessee, but neither work lists the species as introduced. Even though Swift et al. (1986) listed this species as introduced into the Black Warrior drainage in Alabama; neither Boschung (1992) nor Mettee et al. (1996) mentioned it in that drainage. Because of its similarity to *H. etowanum*, specimens south of the Tennessee drainage should be re-examined. If they truly are *H. nigricans*, their presence in these areas is likely a result of introduction (Gilbert, personal communication).

The northern hogsucker may have been introduced into the Chattahoochee drainage where it was recently collected (Lee et al. 1980 et seq.; Swift et al. 1986; Page and Burr 1991). The collection presumably is in the Georgia portion since no Alabama locations were depicted in Mettee et al. (1996). Yerger (1977) did not list this species in the Chattahoochee drainage.

Voucher specimen: Massachusetts (CU 44587).

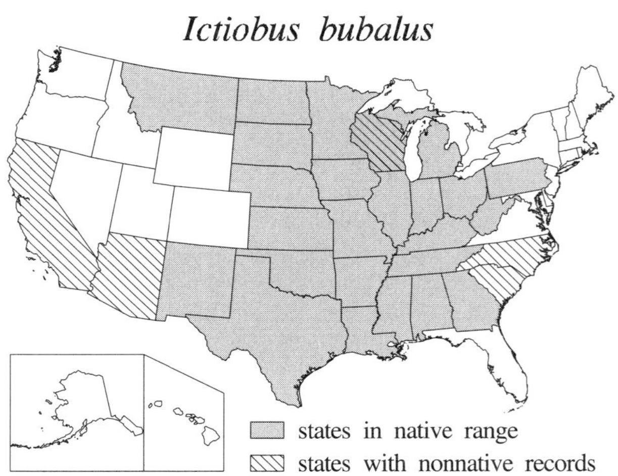

Ictiobus bubalus (Rafinesque 1818)

Common Name: smallmouth buffalo.

Identification: Becker (1983); Hubbs et al. (1991); Page and Burr (1991); Etnier and Starnes (1993). Maximum size: 78 cm.

Native Range: Lake Michigan drainage and Mississippi River basin from Pennsylvania and Michigan to Montana and south to Gulf of Mexico; Gulf Slope drainages from Mobile Bay, Alabama, to Rio Grande, Texas and New Mexico. Also in Mexico (Page and Burr 1991).

Nonindigenous Occurrences: Smallmouth buffalo were stocked in Canyon, Saguaro, and Apache lakes in **Arizona** (Minckley 1973). They were introduced to unspecified area(s) of **California** (Hubbs et al. 1979). They were probably introduced into the Yadkin, Catawba, and Neuse drainages of **North Carolina** (Hocutt et al. 1986; Menhinick 1991), and to the Yadkin and Santee drainages in **South Carolina** (Hocutt et al. 1986; Rohde et al. 1994). Specimens have been taken from the Big Lake and Island Lake region in Vilas County, **Wisconsin** (Becker 1983).

Means of Introduction: Accidental introduction in Arizona in 1918 with bigmouth buffalo *I. cyprinellus* (Minckley 1973; Rinne 1995); unknown means in California, North Carolina, and South Carolina. Stocked in Wisconsin during fish rescue operations from the Mississippi River in the 1930s (Becker 1983).

Status: Unknown in most sites in Arizona. Established in Apache Lake, Arizona, and in North and South Carolina. Extirpated in California. Unknown in Wisconsin; still present in the 1960s (Becker 1983).

Impact of Introduction: Unknown.

Remarks: In the early 1900s all three species of buffalofishes were stocked; *I. bubalus*, *I. cyprinellus*, and *I. velifer* (Leach 1921, 1923). However, when the stockings were reported they were lumped together as "buffalofish" and it is not possible to determine which species were planted. Stocking of buffalofishes occurred outside their native ranges in Lake Erie in Ohio, the Pee Dee and Catawba drainages in North Carolina, and in unknown locations in Massachusetts (Leach 1921, 1923).

Ictiobus cyprinellus

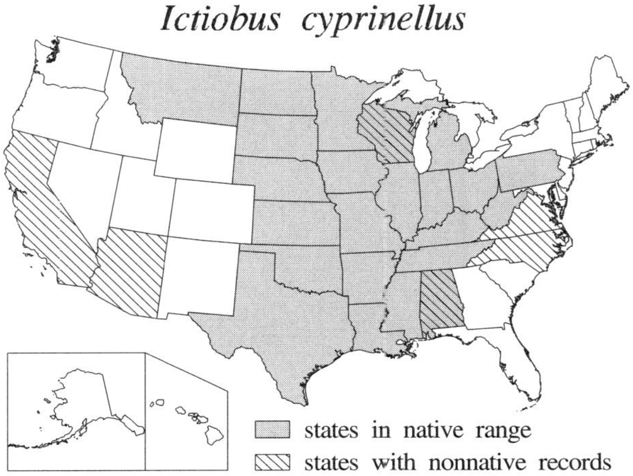

Ictiobus cyprinellus (Valenciennes 1844)

Common Name: bigmouth buffalo.

Identification: Becker (1983); Hubbs et al. (1991); Page and Burr (1991); Etnier and Starnes (1993). Maximum size: 100 cm.

Native Range: Hudson Bay (Nelson River drainage), lower Great Lakes, and Mississippi River basins from Ontario to Saskatchewan and Montana, and south to Louisiana (Page and Burr 1991).

Nonindigenous Occurrences: The bigmouth buffalo was stocked in Perry Lake, Perry County, **Alabama** (Mettee et al. 1996). It was also stocked in **Arizona** in the four in-line reservoirs of the Salt River west of Phoenix in 1918, and is now established. These fish were also stocked in the Colorado River at Yuma (Minckley 1973). In **California**, populations were discovered in the Los Angeles aqueduct system in the late 1940s. Although the populations declined in the 1950s and are no longer commercially harvested, they are still established (Evans 1950; Moyle 1976a). In **North Carolina**, these fish were introduced to the Yadkin and Catawba drainages (Menhinick 1991). Bigmouth buffalo may have escaped from a Virginia Department of Game and Inland Fish hatchery and entered the York drainage, **Virginia** (Jenkins and Burkhead 1994). There are several published reports of this species from the Lake Michigan drainage in **Wisconsin**; however, there are no voucher specimens and the one photograph that is available has been identified by experts as this genus, but the species identification is uncertain (Becker 1983). Another collection of this species in Wisconsin is of a single fish from Big Lake in Vilas County. It is supported by a voucher specimen (Becker 1983).

Means of Introduction: Intentional, authorized stocking for sport fishing in Arizona in 1918 (Minckley 1973); unknown in North Carolina. It is speculated that commercial fishermen transplanted this species from Arizona to California to provide a source closer to the Los Angeles Fish Market (Moyle 1976a). Escaped from an aquaculture facility in Virginia. Population in Alabama was intentionally stocked by a federal fish hatchery during studies (Mettee et al. 1996). The introduction in Big Lake, Wisconsin, is probably a result of a transplant associated with fish rescue operations from the Mississippi River in the 1930s (Becker 1983). The Lake Michigan drainage records may be the result of movement through the Wisconsin-Fox Canal.

Status: Reported from Virginia, but not seen since escape. Established in all other areas with the possible exception of Yuma, Arizona and Perry Lake, Alabama. A few large individuals were collected from Perry Lake, Alabama, in 1992, however, they are probably not reproducing (Mettee et al. 1996).

Impact of Introduction: This species may compete with native minnows and suckers, as well as with juvenile sport fishes, for food and space (Moyle 1976a).

Remarks: In the early 1900s all three species of buffalofishes were stocked; *I. bubalus*, *I. cyprinellus*, and *I. velifer* (Leach 1921, 1923). However, when the stockings were reported they were lumped together as "buffalofish" and it is not possible to determine which species were planted. Stocking of buffalofishes occurred outside their native ranges in Lake Erie in Ohio, the Pee Dee and Catawba drainages in North Carolina, and in unknown locations in Massachusetts (Leach 1921, 1923).

Ictiobus niger

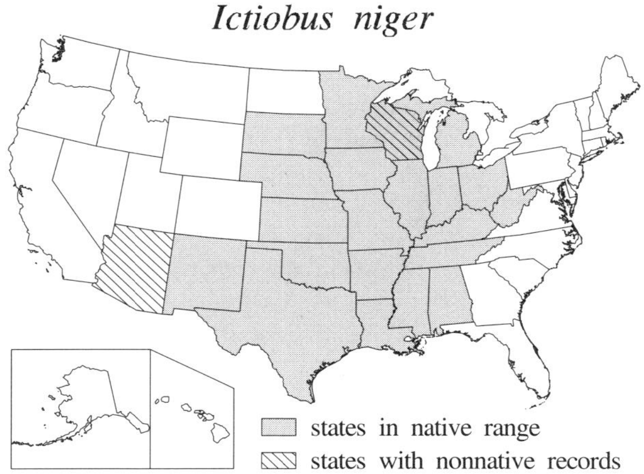

Ictiobus niger (Rafinesque 1819)

Common Name: black buffalo.

Identification: Becker (1983); Hubbs et al. (1991); Page and Burr (1991); Etnier and Starnes (1993). Maximum size: 93 cm.

Native Range: Lower Great Lakes and Mississippi River basins from Michigan and Ohio to South Dakota and south to Louisiana; on Gulf Slope in Sabine Lake, Brazos River, and Rio Grande drainages in Louisiana, Texas, New Mexico, and Mexico (Page and Burr 1991).

Nonindigenous Occurrences: The black buffalo has been introduced to several **Arizona** lakes, including Saguaro Lake, Canyon Lake, Apache Lake, and Roosevelt Reservoir (Minckley 1973). This species was reported from Lac La Belle, Waukesha County, **Wisconsin**, prior to 1927 (Becker 1983). Underhill (1986) reported it as introduced into Lake Michigan and Lake Huron drainages. Presumably the Lake Huron record refers to the Lac La Belle collection in Wisconsin.

Means of Introduction: Accidental introduction in Arizona in 1918 as stock contamination with bigmouth buffalo *I. cyprinellus* (Minckley 1973; Rinne 1995). Wisconsin introduction likely the result of transplant associated with fish rescue operations from the Mississippi River in the 1930s (Becker 1983).

Status: The status of the Saguaro Lake and Canyon Lake populations is unknown. They are established in Apache Lake, and were extirpated from Roosevelt Reservoir by a drought (Minckley 1973). Extirpated from Lac La Belle, Wisconsin (Becker 1983).

Impact of Introduction: Unknown.

Remarks: In the early 1900s all three species of buffalofishes were stocked; *I. bubalus*, *I. cyprinellus*, and *I. velifer* (Leach 1921, 1923). However, when the stockings were reported they were lumped together as "buffalofish" and it is not possible to determine which species were planted. Stocking of buffalofishes occurred outside their native ranges in Lake Erie in Ohio, the Pee Dee and Catawba drainages in North Carolina, and in unknown locations in Massachusetts (Leach 1921, 1923).

Minytrema melanops

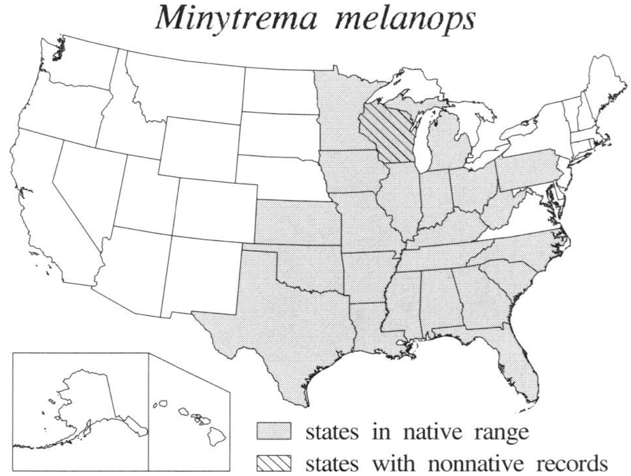

Minytrema melanops (Rafinesque 1820)

Common Name: spotted sucker.

Identification: Becker (1983); Page and Burr (1991); Etnier and Starnes (1993). Maximum size: 50 cm.

Native Range: Lower Great Lakes and Mississippi River basins from Pennsylvania to Minnesota and south to Gulf; Atlantic and Gulf Slope drainages from Cape Fear River, North Carolina, to Colorado River, Texas. Absent from most of peninsular Florida (Page and Burr 1991).

Nonindigenous Occurrences: A single fish collected from Lake Wingra, Dane County, **Wisconsin**, in 1944, was introduced (Becker 1983).

Means of Introduction: This fish was likely stocked in conjunction with fish rescue and transfer efforts from the Mississippi River in the 1930s (Becker 1983).

Status: Reported from Wisconsin.

Impact of Introduction: Unknown.

Remarks: None.

Moxostoma erythrurum

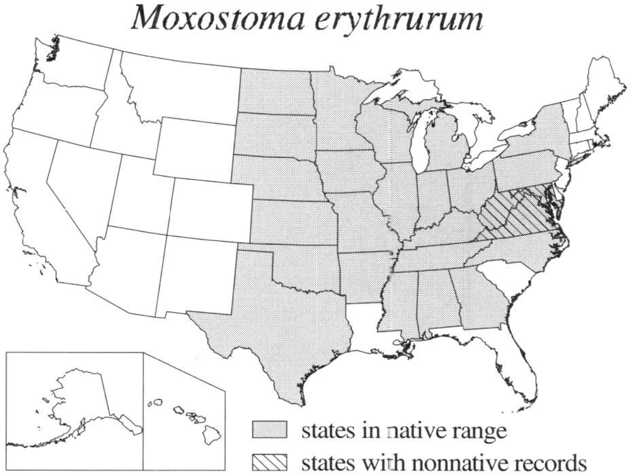

Moxostoma erythrurum (Rafinesque 1818)

Common Names: golden redhorse.

Identification: Becker (1983); Page and Burr (1991); Etnier and Starnes (1993); Jenkins and Burkhead (1994). Maximum size: 78 cm.

Native Range: Great Lakes, Hudson Bay (Red River), and Mississippi River basins from New York and southern Ontario to North Dakota, and south to northern Alabama and southern Oklahoma, with an isolated population in southwestern Mississippi; Atlantic Slope from Potomac River, Maryland, to Roanoke River, North Carolina (absent in Rappahannock and York River drainages), and Mobile Bay drainage in Georgia, Alabama, and southeastern Tennessee (Page and Burr 1991).

Nonindigenous Occurrences: Golden redhorse have been introduced into the Potomac River in **Maryland** and **Virginia** (Hocutt et al. 1986; Jenkins and Burkhead 1994), and into the South Branch Potomac drainage and the Gauley system, of the New drainage, **West Virginia** (Jenkins and Burkhead 1994; Stauffer et al. 1995).

Means of Introduction: Unknown. First appeared in the South Branch Potomac, West Virginia, in two tributaries in 1953. Not collected again until 1971. The first record from the Gauley system was in 1976 (Jenkins and Burkhead 1994).

Status: Established in Maryland, Virginia, and West Virginia.

Impact of Introduction: Unknown.

Remarks: None.

Moxostoma macrolepidotum

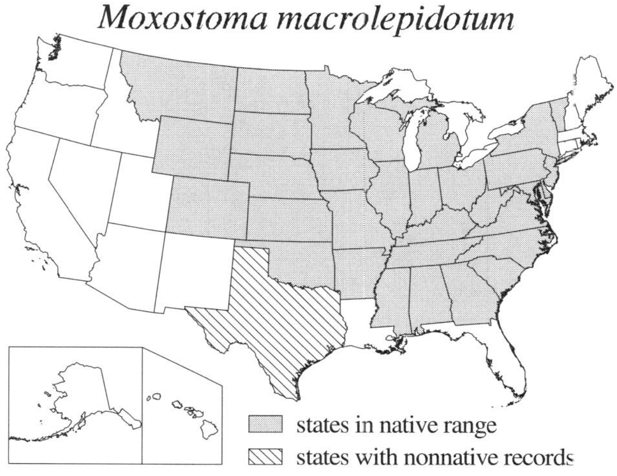

Moxostoma macrolepidotum (Lesueur 1817)

Common Name: shorthead redhorse.

Identification: Page and Burr (1991); Jenkins and Burkhead (1994). Maximum size: 75 cm.

Native Range: Great Lakes–St. Lawrence River, Hudson Bay, and Mississippi River basins from Quebec to Alberta and south to northern Alabama and Oklahoma; Atlantic Slope drainages from Hudson River, New York to Santee River, South Carolina (Page and Burr 1991).

Nonindigenous Occurrences: A single juvenile fish was collected in the Red River on the Oklahoma/Texas border, below the Lake Texoma dam in Grayson County, **Texas**, in 1961 (Riggs and Moore 1963; Cross et al. 1986).

Means of Introduction: Unknown.

Status: Reported from Texas.

Impact of Introduction: Unknown.

Remarks: Page and Burr (1991) refer to an isolated "population" in southern Oklahoma, implying the species is established there. Miller and Robison (1973) reported a single collection from the Red River basin but mapped it on the Oklahoma side of the river.

Scartomyzon cervinus

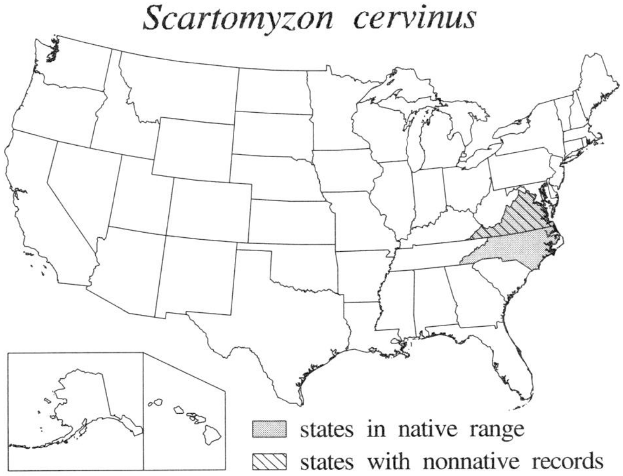

Scartomyzon cervinus (Cope 1868)

Common Name: black jumprock.

Identification: Page and Burr (1991); Jenkins and Burkhead (1994). The genus was revised by Smith (1992). *Scartomyzon* was a subgenus of *Moxostoma*, now elevated to genus. Synonym is *Moxostoma cervinum*. Maximum size: 19 cm.

Native Range: Atlantic Slope drainages from Chowan drainage, Virginia, to Neuse River, North Carolina (Page and Burr 1991; Jenkins and Burkhead 1994).

Nonindigenous Occurrences: First records for the James River drainage are from 1951 just above the mouth of Craig Creek, **Virginia** (Lee et al. 1980 et seq.; Jenkins and Burkhead 1994). The species spread rapidly within and from Craig Creek in the following years. Jenkins and Burkhead (1994) discussed the dispersal from the initial site. Black jumprocks have also been collected recently from the New River drainage in Virginia (Page and Burr 1991; Jenkins and Burkhead 1994). Juveniles and adults were collected in Back Creek in 1987, and one adult was collected from Toms Creek in 1988 (Jenkins and Burkhead 1994). Hocutt et al. (1986) stated black jumprocks may be introduced into the James drainage.

Means of Introduction: The population in the New drainage is thought to be a result of bait bucket introduction from the Roanoke or James drainages. This population was originally thought to have reached Craig Creek by stream capture, but more recent records of collections above and below the creek suggest human introduction (Lee et al. 1980 et seq.). Jenkins and Burkhead (1994) give evidence for introduced status, but suggest no means of introduction for the Craig Creek population.

Status: Established in Virginia.

Impact of Introduction: Unknown.

Remarks: None.

Scartomyzon lachneri

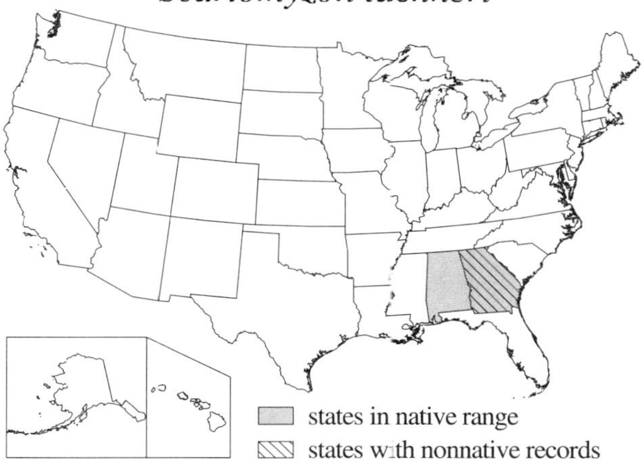

Scartomyzon rupiscartes

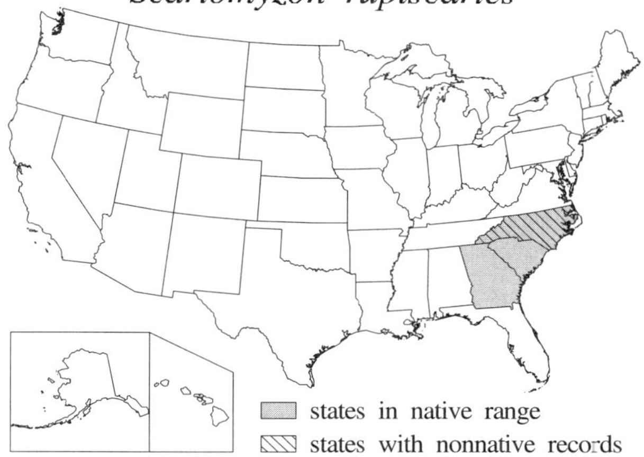

Scartomyzon lachneri Robins and Raney 1956

Common Name: greater jumprock.

Identification: Robins and Raney (1956); Page and Burr (1991); Mettee et al. (1996) Another commonly used name is *Moxostoma lachneri*. Maximum size: 44 cm.

Native Range: Apalachicola basin, Georgia and Alabama (Page and Burr 1991).

Nonindigneous Occurrences: Three individuals were collected in December 1997, from Beach Creek, in the Tallapoosa drainage, Haralson County, **Georgia** (E. Irwin and G. Kowalski, personal communication).

Means of Introduction: Unknown.

Status: Collected in Georgia.

Impact of Introduction: Unknown.

Remarks: None.

Scartomyzon rupiscartes (Jordan and Jenkins 1889)

Common Name: striped jumprock.

Identification: Robins and Raney (1956); Menhinick (1991); Page and Burr (1991). The genus was revised by Smith (1992). *Scartomyzon*, originally a subgenus of *Moxostoma*, has now been elevated to a genus. Another commonly used name is *Moxostoma rupiscartes*. Maximum size: 28 cm.

Native Range: Atlantic Slope from Santee River drainage, North Carolina, to Altamaha River drainage, Georgia; Gulf Slope in upper Chattahoochee River drainage, Georgia. Recently discovered in and possibly introduced to the extreme upper Pee Dee River drainage, North Carolina (Page and Burr 1991).

Nonindigenous Occurrences: Striped jumprock were probably introduced into the Yadkin (Pee Dee) drainage in **North Carolina** (Hocutt et al. 1986; Menhinick 1991).

Means of Introduction: Unknown.

Status: Established in North Carolina.

Impact of Introduction: Unknown.

Remarks: None.

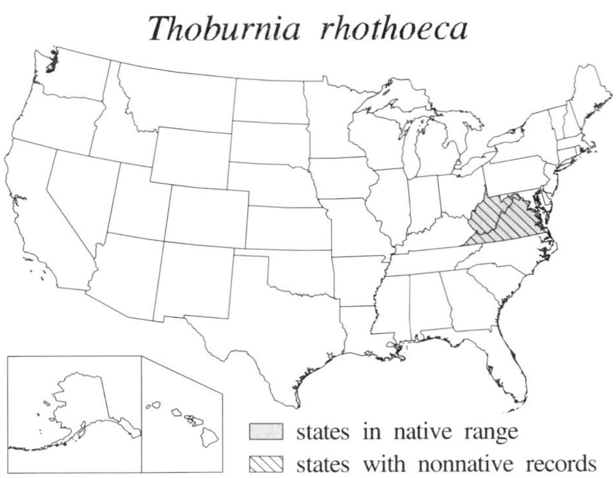

Thoburnia rhothoeca (Thoburn 1896)

Common Name: torrent sucker.

Identification: Page and Burr (1991); Jenkins and Burkhead (1994); Stauffer et al. (1995). Maximum size: 18 cm.

Native Range: Atlantic Slope from upper Potomac River drainage, Virginia and West Virginia, to Roanoke River drainage, northcentral Virginia and adjacent West Virginia (absent in York River system) (Page and Burr 1991).

Nonindigenous Occurrences: Introduced into the upper Rappahannock drainage, **Virginia**, where it was first discovered in 1958 (Lee et al. 1980 et seq.; Page and Burr 1991; Jenkins and Burkhead 1994), and possibly to the New drainage (see Remarks). Also probably introduced into the upper North Fork system of the South Branch Potomac in **West Virginia**. The fish was first discovered in this system in 1973 and is very localized (inhabits only 12 km of the river) (Jenkins and Burkhead 1994).

Means of Introduction: Probably due to a bait bucket release. Jenkins and Burkhead (1994) suggested the Rappahannock introduction took place in the upper Rapidian and spread from there.

Status: Established in Virginia and West Virginia.

Impact of Introduction: Unknown.

Remarks: While there is some question regarding this species' status in the Shenandoah system, Jenkins and Burkhead (1994) treated it as native.

This species was formerly present in the upper New River drainage, Virginia, but is now apparently extirpated (Page and Burr 1991). Its presence in that drainage is somewhat confounding. Jenkins and Burkhead (1994) stated that all of the records from this drainage are suspect and, indeed, rejected one of the records. However, they concede that three of the records from that area cannot be clearly refuted and may be valid. In the end, they "grudgingly agree that *T. rhothoeca* is probably native to the New drainage". Stauffer et al. (1995) also treated it as native to the New drainage in West Virginia but did not discuss the matter. Conversely, the torrent sucker's presence in the New may have been the result of an early introduction (Gilbert, personal communication). Gilbert believes that this is an Atlantic Slope species, not an Ohio drainage species that entered the Atlantic Slope via stream capture. He believes that given the pronounced elevational differences between the plateau on which the New River is situated and the eastern slope, stream capture and faunal transfer in this region could only have occurred in a a one-way direction (west to east). The fact that this species was localized and rare in the drainage (Jenkins and Burkhead 1994) lends support to the hypothesis that it may have been introduced.

Cobitidae - Loaches

The ostariophysan family Cobitidae, whose members are known as loaches, is represented by about 18 genera and about 110 species (Nelson 1994). This family was formerly called Cobitididae. Certain genera formerly placed in the Cobitidae (the most prominent being the genus *Nemacheilus*) are now considered to belong to the Balitoridae. Loaches are widely distributed in Eurasia, with the greatest diversity in Southeast and southern Asia. A few species are native to Morocco in extreme Northwest Africa (Berra 1981; Nelson 1994). In Eurasia, this family is second only to Cyprinidae in number of species (Roberts 1989). Most loaches are small, less than 40 cm long, with slender, elongate bodies. Loaches are typically bottom-dwelling fishes that feed on invertebrates or algae. Several species, such as clown loaches, coolie loaches, botias, and weather loaches, are popular in the aquarium trade. The family is restricted to freshwater.

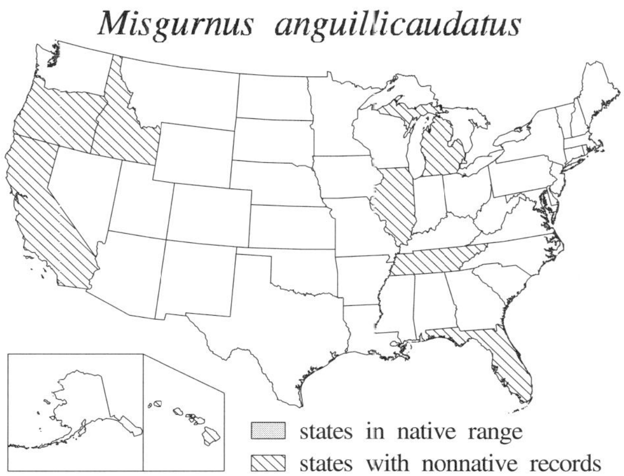

Misgurnus anguillicaudatus (Cantor 1842)

Common Names: oriental weatherfish (dojo, weather loach, Japanese weatherfish, Amur weatherfish).

Identification: Distinguishing characteristics were given by Berg (1949), Sterba (1973), Masuda et al. (1984), Page and Burr (1991), and Talwar and Jhingran (1992). The species is included in identification keys provided by Nichols (1943), Berg (1949), Moyle (1976a), Talwar and Jhingran (1992), and Laird and Page (1996). Illustrations or photographs appeared in Nichols (1943), Berg (1949), Masuda et al. (1984), and Axelrod et al. (1985). Maximum size: 25 cm.

Native Range: Eastern Asia (Berg 1949; Masuda et al. 1984; Talwar and Jhingran 1992).

Nonindigenous Occurrences: Specimens were collected from a three-mile reach of the Westminster flood control channel in Orange County, **California**, in 1968 (St. Amant and Hoover 1969). Additional established populations were discovered upstream from the original collection sites in 1977 and in the adjacent Bolsa Chica Channel in 1979 (Shapovalov et al. 1981). It has also been recorded for Huntington Beach, Orange County, California (e.g., Courtenay et al. 1986; Page and Burr 1991). In **Florida**, this species has been found in the Little Manatee River drainage, since about 1988 (L. G. Nico, unpublished data). The species was introduced into **Hawaii** prior to 1900 (Brock 1960); it now occupies primarily stream habitats in Kauai, Maui, and Oahu (Maciolek 1984; Devick 1991a, 1991b). Specimens have been found in the Harton Davis Canal, in an irrigation ditch at Eagle State Park, in the Boise River system, Ada County, **Idaho**, since the mid 1980s (Courtenay et al. 1988; Idaho Fish and Game 1990). It has been found in the North Shore Channel, Cook County, Lake Michigan drainage, **Illinois**, since 1987, and in the Chicago Sanitary and Ship Canal in 1994 (Page and Laird 1993; Burr et al. 1996; Laird and Page 1996). This species was collected in headwaters of the Shiawassee River, in Oakland and Genesee counties, **Michigan**, in 1958 and 1959 where it was considered to be established (Schultz 1960). The species has been reported from a diked secondary channel of the Clackamas River, **Oregon**, since the mid 1980s; it was found in the Malheur, Owyhee, and Snake River systems in 1995 (Logan et al. 1996). Several specimens were also collected in 1997 from Burlington Bottoms near the Multnomah Channel of the Columbia River (N. Bowers, personal communication). This species is known from **Tennessee** where it was collected in a tributary of Coffee Creek, near the Hiwassee River, Polk County, in April 1995 (D. Etnier, personal communication).

Means of Introduction: California populations were apparently descended from individuals that had escaped from a local goldfish farm, possibly as early as the 1930s (St. Amant and Hoover 1969). Similarly, populations in the upper Shiawassee River system, Michigan, were believed to have descended from individuals imported from Kobe, Japan, that had escaped from a nearly aquarium supply company in Oakland County before 1951 and possibly as early as about 1939 (Schultz 1960; Courtenay et al. 1984). In Hawaii, the species was presumably introduced by Asian immigrants during the 1800s, probably to serve as a source of food (Maciolek 1984; Devick 1991b). Brock (1960) stated that its use as bait for fish was apparently instrumental in its spread in Hawaii. Introductions into other states were presumably the result of aquarium releases.

Status: Established in California, Florida, Hawaii, Idaho, Illinois, Michigan, and Oregon. Reported from Tennessee.

Impact of Introduction: Unknown.

Remarks: In addition to appearing in the aquarium trade, *M. anguillicaudatus* has been introduced into several parts of the world for aquaculture purposes and as a bait fish (Welcomme 1988). It is considered a cool-water species (Welcomme 1988); individuals burrow into soft substrates and tolerate low oxygen conditions by breathing atmospheric air (Sterba 1973). As its name suggests, this fish becomes more active in response to changes in barometric pressure; thus it frequently has been cited as a harbinger of storms (Sterba 1973; Axelrod et al. 1985). There is concern that if *M. anguillicaudatus* becomes more abundant and spreads, it will reduce populations of aquatic insects important as food to native fishes (Page and Laird 1993). Maciolek (1984) categorized *Misgurnus* along with several other introduced fishes as species having an intermediate impact on Hawaiian streams, this conclusion based on their preferred habitat and diet and on their numbers. The Bolsa Chica Channel, California, cited by Shapovalov et al. (1981), likely represents the Huntington Beach, Orange County site reported by other authors (e.g., Courtenay et al. 1986; Page and Burr 1991).

Voucher specimens: California (CAS 24192, LACM; SIO 68-213, UF 87981); Florida (UF 98633, 98634); Hawaii (BPBM 3624, discarded in 1969); Idaho (UNLV 1951); Illinois (INHS 61129, 61130, 61131); Oregon (OS 13196, 13215, 13216, 15467, 15473, 15474; Portland State University); Tennessee (UT 46.17).

Misgurnus mizolepis

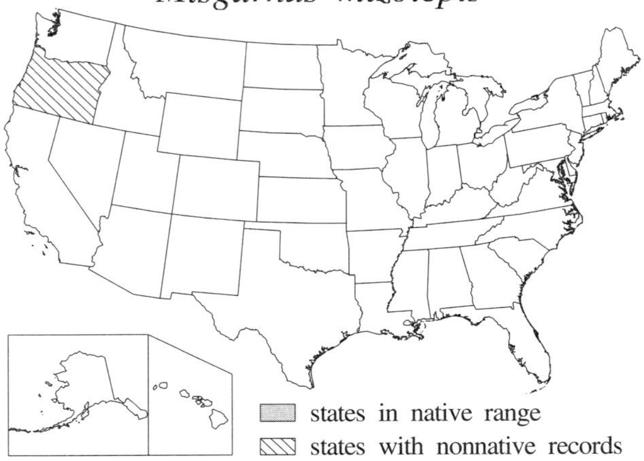

Misgurnus mizolepis (Günther 1888)

Common Name: Chinese fine scaled loach.

Identification: A few distinguishing characteristics were given by Nichols (1943). The genus *Misgurnus* is in need of systematic review. Many, possibly all, of the subspecies of *M. mizolepis* recognized by Nichols should be considered suspect. Illustrations or photographs appeared in Nichols (1943) and Axelrod et al. (1985). Maximum size: 21 cm.

Native Range: Eastern Asia (Nichols 1943).

Nonindigenous Occurrences: A single specimen was taken from the Day Irrigation Canal in the Powder-Snake River drainage, Baker County, **Oregon**, on 25 May 1977 (D. Logan, personal communication).

Method of Introduction: Probable aquarium release (Logan et al. 1996).

Status: Reported from Oregon. Subsequent sampling in the Powder River drainage has yielded no more specimens (Logan, personal communication).

Impact of Introduction: Unknown.

Remarks: Identification of the Oregon specimen was tentatively verified by M. Kottelat (Logan, personal communication).

Voucher specimen: Oregon (OS 8671).

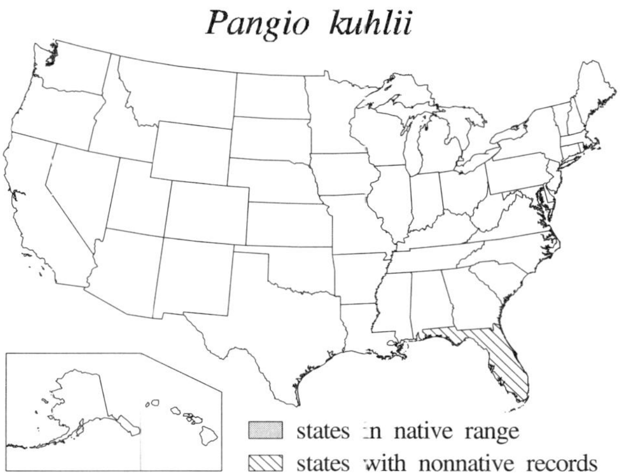

Pangio kuhlii (Valenciennes 1846)

Common Names: coolie loach (giant coolie loach, leopard eel, prickly eye, striped loach, leopard loach).

Identification: A few distinguishing characteristics were given by Sterba (1973) and Kottelat et al. (1993). According to Kottelat et al. (1993), *Pangio kuhlii* may represent a species complex; commonly used synonyms include *Acanthophthalmus kuhli*, *A. semicinctus*, and *Pangio semicincta*. Illustrations or photographs appeared in Sterba (1973), Axelrod et al. (1985), Kottelat et al. (1993), and Sakurai et al. (1993). Maximum size: 8 cm.

Native Range: Sumatra, Borneo, Java, Malaya (Kottelat et al. 1993).

Nonindigenous Occurrences: A single specimen was taken in **Florida** from a ditch adjacent to the Tampa Bypass Canal, Hillsborough County, in the vicinity of an ornamental fish farm in November 1993 (museum specimen).

Method of Introduction: Probable escape from an ornamental fish farm.

Status: Reported from Florida.

Impact of Introduction: Unknown.

Remarks: The coolie loach is common in the aquarium trade. It is a nocturnal scavenger.

Voucher specimen: Florida (UF 98930).

ORDER CHARACIFORMES

Anostomidae - Headstanders

This neotropical characiform family comprises some 12 genera, many of which are monotypic, and about 110 nominal species (Nelson 1994). Anostomids are fusiform, somewhat elongate fishes and have conical heads with small, nonprotractile mouths. Most species are medium- to large-sized omnivores or vegetarians that use their incisorlike teeth to clip aquatic vegetation, or as forceps to remove prey from crevices (Taphorn 1992). The family is restricted to freshwater.

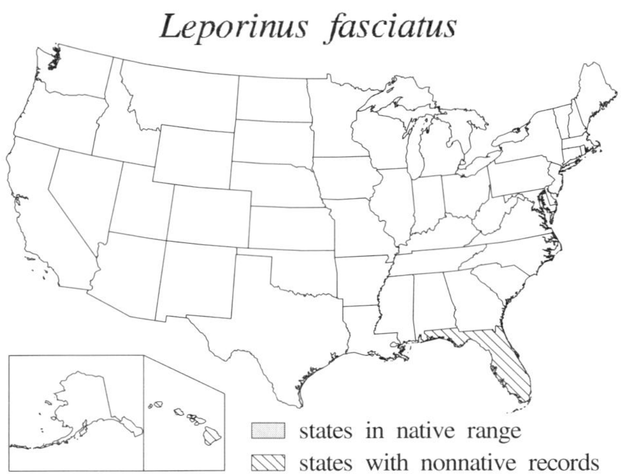

Leporinus fasciatus (Bloch 1794)

Common Names: banded leporinus (black-banded leporinus).

Identification: The genus *Leporinus* comprises about 70 nominal species, many needing additional taxonomic study (Vari 1983). *Leporinus fasciatus* likely represents a species complex. At least ten *Leporinus* species have been described with color patterns that consist of vertical bars along the body (Géry 1977). See Géry (1977) for color photographs and distinguishing characteristics as well as a general discussion of the *L. fasciatus* group. Maximum size: 30 cm SL.

Native Range: Tropical America. This species complex is widely distributed in South America, from Guiana and the Orinoco region in Colombia and Venezuela to the La Plata basin (Géry 1977).

Nonindigenous Occurrences: Single record from an unspecified location in **Florida** between 1984 and 1991 (Courtenay et al. 1991). Reported from reservoir on Oahu, **Hawaii**, in 1984 (Devick 1991a, 1991b).

Means of Introduction: Probable aquarium releases.

Status: Reported from Florida and Hawaii.

Impact of Introduction: Unknown.

Remarks: *Leporinus fasciatus* was listed as having been reported from Colorado by Courtenay and Stauffer (1990); however, that record was the result of a publishing error and thus not valid (Courtenay, personal communication).

There are no known voucher specimens.

Erythrinidae - Trahiras

The neotropical characiform family Erythrinidae comprises 3 genera with at least 10 species (Nelson 1994). These primitive characoids are widely distributed, ranging from the Parana–La Plata basin in south central South America north to Panama (Berra 1981). Erythrinids are cylindrically shaped and in certain ways are similar in appearance to our native bowfin *Amia calva* (Amiidae). Most species are medium-sized fish, but one reaches about one meter. Members of this family are highly predacious and have sharp teeth; all or most species have a swimbladder modified for air breathing. The family is restricted to freshwater.

Hoplias malabaricus

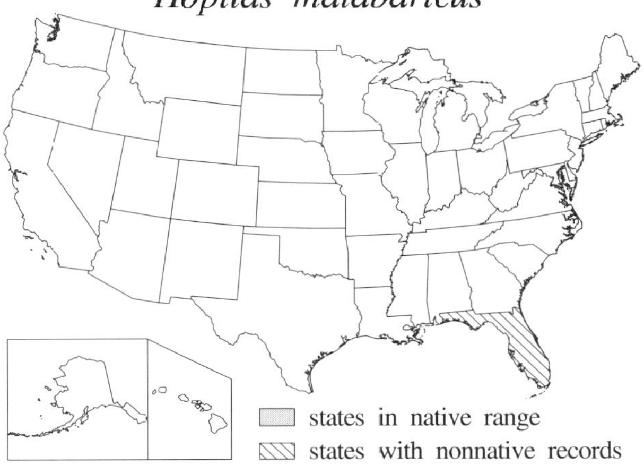

Hoplias malabaricus (Bloch 1794)

Common Names: trahira (guabina, traira, South American snakehead, tigerfish, tiger characin, tararura, haimara).

Identification: Distinguishing characteristics, key, and photographs appeared in Géry (1977). A color photograph of live fish was given in Axelrod et al. (1985). Given its wide native distribution and geographic variation, it is likely that *Hoplias malabaricus* represents a species complex. Maximum size: 50 cm SL.

Native Range: Tropical America. Lowland areas of Central America, and in tropical and subtropical South America including the island of Trinidad (Taphorn 1992).

Nonindigenous Occurrences: All records are from the Little Manatee River drainage in the lower southeastern Tampa Bay area of Hillsborough County, **Florida**. A large number of specimens (about 60) were collected from a small system of drainage ditches and ponds 1.9 km south of the Little Manatee River on U.S. Highway 301, with several samples taken between December 1974 and February 1975 (Hensley and Moody 1975). Postlarvae and juveniles were taken from the same site in August and September 1975 (Hensley 1976).

Means of Introduction: These introductions represented either releases or escapes from a fish farm (Courtenay, personal communication).

Status: Formerly established in Hillsborough County, Florida. No specimens collected since January 1977; presumably species died out as result of extremely cold temperatures during that month (Courtenay and Hensley 1979a).

Impact of Introduction: According to Courtenay (personal communication), Florida *Hoplias* were causing severe injuries to native centrarchids, especially *Lepomis* species.

Remarks: In its native range, this predacious fish has been reported to move overland through wet vegetation or mud at night or during rainstorms (Taphorn 1992). Live fish are difficult to handle and potentially dangerous because of their sharp teeth, strong jaws, and slippery bodies.

Voucher specimens: Florida (UF 97060; FDNR 8842, 8843, 8844, 8962, 8963, 9593).

Characidae - Characins

This diverse family consists of at least 170 genera and more than 880 species (Nelson 1994). Except for the genus *Astyanax,* native to the southwestern United States (Texas), the family Characidae is restricted to South America, Central America into Mexico, and Africa, with most species found in tropical South America. Ideas on classification of the Characidae have fluctuated considerably down through the years, ranging from a very conservative classification involving a single broad family, or (as currently treated) to comprise only the more generalized characoids, with the more specialized groups included in 15 or 16 other separate families. Although not widely accepted, a few authors place the piranhas and their close relatives (e.g., *Colossoma* and *Piaractus*) in a separate family, the Serrasalmidae. Characids are very diverse in terms of morphology, size, and ecology. Many species are omnivorous or carnivorous. The family includes such groups as the piranhas, pacus, and many smaller fishes such as the tetras important in the aquarium trade. The family is restricted to freshwater.

Aphyocharax anisitsi

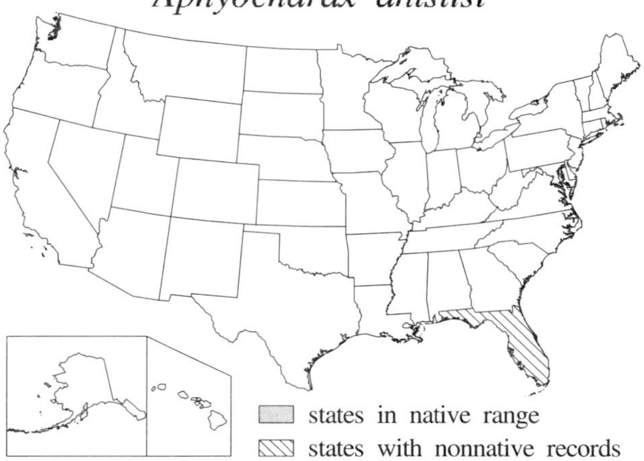

Aphyocharax anisitsi Eigenmann and Kennedy 1903

Common Names: bloodfin tetra (true bloodfin, bloodfin).

Identification: Géry (1977) discussed the genus, provided a key (probably outdated) and considered *Aphyocharax rubripinnis* to be a synonym. The genus *Aphyocharax* contains at least 10 species, all of which share the same common name, bloodfin tetras (Géry 1977). Maximum size: 4 cm.

Native Range: Tropical America. Paraguay River in northern Argentina and Paraguay, South America (Géry 1977; S. H. Weitzman, personal communication).

Nonindigenous Occurrences: Several specimens were taken in **Florida** from two separate stations in the lower Little Manatee River, Hillsborough County, on 15 September 1988 (museum specimens).

Means of Introduction: Records likely the result of escapes from local ornamental fish farms.

Status: Reported from Florida.

Impact of Introduction: Unknown.

Remarks: *Aphyocharax anisitsi* is a well-known aquarium fish. Florida specimens were identified by S.H. Weitzman.

Voucher specimens: Florida (FDNR uncatalogued).

Astyanax mexicanus

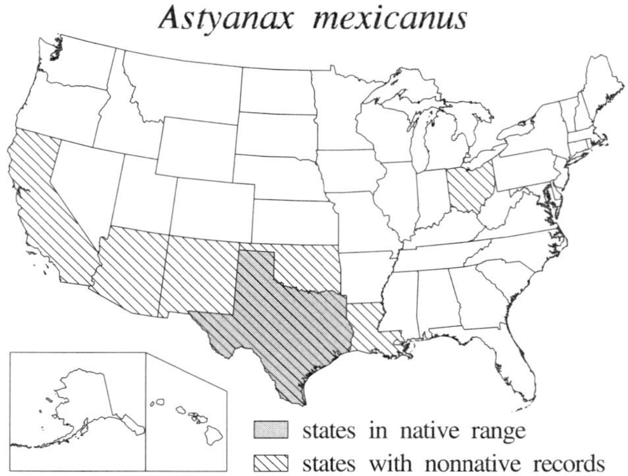

Astyanax mexicanus (Filippi 1853)

Common Names: Mexican tetra (sardina de México).

Identification: Although previously considered a subspecies of *Astyanax fasciatus* (Lee et al. 1980 et seq.), *A. mexicanus* more likely represents a species complex (Mayden et al. 1992). Sublette et al. (1990) and Page and Burr (1991) gave distinguishing characteristics. The species was included in a state key by Miller (1952). Illustrations and photographs were provided by Miller (1952), Miller and Robison (1973), Minckley (1973), Sublette et al. (1990), and Page and Burr (1991). Maximum size: 12 cm.

Native Range: Subtropical America. Mostly eastern and central Mexico and also the lower Rio Grande, lower Pecos, and Nueces rivers in Texas (Lee et al. 1980 et seq.; Page and Burr 1991). Historically, found in the Pecos River drainage and lower Rio Grande in New Mexico, but now restricted to the Pecos drainage downstream of Dexter (Sublette et al. 1990).

Nonindigenous Occurrences: The species was first mentioned as a baitfish in the lower Colorado River at Yuma, **Arizona,** in the early 1950s, but no specimens had been collected up to that time from open waters (Evans and Douglas 1950; Miller 1952). It was first recorded in open waters, when several juvenile to subadult specimens were taken from the Colorado River just below Morelos Dam in 1966 (Minckley 1973). There were several reports from the lower Colorado River in **California** in 1950 (Minckley 1973; Swift et al. 1993). According to Hubbs et al. (1979), the fish was introduced but never became established in the state. There were two records of fish taken from lakes in the Red River drainage, northwestern part of **Louisiana**. The first was based on a single specimen taken from Lake Bistineau Reservoir in October 1954 (Lambou 1962). The species was also recorded from Cross Lake, near Shreveport (Lee et al. 1980 et seq.). The species has been recorded from the Rio Grande drainage in southern **New Mexico** (Lee et al. 1980 et seq.; Page and Burr 1991). A single fish was taken from Big Darby Creek, Union County, **Ohio**, on 22 October 1966 (Trautman 1981). A single fish was taken in 1953 from Lake Texoma, **Oklahoma**, a reservoir of the Red River bordering Texas (Riggs 1954); large numbers of additional specimens were taken from the same reservoir in 1955, suggesting that the species was reproducing (Dowell and Riggs 1958). Miller and Robison (1973) repeated that a breeding population may be established in Lake Texoma. The species was reported as having been introduced, but probably not established in several other lakes including Lake Spavinaw in the Neosho drainage in the northeastern part of state; Lake Murray in Love and Carter counties; Lake Tenkiller in Cherokee and Sequoya counties; Ft. Gibson Lake; and perhaps elsewhere (Miller and Robison 1973; Lee et al. 1980 et seq.). The species has been reported at many sites, especially springs, in central and northeastern parts of **Texas** (Hubbs et al. 1978; Lee et al. 1980 et seq.; Hubbs et al. 1991). Specifically, introduced populations have been found in various parts of the Edwards Plateau and in the Red River drainage above Denton (Lee et al. 1980 et seq.), and reported in drainages from San Antonio Bay to Galveston Bay (Conner and Suttkus 1986). Recently, this species was recorded for the Sabine Lake watershed in the southeastern part of the state (Bechler and Harrel 1994).

Means of Introduction: All or most introductions have probably resulted from bait bucket releases (e.g., Miller and Robison 1973). The first *A. mexicanus* observed outside their native range were seen in a bait shop in Yuma, Arizona, along the Colorado River in 1950; presumably, these fish were a species contaminant that arrived in a minnow shipment (Evans and Douglas 1950; Miller 1952).

Status: Established in Texas (outside native range). Formerly found in the Rio Grande valley in New Mexico, but now possibly extirpated (although these likely represented native records and not introductions). Previously established in southern Oklahoma; the species has not been taken in recent years and is now considered extirpated (Cashner and Matthews 1988). Reported from Arizona, California, and Louisiana.

Impact of Introduction: Unknown.

Remarks: This is the only characid species native to the United States. Because it is predacious and highly aggressive, there is concern that this species could impact native fishes by preying on their young (Miller 1952; Minckley 1973). Lee et al. (1980 et seq.) and Page and Burr (1991) provided distribution maps; these authors considered Mexican tetra populations recorded from the Rio Grande drainage in southern New Mexico, and in the associated nearby border area of Texas, to be introduced. In contrast, Sublette et al. (1990) believed that this area constituted part of the species' natural or historical distribution. It now appears to be extirpated from the Rio Grande in New Mexico and is on the decline in many portions of the Pecos River drainage; as such, the species is classified as "state endangered" by New Mexico (Sublette et al. 1990).

Voucher specimens: Ohio (OSM 14147); Texas (INHS 75684).

Colossoma macropomum

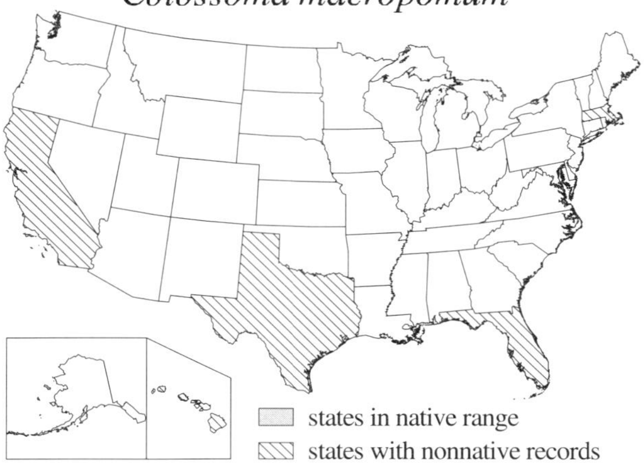

Colossoma macropomum (Cuvier 1818)

Common Names: black pacu (tambaqui, cachama, blackfin pacu).

Identification: Machado-Allison (1982) revised the group and provided information for identification. Many synonyms exist; *C. oculus* and *C. nigripinne* are junior synonyms commonly used in the aquarium literature and in past reports on United States nonindigenous fishes. *Colossoma* taken in United States waters also have been frequently misidentified as red piranhas *Pygocentrus nattereri* or as one of the other pacus *Piaractus* spp. An unpublished identification guide by Taylor (1985) is mostly a direct translation of sections from Machado-Allison (1982) and Britski (1977). For photographs, see Géry (1977, p. 255, misidentified as *C. brachypomum* = *Piaractus brachypomus*), Goulding (1980), Machado-Allison (1982), and Araujo-Lima and Goulding (1997). Maximum size: To 90 cm SL and 30 kg.

Native Range: Tropical America. Native to the Amazon and Orinoco river basins of South America (Machado-Allison 1982; Araujo-Lima and Goulding 1997).

Nonindigenous Occurrences: A single fish, tentatively identified as *C. nigripinnis*, was taken from the Sacramento River in **California**, near Elkhorn Ferry above Sacramento in Yolo County, on 10 October 1977 (Brittan and Grossman 1979). A single fish was taken from Lake Maggiore in St. Petersburg, Pinellas County, **Florida,** on 12 January 1983 (museum specimen); one fish (identified as *Colossoma nigripinne*) was collected from Waverly Pond in Tallahassee, Leon County, on 13 October 1968 (Courtenay et al. 1974; Courtenay and Hensley 1979a; museum specimen); one fish (270 mm TL) was collected from Diversion Canal C-24, near Fort Pierce, St. Lucie County, on 11 December 1988 (J. D. Williams, personal observation; specimen not retained). The first report in **Hawaii** was of a 3.2-kg adult fish taken in Wahiawa Reservoir, a 350-acre, privately owned irrigation reservoir on Oahu, in 1987; several other large adult pacus, possibly members of this same species, have been caught or seen in recent years, but no juveniles have been taken (Devick 1991b, 1992). One fish was sighted in Nuuanu #3 Reservoir, Oahu (Devick 1991b). A single fish was taken with a hook and line from Triphammer Pond in Hingham, Plymouth County, **Massachusetts**, on 22 August 1993 (Cardoza et al. 1993; Hartel et al. 1996). There are several records of single fish taken from various sites in **Texas** including Forests Lake near Longview, Gregg County, on 30 July 1988; Eagle Point, a coastal area (with water salinity of about 10 ppt) near Houston, Harris County, in August 1988; and Branch Creek near Port Arthur, Jefferson County, on 25 May 1990 (Howells et al. 1991a).

Means of Introduction: All introductions likely represent aquarium releases.

Status: Reported in five states, including California, Florida, Hawaii, Massachusetts, and Texas.

Impact of Introduction: Unknown.

Remarks: Pacus are popular as aquarium fish. They are esteemed as food fish in South America. Most U.S. records represent collections of single fish taken by anglers. The Massachusetts fish was originally reported to be a red-bellied pacu in a newspaper account and as a *Colossoma* sp. by Cardoza et al. (1993); the fish was later determined to be *C. macropomum* by W. Fink, based on a photograph of the specimen (Hartel, personal communication). Some specimens previously identified as *C. macropomum* and taken in Florida waters may very well represent hybrid forms produced by artificial means using *Colossoma* and *Piaractus* parents (L. G. Nico, personal observations).

Voucher specimens: Florida (UF 65748, UF 97061). Voucher photographs: California (see Brittan and Grossman 1979).

Gymnocorymbus ternetzi

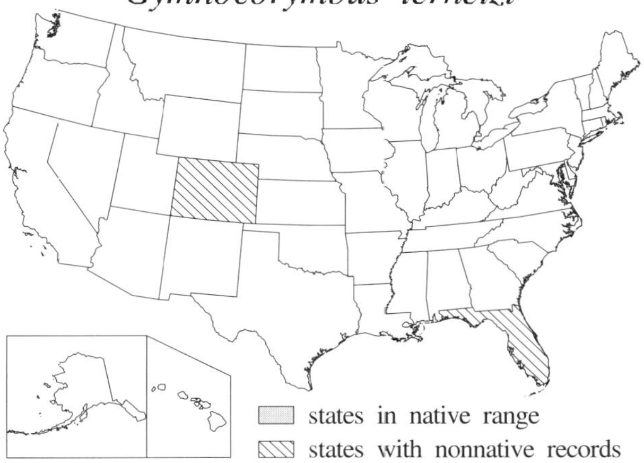

Gymnocorymbus ternetzi (Boulenger 1895)

Common Names: black tetra (black widow, black amoor, butterfly tetra, petticoat tetra).

Identification: Géry (1977) provided an identification key and color photographs. Color photographs were also provided in Axelrod et al. (1985) and Sakurai et al. (1993). Maximum size: 5 cm.

Native Range: Tropical America. Amazon (Guaporé drainage) and La Plata (Paraguay River) basins in southern Brazil, Argentina, and Bolivia, South America (Géry 1977).

Nonindigenous Occurrences: It has been recorded from geothermal waters (hot springs), in the San Luis Valley, Rio Grande basin, **Colorado**, during the period 1980 to 1984 (Zuckerman and Behnke 1986). There are two records from Ruskin, Little Manatee River drainage, Hillsborough County, **Florida**. The first of these fish was taken from a canal on 30 September 1978, the second from a drainage ditch on 26 October 1979 (museum specimens). In probable reference to one of these collections, Courtenay and Hensley (1979a) reported that this species was collected in large numbers from a roadside ditch near a fish farm in Ruskin.

Means of Introduction: The Colorado introduction resulted from an escape from a local tropical fish farm, and the Florida record may represent either an aquarium release or a fish farm escape.

Status: Reported from Colorado and Florida.

Impact of Introduction: Unknown.

Remarks: This is a well-known species among aquarists. The hot springs area of Colorado where this fish was found is at an altitude of 8,000 ft and has very cold winters, but L. Zuckerman (personal communication) suggested that some introduced species might spread downstream during warmer months and reach other thermal refugia.

Voucher specimens: Florida (UF 87974, UF 97058). The Colorado specimens were reportedly deposited with the U.S. Fish and Wildlife Service, Fort Collins (Zuckerman, personal communication). Zuckerman's Colorado collections were later transferred to the University of New Mexico, Albuquerque; however, *Gymnocorymbus* did not appear in a 1995 list of deposited fishes provided to us by the collections manager.

Hemigrammus ocellifer

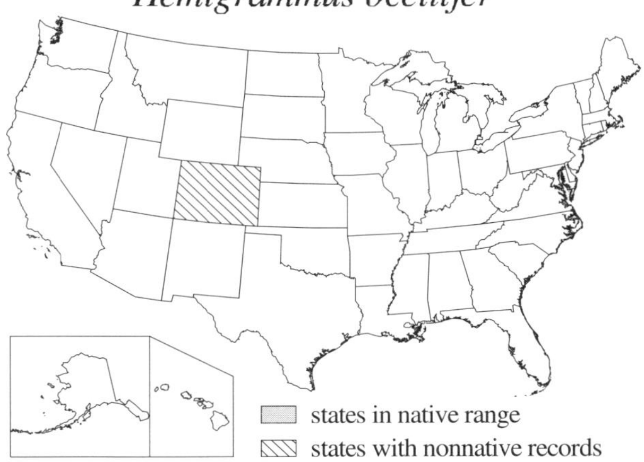

Hemigrammus ocellifer (Steindachner 1882)

Common Names: head-and-taillight tetra (beacon tetra, beacon fish).

Identification: The genus *Hemigrammus* needs systematic revision. See Géry (1977) for identification key and photographs. Color photographs were also provided in Axelrod et al. (1985) and Sakurai et al. (1993). Maximum size: 4 cm.

Native Range: Tropical America. The true *Hemigrammus ocellifer* is native to drainages in northeastern South America, including French Guiana, Guyana, and Venezuela (Géry 1977). Closely related members of the genus occur from Venezuela to Argentina (Géry 1977).

Nonindigenous Occurrences: This fish was taken from geothermal waters (hot springs), in the San Luis Valley, Rio Grande basin, **Colorado**, during the period 1980 to 1984 (Zuckerman and Behnke 1986).

Means of Introduction: Records represent escapes from areas of the hot springs used for culturing ornamental fish.

Status: Reported from Colorado.

Impact of Introduction: Unknown.

Remarks: The hot springs area is located at an altitude of 8,000 ft and has very cold winters, but Zuckerman (personal communication) suggested that some of the introduced species have the potential to spread downstream and reach other thermal refugia during warmer months.

Voucher specimen(s): reportedly were deposited with the U.S. Fish and Wildlife Service, Fort Collins, Colorado (Zuckerman, personal communication); but we have not been able to locate this material.

Hyphessobrycon cf. serpae

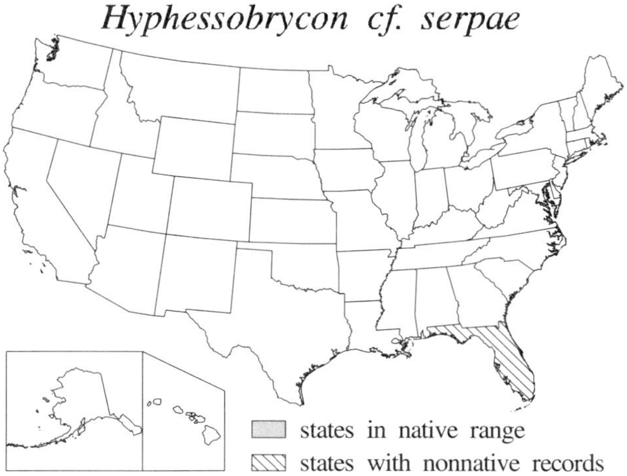

Hyphessobrycon cf. serpae Durbin 1908

Common Names: Serpa tetra (serpae tetra).

Identification: According to Géry (1977), the genus *Hyphessobrycon* contains approximately 65 species and needs systematic revision. *Hyphessobrycon serpae* itself represents a species complex. See Géry for an identification key and a color photograph. A color photograph was also published in Axelrod et al. (1985). Maximum size: 4 cm.

Native Range: Tropical America. Central Amazon basin and Guianas (Géry 1977).

Nonindigenous Occurrences: Two specimens were taken in **Florida** from a canal at the upper end of Mill Bayou, Bay County, in September 1989 (FL DNR, personal communication).

Means of Introduction: These fish were either aquarium releases or fish that escaped from a fish farm.

Status: Reported from Florida.

Impact of Introduction: Unknown.

Remarks: This is a well-known aquarium species.

There are no known voucher specimens.

Metynnis sp.

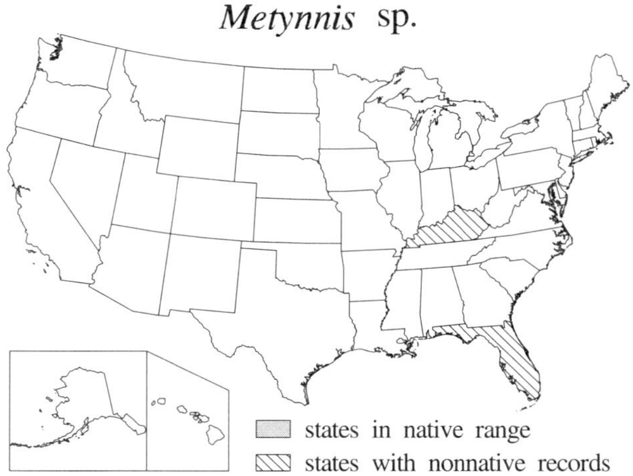

Metynnis spp.

Common Names: spotted metynnis (silver dollar, Guapore metynnis).

Identification: The genus *Metynnis* is in great need of systematic revision. One of the few available keys was given by Géry (1977). According to Géry, the M. *lippincottianus–maculatus* complex seems to comprise several different forms that could represent subspecies or sibling species. A possible synonym is *M. roosevelti*, a name commonly used by aquarists (Géry 1977). Although members of the genus *Metynnis* are characterized by their long-based adipose fins, those belonging to the *M. lippincottianus–maculatus* complex superficially resemble young pacus (i.e., *Colossoma* and *Piaractus* species) as well as certain piranha species. Photographs of several *Metynnis* species appeared in Axelrod et al. (1985) and Sakurai et al. (1993). Maximum size: 12 cm.

Native Range: Tropical America. *Metynnis* species are found throughout much of tropical South America (e.g., in the Orinoco and Amazon basins). The *M. lippincottianus–maculatus* complex is native to Guaporé and Amazon basin (Géry 1977).

Nonindigenous Occurrences: A member of this genus has been reported in **Florida** from an unspecified locality (listed as *Metynnis lippincotianus* in Courtenay et al. 1984; listed as *Metynnis* sp. in Courtenay and Stauffer 1990, and in Courtenay et al. 1991). In **Kentucky**, a single fish (originally identified as a piranha and as *Metynnis roosevelti*) was taken by hook and line from Lighthouse Lake, Louisville, Jefferson County, in the summer of 1981 (Anonymous 1981; Fossett 1981).

Means of Introduction: Records mostly likely represent aquarium releases.

Status: Reported from Florida and Kentucky.

Impact of Introduction: Unknown.

Remarks: Several *Metynnis* species are popular aquarium fishes. There is considerable confusion surrounding the Kentucky record. In original published accounts, the fish was identified as a piranha, but the scientific name provided was *Metynnis roosevelti* (= *Metynnis maculatus*). However, in a photograph of the fish accompanying the newspaper article (Fossett 1981), the specimen actually appears to have a short adipose fin and is probably a pacu, possibly *Piaractus brachypomus*. The collectors gave the live fish to the Louisville Zoo, where it was kept in aquaria; when the fish later died, it was supposedly not preserved. The Kentucky specimen has been the basis for inclusion of the species in published lists of nonestablished foreign species, with earlier listings identifying it as *Metynnis roosevelti* (e.g., Courtenay et al. 1984) and later simply as *Metynnis* sp. (i.e., Courtenay and Stauffer 1990; Courtenay et al. 1991). There are no known voucher specimens.

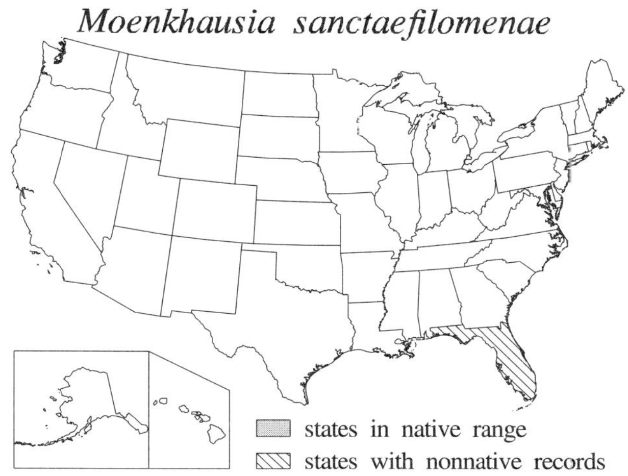

Moenkhausia sanctaefilomenae (Steindachner 1907)

Common Names: redeye tetra (yellowhead characin, red-eye tetra, yellow-banded moenkhausia).

Identification: The genus is in need of systematic revision. Géry (1977) provided a key and photographs. Color photographs appeared in Axelrod et al. (1985) and Sakurai et al. (1993). This species closely resembles the glass tetra *Moenkhausia oligolepis* in body form and coloration but has fewer lateral line scales. Maximum size: 7 cm.

Native Range: Tropical America. Paraguay River basin of South America (Géry 1977).

Nonindigenous Occurrences: Two specimens were taken from a ditch in **Florida** adjacent to Tampa Bypass Canal, near a fish farm east of Tampa in Hillsborough County, on 10 November 1993 (museum specimens).

Means of Introduction: These fish were probably released or escaped from a fish farm, or were aquarium releases.

Status: Reported from one site in Florida.

Impact of Introduction: Unknown.

Remarks: This is species is common in the aquarium trade.

Voucher specimens: Florida (UF 98917).

Myleus cf. rubripinnis

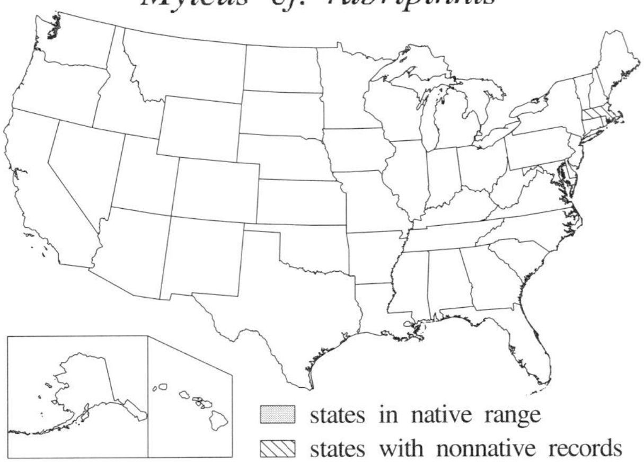

Myleus cf. rubripinnis (Müller and Troschel 1845)

Common Names: redhook myleus (redhook metynnis).

Identification: The genus *Myleus* needs systematic review. See Géry (1977) for key, synonymy, and color photographs. According to Géry, *Myleus rubripinnis* represents a complex and seems to be composed of several different forms that are difficult to distinguish. Photographs of *Myleus* species appeared in Axelrod et al. (1985) and Sakurai et al. (1993). Maximum size: 15 cm SL.

Native Range: Tropical America. This species complex is widely distributed in lowland areas of South America east of the Andes, including the Amazon, Orinoco, and Parana (Paraguay River) basins and various coastal drainages of the Guianas (Géry 1977).

Nonindigenous Occurrences: An angler took a single fish from Cook's Pond, Fall River, Bristol County, **Massachusetts**, in the late summer of 1991 (Hartel 1992; Cardoza et al. 1993; Hartel et al. 1996).

Means of Introduction: Record most likely represents an aquarium release.

Status: Reported in Massachusetts.

Impact of Introduction: Unknown.

Remarks: According to the report the Massachusetts specimen was examined and identified as *Myleus rubripinnis* by Massachusetts Division of Fish and Wildlife biologist Richard Hartley, but was not retained (Hartel 1992). R. Hartley (personal communication) has recently indicated to us that identification of the fish as *M. rubripinnis* should be considered tentative at best. He used an aquarium book to identify the specimen but only had in his possession the specimen's head (the angler and his family had eaten the body) and a sketch prepared by the angler showing a fish with a long, hooked anal fin. In fact, because of taxonomic problems associated with chorused fishes, there remains some doubt that the Massachusetts fish was correctly identified even to genus.

Paracheirodon innesi

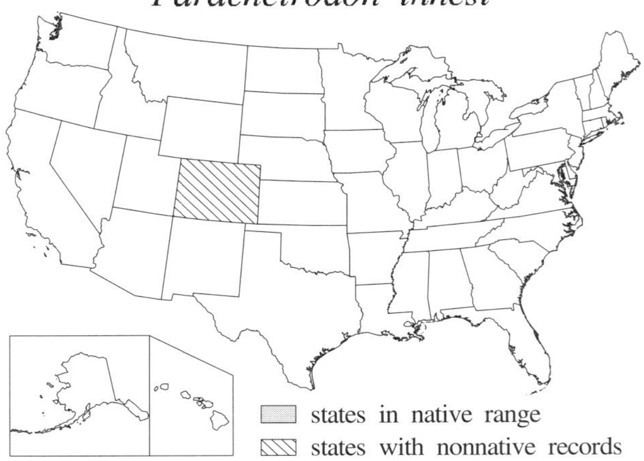

Paracheirodon innesi (Myers 1936)

Common Name: neon tetra.

Identification: Weitzman and Fink (1983) revised the group, and provided distinguishing characteristics, a key, and photographs. Color photographs were provided by Géry (1977), Weitzman and Fink (1987), and Sakurai et al. (1993). Similar species include *Paracheirodon simulans* (green neon) and *P. axelrodi* (cardinal tetra). Maximum size: 4 cm.

Native Range: Tropical America. Upper Amazon basin of Peru, Colombia and Brazil.

Nonindigenous Occurrences: The fish was taken in **Colorado** from geothermal waters (hot springs), San Luis Valley, Rio Grande basin, during the period 1980 to 1984 (Zuckerman and Behnke 1986).

Means of Introduction: Record is most likely the result of fish escapes from local fish farms that use the hot springs for culturing aquarium fish.

Status: Reported from Colorado.

Impact of Introduction: Unknown.

Remarks: The hot springs area is at an altitude of 8,000 ft and has very cold winters, but Zuckerman (personal communication) suggested that some of the introduced species might spread downstream during warmer months and reach other thermal refugia. Specimen(s) were reportedly deposited with the U.S. Fish and Wildlife Service, Fort Collins, Colorado; however, we have been unable to locate the material.

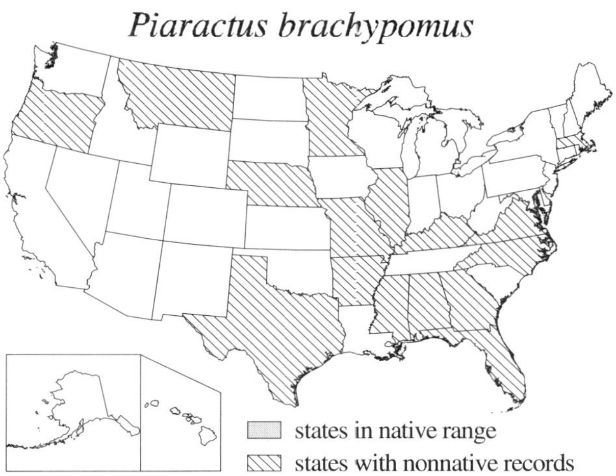

Piaractus brachypomus (Cuvier 1818)

Common Names: redbellied pacu (pirapatinga, morocoto, cachama).

Identification: Machado-Allison (1982) revised the pacus and provided distinguishing characteristics for *Piaractus brachypomus*. *Colossoma bidens* and *Colossoma brachypomum* are junior synonyms that have been widely used in the aquarium literature and in most past reports concerning fish introductions. Many *Piaractus* taken in U.S. waters have been misidentified and reported as the red piranha *Pygocentrus nattereri*. This pacu species is also confused often with the other common pacu species, tambaqui *Colossoma macropomum* and possibly with *Piaractus mesopotamicus*. An unpublished mimeograph by Taylor (1985) used to identify pacus was apparently based on information provided by Britski (1977) and Machado-Allison (1982). For photographs, see Machado-Allison (1982), Géry (1977, identified as *Colossoma* sp. on pages 252 and 256) and Goulding (1980, identified as *Colossoma bidens*). Maximum size: 85 cm SL and 20 kg.

Native Range: Tropical America. Orinoco and Amazon river basins, South America (Machado-Allison 1982).

Nonindigenous Occurrences: A single fish (241 mm long) was taken in **Alabama** by an angler from Elk River in Limestone County on 29 August 1988 (museum specimen); the original newspaper account (Middleton 1988) incorrectly reported the fish as being a red piranha. One fish was taken by an angler from a cemetery pond in **Arkansas** in Fayetteville, Washington County, in June 1995 (Wright 1995a, 1995b). One pacu was captured in **California** at Stevens Creek Reservoir, Santa Clara County, on 4 July 1996 (R. N. Lea, personal communication). There are several records of single fish taken from various sites in **Florida** including a pond near Port Charlotte in DeSoto County in September 1983 (Courtenay, personal communication); Bivens Arm Lake in Gainesville, Alachua County, 5 December 1984 (museum specimen); a lake at St. Lucie West Development, just east of I-95/St. Lucie West interchange in St. Lucie County, 11 June 1991 (museum specimen); a retention pond at the Stoneridge Apartments in Gainesville, Alachua County, 5 September 1993 (museum specimen); Lake Alice, Gainesville, Alachua County, 22 January 1998. There are several records of single fish (identified as *Colossoma bidens*) taken from various sites in **Georgia**, including a private pond in Banks County in 1982, a private pond in Coweta County in 1987, and Big Cotton Indian Creek, east of Stockbridge in Henry County on 6 June 1990 (R. M. Gennings, personal communication). A single fish was taken an by angler from Lake Tara in the Flint River drainage in September 1994 (museum specimen). A single fish taken with a trotline from the Mississippi River in **Illinois**, south of Chester, in or near Randolph County in September 1988; in an erroneous newspaper account (Anonymous 1988), it was identified as a piranha, but the accompanying photograph shows it to be *Piaractus*. Additional state records include a fish taken from Little Grassy Lake, Williamson County on 15 June 1992 (museum specimen); a single fish was taken from a campus lake at Southern Illinois University at Carbondale, Jackson County, on 22 June 1992 (Burr et al. 1996). A single fish was taken from Lake Barkley, Cumberland-Ohio River drainage, **Kentucky**, about river mile 56, in Trigg County, on 16 June 1993 (museum specimen). Another Kentucky pacu (based on a newspaper photograph probably this species) was taken from Lake Cumberland in 1991; it was incorrectly reported as a piranha in original newspaper account (see Lander 1991). One specimen was taken in **Massachusetts** by an angler from Dug Pond in Natick, Middlesex County, in 1988 (Hartel 1992; Hartel et al. 1996). A specimen identified as this species was taken in **Minnesota** from Taft Lake, Hennepin County (K. Schmidt, personal communication). There have been five records from **Mississippi**, all from the Pascagoula River drainage. Two specimens were originally identified as *Colossoma* species, the first collected at Lake Serene, Lamar County, in September 1990, and the second taken from a tributary in the Leaf River system, Forrest County, in July 1991 (Ross and Brenneman 1991; S.T. Ross, personal communication). There are two records of single specimens taken from West Lake in the Leaf River system, Lamar County, during the summer of 1992; another specimen was taken from a pond in the Leaf River system in Hattiesburg, in September 1982 (museum specimens; Ross, personal communication). A single specimen was taken by an angler in southwest **Missouri** from Stockton Lake, Dade County, in August 1995 (T. J. Banek, personal communication). A single specimen was taken by angler in **Montana** from Lake Elmo, Billings, in July 1994 (preserved specimen). This species has been taken from one or more sites in **Nebraska**; Welsch (1996) reported that piranhas were commonly taken in the state but the accompanying photograph of a local 'piranha is that of a *Piaractus brachypomus*. In **North Carolina** a single fish was taken with a hook and line from Elizabeth Lake, Insole County, in October 1991 (Lee 1991). A single specimen was taken in **Oregon** from the Willemite River near Portland in Multinomial County, on 4 July 1988; one or more other collections of pacu from other sites in the state also may represent this species (Logan et al. 1996). Several single fish (identified as *Colossoma brachypomum*) were taken from various sites in **Texas** including Town Lake in Austin, Travis County, in 1980; Lake Bastrop, Bastrop County, during the summer of 1987; the Colorado River near Bastrop, Bastrop County, on 16 July 1987; a private pond near Cleburne, on 12 June 1989; a freshwater canal near Nederland, Jefferson County, on 4 June 1990; Gray's Hisle Camp, Tri Cities Beach Road, near Baytown, Harris County, on 31 July 1990; a creek adjacent to All American Rice Growers Canal, 9.7 km south of Dayton, Liberty County, on 17 October 1990 (an angler reported catching three additional specimens, which he released); the Colorado River 1.6 km upstream from Bastrop, on 26 August 1991 (Howells et al. 1991a). A single fish was collected in Burke Lake in Burke Lake Park, Fairfax County, **Virginia**, in June 1988 (Bohn 1988).

Means of Introduction: Most records likely represent aquarium releases, although some Florida and Georgia records may have resulted from fish farm escapes.

Status: Reported from 16 states including Alabama, Arkansas, Florida, Georgia, Illinois, Kentucky, Massachusetts, Minnesota, Mississippi, Missouri, Montana, Nebraska, North Carolina, Oregon, Texas, and Virginia.

Impact of Introduction: Unknown.

Remarks: This species is a popular aquarium fish. It is a prized food fish in South America. To date, records maintained by the Georgia Department of Natural Resources on introduced foreign fishes show pacus identified as either *Colossoma bidens* (= *Piaractus brachypomus*) or as unidentified pacus. No *Colossoma macropomum* have yet been taken in Georgia. A fish taken from the Snake River in Oregon (OS 13217) was originally identified as *Colossoma macropomum* by Logan (1994), but one of us (LGN) examined this specimen and it more likely represents a hybrid between *C. macropomum* and *Piaractus* (also see Logan et al. 1996). All Texas specimens were taken by anglers and later identified by biologists of the Texas Parks and Wildlife Department.

Selected voucher specimens: Alabama: (UF 98946); Florida (UF 42782, 87879, 94136, 107100); Georgia (UF 98944); Illinois (SIUC 19835); Kentucky (SIUC 22241); Massachusetts (MCZ 96096); Mississippi (USM 14134, 14160, 14637); Montana (uncatalogued); Oregon (OS 11491); Texas (TNHC 11037). Voucher photographs: Arkansas, Florida, Illinois, Minnesota, Missouri, Montana, Nebraska, Texas (video film), and Virginia.

Piaractus mesopotamicus

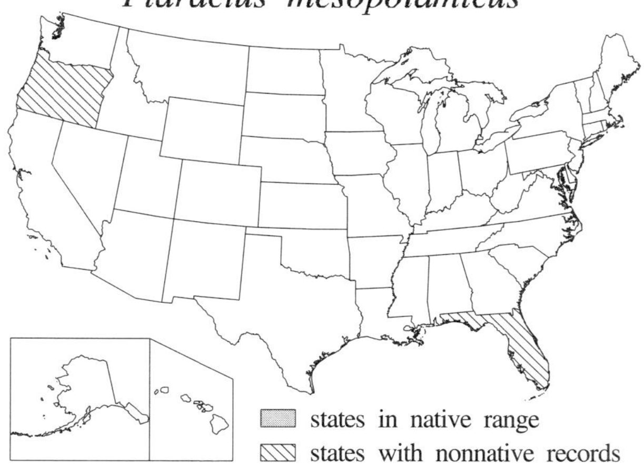

Piaractus mesopotamicus (Holmberg 1887)

Common Name: small-scaled pacu.

Identification: H. A. Britski (personal communication) considers *Piaractus mesopotamicus* one of three valid large pacu species (the others are *P. brachypomus* and *Colossoma macropomum*). Commonly used synonyms include *Piaractus metrei*, *Colossoma metrei*, and *Colossoma mesopotamicus*. In his taxonomic review, Machado-Allison (1982) used the name *P. metrei* (Berg 1895). This, however, is a junior synonym of *P. mesopotamicus*. At least some, and perhaps all, specimens taken in United States waters and identified as this species may actually have been the closely related *Piaractus brachypomus*. In addition, *Piaractus* specimens taken in United States waters have frequently been misidentified as red piranhas *Pygocentrus nattereri*. Maximum size: To about 85 cm SL.

Native Range: Tropical America. Restricted to the La Plata Basin, South America (Machado-Allison 1982).

Nonindigenous Occurrences: A single fish (56 cm and 3515 g) was reported as taken in **Florida** by an angler from a brackish-water canal at Sunshine Parkway bridge near town of Cape Coral, Charlotte Bay drainage, Lee County, on 4 July 1987 (Courtenay, personal communication). A single specimen (identified as *Piaractus mesopotamicus*) was taken in **Oregon** from Herbert's Pond near Canyonville, Douglas County, in the Umpqua River drainage on 21 August 1989 (Logan et al. 1996).

Means of Introduction: All records probably represent aquarium releases.

Status: Reported from Florida and Oregon.

Impact of Introduction: Unknown.

Remarks: The Florida specimen, originally reported as a piranha in newspaper accounts, was not preserved. That specimen was apparently destroyed and the fish was later identified as *Colossoma metrei* (= *Piaractus mesopotamicus*) based on a photograph (Courtenay, personal communication). Because of the close resemblance with other pacus, that identification should be considered tentative.

Voucher specimens: Oregon (OS 11983).

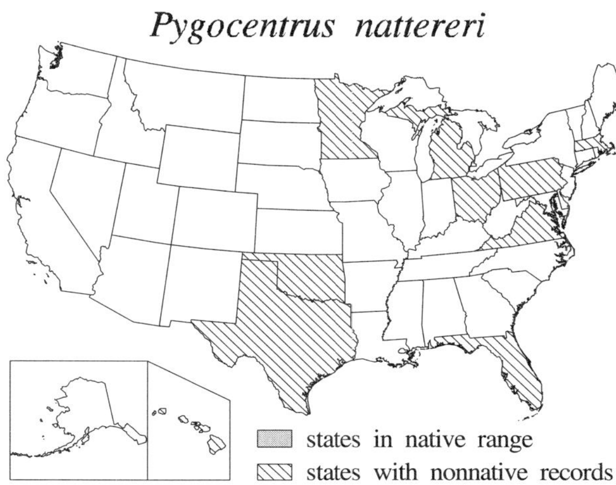

Pygocentrus nattereri Kner 1860

Common Names: red piranha (redbelly piranha).

Identification: Another commonly used name is *Serrasalmus nattereri*. Some United States specimens misidentified as members of this species may actually have been closely related species, the Orinoco piranha (*Pygocentrus notatus*, whose other names include *P. caribe* and *Serrasalmus notatus*). Pacus (*Colossoma* and *Piaractus*) also have been frequently misidentified as piranhas. Fink (1993) recently reviewed the genus *Pygocentrus* and provided a key and photographs. Maximum size: 32 cm SL.

Native Range: Tropical America. Widely distributed in lowland areas of central and southern South America east of the Andes, including the Amazon and Parana basins and various coastal drainages of the Guianas and Brazil (Fink 1993).

Nonindigenous Occurrences: A single fish was taken from a borrow pit connected to Snapper Creek in Miami, Dade County, **Florida**, ca. 1974 (Courtenay et al. 1974). One specimen was taken from Lake Mabo in Boca Raton, Palm Beach County, on 4 April 1979 (as a result, the lake was treated with rotenone to kill all fishes present) (Courtenay and Hensley 1979a: Courtenay, personal communication; museum specimen). There is also a record of one specimen taken from a canal west of Ft. Lauderdale in Broward County (no date given) (Courtenay and Hensley 1979a). The fish was first reported in **Hawaii** in Wahiawa Reservoir, a 350-acre, privately owned irrigation reservoir on Oahu, in June 1992 (Devick 1992); a mature female was taken on 23 February 1993, but there was no indication that the species had become established (Radtke 1995; W. S. Devick, personal communication). There are several records of individual fish taken in **Massachusetts** from various localities: one fish (124 mm SL) from Lexington Reservoir, Middlesex County, on 3 August 1981 (Hartel 1992; Cardoza et al. 1993); one fish (146 mm SL) from Island Grove Pond in Abington, Plymouth County, in August 1984 (Hartel 1992; Cardoza et al. 1993); one fish from a pond near Westminster, Worcester County, on 22 July 1985 (Cardoza et al. 1993); and one fish (tentative identification) from Horn Pond in Woburn, Middlesex County, during the summer of 1993 (Cardoza et al. 1993). One specimen was found dead in **Michigan** on shore of a lake near Ann Arbor, ca. 1977 (Courtenay and Hensley 1979a; W. C. Latta, personal communication). Single specimens have been taken in **Minnesota** from Duban Lake in Rice County in July 1993 (museum specimen) and from Simley Pond in Dakota County in August 1998 (museum specimen). In **Ohio**, a single fish was taken from Rocky River near London, Madison County(?), in early August 1975; a second specimen was apparently observed at the same site during the same period as was the first (Stroud 1976; D. Moreno, personal communication). In **Oklahoma**, one fish (170 mm SL) was found dead in Theta pond on the campus of Oklahoma State University in Stillwater, Payne County, on 12 November 1993 (museum specimen). One, or possibly more reports came from an unspecified locality or localities in **Pennsylvania** (Courtenay et al. 1984, 1991; Courtenay and Stauffer 1990). A single piranha (119 mm TL, 119 g) taken in **Texas** from Boerne City Reservoir in Kendall County (Howells et al. 1991a). A single fish (150 mm) was taken in **Virginia** from Indian Lake (borrow pit) in the Virginia Beach area, on 22 August 1987 (Stone 1987; Jenkins and Burkhead 1994 (identified as *Pygocentrus* sp.); Southwick, personal communication).

Means of Introduction: Probable aquarium releases.

Status: Reported from ten states, including Florida, Hawaii, Massachusetts, Michigan, Minnesota, Ohio, Oklahoma, Pennsylvania, Texas, and Virginia.

Impact of Introduction: Unknown.

Remarks: This species is the most commonly seen piranha in the aquarium trade. Although *P. nattereri* is considered one of the more aggressive piranha species, many experts consider its danger to humans greatly exaggerated. Nevertheless, considerable care must be taken in handling live individuals. Import is prohibited in many southern states (e.g., Florida and Texas). Single or several *P. nattereri* have been found in ponds, lakes, rivers, and borrow pits. In northern states, reports are typically of fish taken by hook and line during the summer or found dead during the cooler months. The reaction by some state agencies has been to rotenone the entire lake where the piranha was taken. This fish has small chance of surviving cold periods. A fish taken from the Elk River, Alabama, previously reported in a newspaper account as a red piranha (Middleton 1988), has been shown to be the pacu *Piaractus brachypomus* (museum specimen). The piranha taken from Duban Lake, Minnesota, was originally misidentified by local biologists as being a pacu. Reports of piranhas taken from the Tamiami Canal in south Florida during the period 1969–1979 were unconfirmed (Courtenay and Hensley 1979a).

Voucher specimens: Florida (UF 97062), Hawaii (UMMZ 220369), Massachusetts (MCZ 57904, NUVC uncatalogued), Minnesota (UF 98887; JFBM 30193), Oklahoma (UF 96188), Texas (uncatalogued, Texas Parks and Wildlife Department), and Virginia (uncatalogued, Virginia Department of Game and Inland Fisheries, Virginia Beach).

Serrasalmus rhombeus

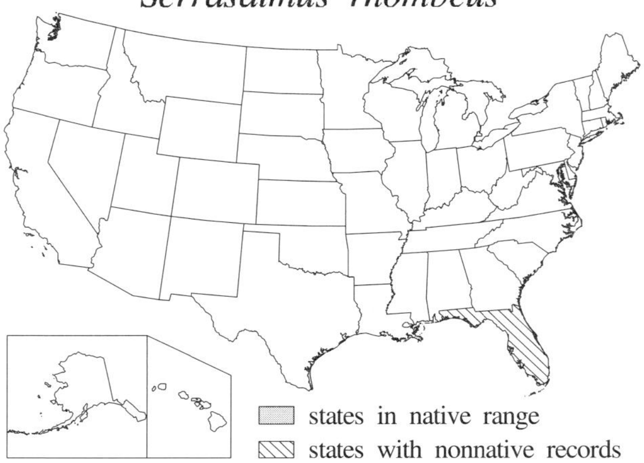

Serrasalmus rhombeus (Linnaeus 1766)

Common Names: redeye piranha (black piranha, white piranha, caribe amarillo, spotted piranha, yellow piranha).

Identification: *Serrasalmus rhombeus* represents a widespread complex of closely related species, a group in need of systematic review. A key covering piranhas from many regions of South America was given by Géry (1977); however, that work is dated and should be considered incomplete. More recently published keys cover only fishes from single regions or drainages (e.g., Nico and Taphorn 1986). Photographs of *S. rhombeus* have been provided by Géry (1977), Goulding (1980), and Nico and Taphorn (1986). A few distinguishing characteristics were given by Fink and Machado-Allison (1992) and Taphorn (1992). Live *Serrasalmus rhombeus* have red eyes; body color changes from silver in juveniles and young adults to black or dark purple in large adults. Maximum size: 40 cm SL.

Native Range: Tropical America. The species complex is widely distributed in lowland areas of South America east of the Andes, including the Amazon, Orinoco, and Essequibo basins and various coastal drainages of the Guianas.

Nonindigenous Occurrences: This species was stocked in **Florida** in an isolated sinkhole pool at Monkey Jungle, a tourist attraction in the Miami area of Dade County in 1963 or 1964. These fish reproduced and the population survived in the pool for 13 or 14 years, until all were killed or removed alive by state personnel with rotenone in September 1977 (Shafland and Foote 1979; identified as *Serrasalmus humeralis*). Two specimens were taken from an abandoned swimming pool in South Miami, Dade County (Courtenay et al. 1974). A single fish was taken from a pond in Gainesville, Alachua County in 1985 (museum specimen).

Means of Introduction: All fish were probably aquarium releases. A small number of live fish were intentionally stocked at one locality in Florida in the early 1960s (Shafland and Foote 1979).

Status: Several reports from Florida. A locally established population in the southern part of state was exterminated in 1977.

Impact of Introduction: Unknown.

Remarks: Similar to *Pygocentrus nattereri*, *Serrasalmus rhombeus* is one of the more aggressive piranha species; however, most experts consider its danger to humans to be greatly exaggerated. The Florida population at Monkey Jungle survived the winter of 1976–1977, one of the areas' coldest periods on record for this area (Shafland and Foote 1979). Preserved specimens from the Miami area deposited at UF were determined by W. L. Fink (personal communication) to be *S. rhombeus*. Import of piranhas is prohibited in many southern states (e.g., Florida, Texas).

Voucher specimens: Florida (UF 87975, UF 97059).

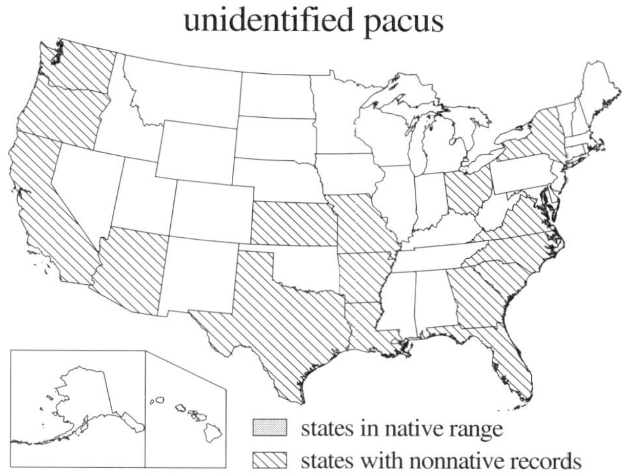

Unidentified pacus

Common Name: pacu.

Identification: Machado-Allison (1982), who provided most recent revision, recognized two genera with three valid species: *Colossoma macropomum*, *Piaractus brachypomus*, and *P. metrei*. Many synonyms exist (Machado-Allison 1982). Britski (personal communication) considered *Piaractus metrei* to be a junior synonym of *P. mesopotamicus*. Maximum size: 90 cm SL.

Native Range: Tropical America. Amazon, Orinoco, and La Plata river basins, South America (Machado-Allison 1982).

Nonindigenous Occurrences: A pacu was reported from an unspecified site in **Arizona** (no date was given) (Courtenay et al. 1984, 1991). There have been several reports of single pacus taken in **Arkansas** including one specimen taken from Lake Valencia, Maumelle in Pulaski County in June, 1995 (T. Horton, personal communication); a single specimen collected in June 1992 from the Arkansas River near Little Rock in 1992 (Wright 1995b); and another from Beaver Reservoir in the early 1990s (Wright 1995b). A pacu was taken in **California** from the San Joaquin River in Fresno County during the summer of 1987 (Courtenay, personal communication); that fish and a second possible pacu from the same area were originally reported to be piranhas in newspaper articles (e.g., Anonymous 1987b; Gordon 1987). Several specimens were collected in **Florida** from ponds in Daytona Beach, Volusia County (Courtenay et al. 1974; Courtenay and Hensley 1979a). There is an unconfirmed report of a pacu caught near Vero Beach, Indian River County, in 1979 (Courtenay and Hensley 1979a);

a single specimen was taken in Crescent Lake, St. Petersburg, Pinellas County (Courtenay and Hensley 1979a); Newnan's Lake, Gainesville, Alachua County, in July 1984 (J.B. Miller, personal communication) (this record may be a reporting error of a *Piaractus brachypomus* taken from the same site in May 1994); Crystal Lake, Pompano Beach, Broward County, ca. 1986 (Courtenay, personal communication); one fish (380 mm TL) was taken from a borrow pit near Vero Beach in Indian River County on 27 November 1988 (B. Barnett, personal communication); Pottsburg Creek on the Arlington River, Duval County, on 10 September 1989 (USGS/BRD-G database); Peace River (Lettuce Lake?), Fort Ogden, DeSoto County, in October 1989 (USGS/BRD-G database); reported from a pond in Stuart, Martin County, on 29 April 1991 (G. Gilmore, personal communication). Additional Florida records include collections or sightings of pacu in Citrus, Dade, Duval, Holmes, and Escambia counties (Massette 1993). There are several records of single fish taken from various sites in **Georgia** including a private pond in Cobb County, in 1987; Jesup or vicinity, Wayne County, in 1989; The Barnes Mill Lake subdivision, Cobb County, in June 1991; Stone Mountain Park Lake, Dekalb County, in July 1991; Lake Lanier, Forsyth County, in September 1991 (Gennings, personal communication). Two fish, identified as pacus, were taken in **Kansas** from a pond in Chatauqua County in 1984 (J. Beam, personal communication). Unidentified pacus have been reported from an unspecified locality in **Louisiana** (G.A. Tilyou, personal communication). Unidentified pacus have been reported from **Missouri** (Courtenay et al. 1991), but the only specific account is that of a single fish taken by an angler from Stockton Lake, Dade County, in 1994 (Banek, personal communication). An unidentified pacu was reported from **New York** in Onondaga Creek in Syracuse, Onondaga County, in August 1991 (R. Werner, personal communication 1991). There is a record of an unidentified pacu taken in **North Carolina** from the Pamlico River in or around 1989 (Lee 1991). A single specimen (about 25 cm long) taken with hook and line from Lake Erie in **Ohio** near an electric plant at Painesville, during the summer of 1975 (Moreno, personal communication); according to Moreno (personal communication), the fish was identified by Cleveland Aquarium staff as *C. nigripinnis* based on its black fins; this same fish was referenced as *Colossoma* sp. in Stroud (1976), and was likely the basis for Ohio *Colossoma* sp. listed in Courtenay et al. (1984, 1991). Several pacu specimens taken in **Oregon** were not identified reliably to species; these included a fish taken by gill net from the Columbia River in September 1990; a specimen taken from a pond in Salem in 1991; a fish taken from the Long Tom River, Willamette River drainage, in August 1992; and a fish taken from Sellwood Park Pond on the Willamette River in Portland, Oregon, in August 1995 (Logan et al. 1996). A single fish identified as a pacu was taken by an angler in **South Carolina** from the Sandy River, Broad River drainage, Chester County, in August 1993 (R. M. Stroud, personal communication); more recently pacus have been reported from Lake Hartwell, near Charleston (Marchand 1998). There are several records of single fish from **Texas** reported as *Colossoma* sp. by Howells et al. (1991a), including ones found in Lake Lavon, Collin County, on 13 June 1987; Medina River in Medina County, in July 1988; the confluence of the San Marcos and Blanco rivers, Hays County, in July 1990; the Rhodair Gully near Taylor Bayou, on 30 September 1990; and a marina at Sam Rayburn Reservoir, Angelina County, on 24 July 1991. Two specimens were taken in **Virginia** from the Roanoke River near Roanoke (Cochran 1987; Jenkins and Burkhead 1994, and a single specimen was caught in the Virginia Beach area during the spring of 1987 (Jenkins and Burkhead 1994). The species has been reported from an unspecified locality in **Washington** (Fletcher, personal communication).

Means of Introduction: These fish were probably aquarium releases, although some records may represent escapes from fish farms.

Status: Reported from at least 16 states including Arizona, Arkansas, California, Florida, Georgia, Kansas, Louisiana, Missouri, New York, North Carolina, Ohio, Oregon, South Carolina, Texas, Virginia, and Washington.

Impact of Introduction: Unknown.

Remarks: All reports represent collections of single fish by anglers except where indicated. All Texas pacus but one (the Medina River fish) were seen and identified by Texas Parks and Wildlife Department fisheries biologists (Howells et al. 1991a). Because most state fish and wildlife agencies do not maintain records of foreign fishes, many additional captures probably have gone unreported. A newspaper account of a piranha taken from the Lehigh River near Easton, Pennsylvania (Anonymous 1995b), was most likely a pacu based on the reported fish size. In addition to the above unidentified pacus, there are at least two cases of what are apparently hybrid pacu, *Colossoma macropomum* × *Piaractus* sp. (L. G. Nico, unpublished data; also see Logan et al. 1996). One of these was taken in Oregon from the Snake River near Ontario State Park in Malheur County, on 19 July 1991 (Logan et al. 1996; museum voucher OS 13217). The other was taken in Florida from the Pompano Canal in Pompano Beach, Broward County, in December 1991 (museum voucher UF 96235). A pacu taken with a gill net from the Columbia River in Oregon in 1990 was originally identified as *Colossoma macropomum* by Logan (personal communication), but that fish was later treated as an unidentified *Piaractus*, possibly *P. brachypomus* by Logan et al. (1996). Brazilian aquaculturists have successfully produced hybrid pacus in captivity and a similar practise is likely taking place in some U.S. ornamental fish farms (M. Goulding, personal communication)

unidentified piranahs

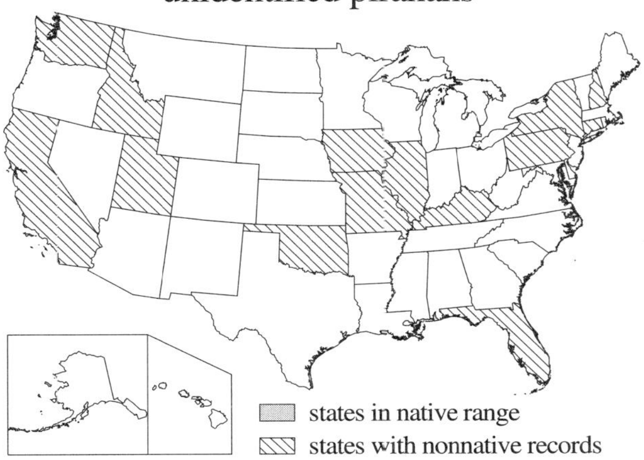

Unidentified piranhas

Common Name: piranha.

Identification: Four piranha genera are recognized, *Pygocentrus*, *Serrasalmus, Pristobrycon*, and *Pygopristis*. Of these, *Pygocentrus* and *Serrasalmus* are the only confirmed groups documented as taken from United States waters to date. Pacus (*Colossoma* and *Piaractus*) are frequently misidentified as piranhas. Maximum size: Depending on species, from 20 to 40 cm.

Native Range: Tropical America. Widely distributed in lowland areas of South America east of the Andes, including the Amazon, Orinoco, São Francisco, and La Plata basins and various coastal drainages of the Guianas and Brazil.

Nonindigenous Occurrences: Two piranhas, one captured, were reportedly occupying a park pond in Santa Barbara, **California** (Anonymous 1989). Piranhas were also reportedly taken from the San Joaquin River (Anonymous 1987b; Gordon 1987), but based on published photographs, these fish were later determined to be pacus. Two characin specimens 5-8 cm long, reportedly piranhas, were taken from the Thames River drainage in **Connecticut** (Whitworth 1996). There have been various reports of unidentified piranhas sighted in southern **Florida** (Courtenay and Hensley 1979a). Piranhas have been reported from unspecified sites in both **Idaho** (Courtenay et al. 1984, 1991; Courtenay and Stauffer 1990) and **Illinois** (Courtenay et al. 1984, 1991; Courtenay and Stauffer 1990). A single fish was taken in **Iowa** by an angler from the backwaters of the Cedar River in Linn County, in August 1983 (Anonymous 1983b). Piranhas have been reported from unspecified sites in both **Kentucky** (Courtenay et al. 1984, 1991; Courtenay and Stauffer 1990) and **Missouri** (Courtenay et al. 1984, 1991; Courtenay and Stauffer 1990). A single fish was caught in **New Hampshire** by an angler in the Merrimack River in Manchester, during the summer of 1990 (Anonymous 1990; Kennedy 1990). There are reports of one or more piranhas being captured at unspecified site(s) and dates from **New York** (Courtenay et al. 1984, 1991; Courtenay and Stauffer 1990), **Oklahoma** (Courtenay et al. 1984, 1991; Courtenay and Stauffer 1990), and **Pennsylvania** (Courtenay and Hensley 1979a; Courtenay et al. 1984, 1991; Courtenay and Stauffer 1990). One fish was taken by an angler in **Utah** from the Ogden River (Courtenay and Stauffer 1990; Courtenay, personal communication). One or more piranhas were reported from the Columbia River in **Washington** (D. Fletch, personal communication); another report is of an unspecified site and date (Courtenay and Stauffer 1990).

Means of Introduction: All were probably aquarium releases.

Status: Reported from 14 states including California, Connecticut, Florida, Idaho, Illinois, Iowa, Kentucky, Missouri, New Hampshire, New York, Oklahoma, Pennsylvania, Utah, and Washington.

Impact of Introduction: Unknown.

Remarks: The above records are not known to be supported by voucher specimens. Most or all records apparently are based on single fish taken by anglers. It is possible that some or many of these reports were based on misidentifications of other fish (e.g., pacus). For example, Welsch (1996) indicated that piranhas had been commonly collected in Nebraska but the photograph accompanying his article, although captioned as being a piranha, actually shows a *Piaractus brachypomus* specimen. The Utah fish was reported as being 15-inches long (Courtenay, personal communication). Given the large size, it also likely was a pacu, not a piranha.

ORDER SILURIFORMES

Ictaluridae - Bullhead catfishes

The freshwater catfish family Ictaluridae is composed of 7 genera and about 45 species (Nelson 1994). The family is restricted to the North American continent and ranges from southern Canada south to Guatemala (Berra 1981; Page and Burr 1991). Ictalurids range in size from small species such as the madtoms *Noturus* spp., most of which are less than 10 cm long, to large, commercially important food species such as the flathead *Pylodictis olivaris* and blue *Ictalurus furcatus* catfishes, which can reach 1.5 meters and 60 kg (Page and Burr 1991). Several other species are important as sport fishes or in aquaculture. Most ictalurid catfishes feed on aquatic invertebrates or fish. The true identity of several shipments of ictalurids introduced into California in the 1800s is quite confusing. Dill and Cordone (1997) provided a summary of information published by several different authors and speculated on which species they may have referred to. They also point out several errors that have been perpetuated in the literature.

Ameiurus brunneus

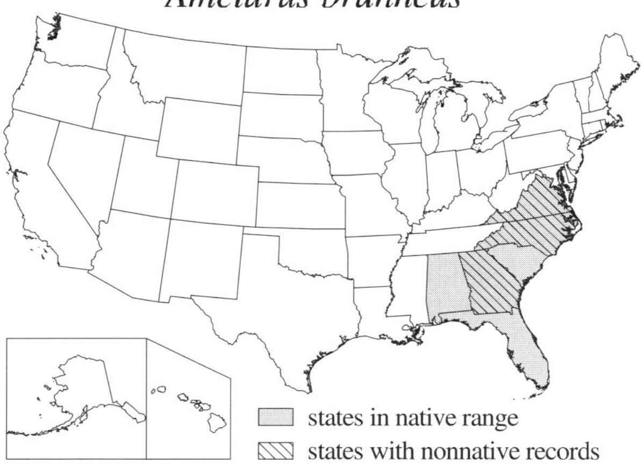

Ameiurus brunneus Jordan 1877

Common Name: snail bullhead.

Identification: Page and Burr (1991); Jenkins and Burkhead (1994); Mettee et al. (1996). Another commonly used name is *Ictalurus brunneus*. Maximum size: 29 cm.

Native Range: Atlantic Slope from the Pee Dee River drainage, southern Virginia, south to Altamaha River drainage, Georgia, and middle St. Johns River drainage, Florida; Gulf Slope in Apalachicola River basin, Georgia, Alabama, and Florida (Page and Burr 1991).

Nonindigenous Occurrences: The snail bullhead was collected in 1990 from the Etowah River in Cherokee County, **Georgia**, and from Amicalola Creek (Etowah system) in Dawson County, Georgia, in 1997 (Etnier, personal communication). This species was introduced into the upper Roanoke drainage in the Dan River above Kerr Reservoir, **Virginia** (Burkhead et al. 1980; Jenkins and Burkhead 1994). The first record from the Dan River was in 1976 (Burkhead et al. 1980; Jenkins and Burkhead 1994). In 1981, specimens were taken from the Smith River (a Dan River tributary) in Rockingham County, **North Carolina** (Menhinick 1991; Jenkins and Burkhead 1994). Menhinick (1991) lists it as native and introduced in the Roanoke drainage. Hocutt et al. (1986) list it as probably introduced in the Roanoke drainage. More recently, specimens have been collected from the Little Tennessee River in Franklin County, North Carolina (Jenkins, personal communication), and from Chatauga Reservoir (Hiwassee drainage), in Clay County (Etnier, personal communication).

Means of Introduction: Introduction of snail bullheads may have resulted from stock contamination of black bullheads (*Ameiurus melas*) that were stocked in Belews Lake, North Carolina, on the upper Dan River (Burkhead et al. 1980). Unknown means in Georgia.

Status: Established in Georgia, North Carolina, and Virginia.

Impact of Introduction: Unknown.

Remarks: Burkhead et al. (1980) provided a map and a discussion of this species' distribution in Virginia. The appearance of this species in the headwaters of the Coosa River system in northern Georgia was likely a result of natural stream capture (Bryant et al. 1979; Lee et al. 1980 et seq.).

Ameiurus catus

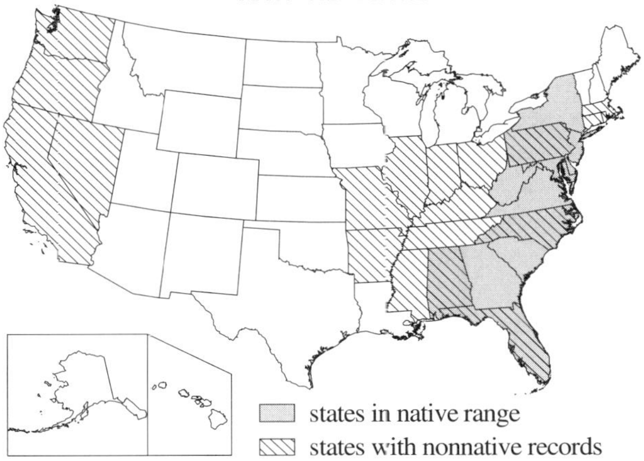

Ameiurus catus (Linnaeus 1758)

Common Name: white catfish.

Identification: Page and Burr (1991); Jenkins and Burkhead (1994); Mettee et al. (1996). Another commonly used name is *Ictalurus catus*. Maximum size: 62 cm.

Native Range: Atlantic and Gulf Slope drainages from lower Hudson River, New York, to Apalachicola basin in Florida, Georgia, and Alabama; south in peninsular Florida to Peace River drainage (modified from Page and Burr 1991).

Nonindigenous Occurrences: The white catfish was introduced into the Choctawhatchee, Tennessee, Cahaba, Coosa and Tallapoosa drainages, and the Mobile Delta, **Alabama** (Mettee et al. 1996); the lower White, lower Ouachita, lower St. Francis, lower Red (Lake Erling), and Arkansas drainages, and the Illinois system, **Arkansas** (Buchanan 1973; Cross et al. 1986; Robison and Buchanan 1988); at least 22 counties in **California** including sites in the Sacramento, San Joaquin, central California coastal, Tulare–Buena Vista lakes, and San Francisco Bay drainages (Smith 1896; Shebley 1917; Neale 1931; Moyle et al. 1976a; Swift et al. 1993; Dill and Cordone 1997); most basins in **Connecticut** (Behnke and Wetzel 1960; Whitworth et al. 1968; Schmidt 1986; Whitworth 1996); the panhandle west of the Apalachicola basin, **Florida** (Boschung 1992) (FLMNH has records from the Yellow drainage); the Illinois, Mississippi, and Kaskaskia rivers in **Illinois** (Smith 1979; Burr and Page 1986; Burr 1991); northern and southern **Indiana** (Nelson and Gerking 1968); the Ohio River, Kinniconick and Tygarts creeks, and Greenbo Lake in **Kentucky** (Clay 1975; Burr 1980; Burr and Warren 1986); the Charles, Connecticut, and Merrimack drainages in **Massachusetts** (Schmidt 1986; Hartel 1992; Cardoza et al. 1993; Hartel et al. 1996); the Tombigbee River system, **Mississippi** (Ross and Brenneman 1991); Missouri River drainage in **Missouri** (Pflieger 1997); the Truckee, Carson, and Humboldt rivers in **Nevada** (Smith 1896; Miller and Alcorn 1946; La Rivers 1962; Deacon and Williams 1984; Sigler and Sigler 1996); the French Broad River (Menhinick 1991; Etnier and Starnes 1993), the Pigeon and Hiawassee systems in the Tennessee drainage (Starnes and Etnier 1986; Menhinick 1991), **North Carolina**; Lake Erie, Sandusky Bay, the mouth of the Portage River, Sippo Lake east of Canton, Springdale Lake near Cincinnati, the Blanchard system, the Muskingum drainage, and the Ohio River in **Ohio** (Trautman 1981; Emery 1985; Hocutt et al. 1986); the Willamette River and other locations in western **Oregon** (Smith 1896; Wydoski and Whitney 1979; Bond 1994); Lake Erie and the Ohio River basin in **Pennsylvania** (Hendricks et al. 1979; Cooper 1983; Emery 1985); **Rhode Island** (Lapin, personal communication); the Clinch River in **Tennessee** (Burkhead, personal communication), and the Cowlitz River and Silver Lake in **Washington** (Smith 1896; Lampman 1946). Underhill (1986) reports this species as introduced into the Lake Erie drainage (state unlisted).

Means of Introduction: Intentional stocking for sport and food. Whitworth (1996) points out that because of this species' high tolerance for salinity, the original movement to large rivers in Connecticut could have been a natural range extension. Primary source of stocked fish in Kentucky is the James River, Virginia (Clay 1975). Stock planted in Indiana is from southeastern Virginia (Nelson and Gerking 1968). It is frequently stocked in fee-fishing lakes and other private waters. Pflieger (1997) indicated that specimens recorded from natural waters in Missouri may represent escapes from such situations.

Status: Many of these introductions have led to established populations. The Mississippi report is of a single individual presumably taken after 1989, because it was not reported in two publications in 1989. In apparent reference to its occurrence in the Willamette drainage and Columbia River of Oregon, Bond (1994) noted establishment as uncertain. A single individual was reported from Tennessee on an angler's stringer in 1993 (Burkhead, personal communication). Apparently not established in Arkansas (Robison and Buchanan 1988). Status in Tennessee is unknown.

Impact of Introduction: White catfish were apparently responsible for the disappearance of Sacramento perch

Archoplites interruptus in Thurston Lake, California (McCarraher and Gregory 1970).

Remarks: Yerger (1977) and Dahlberg and Scott (1971a) considered the white catfish as introduced into the Chattahoochee River. However, Mettee et al. (1996) believe it is native to the Chattahoochee based on its abundance; Boschung (1992) listed it as introduced only west of the Apalachicola; and Lee et al. (1980 et seq.) reported it as native. We have chosen to regard it as native to the Apalachicola–Chattahoochee drainage based on early records and zoogeographic pattern. The Apalachicola–Chattahoochee drainage is commonly the western edge of the distributional range of upland species that occur further to the east, as white catfish do. This species was introduced into California under the common name of "Schuylkill catfish" (Dill and Cordone 1997).

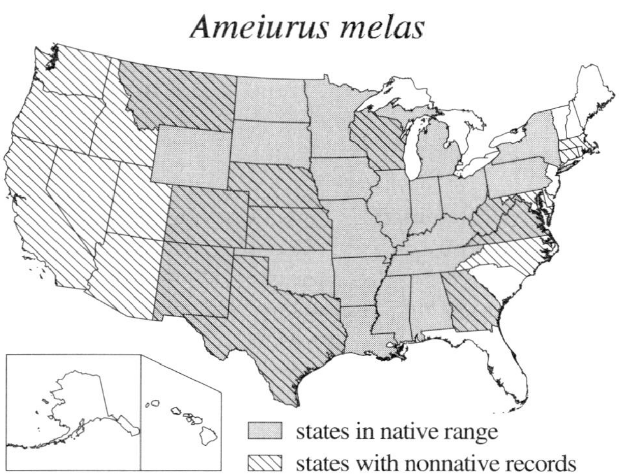

Ameiurus melas (Rafinesque 1820)

Common Name: black bullhead.

Identification: Becker (1983); Page and Burr (1991); Etnier and Starnes (1993); Jenkins and Burkhead (1994). Another commonly used name is *Ictalurus melas*. Maximum size: 46 cm.

Native Range: Great Lakes, Hudson Bay, and Mississippi River basins from New York to southern Saskatchewan and Montana, south to Gulf; Gulf Slope drainages from Mobile Bay, Georgia and Alabama, to northern Mexico. Apparently not native to Atlantic Slope (Page and Burr 1991).

Nonindigenous Occurrences: The black bullhead has been introduced into, or recorded from, rivers and reservoirs in the Salt, Gila, Verde, San Pedro, Santa Cruz, Colorado, and Little Colorado drainages in **Arizona** (Minckley 1973; Hendrickson et al. 1980; Lee et al. 1980 et seq.); the Klamath, Central Valley, Pajaro–Salinas, Lahontan, Colorado and southern California drainages in **California** (Moyle et al. 1976a; Bell 1978; Wydoski and Whitney 1979; Swift et al. 1993; Dill and Cordone 1997); the western slope (Colorado River drainage) in **Colorado** (Vanicek et al. 1970; Everhart and Seaman 1971; Beckman 1974; Tyus et al. 1982); the Connecticut River and Housatonic drainage, **Connecticut** (Behnke and Wetzel 1960; Whitworth et al. 1968; Schmidt 1986; Whitworth 1996); probably introduced into nonnative areas of **Georgia** (Dahlberg and Scott 1971b); the Spokane and middle Snake drainages in western **Idaho** (Linder 1963; Simpson and Wallace 1978; Idaho Fish and Game 1990); fee-fishing lakes in eastern **Kansas** (Cross 1967); upper Chesapeake drainage in **Maryland** (Lee et al. 1980 et seq.); Mill River in Hampden County, **Massachusetts** (Hartel 1992; Cardoza et al. 1993); the Kootenai, Clark Fork, Yellowstone, and Missouri drainages in **Montana** (Brown 1971; Cross et al. 1986; Holton 1990); in the Sand Hills lakes and other areas in **Nebraska** (Jones 1963); the Rio Grande, Pecos, San Juan, San Francisco, and Gila drainages in **New Mexico** (Koster 1957; Tyus et al. 1982; Sublette et al. 1990; Platania 1991); the Carson, Humboldt, Walker, and Reese rivers and the Colorado drainage in **Nevada** (Miller and Alcorn 1946; La Rivers 1962; Hubbs et al. 1974; Deacon and Williams 1984); possibly into the Catawba, Pee Dee (Yadkin), and Roanoke drainages in **North Carolina** (Burkhead et al. 1980; Hocutt et al. 1986; Menhinick 1991); the Willamette (Smith 1896; Lampman 1946), Snake, Columbia, and Chewaucan rivers (Bond 1973, 1994), and Benton County (Logan 1994), **Oregon**; the Trans-Pecos drainages, **Texas** (Hubbs et al. 1991); the Logan, Jordan, Bear (Sigler and Miller 1963), Colorado, Green, Duchesne, Strawberry, White (Vanicek et al. 1970; Tyus et al. 1982), Jordan, Bear, and Logan rivers (Sigler and Miller 1963), and Utah Lake and Willard Bay Reservoir (Sigler and Sigler 1987) in **Utah**; the lower Dan River near Kerr Reservoir, Buffalo Creek at Kerr Reservoir, and the New River in Grayson County (1939), **Virginia** (Burkhead et al. 1980; Lee et al. 1980 et seq.; Hocutt et al. 1986; Jenkins and Burkhead 1994); the Hanford reach of the Columbia River, perhaps Lake Washington, and other non-specific localities in **Washington** (Gray and Dauble 1977; Wydoski and Whitney 1979; Fletcher, personal communication); Sherwood Lake in the Greenbrier River system, **West Virginia** (Burkhead et al. 1980); northern **Wisconsin** (Becker 1983).

Means of Introduction: Intentionally stocked for sport and as a food fish. In Idaho, it was probably accidentally stocked with brown bullhead *A. nebulosus* or with channel catfish *Ictalurus punctatus* (Linder 1963).

Status: Established in most locations where introduced. Apparently not established in Connecticut (Whitworth 1996). A single report from Massachusetts in the 1940s (Hartel 1992).

Impact of Introduction: Introduced black bullhead eat endangered humpback chubs *Gila cypha* in the Little Colorado River, and may exert a major negative effect on the population there (Marsh and Douglas 1997). Minckley (1973) reported that this species is generally considered a pest in Arizona as it forms large stunted populations that compete with more desirable fishes for space and food. Black bullheads are voracious predators of newly hatched gamefish (Whitmore 1997). Introduced predatory fishes, including the black bullhead, are likely at least partially responsible for the decline of the Chiricahua leopard frog *Rana chiricahuensis* in southeastern Arizona (Rosen et al. 1995).

Remarks: Although the black bullhead has not been reported from South Carolina, it is probable that it has been introduced into the state. Tyus et al. (1982) gave a distribution map of the this species in the upper Colorado basin. Occurrence in Maryland is not listed by Lee et al. (1976, 1981) or Rohde et al. (1994). Becker (1983) states that in Wisconsin the black bullhead's distribution is much more extensive now than it was in the mid-1920s, especially in northern Wisconsin.

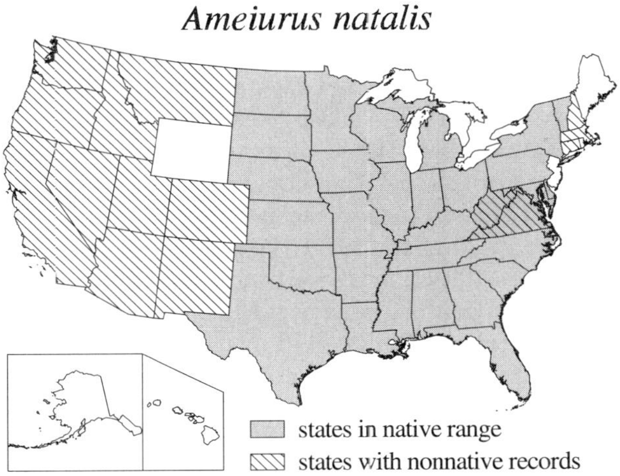

Ameiurus natalis (Lesueur 1819)

Common Name: yellow bullhead.

Identification: Becker (1983); Page and Burr (1991); Etnier and Starnes (1993); Jenkins and Burkhead (1994). Another commonly used name is *Ictalurus natalis*. Maximum size: 47 cm.

Native Range: Atlantic and Gulf Slope drainages from New York to northern Mexico, and St. Lawrence-Great Lakes and Mississippi River basins from southern Quebec west to central North Dakota, and south to Gulf (Page and Burr 1991).

Nonindigenous Occurrences: The yellow bullhead has been introduced into all the major drainages in **Arizona** (Minckley 1973; Wydoski and Whitney 1979; Deacon and Williams 1984), including the Colorado, Gila, Salt, Little Colorado, San Juan, and Bill Williams; the Lost, Colorado, Sacramento, San Joaquin, and Klamath drainages, and drainages in southern **California** (Moyle 1976a; Wydoski and Whitney 1979); the Colorado and Rio Grande drainages, **Colorado** (Zuckerman and Behnke 1986; Walker 1993); the Connecticut, Housatonic and Thames drainages of **Connecticut** (Whitworth 1996); the Snake River, above and below the falls, in **Idaho** (Idaho Department of Fish and Game 1996); drainages in eastern and central **Massachusetts** (Hartel 1992; Cardoza et al. 1993; Hartel et al. 1996); the Beaver, Tongue, Yellowstone, and Powder rivers, two locations in the Clark's Fork drainage in **Montana** (Brown 1971; Cross et al. 1986; Holton 1990); the Colorado River, **Nevada** (Deacon and Williams 1984); the Merrimack River and several other non-specific areas in **New Hampshire** (Bailey and Oliver 1939; Scarola 1973; Schmidt 1986); the Rio Grande, Canadian, Gila, San Francisco, Pecos, Dry Cimarron, and Arkansas drainages in **New Mexico** (Koster 1957; Sublette et al. 1990; Platania 1991); the Willamette River (Smith 1896; Lampman 1946; Bond 1973, 1994), Benton County (Logan 1994), and Warner Valley (Sigler and Sigler 1987), **Oregon**; upper Lake Powell, the Colorado River, and the lower San Juan River, **Utah** (Tyus et al. 1982; Sigler and Sigler 1996); possibly the upper Levisa system in the Big Sandy drainage, **Virginia** (Jenkins and Burkhead 1994); Hanford reach of the Columbia River, and other non-specific locations in **Washington** (Lampman 1946; Gray and Dauble 1977; Fletcher, personal communication); and the New River, **West Virginia** (Burkhead et al. 1980; Jenkins and Burkhead 1994; Stauffer et al. 1995).

Means of Introduction: Intentionally stocked.

Status: Established in most introduced waters. Recently discovered in West Virginia (Jenkins and Burkhead 1994). First found in Connecticut in the late 1970s (Whitworth 1996). Extirpated from the Pecos drainage in New Mexico (Sublette et al. 1990).

Impact of Introduction: Introduced predatory fishes, including the yellow bullhead, are likely at least partially responsible for the decline of the Chiricahua leopard frog *Rana chiricahuensis* in southeastern Arizona (Rosen et al. 1995).

Remarks: Neither Everhart and Seaman (1971) nor Beckman (1974) included this species in the listings of Colorado fishes.

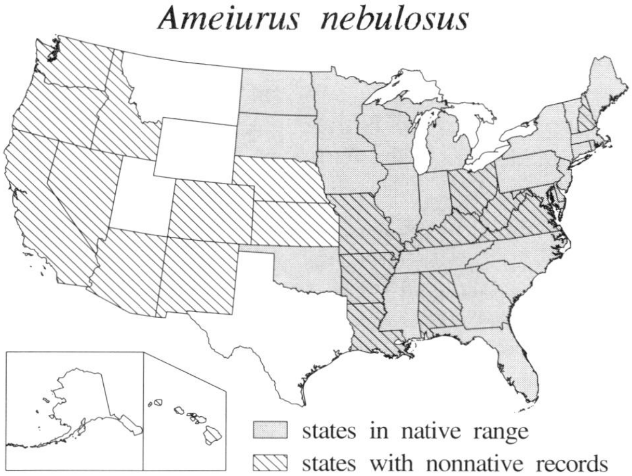

Ameiurus nebulosus (Lesueur 1819)

Common Name: brown bullhead.

Identification: Becker (1983); Page and Burr (1991); Etnier and Starnes (1993); Jenkins and Burkhead (1994). Another commonly used name is *Ictalurus nebulosus*. Maximum size: 50 cm.

Native Range: Atlantic and Gulf Slope drainages from Nova Scotia and New Brunswick to Mobile Bay, Alabama, and St. Lawrence–Great Lakes, Hudson Bay, and Mississippi River basins from Quebec west to southeastern Saskatchewan, and south to Louisiana (Page and Burr 1991). This species may have been originally absent from all or part of the Gulf Coast west of the Apalachicola and east of the Mississippi River (Gilbert, personal communication). This speculation is based on the very spotted distribution of the species both in panhandle Florida and Alabama and the fact that it appears to be largely confined to reservoirs in Alabama. In its native range in peninsular Florida it is found primarily in larger bodies of water; whereas, on the Atlantic Slope in Florida, this species is found in both streams and sloughs (Gilbert, personal communication).

Nonindigenous Occurrences: The brown bullhead has been introduced into the lower Tombigbee in **Alabama** (Lee et al. 1980 et seq.; Mettee et al. 1996); non-specific areas of **Arizona** (Minckley 1973); several impoundments in **Arkansas** (Robison and Buchanan 1988); numerous drainages in **California** (Smith 1896; Shebley 1917; Neale 1931; Lampman 1946; Moyle 1976a; Swift et al. 1993; Dill and Cordone 1997); the South Platte drainage (Propst and Carlson 1986; Walker 1993), and possibly other locations in **Colorado** (Everhart and Seaman 1971; Beckman 1974); the islands of Hawaii and Oahu in **Hawaii** (Cobb 1902; Jordan and Evermann 1902, 1905; Brock 1960; Maciolek 1984); the Snake, Kootenai, Palouse, Spokane, Pend Orielle, Clearwater, and Bear drainages in **Idaho** (Linder 1963; Simpson and Wallace 1978; Idaho Fish and Game 1990); fee-fishing lakes in eastern **Kansas**, including Cherokee, Crawford and Linn counties (Cross 1967; Cross et al. 1986; Cross and Collins 1995); the Licking, Rolling Fork, Salt, and Kentucky drainages in **Kentucky** (Burr and Warren 1986); Bayou Lafourche in the Boeuf drainage in **Louisiana** (Douglas 1974); in sites in the lower Missouri drainage of **Missouri** including the Chariton-Nishnabotna drainage (Pflieger 1971, 1975; Cross et al. 1986) and Truman Reservoir on the Osage River (Pflieger 1997); private ponds in eastern **Nebraska** (Jones 1963; Cross et al. 1986) (Morris et al. (1974) show two locations along the lower Platte River); the Carson, Truckee, Humboldt, and Muddy drainages in **Nevada** (Miller and Alcorn 1946; La Rivers 1962; Hubbs et al. 1974; Deacon and Williams 1984; Sigler and Sigler 1987); possibly the Androscoggin drainage in **New Hampshire** (Scarola 1973); non-specific in **New Mexico** (Koster 1957; Sublette et al. 1990); southern **Ohio** (Trautman 1981); the Willamette, Klamath, and Coos drainages, Devil's Lake and Benton County, **Oregon** (Smith 1896; Chapman 1942; Lampman 1946; Bond 1973, 1994; U.S. Fish and Wildlife Service 1993e); the New River below Claytor Lake, North Fork Holston, and impoundments on the Powell River, **Virginia** (Burkhead et al. 1980; Jenkins and Burkhead 1994); Silver Lake near Castlerock (Lampman 1946) and probably other localities in **Washington** (Smith 1896; Chapman 1942; Wydoski and Whitney 1979; Fletcher, personal communication); and the New River, **West Virginia** (Burkhead et al. 1980; Jenkins and Burkhead 1994; Stauffer et al. 1995).

Means of Introduction: Intentionally stocked for food and sport. Pflieger (1997) suspected that brown bullheads observed in Truman Reservoir, Missouri, may have spread from populations introduced into upstream sites in Kansas.

Status: Generally established in most waters where introduced. Reported in Virginia and West Virginia. Extirpated in New Mexico (Sublette et al. 1990). It was reportedly stocked in 1893 in Hawaii but the fish failed to survive (Jordan and Evermann 1905; Brock 1960).

Impact of Introduction: Unknown.

Remarks: The brown bullhead was among the first fish introduced to the West Coast, as early as 1874 (Smith 1896). Early accounts of "bullhead" or "hornpout" introductions in the West referred to this species (Dill and Cordone 1997). Stauffer et al. (1995) listed this species as native but possibly introduced to the Kanawha drainage above the falls (New). Pflieger (1997) indicated that brown bullhead had been stocked in ponds in various parts of Missouri in the past, but the status of these stockings is not known. He also stated that some reports of a bullhead species may be based on misidentifications based on the fact that collectors are often unable to correctly distinguish the different species in this closely related group of catfishes.

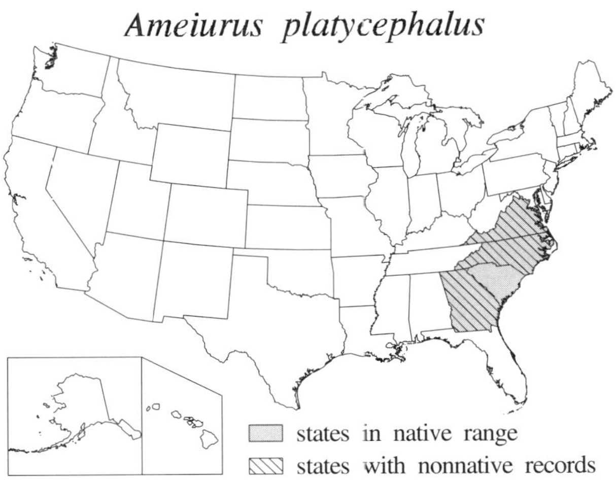

Ameiurus platycephalus (Girard 1859)

Common Name: flat bullhead.

Identification: Page and Burr (1991); Jenkins and Burkhead (1994). Another commonly used name is *Ictalurus platycephalus*. Maximum size: 29 cm.

Native Range: Atlantic Piedmont and Coastal Plain from Roanoke River drainage, Virginia, to Altamaha River drainage, Georgia (Page and Burr 1991).

Nonindigenous Occurrences: This species has been taken from the Chattahoochee River drainage, **Georgia** (Couch et al. 1995). It has been introduced into the French Broad, Nolichucky, Little Tennessee, Toe, and Hiwassee systems of the Tennessee drainage, **North Carolina** (Menhinick 1991; Jenkins and Burkhead 1994). Two individuals collected from the upper James River in Botetourt County, **Virginia**, in 1984 may have escaped from a stocked pond (Jenkins and Burkhead 1994).

Means of Introduction: This species was intentionally stocked in North Carolina. Although tentatively considered native to the James River, Virginia, Jenkins and Burkhead (1994) reported that the species may have escaped from a stocked farm pond. Possibly introduced as a sportfish in drainages of Georgia outside its native range.

Status: Established in North Carolina; probably established in Georgia. Reported in Virginia.

Impact of Introduction: Unknown.

Remarks: The two Virginia specimens taken in the upper James were lost (Jenkins and Burkhead 1994). This species closely resembles the snail bullhead *Ameiurus brunneus*, and some flathead records may be based on misidentifications. Dahlberg and Scott (1971a) did not list this species from the Chattahoochee drainage in Georgia.

Ictalurus furcatus

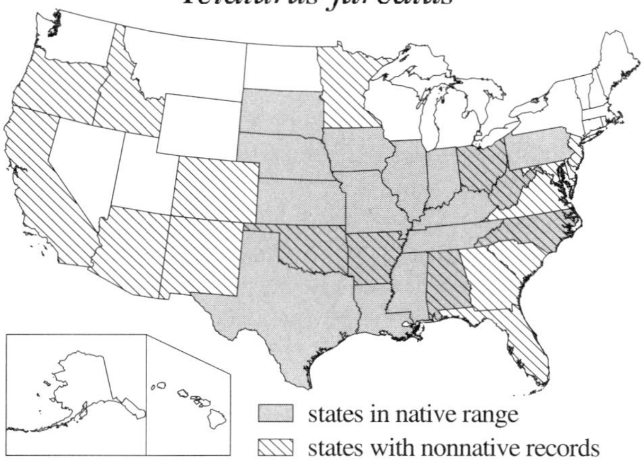

Ictalurus furcatus (Lesueur 1840)

Common Name: blue catfish.

Identification: Smith (1979); Page and Burr (1991); Etnier and Starnes (1993); Jenkins and Burkhead (1994). Maximum size: 165 cm.

Native Range: Mississippi River basin from western Pennsylvania to southern South Dakota and Platte River, southwestern Nebraska, south to Gulf; Gulf Slope from Mobile Bay basin, Alabama, to Rio Grande drainage, Texas and New Mexico. Also Atlantic Slope of Mexico (Page and Burr 1991) if not distinguished separately from *I. medionalis* (Gilbert 1998).

Nonindigenous Occurrences: The blue catfish has been stocked in the Chattahoochee River (Dahlberg and Scott 1971b), the Choctawhatchee River and perhaps the Conecuh River (Mettee et al. 1996) in **Alabama**; in the Colorado River in **Arizona** (Minckley 1973); the Ouachita, White, St. Francis, and Red drainages in **Arkansas** (Robison and Buchanan 1988) (based on native range in Lee et al. 1980 et seq.); several reservoirs in southern **California** drainages (Richardson et al. 1970; Moyle 1976a); the Arkansas, upper Rio Grande, and Platte drainages in **Colorado** (Everhart and Seaman 1971; Barkuloo 1967; Zuckerman and Behnke 1986; Walker 1993); ponds in the **Florida** panhandle (Eglin Air Force Base 1968), and possibly in the Choctawhatchee River downstream from the Alabama collections; the Savannah and Chattahoochee rivers in **Georgia** (Dahlberg and Scott 1971b; Ober, personal communication); the Snake River, **Idaho** (Idaho Fish and Game 1990); the Potomac River, **Maryland** (Lee et al. 1976, 1981; Rohde et al. 1994; museum specimen); Lake St. Croix and Lake Pepin, **Minnesota** (Phillips et al. 1982); Morris and Passaic counties, **New Jersey** (Fowler 1952; Stiles 1978); the San Juan and Canadian rivers, **New Mexico** (Minckley 1973; Sublette et al. 1990); the Cape Fear, Catawba, Neuse, and Yadkin drainages, **North Carolina** (Guire et al. 1984; Hocutt et al. 1986; Menhinick 1991; Rohde et al. 1994); Indian and Buckeye lakes, and the Great Miami and Muskingum drainages, **Ohio** (Trautman 1981; Hocutt et al. 1986; Burr and Page 1986); impoundments in **Oklahoma** (Miller and Robison 1973); the Columbia River and Willamette River, **Oregon** (Lampman 1946; Bond 1994); the Savannah River, **South Carolina** (Dahlberg and Scott 1971b; Rohde et al. 1994); and the Potomac, lower Rappahannock, and lower James drainages, and Lake Anna in the upper York drainage, **Virginia** (Hocutt et al. 1986; Burkhead and Jenkins 1994).

Means of Introduction: Intentionally stocked for food and sport. Stocked in the Cape Fear River, North Carolina, in 1966 (Guire et al. 1984). Introductions in the Choctawhatchee River, Alabama, were due to flooding of a private lake in 1993 (Mettee et al. 1996). Recent introductions into the Chattahoochee River in Alabama and Georgia were due to flooding of catfish farms in Alabama during a storm in March 1990 (Ober, Personal communication).

Status: Established in most locations. Probably extirpated from the San Juan and Canadian drainages in New Mexico (Sublette et al. 1990). Established in the Chattahoochee River in Alabama and Georgia (Ober, personal communication).

Impact of Introduction: Hybridizes with threatened Yaqui catfish *I. pricei* in Mexico (U.S. Fish and Wildlife Service 1994b).

Remarks: Blue catfish has been stocked to feed on the introduced Asian clam *Corbiucula fluminea*. Although the species may not actually control clam populations, it is hoped that clam biomass could be converted to fish biomass and create trophy-sized catfish to catch (Dill and Cordone 1997). Blue catfish in the Cape Fear River, North Carolina, feed almost exclusively on *Corbicula* (M. Moser, personal communication). Not listed as occurring in South Carolina by Loyacano (1975). Not listed as occurring in Idaho by Simpson and Wallace (1978). Reports of *I. furcatus* in the New drainage in West Virginia and Virginia are more likely misidentified *I. punctatus* (Burkhead et al. 1980). See Burkhead et al. (1980) for discussion of these reports. Stauffer et al. (1995) do not list this species for the Kanawha (including the New) drainage of West Virginia.

Ictalurus pricei

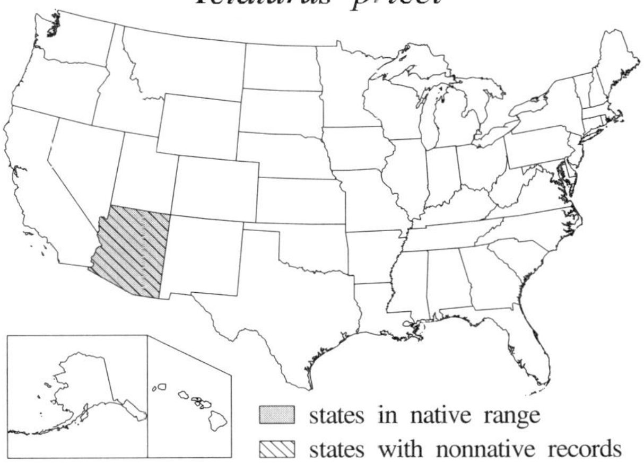

Ictalurus punctatus

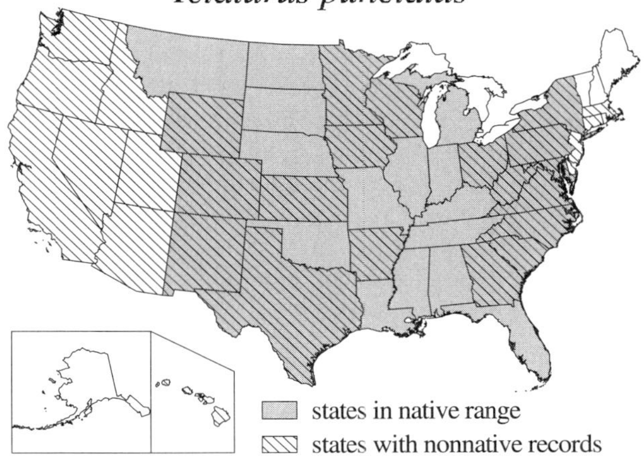

Ictalurus pricei (Rutter 1896)

Common Name: Yaqui catfish.

Identification: Minckley (1973); Page and Burr (1991). Maximum size: 57 cm.

Native Range: Rio Yaqui and Rio Casas Grandes drainages, northwestern Mexico (Page and Burr 1991).

Nonindigenous Occurrences: The Yaqui catfish was introduced in 1899 into the Santa Cruz River system, **Arizona**, in a reservoir fed by Monkey Spring, where the species persisted until the 1950s (Minckley 1973; Hendrickson et al. 1980; Lee et al. 1980 et seq.; U.S. Fish and Wildlife Service 1994b).

Means of Introduction: Unknown.

Status: Extirpated in Arizona and the United States (Minckley 1973; Lee et al. 1980 et seq.).

Impact of Introduction: Unknown.

Remarks: The Yaqui catfish is listed as a federally threatened species (Page and Burr 1991; U.S. Fish and Wildlife Service 1993a). It hybridizes with introduced channel and blue catfish (U.S. Fish and Wildlife Service 1994b).

Ictalurus punctatus (Rafinesque 1818)

Common Name: channel catfish.

Identification: Becker (1983); Page and Burr (1991); Etnier and Starnes (1993); Jenkins and Burkhead (1994). Maximum size: 127 cm.

Native Range: St. Lawrence–Great Lakes, Hudson Bay (Red River drainage), and Missouri-Mississippi River basins from southern Quebec to southern Manitoba and Montana south to Gulf. Possibly also native on Atlantic and Gulf Slopes from Susquehanna River to Neuse River, and from Savannah River to Lake Okeechobee, Florida, and west to northern Mexico and eastern New Mexico (Page and Burr 1991).

Nonindigenous Occurrences: The channel catfish has been extensively introduced into states including **Arkansas** (Pritchard et al. 1978; Robison and Buchanan 1988); the Colorado, Gila, Bill Williams, Little Colorado, and Rio Magdalena drainages in **Arizona** (Minckley 1973; Tyus et al. 1982); Lake Cuyamaca in San Diego County, **California**, in 1891 (Smith 1896), and widely stocked since then throughout the state (Shebley 1917; Lampman 1946; Moyle 1976a; Moyle and Daniels 1982; Dill and Cordone 1997); the Colorado, Green, San Juan, Yampa, White, Republican and Rio Grande drainages in **Colorado** (Vanicek et al. 1970; Everhart and Seaman 1971; Holden and Stalnaker 1975; Tyus et al. 1982; Walker 1993; Tyus 1998); the Connecticut and Housatonic rivers, **Connecticut** (Behnke and Wetzel 1960; Whitworth et al. 1968; Schmidt 1986; Whitworth 1996); the Delaware and Nanticoke rivers in **Delaware** (Lee et al. 1976, 1981; Schmidt 1986; Raasch and Altemus 1991); the Altamaha, Ocmulgee, Oconee, Satilla,

Ogeechee, Savannah, and St. Mary's rivers in **Georgia** (Dahlberg and Scott 1971a, 1971b); the islands of Hawaii, Kaui, Maui, and Oahu, **Hawaii** (Brock 1960; Maciolek 1984); the Boise River, **Idaho**, in 1893 and later the Snake and Little Wood rivers (Linder 1963; Simpson and Wallace 1978); non-specific areas of **Iowa** (Harlan et al. 1987); non-specific areas in **Kansas** (Cross 1967; Cross and Collins 1995); throughout **Maryland** (Lee et al. 1976, 1981); the Connecticut, Merrimack, Chicopee, Manhan, Quaboag, Agawam and Charles rivers, five ponds in Hampden County, and Silver Lake in Essex County, **Massachusetts** (Lee et al. 1980 et seq.; Schmidt 1986; Hartel 1992; Cardoza et al. 1993; Hartel et al. 1996); southwestern **Minnesota** (Phillips et al. 1982); Lake Mead, Pyramid Lake, the lower Humboldt River, and Lahontan Reservoir in **Nevada** (La Rivers 1962; Deacon and Williams 1984); many areas of **New Jersey**, including Greenwood Lake and Lake Hopatcong (Morse 1905; Fowler 1906, 1952; Stiles 1978; R. Soldwedel, personal communication); the Mimbres, Rio Grande, Pecos, San Juan, Little Colorado, San Francisco, and Gila drainages in **New Mexico** (Koster 1957; Tyus et al. 1982; Sublette et al. 1990; Platania 1991); the Susquehanna drainage in **New York** (T. Sinnott, personal communication); the Broad, Catawba, Yadkin, Neuse, Tar, Roanoke, Chowan, Cape Fear, and Northeast Cape Fear drainages in **North Carolina** (Hocutt et al. 1986; Menhinick 1991); Indian Lake, **Ohio** (Trautman 1981); the Columbia, Willamette, Umpqua, and Walla Walla rivers, **Oregon** (Smith 1896; Lampman 1946; Bond 1973; Wydoski and Whitney 1979; Bond 1994); the Susquehanna drainage, **Pennsylvania** (Bean 1892b; Cooper 1983 (for locations); Hocutt et al. 1986); the Pee Dee, Santee-Cooper, Edisto, Combahee, Broad, and Savannah rivers in **South Carolina** (Dahlberg and Scott 1971b; Loyacano 1975; Hocutt et al. 1986); Sanctuary Pond in Pflugerville, ponds and reservoirs in **Texas** (Knapp 1953; Howells and Prentice 1991); Utah Lake and the Green, Colorado, Bear, Jordan, San Juan, Price, San Rafael, Duchesne, and Sevier rivers, **Utah** (Sigler and Miller 1963; Vanicek et al. 1970; Tyus et al. 1982; Sigler and Sigler 1987); the Potomac, Rappahannock, York, James, Chowan and Roanoke drainages in **Virginia** (Hocutt et al. 1986; Jenkins and Burkhead 1994); the Columbia River and Lake Washington, **Washington** (Smith 1896; Lampman 1946; Gray and Dauble 1977; Wydoski and Whitney 1979); the Potomac River, **West Virginia** (Stauffer et al. 1995); the Rock River, Waukesha County lakes, Lake Helen in Portage County, and several rivers in the Manitowoc Sheboygan watershed, **Wisconsin** (Becker 1983); and the North Platte and Bighorn drainages, the Little Snake Connor River, and Flaming Gorge Reservoir in **Wyoming** (Baxter and Simon 1970; Hubert 1994).

Means of Introduction: Intentionally stocked for sport fishing and food. The first introductions in the Colorado River took place in 1892–1893 or in 1906 (Miller and Alcorn 1946). They had become established throughout the Colorado basin by the early 1900s (Holden and Stalnaker 1975). The earliest stocking record for the Yampa River is from 1944 and involved 34,200 fingerling catfish (Tyus 1998). The introductions into Silver Lake and the Charles River in Massachusetts involved albino fish from the aquarium trade (Cardoza et al. 1993).

Status: Established in most waters where introduced.

Impact of Introduction: The channel catfish hybridizes with the threatened Yaqui catfish *I. pricei* in Mexico (U.S. Fish and Wildlife Service 1994b). Colorado squawfish *Ptychocheilus lucius*, an endangered species, have been documented to choke on introduced channel catfish when attempting to eat them (McAda 1983; Pimental et al. 1985; U.S. Fish and Wildlife Service 1990b). Jenkins and Burkhead (1994) speculated that introduced channel catfish may have contributed to the demise of an isolated population of trout-perch *Percopsis omiscomaycus* in the Potomac River in Virginia and Maryland. Introduced channel catfish may exert a major negative effect on populations of various endangered species. For instance, this species is known to prey on small and large endangered humpback chub *Gila cypha* in the Little Colorado River thereby limiting recruitment and also increasing adult mortality (Marsh and Douglas 1997). There is also evidence that this introduced catfish preys heavily on juveniles of razorback sucker *Xyrauchen texanus* that had been reintroduced into the Gila River of Arizona (Marsh and Brooks 1989). Introduced predatory fishes, including the channel catfish, may be partially responsible for the decline of the Chiricahua leopard frog *Rana chiricahuensis* in southeastern Arizona (Rosen et al. 1995).

Remarks: Tyus et al. (1982) gave a distribution map for this species in the upper Colorado basin. Channel catfish have also been stocked in many native areas including Arkansas (Robison and Buchanan 1988); Illinois (Burr, personal communication); Nebraska (Jones 1963). Harlan et al. (1987) stated that stocking in Iowa has widened this species' distribution. Cross and Collins (1995) mapped the species in every county in Kansas. Cross (1967) indicated a much more restricted distribution in the state and did not include every county. Presumably the more recent map indicates the species had been introduced to new locations since the 1967 publication. Cross (1967) also stated that it had been stocked in many lakes and ponds in the state.

Noturus exilis

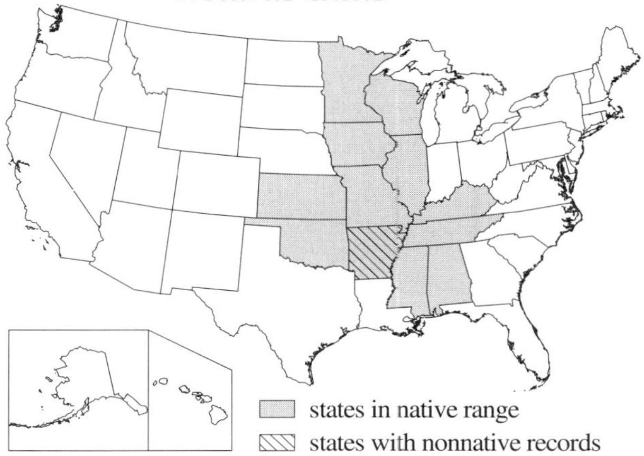

Noturus exilis (Nelson 1876)

Common Name: slender madtom.

Identification: Becker (1983); Robison and Buchanan (1988); Page and Burr (1991); Etnier and Starnes (1993); Pflieger (1997). Maximum size: 15 cm.

Native Range: Green, Cumberland, and Tennessee River drainages from central Kentucky to nothern Alabama; and the upper Mississippi River basin from southern Wisconsin and southern Minnesota to Ozark an Ouachia Highlands of Arkansas, Kansas, and Oklahoma (Page and Burr 1991).

Nonindigenous Occurrences: Two specimens were taken from the Mountain Fork River (Little River system, Red River drainage) in Polk County, **Arkansas**, near the Oklahoma border in 1976 (Robison and Winters 1978; Robison and Buchanan 1988).

Means of Introduction: Possible bait release; alternatively, this collection could represent an unusual natural distribution pattern that may give clues to past river connections (Robison and Winters 1978; Robison and Buchanan 1988).

Status: Reported in Arkansas.

Impact of Introduction: Unknown.

Remarks: This species had never been collected in this area previously, despite intensive sampling, making it unlikely that this species was native but overlooked. In addition, this species is known to be used as bait in the area (Robison and Winters 1978). An additional five specimens were taken from Mill Creek (tributary of Mt. Fork River, Red River drainage) by C.R. Gilbert in 1971 (UF 22742) (Robison and Buchanan 1988).

Noturus flavus

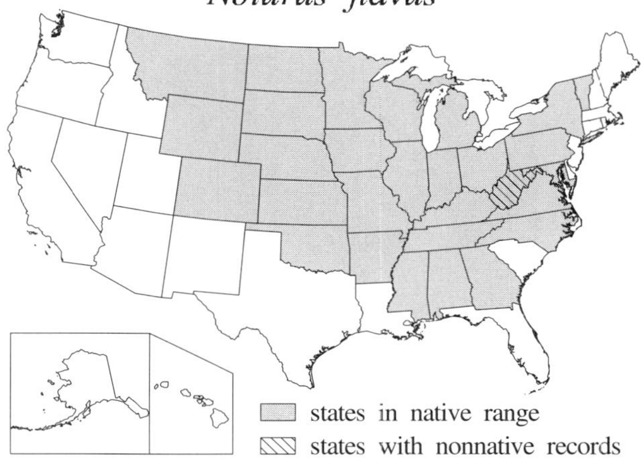

Noturus flavus Rafinesque 1818

Common Name: stonecat.

Identification: Becker (1983); Page and Burr (1991); Etnier and Starnes (1993); Jenkins and Burkhead (1994). Maximum size: 31 cm.

Native Range: St. Lawrence–Great Lakes, Hudson Bay (Red River), and Mississippi River basins from Quebec to Alberta, and southern to northern Alabama, northern Mississippi, and northeastern Oklahoma; Hudson River drainage, New York (Page and Burr 1991).

Nonindigenous Occurrences: The stonecat was possibly introduced into the New drainage, **West Virginia**, where it is known from three sites in the Gauley system. It was first reported from this drainage in 1951 (Hocutt et al. 1986; Jenkins and Burkhead 1994; Stauffer et al. 1995).

Means of Introduction: Unknown; possible bait bucket introduction or stock contamination.

Status: Reported from West Virginia.

Impact of Introduction: Unknown.

Remarks: According to Taylor (1969), it may have entered the Mohawk and upper Hudson River systems from the Great Lakes as a canal immigrant.

Noturus gilberti

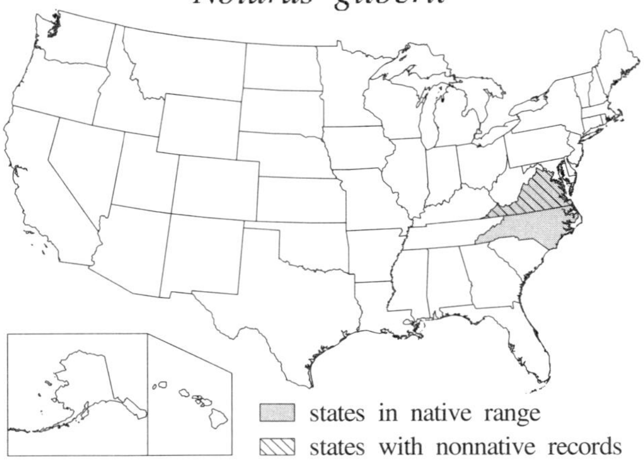

Noturus gilberti Jordan and Evermann 1889

Common Name: orangefin madtom.

Identification: Page and Burr (1991); Jenkins and Burkhead (1994). Maximum size: 10 cm.

Native Range: Upper Roanoke (including Dan River) drainage, Virginia and North Carolina (Page and Burr 1991).

Nonindigenous Occurrences: The orangefin madtom was probably introduced into the middle section of the Craig Creek system in the upper James drainage, **Virginia** (Lee et al. 1980 et seq.; Hocutt et al. 1986; Page and Burr 1991; Jenkins and Burkhead 1994). The first record for this area was from 1951 (Jenkins and Burkhead 1994). See Jenkins and Burkhead for a discussion of collection locations and dates.

Means of Introduction: Probable bait bucket introduction (Burkhead, personal communication).

Status: Established in Virginia.

Impact of Introduction: Unknown.

Remarks: Jenkins and Burkhead (1994) believe the orangefin madtom was introduced into the upper James drainage, Virginia, based on several factors: the species has only recently been recorded from the area, it is absent from good habitat elsewhere in the drainage, and it is a popular bait fish. In addition, Craig Creek is well used by anglers, and the site is near the upper Roanoke, a possible source of bait. The orangefin madtom is proposed for federal listing (U.S. Fish and Wildlife Service 1994a).

Noturus gyrinus

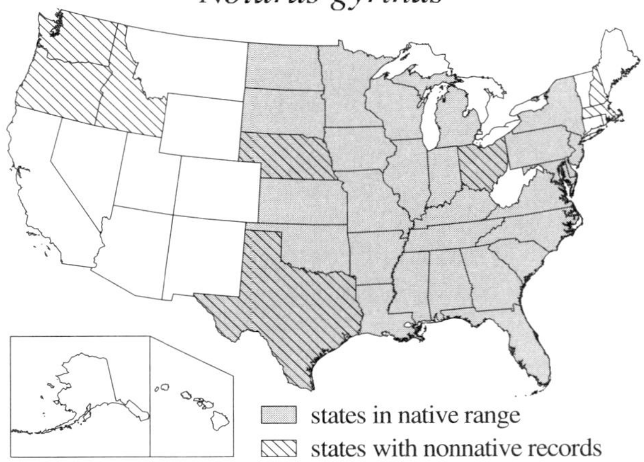

Noturus gyrinus (Mitchill 1817)

Common Name: tadpole madtom.

Identification: Becker (1983); Page and Burr (1991); Etnier and Starnes (1993); Jenkins and Burkhead (1994). Maximum size: 13 cm.

Native Range: Atlantic and Gulf Slope drainages from New Hampshire to Nueces River, Texas; St. Lawrence–Great Lakes, Hudson Bay (Red River), and Mississippi River basins from southern Quebec to southern Saskatchewan, and south to Gulf. Absent from Appalachian and Ozark Highlands (Page and Burr 1991).

Nonindigenous Occurrences: The tadpole madtom is known from the Snake and Boise rivers in **Idaho** (Linder 1963;Taylor 1969; Simpson and Wallace 1978; Wydoski and Whitney 1979; Idaho Fish and Game 1990); the upper Chicopee and Thames rivers in **Massachusetts** (Hartel 1992; Cardoza et al. 1993; Hartel et al. 1996); Frenchman's Creek, Hitchcock County in **Nebraska** (Jones 1963);the Merrimack River drainage in **New Hampshire** (Taylor 1969; Scarola 1973); the Muskingum drainage in **Ohio** (Hocutt et al. 1986); the Snake River drainage in **Oregon** (Taylor 1969; Bond 1973, 1994; Wydoski and Whitney 1979); the Rio Grande River in **Texas** (Hubbs et al. 1991); and the Columbia River in **Washington**, near the mouth of the Walla Walla River (Wydoski and Whitney 1979).

Means of Introduction: Unintentional in most locations. It is believed that introduced channel catfish and bullhead stock was contaminated with tadpole madtoms (Linder 1963; Taylor 1969; Simpson and Wallace 1978; Wydoski and Whitney 1979; Hartel 1992).

Status: Established in Connecticut, Idaho, Massachusetts, New Hampshire, Ohio, Oregon, and Washington; reported in Nebraska and Texas.

Impact of Introduction: Unknown.

Remarks: Introduced populations in Washington resulted from the downriver dispersal of fish introduced into Idaho rivers (Wydoski and Whitney 1979). The tadpole madtom has not been collected from the Thames drainage in Connecticut (B. Gerrish, personal communication). Record from a lake near Poteau, Oklahoma, may be the result of an introduction (Taylor 1969).

Noturus insignis

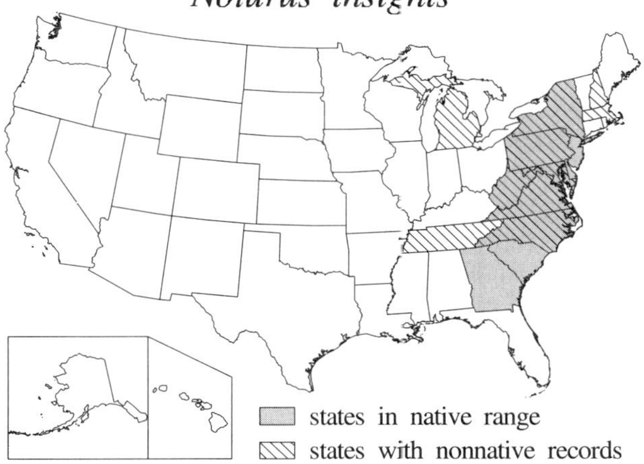

Noturus insignis (Richardson 1836)

Common Name: margined madtom.

Identification: Smith (1985); Page and Burr (1991); Jenkins and Burkhead (1994). Maximum size: 15 cm.

Native Range: Atlantic Slope from St. Lawrence River and southeastern Lake Ontario drainages, New York, to upper Altamaha River drainage, Georgia; upper Kanawha (New) River system, Virginia and North Carolina; upper Monongahela River system, Pennsylvania and Maryland (Page and Burr 1991).

Nonindigenous Occurrences: This catfish has been introduced outside, or possibly outside, its native range, in the Monongahela drainage, **Maryland** (Lee et al. 1984; Jenkins and Burkhead 1994); Crooked Springs Brook in the Merrimack drainage, **Massachusetts** (Hartel 1992; Hartel et al. 1996); Lake Gogebic (=Lake Clark), **Michigan** (Taylor 1969; Emery 1985); the Merrimack River drainage, **New Hampshire** (Taylor 1969; Scarola 1973; Schmidt 1986; Hartel 1992); the Finger Lakes, Mohawk, Oswego, and St. Lawrence drainages, **New York** (Emery 1985; Jenkins and Burkhead 1994); the Watauga River, Tennessee drainage, **North Carolina** (Starnes and Etnier 1986; Menhinick 1991; Page and Burr 1991; Etnier and Starnes 1993; Jenkins and Burkhead 1994); the Monongahela drainage, **Pennsylvania** (Cooper 1983; Hocutt et al. 1986; Jenkins and Burkhead 1994); the Watauga River above Watauga Reservoir, **Tennessee** (Etnier and Starnes 1993); the North Fork Holston River, **Virginia** (Page and Burr 1991; Etnier and Starnes 1993; Jenkins and Burkhead 1994); and the Monongahela River, **West Virginia** (Hocutt et al. 1986; Jenkins and Burkhead 1994; Stauffer et al. 1995).

Means of Introductions: Many of these occurrences may have been the result of bait bucket introductions (Taylor 1969; Jenkins and Burkhead 1994).

Status: Established in most waters where introduced

Impact of Introduction: Unknown.

Remarks: Taylor (1969) and Jenkins and Burkhead (1994) discussed native and possible nonnative distribution. Hocutt et al. (1986) listed this species as native but possibly introduced into the Monongahela drainage.

Noturus miurus

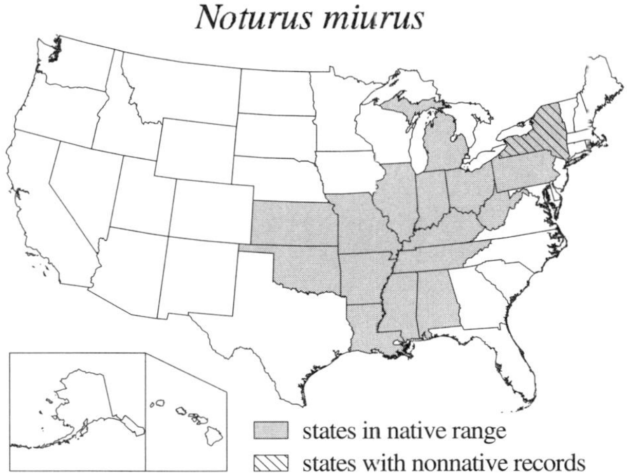

Noturus miurus Jordan 1877

Common Name: brindled madtom.

Identification: Trautman (1981); Smith (1985); Page and Burr (1991); Etnier and Starnes (1993); Pflieger (1997). Maximum size: 13 cm.

Native Range: Lower Great Lakes drainages in Ontario and New York, southwest through most of the Ohio River basin and lower Mississippi River basin (west to eastern Kansas and Oklahoma; Pearl River and Lake Pontchartrain drainages on central Gulf Slope in Mississippi and Louisiana (Page and Burr 1991).

Nonindigenous Occurrences: Introduced into the Mohawk River in **New York** (Mills et al. 1997). Specimens have been taken at Lock 7 near Schenectady and at Lock 9 at Rotterdam Junction (Smith 1985; Mills et al. 1997). Smith (1985) mapped another location on the Mohawk near Little Falls, where it is presumably also introduced. The eastern range expansion is relatively recent, Lee et al. (1980 et seq.) reported it only as far east as Oneida Lake.

Means of Introduction: Canal connection. Probably gained access from Lake Ontario systems in western New York to the Mohawk River via the Erie Canal (Smith 1985; Mills et al. 1997).

Status: Reported in New York.

Impact of Introduction: Unknown.

Remarks: None.

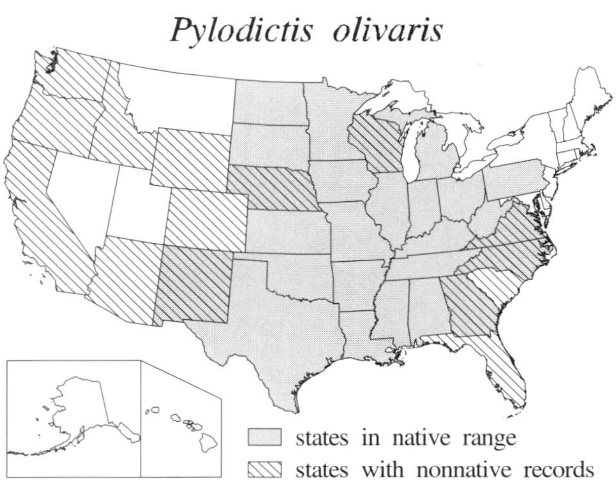

Pylodictis olivaris (Rafinesque 1818)

Common Name: flathead catfish.

Identification: Becker (1983); Page and Burr (1991); Etnier and Starnes (1993); Jenkins and Burkhead (1994). Maximum size: 155 cm.

Native Range: Lower Great Lakes and Mississippi River basins from western Pennsylvania to White-Little Missouri River system, North Dakota, and south to Louisiana; Gulf Slope from Mobile Bay drainage, Georgia and Alabama, to Mexico (Page and Burr 1991).

Nonindigenous Occurrences: Flathead catfish have been introduced to the Gila basin in the Salt, Verde, and San Pedro rivers in **Arizona** (Bottroff et al. 1969; Minckley 1973; S. Stefferud, personal communication); the Colorado River basin in **California** (Bottroff et al. 1969; Minckley 1973; Moyle 1976a; Dill and Cordone 1997); the Arkansas and Platte river drainages in **Colorado** (Ellis 1974; Walker 1993); the Apalachicola, Ochlockonee (Ober, personal communication), Escambia, and Perdido rivers in western **Florida** (Baker 1993), and the Suwannee River in central Florida (museum specimen); the Altamaha River drainage (Bart et al. 1994; C. Jennings, personal communication), the Flint River (Dahlberg and Scott 1971a, 1971b; Yerger 1977), and Lake Hartwell (1968) on the Savannah River (Dahlberg and Scott 1971a, 1971b) and Ochlockonee River (Ober, personal communication) in **Georgia**; the Snake River in **Idaho**, between Lewiston and Swan Falls (Linder 1963; Simpson and Wallace 1978; Idaho Fish and Game 1990); Lake Maloney in the lower South Platte drainage in **Nebraska** (Jones 1963); the Gila River drainage in **New Mexico** (Koster 1957; Minckley 1973; Sublette et al. 1990); the Cape Fear, Tennessee, Yadkin, and Catawba river drainages in **North Carolina** (Hocutt et al. 1986; Menhinick 1991; Moser and Ross 1993; Jenkins and Burkhead 1994); Snake River, **Oregon** (Bond 1973, 1994; Wydoski and Whitney 1979); Lake Hartwell on the Savannah River in **South Carolina** (Dahlberg and Scott 1971b); Occoquan Reservoir, Smith Mountain Reservoir, the lower James, middle Roanoke, and upper James (Botetourt County) drainages in **Virginia** (Hocutt et al. 1986; Jenkins and Burkhead 1994); **Washington** (Fletcher, personal communication); Oconomowoc and Nagawicka lakes, and the Wolf and Fox drainages, **Wisconsin** (Becker 1983); and in the lower North Platte drainage in **Wyoming** (Hubert 1994).

Means of Introduction: The flathead catfish has been intentionally stocked in most cases. In Idaho, however, flatheads were accidentally stocked instead of blue catfish (Simpson and Wallace 1978). Populations in the Apalachicola River, Florida, probably spread from introductions upstream in the Flint River, Georgia. It is believed that flatheads were stocked by anglers circa 1950 in the vicinity of Potato Creek in Upson County, Georgia, with stock from the Tennessee drainage (Quinn 1988). They were recorded in the Flint River below the Warwick Dam at Lake Blackshear in the early 1960s, and at Albany in the early 1970s (Quinn 1988). Although there have been rumors that flatheads were stocked in the Suwannee River by the Florida Game and Freshwater Fish Commission in the 1960s (G. Burgess, personal communication), the rumors are not true (J. Krummrich, personal communication). The species was apparently first stocked in the Cape Fear River in 1966 when 11 sexually mature fish were released near Fayetteville, North Carolina, by North Carolina Wildlife Resources Commission biologists (Guire et al. 1984; Ashley and Buff 1986). Flatheads were stocked in Atlantic drainages (Savannah and Altamaha) in Georgia in the 1970s (Bart et al. 1994; C. Jennings, personal communication). According to Bart et al. (1994), at least some of these were the result of stocking by Georgia Department of Natural Resource personnel. The first known reports of this fish in California were recorded catches made in the lower Colorado River near Yuma in 1966 (Dill and Cordone 1997). The Colorado River populations in California and Arizona resulted, at least in part, from a stocking of about 600 flathead catfish above Imperial Dam made by the Arizona Game and Fish Department in 1962 (Dill and Cordone 1997). According to Dill and Cordone (1997), the believed route of the flathead catfish was downstream to Imperial Dam and subsequently into the All American Canal system to the Imperial Valley. Minckley (1973) reported that the species was introduced prior to 1950 into the Gila River system, a tributary of the Colorado River; however, Dill and Cordone (1997) indicated that, as far as is known, the flathead catfish was not taken in the lower Colorado River basin until after 1962. A single fish was taken 20 November 1995 in Arizona from the upper San Pedro River, about 32 kilometers from the Mexican border (Stefferud, personal communication). It is not known how the species gained access to the upper reach of this river. In Wisconsin, flatheads probably entered the Wolf and Fox drainages via the canal at Portage (Becker 1983). The Ochlockonee River introduction in Florida and Georgia was probably due to illegal stocking by anglers with fish from the nearby Apalachicola River, where the fish had also been introduced.

Status: The flathead catfish has become established in most waters where introduced. For instance, it is widespread and reproducing in the lower Colorado River basin (Dill and Cordone 1997). As of about 1980, the Cape Fear River population had expanded from the site of its initial release near Fayetteville, North Carolina, and was found to inhabit a 201-kilometer stretch of the river (Guire et al. 1984). In samples taken by Guire et al. (1984) from the Cape Fear River, flathead accounted for 10.52% of total fish numbers and 64.7% of total fish weight. Establishment in Oregon is uncertain (Bond 1994). The species does not appear to have survived to reproduce in Wyoming (Hubert 1994). It has been reported from the San Pedro River, Arizona, and from the Suwannee River, Florida.

Impact of Introduction: Many feeding studies have found that flathead catfish prey heavily on sunfish *Lepomis* spp. (Quinn 1988). One study found that they reduced the number of common carp *C. carpio* and bullheads *Ameiurus* spp. (Quinn 1988). However, the introduced population in the Flint River system was found to prey largely on crayfish, and that young-of-the-year flatheads fed on darters *Etheostoma* spp. Clupeids, Catostomids, Ictalurids (including other flatheads), and Centrarchids were also consumed (Quinn 1988). According to Quinn (1988), introduced flatheads in the Flint River rely more on crayfish than any other catfish population yet described. A severe decline in native fish species, particularly native bullhead species, was observed in the Cape Fear River within 15 years of the first flathead catfish introduction (Guire et al. 1984; Jenkins and Burkhead 1994). Feeding studies conducted in the Cape Fear River showed that flatheads consume mainly bullheads, catfishes, shad, and sunfishes (Guire et al. 1984; Ashley and Buff 1986). In 1979, flatheads in the Cape Fear River fed primarily on bullheads. However, by 1986, bullhead populations had declined and flathead catfish had switched to preying on shad (Ashley and Buff 1986). Diet studies also have been conducted in the Oconee River in Georgia, where this catfish had been implicated in causing declines of native bullheads and sunfishes (especially redbreast sunfish *Lepomis auritus*). However, findings of that initial study were inconclusive since most of the flathead catfish examined had empty stomachs (C. Jennings, personal communication). Flathead catfish also may be responsible for declines in other native species in the Altamaha drainage (C. Jennings, personal communication). In the Ocmulgee River, Georgia, abundances of silver redhorse *Moxostoma anisurum*, robust redhorse *M. robustum*, snail bullhead *Ameiurus brunneus*, flat bullhead *A. platycephalus*, and redbreast sunfish *Lepomis auritus*, were negatively correlated with flathead catfish occurrence and abundance (Bart et al. 1994). This correlation may be due to direct predation. Several authors have reported suckers and catfish as common prey items of flatheads (Bart et al. 1994). The snail bullhead and flat bullhead appear to be most affected by the presence of flathead catfish in the Ocmulgee drainage (Bart et al. 1994). One of us (Jim Williams) suspects flatheads may be contributing to the decline of the federally threatened Gulf sturgeon *Acipenser oxyrinchus desotoi* by consuming the young benthic fish in the Apalachicola River. The flathead catfish is thought to be contributing to the decline of the razorback sucker *Xyrauchen texanus*. For instance, Marsh and Brooks (1989) found that intensive predation by flathead catfish and channel catfish on juvenile razorback suckers is likely to prevent hatchery transplants of this southwestern endangered sucker from becoming re-established in portions of its natural range. If the flathead catfish becomes established in the San Pedro River, it could mean a major loss for recovery of several species (Stefferud, personal communication). Gilbert (personal communication) considers that introductions of flathead catfish are probably the most biologically harmful of all fish introductions in North America.

Remarks: The flathead catfish became the dominant predator in the Cape Fear drainage, North Carolina, within 15 years of the introduction (Guire et al. 1984). The species may actually be native to the upper Tennessee drainage in North Carolina (Jenkins and Starnes, personal communication). In their book on Alabama fishes, Mettee et al. (1996) presented conflicting information regarding native versus introduced ranges. These researchers stated, in the species account, that *Pylodictis olivaris* is introduced to the Conecuh and Escatawpa river systems, but they listed the species as "native" in their summary table.

Clariidae - Labyrinth catfishes

The family Clariidae, whose members are known as the air-breathing or labyrinth catfishes, is represented by about 13 genera with about 100 species (Nelson 1994). These fishes are distributed throughout Africa into Syria, and occur in southern and eastern Asia including China, India, Java, Borneo, Taiwan, Japan, and the Philippines (Berra 1981), with the greatest diversity in Africa (Nelson 1994). Most are medium-sized fishes; the largest grow to over one meter long and are characterized by an elongate, cylindrically shaped body. Clariids have an accessory respiratory organ that permits survival in stagnant waters, and some species can leave water and move short distances over land. In Africa and Asia, selected clariids are important food fishes and some are used in aquaculture. They are voracious feeders and many species are aggressive predators (Burgess 1989). Many members of this family are considered potentially harmful and have therefore been placed on the restricted fish list of certain states (Howells 1992b; Dill and Cordone 1997). Like most catfishes, clariids are restricted to freshwater.

Clarias batrachus

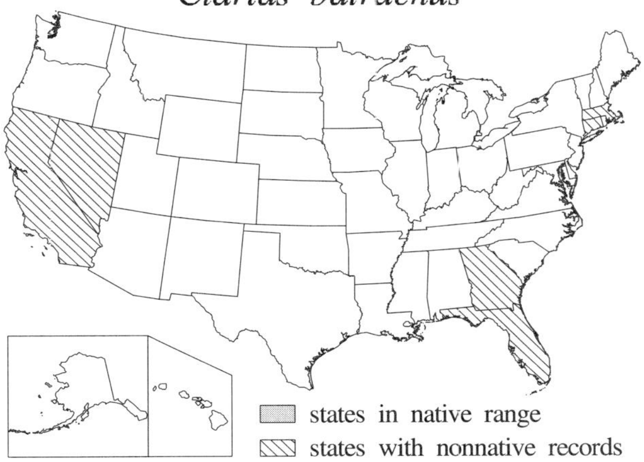

Clarias batrachus (Linnaeus 1758)

Common Names: walking catfish (magur, pla duk dam).

Identification: Page and Burr (1991). Keys and distinguishing characteristics were given by Smith (1945), Talwar and Jhingran (1992), and Kottelat et al. (1993). A photograph appeared in Kottelat et al. (1993). Maximum size: 61 cm in native range; rarely over 36 cm in Florida.

Native Range: Southeastern Asia including eastern India, Sri Lanka, Bangladesh, Burma, Indonesia, Singapore, and Borneo (Lee et al. 1980 et seq.; Talwar and Jhingran 1992; Kottelat et al. 1993). Probably introduced into the Philippines.

Nonindigenous Occurrences: This species has been collected in **California** from the All American Canal west of Yuma, Arizona (Minckley 1973; Courtenay et al. 1984); from the San Joaquin River, Sacramento County (Courtenay and Hensley 1979a; Courtenay et al. 1984, 1986); and from Legg Lake, Los Angeles County (Shapovalov et al. 1981). Specimens have been captured in widely separated water bodies in **Connecticut** (Whitworth 1996). It has been firmly established in southern peninsular **Florida** since the late 1960s (Courtenay 1970, 1978, 1979a; Courtenay et al. 1974; Courtenay and Miley 1975; Courtenay and Hensley 1979a; Loftus and Kushlan 1987; Anonymous 1983a; Lorenz et al. 1997). A specimen has been taken from the Flint River in **Georgia** (Courtenay and Miley 1975; Courtenay and Hensley 1979a; Courtenay et al. 1984, 1991; Gennings, personal communication). A single fish was taken by an angler from Waldo Lake, Brockton in Plymouth County, **Massachusetts**, in August 1971; three or four additional fish were reportedly taken from ponds in the eastern part of the state in the mid-1970s, but the specimens were not retained (Hartel 1992; Cardoza et al. 1993; Hartel et al. 1996). Two specimens were taken from Rogers Spring, Clark County, **Nevada** (Courtenay and Deacon 1983; Deacon and Williams 1984).

Means of Introduction: The walking catfish was imported to Florida, reportedly from Thailand, in the early 1960s for the aquarium trade (Courtenay et al. 1986). The first introductions apparently occurred in the mid-1960s when adult fish imported as brood stock escaped, either from a fish farm in northeastern Broward County or from a truck transporting brood fish between Dade and Broward counties (Courtenay and Miley 1975; Courtenay 1979a; Courtenay and Hensley 1979a; Courtenay et al. 1986). Additional introductions in Florida, supposedly purposeful releases, were made by fish farmers in the Tampa Bay area, Hillsborough County in late 1967 or early 1968, after the state banned the importation and possession of walking catfish (Courtenay and Stauffer 1990). Aquarium releases likely are responsible for introductions in other states (Courtenay and Hensley

1979a; Courtenay and Stauffer 1990; Hartel 1992). Dill and Cordone (1997) reported that this species has been sold by tropical fish dealers in California for some time.

Status: Established in Florida; reported from California, Connecticut, Georgia, Massachusetts, and Nevada.

Impact of Introduction: Largely unknown. In Florida, walking catfish are known to have invaded aquaculture farms, entering ponds where these predators prey on fish stocks. In response, fish farmers have had to erect protective fences to protect ponds (Courtenay and Stauffer 1990).

Remarks: In 1968, this species was confined to three south Florida counties; by 1978, it had spread to 20 counties in the southern half of the peninsula (Courtenay 1979a; Courtenay et al. 1986). Dispersal apparently has occurred by way of the interconnected network of canals along the southeastern coastal region; however, spread was accelerated by overland migration, typically during rainy nights. The walking catfish has been established in Everglades National Park and in Big Cypress National Preserve since the mid-1970s (Courtenay 1989). Populations suffer periodic die-offs due to cold temperatures; consequently, northward dispersal is limited (Courtenay 1978, Courtenay and Stauffer 1990). Although all Florida imports were originally albinos, albinos in the wild are now rare and descendants have reverted to the dominant, dark-color phase (dark brown to gray) probably a result of natural selection by predators (Courtenay et al. 1974; Courtenay and Stauffer 1990). Guarding of free-swimming young may enhance survivorship over that of native species with less advanced (or no) parental care (Taylor et al. 1984).

Voucher specimens: Florida (USNM 203889, 317282, 317283, UF 16424, 93561, TU 71116, 12099), Massachusetts (MCZ 69534). Minckley (1973) indicated that the specimen taken from the All American Canal, California, was deposited at Arizona State University.

Clarias fuscus

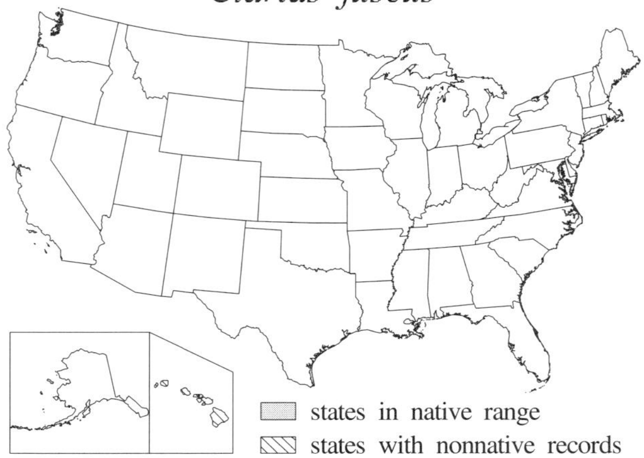

Clarias fuscus (Lacepède 1803)

Common Names: whitespotted clarias (Chinese catfish, puntat).

Identification: A few distinguishing characteristics and illustrations or photographs were given by Nichols (1943) and Masuda et al. (1984). Jordan and Evermann (1905) provided a detailed description of a specimen collected in Hawaii. Maximum size: 40 cm.

Native Range: Eastern Asia in south China, Taiwan, and the island of Ishigakijima, Japan (Masuda et al. 1984).

Nonindigenous Occurrences: This species has been known from streams on the major islands of **Hawaii** since the late 1800s (Jordan and Evermann 1905; Brock 1960; Maciolek 1984; Devick 1991a, 1991b).

Means of Introduction: This catfish is believed to have been introduced to Hawaii from Asia by Chinese immigrants, prior to 1900 (Cobb 1902; Jordan and Evermann 1902, 1905; Maciolek 1984). Jordan and Evermann (1905) stated that the species "was introduced from China a few years ago . . ." The fish was introduced apparently as a food source (Devick 1991b). Cobb (1902) and Jordan and Evermann (1905) stated that the catfish was commonly raised in the irrigation ditches and freshwater ponds in Hawaii and typically sold alive to the Chinese.

Status: Cobb (1902) indicated that that this catfish was fairly common in the vicinity of Honolulu; Brock (1960) stated that it was established in the waters of Kauai, Oahu, and Maui prior to the turn of the century. Based on recent information, the species is considered established on the major islands of Hawaii, including Hawaii, Kauai, Maui, Molakai, and Oahu; it is most common in stream habitats (Brock 1960; Maciolek 1984; Devick 1991a, 1991b).

Impact of Introduction: Unknown.

Remarks: In Hawaii, this catfish inhabits the muddy bottoms of taro patches, streams, and ditches (Morita 1981). Cobb (1902) and Jordan and Evermann (1902) recorded the species found in Hawaii as *Macropternotus maqur*; Jordan and Evermann (1905) recognized it as *Clarias fuscus* (although in obvious error they use the name *Clarias magur* in one part of their report).

Voucher specimens: Hawaii (BPBM 3626, 3627, 30618).

Doradidae - Thorny catfishes

This neotropical catfish family comprises about 35 genera and about 90 species (Nelson 1994). The family is restricted to South America, ranging from the La Plata–Parana basin north to Colombia (Berra 1981; Burgess 1989; Higuchi 1992). Doradids have a single row or series of plates along their flanks, each with a strong spine, hook, or similar ornamentation. They are typically heavy-built benthic fishes that are opportunistic omnivores; they vary in size from the dwarf *Physopyxis*, never more than a few centimeters, to the huge *Oxydoras*, *Rhinodoras*, *Lithodoras*, and *Megalodoras*, reaching about a meter (Higuchi 1992). Some species are used for food in South America and a few are popular aquarium fishes. They are restricted to freshwater.

Oxydoras niger

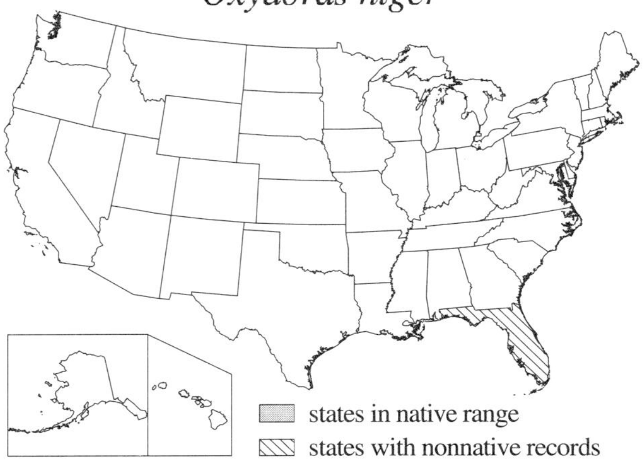

Oxydoras niger (Valenciennes 1821)

Common Names: ripsaw catfish (black doradid, black shielded catfish, blue dolphin, dolphin catfish, fork-snout catfish, terushuki, cuiu-cuiu, sierra negra).

Identification: Higuchi (1992) gave diagnostic characters of the genus. Somewhat dated keys to the genera and species of Doradidae, including *O. niger*, were provided by Eigenmann (1925) and Schultz (1944). Distinguishing characteristics, color photographs of live fish, and a key to doradid genera were given by Burgess (1989). Maximum size: 120-cm fork length and 20 kg.

Native Range: Tropical America. Very common in the Guiana region and entire Amazon basin of South America from Peru to the Brazilian state of Para (Higuchi 1992).

Nonindigenous Occurrences: This species was reported from an unspecified locality in **Florida**, apparently prior to 1984 (Courtenay et al. 1984; 1991; Courtenay and Stauffer 1990; Robins et al. 1991b). Based on information included with an existing photograph of the specimen (labeled as *Oxydoras* sp.), it was taken from Biscayne Bay (or possibly one of its drainages), Dade County.

Means of Introduction: This species was probably an aquarium release.

Status: Reported from Florida.

Impact of Introduction: Unknown.

Remarks: There is some confusion in the literature concerning the Florida record and the positive identification of the Florida specimen. The listing for Florida of *Oxydoras* sp. (Courtenay et al. 1986) and Florida listings of *Oxydoras niger* or *Pseudodoras niger* (Courtenay et al. 1984; 1991; Courtenay and Stauffer 1990; Robins et al. 1991b; Courtenay and Williams 1992) most likely represent one and the same record. *Pseudodoras niger* is a name sometimes used for *Oxydoras knerii*, a species distinct from *O. niger* (C. Ferraris, personal communication). The whereabouts of the Florida specimen is not known. There is a photograph of a large specimen on file with USGS Biological Resources Division center in Gainesville. Information provided with the photograph suggest the fish was taken from Biscayne Bay or one of its tributaries. No date was provided with the photograph.

Platydoras costatus

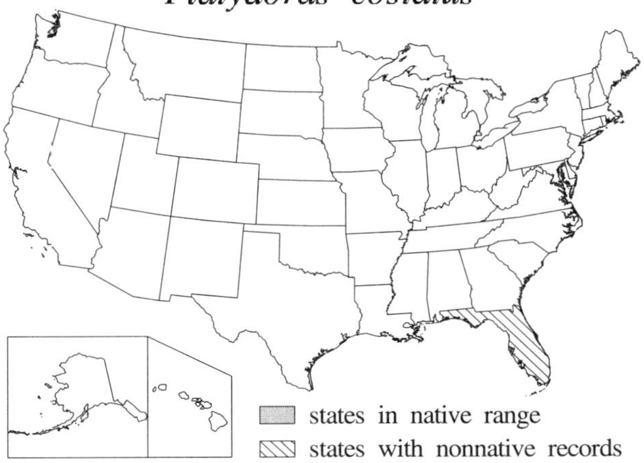

Pterodoras granulosus

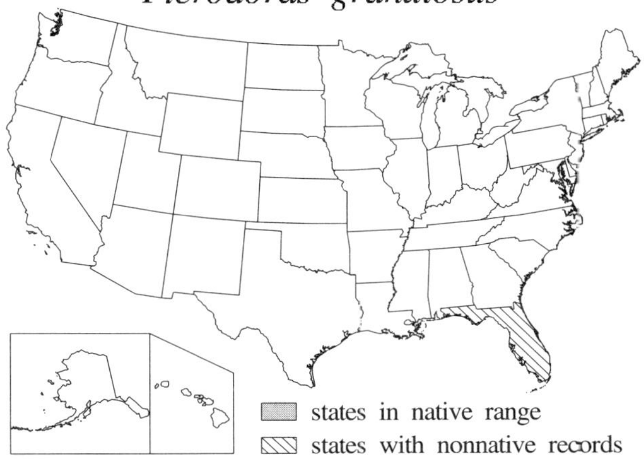

Platydoras costatus (Linnaeus 1766)

Common Names: Raphael catfish (striped raphael, chocolate catfish, bacu).

Identification: Higuchi (1992) gave a brief diagnosis of the genus. Keys to doradid genera and species were given by Eigenmann (1925) and a key to doradid genera appeared in Burgess (1989); however, the information should be considered dated or suspect. A few distinguishing characteristics and color photographs of live fish appeared in Burgess (1989). A photograph of the species also appeared in Sands (1984) and Ferraris (1991). Maximum size: To about 15 cm SL.

Native Range: Tropical America. South America, from Venezuela and Guianas to Argentina (Robins et al. 1991b).

Nonindigenous Occurrences: This species was reported from an unspecified locality in **Florida** (Courtenay et al. 1984; Courtenay and Stauffer 1990; Courtenay et al. 1991; Robins et al. 1991b). The record was probably from the period 1979 to 1984.

Means of Introduction: Record probably based on an aquarium release.

Status: Reported from Florida.

Impact of Introduction: Unknown.

Remarks: This species is one of the more common doradids in the aquarium trade.

There are no known voucher specimens.

Pterodoras granulosus (Valenciennes 1821)

Common Names: granulated catfish (bacu, bacu comum).

Identification: Distinguishing characteristics of the genus *Pterodoras* were given by Higuchi (1992). A few distinguishing characteristics and a key to doradid genera were given by Burgess (1989), although that key should be considered somewhat suspect. The species was also included in illustrations and identification keys of Eigenmann (1925). For photographs of species see Goulding (1980), Sands (1984, incorrectly spelled as *Pterydoras granulosus*), and Burgess (1989). Maximum size: 70 cm.

Native Range: Tropical America. La Plata and Amazon basins of South America.

Nonindigenous Occurrences: A single specimen (55 cm SL) was found dead in a canal north of Brooker Creek, near Lake Tarpon in Pinellas County, **Florida**, in early 1977 (Courtenay and Hensley 1979a; museum specimen).

Means of Introduction: This fish was probably an aquarium release.

Status: Reported from Florida.

Impact of Introduction: Unknown.

Remarks: The Pinellas County, Florida, record is likely the basis for inclusion of this species in several published lists of nonestablished, nonindigenous fishes (e.g.,

Courtenay and Hensley 1980; Courtenay et al. 1984, 1986; Courtenay and Stauffer 1990; Courtenay et al. 1991; Robins et al. 1991b).

Voucher specimens: Florida (FDNR 11967).

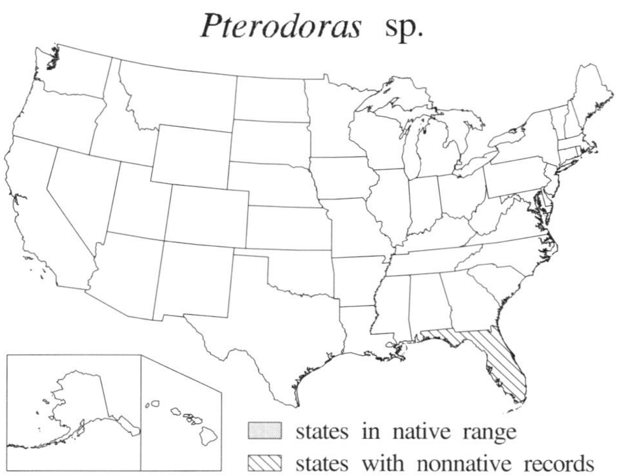

Pterodoras sp.

Common Names: thorny catfish (bacu).

Identification: Distinguishing characteristics of the genus *Pterodoras* were given by Higuchi (1992). A key to doradid genera, and photographs, appeared in Burgess (1989). Illustrations and identification keys were also provided by Eigenmann (1925). Maximum size: 60-cm fork length.

Native Range: Tropical America. Native to Orinoco, Amazon, and La Plata basins of South America.

Nonindigenous Occurrences: This species was reported from an unspecified locality in **Florida** prior to 1984 (Courtenay et al. 1984, 1991).

Means of Introduction: This species was probably an aquarium release.

Status: Reported from Florida.

Impact of Introduction: Unknown.

Remarks: The Florida report of *Pterodoras* sp. (Courtenay et al. 1984, 1991) and that of *Pterodoras granulosus* (Courtenay and Hensley 1979a, 1980; Courtenay et al. 1984, 1986; Courtenay and Stauffer 1990; Courtenay et al. 1991; Robins et al. 1991b) possibly represent the same record.

There is no known voucher specimen.

NOTE: According to W. Minckley (personal communication), the remains of what was likely a doradid catfish were found in Arizona in the Phoenix Canal system in the vicinity of the Phoenix Zoo in early 1998. The fish, about 70 cm total length, had bony plates, and a small mouth with three pairs of barbels (no basal barbels evident). The dead fish was partially decomposed so only the skeleton was preserved and deposited in the ASU fish collection.

Auchenipteridae - Driftwood catfishes

This neotropical catfish family comprises 21 genera and about 60 species (Nelson 1994). Auchenipterids range from the Parana–La Plata basin in Argentina north to Panama (Berra 1981). They are small- to medium-sized fishes without armor or plates. Most species are nocturnal and feed on invertebrates or small fishes. The family is restricted to freshwater with the exception of *Pseudauchenipterus nodosus*, a species reported from brackish waters.

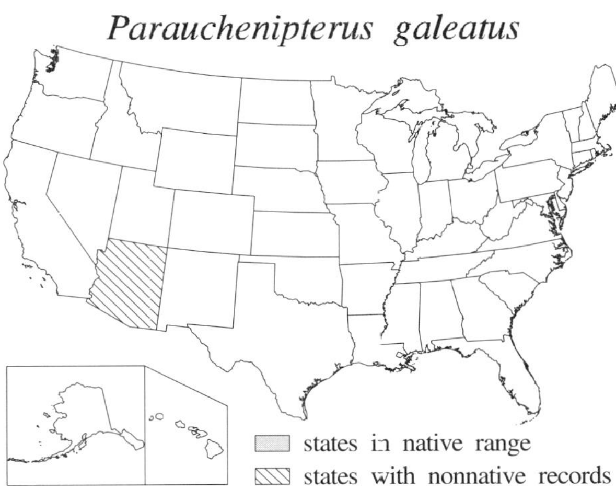

Parauchenipterus galeatus (Linnaeus 1766)

Common Names: driftwood catfish (cachorro-do-padre).

Identification: The genus is in need of taxonomic review. Considering its wide native distribution, *P. galeatus* may very well represent a species complex. A key to genera and distinguishing characteristics of the group were provided by Burgess (1989). Color photographs were given in Burgess (1989) and Ferraris (1991). Maximum size: 18 cm SL.

Native Range: Tropical America. Widely distributed in Atlantic drainages of northern and central South America including the Orinoco and Amazon river basins, drainages in the Guianas, and possibly elsewhere.

Nonindigenous Occurrences: In 1989, a single specimen was taken with a minnow trap from Lake B of Arizona State University's Research Park in Tempe, **Arizona** (P. C. Marsh, personal communication; S. Vives, personal communication).

Means of Introduction: Probable aquarium release.

Status: Reported from Arizona.

Impact of Introduction: Unknown.

Remarks: Lake B is an artificial 62.5-ha, closed-system lake. The voucher specimen, reportedly preserved and frozen, has not been located (Marsh, personal communication). *Parauchenipterus galeatus*, and closely related species (which are nearly impossible to tell apart) are commonly imported as aquarium fishes (Ferraris, personal communication).

Pimelodidae - Longwhiskered catfishes

The family Pimelodidae is a large and very diverse group of neotropical catfishes with about 56 genera and about 300 species (Nelson 1994). These fishes range from the Parana–La Plata Basin north to southern Mexico (Berra 1981; Burgess 1989). Although diverse in terms of morphology, in general these fish are elongate, have a smooth, naked skin, and possess three pairs of barbels. Pimelodids range from relatively small species less than 100 mm long to giants well over 1 m. Most are predators or scavengers and many are popular as aquarium fish. Many of the larger species are commercially important food fishes. Pimelodids are restricted to freshwater.

Perrunichthys perruno

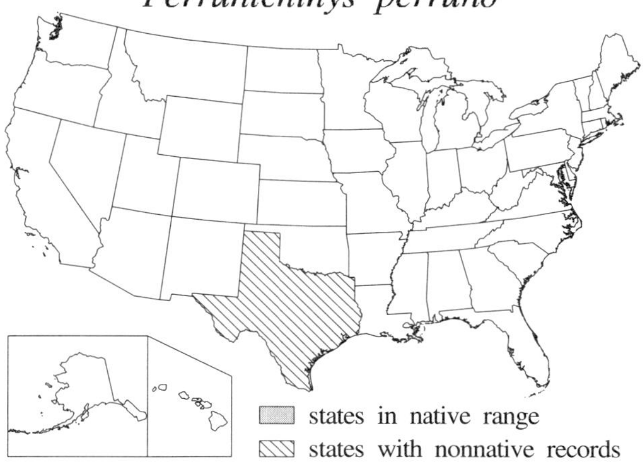

Perrunichthys perruno Schultz 1944

Common Names: leopard catfish (perruno).

Identification: This genus contains a single species. Keys were provided by Schultz (1944) and Taphorn and Lilyestrom (1984). As well as giving diagnostic characteristics in his original description, Schultz (1944) provided a representative illustration of the species. Distinguishing characteristics and color photographs of live fish were given by Burgess (1989). Adults are large and colorful, with a color pattern similar to that of another large pimelodid catfish, *Leiarius marmoratus*. Maximum size: To over 62 cm SL.

Native Range: Tropical America. Restricted to Maracaibo Basin of Venezuela, South America (Schultz 1944).

Nonindigenous Occurrences: A single specimen (about 1.4 kg, greater than 50 cm SL) was taken from Cedar Bayou near Galveston Bay, **Texas,** by an angler in March 1992 (R. G. Howells, personal communication).

Means of Introduction: Probable aquarium release.

Status: Reported from Texas.

Impact of Introduction: Unknown.

Remarks: The Texas record is supported by a photograph of rather poor quality (on file at the USGS Biological Resources Division center in Gainesville). The angler kept the specimen to make it into a trophy mount (Howells, personal communication). Some question still remains as to correct identity of the Texas specimen. According to J. G. Lundberg (personal communication), the dorsal ray count of 10, if correctly reported, makes the case good for *Leiarius marmoratus* even though the color of the fish in the photograph is not exactly correct (*Leiarius* and *Perrunichthys* can be distinguished easily by superficial skull characteristics). Although neither species is apparently common in the aquarium trade, Ferraris (1991) reported that both species have become more popular in recent years. Commercial airline flights from Maracaibo, Venezuela, to the Houston area have been common during the recent past and indicate that *P. perruno* could be easily brought into Texas along with other ornamental fishes.

Phractocephalus hemioliopterus

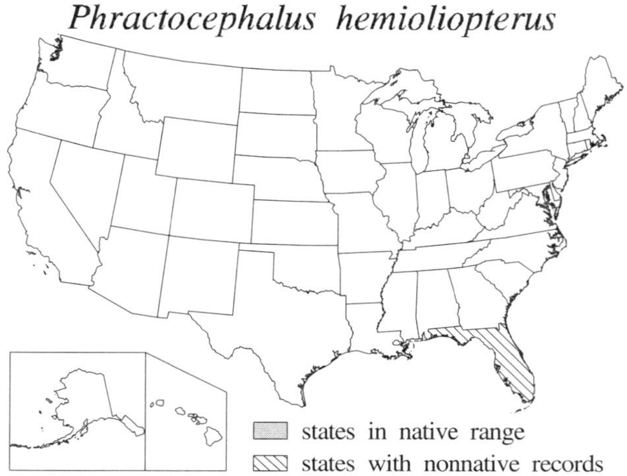

Phractocephalus hemiolipterus (Bloch and Schneider 1801)

Common Names: redtail catfish (pirarara, cajaro, guacamayo, bigorilo, pez torre, laitu, parabepre).

Identification: A somewhat dated key to pimelodid genera was presented by Schultz (1944). A few distinguishing characteristics and photographs were given by Goulding (1980), Burgess (1989), and Barthem and Goulding (1997). Color photographs, including one of a small juvenile, appeared in Ferraris (1991). The color pattern is quite distinct and photographs of this species are common in aquarium literature. Many references to *Phractocephalus hemiolipterus* give incorrect spellings of the scientific name (e.g., *P. hemiliopterus*). Maximum size: 130 cm and 80 kg.

Native Range: Tropical America. Widespread in Amazon and Orinoco basins of South America (Barthem and Goulding 1997).

Nonindigenous Occurrences: There is a single record from an unspecified locality in **Florida** (Courtenay et al. 1991); the date of the record is presumably some time between 1979 and 1991.

Means of Introduction: Probable aquarium release.

Status: Reported from Florida.

Impact of Introduction: Unknown.

Remarks: This large predacious catfish is somwhat common in the ornamental fish trade. There are no known voucher specimens.

Rhamdia quelen

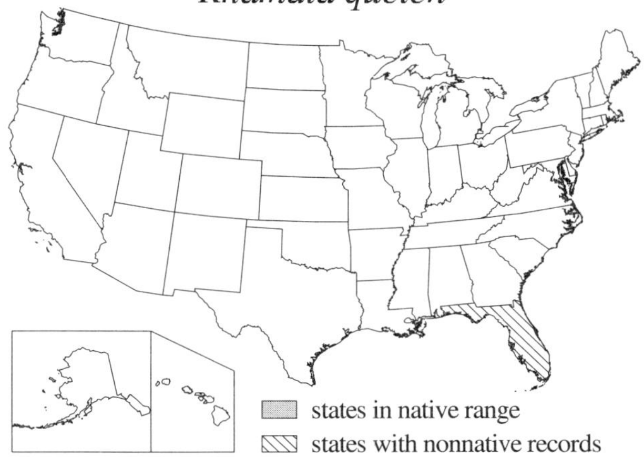

Rhamdia quelen (Quoy and Gaimard 1824)

Common Names: (bagre, bagre negro, bagre-guarié, jandiá, jundiá, jundiá-tinga, mandi, mandi-guaru, sapipoca).

Identification: Silfvergrip (1996) reviewed and revised the genus *Rhamdia*. He included diagnostic characteristics for the genus and for *R. quelen*. Also included were illustrations and photographs of this species. This catfish, along with illustrations, is included in identification keys of Britski et al. (1984). Maximum size: 30 cm.

Native Range: Tropical America. From southern Mexico in the north to central Argentina (Silfvergrip 1996).

Nonindigenous Occurrences: A single specimen, tentatively identified as this species, was taken in **Florida** from a water body west of Miami, Dade County in June 1995 (museum specimen).

Means of Introduction: Probable aquarium release.

Status: Reported from Florida.

Impact of Introduction: Unknown.

Remarks: The Florida specimen was tentatively identified as *R. quelen* based on information given in Silfvergrip (1996). Several members of the genus *Rhamdia* are found in the ornamental fish trade (Burgess 1989).

Voucher specimen: Florida (UF 103762).

Callichthyidae - Plated catfishes

The neotropical catfish family Callichthyidae, whose members are also known as armored or plated catfishes, comprises seven genera and about 130 species (Nelson 1994). They range north from the La Plata–Parana River basin in South America, to Panama and Trinidad (Berra 1981; Burgess 1989). Callichthyids are heavily armored with two lateral rows of bony plates arranged in two rows along the flanks. Most species are small bottom dwellers, and many are popular aquarium fishes. Most members of the family are restricted to freshwater, although some species occur in brackish water habitats.

Callichthys callichthys

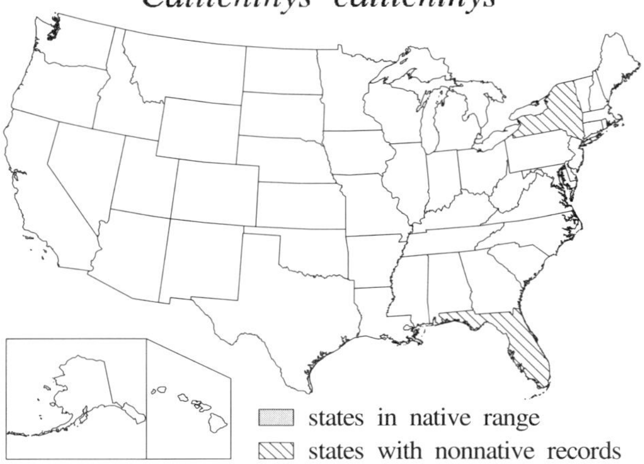

Callichthys callichthys (Linnaeus 1758)

Common Names: cascarudo (mailed catfish, bubblenest catfish, curito, callichthys).

Identification: *Callichthys callichthys* belongs to the subfamily Callichthyinae. Reis (1997) provided a key to the genera of that group. Distinguishing characteristics and identification keys provided by Gosline (1940) and Burgess (1989), but these latter keys are outdated and should be used with caution. Color photographs are given in Burgess (1989). Maximum size: 20 cm.

Native Range: Tropical America. Widespread in South America including, but not limited to, the Amazon, Orinoco, and La Plata–Parana basins, the Guianas, and Trinidad (Burgess 1989).

Nonindigenous Occurrences: Fish tentatively identified as members of this species were collected in a canal along Jog Road in Palm Beach County, **Florida**, in the early 1970s (Courtenay and Hensley 1979a). A single fish (102 mm TL) allegedly was caught by an angler in Great South Bay off Copiague in western Suffolk County, Long Island, **New York**, in June 1978 (P. T. Briggs, personal communication; Courtenay and Hensley 1979a).

Means of Introduction: Florida record may represent escape from fish farm or aquarium release. New York record, if valid, is a probable aquarium release.

Status: Reported from Florida and New York.

Impact of Introduction: Unknown.

Remarks: The identity of the New York fish was confirmed by James Atz of the American Museum of Natural History; the specimen was later returned to the angler for mounting (Briggs, personal communication). Nevertheless, some doubts remain about the actual capture of this tropical freshwater fish, since it was reportedly taken in Great South Bay, which has relatively cool (about 13°C), saline waters.

There are no known voucher specimens.

Corydoras aeneus

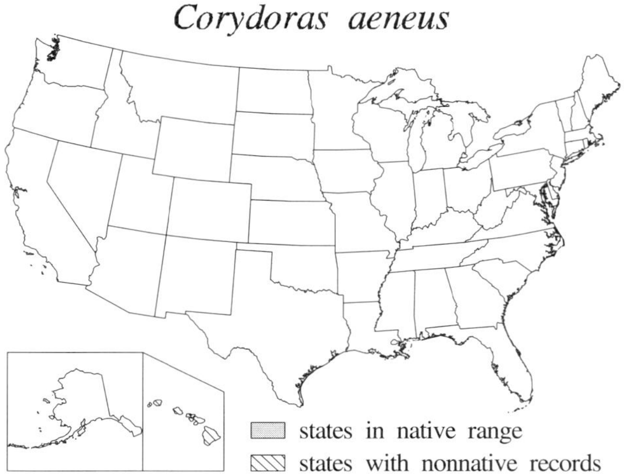

Corydoras aeneus (Gill 1858)

Common Names: bronze corydoras (green corydoras).

Identification: Nijssen and Isbrücker (1980) treated *C. aeneus* as a valid species, but also recognized a *C. aeneus* species complex with 25 species. The key in Gosline (1940) is useful but somewhat dated. A key to a limited number of species, including distinguishing characteristics for *C. aeneus*, was given by Taphorn (1989). A few distinguishing characteristics and photographs appeared in Burgess (1992). Color photographs also appeared in Burgess (1989), although it is uncertain whether these represent the true *C. aeneus*. Maximum size: In aquaria, to about 7.5 cm; in nature, rarely exceeds 4 cm SL.

Native Range: Tropical America. Widespread in South America including, but not limited to, Trinidad (type locality), and the Amazon and Orinoco basins; from Venezuela and the Guianas in the north to São Paulo and Mato Grosso, Brazil, in the south (Nijssen and Isbrücker 1980).

Nonindigenous Occurrences: Established in streams on Oahu and Kauai, **Hawaii**, since about 1984 (Devick 1991a, 1991b).

Means of Introduction: Probable aquarium release or escape.

Status: Established on Oahu and Kauai, Hawaii (Devick 1991a, 1991b).

Impact of Introduction: Unknown.

Remarks: This species is one of the most common *Corydoras* in the aquarium trade (Burgess 1989).

There are no known voucher specimens, although preserved specimens are likely in the possession of Hawaii Department of Land and Natural Resources (M. Yamamoto, personal communication).

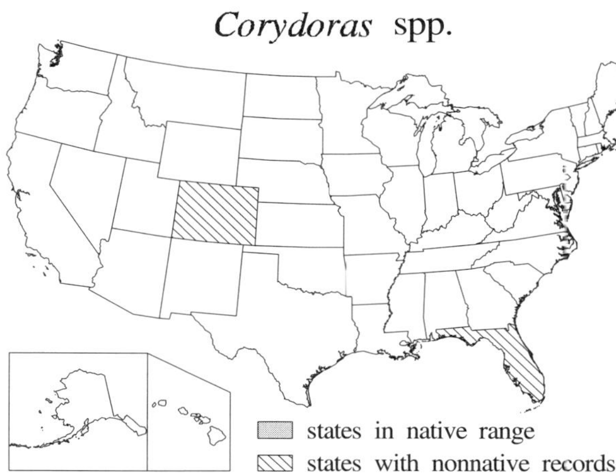

Corydoras spp.

Common Names: corydoras (corys).

Identification: The genus *Corydoras*, as reviewed by Nijssen and Isbrücker (1980), includes approximately 100 species and subspecies. Distinguishing characteristics were provided in Nijssen and Isbrücker (1980) and Burgess (1989, 1992). Color photographs appeared in Burgess (1989, 1992) and Ferraris (1991). Maximum size: 2.5 to 12 cm; most species less than 7 cm.

Native Range: Tropical America. Widespread in South America including, but not limited to, the Amazon, Orinoco, and La Plata–Parana basins, the Guianas, and Trinidad (Burgess 1989).

Nonindigenous Occurrences: Taken from open waters of geothermal springs in the San Luis Valley, Rio Grande basin, **Colorado**, during the period 1980 to 1984 (Zuckerman and Behnke 1986). There is a single report of fish collected from a small roadside borrow pit south of Miami, **Florida** (Shafland 1976).

Means of Introduction: The Colorado hot springs area is used for culturing aquarium fish, and the record represents escape or release from a local fish farm. The Florida record may be result of escape from fish farm or aquarium release.

Status: Reported from Colorado and Florida.

Impact of Introduction: Unknown.

Remarks: The first imports of *Corydoras* catfishes into North America as aquarium fish were around 1915 (Burgess 1989). The Florida record reported by Shafland (1976) is most likely the source for the listing of a *Corydoras* species by Courtenay et al. (1984, 1991). Colorado specimens reportedly were deposited with the U.S. Fish and Wildlife Service in Fort Collins, Colorado (Zuckerman, personal communication). Many of the Colorado fish collected by Zuckerman are now part of the MSB collection; however, *Corydoras* did not appear in a 1995 listing of fish species being transferred to MSB. As such, the voucher specimen has disappeared and is presumed lost.

Hoplosternum littorale

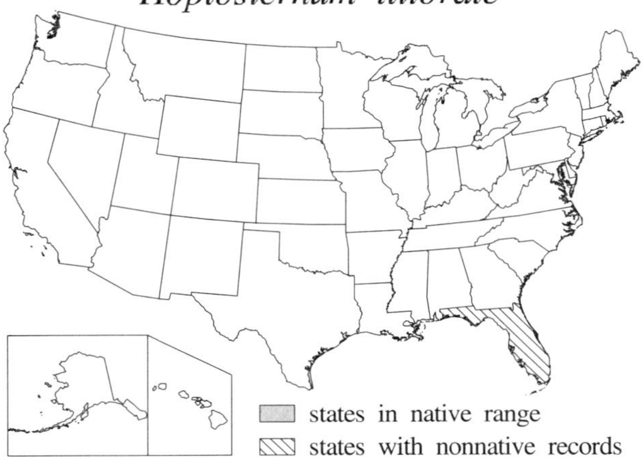

Hoplosternum littorale (Hancock 1828)

Common Names: brown hoplo (tamoata, curito, atipa, cascudo, curito, cascadura).

Identification: The genus *Hoplosternum* was reviewed and revised by Reis (1997). Reis also provided keys, distinguishing characteristics, and illustrations. A photograph also appeared in Nico et al. (1996). Maximum size: 22 cm SL.

Native Range: Tropical America. Widespread in South America east of the Andes and north of Buenos Aires, including the Rio Orinoco, Trinidad, coastal rivers of the Guianas, Rio Amazonas basin, Rio Paraguay, lower Rio Parana and coastal systems in southern Brazil (Reis 1997).

Nonindigenous Occurrences: A population was discovered in ditches of the Indian River lagoon system of **Florida** in late 1995 (Nico et al. 1996). Recently, populations have been found in the St. Johns and Kissimmee River drainages (L. G. Nico, unpublished data).

Means of Introduction: Unknown.

Status: Established in peninsular Florida (Nico et al. 1996); the species has recently expanded its range into several new drainages within the state (Nico, unpublished data).

Impact of Introduction: Largely unknown. In several Florida water bodies this species is locally abundant; because it feeds heavily on benthic invertebrates and detritus, the species may be causing significant changes in food web structure, negatively impacting native invertebrates, and competing with native fishes for food (Nico, unpublished data). This species can breathe air and can tolerate a wide range of environmental conditions (e.g., fresh and brackish waters). There is concern that it will eventually invade many natural freshwater wetlands, and coastal marshes in Florida (Nico et al. 1996).

Remarks: The identity of specimens caught in Florida was confirmed by R. Reis. The distribution, biology, and impacts of introduced populations in Florida are being investigated by U.S. Geological Survey researchers.

Voucher specimens: Florida (UF 101746, 101747, 101748; UMMZ 231955).

Loricariidae - Suckermouth catfishes

The Loricariidae, whose members are known as suckermouth catfishes, is an extremely large and diverse family of neotropical fishes with about 80 genera (Nelson 1994) and more than 600 species (Burgess 1989). These catfishes range from the La Plata–Parana basin of South America north to Costa Rica and Panama in Central America (Berra 1981). Their bodies are covered with bony plates and their mouths are subterminal, typically in the form of a sucker. Most species are bottom-feeding vegetarians or detritivores. Members of the family range in size from relatively small species, less than one or a few centimeters long, to species that exceed one meter. Many are popular aquarium fishes used for controlling algae in tanks. Loricariids are restricted to freshwater.

Ancistrus spp.

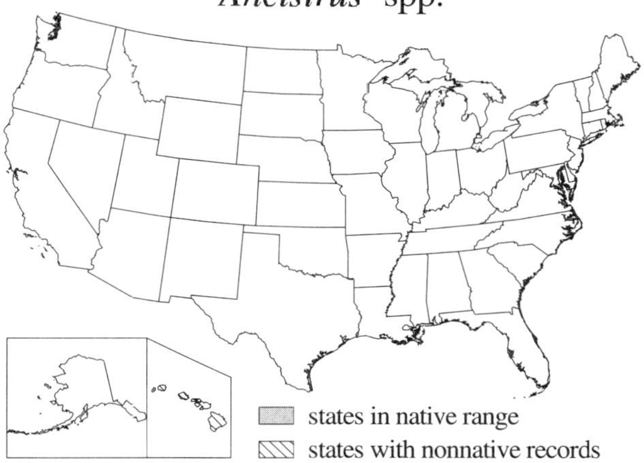

Ancistrus spp.

Common Names: bristle-nosed catfish (bushy-faced catfish, bristle-nosed pleco).

Identification: The genus *Ancistrus* contains about 50 or more described species (Burgess 1989; Armbruster 1997). Members of this genus exhibit marked sexual dimorphism (Ferraris 1991). Burgess (1989) and Armbruster (1997) gave distinguishing characteristics of the genus and a key to loricariid genera; Burgess (1989) also provided key to selected species. Photographs were given in Burgess (1989) and Ferraris (1991). Maximum size: 15 cm (Burgess 1989).

Native Range: Tropical America. Central and South America (Armbruster 1997).

Nonindigenous Occurrences: A member of this genus was first encountered in small numbers in the Nuuanu #3 Reservoir in Oahu, **Hawaii**, in 1984; it is now abundant in that reservoir and in many other Oahu reservoirs and streams (Devick 1991b). An unidentified *Ancistrus* has been reported from Wahiawa Reservoir in Oahu since about 1987 (Devick 1988) and also was collected from Manoa Stream, Oahu, in 1989 (museum specimens).

Means of Introduction: Probable aquarium release.

Status: Established on island of Oahu, Hawaii (Devick 1991b).

Impact of Introduction: Unknown.

Remarks: According to Devick (1991b), at least two different *Ancistrus* species have been found in Hawaii, but identifications to species level have not been made. All or most species feed on attached algae. Devick (1991b) cited *Ancistrus* as a suspected major hazard to native Hawaiian species, apparently because of its abundance.

Voucher specimens: Hawaii (UF 91911, UF 91913).

Hypostomus spp.

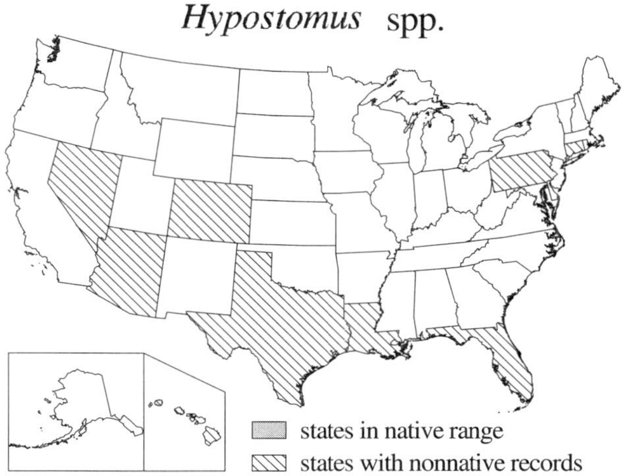

Hypostomus spp.

Common Names: suckermouth catfishes (armored catfishes, plecostomus, plecos, common plecos, armadillo del río).

Identification: The genus *Hypostomus* contains about 116 species (Burgess 1989). Highlighting the serious need for additional taxonomic and systematic work, Armbruster (1997) concluded that it is currently impossible to identify most species in the genus. Several apparently different *Hypostomus* species have been collected in the United States but not definitively identified to species level (Page and Burr 1991; Courtenay and Stauffer 1990). Distinguishing characteristics of the genus and a key to loricariid genera were provided by Burgess (1989) and Armbruster (1997). Photographs appeared in Burgess (1989) and Ferraris (1991). *Hypostomus* has officially replaced the generic name *Plecostomus*. The genus was included in the key to Texas fishes of Hubbs et al. (1991) and several identifying traits were also given by Page and Burr (1991). Maximum size: Depending on species, maximum from 14 to about 50 cm (Burgess 1989).

Native Range: Tropical America. South and Central America from Uruguay north to Panama (Burgess 1989).

Nonindigenous Occurrences: A single specimen was taken from the lower Colorado River near Headgate Rock Dam above Parker, **Arizona**, in 1986 (Lea, personal communication to Courtenay). An unidentified *Hypostomus* was collected in geothermal waters of Upper Rio Grande drainage in the San Luis Valley, Conejos County, **Colorado**, during the period 1980 to 1984 (Zuckerman and Behnke 1986). Several specimens, reportedly members of this genus, were taken from the Thames River drainage, **Connecticut** (Whitworth 1996). There are several reports of *Hypostomus* from **Florida.** Burgess (1958) indicated that *Plecostomus* sp. (= *Hypostomus*) had been reported in the Hillsborough County area, but none were taken during his 1958 fish survey of Six Mile Creek. Later, in the 1970s, collections verified its presence in Six Mile Creek (museum specimens). There also are reports of *Hypostomus* in south Florida. Courtenay et al. (1974) reported that a member of this genus had inhabited the Snapper Creek Canal system since 1959. There also was a record from a rock pit in West Miami, Dade County, in the early 1960s (Rivas 1965). Other south Florida records include Westwood Lake, southwestern Miami in Dade County; Snake Creek Canal on the border of Dade and Broward counties; and Conservation Area No. 3 (Courtenay et al. 1974; Courtenay and Hensley 1979a; Courtenay and McCann 1981). However, the only south Florida record based on a specimen involves a fish taken from Coral Gables Canal, Dade County, in 1960 (museum specimen). There is an unconfirmed report of a specimen taken in Lake Jessup on the St. John's River (Courtenay et al. 1974). An unidentified *Hypostomus* is established and abundant in Manoa Stream, Oahu, **Hawaii**; the catfish first was discovered there in 1988 (Devick 1991a, 1991b). A *Hypostomus* also was collected from an unspecified water body in **Louisiana** (Tilyou, personal communication). A *Hypostomus* species has been established in Indian Spring, a thermal spring in Clark County, **Nevada**, since at least 1966 (Minckley 1973; Courtenay and Deacon 1982; Deacon and Williams 1984; Courtenay and Stauffer 1990). A single member of this genus was taken from Muncy Creek in Lycoming County, **Pennsylvania**, from the Susquehanna River drainage on 8 August 1980 (Courtenay, personal communication; museum specimen). The first specimens collected in open waters of **Texas** were from a headwater stream of the San Antonio River near the San Antonio Zoo in Brackenridge Park, San Antonio, in the spring of 1962; reproduction and apparent overwintering of the population was documented in 1964 (Barron 1964). Subsequent collections and records have confirmed its survival in the same locality (Hubbs et al. 1978; Hubbs 1982; Howells 1992a). Two specimens were collected from Landa Lake at the headwaters of the Comal River in Comal County on 1 August 1990 (Whiteside and Berkhouse 1992). A *Hypostomus* also has been reported from specimens taken from Comal Springs, Comal County, Texas, in 1991 (Hubbs et al. 1991; Howells 1992a).

Means of Introduction: Members of this genus have been introduced through a combination of fish farm escapes or releases, and aquarium releases (Courtenay and Stauffer 1990; Courtenay and Williams 1992). In Texas, the initial introduction occurred when *Hypostomus* entered local streams after escaping from pool and canal systems of the San Antonio Zoological Gardens in or before 1962 (Barron 1964); the Comal County introduction was probably due to an aquarium release (Whiteside and Berkhouse 1992).

Status: Several morphologically distinct but unidentified *Hypostomus* species have been recorded as established in the United States; these included populations in Indian Springs in Nevada; Hillsborough County in Florida; and the San Antonio River in Texas (Courtenay and Deacon 1982; Courtenay et al. 1984, 1986; Courtenay and Stauffer 1990; Page and Burr 1991). A population of an unidentified *Hypostomus* species is firmly established in Hawaii (Devick 1991a, 1991b). There have been no recent records of *Hypostomus* in Hillsborough County, Florida, although another loricariid, *Pterygopli-*

chthys multiradiatus, is locally common (L. G. Nico, personal observation). Reported from Arizona, Colorado, Connecticut, Louisiana, and Pennsylvania,

Impact of Introduction: The effects of these loricariid catfish is largely unknown. In Texas, Hubbs et al. (1978) reported possible local displacement of algae-feeding native fishes such as *Campostoma anomalum* by *Hypostomus*. Because of their abundance in Hawaii, introduced *Hypostomus* and *Pterygoplichthys* may compete for food and space with native stream species (Devick 1989). Once established, armored catfishes are difficult or impossible to eradicate. For example, the use of electroshocking as a control method in Hawaii has proven unsuccessful (Devick 1991a).

Remarks: The Nevada population was reported originally as *Plecostomus punctatus* by Minckley (1973) and as *Hypostomus plecostomus* by Deacon and Williams (1984). Populations from Texas (e.g., Hubbs et al. 1978; Whiteside and Berkhouse 1992) and Florida (e.g., Rivas 1965) occasionally have been reported as *Hypostomus plecostomus*. According to Courtenay et al. (1974), the Florida *Hypostomus* species in the Hillsborough County area was probably different than that reported from the southern part of the state. In addition, most early reports from south Florida, and possibly elsewhere in the state, probably were based on incorrect identifications of *Pterygoplichthys* (Loftus and Kushlan 1987; Ludlow and Walsh 1991; Nico, personal observation). Courtenay (personal communication) reviewed records of loricariid catfishes from southeastern Florida and located only one specimen of the genus *Hypostomus* (now deposited at UF - UF 98938), collected from Coral Gables Canal at Red Road, Dade County, in 1960; he concluded that all other loricariids from Dade County were *Pterygoplichthys*. The *Hypostomus* inhabiting the Tampa area was reported as expanding its range into the Hillsborough River from Six Mile Creek (Courtenay and Stauffer 1990), but there are no supporting specimens, and these also may be based on misidentifications of *Pterygoplichthys*. Whitworth (1996) recorded the capture of specimens of an unidentified loricariid from the Thames River drainage, Connecticut, and listed it as *Hypostomus*. Unfortunately, he does not provide any information that might be useful in its positive identification. In his book, Whitworth included an illustration of a *Hypostomus*, but the drawing is from an old plate and not of the Connecticut fish. Distribution maps for *Hypostomus* found in the United States were given in Courtenay and Hensley (1979a), Lee et al. (1980 et seq.), and Courtenay and McCann (1981), but these maps most likely include records based on what is now recognized to be *Pterygoplichthys*. Members of this genus are popular aquarium fishes.

Voucher specimens: Florida (UF 30864, 98938, 98939, FDNR 08920), Hawaii (UF 91912), Nevada (UF 91914), Pennsylvania (PSU 1433?), Texas (UF 91915).

Otocinclus sp.

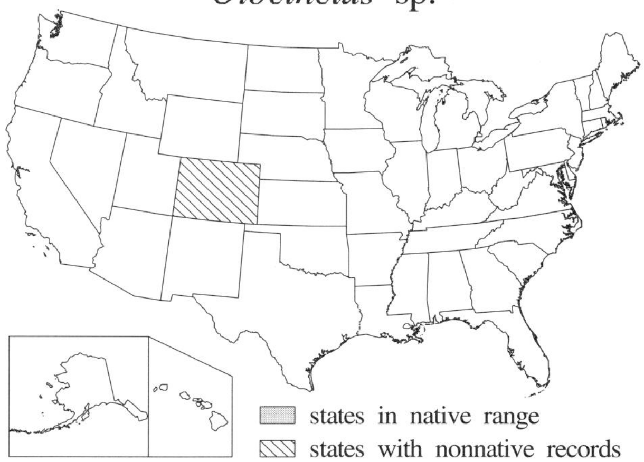

Otocinclus sp.

Common Names: sucker catfish (midget sucker catfish, cascudinhos).

Identification: Schaefer (1997) revised the group and recognized 13 species in the genus *Otocinclus*; he also provided identification keys along with photographs and illustrations. For color photographs see Sands (1984), Burgess (1989), and Ferraris (1991). Similar species include members of the loricariid genera *Microlepidogaster*, *Parotocinclus, Pseudotocinclus*, and *Schizolecis* (Schaefer 1997), and to a lesser extent *Hypoptopoma*. Maximum size: about 4.5 cm SL.

Native Range: Tropical America. South America including the Amazon, Orinoco, La Plata–Parana, and Sao Francisco river basins, and also coastal drainages of southeastern Brazil (Schaefer 1997).

Nonindigenous Occurrences: One or more specimens identified as belonging to this genus were collected in geothermal waters of the Upper Rio Grande drainage in the San Luis Valley, Conejos County, **Colorado**, during the period 1980 to 1984 (Zuckerman and Behnke 1986).

Means of Introduction: Probable escape from local fish farm (Zuckerman and Behnke 1986).

Status: Reported from Colorado.

Impact of Introduction: Unknown.

Remarks: Several *Otocinclus* species occur in the aquarium trade (Sands 1984; Burgess 1989). According to Zuckerman (personal communication), voucher specimens from his San Luis Valley collections were deposited with the U.S. Fish and Wildlife Service in Fort Collins. As of 1996, all, or most of, the preserved material of Zuckerman was being incorporated into the MSB collection in Albuquerque. However, a recent listing of the MSB holdings did not include *Otocinclus*.

Peckoltia sp.

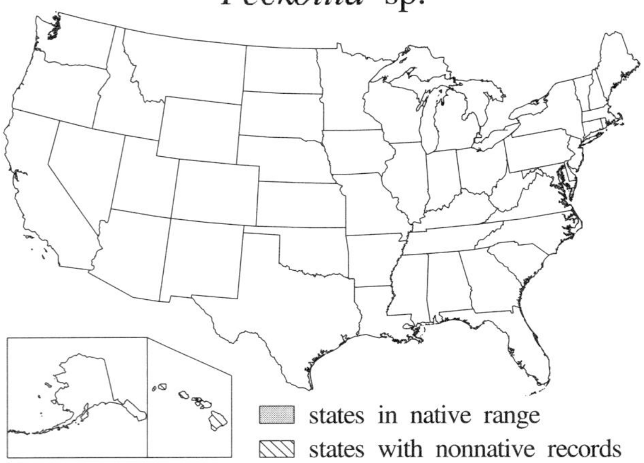

Peckoltia sp.

Common Name: clown pleco.

Identification: *Peckoltia* includes about 12 species; the group is in need of taxonomic revision (Armbruster 1997). Distinguishing characteristics of the genus and a key to loricariid genera were provided by Burgess (1989) and Armbruster (1997). Color photographs appeared in Sands (1984), Burgess (1989), and Ferraris (1991). Armbruster (1997) noted that it is commonly confused with some *Panaque*; as such, he concluded that the aquarium literature should be viewed with skepticism. Maximum size: 22 cm.

Native Range: Tropical America. The Amazon and Orinoco river basins (Armbruster 1997).

Nonindigenous Occurrences: A species identified as *Peckoltia* was taken from Manoa Stream, Oahu, **Hawaii**, during 1990–1991 surveys by the state; there was also an earlier unconfirmed report from Kaukonahua Stream, Oahu, in 1989 (Devick 1991b).

Means of Introduction: Probable aquarium release.

Status: Reported from Hawaii.

Impact of Introduction: Unknown.

Remarks: Many members of this genus are popular aquarium fishes. However, some *Peckoltia* species are difficult to separate from members of the loricariid genus *Panaque*. As such, the Hawaiian record may actually be a *Panaque* species. *Panaque maccus* is the common clown pleco in the ornamental fish trade (J. Armbruster, personal communication).

There are no known voucher specimens.

Pterygoplichthys disjunctivus

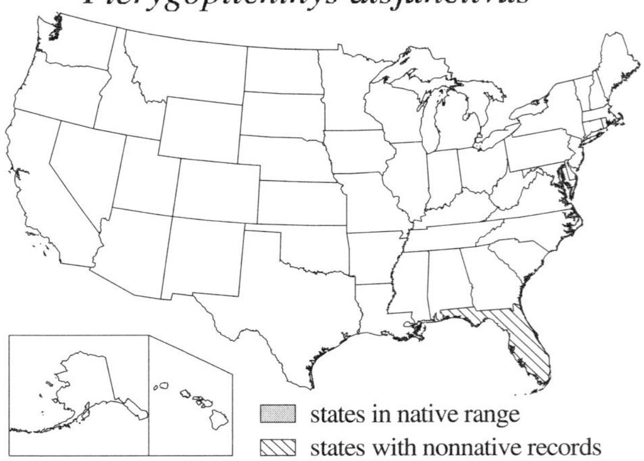

Pterygoplichthys disjunctivus (Weber 1991)

Common Names: vermiculated sailfin catfish (radiated ptero).

Identification: Weber (1991, 1992) assigned sailfin catfishes to three genera and described this species as new under the name *Liposarcus disjunctivus*. Armbruster (1997), after a detailed systematic review, placed the genus *Liposarcus* into the synonymy of *Pterygoplichthys*. Weber (1992) provided a key and distinguishing characteristics and photographs of specimens. Page (1994) also provided a few distinguishing characteristics. Measurements and counts for several Florida specimens were given by Ludlow and Walsh (1991). Maximum size: 70 cm.

Native Range: Tropical America. The Madeira River drainage, Amazon Basin, South America (Weber 1992).

Nonindigenous Occurrences: This species is known from streams, canals and other water bodies, in **Florida**. Published records document its presence in several sites in Hillsborough County, including the Hillsborough River drainage, Hillsborough County, Florida (Ludlow and Walsh 1991, referenced as *Pterygoplichthys cf. multiradiatus*; Page 1994) and Lake Thonotosassa and Baker Creek (Page 1994; museum specimens). More recent records include the Little Manatee, Alafia, Peace, Kissimmee, and Withlacoochee river drainages, and several other streams and lakes in the central part of the state (L. G. Nico, unpublished data). Several armored catfish were taken in **Texas** from a water body in the Galveston Bay basin near Houston in 1998 (R. T. Martin, personal communication; museum specimen); a single small specimen was preserved and we tentatively identified it as *P. disjunctivus*.

Means of Introduction: In Florida, probably originally introduced through fish farm escapes or releases, although aquarium releases cannot be ruled out. Texas record is probably the result of aquarium release.

Status: Established in Florida, the species has recently expanded its range into several new drainages within the state (Nico, unpublished data). Reported, possibly established, in a small drainage in Texas.

Impact of Introduction: Largely unknown. In several natural streams in this species is relatively abundant. Because it grazes and removes attached algae, and also feeds on benthic organisms and detritous, this introduced catfish may be having a significant impact on the aquatic food base and, therefore, negatively effecting native invertebrate and vertebrate species (Nico, personal observations).

Remarks: Prior to its description, specimens of this species taken in Florida (Tampa Bay area) were reported as *Pterygoplichthys multiradiatus* or, in some earlier cases, as *Hypostomus* species. *Pterygoplichthys disjunctivus* specimens from the Hillsborough River drainage of Florida were tentatively identified as *Pterygoplichthys cf. multiradiatus* by Ludlow and Walsh (1991). Page (1994) followed Weber (1992) in using the name *Liposarcus disjunctivus*. The distribution and impact of this species in Florida, and that of other introduced armored catfishes, is currently being investigated by U.S. Geological Survey researchers. Note: as this book was going to press, J. Armbruster examined the Texas specimen and determined it to be *Pterygoplichthys anisitsi*.

Voucher specimens: Florida (UF 79623, 89896; many uncatalogued; FDNR 18185); Texas (UF uncatalogued).

Pterygoplichthys multiradiatus

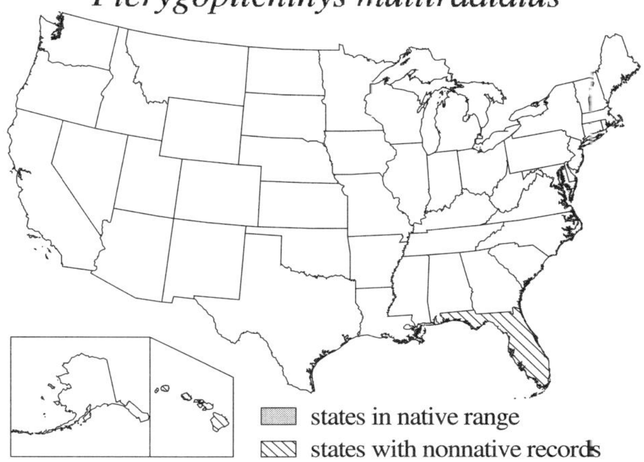

Pterygoplichthys multiradiatus (Hancock 1828)

Common Names: Orinoco sailfin catfish (radiated ptero, butterfly pleco).

Identification: Weber (1991, 1992) assigned sailfin catfishes to three genera and used the name *Liposarcus multiradiatus* for this species (a name used in some of the more recent literature). Armbruster (1997), after a detailed systematic review, placed the genus *Liposarcus* into the synonymy of *Pterygoplichthys*. Weber (1992) provided a key and distinguishing characteristics and photographs of specimens. A few distinguishing characteristics also were given by Page and Burr (1991) and Page (1994). Maximum size: 70 cm.

Native Range: Tropical America. Orinoco River basin in northern South America.

Nonindigenous Occurrences: This catfish is known from various canals and water bodies in south **Florida** within Dade, Palm Beach, and Broward counties (Courtenay and Stauffer 1990; Nico, unpublished data). The species was first reported from **Hawaii** in Wahiawa Reservoir, Oahu, in January 1987; it had become one of the most abundant fish in the reservoir by 1989; it also is established in Kaukonahua Stream, Oahu (Devick 1988, 1991b). The fish is widely distributed in lower elevation reservoirs and streams on northern Oahu (Devick 1988, 1989).

Means of Introduction: This armored catfish has been collected in southeastern Florida since about 1971 (Courtenay et al. 1984). Its presence is most likely the result of escapes or releases from aquarium fish farms (Courtenay and Stauffer 1990). In Hawaii, introductions are presumably the result of aquarium releases that occurred in the 1980s, possibly as early as 1982 (Devick 1991b).

Status: Established in Florida and Hawaii. Recent surveys in Florida indicate its range may be expanding (Nico, unpublished data).

Impact of Introduction: Largely unknown. In Hawaii, the thousands of nesting tunnels excavated by male *Pterygoplichthys* in reservoir and stream banks have contributed to siltation problems (Devick 1989). Because of their abundance in Hawaii, *Pterygoplichthys* and other armored catfishes have the potential to affect native stream species negatively through competition for food and space (Devick 1989). In Florida, this species occupies waters adjacent to Everglades National Park and is considered a threat to the park (Courtenay 1989).

Remarks: Many early reports of *Hypostomus* from south Florida, and some from the Tampa area, may have been based on misidentifications of *Pterygoplichthys* (Loftus and Kushlan 1987; Ludlow and Walsh 1991).

Voucher specimens: Florida (UF many specimens).

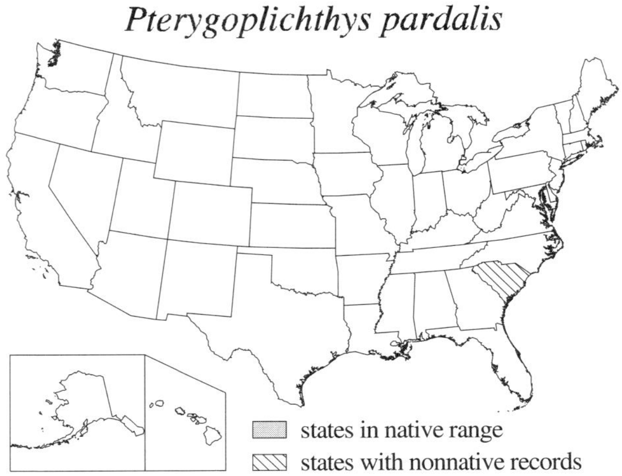

Pterygoplichthys pardalis (Castelnau 1855)

Common Names: Amazon sailfin catfish

Identification: Weber (1991, 1992) assigned sailfin catfishes to three genera and used the name *Liposarcus pardalis* for this species. Armbruster (1997), after a detailed systematic review, placed the genus *Liposarcus* into the synonymy of *Pterygoplichthys*. Weber (1992) provided a key and distinguishing characteristics and photographs of specimens. Maximum size: probably about 70 cm.

Native Range: Tropical America. Amazon River basin.

Nonindigenous Occurrences: A single specimen was taken in Cherokee County, **South Carolina,** from the Broad River at 99 Island Dam in 1992 (museum specimen).

Means of Introduction: Probable aquarium release.

Status: Reported from South Carolina.

Impact of Introduction: Unknown.

Remarks: L. Page examined the specimen and confirmed its identification as *Pterygoplichthys pardalis*.

Voucher specimen: Florida (UF 93286).

ORDER ESOCIFORMES

Esocidae - Pikes

The family Esocidae includes one genus and five species (Nelson 1994). They are distributed holarctically in northern and eastern North America, Europe, and northern Asia (Berra 1981). A single genus, *Esox*, and four species are native to the United States (Page and Burr 1991). Esocids are large fish, cylindrical in shape, have a duckbill-like snout, and (with the exception of *E. americanus*) are large fish. Because of their large size and active predatory behavior, pikes are very popular sport and trophy fishes, and therefore frequently stocked. However, because they are mainly piscivorous, introduced pikes pose a potential threat to smaller native species. The family is restricted to freshwater.

Esox americanus

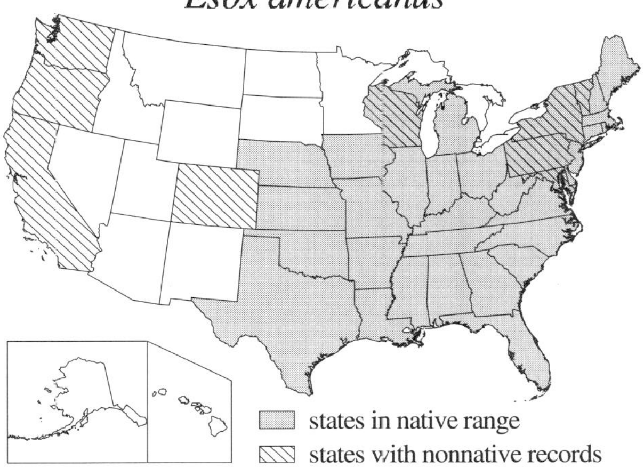

Esox americanus Gmelin 1789

Common Names: grass pickerel, redfin pickerel.

Identification: Becker (1983); Page and Burr (1991); Etnier and Starnes (1993); Jenkins and Burkhead (1994). Two recognized subspecies in this genus; *E. a. americanus*, the redfin pickerel, and *E. a. vermiculatus*, the grass pickerel. Crossman (1978) described and mapped the range of each. Maximum size: 38 cm.

Native Range: Atlantic Slope from the St. Lawrence River drainage, Quebec, to Lake Okeechobee, Florida; Gulf drainages east to the Brazos River, Texas; Mississippi River and Great Lakes basins north to Nebraska, Wisconsin, Michigan, and southern Ontario (range for the species). Redfin pickerel *E. a. americanus* occur on the Atlantic Slope to southern Georgia and intergrade with grass pickerel *E. a. vermiculatus* in the Gulf Slope drainages west to the Pascagoula River. Grass pickerel occupy the rest of the range (Page and Burr 1991). Crossman (1978) gave a distribution map of the subspecies and intergrades.

Nonindigenous Occurrences: Grass pickerel *E. a. vermiculatus* were stocked in Lake Cuyamaca near San Diego, **California** (Smith 1896; Crossman 1978) in the late 1800s; in Palmer Lake, **Colorado** (Ellis 1974); in unspecified areas of **Oregon** (Li, personal communication); in Finnel and Cow lakes, in streams in the Palouse River drainage (Wydoski and Whitney 1979), in lakes near Spokane, and streams of southeastern **Washington** (Chapman 1942; Lampman 1946; Crossman 1978); and the Wisconsin and Chippewa drainages (Burr and Page 1986), and northern **Wisconsin** (Crossman 1978). Redfin pickerel *E. a. americanus* were stocked in Deep Creek, **Maryland** (Crossman 1978); the Lake Champlain drainage in **New York** (Underhill 1986), in the Youghiogheny River and Monongahela drainage, **Pennsylvania** (Hendricks et al. 1979; Hocutt et al. 1986) and the Lake Champlain drainage in **Vermont** (Underhill 1986; museum specimen).

Means of Introduction: This species was stocked intentionally and possibly accidentally; it was stocked illegally in Oregon. Probably introduced into Washington as misidentified northern pike *E. lucius* (Chapman 1942); it is also possible that grass pickerel were introduced into Idaho along with northern pike, and that the former species later reached Washington by "some devious means" (Wydoski and Whitney 1979).

Status: Extirpated in California; established in Colorado, Pennsylvania, and Washington; unknown in Maryland, Oregon, and Wisconsin.

Impact of Introduction: Unknown.

Remarks: This species is sometimes mistaken for the young of other, larger *Esox* species. Smith (1985) mapped several records for the Lake Champlain drainage in New York but did not indicate that these were the result of introductions.

Voucher specimens: Vermont (CU 54844).

Esox lucius

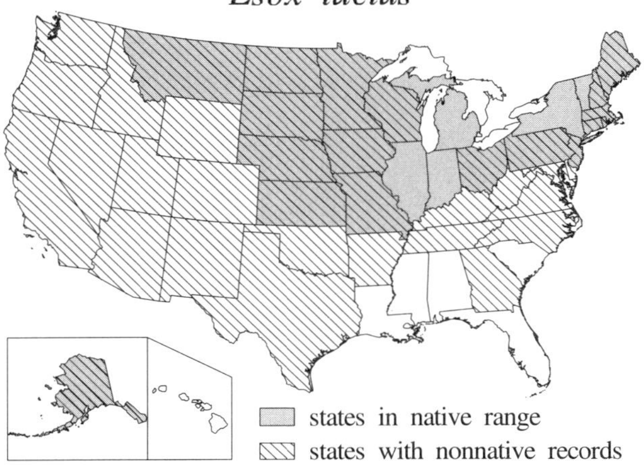

Esox lucius Linnaeus 1758

Common Name: northern pike.

Identification: Scott and Crossman (1973); Pflieger (1975); Becker (1983); Page and Burr (1991). Maximum size: 133 cm.

Native Range: Atlantic, Arctic, Pacific, Great Lakes, and Mississippi River basins from Labrador to Alaska and south to Pennsylvania, Missouri, and Nebraska (Page and Burr 1991). Crossman (1978) gave a distribution map.

Nonindigenous Occurrences: This popular sport fish has been stocked in **Alaska** (Morrow 1980; M. Bell, personal communication); **Arkansas** (Buchanan 1973; Pritchard et al. 1978; Cross et al. 1986); **Arizona** (Minckley 1973; Rinne 1995); **California** (Smith 1896; Shebley 1917; Lampman 1946; Hubbs et al. 1979; Dill and Cordone 1997); **Colorado** (Everhart and Seaman 1971; Propst and Carlson 1986); **Connecticut** (Webster 1942; Behnke and Wetzel 1960; Schmidt 1986; Whitworth 1996); **Georgia** (Dahlberg and Scott 1971a); **Iowa** (Harlan et al. 1987); **Idaho** (Smith 1896; Chapman 1942; Lampman 1946; Linder 1963; Simpson and Wallace 1978; Wydoski and Whitney 1979; McMahon and Bennett 1996); **Kansas** (Todd 1962; Polson 1964; Cross 1967; Cross et al. 1986); **Kentucky** (Clay 1975; Trautman 1981; Burr and Page 1986; Burr and Warren 1986; Hocutt et al. 1986); **Maine** (Kircheis 1994); **Massachusetts** (Hartel 1992; Hartel et al. 1996); **Maryland** (Pritchard et al. 1978); **Minnesota** (Phillips et al. 1982); **Missouri** (Pflieger 1971, 1975, 1997; Cross et al. 1986); **Montana** (Brown 1971; Cross et al. 1986; Holton 1990; McMahon and Bennett 1996; Whitmore 1997); **Nebraska** (Jones 1963); **Nevada** (Deacon and Williams 1984); **New Hampshire** (Scarola 1973; Hartel et al. 1996); **New Jersey** (Fowler 1952; Stiles 1978; Soldwedel, personal communication); **New Mexico** (Sublette et al. 1990); **North Carolina** (Pritchard et al. 1978; Menhinick 1991); **North Dakota** (Owen et al. 1981; Cross et al. 1986); **Ohio** (Trautman 1981; Hocutt et al. 1986; Burr and Page 1986); **Oklahoma** (Miller and Robison 1973; Pritchard et al. 1978; Cross et al. 1986); **Oregon** (Li, personal communication); **Pennsylvania** (Eddy and Underhill 1974; Denoncourt et al. 1975a; Hendricks et al. 1979; Cooper 1983; Hocutt et al. 1986); **Rhode Island** (Lapin, personal communication); **South Dakota** (Underhill 1959; Bailey and Allum 1962; Cross et al. 1986); **Tennessee** (Kuhne 1939; Etnier and Starnes 1993); **Texas** (Luebke 1978; Cross et al. 1986; Howells 1992a); **Utah** (B. Schmidt, personal communication); **Virginia** (Hocutt et al. 1986; Etnier and Starnes 1993; Stauffer et al. 1995; Jenkins and Burkhead 1994; Southwick, personal communication); **Washington** (McMahon and Bennett 1996; Fletcher, personal communication); **West Virginia** (Hocutt et al. 1986; Stauffer et al. 1995); **Wisconsin** (Becker 1983); and **Wyoming** (Baxter and Simon 1970; Hubert 1994; Stone 1995; McMahon and Bennett 1996). McMahon and Bennett (1996) gave a map showing introduced populations in the northwest.

Means of Introduction: This species has been intentionally stocked as a sport fish in most areas. In some cases, introductions were illegal, and these include such sites as Coeur d'Alene Lake, Idaho; Keyhole Reservoir, Wyoming; and Beaver Creek Reservoir, Bitterroot River, and Flathead River, Montana (McMahon and Bennett 1996), and lakes in Alaska (Bell, personal communication). McMahon and Bennett (1996) gave a table of western reservoirs with introduced populations and the method of introduction for each one. First stocked in Arizona in 1967 (Rinne 1995). In addition to being stocked as a sport fish, Pflieger (1997) stated that *Esox lucius* was stocked in Missouri reservoirs to introduce a large predator that could more effectively prey on the large populations of carp and gizzard shad present in such artificial environments.

Status: Established in many localities. Extirpated in California (Hubbs et al. 1979).

Impact of Introduction: The piscivorous northern pike has been shown to significantly reduce prey density and has the potential to cause large-scale changes in fish communities, even resulting in species elimination (He and Kitchell 1990). A study conducted in Wisconsin showed introduced pike mostly affected four minnow species; redbelly dace *Phoxinus eos*, finescale dace *P. neogaeus*, fathead minnow *Pimephales promelas*, and brassy minnow *Hybognathus hankinsoni*. Pike apparently had less effect on other species in the pond (He and Kitchell 1990). Impacts can be either direct, such as by predation, or indirect, such as by causing prey fish to alter their behavior (He and Kitchell 1990). In Montana, northern pike commonly deplete their prey and become stunted (McMahon and Bennett 1996). A study conducted by T. Jones (University of Montana) in 1990, found northern pike eliminated most other fishes except for the pumpkinseed *Lepomis gibbosus*, which was likely protected by its deep body shape and stiff spines, making it difficult prey (McMahon and Bennett 1996). Northern pike may be responsible for declines of native westslope cutthroat trout *Oncorhynchus clarki lewisi* and bull trout *Salvelinus confluentus* in the Stillwater lakes in Montana (McMahon and Bennett 1996). Northern pike are reported to be "a problem" in the Yampa River in Colorado (Whitmore 1997). Illegally stocked lakes in Alaska are home to unique populations of threespine stickleback *Gasterosteus aculeatus*. These populations have been isolated for 22,000 years and have evolved a number of striking phenotypic traits. The presence of nonindigenous pike in these same lakes threatens the stickleback's very existence (Bell, personal communication). Trapping in lakes where pike had not been stocked yielded sticklebacks and native salmonids. Trapping in adjacent lakes where pike had been stocked yielded nothing but pike (Bell, personal communication).

When northern pike are stocked in lakes containing native muskellunge *E. masquinongy*, the two species may hybridize. Although the male tiger muskellunge are sterile, females are often fertile and are capable of backcrossing (Becker 1983). Northern pike are replacing native muskellunge in many Wisconsin lakes (Becker 1983). It is also believed that because northern pike generally spawn a month earlier than muskellunge, the older pike may prey on younger muskellunge (Gilbert, personal communication).

Remarks: Tyus et al. (1982) gave a distribution map of this species in the upper Colorado basin.

Esox masquinongy

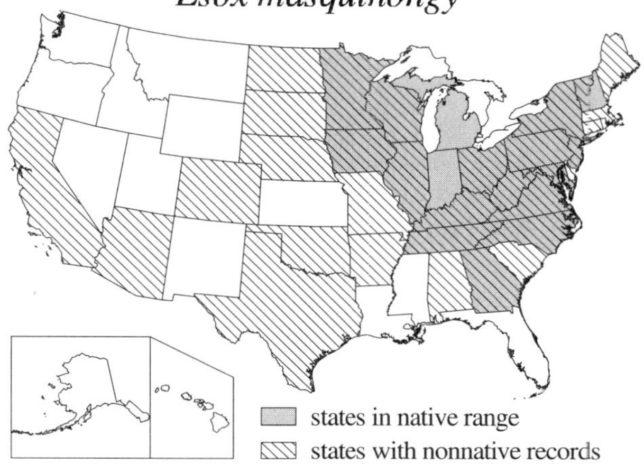

Esox masquinongy Mitchill 1824

Common Names: muskellunge (muskie, musky).

Identification: Becker (1983); Page and Burr (1991); Etnier and Starnes (1993); Jenkins and Burkhead (1994). Three subspecies are sometimes recognized: *E. m. masquinongy*, a spotted form which occurs in the St. Lawrence River and the Great Lakes and their tributaries; *E. m. immaculatus*, a form with either no pattern or barring which occurs in Wisconsin, Minnesota, northwestern Ontario, and southeastern Manitoba; and *E. m. ohioensis*, a form with bars or diffuse spots and blotches which occurs in the Ohio River and its tributaries (Becker 1983). Maximum size: 183 cm.

Native Range: St. Lawrence River–Great Lakes, Hudson (Red River), and Mississippi River basins, from Quebec to southeastern Manitoba; south in the Appalachians to Georgia and in the west to Iowa (Page and Burr 1991). Crossman (1978) gave a distribution map. Although never reported from Mississippi, considering the fact that muskellunge are (or were) native to the main Tennessee River, the species almost certainly historically entered the extreme northeastern part of that state (Gilbert, personal communication).

Nonindigenous Occurrences: This species was introduced into **Alabama** (Boschung 1992; Mettee et al. 1996); **Arizona** (Rinne 1995); the White and Ouachita drainages in **Arkansas** (Buchanan 1973; Cross et al. 1986); Lake Merced, **California** (Smith 1896; Shebley 1917; Lampman 1946; Hubbs et al. 1979); the Connecticut and Thames drainages, **Connecticut** (Milner 1874b; Schmidt 1986; Whitworth 1996); a pond on the Colorado State University campus, **Colorado** (Walker 1993; D. Horak, personal communication); Brandywine Creek,

Delaware (Raasch and Altemus 1991); Lake Blueridge, **Georgia** (Dahlberg and Scott 1971a, 1971b); the Skunk River, Iowa-Cedar River, and several lakes in **Iowa** (Burr and Page 1986; Harlan et al. 1987); reservoirs in **Illinois** (Smith 1979); several areas of **Kentucky** (Axon 1981; Burr and Page 1986; Burr and Warren 1986); the St. John River drainage, **Maine** (Kircheis 1994); Deep Creek, **Maryland** (Lee et al. 1976; Pritchard et al. 1978); **Massachusetts** (Fowler 1907; Hartel 1992; Hartel et al. 1996); **Minnesota** (Phillips et al. 1982; Burr and Page 1986); Lake of the Ozarks, Pony Express Lake, Pomme de Terre Reservoir, below Indian Trail Hatchery, Hazel Creek Lake, and Lake Girardeau in **Missouri** (Pflieger 1971, 1975, 1997; Cross et al. 1986); Cherry County, the Missouri River, and Elwood Reservoir in **Nebraska** (Morris et al. 1974; Bouc 1987); many areas of **New Jersey** (Fowler 1906, 1920, 1952; Stiles 1978; Soldwedel, personal communication); Lake Canadarago, **New York** (Smith 1985); the Roanoke, Broad, New, and Tennessee drainages in **North Carolina** (Pritchard et al. 1978; Menhinick 1991; Jenkins and Burkhead 1994); **North Dakota** (North Dakota Game and Fish Department 1994); many areas of **Ohio** (Trautman 1981); Lake Carl Etling, **Oklahoma** (J. Pigg, personal communication); Brandywine Creek and new impoundments in **Pennsylvania** (Kendall 1917; Fowler 1919; Cooper 1983; Raasch and Altemus 1991); Broad River, **South Carolina** (South Carolina Department of Natural Resources, Division of Freshwater Fisheries 1997); **South Dakota** (North Dakota Game and Fish Department 1994; R. L. Hanten, personal communication); Watauga Lake (Pritchard et al. 1978), Dale Hollow Lake (Burr and Warren 1986), and Norris Reservoir (Etnier and Starnes 1993), **Tennessee**; Inks Lake, Copper Breaks State Park, and Amistad Reservoir, **Texas** (Howells 1992a); the Connecticut River, **Vermont** (Milner 1874b); the Potomac, Rappahannock, James, Roanoke, New, Holston and Clinch-Powell drainages, **Virginia** (Hocutt et al. 1986; Jenkins and Burkhead 1994; Stauffer et al. 1995); the Potomac, Guyandotte, and New rivers, **West Virginia** (Hocutt et al. 1986; Jenkins and Burkhead 1994; Stauffer et al. 1995); and numerous rivers and lakes in **Wisconsin** (Becker 1983; Burr and Page 1986).

Means of Introduction: Intentionally stocked for sportfishing. According to Pflieger (1997) this species was first stocked in Missouri reservoirs in 1966 for the purpose of providing another trophy-sized fish and a large predator capable of preying on the many gizzard shad and other forage fish too large to be eaten by largemouth bass. Muskellunge found in one Missouri creek had escaped from hatchery ponds (Pflieger 1997).

Status: Reported from above areas, some of which may have established populations. Extirpated in California (Hubbs et al. 1979) and in Georgia, where it has not been seen since 1957 (Dahlberg and Scott 1971b). About 25% of all muskellunge populations in Wisconsin are the result of stocking (Becker 1983).

Impact of Introduction: This species likely has impacts similar to that of northern pike, namely reducing the numbers of small fish by way of predation. Muskellunge and young largemouth bass *M. salmoides* frequent the same habitats, brush piles and beds of vegetation. In two Wisconsin lakes, this resulted in the decline of the largemouth bass population (Becker 1983).

Remarks: See Becker (1983) for detailed summary on the biology of this species.

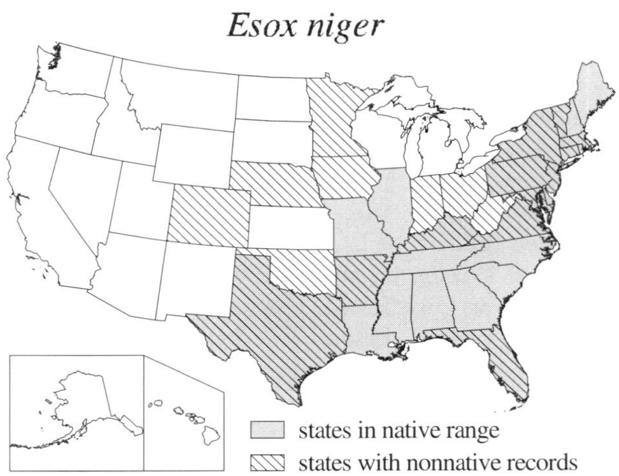

Esox niger Lesueur 1818

Common Name: chain pickerel.

Identification: Pflieger (1975); Page and Burr (1991); Etnier and Starnes (1993); Jenkins and Burkhead (1994). Maximum size: 99 cm.

Native Range: Atlantic Slope from Nova Scotia (where introduced) to southern Florida; Gulf Coast west to Sabine and Red drainages, Texas; Mississippi River basin north to Kentucky and Missouri (mostly Former Mississippi Embayment, but also upland streams in southeastern Missouri) (Page and Burr 1991). Hubbs et al. (1991) list as native to eastern Texas. Crossman (1978) gave a distribution map.

Nonindigenous Occurrences: Chain pickerel have been introduced into several lakes in **Arkansas**, including Lake Wilhelmina in Polk County (Robison and Buchanan 1988); unspecified areas of **Colorado** (Crossman 1978; Page and Burr 1991); the Connecticut River, **Connecticut** (Webster 1942; Whitworth et al. 1968); areas of **Florida** south of Lake Okeechobee (Kushlan and Lodge 1974; Lee et al. 1980 et seq.); eastern **Indiana** (Crossman 1978); eastern **Iowa** (Lee et al. 1980 et seq.); the Ohio River near Louisville, **Kentucky** (unsubstantiated) (Lee et al. 1980 et seq.; Burr and Page 1986), northern Kentucky (Crossman 1978); Chesapeake Bay drainage, the ridge and valley area, and the Ohio basin of **Maryland** (Lee et al. 1976); nonnative areas of **Massachusetts** (Hartel 1992); eastern **Minnesota** (Crossman 1978; Lee et al. 1980 et seq.); Three Mile Lake and Sutherland Reservoir, **Nebraska** (Morris et al. 1974; Crossman 1978); **New Jersey** (Morse 1905); the Lake Erie drainage and the Adirondacks in **New York** (Crossman 1978; Emery 1985; Smith 1985); many areas of **Ohio** (Trautman 1981; Burr and Page 1986; Hocutt et al. 1986), including Long Lake (Crossman 1978); possibly the **Oklahoma-Texas** border in the Red River (Lee et al. 1980 et seq.); the Youghioghenay River and the Lake Erie drainage, **Pennsylvania** (Hendricks et al. 1979; Emery 1985); **Vermont** (Lee et al. 1980 et seq.); the New drainage and Fairystone Reservoir, Smith River, **Virginia** (Hocutt et al. 1986; Jenkins and Burkhead 1994); the New drainage and the main channel of the Ohio River in **West Virginia** (Hocutt et al. 1986; Stauffer et al. 1995; Jenkins and Burkhead 1994).

Means of Introduction: Intentional stocking in most areas; may have gained access to south Florida through artificial canals.

Status: Established in many areas. Reported from Kentucky.

Impact of Introduction: Unknown.

Remarks: A single collection from the Ohio River, Kentucky, is unsubstantiated (Burr, personal communication).

Esox reicherti

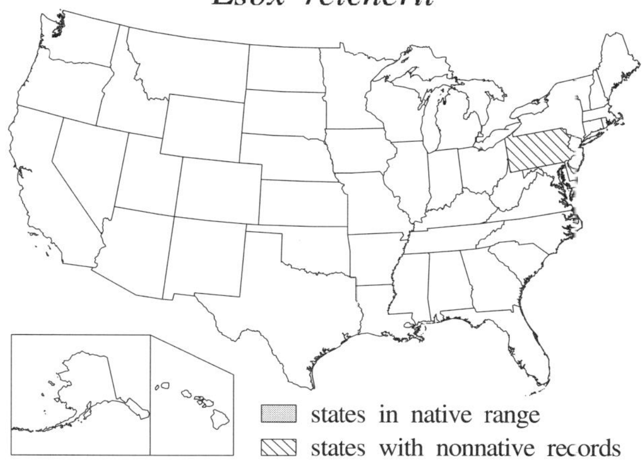

Esox reicherti Dybowski 1869

Common Names: Amur pike (blackspotted pike).

Identification: Distinguishing characteristics, a key, and an illustration were provided by Berg (1948). Maximum size: 110 cm and 16 kg or more.

Native Range: Northeast Asia in the Amur River basin (Berg 1948; Nikolski 1956; Robins et al. 1991b).

Nonindigenous Occurrences: This species was introduced to Glendale Lake, Cambria County, **Pennsylvania**, by the Pennsylvania Fish Commission in 1968 (Denoncourt et al. 1975a; Crossman and Meade 1977; Cooper 1983). Several fish also may have escaped a fish hatchery into Spring Creek (Denoncourt et al. 1975a).

Means of Introduction: Intentionally stocked for sportfishing.

Status: Cooper (1983) reported Amur pike as established; however, Robins et al. (1991b) listed them as not established. T. Bender (personal communication) believes pure Amur pike are extirpated in Pennsylvania.

Impact of Introduction: Unknown.

Remarks: Pure Amur pike were last spawned in 1971. All of the Pennsylvania Fish Commission's brood stock was lost in the summer of 1976 (Bender, personal communication).

HYBRIDS

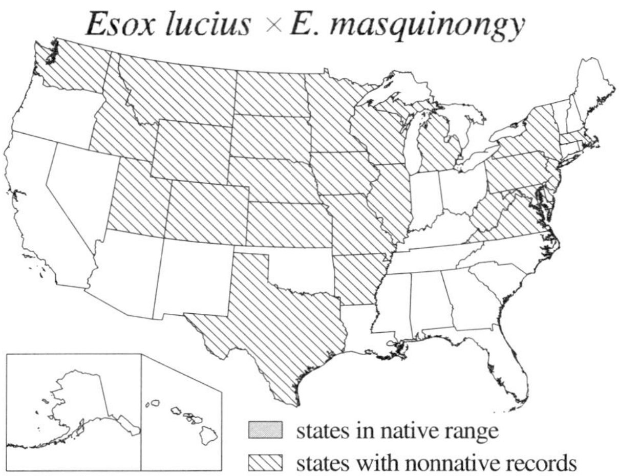

Esox lucius × E. masquinongy

Common Names: tiger muskellunge (tiger musky).

Identification: See the account for *Esox lucius* in Page and Burr (1991); characteristics and photographs of this hybrid were given in Crossman and Buss (1965) and in Buss and Miller (1967). Maximum size: 16.1 kg.

Native Range: A natural and artificial hybrid. Hybridization occurs in nature where the two species' ranges overlap, in the northeastern and western Great Lakes regions (Lee et al. 1980 et seq.).

Nonindigenous Occurrences: This hybrid has been bred artificially and stocked by state fish and game agencies for sport fishing in Spring River and Lake Ashaugh in **Arkansas** (A. Carter, personal communication); the Arkansas drainage and the North Platte River in the Denver, **Colorado** area (Walker 1993; Horak, personal communication); areas in the northern half of **Delaware** (Raasch and Altemus 1991; Martin, personal communication); several drainages in **Idaho** (Idaho Fish and Game 1990); Summerset, Argyle, Braidwood, Clinton, Forbes, Glenn Shoals, Heidecke, Springfield, and Wolf lakes, **Illinois** (International Game Fish Association 1994; Illinois Department of Natural Resources 1997a, 1997b, 1997c); Ohio River, **Indiana** (Indiana Department of Natural Resources 1997a, 1997b); many areas in **Iowa** (Harlan et al. 1987); **Kansas** (B. Hartman, personal communication); Ohio River, **Kentucky** (Indiana Department of Natural Resources 1997a); Potomac River, and two lakes in the Youghiogheny system in **Maryland** (DeLorme Mapping 1993a); ponds statewide in **Massachusetts** (Hartel 1992); Lake Rolard and Lac Vieux Desert, **Michigan** (International Game Fish Association 1994; Michigan Department of Natural Resources 1997); Leech Lake, **Minnesota** (International Game Fish Association 1994); Pony Express Lake, **Missouri** (Pflieger 1975); the eastern half of the Yellowstone drainage, the Little Missouri River drainage, and the Missouri drainage east of the Continental Divide in **Montana** (Holton 1990); Elwood Reservoir, **Nebraska** (Bouc 1987); the Tioughnioga River, Catskill Creek, and the St. Lawrence River, **New York** (International Game Fish Association 1994); **North Dakota** (North Dakota Game and Fish Department 1994, 1997); the Susquehanna River, and numerous other localities in **Pennsylvania** (International Game Fish Association 1994; Pennsylvania Fish and Boat Commission 1997); **South Dakota** (Hanten, personal communication; North Dakota Game and Fish Department 1994); Lake Bastrop and Alcoa Lake, **Texas** (Luebke 1978); Pineview Reservoir, **Utah** (DeLorme Mapping 1993b); the New, Powell (Jenkins and Burkhead 1994), and Potomac rivers (DeLorme Mapping 1993a), **Virginia**; **Washington** (Fletcher, personal communication); the Potomac River, **West Virginia** (DeLorme Mapping 1993a); numerous lakes in **Wisconsin** (Becker 1983; International Game Fish Association 1994), and Grayrocks Reservoir, **Wyoming** (DeLorme Mapping 1992b; Wyoming Game and Fish, Fish Division 1997).

Means of Introduction: Intentionally stocked for sport fishing in most locations. In some areas the hybrid occurs "naturally" after stocking northern pike in native muskellunge waters (Becker 1983). Although the male tiger muskellunge are sterile, females are often fertile and are capable of backcrossing (Becker 1983). Hartel et al. (1996) suggested individuals collected in Massachusetts were a result of stock contamination.

Status: Maintained by stocking in most cases. In some areas, such as many lakes in Wisconsin, the population is maintained by natural hybridization when northern pike are stocked in native muskellunge waters (Becker 1983). Hartel et al. (1996) report only one or two individuals collected in Massachusetts.

Impact of Introduction: This hybrid probably affects smaller fish species through predation. It also is capable of backcrossing. Male tiger muskellunge are always sterile, but females are often fertile (Becker 1983).

Remarks: Tiger muskellunge have not been stocked in Oklahoma (Pigg, personal communication) or Tennessee (T. Churchill, personal communication).

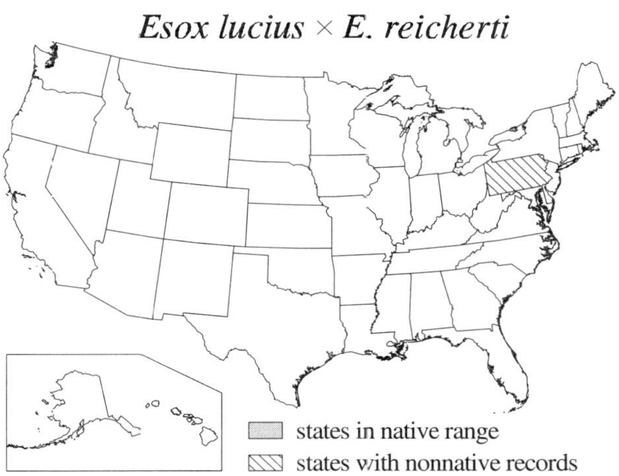

Esox lucius × *E. reicherti*

Common Name: northern pike × Amur pike hybrid.

Identification: Crossman and Meade (1977) provided photographs and characteristics of Amur pike hybrids. Maximum size: Probably similar to parent species.

Native Range: None; artificial hybrid.

Nonindigenous Occurrences: The Pennsylvania Fish Commission stocked these hybrids in Glendale Lake, Cambria County, **Pennsylvania**, in 1976 (Crossman and Meade 1977; Bender, personal communication).

Means of Introduction: One hundred eighty-six fish were intentionally stocked for sport fishing.

Status: A few hybrids may still exist in Glendale Lake, although none have been reported for several years (Bender, personal communication).

Impact of Introduction: Unknown.

Remarks: The captive brood stock was destroyed some years after introduction because of lack of success. Experimental backcrosses of this hybrid with Amur pikes yielded fertile offspring (Crossman and Meade 1977).

Umbridae - Mudminnows

The family Umbridae comprises three genera and at least five species (Nelson 1994). These fishes are distributed in several disjunct sections of the Northern Hemisphere: North America including Alaska, Washington state, and the central and east central parts of the continent; Eastern Europe; and northeastern Siberia in Asia (Berra 1981; Nelson 1994). Mudminnows are small fishes less than 20 cm long, are capable of using atmospheric oxygen, and can survive in habitats unsuitable for many other fishes. In addition to low oxygen levels, they can tolerate cold and drought. The family is restricted to freshwater.

Dallia pectoralis

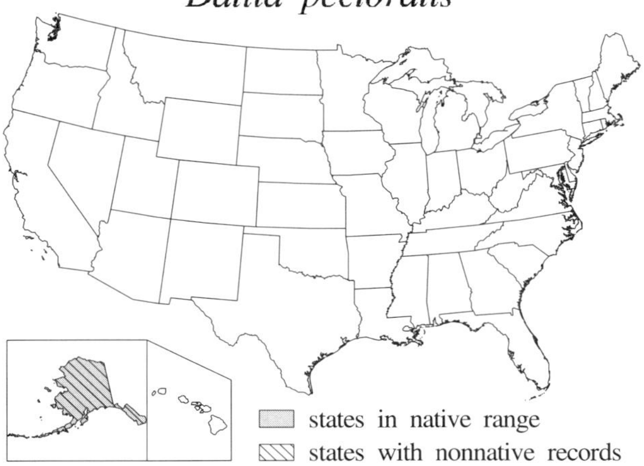

Dallia pectoralis Bean 1880

Common Name: Alaska blackfish.

Identification: Morrow (1980); Page and Burr (1991). Maximum size: 33 cm.

Native Range: Alaska, Colville River delta south to central Alaska Peninsula near Chignik and upstream in Yukon-Tanana drainage to Fairbanks vicinity. Also Bering Sea islands and northeastern Siberia (Page and Burr 1991).

Nonindigenous Occurrences: Alaska blackfish have been introduced into Hood and Spenard Lakes at the Anchorage Airport, **Alaska** (Morrow 1980; Page and Burr 1991).

Means of Introduction: Accidental.

Status: Established in Alaska.

Impact of Introduction: Unknown.

Remarks: The Alaska blackfish was formerly a staple part of the diet of indigenous peoples of Alaska (Scott and Crossman 1973).

Novumbra hubbsi

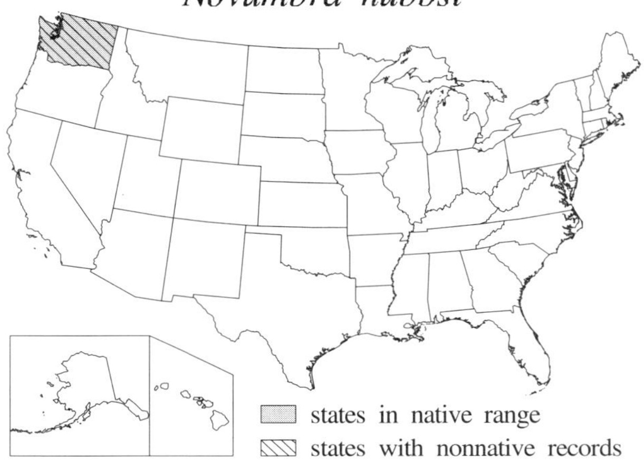

Novumbra hubbsi Schultz 1929

Common Name: Olympic mudminnow.

Identification: Wydoski and Whitney (1979); Page and Burr (1991). Maximum size 8 cm.

Native Range: Coastal lowlands of Olympic Peninsula, Washington, from Ozette Lake and Queets River drainage (where possibly introduced), to upper Chehalis River drainage. Occasionally found in lower Deschutes River (Puget Sound drainage), apparently as a result of floodwater exchange with Chehalis River (Wydoski and Whitney 1979; Page and Burr 1991).

Nonindigenous Occurrences: It is suspected that populations in the Queets River, including the Lake Ozette system, **Washington**, are the result of an introduction (Wydoski and Whitney 1979).

Means of Introduction: Unknown.

Status: Established in Washington.

Impact of Introduction: Unknown.

Remarks: None.

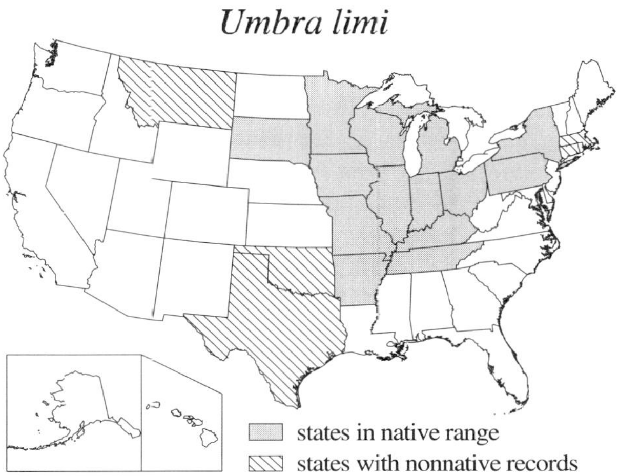

Umbra limi (Kirtland 1840)

Common Names: central mudminnow (western mudminnow, mud minnow, Mississippi mud minnow).

Identification: Becker (1983); Smith (1985); Page and Burr (1991); Etnier and Starnes (1993). Maximum size: 14 cm.

Native Range: St. Lawrence–Great Lakes, Hudson Bay (Red River), and Mississippi River basins from Quebec to Manitoba and south to central Ohio, western Tennessee, and northeastern Arkansas; Hudson River drainage (Atlantic Slope), New York. Isolated populations in Missouri River drainage of east central South Dakota and western Iowa (Page and Burr 1991).

Nonindigenous Occurrences: The central mudminnow has been introduced to the Connecticut River and is found throughout most of its length in **Connecticut** (Schmidt 1986; Whitworth 1996) and in three or four localities in the Connecticut River Valley in **Massachusetts** (Schmidt 1986; Hartel 1992; Hartel et al. 1996). It was reported from Blacktail Creek in Silverbow County, Butte, **Montana**, in 1990 (Holton 1990). It was also listed as introduced to nonspecific area(s) in **Oklahoma** (Moore and Riggs 1963). A single specimen was taken from Keystone Lake, north of Cleveland, Oklahoma, in September 1971 (Pigg et al. 1996). This species was collected from Fish Lake, a pond on the Texas A&M range area in College Station, **Texas**, during treatment of the pond (Champ et al. 1973).

Means of Introduction: Released lab animals in Massachusetts. First collected in 1975 in Massachusetts (Hartel et al. 1996). Unknown in Montana, Oklahoma, and Texas. Biologists of the Oklahoma Department of Wildlife Conservation found a single specimen in a shipment of fathead minnows from a fish farm in Pontotoc County, Oklahoma (Pigg et al. 1996).

Status: Established in Connecticut and Massachusetts. Reported in Montana, Oklahoma, and Texas.

Impact of Introduction: Unknown.

Remarks: This species is sometimes used as a baitfish by anglers (Scott and Crossman 1973). Although Stauffer et al. (1995) had no records from West Virginia, they expected it to occur in the Ohio River tributaries in the northern portion of the state.

Voucher specimens: Oklahoma (OSUS 23256).

ORDER OSMERIFORMES

Osmeridae - Smelts

This family is composed of seven genera and 13 species (Nelson 1994). Smelt are circumpolar in cold and temperate waters of the Northern Hemisphere (Berra 1981). These fishes are small, mostly carnivorous, and are important commercially as food and bait. Most osmerids are marine; however, some coastal species enter rivers during spawning runs, and a few species are established in inland waters.

Hypomesus nipponensis

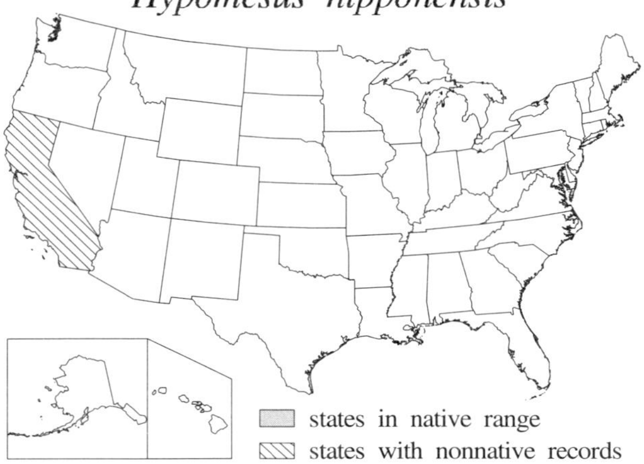

Hypomesus nipponensis McAllister 1963

Common Name: wakasagi.

Identification: Moyle (1976a); Page and Burr (1991). Maximum size: 17 cm.

Native Range: Japan (Page and Burr 1991).

Nonindigenous Occurrences: The wakasagi was stocked in **California** in Sly Park Reservoir (El Dorado County), Dodge Reservoir (Lassen County), Spaulding Reservoir (Nevada County), Big Bear Lake (San Bernardino County), Dwinnel Reservoir (Siskiyou County), Shastina Reservoir (Siskiyou County), and Freshwater Lagoon (Humboldt County) in 1959 (Wales 1962; Moyle 1976a; Courtenay et al. 1984; Dill and Cordone 1997). It was then stocked in Lake Almanor (Plumas County) on the North Fork Feather River in 1972 (Dill and Cordone 1997). It migrated downstream from Lake Almanor to Lake Oroville (Dill and Cordone 1997). It has recently been observed in the lower American River, Cache Slough off the Sacramento River and the Mokelumne River, and at the Central Valley Project and State Water Project fish salvage facilities in the south Delta (Dill and Cordone 1997). The species can be expected to occur in the lower Klamath, the Sacramento, and possibly other drainages (Moyle 1976a). Most recently, it was found in the Sacramento-San Joaquin Estuary (Aasen et al. 1998).

Means of Introduction: Wakasagi were intentionally introduced in 1959 from Japan by the California Department of Fish and Game as an experimental forage fish for trout (Wales 1962; Moyle 1976b; Dill and Cordone 1997).

Status: This species is established in several reservoirs and associated tributaries in California (Moyle 1976a; Shapovalov et al. 1981; Courtenay et al. 1986). It has not been recorded in Big Bear Lake since 1960 (Swift et al. 1993).

Impact of Introduction: This species has been found to negatively impact kokanee *Oncorhynchus nerka* and threadfin shad *D. petenense* (Dill and Cordone 1997). It also is known to hybridize with the native and federally endangered delta smelt *Hypomesus transpacificus*. Hybridization between the two species was suspected by Courtenay et al. (1986), and was later confirmed (Dill and Cordone 1997; P. Moyle, personal communication).

Remarks: Dill and Cordone (1997) reviewed its introduction history in California. In documenting the original introduction, Wales (1962) incorrectly identified the species as *Hypomesus olidus*. Several authors (e.g., Moyle 1976a; Lee et al. 1980 et seq.) treated the introduced wakasagi as a subspecies of *H. transpacificus* (i.e., as *H. t. nipponensis*). In California the wakasagi is generally considered a freshwater species, hence its often-used name "freshwater smelt" in that state; however, it has recently been discovered in brackish waters, further threatening the continued survival of the imperiled delta smelt (Dill and Cordone 1997).

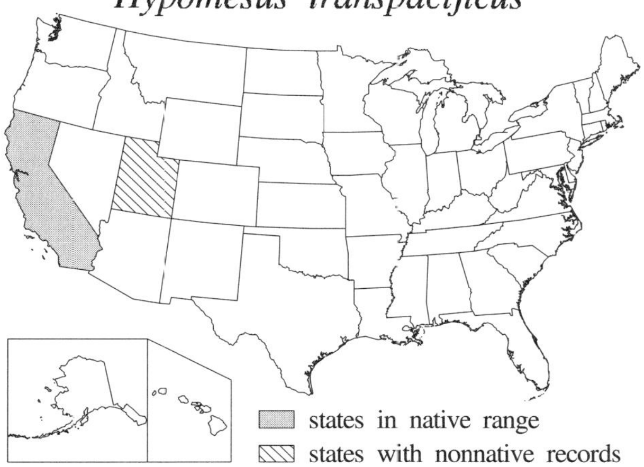

Hypomesus transpacificus McAllister 1963

Common Name: delta smelt.

Identification: Moyle (1976a); Page and Burr (1991). Maximum size: 20 cm.

Native Range: Sacramento-San Joaquin delta region of central California (Page and Burr 1991).

Nonindigenous Occurrences: Introduced into Willard Bay Reservoir, **Utah**, in 1982 (Sigler and Sigler 1987).

Means of Introduction: Intentionally stocked as forage (Sigler and Sigler 1987).

Status: Extirpated (Sigler and Sigler 1996).

Impact of Introduction: Unknown.

Remarks: None.

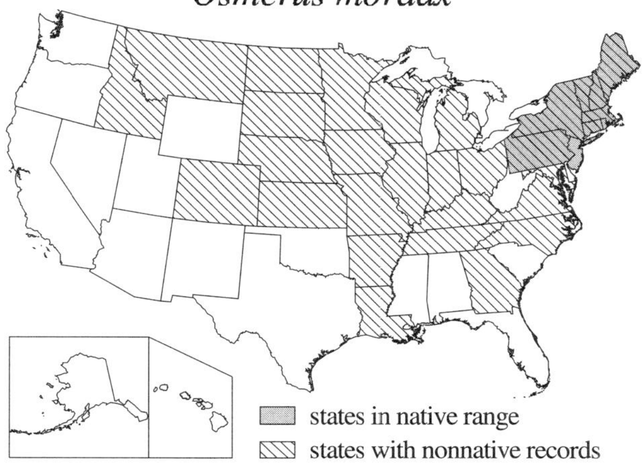

Osmerus mordax (Mitchill 1814)

Common Names: rainbow smelt.

Identification: Scott and Crossman (1973); Becker (1983); Smith (1985); Page and Burr (1991). Maximum size: 33 cm.

Native Range: Atlantic drainages from Lake Melville, Newfoundland, to Delaware River, Pennsylvania, and west through Great Lakes; Arctic and Pacific drainages from Bathurst Inlet, Northwest Territories, to Vancouver Island, British Columbia. Also Pacific drainages of Asia (Page and Burr 1991).

Nonindigenous Occurrences: Rainbow smelt occur in all five Great Lakes and also have been introduced or dispersed after introduction into several large rivers. Areas with introduced smelt include the Mississippi River, **Arkansas** (Pennington et al. 1982; Mayden et al. 1987); reservoirs in the South Platte and Arkansas drainages and headwaters of the Colorado basin in **Colorado** (Woodling 1985; Propst and Carlson 1986); several lakes in **Connecticut** (Webster 1942); the Chattahoochee River below Lake Lanier, **Georgia** (Dahlberg and Scott 1971a, 1971b); lakes in the Sawtooth Mountains of **Idaho** (Linder 1963; Simpson and Wallace 1978); Lake Michigan (Emery 1985; Burr 1991), the Mississippi River, the Illinois River (Burr and Mayden 1980; Mayden et al. 1987; Burr 1991, Burr et al. 1996), and Ohio River (Burr 1991) in **Illinois** (Smith 1979; Burr and Page 1986); Lake Michigan and the Ohio River near Madison, **Indiana** (Emery 1985; Mayden et al. 1987); the Missouri River, **Iowa** (Harlan et al. 1987; Mayden et al. 1987); the Missouri River, **Kansas** (Mayden et al. 1987); the Mississippi River, **Kentucky** (Burr and Warren 1986; Mayden et al. 1987); the Mississippi River, **Louisiana** (Suttkus and Conner 1980; Mayden et al. 1987); Schoodic Lake, **Maine** (Havey 1973); **Maryland** (Ferguson 1876; Jenkins and Burkhead 1994); nonnative waters of **Massachusetts** (Smith 1833; Hartel 1992; Hartel et al. 1996); the Great Lakes in **Michigan** (Emery 1985); Lake Superior, **Minnesota** (Emery 1985; Burr and Page 1986); the Missouri and Mississippi rivers, **Missouri** (Cross et al. 1986; Mayden et al. 1987; Pflieger 1997); the Missouri and Yellowstone rivers, **Montana** (Gould 1981; Cross et al. 1986; Mayden et al. 1987; Holton 1990); the Missouri River in **Nebraska** (Cross et al. 1986; Bouc 1987; Mayden et al. 1987); several dozen lakes in **New Hampshire** (Scarola 1973); Lake Erie (Emery 1985), Lake Ontario, the Finger Lakes, the Adirondack lakes, Neversink Reservoir, and Lake Champlain in **New York** (Werner 1980); Tennessee drainage, **North Carolina** (Menhinick 1991); Lake Sakakawea, **North Dakota** (Gould 1981; Bouc 1987; Harlan et al. 1987; Mayden et al. 1987; Holton 1990); Lake Erie, **Ohio** (Emery 1985); Lake Erie (Emery 1985) and Harvey's Lake (Susquehanna drainage), **Pennsylvania** (Denoncourt et al. 1975a; Hendricks et al. 1979; Cooper 1983; Hocutt et al. 1986); reservoirs on the Missouri River, **South Dakota** (Mayden et al. 1987; Hanten, personal communication); the Mississippi River, Watauga Reservoir, and South Fork Holston River, **Tennessee** (Mayden et al. 1987; Etnier and Starnes 1993); Lake Champlain, Vermont (Werner 1980); the Potomac River and Occoquan Reservoir, **Virginia** (Hocutt et al. 1986; Jenkins and Burkhead 1994); and Lake Superior, **Wisconsin** (Emery 1985; Burr and Page 1986).

Means of Introduction: The earliest known record is from 1912, when eggs were stocked in Crystal Lake, Michigan, which drains into Lake Michigan (Van Oosten 1937). Fish escaped into Lake Michigan and spread quickly throughout the Great Lakes and their tributaries (Creaser 1926; Gerking 1945; Hubbs and Lagler 1947; Nelson and Gerking 1968; Christie 1974; Eddy and Underhill 1974; Smith 1979; Morrow 1980; Phillips et al. 1982; Cooper 1983; Emery 1985). Early records documenting the smelt's range expansion in the Great Lakes include Lake Michigan, 1923 (Christie 1974; Emery 1985), Lake Erie, 1935 (Cooper 1983; Smith 1985), Lake Huron, 1925 (Christie 1974; Eddy and Underhill 1974), Lake Ontario, 1929 (Christie 1974; Smith 1985), and Lake Superior, 1923 (Emery 1985). The Lake Ontario population may be either native to this lake or gained access when the Welland Canal was built (Emery 1985; Smith 1985). Another possibility is that the species was introduced from the Finger Lakes via the Seneca-Cayuga, Erie and Oswego canals (Smith 1985).

Two means have been proposed to explain the introduction of rainbow smelt into the Missouri and Mississippi rivers. It may have spread from Lake Michigan via the Chicago sanitary canal to the Illinois River and then to the Mississippi and Missouri rivers (Burr and Mayden 1980). Alternatively, the species may have gained access to these rivers as a result of a stocking at Lake Sakakawea, North Dakota, in 1971 (Bouc 1987; Mayden et al. 1987; Holton 1990). The second explanation seems more plausible because of a lack of records from the Illinois River. Records of first occurrence in other areas include the Mississippi River, Illinois and Kentucky, 1978; Mississippi River, Louisiana, 1979; Mississippi River, Tennessee and Arkansas, 1980; Missouri River, Missouri, 1980; Missouri River, Kansas, 1982 (Mayden et al. 1987). Mayden et al. (1987) provided a map of the species' distribution, dates of first observation in new areas, and possible introduction pathways. The species was originally introduced into Lake Sakakawea, North Dakota, as forage for salmonids (Mayden et al. 1987).

Status: Introduced populations of this species have been very successful and the rainbow smelt is now established in the Great Lakes and in most rivers and lakes where introduced. This species has done so well in the Great Lakes that a commercial fishery targeting smelt has been operating there for many years (Smith 1985). It is the most abundant fish in some samples taken from the Mississippi River (Pflieger 1997). Nevertheless, no adults of the rainbow smelt have been found in either Missouri (Pflieger 1997) or Tennessee (Etnier and Starnes 1993). As such, Pflieger (1997) concluded that populations in Missouri are maintained by continued escape of fish from upstream reservoirs on the Missouri River. As of 1987, only one specimen had been taken from the Ohio River (Mayden et al. 1987). It is considered extirpated in Georgia; the species has not been observed in that state since its release (Dahlberg and Scott 1971b).

Impact of Introduction: In the Great Lakes, rainbow smelt compete with lake herring *Coregonus artedii* for food (Becker 1983). Christie (1974) supplied some evidence to support this, correlating lake herring decline with smelt increases in most of the lakes. Todd (1986b) also reported that smelt may be partially responsible for the decline of whitefish *Coregonus* spp. in the Great Lakes. Havey (1973) reported increased growth of landlocked Atlantic salmon following the introduction of smelt as a forage species in a lake in Maine. Hrabik et al. (1998) found evidence of competition for food between introduced rainbow smelt and native yellow perch *Perca flavescens* in Wisconsin lake habitats.

Remarks: This species is eaten by humans and used as bait for salmonids and walleye (Pflieger 1997).

Plecoglossus altivelis

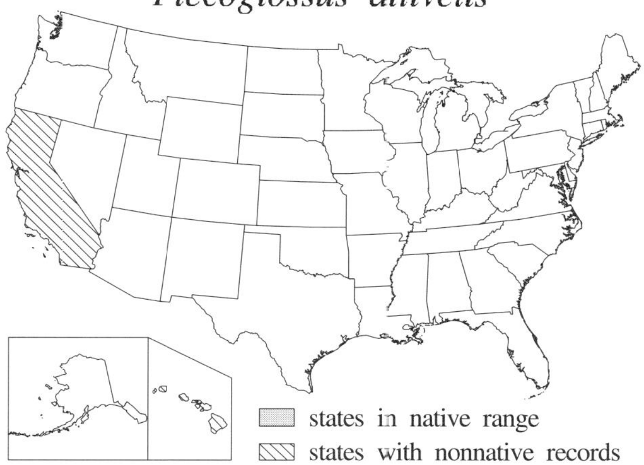

Plecoglossus altivelis Temminck and Schlegel 1846

Common Names: ayu (sweetfish, ko-ayu [land-locked form]).

Identification: This species is characterized by the presence of an adipose fin; 10–12 dorsal rays; 9–17 anal rays; 5 or 6 branchiostegal rays; usually 59–64 vertebrae; no pelvic axillary process; and more than 300 pyloric caeca (Nelson 1984). Masuda et al. (1984) also gave characteristics. The species was formerly placed in its own family, Plecoglossidae (Nelson 1984). Maximum size: 30 cm.

Native Range: Marine and freshwater, anadromous. Japan, Korea, Taiwan, and China (Nelson 1984).

Nonindigenous Occurrences: The ayu was stocked in **California** in 1961–1965 in Morris Lake, Mendocino County; Ruth Reservoir, Trinity County; and Eel River below Fortuna, Humboldt County (Hubbs et al. 1979; Dill and Cordone 1997). Earlier shipments of eggs to California did not survive and were not stocked (Dill and Cordone 1997). The ayu was introduced into **Hawaii** on the islands of Hawaii, Kauai, Maui, and Oahu (Maciolek 1984; Randall 1987).

Means of Introduction: With approval from the California Fish and Game Commission, about 3,845,000 egs and fry were stocked in California waters from 1961 through 1965 (Dill and Cordone 1997). Relished as a food fish by Asians, Dill and Cordone (1997) stated that it was introduced to the state because it was felt to be a plant-eating fish expected to thrive in coastal streams considered marginal trout habitat. An estimated 250,000 fish were imported from Japan and intentionally stocked in Hawaii in 1925 and subsequent years, but failed to establish a reproducing population (Maciolek 1984).

Status: None of the plants in California survived (Shapovalov et al. 1981; Dill and Cordone 1997); it also failed to become established in Hawaii.

Impact of Introduction: Unknown.

Remarks: Dill and Cordone (1997) provided a detailed history of its introduction in California. Adults live and spawn in freshwater. Larvae live in salt water and ascend streams when they reach 5–8 cm TL (Randall 1987).

ORDER SALMONIFORMES

Salmonidae - Trouts

The family Salmonidae includes 11 genera and about 66 species (Nelson 1994). Within this family are salmon, trout, ciscoes, grayling, and whitefishes. Some authors place the whitefishes in their own family (i.e., Coregonidae). Salmonids are distributed throughout most of the cool and cold waters of the Northern Hemisphere (Berra 1981). They are very valuable sport and commercial fishes and have been introduced into cool waters in Africa, South America, Australia, New Zealand, New Guinea, and India. The family Salmonidae contains both freshwater and anadromous species, which make long migrations into freshwater streams to spawn.

Coregonus albula

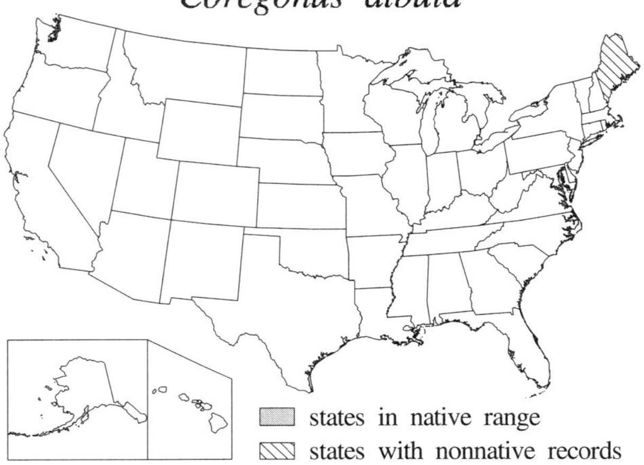

Coregonus albula (Linneaus 1758)

Common Names: vendace (kleine maräne, dwarf white fish, coregoni piccoli).

Identification: Wheeler (1969); Ladiges and Vogt (1986). The nomenclature of the genus *Coregonus* is particularly problematic. Some of the confusion is the result of the many early introductions and transplants of the various species and populations within Europe and elsewhere (e.g., Leviton 1996). Furthermore, the reportedly frequent hybridization between related *Coregonus* species also has contributed to the taxonomic confusion. Maximum size: 35 cm.

Native Range: Northern Europe, northern parts of the former USSR, the Baltic and North Sea basins, and the Swiss Alps (Ladiges and Vogt 1986; Leviton 1996).

Nonindigenous Occurrences: Thousands of young fish or fry, purported to be this species, were stocked in **Maine** in Heart Pond (Penobscot River drainage) near Bucksport, Hancock County, and Lake Hebron, Piscataquis County, during the 1880s (Smiley 1885b; Baird 1889; Bean 1892a; Todd 1983).

Means of Introduction: Several shipments of eggs, originally identified as coming from *Coregonus albula,* were imported from Germany by the U.S. Fish Commission in the 1880s. The Commission instructed C. G. Atkins to hatch the eggs in captivity; all or most of the surviving young were later released into the two Maine lakes (e.g., Smiley 1885b). Intentionally stocked, this species was apparently seen as a potential food fish.

Status: The introductions failed (Kendall 1914a).

Impact of Introduction: Unknown.

Remarks: There is confusion over whether or not the true *Coregonus albula* was actually the one imported and stocked in this country. Todd (1983) investigated the history of the early *Coregonus* transfers and summarized much of the information. He was informed by one German contact that the fish imported to the United States from hatcheries on the Bodensee (such as the 1885 shipment and possibly others) were more likely *C. wartmanni* (a species now considered a junior synonym of *C. lavaretus*). Kendall (1914a) used the name *Leucichthys albula* for this species.

Coregonus artedi

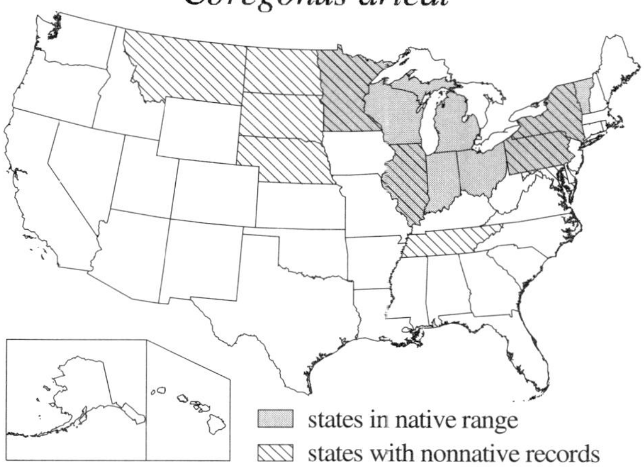

Coregonus artedi Lesueur 1818

Common Names: cisco (lake herring).

Identification: Scott and Crossman (1973); Trautman (1981); Becker (1983); Page and Burr (1991). Commonly used spelling is *C. artedii*. Maximum size: 57 cm.

Native Range: Widespread through much of Canada and northern United States in St. Lawrence–Great Lakes, Arctic, and upper Mississippi River basins from Quebec to Northwest Territories and Alberta, and south to northern Ohio, Illinois, and Minnesota (Page and Burr 1991).

Nonindigenous Occurrences: The cisco was introduced into the Illinois River in **Illinois** (Burr and Page 1986); Susquehanna River in **Maryland** (Lee et al. 1980 et seq.); several lakes in northern **Minnesota** (Eddy and Underhill 1974); the Missouri River in **Montana** in the mid-1980s (Holton 1990); Santee Lake in Sarpy County, and a lake at Nebraska City, **Nebraska** (Morris et al. 1974); Rushford Lake in the Genesee drainage, **New York** (Smith 1985); Conowingo Reservoir and Harvey's Lake, Luzerne County, **Pennsylvania** (Denoncourt et al. 1975a; Lee et al. 1976, 1981; Cooper 1983); the Missouri River drainage in **North Dakota** (North Dakota Game and Fish Department 1994; Power and Ryckman 1998); unspecified location(s) in **South Dakota** (North Dakota Game and Fish Department 1994; Hanten, personal communication); and Calderwood Reservoir on the Little Tennessee River, **Tennessee** (Etnier and Starnes 1993).

Means of Introduction: This species was intentionally stocked either as a food fish or as forage for larger predatory sport fish. Power and Ryckman (1998) noted that the species migrated downstream into the Missouri River drainage of North Dakota from Fort Peck Reservoir, Montana. It was stocked in Tennessee in 1960 (Etnier and Starnes 1993); in Harvey's Lake, Pennsylvania from 1969 to 1972 (Denoncourt et al. 1975a); and in Nebraska in 1888 (Morris et al. 1974).

Status: Some populations in Minnesota and Montana became established. Reproduction has been noted in North Dakota. Stockings that took place in many of the other states reportedly failed.

Impact of Introduction: Unknown.

Remarks: None.

Coregonus clupeaformis

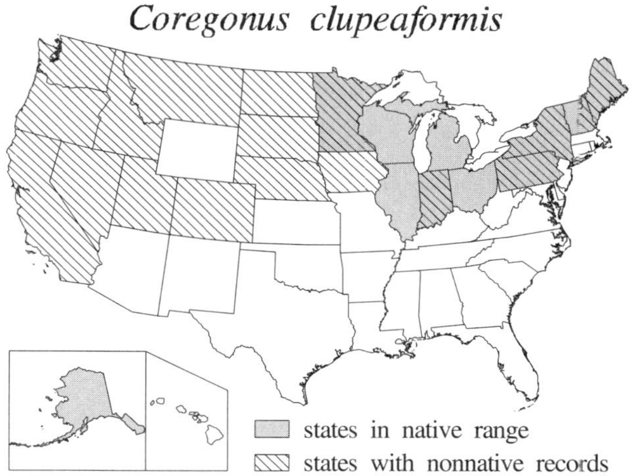

Coregonus clupeaformis (Mitchill 1818)

Common Names: lake whitefish.

Identification: Scott and Crossman (1973); Trautman (1981); Becker (1983); Page and Burr (1991). Maximum size: 80 cm.

Native Range: Throughout Alaska and most of Canada south into New England, the Great Lakes basin, and central Minnesota (Page and Burr 1991).

Nonindigenous Occurrences: This species has been introduced into many lakes in **California** (Smith 1896; Shebley 1917; Hubbs et al. 1979; Dill and Cordone 1997); **Colorado** (Wiltzius 1985; Horak, personal communication); **Idaho** (Smith 1896; Chapman 1942; Linder 1963; Simpson and Wallace 1978; Lee et al. 1980 et seq.; Idaho Fish and Game 1996); **Indiana** (Sweeney 1902); **Iowa** (museum specimen); **Maine** (Kendall 1914a); **Minnesota** (Eddy and Underhill 1974); **Montana** (Brown 1971; Lee et al. 1980 et seq.; Cross et al. 1986; Holton 1990); **Nebraska** (Morris et al. 1974); **Nevada** (Smith 1896; Miller and Alcorn 1946; Deacon and Williams 1984); **New Hampshire** (Scarola 1973); **New York** (Lee et al. 1980 et seq.; Smith 1985); **North Dakota** (Cross et al. 1986; North Dakota Game and Fish Department 1994); **Oregon** (Smith 1896; Chapman 1942); **Pennsylvania** (Ford et al. 1892); **South Dakota** (Todd 1983; Cross et al. 1986; North Dakota Game and Fish Department 1994; Hanten, personal communication); **Utah** (Sigler and Miller 1963; Sigler and Sigler 1996); and **Washington** (Smith 1896; Chapman 1942; Wydoski and Whitney 1979; Lee et al. 1980 et seq.; Washington Department of Fish and Wildlife 1997).

Means of Introduction: The U.S. Fish Commission was active in stocking lake whitefish during the late 1800s and early 1900s (e.g., Smith 1896; Chapman 1942).

Status: Most early attempts to establish populations failed, for instance, all attempts in California, Nebraska, Nevada, Oregon, and Utah. However, populations became established locally in a few lakes in several states.

Impact of Introduction: Unknown.

Remarks: According to Miller and Alcorn (1946), in 1880 25,000 eggs of this species were sent to Eureka, Nevada, but no records of plantings could be found.

Voucher specimens: California (USNM 27198), Iowa (USNM 63643).

Coregonus lavaretus

Coregonus lavaretus (Linnaeus 1758)

Common Names: powan (lavaret, common whitefish, schelly, gwyniad, lavarello).

Identification: *Coregonus lavaretus* presumably represents a species complex and is need of systematic review. Berg (1948) recognized many different subspecies or variants. *Coregonus wartmanni* (T. Todd and R. J. Behnke, personal communication) and *Coregonus maraena* are considered junior synonyms. Ladiges and Vogt (1986) placed *C. maraena* in the *C. lavaretus* group and provided illustrations of *C. maraena* and other group members. Illustrations of *C. lavaretus* were also provided by Berg (1948), Maitland (1977), and Wheeler (1978). See also Wheeler (1969). Maximum size: 130 cm.

Native Range: Northern Europe and Siberia (Berg 1948), also the Baltic and North Seas basin, Britain, and the Swiss Alps (Leviton 1996).

Nonindigenous Occurrences: Several shipments of "*C. albula*" eggs were received between 1884 and 1886 (Smiley 1885b; Baird 1889; Bean 1892a). Some, or all, of these fish actually may have been *C. lavaretus* (Todd 1983). Thousands of hatched young from these eggs were stocked in Heart Pond and Lake Hebron, **Maine** (Smiley 1885b; Baird 1889; Bean 1892a; Todd 1983). Kendall (1914a) referred to the stocking of "*C. albula*" in Maine under the name *Leucichthys albula*. Several hundred young fish, originally identified as *Coregonus maraena*, were stocked into Gardner's Lake, Otsego County, **Michigan**, in 1877 (Baird 1878; Todd 1983).

Means of Introduction: *Coregonus lavaretus*, or a related species, was intentionally stocked as a food fish. Shipments were imported by the U.S. Fish Commission from Germany in the 1880s (Todd 1983). According to Baird (1878), 1,000 eggs of what was identified as *C. maraena* were shipped from Poland to Michigan in 1877. The eggs were provided to the Michigan State Hatching House in Detroit and hatched in captivity. A total of 409 of the young fish were stocked in Gardner's Lake (Baird 1878; Todd 1983). Baird (1878) considered the stocking an experimental introduction of a European food fish.

Status: Reported from Michigan (Baird 1878). The introductions in Maine failed (Kendall 1914a)

Impact of Introduction: Unknown.

Remarks: There is much confusion surrounding the positive identification of whitefish imported from Germany in the late 1800s For instance, Baird (1878) reported that the eggs brought to the United States from Europe and introduced into Michigan represented the 'Madue maraena' whitefish *Coregonus maraena*. Berg (1948) treated that form as a subspecies of *C. laventer*, (i.e., *C. laventer maraena*) indicating the subspecies was native to Lake Madue in northwest Poland. A large number of whitefish eggs imported from Germany during the 1880s were reported as *C. albula* (e.g., Smiley 1885b; Baird 1889; Bean 1892a). However, the fish imported from the hatcheries on the Bodensee, a lake bordering Germany (such as the 1885 shipment and possibly others) were more likely *C. lavaretus* (Todd 1983).

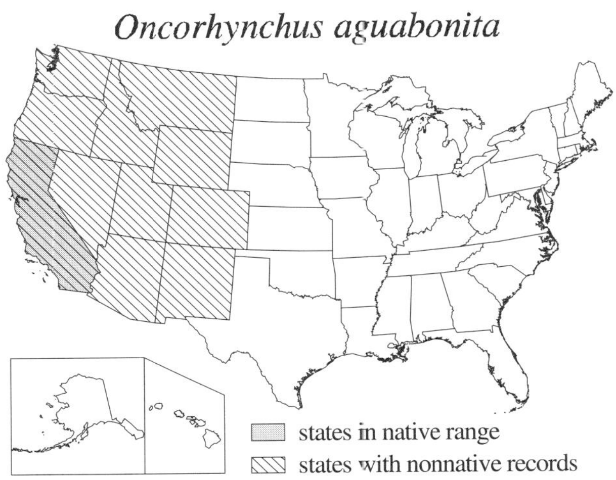

Oncorhynchus aguabonita (Jordan 1893)

Common Names: golden trout.

Identification: Moyle (1976a); Wydoski and Whitney (1979); Sigler and Sigler (1987); Page and Burr (1991); Behnke (1992). Another name for this species is *Salmo aguabonita*. Maximum size: 71 cm.

Native Range: Endemic to upper Kern River basin, Tulare and Kern counties, California (Page and Burr 1991).

Nonindigenous Occurrences: The golden trout has been introduced into many western areas including Santa Cruz County in **Arizona** (Minckley 1973; Rinne 1995); over 300 mountain streams, Lake Tahoe, and lakes in the Central Valley, Mono, Owens and Lahontan drainages in **California** (Miller and Alcorn 1946; Moyle 1976a; Sigler and Sigler 1987); lakes in the North Platte, Rio Grande, and Gunnison drainages in **Colorado** (Everhart and Seaman 1971; Wiltzius 1985; Zuckerman and Behnke 1986; Walker 1993); Kootenai, Snake, and Spokane River drainages and several lakes in independent drainages in **Idaho** (Linder 1963; Simpson and Wallace 1978; Idaho Fish and Game 1990, 1996, 1997); the Clark Fork, upper Yellowstone and upper Missouri drainages and many lakes in **Montana** (Brown 1971; Cross et al. 1986; Holton 1990); lakes in the Ruby Mountains, the Humboldt drainage, and Lake Tahoe in **Nevada** (Miller and Alcorn 1946; La Rivers 1962; Sigler and Sigler 1987); non-specific areas in **New Mexico** (Sublette et al. 1990); high lakes and streams in Wallowa County, and a few other locations in **Oregon** (Bond 1973, 1994; Li, personal communication); Echo Lake, Corn Creek, and Atwood Creek in **Utah** (Sigler and Miller 1963; Sigler and Sigler 1996); several small lakes in the Skykomish system of **Washington** (Wydoski and Whitney 1979; Washington Department of Fish and Wildlife 1997); and Cooks Lake in Sublette County, and lakes in the Wind River, Big Horn, and Medicine Bow mountains in **Wyoming** (Brown 1971; Sigler and Sigler 1987; Hubert 1994).

Means of Introduction: Authorized introductions for sportfishing. First introduced in Arizona in 1971 (Rinne 1995) and in Idaho in 1939 (Linder 1963).

Status: Established populations recorded for California, Utah, and Washington (Fletcher, personal communication). Populations in the high-elevation lakes in the Ruby Mountains, Nevada, no longer exist (Deacon and Williams 1984). Also extirpated in New Mexico.

Impact of Introduction: Unknown.

Remarks: The introduction into New Jersey (Stiles 1978) may have actually been based on stocking of "golden trout", a mutant form of rainbow trout raised in hatcheries (Behnke, personal communication).

Oncorhynchus clarki

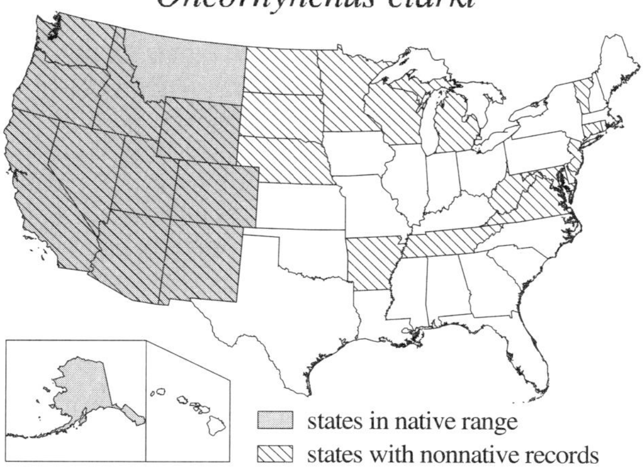

Oncorhynchus clarki (Richardson 1836)

Common Names: cutthroat trout (Yellowstone cutthroat *O. c. bouvieri*, Snake River cutthroat, fine-spotted Snake River cutthroat *O. c.* subsp. or *O. c. carmichaeli* (proposed) (Trotter 1987), Lahontan cutthroat *O. c. henshawi*, coastal cutthroat *O. c. clarki*, Colorado cutthroat *O. c. pleuriticus*, Paiute cutthroat *O. c. selenis*, westslope cutthroat *O. c. lewisi*, greenback cutthroat *O. c. stomias*, Humboldt cutthroat *O. c.* subsp., Alvord cutthroat *O. c.* subsp., Whitehorse cutthroat *O. c.* subsp., Bonneville cutthroat *O. c. utah*, Rio Grande cutthroat *O. c. virginalis*, yellowfin cutthroat *O. c. macdonaldi*, Bear Lake cutthroat *O. c.* subsp.).

Identification: Scott and Crossman (1973); Moyle (1976a); Morrow (1980); Sublette et al. (1990); Page and Burr (1991); Behnke (1992). Behnke (1992) recognized 14 subspecies and provided information on each one. This species was formerly known as *Salmo clarki*. Maximum size: 99 cm.

Native Range: Pacific Coast drainages from Prince William Sound, Alaska, to Eel River, northern California. Freshwater populations range through Rocky Mountains to Hudson Bay, Mississippi River, Great (including Lahontan, Bonneville, and Alvord basins), and Pacific basins from southern Alberta to Rio Grande drainage, New Mexico (Page and Burr 1991).

Nonindigenous Occurrences: Cutthroat trout have been stocked in the White River below Bull Shoals Dam and in the North Fork River below Norfork Dam in **Arkansas** (Robison and Buchanan 1988; Carter, personal communication); Lake Mead, **Arizona** (Minckley 1973; Rinne 1995); the San Joaquin, Surprise Valley, Death Valley, Mono, and Owens lake drainages in **California** (Moyle 1976a; Dill and Cordone 1997); the South Platte, Arkansas, Rio Grande headwaters, Colorado headwaters, Gunnison, Green, Yampa, White and San Juan drainages in **Colorado** (Ellis 1974; Tyus et al. 1982; Wiltzius 1985; Zuckerman and Behnke 1986; Walker 1993); the Connecticut and Housatonic drainages in **Connecticut** (Whitworth 1996); the Snake River above and below Shoshone Falls, and independent drainages in **Idaho** (Simpson and Wallace 1978; Idaho Fish and Game 1990, 1997); non-specific areas in **Kansas** (Cross 1967); non-specific areas in **Maryland** (R. M. Davis, personal communication); south branch of the Pere Marquette River in **Michigan** (Hubbs and Lagler 1947, 1964; Becker 1983; Emery 1985); non-specific areas in **Minnesota** (La Rivers 1962; Eddy and Underhill 1974); Pine, White Clay, and Larrabee creeks in Sheridan County, White River in Sioux County, and Beaver, Chadron, Dead Horse, Indian, Bordeaux, Little Bordeaux, and Big Bordeaux creeks in Dawes County, **Nebraska** (Morris et al. 1974); the White River (Colorado basin), many areas in the Great Basin, and the Guano system (Oregon drainage) in **Nevada** (Miller and Alcorn 1946; La Rivers 1962; Deacon and Williams 1984; Miller et al. 1991); the Cooper River in the lower Delaware drainage in **New Jersey** (Fowler 1952; Stiles 1978); the Canadian, Pecos, Rio Grande, San Juan, and Tularosa Valley drainages in **New Mexico** (Koster 1957; Tyus et al. 1982; Sublette et al. 1990); non-specific areas in **North Dakota** (North Dakota Game and Fish Department 1994, 1997); the John Day River in Grant County, and other locations in **Oregon** (Bond 1973); non-specific areas in **South Dakota** (North Dakota Game and Fish Department 1994; Hanten, personal communication); Dale Hollow and Center Hill reservoirs, and Wilbur tailwaters on the Tennessee River in **Tennessee** (Etnier and Starnes 1993); the Green and Strawberry rivers in **Utah** (Sigler and Miller 1963; Tyus et al. 1982; Sigler and Sigler 1996); non-specific areas in **Vermont** (Cox, personal communication); the upper New drainage in **Virginia**

(Jenkins and Burkhead 1994); western **Washington** (Schultz and DeLacy 1935); non-specific areas in **West Virginia** (D. Cincotta, personal communication); a lake in Washington County, **Wisconsin** (Becker 1983); and the Laramie system in the North Platte, and Flaming Gorge Reservoir on the Green River in **Wyoming** (La Rivers 1962; Baxter and Simon 1970; Tyus et al. 1982; Gorges 1994; Hubert 1994).

Means of Introduction: Many introductions resulted from authorized stocking for sportfishing. Escaped from a fish hatchery in Virginia. Fine-spotted Snake River cutthroat from Wyoming were stocked in Arkansas in 1984 (Robison and Buchanan 1988). Yellowstone, Lahantan and perhaps other subspecies have been stocked in Arizona (Minckley 1973). Paiute (Moyle 1976a), Yellowstone (Dill and Cordone 1997), and probably coastal and Lahontan cutthroat have been transplanted outside their native ranges in California. The state of Colorado planted Yellowstone cutthroats exclusively until sometime after 1974 (Beckman 1974). Since then, Snake River cutthroats have also been stocked in the state (Walker 1993). Both species have been stocked in the same drainages (Walker 1993). Greenback, Colorado River, and Rio Grande cutthroat are native to the state (Walker 1993). Yellowfin cutthroat, formerly native to Twin Lakes, Colorado, is now extinct (Walker 1993). Stocked numerous times since the late 1800s in Connecticut (Whitworth 1996). Lahontan, Bear Lake (Idaho Fish and Game 1997), and perhaps westslope cutthroats (Simpson and Wallace 1978) have been introduced into nonnative areas of Idaho. The Yellowstone subspecies was introduced into Michigan (Hubbs and Lagler 1947). Westslope (Eddy and Underhill 1974) and Lahontan cutthroat (La Rivers 1962) were introduced into Minnesota. Bonneville, and Lahontan (La Rivers 1962), Yellowstone, greenback (Deacon and Williams 1984), Rio Grande, fine-spotted Snake River, and Yellowstone (Sublette et al. 1990) cutthroat subspecies have all been introduced in nonnative areas of Nevada. Yellowstone cutthroat were introduced in Oregon (Bond 1973) and Washington (Schultz and DeLacy 1935). Lahontan cutthroat were recently introduced into Washington (Washington Department of Fish and Wildlife 1997). Westslope (Sigler and Miller 1963), greenback (Tyus et al. 1982), and Yellowstone (Sigler and Sigler 1996) cutthroat were introduced to nonnative areas in Utah. The Lahontan (La Rivers 1962) and Colorado River (Gorges 1994) cutthroat were introduced in Wyoming. Unknown subspecies were stocked in Connecticut, Nebraska, New Jersey, Tennessee, and Wisconsin. Miller et al. (1991) recounted a story of an early trout introduction told to Carl Hubbs by a Nevada pioneer George Schmidtlein. In 1873, George gathered together family and neighbors, including a native American, to transfer cutthroat trout *O. clarki henshawi* from the Reese River system to his pond in Big Smoky Valley. The group, with the use of a hired mule train, spent more than four days transporting the fish live in vinegar kegs over several mountain summits. These fish, or their progeny, eventually spread to many of the area streams.

Status: Established in Arizona, California, Colorado, Nevada, and Utah. Most stocking attempts in other states failed (e.g., Tennessee, Connecticut, Minnesota) or populations replenished with repeated stockings. Established for a time in Michigan, then died out (Hubbs and Lagler 1947, 1964; Becker 1983)

Impact of Introduction: Indiscriminate stocking has obscured recognition of many subspecies (Page and Burr 1991; Behnke 1992). Yellowstone cutthroat *O. c. lewisi* have hybridized extensively with Lahontan cutthroat *O. c. henshawi* in California (McAffee 1966c). Cutthroat trout also hybridize with rainbow trout in situations where one or both species have been introduced (Behnke 1992).

Remarks: Interior (noncoastal) cutthroat (all subspecies except *O. clarki clarki*) have declined dramatically since the 19th century. Brown trout have commonly replaced cutthroat trout in large rivers (Behnke 1992). Introduced brook trout *Salvelinus fontinalis* quickly replaced greenback cutthroat trout in the Arkansas and South Platte drainages (U.S. Fish and Wildlife Service 1995). Recovery plans have been written for greenback cutthroat trout *O. c. stomias* (U.S. Fish and Wildlife Service 1977a, 1995), Paiute cutthroat trout *O. c. selenis* (U.S. Fish and Wildlife Service 1985a), and Lahontan cutthroat trout *O. c. henshawi* (U.S. Fish and Wildlife Service 1994d). Paiute cutthroat trout have hybridized with introduced rainbow trout and Lahontan cutthroat trout, resulting in a loss of distinctiveness of this subspecies in many areas (U.S. Fish and Wildlife Service 1985a). However, some pure populations do still exist (Behnke, personal communication). The Paiute, greenback, and Lahontan subspecies of cutthroat are listed as federally threatened species (U.S. Fish and Wildlife Service 1993a). Prior to the elimination of the U.S. Fish and Wildlife Service candidate species list, the Snake River fine-spotted, Willow or Whitehorse, westslope, Colorado River, and Bonneville subspecies were under review for Federal listing as an endangered or threatened taxa (U.S. Fish and Wildlife Service 1994a). Since 1963, 24 introductions of Lahontan cutthroat were made outside of its native range; 14 of these have established populations (U.S. Fish and Wildlife Service 1994d). Lahontan cutthroats are now established outside their native range in eleven locations in Nevada, nine in Oregon, four in Utah, and nine in California (U.S. Fish and Wildlife Service 1994d).

Oncorhynchus gilae

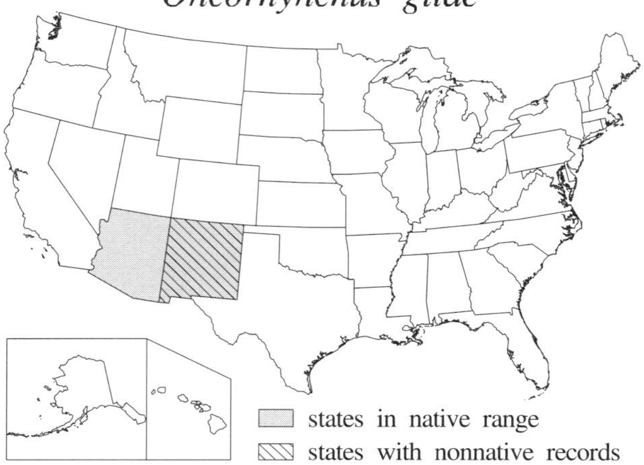

Oncorhynchus gilae (Miller 1950)

Common Name: Gila trout.

Identification: Minckley (1973); Sublette et al. (1990); Page and Burr (1991); Behnke (1992). Formerly known as *Salmo gilae*. Behnke (1992) treated the Gila and Apache trout as two subspecies of one species (i.e., *O. g. gilae* and *O. g. apache*). Maximum size: 32 cm.

Native Range: Gila River system, New Mexico and Arizona, including the San Francisco drainage in Arizona and New Mexico, and the Verde and Agua Fria drainages in Arizona (Page and Burr 1991; U.S. Fish and Wildlife Service 1993c).

Nonindigenous Occurrences: The Gila trout was introduced into McKnight Creek, **New Mexico**, a tributary of the Mimbres River in the Guzman basin (Lee et al. 1980 et seq.; Sublette et al. 1990; Behnke 1992; U.S. Fish and Wildlife Service 1993c). Other transplants took place within the species' native range (Sublette et al. 1990). McKnight Creek is definitely outside the native range, because it is east of the Continental Divide (U.S. Fish and Wildlife Service 1983c).

Means of Introduction: This endangered species was intentionally stocked to increase its distribution. McKnight Creek was stocked in November 1970 with 307 fish from Main Diamond Creek. The creek was restocked in April 1972 with 110 fish from Main Diamond Creek to supplement the population after a severe drought in 1971 reduced its numbers to less than 20. Flooding in 1988 resulted in a loss of greater than 90 percent of the population. The creek was restocked in October 1989 with 200 trout from Main Diamond Creek (U.S. Fish and Wildlife Service 1993c).

Status: Established in McNight Creek, New Mexico (U.S. Fish and Wildlife Service 1993c).

Impact of Introduction: Unknown.

Remarks: According to Behnke (1992), the distribution of this species before European settlement is not known with certainty. He discussed various aspects of its distribution. The Gila trout has been listed as a federally endangered species since 1966 (U.S. Fish and Wildlife Service 1993c, 1994a). This species readily hybridizes with rainbow trout *O. mykiss* and therefore has lost some of its genetic integrity at some sites (Lee et al. 1980 et seq.; U.S. Fish and Wildlife Service 1993c). The Gila trout was also stocked within its historic native range in the Gila and San Francisco drainages (Koster 1957; Behnke 1992; U.S. Fish and Wildlife Service 1993c).

Oncorhynchus gorbuscha

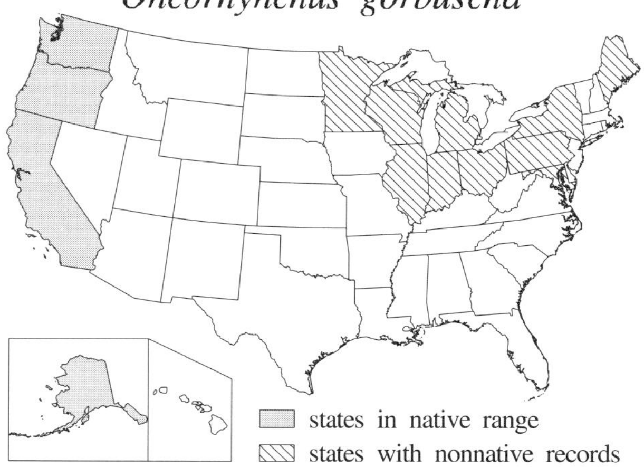

Oncorhynchus gorbuscha (Walbaum 1792)

Common Names: pink salmon.

Identification: Moyle (1976a); Scott and Crossman (1973); Wydoski and Whitney (1979); Morrow (1980);

Eschmeyer et al. (1983); Page and Burr (1991). Maximum size: 76 cm.

Native Range: Arctic and Pacific drainages from Mackenzie River delta, Northwest Territories, to Sacramento River drainage, California; occasionally as far south as La Jolla, southern California. Also in northeastern Asia (Page and Burr 1991).

Nonindigenous Occurrences: Pink salmon have been introduced into the Great Lakes and the surrounding states of **Illinois**, **Indiana**, **Michigan**, **Minnesota**, **New York**, **Ohio**, **Pennsylvania**, and **Wisconsin** (Eddy and Underhill 1974; Becker 1983; Cooper 1983; Underhill 1986), and to the coasts of **Maine** (Kendall 1914a; Morrow 1980) and **Maryland** (Morrow 1980).

Means of Introduction: This species was introduced accidentally into the Canadian side of Lake Superior in Thunder Bay in 1956, when about 100 salmon escaped while being transferred from a hatchery to a seaplane (Schumacher and Eddy 1960; Eddy and Underhill 1974). Another reference states that about 21,000 surplus fingerlings from the hatchery at Port Arthur, Ontario, were disposed of into the Current River, a tributary of Lake Superior (Becker 1983). The hatchery was raising them to stock in Hudson Bay (Eddy and Underhill 1974). The first evidence of reproduction was noted in 1959, when two second-generation fish were caught in Lake Superior, Minnesota (Schumacher and Eddy 1960; Schumacher and Hale 1962; Eddy and Underhill 1974). The species was first observed in a northern tributary of Lake Huron in 1969 (Parsons 1973). By 1979, it had spread throughout the Great Lakes (Cooper 1983). Emery (1981) gave an account of their spread through the Great Lakes and included a map showing the dispersal. The species was also intentionally stocked on the coasts of Maine and Maryland (Morrow 1980). Also see Schumacher and Eddy (1960), Schumacher and Hale (1962), Parsons (1973), Wagner (1976), Emery (1981), and Kwain and Lawrie (1981).

Status: Established population present throughout the Great Lakes (Emery 1981; Trautman 1981; Phillips et al. 1982; Cooper 1983). Populations in Maine and Maryland survived until about 1973, then disappeared (Morrow 1980).

Impact of Introduction: Pink salmon may displace native chubs by way of food competition. The introduced salmon also may compete with lake herring (Becker 1983).

Remarks: Voucher specimens: Maine (USNM 125907, 125936), Minnesota (University of Minnesota), Wisconsin (UWSP 5049).

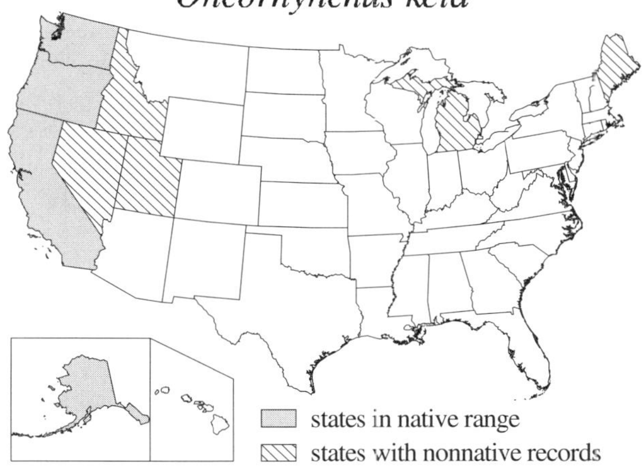

Oncorhynchus keta (Walbaum 1792)

Common Names: chum salmon.

Identification: Moyle (1976a); Scott and Crossman (1973); Wydoski and Whitney (1979); Morrow (1980); Eschmeyer et al. (1983); Page and Burr (1991). Maximum size: 102 cm.

Native Range: Arctic and Pacific drainages from Anderson and Mackenzie rivers, Northwest Territories, through much of Alaska south to Sacramento River drainage, California; rarely to San Lorenzo River, southern California. Also in northeastern Asia (Page and Burr 1991).

Nonindigenous Occurrences: Chum salmon have been introduced into Bannock County, **Idaho** (Linder 1963; Simpson and Wallace 1978), Churchill County (probably Lahontan Reservoir); Damariscotta, Megunticook, and Swan lakes, **Maine** (Everhart 1950); Deep Lake in Oakland County, **Michigan** (Emery 1985); Mineral County (probably Walker Lake), and Washoe County (probably Truckee River), all in the Lahontan drainage, **Nevada** (Miller and Alcorn 1946; Deacon and Williams 1984); and Strawberry Reservoir and Fish Lake in **Utah** (Sigler and Miller 1963; Sigler and Sigler 1996).

Means of Introduction: Authorized stockings took place in the 1930s and 1940s. The stocking in Idaho took place in 1939 (Linder 1963). The stocking in Maine took place in 1938 (Everhart 1950). The Michigan stocking occurred in 1945 (Emery 1985).

Status: Stocking attempts failed to generate reproducing populations.

Impact of Introduction: Unknown.

Remarks: There is some question concerning the identity of the salmon stocked into Michigan. Emery (1985) reported that they may have been misidentified chinook or coho salmon. Parsons (1973) made no mention of this species being stocked in the Great Lakes.

Oncorhynchus kisutch

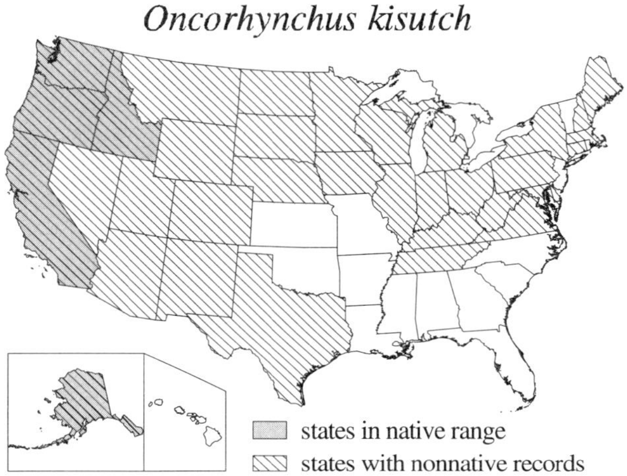

Oncorhynchus kisutch (Walbaum 1792)

Common Names: coho salmon (silver salmon).

Identification: Moyle (1976a); Scott and Crossman (1973); Wydoski and Whitney (1979); Morrow (1980); Eschmeyer et al. (1983); Page and Burr (1991). Maximum size: 98 cm.

Native Range: Arctic and Pacific drainages from Point Hope, Alaska, to Monterey Bay, California, infrequently as far south as Chamalu Bay, Baja California. Also in northeastern Asia (Page and Burr 1991).

Nonindigenous Occurrences: Coho have been stocked outside their native range into many lakes in **Alaska** (Morrow 1980); the Colorado River in **Arizona** (Minckley 1973; Rinne 1995); the Salton Sea and reservoirs in the Sacramento drainage in **California** (Walker et al. 1961; Moyle 1976a; Moyle and Daniels 1982); Lake Granby in the Colorado headwaters, Blue Mesa Reservoir on the Gunnison River, and Navajo Reservoir on the San Juan River, **Colorado** (Everhart and Seaman 1971; Tyus et al. 1982; Wiltzius 1985); all drainages of **Connecticut** (Whitworth 1996); Delaware Bay, **Delaware** (Raasch and Altemus 1991); the Snake River, Cascade Reservoir, Salmon River, and Coeur d'Alene Lake in **Idaho** (Idaho Fish and Game 1990, 1997); Lake Michigan, **Illinois** (Parsons 1973; Smith 1979; Burr 1991); Lake Michigan, **Indiana** (Nelson and Gerking 1968; Parsons 1973; Simon et al. 1992); West Lake Okoboji in the Little Sioux system in **Iowa** (Harlan et al. 1987); the Ohio River in **Kentucky** (Burr and Page 1986; Burr and Warren 1986); numerous rivers in **Maine** (Kendall 1914a; Stolte 1974; Everhart 1976; Schmidt 1986); the coastal waters of **Maryland** (Morrow 1980); the North River, and Merrimack drainage in **Massachusetts** (Stolte 1974; Schmidt 1986; Raasch and Altemus 1991; Hartel 1992; Hartel et al. 1996); the Great Lakes surrounding **Michigan** and their tributaries (Parsons 1973; Eddy and Underhill 1974; Phillips et al. 1982; Emery 1985); Lake Superior and tributaries in **Minnesota** (Parsons 1973; Eddy and Underhill 1974; Phillips et al. 1982); many lakes in **Montana**, including Deadman (Red Rock system), Flathead, Browns (Blackfoot system), Bean (upper Missouri drainage), Fort Peck, Cooney (Clarks Fork drainage), and Georgetown (Flint-Rock system) (Brown 1971; Cross et al. 1986; Holton 1990); Merrit Reservoir in the Snake drainage, and Lake McConaughy in the North Platte drainage **Nebraska** (Morris et al. 1974); the Colorado River in **Nevada** (Miller and Alcorn 1946; Deacon and Williams 1984); inland lakes and coastal **New Hampshire** (Hoover 1936; Scarola 1973; Stolte 1974; Schmidt 1986; Raasch and Altemus 1991); Navajo Reservoir, and Heron, Maloya, Eagle Nest, and Bluewater lakes, in the Canadian, upper Rio Grande, and upper San Juan drainages in **New Mexico** (Tyus et al. 1982; Sublette et al. 1990); Lake Ontario and tributaries **New York** (Parsons 1973; Smith 1985; Hocutt et al. 1986); unspecified areas in **North Dakota** (Cross et al. 1986; North Dakota Game and Fish Department 1994, 1997); Lake Erie and the Ohio and Scioto rivers in **Ohio** (Trautman 1981; Emery 1985; Burr and Page 1986; Burr and Warren 1986); unspecified inland lakes in nonnative areas in **Oregon** (Bond 1973, 1994; Morrow 1980); tributaries of the Delaware River, and Harvey's Lake, Luzurne County in **Pennsylvania** (Parsons 1973; Denoncourt et al. 1975a; Cooper 1983; Hocutt et al. 1986; Raasch and Altemus 1991); unspecified areas in **South Dakota** (Cross et al. 1986; North Dakota Game and Fish Department 1994); Inks Lake in the Middle Colorado drainage in **Texas** (Howells 1992a);

Watauga Reservoir, **Tennessee** (Etnier and Starnes 1993); Smith Mountain Lake and Philpott Reservoir in **Virginia** (Jenkins and Burkhead 1994); unspecified areas in **Utah** (Sigler and Miller 1963; Sigler and Sigler 1996); nonnative areas of **Washington** (Morrow 1980); non-specific locations in **West Virginia** (Cincotta, personal communication); Riley Lake in Chippewa County, Stormy and Pallette lakes in Vilas County and lakes Superior and Michigan in **Wisconsin** (Parsons 1973; Becker 1983); and McDonald Reservoir, Deep Lakes, Christina Lake, and a few other lakes in **Wyoming** (Baxter and Simon 1970).

Means of Introduction: Introductions of coho salmon as sport fish began as early as the 1920s. The coho was introduced into the Great Lakes to control the alewife population (Eddy and Underhill 1974). The first stocking in the Great Lakes was in Lake Erie in 1933 by the Ohio Division of Conservation (Parsons 1973). The first large planting in the Great Lakes occurred in 1966 in Lakes Michigan and Superior. The species had been stocked in all the Great Lakes by 1970 (Parsons 1973). A total of 16 million fish had been stocked in the Great Lakes between 1966 and 1970 (Parsons 1973). Fish intentionally stocked into coastal New Hampshire in the late 1960s dispersed to Maine and Massachusetts (Stolte 1974). It was introduced into the Colorado River in 1967 (Rinne 1995). Colorado first began stocking cohos in the early 1900s (Everhart and Seaman 1971). Originally, the stock used for Colorado came from Oregon; more recently stock from Lake Michigan has been used (Everhart and Seaman 1971). Introduced into Connecticut since the late 1800s (Whitworth 1996). Most attempts to establish populations failed; therefore, stocking was discontinued.

Status: Some reproduction reported in Lake Erie, Lake Michigan, Lake Superior, Manistee River in Michigan, coastal waters of New Hampshire, Maine, and Massachusetts, Salmon River in New York, and Colorado River in Arizona. Stockings in Tennessee failed. Rinne (1994) lists coho salmon as not established in the Colorado River in Arizona. The first evidence of reproduction in the Great Lakes came in 1968 from tributaries of Lakes Michigan and Superior (Parsons 1973). There are no known established populations in Connecticut (Whitworth 1996) or Delaware (Raasch and Altemus 1991).

Impact of Introduction: Coho salmon compete with native lake trout *Salvelinus namaycush* (Page and Laird 1993). Fausch and White (1986) found that coho salmon may compete with brook trout *S. fontinalis* and brown trout *Salmo trutta* for food an space in the Great Lakes if resources become scarce. Coho have an advantage over brook and brown trout because of an earlier emergence and a larger size at emergence (Fausch and White 1986).

Remarks: Coho salmon have not been stocked in Oklahoma (Pigg, personal communication). Parsons (1973) gave detailed stocking information for the Great Lakes.

Voucher specimens: Montana (USNM 104701).

Oncorhynchus masou

Oncorhynchus masou (Brevoort 1856)

Common Names: cherry salmon.

Identification: Distinguishing characteristics and color photographs were given by Masuda et al. (1984). Several subspecies or forms are recognized, some of them previously considered distinct species (Masuda et al. 1984). Behnke et al. (1962) provided a list of taxonomic characters. Other names for this species are *O. macrostomus*, *O. rhodorus*, *Salmo fomosanus*. Maximum size: 78 cm FL.

Native Range: Marine and freshwaters of Japan, Korea, and Russia (Masuda et al. 1984; Courtenay and Stauffer 1984).

Nonindigenous Occurrences: This species was planted in the North Branch of the Boyne River, a tributary of Lake Michigan in Charlevoix County, **Michigan**, by the Michigan Department of Conservation in 1929 (Parsons 1973). Cherry salmon were stocked into Hewitt Lake and Puget Sound in the state of **Washington** circa 1974 (University of Washington Fish Collection 1996; Courtenay et al. 1991).

Means of Introduction: Two hundred 18-month old fingerlings were intentionally stocked in Michigan by the Michigan Department of Conservation. Stocked in Washington.

Status: Failed in Michigan. Presumably failed in Washington.

Impact of Introduction: Unknown.

Remarks: Voucher specimens: Washington (UW 21588).

Oncorhynchus mykiss

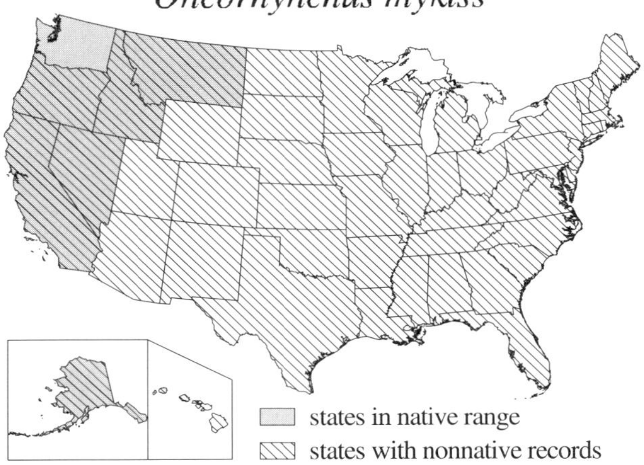

Oncorhynchus mykiss (Walbaum 1792)

Common Names: rainbow trout (steelhead [anadromous form], coastal rainbow trout *O. m. irideus*, Kamloops, redband trout, Columbia River redband trout *O. m. gairdneri*, Eagle Lake trout, Kern River trout *O. m. gilberti*, Shasta trout, Sacramento redband trout *O. m. stonei*, Kamchatkan trout *O. m. mykiss*).

Identification: Moyle (1976a); Scott and Crossman (1973); Wydoski and Whitney (1979); Morrow (1980); Eschmeyer et al. (1983); Page and Burr (1991); Behnke (1992). Behnke (1992) gave accounts and drawings for several subspecies. A commonly used named for this species is *Salmo gairdnerii*, sometimes given as *S. gairdneri*. Maximum size: 114 cm.

Native Range: Pacific Slope from Kuskokwim River, Alaska, to (at least) Rio Santa Domingo, Baja California; upper Mackenzie River drainage (Arctic basin), Alberta and British Columbia; endorheic basins of southern Oregon (Page and Burr 1991).

Nonindigenous Occurrences: The rainbow trout has been extensively stocked throughout the United States, in states including **Alaska** (Morrow 1980); **Alabama** (Mettee 1978; Boschung 1992; Mettee et al. 1996); **Arizona** (Tyus et al. 1982; Minckley 1973); **Arkansas** (Pritchard et al. 1976; Cross et al. 1986); **California** (Moyle 1976a; Moyle and Daniels 1982; Dill and Cordone 1997); **Colorado** (Vanicek et al. 1970; Everhart and Seaman 1971; Tyus et al. 1982; Wiltzius 1985; Propst and Carlson 1986); **Connecticut** (Behnke and Wetzel 1960; Schmidt 1986; Whitworth 1996); **Delaware** (Raasch and Altemus 1991; Rohde et al. 1994); **Florida** (Eglin Air Force Base 1968); **Georgia** (Dahlberg and Scott 1971a, 1971b; Yerger 1977); **Hawaii** (Brock 1960; Maciolek 1984); **Idaho** (Idaho Fish and Game 1996); **Illinois** (Smith 1979; Burr and Page 1986; Underhill 1986; Burr 1991); **Indiana** (Sweeney 1902; Underhill 1986); **Iowa** (Harlan and Speaker 1951; Smith 1979; Harlan et al. 1987); **Kansas** (Cross 1967; Cross et al. 1986); **Kentucky** (Clay 1962, 1975; Burr and Page 1986; Burr and Warren 1986); **Louisiana** (J. Forrester, personal communication); **Maine** (Kendall 1914a; Everhart 1950, 1976; Schmidt 1986); **Maryland** (Lee et al. 1976, 1981; Rohde et al. 1994); **Massachusetts** (Hartel 1992; Cardoza et al. 1993; Hartel et al. 1996); **Michigan** (Hubbs and Lagler 1947; Emery 1985; Underhill 1986); **Minnesota** (Phillips et al. 1982; Burr and Page 1986; Underhill 1986); **Mississippi** (Ross and Brenneman 1991); **Missouri** (Bean 1903; Pflieger 1971, 1975, 1997; Cross et al. 1986); **Montana** (Brown 1971; Cross et al. 1986; Holton 1990); **Nebraska** (Jones 1963; Morris et al. 1974); **Nevada** (Smith 1896; Miller and Alcorn 1946; La Rivers 1962; Deacon and Williams 1984; Sigler and Sigler 1987); **New Hampshire** (Hoover 1936; Bailey and Oliver 1939; Scarola 1973; Schmidt 1986); **New Jersey** (Fowler 1952; Stiles 1978); **New Mexico** (Koster 1957; Tyus et al. 1982; Sublette et al. 1990); **New York** (Bean 1903; Werner 1980; Smith 1985; Schmidt 1986; Underhill 1986); **North Carolina** (Hocutt et al. 1986; Menhinick 1991; Rohde et al. 1994); **North Dakota** (Lee et al. 1980 et seq.; Cross et al. 1986; North Dakota Game and Fish Department 1994); **Ohio** (Trautman 1981; Burr and Page 1986; Hocutt et al. 1986; Underhill 1986); **Oklahoma** (Hall 1956; Miller and Robison 1973; Cross et al. 1986); **Oregon** (Li, personal communication); **Pennsylvania** (Bean 1892b; Denoncourt et al. 1975a; Hendricks et al. 1979; Cooper 1983; Hocutt et al. 1986; Underhill 1986); **South Carolina** (Dahlberg and Scott 1971a, 1971b; Loyacano 1975; Hocutt et al. 1986; Rohde et al. 1994); **South Dakota** (Bailey and Allum 1962; North Dakota Game and Fish Department 1994);

Tennessee (Kuhne 1939; Starnes and Etnier 1986; Etnier and Starnes 1993); **Texas** (Knapp 1953; Conner and Suttkus 1986; Cross et al. 1986; Howells 1992a); **Utah** (Sigler and Miller 1963; Vanicek et al. 1970; Tyus et al. 1982; Sigler and Sigler 1987, 1996); **Vermont** (Lee et al. 1980 et seq.; Underhill 1986); **Virginia** (Hocutt et al. 1986; Jenkins and Burkhead 1994; Rohde et al. 1994); **West Virginia** (Hocutt et al. 1986; Stauffer et al. 1995); **Wisconsin** (Eddy and Underhill 1974; Burr and Page 1986; Underhill 1986); and **Wyoming** (Baxter and Simon 1970; Tyus et al. 1982; Hubert 1994).

Means of Introduction: Beginning in the late 1800s, there have been many stockings of this species for sportfishing purposes by state and federal agencies and by private individuals, mostly into streams and spring branches. Some states stock on an annual basis.

Status: Established in many states, including Hawaii. Also frequently stocked in most states to replenish populations harvested by fishing pressures or in other areas where populations are not self sustaining. One specimen collected from Mississippi (Ross and Brenneman 1991). Stocked once, in 1991, in Louisiana. The stocking failed.

Impact of Introduction: The rainbow trout hybridizes with other, more rare trout species, thereby affecting their genetic integrity (Lee et al. 1980 et seq.; Rinne and Minckley 1985; Page and Burr 1991). In California, rainbow trout have hybridized with Lahontan cutthroat trout *O. clarki henshawi*, golden trout *O. aguabonita*, and redband trout *O. mykiss* subsp. to the point that all three are included in the threatened trout management program of the California Department of Fish and Game (McAffee 1966c; Moyle 1976b; Behnke 1992). In the Lahontan drainage and various Rocky Mountain rivers, hybridization with rainbow trout has been a major factor in the decline of native cutthroat trout (McAffee 1966a). In Nevada, this species is also held responsible for the virtual extinction of Alvord cutthroat *O. mykiss* subsp. (Behnke 1992). In Arizona, the species has hybridized with native Gila trout *O. gilae* and Arizona trout *O. apache* (Minckley 1973; U.S. Fish and Wildlife Service 1979b), the latter species also known as the Apache trout and considered a subspecies of *O. gilae* (i.e., *O. g. apache*). Rainbow trout have replaced Lahontan cutthroat trout in areas where the cutthroat is native and rainbow trout have been introduced (McAffee 1966c). Introduced rainbow trout, and other trout species, were likely responsible for the near-extinction of Lahontan cutthroat in Lake Tahoe in the 1940s (McAffee 1966c). Rainbow trout have been found to negatively affect Little Colorado spinedace *Lepidomeda vittata* through predation and by affecting spinedace behavior. The trout occupied undercut banks that the spinedace normally used for refuge. As a result, spinedace were displaced from preferred microhabitats and pushed into open water, making them vulnerable to predation (Blinn et al. 1993).

Stocking of hatchery rainbow trout in rivers has led to introduction of whirling disease into open waters in approximately 20 states including, most recently, the Madison River and its tributaries in Montana (B. Nehring and R. White, personal communication). In the Madison River, the disease has reduced the rainbow trout population by 90% (White, personal communication). Rainbow trout have the potential to consume native fishes and compete with native salmonids (Page and Laird 1993). Introduced rainbow trout eat endangered humpback chubs *Gila cypha* in the Little Colorado River, and may exert a major negative effect on the population there (Marsh and Douglas 1997). Fausch (1988), Clark and Rose (1997), and numerous papers cited in both, discussed several factors affecting competitive interactions between rainbow and brook trout. Rainbow trout drive nongame fishes such as suckers and squawfish from feeding territories (Li, personal communication to P. Moyle in Moyle 1976a). Introduced predatory fishes, including the rainbow trout, are likely at least partially responsible for the decline of the Chiricahua leopard frog *Rana chiricahuensis* in southeastern Arizona (Rosen et al. 1995).

Remarks: Tyus et al. (1982) mapped the distribution of rainbow trout in the upper Colorado basin.

Oncorhynchus nerka

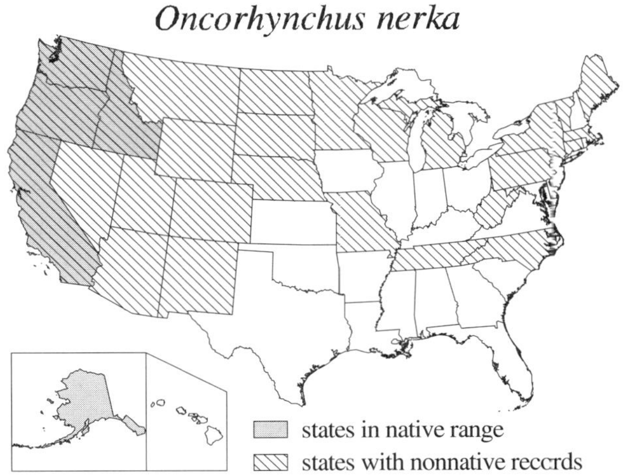

Oncorhynchus nerka (Walbaum 1792)

Common Names: sockeye salmon (anadromous form) (kokanee [landlocked]).

Identification: Moyle (1976a); Scott and Crossman (1973); Wydoski and Whitney (1979); Morrow (1980); Eschmeyer et al. (1983); Page and Burr (1991). Maximum size: 84 cm.

Native Range: Arctic and Pacific drainages from Point Hope, Alaska, to Sacramento River drainage, California. Landlocked populations in Alaska, Yukon, British Columbia, Washington, and Oregon. Also in northeastern Asia (Page and Burr 1991).

Nonindigenous Occurrences: This species was stocked in Ashurst, Luna, and Mojave lakes, **Arizona** (Minckley 1973; Rinne 1995); many lakes in **California** in the Sacramento, Klamath, Lahontan, Mojave, and San Joaquin drainages (La Rivers 1962; Cordone et al. 1971; Moyle 1976a, 1976b); more than 24 coldwater lakes and reservoirs in **Colorado**, including Blue Mesa, Dillon, Shadow Mountain, Navajo, and Lake Granby (Everhart and Seaman 1971; Ellis 1974; Tyus et al. 1982; Wiltzius 1985; Walker 1993); East Twin Lakes in Litchfield, Housatonic drainage, **Connecticut** (Behnke and Wetzel 1960; Whitworth et al. 1968; Webster 1940; Schmidt 1986; CU 10587); Bear Lake, Redfish Lake, and Priest Lake, **Idaho** (Sigler and Sigler 1987; Idaho Fish and Game 1990, 1997); unspecified areas of **Maine** (Buss 1957 cited in Scott and Crossman 1973); the Connecticut, Housatonic, and Thames drainages, **Massachusetts** (Schmidt 1986; Hartel 1992; Cardoza et al. 1993; Hartel et al. 1996); lakes Huron, Michigan, and Erie in **Michigan** (Buss 1957 cited in Scott and Crossman 1973; Parsons 1973; Emery 1985); several lakes in Cook County and Larson Lake in George Washington State Forest, **Minnesota** (Eddy and Underhill 1974); Lake Taneycomo in **Missouri** (Pflieger 1971); many areas of **Montana**, including the Pend Orielle, Kootenai, St. Mary, Yellowstone, and Missouri drainages (Brown 1971; Cross et al. 1986); Lake Ogallala and Lake McConaughy, **Nebraska** (Jones 1963; Morris et al. 1974; Cross et al. 1986); lakes Tahoe and Mojave, Walker, Pyramid, Topaz, and Virginia lakes, Wildhorse Reservoir, and Lahontan Reservoir, **Nevada** (Miller and Alcorn 1946; La Rivers 1962; Cordone et al. 1971; Deacon and Williams 1984; Sigler and Sigler 1987); Canadian, Rio Grande, and San Juan drainages, **New Mexico** (Tyus et al. 1982; Sublette et al. 1990); several areas of **New York** including Green Lake in Onondaga County, lakes and ponds in the Black and Richlieu drainages, the lower Hudson River, Adirondack Mountains, Lake Champlain, and Lake Ontario (Parsons 1973; Werner 1980; Emery 1985; Smith 1985; Underhill 1986; museum specimens at Cornell University); the Tennessee drainage, **North Carolina** (Menhinick 1991); unspecified areas of **North Dakota** (Buss 1957 cited in Scott and Crossman 1973; North Dakota Game and Fish Department 1994); nonindigenous areas of **Oregon** (Bond 1973, 1994; Behnke, personal communication); Upper Woods Pond and Lake Winola, both near Scranton, the Monongahela drainage, and in Harvey's Lake, Luzurne County, **Pennsylvania** (Denoncourt et al. 1975a; Cooper 1983; Hocutt et al. 1986; Stauffer et al. 1995); unspecified areas of **South Dakota** (Hanten, personal communication; North Dakota Game and Fish Department 1994); the Little Tennessee River and Watauga Reservoir, **Tennessee** (Etnier and Starnes 1993); Scofield, East Canyon, Deer Creek, Flaming Gorge, Porcupine, and Strawberry reservoirs; the Uinta Mountains; and Bear, Navajo and Panguitch lakes, **Utah** (Buss 1957 cited in Scott and Crossman 1973; Sigler and Miller 1963; Tyus et al. 1982; Sigler and Sigler 1987, 1996; Teuscher and Luecke 1996); unspecified areas of **Vermont** (Buss 1957 cited in Scott and Crossman 1973; Cox and T. Hess, personal communication); Banks Lake (Grant County) on the Columbia River, and in Loon Lake, Stevens County, **Washington** (Wydoski and Whitney 1979); Upper Bass Lake in Menominee County, a lake in Langlade County, and probably other lakes in **Wisconsin** (Becker 1983; Fago 1992; Wisconsin Department of Natural Resources 1997); several high mountain impoundments in the Potomac drainage in **West Virginia** (Cincotta, personal communication); and Crystal Reservoir, Freemont Lake and Flaming Gorge Reservoir, **Wyoming** (Buss 1957 cited in Scott and Crossman 1973; Baxter and Simon 1970; Tyus et al. 1982; Hubert 1994; Teuscher and Luecke 1996). Hocutt et al. (1986) listed this species as introduced into the Potomac drainage but did not give a state. Jenkins and Burkhead (1994) did not list it in Virginia; Stauffer et al. (1995) did not list it in West Virginia; and Rohde et al. (1993) did not list it in Maryland. On-line searches of the Philadelphia Academy of Sciences, the National Museum of Natural History, Cornell University, Museum of Comparative Zoology, and University of Michigan also returned no specimens for the Potomac drainage, nor for the three states included in that drainage. Presumably if it occurred there, it no longer does.

Means of Introduction: Authorized stocking for sportfishing in most states. Stocked as forage in California (Moyle 1976b). Stocked in Arizona in 1967 (Rinne 1995). Cordone et al. (1971) gave the stocking history for Lake Tahoe. The first stocking in the Great Lakes was in 1950 in Lake Ontario tributaries by the New York Department of Environmental Conservation (Parsons

1973). Between 1950 and 1970, 19 million kokanee were planted in the Great Lakes, primarily in Lakes Ontario and Huron (Parsons 1973). Parsons (1973) gave a detailed account of the Great Lakes stockings.

Status: Established locally in some areas; failed in others. Failed in Tennessee. Rinne (1994) listed it as not established in Arizona. First found to be reproducing in the Great Lakes in Lake Huron in 1968 (Parsons 1973). No longer present in Vermont (Hess, personal communication).

Impact of Introduction: Kokanee was introduced to compete with, and reduce native Utah chub *Gila atraria* populations in Flaming Gorge Reservoir. However, it was found that under food-limiting conditions, the Utah chub was the superior competitor and was not affected by the kokanee (Teuscher and Luecke 1996). Seeley and McCammon (1966) gave a detailed discussion of impacts of kokanee on other species. Some of the impacts listed included: large kokanee populations can inhibit trout production because of competition for zooplankton; introductions of kokanee into small lakes can adversely affect even strongly established populations of rainbow trout; and late-run kokanee and brown trout may compete for the same spawning areas, especially in lower stream reaches. Introduced kokanee were at least partially responsible for the disappearance of three cladoceran species in Lake Tahoe (Morgan et al. 1978). Sometimes stocked kokanee do not have an apparent negative effect. Some of the determining factors which may influence whether this species has negative impacts on a receiving ecosystem or not are waterbody size and depth, the forage base, stocking densities, other species present, and if early or late-run strains are stocked. Seeley and McCammon (1966) cautioned that kokanee's value as a forage fish may be overrated–the species may depress rather than bolster trout populations. Its presence may not enhance a fishery, especially in waters without a deep, cool, well-aerated hypolimnion. The species is not long-lived (2 years) and in many areas the fish is only accessible for fishing for short periods in the spring and fall. Therefore, the entire population may die off after only being briefly available to anglers.

Remarks: Tyus et al. (1982) mapped the distribution the kokanee in the upper Colorado basin.

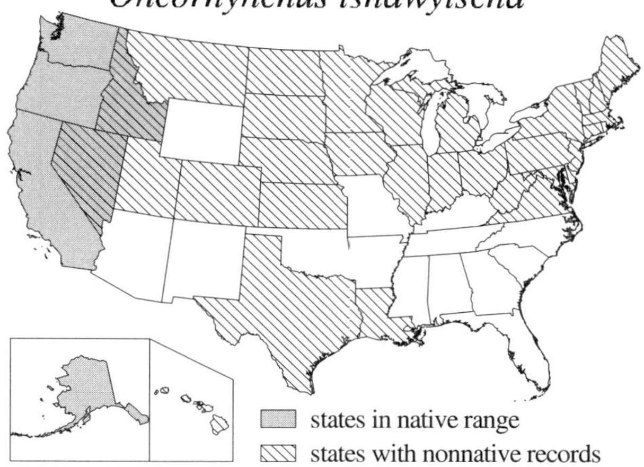

Oncorhynchus tshawytscha (Walbaum 1792)

Common Names: chinook salmon (king salmon).

Identification: Moyle (1976a); Scott and Crossman (1973); Wydoski and Whitney (1979); Morrow (1980); Eschmeyer et al. (1983); Page and Burr (1991). Maximum size: 147 cm.

Native Range: Arctic and Pacific drainages from Point Hope, Alaska, to Ventura River, California. Occasionally strays south to San Diego, California. Also in northeastern Asia (Page and Burr 1991).

Nonindigenous Occurrences: Chinooks have been stocked in the Colorado basin in **Colorado** (Wiltzius 1985; Horak, personal communication); unspecified areas in **Connecticut** (Whitworth 1996); the Delaware and Nanticoke rivers in **Delaware** (Raasch and Altemus 1991); Kaui, **Hawaii** (Brock 1960; Maciolek 1984); the Spokane drainage in **Idaho** (Idaho Fish and Game 1990, 1996, 1997); Lake Michigan, **Illinois** (Parsons 1973; Smith 1979; Emery 1985; Underhill 1986; Burr 1991); Lake Michigan, **Indiana** (Nelson and Gerking 1968; Parsons 1973; Emery 1985; Underhill 1986); the Missouri River, **Iowa** (Baird 1876 in Jones 1963); unspecified areas in **Kansas** (Cross 1967); Tangipahoa and Notalbany rivers, **Louisiana** (Bean 1882); Swan, Damariscotta, Cobboseecontee, and Long lakes and Pemaquid River, **Maine** (Kendall 1914a; Everhart 1950); lower Susquehanna, upper Chesapeake, Gunpowder, Patapsco, Patuxent, Potomac, Monocacy, and Conococheague drainages in **Maryland** (Ferguson 1876); the Shawsheen, Saugus, North, Ipswich, Merrimack, and Nashua rivers, Lake Quinsigamond in Worcester County, and an unnamed river in Bridgewater, Plymouth County, **Massachusetts**, in historic times and in the North River in recent times (1989) (Hartel 1992; Cardoza et al. 1993); the Great Lakes surrounding **Michigan** (Parsons 1973; Phillips et al. 1982; Underhill 1986; Michigan Department of Natural Resources 1997); Lake Superior, and numerous inland lakes in **Minnesota** (Parsons 1973; Eddy and Underhill 1974; Phillips et al. 1982; Emery 1985; Underhill 1986); Fort Peck Reservoir, **Montana** (Holton 1990); the Missouri, Big Blue, Platte, North Platte, Loup, and Elkhorn rivers, Salt Creek, and Birdwood Creek, **Nebraska** (Jones 1963; Morris et al. 1974); lakes and rivers in the central Lahontan basin, **Nevada** (Smith 1896; Miller and Alcorn 1946; La Rivers 1962); two unspecified rivers and large oligotrophic lakes in **New Hampshire**, including Sunapee, First Connecticut, Squam and Ossippee lakes, Round Pond in Pittsburg (Hoover 1936; Bailey and Oliver 1939; McAffee 1966d; Cooper 1983; Schmidt 1986); the Raritan and Delaware rivers in **New Jersey** (Nelson 1890; Fowler 1906, 1952; Raasch and Altemus 1991); Lake Ontario, Lake Erie, Little Moose Lake in Herkimer County (CU 73659), and Green Lake State Park in Onondaga County (CU 72033), **New York** (Parsons 1973; Smith 1985); the Missouri River in **North Dakota** (Cross et al. 1986; North Dakota Game and Fish Department 1994, 1997); Lake Erie and its tributaries, and the Tuscarwas, Muskingum, Scioto, and Great Miami drainages in **Ohio** (Trautman 1957; Parsons 1973; Trautman 1981; Underhill 1986); the Delaware and Susquehanna rivers, and Lake Erie, **Pennsylvania** (Bean 1892b; Parsons 1973; Cooper 1983; Underhill 1986); unspecified locations in **South Dakota** (Cross et al. 1986; North Dakota Game and Fish Department 1994; Hanten, personal communication); unspecified locations in **Texas** (Howells 1992a); Fish Lake, **Utah** (Sigler and Miller 1963; Sigler and Sigler 1996); the Potomac, James, Dan, New, and North and South Fork Holston rivers, **Virginia** (Ferguson 1876; Jenkins and Burkhead 1994); unspecified locations in **Vermont** (Cox, personal communication); the Potomac drainage in **West Virginia** (Cincotta, personal communication); Riley Lake in Chippewa County, Stormy and Pallette lakes in Vilas County, and Lakes Michigan and Superior, **Wisconsin** (Parsons 1973; Phillips et al. 1982; Becker 1983; Emery 1985; Underhill 1986).

Means of Introduction: Authorized introductions for sportfishing. Stocking began as early as 1874 in several states. Parsons (1973) give detailed accounts of stockings in the Great Lakes. The first stocking of large numbers of chinooks in the Great Lakes occurred in 1967 in Lake Michigan and Superior. By 1970 the species had been planted in all the Great Lakes (Parsons 1973). Between 1873 and 1933, about 11 million chinook were stocked in the Great Lakes basin (Parsons 1973). In a second attempt to establish chinooks, another six million were stocked 1967–1970. Chinook were stocked in West Virginia in 1874 (Cincotta, personal communication).

Status: Most attempts to establish populations failed. However, there may be some reproduction in the Great Lakes (Phillips et al. 1982). Chinook are no longer present in Utah (Sigler and Sigler 1996).

Impact of Introduction: Competes with native lake trout *Salvelinus namaycush* (Page and Laird 1993). Chinook salmon are predatory fish and as such may impact populations of smaller fish. The species had totally eliminated rainbow smelt *Osmerus mordax* in two small New Hampshire lakes where the salmon was stocked to control the smelt (McAffee 1966d).

Remarks: Chinook salmon have not been stocked in Oklahoma (Pigg, personal communication). Parsons (1973) gave detailed stocking information for the Great Lakes.

Prosopium cylindraceum

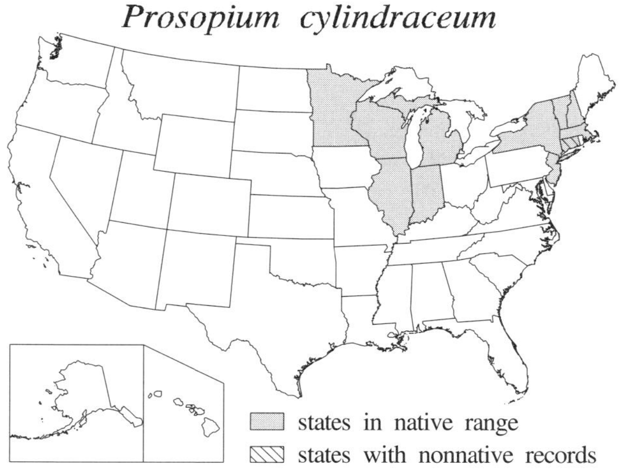

Prosopium cylindraceum (Pallas 1784)

Common Name: round whitefish.

Identification: Page and Burr (1991). Maximum size: 56 cm.

Native Range: Disjunct range. Arctic and Pacific drainages from western shore of Hudson Bay to Alaska and northern British Columbia; Arctic and Atlantic drainages from Labrador to Connecticut and west through St. Lawrence-Great Lakes basin. Also in Arctic drainages of Asia (Page and Burr 1991).

Nonindigenous Occurrences: This species was introduced into East Twin Lakes (i.e., Washinee Lake) in the Housatonic drainage, **Connecticut**, in the 1870s. Individuals are still periodically taken from the lake, indicating an established population (Whitworth 1996).

Means of Introduction: Intentionally stocked.

Status: Established in Connecticut.

Impact of Introduction: Unknown.

Remarks: Schmidt (1986) listed this species as native to Housatonic River drainage.

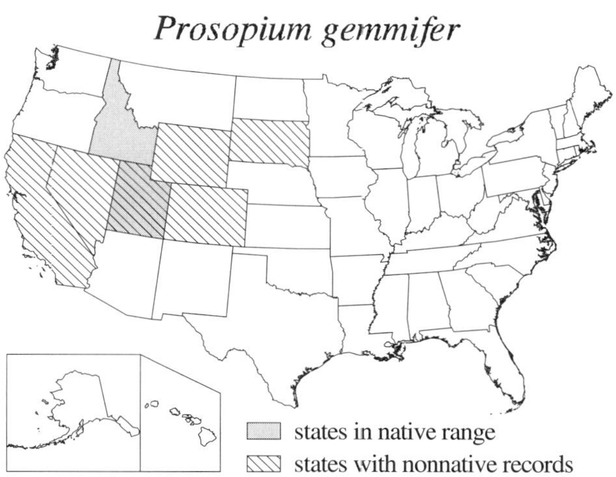

Prosopium gemmifer (Snyder 1919)

Common Name: Bonneville cisco.

Identification: Sigler and Sigler (1987); Page and Burr (1991). Another name for this species is *P. gemmiferum* (Page and Burr 1991). Maximum size: 22 cm.

Native Range: Endemic to Bear Lake in southeastern Idaho and northern Utah (Page and Burr 1991).

Nonindigenous Occurrences: The Bonneville cisco was stocked in Lake Tahoe, **California** and **Nevada**, in 1964, 1965, and 1966 (Frantz and Cordone 1965, 1967; Sigler and Sigler 1996; Dill and Cordone 1997). It was also introduced to Twin Lakes in Lake County, **Colorado** (Wiltzius 1985; Sigler and Workman 1978; Sigler and Sigler 1987, 1996), high mountain areas of **South Dakota** (Sigler and Sigler 1987; North Dakota Game and Fish Department 1994; Sigler and Sigler 1996), Flaming Gorge Reservoir, **Utah** and **Wyoming** (Sigler and Sigler 1987, 1996). Page and Burr (1991) reported this species has been unsuccessfully introduced into several western states.

Means of Introduction: Intentionally stocked. A total of 15,888 cisco were released over a three-year period in Lake Tahoe (Frantz and Cordone 1967). The species was stocked to provide forage for lake trout *Salvelinus namaycush* and other trouts species (Frantz and Cordone 1965). It was stocked by the state of Colorado in 1971.

Status: Extirpated in South Dakota and presumably extirpated in Lake Tahoe because no mention was made of a population by La Rivers (1962), Lee et al. (1980 et seq.), or Page and Burr (1991). Moyle (1976a) listed the introduction as a failure in California. The Colorado introduction also failed to produce an established population. Sigler and Sigler (1987) reported that no survival has been documented from any of the stockings.

Impact of Introduction: Unknown.

Remarks: None.

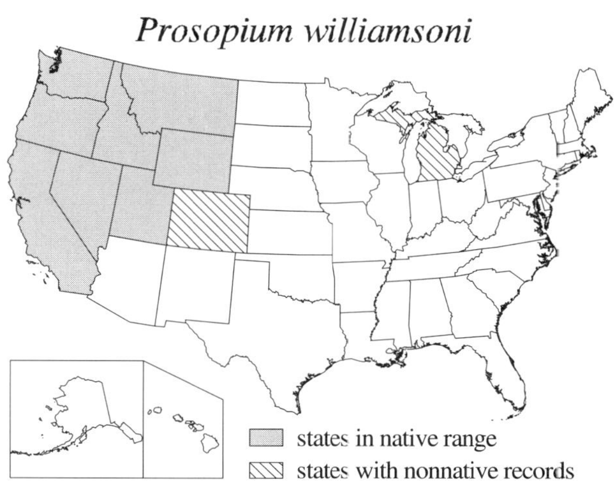

Prosopium williamsoni (Girard 1856)

Common Name: mountain whitefish.

Identification: Scott and Crossman (1973); Wydoski and Whitney (1979); Sigler and Sigler (1987); Page and Burr (1991). Maximum size: 57 cm.

Native Range: Mackenzie River drainage (Arctic basin), Northern Territories, south through western Canada and northwestern United States in Pacific, Hudson Bay, and upper Missouri River basins, to Truckee River drainage, Nevada, and Sevier River drainage, Utah (Page and Burr 1991).

Nonindigenous Occurrences: This species was introduced into the Cache la Poudre drainage (Wiltzius 1985; P. Walker and Behnke, personal communication) and the Roaring Fork River, a tributary to the Colorado River in **Colorado**, in 1955. From the Roaring Fork River, the mountain whitefish spread to the upper Colorado River in the state of Colorado (Behnke, personal communication). It was also introduced into the Great Lakes Basin in **Michigan** in 1920 (Emery 1985).

Method of Introduction: Stocked for sportfishing in Colorado and as a game and food fish in Michigan.

Status: Established in Colorado. Extirpated in Michigan.

Impact of Introduction: Whitefish compete with juvenile and adult trout (McAffee 1966f).

Remarks: None.

Salmo letnica

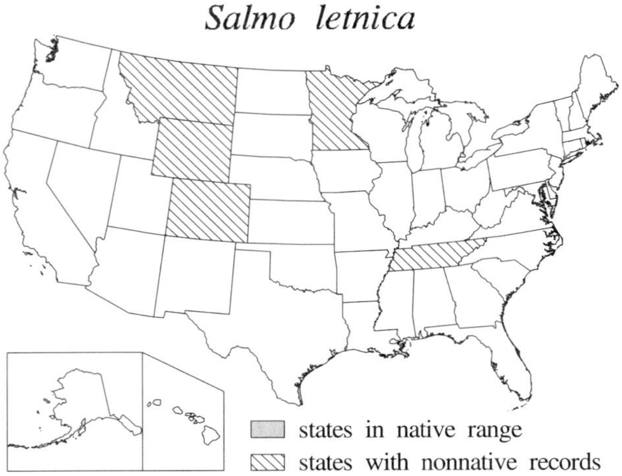

Salmo letnica (Karaman 1924)

Common Name: Ohrid trout.

Identification: Stefanovic (1966). Maximum size: Not reported.

Native Range: The Ohrid trout is native to Yugoslavia in Europe (Robins et al. 1991b).

Nonindigenous Occurrences: The Ohrid trout was stocked in Parvin Lake, Big Creek Reservoir, and Turquoise Lake in **Colorado**, in 1969 (Courtenay and Hensley 1979a; Wiltzius 1985; Courtenay et al. 1991), seven lakes in northern **Minnesota**, including Strawberry Lake, Chester Lake, and Big Trout Lake (Stroud 1969; Courtenay and Hensley 1979a; Radonski et al. 1984); an unspecified locality in **Montana** (Courtenay et al. 1991); Watauga Reservoir (= Ripshin Lake) (Courtenay and Hensley 1979a; Etnier and Starnes 1993) and South Holston Reservoir (Starnes, personal communication), **Tennessee**; and a few reservoirs in **Wyoming**, including Viva Naughton Reservoir on the Green River, lakes near Casper, and the North Platte River (Courtenay and Hensley 1979a; Courtenay et al. 1991; Hubert 1994; Wyoming Game and Fish, Fish Division 1997).

Means of Introduction: The Ohrid trout was brought into the United States by the U.S. Fish and Wildlife Service and shipped to a federal hatchery in Iowa and a state hatchery in Minnesota (Courtenay and Hensley 1979a). In 1965, eggs from Yugoslavia were hatched in the United States; young fish were released into several small lakes in northern Minnesota in 1968 and 1969 (Stroud 1969). Ohrid trout have been stocked in Tennessee since 1971 (Etnier and Starnes 1993). The species was first stocked in Colorado in 1969 (Wiltzius 1985).

Status: Introductions into most of these states failed. Repeatedly stocked in Tennessee with no evidence of reproduction yet (Etnier and Starnes 1993). Courtenay and Hensley (1979a) report that even though there has been no reproduction, spawning has been observed in Tennessee.

Impact of Introduction: Unknown.

Remarks: A state record-sized fish was taken from the North Platte River in Wyoming in 1986 (Wyoming Game and Fish, Fish Division 1997).

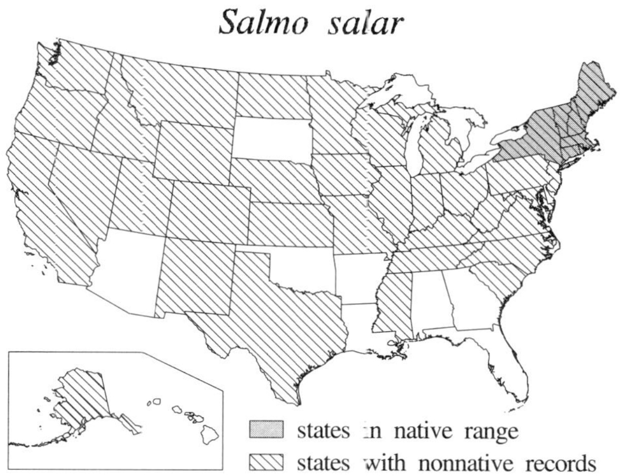

Salmo salar Linnaeus 1758

Common Names: Atlantic salmon.

Identification: Scott and Crossman (1973); Smith (1985); Page and Burr (1991). Maximum size: 140 cm.

Native Range: Atlantic Coast drainages from northern Quebec to Housatonic River, Connecticut (possibly formerly to Delaware); inland to Lake Ontario (where now extirpated). Also eastern Atlantic drainages from the Arctic Circle to Portugal (Page and Burr 1991).

Nonindigenous Occurrences: MacCrimmon and Gots (1979) gave a review of worldwide introductions of this species, including dates of introduction and source of stock. In the United States Atlantic salmon have been introduced into offshore waters of **Alaska** (Wing et al. 1992; Thomson and McKinnell 1994; B. Heard, personal communication); the Lake Tahoe, Truckee, Trinity, Sacramento, San Francisco Bay, Tulare Lake, Mokelumne, Russian, and Klamath drainages in **California** (Smith 1896; Shebley 1917; McAffee 1966e; Dill and Cordone 1997); Twin Lakes near Leadville, **Colorado** (Ellis 1974; Wiltzius 1985); inland areas in **Connecticut** (MacCrimmon and Gots 1979; Whitworth 1996); the Snake River, lakes in the Sawtooth Mountains, Payette Reservoir, and Deadwood Lake, **Idaho** (Linder 1963; Simpson and Wallace 1978; Idaho Fish and Game 1990, 1996); the Mississippi River and Lake Michigan, **Illinois** (MacCrimmon and Gots 1979; Emery 1985; Underhill 1986; Burr 1991); Lake Michigan, **Indiana** (Emery 1985; Underhill 1986; Simon et al. 1992); **Iowa** (Harlan et al. 1987); unspecified areas in **Kansas** (Cross 1967; MacCrimmon and Gots 1979); nonnative areas in **Maine** (Kendall 1914a; Everhart 1976); unspecified areas in **Kentucky** (MacCrimmon and Gots 1979); the Potomac River and inland areas of **Maryland** (Ferguson 1876; Bean 1892b; MacCrimmon and Gots 1979) and **Massachusetts** (Hartel 1992; Cardoza et al. 1993); the Great Lakes surrounding **Michigan** (Parsons 1973; Emery 1985; Underhill 1986; Latta, personal communication); Lake Superior, and lakes in Rice, Hennepin, and Ramsey counties, **Minnesota** (Parsons 1973; Eddy and Underhill 1974; Phillips et al. 1982; Emery 1985; Underhill 1986); unspecified areas in **Mississippi** (MacCrimmon and Gots 1979); unspecified areas in **Missouri** (Ferguson 1876; MacCrimmon and Gots 1979); unspecified areas in **Montana** (MacCrimmon and Gots 1979); the Missouri River in Douglas County and Spring Creek in Cherry County, **Nebraska** (Morris et al. 1974); the Humboldt, Truckee, and Carson rivers and Lake Tahoe, **Nevada** (Smith 1896; Miller and Alcorn 1946; Deacon and Williams 1984); 13 lakes including Squam, Newfound, Winnipesaukee, and Sunapee lakes in **New Hampshire** (Hoover 1936; Scarola 1973); the Passaic, Raritan, and Delaware rivers, and lakes in northern **New Jersey** (Nelson 1890; Bean 1892b; Fowler 1906, 1952; Stiles 1978); unspecified areas **in New Mexico** (MacCrimmon and Gots 1979); Lakes Erie and Ontario, the Hudson River, and numerous inland lakes in **New York** (Bean 1892b, 1903; Fowler 1906; Parsons 1973; Emery 1985; Smith 1985); unspecified areas in **North Carolina** (MacCrimmon and Gots 1979); unspecified areas in **North Dakota** (MacCrimmon and Gots 1979; Idaho Fish and Game 1997); Lake Erie, Sandusky and Maumee rivers in **Ohio** (Jordan 1882; Parsons 1973; Werner 1980; Trautman 1981; Emery 1985); a few landlocked waters of the Deschutes drainage and coastal areas of **Oregon** (Bond 1973, 1994; Li, personal communication); the Susquehanna and upper Delaware drainages and Lake Erie, **Pennsylvania** (Bean 1892b; Ford et al. 1892; Denoncourt et al. 1975a; Hendricks et al. 1979; MacCrimmon and Gots 1979; Cooper 1983; Emery 1985); unspecified areas in **Rhode Island** (MacCrimmon and Gots 1979); unspecified areas in **South Carolina** (MacCrimmon and Gots 1979); unspecified areas in **Tennessee** (MacCrimmon and Gots 1979); unspecified areas in **Texas** (MacCrimmon and Gots 1979); unspecified areas in **Utah** (Sigler and Miller 1963; Sigler and Sigler 1996); unspecified areas in **Vermont** (MacCrimmon and Gots 1979); the Potomac River, **Virginia** (Bean 1892b; MacCrimmon and Gots 1979; Jenkins and Burkhead 1994); Puget Sound and the Chehalis River, **Washington** (Wing et al. 1992; Thomson and McKinnell 1994); unspecified areas in **West Virginia** (MacCrimmon and Gots 1979); Lake Superior and Michigan, **Wisconsin** (Parsons 1973; Eddy and Underhill 1974; MacCrimmon and Gots 1979; Becker 1983; Emery 1985; Underhill 1986); and unspecified areas in **Wyoming** (Baxter and Simon 1970; MacCrimmon and Gots 1979).

Means of Introduction: Attempts were made to restore declining salmon populations in the species' native range. Soon this led to introductions outside its range for sportfishing. In Connecticut, only anadromous salmon were historically recorded. In the late 1800s, both anadromous and landlocked salmon were stocked in the state (Whitworth 1996), and in Massachusetts (Cardoza et al. 1993). Landlocked salmon were first stocked in the Great Lakes in 1874 (Parsons 1973). A total of 743,000 anadromous Atlantic salmon was stocked in the Great Lakes between 1873 and 1947 (Parsons 1973). The Atlantic salmon's popularity has led to unauthorized stocking in some areas, such as the Mississippi River in Illinois, where a specimen was collected in 1978 (Burr 1991). In Alaska and Washington, introduction resulted from escape of cultured fish from aquaculture in British Columbia and Washington. Although most of the releases are a result of small "leakage", occasionally large releases occur. More than 50,000 fish were released in British Columbia in March of 1997 when someone cut the fish pens open (Vovscko 1997). In 1996, 100,000 fish escaped from pens in Cypress Island, Washington (D. Seiler, personal communication). In July 1997, approximately 300,000 fish escaped while pens were being moved in Puget Sound (Associated Press 1997). Atlantic salmon have been cultured in Puget Sound since 1985 (Dodge 1997). Culture of Atlantic salmon is illegal in Alaska (Heard, personal communication). Courtenay (1993) reports that culture pens on the west coast have been ripped apart by pinnipeds, freeing the fish.

Status: Most attempts to establish populations have failed. Atlantic salmon are common and widely dispersed in Alaska, although there is no evidence of reproduction. All captures have been from marine waters. The salmon are found west to the Kodiak peninsula (B. Heard, personal communication). Atlantic salmon have been reported in the commercial fisheries of Washington since 1988 and in Alaska since 1990 (A. J. Thomson, personal communication). Although many hundreds of thousands of Atlantic salmon have escaped from west coast aquaculture facilities there had been no evidence of reproduction (D. Seiler, personal communication). However, in 1998, evidence of reproduction was found in British Columbia (Howard 1998). Most salmon that are collected are emaciated and apparently do not feed well in the wild (Seiler, personal communication).

Impact of Introduction: Infected Atlantic salmon stocked into Puget Sound were apparently responsible for introducing a new disease to the west coast, viral hemorraghic septicemia (VHS). This disease has since been found in two Puget Sound salmon hatcheries (Dentler 1993).

Remarks: Atlantic salmon have been stocked in native areas in Connecticut and Massachusetts to restore extirpated populations there (Cardoza et al. 1993; Whitworth 1996). States where "inland" locations are given, "inland" is assumed because the landlocked form was stocked. Even if the state is in the native range, the species was not native to inland areas in these states (Webster 1942; Cardoza et al. 1993).

Salmo trutta

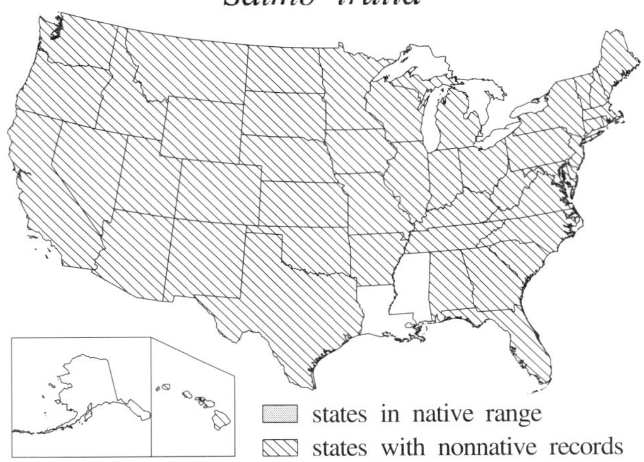

Salmo trutta Linneaus 1758

Common Names: brown trout (von Behr trout, Loch Leven trout, German brown trout).

Identification: Scott and Crossman (1973); Page and Burr (1991); Jenkins and Burkhead (1994). Other names for this species include *S. fario* and *S. lacustris*. Maximum size: 103 cm.

Native Range: Europe, northern Africa, and western Asia (Page and Burr 1991).

Nonindigenous Occurrences: This trout is documented for nearly every state, including **Alabama** (Boschung 1992; Mettee et al. 1996); **Arizona** (MacCrimmon et al. 1970; Lee et al. 1980 et seq.; Tyus et al. 1982); **Arkansas** (MacCrimmon et al. 1970; Buchanan 1973; Cross et al. 1986); **California** (Smith 1896; Shebley 1917; MacCrimmon et al. 1970; Moyle and Daniels 1982; U.S. Fish and Wildlife Service 1993e; Dill and Cordone

1997); **Colorado** (Bean 1903; MacCrimmon et al. 1970; Vanicek et al. 1970; Everhart and Seaman 1971; Tyus et al. 1982; Wiltzius 1985; Propst and Carlson 1986; Walker 1993); **Connecticut** (Webster 1942; Behnke and Wetzel 1960; Whitworth et al. 1968; MacCrimmon et al. 1970; Schmidt 1986; Whitworth 1996); **Delaware** (MacCrimmon and Marshall 1968; MacCrimmon et al. 1970; Lee et al. 1980 et seq.; Raasch and Altemus 1991; Rohde et al. 1994); **Florida** (MacCrimmon and Marshall 1968; MacCrimmon et al. 1970; Courtenay et al. 1984); **Georgia** (MacCrimmon et al. 1970; Dahlberg and Scott 1971a, 1971b; Yerger 1977; Mettee et al. 1996); **Hawaii** (Brock 1960; MacCrimmon and Marshall 1968; MacCrimmon et al. 1970; Maciolek 1984); **Idaho** (Smith 1896; Linder 1963; MacCrimmon et al. 1970; Simpson and Wallace 1978; Idaho Fish and Game 1996); **Illinois** (Smith 1965; MacCrimmon and Marshall 1968; MacCrimmon et al. 1970; Smith 1979; Emery 1985; Underhill 1986; Burr 1991); **Indiana** (MacCrimmon and Marshall 1968; MacCrimmon et al. 1970; Nelson and Gerking 1968; Emery 1985; Underhill 1986; Simon et al. 1992); **Iowa** (MacCrimmon et al. 1970; Lee et al. 1980 et seq.; Burr and Page 1986; Harlan et al. 1987); **Kansas** (MacCrimmon and Marshall 1968; MacCrimmon et al. 1970; Cross et al. 1986; Cross and Collins 1995); **Kentucky** (Clay 1962; MacCrimmon and Marshall 1968; MacCrimmon et al. 1970; Clay 1975; Burr 1980; Burr and Page 1986; Burr and Warren 1986; Hocutt et al. 1986); **Maine** (Kendall 1914a; Everhart 1950; MacCrimmon et al. 1970; Everhart 1976; Lee et al. 1980 et seq.; Schmidt 1986); **Maryland** (Bean 1903; MacCrimmon et al. 1970; Lee et al. 1976; Pritchard et al. 1976; Lee et al. 1981; Rohde et al. 1994); **Massachusetts** (MacCrimmon et al. 1970; Hartel 1992; Cardoza et al. 1993; Hartel et al. 1996); **Michigan** (Bean 1903; MacCrimmon et al. 1970; Emery 1985; Underhill 1986); **Minnesota** (MacCrimmon et al. 1970; Eddy and Underhill 1974; Phillips et al. 1982; Emery 1985; Burr and Page 1986; Underhill 1986); **Missouri** (MacCrimmon and Marshall 1968; MacCrimmon et al. 1970; Pflieger 1971, 1975, 1997; Cross et al. 1986); **Montana** (MacCrimmon et al. 1970; Cross et al. 1986; Holton 1990); **Nebraska** (Bean 1903; Jones 1963; MacCrimmon et al. 1970; Morris et al. 1974; Cross et al. 1986); **Nevada** (Miller and Alcorn 1946; La Rivers 1962; MacCrimmon et al. 1970; Deacon and Williams 1984); **New Hampshire** (Hoover 1936; Bailey and Oliver 1939; MacCrimmon et al. 1970; Scarola 1973); **New Jersey** (Fowler 1952; MacCrimmon et al. 1970); **New Mexico** (Koster 1957; MacCrimmon et al. 1970; Tyus et al. 1982; Sublette et al. 1990); **New York** (Bean 1903; MacCrimmon et al. 1970; Werner 1980; Emery 1985; Smith 1985; Schmidt 1986; Underhill 1986); **North Carolina** (MacCrimmon et al. 1970; Hocutt et al. 1986; Menhinick 1991; Rohde et al. 1994); **North Dakota** (MacCrimmon and Marshall 1968; MacCrimmon et al. 1970; Cross et al. 1986; North Dakota Game and Fish Department 1994, 1997); **Ohio** (MacCrimmon and Marshall 1968; MacCrimmon et al. 1970; Trautman 1981; Emery 1985; Burr and Page 1986; Underhill 1986); **Oklahoma** (Hall 1956; MacCrimmon and Marshall 1968; MacCrimmon et al. 1970; Cross et al. 1986; Pigg, personal communication); **Oregon** (Chapman 1942; MacCrimmon et al. 1970; U.S. Fish and Wildlife Service 1993e; Bond 1994); **Pennsylvania** (Bean 1892b, 1903; MacCrimmon et al. 1970; Hendricks et al. 1979; Cooper 1983; Emery 1985; Hocutt et al. 1986; Underhill 1986); **Rhode Island** (MacCrimmon et al. 1970; Lapin, personal communication); **South Carolina** (MacCrimmon et al. 1970; Dahlberg and Scott 1971a, 1971b; Loyacano 1975; Hocutt et al. 1986; Rohde et al. 1994); **South Dakota** (Bailey and Allum 1962; MacCrimmon et al. 1970; Lee et al. 1980 et seq.; Cross et al. 1986; North Dakota Game and Fish Department 1994); **Tennessee** (Kuhne 1939; MacCrimmon et al. 1970; Starnes and Etnier 1986; Etnier and Starnes 1993); **Texas** (MacCrimmon et al. 1970; Howells 1992a); **Utah** (Sigler and Miller 1963; MacCrimmon et al. 1970; Vanicek et al. 1970; Tyus et al. 1982); **Vermont** (MacCrimmon et al. 1970; Lee et al. 1980 et seq.); **Virginia** (MacCrimmon et al. 1970; Lee et al. 1980 et seq.; Hocutt et al. 1986; Jenkins and Burkhead 1994; Rohde et al. 1994); **Washington** (Chapman 1942; MacCrimmon et al. 1970; Wydoski and Whitney 1979; Washington Department of Fish and Wildlife 1997); **West Virginia** (MacCrimmon et al. 1970; Lee et al. 1980 et seq.; Hocutt et al. 1986; Jenkins and Burkhead 1994; Stauffer et al. 1995); **Wisconsin** (Bean 1903; MacCrimmon et al. 1970; Emery 1985; Burr and Page 1986; Underhill 1986); and **Wyoming** (MacCrimmon et al. 1970; Baxter and Simon 1970; Tyus et al. 1982; Hubert 1994). As of 1970, brown trout had not been introduced into Louisiana, Mississippi, Alabama, and Alaska.

Means of Introduction: The brown trout was first imported to the United States in 1883 from Germany and stocked in the Pere Marquette River, Michigan, by the U.S. Fish Commission (Mather 1889; Courtenay et al. 1984). Since then, the species has been stocked in virtually every states. MacCrimmon et al. (1970) gave dates of first stocking in each state. In most regions the species was first stocked in the late 1800s or early 1900s. Texas and Alabama introductions took place after 1970; MacCrimmon et al. (1970) listed no introductions for those states.

Status: Natural reproduction is low or nonexistent in most state, as such, many states maintain brown trout populations by periodic stocking. Rinne (1995) listed this species as established in Arizona, but the species may not be reproducing in open waters; Rinne apparently used the term 'established' for species that maintain long-term populations through continual or periodic stockings. Courtenay et al. (1984) indicated that introductions failed to establish populations in Florida, Kansas, Ohio, and Oklahoma.

Impact of Introduction: Brown trout have been implicated in reducing native fish populations (especially other salmonids) through predation, displacement, and food competition (Taylor et al. 1984). Many studies have been conducted looking at the effects of brown trout on brook trout including Nyman (1970b), Fausch and White (1981), Waters (1983), Fausch and White (1986), and DeWald and Wilzbach (1992), to name a few. Taylor et al. (1984) list a number of papers citing the effects of brown trout on native fishes. Fausch and White (1981) stated adult brown trout displaced adult native brook trout *Salvelinus fontinalis* from the best habitats in a Michigan stream, and in the northeast in general. Brook trout are also more susceptible to angling and predation. Conversely, juvenile brook trout are dominant over juvenile brown trout of the same size in an artificial stream (Fausch and White 1986). The competitive advantage of the two species may change with size, age, temperature, stream size, or environmental adaptations of different populations (Fausch and White 1986). Although it rarely occurs, the brown trout is one of the few foreign species able to hybridize with natives (e.g., *S. fontinalis*) (Brown 1966; Taylor et al. 1984). In California, competition and predation from brown trout may have contributed to the decline of the Dolly Varden *S. malma* in the McCloud River (Moyle 1976a), and of the golden trout *Oncorhynchus aguabonita* in the Kern River (Krueger and May 1991; Courtenay and Williams 1992). Brown trout may have also depleted the Modoc sucker *Catostomus microps*, an endangered species, in Rush Creek, Modoc County (Moyle and Marciochi 1975). Brown trout have commonly replaced cutthroat trout *O. clarki* in large rivers (Behnke 1992). McAffee (1966c) specifically lists Lahontan cutthroats *O. c. henshawi* as being replaced by brown trout. Introduced brown trout, and other trout species, were likely responsible for the near-extinction of Lahontan cutthroat in Lake Tahoe in the 1940s (McAffee 1966c). Because of their predatory nature, brown trout were introduced into Flaming Gorge Reservoir to reduce populations of the Utah chub *Gila atraria* (Teuscher and Luecke 1996).

Remarks: The state of California has attempted to eradicate brown trout in some areas in order to preserve native golden trout *O. aguabonita* (Taylor et al. 1984; Moyle, personal communication). Tyus et al. (1982) mapped the distribution of the brown trout in the upper Colorado basin. MacCrimmon and Marshall (1968) and MacCrimmon et al. (1970) summarized information on world-wide distribution and introductions.

Salvelinus alpinus

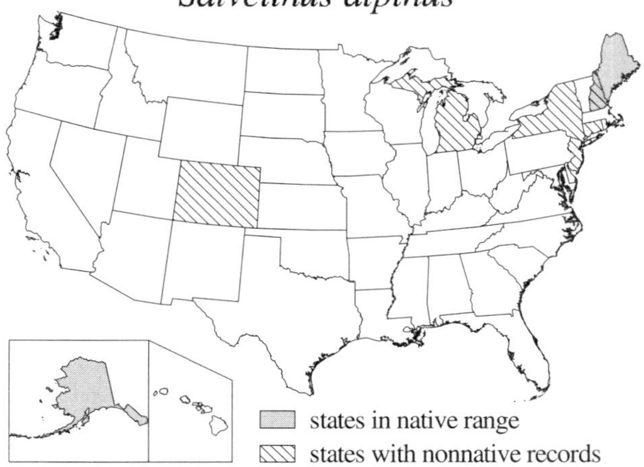

Salvelinus alpinus (Linneaus 1758)

Common Names: Arctic char (formerly included blueback char, blueback trout, Sunapee trout).

Identification: Everhart (1976); Page and Burr (1991); closely resembles the Dolly Varden *S. malma*. Maximum size: 96 cm.

Native Range: Circumpolar; coastal areas in Atlantic, Arctic, and Pacific drainages from Newfoundland to Alaska; south along Atlantic Slope to New England. Also in Eurasia (Page and Burr 1991).

Nonindigenous Occurrences: Arctic char from Alaska were introduced many times since the 1800s into **Connecticut** (Whitworth 1996). The species has also been stocked in Dillon Reservoir in Summit County, and likely many other reservoirs in **Colorado** (Colorado Division of Wildlife 1993; Walker 1993; Horak, personal communication). Arctic char (reported as Swiss lake trout) were stocked into an inland lake in **Michigan** (Emery 1985) and **New Hampshire** (Hoover 1936) in 1890. European saibling (= Arctic char) were stocked in New Hampshire in 1881, 1887, 1888, 1890 (Hoover 1936). They were introduced into Sterling Lake (Somerset County) in **New Jersey** (Fowler 1906, 1952; Stiles 1978) and Sterling Lake (Orange County) (Bean 1903), and the Finger Lakes in **New York** (Emery 1985).

Means of Introduction: Most of these introductions took place a hundred years ago and were intended to enhance sportfishing. Stock for New Jersey (and possibly New York) came from Europe.

Status: Presumably extirpated in New Jersey and New York. Extirpated in Michigan (Emery 1985). Reported in Colorado and Connecticut.

Impact of Introduction: Unknown.

Remarks: None.

Salvelinus aureolus

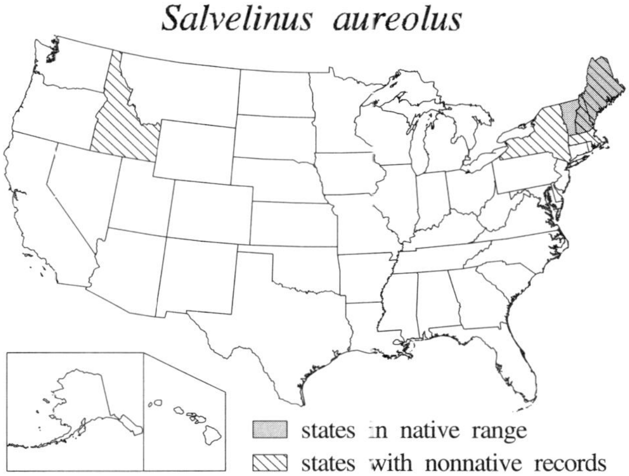

Salvelinus aureolus Bean 1887

Common Names: Sunapee trout, blueback char, blueback trout.

Identification: There has been considerable confusion and disagreement among taxonomists concerning the status of the Sunapee and blueback. For instance, some authors have treated each as distinct species, the blueback as *Salvelinus oquassa* and the Sunapee as *S. aureolus*. Quadri (1974) suggested that the Sunapee and blueback, along with the Quebec red (sometimes listed as *S. marstoni*), are conspecific and should be synonymized as a subspecies of the Arctic char (i.e., *S. alpinus oquassa*). According to Behnke (personal communication) both the Sunapee and blueback trout should be recognized as *S. aureolus oquassa.* Kendall (1914b) provided color plates showing both the Sunapee and blueback trout and gave a table to distinguish between the two forms. He also provided detailed descriptions and morphometrics. Scarola (1973) provided a key for the Sunapee trout. Everhart (1976) gave a photo of blueback trout and a key to both Sunapee and blueback trout. Maximum size: blueback - 36 cm; Sunapee - 58 cm (Kircheis 1976).

Native Range: Sunapee trout were native to Sunapee Lake, New Hampshire; Averill Pond, Vermont; Big Dan Pond, New Hampshire; and Floods Pond, Maine (Behnke, personal communication). Three of the four populations have become extinct and the Sunapee exists only in Floods Pond, Maine, near Bangor (Behnke, personal communication). The Sunapee populations have suffered due to hybridization with introduced lake trout *S. namaycush (*Behnke, personal communication). The blueback trout is native to northwestern Maine in the headwaters of the St. John and Penobscot rivers, specifically in Black Lake, Deboullie Lake, Gardner Lake. Purshineer pond in Arrostook County; Big Reed Pond, Rainbow Lake and Wadleigh Pond in Piscataquis County; Penobscot Lake in Somerset County (erroneously reported as in Piscataquis County); and in Bald Mountain Pond, also in Somerset County (Everhart and Waters 1965). Formerly existed in the Rangeley Lakes but extirpated from there circa 1904 (Kendall 1914b; Everhart and Waters 1965).

Nonindigenous Occurrences: This species was introduced into several lakes at the headwaters of the Salmon River in **Idaho**, including Alice, Vernon, Big Redfish, and Sawtooth lakes (Linder 1963; Deacon et al. 1979; Idaho Fish and Game 1990). It was also stocked in four areas in **Maine**: Upper and Lower South Branch Pond in Piscataquis County, Echo Lake in Hancock County and Coffee Pond in Cumberland County (Maine Department of Inland Fisheries and Game 1974; Kircheis 1976). Each location received 5,000 fish annually from 1969 until at least 1974, except for Echo Lake which began stocking in 1974. This stocking was conducted as part of a management plan to increase the range of the Sunapee trout which has become restricted to one lake (Maine Department of Inland Fisheries and Game 1974). The Sunapee trout was introduced into unspecified area(s) of **Massachusetts** (Fowler 1907). The species was stocked in several lakes **New Hampshire**, including Tewskburry Pond in Grafton and Connor Pond in Ossippee (Scarola 1973; Kircheis 1976). Lake George (Warren County) in **New York** was stocked with trout from Sunapee Lake prior to 1903 (Bean 1903).

The Blueback trout was introduced into the Snake River drainage below Shoshone Falls, **Idaho** (Idaho Fish and Game 1990); unspecified areas of **Maine** shortly after the species discovery in 1874 (Kendall 1914b; Kircheis 1975). Although presumed native, the population at Wadleigh Pond, Maine, might have resulted from these early introductions. An experimental translocation from Wadleigh Pond to Basin Pond in Kennebec County, Fayette Township, Maine, was carried out in 1969 (Kircheis 1975). The blueback trout was also stocked in Newfound, Squam, Connor, and Sunapee lakes, **New Hampshire**, in 1879, and in unknown locations the previous year (Kendall 1914b; Hoover 1936; Kircheis 1975). Stocking in Sunapee Lake took place several years before the Sunapee trout was identified (Kendall 1914b).

Means of Introduction: Most of these introductions took place a hundred years ago and were intended to enhance sportfishing. Blueback stockings in Maine and New Hampshire were mainly as forage for landlocked salmon (Kircheis 1975).

Status: Established in Idaho (in Sawtooth Lake) (Behnke, personal communication) and Massachusetts; unknown in New Hampshire. Early blueback stockings in Maine and New Hampshire failed (with the possible exception of Wadleigh Pond (Kircheis 1975). Experimental stockings of bluebacks in Basin Pond failed (Kircheis 1975).

Impact of Introduction: Unknown.

Remarks: Sunapee trout hybridize with introduced brook trout *S. fontinalis* in Idaho (Behnke, personal communication).

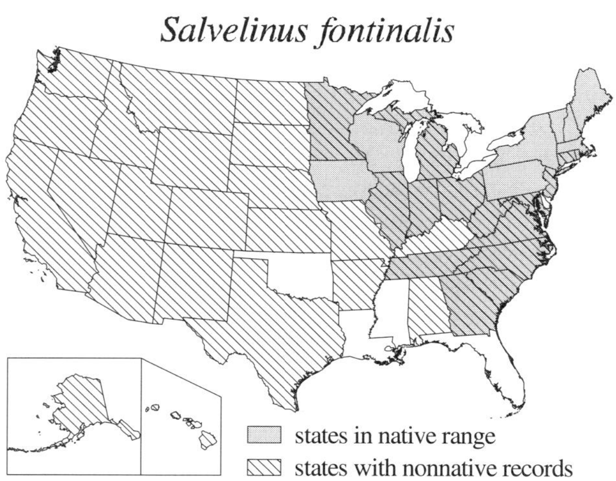

Salvelinus fontinalis (Mitchill 1814)

Common Names: brook trout.

Identification: Scott and Crossman (1973); Becker (1983); Page and Burr (1991); Etnier and Starnes (1993); Jenkins and Burkhead (1994). Maximum size: 70 cm.

Native Range: Most of eastern Canada from Newfoundland to western side of Hudson Bay; south in Atlantic, Great Lakes, and Mississippi River basins to Minnesota and (in Appalachian Mountains) northern Georgia (Page and Burr 1991).

Nonindigenous Occurrences: Since the late 1800s, brook trout have been introduced successfully outside their native range. Introductions have been made into southeastern **Alaska** (Morrow 1980); the White and North Fork rivers, **Arkansas** (Robison and Buchanan 1988); unspecified lakes and streams in **Arizona** (Minckley 1973; Rinne 1995); Lake Tahoe, Truckee, Walker, Central Valley, Klamath, Trinity, Pit, Goose Lake, McCloud, Feather, Yuba, Kern, King, Merced, San Francisco Bay, St. Maria–St. Inez, San Diego, Surprise, Mono Lake, Owens, and Mojave systems in **California** (Smith 1896; Moyle 1976a; Dill and Cordone 1997); South Platte, Arkansas, Rio Grande, Blue, Gunnison, Green, Yampa, and San Juan drainages in **Colorado** (Everhart and Seaman 1971; Wiltzius 1985; Propst and Carlson 1986); Moriority's Pond and West Hill Pond, **Connecticut** (Webster 1942); Brandywine Creek, **Delaware** (Raasch and Altemus 1991); the Savannah, Apalachicola (Chattahoochee), Coosa, Etowah, and Tennessee drainages in **Georgia** (Dahlberg and Scott 1971a, 1971b) (see remarks); Hawaii and Kauai, **Hawaii** (Cobb 1902; Jordan and Evermann 1902, 1905; Brock 1960; Maciolek 1984); the Palouse, Spokane, Pend Orielle, Kootenai, Snake, Bear, and other independent drainages in **Idaho** (Linder 1963; Simpson and Wallace 1978; Idaho Fish and Game 1990, 1996); streams in northern **Illinois** (Smith 1979); streams in the St. Joseph, Eel, Sugar, East Fork White, and Kankakee drainages, **Indiana** (Sweeney 1902; Gerking 1945; Nelson and Gerking 1968); unspecified locations in **Kansas** (Cross 1967); Bad Branch in Letcher County, Shillalah Creek and Martin's Fork in Bell and Harlan counties, and Station Camp Creek in Wolfe County, **Kentucky** (Clay 1975; Burr 1980; Burr and Warren 1986); Deep Creek Lake in the Youghiogheny drainage, and other locations in **Maryland** (Lee et al. 1976; Pritchard et al. 1976; Lee et al. 1981); the northern part of the lower peninsula, the Au Sable River system, Isle Royal in Lake Superior (Hubbs and Lagler 1949), and the southern 3/4 of the lower peninsula of **Michigan** (Lee et al. 1980 et seq.); lakes in Cook, Lake, and St. Louis counties, the Minnesota River, the Vermilion, St. Louis, and Cloquet systems, and the St. Croix drainage in **Minnesota** (Eddy and Underhill 1974; Phillips et al. 1982; Burr and Page

1986); unspecified locations in **Missouri** (Pflieger 1971); the Little Missouri, Missouri, Yellowstone, Belly, St. Mary, Kootenai, Pend Orielle (Clark Fork) drainages in **Montana** (Brown 1971; Cross et al. 1986; Holton 1990); Brown County, a pond north of Omaha, the North Platte drainage, Hat Creek, White River, West Fork Big Blue, and the Elkhorn River in **Nebraska** (Morris et al. 1974; Cross et al. 1986; Bouc 1987); the White, Truckee, Lake Tahoe, Humboldt, Carson, Walker, Pyramid Lake, and Snake drainages in **Nevada** (Miller and Alcorn 1946; La Rivers 1962; Deacon and Williams 1984); nonnative streams in **New Jersey** (Morse 1905; Fowler 1906, 1952); the Rio Grande, San Francisco, upper San Juan, upper Pecos, Tularosa, and Rio San Jose drainages in **New Mexico** (Koster 1957; Tyus et al. 1982; Sublette et al. 1990); the Yadkin, Catawba, Broad, and Savannah drainages in **North Carolina** (Hocutt et al. 1986; Menhinick 1991); the White–Little Missouri drainage in western **North Dakota** (Cross et al. 1986; North Dakota Game and Fish Department 1994); the Sandusky, upper Great Miami (Mad River), and Muskingum drainages in **Ohio** (Trautman 1981; Burr and Page 1986; Hocutt et al. 1986; Underhill 1986); the Rogue, Umatilla, and middle Columbia (near Hood River Station) drainages in **Oregon** (Smith 1896; Chapman 1942; Moyle 1976a; also see Bond 1994); the Savannah drainage in **South Carolina** (Dahlberg and Scott 1971a) (see remarks); the Black Hills in the Cheyenne drainage, and the Minnesota drainage in **South Dakota** (Bailey and Allum 1962; Cross et al. 1986; North Dakota Game and Fish Department 1994; Hanten, personal communication); the Little Tennessee and Hiwassee systems, and the Cumberland drainage in **Tennessee** (Kuhne 1939; Dahlberg and Scott 1971a; Starnes and Etnier 1986; Etnier and Starnes 1993); the Guadalupe River, **Texas** (Howells 1992a); the Wasatch and Uinta mountains, southwestern Utah, and the Bear, upper Green, Strawberry, Utah Lake, and Jordan drainages in **Utah** (Tyus et al. 1982; Sigler and Sigler 1996); the Pee Dee, Clinch, Powell, North and Middle Fork Holston, and possibly the Roanoke drainages in **Virginia** (Hocutt et al. 1986; Jenkins and Burkhead 1994); mountain lakes, lakes in the Puget Sound drainage, and northeastern **Washington** (Smith 1896; Chapman 1942; Wydoski and Whitney 1979; Fletcher, personal communication); and Flaming Gorge Reservoir, the Green River, and streams of major mountain ranges in **Wyoming** (Baxter and Simon 1970; Tyus et al. 1982; Hubert 1994); and the Little Kanawha and probably the Kanawha drainage below the falls in **West Virginia** (Hocutt et al. 1986).

Means of Introduction: This species has been stocked for sportfishing since the 1800s. It was first stocked in Arizona in 1920 (Rinne 1995). It was stocked in Missouri from 1879 to 1914 (Pflieger 1971).

Status: Established locally in many areas.

Impact of Introduction: In Black Hollow Creek, near Fort Collins, Colorado, stocked brook trout completely replaced greenback cutthroat trout *O. clarki stomias* within a period of five years (Behnke 1992, U.S. Fish and Wildlife Service 1995). Brook trout also have replaced Lahontan cutthroat trout *O. c. henshawi* in areas where the cutthroat is native and brook trout have been introduced. Introduced brook trout, and other trout species, were likely responsible for the near-extinction of Lahontan cutthroat in Lake Tahoe in the 1940s (McAffee 1966c). Becker (1983) hypothesized that brook trout may have been responsible for lack of success in establishing grayling populations by stocking because of the brook trout's more aggressive nature. Introduced brook trout inhabit approximately 40% of bull trout *S. confluentus* streams in Montana and may be replacing native bull trout in that state through hybridization. Almost all the resulting hybrids are sterile. Bull trout are now a species of special concern in Montana (Holton 1990). Introduced brook trout are known to replace native golden trout *O. aguabonita* (McAffee 1966b; Moyle 1976a). Introduced predatory fishes, including the brook trout, are likely at least partially responsible for the decline of the Chiricahua leopard frog *Rana chiricahuensis* in southeastern Arizona (Rosen et al. 1995).

Remarks: A popular sport fish, especially with fly fishermen. Tyus et al. (1982) mapped the distribution of brook trout in the upper Colorado basin. Loyacano (1975) lists brook trout in the Santee-Cooper and Savannah drainages in South Carolina but does not indicate that it is introduced. Rohde et al. (1994) do not show it in the Santee-Cooper drainage in South Carolina. Swift et al. (1986) listed it as native to the Savannah (South Carolina/Georgia), Chattahoochee (Georgia), and Coosa (Georgia) drainages. Hocutt et al. (1986) listed it as native but possibly introduced to the Santee drainage (SC). Etnier and Starnes (1993) listed the native status of this species as uncertain in the Conasauga (Coosa) system. Underhill (1986) listed it as native to Isle Royal in Lake Superior (Michigan). Stauffer et al. (1995) listed this species as native to the Kanawha below the falls, and as probably present and native to the Little Kanawha drainage in West Virginia.

Salvelinus malma

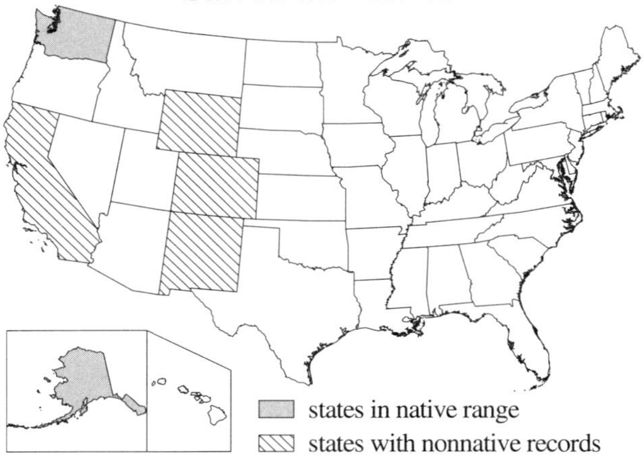

Salvelinus namaycush

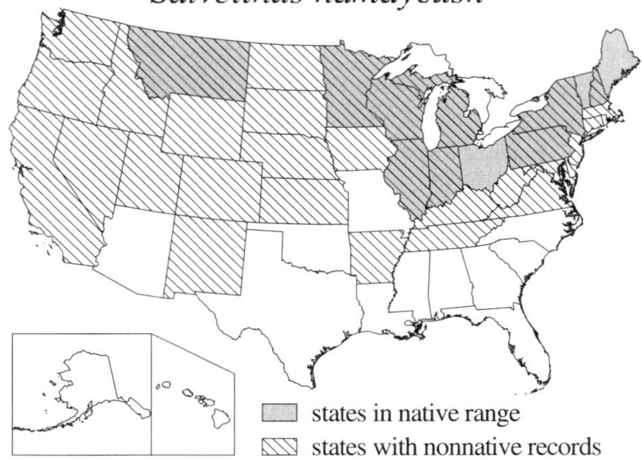

Salvelinus malma (Walbaum 1792)

Common Name: Dolly Varden.

Identification: Scott and Crossman (1973); Wydoski and Whitney (1979); Morrow (1980); Page and Burr (1991). Maximum 63 cm.

Native Range: Arctic and Pacific drainages from Alaska to Puget Sound, Washington. Also in Asia (Page and Burr 1991).

Nonindigenous Occurrences: The Dolly Varden has been introduced into the Central Valley drainage in **California** (Moyle 1976a) and to Fallen Leaf Creek near Lake Tahoe (museum specimen: USNM 75715); both the east and west slope of **Colorado** (Wiltzius 1985; Horak, personal communication); unspecified locations in **New Mexico** (Sublette et al. 1990); and Lake Hattie Reservoir in **Wyoming** (Baxter and Simon 1970).

Means of Introduction: Intentional introductions for sportfishing. Introductions in Colorado took place in the 1890s and in 1968–1970 (Wiltzius 1985). Dolly Varden were introduced once into Wyoming, in 1968 (Baxter and Simon 1970).

Status: Introductions have failed to establish populations.

Impact of Introduction: Unknown.

Remarks: USNM specimen erroneously labeled from Nevada.

Salvelinus namaycush (Walbaum 1792)

Common Names: lake trout (mackinaw, siscowet).

Identification: Scott and Crossman (1973); Becker (1983); Smith (1985); Page and Burr (1991). Maximum size: 126 cm.

Native Range: Widely distributed from northern Canada and Alaska (missing in southern prairie provinces) south to New England and Great Lakes basin (Page and Burr 1991).

Nonindigenous Occurrences: The lake trout has been stocked in Bull Shoals Lake, Greer's Ferry Lake, and the Little Red River below Greer's Ferry Lake in **Arkansas** (Robison and Buchanan 1988); Lake Tahoe, and Fallen Leaf, Stony Ridge, Donner, Crystal, Eagle, and Clear lakes, Golden Gate Park, and several lakes in the Sierra Nevada Mountains in **California** (Smith 1896; Shebley 1917; Cordone and Frantz 1966; Moyle 1976a; Lee et al. 1980 et seq.; Sigler and Sigler 1987; Dill and Cordone 1997); lakes in the Flattop Mountains (Colorado headwaters), and in the Clear, Arkansas headwaters, Rio Grande headwaters, Blue, Frying Pan, and upper Gunnison drainages in **Colorado** (Everhart and Seaman 1971; Lee et al. 1980 et seq.; Tyus et al. 1982; Wiltzius 1985); the Connecticut and Thames drainage, and East Twin Lakes and Wononskopomuc Lake in the Housatonic drainage, **Connecticut** (Hubbs and Lagler 1947; Webster 1942; Behnke and Wetzel 1960; Whitworth et al. 1968; Lee et al. 1980 et seq.; Schmidt 1986); Delaware River and Delaware Bay, **Delaware** (Raasch and Altemus 1991); the Bear, Snake River above and below Shoshone falls, Pend Orielle, Palisades, and North Fork

Payette drainages in **Idaho** (Linder 1963; Simpson and Wallace 1978; Lee et al. 1980 et seq.; Idaho Fish and Game 1996); Rock drainage, **Illinois** (Smith 1979; Burr and Page 1986); a lake in the St. Joseph drainage, **Indiana** (Sweeney 1902); West Lake Okoboji, **Iowa** (Harlan et al. 1987); unspecified areas in **Kansas** (Breukelman 1946; Cross 1967); Dale Hollow Reservoir and Lake Cumberland, **Kentucky** (Burr and Warren 1986); Jennings Randolph Lake in the Potomac drainage (Cincotta, personal communication) and the Gunpowder system in **Maryland** (Ferguson 1876; Pritchard et al. 1976); Wachusett and Quabbin reservoirs in **Massachusetts** (Lee et al. 1980 et seq.; Hartel 1992; Cardoza et al. 1993; Hartel et al. 1996); numerous Great Lake tributaries in **Michigan** (Lee et al. 1980 et seq.); the Minnesota drainage, and lakes in central and southern **Minnesota** in the Mississippi drainage (Eddy and Underhill 1974; Burr and Page 1986); Fort Peck Reservoir, and the Pend Orielle, Missouri headwaters, upper Yellowstone, Bighorn, Marais, and upper Kootenai drainages in **Montana** (Lee et al. 1980 et seq.; Cross et al. 1986; Holton 1990); lakes in the North Platte, upper White, Niobrara, Red Willow Creek in Morrill County, McConaughy Lake and lakes in eastern and northern, **Nebraska** (Jones 1963; Morris et al. 1974; Cross et al. 1986); Lake Tahoe, Truckee River, Cascade Lake, and Walker, Fallen Leaf, and Tallac lakes, **Nevada** (Smith 1896; La Rivers 1962; Cordone and Frantz 1966; Deacon and Williams 1984; Sigler and Sigler 1987); many lakes in **New Hampshire** (Hoover 1936; Scarola 1973; Lee et al. 1980 et seq.); many lakes statewide including those in the Raritan and upper Delaware drainages in **New Jersey** (Fowler 1952; Stiles 1978; Soldwedel, personal communication); upper Canadian, Cimarron, Rio Grande, and Rio Chama drainages in **New Mexico** (Sublette et al. 1990); lakes in Putnam, Sullivan, Westchester, and possibly Jefferson and St. Lawrence counties, **New York** (Smith 1985); unspecified locations in **North Dakota** (North Dakota Game and Fish Department 1994; T. Steinwand, personal communication); Odell, Crescent, and Summit lakes in Klamath County, and Big Cultus Lake in Deschutes County, **Oregon** (Bond 1973, 1994; Lee et al. 1980 et seq.; Li, personal communication); Harvey's Lake in the Susquehanna drainage in **Pennsylvania** (Denoncourt et al. 1975a; Hendricks et al. 1979; Lee et al. 1980 et seq.; Cooper 1983; Hocutt et al. 1986); unspecified locations in **South Dakota** (North Dakota Game and Fish Department 1994; Hanten, personal communication); Dale Hollow Reservoir, **Tennessee** (Burr and Warren 1986; Etnier and Starnes 1993); Flaming Gorge Reservoir, Starvation Reservoir, and Utah, Bear, Provo, St. Mary's, and Fish lakes in **Utah** (Sigler and Miller 1963; Lee et al. 1980 et seq.; Tyus et al. 1982; Sigler and Sigler 1987, 1996); the New and upper Roanoke (Dan) drainages in **Virginia** (Jenkins and Burkhead 1994); lakes in the upper Columbia, Puget Sound, and Yakima drainages in **Washington** (Chapman 1942; Wydoski and Whitney 1979; Washington Department of Fish and Wildlife 1997); Jennings Randolph Lake in the Potomac drainage in **West Virginia** (Cincotta, personal communication); the Fox, Chippewa, Wisconsin, and Rock drainages in **Wisconsin** (Lee et al. 1980 et seq.; Becker 1983); and Flaming Gorge Reservoir, the Green River, Fremont Lake in the New Fork system, and lakes in the Snake headwaters **Wyoming** (Sigler and Miller 1963; Baxter and Simon 1970; Lee et al. 1980 et seq.; Tyus et al. 1982; Hubert 1994; Kaeding et al. 1996).

Means of Introduction: Lake trout have been intentionally stocked as a sport fish. In the Great Lakes, the species was stocked to restore populations within its native range that had been decimated by the sea lamprey. Kaeding et al. (1996) summarized what is known about the illegal introduction into Yellowstone Lake.

Status: Reported as established locally in California, Colorado, Idaho, Massachusetts, North Dakota, South Dakota, Washington, and Wyoming. Extirpated in Connecticut (Whitworth 1996). Scientists have concluded lake trout are established in Yellowstone Lake and are present at such high numbers eradication is probably not possible (Kaeding et al. 1996).

Impact of Introduction: Lake trout often lead to the demise of other trout species where it is introduced. For instance, introductions into Lake Tahoe led to the elimination of the native Lahontan cutthroat *O. c. henshawi* (McAffee 1966c; Moyle 1976a; Behnke, personal communication). Cordone and Frantz (1966) reported a drastic decline and eventual extinction of native cutthroat in Lake Tahoe after the introduction of lake trout. Lahontan cutthroat were abundant in the lake in 1907, with only an occasional lake trout reported. By 1938, the cutthroat has virtually disappeared, and by 1966, lake trout comprised 70% of angler catch in the lake (Cordone and Frantz 1966). Stocked lake trout have replaced native cutthroats in deep Rocky Mountain lakes (Benson et al. 1961). This trout has also virtually eliminated cutthroat and bull trout in Flathead Lake, Montana, and Pend Orielle Lake, Idaho. The recent introduction into Yellowstone Lake poses a similar threat to the native Yellowstone cutthroat trout *O. c. bouvieri*. The introduction of lake trout led to the extirpation of three of the four populations of Sunapee trout (Behnke, personal communication). Lake trout introduced into Flaming Gorge Reservoir were found to prey heavily on the Utah chub *Gila atraria* (Teuscher and Luecke 1996).

Remarks: A recent illegal introduction in Yellowstone Lake, Wyoming, has been of great concern there. Biologists are worried about effects on native Yellowstone cutthroat trout (McCullen 1994; Anonymous 1994a; Kaeding et al. 1996). Although it is found in some surrounding states, the lake trout has not been stocked in Oklahoma (Pigg, personal communication).

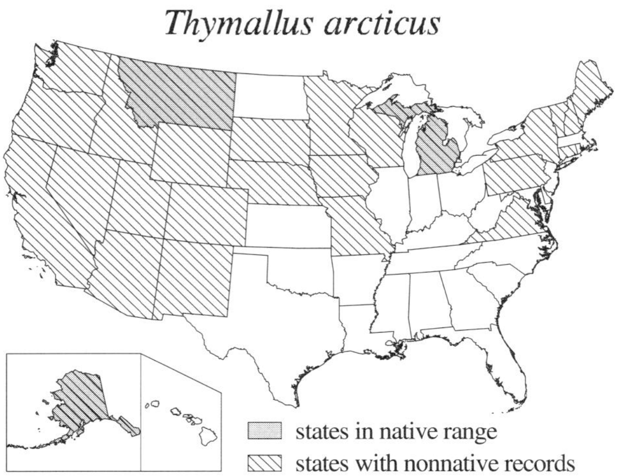

Thymallus arcticus (Pallas 1776)

Common Names: Arctic grayling.

Identification: Scott and Crossman (1973); Morrow (1980); Page and Burr (1991). At one time *T. arcticus* was divided into four separate species: *T. signifer*, *T. montanus*, *T. tricolor*, and *T. ontariensis*. *Thymallus signifer* is now treated as synonymous with *T. arcticus*, and the others are considered subspecies of *T. arcticus* (Lee et al. 1980 et seq.). Maximum size: 76 cm.

Native Range: Widespread in Arctic drainages from Hudson Bay to Alaska and in Arctic and Pacific drainages to central Alberta and British Columbia; upper Missouri River drainage, Montana. Formerly in Great Lakes basin, Michigan. Also in Asia (Page and Burr 1991).

Nonindigenous Occurrences: Grayling have been introduced into **Alaska** (Morrow 1980); the Salt and Verde drainages in **Arizona** (Minckley 1973; Lee et al. 1980 et seq.; Rinne 1995); at least 26 waterbodies in 11 counties, including the lower Klamath, Sacramento (Pit), San Joaquin, and northern Mojave drainages in **California** (Shebley 1917; Moyle 1976a; Dill and Cordone 1997); Zimmerman Lake in the Cache la Poudre system, the South Platte River, the San Luis Valley in the Rio Grande headwaters, the Eagle and Frying Pan rivers on the west slope, and the Arkansas drainage in **Colorado** (Bowers 1901; Everhart and Seaman 1971; Ellis 1974; Morrow 1980; Wiltzius 1985; Zuckerman and Behnke 1986; Walker 1993); Housatonic drainage in **Connecticut** (Whitworth 1996); the Maquoketa and Coon-Yellow systems in **Iowa** (Bowers 1901); the Kootenai, Spokane, and Snake (above and below the falls) drainages, and other independent drainages in **Idaho** (Bowers 1901; Linder 1963; Simpson and Wallace 1978; Lee et al. 1980 et seq.; Idaho Fish and Game 1990, 1996); unspecified locations in **Maine** (Kendall 1914a); Ford Lake, **Michigan** (Hubbs and Lagler 1947); Twin Lake near Isabella in the Rainey headwaters, and several lakes in the Arrowhead region of the Baptism-Brule drainage, including Musquash Lake, **Minnesota** (Bowers 1901; Eddy and Underhill 1974; Phillips et al. 1982); unspecified locations in **Missouri** (Pflieger 1971); the Belly, Red Rock, Madison, Gallatin, upper Missouri-Dearborn, Sun, Clarks Fork Yellowstone, Pend Orielle, and Rock systems in **Montana** (Bowers 1901; Lee et al. 1980 et seq.; Cross et al. 1986; Holton 1990); several unspecified streams in **Nebraska** (Jones 1963; Morris et al. 1974; Cross et al. 1986); high elevation lakes in the northern part of the state, Ruby Valley, Steele Lake in Elko County, **Nevada** (Miller and Alcorn 1946; La Rivers 1962; Deacon and Williams 1984; Sigler and Sigler 1987); Long Pond in Benton (Hoover 1936), and Sunapee Lake (Bickford 1914), **New Hampshire**; the Canjilon area in **New Mexico** (Koster 1957; Sublette et al. 1990); unspecified areas of **New York** (Bickford 1914); upper Deschutes (Lee et al. 1980 et seq.) and Umatilla (Bowers 1901) systems, **Oregon**; the Atlantic basin (Susquehanna and/or Delaware drainage), **Pennsylvania** (Bean 1892b); unspecified areas of **South Dakota** (Johnson 1937); lakes and streams in Duchesne, Summitt, Cache, and Daggett counties, and in the Uintah and Wasatch mountains, streams near Salt Lake City, Blind, Navajo, Lockawaxen, Caroly, Round, Sand, and Blue lakes, and Red Creek, Big Ells and Labaron reservoirs in **Utah** (Sigler and Miller 1963; Lee et al. 1980 et seq.; Sigler and Sigler 1987, 1996); Caspian Lake (Bowers 1901) and other unspecified mountainous areas in **Vermont** (Lee et al. 1980 et seq.; Morrow 1980; Cox, personal communication); below Philpott Reservoir, **Virginia** (Jenkins and Burkhead 1994); one mountain lake in **Washington** (Wydoski and Whitney 1979; Fletcher, personal communication); the Namekagon, Wolf, Brule, and Beartrap-Nemadji systems in **Wisconsin** (Bowers 1901; Becker 1983); and Yellowstone Lake, Bighorn, Wind, and Medicine Bow rivers, Big Sandy Reservoir, Jackson Lake, Babione Creek (upper Tongue), Beartooth Lake (Clarks Fork Yellowstone), and Frye Lake (Popo Agie),

Wyoming (Johnson 1937; Baxter and Simon 1970; Lee et al. 1980 et seq.; Cross et al. 1986; DeLorme Mapping 1992b; Gorges 1994; Hubert 1994).

Means of Introduction: Authorized stocking for sportfishing. First stocked in Arizona in 1943 (Rinne 1995). First stocked in Connecticut in the 1870s (Whitworth 1996). Stocked in Nebraska in 1939 (Jones 1963).

Status: Established in several states, including Arizona (Rinne 1995). Reported from Colorado, Montana, Nevada, Oregon, and Vermont. Failed in Connecticut, Maine, Missouri, Nebraska, New Mexico, and Virginia. Extirpated from Pennsylvania.

Impact of Introduction: Unknown.

Remarks: None.

HYBRIDS

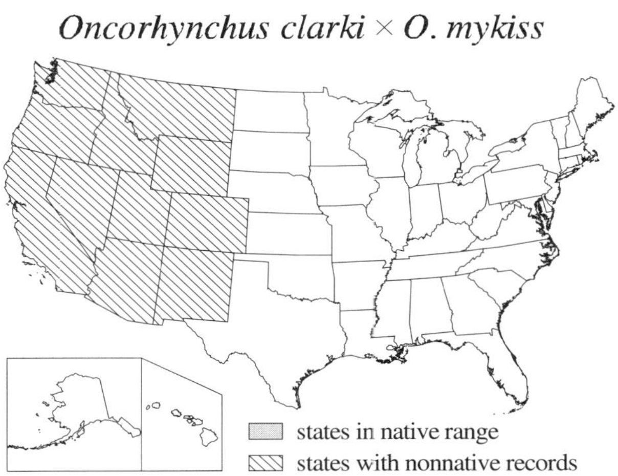

Oncorhynchus clarki × *O. mykiss*

Common Name: cutbow trout.

Identification: Rourke and Wallace (1978); Behnke (1992). Maximum size: 2.46 kg.

Native Range: Not applicable; artificial hybrid. Can occur "naturally" where both species come in contact through stocking.

Nonindigenous Occurrences: According to Behnke (personal communication), these hybrids occur throughout the native range of the cutthroat trout including the states of **Arizona**, **California**, **Colorado**, **Idaho**, **Montana**, **New Mexico**, **Nevada**, **Oregon**, **Utah**, **Washington**, and **Wyoming**. A record-sized fish was caught in Spinney Mt. Reservoir in Hartzel, Park County, **Colorado**, in 1993 (International Game Fish Association 1994). Cutbow trout have been collected in the Arkansas, Colorado, Rio Grande, and Platte river drainages in Colorado (Walker 1993); Henry's Lake and Pend Orielle, **Idaho** (DeLorme Mapping 1992a; Idaho Fish and Game 1997); and in Hobart Creek Reservoir, Spooner Lake, Big Spring Reservoir, Blue Lakes, Squaw Creek Reservoir (DeLorme Mapping 1996b), and Pyramid Lake, **Nevada** (Behnke, personal communication). Miller and Alcorn (1946) reported stockings of this hybrid in Clear Creek and Stoney Lake in Ormsby County, and in the Truckee River, Washoe county, Nevada, in the early 1900s. One fly fishing web page lists them as present in the Rio Grande Gorge in **New Mexico** (Martin 1995). This hybrid has been reported as common in **Utah** (Sigler and Miller 1963).

Means of Introduction: Intentionally stocked as sport fish.

Status: Maintained by stocking either the hybrid or a parent species where the other parent species naturally occurs (usually stocking the rainbow in cutthroat native range).

Impact of Introduction: Native cutthroat are being replaced by introduced rainbow trout through hybridization and competition. Where the two species naturally co-occur, they rarely hybridize (Sigler and Miller 1963; Behnke, personal communication).

Remarks: Recorded from the Southwest as early as 1918 (Sigler and Miller 1963). Private hatcheries sell hybrids between rainbows and Snake River cutthroats (Behnke, personal communication). These two species do not hybridize in areas where both are native (Sigler and Miller 1963). Hybrid trout have not been stocked in Oklahoma (Pigg, personal communication).

Salmo salar × S. trutta

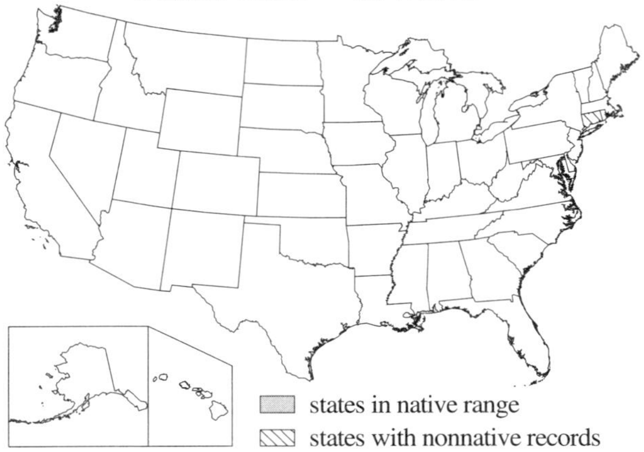

Salmo salar × S. trutta

Common Name: Sambrown.

Identification: Nyman (1970a). Maximum size: Probably similar to parent species: 103–140 cm.

Native Range: These hybrids have been found naturally occurring in Europe and in Canada (Beland et al. 1981), where *S. trutta* has been introduced. This account deals with artificial hybrids.

Nonindigenous Occurrences: This hybrid was stocked in **Connecticut** from 1971 until at least 1979 (MacCrimmon and Gots 1979).

Means of Introduction: Intentionally stocked as sport fish.

Status: Maintained by stocking.

Impact of Introduction: Unknown.

Remarks: None.

Salmo trutta × Salvelinus fontinalis

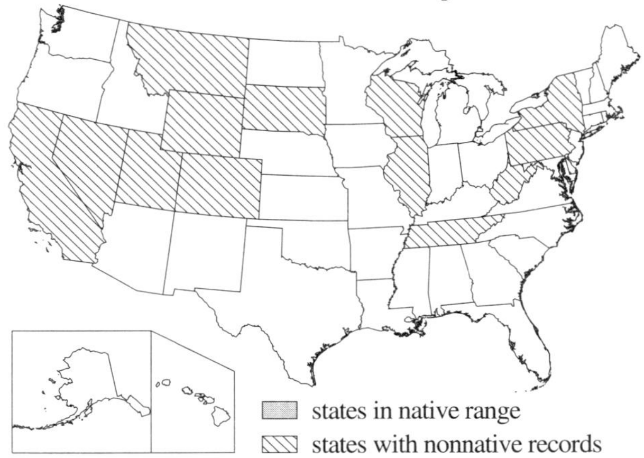

Salmo trutta × Salvelinus fontinalis

Common Name: tiger trout.

Identification: Buss and Wright (1958); Brown (1966); Suzuki (1974); Becker (1983); International Game Fish Association (1994). Maximum size: Probably similar to parent species: 70–103 cm.

Native Range: Mainly an artificial hybrid, but has been known to occur in nature (Brown 1966; Allan 1977; International Game Fish Association 1994). (Even where it occurs "naturally," it is the result of stocking a nonindigenous species, the brown trout).

Nonindigenous Occurrences: The tiger trout has been introduced into, or been recorded from, an unspecified area in **California** (Brown 1966); the Arkansas drainage in **Colorado**; Lake Michigan, **Illinois** (Illinois Department of Natural Resources 1997a); Red Clay Creek in northern New Castle County, **Delaware**, in 1972–1973 (Raasch and Altemus 1991); waters west of the Continental Divide, and throughout much of the Missouri drainage in **Montana** (Brown 1966; Holton 1990); Blue Lakes, Knott Creek Reservoir, Onion Valley Reservoir (Humboldt County), and Ruby Reservoir, **Nevada** (DeLorme Mapping 1996b); the Connetquat River, Oakdale, **New York**, where record-sized fish were caught in 1988 and 1989 (International Game Fish Association 1994); an unspecified area in **Pennsylvania** (Brown 1966); Whitewood Creek in the lower Belle Fourche drainage, **South Dakota** (South Dakota Game, Fish and Parks 1997); unspecified areas of **Tennessee** (Etnier and Starnes 1993); Woodruff Reservoir (Rich County) and Huntington Reservoir (Emery County), **Utah** (DeLorme Mapping 1993b); the Gauley, Greenbrier, New, Twelvepole, Tug, Elk, Tygart Valley, Little Kanawha, and upper Kanawha systems, **West Virginia** (Stauffer et al. 1995); Lake Michigan and numerous creeks, **Wisconsin**, in 1978 (International Game Fish Association 1994; Illinois Department of Natural Resources 1997a; Wisconsin Department of Natural Resources 1997); and Soda Lake in Sublette County, Jakeys Fork in Fremont County, and Sybille Creek in Platte County, **Wyoming** (Baxter and Simon 1970).

Means of Introduction: Some stocked intentionally as sport fish; others occur "naturally" as a result of brown trout stockings in native brook trout range (Holton 1990).

Status: Maintained by stocking. In Montana, some hybridization occurs "naturally" because both species are stocked in the same rivers (Holton 1990).

Impact of Introduction: Unknown.

Remarks: This hybrid is the result of a cross between a female brown trout and a male brook trout. Mortality of eggs and alevin is high because of lack of relatedness between parent species. Survivors are not reproductively viable (International Game Fish Association 1994). This hybrid is more resistant to fungus mortality related to spawning stress than are the parental species (Becker 1983).

Salvelinus fontinalis × *S. namaycush*

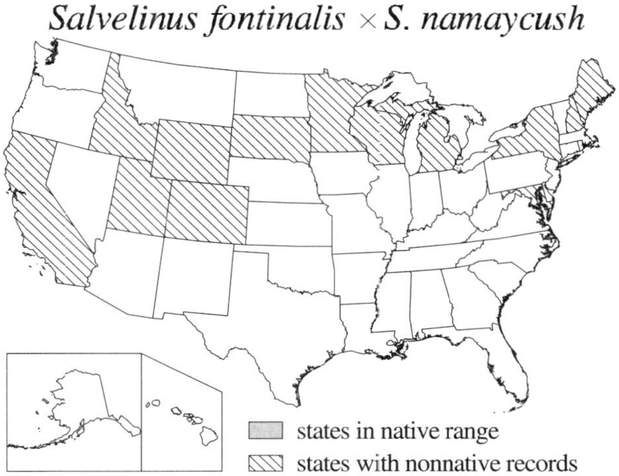

Salvelinus fontinalis × *S. namaycush*

Common Names: splake (brookinaw, wendigo).

Identification: Buss and Wright (1958); Becker (1983); Smith (1985); International Game Fish Association (1994). Maximum size: 9.39 kg.

Native Range: Not applicable; artificial hybrid.

Nonindigenous Occurrences: Splake have been stocked in Lower Salmon and Deer lakes in Sierra County, **California** (Dill and Cordone 1997); several lakes and reservoirs in the Arkansas drainage of **Colorado** (Wiltzius 1985; Colorado Division of Wildlife 1993; Walker 1993; Horak, personal communication); Hayden Lake Outlet, **Idaho** (Idaho Fish and Game 1997); Basin Pond, **Maine** (Internet Waterway, Inc. 1998); Deep Creek Lake (Garrett County) in **Maryland** (Pritchard et al. 1976); Lake Huron, **Michigan** (Becker 1983; Michigan Department of Natural Resources 1997); Pierz Lake (Morrison County) in **Minnesota** (Eddy and Underhill 1974); Crystal Lake in Carroll County and Long Pond in Sullivan County, **New Hampshire** (DeLorme Mapping 1988); unspecified areas of **New York** (Smith 1985); Deerfield Lake in **South Dakota** (North Dakota Game and Fish Department 1994; South Dakota Game, Fish and Parks 1997); Fish Lake, Joe's Valley Reservoir, Seeley Creek, and Blind Lake, **Utah** (DeLorme Mapping 1993b); in at least eight inland lakes in **Wisconsin**, as well as in Lake Superior (Becker 1983); and Brooks, Torrey, and Louis lakes in Fremont County, and Rob Roy Reservoir in Albany County, **Wyoming** (DeLorme Mapping 1992b; Wyoming Game and Fish, Fish Division 1997).

Means of Introduction: Intentionally stocked for sportfishing. According to Krueger and May (1991), splake were first stocked in the 1950s as a replacement for the lake trout which had greatly declined as a result of sea lamprey predation.

Status: Reported in states listed above. Natural spawning of this hybrid has occurred in Lake Huron (Krueger and May 1991).

Impact of Introduction: Introduction of hybrid could affect the genetics of co-occurring parent species.

Remarks: This is a fertile hybrid capable of reproducing and of back-crossing (Krueger and May 1991; International Game Fish Association 1994). Berst et al. (1981) reviewed spawning behavior. In 1969, splake selectively bred for early maturity and retention of swimbladder gas were begun to be stocked in the Great Lakes (Scott and Crossman 1973; Krueger and May 1991). In addition to hybrids, backcrosses between selected splake males and lake trout female are known from portions of the Great Lakes (Krueger and May 1991).

ORDER PERCOPSIFORMES

Percopsidae - Trout-perches

The family Percopsidae comprises a single genus with two extant species. It is distributed in northern North America, primarily from Alaska to Quebec and southward to Missouri and Kentucky (Nelson 1994). Trout-perches are small fish, less than about 20 cm, and are characterized by a large scale-less head, both cycloid and ctenoid scales, subthoracic pelvic fins, a single large dorsal fin and an adipose fine, and spines in the dorsal, anal, and pelvic fins (Page and Burr 1991; Jenkins and Burkhead 1994; Nelson 1994). They are restricted to freshwater, but the group may have had a marine origin.

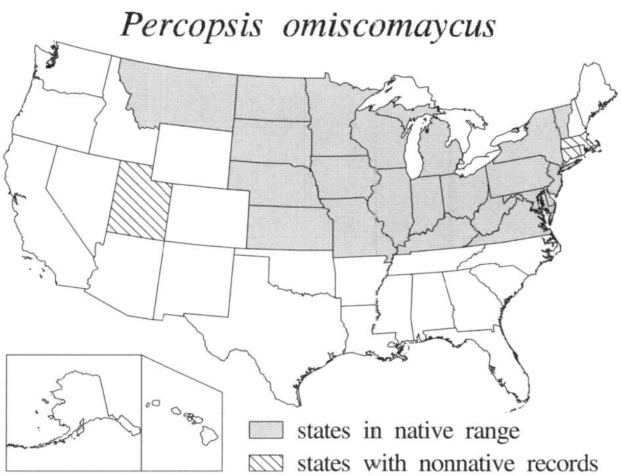

Percopsis omiscomaycus (Walbaum 1792)

Common Name: trout-perch.

Identification: Trautman (1981); Smith (1985); Page and Burr (1991); Jenkins and Burkhead (1994). Maximum size: 20 cm.

Native Range: Atlantic and Arctic basins throughout most of Canada from Quebec to Yukon and British Columbia, and south to Potomac River drainage, Virginia; Yukon River drainage, Yukon and Alaska; Great Lakes and Mississippi River basins south to West Virginia, eastern Kentucky, southern Illinois, central Missouri, North Dakota, and northern Montana (Page and Burr 1991).

Nonindigenous Occurrences: Specimens were taken from the Housatonic River drainage, **Connecticut**, in 1879, and from the Housatonic drainage in **Massachusetts** between 1940 and 1942 (Whitworth 1996; but see Remarks section). It was introduced into Willard Bay Reservoir and Utah Lake, **Utah**, in 1983 (Sigler and Sigler 1987).

Means of Introduction: Unknown.

Status: Reported from Connecticut and Massachusetts. Possibly established in Utah.

Impact of Introduction: Unknown.

Remarks: There exist conflicting views, and some confusion, in the literature concerning the native versus nonnative distribution of this species in the eastern United States. Lee et al. (1981) stated that early records of this species in the Potomac and Susquehanna river drainages may represent introduced fish from the Ohio River drainage that were transplanted by man during the late 19th and early 20th centuries. However, Jenkins and Burkhead argued that the species was native to the two drainages. They believed introduction to be unlikely because most early stockings were of game fishes, and government agencies in charge (of stocking fish) were infant when the trout-perch was first reported. The species apparently was common in some parts of Chesapeake and in the Ohio Canal in the early 1900s (Radcliffe and Welsh 1916). Jenkins and Burkhead (1994) suggested that early stockings of predatory sportfishes may have contributed to the decline or disappearance of *Percopsis omiscomaycus* in the Potomac and Susquehanna drainages. As an aside, it should be noted that Jenkins and Burkhead (1994) apparently cited Lee et al. (1976) by mistake (most likely intending to cite Lee et al. 1981). Other researchers also apparently agree with the conclusion that it is native. In fact, Hocutt et al. (1986) listed the trout-perch as native to the Susquehanna and extirpated from the Potomac. Similarly, Page and Burr (1996) noted that the species ranged south to the Potomac River drainage without discussing the possibility of introduction. On the other hand, the species may have been introduced to another Atlantic drainage. Whitworth (1996) included trout-perch in his table of species that were probably transported to the fresh waters of Connecticut by humans (and that do not have at least one recently verified established population). He noted that the only report from Connecticut was a specimen taken from the Housatonic River drainage in 1879. However, Schmidt (1986) listed it as native to the Housatonic and Hartel et al. (1996) also apparently considered this species to be native to the drainage. Based on Sigler and Sigler (1987), the trout-perch was introduced into two water bodies in Utah. However, there is some uncertainty concerning the status of status of the species in that state. In their summary list of Utah fishes, Sigler and Sigler (1996) included trout-perch in such a way as to suggest it was introduced and still present; however, these authors did not include additional details, nor did they include trout-perch in a species account or in their fish identification keys. To date, we have been unable to find additional data documenting the continued presence of this species in Utah. In fact, we recently contacted several fishery biologists working in Utah and they were unaware of trout-perch having ever been present in the state.

Voucher specimens: Connecticut (MCZ).

Aphredoderidae - Pirate perches

The family Aphredoderidae consists of a single living species, the pirate perch (Nelson 1994). The pirate perch is distributed in the eastern United States, primarily in the lowlands of Atlantic drainages from Long Island southward, the Gulf of Mexico Slope, the Mississippi Valley, and part of the Great Lakes drainage (Nelson 1994). The family is characterized by having a large mouth, ctenoid scales, a single dorsal fin with both spines and rays, thoracic pelvic fins, and the anus and urogenital openings between the gill membranes. The anus and urogenital openings are positioned posteriorly in juveniles, and migrate forward to the throat during development. The diet consists mainly of small invertebrates, particularly aquatic insects, and small fishes (Becker 1983). The pirate perch is restricted to freshwater.

Aphredoderus sayanus

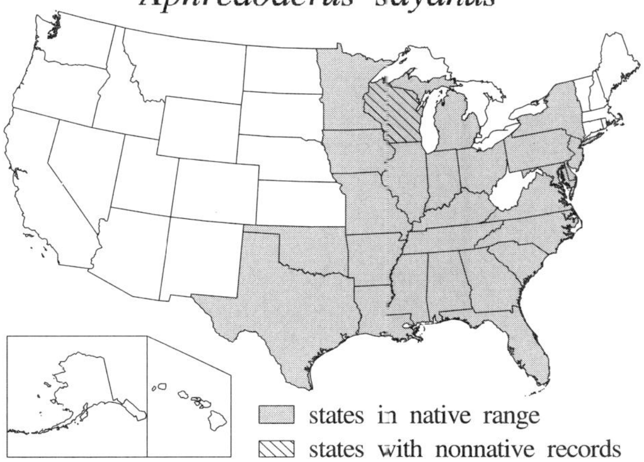

Aphredoderus sayanus (Gilliams 1824)

Common Name: pirate perch.

Identification: Becker (1983); Smith (1985); Page and Burr (1991); Etnier and Starnes (1993); Jenkins and Burkhead (1994); Pflieger (1997). Two subspecies: *Aphredoderus s. sayanus* and *A. s. gibbosus* (Smith 1985; Boltz and Stauffer 1993; Etnier and Starnes 1993). Maximum size: 14 cm.

Native Range: Atlantic and Gulf Slopes from Long Island, New York, to Colorado river drainage, Texas; Great Lakes and Mississippi River basins from Michigan, Wisconsin, and southern Minnesota to the Gulf. Isolated populations in Lakes Ontario and Erie drainages, New York (Page and Burr 1991). *Aphredoderus s. saynus* is native to the Atlantic Coast drainages; *A. s. gibbosus* is native to the Mississippi Valley; and eastern Gulf Coast, Florida and southern Georgia populations are considered to be intergrades (Boltz and Stauffer 1993).

Nonindigenous Occurrences: This species is known from the Embarrass River (Marathon County) in the Lake Michigan basin, and from the upper Wisconsin River system (Oneida County) in **Wisconsin** (Becker 1983).

Means of Introduction: Unknown; possible bait bucket introduction.

Status: Reported in Wisconsin outside what is considered its native range.

Impact of Introduction: Unknown.

Remarks: Becker (1983) noted that populations are widely separated and stated that it seems probable that the Wisconsin records from the Embarrass River and the upper Wisconsin system, resulted from introductions. He did not suggest a possible means of introduction.

Amblyopsidae - Cavefishes

The family Amblyopsidae, whose members are known as cavefishes, is composed of four or five genera, and six species (Page and Burr 1991; Nelson 1994). The family is restricted to North America, distributed in the southern and eastern sections of the United States. Most of these fishes inhabit caves, or springs and swamps associated with caves, in limestone formations. They are small fishes, 6 to 11 cm long, and have small or rudimentary eyes. Four of the six species are white and blind. The family is restricted to freshwater.

Forbesichthys agassizi

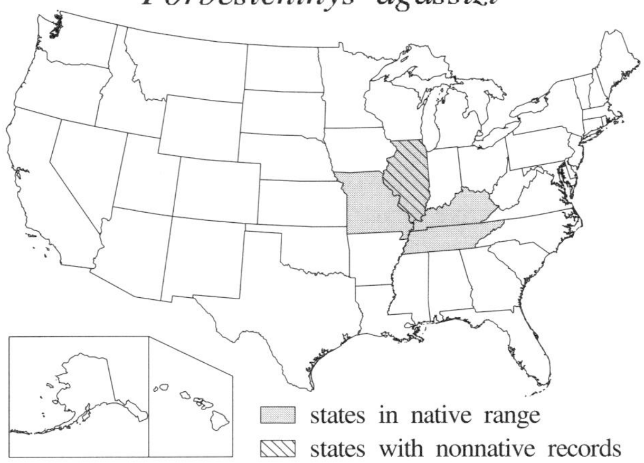

Forbesichthys agassizi (Putnam 1872)

Common Name: spring cavefish.

Identification: Pflieger (1975); Page and Burr (1991); Etnier and Starnes (1993). Another commonly used name is *Chologaster agassizi* (Nelson 1984; Robins et al. 1991a). The names *Forbesella agassizi* and *Chologaster papilliferus* have also been used in the literature (Weise 1957; Lee et al. 1980 et seq.; Page and Burr 1991). *Forbesichthys agassizi* is one of two species in this family with functional eyes (Nelson 1984). Maximum size: 8.4 cm.

Native Range: Springs and caves from central and western Kentucky (west to the Tennessee River) to south central Tennessee; west across southern Illinois to southeastern Missouri (Page and Burr 1991).

Nonindigenous Occurrences: In March 1954, 36 fish were introduced into a cave spring in Adams County, **Illinois** (Weise 1957; Lee et al. 1980 et seq.).

Means of Introduction: Intentionally stocked from southern Illinois sites to establish a population near Quincy College. This population was intended to serve as a nearby source of fish for research (Weise 1957).

Status: The population apparently failed to reproduce and no longer exists. Weise (1957) suggested that the cavefish were swept downstream and subsequently taken by predators.

Impact of Introduction: Unknown.

Remarks: This species is listed as state endangered in Missouri.

There are no known voucher specimens.

ORDER GADIFORMES

Gadidae - Cods

The family Gadidae, including cod, haddock, and cuskfishes, is composed of 15 genera and about 30 species (Nelson 1994). The family is found in the Northern Hemisphere, including the Arctic, Atlantic, and Pacific oceans (Berra 1981; Nelson 1994). Cod are commercially important fishes, and some species reach a size of one meter or more. All members of this family are marine except for the burbot *Lota lota*, a freshwater species found in the northern parts of Eurasia and North America (Page and Burr 1991; Nelson 1994). A second species, the Atlantic tomcod *Microgadus tomcod* frequently enters Atlantic coastal drainages of North America and is known to populate and spawn in freshwater (Page and Burr 1991; Nelson 1994).

Lota lota

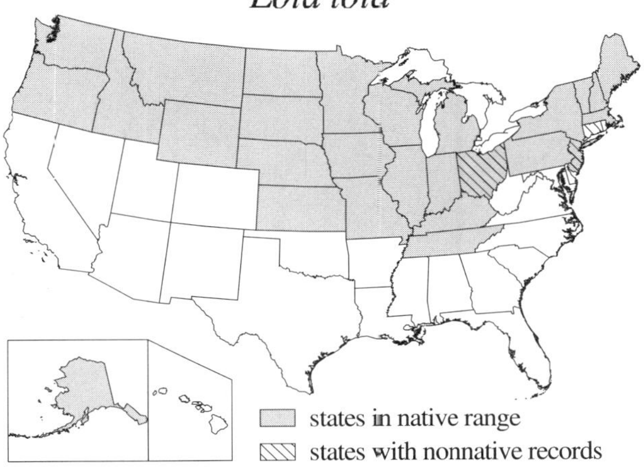

Lota lota (Linnaeus 1758)

Common Names: burbot (ling, freshwater codfish, eelpout).

Identification: Scott and Crossman (1973); Becker (1983); Smith (1985); Page and Burr (1991). Maximum size: 84 cm.

Native Range: Throughout Canada, Alaska, and northern United States (south to Pennsylvania, Kentucky, Missouri, Wyoming, and Oregon). Also in northern Eurasia (Page and Burr 1991). Etnier and Starnes (1993) reported this species from Tennessee.

Nonindigenous Occurrences: Whitworth (1996) refers to burbots as introduced in **Connecticut** and reported introductions as early as 1844. Goode (1884) reported the burbot from the Connecticut River; however, Webster (1942) did not list the species in the state. Collections made in recent years have been from only two drainages. He noted that the only viable population is in the Hollenbeck River of the Housatonic drainage. Lee et al. (1980 et seq.) showed collections both here and in the Connecticut River, which is presumably the second drainage to which Whitworth was referring. Burbots have been stocked and are established in **New Jersey**; they were first listed for the state in 1920, although the date of initial introduction is not known (Fowler 1920, 1952; Stiles 1978). Burbots were also introduced into **Ohio** outside their native range, including Buckeye Lake, stocked during the period 1915 to 1930 with fish caught in Lake Erie, and at least one pay-fishing lake in Tuscarawas County, apparently stocked during the 1960s (Trautman 1981). Specimens were collected from several sites in Ohio during the 1960s, including the Little Miami River in Hamilton County, the Stillwater River in Miami County, and the Great Miami River; these fish may have been introduced into or possibly represent strays from the upper Mississippi River (Trautman 1981). Hocutt et al. (1986) reported this species as introduced into the Muskingum drainage in Ohio, and possibly introduced into the Susquehanna drainage. Burr and Page (1986) reported it as possibly introduced into the Ohio River proper, the Little Miami, Licking, Great Miami, Kentucky, and Wabash drainages in Ohio, Kentucky, Illinois and Indiana.

Means of Introduction: Intentionally stocked for sportfishing.

Status: Established in Connecticut, New Jersey and Ohio. Extirpated in Pennsylvania.

Impact of Introduction: Unknown.

Remarks: Goode (1884) reported a single collection from the Susquehanna River in Pennsylvania. Cooper (1983) and Smith (1985) mention relict populations in the upper Susquehanna in New York. The Pennsylvania record may be a native.

ORDER MUGILIFORMES

Mugilidae - Mullets

The family Mugilidae, also known as mullets, consists of 17 genera with between 66 and 80 species (Nelson 1994). Mullets are found worldwide in tropical and temperate seas; a few spend their lives in freshwater (Robins and Ray 1986; Nelson 1994). They are small to fairly large fishes characterized by a torpedo-shaped body that is rounded in cross-section near the front, and compressed toward the rear. They have widely separated spiny and rayed dorsal fins, ventral pelvic fins located much farther back than the pectoral fins (which are high on the side), an exceedingly long intestine; large- to medium-sized cycloid scales (usually), fatty tissue covering part of the eye, and no lateral line (Eschmeyer et al. 1983; Robins and Ray 1986; Nelson 1994). Most species are green, blue, or olive on the back and silvery on the side and belly (Eschmeyer et al. 1983). Adults travel in schools (Robins and Ray 1986).

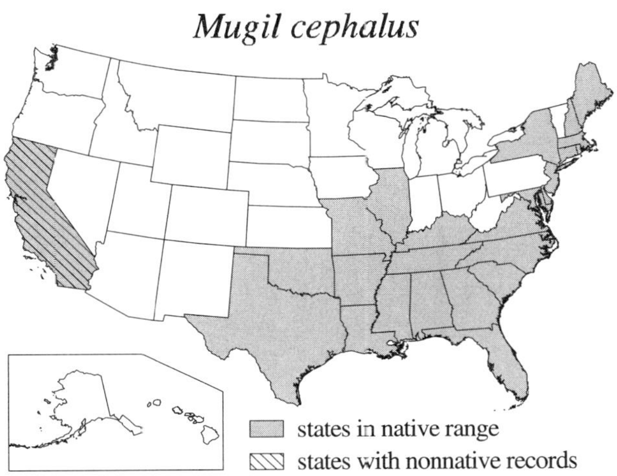

Mugil cephalus Linnaeus 1758

Common Name: striped mullet.

Identification: Miller and Lea (1972); Eschmeyer et al. (1983); Robins and Ray (1986); Etnier and Starnes (1993); Fischer et al. (1995b). Maximum size: 91 cm.

Native Range: Marine ascending into freshwaters. Atlantic and Pacific coasts. Native range generalized as world-wide in coastal waters from 42°N to 42°S (Collins 1985 in Gilbert 1993). Gilbert (1993) gave an update on the distribution of the striped mullet in the Atlantic and eastern Pacific oceans. San Francisco Bay to Chile and Galapagos Islands; rare north of southern California (Eschmeyer et al. 1983; Gilbert 1993) in the Pacific. Nova Scotia to Tampico, Mexico, but absent from Bahamas and most of the West Indies and Caribbean in the western Atlantic (Gilbert 1993). Throughout the Mediterranean and Black seas, the islands of Madeira and the Azores, and the west coast of Africa to Ghana or possibly Togo, in the eastern Pacific. Southwestern Indian Ocean (Gilbert 1993). Previous reports from the Atlantic side of the Cape of Good Hope in Africa, the West Indies, and the Atlantic coasts of Central and South America are erroneous (Gilbert 1993). Etnier and Starnes (1993) mapped inland range in the United States including Mississippi River as far north as Illinois/Kentucky, and Kentucky Reservoir, Tennessee; also the Red River in Texas/Oklahoma.

Nonindigenous Occurrences: Introduced into Salton Sea, **California**, in 1950 (Walker et al. 1961; Dill and Cordone 1997).

Means of Introduction: Intentionally stocked. Fifteen fish from San Felipe, Mexico, were stocked (Walker et al. 1961).

Status: The introduction failed to produce a viable population and the species is extirpated in the Salton Sea.

Impact of Introduction: Unknown.

Remarks: None.

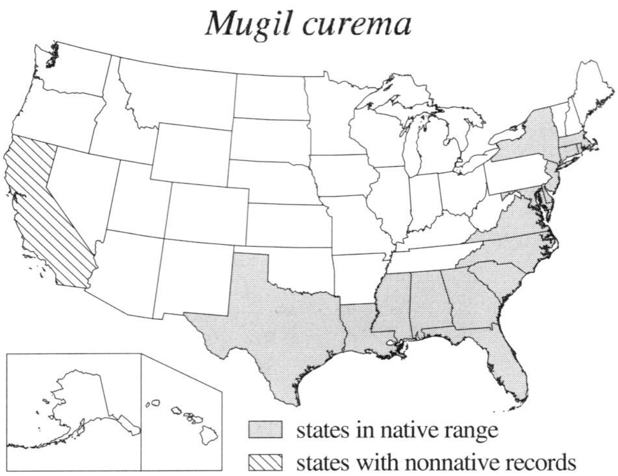

Mugil curema Valenciennes 1836

Common Name: white mullet.

Identification: Robins and Ray (1986); Fischer et al. (1995b). Maximum size: 20 cm.

Native Range: Marine, ascending into freshwaters. Atlantic and Pacific coasts. Massachusetts, Bermuda, and northern Gulf of Mexico to southeastern Brazil in the Atlantic; also in the West Indies, the eastern Atlantic and the eastern Pacific (Ramirez-Hernández and Gonzales Pages 1976; Robins and Ray 1986). Gulf of California to Chile in the Pacific (Ramirez-Hernández and Gonzales Pages 1976; Fischer et al. 1995b).

Nonindigenous Occurrences: Introduced into Salton Sea, **California**, in 1950 and 1951 (Walker et al. 1961; Dill and Cordone 1997).

Means of Introduction: Intentionally stocked. A total of 90 fish was stocked in 1950–1951(Walker et al. 1961).

Status: The introduction failed to produce a viable population and the species is extirpated in the Salton Sea.

Impact of Introduction: Unknown.

Remarks: None.

ORDER ATHERINIFORMES

Melanotaeniidae - Rainbowfishes

This family includes 6 genera and about 53 species restricted to tropical and subtropical regions of eastern and northern Australia, New Guinea, and a few nearby islands (Allen 1991; Nelson 1994). Rainbowfishes are small, usually less than 12 cm in length, and are characterized by their compressed bodies, somewhat pointed heads, flattened snouts and foreheads, two dorsal fins set close together, and large scales (Merrick and Schmida 1984; Allen 1991; Nelson 1994). The lateral line is absent or weakly developed and the mouth is small, usually terminal. They are typically schooling fishes that feed on a variety of plants and small invertebrates (Allen 1991). Most species reach maturity within the first year, and most also exhibit some sexual dimorphism (Allen 1991). Rainbowfishes have been used as mosquito control agents (Berra 1981), and several species are common in the ornamental fish trade. Most rainbowfishes inhabit freshwaters, although they are sometimes found in brackish habitats (Nelson 1994).

Melanotaenia nigrans

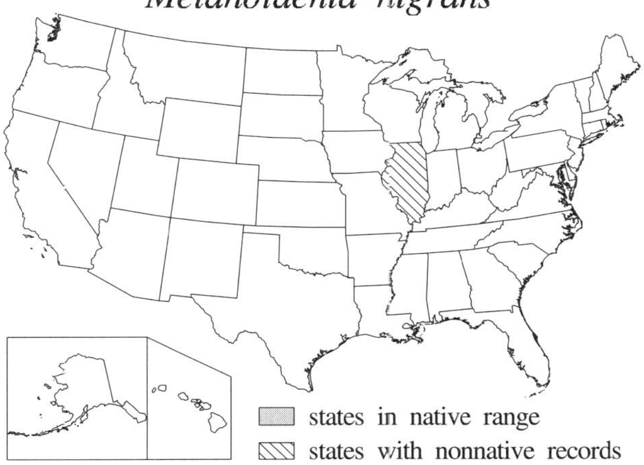

Melanotaenia nigrans (Richardson 1843)

Common Names: Black-banded rainbowfish (Australian rainbowfish, dark Australian rainbowfish, rainbowfish, Australian red-tailed rainbowfish).

Identification: Allen (1989); Allen and Cross (1982) provided a summary of most of the known species, including *Melanotaenia nigrans*. Distinguishing characteristics were also presented by Sterba (1973) and Merrick and Schmida (1984). Color photographs appeared in Merrick and Schmida (1984) and in Axelrod et al. (1985). Maximum size: 7 cm.

Native Range: Tropical and subtropical Australia. Discontinuous distribution in northeastern drainages of the Timor Sea, Gulf of Carpentaria, and Northeast Coast drainage divisions (Merrick and Schmida 1984).

Nonindigenous Occurrences: Three specimens were taken with a small seine from the edge of a sandbar of the Mississippi River in Randolph County, **Illinois**, in July 1930 (O'Donnell 1935; Smith 1965).

Means of Introduction: It was supposed that these fish were escapes from a tropical fish establishment (store?) in the St. Louis area, about 40 miles upstream (O'Donnell 1935; Smith 1965). However, an aquarium release cannot be ruled out.

Status: Reported from Illinois.

Impact of Introduction: Unknown.

Remarks: The record of O'Donnell (1935) represents one of the earliest accounts of an introduced ornamental fish taken in United States open waters. Specimens were identified by Carl Hubbs (O'Donnell 1935). There are no known voucher specimens.

Atherinidae - Silversides

The family Atherinidae contains about 25 genera and about 165 species (Nelson 1994). The group has a worldwide distribution (Berra 1981; Nelson 1994). Only three species occur naturally in freshwaters of the United States (Page and Burr 1991). Atherinids are small fishes, ranging in length from 5 to 70 cm, but most are less than 15 cm in length (Berra 1981). They are characterized by a silvery, elongate, compressed body with two widely separated dorsal fins, large eyes, and a long snout with a terminal mouth. Atherinids are schooling fishes and important as prey species. As such, they are occasionally stocked as forage fishes or used as bait (Robins and Ray 1986, Robins et al. 1991b). Most members of the family are marine; however, about 50 species are confined to freshwater (Nelson 1994), and others inhabit brackish waters or enter coastal rivers (Berra 1981).

Atherinops regius

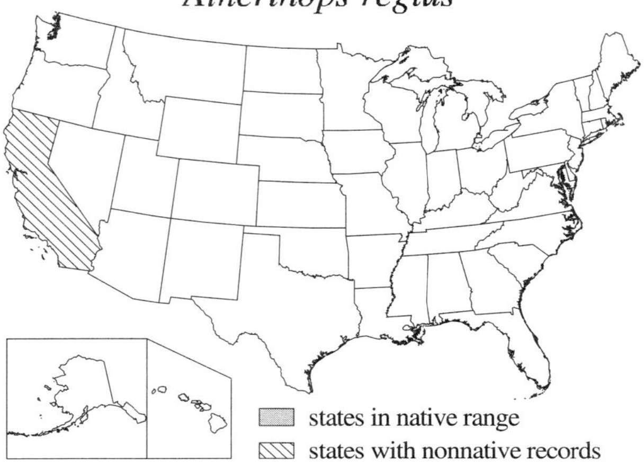

Atherinops regius Jenkins and Evermann 1889

Common Name: (silverside, false grunion).

Identification: Fischer et al. (1995a) Reported as *Colpichthys regis* by Walker (1961) and Dill and Cordone (1997). The name has been changed to *Atherinops regius* (W. Eschmeyer, personal communication). Another spelling is *Atherinops regis*.

Native Range: Marine; Gulf of California (Fischer et al. 1995a).

Nonindigenous Occurrences: Introduced into Salton Sea, **California**, in 1950 and 1951 (Walker et al. 1961; Dill and Cordone 1997).

Means of Introduction: Intentionally stocked as forage on five different occasions. Stock was obtained from San Felipe, Mexico (Walker et al. 1961).

Status: The introduction failed to produced a viable population and the species is extirpated from the Salton Sea.

Impact of Introduction: Unknown.

Remarks: None.

Chirostoma jordani

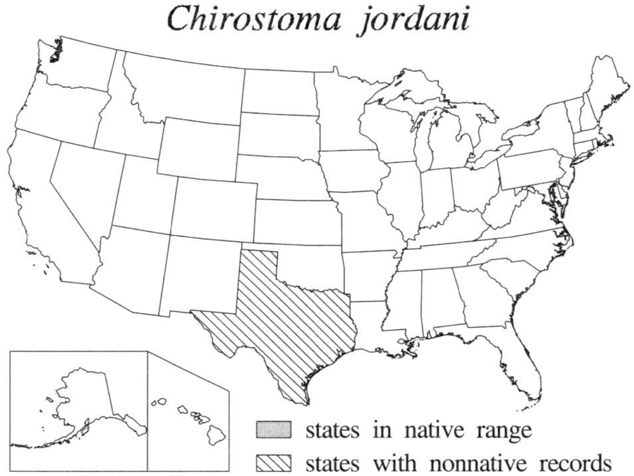

Chirostoma jordani Woolman 1894

Common Names: mesa silverside (charal, charal comun, Mexican silverside, charale).

Identification: Barbour (1973) provided a key to members of the genus. Maximum size: 9.1 cm SL.

Native Range: Central Mexico, Lago de Chapala and the Valley of Mexico (Barbour 1973; Contreras and Escalante 1984).

Nonindigenous Occurrences: This species, or a closely related form, was stocked into Amistad and Falcon reservoirs on the Rio Grande River, on the Mexico-**Texas** border, before June 1975 (Courtenay and Hensley 1979a; Contreras and Escalante 1984; Robins et al. 1991b; Howells 1992a).

Means of Introduction: Intentionally stocked by the Mexican federal government into two Rio Grande reservoirs, apparently with the hope of establishing a population that would serve as a source of food for local human residents (Courtenay and Hensley 1979a; Contreras and Escalante 1984). Introductions were made in the form of millions of fertilized eggs taken from Lake Chapala, Mexico (Courtenay and Hensley 1979a; C. Hubbs, personal communication to Courtenay; Courtenay, personal communication).

Status: The species failed to become established (Courtenay and Hensley 1979a; Contreras and Escalante 1984).

Impact of Introduction: Unknown (Contreras and Escalante 1984).

Remarks: There is some confusion concerning the Texas *Chirostoma* record and the species' true identity. The original information on the stocking apparently came from a Mexican news report that mentioned the release of 18 million fertilized *Chirostoma* eggs from Lake Chapala into Amistad and other Mexican reservoirs (Hubbs, personal communication to Courtenay, 1975). Lake Chapala is inhabited by several *Chirostoma* species. Consequently, *Chirostoma* eggs taken from the lake and used for stocking may have been from any one, or several species (Courtenay and Hensley 1979a; Courtenay, personal communication). The statement by Howells (1991b) that the *C. jordani* record was "probably based on collections by C. Hubbs" is false. There are no known voucher specimens from Lake Amistad and no known collections of *Chirostoma* in Texas waters (D. Hendrickson, personal communication; Hubbs to Hendrickson, personal communication). The Mexican stocking has been the basis for inclusion of *Chirostoma jordani* in several published lists of nonestablished foreign species of the United States (e.g., Courtenay et al. 1984, 1986).

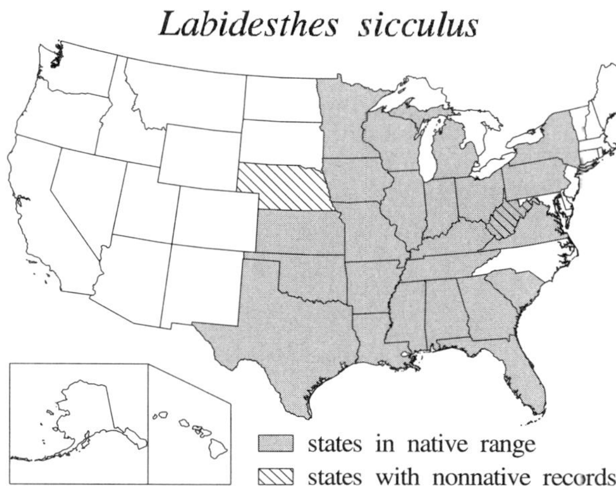

Labidesthes sicculus (Cope 1865)

Common Name: brook silverside.

Identification: Becker (1983); Page and Burr (1991); Etnier and Starnes (1993); Jenkins and Burkhead (1994). Maximum size: 13 cm.

Native Range: Freshwaters of eastern North America including the St. Lawrence–Great Lakes (except Lake Superior); the Mississippi River basins from southern Quebec to eastern Minnesota and south to Louisiana; and the Atlantic and Gulf slopes from Santee River drainage, South Carolina, to Galveston Bay drainage, Texas (Page and Burr 1991).

Nonindigenous Occurrences: The brook silverside was introduced into the Platte River, **Nebraska**, as well as reservoirs and several lakes in the drainage (Rowe 1992). It was also introduced into the New River below Bluestone Lake and the Potomac River in **West Virginia** (Hocutt et al. 1986; Jenkins and Burkhead 1994; Stauffer et al. 1995).

Means of Introduction: Page and Burr (1991) reported that the species had been introduced, usually into impoundments, as forage for sport fishes. Apparently, therefore, most introductions have been intentional (e.g., Jenkins and Burkhead 1994, and references cited therein). The species was accidentally introduced in Nebraska at the Sutherland cooling pond in July 1979 (Rowe 1992).

Status: Established in Nebraska and in the New River, West Virginia. Reported from the Potomac River, West Virginia.

Impact of Introduction: Unknown.

Remarks: Large specimens reportedly are used for bait by anglers (Scott and Crossman 1973), but there are no documented records of bait bucket releases.
Voucher specimens: Nebraska (UN 3314, 01498, 3291, 3292, 3293, 5603).

Leuresthes sardina

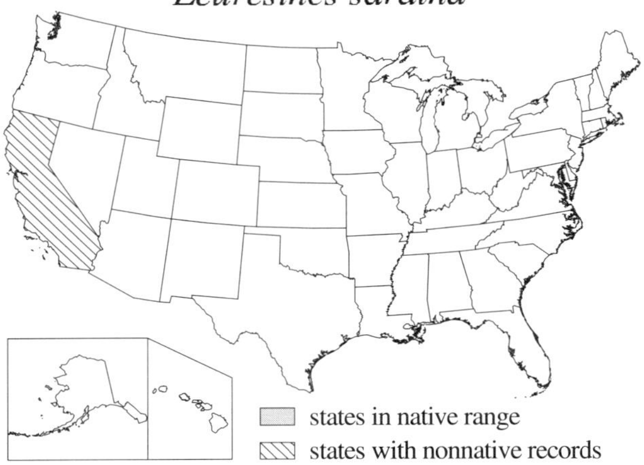

Leuresthes sardina (Jenkins and Evermann 1889)

Common Name: (grunion, Gulf grunion).

Identification: Fischer et al. (1995a). Maximum size: 25 cm.

Native Range: Marine; Gulf of California (Fischer et al. 1995a).

Nonindigenous Occurrences: Introduced into Salton Sea, **California**, in 1951 (Walker et al. 1961; Dill and Cordone 1997).

Means of Introduction: Intentionally stocked as forage. This species was only stocked once; 300 fish were stocked in March 1951. Stock was obtained from San Felipe, Mexico (Walker et al. 1961).

Status: The introduction failed to produced a viable population and the species is extirpated in the Salton Sea.

Impact of Introduction: Unknown.

Remarks: None.

Membras martinica

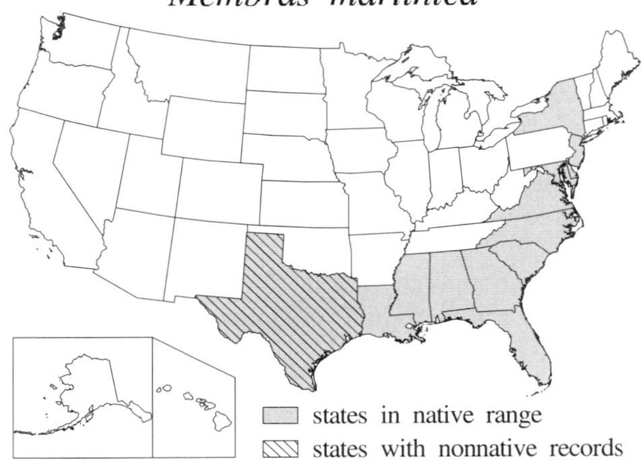

Membras martinica (Valenciennes 1835)

Common Name: rough silverside.

Identification: Smith (1985); Robins and Ray (1986); Hubbs et al. (1991); Menhinick (1991). Maximum size: 12.5 cm.

Native Range: Marine; eastern North America along the Atlantic and Gulf coasts from New York to northern Mexico (Robins and Ray 1986).

Nonindigenous Occurrences: The rough silverside has been stocked as a forage fish in the Amistad and Falcon reservoirs on the Rio Grande River, **Texas** (Hubbs et al. 1991).

Means of Introduction: Intentionally stocked as forage for sport fish.

Status: Possibly established in Texas.

Impact of Introduction: Unknown.

Remarks: None.

Menidia beryllina

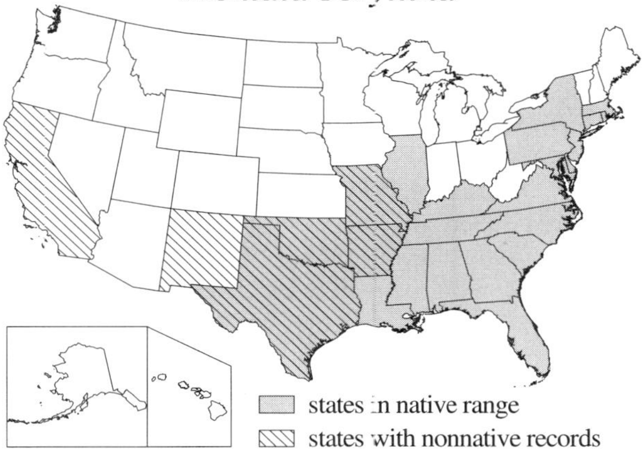

Menidia beryllina (Cope 1867)

Common Names: inland silverside (Mississippi silverside).

Identification: Smith (1985); Hubbs et al. (1991); Page and Burr (1991); Etnier and Starnes (1993); Chernoff et al. (1981) concluded that the Mississippi silverside *M. audens* is conspecific with *M. beryllina*; they also recognized that peninsular Florida populations living in marine environs, formerly considered *M. beryllina*, to be a distinct species (i.e., *M. peninsulae*). Maximum size: 15 cm.

Native Range: Eastern North America including Atlantic and Gulf slopes (mostly near the coast) from Massachusetts to the Rio Grande drainage, Texas, and southeastern New Mexico; north from the Mississippi River and major tributaries (mainly Arkansas and Red rivers) to southern Illinois and eastern Oklahoma. Also in Mexico (Page and Burr 1991).

Nonindigenous Occurrences: This species was introduced into, or recorded from, large reservoirs and other areas of **Arkansas** (Robison and Buchanan 1988; Cross et al. 1986), as well as several locations and drainages in **California**, including Blue Lakes and Clear Lake, Lake County; Putah and Cache creeks, Yolo County; and reservoirs in the San Francisco Bay area (Cook and Moore 1970; Moyle et al. 1974a; Li et al. 1976; Moyle 1976a, 1976b; Shapovalov et al. 1981; Swift et al. 1993). It has also been introduced into Stockton Reservoir in the lower Missouri River drainage and to several other lakes and ponds in **Missouri** (Pflieger 1975, 1997; Lee et al. 1980 et seq.; Cross et al. 1986); the lower Pecos River drainage in **New Mexico** (Smith and Miller 1986; Sublette et al. 1990); Boomer Lake on the lower Cimarron (Miller and Robison 1973; Cross et al. 1986) and a tributary of the Canadian River, Pittsburgh County, **Oklahoma** (Li et al. 1976; Lee et al. 1980 et seq.; Cross et al. 1986; Smith and Miller 1986); and the Pecos and Cochos drainages, and White River and Greenbelt reservoirs in **Texas** (Kraai et al. 1983; Conner and Suttkus 1986; Smith and Miller 1986).

Means of Introduction: Intentionally stocked as forage for sport fish in most locations, but unknown in the Canadian River system. Unauthorized introduction into Clear Lake, California, in 1967 to control the Clear Lake gnat *Chaoborus astictopus* and chironomid midges, and to serve as a nutrient reservoir to control bluegreen algae blooms (Moyle et al. 1974a; Moyle 1976a). A few introductions in California were authorized; however, many additional stockings were unauthorized and help to spread this fish (Moyle et al. 1974a). Introduction to the upper Mississippi and Ohio Rivers may be the result of dispersal through the Tennessee-Tombigbee waterway (Etnier and Starnes 1993) or natural dispersal up the Mississippi made possible by a change in water quality (Burr et al. 1996). Brood stock for Missouri introductions came from Lake Texoma, Oklahoma (Pflieger 1997).

Status: Established in Arkansas, California, Illinois, Kentucky, possibly Missouri, Oklahoma, New Mexico, and Texas. Two years after these fish were introduced into Clear Lake, California, this species became the most abundant fish in the littoral zone (Moyle 1976a). Pflieger (1997) indicated that none of the Missouri impoundments currently support populations; for instance, the Stockton Reservoir population in Missouri died out during the severe winter of 1976 to 1977.

Impact of Introduction: In Oklahoma, the inland silverside has almost completely replaced the brook silverside *Labidesthes sicculus*, an ecologically similar species (Gomez and Lindsay 1972; Moyle 1976a). Cook and Moore (1970) gave details of the stocking in Clear Lake, California. In Clear Lake, inland silversides were reported as having displaced native fishes, including the hitch *Lavinia exilicauda*, the Sacramento blackfish *Orthodon microlepidotus*, and the now extinct Clear Lake splittail *Pogonichthys ciscoides*, apparently through competition for food. Inland silversides also may have replaced or diminished introduced bluegill (Moyle 1976a). Li et al. (1976) found the introduction of Mississippi silversides inhibited growth of black and white crappies *Pomoxis* spp. for the first two years of life and enhanced growth after that time. Moyle and Holzhauser (1978) discussed effects on feeding habits of introduced largemouth bass. Moyle et al. (1974a) speculated that if this silverside reaches the delta area, it could negatively impact the endangered Delta smelt *Hypomesus t. transpacificus* and perhaps juvenile striped bass. Moyle (1976a) and Dill and Cordone (1997) provided detailed discussion of the history and reasons for this species' introduction into California and of its impacts.

Remarks: The Oklahoma records are from upstream of the native range of the inland silverside (Lee et al. 1980 et seq.). Moyle et al. (1974a) showed a map of the distribution of silversides in California. There is some disagreement concerning the records of *Menidia beryllina* in the Mississippi and Ohio rivers in the Illinois and Kentucky area. According to Burr et al. (1996), the species may have either "expanded its range after the low water levels of the late 1980s created water-quality conditions (high dissolved solids) favorable for this species to disperse" or "it may have entered the Ohio River via the Tennessee-Tombigbee waterway which now connects Gulf Coast drainages to the Ohio River, especially because of recent (1991) records from both Kentucky and Barkley reservoirs (Etnier and Starnes 1993)." Burr considered either one equally possible. However, Mettee et al. (1996) show no records of this species from the Tennessee-Tombigbee waterway. Therefore, we believe that the presence of the species in the Mississippi River in southern Illinois and in the lower Ohio River in Illinois and Kentucky are a result of natural dispersal.

ORDER BELONIFORMES

Adrianichthyidae - Ricefishes

The beloniform family Adrianichthyidae comprises 4 genera and 18 species (Nelson 1994). Most members of the family, 13 species, are in the ricefish genus, *Oryzias*. The native range of adrianichthyids extends from India to Japan and to the Indo-Australian archipelago (Nelson 1994). Prior to recent systematic revisions, the group was divided into three families, Oryziidae (ricefishes or medakas), Horaichthyidae (thready killifishes), and Adrianichthyidae, and all were placed in the order Cyprinodontiformes. These are small fishes, ranging in length from about 2 to 20 cm, with most species less than 10 cm long (Berra 1981; Nelson 1994). Most or all apparently feed primarily on invertebrates. Because some species prey heavily on mosquito larvae, they are considered useful as biological control agents (Talwar and Jhingran 1992). Adrianichthyids inhabit fresh and brackish waters (Nelson 1994).

Oryzias latipes

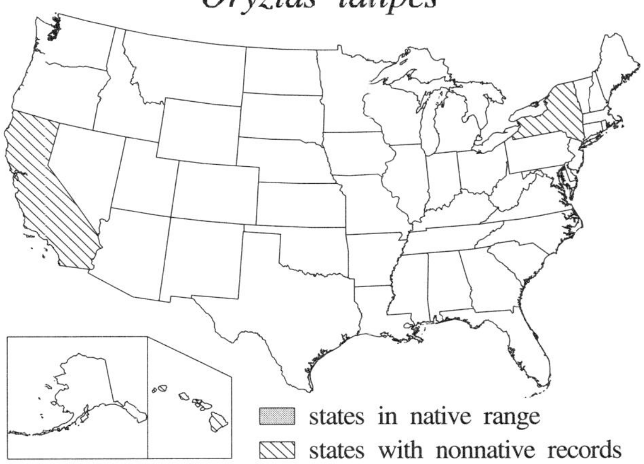

Oryzias latipes (Temminck and Schlegel 1846)

Common Names: Japanese medaka (medaka, ricefish).

Identification: Distinguishing characteristics were provided by Masuda et al. (1984). Photographs or illustrations appeared in Lee et al. (1980 et seq.), Masuda et al. (1984), and Axelrod et al. (1985). Maximum size: 4 cm.

Native Range: Eastern Asia. Japan, Korea, and eastern China (Lee et al. 1980 et seq.; Sakaizumi et al. 1983; Masuda et al. 1984).

Nonindigenous Occurrences: This species was discovered in San Francisquito Creek, Santa Clara County, **California** (Snyder 1935; also see Shapovalov et al. 1959, 1981; Lee et al. 1980 et seq.). It was introduced to Oahu, **Hawaii**, in 1922 (Brock 1960; Maciolek 1984), and to ponds in Long Island, Suffolk County, **New York**, where a population was temporarily established during the 1970s and possibly before (Courtenay and Hensley 1979a; Courtenay et al. 1984).

Means of Introduction: The California introduction reportedly resulted from an aquarium release (Snyder 1935). It was intentionally stocked in unknown numbers in Hawaii (Brock 1960; Maciolek 1984), possibly as a potential biological control for mosquitoes. Sources of and reasons for introductions in New York are not known, but may be associated with research or aquarium releases (Lee et al. 1980 et seq.).

Status: The New York population was extirpated as a result of the cold winter of 1978; the California population disappeared for reasons unknown (Courtenay et al. 1984). Contrary to information provided by some authors, the species was reported but apparently never established in California (Dill and Cordone 1997). In Hawaii, the species never became established (Brock 1960; Maciolek 1984).

Impact of Introduction: Unknown.

Remarks: According to Coates (1950), the California populations were thriving. Shapovalov et al. (1959, 1981) stated that there have been no records of *Oryzias* being taken from open waters of that state since about 1935. It was not mentioned by Moyle (1976a) or Swift et al. (1993) in their reports on freshwater fishes of California. Dill and Cordone (1997) referenced an early letter by Snyder and indicated that the first (and possibly only) record of this fish in California is based on a single specimen taken from San Francisquito Creek by Stanford University students. According to these authors, that single fish may be the only basis for Snyder (1935) and all subsequent mentions of this fish recorded for that state. This freshwater species shows broad temperature and salinity tolerances. Various color varieties or mutants are known and some of these are in the aquarium trade (Sakaizumi et al. 1983).

There are no known voucher specimens.

Belonidae - Needlefishes

The family Belonidae, known as needlefishes, includes 10 genera and 32 species (Nelson 1994). Belonids are found throughout much of the world in marine, coastal, and some freshwater regions (Berra 1981). They are fast-swimming predators characterized by thin, elongate bodies, long, narrow snouts, and numerous jaw teeth. A few species attain a length of about one meter. About 11 species are restricted to freshwater; these are found mainly in tropical South America, southern Asia, and Australia (Berra 1981; Nelson 1994). There are no strictly freshwater belonids native to North America, although one primarily marine species, *Strongylura marina*, enters streams along the northern Gulf of Mexico and the eastern United States (Lee et al. 1980 et seq.).

Strongylura marina

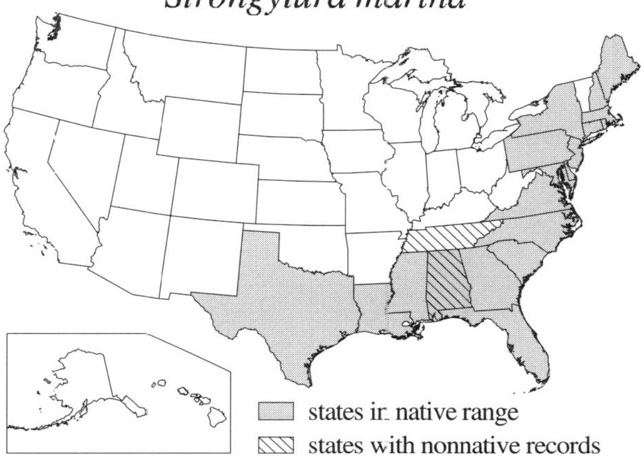

Strongylura marina (Walbaum 1792)

Common Name: Atlantic needlefish.

Identification: Smith (1985); Robins and Ray (1986); Mettee et al. (1996). Maximum size: 61 cm.

Native Range: Marine and freshwater; western Atlantic from Massachusetts to Rio de Janeiro, including coasts of the Gulf of Mexico and the western Caribbean Sea (Boschung 1992).

Nonindigenous Occurrences: This species has been recorded from the Tennessee River drainage, including Pickwick Pool on the Tennessee River, **Alabama** (Boschung 1992; Etnier and Starnes 1993; Mettee et al. 1996), and the Kentucky Reservoir, **Tennessee**, in 1992 (Etnier and Starnes 1993).

Means of Introduction: Gained access to the Tennessee River drainage in Alabama and Tennessee via the Tennessee-Tombigbee Waterway, an artificial canal completed in 1985 (Boschung 1992; Etnier and Starnes 1993).

Status: Reported in Alabama and Tennessee.

Impact of Introduction: Unknown.

Remarks: This belonid occurs naturally in many coastal streams of the southern United States. Mettee et al. (1996) reported evidence that the adults live and spawn in the Alabama, Black Warrior, Coosa, and lower Tombigbee rivers.

Xenentodon cancila

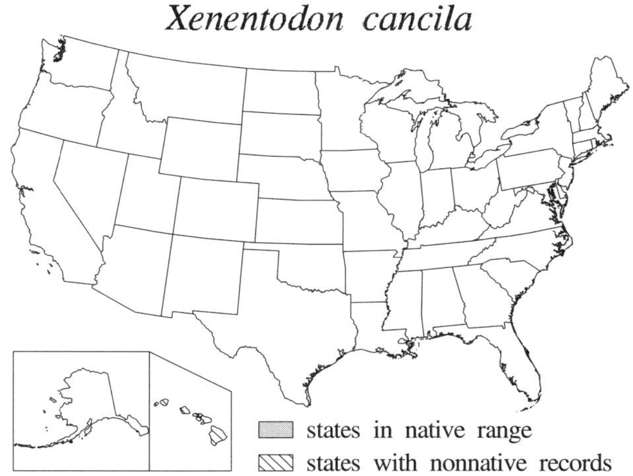

Xenentodon cancila (Hamilton 1822)

Common Names: Asian needlefish (stickfish, freshwater garfish).

Identification: Key and distinguishing characteristics appeared in Smith (1945) and Talwar and Jhingran (1992). Maximum size: 40 cm.

Native Range: Fresh water, tropical Asia. Pakistan, India, Bangladesh, Sri Lanka, Burma, and Thailand (Talwar and Jhingran 1992).

Nonindigenous Occurrences: This species was introduced into the Wahiawa Reservoir (Lake Wilson), Oahu, **Hawaii**, where it was first reported in the spring of 1988 (Devick 1991b).

Means of Introduction: Probable aquarium release.

Status: Established in Hawaii.

Impact of Introduction: During the summer of 1989, *Xenentodon* in Hawaii were preying on largemouth bass (also introduced), to the extent that the bass population appeared to be threatened (Devick 1991b).

Remarks: This species displays sexual dimorphism and spawns throughout the year in Hawaii. However, it apparently cannot survive outside of the reservoir and the species is unlikely to do well during an extended reservoir drawdown (Devick 1991b). Courtenay et al. (1991) and Robins et al. (1991b) incorrectly reported the long tom *Strongylura kreffti* as having been introduced into Hawaii, based on an early tentative identification of *X. cancila* (Devick, personal communication).

Voucher specimens: Hawaii (USNM 308225).

ORDER CYPRINODONTIFORMES

Aplocheilidae - Rivulins

This cyprinodontiform family includes eight genera and about 185 species found in tropical Africa and southern Asia to Java (Parenti 1981; Nelson 1994). Aplocheilids are small, fusiform fishes that exhibit marked sexual dimorphism. Some authors (e.g., Nelson 1994) treat the South American rivulids (Rivulidae) and Aplocheilidae as a single family. We follow Parenti (1981) in recognizing the two as separate families. Some species are popular in the aquarium trade. Species occur in fresh and brackish waters.

Aplocheilus lineatus

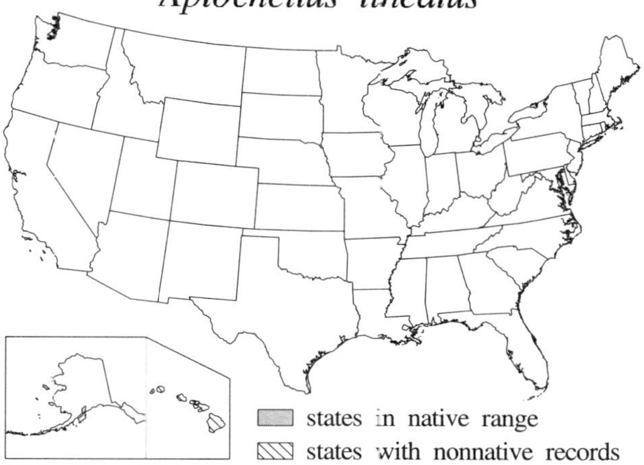

Aplocheilus lineatus (Valenciennes 1846)

Common Names: striped panchax (Malabar killie, panchax lineatus).

Identification: Key and distinguishing characteristics were given in Talwar and Jhingran (1992). Diagnostic characteristics and definition of genus were presented by Parenti (1981). Color photographs appeared in Axelrod et al. (1985), Scheel (1990), and Rosler (1994). Maximum size: 10 cm.

Native Range: Tropical Asia. Western and southeastern regions of India (Talwar and Jhingran (1992).

Nonindigenous Occurrences: This species was released or reported sometime prior to 1960 from **Hawaii**, possibly from Oahu (Maciolek 1984).

Means of Introduction: Unknown. Maciolek (1984) is unclear as to whether this is one of the cyprinodonts imported by the Hawaii Department of Health and, if so, whether the species was intentionally stocked for mosquito control.

Status: Reported from Hawaii.

Impact of Introduction: Unknown.

Remarks: The striped panchax is found in the aquarium trade.

There are no known voucher specimens.

Nothobranchius guentheri

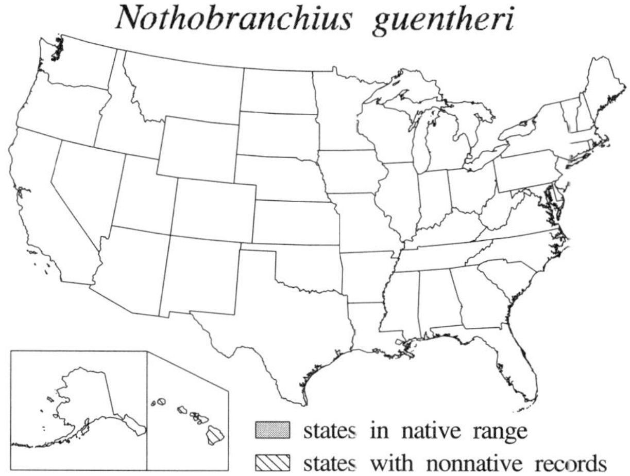

Nothobranchius guentheri (Pfeffer 1893)

Common Name: redtail notho.

Identification: Diagnostic characteristics and definition of genus were given by Parenti (1981). Color photographs appeared in Axelrod et al. (1985) and Scheel (1990). Maximum size: 5 cm.

Native Range: Tropical Africa. Type locality not clearly known; may be island of Zanzibar (Scheel 1990).

Nonindigenous Occurrences: This species was introduced into Oahu, **Hawaii,** ca. 1967 (Kanayama 1968; Maciolek 1984).

Means of Introduction: Imported and released by the Hawaii State Department of Health for possible use as mosquito control agents (Kanayama 1968).

Status: Reported from Hawaii. Kanayama (1968) listed this species as one that may become established in Hawaii; however, later accounts (e.g., Devick 1991b) indicated that it did not persist.

Impact of Introduction: Unknown.

Remarks: *Nothobranchius* are annual fishes; they are able to maintain permanent populations in temporary habitats by combining rapid growth and development with diapausing eggs that survive the dry season buried in the mud. *Nothobranchius guentheri* was used for experimental purposes at the Agricultural Experiment Station at the University of California at Riverside; however, the experiment lasted only one year (1964–1965) and was terminated, supposedly because the species was intolerant of low water temperatures (Dill and Cordone 1997). There is no evidence that the species was ever released into California open waters.

There are no known voucher specimens.

Rivulidae - New World rivulins

This family includes 13 genera and more than 150 species found in tropical and subtropical America including southern Florida, Central America, the Bahamas, the Antilles, and South America from the La Plata River basin of Argentina to northern Colombia (Costa 1990). Some authors (e.g., Nelson 1994) treat the South American rivulids (Rivulidae) and Old World Aplocheilidae as one family. We follow Parenti (1981) and many others in recognizing the two as separate families. Rivulids are small, fusiform fishes and most species exhibit marked sexual dimorphism. They are commonly found in shallow waters and a number of species (e.g., *Cynolebias* spp.) are adapted to survive in temporary habitats. Some species are popular as aquarium fishes. Most species inhabit freshwaters, although some are found in brackish habitats.

Cynolebias bellottii

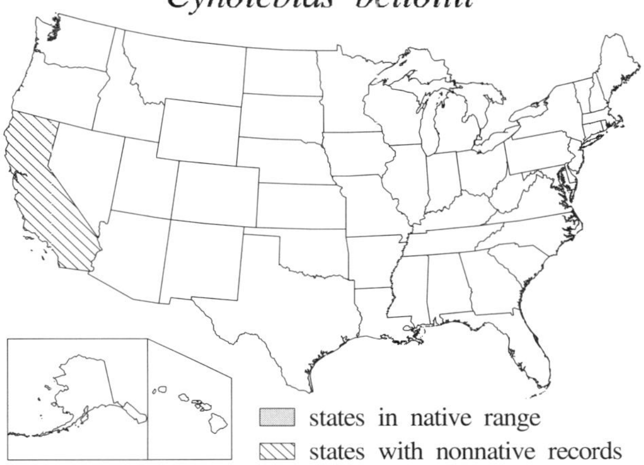

Cynolebias bellottii Steindachner 1881

Common Names: Argentine pearlfish (pavito).

Identification: Costa (1995) provided diagnostic characters and included this species in his identification keys to the genera and species of the subfamily Cynolebiatinae. A few distinguishing characteristics were also provided by Boschi (1957) and Sterba (1973). Moyle (1976a) included this species in his keys for California fishes. Color photographs appeared in Mills and Vevers (1982), Axelrod et al. (1985), and Costa (1995). The spelling of this species is sometimes given as *Cynolebias bellotti*. Maximum size: 6.5 cm.

Native Range: Tropical America. Lower Parana basin, around Rio de la Plata, in Argentina and Uruguay, South America (W. Costa, personal communication).

Nonindigenous Occurrences: During the period 1964-1967, small numbers of adults were stocked in southern **California** in study ponds at the Riverside campus of the University of California, Riverside County; at the Biggs Rice Experiment Station, Butte County; and at two duck clubs, Kern County (Bay 1966; Moyle 1976a; Shapovalov et al. 1981; Dill and Cordone 1997). In addition to experimental ponds of the University of California., Moyle (1976a) recorded it as introduced into southern California ponds and rice fields in the Central Valley in the late 1960s or early 1970s (he included it in his table as introduced to the Los Angeles basin). Additional releases, poorly documented, may have taken place at one or more sites in Los Angeles County, and possibly Orange County, California (Moyle 1976a; Courtenay et al. 1991; Dill and Cordone 1997).

Means of Introduction: Intentionally stocked to assess its ability as a mosquito control agent, especially in rice-fields (Moyle 1976a; Dill and Cordone 1997). These introductions were with the approval of the Fish and Game Commission (Dill and Cordone 1997).

Status: Reported from California where all open water introductions apparently failed. Contrary to Courtenay et al. (1986, 1991), Dill and Cordone (1997) concluded that there is no evidence of past survival and establishment of this species in California open waters. However, it was temporarily established in experimental ponds at the University of California (Bay 1966; Moyle 1976a; Dill and Cordone 1997); that population reportedly survived for five or six years despite repeated floodings and dryings before dying out (Legner and Medved 1972; Shapovalov et al. 1981).

Impact of Introduction: Unknown. Moyle (1976a) expressed concern about the possible impact introduced annual fishes might have on the invertebrate fauna inhabiting California temporary pools.

Remarks: Similar to many tropical killifish species, *Cynolebias bellottii* is an annual fish able to maintain permanent populations in temporary habitats by combining rapid growth and development with diapausing eggs that survive the dry season buried in the mud. Bay (1966) demonstrated that *Cynolebias bellottii* could survive seasonal conditions typical of southern California, although egg survival was low. *Cynolebias bellottii* was reported as an introduction into Hawaii but apparently never was stocked in open water (Maciolek 1984). The literature is fragmentary and somewhat contradictory concerning the introduction, and possible establishment, of this species in open waters of California. Courtenay and Hensley (1979a), citing Moyle (1976a), stated that the species probably was established in Riverside and possibly in Los Angeles counties. Although unclear, the Los Angeles County report by Moyle (1976a) may represent an open water record, whereas his Riverside record (Moyle 1976a), apparently refers to completely confined populations restricted to experimental ponds at the University of California, Riverside, Riverside County. The literature does not provide information on the outflow, if any, of the experimental ponds nor of the ricefield areas. Courtenay et al. (1986, 1991) listed this species as an exotic formerly established in California but now extirpated. Hubbs et al. (1979) cited it as not known to occur in California. Swift et al. (1993) did not mention *C. bellottii* in their work on freshwater fishes of southern California. Recently, Dill and Cordone (1997) provided details of the history of this species in California. They verified that a small number of adult Argentine pearlfish were stocked in study plots at the Riverside campus in Riverside County, at the Biggs Rice Experiment Station in Butte County, and at two duck clubs in Kern County. Although possibly introduced to sites in Orange and Los Angeles counties, Dill and Cordone could not find any information on such releases.

There are no known voucher specimens.

Cynolebias nigripinnis

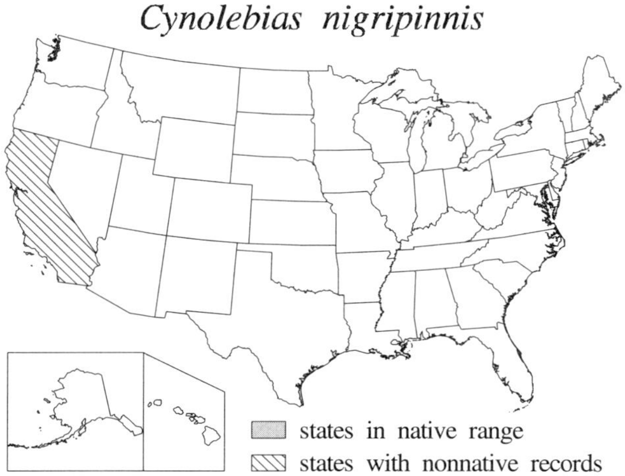

Cynolebias nigripinnis Regan 1912

Common Names: blackfin pearlfish (black-finned pearl fish).

Identification: Costa (1995) provided diagnostic characters and included this species in his identification keys to the genera and species of the subfamily Cynolebiatinae. A few distinguishing characteristics were also given in Sterba (1973). Color photographs appeared in Mills and Vevers (1982), Axelrod et al. (1985), and Costa (1995). Maximum size: 5 cm.

Native Range: Tropical America. Lower Parana basin, around Rio de la Plata, in Argentina and Uruguay, and an area along the Rio Uruguay floodplains, South America (Costa, personal communication).

Nonindigenous Occurrences: This pearlfish has been listed as a nonestablished species known from open waters in **California** (Courtenay et al. 1986, 1991; Williams and Jennings 1991). That listing likely is based on its introduction into experimental rice plots and ponds on lands of the Butte County Mosquito Abatement District in 1973 and 1974 (e.g., Shapovalov et al. 1981; Dill and Cordone 1997).

Means of Introduction: Intentionally stocked to assess its ability as a mosquito control agent, especially in ricefields (Dill and Cordone 1997). These studies were with the approval of the Fish and Game Commission (Dill and Cordone 1997).

Status: Reported from California. However, Dill and Cordone (1997) concluded that there is no evidence that the species was ever an inhabitant of open waters (see Remarks section).

Impact of Introduction: Unknown.

Remarks: Like other annual killifishes, this species is capable of surviving in temporary pool habitats. The literature is fragmentary and somewhat contradictory concerning the introduction of this species in open waters of California. Dill and Cordone (1997) provided details on the history of this fish in California. They related that the species was used for experimental purposes at the Agricultural Experiment Station at the University of California at Riverside, either carried out in the laboratory or in outdoor plots. These experiments were terminated in 1965 after one year because the species did not reproduce adequately, was prone to disease, and did not tolerate cold temperatures. Dill and Cordone also reported that the Butte County Mosquito Abatement District used the blackfin pearlfish in tests at the Biggs Rice Experiment Station in Butte County, California. As part of these tests, young blackfin pearlfish were introduced to ponds and rice paddies at the Station and in an adjacent temporary pool. None of the fish survived and tests were terminated. In contrast to other authors (e.g., Courtenay et al. 1986, 1991; Williams and Jennings 1991), Dill and Cordone (1997) did not consider these California sites to be open waters. Unfortunately, the literature does not provide details on the outflow connections, if any, of the experimental ponds or of the ricefield areas where this species was used. In addition, if these sites were subject to periodic flooding from adjacent areas, then the introductions may be considered open water.

There are no known voucher specimens.

Rivulus hartii

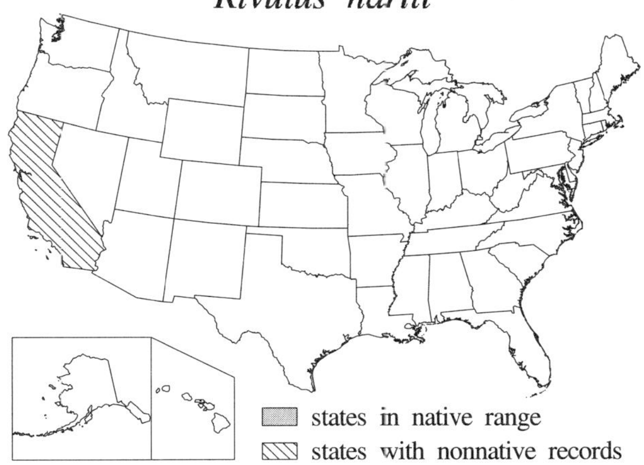

Rivulus hartii (Boulenger 1890)

Common Names: giant rivulus (Hart's rivulus, Trinidad rivulus).

Identification: Distinguishing characteristics and photographs were given by Huber (1992). The spelling for this species is sometimes given as *Rivulus harti*. Maximum size: 10 cm.

Native Range: Tropical America. Caribbean Slope drainages of Venezuela, South America, from Rio Aroa east to Rio San Juan; islands of Trinidad, Tobago, Grenada, and Margarita (J. E. Thomerson, personal communication).

Nonindigenous Occurrences: A population was discovered in a small ditch (flowing into the Salton Sea) draining a tropical fish farm in Imperial County, **California** (St. Amant 1970; Moyle 1976a; Dill and Cordone 1997).

Means of Introduction: The California fish were first found in January 1967 (St. Amant 1970). This introduction was probably due to escape or release from a nearby fish farm.

Status: Locally established in California. The species was thought to have disappeared after 1969 (Shapavalov et al. 1981), but additional specimens were taken in 1990 (Courtenay et al. 1991; Swift et al. 1993).

Impact of Introduction: Unknown.

Remarks: The first specimens from Imperial County were identified by C. Hubbs. The tropical fish farm, mentioned as the possible source of the introduction, has been defunct for many years (Moyle 1976a; Shapovalov et al. 1981; Dill and Cordone 1997).

There are no known voucher specimens.

Simpsonichthys whitei

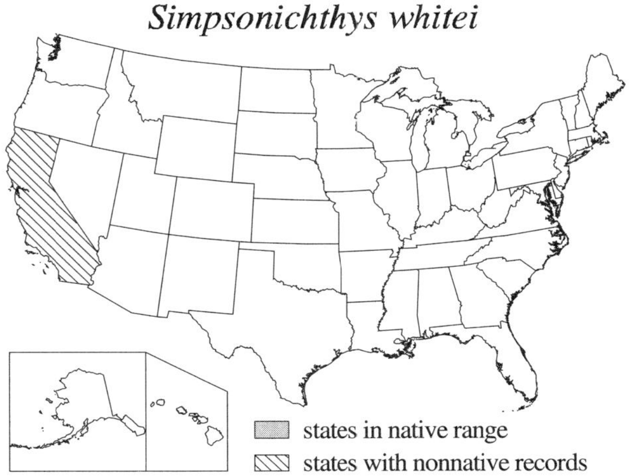

Simpsonichthys whitei (Myers 1942)

Common Names: Rio pearlfish (White's pearlfish).

Identification: Costa (1995) provided diagnostic characters and included this species in his identification keys to the genera and species of the subfamily Cynolebiatinae. Color photographs appeared in Axelrod et al. (1985) and Costa (1995). This species, formerly known by the name *Cynolebias whitei*, was recently assigned to the genus *Simpsonichthys* by Costa (1996). Maximum size: 9 cm.

Native Range: Tropical America. South America in the coastal plains of Rio de Janeiro State, between Rio das Ostras and Marica, southeastern Brazil (Costa, personal communication).

Nonindigenous Occurrences: This pearlfish has been listed as a nonestablished species known from open waters of **California** (Courtenay et al. 1986, 1991; Courtenay and Williams 1991; Williams and Jennings 1991). That listing likely based on its introduction into experimental rice plots and ponds on lands of the Butte County Mosquito Abatement District in 1973 and 1974 (e.g., Shapovalov et al. 1981; Dill and Cordone 1997).

It was introduced as a mosquito control agent into ponds in southern **California** and rice fields in the Central Valley, in the late 1960s or early 1970s, but the population did not survive (Moyle 1976a).

Means of Introduction: Intentionally stocked to assess its ability as a mosquito control agent, especially in ricefields (Dill and Cordone 1997). These studies were with the approval of the Fish and Game Commission (Dill and Cordone 1997).

Status: Reported from California. However, as in the case of the black pearlfish *Cynolebias nigripinnis*, Dill and Cordone (1997) concluded that there is no evidence that this pearlfish species was ever an inhabitant of open waters (see Remarks section).

Introductions into California failed (Moyle 1976a).

Impact of Introduction: Unknown.

Remarks: Like other annual killifishes, this species is capable of surviving in temporary pool habitats. Dill and Cordone (1997) gave details on the history of this species in California. It is essentially the same story as that of the blackfin pearlfish *C. nigripinnis*. Dill and Cordone argued that the ponds and rice paddies where these fish were released should not be considered open waters. Unfortunately, the literature does not provide details on the outflow connections, if any, of the experimental ponds or of the ricefield areas where this species was used. In addition, if these sites were subject to periodic flooding from adjacent areas, then the introductions may be considered open water.

There are no known voucher specimens.

Fundulidae - Topminnows

The family Fundulidae comprises 5 genera and about 48 species in North and Central America from southeastern Canada to the Yucatan peninsula, including the Mississippi River basin, and in Bermuda and Cuba (Page and Burr 1991; Nelson 1994). About 24 species are native to freshwaters of the United States (Page and Burr 1991). Topminnows are small fishes; most species are less than 10 cm long, characterized by a cylindrical body, a flattened head and back, and an upturned mouth (Page and Burr 1991). They feed primarily on small invertebrates. Fundulids are typically surface-dwelling fishes that inhabit fresh and brackish waters, and coastal marine areas.

Fundulus catenatus

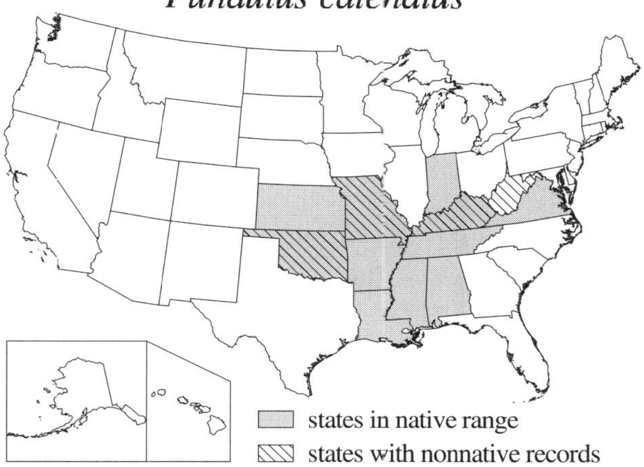

Fundulus catenatus (Storer 1846)

Common Name: northern studfish.

Identification: Page and Burr (1991); Etnier and Starnes (1993); Jenkins and Burkhead (1994); Pflieger (1997). Maximum size: 18 cm.

Native Range: Upper East Fork White River system, Indiana; upper Salt and Kentucky River drainages, Kentucky; upper Green, middle, and lower Cumberland, and Tennessee River drainages, Virginia, Kentucky, Tennessee, Alabama, and Mississippi; west of Mississippi River (primarily Ozark and Ouachita uplands) in central and southern Missouri, southeastern Kansas, eastern Oklahoma, and southern Arkansas; southwestern Mississippi in Mississippi (Coles Creek, Homochitto River, and Buffalo Bayou) and Gulf Slope drainages (Amite River and Pearl River) (Page and Burr 1991). Recently reported from the Amity River in Louisiana (Warren and Denette 1992).

Nonindigenous Occurrences: This species was introduced into several streams in **Kentucky** including the Big Sandy, Beech Fork, Rolling Fork, Dix, Middle Creek, Rockhouse Fork, and Licking rivers, and Cave Run Lake (Branson and Batch 1971; Burr and Page 1986; Burr and Warren 1986). Possibly introduced into several sites in **Missouri** including the Elk River system in extreme southwest part of state (Pflieger 1971, 1975); it may also have been introduced to selected streams along the northern periphery of the Missouri Ozarks (Pflieger 1997). Hall (1956) suggested that the presence of this species in streams of northeastern **Oklahoma** in the Neosho drainage may be the result of introductions. This species was taken from Middle Grave Creek, Marshall County, **West Virginia** in 1987 (Cincotta et al. 1990).

Means of Introduction: Bait bucket transfer may be responsible for most of the introductions in Kentucky, although some may be a result of stream capture (Burr and Warren 1986). The Oklahoma introductions is believed to be the result of a bait introduction prior to 1954 (Hall 1956). This species was not taken from the Elk River drainage of Missouri until the 1960s, and Pflieger (1997) concluded that these populations represent range extensions and are descended from northern studfish originally introduced into northeastern Oklahoma streams. In West Virginia, the introductions may have been a result of bait bucket release, aquarium release, or pond escape (Cincotta et al. 1990).

Status: Established or possibly established in Middle Creek, Rockhouse Fork, and Licking rivers, Kentucky. Collected elsewhere in Kentucky (Burr and Warren 1986). Established in Missouri, Oklahoma, and West Virginia. Seven individuals were collected in West Virginia and the size distribution of these specimens indicate reproduction had occurred (Cincotta et al. 1990).

Impact of Introduction: Unknown.

Remarks: Stauffer et al. (1995) listed species as native to Middle Grave Creek in West Virginia. Although it was present in 1987, surveys in this same area in 1982 failed to find this species (Cincotta et al. 1990), suggesting an introduction.

Fundulus chrysotus

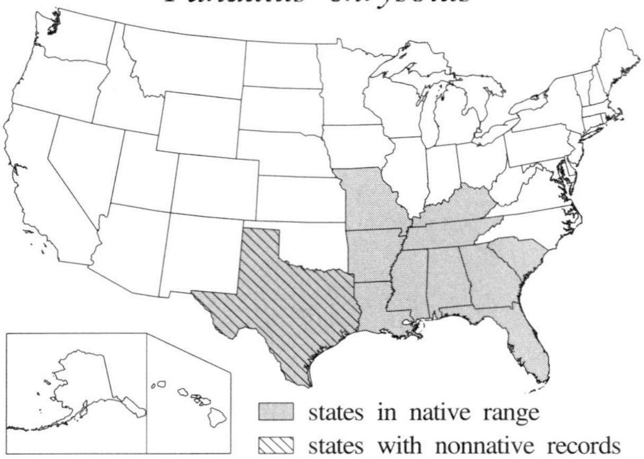

Fundulus chrysotus (Günther 1866)

Common Name: golden topminnow.

Identification: Hubbs et al. (1991); Page and Burr (1991); Etnier and Starnes (1993); Mettee et al. (1996). Maximum size: 7.5 cm.

Native Range: Atlantic and Gulf Coastal Plain from Santee River drainage, South Carolina, to Trinity River drainage, Texas; Former Mississippi Embayment north to Kentucky and Missouri. East of Mississippi River, mostly restricted to lower Coastal Plain (Page and Burr 1991).

Nonindigenous Occurrences: This species was collected twice in 1991 from Birds Creek (a tributary of the Guadalupe River) in Dewitt County, **Texas** (Whiteside and Berkhouse 1992).

Means of Introduction: Unknown.

Status: Probably established in Texas.

Impact of Introduction: Unknown.

Remarks: Whiteside and Berkhouse (1992) concluded that the Texas stream population was reproducing based on the number of specimens captured (n = 36) and the range of sizes (14–48 mm).

Fundulus diaphanus

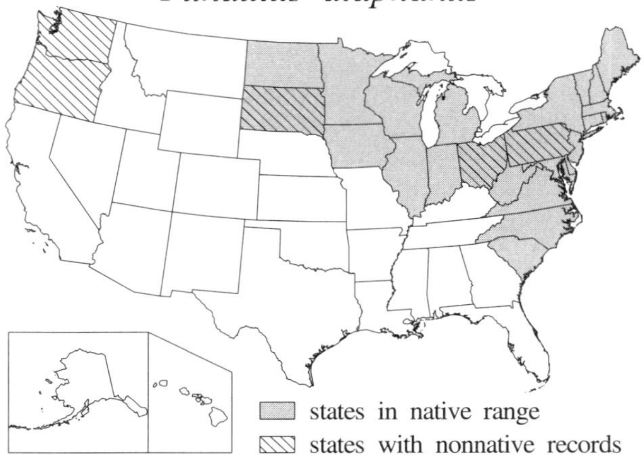

Fundulus diaphanus (Lesueur 1817)

Common Names: banded killifish (eastern banded killifish, western banded killifish).

Identification: Becker (1983); Smith (1985); Page and Burr (1991); Jenkins and Burkhead (1994). This species is represented by two subspecies, *F. d. diaphanus* and *F. d. menona*. Maximum size: 13 cm.

Native Range: Atlantic Slope drainages from Newfoundland to Pee Dee River, South Carolina; St. Lawrence–Great Lakes and Mississippi River basins from Quebec to Manitoba, and south to southern Pennsylvania, northern Illinois, and northeastern Nebraska. Two subspecies with intergrades in St. Lawrence and Lake Erie drainages: *F. d. diaphanus* on the Atlantic Slope, *F. d. menona* in the rest of the range (Page and Burr 1991).

Nonindigenous Occurrences: The eastern subspecies *F. d. diaphanus* was introduced into Lake Erie, the Little Miami drainage, and several areas in the upper Ohio River basin in **Ohio,** and into the Ohio and Beaver River drainages (Raney 1938, cited in Trautman 1981), the Monongahela drainage (Hocutt et al. 1986), and the Youghiogheny River (Hendricks et al. 1979) in **Pennsylvania**. This species was accidentally stocked in Lake Andes, **South Dakota**, by a federal hatchery stocking of largemouth bass (Bailey and Allum 1962). *Fundulus diaphanus* is known from the Columbia and Willamette rivers, **Oregon** (Bond, 1994). It also has been reported from **Washington** (Fletcher, personal communication).

Means of Introduction: The fish was transplanted to parts of eastern Ohio and western Pennsylvania from the

Delaware River drainage of eastern Pennsylvania (Raney 1938, cited in Trautman 1981). A population in Clough Creek, Ohio, was likely the result of an aquarium release (Trautman 1981). One lake in South Dakota was stocked accidentally with largemouth bass (Bailey and Allum 1962). The species was introduced into other areas through unknown means; possibly bait bucket releases.

Status: Established in Ohio, Pennsylvania, and South Dakota. Reported from Oregon and Washington. In Pennsylvania, known from the Ohio and Beaver drainages and established in the Youghiogheny.

Impact of Introduction: Unknown.

Remarks: According to Bond (1994), the Oregon introduction involved the subspecies *Fundulus diaphanus diaphanus*; however, Li (personal communication) reported it as being the western subspecies *F. d. menona*. The Washington record was also reported as *F. d. menona* (Fletcher, personal communication).

Voucher specimens: Ohio (OSM 33514, 13865).

Fundulus grandis

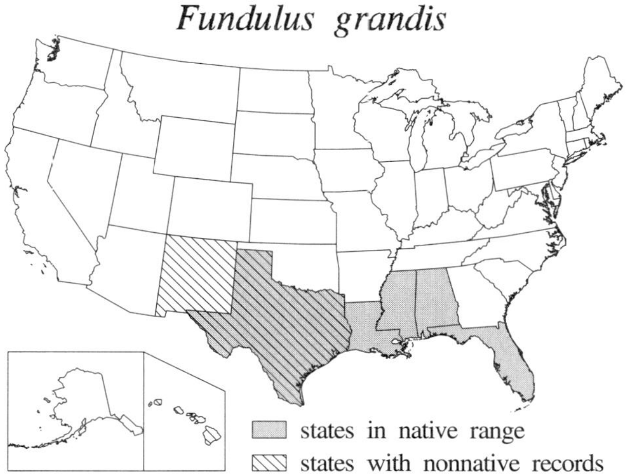

Fundulus grandis Baird and Girard 1853

Common Name: gulf killifish.

Identification: Hoese and Moore (1977); Robins and Ray (1986). Maximum size: 18 cm.

Native Range: Fresh and brackish waters from northeastern Florida to Key West and northern Gulf of Mexico to Cuba (Lee et al. 1980 et seq.; Robins and Ray 1986).

Nonindigenous Occurrences: This species has been introduced into all islands of **Hawaii** (Brock 1960; Maciolek 1984; Randall 1987), the Pecos River, **New Mexico** (Sublette et al. 1990), and the Pecos, Rio Grande, and Brazos river drainages in **Texas** (Hubbs et al. 1991).

Means of Introduction: The fish was introduced through bait bucket release into New Mexico and Texas. It was stocked intentionally but unsuccessfully in Hawaii for mosquito control in 1905, with specimens from Texas (Brock 1960; Maciolek 1984; Randall 1987).

Status: Established in Texas; extirpated in Hawaii. Sublette et al. (1990) reported that it was extirpated in New Mexico, but Burr (personal communication) believes species still established in lower Pecos River as of 1992.

Impact of Introduction: Unknown.

Remarks: None.

Fundulus heteroclitus

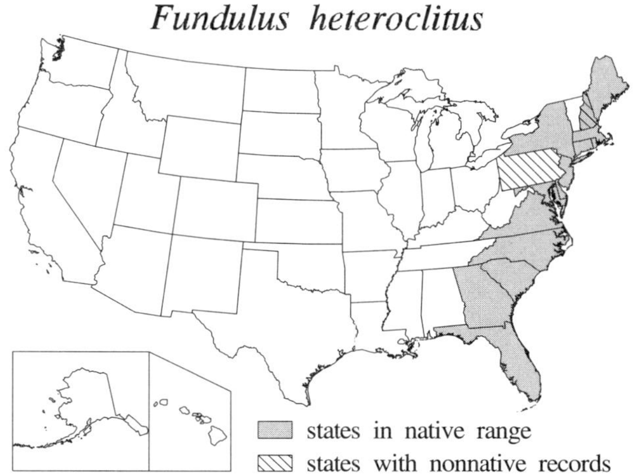

Fundulus heteroclitus (Linnaeus 1766)

Common Name: mummichog.

Identification: Smith (1985); Robins and Ray (1986); Menhinick (1991). Maximum size: 12.5 cm.

Native Range: Marine, brackish, and occasionally freshwaters from the Gulf of St. Lawrence to northeastern Florida (Robins and Ray 1986).

Nonindigenous Occurrences: This species was stocked in ponds in Windham, **New Hampshire** (Scarola 1973). It was collected from the upper Ohio and Beaver river systems in far western **Pennsylvania,** apparently in the late 1930s and early 1940s (Raney 1938, cited in Trautman 1981). It has also been collected in Pennsylvania from the Juniata River in Blair County; Sandy Run, a tributary of the Little Juniata; and in Conowingo Pond in Lancaster County (Denoncourt et al. 1975a).

Means of Introduction: This species was introduced into ponds in New Hampshire, apparently via bait bucket release (Scarola 1973). It was transferred to far western Pennsylvania from the Delaware River drainage of the eastern part of the state (Raney 1938, cited in Trautman 1981), possibly as a baitfish. The other collections in Pennsylvania are believed to be bait bucket introductions (Denoncourt et al. 1975a).

Status: Previously established or locally established in New Hampshire and Pennsylvania (Scarola 1973; Trautman 1981); present status unknown. Denoncourt et al. (1975a) reported only two specimens from Sandy Run, and one specimen from the Juniata River.

Impact of Introduction: Unknown.

Remarks: None.

Fundulus lineolatus

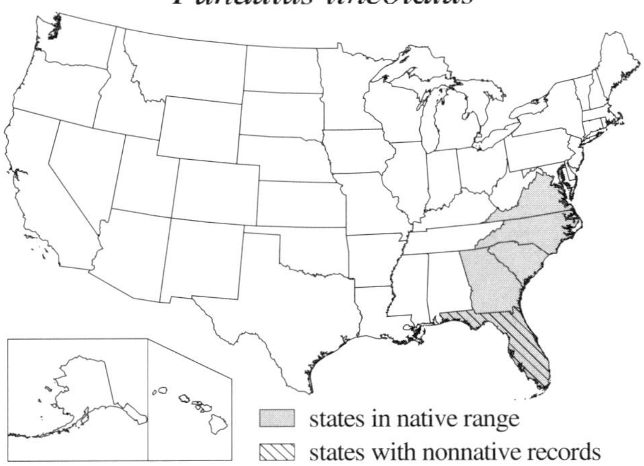

Fundulus lineolatus (Agassiz 1854)

Common Name: lined topminnow.

Identification: Page and Burr (1991); Jenkins and Burkhead (1994); Rohde et al. (1994). Maximum size: 8.4 cm.

Native Range: Atlantic and Gulf Coastal Plains from southern Virginia to Ochlockonee River, Georgia and Florida; south in Florida to Lake Okeechobee (Page and Burr 1991).

Nonindigenous Occurrences: A single specimen was collected from the Apalachicola River on the border of Jackson and Gadsden counties, **Florida**, in March 1958 (Gilbert 1987; Gilbert and J. D. Williams, personal communication).

Means of Introduction: Probable bait-bucket release (Gilbert 1987; Gilbert and J. D. Williams, personal communication).

Status: Reported in Florida.

Impact of Introduction: Unknown.

Remarks: Voucher specimens: Florida (UF 54521).

Fundulus notatus

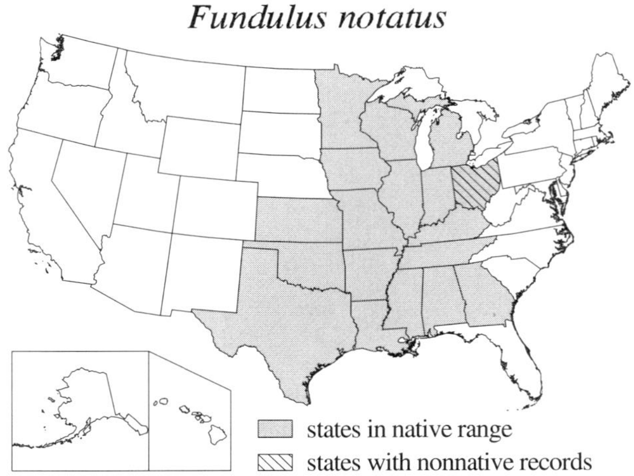

Fundulus notatus (Rafinesque 1820)

Common Name: blackstripe topminnow.

Identification: Page and Burr (1991). Maximum size: 7.4 cm.

Native Range: Lake Erie, Lake Michigan, and Mississippi River basins from southern Ontario, Michigan,

Wisconsin, and northern Iowa south to the Gulf (west to central Kansas and Oklahoma; Gulf Slope drainages from Mobile Bay, Alabama, to San Antonio Bay, Texas.

Nonindigenous Occurrences: The blackstripe topminnow was introduced into Burr Oak Lake in the Hocking system, Buckeye Lake in the Licking system, both of the Muskingum drainage in **Ohio** (Trautman 1981; Hocutt et al. 1986).

Means of Introduction: Probably inadvertently stocked into Burr Oak Reservoir (Trautman 1981). Access to Buckeye Lake via the Ohio and Erie canals (Trautman 1981).

Status: A population was discovered in Burr Oak Reservoir in 1962 (Trautman 1981). Established in Ohio.

Impact of Introduction: Unknown.

Remarks: None.

Fundulus sciadicus

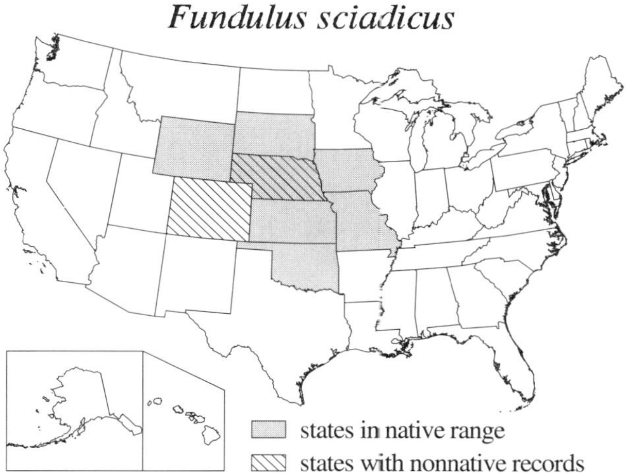

Fundulus sciadicus Cope 1865

Common Name: plains topminnow.

Identification: Bouc (1987); Page and Burr (1991); Pflieger (1997). Maximum size: 7 cm.

Native Range: Disjunct distribution, Missouri River basin from western Iowa to eastern Wyoming; Missouri drainage, central Missouri; and Neosho River system, southwestern Missouri, southeastern Kansas, and northeastern Oklahoma (Page and Burr 1991).

Nonindigenous Occurrences: This species was collected from the White River (Rio Blanco County) in **Colorado** (Wick et al. 1981; Tyus et al. 1982; Woodling 1985). It also was introduced into the San Luis Valley (Rio Grande headwaters) in Colorado (Zuckerman and Behnke 1986). The plains topminnow has also been introduced into the Kansas drainage in Frenchman Creek, a tributary of the Republican River in Hitchcock County, **Nebraska** (Cross 1967; Cross et al. 1986).

Means of Introduction: The species was introduced accidentally into the San Luis Valley by the Colorado Department of Wildlife (Zuckerman and Behnke 1986). It was introduced into Rio Blanco County via unknown means, probably bait bucket release. Probable bait bucket introduction in Nebraska (Cross 1967). This species was first captured below Enders Reservoir, an area commonly fished with bait minnows, in 1959. The fish likely came from the nearby Platte River (Cross 1967).

Status: Reported in White River drainage (Woodling 1985); extirpated in Rio Grande drainage, Colorado (Zuckerman and Behnke 1986). Established in Nebraska.

Impact of Introduction: Unknown.

Remarks: Voucher specimens: Nebraska (KU 4856).

Fundulus seminolis

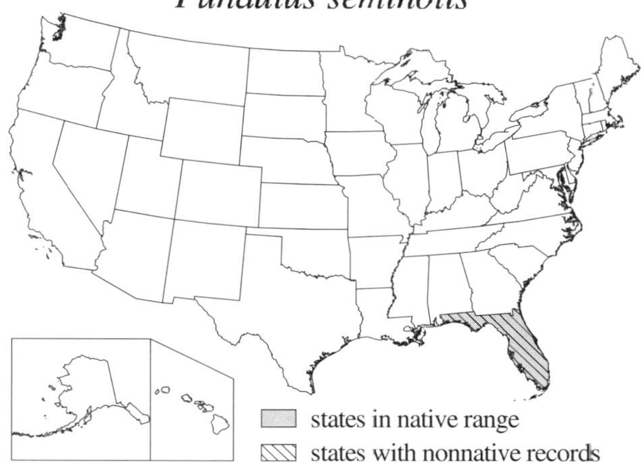

Fundulus seminolis Girard 1959

Common Name: Seminole killifish.

Identification: Page and Burr (1991). Maximum size: 16 cm.

Native Range: Peninsular Florida from St. Johns drainage on the Atlantic to New drainage in the Gulf, south to just below Lake Okeechobee (Page and Burr 1991).

Nonindigenous Occurrences: A single individual was collected from North Bay in St. Andrew Bay, Bay County, **Florida**, in 1966 (Swift et al. 1986; museum specimen).

Means of Introduction: Unknown. Possible bait bucket release.

Status: Reported from Florida.

Impact of Introduction: Unknown.

Remarks: Voucher specimen: UF 64680 (=FSU 14680).

Fundulus stellifer

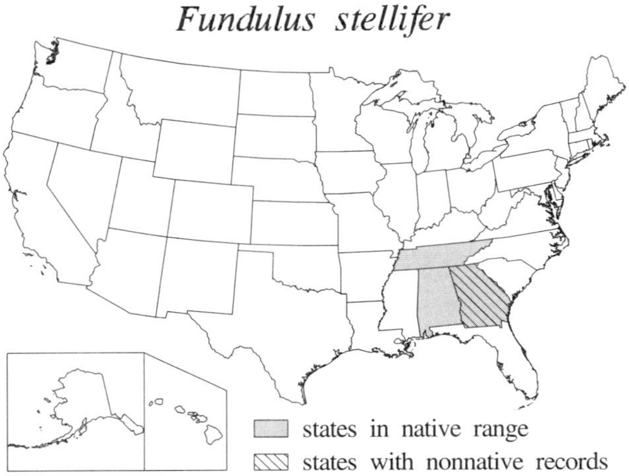

Fundulus stellifer (Jordan 1877)

Common Name: southern studfish.

Identification: Page and Burr (1991); Etnier and Starnes (1993); Mettee et al. (1996). Maximum size: 12 cm.

Native Range: Alabama River drainage of Alabama, Georgia, and Tennessee (except Tallapoosa River system), and upper Chattahoochee River drainage, Georgia (Thomerson 1969).

Nonindigenous Occurrences: A single specimen was taken from West Fork Chickamauga Creek, a tributary of the Tennessee River in Walker County, northwestern **Georgia** in 1964 (Thomerson 1969; Dahlberg and Scott 1971a; Lee et al. 1980 et seq.; Page and Burr 1991).

Means of Introduction: This species was introduced through unknown means; possibly bait bucket release (Thomerson 1969).

Status: Reported from Georgia.

Impact of Introduction: Unknown.

Remarks: Thomerson (1969) was uncertain whether the Georgia specimen represented an introduction or a native record. The fish was taken with a northern studfish *F. catenatus*, a rare and possibly unnatural co-occurrence.

Voucher specimens: Georgia (TU 34975).

Fundulus zebrinus

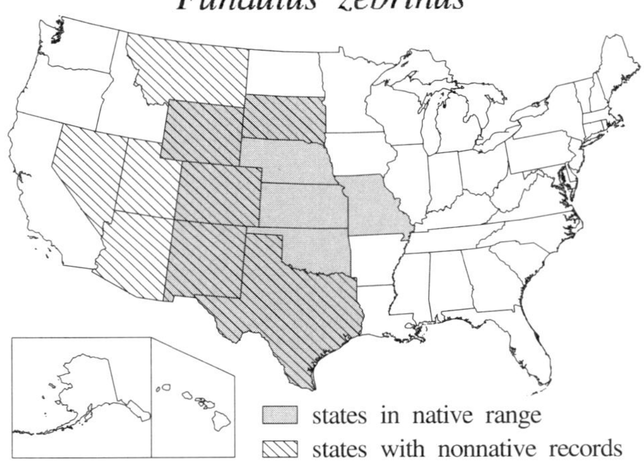

Fundulus zebrinus Jordan and Gilbert 1883

Common Name: plains killifish.

Identification: Sigler and Sigler (1987); Hubbs et al. (1991); Page and Burr (1991); Pflieger (1997). *Fundulus kansae* is considered a junior synonym (Poss and Miller 1983). Maximum size: 10 cm.

Native Range: Mississippi River and Gulf Slope basins from north central Missouri to central Wyoming, and south to Colorado River, Brazos River, Galveston Bay, and Rio Grande (primarily Pecos River) drainages, Texas. Mostly on Great Plains (Page and Burr 1991).

Nonindigenous Occurrences: This species is known from the Colorado and Little Colorado river drainages in **Arizona** (Miller and Lowe 1967; Minckley 1973; Hughes 1981; Poss and Miller 1983); the Colorado River and several tributaries (e.g., Gunnison, Green, Yampa, San Juan) in **Colorado** (Holden and Stalnaker 1975; Hughes 1981; Tyus et al. 1982; Poss and Miller 1983; Woodling 1985); Fort Peck Reservoir, O'Fallon Creek (a tributary of the Yellowstone), and the Little Missouri, Big Horn, and Yellowstone rivers in **Montana** (Brown 1971; Poss and Miller 1983; Cross et al. 1986; Holton 1990); tributaries of Lake Mead and along the Colorado River, **Nevada** (Deacon and Williams 1984); the Zuni and San Juan rivers in **New Mexico** (Hughes 1981; Tyus et al. 1982; Poss and Miller 1983; Sublette et al. 1990); the Cheyenne River drainage in **South Dakota** (Bailey and Allum 1962; Poss and Miller 1983); the Rio Grande and a few of its tributaries in and near Big Bend National Park, and possibly elsewhere (e.g., Trinity River) in **Texas** (Poss and Miller 1983; Hubbs et al. 1991); Juab County near Mona Reservoir and the Colorado River in **Utah** (Sigler and Miller 1963; Minckley 1973; Hughes 1981; Tyus et al. 1982; Poss and Miller 1983; Woodling 1985; Sigler and Sigler 1987); and the Bighorn and Cheyenne river drainages, **Wyoming** (Poss and Miller 1983).

Means of Introduction: Most introductions apparently originated from bait bucket releases. In western Colorado, the species may have been introduced by bait bucket transfers or accidentally stocked as a contaminant with other species, although upstream expansion from Utah cannot be ruled out (Woodling 1985). It possibly spread downstream into the Big Horn River of Montana from Wyoming (Brown 1971). The species has been introduced widely since about the 1930s (Poss and Miller 1983).

Status: Established in many areas of the Colorado River above and below Glen Canyon dam (Poss and Miller 1983). Presumably established in other localities mentioned.

Impact of Introduction: Unknown.

Remarks: Poss and Miller (1983) discussed the introduction history of this species and provided a dot distribution map distinguishing native and introduced records. According to Lee et al. (1980 et seq.), many records represent natural occurrences (e.g., the Yellowstone River drainage records in Montana and Wyoming, the Cheyenne River records in Wyoming and South Dakota, the non-Pecos records in the Rio Grande Basin in Texas); however, Poss and Miller (1983) concluded these populations were introduced. Holton (1990) found the species to be more widespread in Montana than previously thought, inhabiting the Little Missouri and Yellowstone drainages and the Fort Peck Reservoir area; thus, he suggested that the species may be native to the state. Tyus et al. (1982) gave a map showing its distribution in the upper Colorado basin. The plains killifish has never been collected in North Dakota (Steinwand, personal communication) even though it has been collected in the Yellowstone River in Montana near the Montana/North Dakota state line (Elser et al. 1980). That area of western North Dakota has been intensively surveyed (Steinwand, personal communication), so it is not likely that the species was overlooked.

Lucania goodei

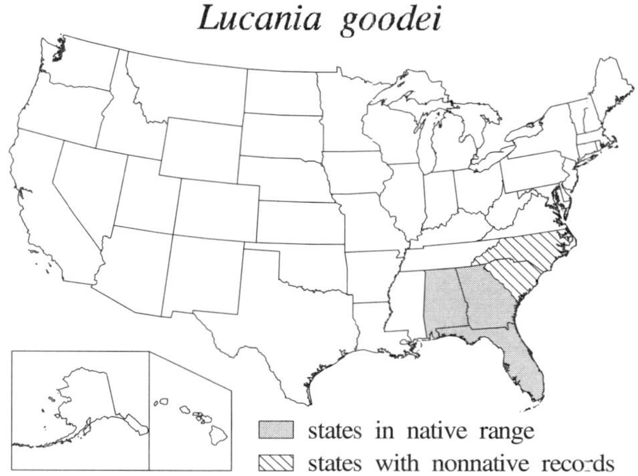

Lucania goodei Jordan 1880

Common Name: bluefin killifish.

Identification: Menhinick (1991); Page and Burr (1991); Mettee et al. (1996). Maximum size: 5 cm.

Native Range: Throughout Florida, except in panhandle only as far west as Choctawhatchee River drainage and southeastern Alabama in Chipola River drainage (Page and Burr 1991); sporadically along Atlantic Coast as far north as central South Carolina (where possibly introduced).

Nonindigenous Occurrences: This species was probably introduced into both the Northeast Cape Fear drainage near Wilmington, **North Carolina** (Menhinick 1991) and the Cooper River on the central **South Carolina** coast (Loyacano 1975; Lee et al. 1980 et seq.;

Christie and Curtis 1983; Page and Burr 1991; Rhode et al. 1994).

Means of Introduction: Unknown. Based on meristics, the South Carolina population came from Florida (Christie and Curtis 1983).

Status: Established in North Carolina and South Carolina.

Impact of Introduction: Unknown.

Remarks: The isolated population near Wilmington, North Carolina, is similar to the isolated population of *Heterandria formosa*, another species listed by Menhinick (1991) as introduced. These similar distributions could indicate a shared transport event, or alternatively, a shared refugium (Starnes, personal communication).

Lucania parva

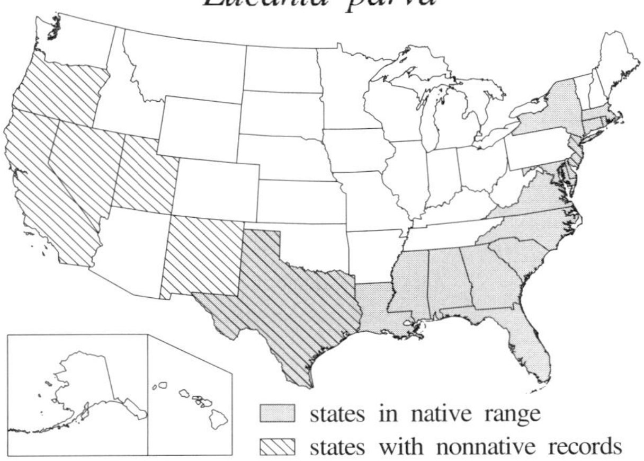

Lucania parva (Baird and Girard 1855)

Common Name: rainwater killifish.

Identification: Smith (1985); Menhinick (1991); Page and Burr (1991); Mettee et al. (1996). Maximum size: 7 cm.

Native Range: Marine, but enters freshwater from Massachusetts to Mexico. Ascends Rio Grande and Pecos rivers, Texas and New Mexico (Hubbs and Miller 1965; Page and Burr 1991).

Nonindigenous Occurrences: The species was introduced into Lake Irvine in the Los Angeles basin and into the San Francisco Bay area, **California** (Sigler and Miller 1963; Hubbs and Miller 1965; Moyle 1976a; Dill and Cordone 1997); the Blue Lake area, **Nevada**, near the Utah border (Deacon and Williams 1984); the Great Swamp, **New Jersey** (Stiles 1978); the lower Pecos and Rio Grande drainages, **New Mexico** (Sublette et al. 1990; Platania 1991); the coastal region of **Oregon**, including Yaquina Bay drainage (Sigler and Miller 1963; Hubbs and Miller 1965; Moyle 1976a; Bond 1994; Dill and Cordone 1997); Clear Creek (San Saba River drainage), **Texas** (Hubbs et al. 1991); and Timpie Springs and Blue Lake, **Utah** (Sigler and Miller 1963; Hubbs and Miller 1965; Moyle 1976a; Dill and Cordone 1997).

Means of Introduction: Many introductions likely were the result of contaminated stock (Sigler and Miller 1963; Hubbs and Miller 1965; Moyle 1976a, 1976b). In Utah, rainwater killifish arrived as a stock contaminate in a shipment of largemouth bass (Sigler and Sigler 1987). In northern California and Oregon, rainwater killifish may have been transported as eggs on planted oysters (Moyle 1976a, 1976b) or with ballast water from the Atlantic Coast (Hubbs and Miller 1965). In southern California it was a contaminant in a shipment of largemouth bass (Moyle 1976b). Hubbs and Miller (1965) indicated some introductions may have resulted when the species was stocked along with *Gambusia* during mosquito control operations, or from aquarium release. Most of the original introductions in the western United States apparently occurred during the 1940s and 1950s (Hubbs and Miller 1965). The species was introduced into the Great Swamp in New Jersey in 1969 (Stiles 1978) and into Texas in about 1980 (Hubbs et al. 1991), but the reasons for these introductions were not given.

Status: Presumably established in California, Nevada, New Jersey, New Mexico, Oregon, Texas, and Utah.

Impact of Introduction: Unknown.

Remarks: Hubbs and Miller (1965) reviewed the introduction history of this species and provided a map distinguishing introduced and native records.

Poeciliidae - Live bearers

The family Poeciliidae includes 30 genera and about 293 species found in the eastern United States south to South America, and in Africa (Nelson 1994). Some 12 species are native to the United States (Page and Burr 1991). Poeciliids are small fishes; most species are less than 10 cm long. They closely resemble killifish and topminnows, but males of the live-bearing poeciliids have elongated anterior anal fin rays that are used for internal fertilization. In most species females attain a larger size than males. Poeciliids are live bearers, except for two groups (formerly placed in the family Cyprinodontidae) that have external fertilization (Parenti 1981; Nelson 1994). Although a few species are piscivores, most poeciliids are omnivores and feed on mixtures of invertebrates, detritus, algae, and vascular plants (Meffe and Snelson 1989). The mosquitofishes, *Gambusia affinis* and *G. holbrooki*, have been introduced widely and indiscriminately into many parts of the world as mosquito control agents. Many poeciliids are popular in the aquarium trade, such as the guppy and mollies of the genus *Poecilia*, and swordtails and platies of the genus *Xiphophorus*. Many of these species have been backcrossed and crossbred to produce an incredible variety of offspring, many of which are fertile. Given the natural diversity of the group, coupled with frequent release of artificially-produced strains and hybrids, the positive identification of introduced populations is often difficult to determine. Poeciliids inhabit fresh and brackish waters (Nelson 1994).

Belonesox belizanus

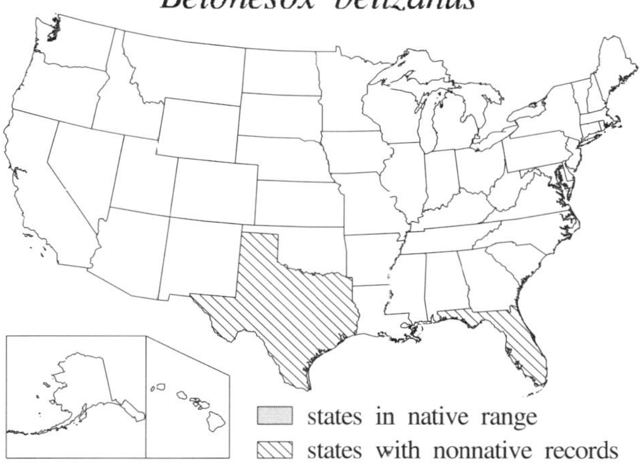

Belonesox belizanus Kner 1860

Common Names: pike killifish (pike top live bearer, pike topminnow, picudito).

Identification: Distinguishing characteristics and illustrations were given by Rosen and Bailey (1963), Page and Burr (1991), and Howells (1992a). The species is included in identification keys of Greenfield and Thomerson (1997). Photographs also appeared in Bussing (1987), Dawes (1991), Sakurai et al. (1993), and Wischnath (1993). Maximum size: Females 15 cm; males 11 cm.

Native Range: Atlantic Slope from the Río Antigua system, Veracruz, Mexico, south to the Yucatan Peninsula, Guatemala, Belize, Honduras, Nicaragua, and northern Costa Rica (Rosen and Bailey 1963; Miller 1966; Greenfield and Thomerson 1997).

Nonindigenous Occurrences: This species was first documented in **Florida** in canals in southeastern Dade County in 1957 (Belshe 1961). It is now established or has been taken in canals, ditches, mangrove swamps, and other habitats in Dade and Collier counties, including parts of Everglades National Park and Big Cypress National Preserve (Belshe 1961; Courtenay et al. 1974; Miley 1978; Turner and Snelson 1984; Loftus and Kushlan 1987; Lorenz et al. 1997; museum specimens). A separate population was found in a borrow pit on North Key Largo, Monroe County, Florida, in 1984 (Courtenay and Meffe 1989). A few specimens were taken from a roadside ditch in Hillsborough County, Florida, in 1997 (Nico, unpublished data). A population was found in **Texas** in a spring run of the San Antonio River, San Antonio, in the early 1960s (Barron 1964), but that population apparently has died out (Hubbs et al. 1978; Howells 1992a).

Means of Introduction: This species was introduced into south Florida in 1957 when individuals reared for medical research purposes were released into a local canal (Belshe 1961). Those found in Hillsborough County in 1997 were escapees from a nearby ornamental fish farm (Nico, personal communication). The Texas introduction was probably an aquarium release (Howells 1992a).

Status: Established in southern Florida. The Key Largo, Florida, population apparently did not become established (Courtenay and Meffe 1989). A population was possibly established temporarily in the San Antonio area of Texas. However, the species was absent in later samples and the population was assumed to have been extirpated (Hubbs et al. 1978; Howells 1992a).

Impact of Introduction: The pike killifish is a voracious predator and has been known to reduce populations of eastern mosquitofish *Gambusia holbrooki* and other native poeciliid and cyprinodontid populations in areas where it has been introduced (Belshe 1961; Loftus and Kushlan 1987; Courtenay and Meffe 1989).

Remarks: The Florida population is descended from stock that is thought to have come from the Yucatan peninsula (Belshe 1961); it is the subspecies *B. b. maxillosus* (Rosen and Bailey 1963; Page and Burr 1991). In Florida, reproduction has been recorded in salinities up to at least 35 parts per thousand (Turner and Snelson 1984).

Voucher specimens: Florida (UMMZ 178973, UF many collections).

Gambusia affinis and *G. holbrooki*

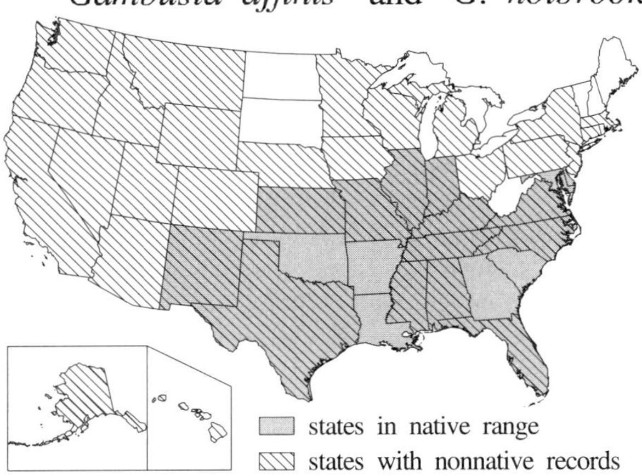

Gambusia affinis (Baird and Girard 1853) and *G. holbrooki* Girard 1859

Common Names: western and eastern mosquitofishes.

Identification: Distinguishing characteristics were provided by Rauchenberger (1989a) and Page and Burr (1991) (although the latter authors treated the two forms as subspecies). These two species were long considered subspecies of *G. affinis*, and were only recently recognized as separate species (Wooten et al. 1988; Rauchenberger 1989a; Robins et al. 1991a). Complicating matters of identification, most introductions occurred before the recent taxonomic change; furthermore, the origins of introduced stocks were usually unknown or unreported. In addition, both forms were widely available and thought to have been dispersed widely by humans. As a consequence, it often is not possible to determine if many of the earlier records represent introductions of *G. affinis* or of *G. holbrooki*. Maximum size: 6.5 cm.

Native Range: Atlantic and Gulf Slope drainages from southern New Jersey to Mexico; Mississippi River basin from central Indiana and Illinois south to Gulf. *Gambusia holbrooki* is native to Atlantic and Gulf Slope drainages as far west as southern Alabama; *G. affinis* occurs throughout rest of the range (Rauchenberger 1989a; Page and Burr 1991).

Nonindigenous Occurrences: Mosquitofish have been stocked in **Alabama** (Boschung 1992); **Alaska** (Krumholz 1948); **Arizona** (Dees 1961; Miller and Lowe 1967; Minckley 1973; Lee et al. 1980 et seq.); **California** (Dees 1961; La Rivers 1962; Minckley 1973; Moyle 1976a; Lee et al. 1980 et seq.; Moyle and Daniels 1982; Smith 1982; Taylor et al. 1982); **Colorado** (Woodling 1985; Tyus et al. 1982; Zuckerman and Behnke 1986; Dill and Cordone 1997); **Connecticut** (Whitworth 1996); **Florida** (J. Chick, personal communication; J. D. Williams, personal observation); **Hawaii** (Brock 1960; Maciolek 1984); **Idaho** (Simpson and Wallace 1978; Lee et al. 1980 et seq.; Idaho Fish and Game 1990); **Illinois** (Dees 1961; Smith 1979); **Indiana** (Dees 1961; Simon et al. 1992); **Iowa** (Harlan et al. 1987); **Kansas** (Cross 1954, 1967; Clarke et al. 1958; Dees 1961); **Kentucky** (Clay 1975; Burr and Warren 1986); **Massachusetts** (Krumholz 1948; Dees 1961); **Michigan** (Krumholz 1948; Hubbs and Lagler 1958; Dees 1961; Lee et al. 1980 et seq.); **Minnesota** (Eddy and Underhill 1974; Phillips et al. 1982); **Mississippi** (Krumholz 1948); **Missouri** (Dees 1961; Cross 1967; Pflieger 1997); **Montana** (Brown 1971; Lee et al. 1980 et seq.; Holton 1990); **Nebraska** (Haynes 1983; Lynch 1988a, 1988b, 1991); **Nevada** (Miller and Alcorn 1946; Dees 1961; La Rivers 1962; Lee et al. 1980 et seq.; Deacon and Williams 1984); **New Jersey** (Krumholz 1948; Fowler 1952; Dees 1961); **New Mexico** (Barber et al. 1929; Koster 1957; Dees 1961; Lee et al. 1980 et seq.; Tyus et al. 1982; Sublette et al. 1990; Plantania 1991); **New York** (Dees 1961; Lee et al. 1980 et seq.; Smith 1985; Schmidt 1986); **North Carolina** (Menhinick 1991); **Ohio** (Dees 1961; Lee et al. 1980 et seq.; Trautman 1981; Hocutt et al. 1986; Burr and Page 1986); **Oregon** (Bond 1961, 1973, 1994); **Pennsylvania** (Dees 1961; Cooper 1983); **Tennessee** (Kuhne 1939; Etnier and Starnes 1993); **Texas** (Hubbs, personal communication); **Utah** (Rees 1934, 1945; Dees 1961; Sigler and Miller 1963; Minckley

1973; Lee et al. 1980 et seq.; Tyus et al. 1982); **Virginia** (Jenkins and Burkhead 1994); **Washington** (Dees 1961; Wydoski and Whitney 1979; Fletcher, personal communication); **West Virginia** (Cincotta, personal communication); **Wisconsin** (Krumholz 1948; Dees 1961); **Wyoming** (Hubert 1994; Stone 1995), and probably other states.

Means of Introduction: Because of their reputation as mosquito-control agents, both *G. holbrooki* and *G. affinis* have been stocked routinely and indiscriminately in temperate and tropical areas around the world. In the United States the first known introductions of mosquitofish took place in the early 1900s (Krumholz 1948). In 1905 about 150 *G. affinis* were introduced into Hawaii from Texas to test their effectiveness in preying on mosquito larvae (Seale 1905), and by 1910 their descendants had been released into parts of Oahu, Hawaii, Maui, Kauai, and Molokai (Van Dine 1907; Stearns 1983). Also, in 1905 *Gambusia*, reportedly from North Carolina, were released into New Jersey waters for the purpose of controlling mosquitoes (Seal 1910; Krumholz 1948). Mosquitofish were commonly and widely introduced during the following decades by such organizations as the former U.S. Public Health Service, in large part because they were thought of as an effective and inexpensive means of combating malaria (Krumholz 1948). In more recent years, employees of many state and local health departments apparently view the use of mosquitofish to control mosquito larvae as an attractive alternative to the use of insecticides. In some areas range extensions have occurred through natural dispersal far from sites where originally introduced (e.g., Pflieger 1997).

Status: Established in most states where stocked outside its native range. Its establishment and spread in northern states is greatly restricted because the species are not, in general, cold tolerant. In most cases, overwintering in colder regions requires surfacing groundwater springs (e.g., Woodling 1985; but see Lynch 1988b). Established in Nebraska, although the populations suffer heavy (up to 99%) winter mortality (Haynes 1983). Pflieger (1997) noted that *Gambusia affinis* is more widespread and abundant in Missouri now than it was half a century ago. For instance, Pflieger indicated that, by the early 1980s, it had become established northward along the Mississippi River to Clark County, Missouri, and westward near the Missouri River to Andrew County, a range expansion attributed to a combination of natural dispersal and undocumented introductions.

Impact of Introduction: According to Courtenay and Meffe (1989), mosquitofish have had the greatest ecological impact by far of any of the introduced poeciliids. Although widely introduced as mosquito control agents, recent critical reviews of the world literature on mosquito control have not supported the view that *Gambusia* are particularly effective in reducing mosquito populations or in reducing the incidence of mosquito-borne diseases (Courtenay and Meffe 1989; Arthington and Lloyd 1989). Because of their aggressive and predatory behavior, mosquitofish may negatively affect populations of small fish through predation and competition (Myers 1967; Courtenay and Meffe 1989). In some habitats, introduced mosquitofish reportedly displaced select native fish species regarded as better or more efficient mosquito control agents (Danielsen 1968; Courtenay and Meffe 1989). Introduced mosquitofish have been particularly destructive in the American West where they have contributed to the elimination or decline of populations of federally endangered and threatened species (Courtenay and Meffe 1989). Specific examples of their negative effects include a habitat shift and a reduction in numbers of the threatened Railroad Valley springfish *Crenichthys baileyi* in springs in Nevada (Deacon et al. 1964) and the local elimination of the endangered Sonoran topminnow *Poeciliopsis occidentalis* in Arizona (Moyle 1976a; Meffe et al. 1983, Meffe 1985). Western mosquitofish use the same habitat as the plains topminnow *Fundulus sciadicus* and have displaced these topminnows and other species with their aggressive behavior (Whitmore 1997). The mosquitofish is also responsible for the elimination of the least chub *Iotichthys phlegethontis* in several areas of Utah (Whitmore 1997). Meffe (1983, 1985) found that mosquitofish are very aggressive, even toward larger fish. They often attack, shred fins, and sometimes kill other species. Mosquitofish are known to prey on eggs, larvae, and juveniles of various fishes, including those of largemouth bass and common carp; they are also known to prey on adults of smaller species (Meffe 1985; Courtenay and Meffe 1989). Courtenay and Meffe (1989) listed impacts on a variety of native fishes. Introducing mosquitofish also can precipitate algal blooms when the fish eat the zooplankton grazers (Hurlbert et al. 1972), or in an increase in the number of mosquitoes if the fish eat the invertebrate predators (Hoy et al. 1972). Introduced fishes, including mosquitofish, are likely at least partially responsible for the decline of the Chiricahua leopard frog *Rana chiricahuensis* in southeastern Arizona (Rosen et al. 1995).

Remarks: Summaries and reviews of mosquitofish introductions were provided by Krumholz (1948), Hardy (1978), Courtenay and Meffe (1989), and Dill and Cordone (1997). The identity of some mosquitofish populations introduced into selected areas is correctly known. In most cases this is because the source of the stock was reported. The western mosquitofish, *G. affinis*, has been documented as introduced into Arizona (Dees 1961; Miller and Lowe 1967); California (Dees 1961; Moyle 1976a; Swift et al. 1993; Dill and Cordone 1997); Florida (Chick, personal communication; Williams, personal observation); Hawaii (Stearns 1983); Illinois (Krumholz 1948; Dees 1961); Indiana (Dees 1961); Kansas (Dees 1961; Cross 1967); Massachusetts (Dees 1961); Michigan (Dees 1961); Missouri (Dees 1961; Pflieger 1997); Nebraska (Lynch 1988a, 1988b); Nevada (Miller and Acorn 1946; Dees 1961); New Jersey (Dees 1961); New Mexico (Dees 1961); New York (Dees 1961); Ohio (Dees 1961); Pennsylvania (Dees 1961); Utah (Rees 1945; Dees 1961); Washington (Dees 1961); and Wisconsin (Dees 1961). *Gambusia holbrooki* was introduced into New Jersey (Fowler 1952) and into Tennessee near Knoxville and maybe to other locations as well (Starnes, personal communication). Both species have been introduced into Alabama (Boschung 1992). Shapovalov et al. (1981) indicated that both species were introduced into California, but Swift et al. (1993) argued that *G. holbrooki* never has been taken in the state and probably never was stocked. There was even mention that a hybrid between the two species was released into California waters (Dill and Cordone 1997). In their recent tome on fishes introduced into California, Dill and Cordone (1997) related their strong suspicions that pure *Gambusia holbrooki* had been introduced into that state. They based their conclusion, in part, on the importance and size of the mosquito control program in California, and the central role mosquitofish played in those attempts. However, Dill and Cordone did admit that there was no real proof that *G. holbrooki* became established in the state. In some cases *Gambusia* stocks native to a particular region of a state were moved within the same state, in Virginia for example (Jenkins and Burkhead 1994). In contrast, Krumholz (1948) reported that mosquitofish from southern Illinois, where the species is native, were introduced into northern Illinois, an area outside its native range. Hubbs and Lagler (1958) reported that intergrades between *G. affinis* and *G. holbrooki* have been introduced into southern Michigan, but the stock did not become established. Galat and Robertson (1992) found that the Yaqui topminnow *P. occidentalis sonoriensis* occurring in some sites increased their fecundity in response to the presence of introduced *Gambusia*; however, the researchers noted that such habitats must also have certain environmental conditions (e.g., uniform temperatures) for maintenance of vigourous *P. o. sonoriensis* populations. Galat and Robertson concluded that conservation of some extant populations of *P. occidentalis* depends primarily on control of *Gambusia*.

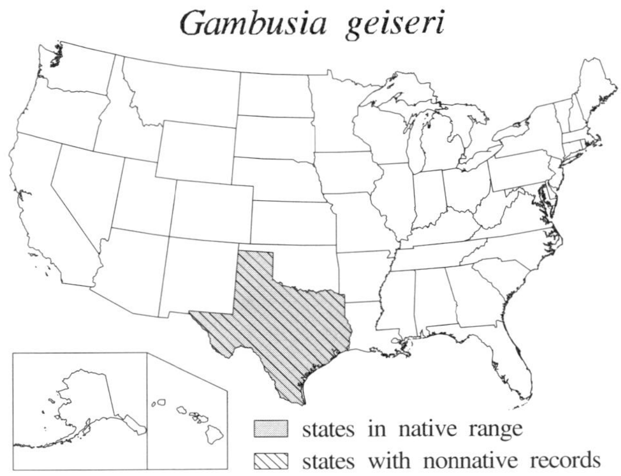

Gambusia geiseri Hubbs and Hubbs 1957

Common Name: largespring gambusia.

Identification: Hubbs et al. (1991); Page and Burr (1991). Maximum size: 4.4 cm.

Native Range: Springs in the San Marcos and Guadalupe River systems, central Texas (Lee et al. 1980 et seq.; Hubbs et al. 1991; Page and Burr 1991).

Nonindigenous Occurrences: The largespring gambusia has been introduced into several localities in **Texas** including Diamond Y Springs on Leon Creek (Pecos County); the Balmorhea springs complex including San Solomon, Giffin, East and West Sandia, and Phantom Cave springs, and irrigations ditches in that vicinity (Reeves County); Lazy Pond on Independence Creek, and nearby Chandler Springs (Terrel County); the headwaters of the Concho River at Head of the Run Ranch (formerly known as Anson Spring) south of Christoval (Tom Green County) (Hubbs et al. 1991; Hubbs, personal communication); and Comanche Springs (now desiccated) in Fort Stockton (Reeves County) (U.S. Fish and Wildlife Service 1982b).

Means of Introduction: Stocks from San Marcos, Texas, were introduced into Comanche Springs and Balmorhea area in 1937 and 1956, respectively (U.S. Fish and Wildlife Service 1982b). Dates of other introductions are not known. Introductions into the springs at Balmorhea were separate introductions as the springs are not connected (Hubbs, personal communication). Hendrickson and Brooks (1991) indicated that *G. geiseri* was stocked into at least one site for conservation purposes, but the authors did not provide details or information on the locality.

Status: Locally established in Texas in the Balmorhea area springs. Extirpated at Comanche Springs which periodically becomes dry (Hubbs, personal communication).

Impact of Introduction: This species reportedly competes with another spring-dwelling species, the endangered Pecos gambusia *G. nobilis* and is considered a potential threat to that species' existence (U.S. Fish and Wildlife Service 1982b).

Remarks: None.

Gambusia nobilis

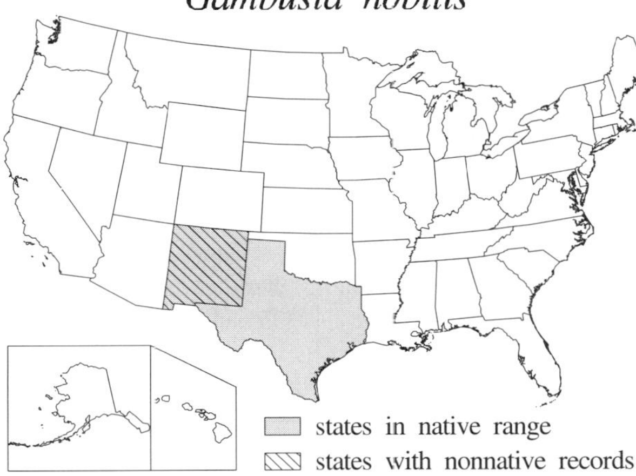

Gambusia nobilis (Baird and Girard 1853)

Common Name: Pecos gambusia.

Identification: Sublette et al. (1990); Hubbs et al. (1991); Page and Burr (1991). Maximum size: 4.8 cm.

Native Range: Endemic to springs and sinkholes in the Pecos River system, New Mexico and Texas (Hubbs et al. 1991; Page and Burr 1991).

Nonindigenous Occurrences: The Pecos gambusia was introduced into **New Mexico** in Lake St. Francis and in sinkholes at Bitter Lake National Wildlife Refuge, Chaves County, that did not contain native populations of this species (other sinkholes where it was introduced contained native populations); Ink Pot, Salt Creek Wilderness Area, Chaves County; Geyser Spring; and artificial ponds at the Living Desert State Park near Carlsbad, Eddy County (U.S. Fish and Wildlife Service 1982b; Sublette et al. 1990).

Means of Introduction: This endangered species was intentionally stocked to create additional populations. Ink Pot and one sinkhole at Bitter Lake were stocked in 1973. The source of the Living Desert State Park population stocked in 1975 is apparently Blue Spring, near Black River Village, Eddy County (U.S. Fish and Wildlife Service 1982b). Pecos gambusia were transplanted from waters near the north end of the Bitter Lake National Wildlife Refuge into 19 new localities within the same refuge and within the Salt Creek Wilderness Areas in 1972 and 1973. Other sinkholes at Bitter Lake were stocked in July and August 1981 (U.S. Fish and Wildlife Service 1982b). As part of recovery efforts, health officials in New Mexico investigated the use of *G. nobilis* in stock ponds within the Pecos River drainage to control mosquitoes, but the project was stopped when stocked fish failed to survive (Johnson and Hubbs 1989).

Status: Extirpated at Lake St. Francis and Geyser Spring, Bitter Lake National Wildlife Refuge, probably due to predation by introduced green sunfish *Lepomis cyanellus* and largemouth bass *Micropterus salmoides* (U.S. Fish and Wildlife Service 1982b). New populations became established in two sinkholes and at Ink Pot as result of 1972–1973 stockings. The other 16 transplants made at that time failed. Status of 1981 populations not known (U.S. Fish and Wildlife Service 1982b).

Impact of Introduction: Unknown.

Remarks: The Pecos gambusia was listed as a federally endangered species in 1970 (U.S. Fish and Wildlife Service 1982b, 1993a). It has been affected negatively by hybridization, predation, and competition with two introduced species: the western mosquitofish *G. affinis* and the largespring gambusia *G. geiseri* (U.S. Fish and Wildlife Service 1982b; Hendrickson and Brooks 1991; Minckley et al. 1991). Hendrickson and Brooks (1991) indicated that *G. nobilis* had been stocked into 33 different sites for conservation purposes, but they did not distinguish between historic native habitats and sites outside its former native range.

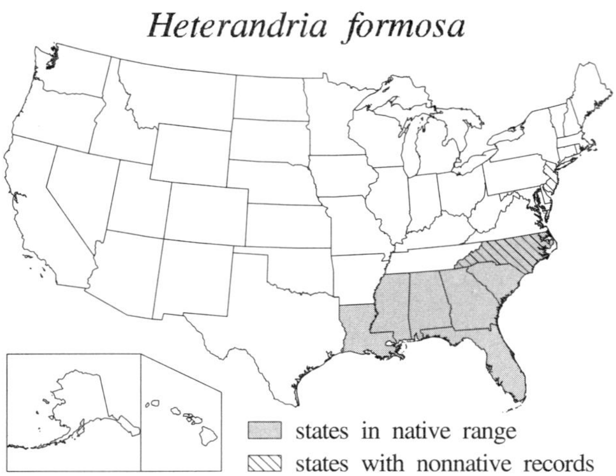

Heterandria formosa (Agassiz 1855)

Common Name: least killifish.

Identification: Menhinick (1991); Page and Burr (1991); Mettee et al. (1996). Maximum size: 3.6 cm.

Native Range: Lower Coastal Plain from South Carolina to southern Louisiana (Rosen 1979; Page and Burr 1991).

Nonindigenous Occurrences: The species was stocked into **New Jersey** (Seal 1910) and introduced into Cape Fear and Northeast Cape Fear drainages near Wilmington, **North Carolina** (Menhinick 1991; Page and Burr 1991).

Means of Introduction: In 1905 about 10,000 *Heterandria* and *Gambusia* were stocked into New Jersey for mosquito control purposes (Seal 1910). The reason for the North Carolina stocking is not known.

Status: The New Jersey population apparently did not survive; the species is established in North Carolina.

Impact of Introduction: Unknown.

Remarks: Seal (1910) did not provide the locality or localities in New Jersey where he stocked fish, although he did mention that at least one of the sites was some 90 miles from Cape May. Krumholz (1948) indicated that the fish stocked in New Jersey came from North Carolina, but that information does not appear in Seal (1910). The isolated population near Wilmington, North Carolina, is similar to the distribution of *Lucania goodei*, another species listed by Menhinick (1991) as introduced. These similar distributions could indicate a shared transport event, or conversely, a shared refugium (Starnes, personal communication).

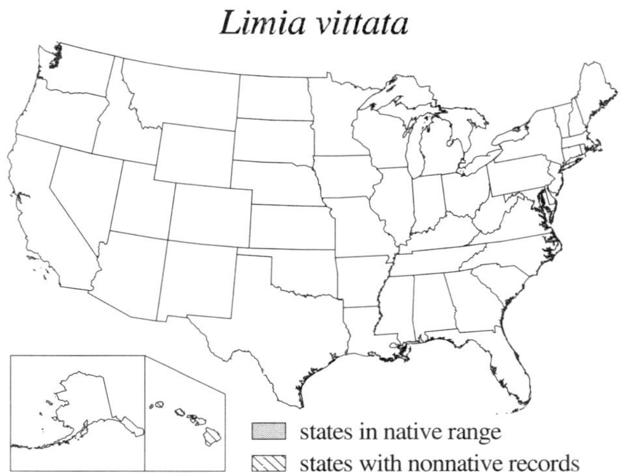

Limia vittata (Guichenot 1853)

Common Names: Cuban limia (banded limia).

Identification: Distinguishing characteristics were provided by Rosen and Bailey (1963) and Rivas and Fink (1970). For photographs see Dawes (1991) and Wischnath (1993). A commonly used name for this species is *Poecilia vittata* (Rosen and Bailey 1963; Rivas 1980). Maximum size: Males to 50 mm SL; females to 70 mm SL.

Native Range: Streams, lakes, estuaries, coastal lagoons, and mangrove swamps throughout Cuba and Isle of Pines (Lee et al. 1980 et seq.).

Nonindigenous Occurrences: The only known occurrence of this species in the United States is in **Hawaii** on Oahu (Brock 1960; Maciolek 1984; Devick 1991b).

Means of Introduction: According to Brock (1960), the origin and date of introduction into Hawaii are not known. Devick (1991b) listed the species as an aquarium release some time between 1900 and 1940; although its presence in Hawaii was not known until about 1950.

Status: Established in Hawaii by 1950 (Devick 1991b).

Impact of Introduction: Unknown.

Remarks: Brock (1960) stated that the species inhabits several streams flowing through Honolulu, occupying the lower portions of the streams and in salt water. There are no known voucher specimens.

Poecilia formosa

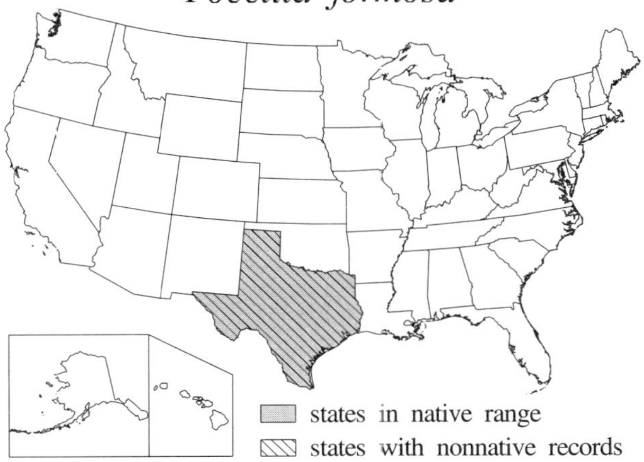

Poecilia formosa (Girard 1859)

Common Name: Amazon molly.

Identification: Distinguishing characteristics were given in Miller (1983), Hubbs et al. (1991), and Page and Burr (1991); other names used in the past for this species include *Mollienesia formosa* and *Limia formosa* (Hubbs 1955; Miller 1983). It was included in Miller's (1983) key to the Mexican *Poecilia*. A photograph appeared in Dawes (1991). Maximum size: 9.6 cm.

Native Range: Rio Grande drainage, extreme southern Texas; also in Mexico south to Veracruz (Miller 1983; Page and Burr 1991).

Nonindigenous Occurrences: Introduced populations have been recorded from the San Antonio River at San Antonio, the San Marcos River at San Marcos, and the lower Neuces River, **Texas** (Drewry et al. 1958; Lee et al. 1980 et seq.; Conner and Suttkus 1986; Hubbs et al. 1991; Page and Burr 1991).

Means of Introduction: Unknown.

Status: Several introduced populations have been documented in Texas (e.g, Hubbs et al. 1991; Page and Burr 1991); as such, it can be assumed that the species is, at a minimum, locally established.

Impact of Introduction: Unknown.

Remarks: *Poecilia formosa* is a gynogenetic, unisexual (all-female) molly of hybrid origin (Hubbs 1955; Miller 1983). The spermatozoa from a related species stimulate egg development but do not contribute any genetic material (Meffe and Snelson 1989). There is some doubt concerning the positive identification of this and some of the other molly species taken in Texas and as to whether certain populations are actually nonnative or possibly hybrids (Howells 1992a).

Poecilia latipinna

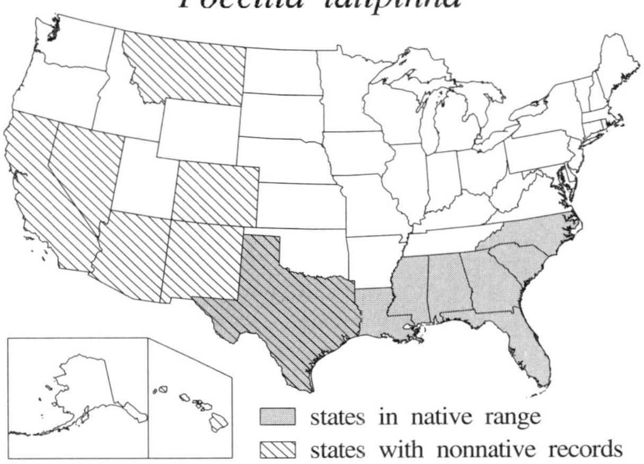

Poecilia latipinna (Lesueur 1821)

Common Name: sailfin molly.

Identification: Distinguishing characteristics given by Miller (1983), Hubbs et al. (1991), Menhinick (1991), and Page and Burr (1991); a formerly used name is *Mollienesia latipinna* (Miller 1983). Miller (1983) gave a key to this species and other Mexican *Poecilia*. Photographs of wild-caught and aquarium strains appeared in Wischnath (1993). Maximum size: 15 cm.

Native Range: Atlantic and Gulf Coast drainages from Cape Fear drainage, North Carolina, to Veracruz, Mexico. Restricted to coastal areas in most of range; found farther inland in Florida, Louisiana, and Texas (Page and Burr 1991).

Nonindigenous Occurrences: This species has been introduced into several areas of **Arizona** (Minckley 1973; Lee et al. 1980 et seq.); parts of southern **California** (Shapovalov et al. 1959, 1981; St. Amant 1966; Minckley 1973; Mearns 1975; Moyle 1976a; Schoenherr 1979; Lee et al. 1980 et seq.; U.S. Fish and Wildlife Service 1983a; Swift et al. 1993; Williams et al. 1998); waters in Conejos County, **Colorado** (Woodling 1985; Zuckerman and Behnke 1986); all major islands of **Hawaii** (Seale 1905; Van Dine 1907; Brock 1960; Maciolek 1984; Devick 1991b); Trudau Pond, Madison County, **Montana** (Holton 1990); several springs in **Nevada** (La Rivers 1962; Deacon et al. 1964; Hubbs and Deacon 1964; Minckley 1973; Lee et al. 1980 et seq.; Courtenay and Deacon 1982; Deacon and Williams 1984); Dona Ana County, **New Mexico** (Sublette et al. 1990); and spring-influenced headwaters in central **Texas** (Brown 1953; Hubbs et al. 1991).

Means of Introduction: In most areas this species probably was introduced by way of aquarium releases. It was first brought to Hawaii from Texas in 1905 to test its effectiveness in controlling mosquitoes (Seale 1905; Van Dine 1907; Brock 1960). A failure in Hawaii at mosquito control, this fish has on occasion been used as a tuna baitfish in that state (Randall 1987).

Status: Established or locally established in Arizona (Minckley 1973), California (Swift et al. 1993), Colorado (Zuckerman and Behnke 1986), Hawaii (Devick 1991b), Montana (Holton 1990), Nevada (Deacon and Williams 1984), and Texas (Hubbs et al. 1991).

Impact of Introduction: The sailfin molly is responsible for the decline of the desert pupfish *Cyprinodon macularius* in California (U.S. Fish and Wildlife Service 1983a). Sigler and Sigler (1987) stated that the sailfin molly has probably impacted native species adversely.

Remarks: The sailfin molly has been stocked for mosquito control, even though it is largely or completely herbivorous (Courtenay and Meffe 1989). Records of this species in a few areas apparently are based on reports of the black molly, a hybrid, and not pure *P. latipinna* (Courtenay and Meffe 1989). Wischnath (1993) stated that U.S. commercial breeders have released various domestically bred forms, including *P. latipinna* hybrids, into natural waters. Contrary to Brown (1953) and Hubbs et al. (1991), Lee et al. (1980 et seq.) argued that *P. latipinna* found in inland waters of Texas were native. Improperly citing Van Dine (1907), Kanayama (1968) incorrectly used the name *Mollienesia latipes* for the species introduced to Hawaii.

Voucher specimens: California (LACM 36402-3), Montana (UMMZ 188981).

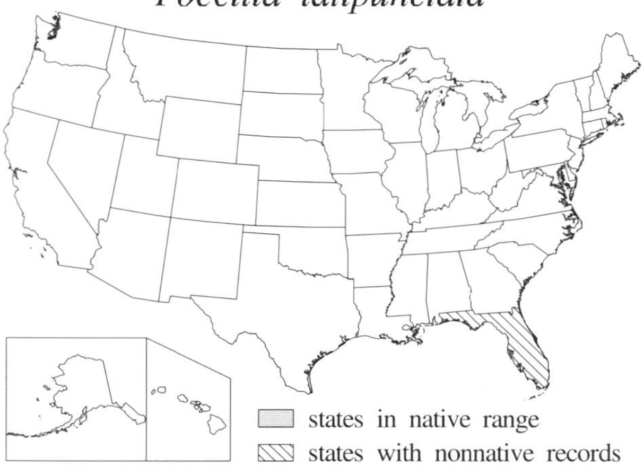

Poecilia latipunctata Meek 1904

Common Name: broadspotted molly (Tamesi molly, plain molly).

Identification: Distinguishing characteristics and an identification key that includes this species were given by Miller (1983). This species is part of the *P. sphenops* species complex (Lee et al. 1980 et seq.; Miller 1983). For illustration or photograph of the species see Meek (1904), Axelrod et al. (1985), and Wischnath (1993). Maximum size: 10 cm SL.

Native Range: Río Tamesí drainage of eastern Mexico (Miller 1983).

Nonindigenous Occurrences: Collected in Lake Worth Drainage Canal L-15, Palm Beach County, **Florida**, between 1969 and 1970 (Courtenay et al. 1974; Courtenay, personal communication).

Means of Introduction: Introductions are due to escapes and intentional releases from fish farms, as well as releases by aquarists (Lee et al. 1980 et seq.).

Status: Reported from Palm Beach County, Florida.

Impact of Introduction: Unknown.

Remarks: Wischnath (1993) gave a brief discussion on coloration and reported that intensely black piebald fish and nearly black males are seen on rare occasions.

There are no known voucher specimens.

Poecilia mexicana

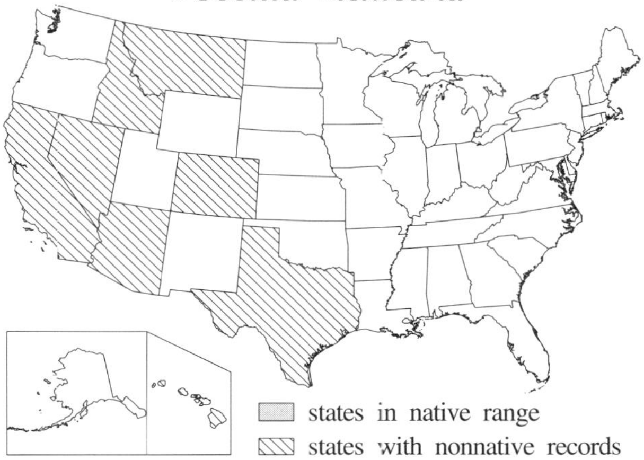

Poecilia mexicana Steindachner 1863

Common Name: shortfin molly (Mexican molly, Atlantic molly, orangefin molly).

Identification: Distinguishing characteristics were given by Miller (1983) and Page and Burr (1991); the species is included in identification keys of Miller (1983) and Greenfield and Thomerson (1997). It is part of the *P. sphenops* complex. Photographs appeared in Bussing (1987), Dawes (1991), and Wischnath (1993). Maximum size: 11 cm.

Native Range: Atlantic Slope from Rio San Juan, Mexico, to Guatemala, possibly south to Nicaragua (Miller 1983; Page and Burr 1991).

Nonindigenous Occurrences: This species has been recorded from Phoenix canals, **Arizona** (Minckley and Deacon 1968; Minckley 1973; W. Silvey, personal communication); southern **California** (St. Amant 1966; Mearns 1975; Moyle 1976a; Hubbs et al. 1979; Schoenherr 1979; Shapovalov et al. 1981; Courtenay et al. 1984, 1991; Swift et al. 1993); Conejos and Sagauche counties, **Colorado** (Hahn 1966; Woodling 1985; Zuckerman and Behnke 1986); Oahu, **Hawaii** (Maciolek 1984; Devick 1991b); the Brunneau River at Brunneau Hot Springs, **Idaho** (Courtenay et al. 1987); three locations in Madison County, **Montana** (Brown 1971; Holton 1990; Courtenay, personal communication); the Moapa River (U.S. Fish and Wildlife Service 1983a) and several springs in **Nevada** (Deacon et al. 1964; Hubbs and Deacon 1964; Deacon and Bradley 1972; Cross 1976; Courtenay and Deacon 1983; Deacon and Williams 1984; Scoppottone et al. 1998); and **Texas** (Courtenay et al. 1984, 1991; Howells 1992a).

Means of Introduction: Most introductions presumably were due to fish farm escapes or aquarium releases.

Status: Locally established in California (Swift et al. 1993), Colorado (Zuckerman and Behnke 1986), Hawaii (Devick 1991b), Idaho (Courtenay et al. 1988), and Nevada (Deacon and Williams 1984; Scoppottone et al. 1998). Reported from Arizona (Courtenay and Meffe 1989) and Texas (Howells 1992a); extirpated from Montana (Courtenay and Meffe 1989). It has become the most abundant fish species in Ash Springs, Nevada (Sigler and Sigler 1987).

Impact of Introduction: This species may prey upon native larval fishes (Scoppettone 1993). Species from the *P. sphenops* complex (specifically the Mexican molly) have negatively affected Moapa dace *Moapa coriacea* and White River springfish *Crenichthys baileyi* in Nevada (Lee et al. 1980 et seq.; U.S. Fish and Wildlife Service 1983a). The Moapa dace is listed as federally endangered, and the Moapa White River springfish is being considered for listing (Scoppettone 1993; U.S. Fish and Wildlife Service 1994a). Hoffman and Schubert (1984) reported that the parasitic copepod of the genus *Pseudocaligus* was transferred to Hawaii by way of *P. mexicana*.

Remarks: There is considerable confusion surrounding U.S. records of this and other members of the *P. sphenops* complex. For instance, Howells (1992a) expressed doubt as to the authenticity of the report of this species in Texas because of the common difficulty in distinguishing it from other members of the species complex, as well as from various molly hybrids. There is a file report indicating that this species was collected in July 1984 in Kelly Warm Springs, Teton County, Wyoming; however, that record is apparently in error (Courtenay, personal communication). There is no mention of this species being found in Kelly Warm Springs by Courtenay et al. (1988) although another introduced poeciliid, *P. reticulata*, was reported from the site. Wischnath (1993) reported that natural hybrids with *P. latipinna* are known from the Rio Mante Canal in Mexico.

Voucher specimens: Colorado (MSB, uncataloged); Nevada (UF 91917).

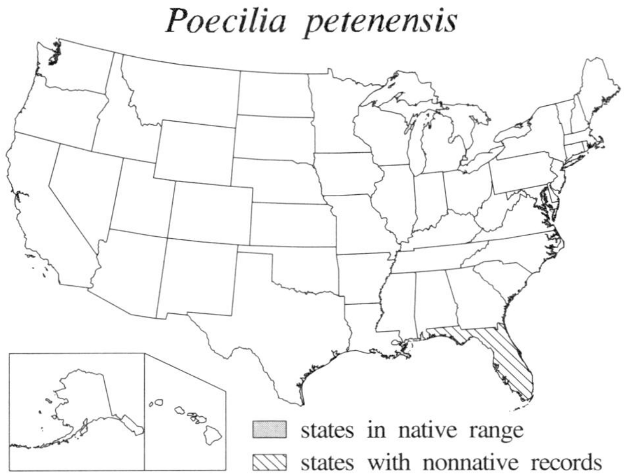

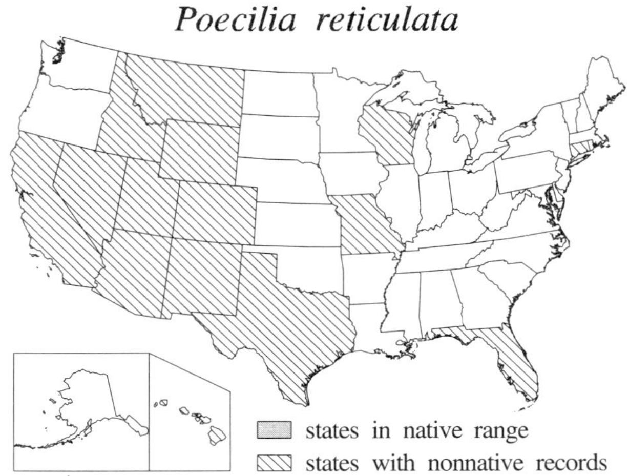

Poecilia petenensis (Günther 1866)

Common Names: Peten molly (swordtail molly, spiketail molly).

Identification: Distinguishing characteristics and identification keys that includes this species were given by Miller (1983) and Greenfield and Thomerson (1997). This species is part of the *Poecilia latipinna* complex. Another name is *Mollienesia petenensis*. A photograph appeared of this species appeared in Wischnath (1993). Maximum size: Females 10 cm; males 13 cm.

Native Range: Middle America from southeastern Mexico to Belize and Guatemala (Miller 1983; Greenfield and Thomerson 1997).

Nonindigenous Occurrences: Reported from Hillsborough and Palm Beach Counties, **Florida**, prior to about 1974 (Courtenay et al. 1974).

Means of Introduction: Probably due to aquarium releases.

Status: This molly has been reported from Florida. It has been suggested that introduced populations have survived for brief periods and possibly reproduced in open waters of that state, but no established populations are known (Courtenay et al. 1974).

Impact of Introduction: Unknown.

Remarks: Males of this species are characterized by a sail-like dorsal fin, similar to that of *P. latipinna*, and a short sword in the caudal fin (Dawes 1991).

There are no known voucher specimens.

Poecilia reticulata Peters 1859

Common Name: guppy.

Identification: Distinguishing characteristics were provided by Rosen and Bailey (1963); Sigler and Sigler 1987; and Page and Burr (1991). It is included in keys of Minckley (1973) and Sublette et al. (1990). Photographs or illustrations appeared in Minckley (1973), Axelrod et al. (1985), Petrovicky (1988), Sublette et al. (1990), Dawes (1991), Sakurai et al. (1993), and Wischnath (1993). This species was previously known by the name *Lebistes reticulatus*. Maximum size: Males to 2.5 cm; females to 5.1 cm.

Native Range: West Indies and northern South America from western Venezuela to Guyana (Rosen and Bailey 1963; Page and Burr 1991).

Nonindigenous Occurrences: This species is known from **Arizona** (Minckley 1973; Courtenay et al. 1984, 1991); **California** (St. Amant and Hoover 1969; Moyle 1976a; Shapovalov et al. 1981; Courtenay et al. 1984, 1991; Dill and Cordone 1997); **Colorado** (Zuckerman and Behnke 1986); **Connecticut** (Whitworth 1996); **Florida** (Courtenay et al. 1974; J. D. Williams, personal observation); **Hawaii** (Brock 1960; Kanayama 1968; Maciolek 1984; Devick 1991b); **Idaho** (Linder 1964; Baxter and Simon 1970; Simpson and Wallace 1978; Courtenay 1985; Courtenay et al. 1987; Idaho Fish and Game 1990); **Missouri** (Keevin 1978); **Montana** (Moyle 1976a); **Nevada** (Deacon et al. 1964; Moyle 1976a; Williams and Williams 1981; Courtenay and Deacon 1982, 1983; Deacon and Williams 1984); **New Mexico** (Sublette et al. 1990); **Texas** (Edwards 1976; Hubbs et al. 1978; Courtenay et al. 1984; Conner and

Suttkus 1986; Courtenay et al. 1991; Hubbs et al. 1991; Howells 1992a); **Utah** (F. B. Guttermuth, personal communication); **Wisconsin** (Becker 1983); and **Wyoming** (Baxter and Simon 1970; Courtenay et al. 1987; Hubert 1994; Stone 1995).

Means of Introduction: Most introductions probably due to fish farm or aquarium releases (e.g., Zuckerman and Behnke 1986; Courtenay and Meffe 1989; Howells 1992a; Dill and Cordone 1997). Some of the California introductions were for mosquito control (Shapovalov et al. 1981; Dill and Cordone 1997).

Status: Locally established in warmwater sites in Hawaii (Devick 1991b), Idaho (Courtenay et al. 1988), New Mexico (Sublette et al. 1990), Nevada (Courtenay and Deacons 1983), Texas (Hubbs et al. 1991), and Wyoming (Courtenay et al. 1988); reported from California (Shapovalov et al. 1981), Colorado (Zuckerman and Behnke 1986), Connecticut (Whitworth 1996), Missouri (Keevin 1978), Montana (Moyle 1976a), and Wisconsin (Becker 1983). Considered as established in Arizona by Minckley (1973); however, according to Courtenay and Meffe (1989), populations in Arizona and Florida do not appear to be self-sustaining. Although introduced widely in Texas, the only established population is one found in the San Antonio River near Brackenridge Park (Hubbs et al. 1991).

Impact of Introduction: The guppy is considered a hazard to native cyprinids and killifishes (Sigler and Sigler 1987). For instance, it has adversely affected the White River springfish *Crenichthys baileyi* in a Nevada spring (Deacon et al. 1964). Two subspecies of this springfish are now listed as federally endangered (U.S. Fish and Wildlife Service 1993a), and three others are proposed for listing (U.S. Fish and Wildlife Service 1994). The guppy has also been implicated in the decline of the Utah sucker *Catostomus ardens* in a thermal spring in Wyoming (Courtenay et al. 1988). The guppy has become the dominant species in some warm water springs in the west (Courtenay, personal communication). Courtenay and Meffe (1989) summarized the effect of this species on native fishes. This species also presents a threat because it is a known carrier of certain exotic trematode parasites (Leberg and Vrijenhoek 1994).

Remarks: This species has had a long and popular history as an ornamental fish. A wide variety of strains differing in color and fin shape have been cultivated by aquarists (Axelrod et al. 1985; Sakurai et al. 1993; Wischnath 1993). Some fish reported from the United States as *P. reticulata* actually may represent one of the other species in the genus. In 1989, M. Rauchenberger (personal communication) examined the *P. reticulata* voucher specimens (UF 91918) taken from Kelly Warm Springs, Wyoming, in 1984 by Courtenay et al. (1988), but she could not confirm that identification and labeled them as *Poecilia* species.

Voucher specimens: Nevada (TU 94347, 123712, 123718), Texas (TNHC 8958).

Poecilia sphenops

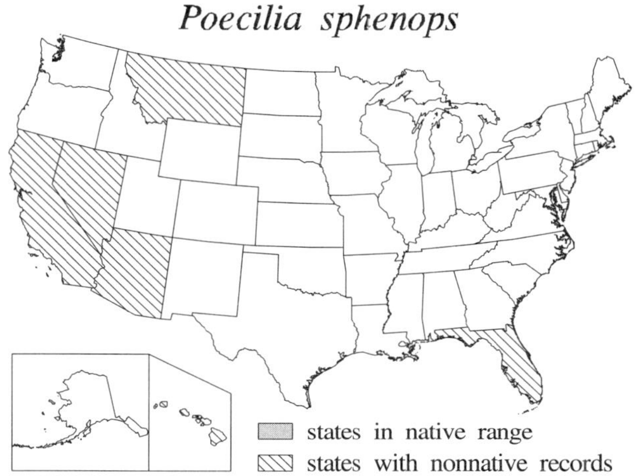

Poecilia sphenops Valenciennes 1846

Common Names: Mexican molly (liberty molly).

Identification: Distinguishing characteristics were given by Miller (1983). Miller (1983) provided a key to this species and other *Poecilia* found in Mexico. Photographs or illustrations appeared in Schultz and Miller (1971), Mills and Vevers (1982), Axelrod et al. (1985), Petrovicky (1988), Dawes (1991), Sakurai et al. (1993), and Wischnath (1993). This species is part of the *P. sphenops* complex (Rosen and Bailey 1963; Schultz and Miller 1971; Miller 1983). Formerly known as *Mollienesia sphenops*. Maximum size: 10 cm SL.

Native Range: Atlantic Slope of eastern Mexico from north of Veracruz City to Guatemala (Miller 1983).

Nonindigenous Occurrences: This species is known from southern **California** (St. Amant and Hoover 1969; Lee et al. 1980 et seq.; Dill and Cordone 1997); **Florida** (Courtenay et al. 1974); **Hawaii** (Maciolek 1984); the Ruby River, Madison County, **Montana** (Lee et al. 1980 et seq.); and Rogers Spring, **Nevada** (Courtenay, personal communication).

Means of Introduction: Introductions are due to escapes, intentional releases from fish farms, and releases by aquarists (Lee et al. 1980 et seq.; Dill and Cordone 1997).

Status: Locally established in Hawaii, Montana, and Nevada, and possibly Florida; reported from California.

Impact of Introduction: Introduced populations have adversely affected the Moapa dace *Moapa coriacea* and the White River springfish *Crenichthys baileyi* (both endangered species) and are a potential threat to other native fishes in the Pahranagat Valley, Nevada (Lee et al. 1980 et seq.).

Remarks: Some, conceivably all, of the records of this species in the United States may turn out to represent one of the other members of the *P. sphenops* species complex and not *P. sphenops*. For instance, the Hawaiian records of *P. sphenops* (Maciolek 1984) and that of *P. mexicana* (Devick 1991b) are possibly the same fish. Similarly, Schoenherr (1979) reported taking *P. sphenops* from a canal northwest of the Salton Sea in California; however, according to Dill and Cordone (1997), later authors questioned the identification and therefore did not include *P. sphenops* in their listings of California fishes (e.g., Hubbs et al. 1979; Shapovalov et al. 1981). At least five members of the *P. sphenops* species complex have been reported from the United States (Courtenay and Hensley 1979a).

We are unaware of any voucher specimens.

Poeciliopsis gracilis

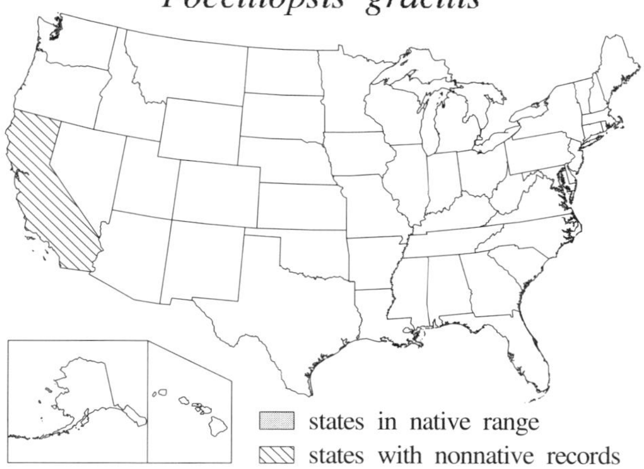

Poeciliopsis gracilis (Heckel 1848)

Common Name: porthole livebearer.

Identification: A few distinguishing characteristics provided by Rosen and Bailey (1963), Page and Burr (1991), and Wischnath (1993). For photographs of species see Dawes (1991) and Wischnath (1993). Maximum size: Females 5 cm SL; males 4 cm SL.

Native Range: Southern Mexico to Honduras on Atlantic and Pacific slopes (Rosen and Bailey 1963; Page and Burr 1991).

Nonindigenous Occurrences: This species was first discovered in 1974 in an irrigation canal near Mecca, Riverside County, **California** (Mearns 1975; Hubbs et al. 1979; Shapovalov et al. 1981; Courtenay et al. 1984, 1991; Swift et al. 1993).

Means of Introduction: Probably due to an aquarium release or escape from a local fish farm (Mearns 1975; Lee et al. 1980 et seq.).

Status: Population in Riverside County, California, was locally established (Mearns 1975; Shapovalov et al. 1981); however, it may no longer exist. Canal reportedly filled in as of 1987 (Courtenay, personal communication).

Impact of Introduction: Unknown.

Remarks: Swift et al. (1993), incorrectly citing Mearns, mistakenly reported that the species was first found in California in 1965. This species exhibits a broad range in body build and coloration, especially in terms of spotting pattern (Wischnath 1993).

Voucher specimens: California (SIO 51074-126).

Poeciliopsis occidentalis

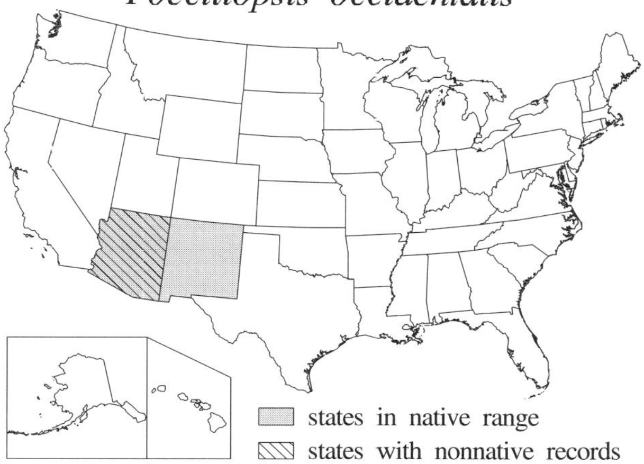

Poeciliopsis occidentalis (Baird and Girard 1853)

Common Names: Sonoran topminnow (Gila topminnow, Yaqui topminnow).

Identification: Minckley (1973); Sublette et al. (1990); Page and Burr (1991). Two subspecies: *P. o. occidentalis* - Gila topminnow; *P. o. sonoriensis* - Yaqui topminnow. For a photograph of the species see Dawes (1991). Maximum size: Females 5 cm SL; males to about 3 cm.

Native Range: *P. o. occidentalis* endemic to Gila River system of Arizona, New Mexico, and northern Sonora, Mexico. *P. o. sonoriensis* native to Yaqui River system in southeastern Arizona and northern Mexico (U.S. Fish and Wildlife Service 1983b). Now extirpated in New Mexico and rare in Arizona (Page and Burr 1991).

Nonindigenous Occurrences: Species was stocked in several sites in **Arizona** including Hidden Waters, a tributary of the Salt River; the Boyce Thompson Arboretum, near Superior, Pinal County; Seven Springs, a tributary to the Gila River; Tule Creek, a tributary to the Agua Fria River; Arivaca Creek; and Cow Creek, a tributary to the Agua Fria River (U.S. Fish and Wildlife Service 1983b).

Means of Introduction: This endangered species was intentionally stocked to create additional populations. It was introduced into Arivaca Creek in 1936. Populations were abundant by 1957, but were extirpated in less than two years by introduced mosquitofish (U.S. Fish and Wildlife Service 1983b). The species is established in Hidden Waters, where it was first stocked in 1976. It was stocked in Tule Creek in 1968, but extirpated by flooding in 1978. It was restocked in October 1981 and was still established as of 1983. The fish first was stocked at Seven Springs and Cave Creek in the late 1960s, but the population was lost to flooding several times. It was restocked in 1975, but extirpated by flooding again in 1978. It was stocked again in February 1980 and became established. It was stocked in Cow Creek in September 1981 and became established (U.S. Fish and Wildlife Service 1983b). Simons (1987, cited in Williams 1991) documented more than a hundred introductions of *Poeciliopsis o. occidentalis* into isolated Arizona waters, many of which were created for livestock watering and therefore, were historically fishless.

Status: Stocked populations at Hidden Waters and Arivaca Creek were extirpated because of introduction of nonnative fish and habitat alteration. Established at Tule Creek, Seven Springs, and Cow Creek (U.S. Fish and Wildlife Service 1983b).

Impact of Introduction: Unknown.

Remarks: Wischnath (1993) considered *P. occidentalis* to be the most attractive member of the genus. It is listed as federally endangered (U.S. Fish and Wildlife Service 1993a). The Yaqui topminnow also has been stocked commonly in portions of Yaqui River system where it is native (U.S. Fish and Wildlife Service 1983b, 1994b). Likewise, Minckley (1969) listed locations within the natural range where reintroduction of the Gila topminnow was attempted. Both subspecies are affected by introduced western mosquitofish *Gambusia affinis* which prey on juvenile topminnows and harass adults (U.S. Fish and Wildlife Service 1983b, 1994b).

Poeciliopsis sp.

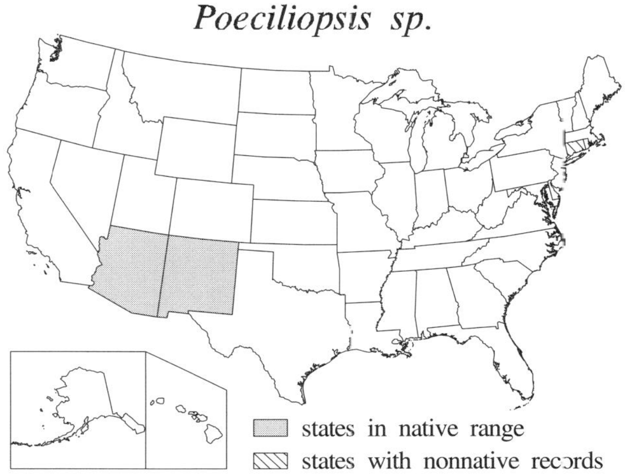

Poeciliopsis sp.

Common Name: livebearer, topminnow.

Identification: Rauchenberger (1989b) listed 20 extant species in the genus *Poeciliopsis*. Rosen and Bailey (1963) provided diagnostic characters of the genus and a limited amount of information on the species known at that time. Distinguishing characteristics for species previously found in the United States were given by Page and Burr (1991). Photographs of several species appeared in Dawes (1991) and Wischnath (1993). Maximum size: 5.1-6 cm.

Native Range: Members of the genus *Poeciliopsis* occur in Pacific drainages from the southwestern United States to Colombia and in some Atlantic streams of southeastern Mexico, Guatemala, and Honduras (Rosen and Bailey 1963). Only one species, *P. occidentalis*, is native to the United States where it occurs in the Gila River system, New Mexico and Arizona, and in western Mexico; now extirpated in New Mexico (Page and Burr 1991).

Nonindigenous Occurrences: A few specimens, identified as belonging to this genus, were taken from the Thames River drainage in **Connecticut** (Whitworth 1996).

Means of Introduction: Unknown. Possible aquarium release.

Status: Reported in Connecticut.

Impact of Introduction: Unknown.

Remarks: Males of this genus have very long, pointed gonopodia which exhibit sinistral asymmetry (Dawes 1991).

Xiphophorus hellerii

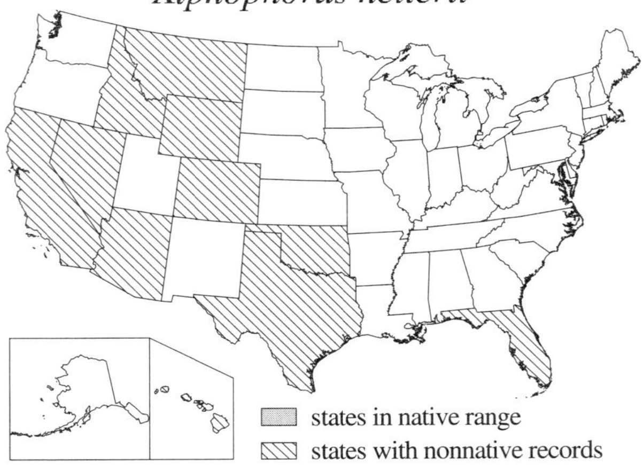

Xiphophorus hellerii Heckel 1848

Common Names: green swordtail (red swordtail, swordtail).

Identification: Distinguishing characteristics were given by Rosen (1960, 1979) and Page and Burr (1991). This species is included in keys of Rosen (1960, 1979), Brown (1971), Minckley (1973), and Greenfield and Thomerson (1997). Photographs or illustrations appeared in Rosen (1979), Mills and Vevers (1982), Axelrod et al. (1985), Petrovicky (1988), Dawes (1991), Sakurai et al. (1993), and Wischnath (1993). Name often spelled as *Xiphophorus helleri.* Maximum size: Females up to 16 cm; males up to 14 cm, excluding sword.

Native Range: Middle America from Rio Nautla, Veracruz, Mexico, to northwestern Honduras (Rosen 1960, 1979; Page and Burr 1991; Greenfield and Thomerson 1997).

Nonindigenous Occurrences: This species has been recorded from Rock Spring in Maricopa County, **Arizona** (Minckley 1973); several counties in **California** (Coots 1956; St. Amant and Hoover 1969; St. Amant 1970; Mearns 1975; Courtenay et al. 1984, 1986, 1991; Swift et al. 1993; Dill and Cordone 1997); Conejos and Saguache counties in **Colorado** (Woodling 1985, Zuckerman and Behnke 1986; S. Platania, personal communication); several counties in **Florida** (Courtenay et al. 1974; Courtenay and Hensley 1979a; Dial and Wainright 1983; museum specimens); **Hawaii** (Brock 1960; Maciolek 1984; Devick 1991b); several geothermal waters in **Idaho** (Courtenay 1985; Courtenay et al. 1987; Idaho Fish and Game 1996); Madison County, **Montana** (Brown 1971; Courtenay 1985; Courtenay and Meffe

1989; Holton 1990); Indian Spring and Rogers Spring, Clark County, **Nevada** (La Rivers 1962; Deacon et al. 1964; Minckley 1973; Courtenay and Deacon 1982; Deacon and Williams 1984); the Verdigris River near Catoosa, **Oklahoma** (Pigg et al. 1996); Bexar County, **Texas** (Howells 1992a); and Teton County, **Wyoming** (Courtenay et al. 1987; Hubert 1994; Stone 1995).

Means of Introduction: Most introductions probably due to aquarium releases. Its origin in the Westminster flood control channel in California was believed to be a goldfish farm (St. Amant and Hoover 1969). Florida records may also be due to fish farm escapes or releases.

Status: It is locally established, or possibly established locally, in Colorado (Zuckerman and Behnke 1986), Florida (Nico, personal communication), Hawaii (Devick 1991b), Idaho (?) (Courtenay and Meffe 1989), Montana (Courtenay and Meffe 1989), Nevada (Courtenay and Meffe 1989), Texas (Howells 1992a), and Wyoming (Stone 1995). A breeding population existed in Arizona at Rock Spring, but it disappeared after a flood in 1965 (Minckley 1973). It has been reported from several sites in California (Dill and Cordone 1997) and a single locality in Oklahoma (Pigg et al. 1996).

Impact of Introduction: Largely unknown. The green swordtail has been implicated in the decline of the Utah sucker *Catostomus ardens* in a thermal spring in Wyoming (Courtenay et al. 1988).

Remarks: This species exhibits a wide natural range of body form and color patterns (Rosen 1960, 1979; Dawes 1991; Wischnath 1993). It has long been popular as an ornamental fish. The green swordtail was recorded from Indian Spring, Clark County, Nevada, as early as 1975 (Courtenay and Deacon 1982). However, the *Xiphophorus* taken from Indian Spring during later collections were identified as hybrids of *X. hellerii* × *X. maculatus* (Courtenay and Deacon 1982; Deacon and Williams 1984; Page and Burr 1991). That conclusion apparently was based on the fact that the live fish were yellow to pale orange (Courtenay and Deacon 1982). In 1989, Rauchenberger (personal communication) examined the preserved voucher material (UF 91919) and determined that all specimens were *X. hellerii*. Nevertheless, some of the other above records actually may represent hybrids rather than pure *X. hellerii*. Myers (1940) mentioned an unconfirmed report of *X. hellerii* breeding in the Florida Everglades.

Voucher specimens: Colorado (MSB, uncataloged); Florida (USNM 89437; UF 30867, 86323, 91920, 92139, 97844, 98928), Hawaii (BPBM 35798), Nevada (UF 91919), Oklahoma (OSU 27459).

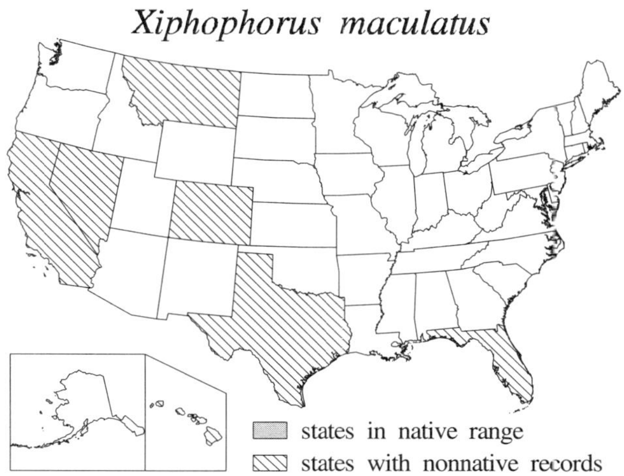

Xiphophorus maculatus (Günther 1866)

Common Names: southern platyfish (moonfish, platy).

Identification: Distinguishing characteristics were given by Rosen (1960, 1979) and Page and Burr (1991). It is included in identification keys of Rosen (1960, 1979) and Greenfield and Thomerson (1997). Photographs or illustrations appeared in Rosen (1960), Mills and Vevers (1982), Petrovicky (1988), Dawes (1991), Sakurai et al. (1993), and Wischnath (1993). Formerly known as *Platypoecilus maculatus*. Maximum size: Females to about 6 cm; males to about 3.5 cm.

Native Range: Atlantic Slope of Middle America from Ciudad Veracruz, Mexico, to Belize (Rosen 1960, 1979; Page and Burr 1991; Greenfield and Thomerson 1997).

Nonindigenous Occurrences: This species has been recorded from Orange County, **California**, near Westminster (St. Amant and Hoover 1969; Courtenay et al. 1984, 1991; Swift et al. 1993); near a fish farm in Conejos County, **Colorado** (Woodling 1985; Zuckerman and Behnke 1986); several counties in **Florida** (Courtenay et al. 1974; Courtenay and Hensley 1979a; Dial and Wainright 1983; museum specimens); **Hawaii** (Brock 1960; Maciolek 1984; Devick 1991b); Beaverhead Rock Pond (Madison County), **Montana** (Page and Burr 1991); Clark County, **Nevada** (La Rivers 1962; Courtenay and Deacon 1982; Courtenay et al. 1984, 1991; Deacon and Williams 1984); and **Texas** (Hubbs 1982; Courtenay et al. 1984, 1991).

Means of Introduction: Probably due to fish farm or aquarium releases.

Status: It is locally established, or possibly established locally, in several states, including Florida (Courtenay and Meffe 1989; Nico, personal communication), Colorado (Zuckerman and Behnke 1986), Hawaii (Kanayama 1968; Devick 1991b), and Montana (Page and Burr 1991). It was reported from Texas (Courtenay et al. 1984, 1991). The species was mentioned as being extirpated from California (Swift et al. 1993); however, there is no evidence that it was ever established in that state (Courtenay and Meffe 1989). The species was introduced into Blue Point Springs near Lake Mead, in Nevada (Deacon and Williams 1984); the status of that population is assumed to be locally established. A second population, inhabiting Indian Springs, Clark County, Nevada, was considered to be a hybrid with *X. hellerii* (Deacon and Williams 1984; Courtenay and Meffe 1989); however, in 1989 M. Rauchenberger examined specimens (UF 91919) taken from Indian Spring and determined that they represented *X. hellerii* and not a hybrid.

Impact of Introduction: Unknown.

Remarks: This species exhibits a wide range of color patterns (Rosen 1960, 1979; Dawes 1991; Wischnath 1993). Kallman and Brunetti (1983) examined the genetic basis of several mutant color varieties of this species. It has long been a popular ornamental fish. Myers (1940) received unconfirmed reports that *X. maculatus*, along with several other introduced species, had breeding populations in the Florida Everglades. There is some confusion surrounding the positive identification of a Nevada population (see account for *X. hellerii*). Courtenay and Meffe (1989) indicated that the Nevada record of Courtenay and Deacon (1982) and Deacon and Williams (1984) was actually a hybrid with *X. hellerii*. As such, the listing of pure *X. maculatus* for Nevada is obviously tentative. Nevada records are also covered in the hybrid accounts.

Voucher specimens: Florida (UF 91921, 91922, 91923, 91924, 91925, 92140, 97845, 98929, uncataloged).

Xiphophorus variatus

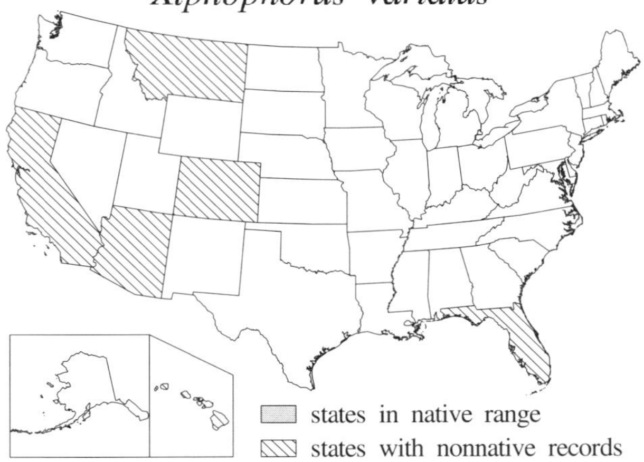

Xiphophorus variatus (Meek 1904)

Common Names: variable platyfish (sunset platy, shortblade swordtail).

Identification: Distinguishing characteristics were given by Rosen (1960, 1979), Page and Burr (1991), and Wischnath (1993). It was included in keys of Rosen (1960, 1979). This species was formerly known as *Platypoecilus variatus*. Photographs or illustrations appeared in Rosen (1960), Mills and Vevers (1982), Petrovicky (1988), Dawes (1991), Sakurai et al. (1993), and Wischnath (1993). Maximum size: Females up to 7 cm; males to about 5.5 cm.

Native Range: Atlantic Slope of Mexico from southern Tamaulipas to northern Veracruz (Rosen 1960, 1979; Page and Burr 1991).

Nonindigenous Occurrences: This species has been recorded from Yuma and Tempe, **Arizona** (Minckley 1973); several sites in southern **California** (St. Amant and Hoover 1969; St. Amant and Sharp 1971; Moyle 1976a; Hubbs et al. 1979; Shapovalov et al. 1981; Courtenay et al. 1984, 1991; Swift et al. 1993); **Colorado** (Zuckerman and Behnke 1986); as many as six counties in **Florida** (Courtenay et al. 1974; Burgess et al. 1977; J. D. Williams, personal observation; museum specimens); Oahu, **Hawaii** (Maciolek 1984); and Beaverhead, Broadwater, Granite, and Madison counties, **Montana** (Brown 1971; Courtenay 1985; Holton 1990).

Means of Introduction: Probably due to aquarium releases.

Status: Locally established in Florida (Courtenay and Meffe 1989; museum specimens), and Montana (Courtenay and Meffe 1989); possibly established locally, in California (Swift et al. 1993) and Hawaii (Maciolek 1984). It was established locally in Arizona from 1963 to 1965, but that population was presumably destroyed by a flood in late 1965 (Minckley 1973). The species was reported from Colorado (Zuckerman and Behnke 1986). Although listed as established in Hawaii by Maciolek (1984), there is no mention of this species lists of introduced fishes provided by Devick (1991b). A population in Alachua County, Florida, first recorded in the 1970s (Burgess et al. 1977), still survives locally.

Impact of Introduction: Unknown.

Remarks: Similar to some of the other poeciliids, this species exhibits considerable natural variation (e.g., Wischnath 1993). It has been popular as an ornamental fish for many years.

Voucher specimens: Florida (UF 16992, 30868, 91926, 91927, 91928).

HYBRIDS

Poecilia latipinna × *P. velifera*

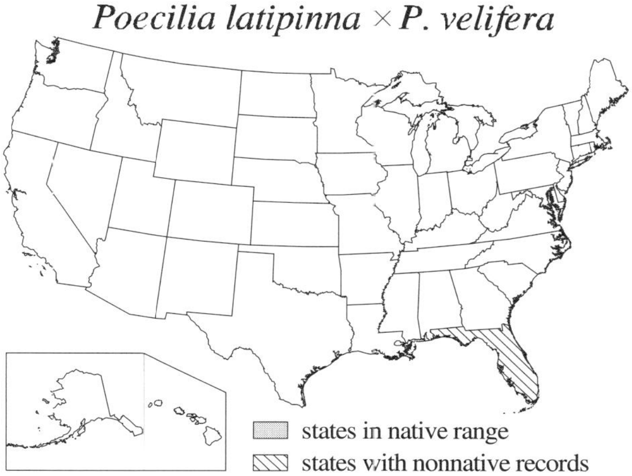

Poecilia latipinna × *P. velifera*

Common Name: black molly (sailfin molly × Yucatan molly)

Identification: Both *P. latipinna* and *P. velifera* are part of the *Poecilia latipinna* complex (Miller 1983). Illustrations of black molly forms appeared in Mills and Vevers (1982), Axelrod et al. (1985), Petrovicky (1988), and Wischnath (1993). Maximum size: Probably similar to *P. latipinna* (i.e., 15 cm).

Native Range: Not applicable.

Nonindigenous Occurrences: The black molly has been recorded from Brevard, Hillsborough, Palm Beach, and Polk counties in **Florida** (Courtenay et al. 1974); Madison County, **Montana** (Holton 1990); and Big Spring, Ash Meadows, **Nevada**, in 1981 and 1986 (Courtenay, personal communication).

Means of Introduction: Aquarium release or escape from fish farm (Florida).

Status: Locally established in Florida and Montana; reported from Nevada.

Impact of Introduction: Unknown.

Remarks: Our review of the literature suggests that natural hybrids between these two species are unknown. As such, the correct identification of the introduced forms is somewhat questionable. Several black molly forms are in the aquarium trade. According to Mills and Vevers (1982), the exact origin of some of the black mollies is not clear, but it is likely the original male parent was *Poecilia sphenops*. Mills and Vevers also reported that a hybrid, known as the crescenty black molly, was produced by crossing the original black molly with the sailfin molly *Poecilia latipinna*. Petrovicky (1988) illustrated several black forms. Two variants were identified as variants of *Poecilia sphenops* (the marbled molly and the lyre molly), another was recognized as *Poecilia velifera* (according to Petrovicky, the black form of the latter species possibly also occurs in nature). However, the identification of these fish as pure species should be considered highly suspect since aquarists have made a long and successful practice of artificially crossing various combinations of poeciliids.

There are no known voucher specimens.

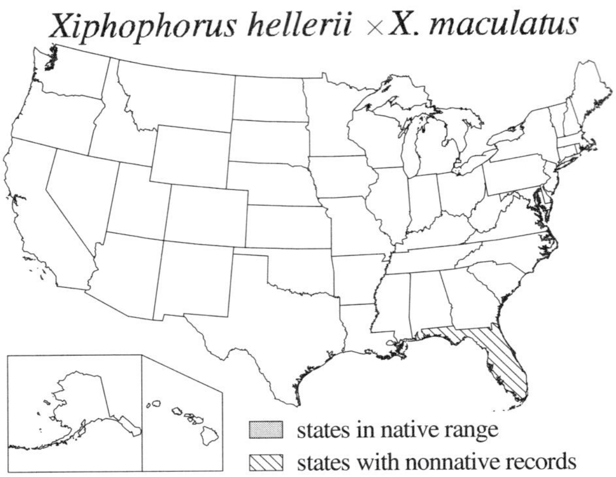

Xiphophorus hellerii × *X. maculatus*

Common Names: green swordtail × southern platyfish (red swordtail × sourthern platyfish).

Identification: For information on hybridization between these species and their offspring see Gordon and Rosen (1951), Gordon and Gordon (1957), and Atz (1962). Distinguishing characteristics of the parent species were given by Rosen (1960, 1979) and Page and Burr (1991). Maximum size: Probably similar to parent species: 6–13 cm.

Native Range: Not applicable.

Nonindigenous Occurrences: This hybrid combination has been recorded from Brevard and Hillsborough counties in **Florida** (Courtenay et al. 1974).

Means of Introduction: Most likely introduced through aquarium releases or fish farm escapes or releases.

Status: Reported from Florida.

Impact of Introduction: Unknown.

Remarks: In captivity, these two species are often crossbred to produce a wide variety of hybrid offspring, many of which are fertile (Dawes 1991). Although easily hybridized experimentally, natural hybrids are unknown (Rosen 1960). A population found in Indian Spring, Clark County, Nevada, was originally reported to be this hybrid (Courtenay and Deacon 1982; Deacon and Williams 1984; Page and Burr 1991); however, in 1989 Rauchenberger (personal communication) examined the Nevada voucher specimens (UF 91919) and concluded that all were *X. hellerii.*

There are no voucher specimens to support the Florida record.

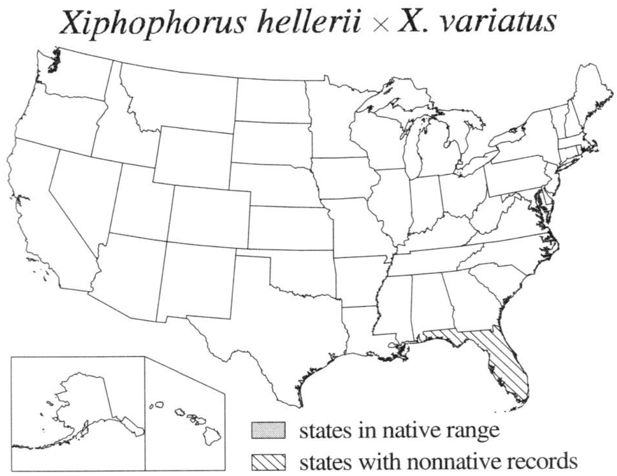

Xiphophorus hellerii × *X. variatus*

Common Names: green swordtail × variable platyfish (papagai platy).

Identification: See Gordon and Rosen (1951), Gordon and Gordon (1957), and Atz (1962). For photographs or illustrations of selected resulting offspring from this cross see Petrovicky (1988) and Dawes (1991). Maximum size: Probably similar to parent species: 7–13 cm.

Native Range: Not applicable.

Nonindigenous Occurrences: Recorded from an unspecified locality in **Florida** (Courtenay et al. 1974).

Means of Introduction: Possibly the result of an aquarium release or fish farm escape or release.

Status: Reported from Florida.

Impact of Introduction: Unknown.

Remarks: In captivity, *X. hellerii* is commonly crossbred with *X. variatus* to produce a wide variety of hybrids, many of which are fertile (Dawes 1991).

There are no known voucher specimens.

Xiphophorus maculatus × *X. variatus*

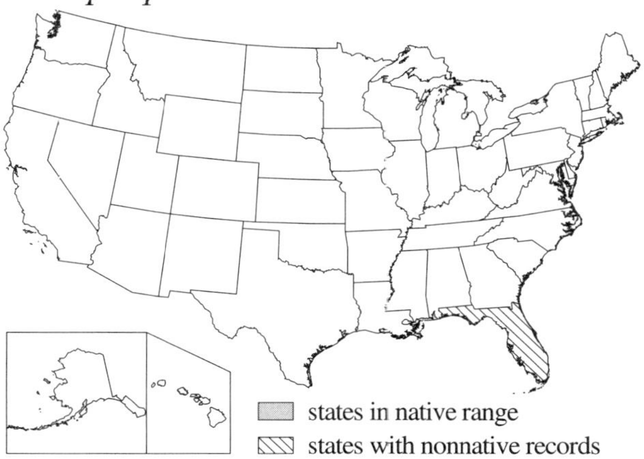

Xiphophorus maculatus × *X. variatus*

Common Name: southern platyfish × variable platyfish.

Identification: See Gordon and Gordon (1957) and Atz (1962). Maximum size: Probably similar to parent species: 6–7 cm.

Native Range: Not applicable.

Nonindigenous Occurrences: This hybrid has been recorded for Brevard County, **Florida** (Courtenay et al. 1974).

Means of Introduction: Most likely introduced through aquarium releases or fish farm escapes or releases.

Status: Reported from Florida.

Impact of Introduction: Unknown.

Remarks: There are no known voucher specimens.

Note: Hubbs (1955) mentioned the discovery of a hybrid between species of "*Mollienisia*" (i.e., *Molliene-sia*) in Hawaii. He added that "In Hawaii, where both genera were introduced, I have also found hybrids between the related genera *Limia* and *Mollienisia*." However, the identity of the two species addressed is not known. As many as five introduced poeciliids in the modern genus *Poecilia* may have reproducing populations in Hawaii. Such species as *P. latipinna* and *P. sphenops*, were formerly placed in the genus *Mollienesia*. *Limia vittata* (formerly *Poecilia vittata*) was already present in Hawaii as early as 1950, and probably before (Maciolek 1984; Devick 1991b). Given the above, it is conceivable that one of the parental species in the Hawaiian hybrid discovered by Hubbs was *Limia vittata*.

Goodeidae - Splitfins

The family Goodeidae, whose members are also called splitfins or springfishes, comprises about 19 genera with about 40 species found in west central Mexico and the southwestern United States (Nelson 1994). Four species are native to the United States, all in Nevada and associated with spring habitats (Page and Burr 1991; Nelson 1994). Goodeids are small fishes; most species are less than 10 cm long (Berra 1981; Page and Burr 1991). Mexican species have internal fertilization and include both deep- and long-bodied forms (Berra 1981; Nelson 1994). Species native to the United States lay eggs and are characterized by thick bodies, lack of pelvic fins, and posteriorly placed dorsal fins (Page and Burr 1991). They occur only in freshwater habitats (Nelson 1994).

Ameca splendens

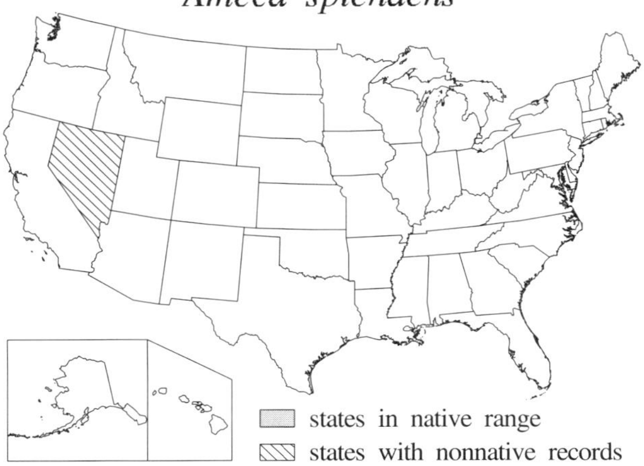

Ameca splendens Miller and Fitzsimons 1971

Common Name: butterfly splitfin (butterfly goodeid).

Identification: Miller and Fitzsimons (1971); Sanford (1995) – photo. Maximum size: 10 cm.

Native Range: Río Ameca basin, Pacific Slope of western Mexico (Miller and Fitzsimons 1971).

Nonindigenous Occurrences: One individual was taken in November 1981 (museum specimen) and another in August 1983 from Rogers Spring, **Nevada** (Courtenay and Deacon 1983; Deacon and Williams 1984). Others were seen but not collected (Courtenay, personal communication).

Means of Introduction: Probably an aquarium release.

Status: Reported from Nevada.

Impact of Introduction: Unknown.

Remarks: This species is raised commercially for the aquarium industry in Florida (Robins et al. 1991b).

Voucher material from Nevada was deposited with Florida Atlantic University, but subsequently lost.

Crenichthys baileyi

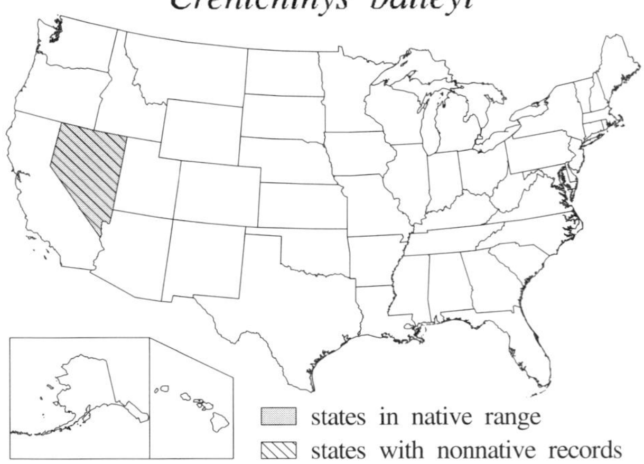

Crenichthys baileyi (Gilbert 1893)

Common Name: White River springfish.

Identification: La Rivers (1962); Sigler and Sigler (1987); Page and Burr (1991). This species consists of five subspecies: *C. b. albivallis*, *C. b. baileyi*, *C. b. grandis*. *C. b. moapae*, and *C. b. thermophilus*. Maximum size: 9 cm.

Native Range: Springs and remnants of the White River system in eastern Nevada (Page and Burr 1991).

Nonindigenous Occurrences: The Hiko White River springfish *C. b. grandis* was transplanted into Blue Link Spring in eastern Mineral County, **Nevada**, in 1985 (Baugh et al. 1985).

Means of Introduction: According to Sevon and Delany (1987), the introduction of tropical aquarium fishes and largemouth bass had decimated the populations of *C. b. grandis* in their native habitats at Hiko Spring and Crystal Spring in the White River drainage of Nevada. Blue Link Spring was selected by federal and state agencies to serve as a refuge for this endangered subspecies; the site is far outside the native range of the Hiko White River springfish and differs substantially from their native habitat (Sevon and Delany 1987). The original transplant took place in early 1985 when 59 fish were planted in Blue Link Spring; a second transplant, in late 1985, involved the stocking of 178 additional fish (Sevon and Delany 1987).

Status: Hiko White River springfish populations fluctuated widely in Blue Link Spring; however, the total number had increased steadily to an estimated 5,500 individuals by 1994 (Lujan 1994). Native populations in Hiko and Crystal springs compete with various nonindigenous fishes, and their numbers remain low in those sites (Clemmer, personal communication).

Impact of Introduction: Unknown.

Remarks: *Crenichthys b. baileyi* and *C. b. grandis* are listed as federally endangered taxa. Prior to the elimination of the U.S. Fish and Wildlife Service candidate species list, *Crenichthys b. albivallis*, *C. b. moapae*, and *C. b. thermophilus* were under review for Federal listing as an endangered or threatened taxa (U.S. Fish and Wildlife Service 1994a).

Crenichthys nevadae

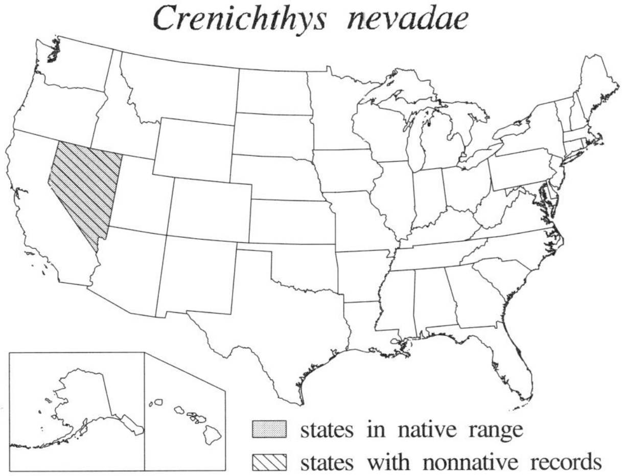

Crenichthys nevadae Hubbs 1932

Common Name: Railroad Valley springfish.
Identification: La Rivers (1962); Sigler and Sigler (1987); Page and Burr (1991). Maximum size: 6 cm.

Native Range: Springs in Railroad Valley, Nye County, Nevada (Page and Burr 1991).

Nonindigenous Occurrences: In 1947, six individuals were stocked into artificial ponds created near a natural seep of warm water, at Sodaville, in the southeastern part of Mineral County, **Nevada** (La Rivers 1962; Hubbs et al. 1974). The species was introduced into several pools along the outflow of Chimney Hot Springs (also called Chimney Springs), Nye County, in 1978 and again in 1982 (Deacon and Williams 1984; Williams and Williams 1989).

Means of Introduction: The Sodaville introduction was performed by Thomas Trelease of the Nevada Fish and Game Commission in response to the possible stocking of bass into the species' native habitat (La Rivers 1962). The U.S. Bureau of Land Management and Nevada Department of Wildlife introduced the species into Chimney Hot Springs in response to deterioration of its native habitats (Williams and Williams 1989). According to Williams and Williams (1989), the purpose of these introductions was to establish a refuge population for conservation of the species.

Status: Transplants have helped secure the long-term viability of this species. The population at Chimney Hot Springs, Nevada, was estimated at 1,881 adults in 1985; however, various habitat perturbations, including low water levels, unstable water temperatures, and trespass by cattle contributed to a population decline. The Sodaville population appeared stable during the late 1980s, although no firm population data was available (Williams and Williams 1989).

Impact of Introduction: Unknown.

Remarks: Williams and Williams (1989) made reference to an introduced population at Old Dugan Ranch, but they did not report details. The Railroad Valley springfish is protected as a threatened species (U.S. Fish and Wildlife Service 1993a). Introductions outside its native range occurred well before the species' 1986 formal listing as threatened (Williams and Williams 1989). In part of the native habitat of *C. nevadae*, introduced channel catfish *Ictalurus punctatus* and guppies *Poecilia reticulata* have large populations (Williams and Williams 1989). According to Clemmer (personal communication), commercial catfish farming in Duckwater Valley has depressed springfish populations and limited them to the fringes of the warm springs and outflows. A photograph of the Chimney Hot Springs Refuge appeared in Williams (1991).

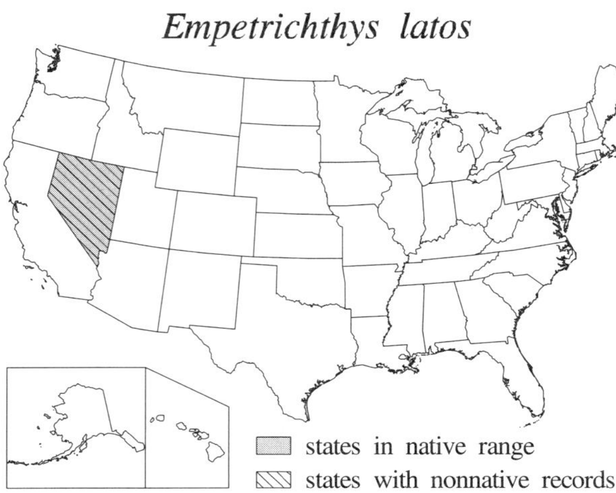

Empetrichthys latos Miller 1948

Common Names: Pahrump poolfish (Pahrump killifish).

Identification: La Rivers (1962); Page and Burr (1991). This species was once represented by three subspecies: *E. l. latos*, *E. l. concavus*, and *E. l. pahrump*; two are extinct, and only *E. l. latos* remains (U.S. Fish and Wildlife Service 1979a). Maximum size: 6 cm.

Native Range: Each subspecies was confined to an isolated spring in Pahrump Valley, Nye County, Nevada, where it was the only native fish. Two of these populations have become extinct. The Pahrump poolfish *E. l. latos* disappeared from Manse Springs in August 1975 and now exists only outside the Pahrump Valley in transplanted locations (U.S. Fish and Wildlife Service 1979a; Page and Burr 1991).

Nonindigenous Occurrences: This species has been introduced to several sites in **Nevada** including a refuge complex, Shoshone Ponds, White Pine County, about 40 miles southwest of Ely (Hubbs et al. 1974; Minckley et al. 1991); Corn Creek Spring northwest of Las Vegas on the Desert National Wildlife Range, Nye County; and to a reservoir at Spring Mountain Ranch State Park, Clark County, in 1983 (Minckley et al. 1991); Chimney Springs and Hot Creek, Nye County; and Sodaville Springs, Mineral County (Clemmer, personal communication).

Means of Introduction: This species was intentionally stocked to establish refuge populations for preservation of the species. The transplant to Corn Creek Spring took place in 1971 and involved 29 fish; Shoshone Ponds complex, a site created by the US Bureau of Land Management, was stocked in 1972 with 16 fish from Corn Creek or Manse Ranch Spring (an additional 50 fish were transplanted to the site in 1976); Spring Mountain Ranch State Park was stocked in 1983 with an unrecorded number of fish (Pister 1974; U.S. Fish and Wildlife Service 1979a; Minckley et al. 1991). The Nevada Department of Fish and Game was instrumental in at least tow of these introductions (Pister 1974).

Status: The species is established at Corn Creek Spring, the Shoshone Ponds complex, and Spring Mountain Ranch State Park (Minckley et al. 1991); populations also are established and stable at Chimney Springs, Hot Creek, and Sodaville Springs as of 1992 (Clemmer, personal communication). In 1989, Shoshone Ponds had approximately 450 fish (Bureau of Land Management, Ely, records). Several transplants have failed (Minckley et al. 1991). Soltz and Naiman (1978) reported that relatively large reproducing populations of Pahrump poolfish have been established in ponds at Corn Creek and in an isolated canyon above the Colorado River. They indicated that a third population was established in an artificial refugium in Ash Meadows but that it died out in 1977.

Impact of Introduction: Unknown.

Remarks: In an earlier report by the U.S. Fish and Wildlife Service (1979a), location of introduction sites was not provided. This species is federally listed as an endangered species (U.S. Fish and Wildlife Service 1993a). The introduction of other fishes into its native spring habitat contributed to the decline of the Pahrump poolfish. For instance, Deacon et al. (1964) reported that the establishment of goldfish *Carassius auratus* in Manse Spring resulted in population depression.

Cyprinodontidae - Pupfishes

The family Cyprinodontidae includes 9 genera and 100 species found in North and Central America, the West Indies, northern South America, northern Africa, and the Mediterranean Anatolian region (Nelson 1994). There are 13 species native to freshwaters of North America north of Mexico, most of which inhabit the southwestern desert regions (Page and Burr 1991). Pupfishes are small, North American species less than 8 cm long, and are characterized by a somewhat deep body, a flattened head and back, and an upturned mouth (Page and Burr 1991). They inhabit fresh and brackish waters, and coastal marine areas (Nelson 1994).

Cyprinodon diabolis

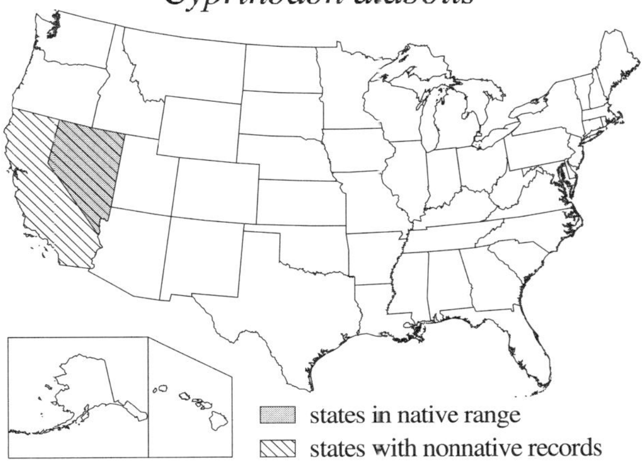

Cyprinodon diabolis Wales 1930

Common Name: Devils Hole pupfish.

Identification: La Rivers (1962); Sigler and Sigler (1987); Page and Burr (1991). Maximum size: 3.4 cm.

Native Range: Devil's Hole in central Ash Meadows, Death Valley National Monument, Nye County, Nevada (U.S. Fish and Wildlife Service 1980b; Page and Burr 1991).

Nonindigenous Occurrences: This species was introduced into five, supposedly suitable, spring habitats widely scattered over the deserts of **California** and **Nevada** (Soltz and Naiman 1978). According to Lee et al. (1980 et seq.) the pupfish was introduced into two natural springs in Nevada. These were identified as School Spring and Point of Rocks Spring, both in Ash Meadows by Clemmer (personal communication).

Means of Introduction: This species was intentionally stocked to create another population of a restricted endangered species. Point of Rocks Springs was stocked with 16 fish in 1992. The population declined to 5 fish and was augmented with another 16.

Status: Early introductions into spring habitats failed to become established (Soltz and Naiman 1978). More recent introductions may have been successful, but current status is not certain and available information is somewhat conflicting. According to J. Heinrich (personal communication), the Nevada population in School Spring numbered 127 individuals in 1993; the population in Point of Rocks Spring was 140. However, A. A. Echelle (personal communication) questioned the continued existence of the *Cyprinodon diabolis* populations in School Spring and Point of Rocks Spring. He noted that in 1989 these springs contained two subspecies of Amargosa pupfish (*Cyprinodon nevadensis mionectes* and *C. n. pectoralis*, respectively), not *C. diabolis*.

Impact of Introduction: Unknown.

Remarks: This species has been listed as a federally endangered species since 1967 (U.S. Fish and Wildlife Service 1980b, 1993a). In its only known native habitat, the species faces several threats to survival including reduction or loss of water, surface water runoff, vandalism, and accidents (U.S. Fish and Wildlife Service 1980b, 1993a). A captive population was created in the 1970s in an artificial refugium consisting of a concrete-lined pond constructed below Hoover Dam, Clark County, Nevada (Lee et al. 1980 et seq.; U.S. Fish and Wildlife Service 1980b; Williams 1991). As a result of a greater food supply, and probably other factors, individuals in the refuge population differed from those found in the native habitat were larger and more brightly colored (Williams et al. 1988). According to Williams (1991), the Hoover Dam refugium population is no longer extant. A second refugium, the Amargosa Pupfish Refugium, was constructed in Ash Meadows and stocked with fish in 1980 (U.S. Fish and Wildlife Service 1980b; Sigler and Sigler 1987; Williams 1991).

Cyprinodon macularius

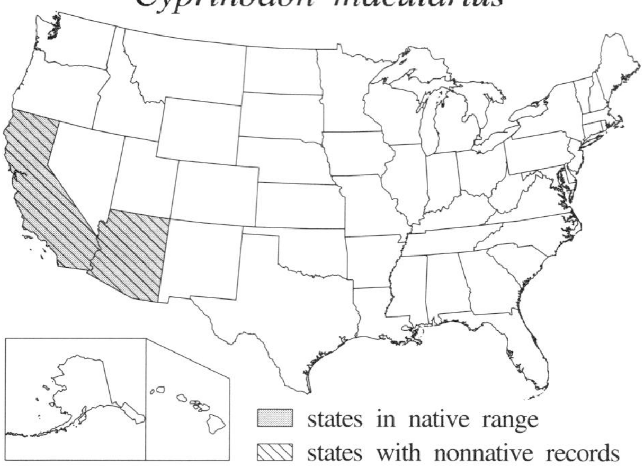

Cyprinodon macularius Baird and Girard 1853

Common Name: desert pupfish.

Identification: Minckley (1973); Moyle (1976a); Page and Burr (1991); two subspecies in the United States, a Colorado River form *C. m. macularius* and a Quitobaquito form *C. m. eremus*. A third undescribed subspecies occurs in Mexico (U.S. Fish and Wildlife Service 1993b). Maximum size: 7.2 cm.

Native Range: This species occurs in the lower Colorado River drainage, including the Gila River system and south through southern Arizona and California (including the Salton Sea) into northern Mexico (Page and Burr 1991). The Colorado River form exists naturally in California in two streams tributary to, and a few shoreline pools and irrigation drainages of, the Salton Sea, and on the Colorado River Delta, and in the Laguna Salada basin. The Quitobaquito form exists only in a single modified spring at Organ Pipe Cactus National Monument, Pima County, Arizona. The Mexican subspecies is found at scattered localities along the Rio Sonoyta (U.S. Fish and Wildlife Service 1993b; Echelle, personal communication).

Nonindigenous Occurrences: This species has been introduced extensively into natural springs, and springs and artificial ponds in public and private parks or educational institutions, and elsewhere, in southern **Arizona** and **California** (Miller 1968; Black 1980; Lee et al. 1980 et seq.; Schoenherr 1988; Hendrickson and Varela-Romero 1989; Minckley et al. 1991; U.S. Fish and Wildlife Service 1993b).

Means of Introduction: Most introductions were for the purpose of establishing refuge populations of this imperiled species. Hendrickson and Varela-Romero (1989) reported on the spread of one population via artificial canals. Miller (1968) reported that six individuals escaped from a trap into Dos Palmas Spring, near the northeastern corner of the Salton Sea, in May 1939. Based on the distribution map for this species in Lee et al. (1980 et seq.), the spring site is apparently in or near its native range. The stocking of a least one site in California was part of a series of experiments to test the effects of changed environment on meristic and morphometric characters (Miller 1968).

Status: Established in Arizona and California.

Impact of Introduction: Unknown. Walters and Legner (1980) looked at the diets of desert pupfish in experimental ponds. Desert pupfish ate mostly benthos, especially chironomid midge larvae, detritus, aquatic vegetation, and snails. Pupfish also ate zooplankters in weedy or benthic habitats. Walters and Legner (1980) concluded that, in the Southwest, pupfish may be a better alternative for mosquito control than stocking mosquitofish because they are less piscivorous than mosquitofish.

Remarks: The desert pupfish was listed as a federally endangered species in 1986 (Minckley et al. 1991; U.S. Fish and Wildlife Service 1993a). Populations within parts of its native range have declined partly as a result of introduction of nonindigenous fishes, especially tilapias (Schoenherr 1988; Minckley et al. 1991; Swift et al. 1993). The recovery plan for this species lists all known populations and transplants (i.e., U.S. Fish and Wildlife Service 1993b). Several authors (Schoenherr 1988: Hendrickson and Varela-Romero 1989; Minckley et al. 1991) reviewed the distribution and introduction of the desert pupfish; unfortunately, these authors did not always clearly distinguish between introductions made to sites outside the species range and transplant of one subspecies into the natural range of another *C. macularius* subspecies. Many introduced populations may be genetically contaminated because of mixing between subspecies (Minckley et al. 1991; U.S. Fish and Wildlife Service 1993b). In addition to the many spring and pond refugia, desert pupfish also are being held in captivity at various institutions (Hendrickson and Varela-Romero 1989; U.S. Fish and Wildlife Service 1993b).

Cyprinodon nevadensis

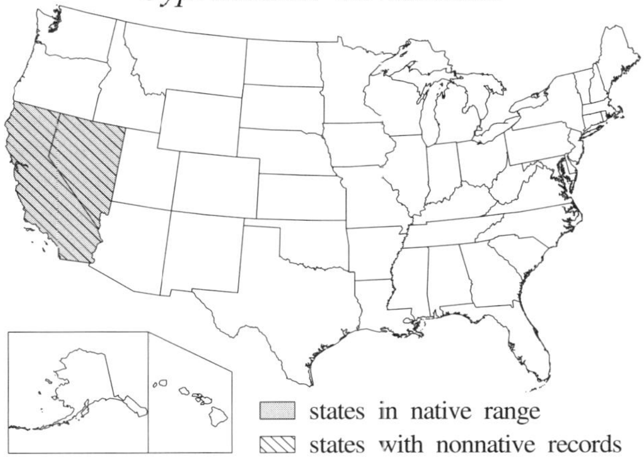

Cyprinodon nevadensis Eigenmann and Eigenmann 1889

Common Names: Amargosa pupfish (Nevada pupfish, Warm Springs pupfish, Tecopa pupfish, Shosone pupfish).

Identification: La Rivers (1962); Moyle (1976a); Page and Burr (1991). Six subspecies. Maximum size: 7.8 cm.

Native Range: Amargosa River basin, California and Nevada (Page and Burr 1991).

Nonindigenous Occurrences: This species was introduced into several counties and drainages in the southern part of **California** including a ranch reservoir in the Lucerne Valley, San Bernadino County in 1939 and again in 1940; a spring and spring-fed ditches at Little Lake, Owens Valley in 1939 and 1940; a spring near the northeast end of the Salton Sea, Riverside County, in 1939 and again in 1940; San Felipe Creek in Sentenac Canyon, in 1939; and a site on the west side of Death Valley in 1940 (Miller 1968; Moyle 1976a). It was also introduced to the head spring at Pahrump Ranch in the Pahrump Valley, Nye County, **Nevada**, in 1940 (Miller 1968). A sample of 20 individuals was taken from a spring seepage area at a golf course in Las Vegas, Nevada, in 1941 (Miller and Alcorn 1946; La Rivers 1962).

Means of Introduction: Most introductions were the result of intentional stockings, of which some were presumably attempts to expand the range of this imperiled species. Fish found in the Nevada golf course were thought to have escaped with overflow from ponds at a nearby US Fish and Wildlife Service facility. Miller and Alcorn (1946) presumed that this small species had been brought to the federal facility from the Amargosa drainage for use as forage for the bass and sunfish populations maintained at the station. A few of the earlier stockings in California were apparently part of a series of experiments to test the effects of changed environment on meristic and morphometric characters (Miller 1968).

Status: Moyle (1976a) listed the occurrence of this species in the Mojave basin of California, but he did not provide details nor indicate its status. His Mojave basin record may very well be based on one or more of the early reports by Miller (1968). Miller (1968) documented the introduction of various subspecies into some seven sites in California and Nevada, but he concluded that all introductions failed to produce any long-term reproducing populations. The Las Vegas golf course site was searched in 1940, but no additional specimens were found (Miller and Alcorn 1946).

Impact of Introduction: Unknown.

Remarks: This species is comprised of six subspecies, of which at least one, *C. n. calidae* is extinct and the others are at various levels of imperilment (Minckley et al. 1991). Two subspecies (*C. n. mionectes* and *C. n. pectoralis*) are federally listed endangered subspecies (U.S. Fish and Wildlife Service 1993a). Seven of the nine stockings reported by Miller (1968) involved the subspecies *C. n. amargosae*. The 1939 stocking into Lucerne Valley was of *C. n. nevadensis*; the stocking into western Death Valley was of *C. n. shoshone* (Miller 1968).

Cyprinodon radiosus

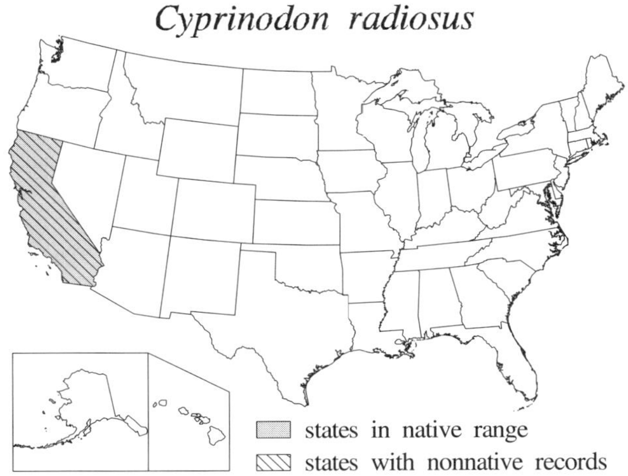

Cyprinodon radiosus Miller 1948

Common Name: Owens pupfish.

Identification: Moyle (1976a); Page and Burr (1991). Maximum size: 7.2 cm.

Native Range: Owens River system in Owens Valley, California, from the springs at Fish Slough in Mono County, to (but not in), Owens Lake, Inyo County, and in the springs around the lake. Currently only found naturally in the Owens Valley Native Fish Sanctuary and in BLM Spring in Fish Slough, a tributary of the Owens River, just north of Bishop (Miller 1968; Miller and Pister 1971; U.S. Fish and Wildlife Service 1984c).

Nonindigenous Occurrences: Introduced to Warm Springs, a protected area about 9 km north of Big Pine at the base of the White Mountains in Inyo County, **California**, in 1970 or shortly thereafter (Miller 1968; U.S. Fish and Wildlife Service 1984c).

Means of Introduction: Intentionally stocked to create an additional population of this endangered species where the pupfish could survive without danger of extermination by predatory, nonindigenous fishes or further threatened by competition with introduced mosquitofish (Miller 1968). According to Miller (1968), this site was also intended to study the interspecific relationships among various fishes native to the Owens Valley. The Warm Springs refuge was created in 1970 under the supervision of the California Department of Fish and Game (Miller 1968) and presumably stocked soon afterwards.

Status: Established at Warm Springs, California (U.S. Fish and Wildlife Service 1984c).

Impact of Introduction: Unknown.

Remarks: This pupfish is listed as a federally endangered species (Page and Burr 1991; U.S. Fish and Wildlife Service 1993a). The Warm Springs refugium is a small pond formed by a small dam and fed by a thermal spring located just outside of what is considered the species' native range (U.S. Fish and Wildlife Service 1984c). At least two populations of this species also survive as result of reintroductions made in 1969 and 1970 into sites within its historical range within the Owens River Valley near Bishop, California; both are in or adjacent to Fish Slough, one at the Owens Valley Native Fish Sanctuary and the other at the BLM Spring refugium (Miller 1968; U.S. Fish and Wildlife Service 1984c; Minckley et al. 1991). Minckley et al. (1991) reported that the California Department of Fish and Game stocked six other potential sites after about 1986, and that the species has survived in at least two of these sites. However, Minckley and his colleagues did not provide details on the site locations; thus it is not known if any of these stockings represent true introductions as opposed to reintroductions. Introduction of nonindigenous species has contributed to the Owen pupfish's decline (e.g., Miller 1968). Species introduced into the Owens River include largemouth bass *Micropterus salmoides*, brown trout *Salmo trutta*, mosquitofish *Gambusia affinis*, and crayfish *Procambarus clarki*. The largemouth bass, brown trout, and crayfish prey on adults. Mosquitofish probably compete with the Owens pupfish for food and space and may prey on juveniles. Crayfish also prey on pupfish eggs (U.S. Fish and Wildlife Service 1993a). Creation of the various refugia typically involved elimination of nonindigenous species and erection of fish barriers preventing access to the site by nonindigenous predators, particularly largemouth bass (Miller 1968; Minckley et al. 1991).

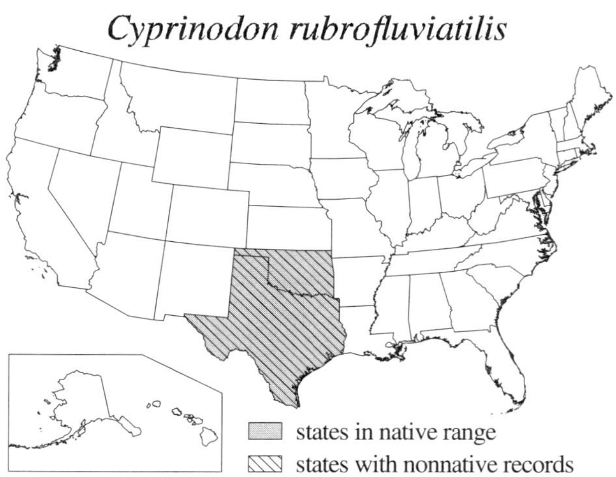

Cyprinodon rubrofluviatilis Fowler 1916

Common Name: Red River pupfish.

Identification: Hubbs et al. (1991); Page and Burr (1991). Maximum size: 5.8 cm.

Native Range: Upper Red and Brazos River drainages, Oklahoma and Texas (Page and Burr 1991).

Nonindigenous Occurrences: Populations of this species exist in the Canadian River system of the Arkansas River drainage in **Oklahoma** and **Texas**, and in the Colorado River drainage in **Texas** (Echelle et al. 1977; Lee et al. 1980 et seq.; Conner and Suttkus 1986; Cross et al. 1986; Hubbs et al. 1991; Pigg et al. 1995).

Means of Introduction: Unknown. Echelle et al. (1977) speculated that the species was transported to the Colorado River drainage by bait-bucket or other artificial means; however, they could not completely discount the slight possibility that it was native to the drainage. Ashbaugh et al. (1994) suggested that introduced populations in the South Canadian and Colorado river drainages appear to have been derived from Red River parent stock. Its spread within the Canadian River drainage may have been aided by lower than normal peak stream flows during recent years (Pigg et al. 1995).

Status: Recorded and presumably established in an Oklahoma drainage and in two Texas drainages outside its native range.

Impact of Introduction: Unknown.

Remarks: Pigg et al. (1995) listed sites where this species was taken in the South Canadian River drainage.

Cyprinodon salinus

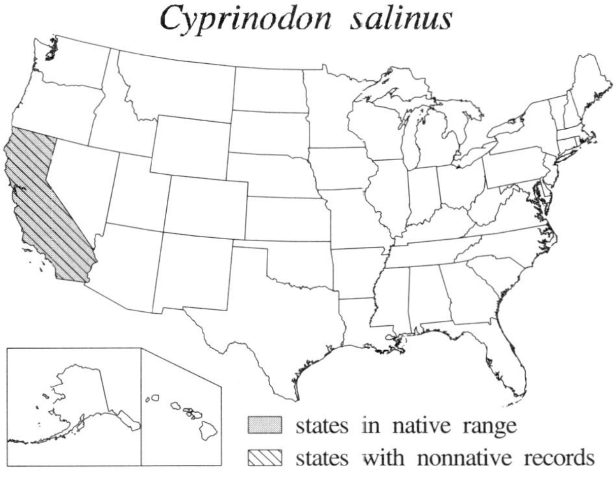

Cyprinodon salinus Miller 1943

Common Name: Salt Creek pupfish.

Identification: Moyle (1976a); Page and Burr (1991). Maximum size: 7.8 cm.

Native Range: Salt Creek, Death Valley, California (Page and Burr 1991).

Nonindigenous Occurrences: This species was introduced into several sites in **California** including Zzyzx Spring, a spring-fed pool on the west side of Soda Lake, San Bernardino County, in 1939 and again in 1940; into spring-fed ditches of Deep Springs Valley, east of Bishop in northern Inyo County in 1939 and again in 1940; one of the spring sources of the slough at Fish Springs in Owens Valley, Inyo County, in 1940; and the main spring pool at River Springs, Adobe Valley, Mono County in 1940 (Miller 1968), and into River Springs, Mono County in **California** (Lee et al. 1980 et seq.; Sigler and Sigler 1987).

Means of Introduction: Many of the early stockings were apparently part of a series of experiments to test the effects of changed environment on meristic and morphometric characters (Miller 1968). Some introductions were presumably for establishing additional populations of a rare species with limited distribution.

Status: According to Miller (1968) the Soda Lake introduction was one of two that proved successful. He found the pupfish to be abundant during a visit to River Springs in 1965. Similarly, Lee et al. (1980 et seq.) recorded the species as being established in both Soda Lake and River Springs. The other introductions apparently failed (Miller 1968).

Impact of Introduction: Unknown.

Remarks: Some authors (e.g., Minckley et al. 1991), recognize two subspecies, *C. s. milleri* and *C. s. salinus*; both are rare due to limited distribution. However, Moyle (1976a) and Page and Burr (1991), among others, treated these as separate species, *C. salinus* and the Cottonball Marsh pupfish *C. milleri*. The Cottonball Marsh pupfish inhabits Cottonball Marsh, adjacent to Salt Creek in Death Valley, Inyo County, California (Minckley et al. 1991; Page and Burr 1991).

Cyprinodon variegatus

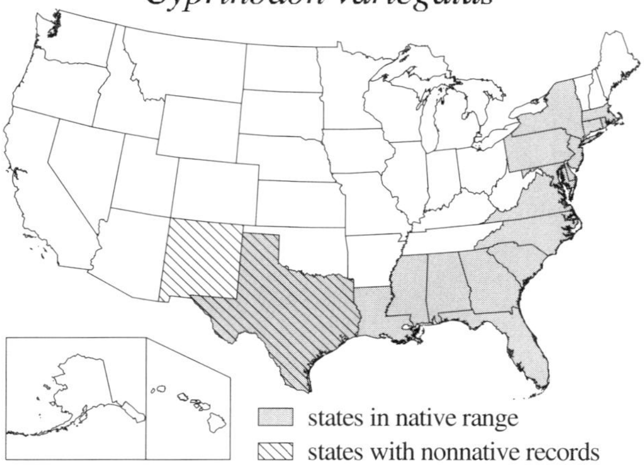

Cyprinodon variegatus Lacepède 1803

Common Name: sheepshead minnow.

Identification: Smith (1985); Hubbs et al. (1991); Menhinick (1991); Page and Burr (1991). Maximum size: 7.5 cm.

Native Range: Coastal waters from Massachusetts to northeastern Mexico. Also in West Indies (Page and Burr 1991).

Nonindigenous Occurrences: This species (or hybrids with other *Cyprinodon*) has been recorded from the Pecos River drainage in **New Mexico** and **Texas** on several occasions over the past few decades including Lake Balmorhea in Texas (Stevenson and Buchanan 1973; Echelle and Echelle 1994), a spring-fed section of Diamond Y Draw in Texas (referred to as Leon Creek in earlier publications) (Kennedy 1977), and the Pecos River proper of Texas (Echelle and Connor 1989; Wilde and Echelle 1992) and, more recently, New Mexico (Echelle et al. 1997). It is also known from the San Antonio drainage of Texas (Hubbs et al. 1991).

Means of Introduction: Initial introductions were presumably the result of bait-bucket releases (Stevenson and Buchanan 1973; Kennedy 1977). The species was first found in Lake Balmorhea in the 1960s (Stevenson and Buchanan 1973) and in Leon Creek in the 1970s (Kennedy 1977). Hubbs (1980) suggested that its appearance in Leon Creek may have resulted from a water change that took place at a highway bridge crossing during the transport of live bait from the Gulf Coast to the Pacific Coast. Kennedy (1977) suggested that it was probably introduced by local fishermen or bait dealers. The species first appeared in the Pecos River proper between 1980 and 1984 (Echelle and Connor 1989). Wilde and Echelle (1992) believed that Lake Balmorhea could have been the source of *C. variegatus* introduced into Leon Creek or the Pecos River proper; however, the researchers could not rule out the possibility that the species was introduced into these waters from elsewhere. Based on results of their electrophoretic study of fish taken from the Pecos River proper, Wilde and Echelle (1992) suggested that hybrids are widespread as a result of intradrainage dispersal and artificial transport of hybrids rather than multiple introductions of *C. variegatus*. Their data also suggested that the initial introduction of this species into the Pecos River probably occurred in the vicinity of Pecos, Reeves County, Texas, and hybrids subsequently dispersed both up- and down stream. However, recent research suggests that Red Bluff Reservoir was the original site of introduction (Childs et al. 1996).

Status: Established in Texas outside of its native range.

Impact of Introduction: Introduced *Cyprinodon variegatus* are known to hybridize with other native *Cyprinodon* species, including the endangered Commanche Springs pupfish *C. elegans* (Stevenson and Buchanan 1973), the Pecos pupfish *C. pecosensis* (Echelle and Connor 1989; Wilde and Echelle 1992, 1997; Childs et al. 1996; Echelle et al. 1997), and the endangered Leon Springs pupfish *C. bovinus* (Minckley et al. 1991; Echelle and Echelle 1997). Hubbs et al. (1991) stated that *C. pecosensis* is nearly extirpated in Texas due to introgressive hybridization with introduced *C. variegatus*. Wilde and Echelle (1992) reported that hybrids between introduced *C. variegatus* and Pecos pupfish *C. pecosensis* have replaced endemic Pecos pupfish in a 430-km stretch of the Pecos River and have dispersed into downstream areas well outside the recorded range of Pecos pupfish. They concluded that hybridization with *C. variegatus* had reduced the range of the Pecos pupfish by 60 percent. Because the appearance of *C. variegatus* was likely the result of a bait bucket introduction involving few individuals, it serves as an example that a small number of individuals can have a dramatic effect on another species (Echelle and Connor 1989). Introduction of *C. variegatus* into a spring-fed watercourse of Diamond Y Draw in the 1970s rapidly led to extensive hybridization with the endemic *C. bovinus* and local genetic swamping of the native species (Hubbs 1980; Minckley et al. 1991). The hybrids were eradicated and the native species was restored to the area (Hubbs 1980). Subsequently, however, *C. variegatus* was again introduced to the area, and this led to genetic introgression of the entire wild population of *C. bovinus* (Echelle and

Echelle 1997). In contrast to the high levels of genetic introgression resulting from introductions of *C. variegatus* into the ranges of the other two species of pupfish endemic to the Pecos River drainage, Echelle and Echelle (1994) found little evidence of genetic introgression after more than two decades of contact between *C. elegans* and *C. variegatus*.

Remarks: As of 1992, the distribution of *C. variegatus* × *C. pecosensis* hybrids in the New Mexico portion of the Pecos River drainage was restricted to headwaters of Red Bluff Reservoir and the mouth of the Delaware River a few kilometers north of the New Mexico–Texas border, Eddy County, New Mexico (Wilde and Echelle 1992). More recently, Echelle et al. (1997) reported that all known New Mexico populations of *C. pecosensis* from Loving downstream to Red Bluff Reservoir now contain genetic material from *C. variegatus*. Childs et al. (1996) suggested that the introduction of *C. variegatus* may have occurred at a time when *C. pecosensis* populations had been catastrophically depleted, and therefore susceptible to genetic swamping. They felt that this would explain the rapid spread of introduced genetic material in the Pecos River and other genetic changes observed among sample localities. Many voucher specimens, presumably a mixture of *C. variegatus*, *C. pecosensis*, and hybrids, are housed at OSUS.

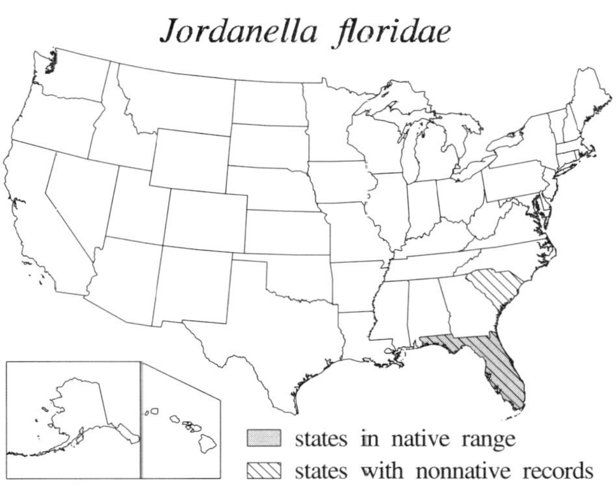

Jordanella floridae Goode and Bean 1879

Common Name: flagfish.

Identification: Boschung et al. (1983); Page and Burr (1991). Maximum size: 6.5 cm.

Native Range: Peninsular Florida north to St. Johns and Ochlockonee River drainages (Page and Burr 1991).

Nonindigenous Occurrences: A single specimen was taken from a small stream in the Ochlockonee River drainage west of Tallahassee in Leon County, **Florida** (Swift et al. 1977; Lee et al. 1980 et seq.). The specimen (UF 56238, formerly FSU 6238) was collected in 1959. There are also unconfirmed records of flagfish being collected in Charleston County, **South Carolina** (Lee et al. 1980 et seq.).

Means of Introduction: Unknown.

Status: Reported from western portion of panhandle of Florida and from South Carolina.

Impact of Introduction: Unknown.

Remarks: Swift et al. (1977) believed the Florida panhandle record was the result of an introduction since all other localities known to them were close to the coast. However, in peninsular Florida *Jordanella floridae* occurs naturally far inland (e.g., the Kissimmee River drainage).

ORDER GASTEROSTEIFORMES

Gasterosteidae - Sticklebacks

The Gasterosteidae comprises five genera and about seven species (Nelson 1994). The family is widely distributed in temperate areas of the Northern Hemisphere (Berra 1981; Nelson 1994). Four species are native to North America north of Mexico (Page and Burr 1991). Sticklebacks are small fish, less than about 15 cm in length, characterized by a scaleless body with 3–16 isolated dorsal spines along the back (Page and Burr 1991). They occupy shallow, cool habitats and are found in marine, fresh, and brackish waters (Berra 1981; Nelson 1994).

Apeltes quadracus

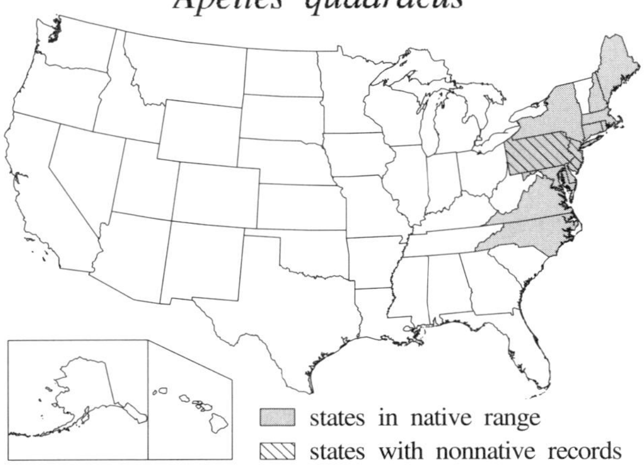

Apeltes quadracus (Mitchill 1815)

Common Name: fourspine stickleback.

Identification: Scott and Crossman (1973); Smith (1985); Robins and Ray (1986); Page and Burr (1991). Maximum size: 6.4 cm.

Native Range: Atlantic Slope from Gulf of St. Lawrence to Trent River system, North Carolina. Mostly a nearshore marine species, but far inland populations occur in lakes in Nova Scotia, and in Hudson, Delaware, and Susquehanna River drainages (Page and Burr 1991).

Nonindigenous Occurrences: This species was introduced to the Great Swamp in **New Jersey** in 1969 (Stiles 1978). It was recorded from the Susquehanna River drainage in **Pennsylvania** (Denoncourt et al. 1975c). The Pennsylvania record is based on two specimens taken in the fall of 1973 from an inlet stream of Harvey's Lake in Luzurne County, and 78 specimens taken in July 1965 from Big Spring Creek, a tributary of Conodquinet Creek, in Cumberland County.

Means of Introduction: Unknown. Denoncourt et al. (1975c) were uncertain if the Susquehanna populations of this species, and that of several other additions to the drainage, represented natural occurrences or were the result of accidental introductions.

Status: Present status in New Jersey is unknown. Apparently established in Pennsylvania.

Impact of Introduction: Unknown.

Remarks: In 1986, an introduced population was discovered on the Canadian side of Lake Superior in Thunder Bay; the suggested means of introduction in that case was ballast water (Holm and Hamilton 1988).

Culaea inconstans

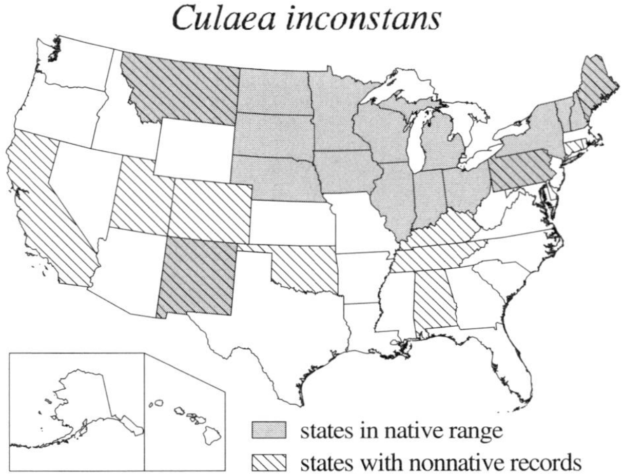

Culaea inconstans (Kirtland 1840)

Common Names: brook stickleback (five-spined stickleback).

Identification: Scott and Crossman (1973); Becker (1983); Smith (1985); Page and Burr (1991). Maximum size: 8.7 cm.

Native Range: Atlantic and Arctic drainages from Nova Scotia to Northwest Territories; Great Lakes–Mississippi River basins south to southern Ohio and New Brunswick, and west to Manitoba and eastern British Columbia. Isolated population in Canadian River system, northeastern New Mexico (Page and Burr 1991).

Nonindigenous Occurrences: This species has been recorded from Butler Creek, Tennessee River drainage, in Lauderdale County, **Alabama** (Smith-Vaniz 1968; Boschung 1992); the Scott River proper and several of its tributaries in Siskiyou County, **California** (Dill and Cordone 1997); the San Luis Valley of the Rio Grande drainage, **Colorado** (Zuckerman and Behnke 1986); the Connecticut River drainage in **Connecticut** (Whitworth et al. 1968; Lee et al. 1980 et seq.; Schmidt 1986); several streams in eastern **Kentucky** (Burr and Warren 1986); an area in coastal **Maine** (Lee et al. 1980 et seq.); Swan and Flathead lakes, **Montana** (W. Fredenberg, personal communication); several sites in the Canadian River drainage of **New Mexico** (Nelson 1969; Sublette et al. 1990; Gach 1996); unspecified areas of **Oklahoma** (Moore and Riggs 1963); Sandy Run, a tributary of the Susquehanna River, in **Pennsylvania** (Denoncourt et al. 1975c; Hocutt et al. 1986); the Obey River near Celina, and ponds and creeks in the White Oak Creek system in Roane County, **Tennessee** (Etnier and Starnes 1993); and Willard Bay Reservoir and Utah Lake, **Utah** (Sigler and Sigler 1987).

Means of Introduction: Many introductions have been attributed to bait bucket releases. According to Smith-Vaniz (1968), the Alabama record was based on a single specimen, and was likely the result of an escape or release from a minnow hatchery located several miles downstream. Holding tanks at the hatchery contained stickleback, and the species was frequently observed in shipments of commercial bait minnows brought to Alabama from Wisconsin. Nelson (1969) also reported the species as commonly found with *Pimephales promelas* in baitfish shipments from Minnesota and Wisconsin. In certain southwestern localities, the species may have been introduced accidentally as a contaminant in stockings of sport fishes. For instance, Hatch (1985, cited by Sublette et al. 1990) believed that the presence of this species in New Mexico was the result of baitbucket introductions or introduced incidental to stocking of salmonids. He based his conclusions on the fact that this stickleback had not been found in New Mexico prior to 1954 and that headwaters of the Canadian River drainage are popular with trout anglers. Zuckerman and Behnke (1986) suggested that the Colorado Division of Wildlife may have accidentally introduced the species into the Rio Grande drainage of Colorado waters along with northern pike brought in from Minnesota. Pigg et al. (1993) discovered brook stickleback in a bait shipment of fathead minnows from Minnesota and suggested a possible link between its introduction into Oklahoma and the bait fish industry. Bait bucket release is also thought to be the source of the introduced populations in Montana (Fredenberg, personal communication).

It was introduced to two sites in Utah in 1983 (Sigler and Sigler 1987), possibly intentional stockings as forage. The species was first discovered in California in 1991, but the means of introduction remains unknown (Dill and Cordone 1997). Denoncourt et al. (1975c) were uncertain as to the origin of this species in the Susquehanna River drainage of Pennsylvania; they stated that the populations may be natural or the result of accidental introductions.

Status: Reported from Alabama, Kentucky, Minnesota, and Oklahoma. The species is no longer present in Utah (Sigler and Sigler 1996). According to Whitworth (1996), it has not been found in Connecticut since the report by Whitworth et al. (1968). It has been introduced to Tennessee on several occasions resulting in populations of varied persistence; the White Oak Creek system population is considered to be locally established in that state (Etnier and Starnes 1993). Koster (1957) reported that it is likely extirpated in New Mexico because its only known locality was poisoned. In Montana, it was introduced into Swan Lake within the past five years or so and has become established there. It was first taken in nearby Flathead Lake in 1997. A single individual was collected in 1997 and four more in 1998 (Fredenberg, personal communication). The species is established or possibly locally established in other states mentioned.

Impact of Introduction: Largely unknown. Woodling (1985) stated that the species is pugnacious and preys on eggs. As such, he said that its use as a bait fish is discouraged so as to prevent accidental range expansion.

Remarks: Zuckerman and Behnke (1986) indicated that there is uncertainty as to whether *Culaea inconstans* is introduced or native to the Rio Grande drainage. Stickleback found in some New Mexico minnow ponds may have been introduced; however, the population inhabiting the headwaters of the Canadian River drainage, New Mexico, may represent a glacial relict (Nelson 1969). Page and Burr (1991) clearly agreed with that conclusion. However, Sublette et al. (1990) presented strong evidence to support the hypothesis that the species' presence in New Mexico is the result of introduction. In a recent study, Gach (1996) found that the disjunct population in New Mexico shares identical mitochondrial DNA restriction profiles with Michigan populations and likely repreents an accidental introduction rather than a Pleistocene relict. She also found evidence indicating that *Culaea* had been introduced into parts of Michigan including the Saline River and the Huron River drainage. Apparently, these were sites where the species is already native, but the introduction caused a change in the genetic composition of existing populations. The Saline River introduction reportedly occurred in 1967 and involved *Culaea* that had contaminated a commercial shipment of *Pimephales promelas* used in a toxicity study. Woodling (1985) recorded the presence of *C. inconstans* from the South Platte drainage and Arkansas River system in eastern Colorado. Based on the conclusions of others, he suggested its occurrence in these drainages represented native records. Similarly, Propst and Carlson (1986) provisionally listed *C. inconstans* as indigenous to the Platte River drainage of Colorado. However, Page and Burr (1991) did not include Colorado within the species' native range. Nelson (1969) indicated that a few early workers included Kansas in the range limits of *Culaea*; however, he concluded that such a southern extreme was doubtful. According to Etnier and Starnes (1993), Tennessee formerly prohibited import of this species into the state based on the unlikely notion that their sharp spines posed a potential hazard to sport fishes. Because sticklebacks are highly aggressive, Zuckerman and Behnke (1986) felt there was a need to address the impact of their introduction on native fish communities. Woodling (1985) felt that the species is not a very good bait fish. Nevertheless, it is occasionally found in bait buckets.

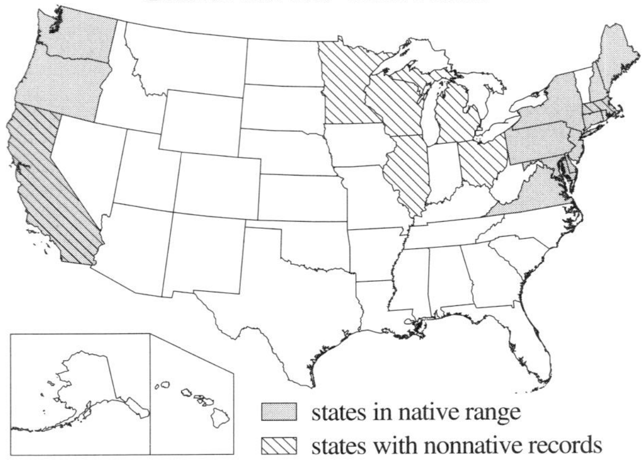

Gasterosteus aculeatus Linnaeus 1758

Common Name: threespine stickleback.

Identification: Scott and Crossman (1973); Smith (1985); Robins and Ray (1986); Page and Burr (1991); Jenkins and Burkhead (1994). *Gasterosteus aculeatus* actually may represent a complex of two or more distinct species; subspecies have been proposed but their ranges are poorly defined (Miller and Hubbs 1969; Page and Burr 1991). Because the taxonomy of this group is so complex it is probably irresolvable (Gilbert, personal communication). Maximum size: 10 cm.

Native Range: Marine and freshwater. Arctic and Atlantic drainages from Baffin Island and western side of Hudson Bay to Cape Fear Estuary, North Carolina (Schwartz et al. 1981; Page and Burr 1991); Pacific drainages from Alaska to Baja California. Eastern freshwater populations found far inland, including Lake Ontario. Also in Europe, Iceland, Greenland, and Pacific coast of Asia (Page and Burr 1991). In the Great Lakes native only below Niagara Falls (Smith 1985).

Nonindigenous Occurrences: This species is known from the Santa Maria River system, Gull and June lakes in Mono County, Mojave River drainage, and Big Bear Lake in San Bernardino County, **California** (Miller and Hubbs 1969; Moyle 1976a); Lake Michigan, **Illinois** (Johnston 1991; Burr 1991; Page and Laird 1993); a park in Boston, **Massachusetts** (Hartel et al. 1996); Lake Michigan and Lake Huron, **Michigan** (Stedman and Bowen 1985; Bailey and Smith 1991; Johnston 1991; Hirsch 1998); several areas of Lake Superior in **Minnesota** including Silver Bay, an unspecified tributary on the north shore near Grand Marais, southwest of the Knife

River, offshore from the Duluth ship canal, at Hovland, the St. Louis estuary (Hirsch 1998; S. A. Geving, personal communication), Skunk Creek and the mouth of the Baptism River (K. Schmidt, personal communication); Maumee Bay, Lake Erie, **Ohio** (T. Cavender, personal communication); and Lake Michigan, **Wisconsin** (Johnston 1991; K. Scheidegger, personal communication).

Means of Introduction: The threespine stickleback was not known from above Niagara Falls before 1979. The first specimens collected from above the falls were taken in 1980 from South Bay in Lake Huron (Stedman and Bowman 1985). Smith (1985) stated that the threespine stickleback gained access to the Upper Great Lakes from the Ottawa River and Lake Ontario through the artificial Nipissing Canal. However, Stedman and Bowman (1985) presented the possibility that they were transported by bait dealers and subsequently released by anglers. Mandrak and Crossman (1992) recorded it from Thunder Bay, Lake Superior (Canada) and attributed its presence there to a ballast water introduction. According to Miller and Hubbs (1969), it was introduced into the Mohave River drainage of California between 1938 and 1940. The species was presumably introduced as a result of escape or release of bait fish brought in by anglers from southern California coastal area. In addition to the earlier releases, the species was probably introduced into the Mohave River with trout from the Fillmore Hatchery in 1947 (Miller and Hubbs 1969). The species was believed to have been introduced into Gull and June lakes by anglers, probably from the Ventura River drainage; and it was introduced into the Santa Maria River system along with trout from the Santa Ynez River in 1940 (Miller and Hubbs 1969). A population in Holcomb Creek in the Mojave River drainage may have been introduced with trout in the late 1800s (Bell 1982; Sigler and Sigler 1987). According to Hartel et al. (1996), the population in Boston's Olmsted Park may have been introduced as part of a museum exhibit. The park contains a pool that Fredrick Law Olmsted had designated for a stickleback exhibit

Status: Reported in Ohio and Wisconsin, established in California, Massachusetts, and Michigan. First recorded in Lake Huron in 1982, where it apparently gained access through the Nipissing Canal from the Ottawa River (Smith 1985). During recent years, species reported as spreading rapidly through the upper Great Lakes (Smith 1985; Burr 1991). Recently this species was reported from Taconite Harbor, Lake Superior, Minnesota, where eight specimens were taken from cooling tower intakes in June 1994 (Hirsh 1998; J. L. Gunderson, personal communication). A single fish was collected from the stomach of a lake trout taken in 150 feet of water southwest of Knife River in May 1996 (Geving, personal communication). Also in May, two were collected in a commercial pound net in the St. Louis estuary under the Blatnik Interstate Bridge (Geving, personal communication). It also has been reported from the Canadian side of Lake Superior (Gunderson, personal communication).

Impact of Introduction: Miller and Hubbs (1969) reported the stockings of armoured threespine stickleback *G. a. microcephalus* into the native range of the unarmoured threespine stickleback *G. a. williamsoni* in certain California drainages. One consequence has been extensive hybridization between the two subspecies (Miller and Hubbs 1969; Moyle 1976b). The threespined stickleback is known to prey on eggs of other species (Page and Laird 1993).

Remarks: The unarmoured threespine stickleback *G. a. williamsoni* has been on the decline in California and has been listed as Federally endangered since 1971. The population in Boston, Massachusetts is unique in several ways. It is the southernmost freshwater population; it contains 3 distinct lateral-plate morphs; and it represents only the fourth record of low plate individuals. Its urban location is another factor in support of an introduction (Hartel et al. 1996). Hubbs (1919) advocated the stocking of artificial and natural waterbodies in California as a biological control against mosquitos.

Voucher specimens: Illinois (INHS 64211, 64481, 64482, 59309), Michigan (UWZM 8269), Minnesota (USGS–Biological Resources center, Ashland, WI; JFBM), Ohio (OSM), Wisconsin (UWZM 9093, 9094, 9727).

Pungitius pungitius

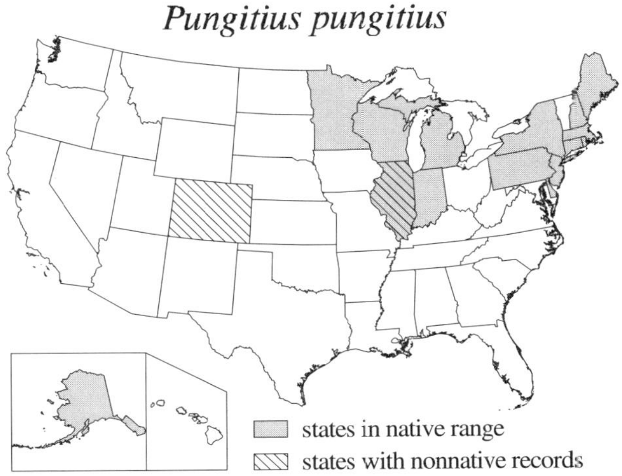

Pungitius pungitius (Linnaeus 1758)

Common Name: ninespine stickleback.

Identification: Scott and Crossman (1973); Becker (1983); Smith (1985); Page and Burr (1991). There are two recognizable forms of this species; a coastal form and an inland form. However, they have not yet been formally recognized as subspecies (Page and Burr 1991). Maximum size: 9 cm.

Native Range: Arctic and Atlantic drainages across Canada and Alaska, and as far south as New Jersey; Pacific Coast of Alaska; Great Lakes basin. Also in Eurasia (Page and Burr 1991).

Nonindigenous Occurrences: This species has been found in live bait buckets in the South Platte drainage of **Colorado** and have presumably been introduced there (Walker 1993). It also has been reported from the Illinois drainage in **Illinois** and the Mississippi drainage (presumably in the Illinois portion) (Burr and Page 1986). Smith (1979) depicted a collection from the Illinois River and stated that it occasionally enters the Des Plaines and upper Illinois rivers during winter floods.

Method of Introduction: Possible bait bucket release in Colorado. Canal connection in Illinois; the Chicago Shipping and Sanitary Canal connects Lake Michigan to the Illinois River.

Status: Reported from Colorado and Illinois. Smith (1979) stated that there was no evidence of reproduction outside of Lake Michigan in Illinois.

Remarks: Many voucher specimens deposited at UMMZ and JFBM.

ORDER SYNBRANCHIFORMES

Synbranchidae - Swamp eels

The family Synbranchidae includes 4 genera and 17 species (Bailey and Gans 1998). These fishes are found in tropical and subtropical regions including Mexico, Central and South America, and west Africa, and from India east to Australia, including Japan (Berra 1981; Nelson 1994). Synbranchids are medium-sized, eel-like fishes, characterized by reduced dorsal and anal fins and the absence of pectoral and pelvic fins. Most species lack scales. Most are also capable of air breathing, which allows them to inhabit stagnant waters. Synbranchids are primarily freshwater fishes, although some species occasionally occur in brackish waters.

Monopterus albus

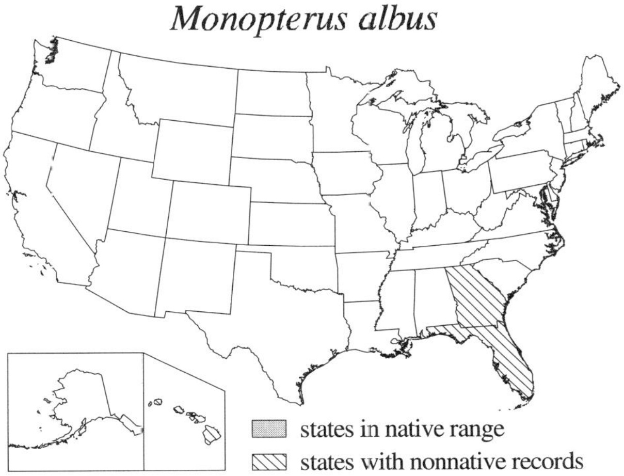

Monopterus albus (Zuiew 1793)

Common Names: Asian swamp eel (rice eel, ricefield eel, belut, rice paddy eel, ta-unagi).

Identification: Distinguishing characteristics and illustrations or photographs were given by Day (1958), Sterba (1973), Rosen and Greenwood (1976), Masuda et al. (1984), Talwar and Jhingran (1992), and Kottelat et al. (1993); characteristics were also provided by Smith (1945) and Merrick and Schmida (1984). This eel was included in keys of Rosen and Greenwood (1976), Talwar and Jhingran (1992), and Kottelat et al. (1993). Other names used for this species include *Monopterus alba*, *Fluta alba*, and *Monopterus javanenesis*. Maximum size: 100 cm.

Native Range: Asia, from northern India and Burma to China, perhaps Asiatic Russia, Japan, and the Indo-Malayan Archipelago (Bailey and Gans 1998); possibly northeastern Australia (Merrick and Schmida 1984).

Nonindigenous Occurrences: A synbranchid tentatively identified as this species was recently discovered in several waterways in peninsular **Florida** (Nico, unpublished data). It also is known from three spring-fed impoundments (Chattahoochee River drainage) at the Chattahoochee Nature Center in Roswell, Fulton County, **Georgia** (Starnes et al. 1998). This species was introduced to Oahu, **Hawaii**, before 1900 (Brock 1960; Maciolek 1984, Devick 1991a, 1991b).

Means of Introduction: Its introduction into Florida was probably the result of either an aquarium release or a fish farm escape or release, or a combination of all of the above (Nico, personal communication). The eel was probably introduced into Georgia as an aquarium release. In Georgia, adults were first collected in 1996, although they were likely present since 1990 or before (Starnes et al. 1998). It was presumably brought to Hawaii by Asian immigrants as a food fish (Devick 1991b). Brock (1960) stated that it was established in Hawaii prior to 1900. Devick (1991b) listed it as one of six fish species that were successfully introduced into Hawaii in the 19th century. However, Cobb (1902) and Jordan and Evermann (1902, 1905) made no mention of finding the species in their reports of their turn-of-the-century surveys of Hawaiian fishes.

Status: Established in Florida (Nico, unpublished data), Georgia (Starnes et al. 1998), and Hawaii (Devick 1991b).

Impact of Introduction: Largely unknown. Because they are generalized predators, this species is a potential threat to native fishes, frogs, and aquatic invertebrates. Based on sampling in 1996, the Georgia population had apparently eliminated local native sunfishes *Lepomis* species (Starnes et al. 1998). In both Georgia and Florida there is concern that the species will spread to adjacent water bodies. The Georgia impoundments where this eel has been taken are connected to the Chattahoochee River.

Remarks: The taxonomy of the genus *Monopterus* is in need of systematic review. This eel is a voracious piscivore capable of living out of water for a considerable length of time (Day 1958). The Georgia population has shown some cold tolerance, as evidenced by having survived air temperatures below freezing and ice cover over their pond habitat (Starnes et al. 1998). The distribution, biology, and impacts of Florida populations are being investigated by U.S. Geological Survey researchers in conjunction with university biologists. Georgia specimens were identified by Starnes.

Voucher specimens: Florida (UF uncatalogued); Hawaii (BPBM 3448, 3546, 5542, 5970, 28361, 30601, 30604, 33499; UF 98934); Georgia (UF 99872, USNM uncatalogued, NCSM).

ORDER SCORPAENIFORMES

Cottidae - Sculpins

Cottidae is a largely marine family, comprising about 70 genera and about 300 species (Nelson 1994). With most species found along the northern Pacific coastline of North America, the family is distributed primarily within the Northern Hemisphere; however, sculpins also occur near the coasts of Australia, New Guinea, and New Zealand (Berra 1981; Nelson 1994). Freshwater sculpins inhabit freshwaters of North America and northern Eurasia (Berra 1981). About 115 species are known from fresh and marine waters in North America (Nelson 1994); of these, 25 *Cottus* and two *Myoxocephalus* species are native to freshwaters of North America north of Mexico (Page and Burr 1991). Some marine species may attain a length of 60 cm, but freshwater forms rarely exceed 17 cm in length (Berra 1981). Sculpins are bottom-dwelling fishes with large heads.

Cottus asper

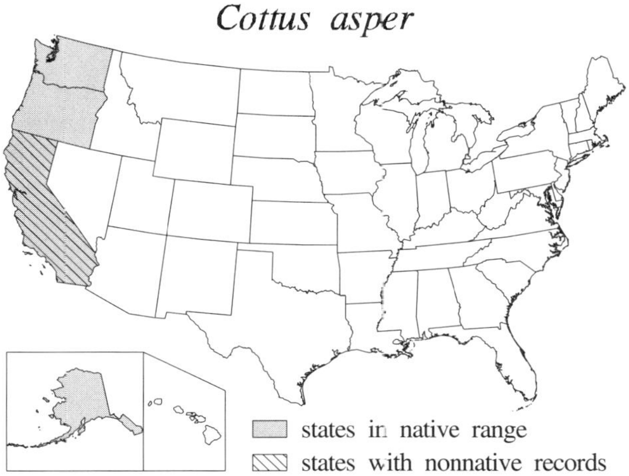

Cottus asper Richardson 1836

Common Name: prickly sculpin.

Identification: Moyle (1976a); Morrow (1980); Page and Burr (1991). Maximum size: 30 cm.

Native Range: Pacific Slope drainages from Seward, Alaska, to Ventura River, California; east of Continental Divide in upper Peace River, British Columbia (Page and Burr 1991).

Nonindigenous Occurrences: This species is known from the Santa Clara River system, **California** (U.S. Fish and Wildlife Service 1977b; Bell 1978).

Means of Introduction: Apparently introduced in imported water, possibly with water released into Piru Creek, a tributary of the Santa Clara River, from Pyramid Reservoir during the 1970s (Bell 1978).

Status: Established in California.

Impact of Introduction: Unknown. The occurrence of this cottid in the Santa Clara River system has been viewed as a possible threat to the survival of the Federally endangered unarmoured threespine stickleback *Gasterosteus aculeatus williamsoni* (U.S. Fish and Wildlife Service 1977b).

Remarks: According to Bell (1978), this species and others would have survived passage through pumping and power plants in route to Pyramid Reservoir before colonizing the Santa Clara River system. *Cottus asper* is capable of surviving in a wide range of habitats including disturbed sites; it is also known to prey on salmon eggs and fry (Moyle 1976a).

Cottus bairdi

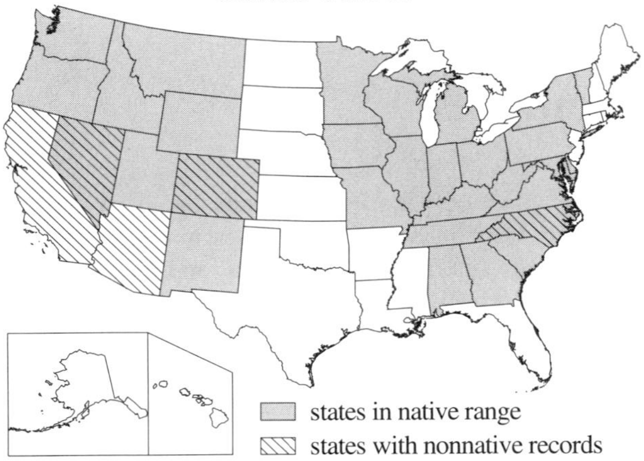

Cottus bairdi Girard 1850

Common Name: mottled sculpin.

Identification: Wydoski and Whitney (1979); Becker (1983); Page and Burr (1991); Jenkins and Burkhead (1994). Maximum size: 15 cm.

Native Range: This species has a broad distribution, with disjunct eastern and western populations. In eastern North America it occurs in the Arctic, Atlantic, and Mississippi River basins from Labrador and northern Quebec west to western Manitoba, and south to the Roanoke River drainage, Virginia, and the Tennessee River drainage, northern Georgia and Alabama, with isolated populations in the extreme upper Santee (North Carolina), Savannah (South Carolina and Georgia), Chattahoochee (Georgia), Coosa (Georgia), and Osage (Missouri) River systems. In western North America populations exist in the upper Missouri River basin, in Alberta, Montana, and Wyoming; in the Columbia River drainage from British Columbia south to Oregon and east to Wyoming; in the upper Colorado River drainage in New Mexico, Colorado, and Wyoming; isolated populations exist in endorheic basins in Utah and Nevada (Page and Burr 1991).

Nonindigenous Occurrences: This species has been observed in the Colorado River near Davis Dam in **Arizona/Nevada** (Miller 1952; Miller and Lowe 1967; Minckley 1973). Miller (1952) also reported a mottled sculpin that was sighted in Lake Havasu just south of the Needles Boat Landing in **Arizona/California**. It has been collected from the San Luis Valley in **Colorado** (Zuckerman and Behnke 1986). The species is known from two localities in the Broad drainage in **North Carolina** (Menhinick 1991).

Means of Introduction: Bait bucket release in the Colorado River in Arizona, California, and Nevada. Miller (1952) and Miller and Lowe (1967) reported the use of this species as a bait fish in the lower Colorado River. The fish was introduced accidently into Colorado by the Colorado Division of Wildlife (Zuckerman and Behnke 1986); the pathway of introduction is unknown for North Carolina.

Status: Reported in Arizona, California, and Nevada. Minckley (1973) speculated it may be established in the lower Colorado River; however, it is not established there (Minckley, personal communication). Established in North Carolina; reported from Colorado.

Impact of Introduction: Possible predator on young trout and eggs (Miller 1952).

Remarks: There is disagreement concerning native and introduced ranges of this species in North Carolina. Menhinick (1991) listed this species as probably introduced to the Broad River drainage. In contrast, Starnes (personal communication) believes that records of this species in the Broad drainage in North Carolina may have been the result of stream capture from the French Broad. Miller (1952) reported observations of an individual who saw a six-inch sculpin in Lake Havasu. This observation, based on the large size, may have been based on *Gillichthys mirabilis*, a species of goby (family Gobiidae), which is also commonly used as bait in that area.

Cottus extensus

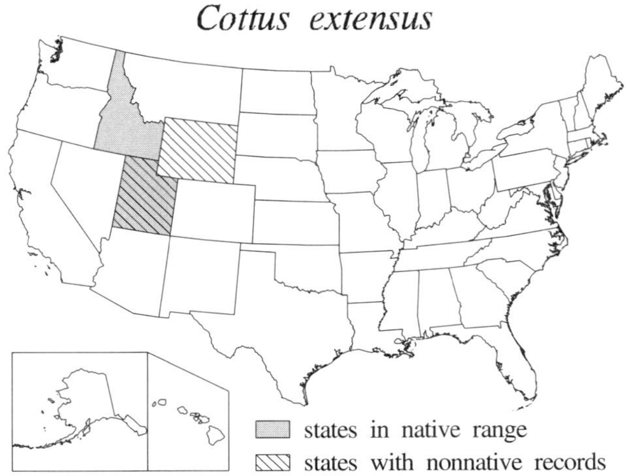

Cottus extensus Bailey and Bond 1963

Common Name: Bear Lake sculpin.

Identification: Page and Burr (1991); Sigler and Sigler (1996). Maximum size: 13 cm.

Native Range: Bear Lake, Utah and Idaho (Page and Burr 1991).

Nonindigenous Occurrences: This species was introduced into Flaming Gorge Reservoir on the Green River, Colorado River basin, **Utah** and **Wyoming** (Sigler and Sigler 1987, 1996).

Means of Introduction: Intentionally stocked into Flaming Gorge Reservoir (Sigler and Sigler 1987, 1996), apparently to serve as forage for large predatory fish species.

Status: Unknown (Sigler and Sigler 1987, 1996).

Impact of Introduction: Unknown.

Remarks: In its native habitat this species is very abundant and considered important forage for large predators such as cutthroat trout and lake trout (Sigler and Sigler 1987, 1996).

Cottus greenei

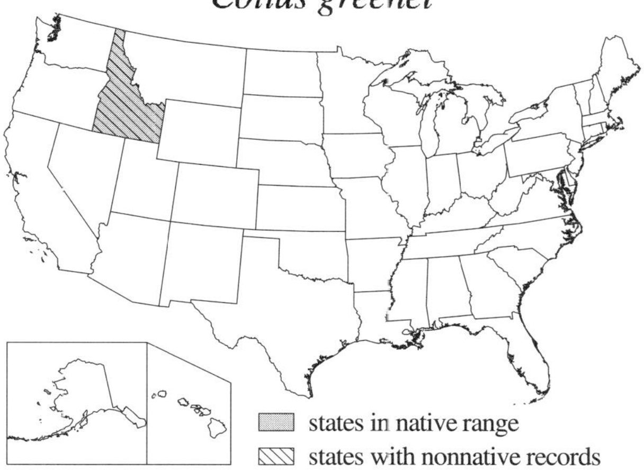

Cottus greenei (Gilbert and Culver 1898)

Common Name: Shoshone sculpin.

Identification: Simpson and Wallace (1978); Page and Burr (1991).

Native Range: Rocky springs and their effluents in Hagerman Valley in the Snake River drainage, Idaho (Wallace et al. 1984; Page and Burr 1991; Kuda and Griffith 1993).

Nonindigenous Occurrences: A total of 419 Shoshone sculpins were transplanted from Bickel Spring at the Hagerman National Fish Hatchery to an unnamed spring, sometimes called "Transplant Spring," in the Thousand Springs formation near Hagerman, **Idaho** in August 1983 (Kuda and Griffith 1993). The site of introduction was at river kilometer 965.7 on the Snake River in Gooding County, about 16 km upstream from springs where it was naturally occurring. In 1991 the species was taken from a second spring, 0.1 km downstream from the original site (Kuda and Griffith 1993).

Means of Introduction: Stocked for conservation purposes in an attempt to expand this species' very limited range. The Shoshone sculpin's native habitat was under immediate threat due to water diversion for irrigation and development of springs for commercial trout production (Simpson and Wallace 1978).

Status: The introduced population became established and was the predominant fish in Transplant Spring within eight years of its introduction (Kuda and Griffith 1993).

Impact of Introduction: Kuda and Griffith (1993) suggested that *C. greenei* displaced native mottled sculpin *C. bairdi*, causing the latter species to move from aquatic vegetation and low water velocity areas into sites with more moderate current.

Remarks: The Transplant Spring plant represents a true introduction as opposed to a reintroduction of a species into its former native habitat because the spring was apparently never occupied by *C. greenei* in the past. Prior to the elimination of the U.S. Fish and Wildlife Service candidate species list, the Shoshone sculpin was under review for Federal listing as an endangered or threatened species (U.S. Fish and Wildlife Service 1994a).

ORDER PERCIFORMES

Centropomidae - Snooks

The family Centropomidae, including the snooks and giant perches, comprises 3 genera and about 22 species (Nelson 1994). The family is widely distributed in coastal and estuarine waters of the Atlantic, Pacific, and Indian oceans, and in certain inland areas, mainly in Africa (Berra 1981; Nelson 1994). Snooks are moderate- to large-sized predatory fishes, some reaching two meters in length. Many are important food and sport fish (Robins and Ray 1986). *Centropomis undecimalis* is an Atlantic species that commonly occurs in canals and rivers of the southeastern United States coast (Robins and Ray 1986). Centropomids inhabit marine, often brackish, and freshwaters; some are catadromous (Nelson 1994). The only introduction records of this family in United States inland waters are from Texas. The Texas Parks and Wildlife Department imported three African species for research in 1975 and live fish were later released in four Texas reservoirs.

Lates angustifrons

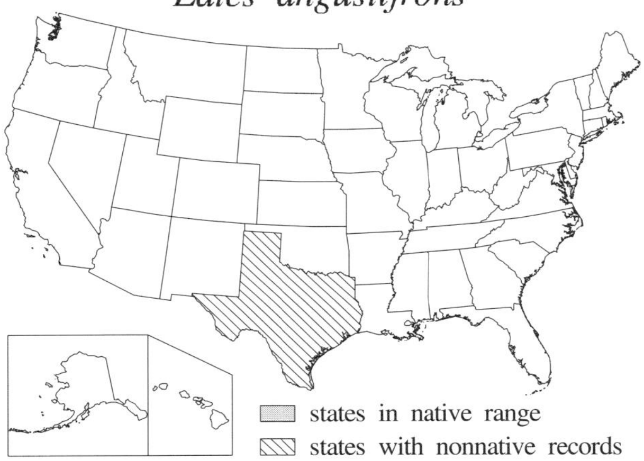

Lates angustifrons Boulenger 1906

Common Names: Tanganyika lates (sangala, sangara, pamba, Tanganyika salmon (market name)).

Identification: Eccles (1992); Howells 1992b. Maximum size: 4.5 kg.

Native Range: Lake Tanganyika, Africa (Robins et al. 1991b).

Nonindigenous Occurrences: Nile perch were imported to **Texas** in 1975 for research and were stocked in Coleto Creek Reservoir in 1981 (Howells 1992a, 1992b) and in Smithers Reservoir in 1985 (Howells 1992a, 1992b; Howells and Garrett 1992) when the research ended.

Means of Introduction: Intentional stocking by the Texas Parks and Wildlife Department for sportfishing.

Status: It is believed that both Texas stockings failed, since there have been no recent recoveries (Howells 1992a, 1992b).

Impact of Introduction: Unknown.

Remarks: None.

Lates mariae

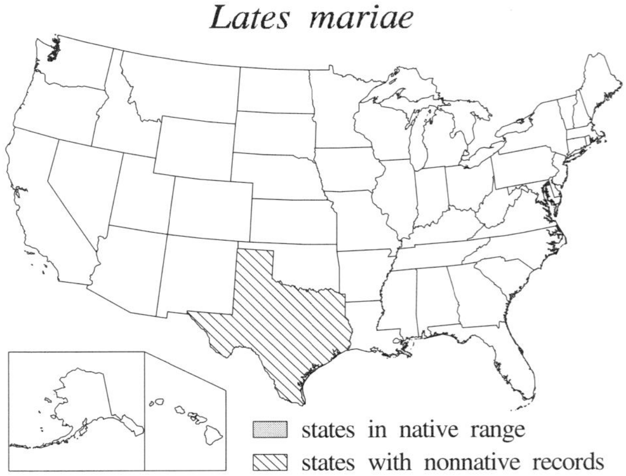

Lates mariae Steindachner 1909

Common Names: bigeye lates (golden perch, Tanganyika salmon [market name]).

Identification: Eccles (1992); Howells (1992b). Maximum size: 500 g.

Native Range: Endemic to Lake Tanganyika, Africa (Robins et al. 1991b).

Nonindigenous Occurrences: This species was stocked in **Texas** in Coleto Creek Reservoir in 1981 (Howells 1992a, 1992b), in Lake Fairfield in 1983, in Victor Braunig Reservoir in 1984 (Howells 1992a, 1992b; Howells and Garrett 1992), and in Smithers Reservoir the following year (Howells 1992a, 1992b; Howells and Garrett 1992).

Means of Introduction: Intentional stocking by the Texas Parks and Wildlife Department for sportfishing.

Status: All populations are now believed to be extirpated (Howells 1992a, 1992b).

Impact of Introduction: Unknown.

Remarks: None.

Lates niloticus

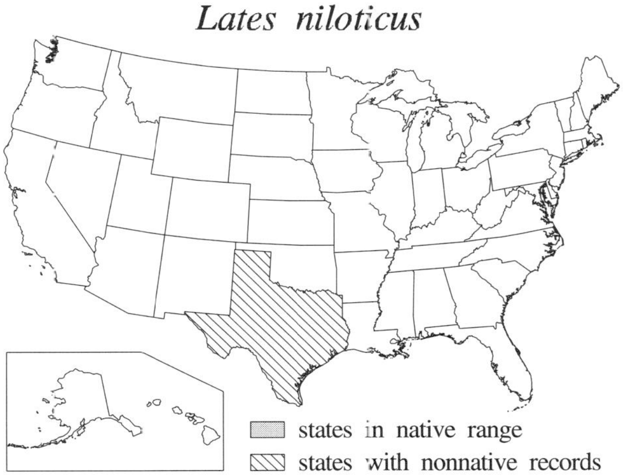

Lates niloticus (Linnaeus 1758)

Common Name: Nile perch.

Identification: Eccles (1992); Howells 1992b. Maximum size: 109 kg.

Native Range: Lake Chad and the Nile, Zaire, and Niger rivers of Africa (Robins et al. 1991b).

Nonindigenous Occurrences: This species was first stocked in **Texas** in Victor Braunig Reservoir in 1978 (Howells 1992a, 1992b; Howells and Garrett 1992) and in Smithers Reservoir in 1985 (14 individuals) (Howells and Garrett 1992; Howells 1992a, 1992b). One individual (115.5 cm, 27.2 kg) was collected from Smithers Reservoir in January 1990 (Howells and Garrett 1992). It is believed that this fish died due to cold water temperatures.

Means of Introduction: Intentional stocking by the Texas Parks and Wildlife Department for sportfishing.

Status: Both populations are probably extirpated (Howells 1992a, 1992b).

Impact of Introduction: Unknown in Texas.

Remarks: Introduction of *Lates* into Lake Victoria, Africa, has contributed to major changes in the local fish community and collapse of the commercial fishery (Goldschmidt et al. 1993). Harrison (1991) found difficulties in separating the different *Lates* species introduced into African lakes. He recommended a reappraisal of Nile perch taxonomy. As such, the positive identification of one or more of the *Lates* species introduced to Texas may eventually also be called into question.

Moronidae - Temperate basses

The family Moronidae (formerly included in Percichthyidae) includes two genera with six species (Nelson 1994) found in eastern North America, Europe, and northern Africa (Page and Burr 1991). Four species occur naturally in North American freshwaters north of Mexico (Page and Burr 1991). Moronids are moderate- to large-sized predatory fishes that are important as sport fish (Page and Burr 1991). Members of this family inhabit fresh and brackish waters, and coastal marine areas (Nelson 1994). One primarily marine species, the striped bass *Morone saxatilis,* migrates far up U.S. rivers to spawn (Page and Burr 1991).

Morone americana

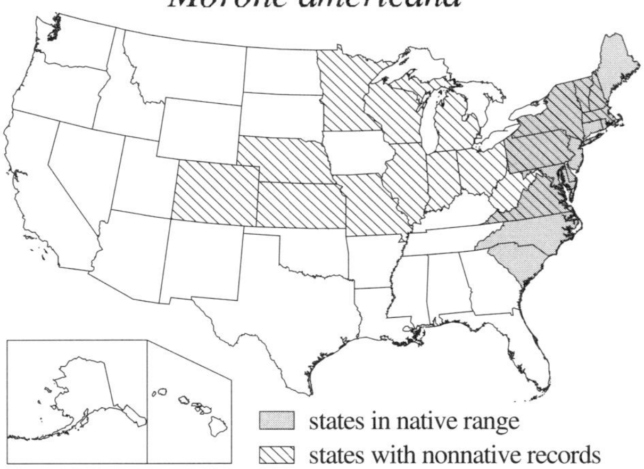

Morone americana (Gmelin 1789)

Common Name: white perch.

Identification: Woolcott (1962); Mansueti (1964); Smith (1985); Page and Burr (1991); Jenkins and Burkhead (1994). Maximum size: 58 cm.

Native Range: Atlantic Slope drainages from St. Lawrence–Lake Ontario drainage, Quebec, south to Pee Dee River, South Carolina (Page and Burr 1991). Populations in Lake Ontario drainage probably became established following construction of the Erie Canal.

Nonindigenous Occurrences: This species recorded for **Colorado** (Everhart and Seaman 1971); Lake Michigan (Savitz et al. 1989a; Mills et al. 1993), the Illinois River (Cochran and Hesse 1994), and the Mississippi River (Cochran and Hesse 1994), **Illinois** (Burr et al. 1996); Lake Michigan, **Indiana** (Mills et al. 1993; R. Horner, personal communication); the Missouri River, **Iowa** (Hergenrader 1980); Cheney and Wilson reservoirs, and Browning Oxbow on the Missouri River, **Kansas** (Whitmore 1997; T. Mosher, personal communication); nonnative, inland waters of **Massachusetts** (Hartel 1992; Hartel et al. 1996); the Great Lakes, **Michigan** (Johnson and Evans 1990; Mills et al. 1993); Duluth Harbor, Lake Superior, **Minnesota** (Johnson and Evans 1990; Mills et al. 1993); Lake Contray in Buchanan County, Big Lake in Holt County, and the Missouri River in Carroll and Howard counties, in **Missouri** (Pflieger 1997); the Missouri River and the Platte River drainage in **Nebraska** (Hergenrader and Bliss 1971; Morris et al. 1974; Hergenrader 1980; Cross et al. 1986; Whitmore 1997); inland lakes in **New Hampshire** (Scarola 1973); Lake Champlain (Plosila and Nashett 1990; Good, personal communication) and the Great Lakes drainage, **New York** (Lee et al. 1980 et seq.; Emery 1985; Smith 1985; Johnson and Evans 1990; Mills et al. 1993), including lakes Ontario and Erie, Oneida Lake, Cross Lake, and the Seneca River (Dence 1952); Lake Erie drainage of **Ohio** (Busch et al. 1977; Trautman 1981; Smith 1985; Johnson and Evans 1990; Mills et al. 1993); Lake Erie, **Pennsylvania** (Larsen 1954; Busch et al. 1977; Johnson and Evans 1990; Page and Burr 1991; Mills et al. 1993); Lake Champlain, **Vermont** (Plosila and Nashett 1990; Good, personal communication). Smith Mountain Lake and Kerr Reservoir, **Virginia** (Jenkins and Burkhead 1994); the upper Potomac drainage, **West Virginia** (Cincotta, personal communication); and Lake Michigan at Green Bay, **Wisconsin** (Savitz et al. 1989a; Johnson and Evans 1990; Mills et al. 1993; Cochran and Hesse 1994; Scheidegger, personal communication).

Means of Introduction: The first report of white perch in the Great Lakes drainage was from Cross Lake, central New York, in 1950 (Dence 1952). The species apparently gained access to the lake via movement through the Erie Barge Canal during the warm weather in the 1930s and 1950s (Lee et al. 1980 et seq.; Johnson and Evans 1990; Mills et al. 1993). Scott and Christie (1963) stated that the white perch most likely gained access to Lake Ontario via the Oswego River, as a result of spread of Hudson River populations northward and westward through the Mohawk River Valley and Erie Barge Canal. Once in Lake Ontario, it gained access to Lake Erie through the Welland Canal in 1953 and continued to spread to the upper Great Lakes (Johnson and Evans 1990; Mills et al. 1993). The first reports of westward movement through the Great Lakes are as follows: Lake Erie in 1953 (Larsen 1954), Lake St. Clair in 1977, Lake Huron in 1987 (Johnson and Evans 1990), Lake Michigan at Green Bay–Fox River, Wisconsin in May 1988 (Cochran and Hesse 1994), and Illinois waters of Lake Michigan off Chicago in September 1988 (Savitz et al. 1989a). One oddity is that the first record from Lake Superior was in 1986 from Duluth Harbor—one year before the fish was found in Lake Huron, and two years before it was seen in Lake Michigan. That population is still restricted to Duluth Harbor, possibly because it is the warmest part of the lake. This population likely represents a separate introduction, because it does not fit the pattern of western dispersal (Johnson and Evans 1990). White perch were brought from New Jersey to Nebraska in 1964, and fry produced that year in a hatchery were accidentally introduced into a reservoir that provided access to the Missouri River (Hergenrader and Bliss 1971). White perch have been stocked intentionally in other areas for sportfishing. In Kansas, fish found at Browning Oxbow on the Missouri River are believed to have come from Nebraska. The species was not recorded from Missouri River in Missouri until the 1990s (Pflieger 1997). The source of the fish in the two Kansas reservoirs is not known (Mosher, personal communication). White perch were stocked in West Virginia in the early 1900s (Cincotta, personal communication).

Status: Established in all five Great Lakes and their surrounding states, as well as in Kentucky, Massachusetts, Missouri, Nebraska, New Hampshire, and Vermont. Current status in Colorado and Kansas unknown.

Impact of Introduction: White perch prey on eggs of walleye *Stizostedion vitreum vitreum*, white bass *Morone chrysops*, eggs of its own species, and possibly eggs of other species (Schaeffer and Margraf 1987). Apparently, fish eggs are an important component of the diet of white perch in the spring months. At times, depending on which fish is spawning, either walleye or white bass eggs make up 100% of the diet. During a three-year study, this diet was found to be unique in that eggs were eaten for a comparatively long time, they were the only significant food item eaten by adults during two of the three years, large volumes were eaten per individual, and most fish were feeding. White perch also feed heavily on minnows *Notropis* spp. (Schaeffer and Margraf 1987). The collapse of the Bay of Quinte (north shore of Lake Ontario) walleye fishery coincided with the increase in the white perch population and may have been a result of egg predation and lack of recruitment (Schaeffer and Margraf 1987).

It was hypothesized that white perch compete with native yellow perch *Perca flavescens* for zooplankton. when it was determined that growth rates of yellow perch had declined since the invasion of white perch in Lake Erie, especially in the western basin. It was also determined that the two species had considerable diet overlap (Parrish and Margraf 1990). Parrish and Margraf (1990) found in one sample, white perch consumed 27 percent more food than yellow perch.

Parrish and Margraf (1994) speculated that competition between white perch and forage fishes, such as emerald shiner *Notropis atherinoides* and spottail shiner *N. hudsonius*, may actually be more complex and may be responsible for the declines of the latter species. Decline of these species could also affect walleye *Stizostedion vitreum*, the top predator in Lake Erie (Parrish and Margraf 1994).

Within three years after being introduced into a Nebraska reservoir, white perch had completely replaced the previously dominant black bullhead *Ameiurus melas*. Species composition changed from 74 percent black bullhead to 70 percent white perch in that timeframe (Hergenrader and Bliss 1971).

Historically allopatric (Todd 1986a), the white perch and white bass *M. chrysops* became sympatric after the white perch invaded the Great Lakes. This sympatric distribution allowed the two species to hybridize. White perch hybridize with native white bass *Morone chrysops* in western Lake Erie, in Ohio and Michigan (Todd 1986a). They have also been reported from the Detroit River and the St. Clair River in Michigan (Todd 1986a). These hybrids were first noted in western Lake Erie in the early 1980s, the same period during which white perch were increasing in this area (Todd 1986a). These hybrids probably occur in other Great Lakes because the two species are sympatric throughout the chain of lakes. However, Todd was not aware of any other locations with these hybrids. His extensive surveys around Saginaw Bay, Lake Huron, and Lake Ontario in the mid–1980s failed to find any (Todd, personal communication).

Todd (1986a) provided photographs of both parent species and the hybrid and gave characteristics of each. Because these hybrids are capable of backcrossing with the parental species, and possibly producing of F_2 hybrids by crossing amongst themselves (Todd 1986a), they dilute the gene pool of each parent species. The white perch/white bass hybrid is the first naturally occurring *Morone* hybrid known (Todd 1986a).

Remarks: Although the white perch was found in the Missouri River in Missouri almost to the Missouri/Iowa state border (Pflieger 1997), as of March 1998, there are no known collections in the state of Iowa (M. Konrad, personal communication).

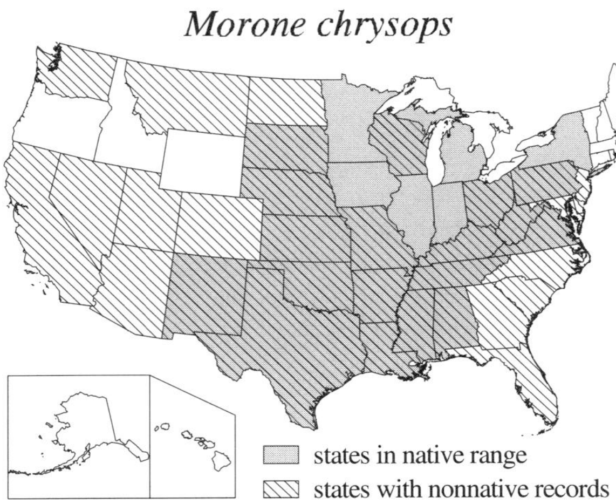

Morone chrysops (Rafinesque 1820)

Common Names: white bass (silver bass).

Identification: Becker (1983); Page and Burr (1991); Etnier and Starnes (1993); Jenkins and Burkhead (1994). Maximum size: 45 cm.

Native Range: St. Lawrence–Great Lakes, Hudson Bay (Red River), and Mississippi River basins from Quebec to Manitoba and south to Louisiana; Gulf Slope drainages from Mississippi River, Louisiana, to Rio Grande, Texas and New Mexico (Page and Burr 1991).

Nonindigenous Occurrences: White bass have been widely introduced into **Alabama** (Bonn et al. 1955; Smith-Vaniz 1968; Lee et al. 1980 et seq.; Boschung 1992; Mettee et al. 1996); **Arizona** (Minckley 1973; Grabowski et al. 1984; Rinne 1995); **Arkansas** (Pritchard et al. 1978); **California** (Smith 1896; von Geldern 1966; Moyle 1976a; Dill and Cordone 1997); **Colorado** (Everhart and Seaman 1971; Ellis 1974; Tyus et al. 1982); **Delaware** (Raasch and Altemus 1991); **Florida** (Yerger 1977; Lee et al. 1980 et seq.); **Georgia** (Bonn et al. 1955; Dahlberg and Scott 1971b; Pritchard et al. 1978; Yerger 1977); **Kansas** (Bonn et al. 1955; Cross 1967; Lee et al. 1980 et seq.); **Kentucky** (Bonn et al. 1955; Burr and Warren 1986); **Louisiana** (Conner and Suttkus 1986); **Maryland** (Hendricks et al. 1983); **Mississippi** (Pritchard et al. 1978); **Missouri** (Pflieger 1975); **Montana** (Gould 1981; Holton 1990); **Nebraska** (Jones 1963; Cross et al. 1986; Bouc 1987); **Nevada** (Deacon and Williams 1984); **New Jersey** (Bean 1903; Fowler 1906, 1952); **New Mexico** (Koster 1957; Sublette et al. 1990); **North Carolina** (Tatum 1958; Pritchard et al. 1978; Hocutt et al. 1986; Menhinick 1991; Jenkins and Burkhead 1994); **North Dakota** (Gould 1981; Cross et al. 1986; Holton 1990; North Dakota Game and Fish Department 1994, 1997); **Ohio** (Trautman 1981; Hocutt et al. 1986); **Oklahoma** (Bonn et al. 1955; Oklahoma Wildlife Commission 1963); **Pennsylvania** (Ford et al. 1892; Denoncourt et al. 1975a; Hendricks et al. 1979; Hocutt et al. 1986); **South Carolina** (Loyacano 1975; Rohde et al. 1994); **South Dakota** (Cross et al. 1986; North Dakota Game and Fish Department 1994); **Tennessee** (Bonn et al. 1955; Lee et al. 1980 et seq.); **Texas** (Bonn et al. 1955; Kraai et al. 1983; Prentice 1985; Conner and Suttkus 1986); **Utah** (Sigler and Miller 1963; B. Schmidt, personal communication); **Virginia** (Lee et al. 1980 et seq.; Hocutt et al. 1986; Jenkins and Burkhead 1994); **Washington** (Lee et al. 1980 et seq.); **West Virginia** (Stauffer et al. 1995); and **Wisconsin** (Becker 1983).

Means of Introduction: Intentionally stocked for sportfishing. In California, white bass were initially intentionally stocked for sport fishing in Lake Nacimiento, an area that was managed to support them and where they would not pose a threat to other vulnerable species. However, the fish mysteriously appeared in Lake Kaweah. Apparently the fish had been transported and stocked illegally. The fish then gained access to the Tulare Lake basin when Lake Kaweah flooded in 1982 and 1983 (Dill and Cordone 1997).

Status: The white bass has been reported as established locally in most areas, except in Delaware, where the species is known from a single record in 1888 and, presumably is extirpated (Raasch and Altemus 1991). Attempts to establish the species in the Youghiogheny River, Maryland and Pennsylvania, have failed (Hendricks et al. 1979, 1983).

Impact of Introduction: When stocked in the native range of the yellow bass *Morone mississippiensis* these two species hybridize (Fries and Harvey 1989). In Utah Lake, Utah, white bass dominate all other fish species; because they are so abundant they severely deplete the food supply (Sigler and Sigler 1987). Once white bass were in the Tulare Lake basin they could potentially reach the Delta region where several threatened and endangered fishes and important sport and commercial species could be at risk from this "voracious predator" (Dill and Cordone 1997). The state opted to conduct a chemical treatment to eradicate the bass from Lake

Kaweah and downstream waters. This was one the most extensive chemical treatments ever conducted in the United States. The cost of the project was about $7.5 million (Dill and Cordone 1997). The state may need to spend several million dollars more to treat other lakes linked to the Delta. This has been a controversial issue and is still on hold due to the controversy and finances (Dill and Cordone 1997). There is now no limit to the number of white bass that may be kept by an angler. Regulations also make it illegal to possess or transport live white bass and require immediate killing of any retained fish or the immediate return of any unwanted fish into the water body from which it was taken (Dill and Cordone 1997).

Remarks: Dill and Cordone (1997) gave a fascinating account of the introduction of this species into California and the resulting fallout. This account is also a good illustration of how fish can be dispersed even when they are thought to be contained.

Morone mississippiensis

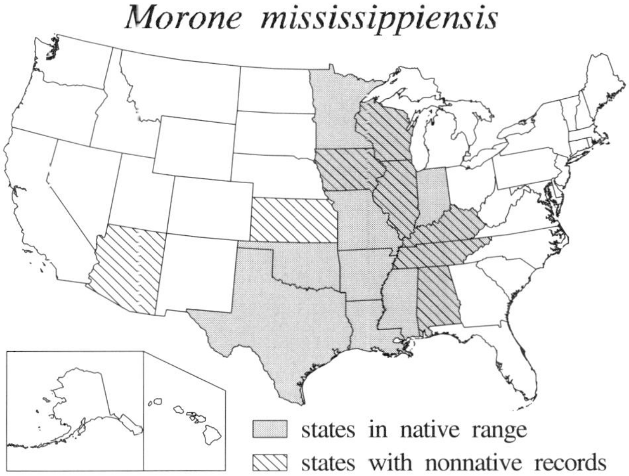

Morone mississippiensis Jordan and Eigenmann 1887

Common Name: yellow bass.

Identification: Pflieger (1975); Becker (1983); Page and Burr (1991); Etnier and Starnes (1993). Maximum size: 46 cm.

Native Range: Lake Michigan and Mississippi River basins from Wisconsin and Minnesota south to Gulf; east to western Indiana and eastern Tennessee, and west to western Iowa and eastern Oklahoma. On Gulf Slope in lower Mobile Bay drainage, Alabama, and from Pearl River drainage, Louisiana, to Galveston Bay drainage, Texas (Page and Burr 1991).

Nonindigenous Occurrences: The yellow bass has been introduced into the Coosa River (Lee et al. 1980 et seq.), the upper Tombigbee River (Boschung 1992), the Cahaba River at Lake Purdy, and Inland Lake in the Black Warrior River (Mettee et al. 1996) in **Alabama**; the Salt River in **Arizona** (Lee et al. 1980 et seq.); **Illinois** (Lee et al. 1980 et seq.); tributaries of the Mississippi and Missouri rivers in **Iowa** (Lee et al. 1980 et seq.; Cross et al. 1986); non-specific areas in **Kansas** before 1885 (Cross 1967), and recently from Browning Oxbow on the Missouri River (Mosher, personal communication); several areas in **Kentucky** (Burr and Warren 1986); the Tennessee River, **Tennessee** (Lee et al. 1980 et seq.); and widely into **Wisconsin** (Helm 1958).

Means of Introduction: Yellow bass are intentionally stocked for sportfishing. They have gained access to the upper Tombigbee River through the Tennessee–Tombigbee Waterway (Boschung 1992). A recent collection from Browning Oxbow, Kansas, may be a result of migration from stocked Lake Icaria, Iowa (Mosher, personal communication).

Status: Reported as established in Arizona (Rinne 1995) and Wisconsin (Helm 1958). Stockings in Kansas in the late 1800s failed to produce sustaining populations (Cross 1967). However, yellow bass have more recently been collected from Browning Oxbow in Kansas (Mosher, personal communication). The status of the other introductions is unknown.

Impact of Introduction: When stocked in the native range of the white bass *M. chrysops* these two species hybridize (Fries and Harvey 1989).

Remarks: None.

Morone saxatilis

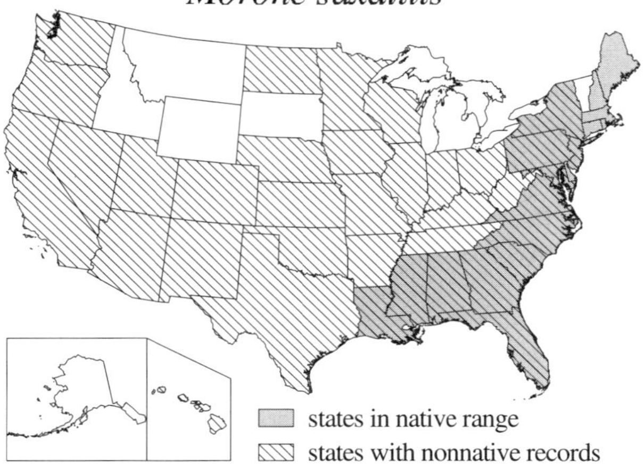

Morone saxatilis (Walbaum 1792)

Common Names: striped bass (striper, rock, rockfish).

Identification: Smith (1985); Page and Burr (1991) Jenkins and Burkhead (1994); Mettee et al. (1996). Maximum size: 2 m.

Native Range: Atlantic drainages from St. Lawrence River, New Brunswick to St. Johns River in Florida; Gulf Slope drainages from Lake Pontchartrain, Louisiana to Suwannee River drainage, Florida (Page and Burr 1991). Apparently, striped bass formerly occurred in coastal areas of eastern Texas (Anonymous 1987c). Higgins and Lord (1926) reported it as one of the chief commercial species in that area.

Nonindigenous Occurrences: Striped bass have been widely stocked in nonindigenous waters from New York to California. Introductions of striped bass have been reported from **Alabama** (Smith-Vaniz 1968; Pritchard et al. 1978; Grabowski et al. 1984; Boschung 1992; Mettee et al. 1996); **Arizona** (St. Amant 1959; Edwards 1974; Tyus et al. 1982); **Arkansas** (Gray 1958; Buchanan 1973; Bailey 1975; Pflieger 1975; Grabowski et al. 1984; Cross et al. 1986); **California** (Smith 1896; Shebley 1917; Surber 1958; St. Amant 1959; Moyle 1976a; Dill and Cordone 1997); **Colorado** (Everhart and Seaman 1971; Grabowski et al. 1984); **Florida** (Ware 1971; Swift et al. 1977; Grabowski et al. 1984); **Georgia** (Dahlberg and Scott 1971b; Yerger 1977); **Hawaii** (Maciolek 1984); **Illinois** (Smith 1979; Burr and Page 1986); **Indiana** (Lee et al. 1980 et seq.); **Iowa** (Burr and Page 1986; Cross et al. 1986; Harlan et al. 1987); **Kansas** (Lee et al. 1980 et seq.; Cross et al. 1986); **Kentucky** (Bailey 1975; Clay 1975; Burr and Page 1986; Burr and Warren 1986; Hocutt et al. 1986); **Louisiana** (Douglas 1974; Grabowski et al. 1984; Conner and Suttkus 1986); **Maryland** (Surber 1958; Bailey 1975; Pritchard et al. 1978; Grabowski et al. 1984); **Minnesota** (Phillips et al. 1982); **Mississippi** (Pritchard et al. 1978); **Missouri** (Pflieger 1975, 1997; Cross et al. 1986); **Nebraska** (Jones 1963; Morris et al. 1974; Pritchard et al. 1978; Cross et al. 1986; Bouc 1987); **Nevada** (Edwards 1974; Deacon and Williams 1984; Grabowski et al. 1984); **New Jersey** (Surber 1958; Grabowski et al. 1984); **New Mexico** (Grabowski et al. 1984; Sublette et al. 1990); **New York** (Grabowski et al. 1984); **North Carolina** (Surber 1958; Bailey 1975; Grabowski et al. 1984; Menhinick 1991; Jenkins and Burkhead 1994); **North Dakota** (North Dakota Game and Fish Department 1994, 1997); **Ohio** (Trautman 1981; Burr and Page 1986); **Oklahoma** (Hall 1956; Surber 1958; Grabowski et al. 1984; Cross et al. 1986); **Oregon** (Moyle 1976a; Wydoski and Whitney 1979; Grabowski et al. 1984; Bond 1994); **Pennsylvania** (Denoncourt et al. 1975a; Hendricks et al. 1979; Cooper 1983); **South Carolina** (Surber 1958; Bailey 1975; Grabowski et al. 1984); **Tennessee** (Smith-Vaniz 1968; Grabowski et al. 1984; Etnier and Starnes 1993); **Texas** (Moczygemba and Morris 1977; Conner and Suttkus 1986; Anonymous 1992a; Howells 1992a); **Utah** (Tyus et al. 1982; Grabowski et al. 1984); **Virginia** (Lee et al. 1980 et seq.; Hocutt et al. 1986; Jenkins and Burkhead 1994); **Washington** (Setzler et al. 1980); **West Virginia** (Trautman 1981; Hocutt et al. 1986; Stauffer et al. 1995); and **Wisconsin** (Wisconsin Department of Natural Resources 1998).

Means of Introduction: Intentionally stocked for sportfishing. Some records involved fish dispersed to other areas far from point of original introduction, for instance, downstream of reservoirs after passage over or through dams (e.g., Pflieger 1997).

Status: Reported as locally established in many states. Annual stocking of lakes and reservoirs maintains the sport fishery. Attempts to establish this species in the Youghiogheny River, Pennsylvania, failed (Hendricks et al. 1979). Not established in Colorado (Walker 1993).

Impact of Introduction: Preys on small fishes. Bailey (1975) summarized what was known about impacts of stocked striped bass on clupeids in lakes and reservoirs in the Southeast. Maumelle Reservoir, Arkansas, had a possible reduction in clupeids; Toledo Bend and D'Arbonne reservoirs, Louisiana, had slight reductions; J. Percy Priest Reservoir, Tennessee, had a significant reduction; and Hunter land Julianna reservoirs suffered drastic reductions. Morgan and Gerlach (1950) reported finding numerous trout and salmon fry and fingerlings in striped bass stomachs at Coos Bay, Oregon.

Remarks: The striped bass is a highly prized sportfish. Bailey (1975) provided a table of locations, numbers, and factors affecting success of striped bass stocked in the Southeast, and the outcome. Setzler et al. (1980) provided a biological synopsis of this species.

HYBRIDS

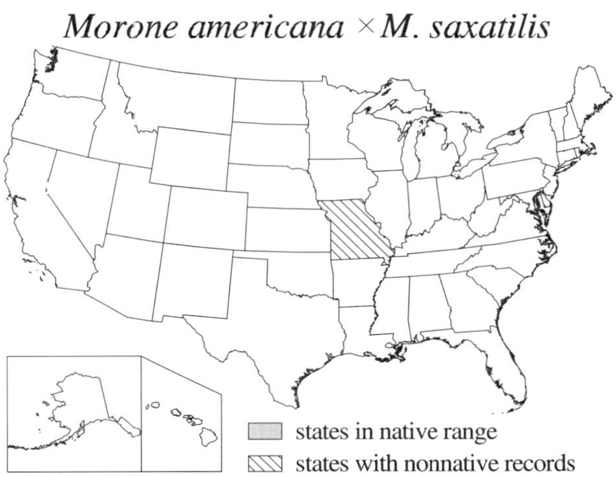

Morone americana × *M. saxatilis*

Common Names: white perch × striped bass hybrid (Virginia bass, Maryland bass).

Identification: Kerby (1979); Setzler et al. (1980); Page and Burr (1991). Maximum size: Unknown.

Native Range: None; artificial hybrid.

Nonindigenous Occurrences: This hybrid was stocked in **Delaware** at Moores Lake and Silver Lake, and in the St. Jones River (all in Kent County) (Raasch and Altemus 1991) and into various impoundments in **Missouri** (Pflieger 1997).

Means of Introduction: Intentionally stocked for sportfishing by the Delaware Division of Fish and Wildlife.

Status: Maintained by stocking.

Impact of Introduction: Unknown.

Remarks: The native ranges of the white perch and striped bass overlap, but we are unaware of any natural hybridization. This hybrid was first cultured in 1966, when female striped bass were crossed with male white perch (Kerby 1979).

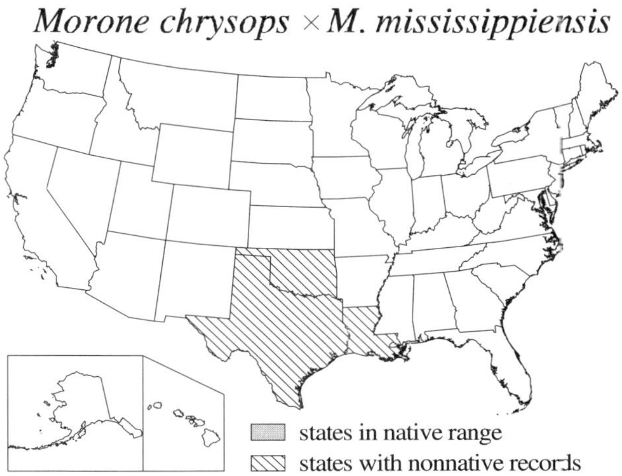

Morone chrysops × *M. mississippiensis*

Common Name: white bass × yellow bass hybrid.

Identification: Hybrid. Maximum size: 1.04 kg.

Native Range: None; artificial hybrid.

Nonindigenous Occurrences: White bass-yellow bass hybrids have been stocked in unspecified areas of **Louisiana** (Louisiana Department of Wildlife and Fisheries 1997; T. Morrison, personal communication). One fish was collected by an angler in March 1991 from the Kiamichi River, **Oklahoma** (International Game Fish Association 1994). These hybrids are found in four eastern **Texas** reservoirs due to stocking of parental species and subsequent "natural" hybridization (Fries and Harvey 1989).

Means of Introduction: Intentionally stocked as a sport fish in Louisiana and Oklahoma. Natural hybridization in Texas.

Status: Maintained by stocking in Louisiana and Oklahoma. Established in Texas.

Impact of Introduction: These hybrids represent a threat to the genetic integrity of the parental species because of potential backcrossing (Fries and Harvey 1989).

Remarks: None.

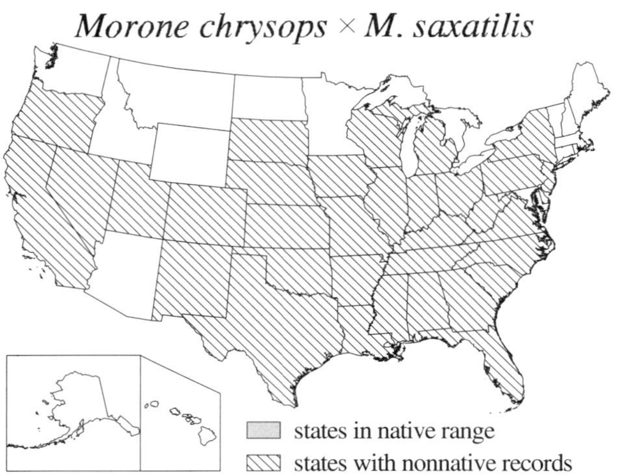

Morone chrysops × *M. saxatilis*

Common Names: white bass × striped bass hybrid, sunshine bass, palmetto bass, wiper, whiterock bass.

Identification: Kerby (1971); Ware (1975); Williams (1976); Kerby (1979); Setzler et al. (1980); Etnier and Starnes (1993); International Game Fish Association (1994); Jenkins and Burkhead (1994); Mettee et al. (1996). Bayless (1972) and Bishop (1968) provided a photograph comparing white bass, striped bass, and their hybrid. Maximum size: 10.97 kg (record size).

Native Range: None; artificial hybrid.

Nonindigenous Occurrences: Many rivers and reservoirs have been stocked with these hybrids including rivers in **Alabama** (Boschung 1992; Mettee et al. 1996); lakes Norfolk, Beaver, DeGray, Greer's Ferry, Greeson, Dequeen, and the Little Red River (Carter, personal communication), Storm Creek Lake and several other reservoirs (Robison and Buchanan 1988) in **Arkansas**; Lake Kaweah, Irvine Lake and Anaheim Lake, **California** (Dill and Cordone 1997); the East Slope drainages (Arkansas, Republican and Platte) of **Colorado** (Walker 1993; Horak, personal communication); the Brandywine-Christina and Broadkill-Smyrna drainages, **Delaware** (Raasch and Altemus 1991); the St. Johns, St. Mary's, Apalachicola, Santa Fe, Manatee, Oklawaha, Peace, Everglades and Escambia drainages in **Florida** (Yerger 1977; Florida Game and Fresh Water Fish Commission, unpublished data); the Savannah River (Dahlberg and Scott 1971b; Williams 1972) and Altoona Reservoir (Mettee et al. 1996) in **Georgia**; nonspecific locations in **Illinois** (Burr, personal communication); the Ohio River, **Indiana** (Indiana Department of Natural Resources 1997a); the Des Moines River, including Saylorville Reservoir, and pool 14 of the Mississippi River, **Iowa** (Harlan et al. 1987; Iowa Department of Natural Resources 1997); many reservoirs in **Kansas** (Cross and Collins 1995; Kansas Department of Wildlife and Parks 1997); Ohio River and perhaps other locations in **Kentucky** (Burr and Warren 1986; Indiana Department of Natural Resources 1997a); non-specific areas of **Louisiana** (Louisiana Department of Wildlife and Fisheries 1997; Morrison, personal communication); Kalamazoo River, **Michigan** (Michigan Department of Natural Resources 1997); the Tennessee-Tombigbee Waterway, **Mississippi** (U.S. Army Corps of Engineers 1997); the Osage River, **Missouri** (International Game Fish Association 1994); Lake McConaughy, Sutherland Reservoir, Bluestem Lake, Lake North, and Freemont Lake, **Nebraska** (Bouc 1987); the lower Humboldt River and Lahontan Reservoir, **Nevada** (DeLorme Mapping 1996b); Lake Hopatcong, **New Jersey** (New Jersey Division of Fish, Game and Wildlife 1997); the Pecos River, **New Mexico** (Sublette et al. 1990); the upper Susquehanna drainage in **New York** (H. Buell and Sinnott, personal communication); many reservoirs in **North Carolina** (Menhinick 1991); the Ohio River and a few lakes and reservoirs in **Ohio** (Stauffer et al. 1995; Ohio Department of Natural Resources, Division of Wildlife 1997); Sooner Lake, **Oklahoma** (International Game Fish Association 1994); Ten Mile lakes and Anna Reservoir, **Oregon** (Bond 1994; International Game Fish Association 1994); Nockamixon Lake, **Pennsylvania** (International Game Fish Association 1994); a reservoir on northern part of the Keowee River (Menhinick 1991), and the Savannah River, **South Carolina** (Dahlberg and Scott 1971b; Williams 1972); **South Dakota** (North Dakota Game and Fish Department 1994); Percy Priest Lake (International Game Fish Association 1994), and Cherokee, Ft. Loudon, Norris, and Chickamauga reservoirs, **Tennessee** (Ware 1975; Etnier and Starnes 1993); numerous reservoirs in **Texas** (Guest 1985; Anonymous 1992a); Mona Reservoir, **Utah** (Sigler and Sigler 1996; Utah Division of Wildlife Resources 1997); Leesville Lake on the Rappahonnock River, **Virginia** (Kerby 1979; International Game Fish Association 1994); the Ohio River and its tributaries, and the Kanawha River, **West Virginia** (Stauffer et al. 1995); and Lake Columbia in Columbia County, **Wisconsin** (Wisconsin Department of Natural Resources 1997).

Means of Introduction: Intentionally stocked for sportfishing.

Status: Populations maintained through stocking programs. Alabama has stocked more than 18 million palmetto bass since 1974 (Mettee et al. 1996).

Impact of Introduction: In the Savannah River, backcrossing to striped bass was observed (Avise and Van den Arvyle 1984). These hybrids also occasionally back cross with white bass (Avise and Van den Arvyle 1984; Etnier and Starnes 1993). The potential outcome of such mating includes loss of genetic integrity of the parent species or even the loss of a native species, subspecies, or of a unique population (Campton 1987). Sunshine bass apparently are backcrossing to parentals in the Ohio River, where this hybrid is very common (Burr, personal communication).

Remarks: The whiterock (palmetto, wiper, Cherokee) is the result of crossing a female striped bass with a male white bass, and was first cultured in 1965. The sunshine bass, first produced in 1973, is the result of the reverse cross (Ware 1975; Kerby 1979; International Game Fish Association 1994). Most introductions are of the whiterock hybrid; however, a few sites have been stocked with sunshine bass, including some lakes in Florida (Ware 1975). These hybrids reportedly grow faster, survive better, and are caught more readily than striped bass (Ware 1975). However, they don't attain as large a size as do striped bass (Gilbert, personal communication). These hybrids occur naturally in Arkansas (Etnier and Starnes 1993).

Morone mississippiensis × M. saxatilis

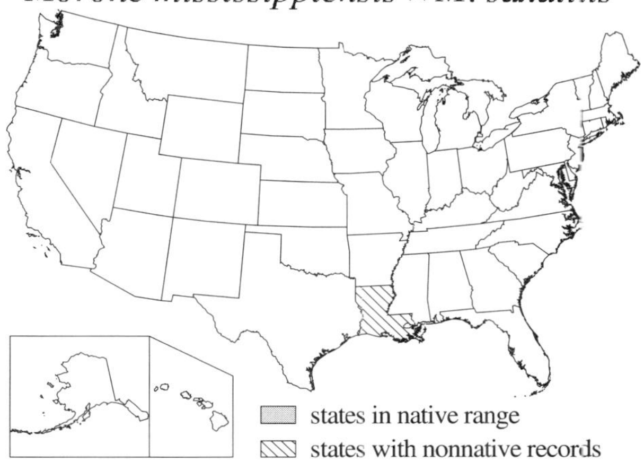

Morone mississippiensis × M. saxatilis

Common Name: yellow bass × striped bass.

Identification: Hybrid.

Native Range: None; artificial hybrid.

Nonindigenous Occurrences: Stocked in non-specific areas of **Louisiana** (Louisiana Department of Wildlife and Fisheries 1997; Morrison, personal communication).

Means of Introduction: Intentionally stocked for sportfishing.

Status: Populations maintained through stocking programs.

Impact of Introduction: Unknown.

Remarks: None.

Serranidae - Sea basses

The family Serranidae, containing sea basses and groupers, is a large family comprised of 69 genera and 449 species (Nelson 1994). They occur in temperate and tropical seas, and a few live in freshwater (Eschmeyer et al. 1983; Nelson 1994). Sea basses have the classical bass shape and are characterized by three spines on the opercule, a complete lateral line not extending onto the caudal fin, a continuous dorsal fin that may be notched, and a rounded, truncate, or lunate caudal fin (Eschmeyer et al. 1983; Nelson 1994). Serranids are small to very large fishes which are usually solitary and territorial (Robins and Ray 1986). Sea basses are hermaphroditic. Usually, the two sexes do not develop at the same time (Nelson 1994), but sometimes they do (Eschmeyer et al. 1983). Many mature as females first and become males as they get larger (Eschmeyer et al. 1983).

Paralabrax maculatofasciatus

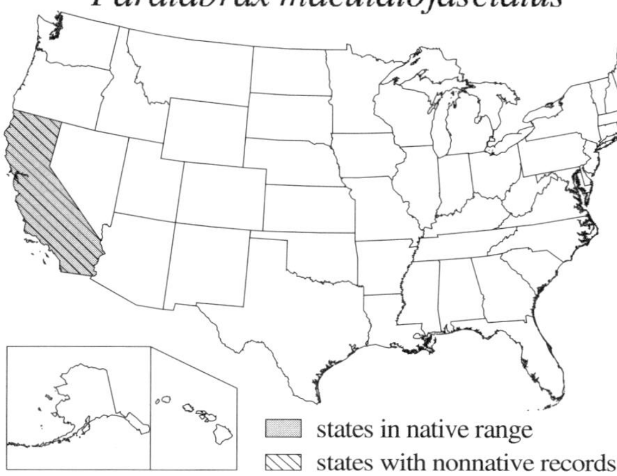

Paralabrax maculatofasciatus (Steindachner 1868)

Common Name: spotted sand bass.

Identification: Miller and Lea (1972); Eschmeyer et al. (1983); Fischer et al. (1995b). Maximum size: 56 cm.

Native Range: Marine; Monterey to Mazatlan, Mexico, including Gulf of California; recorded as far north as San Francisco in the late 1800s (Eschmeyer et al. 1983).

Nonindigenous Occurrences: Introduced into Salton Sea, **California**, in 1951 (Walker et al. 1961; Dill and Cordone 1997).

Means of Introduction: Twelve fish from San Felipe, Mexico were intentionally stocked (Walker et al. 1961).

Status: The above introduction failed to produce a viable population and the species is extirpated in the Salton Sea.

Impact of Introduction: Unknown.

Remarks: None.

Centrarchidae - Sunfishes

The Centrarchidae comprises eight genera and 30 species (Page and Burr 1991). It is naturally found only in North America, mostly east of the Rocky Mountains (Page and Burr 1991). Centrarchids are small- to medium-sized fishes, ranging from about 8 cm for the bluespotted sunfish *Enneacanthus gloriosus* to 83 cm for the largemouth bass *Micropterus salmoides*. They are characterized by deep, laterally compressed bodies and spiny-rayed fins (Page and Burr 1991). Many species, mainly the basses, crappies, and some sunfishes, are popular sport fishes, and therefore have been stocked widely outside and within their native ranges in reservoirs, ponds, and streams of North America. Centrarchids also have been introduced to other continents, mainly South America, Africa, and Europe. The family is restricted to freshwater.

Ambloplites cavifrons

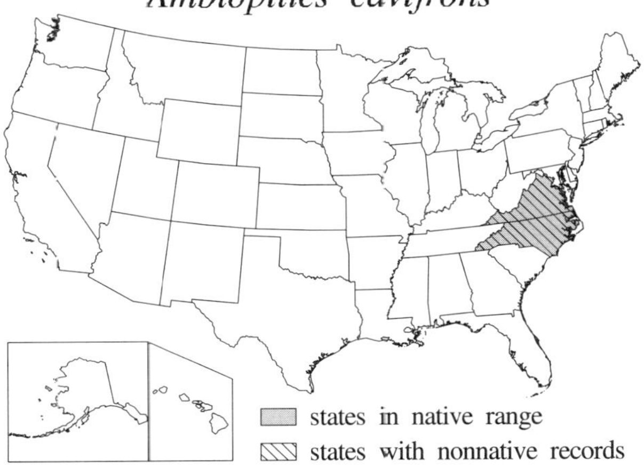

Ambloplites cavifrons Cope 1868

Common Name: Roanoke bass.

Identification: Page and Burr (1991); Jenkins and Burkhead (1994). Maximum size: 36 cm.

Native Range: Endemic to the Chowan, Roanoke, Tar, and Neuse river drainages of Virginia and North Carolina (Page and Burr 1991).

Nonindigenous Occurrences: Stocked in the Cape Fear (Deep system), and Tennessee drainages, and in nonnative areas of the Neuse and Tar drainages in **North Carolina** (Menhinick 1991). Also stocked in the upper and lower James, and middle New drainages of **Virginia** (Jenkins and Burkhead 1994).

Means of Introduction: Intentional stocking for sportfishing. The lower James River was stocked in 1877 with fish from "Stony Creek". There are many "Stony Creeks" in Virginia, but this probably refers to one in the Chowan drainage. One individual was caught six years later. No others have been recorded. The upper James was also stocked in 1877 with fish from the Roanoke. This stocking also failed. In 1879, two adults from the Roanoke River were stocked in the New River at Radford, but they too failed to reproduce (Jenkins and Burkhead 1994).

Status: Extirpated in the upper and lower James and middle New drainages in Virginia. Stockings in North Carolina did not establish populations (McBride et al. 1982; Jenkins and Burkhead 1994).

Impact of Introduction: Unknown.

Remarks: Roanoke bass are nearly extirpated in the upper Roanoke drainage due to competition from the introduced rock bass *A. rupestris* (Lee et al. 1980 et seq.; Jenkins and Burkhead 1994). Both North Carolina and Virginia list Roanoke bass as a species of special concern (Jenkins and Burkhead 1994).

Ambloplites constellatus

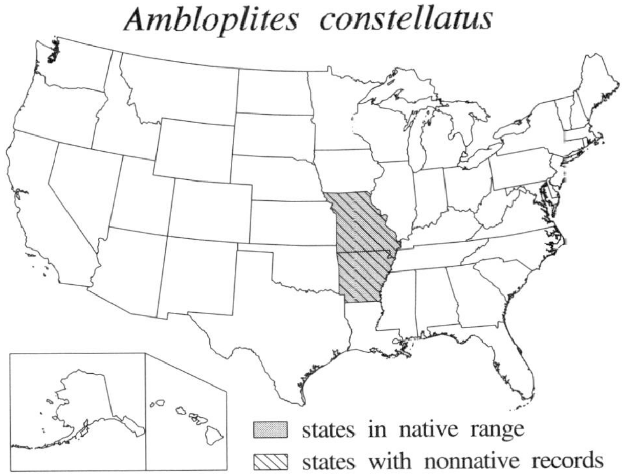

Ambloplites constellatus Cashner and Suttkus 1977

Common Name: Ozark bass.

Identification: Robison and Buchanan (1988); Page and Burr (1991). Maximum size: 19 cm. Prior to its description in 1977, this species was not distinguished from the rock bass *A. rupestris* (Robison and Buchanan 1988).

Native Range: Upper White River drainage in Arkansas and Missouri (Cashner and Suttkus 1977; Robison and Buchanan 1988; Page and Burr 1991).

Nonindigenous Occurrences: This species has been widely introduced into nonnative drainages in **Arkansas** and **Missouri** (Cashner and Suttkus 1977; Lee et al. 1980 et seq.; Robison and Buchanan 1988; Page and Burr 1991). The only area specifically mentioned as being stocked in the above mentioned areas is the Illinois drainage in northwestern Arkansas (Robison and Buchanan 1988). This drainage was stocked in the 1960s. Lee et al. (1980 et seq.) showed a record from what appears to be Stockton Lake, Missouri. Cashner and Suttkus (1977) reported the two known records from the Osage River in Missouri likely represent introductions. Cross et al. (1986) listed the lower Missouri River drainage as an area where this species was introduced.

Means of Introduction: Intentionally stocked for sport. Distributed by the Arkansas state fish hatchery at Centerton (Robison and Buchanan 1988).

Status: There is no evidence of established populations in nonnative areas (Page and Burr 1991).

Impact of Introduction: Unknown.

Remarks: This species does not appear to hybridize with *A. rupestris* or with other adjacent populations of *Ambloplites* (Cashner and Suttkus 1977; Lee et al. 1980 et seq.).

Ambloplites rupestris

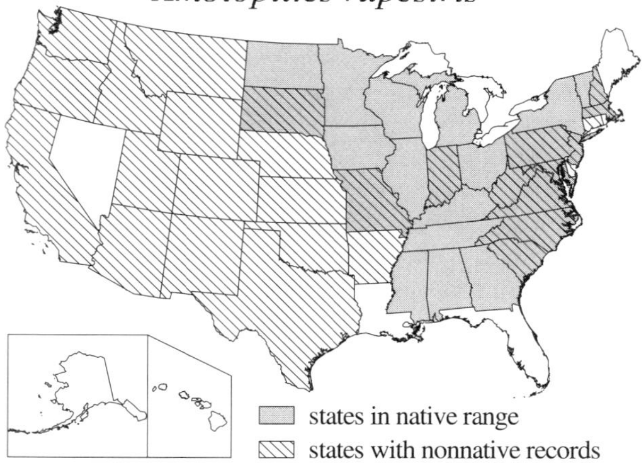

Ambloplites rupestris (Rafinesque 1817)

Common Name: rock bass.

Identification: Becker (1983); Page and Burr (1991); Etnier and Starnes (1993); Jenkins and Burkhead (1994). Maximum size: 43 cm.

Native Range: St. Lawrence River and Great Lakes, Hudson Bay (Red River), and Mississippi River basins, from Quebec to Saskatchewan, south to the Savannah River drainage, Georgia, and northern Alabama, and Missouri (native in Missouri only to the Meramec River) (Page and Burr 1991; Cashner et al. 1992).

Nonindigenous Occurrences: Rock bass have been stocked within their native range, as well as in nonnative areas in the following areas: the lower Neosho and Illinois drainages, and Dardanelle Reservoir in **Arkansas** (Lee et al. 1980 et seq.; Cross et al. 1986; Robison and Buchanan 1988); lakes above the Mogollon Rim, the upper Verde system, and lowermost Clear Creek in the Little Colorado drainage in **Arizona** (Minckley 1973); Napa Creek (a tributary of San Pablo Bay), Feather River, a pond near Lake Morena (San Diego County), and perhaps Lake Cuyamaca (San Diego County), **California** (Smith 1896; Shebley 1917; Dill and Cordone 1997); the vicinity of Las Animas in the upper Arkansas drainage, **Colorado** (Ellis 1974); Connecticut, Thames and Housatonic drainages in **Connecticut** (Behnke and Wetzel 1960; Whitworth et al. 1968; Whitworth 1996); Brandywine system in **Delaware** (Raasch and Altemus 1991); Payette and Troy in the Payette and Clearwater systems in **Idaho** (Linder 1963); non-specific areas in **Indiana** (Sweeney 1902); Shoal Creek in Cherokee

County, **Kansas** (Clarke et al. 1958; Cross 1967; Cross and Collins 1995); most of western **Maryland**, including the Potomac drainage and the western part of the upper Chesapeake drainages (Lee et al. 1976, 1980 et seq.; Jenkins and Burkhead 1994; Rohde et al. 1994); drainages in central and western **Massachusetts** (Hartel 1992), including the Connecticut, Hoosic, and Housatonic drainages (Cardoza et al. 1993; Hartel et al. 1996); the lower Missouri and the Neosho drainage (Lee et al. 1980 et seq.; Cross et al. 1986) and possibly the Osage and Gasconade drainages (Pflieger 1975, 1997; Lee et al. 1980 et seq.) in **Missouri**; the Tongue River, **Montana** (Brown 1971; Cross et al. 1986; Holton 1990); in all major drainages in **Nebraska** (Jones 1963); Connecticut River and Granite Lake in Stoddard, **New Hampshire** (Scarola 1973); West Creek in southern Ocean County and the Delaware River and its tributaries in **New Jersey** (Nelson 1890; Fowler 1906, 1952; Stiles 1978); the Rio Grande and its reservoirs including Elephant Butte and Caballo, the Pecos River including Bataan and Carlsbad reservoirs, and Blue Spring near Carlsbad, **New Mexico** (Koster 1957; Sublette et al. 1990); probably introduced into the New, Broad, Catawba and Yadkin drainages, **North Carolina** (Hocutt et al. 1986; Menhinick 1991); Robert S. Kerr Reservoir and the Neosho drainage in **Oklahoma** (Lee et al. 1980 et seq.; Cashner and Matthews 1988); Willamette River in **Oregon** (Lampman 1946); eastern rivers, mountain lakes and larger streams in **Pennsylvania**, including the Susquehanna River, Delaware River, and Yellow Breeches River (Ford et al. 1892; Lee et al. 1980 et seq.; Hocutt et al. 1986); the Santee drainage, **South Carolina** (Hocutt et al. 1986); Cheyenne drainage, **South Dakota** (Bailey and Allum 1962); upper Guadalupe, Comal, and San Marcos rivers, and the Colorado and San Antonio Bay drainages in **Texas** (Conner and Suttkus 1986; Hubbs et al. 1991; Howells 1992a); Bear River and Utah Lake, **Utah** (Popov and Low 1953; Sigler and Miller 1963; Sigler and Sigler 1996); Potomac, Rappahannock, James, Roanoke, and New drainages in **Virginia** (Lee et al. 1980 et seq.; Hocutt et al. 1986; Jenkins and Burkhead 1994); American and Steilacoom lakes in Pierce County, Saint Clair Lake in Thurston County (Slipp 1943), Skookumchuk River, Silver Lake, Lake Kapowsin, and Long Lake near Olympia, **Washington** (Wydoski and Whitney 1979); Potomac, James, and New drainages in **West Virginia** (Lee et al. 1980 et seq.; Stauffer et al. 1995); and the Tongue River, Lake DeSmet, LAK Reservoir, and ponds in northeastern **Wyoming** (Simon 1946; Baxter and Simon 1970; Brown 1971; Holton 1990; Hubert 1994; Stone 1995).

Means of Introduction: Intentional stocking for sportfishing. Introduced into the Tongue River in Wyoming and moved downstream into Montana (Holton 1990). Rock bass were extensively stocked in Missouri by state personnel during the 1930s and 1940s (Pflieger 1997).

Status: Established in most locations. Extirpated in California (Hubbs et al. 1979; Dill and Cordone 1997). Reports of rock bass in California after circa 1930 are apparently erroneous; see Dill and Cordone (1997) for discussion. Apparently extirpated in Idaho; Simpson and Wallace (1978) do not mention it in their book of Idaho fishes. Extirpated in Utah (Sigler and Sigler 1996). Extirpated in the Rio Grande drainage, New Mexico, and persisting only in Blue Spring in the Pecos drainage.

Impact of Introduction: Rock bass have severely affected Roanoke bass populations in the upper Roanoke drainage through hybridization and competition (Lee et al. 1980 et seq.; Jenkins and Burkhead 1994). Roanoke bass declined after 1965, when rock bass reached high densities (Jenkins and Burkhead 1994). *Ambloplites* in eastern Oklahoma may represent hybrid forms between *A. ariommus* and *A. rupestris* as a result of intensive stocking efforts during the late 1800s and early 1900s (Cashner and Matthews 1988). Jenkins and Burkhead (1994) speculated that introduced rock bass may have contributed to the demise of an isolated population of trout-perch *Percopsis omiscomaycus* in the Potomac River in Virginia and Maryland.

Remarks: Although Loyacano (1975) lists this species in the Santee drainage, South Carolina, he did not distinguish it as introduced there. However, his publication only distinguished species not native to the state rather than to certain drainages.

Archoplites interruptus

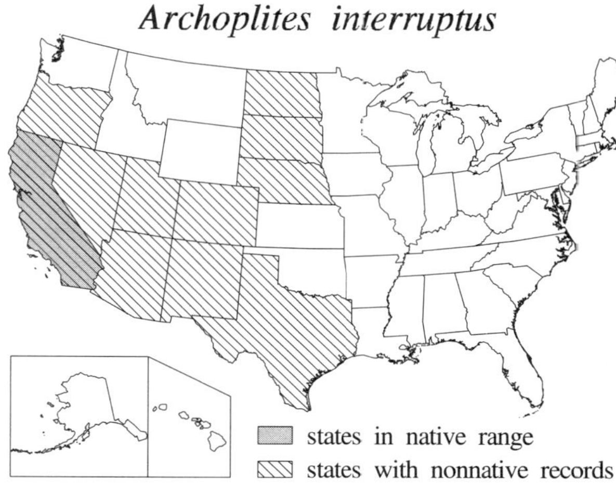

Archoplites interruptus (Girard 1854)

Common Name: Sacramento perch.

Identification: Moyle (1976a); Sigler and Sigler (1987); Page and Burr (1991). Maximum size: 73 cm.

Native Range: Sacramento–San Joaquin, Pajaro, and Salinas River drainages, and Clear Lake in Lake County, California (Moyle 1976a; Page and Burr 1991).

Nonindigenous Occurrences: This species has been introduced to a borrow pit in Maricopa County, **Arizona** (McCarraher and Gregory 1970; Minckley 1973); Clear Lake Reservoir in Modoc County, West Valley Reservoir, San Luis Reservoir, Pit River drainage, Blue Lake in Lake County, Lake Almanor in Plumas County, Russian River, Crowley Lake in Inyo County, and Owens Lake, **California** (McCarraher and Gregory 1970; Moyle 1976a; Moyle and Daniels 1982; U.S. Fish and Wildlife Service 1993e); lakes near Ft. Collins in the Cache la Poudre drainage, Banner Lakes, Newell Lake in northwest Weld County, and Nee Granda Reservoir, **Colorado** (McCarraher and Gregory 1970; Everhart and Seaman 1971; Moyle 1976a); the Sand Hill lakes in **Nebraska** (McCarraher and Gregory 1970, Morris et al. 1974; Moyle 1976a); Rye Patch Reservoir, Washoe Lake, Pyramid Lake, Lahontan Reservoir, Walker Lake, Steptoe and Spring valleys (Bassett and Little Meadow lakes), Indian Lakes, Truckee Meadows, Stillwater Marsh, and the Carson Desert, **Nevada** (Smith 1896; Miller and Alcorn 1946; La Rivers 1962; McCarraher and Gregory 1970; Moyle 1976a; Vigg and Kucera 1981; Deacon and Williams 1984; Sigler and Sigler 1987); Round Lake and Lake George, **North Dakota** (McCarraher and Gregory 1970; Moyle 1976a); Lazy Lagoon, Mirro, and Lea lakes, **New Mexico** (McCarraher and Gregory 1970; Sublette et al. 1990); Klamath Lake drainage, **Oregon** (U.S. Fish and Wildlife Service 1993e; Bond 1994; Li, personal communication); White Lake, **South Dakota** (McCarraher and Gregory 1970; Moyle 1976a); Hamlin Lake, **Texas** (McCarraher and Gregory 1970); and Cutler Reservoir, and Garrison Lake (Pruess) Reservoir, **Utah** (La Rivers 1962; Sigler and Miller 1963; McCarraher and Gregory 1970; Moyle 1976a; Sigler and Sigler 1987). McCarraher and Gregory (1970) gave an account of introductions prior to 1970.

Means of Introduction: Intentional stocking for sportfishing. The earliest introduction of Sacramento perch took place circa 1877, into western Nevada (McCarraher and Gregory 1970). Nebraska, North Dakota, Colorado, New Mexico, Texas, and Arizona were all stocked in the 1960s (McCarraher and Gregory 1970). Sacramento perch were discovered in Crowley Lake, California in the late 1960s, but the exact date and reason for the introduction is not known (McCarraher and Gregory 1970).

Status: According to McCarraher and Gregory (1970), survival has been poor in most states. Small numbers still persist in private lakes near Fort Collins, Colorado (Walker 1993). No longer present in Walker Lake, Nevada (Sigler and Sigler 1987). Probably extirpated in Arizona. It was known to have reproduced at least once in that state after its introduction in 1968 (Minckley 1973). Extirpated in New Mexico (Sublette et al. 1990).

Impact of Introduction: Introduced Sacramento perch, along with many other introduced fishes, may have contributed to the decline of the Lost River sucker *Deltistes luxatus* and the shortnose sucker *Chasmistes brevirostris* in the upper Klamath drainage in Oregon (U.S. Fish and Wildlife Service 1993e).

Remarks: This is the only native centrarchid west of the Rockies. It is declining in its native range due to competition with introduced species (Moore 1968; McCarraher and Gregory 1970; Minckley 1973).

Centrarchus macropterus

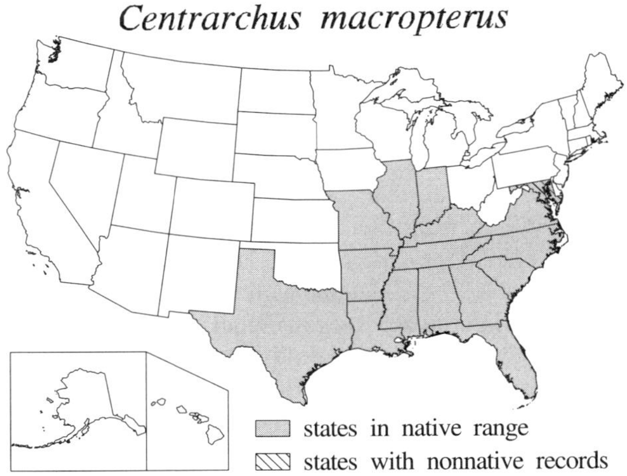

Centrarchus macropterus (Lacepède 1801)

Common Name: flier.

Identification: Hubbs et al. (1991); Page and Burr (1991); Etnier and Starnes (1993); Jenkins and Burkhead (1994). Maximum size: 19 cm.

Native Range: Coastal Plain from Potomac drainage, Maryland (introduced), to central Florida and west to Trinity River, Texas; north in former Mississippi Embayment to southern Illinois and southern Indiana; above Fall Line in southern Indiana, southern Illinois; rarely elsewhere.

Nonindigenous Occurrences: Collected from a farm pond in **Maryland** in St. Marys County in the lower Potomac drainage prior to 1981 (Lee et al. 1981; Jenkins and Burkhead 1994).

Means of Introduction: Probably stocked, either intentionally or as stock contamination. Lee et al. (1981) questioned the flier's native status in Maryland. Jenkins and Burkhead (1994) stated that if it is native to the Potomac River in Maryland, it should have been found in the Virginia portion of that river as well (it was not).

Status: Reported from Maryland.

Impact of Introduction: Unknown.

Remarks: None.

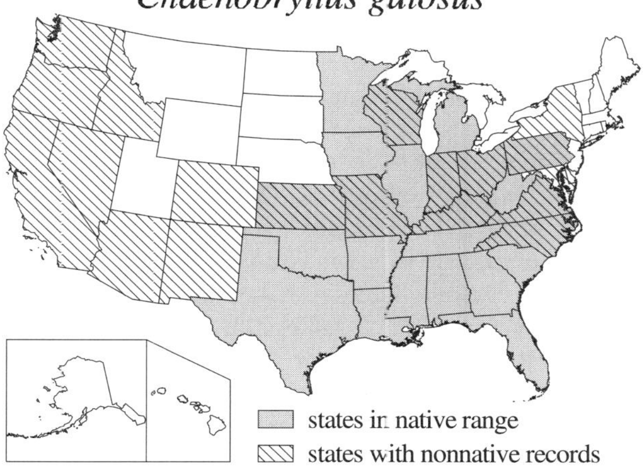

Chaenobryttus gulosus (Cuvier 1829)

Common Name: warmouth.

Identification: Moyle (1976a); Becker (1983); Page and Burr (1991); Etnier and Starnes (1993); Jenkins and Burkhead (1994). Other names: *Lepomis gulosus* and *C. coronarius*. Maximum size: 31 cm.

Native Range: Great Lakes and Mississippi River basins from western Pennsylvania to Minnesota, and south to the Gulf; Atlantic and Gulf slope drainages from Rappahannock River, Virginia, to the Rio Grande, Texas and New Mexico (Page and Burr 1991).

Nonindigenous Occurrences: These fish have been introduced into the Colorado River and reservoirs on the Salt River, **Arizona** (Minckley 1973; Lee et al. 1980 et seq.); the Colorado River, Sacramento-San Joaquin Delta, the Central Valley, the San Diego, Feather, and San Joaquin drainages in **California** (Smith 1896; Shebley 1917; Evermann and Clark 1931; Lampman 1946; Lanse 1965; Moyle 1976a; Dill and Cordone 1997); Rio Grande headwaters (San Luis Valley) and probably other areas of the state in **Colorado** (Ellis 1974; Zuckerman and Behnke 1986); the coastal plain of **Delaware** (Lee et al. 1976, 1981; Raasch and Altemus 1991); the Snake River from C.J. Strike Reservoir to Brownlee Reservoir, and the Boise River, **Idaho** (Smith 1896; Lampman 1946; Linder 1963; Simpson and Wallace 1978; Idaho Fish and Game 1990); central **Indiana** (Nelson and Gerking 1968); possibly in reservoirs in eastern **Kansas** (native status uncertain) (Cross 1967); middle and upper Cumberland, upper Green, and Barren drainages in **Kentucky** (Burr and Warren 1986); probably introduced in the coastal plain, and the Piedmont and the ridge and valley section (Potomac drainage) of **Maryland** (Lee et al. 1976, 1981); the southern and eastern parts of **Missouri** (Pflieger 1997); Colorado River in **Nevada** (Lee et al. 1980 et seq.); Pecos and Rio Grande drainages in **New Mexico** (Sublette et al. 1990); Woodbury Creek in Orange County and the Saw Kill in Dutchess County, **New York** (Smith 1985; Schmidt 1986); probably introduced in the upper New and upper Tennessee drainages in **North Carolina** (Bailey et al. 1954; Menhinick 1991); southern **Ohio** drainages including the Ohio and Scioto (Trautman 1981); the Willamette and Columbia rivers, and Devil's Lake, **Oregon** (Lampman 1946; also see Bond 1994); the lower Delaware drainage, **Pennsylvania** (Cooper 1981; Schmidt 1986); the New and Holston drainages, probably introduced in the York, Rappahonnock, and Potomac drainages, and possibly introduced in the James drainage in **Virginia** (Hocutt et al. 1986; Jenkins and Burkhead 1994); lakes in **Washington** including Loon and Silver (Cowlitz County) lakes (Smith 1896; Lampman 1946; Wydoski and Whitney 1979; Fletcher, personal communication); and the Wolf and Fox systems, and eastern **Wisconsin** drainages (Becker 1983).

Means of Introduction: Intentional stocking for sportfishing in most locations. Presumably accidentally stocked in Arizona with either bluegills *L. macrochirus* or bass *Micropterus* sp. (Minckley 1973). Likely gained access to the Wolf and Fox systems in Wisconsin via the artificial Fox-Wisconsin Canal at Portage which connects the Mississippi and Great Lakes basins (Becker 1983). Introductions into eastern Wisconsin were probably from fish rescue operations in the 1870s to 1930s that transplanted stranded fish from the Mississippi River to other locations (Becker 1983).

Status: Established in most locations where introduced.

Impact of Introduction: Warmouth hybridize with *Lepomis* species (Moyle 1976a). Minckley (1973) described this fish as "pugnacious". Warmouth compete directly with *Lepomis* species for macroinvertebrates (Madsen et al. 1994). Introduced predatory centrarchids are likely responsible for the decline of native ranid frogs in California and for the decline of California tiger salamander *Ambystoma californiense* populations (Hayes and Jennings 1986; Dill and Cordone 1997).

Remarks: There was no mention of this species' introduction in Utah by Sigler and Miller (1963), Sigler and Sigler (1987, 1996).

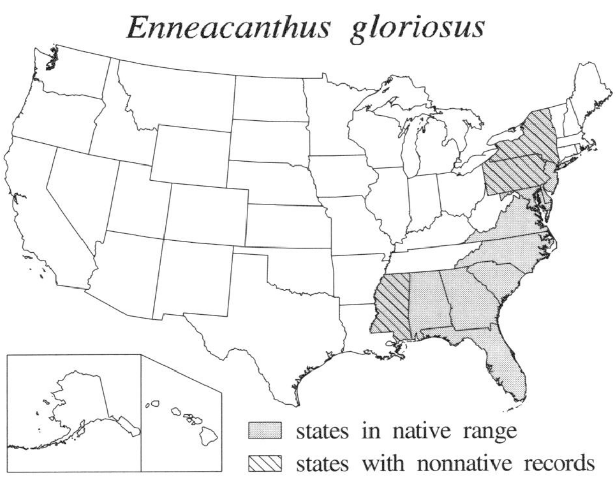

Enneacanthus gloriosus (Holbrook 1855)

Common Name: bluespotted sunfish.

Identification: Smith (1985); Page and Burr (1991); Jenkins and Burkhead (1994); Mettee et al. (1996). Maximum size: 9.5 cm.

Native Range: Atlantic and Gulf Slope drainages below Fall Line from southern New York to lower Tombigbee River, Alabama, south to southern Florida; above Fall Line in New York and Pennsylvania (Page and Burr 1991).

Nonindigenous Occurrences: A population was discovered in the Jamesville Reservoir, south of Syracuse, **New York**, in 1971 (Werner 1972; Lee et al. 1980 et seq.; Smith 1985). Collections also were made in 1916 from Oneida Lake, just north of Syracuse (Smith 1985) and may represent an early introduction or a relict population. The fish also has been introduced into the Big Black River drainage in **Mississippi** (Peterson and Ross 1987; Ross and Brenneman 1991). It has been introduced to the Susquehanna River in **Pennsylvania**, were it has been collected from Mountain Creek in Cumberland County, Stony Creek in Dauphin County, Lake Winola in Wyoming County, and Harvey's Lake in Luzerne County (Denoncourt et al. 1975c).

Means of Introduction: The presence of this species in the Jamesville Reservoir is probably due to an aquarium release sometime between 1951 and 1966 (Werner 1972). However, it is theoretically possible that the species migrated up from the Hudson River, through the Erie or Barge Canal, to Oneida Lake, then up the Chittenango River and Butternut Creek into Jamesville Reservoir. Although Werner (1972) states that no specimens have ever been collected in the intervening area, there is a collection from Oneida Lake dating back to 1916; the only other collection in the Great Lakes basin (Smith 1985). These fish may either represent a relict population or an introduction earlier than what was calculated in Jamesville Reservoir (Smith 1985). One factor against the canal migration hypothesis is that there is a 2–3 m-high weir with which a fish moving upstream from Oneida Lake to Jamesville Reservoir would have to contend (Werner 1972). Jamesville Reservoir was created in 1874, the same year the Erie Canal opened (Werner 1972). The origin of the Pennsylvania fish is unknown. They may be either natural populations or accidental introductions. Because this species is sometime kept as an aquarium species, aquarium release is a possible means of introductions in Pennsylvania (Denoncourt et al. 1975c). Unknown means of introduction for the Big Black River drainage, Mississippi.

Status: Established in Jamesville Reservoir, New York, and Big Black River drainage, Mississippi. Collected from Susquehanna River, Pennsylvania. Extirpated from Oneida Lake, New York.

Impact of Introduction: Unknown.

Remarks: Voucher specimens: New York (College of Environmental Science and Forestry at Syracuse University), Pennsylvania (CU 16429; PSU; University of Maryland, Appalachian Environmental Laboratory, LaVale).

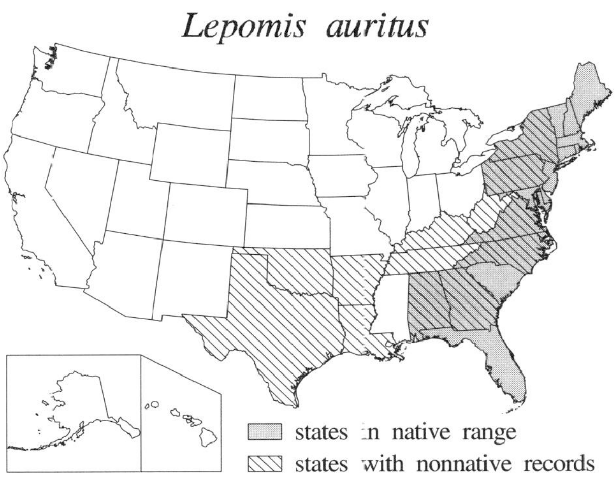

Lepomis auritus (Linnaeus 1758)

Common Name: redbreast sunfish.

Identification: Smith (1985); Hubbs et al. (1991); Page and Burr (1991); Jenkins and Burkhead (1994). Maximum size: 24 cm.

Native Range: Atlantic and Gulf Slope drainages, from New Brunswick to central Florida, and west to the Apalachicola and Choctawhatchee drainages, Georgia and Florida (Page and Burr 1991).

Nonindigenous Occurrences: Introduced in the Black Warrior (Mobile basin), Choctawhatchee, and Tennessee drainages and perhaps also in the Coosa and Tallapoosa drainages in **Alabama** (Smith-Vaniz 1968; Lee et al. 1980 et seq.; Boschung 1992; Mettee et al. 1996); a small lake at Camp Hedges in Stone County (Sylamore Creek, White River drainage), Spring Lake (Petit Jean system, Arkansas drainage) in Yell County, and Spring River near Hardy in Sharp county (White drainage), **Arkansas** (Buchanan 1973; Cross et al. 1986; Robison and Buchanan 1988); the Coosa and Tennessee drainages, **Georgia** (Dahlberg and Scott 1971b); the Big Sandy, Green, Licking, upper and lower Cumberland, and lower Kentucky drainages, **Kentucky** (Burr and Page 1986; Burr and Warren 1986); Sabine drainage, **Louisiana** (Lee et al. 1980 et seq.; Conner and Suttkus 1986); the northern Adirondack Mountains in **New York** (Lee et al. 1980 et seq.); Cape Fear, Chowan, Lumber, Neuse, Pamlico, Pee Dee, Roanoke, Savannah, upper Broad, upper Catawba, upper New, upper Tennessee, and Waccamaw drainages, **North Carolina** (Bailey et al. 1954; Menhinick 1991); Love, McCurtain, and Muskogee counties in southeastern **Oklahoma** in the Red drainage (Miller and Robison 1973; Cross et al. 1986); Susquehanna drainage in southcentral **Pennsylvania**, west of native range (Lee et al. 1980 et seq.); Tennessee drainage down to Chattanooga, and the Conasauga and Middle Fork Forked Deer systems in **Tennessee** (Starnes and Etnier 1986; Etnier and Starnes 1993); eastern and central **Texas** to the Pecos drainage, including the Nueces, San Antonio Bay, Colorado, Brazos, Galveston Bay, Red, and Sabine Lake drainages (Lee et al. 1980 et seq.; Conner and Suttkus 1986; Cross et al. 1986; Hubbs et al. 1991; Howells 1992a); the New and Big Sandy drainages of the Ohio basin, and the Holston and Clinch-Powell systems of the Tennessee drainage in **Virginia** (Lee et al. 1980 et seq.; Hocutt et al. 1986; Jenkins and Burkhead 1994); and the New, Kanawha, and Monongahela drainages in **West Virginia**, where it is probably introduced (Hocutt et al. 1986; Stauffer et al. 1995).

Cross et al. (1986) also listed the redbreast sunfish as introduced into the Ouachita drainage but did not indicate the state. No records were found for this drainage in Arkansas (Lee et al. 1980 et seq.; Robison and Buchanan 1988) or Louisiana (Lee et al. 1980 et seq.).

Means of Introduction: Intentional stocking for sportfishing.

Status: Established in most locations. It has expanded its range westward due to stocking. In Kentucky, only established in the extreme southeastern portion of the state (Jenkins and Burkhead 1994).

Impact of Introduction: There is circumstantial evidence to indicate that redbreast sunfish are displacing native longear sunfish *L. megalotis* in eastern Tennessee through direct competition (Etnier and Starnes 1993).

Remarks: In North Carolina, Menhinick (1991) lists this species as introduced into the New and Tennessee drainages, and native and introduced in the remaining drainages. It is unclear if this sunfish is introduced within or outside its native range in these drainages. There are no records of redbreast sunfish from Mississippi (Scott and Crossman 1973; Lee et al. 1980 et seq.). It is also not mentioned in the manuscript on Mississippi fishes by Ross and Brenneman (1991). This species is not mentioned as occurring in Kentucky by Clay (1975). In their book on Alabama fishes, Mettee et al. (1996)

presented some conflicting information regarding native versus introduced ranges. These researchers stated that *Lepomis auritus* is introduced to the Tennessee, Black Warrior, and Choctawhatchee drainages, but they listed the species as "native" in their summary table.

Lepomis cyanellus

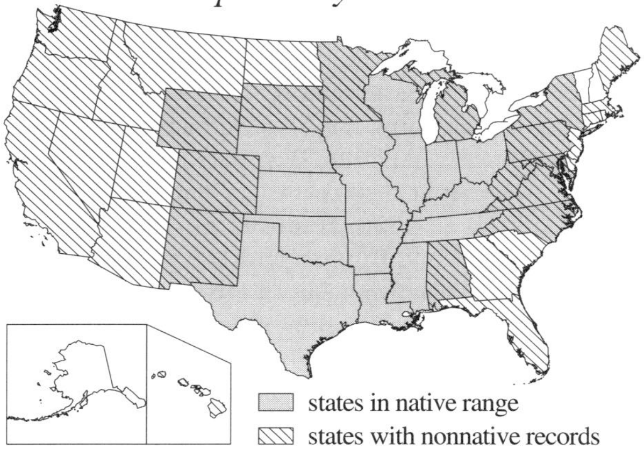

Lepomis cyanellus Rafinesque 1819

Common Name: green sunfish.

Identification: Becker (1983); Page and Burr (1991); Sublette et al. (1990); Etnier and Starnes (1993); Jenkins and Burkhead (1994). Maximum size: 31 cm.

Native Range: Great Lakes, Hudson Bay, and Mississippi River basins from New York and Ontario west to Minnesota, South Dakota, and Wyoming, and south to the Gulf; Gulf Slope drainages from Mobile Bay, Georgia and Alabama, to the Rio Grande, Texas. Also northern Mexico (Page and Burr 1991).

Nonindigenous Occurrences: This sunfish has expanded both its eastward and westward range greatly, mainly as a result of accidental stocking. Green sunfish have been introduced to eastern coastal drainages in **Alabama** (Lee et al. 1980 et seq.; Swift et al. 1986; Boschung 1992); the Colorado River, Gila, Bill Williams, Little Colorado, Salt, Verde, San Carlos, Yaqui, and Santa Cruz drainages in **Arizona** (Miller and Lowe 1967; Minckley 1973; Hendrickson et al. 1980; Tyus et al. 1982); all drainages in **California** (Smith 1896; Shebley 1917; Lampman 1946; Moyle and Nichols 1973; Moyle 1976a; Smith 1982; Taylor et al. 1982; U.S. Fish and Wildlife Service 1993e; Dill and Cordone 1997); the Republican drainage and the Green, Gunnison, White, Yampa, and San Juan drainages, and the Colorado River in **Colorado** (Vanicek et al. 1970; Holden and Stalnaker 1975; Lee et al. 1980 et seq.; Tyus et al. 1982); all drainages in **Connecticut** (Behnke and Wetzel 1960; Whitworth et al. 1968; Schmidt 1986; Whitworth 1996); the Brandywine-Christina drainage in northern **Delaware** (Raasch and Altemus 1991; Rohde et al. 1994); the Apalachicola and other panhandle drainages in **Florida** (Kilby et al. 1959; Yerger 1977; Lee et al. 1980 et seq.; Swift et al. 1986; UF museum specimens); Savannah, Altamaha (Oconee), and Chattahoochee drainages in **Georgia** (Dahlberg and Scott 1971a, 1971b; Yerger 1977; Swift et al. 1986); the island of Kaui, **Hawaii** (Devick 1991a); southeastern **Idaho** (Linder 1963; Simpson and Wallace 1978; Idaho Fish and Game 1990); a pond in the town of Argyle (lower Penobscot drainage) in **Maine** (F. W. Kircheis, personal communication); all areas of **Maryland** except the extreme western portion where it is native (Lee et al. 1976, 1980 et seq.; Rohde et al. 1994); the Merrimack (Assibet) drainage, Buzzards Bay drainage, Bristol County, and possibly in the Taunton, Blackstone, and Quinebaug river drainages, **Massachusetts** (Hartel 1992; Cardoza et al. 1993; Hartel et al. 1996); Houghton County in the upper peninsula (Lake Superior drainage) of **Michigan** (Becker 1983); extreme northeastern **Minnesota** (Lee et al. 1980 et seq.); the Yellowstone, Little Missouri, Tongue, Musselshell, Redwater, and Belle Fourche drainage, in southeastern **Montana** (Brown 1971; Cross et al. 1986; Holton 1990); Lake Mead, the Virgin and Moapa, Truckee and Carson drainages in **Nevada** (Miller and Alcorn 1946; La Rivers 1962; Deacon and Williams 1984); northern **New Jersey** (Lee et al. 1980 et seq.); San Juan, Zuni, Gila, San Francisco, Tularosa, Mimbres, Rio Chama, and Rio Grande drainages in **New Mexico** (Tyus et al. 1982; Sublette et al. 1990; Platania 1991); the upper Susquehanna, Webatuck River in the Housatonic drainage, and New Croton Reservoir, Iron Mine Pond and the Wallkill River in the lower Hudson drainage in **New York** (Smith 1985; Schmidt 1986); the upper Catawba, Lumber, Waccamaw, Yadkin, Cape Fear, Neuse, Tar, Chowan, Roanoke, and perhaps the Tennessee drainages in **North Carolina** (Hocutt et al. 1986; Menhinick 1991); southern **North Dakota** (Lee et al. 1980 et seq.); several sites in **Oregon** including Blue Lake near Portland (lower Willamette drainage) and Klamath, Lost, Rogue, and Umpqua rivers (Wydoski and Whitney 1979; Bond 1994); eastern **Pennsylvania** in the Delaware, Susquehanna, and Potomac drainages (Lee et al. 1980 et seq.; Cooper 1983; Hocutt et al. 1986; Schmidt 1986; Raasch and Altemus 1991); the Santee-Cooper and Savannah drainages in **South Carolina** (Loyacano 1975; Lee et al. 1980 et seq.; Hocutt et al. 1986; Rohde et al. 1994); northwestern **South Dakota** in the Grand Moreau and

Belle Fourche drainages (Bailey and Allum 1962; Lee et al. 1980 et seq.); the upper Colorado basin including the Green, White, Dirty Devil, Dolores, Duchesne, and San Juan drainages, the Colorado River itself, Lake Powell, and the Weber River and Utah Lake in the Great Basin in **Utah** (Sigler and Miller 1963; Vanicek et al. 1970; Tyus et al. 1982; Sigler and Sigler 1987, 1996); the Potomac, Rappahannock, York, James, Pee Dee, Chowan, and Roanoke drainages, and perhaps the Holston, Clinch-Powell, and Big Sandy drainages in **Virginia** (Hocutt et al. 1986; Jenkins and Burkhead 1994); Deer, Colville, and Loon lakes in the Colville drainage, Sacheen and Diamond lakes in the Little Spokane drainage, **Washington** (Smith 1896; Lampman 1946; Wydoski and Whitney 1979; Fletcher, personal communication); the Potomac, James and probably the New drainages in **West Virginia** (Stauffer et al. 1995) (Jenkins and Burkhead [1994] believe it is native to the New); and the Big Horn, Niobrara, and North Platte drainages in **Wyoming** (Simon 1946; Baxter and Simon 1970; Hubert 1994; Stone 1995).

Means of Introduction: Accidentally stocked as bluegill *L. macrochirus* or with other intended species (i.e., stock contaminant). In Hawaii, first discovered in the Western Kauai Lagoons. That site had been previously stocked with "bluegill" from various plantation reservoirs on Kauai; therefore, it is likely that *L. cyanellus* was present in Hawaii some time prior to its collection (Devick 1991a). It has been suggested that green sunfish were planted inadvertently in Lake Mead in 1937 (Holden and Stalnaker 1975). It may have reached the upper part of the Colorado River either by moving upstream from Lake Mead or by being introduced separately in that region (Holden and Stalnaker 1975).

Status: Established in most locations where it has been introduced.

Impact of Introduction: Green sunfish appear to be at least partially responsible for local extinctions of California roach *Hesperoleucus symmetricus* in California (Moyle and Nichols 1974; Moyle 1976a, 1976b; Smith 1982). In California, aggressive green sunfish outcompete native Sacramento perch *Archoplites interruptus* (Moyle 1976a). They may chase Sacramento perch away from spawning areas and out of favored places, such as shallow weedy areas, and into open water (Moyle 1976a). Once in open water, the perch are more vulnerable to predation and have less available food. Introduced green sunfish also compete with and prey on other native fishes (Moyle 1976a; Lemly 1985). Lemly (1985) found that green sunfish reduced populations of native species, and altered population structure, relative dominance, and distribution patterns in North Carolina Piedmont streams. In that study, he found that green sunfish preyed heavily on minnows, and in fact, may have been responsible for the elimination of two cyprinid species in the study area. Green sunfish may compete with and adversely affect young Colorado pikeminnow *Ptychocheilus lucius*, an endangered species (Karp and Tyus 1990). Green sunfish readily hybridize with other *Lepomis* species (Moyle 1976a; Sigler and Sigler 1987). In cold lakes the often overpopulate and reduce trout populations (McKechnie and Tharratt 1966). Their large mouth allows them to compete with larger fish for prey items, and to prey on eggs and young of other fishes (McKechnie and Tharratt 1966). Green sunfish and other introduced predatory centrarchids are likely responsible for the decline of native ranid frogs in California, California tiger salamander *Ambystoma californiense* populations (Hayes and Jennings 1986; Dill and Cordone 1997), and the Chiricahua leopard frog *Rana chiricahuensis* in southeastern Arizona (Rosen et al. 1995).

Remarks: Tyus et al. (1982) gave a distribution map of this species in the upper Colorado basin. The green sunfish has been widely stocked in Ohio (Trautman 1981). But stockings appear to have taken place within its native range.

Lepomis gibbosus

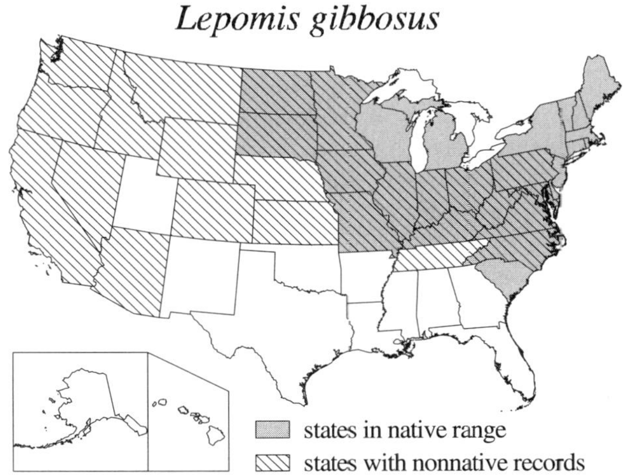

Lepomis gibbosus (Linnaeus 1758)

Common Name: pumpkinseed.

Identification: Moyle (1976a); Becker (1983); Page and Burr (1991); Jenkins and Burkhead (1994). Maximum size: 30 cm.

Native Range: Atlantic Slope drainages from New Brunswick to Edisto River, South Carolina; Great Lakes, Hudson Bay, and upper Mississippi basins from Quebec and New York west to southeastern Manitoba and North Dakota, and south to northern Kentucky and Missouri (Page and Burr 1991).

Nonindigenous Occurrences: Introduced to many areas west of native range. States where stockings have occurred in nonnative areas include: a pond near Safford in Graham County, **Arizona** (Minckley 1973); sloughs of the Susan River in Lassen County, near Mecca in Riverside County, the Klamath, Santa Ana (Big Bear Lake), and Honey Lake drainages in **California** (Moyle 1976a; U.S. Fish and Wildlife Service 1993e; Dill and Cordone 1997); the Platte, Arkansas and Rio Grande (San Luis Valley) drainages in **Colorado** (Everhart and Seaman 1971; Ellis 1974; Zuckerman and Behnke 1986); Kootenai, Pend Orielle, Spokane, Snake, and Palouse drainages in **Idaho** (Linder 1963; Simpson and Wallace 1978; Idaho Fish and Game 1990); the Big Muddy and Little Wabash systems in **Illinois** (Burr and Page 1986); southern artificial lakes including an artificial lake in Brown County and another in Clay County, **Indiana** (Gerking 1945; Nelson and Gerking 1968); southwestern **Iowa** (Lee et al. 1980 et seq.; Harlan et al. 1987); a pond in Johnson County, **Kansas** (Cross 1967; KU 10274); Licking River (Burr and Page 1986) in Rowan County, Greenbo Lake in Greenup County (both locations based on the fact that this species had not been collected here prior to Burr's 1980 paper) (Burr and Warren 1986), and possibly the Ohio River in **Kentucky** (Burr 1980); the Youghiogheny drainage in **Maryland** (Lee et al. 1976, 1981); a few lakes in the Lake Superior drainage in **Minnesota** (Eddy and Underhill 1974; Phillips et al. 1982); possibly the Meramec River in Crawford County and Salt Creek in Chariton County, **Missouri** (Pflieger 1975, 1997; Cross et al. 1986); the upper Missouri River, and the Yellowstone, Milk, Beaver, Tongue, Little Missouri, Kootenai, Flathead, and Clark Fork drainages in **Montana** (Brown 1971; Cross et al. 1986; Holton 1990); the upper Niobrara River and southeastern **Nebraska** (Lee et al. 1980 et seq.; Cross et al. 1986); Dufurrena Ponds in Sheldon National Wildlife Refuge, Humboldt County, **Nevada** (Deacon and Williams 1984; Sigler and Sigler 1987); upper New, upper Tennessee (Menhinick 1991), and perhaps the Pee Dee (Jenkins and Burkhead 1994) drainages in **North Carolina**; extreme eastern **North Dakota** (Lee et al. 1980 et seq.); southern **Ohio** (Trautman 1981); the Klamath, Willamette, lower Columbia, Nehalem, Malheur, Snake, and Rogue drainages in **Oregon** (Lampman 1946; Wydoski and Whitney 1979; Lee et al. 1980 et seq.; U.S. Fish and Wildlife Service 1993e); Youghiogheny drainage, **Pennsylvania** (Hendricks et al. 1979); Big Sioux, James, White, Bad, and Grand drainages in **South Dakota** (Cleary 1956; Bailey and Allum 1962; Lee et al. 1980 et seq.; Cross et al. 1986); Boone and South Holston reservoirs on the South Fork Holston River, Tennessee drainage in **Tennessee** (Starnes and Etnier 1986; Etnier and Starnes 1993); New, Holston, Powell, and perhaps the Delaware drainages in **Virginia** (Hocutt et al. 1986; Jenkins and Burkhead 1994); the Columbia River and lakes in western **Washington** (Lampman 1946; Gray and Dauble 1977; Wydoski and Whitney 1979); the New and possibly the Monongahela drainage in **West Virginia** (Hocutt et al. 1986; Stauffer et al. 1995); and Sloans Lake in Cheyenne, Festo Lake near Wheatland, sloughs near the North Platte River at Guernsey, and sloughs along the Big Horn River near Kane, **Wyoming** (Simon 1946; Baxter and Simon 1970; Lee et al. 1980 et seq.; Cross et al. 1986; Hubert 1994).

Means of Introduction: Intentional stocking for sportfishing. Trautman (1981) reported this species was originally absent from the southern unglaciated part of Ohio. However, he implies construction of canals gave them access to those areas and, furthermore, millions of pumpkinseeds were stocked in southern Ohio between 1920 and 1950.

Status: Established in most locations where introduced. Extirpated in Arizona (Minckley 1973). Reported in Kansas, but failed to become established (Cross 1967).

Impact of Introduction: Pumpkinseeds will hybridize with most other *Lepomis*, especially with bluegill and green sunfish. The result is usually fast-growing, sterile male hybrids (Moyle 1976a).

Remarks: Sigler and Miller (1963), Sigler and Sigler (1987, 1996) make no mention of pumpkinseeds in Utah. Although Burr (1980) lists this species as probably introduced to the Ohio River in Kentucky, resulting from escapees from commercial fishing lakes and farm ponds, Burr and Warren (1986) believed they were more likely strays from the upper Ohio River. Menhinick (1991) reports pumpkinseeds as native and introduced in all remaining drainages (other than New and Tennessee) in the state. It is unclear if these introductions are within their native range in those drainages or outside their native range. Cross et al. (1986) also listed the pumpkinseed as introduced into the Kansas River drainage but did not indicate which state. No records were found for this drainage in Kansas (Lee et al. 1980 et seq.; Cross and Collins 1995) or Nebraska (Lee et al. 1980 et seq.). However, Cross (1967) stated that he believes all previous records of pumpkinseeds in the Kansas drainage are actually *L. megalotis*. He gave a discussion of reasons for the misnomer and the perpetuated error. However, he did give one "recent" (circa 1967) record from Johnson County (KU 10274) which may have been in the Kansas drainage. This could be the basis for the Cross et al. (1986) report.

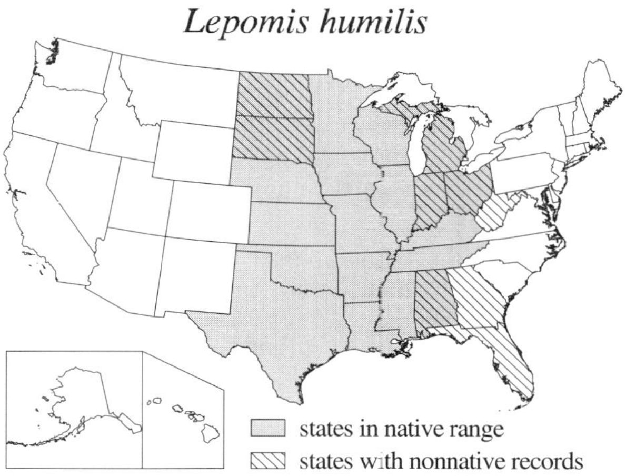

Lepomis humilis (Girard 1858)

Common Name: orangespotted sunfish.

Identification: Becker (1983); Hubbs et al. (1991); Page and Burr (1991); Etnier and Starnes (1993). Maximum size: 15 cm.

Native Range: Lower Great Lakes (southern ends of lakes Erie and Michigan), Hudson Bay (Red River), and Mississippi River basin from Ohio to southern North Dakota, and south to Louisiana; Gulf Slope drainages from Mobile Bay, Alabama, to the Colorado River, Texas (Page and Burr 1991).

Nonindigenous Occurrences: This fish has been introduced into Martin Reservoir, Tallapoosa drainage, **Alabama** (probably) (Boschung 1992); the Apalachicola River, **Florida** (Kilby et al. 1959; Yerger 1977; Swift et al. 1986); the Chattahoochee drainage, **Georgia** (Dahlberg and Scott 1971b; Yerger 1977; Swift et al. 1986); the Tippecanoe River in **Indiana** (Nelson and Gerking 1968); western Lake Erie systems in **Michigan** (Trautman 1981; Bailey and Smith 1992; Mandrak and Crossman 1992; Latta, personal communication); Lake Sakakawea, James and Sheyenne rivers in **North Dakota** (Owen et al. 1981) (based on native range depicted in Lee et al. et seq. 1980); systems of western Lake Erie, and the Scioto, Great Miami, Little Miami, Wabash drainages, other smaller tributaries of the Ohio River, and the river itself in **Ohio** (Trautman 1981; Emery 1985); the Bad (introduced/native?) and Cheyenne drainages in western **South Dakota** (Bailey and Allum 1962); and probably introduced in the Little Kanawha drainage in **West Virginia** (Hocutt et al. 1986).

Means of Introduction: Usually unintentional stocking as stock contaminant with other centrarchids. Probably competes with young bass *Micropterus* spp., bluegill *L. macrochirus*, and crappies *Pomoxis* spp. for food (Cross 1967). Orangespotted sunfish expanded their range eastward across Ohio aided by introductions into farm ponds and reservoirs (Trautman 1981). It gained access to the Lake Erie drainage when it overcame a spillway (probably aided by humans) dividing the Wabash system (Ohio basin) from the St. Marys system (Great Lakes basin). It then invaded the Maumee River and progressed downriver to Lake Erie (Trautman 1981).

Status: Presumably established in all locations.

Impact of Introduction: Orangespotted sunfish probably compete for food with young bass, bluegill, and crappies (Cross 1967).

Remarks: Nelson and Gerking (1968) reported that the orangespotted sunfish was not collected in the Tippecanoe drainage, Indiana, prior to 1945, but do not directly state that it was introduced there. Stauffer et al. (1995) apparently had no definitive records for this species in the Little Kanawha drainage, West Virginia. However, they indicated that it probably occurred there. The lack of specimens may be the reason why Hocutt et al. (1986) reported this species as probably introduced there.

Lepomis macrochirus

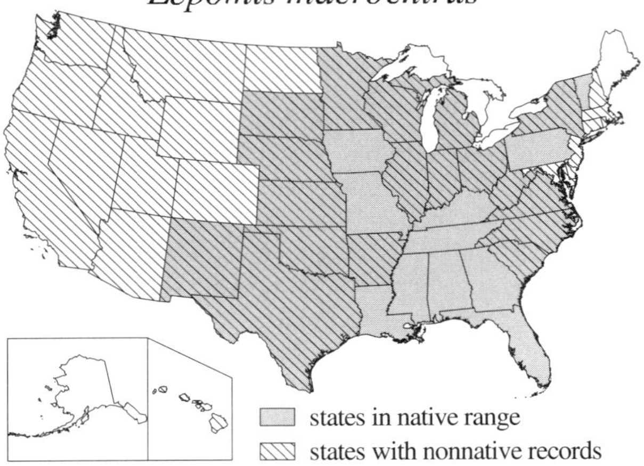

Lepomis macrochirus Rafinesque 1819

Common Names: bluegill (bream, copperbelly).

Identification: Moyle (1976a); Becker (1983); Page and Burr (1991); Jenkins and Burkhead (1994). Two subspecies: *L. macrochirus mystacalis* in Peninsular Florida, and *L. m. macrochirus* throughout the rest of the range (Gilbert, personal communication). Maximum size: 41 cm.

Native Range: St. Lawrence–Great Lakes and Mississippi River basins from Quebec and New York to Minnesota and south to the Gulf; Atlantic and Gulf Slope drainages from the Cape Fear River, Virginia, to the Rio Grande, Texas and New Mexico. Also in northern Mexico (Page and Burr 1991).

Nonindigenous Occurrences: Bluegills have been stocked widely both within and outside their native range. They have been stocked in **Arkansas** (Robison and Buchanan 1988); **Arizona** (Minckley 1973; Hendrickson et al. 1980; Lee et al. 1980 et seq.; Tyus et al. 1982); **California** (Smith 1896; Shebley 1917; Nelae 1931; Lampman 1946; Moyle 1976a; Moyle and Daniels 1982; Smith 1982; Taylor et al. 1982; Dill and Cordone 1997); **Colorado** (Everhart and Seaman 1971; Ellis 1974; Holden and Stalnaker 1975; Tyus et al. 1982; Wiltzius 1985; Propst and Carlson 1986); **Connecticut** (Webster 1942; Behnke and Wetzel 1960; Whitworth et al. 1968; Schmidt 1986; Whitworth 1996); **Delaware** (Lee et al. 1976, 1980 et seq., 1981); **Hawaii** (Brock 1960; Courtenay et al. 1984; Devick 1991b); **Idaho** (Linder 1963; Simpson and Wallace 1978; Idaho Fish and Game 1990, 1996); **Illinois** (Burr, personal communication); **Indiana** (Blatchley 1938); **Kansas** (Cross 1967); **Maryland** (Lee et al. 1976; Lee et al. 1980 et seq.; Lee et al. 1981); **Massachusetts** (Schmidt 1986; Hartel 1992; Cardoza et al. 1993; Hartel et al. 1996); **Michigan** (Hubbs and Lagler 1947); **Minnesota** (Lee et al. 1980 et seq.); **Montana** (Brown 1971; Cross et al. 1986; Holton 1990); **Nebraska** (Jones 1963; Morris et al. 1974); **Nevada** (Moffett 1943; Miller and Alcorn 1946; La Rivers 1962; Deacon and Williams 1984); **New Hampshire** (Scarola 1973; Schmidt 1986); **New Jersey** (Fowler 1920, 1952; Stiles 1978; Soldwedel, personal communication); **New Mexico** (Koster 1957; Tyus et al. 1982; Sublette et al. 1990; Platania 1991); **New York** (Smith 1985; Schmidt 1986); **North Carolina** (Hocutt et al. 1986; Menhinick 1991); **North Dakota** (Owen et al. 1981); **Ohio** (Trautman 1981); **Oklahoma** (Miller and Robison 1973); **Oregon** (Lampman 1946; U.S. Fish and Wildlife Service 1993e; Bond 1994); **Rhode Island** (Lapin, personal communication); **South Carolina** (Rohde, personal communication); **South Dakota** (Underhill 1959; Bailey and Allum 1962); **Texas** (Howells 1987; Howells and Prentice 1991); **Utah** (Sigler and Miller 1963; Tyus et al. 1982); **Virginia** (Hocutt et al. 1986; Jenkins and Burkhead 1994); **Washington** (Smith 1896; Gray and Dauble 1977; Wydoski and Whitney 1979); **West Virginia** (Stauffer et al. 1995); **Wisconsin** (Becker 1983); and **Wyoming** (Baxter and Simon 1970; Hubert 1994).

Means of Introduction: Intentional stocking for sportfishing.

Status: Established in most locations.

Impact of Introduction: In California, aggressive bluegill outcompete native Sacramento perch *Archoplites interruptus* (Moyle et al. 1974b; Moyle 1976a). Bluegill may chase Sacramento perch away from spawning areas and out of favored places, such as shallow weedy areas, and into open water (Moyle 1976a). Once in open water, the perch are more vulnerable to predation and have less available food. Introduced predatory centrarchids are likely responsible for the decline of native ranid frogs in California, California tiger salamander *Ambystoma californiense* populations (Hayes and Jennings 1986; Dill and Cordone 1997), and the Chiricahua leopard frog *Rana chiricahuensis* in southeastern Arizona (Rosen et al. 1995).

Remarks: Bluegill are commonly stocked as forage for largemouth bass in farm ponds. Because introduced California bluegill are typically small, possibly due to a limited genetic background, the California Department of Fish and Game began introducing bluegill from Florida in an effort to obtain a larger, faster-growing fish

(Moyle 1976a). Tyus et al. (1982) gave a distribution map of the this species in the upper Colorado basin.

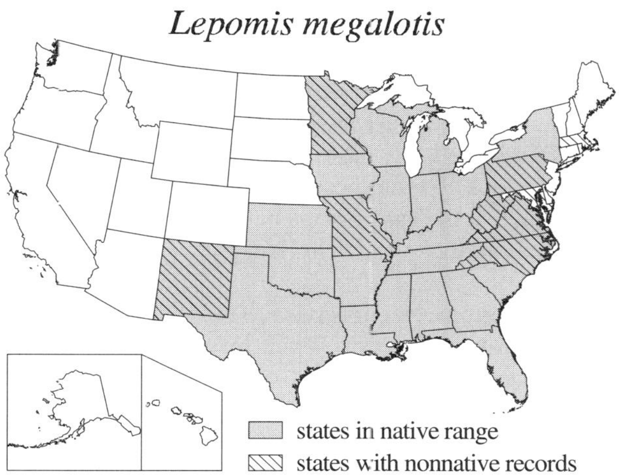

Lepomis megalotis (Rafinesque 1820)

Common Name: longear sunfish.

Identification: Becker (1983); Page and Burr (1991); Etnier and Starnes (1993); Jenkins and Burkhead (1994). Maximum size: 24 cm.

Native Range: St. Lawrence and Great Lakes, Hudson Bay, Mississippi River, and Gulf Slope drainages west of the Appalachian Mountains from southern Quebec west to western Ontario and Minnesota, and south to the Florida panhandle and southern Texas; on the Gulf Slope from the Choctawhatchee River, Florida, to the Rio Grande, Texas and New Mexico. Also in northeastern Mexico (Page and Burr 1991).

Nonindigenous Occurrences: This species has been introduced into the Potomac drainage, the Blue Ridge–Great Valley region, and the coastal plain of southern **Maryland** (Lee et al. 1976, 1981); the Chippewa and Minnesota drainages (Burr and Page 1986), and possibly introduced into Little Rock and Hustler lakes in **Minnesota** (Phillips et al. 1982); Lake St. Louis in St. Charles County, **Missouri** (Pflieger 1997); probably introduced in the Tennessee drainage of **North Carolina** (Menhinick 1991; Gilbert, personal communication); the Rio Grande and Mimbres drainages of **New Mexico** (Sublette et al. 1990); the Potomac and New drainages, **Virginia** (Hocutt et al. 1986; Jenkins and Burkhead 1994); and the Potomac and Kanawha (above the falls) drainages in **West Virginia** (Hocutt et al. 1986; Jenkins and Burkhead 1994; Stauffer et al. 1995). Schmidt (1986) reported this species as introduced into the Connecticut drainage but does not indicate the state. Whitworth (1996) does not list this species in the state of Connecticut. The drainage record may refer to a sunfish tentatively identified as this species collected from a pond in Tewksbury, **Massachusetts**, in 1923 (Cardoza et al. 1993). An undetermined species of *Lepomis* was stocked in Massachusetts through at least 1946 (Cardoza et al. 1993).

Means of Introduction: Accidental stocking as a result of stock contamination with other centrarchids. The record from Missouri refers to a 1985 collection that was the first in Missouri from any Mississippi River tributary above the junction of the Mississippi and Missouri rivers. The longear's presence here may be due to an undocumented introduction (Pflieger 1997).

Status: Established in all locations except the New drainage of Virginia, where it may be extirpated, and Massachusetts, where there are no recent records (Cardoza et al. 1993).

Impact of Introduction: Unknown.

Remarks: Some ichthyologists believe *L. megalotis* may actually be native to the Tennessee drainage in North Carolina. Populations there may be at low levels due to extirpation by introduced *L. auritus* (i.e., Starnes and Jenkins, personal communication). On the other hand, Gilbert (personal communication) points out that there are many specie native to the Tennnessee River drainage of Tennessee that don't enter North Carolina and whose ranges stop abruptly at the state line (which marks the western limits of the Blue Ridge plateau), such as *Luxilus chrysocephalus*, *Etheostoma duryi*, *Etheostoma camurum*, etc. Gilbert states that the state line is characterized by sharp increases in stream gradient of the various tributaries which is most pronounced in the Watauga system and least pronounced in the French Broad River. It is probably not coincidental that several other species otherwise found in North Carolina, range (or once ranged) into that state only in the French Broad system (e.g., *Hiodon tergisus*, *Percina caprodes*, *Etheostoma simoterum*, *Noturus eleutherus*, etc.). Further evidence for introduction is that although found in flowing streams, *L. megalotis* is not partial to the fast-flowing mountain streams characteristic of the Blue Ridge plateau. Stauffer et al. (1995) consider this species native, but possibly introduced to the Potomac drainage. Jenkins and Burkhead (1994) believe it is introduced and reported that the first record from the Potomac drainage in West Virginia was in 1953. The native status of longear sunfish in Minnesota is unclear. Eddy and Underhill (1974) mentioned records from Pomme de Terre River and Big Stone Lake from 1897 but question their validity. They state that many longear sunfish are misidentified. They accepted only a single record for the state, the one at Little Rock Lake. They also noted a population at International Falls, Ontario. Phillips et al. (1982) wrote that only two fish had been collected in the state, one at Little Rock Lake and the other at Hustler Lake, and that these may have been a result of introductions. However, Lee et al. (1980 et seq.) mapped six collection localities in the state. This may be an indication that this species is more widespread and possibly native to Minnesota and earlier authors overlooked these records; or as Eddy and Underhill suggest, they may be based on misidentifications. If the latter is true, then this species is very rare in Minnesota and it's presence there maybe the result of either introductions or isolated relict populations.

Collections made in southeastern New York depicted in Lee et al. (1980 et seq.) were probably based on specimens identified by C. L. Smith. Smith later reidentified these fish as hybrids between the pumpkinseed and redbreast sunfish (Smith 1985).

Lepomis microlophus

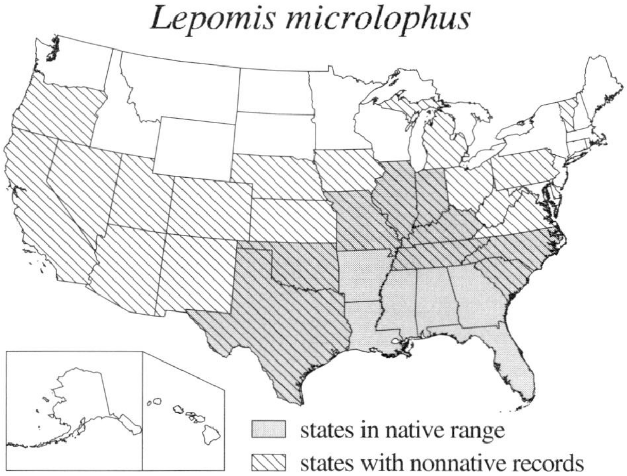

Lepomis microlophus (Günther 1859)

Common Names: redear sunfish (shellcracker).

Identification: Moyle (1976a); Hubbs et al. (1991); Page and Burr (1991); Etnier and Starnes (1993). *Lepomis microlophus* is composed of two unnamed subspecies; one in Florida, Georgia, and southern Alabama, the other throughout the rest of its range. The two subspecies may no longer be distinguishable due to interbreeding caused by stocking programs (Page and Burr 1991). Maximum size: 25 cm.

Native Range: The native range for this species is unclear (Etnier and Starnes 1993; Jenkins and Burkhead 1994). Page and Burr (1991) report the range as Atlantic and Gulf Slope drainages from about the Savannah River, South Carolina, to the Nueces River, Texas; north in the Mississippi River basin to southern Indiana and Illinois. Jenkins and Burkhead (1994) cite Bailey (1938) in reporting that the native range on the Atlantic slope may not be any farther north than Georgia.

Nonindigenous Occurrences: This species is known from the lower Colorado River, Roosevelt Lake on the Salt River, and other reservoirs in central **Arizona** (Minckley 1973; Grabowski et al. 1984; Rinne 1995); many drainages within the Sacramento–San Joaquin, Death Valley, Southern California regions, and the Colorado River in **California** (Smith 1982; Douglas 1974; Moyle 1976a; Moyle and Daniels 1982; Grabowski et al. 1984); the Platte, Republican, and Arkansas drainages (Walker 1993), and College Lake in Larimer County in the Cache La Poudre system in **Colorado** (Horak, personal communication); Wagamons Pond, Sussex County, **Delaware** (Raasch, personal communication); the northern two-thirds of **Illinois**, including the Lake Michigan, Illinois, Kaskaskia, Sangamon, Embarras, and Vermilion drainages (Smith 1979; Burr and Page 1986); southern farm ponds and reservoirs and northern **Indiana**, including the Tippecanoe and lower Wabash drainages, and the Anderson River (Gerking 1945; Nelson and Gerking 1968); artificial recreational lakes and interior streams in **Iowa** (Harlan et al. 1987); the Kansas River drainage and numerous counties in **Kansas** (Cross 1967; Cross and Collins 1995); locations east of the lower Green River in **Kentucky**, including the Licking, Kentucky, Salt and Rolling Fork drainages (Burr and Page 1986; Burr and Warren 1986); Branch, Hardin, Oakland, Hillsdale, and Washtenaw counties, **Michigan** (Bailey and Smith 1992; UMMZ specimens); the lower Missouri drainage, and reservoirs and ponds outside its native range in **Missouri** (Pflieger 1971, 1975, 1997; Cross et al. 1986); the Platte-Niobrara drainage, the Elkhorn River in Dodge and Washington counties, and West Oak Lake in Lancaster County, **Nebraska** (Morris et al. 1974; Cross et al. 1986); lower Colorado River, **Nevada** (Deacon and Williams 1984); the Pecos drainage in **New Mexico** (Koster 1957; Lee et al. 1980 et seq.); Tennessee and Roanoke drainages, probably introduced in the Neuse and Tar drainages, and regarded as native but possibly introduced in the Catawba, Yadkin,

Lumber, and Waccamaw drainages in **North Carolina** (Hocutt et al. 1986; Menhinick 1991); Honey Creek at Buckeye Lake, Greter's Lake in Richland County, Pippen Lake in Portage County, the Scioto and Great Miami drainages, and other locations in **Ohio** (Lee et al. 1980 et seq.; Trautman 1981; Burr and Page 1986; Hocutt et al. 1986; Underhill 1986); non-specific, widely distributed in **Oklahoma** (Hall 1956; Miller and Robison 1973; Douglas 1974); parts of western **Oregon** including ponds in the Willamette Valley (Bond 1994) and Benton County (Logan 1994); at least eight lakes in **Pennsylvania**, seven in the Susquehanna drainage and one in the Allegheny drainage (Denoncourt et al. 1975a; Cooper 1983; Hocutt et al. 1986), Lee et al. (1980 et seq.) also mapped collections in eastern Pennsylvania in the Delaware drainage; all drainages of **South Carolina** including the Pee Dee, Santee–Cooper, Edisto, Combahee, Broad, and Savannah (Loyacano 1975; Hocutt et al. 1986; Rohde, personal communication); non-specific localized populations in **Tennessee** (Etnier and Starnes 1993); western **Texas** (Lee et al. 1980 et seq.); Holmes Creek Reservoir, Davis County, **Utah** (Sigler and Miller 1963); two lakes in Caledonia County, northeastern **Vermont** (Whittier and Hartel 1997); Potomac, Rappahannock, York, James, Chowan, Roanoke, and Big Sandy drainages in **Virginia** (Hocutt et al. 1986; Jenkins and Burkhead 1994); and although Stauffer et al. (1995) listed the redear as native, collections in the Ohio River in **West Virginia** would be based on introductions according to Lee et al. (1980 et seq.) and Jenkins and Burkhead (1994).

Means of Introduction: Intentional stocking for sportfishing. In Iowa, fish found in interior streams are believed to be escapes from stocked lakes (Harlan et al. 1987). The species' recent (1991) discovery in Vermont is somewhat of a mystery. It appears the most likely means of introduction there is stock contamination of privately stocked *Micropterus* (Whittier and Hartel 1997).

Status: Established in most locations, but extirpated in New Mexico (Sublette et al. 1990) and Utah (Sigler and Sigler 1996). The Delaware population at Wagamons Pond was discovered in October 1994 (Raasch, personal communication). Not established in Missouri (Pflieger 1997). Established in at least one of the two lakes in Vermont (Whittier and Hartel 1997).

Impact of Introduction: Redear are highly molluscivorous. In inland lakes of southern Michigan, introduced redear are associated with ecological changes in populations of pumpkinseed *L. gibbosus*, a native molluscivore. Effects of introduced redear on pumpkinseed include reduced consumption of snails and reduced population densities (Huckins 1996). The effects appear to be driven by differences in pharyngeal morphology and competitive ability; redear exert greater crushing forces and consume more snails than sympatric pumpkinseeds (Huckins 1997).

Remarks: Robison and Buchanan (1988) reported redear sunfish as widely stocked in Arkansas in ponds and reservoirs, presumably in native waters. Menhinick (1991) listed this species as "regarded as native but possibly introduced" and "introduced" into the Catawba, Yadkin, Lumber, and Waccamaw drainages. Stauffer et al. (1995) differed with other authors (Lee et al. 1980 et seq. and Jenkins and Burkhead 1994) in reporting this species as native to the Potomac, Big Sandy, Kanawha, and Little Kanawha drainages, and Ohio basin. No records exist for New York (Whittier and Hartel 1997).

Lepomis miniatus

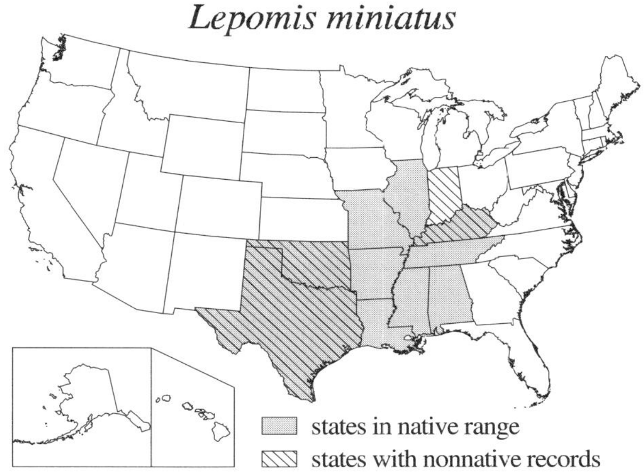

Lepomis miniatus Jordan 1877

Common Names: redspotted sunfish (western spotted sunfish).

Identification: Pflieger (1975); Page and Burr (1991); Warren (1992); Etnier and Starnes (1993); recently distinguished from *L. punctatus* as a separate species (Warren 1992). Maximum size: 20 cm.

Native Range: Western spotted sunfish are native to the Illinois River in Illinois, ranging south through the Mississippi River and along the Gulf Coast from Alabama to Texas (Warren 1992).

Nonindigenous Occurrences: The western spotted sunfish may have been introduced into Round Lake near Columbia City, **Indiana** (Gerking 1955). Lee et al. (1980 et seq.) mapped a collection from the Red River (Cumberland drainage) in **Kentucky**, and indicated it was introduced to this location. Also introduced to the Verdigris River and Eufala Lake in eastern **Oklahoma** (Lee et al. 1980 et seq.); and the Wichita River (Lee et al. 1980 et seq.) and Devils River (Page and Burr 1991) of the Rio Grande drainage in **Texas**. Cross et al. (1986) listed this species as introduced into the middle Arkansas and Canadian drainages but do not give the states where it has been introduced within these drainages.

Means of Introduction: Unknown; possibly stock contamination.

Status: Established in Indiana; reported in Kentucky.

Impact of Introduction: Unknown.

Remarks: These reports were all made before *L. punctatus* was split into *L. punctatus* and *L. miniatus*. We assumed these collections were of *L. miniatus* because that is the species native to three of these states, and closest to the fourth (Oklahoma). The report from Oklahoma is somewhat questionable. Burr and Warren (1986) considered the records of introduced populations in the Red River (Cumberland drainage), Kentucky, depicted in Lee et al. (1980 et seq.), to be unsubstantiated and likely erroneous. However, Warren (1992) did map a collection (with specimens he examined) from the Cumberland just over the border in Tennessee. But, Etnier and Starnes (1993) did not include this location in their distribution map for the species in Tennessee. Therefore, whether it does now or has ever occurred in the Cumberland drainage is not clear. It is not clear whether this species is native or introduced in Indiana.

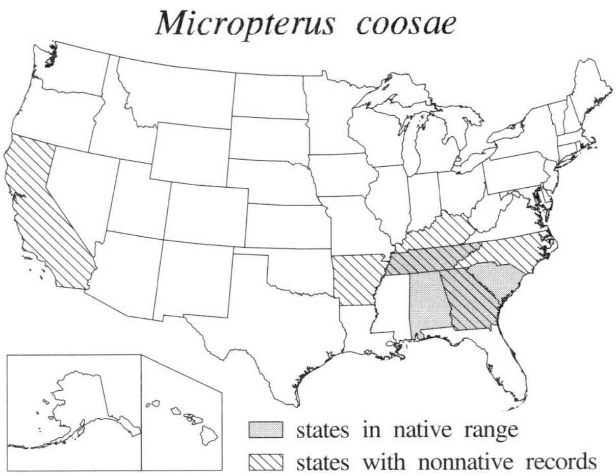

Micropterus coosae Hubbs and Bailey 1940

Common Name: redeye bass (Coosa bass).

Identification: Page and Burr (1991); Etnier and Starnes (1993); Mettee et al. (1996). Maximum size: 47 cm.

Native Range: Above the Fall Line in the Savannah, Chattahoochee, and Mobile Bay basins, of North Carolina, South Carolina, Georgia, Tennessee, and Alabama (Page and Burr 1991).

Nonindigenous Occurrences: Stocked in the Spring River downstream of Mammoth Spring, **Arkansas** (Buchanan 1973; Lee et al. 1980 et seq.; Robison and Buchanan 1988; Cross et al. 1986), Alder Creek in Sacramento County, Stanislaus River in Tuolumne County, Dry Creek in Nevada County, Santa Ana River in Riverside County, San Margarita River in San Diego County, and Sisquoc River in Santa Barbara County, **California** (Goodson 1966b; MacCrimmon and Robbins 1975; Moyle 1976a; Dill and Cordone 1997); the Tennessee (Dahlberg and Scott 1971b; Page and Burr 1991) and the Atlamaha drainage (Page and Burr 1991) in **Georgia**; Martin's Fork and its tributaries in the upper Cumberland River in extreme southeastern **Kentucky** (MacCrimmon and Robbins 1975; Lee et al. 1980 et seq.; Burr and Warren 1986); and the Tennessee (Little Tennessee and Hiawassee systems) and Catawba drainages in **North Carolina** (Menhinick 1991); and the Hiawassee River, Emory River, and Chickamauga Creek of the Tennessee River drainage, and on the Cumberland Plateau of the Cumberland drainage in **Tennessee** (MacCrimmon and Robbins 1975; Starnes and Etnier 1986; Etnier and Starnes 1993).

Means of Introduction: Intentional stocking for sportfishing. Introduced into California between 1962 and 1964 (Moyle 1976a). Goodson (1966b) gave exact dates and numbers of fish introduced. Introduced into Kentucky around 1950 with stock obtained from Georgia (Burr and Warren 1986). Populations in the Hiwassee system in Tennessee are the result of introductions in 1943 (MacCrimmon and Robbins 1975; Etnier and Starnes 1993). Populations in the Cumberland Plateau were introduced in 1953 (MacCrimmon and Robbins 1975; Etnier and Starnes 1993).

Status: Presumably extirpated in Arkansas (Robison and Buchanan 1988). In California, only the Sisquoc River introduction was successful (Moyle 1976a). This location was the only one in California to receive a sizeable number of stocked fish (Goodson 1966b). Established in Georgia (Dahlberg and Scott 1971a, 1971b; Page and Burr 1991), Kentucky (Burr and Warren 1986), and Tennessee (Etnier and Starnes 1993). Unknown in North Carolina.

Impact of Introduction: Introduced predatory centrarchids are likely responsible for the decline of native ranid frogs in California and for the decline of California tiger salamander *Ambystoma californiense* populations (Hayes and Jennings 1986; Dill and Cordone 1997).

The redeye bass was introduced into headwater streams in Tennessee in the 1950s for sportfishing. Smallmouth bass were native to these same streams farther downriver. Eventually the redeye bass extended their range downriver and became sympatric with the smallmouth bass and hybridized (Turner 1989; Turner et al. 1991; Pipas 1996; Pipas and Bulow 1998). The redeye bass has been documented to hybridize with native smallmouth bass *M. dolomieu* in parts of Tennessee, including the Ocoee and Obed systems in the Tennessee drainage and Roaring Fork system in the Cumberland drainage (Turner 1989; Turner et al. 1991; Pipas 1996; Pipas and Bulow 1998). These hybrids are fertile and are also capable of backcrossing. Hybridization and backcrossing has occurred over several generations and resulted in altered genetics of the parental species. These introgressed individuals are phenotypically similar to the parental species but differ genetically (Turner et al. 1991; Pipas and Bulow 1998). It was recently determined that 67% of the *Micropterus* sampled in nine streams in Tennessee were electrophoretic hybrids; 31% sampled from 11 streams were meristic hybrids (Pipas 1996). Meristics of both parental species and the resulting hybrid are given in Turner et al. (1991).

Remarks: Although Page and Burr (1991) reported *M. coosae* as introduced into the Altamaha, Dahlberg and Scott (1971a,1971b) and Etnier and Starnes (1993) reported it as native to that drainage. Redeye bass were originally brought into California in 1953 as broodstock. However, none of these fish were released into the wild and none survived (Kimsey 1957). Although Menhinick (1991) list redeye bass as introduced into the Tennessee drainage in North Carolina, his map shows the collections are actually in the Tennessee portions of the drainage. MacCrimmon and Robbins (1975) showed a map depicting this species' native and introduced range.

Micropterus dolomieu

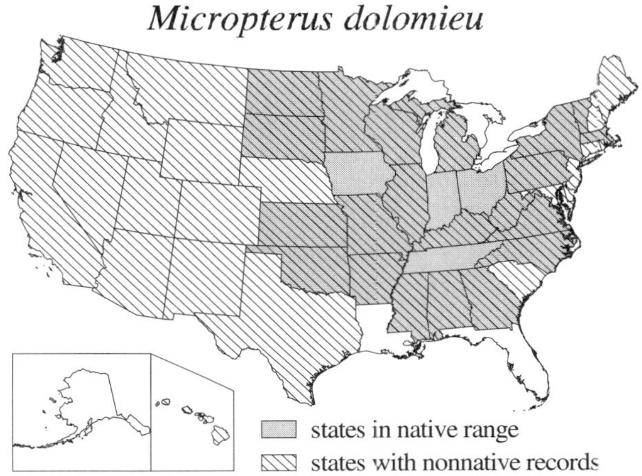

Micropterus dolomieu Lacepède 1802

Common Names: smallmouth bass (black bass).

Identification: Moyle (1976a); Becker (1983); Page and Burr (1991); Etnier and Starnes (1993); Jenkins and Burkhead (1994); other names *M. dolomieui*. Maximum size: 69 cm.

Native Range: St. Lawrence and Great Lakes, Hudson Bay (Red River), and Mississippi River basins from southern Quebec to North Dakota and south to northern Alabama and eastern Oklahoma; Atlantic and Gulf slope drainages from Virginia to central Texas (Page and Burr 1991).

Nonindigenous Occurrences: The smallmouth bass has been stocked extensively and is reported as being stocked in **Alabama** (Boschung 1992; Mettee et al. 1996); **Arkansas** (Lee et al. 1980 et seq.); **Arizona** (Minckley 1973); **California** (Smith 1896; Shebley 1917; Neale 1931; Lampman 1946; Moyle 1976a); **Colorado** (Everhart and Seaman 1971; Tyus et al. 1982); **Connecticut** (Webster 1942; Behnke and Wetzel 1960;

Whitworth et al. 1968; Schmidt 1986; Whitworth 1996); **Delaware** (Raasch and Altemus 1991); **Georgia** (Dahlberg and Scott 1971b; Yerger 1977); **Hawaii** (Brock 1960; Maciolek 1984); **Idaho** (Linder 1963; Idaho Fish and Game 1990); **Illinois** (Burr, personal communication); **Kansas** (Cross 1967; Cross et al. 1986); **Kentucky** (Burr and Warren 1986); **Maine** (Kendall 1914a; Everhart 1976; Schmidt 1986); **Maryland** (Davis, personal communication); **Massachusetts** (Schmidt 1986; Hartel 1992; Hartel et al. 1996); **Michigan** (Hubbs and Lagler 1947); **Minnesota** (Eddy and Underhill 1974); **Mississippi** (Ross and Brenneman 1991); **Missouri** (Pflieger 1971, 1975; Lee et al. 1980 et seq.); **Montana** (Cross et al. 1986; Holton 1990); **Nebraska** (Morris et al. 1974; Cross et al. 1986); **Nevada** (Smith 1896; Miller and Alcorn 1946; La Rivers 1962; Deacon and Williams 1984); **New Hampshire** (Hoover 1936; Scarola 1973; Schmidt 1986); **New Jersey** (Fowler 1906, 1952); **New Mexico** (Sublette et al. 1990); **New York** (Cheney 1897; Schmidt 1986); **North Carolina** (Hocutt et al. 1986; Menhinick 1991); **North Dakota** (Lee et al. 1980 et seq.; Cross et al. 1986); **Oklahoma** (Miller and Robison 1973); **Oregon** (Wydoski and Whitney 1979; Bond 1994); **Pennsylvania** (Cooper 1983; Hocutt et al. 1986); **Rhode Island** (Lapin, personal communication); **South Carolina** (Loyacano 1975; Lee et al. 1980 et seq.; Hocutt et al. 1986); **South Dakota** (Lee et al. 1980 et seq.; Cross et al. 1986); **Texas** (Hubbs and Peden 1968; Kraai et al. 1983; Conner and Suttkus 1986); **Utah** (Sigler and Miller 1963; Tyus et al. 1982; Teuscher and Luecke 1996); **Vermont** (Whittier and Hartel 1997); **Virginia** (Lee et al. 1980 et seq.; Hocutt et al. 1986; Jenkins and Burkhead 1994); **Washington** (Lampman 1946; Gray and Dauble 1977; Wydoski and Whitney 1979); **West Virginia** (Stauffer et al. 1995); **Wisconsin** (Becker 1983); and **Wyoming** (Baxter and Simon 1970; Hubert 1994; Stone 1995).

Means of Introduction: Intentional stocking for sportfishing.

Status: Established in most of the above states. Reported in Mississippi.

Impact of Introduction: In Arizona, smallmouth bass reportedly are responsible for eliminating or reducing some populations of native fishes (Minckley 1973). Smallmouth bass have been shown to eat smolts of Pacific salmonids, therefore posing a threat to these already declining species in the Columbia River (Dentler 1993). Jenkins and Burkhead (1994) speculated that introduced smallmouth bass may have contributed to the demise of an isolated population of trout-perch *Percopsis omiscomaycus* in the Potomac River in Virginia and Maryland. Smallmouth bass were introduced into Flaming Gorge Reservoir to reduce the Utah chub *Gila atraria* (Teuscher and Luecke 1996). Introduced predatory centrarchids are likely responsible for the decline of native ranid frogs in California and for the decline of California tiger salamander *Ambystoma californiense* populations (Hayes and Jennings 1986; Dill and Cordone 1997). Introduction of the smallmouth bass into the native range of Guadalupe bass *M. treculii* in southcentral Texas has resulted in hybridization between the two species (Edwards 1979; Whitmore 1983). This hybrid bass is found in Canyon Lake and the San Marcos system, Guadalupe River drainage, on the Edwards Plateau in Texas (Whitmore 1983; International Game Fish Association 1994). It was first recognized by Edwards (1979) and later verified by Whitmore (1983). The hybrid is fertile and is capable of backcrossing to the parent species. There appears to be more backcrossing to smallmouth bass than to Guadalupe bass (Whitmore 1983). Introgressive hybridization represents another threat to the already depleted Guadalupe bass by compromising its genetic integrity (Whitmore 1983). Although Lake Travis, on the Colorado River, also has been stocked with smallmouth bass, Whitmore (1983) found no evidence of hybridization there. The smallmouth bass also hybridizes with the spotted bass *M. punctulatus* when stocked in the spotted bass's native range or when both species are stocked in the same area. The smallmouth/spotted bass hybrid has been found in the Verde River, Arizona (Minckley 1973); California (Moyle 1976a); the Marmaton River, Barbour County, Kansas (Cross 1967; museum specimen KU 4682); and southeastern Oklahoma (formerly described as *M. punctulatus wichitae*) (Cofer 1995). A third hybrid resulting from stocking smallmouth bass is the smallmouth/largemouth hybrid. Introduced smallmouth bass hybridize with native largemouth bass in Squaw Reservoir in northcentral Texas (Whitmore and Hellier 1988).

Remarks: Introduced bass likely affect small fish populations through predation. Tyus et al. (1982) gave a distribution map of the this species in the upper Colorado basin. MacCrimmon and Robbins (1975) showed a map depicting this species' native and introduced range.

Micropterus punctulatus

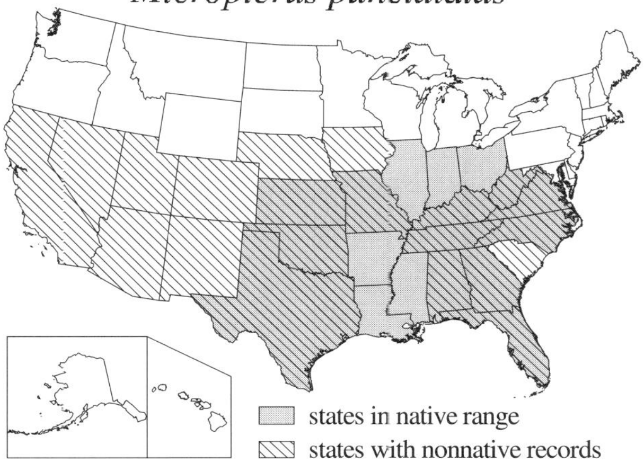

Micropterus punctulatus (Rafinesque 1819)

Common Name: spotted bass (Kentucky bass).

Identification: Moyle (1976a); Hubbs et al. (1991); Page and Burr (1991); Etnier and Starnes (1993); Jenkins and Burkhead (1994). This species was formerly composed of three subspecies: the northern spotted bass – *M. p. punctulatus*, the Alabama spotted bass – *M. p. henshalli* and the Wichita spotted bass – *M. p. wichitae*. Cofer (1995) determined the Wichita subspecies was actually a hybrid with *M. dolomieu* and is therefore invalid. Maximum size: 61 cm.

Native Range: Mississippi River basin from southern Ohio and West Virginia to southeastern Kansas, and south to the Gulf; Gulf Slope drainages from the Chattahoochee River, Georgia (where possibly introduced), to the Guadalupe River, Texas (Page and Burr 1991).

Nonindigenous Occurrences: Introduced into the Chattahoochee River, **Alabama** (Lee et al. 1980 et seq.; Yerger 1977); the Verde system and West Clear Creek, **Arizona** (Miller and Lowe 1967; Minckley 1973; MacCrimmon and Robbins 1975); the Arkansas drainage in **Colorado** (Walker 1993); several areas of **California** (McKechnie 1966b; MacCrimmon and Robbins 1975; Moyle 1976a; Dill and Cordone 1997); the Apalachicola River in **Florida** (Yerger 1977; Williams and Burgess, in prep.; museum specimens); the Chattahoochee, Altamaha (Oconee) (Dahlberg and Scott 1971a, 1971b; Yerger 1977; Lee et al. 1980 et seq.), and Savannah (MacCrimmon and Robbins 1975) drainages in **Georgia**; the Iowa-Cedar River drainage, the Middle Raccoon River, Whitebreast Creek, Coralville Reservoir, and Lake MacBride in **Iowa** (MacCrimmon and Robbins 1975; Burr and Page 1986; Harlan et al. 1987); many rivers in **Kansas** including those in the Osage and Kansas drainages (Cross 1967; MacCrimmon and Robbins 1975; Cross and Collins 1995); the upper Cumberland River, **Kentucky** (Burr and Warren 1986); Salt, Osage, Gasconade, and Moreau drainages, and the Lamine, Perche, Loutre systems in **Missouri** (Pflieger 1971; MacCrimmon and Robbins 1975; Pflieger 1975; Burr and Page 1986; Cross et al. 1986); a pond in Lincoln County (Jones 1963), the Niobrara and Platte rivers, and the Walgreen, Harlan County, and Box Butte reservoirs in **Nebraska** (Morris et al. 1974; MacCrimmon and Robbins 1975); Eagle Valley Creek and Eagle Valley Reservoir, **Nevada** (La Rivers 1962; Deacon and Williams 1984; Sigler and Sigler 1987); the Pecos and Black rivers and Sumner Reservoir in **New Mexico** (MacCrimmon and Robbins 1975; Sublette et al. 1990); the Yadkin, Cape Fear, Catawba, and Roanoke drainages in **North Carolina** (Menhinick 1991); central **Oklahoma** (MacCrimmon and Robbins 1975); the Savannah drainage (MacCrimmon and Robbins 1975) in **South Carolina** (Sigler and Sigler 1987); Reelfoot Lake, **Tennessee** (Kuhne 1939); the Medina River, and the Colorado and Guadalupe drainages in **Texas** (Hubbs 1954; MacCrimmon and Robbins 1975); the York, New, Roanoke, and James drainages in **Virginia** (MacCrimmon and Robbins 1975; Lee et al. 1980 et seq.; Hocutt et al. 1986; Page and Burr 1991; Jenkins and Burkhead 1994); and introduced, but possibly native, to the Kanawha drainage above the falls (New) in **West Virginia** (MacCrimmon and Robbins 1975; Hocutt et al. 1986; Jenkins and Burkhead 1994; Burkhead, personal communication).

Means of Introduction: Intentional stocking for sportfishing. In Missouri, the spotted bass was stocked in the Sac River and spread downstream through the Osage River into the Missouri River (MacCrimmon and Robbins 1975).

Status: Established in most locations. Probably not established in Colorado.

Impact of Introduction: The spotted bass hybridizes with the smallmouth bass when stocked in the smallmouth bass's native range or when both species are stocked in the same area. Spotted bass introduced into central Missouri have hybridized with the native smallmouth bass *M. dolomieu* (Pflieger and Fajen 1970; Pflieger 1975, 1997; Whitmore 1983). Reportedly, the decline of smallmouth populations in the Moreau drainage in Missouri is at least partly attributable to hybridization with introduced spotted bass (Pflieger 1997). The hybrid has been also been found in the Verde River,

Arizona (Minckley 1973); California (Moyle 1976a); and the Marmaton River, Barbour County, Kansas (Cross 1967; museum specimen KU 4682). The form formerly believed to be the subspecies *M. p. wichitae* from southwest Oklahoma was determined to be this hybrid (Cofer 1995).

Introduced predatory centrarchids are likely responsible for the decline of native ranid frogs in California and for the decline of California tiger salamander *Ambystoma californiense* populations (Hayes and Jennings 1986; Dill and Cordone 1997).

Remarks: Introduced bass likely affect populations of small fishes through predation. Two subspecies have been transplanted: *M. p. punctulatus* in Alabama (Lee et al. 1980 et seq.), California (Dill and Cordone 1997), Florida, Georgia (Lee et al. 1980 et seq.), Kansas (Cross and Collins 1995), Kentucky (Burr and Warren 1986), North Carolina (Lee et al. 1980 et seq.), Texas (Hubbs 1954), Virginia (Lee et al. 1980 et seq.; Jenkins and Burkhead 1994) and probably West Virginia; and *M. p. henshalli* in California (Lee et al. 1980 et seq.; Dill and Cordone 1997), and the upper Chattahoochee drainage in Georgia (JDW, personal observation). Subspecies identity of other introductions is unknown. Native and introduced in the Tennessee drainage in North Carolina (Menhinick 1991). Not listed as occurring in South Carolina by Loyacano (1975) or by Rohde et al. (1994). Stauffer et al. (1995) listed this species as native but possibly introduced in the Kanawha (including the New) drainage, West Virginia. MacCrimmon and Robbins (1975) showed a map depicting this species' native and introduced range.

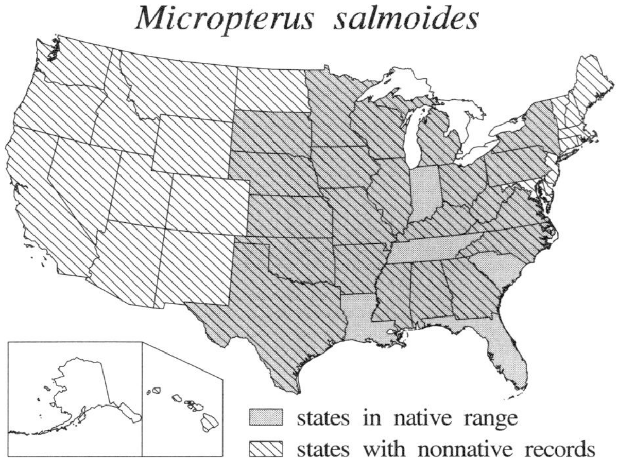

Micropterus salmoides (Lacepède 1802)

Common Names: largemouth bass (black bass, Oswego bass).

Identification: Moyle (1976a); Becker (1983); Page and Burr (1991); Etnier and Starnes (1993); Jenkins and Burkhead (1994). Maximum size: 97 cm.

Native Range: St. Lawrence and Great Lakes, Hudson Bay (Red River), and Mississippi River basins from southern Quebec to Minnesota and south to the Gulf; Atlantic Slope drainages from North Carolina to Florida; Gulf Slope drainages from southern Florida into northern Mexico (Page and Burr 1991).

Nonindigenous Occurrences: The largemouth bass is reported as stocked in **Alabama** (Mettee et al. 1996); **Arkansas** (Pritchard et al. 1978); **Arizona** (Minckley 1973; Hendrickson et al. 1980; Tyus et al. 1982); **California** (Smith 1896; Shebley 1917; Neale 1931; Moyle 1976a; Hubbs et al. 1979; U.S. Fish and Wildlife Service 1993e); **Colorado** (Everhart and Seaman 1971; Holden and Stalnaker 1975; Tyus et al. 1982; Propst and Carlson 1986); **Connecticut** (Behnke and Wetzel 1960; Whitworth et al. 1968; Schmidt 1986; Whitworth 1996); **Delaware** (Raasch and Altemus 1991); **Georgia** (Pritchard et al. 1978); **Hawaii** (Brock 1960; Maciolek 1984); **Idaho** (Smith 1896; Linder 1963; Idaho Fish and Game 1990); **Illinois** (Burr, personal communication); **Iowa** (Bailey and Allum 1962); **Kansas** (Cross 1967); **Kentucky** (Burr and Warren 1986); **Maine** (Kendall 1914a; Everhart 1976; Schmidt 1986); **Maryland** (Ferguson 1876; Truitt et al. 1929); **Massachusetts** (Schmidt 1986; Hartel 1992; Hartel et al. 1996); **Michigan** (Lee et al. 1980 et seq.); **Minnesota** (Phillips et al. 1982); **Mississippi** (Pritchard et al. 1978); **Missouri** (Pflieger 1971, 1975); **Montana** (Cross et al. 1986; Holton 1990); **Nebraska** (Morris et al. 1974); **Nevada** (Miller and Alcorn 1946; La Rivers 1962; Hubbs et al. 1974; Deacon and Williams 1984); **New Hampshire** (Scarola 1973; Schmidt 1986); **New Jersey** (Nelson 1890; Fowler 1952; Soldwedel, personal communication); **New Mexico** (Tyus et al. 1982; Sublette et al. 1990); **New York** (Smith 1985; Schmidt 1986); **North Carolina** (Menhinick 1991); **North Dakota** (Lee et al. 1980 et seq.; Cross et al. 1986); **Ohio** (Trautman 1981); **Oklahoma** (Miller and Robison 1973); **Oregon** (Smith 1896; Wydoski and Whitney 1979; U.S. Fish and Wildlife Service 1993e; Bond 1994); **Pennsylvania** (Lee et al. 1980 et seq.; Hocutt et al. 1986; Schmidt 1986); **Rhode Island** (Lapin, personal communication); **South Dakota** (Bailey and

Allum 1962; Cross et al. 1986). **Texas** (Kraai et al. 1983); **Utah** (Smith 1896; Sigler and Miller 1963; Tyus et al. 1982); **Vermont** (Lee et al. 1980 et seq.); **Virginia** (Lee et al. 1980 et seq.; Hocutt et al. 1986; Jenkins and Burkhead 1994); **Washington** (Smith 1896; Gray and Dauble 1977; Wydoski and Whitney 1979); **West Virginia** (Stauffer et al. 1995); **Wisconsin** (Becker 1983); and **Wyoming** (Baxter and Simon 1970; Tyus et al. 1982; Cross et al. 1986; Hubert 1994).

Means of Introduction: This species has been an important sport fish for many years and as such has been stocked widely in areas where it is nonindigenous. Intentional stocking for sportfishing.

Status: Established in most locations.

Impact of Introduction: Introduced bass usually affect populations of small native fishes through predation, sometimes resulting in the decline or extinction of such species (Minckley 1973). Species that have suffered such effects include relict dace *Relictus solitarius*, Clover Valley speckled dace *Rhinichthys osculus oligoporus*, Independence Valley tui chub *Gila bicolor lethoporus* (U.S. Fish and Wildlife Service 1985b), a distinct population of Gila chub *G. intermedia*, Monkey Spring pupfish *Cyprinodon* sp. (Minckley 1973), White River springfish *Crenichthys baileyi*, Pahranagat roundtail chub *Gila robusta jordani* (U.S. Fish and Wildlife Service 1985b), Owens pupfish *Cyprinodon radiosus* (Miller and Pister 1971), and White River spinedace *Lepidomeda albivallis* (U.S. Fish and Wildlife Service 1994e). Jenkins and Burkhead (1994) speculated that introduced largemouth bass may have contributed to the demise of an isolated population of trout-perch *Percopsis omiscomaycus* in the Potomac River in Virginia and Maryland. Introduced predatory centrarchids are likely responsible for the decline of native ranid frogs in California, California tiger salamander *Ambystoma californiense* populations (Hayes and Jennings 1986; Dill and Cordone 1997), and the Chiricahua leopard frog *Rana chiricahuensis* in southeastern Arizona (Rosen et al. 1995). In Squaw Creek Reservoir in northcentral Texas, introduced Florida largemouth intergrade with native northern largemouth (Whitmore and Hellier 1988).

Remarks: This account includes introductions of both subspecies *M. s. salmoides*, the northern largemouth bass, and *M. s. floridanus*, the Florida largemouth bass. For instance, both subspecies have been introduced into Nevada (Deacon and Williams 1984). Tyus et al. (1982) gave a distribution map of the this species in the upper Colorado basin. MacCrimmon and Robbins (1975) showed a map depicting this species' native and introduced range. Jenkins and Burkhead (1994) reported the largemouth as introduced into the Roanoke drainage in Virginia. Recently prehistoric bones of *M. salmoides* were discovered near the Roanoke River in Roanoke, Virginia, indicating that the species is native there (Jenkins, personal communication).

Micropterus sp. (undescribed)

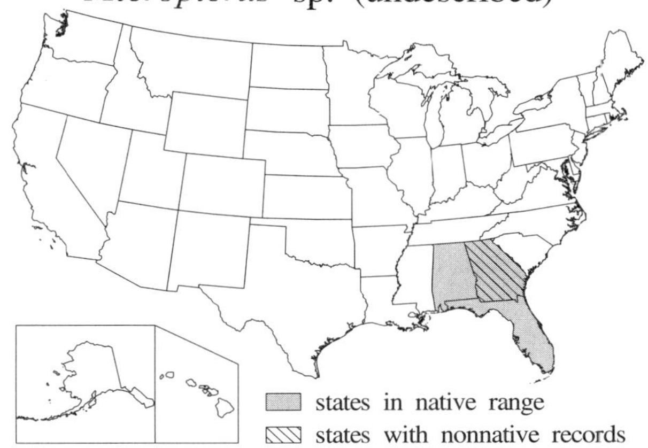

Micropterus sp. (undescribed)

Common Name: shoal bass.

Identification: Page and Burr (1991); Mettee et al. (1996). Maximum size: 39 cm.

Native Range: Apalachicola River drainage of Alabama and Georgia (Page and Burr 1991).

Nonindigenous Occurrences: Stocked in the Ocmulgee River (Altamaha drainage), **Georgia**, in 1975 (Bart et al. 1994; J. D. Williams and G. Burgess, unpublished data).

Means of Introduction: Intentional authorized stocking for sportfishing.

Status: Established from about 35 miles above to 25 miles below the Fall Line.

Impact of Introduction: Unknown.

Remarks: There are concerns regarding hybridization with native redeye bass *M. coosae* (Bart et al. 1994).

Micropterus treculii

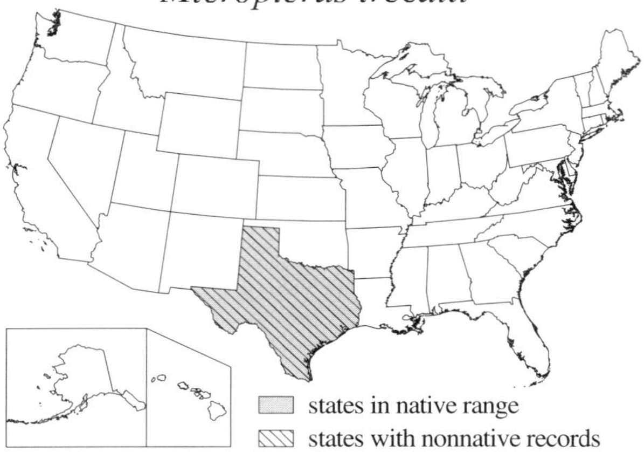

Micropterus treculii (Vaillant and Bocourt 1883)

Common Name: Guadalupe bass.

Identification: Hubbs et al. (1991); Page and Burr (1991). Maximum size: 40 cm.

Native Range: Edwards Plateau in the Brazos, Colorado, Guadalupe, and San Antonio river drainages, Texas (Page and Burr 1991).

Nonindigenous Occurrences: Introduced into the upper Nueces River drainage in **Texas** (Lee et al. 1980 et seq.; Conner and Suttkus 1986; Page and Burr 1991).

Means of Introduction: Unknown.

Status: Established in Texas.

Impact of Introduction: Unknown.

Remarks: None.

Pomoxis annularis

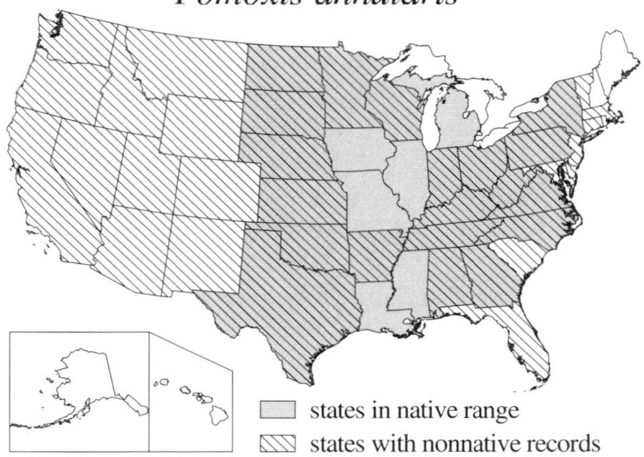

Pomoxis annularis Rafinesque 1818

Common Name: white crappie.

Identification: Moyle (1976a); Becker (1983); Page and Burr (1991); Etnier and Starnes (1993); Jenkins and Burkhead (1994). Maximum size: 53 cm.

Native Range: Great Lakes, Hudson Bay (Red River), and Mississippi River basins from New York and southern Ontario west to Minnesota and South Dakota, and south to the Gulf; Gulf Slope drainages from Mobile Bay, Georgia and Alabama, to the Neuces River, Texas (Page and Burr 1991).

Nonindigenous Occurrences: Introduced into **Alabama** (Smith-Vaniz 1968; Lee et al. 1980 et seq.; Mettee et al. 1996); **Arizona** (Lee et al. 1980 et seq.); **Arkansas** (Robison and Buchanan 1988); **California** (Shebley 1917; Neale 1931; Moyle 1976a); **Colorado** (Everhart and Seaman 1971; Tyus et al. 1982); **Connecticut** (Behnke and Wetzel 1960; Whitworth et al. 1968; Schmidt 1986; Whitworth 1996); **Delaware** (Raasch and Altemus 1991); **Florida** (Kilby et al. 1959; Yerger 1977; Lee et al. 1980 et seq.); **Georgia** (Dahlberg and Scott 1971a, 1971b; Yerger 1977); **Idaho** (Linder 1963; Idaho Fish and Game 1990); **Indiana** (Sweeney 1902; Gerking 1945); **Kansas** (Cross 1967); **Kentucky** (Burr and Warren 1986); **Maryland** (Lee et al. 1980 et seq.); **Massachusetts** (Fowler 1907; Schmidt 1986; Hartel 1992; Hartel et al. 1996); **Minnesota** (Eddy and Underhill 1974; Phillips et al. 1982); **Montana** (Cross et al. 1986; Holton 1990); **Nebraska** (Baxter and Simon 1970), **Nevada** (La Rivers 1962; Deacon and Williams 1984); **New Jersey** (Morse 1905; Fowler 1906, 1952; Stiles 1978; Soldwedel, personal communication); **New Mexico** (Tyus et al. 1982; Sublette et al. 1990); **New York** (Lee et al.

1980 et seq.; Schmidt 1986); **North Carolina** (Hocutt et al. 1986; Menhinick 1991); **North Dakota** (Lee et al. 1980 et seq.; Cross et al. 1986); **Ohio** (Trautman 1981); **Oklahoma** (Miller and Robison 1973); **Oregon** (Lampman 1946; Bond 1994); **Pennsylvania** (Cooper 1983; Hocutt et al. 1986; Schmidt 1986); **South Carolina** (Lee et al. 1980 et seq.; Hocutt et al. 1986); **South Dakota** (Bailey and Allum 1962; Cross et al. 1986); **Tennessee** (Walters 1997); **Texas** (Kraai et al. 1983); **Utah** (Sigler and Miller 1963); **Vermont** (Cox, personal communication); **Virginia** (Lee et al. 1980 et seq.; Hocutt et al. 1986; Jenkins and Burkhead 1994); **Washington** (Gray and Dauble 1977; Fletcher, personal communication); **West Virginia** (Stauffer et al. 1995); **Wisconsin** (Lee et al. 1980 et seq.; Becker 1983); and **Wyoming** (Baxter and Simon 1970; Hubert 1994; Stone 1995).

Means of Introduction: Intentional stocking for sportfishing.

Status: Established in most locations.

Impact of Introduction: Unknown.

Remarks: It is probable that all the white crappie in California are descendents of the original 16 fish planted in the state in 1917 (Goodson 1966a).

Pomoxis nigromaculatus

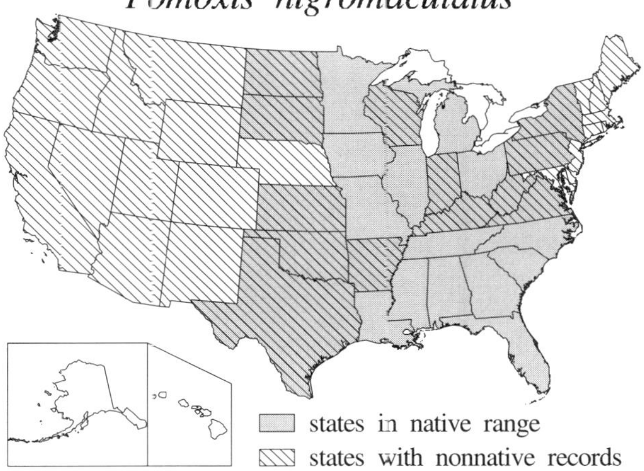

Pomoxis nigromaculatus (Lesueur 1829)

Common Names: black crappie (strawberry bass, calico bass, speck).

Identification: Moyle (1976a); Becker (1983); Page and Burr (1991); Etnier and Starnes (1993); Jenkins and Burkhead (1994). Maximum size: 49 cm.

Native Range: So widely introduced that native range is difficult to determine; presumably Atlantic Slope from Virginia to Florida, Gulf Slope west to Texas, St. Lawrence, Great Lakes and Mississippi River basins from Quebec to Manitoba south to the Gulf (Page and Burr 1991).

Nonindigenous Occurrences: Introduced into **Arizona** (Hendrickson et al. 1980; Lee et al. 1980 et seq.; Tyus et al. 1982); **Arkansas** (Robison and Buchanan 1988); **California** (Smith 1896; Shebley 1917; Neale 1931; Moyle 1976a; Wydoski and Whitney 1979; Smith 1982); **Colorado** (Everhart and Seaman 1971; Ellis 1974; Tyus et al. 1982; Propst and Carlson 1986); **Connecticut** (Behnke and Wetzel 1960; Whitworth et al. 1968; Schmidt 1986; Whitworth 1996); **Delaware** (Raasch and Altemus 1991); **Idaho** (Smith 1896; Linder 1963; Idaho Fish and Game 1990); **Indiana** (Gerking 1945); **Kansas** (Cross 1967; Cross et al. 1986); **Kentucky** (Burr and Warren 1986); **Maine** (Everhart 1976); **Maryland** (Lee et al. 1980 et seq.); **Massachusetts** (Schmidt 1986; Hartel 1992; Hartel et al. 1996); **Montana** (Brown 1971; Cross et al. 1986; Holton 1990); **Nebraska** (Lee et al. 1980 et seq.; Cross et al. 1986); **Nevada** (La Rivers 1962; Deacon and Williams 1984); **New Hampshire** (Scarola 1973; Schmidt 1986); **New Jersey** (Morse 1905; Fowler 1906; Stiles 1978; Soldwedel, personal communication); **New Mexico** (Tyus et al. 1982; Sublette et al. 1990); **New York** (Smith 1985; Schmidt 1986); **North Dakota** (Lee et al. 1980 et seq.; Cross et al. 1986); **Oklahoma** (Miller and Robison 1973; Cross et al. 1986); **Oregon** (Lampman 1946; Bond 1994); **Pennsylvania** (Cooper 1983; Hocutt et al. 1986; Schmidt 1986); **Rhode Island** (Cooper 1983); **South Dakota** (Bailey and Allum 1962; Cross et al. 1986); **Texas** (Conner and Suttkus 1986; Cross et al. 1986; Howells 1991a); **Utah** (Sigler and Miller 1963; Tyus et al. 1982; Sigler and Sigler 1987); **Vermont** (Cox, personal communication); **Virginia** (Lee et al. 1980 et seq.; Hocutt et al. 1986; Jenkins and Burkhead 1994); **Washington** (Lampman 1946; Gray and Dauble 1977; Wydoski and Whitney 1979); **West Virginia** (Stauffer et al. 1995); **Wisconsin** (Becker 1983); and **Wyoming** (Baxter and Simon 1970; Hubert 1994; Stone 1995).

Means of Introduction: Its popularity as a sport fish has led to stockings throughout the west and northeast. Intentional stocking for sportfishing.

Status: Established in most or all states where it has been introduced.

Impact of Introduction: Unknown.

Remarks: Tyus et al. (1982) gave a distribution map of the this species in the upper Colorado basin.

HYBRIDS

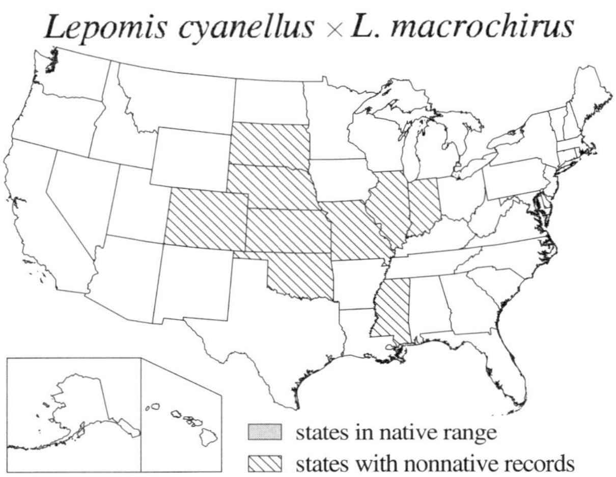

Lepomis cyanellus × *L. macrochirus*

Common Name: green sunfish × bluegill.

Identification: Childers (1967). Maximum size: 0.55 kg.

Native Range: Not applicable.

Nonindigenous Occurrences: Recorded in **Colorado** (Horak, personal communication); **Delaware** (Raasch and Altemus 1991); **Illinois**; **Indiana**; **Kansas** (Burr, personal communication); **Mississippi** (International Game Fish Association 1994); **Missouri** (Burr, personal communication); **Nebraska** (Jones 1963); **Oklahoma** (Burr, personal communication); and **South Dakota** (North Dakota Game and Fish Department 1994).

Means of Introduction: Either intentionally stocked as a sport fish or occurs as the result of stocking one or both parent species in an area.

Status: Established in Delaware. Reported from other states mentioned above.

Impact of Introduction: Unknown.

Remarks: These hybrids are capable of reproducing and backcrossing. Most progeny from such unions are males (92–97%) (Raasch and Altemus 1991).

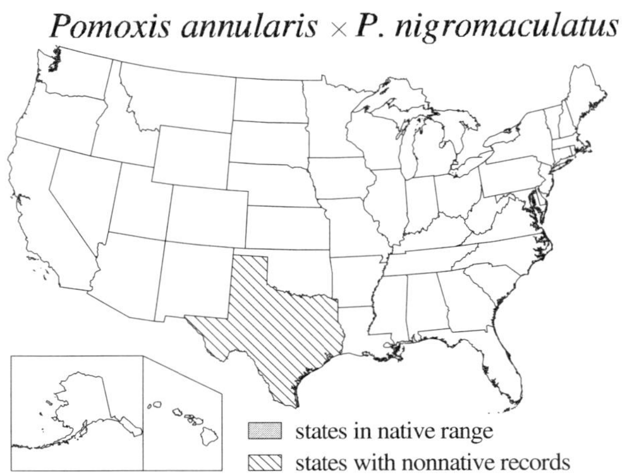

Pomoxis annularis × *P. nigromaculatus*

Common Name: crappie hybrid.

Identification: Hybrid. Maximum size: Probably similar to that of parent species.

Native Range: Not applicable.

Nonindigenous Occurrences: Victor Braunig Reservoir, Bexar County, **Texas**, 1987 (Howells 1991a).

Means of Introduction: Intentionally stocked as a sport fish.

Status: Reported from Texas.

Impact of Introduction: Unknown.

Remarks: None.

Percidae - Perches

The family Percidae, including the darters, perches, walleye, and sauger, comprises 10 genera and 162 species (Nelson 1994). The family is distributed naturally primarily in northern temperate zones, from most of eastern North America through Europe and northern Asia (Berra 1981). All but three of the more than 150 North American species are darters (Page and Burr 1991). Percids are small- to medium-sized fishes, ranging from less than 3 cm in length in the darters to about 90 cm in the largest perches. They are characterized by somewhat elongate, perciform bodies and two dorsal fins. Several species are important as game and food fishes, and many of the larger species have been introduced outside their native ranges. The family is restricted to freshwater.

Ammocrypta bifascia

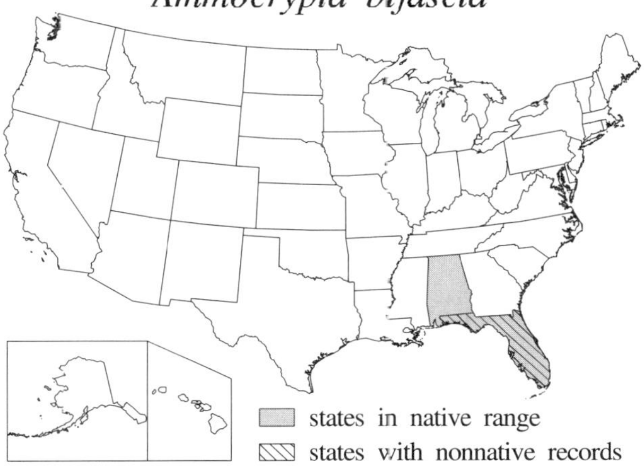

Ammocrypta bifascia Williams 1975

Common Name: Florida sand darter.

Identification: Williams (1975); Page (1983); Page and Burr (1991) (under the name *Etheostoma bifascia*); Mettee et al. (1996). Maximum size: 7.7 cm.

Native Range: Gulf Slope drainages from Choctawhatchee River east to Perdido River, southern Alabama and Florida panhandle (Williams 1975).

Nonindigenous Occurrences: Two specimens have been collected below Jim Woodruff Dam on the Apalachicola River in **Florida** (Lee et al. 1980 et seq.; Swift et al. 1986). One collection was from 1974, the other from 1977 (Starnes and Starnes 1979).

Means of Introduction: The introduction of the Florida sand darter was most likely via a bait bucket (Gilbert and J. D. Williams, unpublished data). While this highly translucent darter may not be used for bait, it easily may have been taken as by-catch by anglers seining for bait. It is common in streams of the Choctawhatchee River drainage, just west of the Apalachicola River, and anyone seining for bait minnows in streams with sandy bottom pools could have captured Florida sand darters.

Status: Established in Florida. Many more collections of Florida sand darter have been reported since the original two. The species is now thought to be fairly common in the Apalachicola River (Gilbert and J. D. Williams, personal communication).

Impact of Introduction: Unknown.

Remarks: Starnes and Starnes (1979) concluded that the species' presence in the Apalachicola was a natural occurrence and resulted from coastal dispersal. They argued against the possibility of a bait bucket release because the two known collections were temporally and geographically separated. However, there are several reasons to support the idea that the Florida sand darter's occurrence in this area resulted from human introduction. The species had never been taken during the many fish samplings conducted in the Apalachicola prior to 1974. The species is known to occur naturally west of the Apalachicola in the upstream reaches of streams draining the upper Coastal Plain. Its absence in the Chattahoochee and Flint rivers in Alabama and Georgia, two upstream tributaries of the Apalachicola River, tends to support a recent introduction.

Voucher specimens: Florida (UF 24415, UF 73263).

Ammocrypta clara

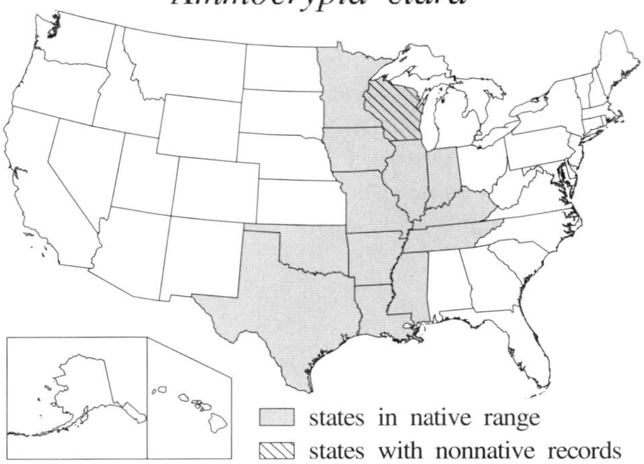

Etheostoma blenniodes

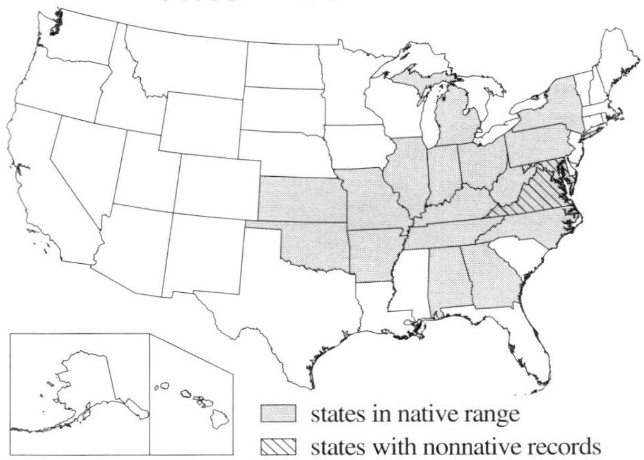

Ammocrypta clara Jordan and Meek 1885

Common Name: western sand darter.

Identification: Williams (1975); Becker (1983); Page and Burr (1991) under the name *Etheostoma clarum*; Etnier and Starnes (1993). Maximum size: 7.1 cm.

Native Range: Mississippi River basin from Wisconsin and Minnesota south to Mississippi and Sabine and Neches river drainages, Texas; also in the Green, Cumberland, and Tennessee river drainages (presumably extinct in the first two); generally sporadic and rare (Williams 1975; Page and Burr 1991).

Nonindigenous Occurrences: Two specimens have been taken from the Wolf system in the Lake Michigan drainage, **Wisconsin**; the first in 1960 from the Waupaca River (museum specimen; Becker 1983), and the second in 1973 from the Wolf River in Outagamie County (Becker 1983).

Means of Introduction: These fish likely gained access to the Wolf system via the Wisconsin-Fox Canal (Becker 1983).

Status: Reported in Wisconsin.

Impact of Introduction: Unknown.

Remarks: Voucher specimens: Wisconsin (UWSP 1155).

Etheostoma blenniodes Rafinesque 1819

Common Name: greenside darter.

Identification: Page and Burr (1991); Etnier and Starnes (1993); Jenkins and Burkhead (1994); Pflieger (1997). Maximum size: 17 cm.

Native Range: Great Lakes and Mississippi River basins from New York and Maryland to eastern Kansas and Oklahoma, and from Ontario south to Georgia, Alabama, and Arkansas; Atlantica Slope in Mohawk, and Susquehanna drainages, New York to Maryland (Page and Burr 1991).

Nonindigenous Occurrences: The greenside darter is probably introduced in the Potomac drainage in **Maryland** and **Virginia** (Jenkins and Burkhead 1994). It was not collected in this drainage until 1954 when the first specimens were taken in Maryland. The first collections from the Virginia portion of the drainage were in 1964. Its recent appearance combined with its rapid range expansion suggest an introduction (Jenkins and Burkhead 1994).

Method of Introduction: Unknown.

Status: Established.

Remarks: None.

Etheostoma caeruleum

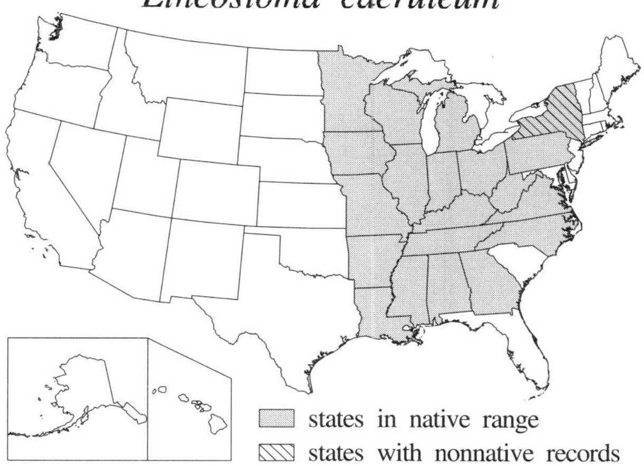

Etheostoma caeruleum Storer 1845

Common Name: rainbow darter.

Identification: Becker (1983); Page (1983); Page and Burr (1991); Etnier and Starnes (1993); Jenkins and Burkhead (1994). Maximum size: 7.7 cm.

Native Range: Great Lakes and Mississippi River basins from southern Ontario and western New York to Minnesota, and south to northern Alabama and Arkansas. Isolated populations in southwestern Mississippi and eastern Louisiana, and in upper Potomac River drainage (Atlantic Slope), West Virginia (Page and Burr 1991).

Nonindigenous Occurrences: The rainbow darter has been introduced into the Genesee River system of **New York**. The species is now common from Portageville to Caneadea and in Angelica and Caneadea creeks (Smith 1985). Hocutt et al. (1986) listed this species only as questionably native in the Potomac and the Kanawah drainage above the falls.

Means of Introduction: Unknown.

Status: The species is established in the Genesee River system, apparently due to a fairly recent introduction (late 1970s). The population was expanding explosively as of the early 1980s (Smith 1985).

Impact of Introduction: Unknown.

Remarks: Cross et al. (1986) reported the rainbow darter as introduced into the Middle Arkansas drainage. No records of this species were found from this drainage in Arkansas (Robison and Buchanan 1988), Oklahoma (Miller and Robison 1973), or Kansas (Cross 1967; Cross and Collins 1995). However, Cross (1967) mentioned that *E. caeruleum* had been reported in Kansas previously, but those records were based on misidentified *E. spectabile*.

Etheostoma cragini

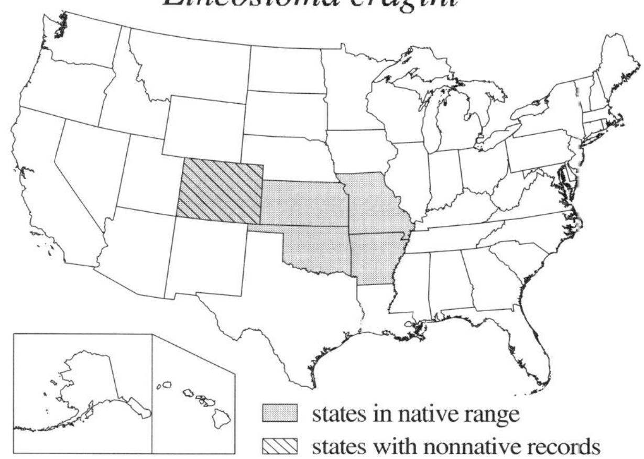

Etheostoma cragini Gilbert 1885

Common Name: Arkansas darter.

Identification: Robison and Buchanan (1988); Page and Burr (1991); Cross and Collins (1995); Pflieger (1997). Maximum size: 6 cm.

Native Range: Arkansas River drainage in southwestern Missouri, northwestern Arkansas, Oklahoma, Kansas, and Colorado (Page and Burr 1991).

Nonindigenous Occurrences: This species was introduced into the spring-fed inlet to Lytle Pond on the Fort Carson Military Reservation, El Paso County, **Colorado** in July 1980 (Miller 1984).

Means of Introduction: According to Miller (1984), the Lytle Pond introduction was carried out by the Colorado Division of Wildlife as part of an experimental transplant, apparently for the preservation of an imperiled or uncommon species. The introduction involved 34 specimens taken from a spring slough along Fountain (about 5 kilometers south of town of Fountain), El Paso County, Colorado.

Status: Apparently established in Colorado in a single site outside its native range. Numerous small darters were observed in Lytle Pond in October 1982; one male and six females, in reproductive condition, were trapped in March 1983. Miller (1984) indicated that a quantitative census has yet to be made.

Impact of Introduction: Unknown.

Remarks: Prior to introduction, the inlet and Lytle Pond were unoccupied by fish (Miller 1984). All, or most of, the Fort Carson Military Reservation in El Paso County is within the Fountain Creek system (Arkansas River drainage).

Etheostoma edwini

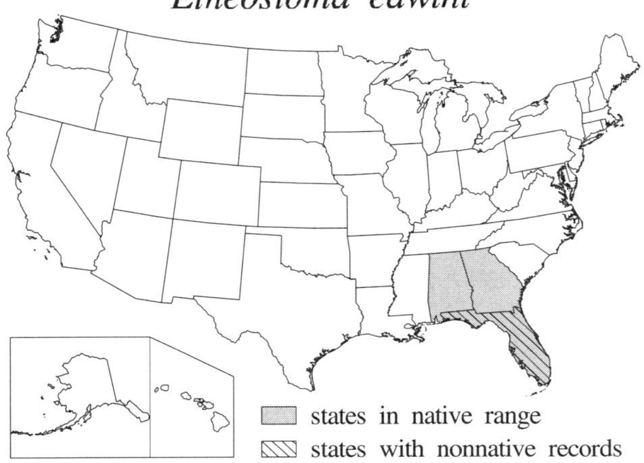

Etheostoma edwini (Hubbs and Cannon 1935)

Common Name: brown darter.

Identification: Page (1983); Page and Burr (1991); Mettee et al. (1996). Maximum size: 5.3 cm.

Native Range: St. Johns drainage, Florida, to Perdido River drainage, Alabama (Page and Burr 1991).

Nonindigenous Occurrences: The brown darter has been found in some tributaries of the Choctawhatchee Bay in Okaloosa and Walton counties, **Florida** (Lee et al. 1980 et seq.). The species was first recorded from this area in October 1964, when four individuals were collected from Swift Creek at the State Route 20 bridge (museum specimen). Since then, many more have been captured from the Rocky Bayou system. Only one individual has been reported from the Boggy Bayou system. On 5 May 1971, one specimen was captured from Juniper Creek at the State Route 85 bridge (museum specimen; H. Jelks, personal communication).

Means of Introduction: It is possible that the brown darter spread by natural dispersal through estuaries into new tributaries of Rocky Bayou. Introduction at Boggy Bayou was probably via bait bucket. Multiple bait bucket introductions may have occurred in this area (Burkhead and Williams 1992). Another possible means of introduction is stock contamination. In the 1950s and 1960s, certain sections of the Rocky Creek watershed were stocked with rainbow trout, largemouth bass, and smallmouth bass from the federal fish hatchery in Welaka, FL. The hatchery is located within the native range of the brown darter, and it is possible that brown darters were accidentally included in the shipment (Burkhead and Williams 1992).

Status: Established in lower reaches of Rocky Bayou stream systems; reported from Boggy Bayou. Populations fluctuate over a period of 3–5 years.

Impact of Introduction: The brown darter is replacing the Okaloosa darter *E. okaloosae* in portions of several creeks tributary to Choctawhatchee Bay (Burkhead, personal communication).

Remarks: It is debated whether the brown darters in Rocky Bayou represent introductions or a species that was overlooked.

Voucher specimens: (Northeast Louisiana University Museum of Zoology - NELUMZ 21364-5, UAIC 1493.56).

Etheostoma exile

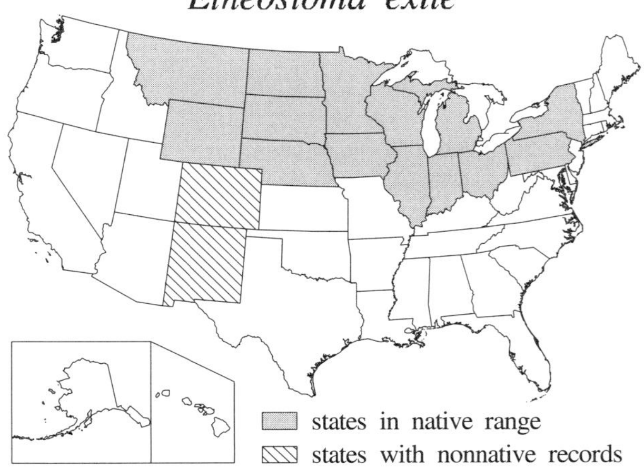

Etheostoma exile (Girard 1859)

Common Name: Iowa darter.

Identification: Becker (1983); Page (1983); Smith (1985); Woodling (1985); Page and Burr (1991). Maximum size: 7.2 cm.

Native Range: St. Lawrence–Great Lakes, Hudson Bay, and Mississippi River basins from southern Quebec to northern Alberta, and south to Ohio, Illinois, and Colorado (Page and Burr 1991).

Nonindigenous Occurrences: The Iowa darter was introduced into the headwaters of the Colorado River in Shadow Mountain Reservoir, **Colorado** (Tyus et al. 1982; Woodling 1985) and to the San Juan River in **New Mexico** (Lee et al. 1980 et seq.; Tyus et al. 1982).

Means of Introduction: Unknown.

Status: Reported in Colorado and New Mexico.

Impact of Introduction: Unknown.

Remarks: Sublette et al. (1990) imply the specimen the New Mexico record is based on may not be this species. They report the specimen as *Etheostoma* sp. and also mention the presence of introduced johnny darters *E. nigrum* upstream in the Animas drainage reported by Woodling (1985) (see *E. nigrum* account).

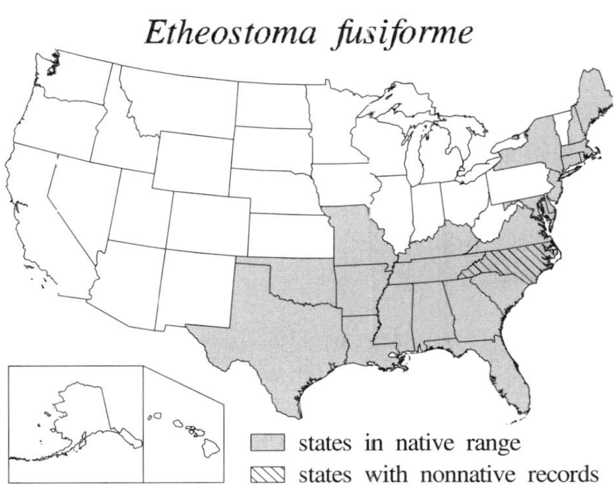

Etheostoma fusiforme (Girard 1854)

Common Name: swamp darter.

Identification: Page (1983); Smith (1985); Page and Burr (1991); Etnier and Starnes (1993); Jenkins and Burkhead (1994). Maximum size: 5.9 cm.

Native Range: Seaboard Lowlands, Atlantic and Gulf coastal plains from southern Maine to Louisiana (Sabine River) and southeastern Oklahoma (Red River); Former Mississippi Embayment north to Kentucky and southeastern Missouri (Page and Burr 1991). Collette (1962) gave a dot distribution map.

Nonindigenous Occurrences: This species was collected in the Tennessee River drainage from a pond near the French Broad River, approximately seven miles south of Asheville, **North Carolina** (Bailey 1950; Bailey et al. 1954; Collette 1962; Menhinick 1991; Page and Burr 1991).

Means of Introduction: Unknown.

Status: Established in North Carolina, at least from 1947–1952 (Bailey 1950; Collette 1962). Present status unknown.

Impact of Introduction: Unknown.

Remarks: This species was incorrectly considered introduced to a coastal stream west of the Thames River, Connecticut, by Schmidt and Whitworth (1979) (Schmidt 1986). Fish collected from this location were incorrectly described as a new subspecies, *Hololepis barratti apalachia* (Bailey 1950; Bailey et al. 1954). *Hololepis* is now considered a subgenus of *Etheostoma* and *barratti*, which was once a subspecies of *fusiforme,* was placed back in the species *fusiforme* (Collette 1962).

Etheostoma nigrum

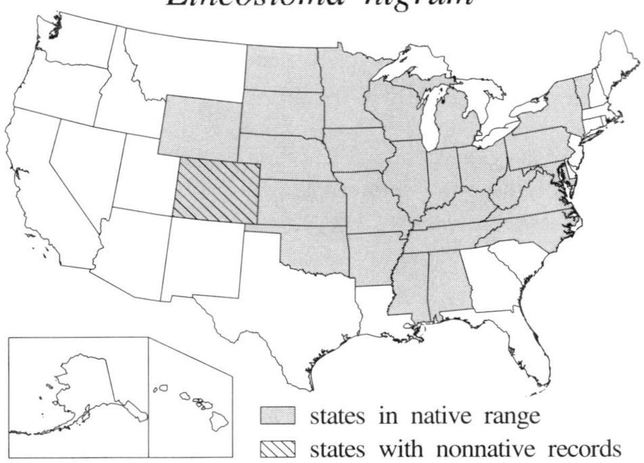

Etheostoma nigrum Rafinesque 1820

Common Name: johnny darter.

Identification: Becker (1983); Page (1983); Page and Burr (1991); Etnier and Starnes (1993); Jenkins and Burkhead (1994). Maximum size: 7.2 cm.

Native Range: St. Lawrence–Great Lakes, Hudson Bay, and Mississippi River basins from Hudson Bay to southern Mississippi, and from Quebec and Virginia to Saskatchewan and Colorado; on Atlantic Slope in James, Roanoke, Tar, and Neuse River drainages, Virginia and North Carolina; on Gulf Slope in Mobile Bay drainage, Alabama and Mississippi. Absent in White River drainage (Arkansas and Missouri), most of Arkansas River drainage, upper Tennessee River drainage, and middle Cumberland River drainage (Page and Burr 1991).

Nonindigenous Occurrences: This species was collected in the Colorado River from Lake Granby and Shadow Mountain Reservoir (Tyus et al. 1982) in Grand County and from Montezuma County, **Colorado** (Woodling 1985; Page and Burr 1991). Propst and Carlson (1986) reported that G. T. Baxter (University of Wyoming) believes that the johnny darter was not native to the North Platte drainage in Colorado.

Means of Introduction: Unknown.

Status: Reported from Colorado.

Impact of Introduction: Unknown.

Remarks: An unidentified introduced darter collected from the San Juan drainage in New Mexico may be this species (Sublette et al. 1990).

Etheostoma nuchale

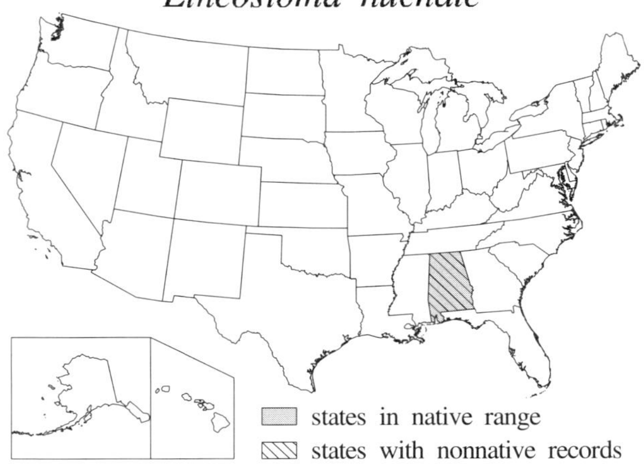

Etheostoma nuchale Howell and Caldwell 1965

Common Name: watercress darter.

Identification: Page (1983); Page and Burr (1991); Mettee et al. (1996). Maximum size: 5.4 cm.

Native Range: Thomas', Roebuck, and Glenn springs in the Black Warrior drainage, Jefferson County, Alabama (U.S. Fish and Wildlife Service 1992).

Nonindigenous Occurrences: The watercress darter was introduced into three springs in Jefferson County, **Alabama**: Prince Springs, a tributary of Valley Creek, approximately one air mile northwest of Glenn Springs; Tapawingo Springs (also known as Penney Springs), a tributary of Turkey Creek; and Avondale Springs (U.S. Fish and Wildlife Service 1992; M. Howell, personal communication).

Means of Introduction: The species was originally stocked to create a second population. At the time of the original introduction, the species was known only from Glenn Springs. That site was immediately adjacent to Jefferson County Route 20, and would have been imperiled by any road expansions. In 1965, 43 specimens from Glenn Springs were stocked in Prince Springs in hope that a second population would afford the species more security (U.S. Fish and Wildlife Service 1980a;

Howell, personal communication). The watercress darter was listed as a federally endangered species in 1970 (U.S. Fish and Wildlife Service 1980a). The species was later discovered at two other sites during a survey conducted between 1977 and 1978. During the period 1976–1977, all three native populations were at risk. Glenn Springs deteriorated because of sewage contamination; Thomas' Spring was to be dredged, the sportfish removed and then replaced with catfish; and the Roebuck Springs population contracted "gas bubble" disease (U.S. Fish and Wildlife Service 1980a). On 19 January 1988, 400 fish were transplanted from Roebuck Springs, 200 to Tapawingo Springs and 200 to Avondale Springs, as part of an endangered species recovery plan designed to save this species from extinction (U.S. Fish and Wildlife Service 1992).

Status: Established at Tapawingo Springs (Howell, personal communication). Failed at Prince Springs, possibly due to competition from redfin darters *E. whipplei* (U.S. Fish and Wildlife Service 1992). Sampling the year after the initial transplant failed to find any individuals. Additional sampling trips in the past ~30 years also have failed to find any individuals (Howell, personal communication). The transplant is believed to also have failed at Avondale Springs (Howell, personal communication). No individuals have been recovered; however, the site is difficult to sample (U.S. Fish and Wildlife Service 1992).

Impact of Introduction: Unknown.

Remarks: The watercress darter is listed as a federally endangered species. The native population at Thomas' Spring was decimated by removal of aquatic vegetation by an introduced grass carp. In 1981, or shortly thereafter, the spring was restocked with watercress darters from Glenn Springs. A new pond, owned by the U.S. Fish and Wildlife Service, was also constructed just downstream of Thomas' Spring in 1988 and stocked with 100 individuals from Thomas' Spring (U.S. Fish and Wildlife Service 1992). The species is established in this pond (Howell, personal communication).

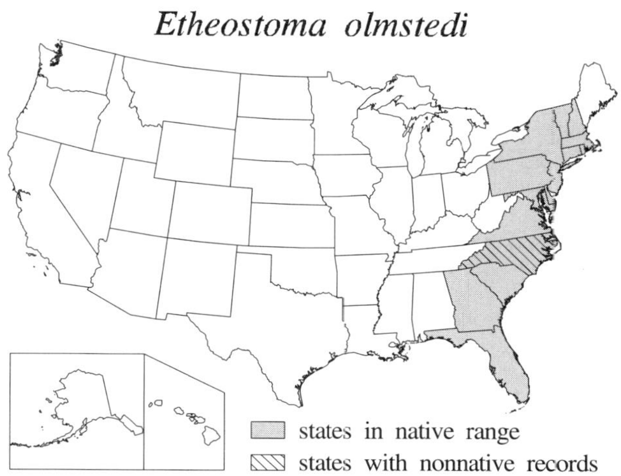

Etheostoma olmstedi Storer 1842

Common Name: tessellated darter.

Identification: Page (1983); Smith (1985); Page and Burr (1991); Jenkins and Burkhead (1994). Maximum size: 11 cm.

Native Range: Atlantic drainages from St. Lawrence River, Quebec and Ontario (absent in Maine) to Altamaha River, Georgia; St. Johns River drainage, Florida; Lake Ontario drainage, New York; mostly restricted to Coastal Plain in Virginia (Page and Burr 1991).

Nonindigenous Occurrences: A population of introduced tessellated darters was discovered in 1986 in the Little River system of the New drainage in **North Carolina**, just south of the Virginia border (Menhinick 1991; Jenkins and Burkhead 1994).

Means of Introduction: Unknown.

Status: Established in North Carolina.

Impact of Introduction: Unknown.

Remarks: None.

Etheostoma simoterum

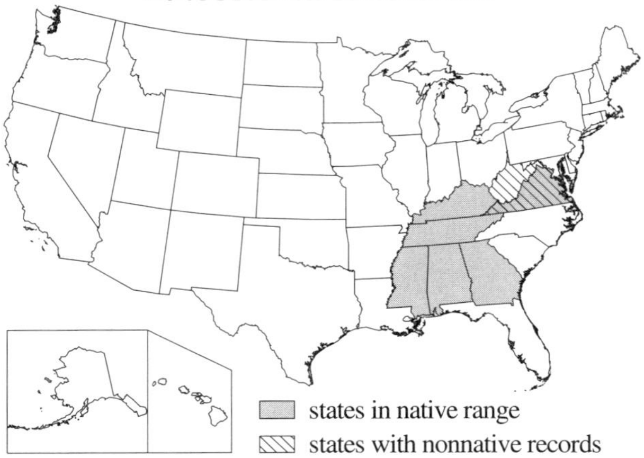

Etheostoma zonale

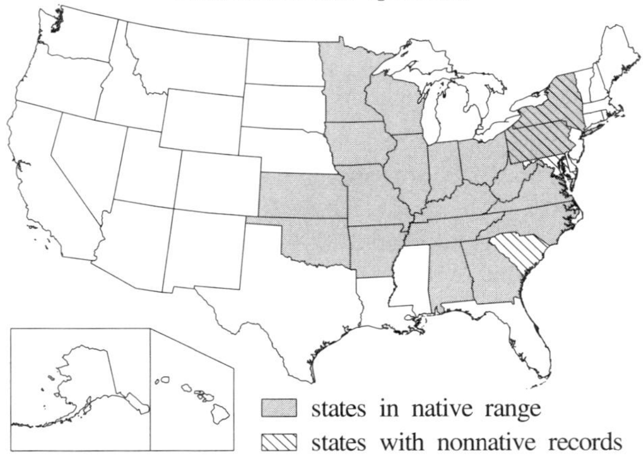

Etheostoma simoterum (Cope 1868)

Common Name: snubnose darter (Tennessee snubnose darter).

Identification: Page and Burr (1991); Etnier and Starnes (1993). Maximum size: 7.7 cm.

Native Range: Cumberland and Tennessee drainages in Virginia, North Carolina, Kentucky, Tennessee, Georgia, and Alabama (Page and Burr 1991). However, Menhinick (1991) did not list in North Carolina. The inclusion of North Carolina is probably based on Cope (1870) who listed the species in the state without comment.

Nonindigenous Occurrences: Hocutt et al. (1986) listed this species as probably introduced into the Kanawha (New) drainage above the falls. This drainage occurrs in **Virginia** and **West Virginia**. However, Jenkins and Burkhead (1994) list this species as native, but possibly introduced into the New drainage and to the Big Sandy drainage in Virginia. Stauffer et al. (1995) listed this species as probably present, expected, and possibly introduced into the Kanawha above the falls in West Virginia. Page and Burr (1991) also do not include the Kanawha drainage (part of the Ohio basin) in their description of the species' range. They restrict the range to only the Tennessee drainage.

Status: Established in Virginia. Unknown in West Virginia.

Impact of Introduction: Unknown.

Remarks: None.

Etheostoma zonale (Cope 1868)

Common Name: banded darter.

Identification: Becker (1983); Page (1983); Page and Burr (1991); Etnier and Starnes (1993); Jenkins and Burkhead (1994). Maximum size: 7.8 cm.

Native Range: Lake Michigan and Mississippi River basins from southwestern New York to Minnesota, and south to northern Georgia, northern Alabama, and southern Arkansas. Absent from Former Mississippi Embayment; Wabash River drainage of Ohio, Indiana, and Illinois; and streams of southern Illinois, southern Iowa, and northern Missouri (Page and Burr 1991).

Nonindigenous Occurrences: The banded darter has been introduced into the Susquehanna drainage (Hocutt et al. 1986) in **Maryland** (Cooper 1983), **New York** (Cooper 1983; Smith 1985), and **Pennsylvania** (Cooper 1983); and the Toxaway River in the Savannah drainage, **South Carolina** (Tsai and Raney 1974; Lee et al. 1980 et seq.; Page and Burr 1991). Swift et al. (1986) listed it as regarded as native, but possibly introduced in the Savannah.

Means of Introduction: Unknown. The species was introduced into the Susquehanna River in the late 1960s and is now the most abundant darter in many parts of the drainage (Raesly et al. 1990). Only a single individual was collected from South Carolina in July 1962, 1.9 mi from the North Carolina state line (M. Taylor, personal communication). Tsai and Raney (1974) stated that *E. zonale* may have entered the Savannah River drainage of the Atlantic Slope via stream capture.

Status: Established in Maryland, New York, and Pennsylvania. Reported from South Carolina.

Impact of Introduction: Introduced banded darters are hybridizing with native tessellated darters *E. olmstedi* in the Susquehanna River, Pennsylvania (Raesly et al. 1990).

Remarks: Menhinick (1991) did not list this species as occurring in North Carolina, and Dahlberg and Scott (1971a) report it was not collected in the Savannah drainage in Georgia.

Voucher specimen: South Carolina (TU 29485).

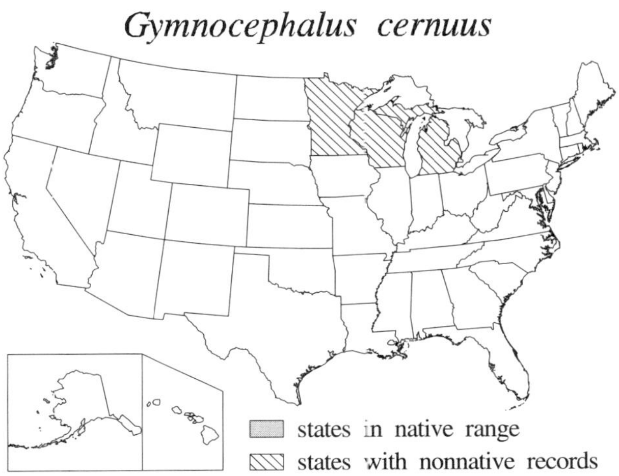

Gymnocephalus cernuus (Linnaeus 1758)

Common Names: ruffe (pope).

Identification: Distinguishing characteristics were provided by Wheeler (1969, 1978), Maitland (1977), Page and Burr (1991), McLean (1993), and Stepien et al. (1998). Detailed traits and an identification key to members of the genus were given by Holcik and Hensel (1974). Name given by some authors is *Acerina cernua* (e.g., Berg 1949), also *Gymnocephalus cernua* (e.g., Holcik and Hensel 1974). Maximum size: 25 cm.

Native Range: Northern Europe and Asia (Berg 1949; Holcik and Hensel 1974; Wheeler 1978; Page and Burr 1991).

Nonindigenous Occurrences: The ruffe was first collected in 1986 from the St. Louis River at the border of **Minnesota** and **Wisconsin** (Pratt et al. 1992). It subsequently spread into Duluth Harbor in Lake Superior and several tributaries of the lake (Underhill 1989; Scheidegger, personal communication; J. Slade, personal communication). In August 1994, it was found in Saxon Harbor, Wisconsin, and in the upper peninsula of **Michigan** at the mouths of the Black and Ontonagon rivers (K. Kindt, personal communication). In the lower Peninsula of Michigan along Lake Huron, the first three specimens were caught at the mouth of the Thunder Bay River in August 1995 (Kindt, personal communication).

Means of Introduction: Unintentional. The ruffe was probably introduced via ship ballast water discharged from a vessel arriving from a Eurasian port, possibly as early as 1982–1983 (Simon and Vondruska 1991; Ruffe Task Force 1992). Within the Great Lakes, the species spread may have been augmented by intra-lake shipping transport (Pratt et al. 1992; Stepien et al. 1998). Recent genetic research has indicated that the origin of ruffe introduced to the Great Lakes was southern Europe, not the Baltic Sea as previously believed (Stepien et al. 1998).

Status: Established in western portion of Lake Superior since about 1988 and expanding in an easterly direction. Reported from Lake Huron at Thunder Bay River, and in Thunder Bay, Lake Superior, Ontario, Canada. It has become the dominant species in the St. Louis River estuary (McLean 1993). Considered the most abundant of the 60 species found in Duluth Harbor (Ruffe Task Force 1992). Population in Duluth Harbor was estimated at two million adult fish in 1991 (Ruffe Task Force 1992).

Impact of Introduction: The ruffe has affected fish populations in other areas where introduced. In Scotland, native perch populations declined, and in Russia whitefish numbers have declined because of egg predation by ruffe (McLean 1993). Ruffe exhibit rapid growth and high reproductive output, and adapt to a wide range of habitat types (McLean 1993); therefore the species may pose a threat to native North American fish. Yellow perch *Perca flavescens*, emerald shiners *Notropis atherinoides*, and trout-perch *Percopsis omiscomaycus* have all declined since the introduction of this fish, although the association is not clear (McLean 1993). There is much concern that ruffe may have a detrimental effect on more desirable species in Lake Superior, such as yellow perch and walleye, by feeding on the young of these species (Raloff 1992), or by competing for food (McLean 1993). Savino and Kolar (1996) conducted a laboratory study to test for competition for food between ruffe and yellow perch. They found that competition could occur between the two species but that the outcome would not always be clear. Each species exhibited competitive advantages and disadvantages. Ogle et al. (1995) studied the diet of introduced ruffe inhabiting the St. Louis estuary. Their findings indicated that the species prey heavily on benthic insects thereby suggesting that ruffe compete for food with yellow perch, trout-perch, and other native benthic-feeding fishes.

Remarks: The ruffe also has been collected in the Canadian waters of Lake Superior at Thunder Bay and in Kaministiquia River estuary, 290 kilometers northeast of Duluth. Seven fish were collected from the latter location in 1991 (Ruffe Task Force 1992). Busiahn (1993) indicated that the potential North American range of ruffe may well extend from the Great Plains to the eastern seaboard and north into Canada. However, early reports that the ruffe was established in Lake Michigan (e.g., Page and Burr 1991) are considered erroneous. In March 1997, an international symposium was held in Ann Arbor, Michigan, to exchange information on the biology and management of ruffe (Jensen 1997). Ogle et al. (1996) found that certain native species preyed on introduced ruffe; however, their study indicated that predation is unlikely to effectively prevent ruffe from colonizing new areas in the Great Lakes.

Perca flavescens

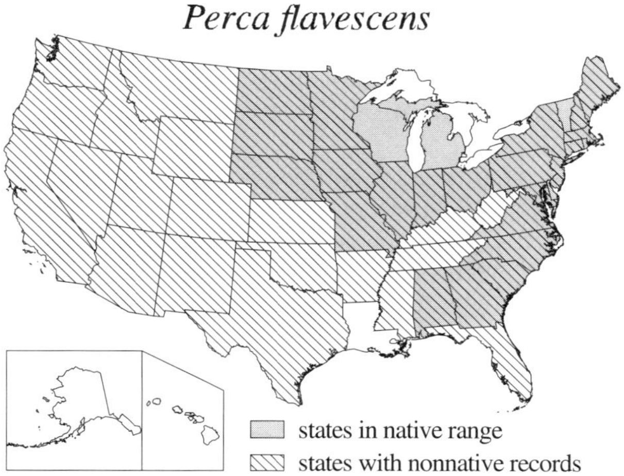

Perca flavescens (Mitchill 1814)

Common Names: yellow perch.

Identification: Becker (1983); Page and Burr (1991); Etnier and Starnes (1993); Jenkins and Burkhead (1994). Maximum size: 40 cm.

Native Range: Atlantic, Arctic, Great Lakes, and Mississippi River basins from Nova Scotia and Quebec west to Great Slave Lake, Northwest Territories, and south to Ohio, Illinois, and Nebraska; south in Atlantic drainages to Santee River, South Carolina (Page and Burr 1991). Goode (1884) reported the species east of the Alleghany Mountains as far south as Georgia.

Nonindigenous Occurrences: The yellow perch has been introduced into **Alabama** (Smith-Vaniz 1968; Boschung 1992; Mettee et al. 1996); **Arizona** (O'Malley 1920; Minckley 1973); **Arkansas** (O'Malley 1920); **California** (Smith 1896; Shebley 1917; Neale 1931; Moyle 1976a; Dill and Cordone 1997); **Colorado** (O'Malley 1919, 1920; Ellis 1974; Tyus et al. 1982; Propst and Carlson 1986); **Connecticut** (Johnson 1915; O'Malley 1919, 1920); **Florida** (Kilby et al. 1959; Yerger 1977; Lee et al. 1980 et seq.); **Georgia** (Dahlberg and Scott 1971a, 1971b; Yerger 1977); **Idaho** (Linder 1963; Idaho Fish and Game 1990); **Illinois** (Bowers 1906; Johnson 1915; O'Malley 1919, 1920; Smith 1979; Burr and Page 1986); **Indiana** (Bowers 1906; Johnson 1915; O'Malley 1919, 1920; Nelson and Gerking 1968); **Iowa** (Titcomb 1904; Bowers 1906; Johnson 1915; O'Malley 1919; O'Malley 1920); **Kansas** (O'Malley 1919; Cross 1967; Cross et al. 1986); **Kentucky** (Johnson 1915; O'Malley 1919, 1920; Burr and Warren 1986); **Maine** (O'Malley 1920); **Maryland** (Bowers 1906; Johnson 1915; O'Malley 1919, 1920; Lee et al. 1976); **Massachusetts** (Bowers 1906; Johnson 1915; O'Malley 1919, 1920); **Minnesota** (O'Malley 1919; Johnson 1915); **Mississippi** (Bowers 1905; O'Malley 1919; Mettee et al. 1996); **Missouri** (Johnson 1915; O'Malley 1919, 1920; Pflieger 1997); **Montana** (Brown 1971; Cross et al. 1986; Whitmore 1997); **Nebraska** (Smith 1896; O'Malley 1919; Lee et al. 1980 et seq.; Cross et al. 1986); **Nevada** (Miller and Alcorn 1946; La Rivers 1962; Deacon and Williams 1984); **New Jersey** (Johnson 1915; O'Malley 1919); **New Mexico** (Johnson 1915; Sublette et al. 1990); **New York** (Bean 1903; Bowers 1906; Johnson 1915; O'Malley 1920); **North Carolina** (Johnson 1915; O'Malley 1919; Menhinick 1991; Jenkins and Burkhead 1994); **North Dakota** (Titcomb 1904; Johnson 1915; O'Malley 1919, 1920); **Ohio** (Trautman 1981; Burr and Page 1986; Hocutt et al. 1986); **Oklahoma** (Johnson 1915; O'Malley 1919, 1920; Hall 1956; Miller and Robison 1973); **Oregon** (Lampman 1946; Bond 1994); **Pennsylvania** (Johnson 1915; O'Malley 1919, 1920); **South Carolina** (Johnson 1915; Dahlberg and Scott 1971a); **South Dakota** (Johnson 1915; O'Malley 1920; Bailey and Allum 1962); **Tennessee** (Lee et al. 1980 et seq.; Starnes and Etnier 1986; Ryon and Loar 1988; Etnier and Starnes 1993); **Texas** (Hubbs et al. 1991); **Utah** (Sigler and

Miller 1963; Tyus et al. 1982); **Vermont** (Bowers 1906; Johnson 1915; O'Malley 1919, 1920); **Virginia** (Bowers 1906; Johnson 1915; O'Malley 1919, 1920; Hocutt et al. 1986; Jenkins and Burkhead 1994); **Washington** (Smith 1896; Johnson 1915; Chapman 1942; Lampman 1946; Gray and Dauble 1977; Wydoski and Whitney 1979); **West Virginia** (Hocutt et al. 1986; Stauffer et al. 1995); and **Wyoming** (Baxter and Simon 1970; Hubert 1994).

Means of Introduction: Intentionally stocked as a food fish and for sportfishing. Some of the earliest introductions were carried out by the U.S. Fish Commission during the late 1800s (Smith 1896). In 1873, Livingston Stone of the Fish Commission attempted to transport yellow perch taken from a Vermont river to California, but the train car carrying the 110 yellow perch, and many other species, was lost when a bridge collapsed crossing the Elkhorn River in Nebraska. The Fish Commission was successful in stocking this species into western waters during the 1890s (Smith 1896; Dill and Cordone 1997, and references therein). A single fish collected from the upper Tombigbee River system probably gained access to the river via the Tennessee-Tombigbee Waterway (Mettee et al. 1996).

Status: Established in most areas where introduced. Extirpated in Arkansas.

Impact of Introduction: Yellow perch compete with trout for food and likely prey on young trout (Coots 1966).

Remarks: Although Yerger (1977) and Lee et al. (1980 et seq.) considered this species to be introduced into the Apalachicola River in Florida, it may actually be native. Because the Apalachicola drainage was not adequately sampled in early years, introduced status may be incorrectly assumed for some species. Yellow perch was first collected in the Mobile basin circa 1850, long before any stocking took place, indicating the species was native to this region. This record was overlooked by Lee et al. (1980 et seq.). Hence, *Perca flavescens* may be native to eastern Gulf drainages in Alabama, Florida, and Georgia, not introduced. Populations in Atlantic coastal drainages of Georgia, such as the Altamaha and Savannah, have been reported (Dahlberg and Scott 1971b) to be introduced. In fact we believe these populations are more likely native. Dill and Cordone (1997) gave a detailed history of this species and its introduction into California.

Percina caprodes

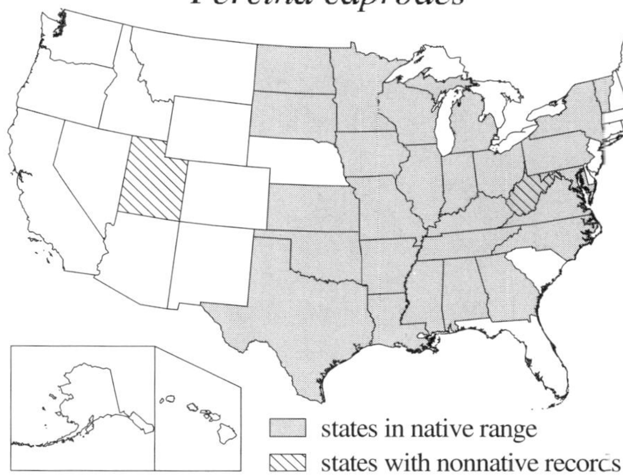

Percina caprodes (Rafinesque 1818)

Common Name: logperch.

Identification: Becker (1983); Page (1983); Page and Burr (1991); Etnier and Starnes (1993); Jenkins and Burkhead (1994); three subspecies are recognized. Maximum size: 18 cm.

Native Range: St. Lawrence–Great Lakes, Hudson Bay, and Mississippi River basins from eastern Quebec to Saskatchewan and south to Gulf; Atlantic Slope drainages from Hudson River, New York, to Potomac River, Maryland; Gulf Slope drainages from Alabama to Mississippi River, Louisiana. *P. c. semifasciata* in Canada and northern states; *P. c. fulvitaenia* in tributaries of Missouri and Arkansas rivers in Missouri, Arkansas, Kansas, and Oklahoma, and in Blue River, Oklahoma; *P. c. caprodes* throughout rest of range (Page and Burr 1991). The Florida population was described by Thompson (1995) as *P. austroperca*.

Nonindigenous Occurrences: The logperch was introduced into Willard Bay Reservoir, **Utah**, in 1983 (Sigler and Sigler 1987). It may have been introduced into **West Virginia** in the Kanawha drainage above the falls (Stauffer et al. 1995).

Means of Introduction: The introduction in Utah was probably the result of stocking for forage.

Status: Established in Utah (Sigler and Sigler 1996), and West Virginia.

Impact of Introduction: Unknown.

Remarks: Reports in the literature of this species being introduced to California (Shapovalov et al. 1959; McKechnie 1966c; Farley 1972; and Moyle et al. 1974a), actually refer to *P. macrolepida*. Although this species was reportedly introduced into Colorado, the species was later re-identified as *P. macrolepida* (Behnke, personal communication via S. Flickinger). The Utah introduction may also be *P. macrolepida*; specimens should be re-examined.

Percina macrolepida

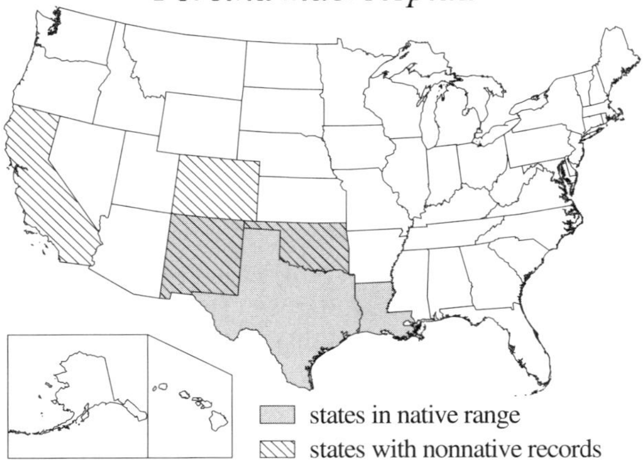

Percina macrolepida Stevenson 1971

Common Name: bigscale logperch.

Identification: Sturgess (1976); Page (1983); Sublette et al. (1990); Hubbs et al. (1991); Page and Burr (1991). Maximum size: 11 cm.

Native Range: From Sabine River, Louisiana, and Red River, Oklahoma, to Rio Grande drainage, Texas, New Mexico, and Mexico (Page and Burr 1991). Stevenson and Thompson (1978) reviewed the distribution of this species.

Nonindigenous Occurrences: This species was originally introduced into the Sacramento River drainage in Miller, Blackweld, and Polk lakes on Beale Air Force Base, Yuba County, **California** in 1953 (McKechnie 1966c). It is established in the Sacramento–San Joaquin basin as well as in Silverwood, Pyramid, Castaic, and Irvine lakes in southern California since 1988 (Moyle 1976a; Page and Burr 1991; Swift et al. 1993). Although the species was reported as *P. caprodes* (Ellis 1974; Wiltzius 1985), it was actually *P. macrolepida* that was introduced into Two Buttes Reservoir, Baca County, **Colorado** (Behnke, personal communication via S. Flickinger). The bigscale logperch has also been introduced into Ute Reservoir in Quar County, **New Mexico** (Sublette et al. 1990). Cashner and Matthews (1988) suspected that this species was introduced into several sites in **Oklahoma**, including the Washita River in Custer County, the North Canadian River in Woodward County, Fort Supply Lake in Woodward County, and possibly into Lake Optima in Texas County (Cashner and Matthews 1988).

Means of Introduction: Unintentional. The bigscale logperch was introduced into California with contaminated stock from the Trinity River, Texas, in 1953 (Moyle 1976a; Dill and Cordone 1997). These introductions were carried out by the U.S. Fish and Wildlife Service during intentional stocking of largemouth bass and bluegill (McKechnie 1966c). This percid subsequently entered the Del Valle Reservoir, Alameda County, California, via the Tracy pumping plant and the South Bay Aqueduct (Farley 1972; Moyle et al. 1974a; Dill and Cordone 1997). It was introduced by bait bucket into New Mexico (Sublette et al. 1990). The Colorado introduction was the result of stock contamination of fishes transplanted from Texas prior to 1952 (Ellis 1974; Wiltzius 1985).

Status: Established and expanding in California. Since its original introduction, the bigscale logperch has extended its range southward into the San Joaquin drainage (Farley 1972; Dill and Cordone 1997); and northward into the Yuba, Sacramento and Feather rivers (Boles 1976). Established in New Mexico, and presumably in Oklahoma. Probably extirpated in Colorado. Two Buttes Reservoir has been periodically dry since 1979 (C. Bennett, personal communication).

Impact of Introduction: Largely unknown. The species reportedly feed on eggs of centrarchids (Moyle et al. 1974a).

Remarks: Lee et al. (1980 et seq.) considered the Washita River, Oklahoma population to be native. They did not map the other Oklahoma sites. Specimens from Lake Optima, Oklahoma have not been positively identified but are reportedly logperch and well outside the native range of both *P. caprodes* and *P. macrolepida* (Cashner and Matthews 1988). Prior to 1976 many authors including Shapovalov et al. (1959), McKechnie (1966c), Farley (1972), and Moyle et al. (1974a), referred to introductions of this species and reported them as *P. caprodes*. Since that time *P. macrolepida* was described and has been split from *P. caprodes*. Sturgess

(1976) determined the species in California was actually *P. macrolepida* and not *P. caprodes*. Moyle et al. (1974a) provided a map showing the range of the bigscale logperch in California and Dill and Cordone (1997) gave an overview of the species' history in that state.

Percina maculata

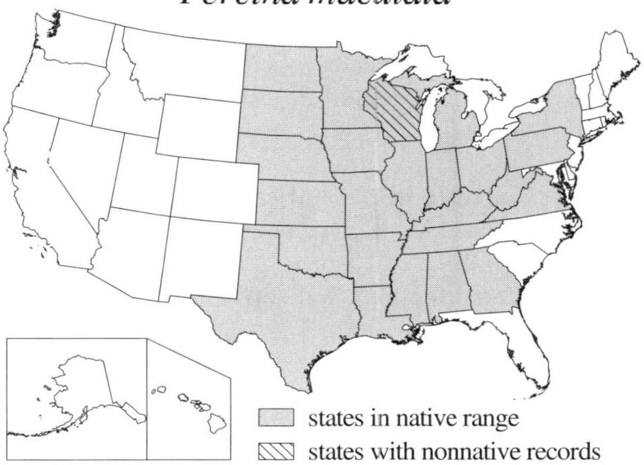

Percina maculata (Girard 1859)

Common Name: blackside darter.

Identification: Becker (1983); Smith (1985); Page and Burr (1991); Etnier and Starnes (1993). Maximum size: 11 cm.

Native Range: Great Lakes, Hudson Bay, and Mississippi River basins from southern Ontario and New York to southeastern Saskatchewan and south to Louisiana; Gulf drainages from Mobile Bay, Alabama to Calcasieu River, Louisiana (Page and Burr 1991). Populations in North Carolina have been described as *P. gymnocephala*.

Nonindigenous Occurrences: Introduced into the upper St. Croix-Namekagon and the upper Red Cedar systems in **Wisconsin** (Becker 1983).

Means of Introduction: Unknown; possible bait bucket release; although Becker (1983) reported it is not a bait fish.

Status: Collected only once from the upper Red Cedar drainage prior to 1935 (Becker 1983). Recently introduced into the upper St. Croix-Namekagon (Becker 1983), where it is apparently established.

Impact of Introduction: Unknown.

Remarks: None.

Percina roanoka

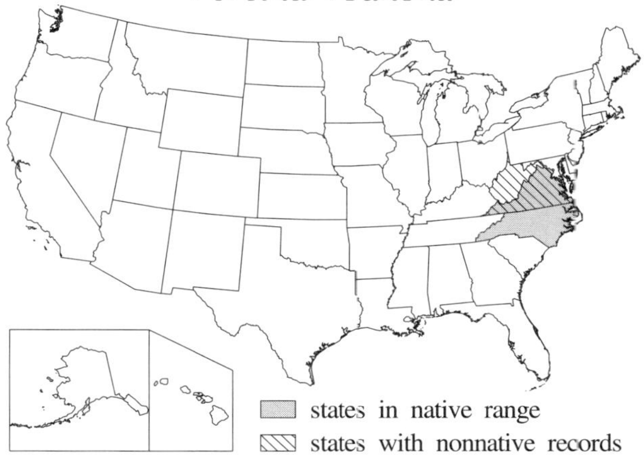

Percina roanoka (Jordan and Jenkins 1889)

Common Name: Roanoke darter.

Identification: Page (1983); Page and Burr (1991); Jenkins and Burkhead (1994). Maximum size: 7.8 cm.

Native Range: Roanoke, Neuse, and Tar River drainages, Virginia and North Carolina (Page and Burr 1991).

Nonindigenous Occurrences: The Roanoke darter was introduced into the James and New River drainages in **Virginia** (Hocutt et al. 1986; Page and Burr 1991; Jenkins and Burkhead 1994; Stauffer et al. 1995). The population in the James River was originally considered to be native; however, due to its range expansion it is now considered to be introduced (Jenkins and Burkhead 1994). Craig Creek is the probable site of introduction. Jenkins and Burkhead (1994) gave a detailed account of the expansion of this species in the James River. The Roanoke darter was first taken in the New River in Virginia in 1963 or 1964 midway between Claytor Dam and Bluestone Reservoir (Jenkins and Burkhead 1994). This species also is introduced in the New River drainage in **West Virginia**, where it spread downstream from Virginia (Jenkins and Burkhead 1994; Stauffer et al. 1995). It was first taken in West Virginia below Bluestone Reservoir in 1970 (Jenkins and Burkhead 1994).

Means of Introduction: Unknown.

Status: Established in Virginia and West Virginia. It is now the most abundant darter for a 20-mi stretch on the New River below Bluestone Reservoir (Jenkins and Burkhead 1994).

Impact of Introduction: Unknown.

Remarks: Voucher specimens: Virginia (USNM 194739, 194784, 230963).

Percina shumardi

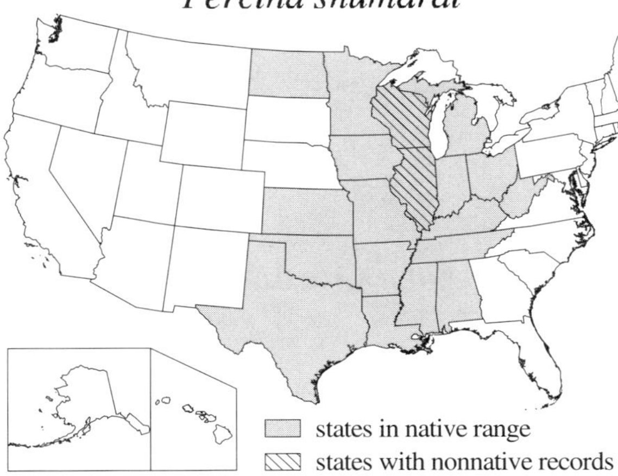

Percina shumardi Girard 1859

Common Name: river darter.

Identification: Becker (1983); Page and Burr (1991); Etnier and Starnes (1993). Maximum size: 7.8 cm.

Native Range: Hudson Bay basin, Ontario, Manitoba, North Dakota, and south in Great Lakes and Mississippi River basins to Louisiana; Gulf drainages from Mobile Bay, Alabama, to Neches River, Texas; isolated population in San Antonio drainage, Texas (Page and Burr 1991). Although the species has been recorded from the Tennessee drainage in northeastern Tennessee, it has not been reported in Virginia (Jenkins and Burkhead 1994), North Carolina (Menhinick 1991), or Georgia (Dahlberg and Scott 1971a).

Nonindigenous Occurrences: The river darter was collected from the city of Chicago's water system prior to 1926. The city's water source, and hence the source of the fish, was off-shore in Lake Michigan, **Illinois** (Becker 1983). It was introduced into the Lake Michigan drainage in **Wisconsin**, in the Wolf and Waupaca rivers and in lakes Winnebago and Poygan (Becker 1983).

Means of Introduction: The river darter may have found its way to Illinois waters of Lake Michigan through the Chicago Sanitary and Ship Canal. The species likely gained access to the Lake Michigan drainage in Wisconsin via the Wisconsin-Fox Canal at Portage, Wisconsin (Becker 1983). It was first reported from this area in 1963 from the western shore of Lake Winnebago (Becker 1983). This species was not found in this drainage in surveys conducted in the late 1920s (Becker 1983).

Status: Collected in Lake Michigan, Illinois. Established in the Lake Michigan drainage, Wisconsin.

Impact of Introduction: Unknown.

Remarks: None.

Percina tanasi

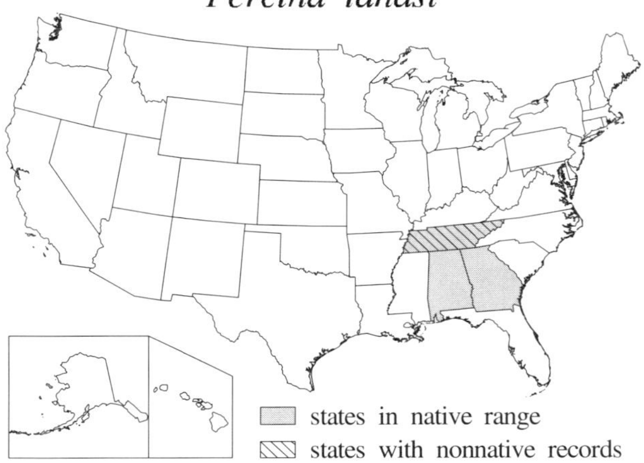

Percina tanasi Etnier 1976

Common Name: snail darter.

Identification: Page (1983); Page and Burr (1991); Etnier and Starnes (1993). Maximum size: 9 cm.

Native Range: Upper Tennessee drainage of eastern Tennessee, northern Georgia, and Alabama, including Little Tennessee River, Watts Bar Reservoir below Fort Loudon Dam, South Chickamauga Creek, lower portion of Big Sewee Creek in Meigs County, lower Sequatchie

River in Marion County, Little River in Blount County, lower French Broad River in Sevier County, and lower Paint Rock River in Madison County, Alabama (Page and Burr 1991; Etnier and Starnes 1993).

Nonindigenous Occurrences: Introduced in **Tennessee** into the Hiwassee River, Polk County; the Nolichucky River; the lower Holston River; and the middle Elk River (Lee et al. 1980 et seq.; U.S. Fish and Wildlife Service 1982a; Etnier and Starnes 1993).

Means of Introduction: The snail darter was intentionally introduced to create an additional population of this endangered fish when its only known habitat was threatened by construction of a dam. Seven hundred and ten snail darters were introduced into the Hiwassee River from June 1975 to February 1976 (U.S. Fish and Wildlife Service 1982a; Etnier and Starnes 1993). In October 1975, 61 were introduced into the Nolichucky River. Introductions into the Nolichucky River were halted when the sharphead darter *Etheostoma acuticeps* was discovered there, for fear the introduction would jeopardize this rare species (U.S. Fish and Wildlife Service 1982a; Etnier and Starnes 1993). The Holston River was stocked with 533 snail darters from the Hiwassee and Little Tennessee rivers during the period 1978 to 1979. The Elk River was stocked in July 1980 with 425 snail darters from the Little Tennessee River (U.S. Fish and Wildlife Service 1982a).

Status: Established in Hiwassee River and range expanding (U.S. Fish and Wildlife Service 1982a). One darter observed in Nolichucky River in 1980. Single individual possibly from small reproducing population or escapee from fish hatchery upstream; none found since. Status of Holston River and Elk River populations uncertain (U.S. Fish and Wildlife Service 1982a; Etnier and Starnes 1993). Snail darters found in lower French Broad and lower Holston rivers in 1988 and 1989 presumably represent progeny of Holston River transplants (Etnier and Starnes 1993).

Impact of Introduction: Unknown.

Remarks: The snail darter was listed as a federally endangered species in 1975 (U.S. Fish and Wildlife Service 1982a). After several more populations were discovered in the early 1980s, the species' status was changed to threatened in 1984 (Etnier and Starnes 1993).

Stizostedion canadense

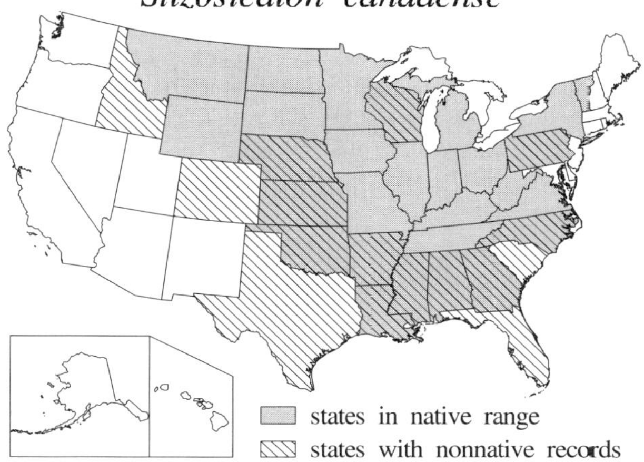

Stizostedion canadense (Smith 1834)

Common Names: sauger.

Identification: Becker (1983); Page and Burr (1991); Etnier and Starnes (1993); Jenkins and Burkhead (1994). Maximum size: 76 cm.

Native Range: St. Lawrence–Great Lakes, Hudson Bay, and Mississippi River basins from Quebec to Alberta, and south to northern Alabama and Louisiana (Page and Burr 1991).

Nonindigenous Occurrences: This species has been recorded from the Alabama and Chattahoochee rivers in **Alabama** (Lee et al. 1980 et seq.); the Arkansas River, lower White River, and Greers Ferry Lake in **Arkansas** (Lee et al. 1980 et seq.); the Arkansas drainage in **Colorado** (Walker 1993; Horak, personal communication); the Apalachicola River, **Florida** (Yerger and Beecher 1975; Yerger 1977); the Chattahoochee and Savannah rivers in **Georgia** (Dahlberg and Scott 1971b; Yerger and Beecher 1975; Yerger 1977; Lee et al. 1980 et seq.); the Bear River drainage in **Idaho** (Idaho Fish and Game 1990); several state lakes and reservoirs in **Kansas** (Cross and Collins 1995); the Mississippi and several other rivers in **Louisiana**; the Mississippi and Pearl rivers in **Mississippi**; the Niobrara and Platte rivers in **Nebraska** (Jones 1963); the Catawba River in **North Carolina** (Menhinick 1991); Arkansas River impoundments in **Oklahoma** (Lee et al. 1980 et seq.); the Youghiogheny, lower Allegheny, and Beaver rivers in **Pennsylvania** (Cooper 1983); the Savannah River in **South Carolina** (Dahlberg and Scott 1971b; Lee et al. 1980 et seq.); Lake Texoma in **Texas** (R. J. Edwards and Hubbs, personal communication); and Waukesha County and the Lake Michigan drainage in **Wisconsin** (Becker 1983).

Means of Introduction: Intentionally stocked for sportfishing in most areas. Introductions in the Lake Michigan drainage of Wisconsin are likely due to an canal connection linking the Wisconsin drainage to the Fox drainage.

Status: Established in Georgia, Oklahoma, Pennsylvania, South Carolina, Texas and Wisconsin. Reported from Florida and Georgia. Extirpated in North Carolina. Unknown in other areas.

Impact of Introduction: Unknown.

Remarks: None.

Stizostedion lucioperca

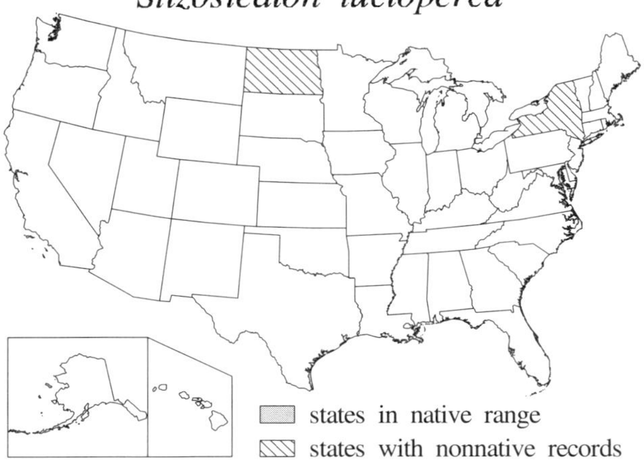

Stizostedion lucioperca (Linnaeus 1758)

Common Name: zander.

Identification: Berg (1949), Wheeler (1969, 1978), Maitland (1977), Howells (1992b). Formerly known as *Lucioperca lucioperca* (e.g., Berg 1949). Maximum size: 1 m (Robins et al. 1991b).

Native Range: Continental Europe to western Siberia (Berg 1949; Robins et al. 1991b).

Nonindigenous Occurrences: This species was reportedly stocked in ponds near Cooperstown, **New York**, in the late 1970s by an individual from Germany (J. Nickum, personal communication). Hatchery-reared zander fingerlings were stocked into Spiritwood Lake (Stutzman County), **North Dakota** in 1989 (Lohman 1989; Anderson 1992).

Means of Introduction: Illegally stocked in New York by an individual carrying fish brought from Germany (Nickum, personal communication). Zander fingerlings were intentionally stocked into one lake by the North Dakota Game and Fish Department biologists in 1989 to research the suitability for stocking zander in this country as a sport fish; the fish were imported as fry from Finland (Lohman 1989; Anderson 1992). However, only one survivor was recovered during a subsequent netting survey (Anderson 1992).

Status: Zander stocked into a North Dakota lake reportedly did not survive (e.g., Anderson 1992). Reported from New York (Courtenay et al. 1988).

Impact of Introduction: Unknown.

Remarks: Courtenay et al. (1986) listed this species from New York, but the record was based on an unconfirmed report. The history of its introduction into North Dakota is not well documented in the scientific literature. Apparently the North Dakota Game and Fish Department had been interested in zander as a sport fish for many years and that agency chose Spirit Lake as the site of an experimental release because the water body was completely enclosed (Anderson 1992). In 1987, prior to the lake introduction, the state had hatched eggs imported from Holland, but the resulting fry were destroyed for fear that they carried pike fry rhobdo virus (Anonymous 1987a; Lohman 1989). Those wanting to introduce zander thought that it would be a boon to the fisheries of North America (e.g., Anderson 1992), whereas others expressed strong reservations (e.g., Wright 1992). Some fisheries personnel in states surrounding North Dakota and nearby Canadian provinces expressed doubts concerning the species' introduction, particularly because its effect on native species was unknown and because of its potential to spread (e.g., Wingate 1992). The zander has been widely introduced into western Europe and the species was illegally introduced into portions of England. According to Hickley (1986), the success of introduced populations seemed to be limited by the availability of the species' preferred habitat, characterized as "eutrophic, turbid, well oxygenated and of low mean depth, and, if a river, slow-flowing rather than turbulent." Zander feed heavily on prey of small size. Because of this, there is concern among European fish resource managers that introduced zander may cause a collapse in resident prey fish stocks (Hickley 1986 and references therein).

There are no known voucher specimens.

Stizostedion vitreum

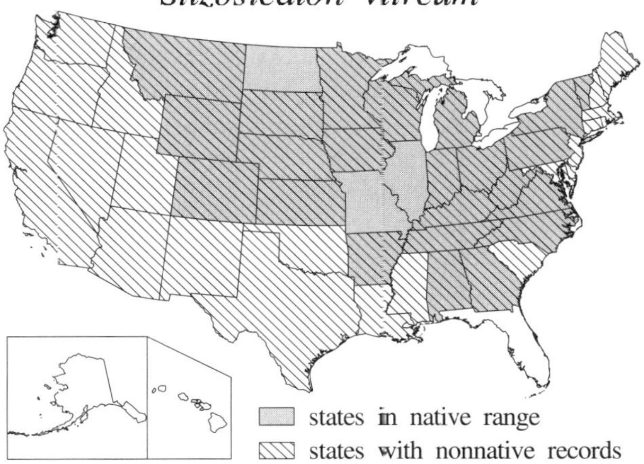

Stizostedion vitreum (Mitchill 1818)

Common Names: walleye (walleyed pike).

Identification: Becker (1983); Page and Burr (1991); Etnier and Starnes (1993); Jenkins and Burkhead (1994). Maximum size: 91 cm.

Native Range: St. Lawrence–Great Lakes, Arctic, and Mississippi River basins from Quebec to Northwest Territories, and south to Alabama and Arkansas (Page and Burr 1991).

Nonindigenous Occurrences: This species has been introduced into **Alabama** (Mettee 1978; Lee et al. 1980 et seq.; Swift et al. 1986; Billington and Maceina 1997); **Arizona** (Minckley 1973; Tyus et al. 1982); **Arkansas** (Robison and Buchanan 1988; Cross et al. 1986); **California** (Smith 1896; Shebley 1917; Hubbs et al. 1979; Dill and Cordone 1997); **Colorado** (Vanicek et al. 1970; Everhart and Seaman 1971; Holden and Stalnaker 1975; Tyus et al. 1982); **Connecticut** (Whitworth et al. 1968; Schmidt 1986; Whitworth 1996; Jones, personal communication); **Delaware** (Raasch, personal communication); **Georgia** (Dahlberg and Scott 1971b; Yerger 1977; Swift et al. 1986); **Idaho** (Linder 1963; Idaho Fish and Game 1990); **Indiana** (Sweeney 1902); **Iowa** (Harlan et al. 1987); **Kansas** (Cross 1967); **Kentucky** (Burr and Warren 1986); **Louisiana** (Douglas 1974); **Maine** (Kendall 1914a); **Massachusetts** (Schmidt 1986; Hartel 1992; Hartel et al. 1996); **Maryland** (Lee et. al. 1976; Davis, personal communication); **Michigan** (Hubbs and Lagler 1947); **Minnesota** (Eddy and Underhill 1974); **Montana** (Brown 1971; Cross et al. 1986); **Nebraska** (Jones 1963); **New Hampshire** (Hoover 1936; Scarola 1973); **New Jersey** (Nelson 1890); **New Mexico** (Sublette et al. 1990); **Nevada** (Deacon and Williams 1984); **North Carolina** (Hocutt et al. 1986; Menhinick 1991); **Ohio** (Trautman 1981); **Oklahoma** (Hall 1956; Miller and Robison 1973; Cross et al. 1986); **Oregon** (Lee et al. 1980 et seq.; Bond 1994; McMahon and Bennett 1996); **Pennsylvania** (Ford et al. 1892); **South Carolina** (Dahlberg and Scott 1971b; Loyacano 1975); **South Dakota** (Bailey and Allum 1962); **Tennessee** (Etnier and Starnes 1993; Jenkins and Burkhead 1994); **Texas** (Prentice 1977; Conner and Suttkus 1986; Cross et al. 1986); **Utah** (Sigler and Miller 1963; Vanicek et al. 1970; Tyus et al. 1982; Sigler and Sigler 1987); **Virginia** (Lee et al. 1980 et seq.; Hocutt et al. 1986; Jenkins and Burkhead 1994); **Vermont** (Lee et al. 1980 et seq.); **Washington** (Gray and Dauble 1977; Wydoski and Whitney 1979; McMahon and Bennett 1996); **West Virginia** (Stauffer et al. 1995); **Wisconsin** (Cahn 1927; Kempinger and Churchill 1972); and **Wyoming** (Baxter and Simon 1970; Hubert 1994).

Means of Introduction: Intentionally stocked as a food fish and for sportfishing. One of the earliest introductions occurred in 1874 when Livingston Stone gathered a small number of adult walleye captured in Vermont and transported them to California where the fish were released into the Sacramento River (Smith 1896). According to Dill and Cordone (1997), in the 1890s the California Fish Commission applied to the U.S. Fish Commission for shipments of walleye for use in controlling carp in Clear and Blue lakes; however, no walleye were imported at the time. These same authors also noted that the state received walleye eggs from Minnesota in 1959 and that these fish were to be used to control bluegill and support other sport fish in southern California reservoirs. In Idaho, walleye may have been stocked accidentally with yellow perch *Perca flavescens* (Linder 1963). McMahon and Bennett (1996) state the first introduction into southern Idaho reservoirs was in 1974. The person or agency responsible for introducing the species into Washington is uncertain. The federal government may have introduced them in the early 1960s (Dentler 1993). A sport fishery had developed in Lake Roosevelt, Washington, by the 1960s (McMahon and Bennett 1996). Walleye was first reported in Wyoming in 1961 from Seminoe Reservoir in the upper North Platte River. The fish were swept downstream and are now established in a 450-km stretch of river (McMahon and Bennett 1996). The walleye was stocked illegally in Canyon Ferry Reservoir, Montana, and was found first circa 1991 (White, personal communication). More recently, the species also was illegally stocked in Noxon Reservoir on the Clark Fork of the Columbia River, Montana (McMahon and Bennett 1996). Illegal introductions seem to be a growing problem in western states (McMahon and Bennett 1996).

Status: Many states have had some success in establishing reproducing populations. Other states have maintained populations with annual stocking. Occurrences in Delaware are due to strays from Pennsylvania stockings (Raasch, personal communication). Extirpated in California (Hubbs et al. 1979; Dill and Cordone 1997). Dentler (1993) indicated that walleye populations were spreading throughout the Columbia River basin.

Impact of Introduction: McMahon and Bennett (1996) recently reviewed the literature and presented a summary of impacts of walleye in the Northwest. Overall, the effects of its introduction were considered complex and varied. The walleye has been shown to prey on smolts of Pacific salmon, and therefore pose a threat to these already declining species in the Columbia River (Dentler 1993; McMahon and Bennett 1996). For instance, it is estimated that walleye consume two million smolts annually in the Columbia River, about one third of total predation loss (McMahon and Bennett 1996). A study in Seminoe Reservoir, Wyoming, found walleye stocking to result in a sharp decline in native minnows *Hybognathus* spp., darters *Etheostoma* spp., suckers *Catostomus* spp., rainbow trout *Oncorhynchus mykiss*, and crayfish *Orconectes obscurus*. For instance, most of the 500,000 trout fingerlings stocked annually were eaten within a few weeks. Consequently, there was a need to stock larger rainbow trout to avoid predation, an action that increased hatchery operation costs (McMahon and Bennett 1996). In their native eastern habitat, walleye and salmonids are able to limit interaction by living at different water temperatures (depths) and in different habitats (McLean and Magnuson 1977). However, in western reservoirs the lack of a strong thermocline and a small littoral area does not permit this separation (McMahon and Bennett 1996). Numbers and health of brown trout *Salmo trutta* were found to decrease after introduced walleyes consumed a large portion of the crayfish population, the brown trout's favorite food (McMahon and Bennett 1996). When the walleye initially was introduced into Salmon Falls Creek Reservoir, Idaho, yellow perch *Perca flavescens* comprised 80% of the sport fish. However, 12 years later, walleye made up 80% and perch only 1% of the fish in the reservoir (McMahon and Bennett 1996). Similar perch collapses also have happened at two other reservoirs in Wyoming (McMahon and Bennett 1996). A crash in the yellow perch population in Canyon Ferry Reservoir may be related to past walleye introduction; studies are being conducted to look at the problem; in addition, it has been predicted that walleye will have a large impact on the trout fishery in the reservoir (White, personal communication). In many cases introduced walleye deplete the forage base. As a consequence, the surviving walleye population consists of stunted individuals and the species no longer serves as a valuable fishery (McMahon and Bennett 1996). Some states now prohibit the introduction of walleye into certain waters. For instance, walleye introductions are banned in the Snake River drainage in Idaho because of concern about predation on anadromous salmonids (McMahon and Bennett 1996). Further introductions in Oregon also are forbidden due to concern about predation on salmonid smolts in the Columbia River (McMahon and Bennett 1996). Although 40 million walleye are stocked annually to maintain an important sport fishery in eastern Montana, the species has been banned from waters west of the Continental Divide in that state due to concern for important native and nonindigenous salmonid stocks (McMahon and Bennett 1996).

Remarks: The walleye is a desirable sport and food fish. Although the species was thought to be native to a few drainages flowing into the Atlantic, Jenkins and Burkhead (1994) reviewed and evaluated the literature on the distribution of eastern populations and concluded that the populations on the Atlantic slope south of the St. Lawrence probably are introduced. As with many of the Virginia species, Jenkins and Burkhead provided extensive detail on the introduction history of walleye in waters of Virginia and surrounding states. McMahon and Bennett (1996) provided a map and a table of walleye introductions in the northwest. The species' distribution in Alabama south of the Tennessee drainage was discussed by Brown (1962), who speculated that they are native to that region. Lee et al. (1980 et seq.) reported them as introduced. Billington and Maceina (1997) investigated the genetic status of walleyes in Alabama, where the southern walleyes are native but northern walleyes from Ohio and Pennsylvania have been stocked. They concluded that transplanted female northern walleyes did not survive to reproduce. However, because of the type of analysis done (mtDNA) they could not tell if any of the transplanted males survived.

One especially problematic record comes from the Escambia drainage in Alabama (Brown 1962; Mettee et al. 1996). Only a single individual has ever been collected from the drainage. None have been taken downstream in the Florida portion of the drainage. Swift et al. (1986) reported it as introduced into the drainage. In discussions with Gilbert (personal communication), he believes the species was introduced to the Escambia based on the fact that only one specimen has been collected, the apparent lack of suitable habitat, and the fact that this sport-fish is more likely to be introduced than less desirable species. He also pointed out that Bailey et al. (1954) failed to include this species in their paper on the Escambia and that Mettee et al. (1996) did not find it in their survey work. He also believes that if the walleye were native to the Escambia, it would be present in the lowermost (Florida) section because that is the stretch with the most suitable habitat. He likens the walleye to *Crystallaria asprella*, which is found only in the lower section of the drainage (Gilbert 1992). However, JDW (author) believes it is native to the drainage because of the presence of several other native large-river fish and mussel species; the collection was before the state began stocking this species, the drainage has never been extensively sampled, and some sections do contain suitable habitat. Much habitat was lost when two dams were constructed on the river in the 1940s. Many large-river mussels suffered from these impoundments (JDW, personal observation) and walleye could have done the same.

HYBRIDS

Stizostedion canadense × *S. vitreum*

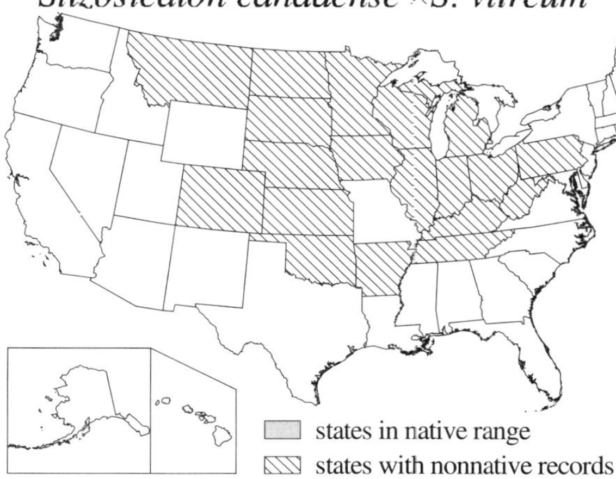

Stizostedion canadense × *S. vitreum*

Common Name: saugeye.

Identification: Etnier and Starnes (1993). Maximum size: 5.63 kg.

Native Range: Not applicable; artificial hybrid.

Nonindigenous Occurrences: This hybrid was reported in Crown and Gilliam lakes, Fulton and Howard counties, **Arkansas** (Carter, personal communication); lakes in the Arkansas drainage in **Colorado** (Walker 1993; Horak, personal communication); Illinois River, **Illinois** (Illinois Department of Natural Resources 1997a); Ohio and St. Joseph rivers, **Indiana** (Indiana Department of Natural Resources 1997a); Rock Creek Lake in Jasper County and Saylorville Lake in Polk County, **Iowa** (Iowa Department of Natural Resources 1997); several state lakes and reservoirs in **Kansas** (Cross and Collins 1995; Kansas Department of Wildlife and Parks 1997); Cumberland Lake (Hearn 1986) and Ohio River (Indiana Department of Natural Resources 1997a), **Kentucky**; St. Joseph River, **Michigan** (Indiana Department of Environmental Resources 1997a); Mississippi River, **Minnesota** (Wisconsin Department of Natural Resources 1997); Fort Peck Reservoir, **Montana** (Illinois Department of Natural Resources 1997a); Bluestem, Conestoga, and Holmes lakes near Lincoln, Buffalo Creek near Lexington, Rockford Lake, and Standing Bear Lake, **Nebraska** (Morris et al. 1974; DeLorme Mapping 1996a); Lake Sakakawea, **North Dakota** (North Dakota Game and Fish Department 1994, 1997); lakes Thunderburg, Boomer, and Stillwater, **Oklahoma** (Pigg, personal communication); at least 46 areas including the Ohio River and its tributaries, and numerous other lakes and reservoirs in **Ohio** (International Game Fish Association 1994; Stauffer et al. 1995; Ohio Department of Natural Resources, Division of Wildlife 1997); several reservoirs in **Pennsylvania** (Pennsylvania Fish and Boat Commission 1997); **South Dakota** (North Dakota Game and Fish Department 1994); the tailwaters of Fort Loudoun Reservoir, **Tennessee** (Etnier and Starnes 1993); the Ohio, Twelvepole, and Little Kanawha rivers in **West Virginia** (Stauffer et al. 1995), and the Mississippi River, **Wisconsin** (Wisconsin Department of Natural Resources 1997).

Means of Introduction: This hybrid or one of the parent species was intentionally stocked for sportfishing. This hybrid can be artificially produced and stocked, or can result from stocking one or both parent species in the same lake.

Status: Maintained by stocking in most areas. Established in West Virginia where it spawns during April (Stauffer et al. 1995), and presumably also established in Ohio.

Impact of Introduction: Unknown.

Remarks: These hybrids and their backcrosses are fertile (Hearn 1986; Thorgaard and Allen 1992).

Nematistiidae - Roosterfish

The family Nematistiidae contains a single species, the roosterfish, *Nematistius pectoralis*. The roosterfish is found in the tropical eastern Pacific (Nelson 1994) and is an important game fish. It has a compressed body and seven very elongate spines in the first dorsal (Nelson 1994). The roosterfish was formerly placed in the family Carangidae. It possesses an adaptation to presumably increase hearing sensitivity: the swim bladder enters the skull and contacts the inner ear (Nelson 1994).

Nematistius pectoralis

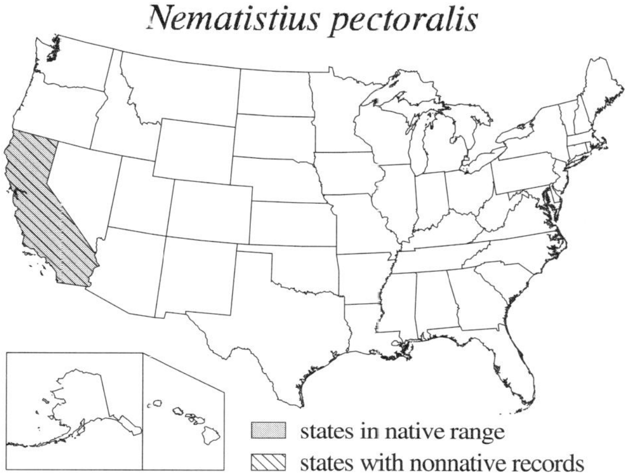

Nematistius pectoralis Gill 1862

Common Name: roosterfish.

Identification: Miller and Lea (1972); Eschmeyer et al. (1983); Fischer et al. (1995b). Maximum size: 122 cm.

Native Range: Marine; San Clemente in southern California to Peru, including Galapagos Islands: rare north of Baja (Eschmeyer et al. 1983; Nelson 1994; Fischer et al. 1995b).

Nonindigenous Occurrences: A single roosterfish was introduced into Salton Sea, **California**, in 1950 (Walker et al. 1961; Dill and Cordone 1997).

Means of Introduction: Intentionally stocked.

Status: The introduction failed (not surprisingly) to produce a viable population and the species is extirpated in the Salton Sea.

Impact of Introduction: Unknown.

Remarks: None.

Carangidae - Jacks and pompanos

The family Carangidae, whose members are also known as jacks and pompanos, includes about 32 genera and 140 species (Nelson 1994; W. F. Smith-Vaniz, personal communication). Carangids are found in the Atlantic, Indian, and Pacific oceans. They are extremely variable in shape, and range from shallow-bodied to extremely thin and deep-bodied. In general, jacks are characterized by a compressed body, silvery color, up to nine detached finlets behind the dorsal and anal fins, two dorsal fins, the soft dorsal and anal fins usually highest near the front and lobed, a narrow caudal peduncle, and a deeply-forked caudal fin (Eschmeyer et al. 1983; Robins and Ray 1986; Nelson 1994). Jacks are predaceous, fast-swimming fishes (Robins and Ray 1986). Many species are important sport and food-fish.

Caranx caballus

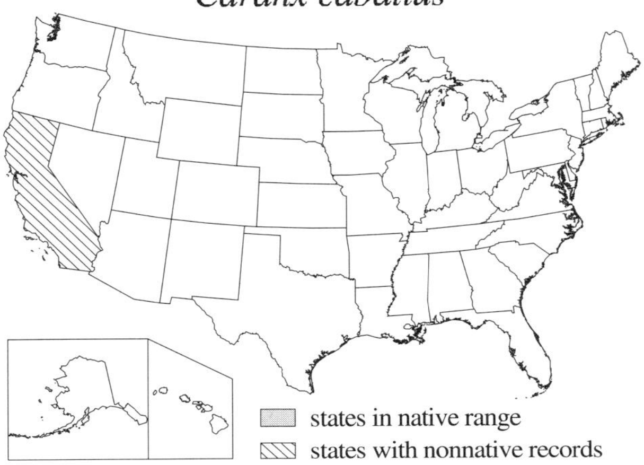

Caranx caballus Günther 1868

Common Name: green jack.

Identification: Miller and Lea (1972); Eschmeyer et al. (1983); Fischer et al. (1995a). Maximum size: 55 cm.

Native Range: Marine; Santa Cruz Island, Northern Baja and Gulf of California south to Peru, including the Galapagos Islands (Ramirez-Hernández and Gonzales Pages 1976; Miller and Lea 1972; Fischer et al. 1995a).

Nonindigenous Occurrences: Introduced into Salton Sea, **California**, in 1948 (Walker et al. 1961; Dill and Cordone 1997).

Means of Introduction: Intentionally stocked as forage. Stock was obtained from San Diego, California (Walker et al. 1961; Dill and Cordone 1997).

Status: The introduction failed to produce a viable population and the species is extirpated in the Salton Sea.

Impact of Introduction: Unknown.

Remarks: None.

Trachinotus paitensis

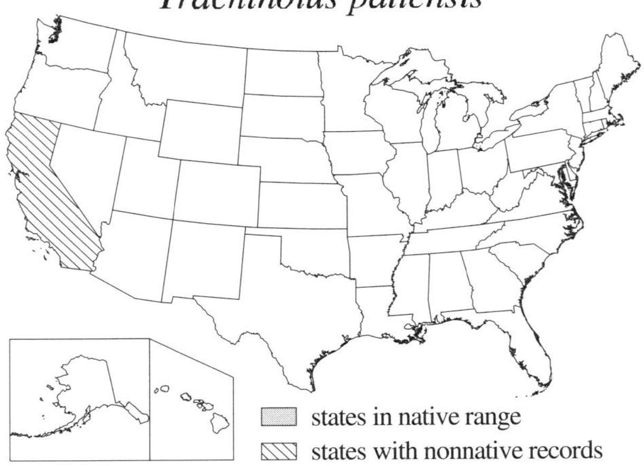

Trachinotus paitensis Cuvier 1832

Common Name: palometa (paloma pompano).

Identification: Miller and Lea (1972); Fischer et al. (1995a). Maximum size: 42 cm.

Native Range: Marine; Redondo Beach, Baja and Gulf of California south to Peru (Miller and Lea 1972; Fischer et al. 1995a).

Nonindigenous Occurrences: Introduced into Salton Sea, **California**, in 1950 and 1953 (Walker et al. 1961; Dill and Cordone 1997).

Means of Introduction: Intentionally stocked as forage. Two fish were stocked in 1950; one in 1953. The fish were obtained from San Felipe, Mexico (Walker et al. 1961).

Status: The introduction failed to produce a viable population and the species is extirpated in the Salton Sea.

Impact of Introduction: Unknown.

Remarks: None.

Gerreidae - Mojarras

The family Gerreidae, also known as mojarras, comprises eight genera with about 40 species (Nelson 1994). Mojarras are found in most warm seas and occasionally in brackish and rarely in freshwater. They are characterized by a pointed snout with a highly protrusible mouth, a deeply forked tail, a single dorsal with spines and rays, and a scaly sheath along the bases of the dorsal and anal fins (Eschmeyer et al. 1983; Nelson 1994). Most species are small (less than 20 cm) and silvery in color (Eschmeyer et al. 1983). This family was formerly spelled Gerridae; however, the name was changed because it had already been used for an order for water striders (Nelson 1994).

Eucinostomus argenteus

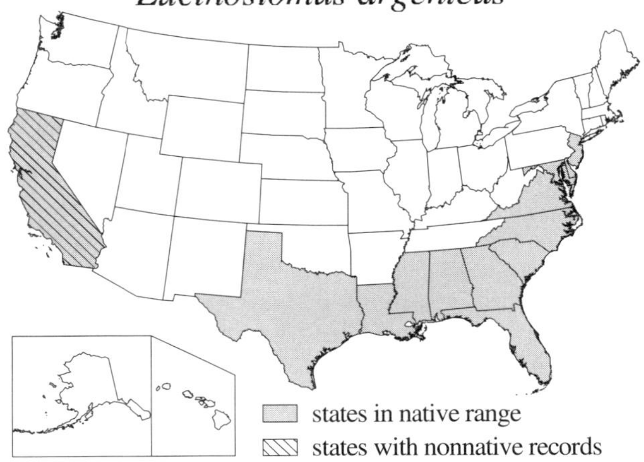

Eucinostomus argenteus Baird and Girard 1855

Common Name: spotfin mojarra (silver mojarra).

Identification: Miller and Lea (1972); Eschmeyer et al. (1983); Fischer et al. (1995a). Maximum size: 20 cm.

Native Range: Marine; Anaheim Bay in southern California to Peru (Miller and Lea 1972; Eschmeyer et al. 1983; Fischer et al. 1995a) in the Pacific; New Jersey, Bermuda, and northern Gulf of Mexico to southeastern Brazil (Ramirez-Hernández and Gonzales Pages 1976; Robins and Ray 1986) in the Atlantic.

Nonindigenous Occurrences: Introduced into Salton Sea, **California**, in 1950 and 1951 (Walker et al. 1961; Dill and Cordone 1997).

Means of Introduction: Intentionally stocked four times during 1950–1951. Two of those stockings contained only one fish; a third one contained less than 12. All stock was obtained from San Felipe, Mexico (Walker et al. 1961).

Status: The introduction failed to produce a viable population and the species is extirpated in the Salton Sea.

Impact of Introduction: Unknown.

Remarks: None.

Eucinostomus gracilis

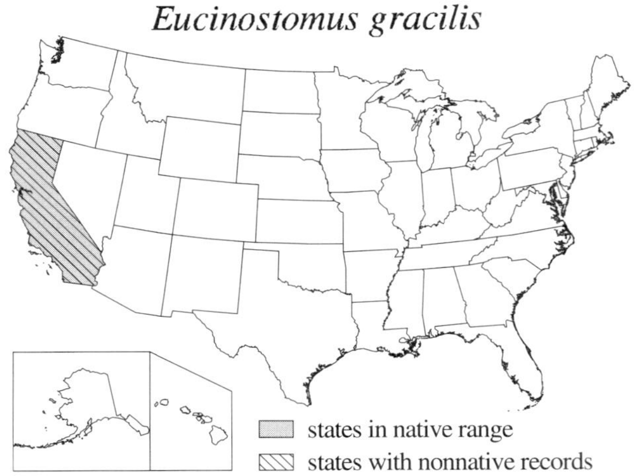

Eucinostomus gracilis (Gill 1862)

Common Name: Pacific flagfin mojarra (graceful mojarra).

Identification: Miller and Lea (1972); Eschmeyer et al. (1983); Fischer et al. (1995a). Maximum size: 21 cm.

Native Range: Marine; Anaheim Bay in southern California to Peru; rare north of Baja (Miller and Lea 1972; Eschmeyer et al. 1983).

Nonindigenous Occurrences: Introduced into Salton Sea, **California**, in 1950 (Walker et al. 1961; Dill and Cordone 1997).

Means of Introduction: Small numbers of fish were stocked twice in December 1950. Stock obtained from San Felipe, Mexico (Walker et al. 1961).

Status: The introduction failed to produce a viable population and the species is extirpated in the Salton Sea.

Impact of Introduction: Unknown.

Remarks: None.

Haemulidae - Grunts

The perciform family Haemulidae (Pomadasyidae) comprises 17 genera with about 150 species (Nelson 1994). Grunts are found in tropical and subtropical coastal waters of the Atlantic, Indian, and Pacific oceans, although a few species extend into temperate waters (Robins and Ray 1986). They are medium-sized, spiny-rayed fishes; most grow to about 30 cm (Eschmeyer et al. 1983). Although variable in body form, grunts superficially resemble snappers or even some basses. Their common name comes from the grunting noise they make by rubbing tooth plates in the throat together (Eschmeyer et al. 1983). These fish are used for food. Most grunts are marine, but a few occur in brackish or freshwater (Eschmeyer et al. 1983).

Anisotremus davidsonii

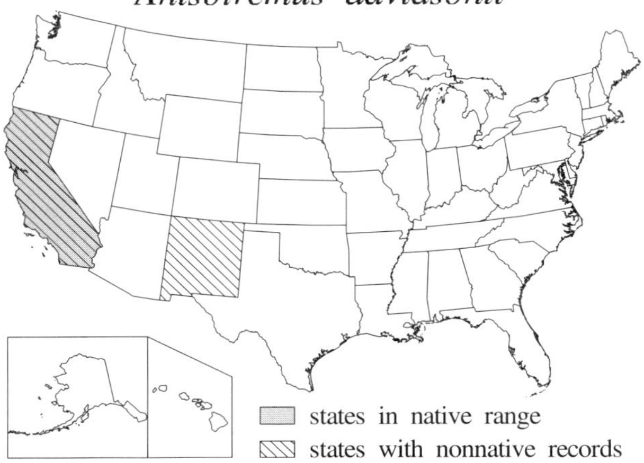

Anisotremus davidsonii (Steindachner 1875)

Common Name: sargo.

Identification: Miller and Lea (1972); Thomson et al. (1979); Eschmeyer et al. (1983). Also appears as *A. davidsoni*. Maximum size: 58 cm.

Native Range: Tropical, marine. The sargo is native to the Pacific Ocean from Santa Cruz, California, to southern Baja California (Eschmeyer et al. 1983).

Nonindigenous Occurrences: This species is now established in the Salton Sea in **California** (Walker et al. 1961; Shapovalov et al. 1981). It was introduced into and later extirpated at an unspecified location in **New Mexico** (Sublette et al. 1990).

Means of Introduction: A total of 65 fish were stocked in the Salton Sea in 1951 by the California Department of Fish and Game (Walker et al. 1961). The means and reason for introduction to New Mexico are unknown.

Status: Established in California; extirpated in New Mexico.

Impact of Introduction: Unknown.

Remarks: This species feeds on crustaceans, mollusks, and bryozoans (Eschmeyer et al. 1983). Its population in the Salton Sea is very large (Walker et al. 1961; Shapovalov et al. 1981).

Sciaenidae - Drums

The family Sciaenidae, whose members are known as drums or croakers, includes about 70 genera and about 270 species (Nelson 1994). Most sciaenids are found on the continental shelves of tropical and temperate areas of the Atlantic, Indian, and Pacific oceans (Berra 1981; Nelson 1994). The freshwater drum *Aplodinotus grunniens* is the only species native to inland waters of the United States, although marine species are established in the Salton Sea, southern California (Page and Burr 1991). Drums range in size from 28 to just over 200 cm, and many are important food fishes (Berra 1981). They get their common name from the sound they make using their swimbladder as a resonating chamber.

Aplodinotus grunniens

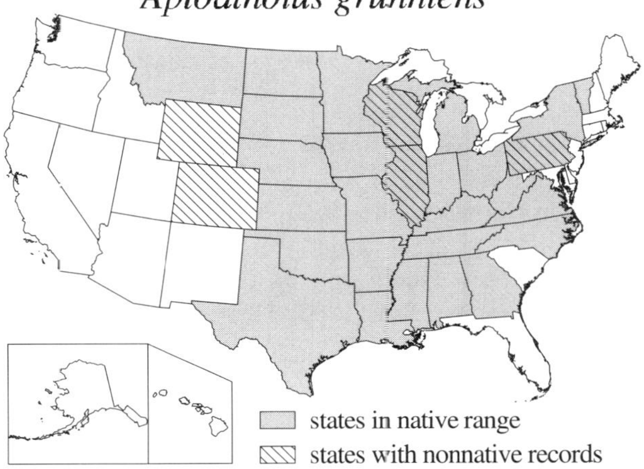

Aplodinotus grunniens Rafinesque 1819

Common Names: freshwater drum (sheepshead).

Identification: Becker (1983); Page and Burr (1991); Etnier and Starnes (1993); Jenkins and Burkhead (1994). Maximum size: 89 cm.

Native Range: East of Rocky Mountains in St. Lawrence–Great Lakes, Hudson Bay, and Mississippi River basins from Quebec to northern Manitoba and southern Saskatchewan, and south to Gulf; Gulf Coast drainages from Mobile Bay, Georgia and Alabama, through eastern Mexico to Rio Usumacinta system, Guatemala (Page and Burr 1991).

Nonindigenous Occurrences: Freshwater drum was stocked in Bonny Reservoir, Yuma County, **Colorado**, with stock from Mead County, Kansas, in 1951 (Everhart and Seaman 1971; Ellis 1974; Walker 1993). Drum introduced and established in Bonny Reservoir became the stock source for many other reservoirs in the Republican and Platte drainages of eastern Colorado (Everhart and Seaman 1971; Walker 1993). The species also was introduced into the Fox drainage in **Illinois** (Burr and Page 1986), and the Youghiogheny system in **Pennsylvania** (Hendricks et al. 1983). Two drum were reported in 1947 from the Moen's Lake chain in Oneida County, **Wisconsin**, in the upper Wisconsin River drainage (Becker 1983). It has been introduced into **Wyoming** (Stone 1995), including Keyhole and Grayrocks reservoirs (Hubert 1994).

Means of Introduction: Freshwater drum was stocked in the early 1950s in Colorado lakes as a sport fish (Everhart and Seaman 1971). It was accidentally stocked in two Wyoming reservoirs (Hubert 1994). The two fish reported from the upper Wisconsin drainage in Wisconsin were likely the result of fish rescue and transfer operations from the Mississippi River in the 1930s (Becker 1983).

Status: Established in Colorado; reported in Illinois, Wisconsin, and Wyoming. Extirpated in Pennsylvania (Hendricks et al. 1983).

Impact of Introduction: Unknown.

Remarks: In Wisconsin, there are no other records of freshwater drum from the upper Wisconsin River; the species is considered native farther downstream near and below the Portage area (Becker 1983).

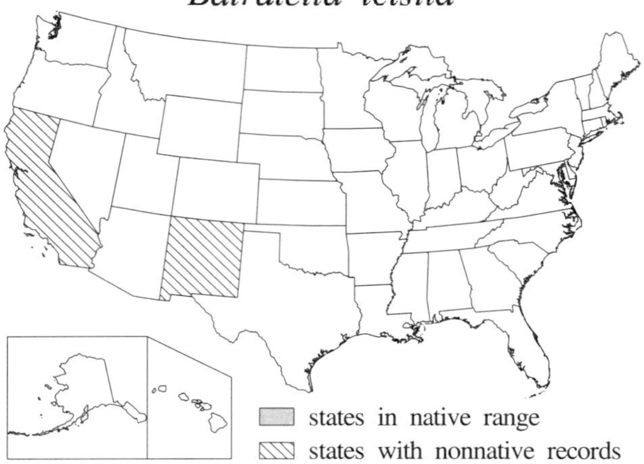

Bairdiella icistia (Jordan and Gilbert 1882)

Common Name: bairdiella.

Identification: Walker et al. (1961); Page and Burr (1991). Maximum size: 30 cm.

Native Range: Marine. Pacific Coast south from Almejas Bay, east of Isla Santa Margarita, Baja California (Page and Burr 1991).

Nonindigenous Occurrences: The bairdiella was introduced to the Salton Sea in southern **California** in 1950 (Walker et al. 1961). It reportedly also was introduced into **New Mexico**, but has since been extirpated (Sublette et al. 1990).

Means of Introduction: In 1950, 57 bairdiella were stocked in the Salton Sea, California; in 1951, an additional 10 fish were stocked (Walker et al. 1961). By 1953, millions were present. This species was introduced as a sport fish and as forage for another introduced sciaenid, the orangemouth corvina *Cynoscion xanthulus* (Walker et al. 1961).

Status: Bairdiella quickly became established as one of the dominant species found in the Salton Sea. It is extirpated in New Mexico (Sublette et al. 1990).

Impact of Introduction: Unknown.

Remarks: Moyle (1976a) listed this species as one of the three most dominant fishes in the Salton Sea. Sublette et al. (1990) did not provide details concerning the New Mexico introduction.

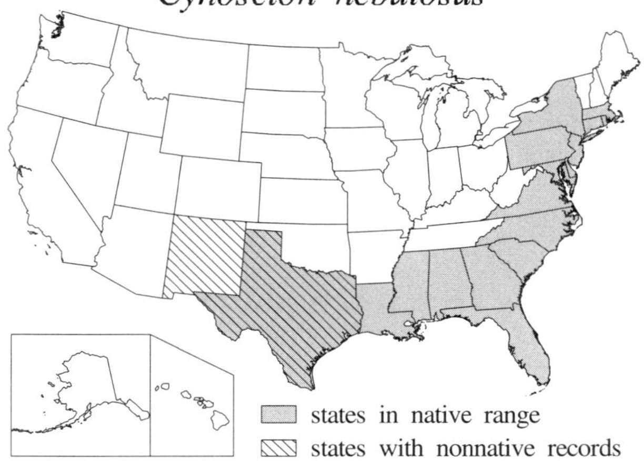

Cynoscion nebulosus (Cuvier 1830)

Common Names: spotted seatrout (speckled seatrout).

Identification: Hoese and Moore (1977); Manooch (1984); Robins and Ray (1986); Howells (1992b). Maximum size: 125 cm.

Native Range: Marine. New York to southern Florida and entire Gulf of Mexico (Robins and Ray 1986).

Nonindigenous Occurrences: This species was introduced into Walter E. Long Reservoir and Victor Braunig Reservoir in Travis and Bexar counties, **Texas** (Lasswell et al. 1979; Howells 1991c; Howells and Garrett 1992). Seatrout also was introduced into **New Mexico** but it has since been extirpated (Sublette et al. 1990).

Means of Introduction: It was stocked by the Texas Parks and Wildlife Department in the 1970s and 1980s as a sportfish (Lasswell et al. 1979; Howells 1991c; Howells and Garrett 1992).

Status: Because there has been no reproduction, it was predicted that the inland population in Texas would die out due to natural mortality and fishing (Howells and Garrett 1992). This species is extirpated in New Mexico (Sublette et al. 1990).

Impact of Introduction: Unknown.

Remarks: The Texas stockings totaled 78,092 fish (Howells and Garrett 1992). Sublette et al. (1990) included this species in their list of "fishes introduced into Mexico, now considered extirpated." However, they provided no additional information.

Cynoscion othonopterus

Cynoscion parvipinnis

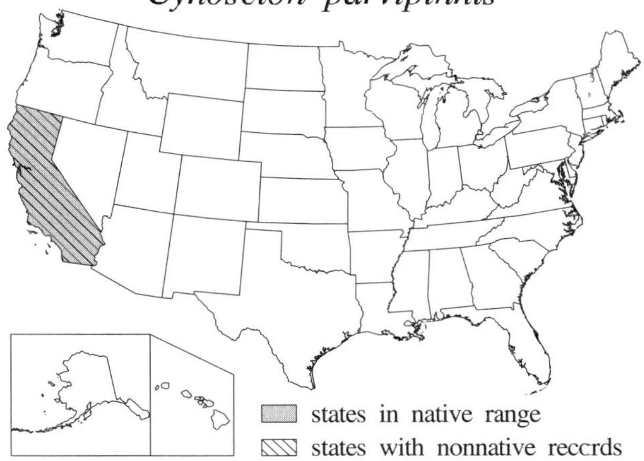

Cynoscion othonopterus (Günther 1868)

Common Name: (scaly corvina, Gulf weakfish).

Identification: Fischer et al. (1995b). Maximum size: 70 cm.

Native Range: Marine; Gulf of California (Fischer et al. 1995b).

Nonindigenous Occurrences: Introduced into Salton Sea, **California**, in 1951 and 1956 (Walker et al. 1961; Dill and Cordone 1997).

Means of Introduction: Intentionally stocked for sportfishing. Less than 200 fish were planted in 1951; and only one fish in 1956. Stock was obtained from San Felipe, Mexico (Walker et al. 1961).

Status: The introduction failed to produce a viable population and the species is extirpated in the Salton Sea.

Impact of Introduction: Unknown.

Remarks: None.

Cynoscion parvipinnis Ayres 1862

Common Name: shortfin corvina.

Identification: Eschmeyer et al. (1983) and Howells (1992b). Maximum size: 50 cm.

Native Range: Marine. Pacific Ocean from Mazatlan, Mexico, to Santa Barbara, California, including the Gulf of California (Howells 1992b).

Nonindigenous Occurrences: The shortfin corvina was introduced into the Salton Sea in southern **California** (Walker et al. 1961).

Means of Introduction: Intentionally introduced as a sport fish during the early 1950s (Walker et al. 1961).

Status: Although individual fish were able to survive for at least one year, the population failed to reproduce and eventually died out (Walker et al. 1961).

Impact of Introduction: Unknown.

Remarks: None.

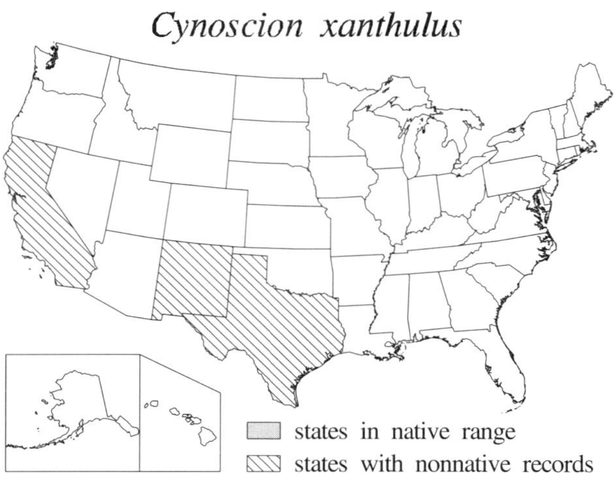

Cynoscion xanthulus Jordan and Gilbert 1882

Common Name: orangemouth corvina.

Identification: Walker et al. (1961); Howells (1991c); Page and Burr (1991). Maximum size: 90 cm.

Native Range: Marine. Pacific Coast from Gulf of California to Acapulco, Mexico (Page and Burr 1991).

Nonindigenous Occurrences: The orangemouth corvina was stocked in the Salton Sea in southern **California** from 1950 to 1955 (Walker et al. 1961; Moyle 1976a). It was introduced into **New Mexico** (Sublette et al. 1990). The Texas Parks and Wildlife Department transferred orangemouth corvina from the Salton Sea to stock Calaveras and Victor Braunig reservoirs in Bexar County, **Texas** (Howells 1992a; Howells and Garrett 1992).

Means of Introduction: In California, stockings were carried out by the California Department of Fish and Game, and in Texas by the Texas Parks and Wildlife Department, for the purpose of introducing a sportfish. Total number of fish introduced into the Salton Sea did not exceed 272 (Walker et al. 1961); the Texas stockings totaled nearly 770,000 fish (Howells and Garrett 1992).

Status: This species is established in the Salton Sea, California, where it was considered an important sportfish (Walker et al. 1961). There has been no evidence of reproduction in Texas, and existing populations were predicted to disappear (Howells 1991c; Howells 1992a; Howells and Garrett 1992). It is extirpated in New Mexico (Sublette et al. 1990).

Impact of Introduction: Largely unknown. Texas biologists expressed concern over the potential for establishment of this species in the Gulf of Mexico (Howells and Garrett 1992).

Remarks: The orangemouth corvina has become a dominant species in the Salton Sea. Sublette et al. (1990) did not provide details for New Mexico introduction.

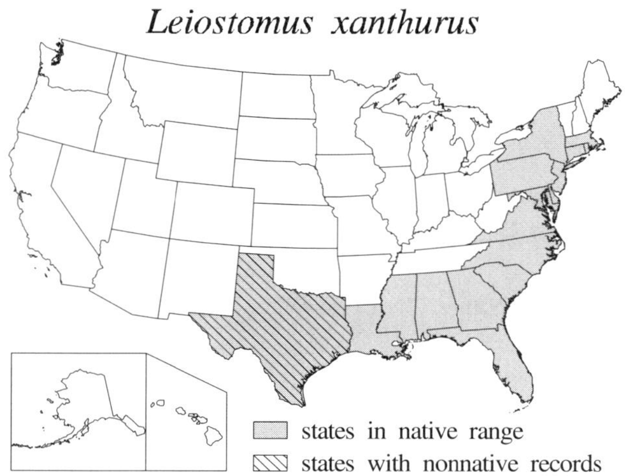

Leiostomus xanthurus Lacepède 1802

Common Name: spot.

Identification: Hoese and Moore (1977); Manooch (1984); Robins and Ray (1986). Maximum size: 36 cm.

Native Range: Marine. Massachusetts to northern Mexico; absent from southern Florida (Robins and Ray 1986).

Nonindigenous Occurrences: Spot from coastal Texas waters reportedly was introduced to inland waters in **Texas** prior to 1992 (probably in the 1950s or 1960s) (Howells and Garrett 1992).

Means of Introduction: Intentionally stocked as a sportfish.

Status: The introduction failed.

Impact of Introduction: Unknown.

Remarks: Howells and Garrett (1992) provided no details concerning the stocking of spot in Texas. Presumably, its introduction took place in the 1950s, or shortly thereafter, when Texas agencies began wholesale capturing of native marine species along the Texas coast for later transport and stocking in reservoirs and rivers within the state.

Menticirrhus nasus

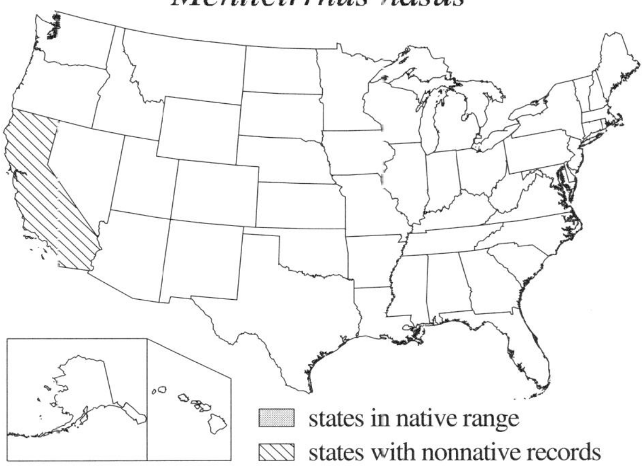

Menticirrhus nasus (Günther 1869)

Common Name: (corbina, highfin king croaker).

Identification: Fischer et al. (1995b). Maximum size: 50 cm.

Native Range: Marine; Baja and Gulf of California (Fischer et al. 1995b).

Nonindigenous Occurrences: This species was introduced into the Salton Sea, **California**, in 1950 and 1951 (Walker et al. 1961; Dill and Cordone 1997).

Means of Introduction: Intentionally stocked. A single specimen was stocked in 1950; two in 1951. Stock was obtained from San Felipe, Mexico (Walker et al. 1961).

Status: The introduction failed to produce a viable population and the species is extirpated in the Salton Sea.

Impact of Introduction: Unknown.

Remarks: None.

Menticirrhus undulatus

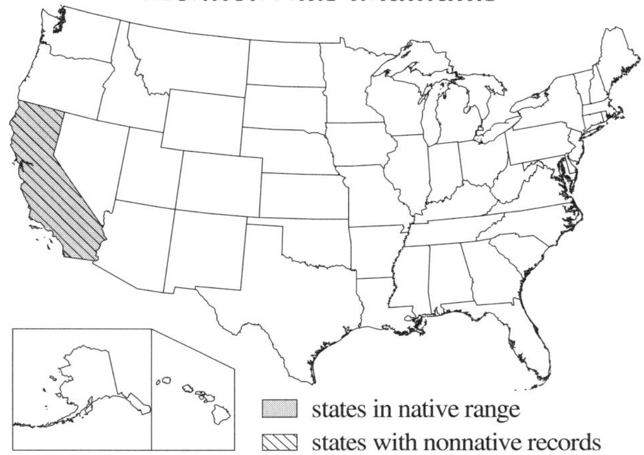

Menticirrhus undulatus (Girard 1854)

Common Name: California corbina (California kingcroaker).

Identification: Miller and Lea (1972); Eschmeyer et al. (1983); Fischer et al. (1995b). Maximum size: at least 71 cm.

Native Range: Marine; Pt. Conception to Gulf of California (Miller and Lea 1972; Eschmeyer et al. 1983; Fischer et al. 1995b).

Nonindigenous Occurrences: This species was introduced into the Salton Sea, **California**, in 1950 and again in 1953 (Walker et al. 1961; Dill and Cordone 1997).

Means of Introduction: Intentionally stocked. This species was introduced on five separate occasions, but the number of individuals totaled only 14 fish (Walker et al. 1961).

Status: The introduction failed to produce a viable population and the species is extirpated in the Salton Sea.

Impact of Introduction: Unknown.

Remarks: None.

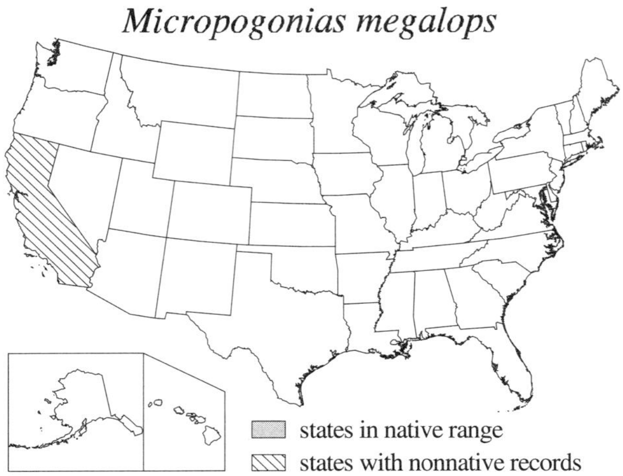

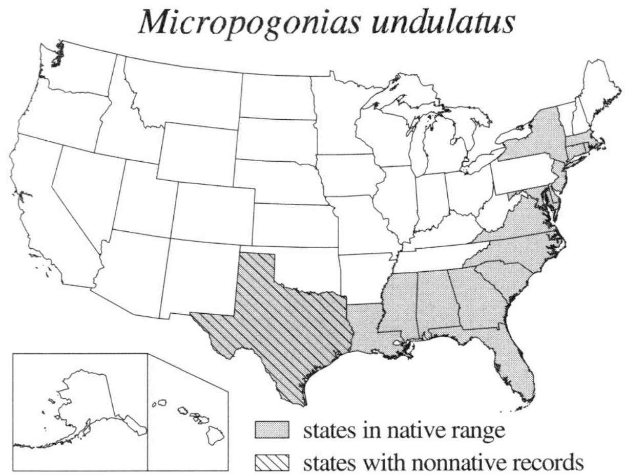

Micropogonias megalops (Gilbert 1890)

Common Name: (croaker, bigeye croaker).

Identification: Fischer et al. (1995b). Reported as *Micropogon megalops*. The genus for this fish has been changed to *Micropogonias*. Maximum size: 40 cm.

Native Range: Marine; Gulf of California south to southern Mexico (Fischer et al. 1995b).

Nonindigenous Occurrences: The croaker was introduced into the Salton Sea, **California**, in 1950 and 1953 (Walker et al. 1961; Dill and Cordone 1997).

Means of Introduction: Intentionally stocked for sportfishing. Fifteen fish were planted in 1950; 35 in 1953. Stock was obtained from San Felipe, Mexico (Walker et al. 1961).

Status: The introduction failed to produce a viable population and the species is extirpated in the Salton Sea.

Impact of Introduction: Unknown.

Remarks: None.

Micropogonias undulatus (Linnaeus 1766)

Common Name: Atlantic croaker.

Identification: Hoese and Moore (1977); Manooch (1984); Robins and Ray (1986). Maximum size: 67 cm.

Native Range: Marine. Atlantic Coast from Massachusetts and the northern Gulf of Mexico to northern Mexico, except for southern Florida. Also possibly from southern Brazil to Argentina (Robins and Ray 1986).

Nonindigenous Occurrences: The Atlantic croakers was introduced into Town Lake Reservoir in Travis County, **Texas** (Howells and Garrett 1992).

Means of Introduction: In 1975, the Texas Parks and Wildlife Department stocked 250 croakers for sport fishing (Howells and Garrett 1992).

Status: Failed to establish a reproducing population.

Impact of Introduction: Unknown.

Remarks: None.

Pogonias cromis

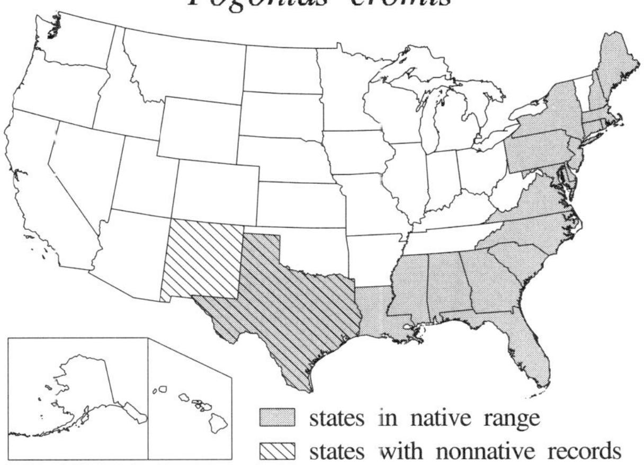

Pogonias cromis (Linnaeus 1766)

Common Name: black drum.

Identification: Hoese and Moore (1977); Manooch (1984); Robins and Ray (1986). Maximum size: 170 cm.

Native Range: Marine. Nova Scotia to northern Mexico, including south Florida; also southern Brazil to Argentina (Robins and Ray 1986).

Nonindigenous Occurrences: Black drum was stocked in four **Texas** reservoirs: Victor Braunig, Lake Creek, Red Bluff, and Walter E. Long (Howells and Garrett 1992). It also was introduced into **New Mexico** (Sublette et al. 1990).

Means of Introduction: Black drum was stocked in 1976 and 1977 by the Texas Parks and Wildlife Department in Texas reservoirs.

Status: The stocking attempts failed to establish reproducing populations in Texas. The species is extirpated in New Mexico.

Impact of Introduction: Unknown.

Remarks: Sublette et al. (1990) included this species in their list of "fishes introduced into Mexico, now considered extirpated." However, the authors provided no additional information.

Sciaenops ocellatus

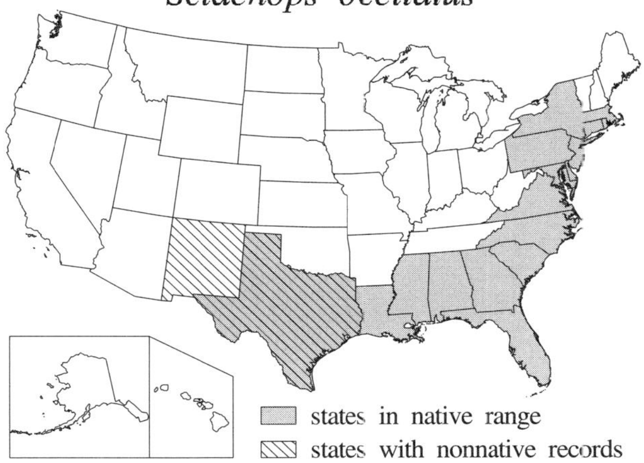

Sciaenops ocellatus (Linnaeus 1766)

Common Names: red drum (redfish).

Identification: Hoese and Moore (1977); Manooch (1984); Robins and Ray (1986). Maximum size: 155 cm.

Native Range: Marine. Massachusetts to northern Mexico, including southern Florida (Robins and Ray 1986).

Nonindigenous Occurrences: This species was introduced to **New Mexico** (Sublette et al. 1990), and into at least 27 **Texas** reservoirs and lakes (Lasswell et al. 1979; Dolman 1983; Howells and Garrett 1992).

Means of Introduction: Beginning in the 1950s, several million red drum have been intentionally stocked by the Texas Parks and Wildlife Department for use as a sportfish (Howells and Garrett 1992).

Status: Because the red drum will not reproduce, Texas populations must be supplemented with repeated stockings.

Impact of Introduction: Freshwater mussels in the family Unionidae have had very poor recruitment over the past 20 or so years in Tradinghouse Creek and Fairfield reservoirs in Texas. Reproduction of these mussels may be affected by red drum feeding on juvenile mussels. Red drum, which feed heavily on mussels in marine waters, were first stocked in Tradinghouse Creek and Fairfield reservoirs in 1975 and 1984, respectively, about the time of the mussel decline. Blue tilapia also was stocked around the same time and may have contributed to the decline by eating juveniles (Howells 1995).

Remarks: Sublette et al. (1990) included this species in their list of "fishes introduced into Mexico, now considered extirpated." However, they provided no additional information.

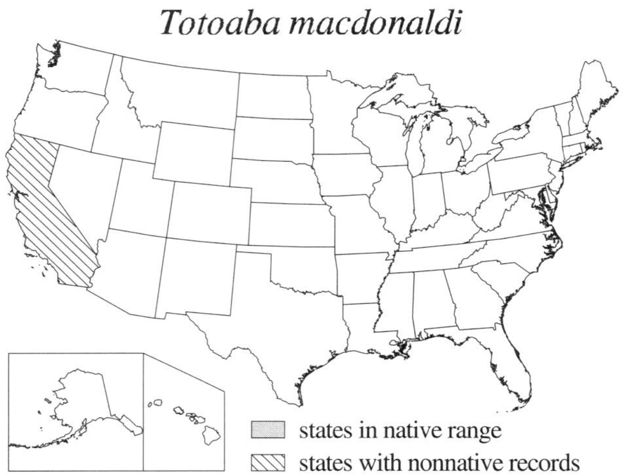

Totoaba macdonaldi (Gilbert 1890)

Common Name: (totuava, totoaba).

Identification: Fischer et al. (1995b). Synonym is *Cynoscion macdonaldi*. Maximum size: 200 cm.

Native Range: Marine; Gulf of California (Fischer et al. 1995b).

Nonindigenous Occurrences: This species was introduced to the Salton Sea, **California**, on three occasions during the period 1950–1956 (Walker et al. 1961; Dill and Cordone 1997).

Means of Introduction: Intentionally stocked for sportfishing. In 1950, a single fish was stocked; in 1951, less than 200; and in 1956, 8 were stocked. Stock was obtained from San Felipe, Mexico (Walker et al. 1961).

Status: The introduction failed to produce a viable population and the species is extirpated in the Salton Sea.

Impact of Introduction: Unknown.

Remarks: Walker (1961) reported this species as *Cynoscion macdonaldi*.

HYBRIDS

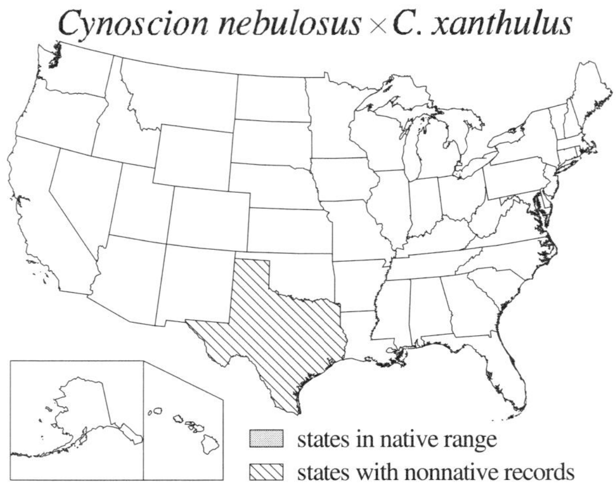

Cynoscion nebulosus × *C. xanthulus*

Common Name: spotted seatrout × orangemouth corvina.

Identification: Howells (1992b). Maximum size: Probably similar to parent species, 90–125 cm.

Native Range: None; artificial hybrid.

Nonindigenous Occurrences: This hybrid combination was stocked in Calaveras Reservoir, Bexar County, **Texas** (Howells 1991c, 1992a; Howells and Garrett 1992).

Means of Introduction: Intentionally stocked by the Texas Parks and Wildlife Department from 1984 through 1986 as a sport fish.

Status: As of 1990, only a few hybrids still survived in Calaveras Reservoir. It was predicted that the populations would soon die out (Howells 1991a, 1991c; Howells and Garrett 1992).

Impact of Introduction: Unknown.

Remarks: The Calaveras Reservoir stocking totaled 75,642 fish (Howells and Garrett 1992).

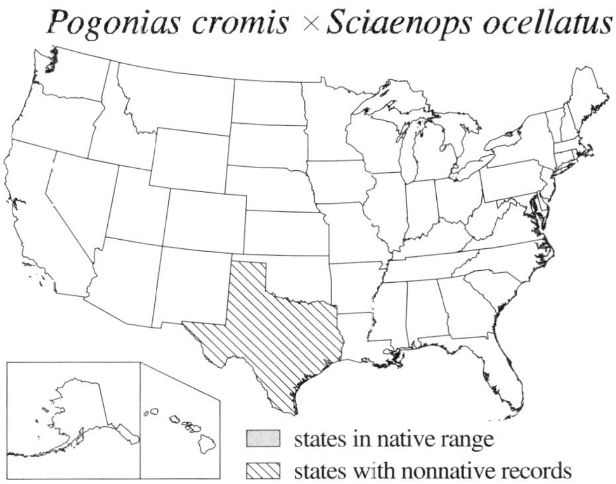

Pogonias cromis × Sciaenops ocellatus

Common Name: black drum × red drum.

Identification: Hybrid. Maximum size: Probably similar to parent species, 155–170 cm.

Native Range: None; artificial hybrid.

Nonindigenous Occurrences: This hybrid was stocked in **Texas** in Victor Braunig and Lake Creek reservoirs in during the period 1983 to 1984 (Howells and Garrett 1992).

Means of Introduction: Intentionally stocked by the Texas Parks and Wildlife Department for sportfishing.

Status: These populations are likely extirpated due to discontinued stocking and lack of reproduction (Howells and Garrett 1992).

Impact of Introduction: Unknown.

Remarks: The Victor Braunig Reservoir stocking totaled 56,346 fish; the Lake Creek Reservoir stocking totaled 86,800 (Howells and Garrett 1992).

Kyphosidae - Sea chubs

The family Kyphosidae, also called sea chubs, consists of 15 genera with about 42 species. Other taxonomists split subfamilies of this family into distinct families (Nelson 1994). Sea chubs are found in the Atlantic, Indian, and Pacific oceans (Nelson 1994). They are medium-sized, schooling fishes that are common in clear water, tropical harbors, reefs, and around small ships (Robins and Ray 1986). Sea chubs are characterized by a fairly stream-lined body that is ovate in side-view; a blunt head with a small, terminal, nibbling mouth; and an indented caudal fin (Eschmeyer et al. 1983; Robins and Ray 1986). They are medium-sized fish (most less than 61 cm) and are usually drab with yellowish or bluish stripes; others are spotted (Eschmeyer et al. 1983).

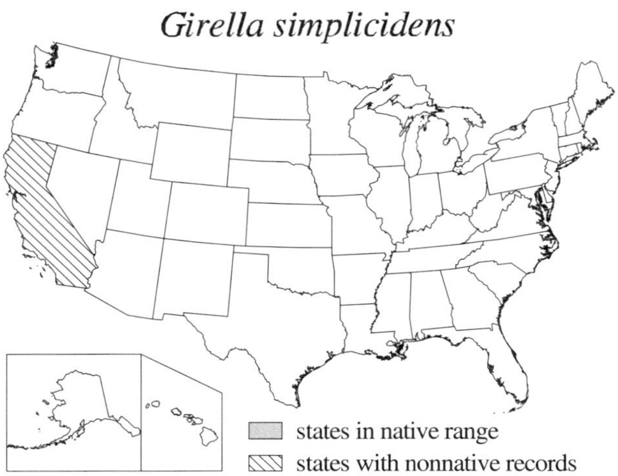

Girella simplicidens Osburn and Nichols 1916

Common Name: (opaleye, Gulf opal eye).

Identification: Miller and Lea (1972); Thomson et al. (1979); Fischer et al. (1995a). Maximum size: 46 cm.

Native Range: Marine; San Francisco to Cape San Lucas, Baja California (Miller and Lea 1972). Gulf of California (Fischer et al. 1995a).

Nonindigenous Occurrences: Introduced into Salton Sea, **California**, in 1951 (Walker et al. 1961; Dill and Cordone 1997).

Means of Introduction: Intentionally stocked. Seven fish obtained from San Felipe, Mexico, were stocked (Walker et al. 1961).

Status: The introduction failed to produce a viable population and the species is extirpated in the Salton Sea.

Impact of Introduction: Unknown.

Remarks: None.

Kuhliidae - Flagtails

The family Kuhliidae, whose members are also known as flagtails or aholeholes, includes a single genus with about eight species (Nelson 1994). Kuhliids are found in the tropical Indo-Pacific region (Berra 1981; Nelson 1994). They are small- to medium-sized silvery fishes, with a maximum size of about 50 cm, that resemble North American sunfishes (Centrarchidae) in body form (Allen 1991). Most species inhabit marine and brackish waters (i.e., shallow coral reefs and estuaries), and some occur in rivers of continents and oceanic islands; *Kuhlia rupestris* is the only species found primarily in freshwater (Allen 1991; Nelson 1994).

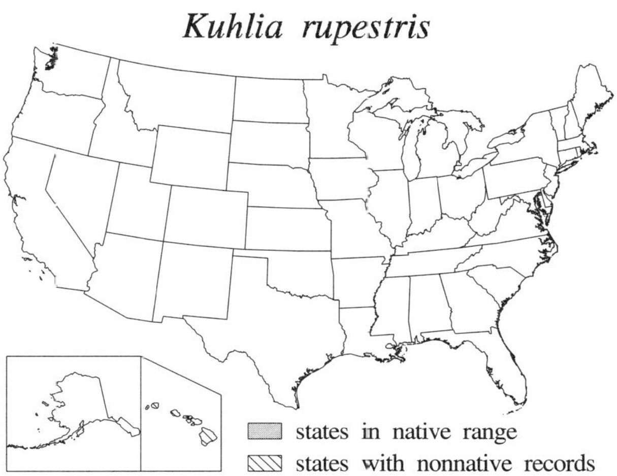

Kuhlia rupestris (Lacepède 1802)

Common Names: nato (rock flagtail, jungle perch).

Identification: Distinguishing characteristics, key, and an illustration were given by Allen (1991). Maximum size: 40 cm SL.

Native Range: Tropical Indo-Pacific; widely distributed in coastal streams from East Africa to Melanesia and eastern Australia (Allen 1991).

Nonindigenous Occurrences: Nato from Guam were imported to **Hawaii**; 104 of these fish were released into a small Oahu stream at Bellows Field in 1958 (Brock 1960; Maciolek 1984; Randall 1987).

Means of Introduction: Intentionally stocked, most likely as a sport fish. According to Brock (1960), a shipment of 43 specimens was brought to Hawaii from Guam in 1957 and held in laboratory tanks; a second shipment, containing 170 fish, was received in 1958. Specimens from the second shipment were subsequently stocked.

Status: Did not survive; extirpated.

Impact of Introduction: Unknown.

Remarks: This species is a popular angling fish in Australia (Allen 1991). It inhabits mainly fast-flowing clear coastal streams, but sometimes travels 200–300 km inland in large rivers (Allen 1991). A second species, *K. sandvicensis*, is endemic to marine and lower stream habitats of Hawaii (Devick 1991b).

Cichlidae - Cichlids

This very diverse perciform family consists of about 105 genera and at least 1,300 species (Nelson 1994). Cichlids are widely distributed in Africa and in Central and South America. A few species are also native to India, Syria, Palestine, and Mexico, and to such islands as Cuba, Hispaniola, Madagascar, and Ceylon (Berra 1981; Nelson 1994). The Rio Grande cichlid *Cichlasoma cyanoguttatum* is the only member of the family native to the United States (Page and Burr 1991). Cichlids are small- to medium-sized fishes that superficially resemble North American sunfishes, but are easily distinguished by their two-part lateral line and by having only a single nostril on each side (Page and Burr 1991). Many cichlids are popular aquarium fish and many introductions are a result of aquarium releases or escapes from fish farms. The larger species are commonly used in aquaculture, and some are stocked as sport and food fish or for aquatic weed control. Most cichlids are found in freshwater; nevertheless, many show some salt tolerance, and many of these, particularly some of the tilapias, are markedly euryhaline.

Aequidens pulcher

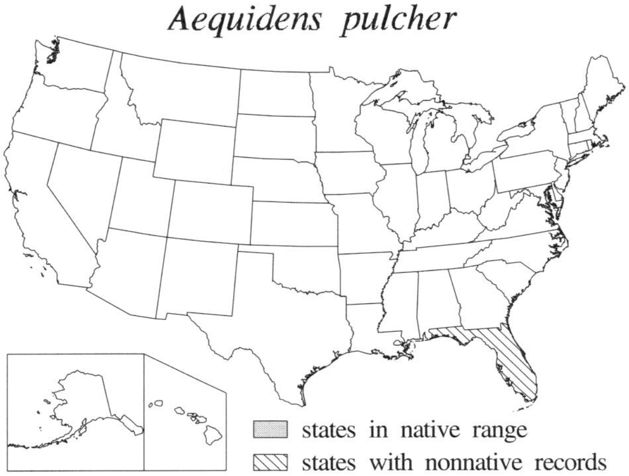

Aequidens pulcher (Gill 1858)

Common Name: blue acara.

Identification: Kullander (1983) considered the genus *Aequidens*, like the genus *Cichlasoma*, a catch-all for several different lineages. According to Kullander (1983, 1986), "*Aequidens*" *pulcher* represents a species complex consisting of "*A.*" *pulcher*, "*A.*" *coeruleopunctatus*, "*A.*" *latifrons*, and "*A.*" *rivulatus* belonging to a genus distinct from *Aequidens* as now restricted by Kullander; however, a modern systematic review of the "*A.*" *pulcher* complex has not been completed. Color photographs of *Aequidens pulcher*, or a closely related form, appeared in Axelrod (1993). Maximum size: 15 cm.

Native Range: Tropical America. Genus native to Orinoco, Amazon, Parnaiba, and Paraguay basins and to the Guianas of South America (Kullander and Nijssen 1989).

Nonindigenous Occurrences: This species was formerly considered common and reportedly was reproducing in canals and ditches around fish farms in the Tampa Bay, **Florida**, area, in Hillsborough and Manatee counties, during the 1960s; it was not collected in that area during 1970–1972 surveys or since, and the species is considered extirpated (Courtenay et al. 1974; Courtenay and Hensley 1979a; Courtenay and Stauffer 1990; Courtenay and Williams 1992).

Means of Introduction: Probable release or escape from local ornamental fish farms.

Status: Formerly established in Florida but now considered extirpated (Courtenay et al. 1974; Courtenay and Hensley 1979a); disappearance possibly a result of low winter temperatures (Courtenay and Stauffer 1990).

Impact of Introduction: Unknown.

Remarks: The past presence of this species in Florida is poorly documented and possibly based on a misidentification. There are no known voucher specimens. W. Smith-Vaniz and other U.S. Geological Survey researchers are finalizing a manuscript on the identification, distribution, and environmental biology of cichlids introduced to Florida. That work treats in detail this species and many others given in the following accounts.

Astronotus ocellatus

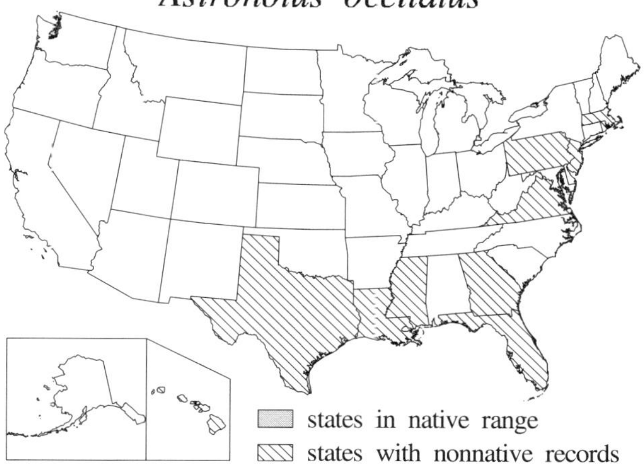

Astronotus ocellatus (Agassiz 1831)

Common Names: oscar (velvet cichlid, red oscar, marble cichlid).

Identification: For distinguishing characteristics see Kullander (1986) and Page and Burr (1991). The genus is in need of systematic review and future study may reveal that the species in the aquarium trade and in U.S. open waters is another species of *Astronotus* and not *A. ocellatus*. Photographs were provided by Kullander (1986) and Axelrod (1993). Maximum size: 40 cm.

Native Range: Tropical America. Genus *Astronotus* is native to South America including Orinoco and Amazon basins; also to French Guiana, and to northern part of Paraguay drainage, Parana basin (Kullander 1986).

Nonindigenous Occurrences: This species is established in ponds, rock pits, and canals in the southern part of **Florida** in Broward, Collier, Dade, Glades, Hendry (possibly), Monroe, and Palm Beach counties, including portions of Everglades National Park and Big Cypress National Preserve (Rivas 1965; Courtenay et al. 1974, 1984; Hogg 1976a, 1976b; Courtenay and Hensley 1979a, 1979b; Loftus and Kushlan 1987; Courtenay and Stauffer 1990). The species has been taken in many other Florida counties but is not known to be established. A single fish was observed in a canal in the Indian River drainage, Brevard County, on 3 July 1990 (J. Snodgrass, personal communication). Thirty-four specimens, 18.5–28 cm TL, were collected by hook and line from canals surrounding water conservation area 3A near Holiday Park, Pasco County, in or before January 1989 (USGS file record). A large specimen was taken by an angler from a river at Hontoon Island in the St. Johns River drainage, west of Orange City, Volusia County, in August 1991 (USGS file record). Records from Alachua County include a single fish taken from a pond near the town of Newberry on 27 October 1983, and one taken from a stream on the University of Florida campus, no date given (museum specimen). A large specimen, 33.7 cm TL, was taken by an angler from Boy's Club Pond in Panama City, Bay County, on 11 August 1991 (Anonymous 1993). In **Georgia**, a single fish was taken by an angler from Gum Creek in the Flint drainage, Crisp County, on 5 June 1991; a single fish was taken by an angler from the Chattahoochee River, Fulton County, on 27 August 1989 (Gennings, personal communication). It was introduced into **Hawaii** in the in the 1950s on the islands of Kauai and Oahu (Brock 1960; Shima 1972; Morita 1981; Maciolek 1984; Devick 1991a, 1991b), water bodies stocked included Nuuanu Reservoir 2 and Wailua Reservoir on Kauai and Wahiawa Reservoir on Oahu (Brock 1960). A fish identified as *Astronotus* sp. reportedly was collected from an unspecified public water body in **Louisiana**; no date was given (Tilyou, personal communication). A single fish was taken by an angler from Congamond Lake in Southwick, Hampden County, **Massachusetts**, in August 1978; a sight record exists from Amherst, Hampshire County, in 1992; other unconfirmed reports from the state are known (Hartel 1992; Cardoza et al. 1993). A fish was taken by hook and line in a local river in **Mississippi**, in 1978 (North American Native Fish Association 1978); it also has been referenced for this same state, but with no additional data, by others (e.g., Courtenay et al. 1984; Courtenay and Stauffer 1990). A single fish was caught in the Millstone River near its confluence with the Raritan River in Somerset County, **New Jersey**, in August 1994 (W. Wakefield, personal communication; photo). Specimens occasionally have been collected from unspecified localities in **Pennsylvania** (Courtenay and Hensley 1979a; Courtenay et al. 1984, 1991; Courtenay and Stauffer 1990). It has been reported from **Rhode Island** (Courtenay and Hensley 1979a; Courtenay et al. 1984; Courtenay and Stauffer 1990) (the inclusion of Rhode Island is apparently based on a report, poorly documented in our files, of two specimens taken by an angler from a pond in northern part of state in the early 1970s). Several **Texas** records are based on specimens taken by anglers, but no spawning or established populations are known (Howells 1992a). Specific Texas sites mentioned in correspondence include Victor Braunig Reservoir near San Antonio, Bexar County (Howells, personal communication), and the San Antonio River near San Antonio, Bexar County (Howells, referenced in USGS file records). A single specimen was taken from a Virginia Beach, **Virginia,** area pond during 1987 (Jenkins and Burkhead 1994, based on Southwick, personal communication).

Means of Introduction: The first Florida records were the result of deliberate stockings from an aquarium fish farm in southeastern Dade County in the late 1950s (Courtenay et al. 1974; Courtenay and Hensley 1979a; Courtenay and Stauffer 1990). During the same period, attempts were made to popularize the species as a sport fish under the name "velvet cichlid." Anglers have aided dispersal in Florida by transporting fish to new locations (Courtenay et al. 1974). It was sighted at the Anhinga Trail in Everglades National Park in the late 1980s (Loftus and Kushlan 1987). It apparently gained access to the park by way of the Southeast Florida Water Management District's Canal L31W, and it is now considered established (Courtenay 1989). The first introduction of this species into Hawaii was in 1952, for purposes of recreational fishing (Devick 1991a, 1991b); Hawaii's Division of Fish and Game made additional reservoir stockings in 1958; the original source of the Hawaii plants was a small stock obtained from Steinhart Aquarium in San Francisco (Brock 1960). Introductions into all other states and in more northern portions of Florida are apparently the result of aquarium releases (Courtenay and Stauffer 1990; Nico, personal communication).

Status: Established in south Florida and Hawaii. Reported from nine other states, including Georgia, Louisiana, Massachusetts, Mississippi, New Jersey, Pennsylvania, Rhode Island, Texas, and Virginia. Species established in Dade County, southeastern Florida, since late 1950s, but fish began to expand range greatly in late 1970s and early 1980s (Hogg 1976a, 1976b; Courtenay and Stauffer 1990).

Impact of Introduction: Largely unknown. In 1993, the state of Florida issued a health advisory warning people about eating freshwater fish contaminated with mercury. For portions of south Florida, the oscar was included in the list of fish affected and for which limited consumption was recommended.

Remarks: Oscars are common in the aquarium trade. Unfortunately, the taxonomy of this group is problematic and future research may determine that some oscars in the aquarium trade, as well as those taken in U.S. waters, are not *A. ocellatus* but another member of the genus (S. O. Kullander, personal communication). To confound the issue, artificial breeding has produced several color variants (Axelrod 1993). Oscars are considered potential competitors with native centrarchids for food and possibly for spawning areas (Courtenay and Hensley 1979a). Based on recent reports of the Florida Game and Freshwater Fish Commission, oscar now makes up a substantial proportion of the recreational fishery catch in some south Florida sectors. The Florida size record was a 1.1-kg, 32-cm TL fish taken from Lake Okeechobee, Palm Beach County, on 16 March 1994 (Ray 1994).

Voucher specimens: Florida (UF 30869, 41396, 44882, 91613, 91898, 91934); New Jersey (uncertain location).

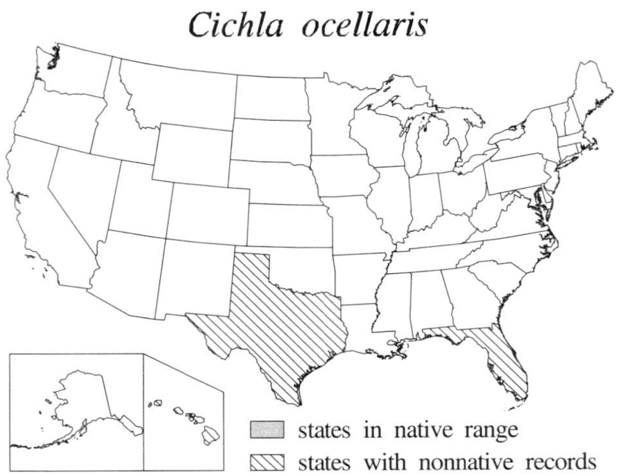

Cichla ocellaris Bloch and Schneider 1801

Common Names: peacock cichlid (pavon, tucunare, tuc, peacock bass, butterfly peacock bass, tucunare comun, eyespot cichlid, lukanani).

Identification: Currently, systematists recognize five *Cichla* species; however, taxonomic problems still exist and several species require description (Kullander and Nijssen 1989). The genus is currently being revised by Sven Kullander of Sweden. Many fish currently called *Cichla ocellaris* by state resource agencies may be members of another *Cichla* species or possibly hybrids. Photographs of *C. ocellaris*, or of closely related forms, appeared in Kullander and Nijssen (1989), Axelrod (1993), and in Larsen (1993). Maximum size: 70 cm TL.

Native Range: Tropical America. Although the genus *Cichla* is widespread in the Amazon and Orinoco basins of South America, the true *Cichla ocellaris* apparently is restricted to the Guianas (Kullander 1986; Kullander and Nijssen 1989).

Nonindigenous Occurrences: Fish identified as this species were introduced into southeastern **Florida** in the early 1960s (Moe 1964; Courtenay et al. 1974; Courtenay and Robins 1989); additional stockings were carried out in canals in Dade and Broward counties, starting in 1984 (Larsen 1993; Shafland 1995a). To date, it has been recorded from more than ten Florida canals and many of their lateral extensions, including Cypress Creek Canal (C-14 Canal), Middle River Canal (C-13 Canal), North New River Canal (G-15 Canal), South New River Canal (C-11 Canal); Snake Creek Canal (C-9 Canal), Biscayne Canal (C-8 Canal), Miami Canal (C-6 Canal), Snapper Creek Canal (C-2 Canal), Cutler Canal (C-100 Canal), Black Creek Canal (C-1 Canal), Princeton Canal (C-102 Canal), Mowry Canal (C-103 Canal), and Tamiami Canal (C-4 Canal), and from lakes near Miami International Airport (Larsen 1993; Hidalgo 1997). It was discovered in Canal C-111 along the eastern border of Everglades National Park in early 1997 (Nico, unpublished data). The species also may be present north to Jupiter Inlet in Palm Beach County (Larsen 1993). In **Hawaii**, there were two or more releases into reservoirs on Kauai and Oahu, starting in 1961 (Maciolek 1984; Larsen 1993). This species is established in Wahiawa Reservoir, a 350-acre, privately owned irrigation reservoir on Oahu (Devick 1991a, 1992; Larsen 1993). It was introduced along with speckled pavon *Cichla temensis* into five electropower plant reservoirs in **Texas** between 1978 and 1984: Alcoa Reservoir, Milam County; Lake Bastrop, Bastrop County; Coleto Creek Reservoir, Goliad County; Wilkes Reservoir, Upshur County; and Tradinghouse Creek Reservoir, McLennan County. All fish ultimately died as a result of temperature extremes (Garrett 1982; Howells and Garrett 1992).

Means of Introduction: Peacock cichlids have been stocked by state agencies as a sport fish. The Florida Game and Fresh Water Fish Commission obtained breeding stock from several regions of South America. The progeny were released into open waters primarily as a sport fish, but also with the hope that it would prey on and thus control other introduced cichlids (Courtenay and Robins 1989; Shafland 1995a). About 10,000 juveniles were released in the Fort Lauderdale area of Dade County, Florida, in 1964 (Moe 1964), but apparently these fish did not survive the cold winter of 1964–1965 (Courtenay et al. 1974; Courtenay and Robins 1989). More recent introductions into canals in Dade and Broward counties, starting in late 1984, resulted in established populations (Larsen 1993; Shafland 1995a). The Hawaiian Division of Fish and Game obtained their broodstock from an aquatic supply dealer in New York, ca. 1957 (Kanayama 1968). These fish reportedly came from Guyana (Larsen 1993). The first Texas populations were released by the Texas Parks and Wildlife Department in 1978. Texas stock came from Colombia and possibly Brazil, and from the Florida Game and Fresh Water Fish Commission.

Status: Established in south Florida (Courtenay and Robins 1989; Shafland 1995a) and Hawaii (Maciolek 1984). Shafland (1995b) indicated that fishable populations of peacock cichlid in Florida exist in more than 500 km of canals, plus numerous urban lakes in the metropolitan Miami-Ft. Lauderdale area. Some *Cichla* populations in Texas survived and reproduced for a brief period, but by 1992 all fish had died. For the most part, *Cichla* species are unable to survive cold winters, although evidence indicated fish in one Texas reservoir succumbed to high summer temperatures (Garrett 1982; Courtenay and Robins 1989; Howells and Garrett 1992).

Impact of Introduction: Largely unknown. Introduced *Cichla* in Florida have been found to include native fishes in their diets (Nico, unpublished data). In their native habitats, members of this genus are diurnal piscivores that consume a variety of prey (Winemiller et al. 1997). The introduction of peacock cichlids into Lake Gatun, Panama, was followed by largescale changes in food web structure and aquatic community composition (Zaret and Paine 1973).

Remarks: Research by Texas Parks and Wildlife Department on *Cichla* species was discontinued in 1986 (Howells and Garrett 1992). Established populations in Florida first were bred in captivity by the Florida Game and Fresh Water Fish Commission. The original breeding stock included fish reportedly imported from Brazil, Guyana, and Peru. Populations in Hawaiian reservoirs are supposedly descended from a single breeding pair (Kanayama 1968; Maciolek 1984). Courtenay and Robins (1989) and Shafland (1995a) reviewed the history of *Cichla* introductions into Florida and Texas. The taxonomic status of *Cichla* introduced into the United States is being reviewed by U.S. Geological Survey researchers in collaboration with S. Kullander.

Voucher specimens: Florida (UF 92166, 92181, 96389).

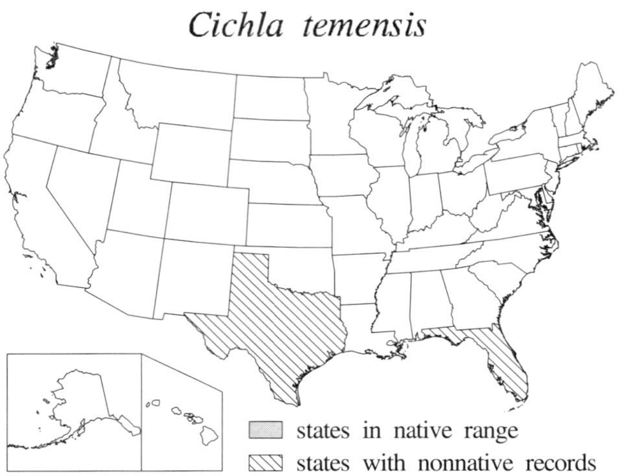

Cichla temensis Humboldt 1833

Common Names: speckled pavon (banded peacock cichlid, speckled peacock bass, tucunare pinema, striped tucunare, royal pavon, painted pavon, pavon lapa).

Identification: Taphorn and Barbarino-Duque (1993) provided distinguishing characteristics and a key to a few *Cichla* species, including *C. temensis*. A color photograph of *C. temensis*, or of a closely related form, appeared in Axelrod (1993). The genus currently is being revised by Sven Kullander, of Sweden. He has tentatively distinguished at least eight different species (Kullander 1986). Maximum size: 70 cm TL, possibly larger.

Native Range: Tropical America. Widespread in clear and black waters of Rio Negro and Orinoco basin, and in Central Amazon of South America (Kullander 1986).

Nonindigenous Occurrences: The speckled pavon was stocked by the Florida Game and Fresh Water Fish Commission in canals in southeastern **Florida** during the middle to late 1980s (Larsen 1993). It was stocked along with peacock cichlid *Cichla ocellaris* in five electropower plant reservoirs in **Texas** between 1978 and 1984: Alcoa Reservoir, Milam County; Lake Bastrop, Bastrop County; Coleto Creek Reservoir, Goliad County; Wilkes Reservoir, Upshur County; and Tradinghouse Creek Reservoir, McLennan County. All fish ultimately died as a result of temperature extremes (Garrett 1982; Howells and Garrett 1992).

Means of Introduction: Stocked by state agencies for use as sport fish. The Florida stocking reportedly involved only 110 fish, the last of which were released in 1988 (Larsen 1993).

Status: Species successfully overwintered at least two years in south Florida canals, but it is not yet considered to be established (Larsen 1993). Some *Cichla* populations in Texas survived and reproduced for a brief period, but by 1992 all fish had died. For the most part, *Cichla* species are unable to survive cold winters, and evidence indicates fish in one Texas reservoir succumbed to high summer temperatures (Garrett 1982; Howells and Garrett 1992).

Impact of Introduction: Unknown. In their native habitats, members of this genus are diurnal piscivores that consume a variety of prey (Winemiller et al. 1997).

Remarks: There are no known voucher specimens.

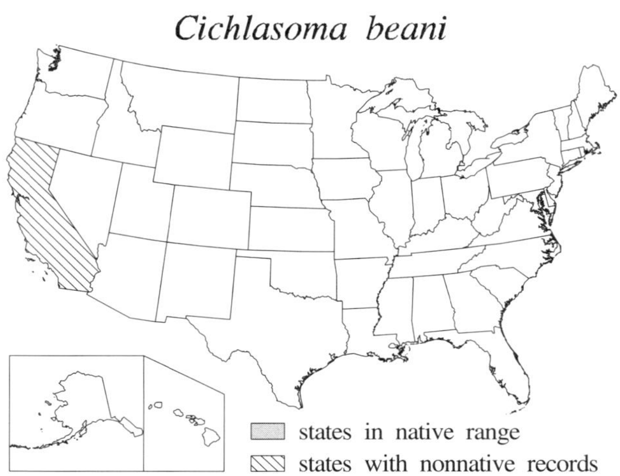

Cichlasoma beani (Jordan 1889)

Common Names: Sinaloan cichlid(green guapote, mojarra de Sinaloa, mojarra verde).

Identification: Another name is *Herichthys beani*. Identifying characteristics were given in Conkel (1993). Color photographs appeared in Konings (1989), Axelrod (1993), and Conkel (1993). Maximum size: 30 cm SL.

Native Range: Tropical and subtropical America. Pacific Slope drainages of Mexico from the Río Presidio to Río Grande de Santiago (Conkel 1993).

Nonindigenous Occurrences: The species was formerly established in several ponds adjacent to Putah Creek, Solano County, **California**, ca. 1975 (Shapovalov et al. 1981; Courtenay and Stauffer 1990).

Means of Introduction: Probable aquarium release.

Status: The California population disappeared by 1979 because some of pond habitat had dried up (Shapovalov et al. 1981), or for some other unknown reasons (Courtenay and Stauffer 1990).

Impact of Introduction: Unknown.

Remarks: There are no known voucher specimens.

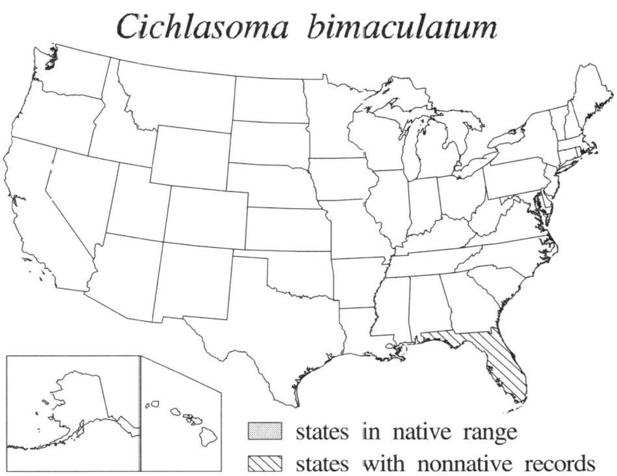

Cichlasoma bimaculatum (Linnaeus 1758)

Common Names: black acara (port cichlid, brown acara, twospot cichlid, coscorob).

Identification: This is a somewhat variable species that is very similar to *Cichlasoma amazonarum* (Kullander 1983). Distinguishing characteristics and photographs were given in Kullander (1983) and Kullander and Nijssen (1989); a few identifying characteristics also were provided by Page and Burr (1991). Kullander (1983) restricted the genus and provided a key to the 12 species, including *C. bimaculatum*, that he recognized as true *Cichlasoma*. Photographs of *C. bimaculatum* were given in Courtenay (1980). Many photographs in the aquarium literature are incorrectly captioned as *C. bimaculatum* (Kullander, personal communication). Maximum size: 20 cm SL.

Native Range: Tropical America. South America in the Guianas from Moruka River in Guyana east to about Sinnamary in French Guiana; also in Essequibo basin, and in adjacent Rio Branco drainage of Amazon basin (Kullander and Nijssen 1989).

Nonindigenous Occurrences: The species has been established in **Florida**, since the early 1960s; it was first discovered in Dade County (Rivas 1965). The expanded geographic range of the species includes the counties of Broward (Courtenay et al. 1974; Courtenay and Hensley 1979a), Collier (Courtenay and Hensley 1979a; Courtenay et al. 1986), Dade (Kushlan 1972; Courtenay et al. 1974; Hogg 1976a; Courtenay and Hensley 1979a; Loftus and Kushlan 1987), Hendry (Courtenay and Hensley 1979a), Lee (Nico, unpublished data), Martin (museum specimens), Monroe (Kushlan 1972; Courtenay et al. 1974; Courtenay and Hensley 1979a; Loftus and Kushlan 1987), and Palm Beach (Courtenay et al. 1974; Courtenay and Hensley 1979a). There are unconfirmed reports or possible occurrences of this species in Glades and Okeechobee counties (Courtenay et al. 1986; Courtenay and Stauffer 1990). It is established in Big Cypress National Preserve and in Everglades National Park (Kushlan 1972; Loftus and Kushlan 1987; Lorenz et al. 1997).

Means of Introduction: In southeastern Florida, introduction was the result of escapes and intentional releases from fish farms, probably during the mid- to late 1950s (Courtenay and Hensley 1979a). Some releases were deliberate attempts to dispose of unwanted and sometimes illegal fish stocks (Courtenay and Stauffer 1990).

Status: Established in Florida.

Impact of Introduction: Considered a competitor with native sunfish for spawning areas; juvenile bands of *C. bimaculatum* have been observed to surround spawning bluegill *Lepomis macrochirus* aggressively in attempts to feed on eggs (Hogg 1976a).

Remarks: The first Florida specimens of *C. bimaculatum* were reported and identified as *Aequidens portalegrensis* (= *Cichlasoma portalegrense*) by Rivas (1965), Bailey et al. (1970), Kushlan (1972), and others. Cultured in Florida possibly since the 1930s, *Cichlasoma bimaculatum* remained an important part of the aquarium trade until the late 1950s, and probably was the first aquarium fish to become established in open waters of Florida (Courtenay and Stauffer 1990). One of the most abundant introduced species in south Florida canal systems during the early 1970s, *C. bimaculatum* has been gradually replaced by *Tilapia mariae* as the most abundant cichlid, possibly through competition for space (Courtenay and Hensley 1979a, 1979b; Loftus and Kushlan 1987). Loftus and Kushlan (1987) provided a map showing the species' distribution in south Florida. Throughout its south Florida range, it is much more common in disturbed habitats, mainly canals, than in natural habitats (Loftus and Kushlan 1987). Spawning has been observed during every month of the year in southeastern Florida (Courtenay 1989).

Voucher specimens: Florida (UF 9896, 16981, 34906, 77572, 90574, 91855, 91861, 91865, many others).

Cichlasoma citrinellum

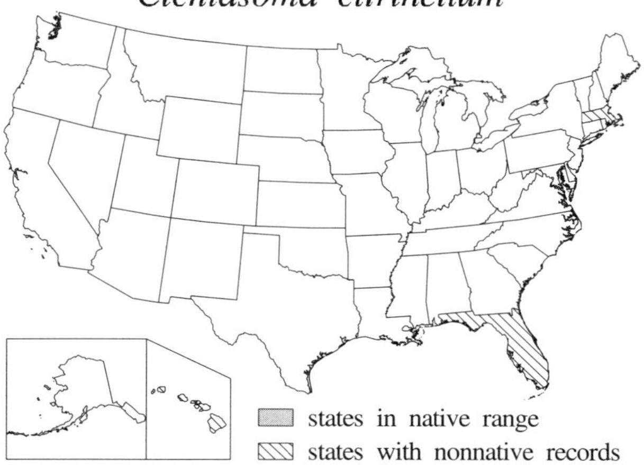

Cichlasoma citrinellum (Günther 1864)

Common Names: midas cichlid (red devil cichlid, mojarra rayada).

Identification: Synonyms and other names include *Cichlasoma granadense* and *Heros citrinellus*. Distinguishing characteristics were provided by Loiselle (1980a) and Page and Burr (1991). Color photographs appeared in Loiselle (1980a), Konings (1989) and Conkel (1993). Although morphologically very similar to *Cichlasoma labiatum*, the species is considered distinct (Barlow and Munsey 1976; Villa 1976). Maximum size: 24 cm.

Native Range: Tropical America. Atlantic Slope drainages in Central America from Great Lakes and crater lakes of Nicaragua to Río Coto in southeastern Costa Rica (Bussing 1987). Distribution maps given in Bussing (1987) and in Conkel (1993).

Nonindigenous Occurrences: First discovered in July 1980, the species is now established in the Black Creek Canal system, Dade County, **Florida** (Anderson et al. 1984; Loftus and Kushlan 1987). It was taken from the Mowry Canal C-103, Dade County, in July 1992 (Smith-Vaniz, personal communication). There is an unconfirmed report from Bivens Arm Lake in Gainesville, Alachua County, March 1985 (Miller, personal communication); this record is not supported by a voucher specimen and the report is considered somewhat suspect. The first report from **Hawaii** was of six specimens found dead, possibly from fungal infections, in Wahiawa Reservoir on Oahu, in February 1991 (Devick 1991a); the species is now documented as being firmly established in Wahiawa Reservoir and it also may be in other localities on Oahu (Yamamoto, personal communication). One fish was taken from Dodgeville Pond in Attleboro, Bristol County, **Massachusetts**, in July 1990 (Hartel 1992; Cardoza et al. 1993).

Means of Introduction: Introductions into Florida were via aquarium or fish farm releases. In other states, introductions most likely represent aquarium releases.

Status: Established in Florida and Hawaii. Reported from Massachusetts.

Impact of Introduction: Unknown.

Remarks: This species exhibits two basic color morphs, gray and orange. Most Florida fish are the orange morph (Anderson et al. 1984). A small specimen taken in Hillsborough County, Florida, in 1976 was reported as *C. labiatum* (Courtenay and Stauffer 1990), but was later tentatively reidentified as *C. citrinellum* (Smith-Vaniz, personal communication).

Voucher specimens: Florida (UF 31651, 77515, 91875, 92165).

Cichlasoma cyanoguttatum

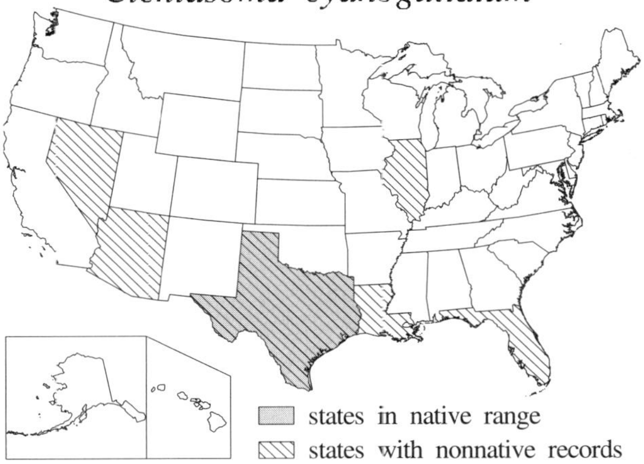

Cichlasoma cyanoguttatum (Baird and Girard 1854)

Common Names: Rio Grande cichlid (Texas cichlid, Rio Grande perch, mojarra de Norte).

Identification: Synonyms and other names include *Cichlasoma pavonaceum*, *C. laurae*, and *Herichthys cyanoguttatus*. For distinguishing characteristics and figure see Hubbs et al. (1991) and Page and Burr (1991). Color photographs were given in Conkel (1993). Maximum size: 30 cm.

Native Range: Tropical and subtropical America. Northeast Mexico and southern Texas, on the Atlantic Coast from the Río Conchos to the Río Grande basin. Native Texas range limited to the Nueces and Rio Grande drainages (Hubbs et al. 1978). Distribution maps (incomplete) given in Lee et al. (1980 et seq.), Conkel (1993), and Page and Burr (1991).

Nonindigenous Occurrences: Two young cichlids (tentatively identified as *C. cyanoguttatum*) were taken from a canal in Tempe, Maricopa County, **Arizona**, prior to 1973 (Minckley 1973). Specimens were taken from several locations in the Tampa, **Florida**, area, some as early as the 1940s. It is established in several areas in Florida including Six Mile Creek in Hillsborough County (Burgess 1958), a phosphate pit in Mulberry in Polk County (Courtenay et al. 1974), Lake Seminole in Pinellas County (Courtenay and Hensley 1979a), and in canals near fish hatcheries in Brevard (died out) and Indian River counties (Smith-Vaniz and J. D. Williams, personal communication). Conkel (1993) reported on a population in the Alafia River in Tampa. Populations have been established locally in Blue Hole at Key Deer National Refuge on Big Pine Key, Monroe County, since about 1971 (Smith-Vaniz, personal communication). This cichlid was recorded as present and presumably reproducing in heated power plant effluents in Powerton Lake near Pekin, **Illinois** (Illinois River drainage), Tazewell County, after having been released accidentally in the mid-1980s (Burr 1991; Page and Laird 1993). A single specimen was taken from Irish Bayou, Lake Pontchartrain, Orleans Parish, **Louisiana**, in June 1996 (Burr, personal communication). Several more were taken from the lake area in 1997 (R. Cashner, personal communication). It was in Roger's Spring near Lake Mead, Clark County, **Nevada**, in 1983 (Courtenay and Deacon 1983; Deacon and Williams 1984). It has been widely introduced and reported as established in the Colorado, San Antonio, San Marcos, Guadalupe, and Comal rivers in the Edwards Plateau region of central **Texas** (Brown 1953; Hubbs et al. 1978). The San Gabriel River, in the Brazos River system, contains the state's northernmost established population (Hubbs et al. 1991).

Means of Introduction: *Cichlasoma cyanoguttatum* was brought to the Guadalupe River drainage by the Fish Cultural Station, part of the U.S. Fish and Wildlife Service, at San Marcos, Texas, in 1928 and released into waters on the Edwards Plateau between 1928 and 1941 or 1943 (Brown 1953; Hubbs et al. 1978). Presumably, San Antonio River stocks came from near Mission, Texas (Hubbs et al. 1978). The species reportedly was introduced first into Florida from Texas stocks around 1941 by a private individual from Mulberry, Polk County. The same individual is reported as having made additional introductions at Mulberry over a period of several years (Courtenay and Hensley 1979a). A fish farm may have been the source of introduction into Six Mile Creek, Hillsborough County, Florida, because this species was cultured as an aquarium fish during the 1940s and early 1950s under the trade name "Texas bluespot" (Burgess 1958; Courtenay and Hensley 1979a). Some Florida introductions supposedly were due to flooding of resident fish farms (Conkel 1993). Introductions elsewhere were most likely the result of aquarium releases or fish farm escapes.

Status: Introduced and established or locally established in parts of Florida and Texas. It was established locally in a power plant lake in Illinois but the species no longer is thought to be reproducing in that state (Laird and Page 1996). It was reported from Arizona and Louisiana. Formerly established in Nevada, but now considered extirpated in that state.

Impact of Introduction: Unknown.

Remarks: *Cichlasoma cyanoguttatum*, naturally occurring in parts of southern Texas, represents the only cichlid native to the United States. According to Courtenay et al. (1974), reports of this species at Gibsonton, Riverview, and Ruskin in Hillsborough County, Florida, by Buntz and Chapman (1970) were based on misidentifications of Jack Dempsey *C. octofasciatum*. Introduced Texas populations mainly occur in larger springs and spring outflows with favorable water temperatures (Hubbs et al. 1978; Lee et al. 1980 et seq.).

Voucher specimens: Florida (UF 40190, 56196, 91892, 92197), Illinois (SIUC 10805, INHS 68039), Louisiana (SIUC 26460; UNO 9487), Texas (INHS 74255).

Cichlasoma labiatum

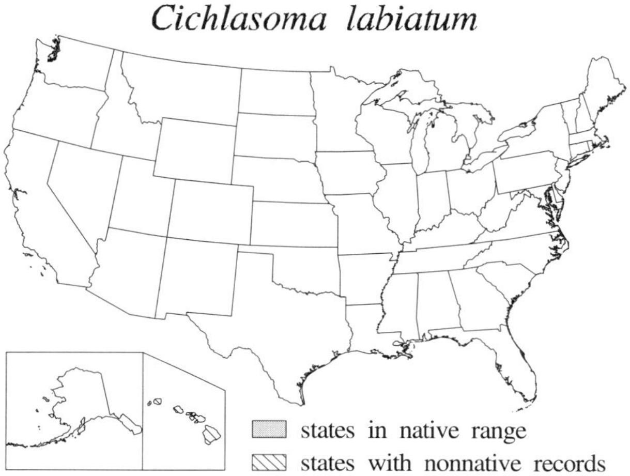

Cichlasoma labiatum (Günther 1864)

Common Names: red devil (mojarra picuda).

Identification: For identification see Barlow and Munsey (1976) and Villa (1976). Distinguishing characteristics and color photographs were provided by Loiselle (1980a) and Conkel (1993). Photographs also appeared in Konings (1989). Although morphologically very similar to the midas cichlid *C. citrinellum*, the species is considered distinct (Barlow and Munsey 1976; Villa 1976). Maximum size: 24 cm SL (Conkel 1993).

Native Range: Tropical America. Atlantic Slope drainages in Central America, including Great Lakes of Nicaragua (Conkel 1993).

Nonindigenous Occurrences: A single live fish was taken from Wahiawa Reservoir on Oahu, **Hawaii**, in February 1991 (Devick 1991a).

Means of Introduction: Probable aquarium release.

Status: Reported from Hawaii.

Impact of Introduction: Unknown.

Remarks: A fish taken from Hillsborough County, Florida, in 1976, originally was reported to be this species (Courtenay and Hensley 1979a; Courtenay and Stauffer 1990), but that specimen later was reidentified tentatively as *Cichlasoma citrinellum* (Smith-Vaniz, personal communication).

There are no known voucher specimens.

Cichlasoma managuense

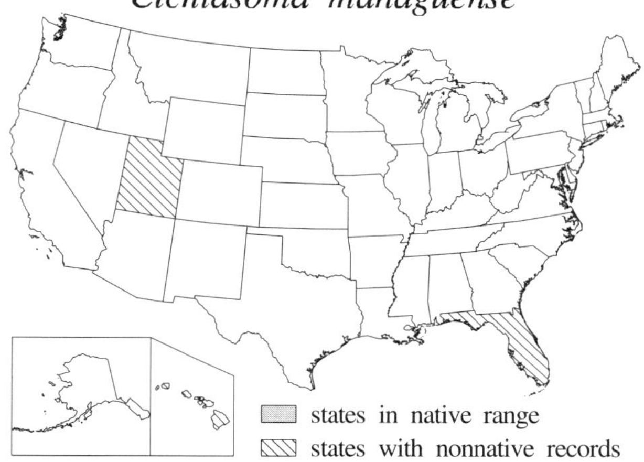

Cichlasoma managuense (Günther 1867)

Common Names: jaguar guapote (managuense, tiger guapote, Aztec cichlid).

Identification: For distinguishing characteristics see Page and Burr (1991) and Kullander and Hartel (1997). Color photographs appeared in Bussing (1987), Konings (1989), and Conkel (1993). Recently, Kullander and Hartel (1997) concluded that the more valid scientific name for this species is *Parachromis managuensis*. Maximum size: 25 cm SL.

Native Range: Tropical America. Atlantic Slope drainages in Central America from Río Ulúa in Honduras to Río Matina in Costa Rica (Bussing 1987). Distribution maps in Bussing (1987) and Conkel (1993).

Nonindigenous Occurrences: The first records of this species in **Florida** were based on specimens taken from canals and other water bodies in the Miami area, Dade County, in the early 1990s. More recently, the species has appeared in other canals in the southeastern part of the state including sites west of Miami in canals located in the eastern portions of the Everglades (Gestring and Shafland 1997; Nico, unpublished data). The species is locally established in a pond on the campus of University of Hawaii on Oahu, **Hawaii**; it was first found at that site ca. 1992 (Yamamoto, personal communication). It is established in Boiler Spring, a thermal spring pool in the Virgin-Colorado Rivers drainage of southwestern **Utah**, in the community of Washington, Washington County (Marsh et al. 1989; Courtenay and Stauffer 1990). Boiler Spring fish first were discovered in April 1988 during a rotenone operation by state personnel; the population consisted of 3–4 different size classes totaling an estimated 500 fish (Minckley and C. Knight, personal communication). Voucher specimens were taken in September 1988; at that time, efforts to eliminate the fish with rotenone and explosives failed (Marsh et al. 1989). There were additional unconfirmed reports of fish in other habitats, including at least one breeding population in a local private pond (Marsh et al. 1989). State personnel more recently indicated that the species has been eradicated from Utah (B. Schmidt, personal communication).

Means of Introduction: Introductions are presumably due to aquarium releases, although escapes from fish farms cannot be completely ruled out in southern Florida. The first documented U.S. introduction record was from Utah in 1988 (Marsh et al. 1989).

Status: Established in Florida; formerly established in Utah, presumably eradicated; locally established in Hawaii.

Impact of Introduction: Largely unknown. In some south Florida localities this species apparently is displacing native centrarchids (Nico, unpublished data).

Remarks: On 14 September 1988, Utah Division of Wildlife personnel killed several hundred *C. managuense* at Boiler Spring with rotenone and explosives, but they were unsuccessful in eradicating the fish (Marsh et al. 1989); however, the most recent report indicates the species has been eradicated throughout the state (B. Schmidt, personal communication). Boiler Spring flows to Virgin River via Mill Creek and irrigation systems. This aggressive cichlid was considered a potential threat as a predator of native fish fauna of the Virgin River (Marsh et al. 1989). The taxonomy of many Central American cichlids is problematic and future research may determine that the "jaguar guapote" introduced to U.S. waters is not *C. managuensis* but another closely-related species.

Voucher specimens: Florida (UF uncatalogued), Hawaii (UF 98933), Utah (ASU 11861).

Cichlasoma meeki

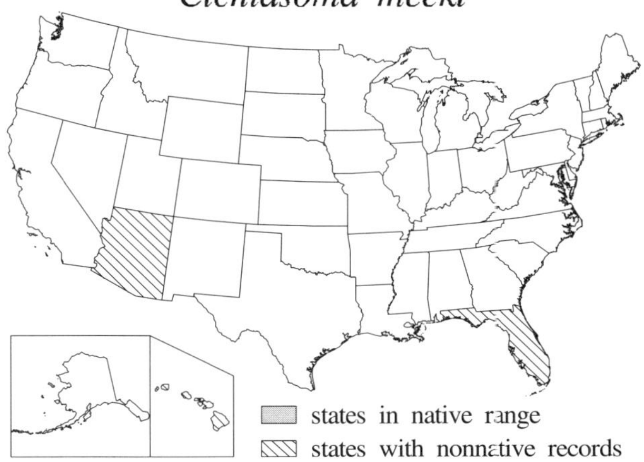

Cichlasoma meeki (Brind 1918)

Common Names: firemouth cichlid (firemouth, red-breasted cichlid).

Identification: Hasse (1981) redescribed the species. Possible synonyms include *Cichlasoma hyorhynchum* and *C. champotonis* (Hasse 1981; Conkel 1993). For distinguishing characteristics and figures see Page and Burr (1991) and Greenfield and Thomerson (1997). Color photographs were given in Konings (1989) and in Conkel (1993). Maximum size: 17 cm.

Native Range: Tropical America. Atlantic Slope drainages in Middle America from the Ríos Grijalva and Usumacinta basins of Mexico and Guatemala, and the Yucatan Peninsula south to southern Belize (Conkel 1993; Greenfield and Thomerson 1997). Distribution maps (incomplete) given in Lee et al. (1980 et seq.), in Conkel (1993), and in Greenfield and Thomerson (1997).

Nonindigenous Occurrences: A single fish was collected from a canal in Mesa, Maricopa County, **Arizona**, in 1973 (Minckley 1973; also see Courtenay and Stauffer 1990). The first record from **Florida** was a report of an established population in a rock pit in northwest Miami, Dade County (Rivas 1965). The species subsequently was found in several canals, ditches, borrow pits, and sinkhole and quarry ponds in various parts of southeastern Florida in Dade, Palm Beach, Brevard, Broward, and Monroe counties during the late 1960s and 1970s (Ogilvie 1969; Courtenay et al. 1974; Courtenay and Hensley 1979a; Courtenay and Stauffer 1990; Hogg 1976a, 1976b; museum specimens). Apparently introduced into a Miami area canal system subsequent to 1972 (Hogg 1976a), it was considered established in Comfort Canal, Miami, with a range extending west to several small canals south of the Tamiami Canal in Dade County as of 1976 (Hogg 1976a, 1976b; Courtenay and Hensley 1979a). A large population was established in an isolated borrow pit adjacent to a former amusement park in Dania, Broward County; that population was eradicated by state personnel in July 1981 (Courtenay et al. 1984; Courtenay and Stauffer 1990). First released in **Hawaii** in Nuuanu Reservoir in 1940 (Brock 1960), this species is now established on Oahu (Brock 1960; Maciolek 1984; Devick 1991a, 1991b). The species was also reported as occurring in a drainage canal in the McCully District of Honolulu (Brock 1960).

Means of Introduction: The Arizona and Hawaii introductions likely represent aquarium releases. Hawaii fish were listed as a deliberate introduction by Devick (1991a, 1991b). Some Florida introductions were likely a result of escapes or releases from former fish farms (Hogg 1976a, 1976b; Courtenay and Hensley 1979a; Lee et al. 1980 et seq.; Courtenay and Stauffer 1990).

Status: Established in Florida since about 1976 and in Hawaii since 1940; reported from Arizona.

Impact of Introduction: Unknown.

Remarks: As with other cichlids, there is the potential that *C. meeki* will compete with native centrarchids. Loftus and Kushlan (1987) did not collect this species south of the Tamiami Canal during their 1976–1983 fish surveys of southern Florida and concluded that *C. meeki* was either very localized in distribution or reduced in numbers and range since Hogg's (1976a) work. The species appears to tolerate a wide range in salinity. The dominant habitat in Hawaii was listed as reservoirs by Maciolek (1984), but as streams by Devick (1991a, 1991b).

Voucher specimens: Florida (UF 91892, 91893, 92200, 92203, 93537).

Cichlasoma nigrofasciatum

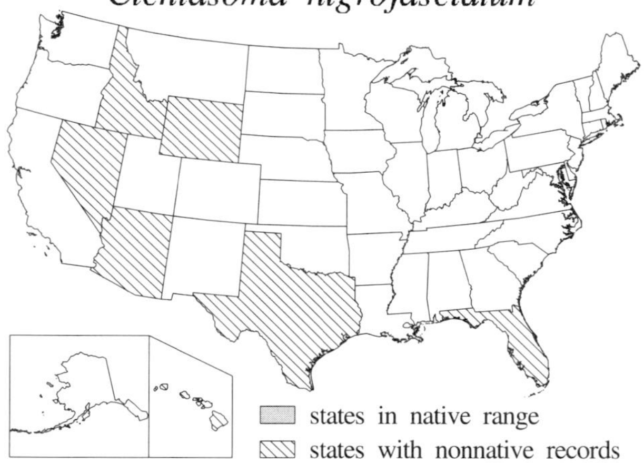

Cichlasoma nigrofasciatum (Günther 1867)

Common Names: convict cichlid (zebra cichlid, zebra chanchito, punto rojo, punto naranja).

Identification: For distinguishing characteristics and figure see Page and Burr (1991). Color photographs were given in Bussing (1987), Konings (1989), and Conkel (1993). Maximum size: 12 cm.

Native Range: Central America, including Pacific Slope drainages of Guatemala, from Río Suchiate to Río Grande de Taracoles in northwestern Costa Rica. Atlantic Slope drainages in Central America, including Honduras, from Río Aguan to Río Guarumo in Panama (Conkel 1993). Distribution maps in Bussing (1987) and in Conkel (1993).

Nonindigenous Occurrences: A population occupied a borrow pit of the often dry Salt River, at the eastern edge of Phoenix, Maricopa County, **Arizona**, from 1969 through 1972, but apparently was eliminated by 1973 floods; a reproducing population found in Mesa, Maricopa County, during the summer of 1970, did not overwinter (Minckley 1973). Single specimens are not uncommon. Reproducing populations may still survive in canals of the Phoenix metropolitan area and also may be established in warmer areas of the southwestern part of the state (Minckley 1973; Lee et al. 1980 et seq.). The first record from **Florida** was an unconfirmed report of an established population in a rock pit in northwest Miami, Dade County (Rivas 1965); however, the site's location is not known, and there are no known voucher specimens. A population was discovered in Green Pond on the University of Florida campus in Gainesville, Alachua County, in 1997 (G. T. Bergman, personal communication). In **Hawaii** the species first was reported in an irrigation ditch and a reservoir near Haleiwa, on Oahu, in 1983 (Devick 1991b). It was established in Nuuanu #4 Reservoir and found in the lower reaches of several windward streams on Oahu (Devick 1991b). It recently was established in Opaek'a and Kalama streams (adjacent to the North Fork of the Wailua River), on Kauai, ca. 1990 (Devick 1991a). The only documented case in **Idaho** is an established population of an amelanic form in Barney Hot Spring and the upper end of Barney Creek in Little Lost River Valley, Custer County; introduction into the site first was reported in 1985, when a sample of 19 fish (15–92 mm SL) was taken on 7 September 1985 (Courtenay et al. 1987). A report produced by the Idaho Game and Fish (Fisheries Management Plan 1991–1995) listed this species as being confined to one or a few geothermal waters in the region known as Snake River drainage below Shoshone Falls, in the south central part of the state; however, that report may contain erroneous information. The species is established in several warm springs along the White and Moapa (Muddy) rivers in southeastern **Nevada**. The earliest records are from Rogers Spring, near the Overton arm of Lake Mead, Clark County, dating to March 1963 (Deacon et al. 1964; Courtenay and Deacon 1983). There are three springs along the White River with reported established populations, all in Lincoln County: Ash Spring and its outflow near Alamo, since June 1964; Crystal Springs since the 1970s; and Hiko Spring since about 1984 (Hubbs and Deacon 1964; Courtenay and Hensley 1979a; Courtenay and Deacon 1982; Deacon and Williams 1984; Courtenay et al. 1985). One specimen was seined from the San Marcos River below the Spring Lake dam in Hays County, **Texas**, on 19 September 1989 (Whiteside and Berkhouse 1992). Several fish were taken from Kelly Warm Springs in Teton County, **Wyoming**, on 10 October 1991 (G. Baxter, personal communication), but the species is not known to be established (M. Stone, personal communication). This species was likely introduced into that site after the mid-1980s because it was not found when the site was sampled by Courtenay on 23 July 1984 (Courtenay et al. 1987).

Means of Introduction: Most introductions probably represent aquarium releases. The species reportedly was introduced into Nuuanu #4 Reservoir, Oahu, Hawaii, by poachers who were using the cichlid as bait for channel catfish; other sites represent separate releases, most likely of aquarium fish (Devick 1991b). The first Hawaii record is from 1983 (Devick 1991b). Discovery of a population in Rogers Spring, Nevada, on 7 March 1963, represents the first record for U.S. open waters (Deacon et al. 1964; Hubbs and Deacon 1964). The first Idaho record dates to 1985 (Courtenay et al. 1987).

Status: Established or locally established in Hawaii, Idaho, and Nevada; possibly established in Arizona, but current status unknown (Courtenay and Hensley 1979a); apparently established locally in Florida; reported from Wyoming and Texas.

Impact of Introduction: Mostly unknown. According to Deacon et al. (1964), *C. nigrofasciatum*, in combination with other foreign fishes, apparently caused the decline and extermination of a population of the native speckled dace *Rhinichthys osculus* near the Overton arm of Lake Mead, Nevada. Along with other foreign species, it has been implicated as a threat to the survival of the endangered White River springfish *Crenichthys baileyi* and other native White River fish in southeastern Nevada (Deacon and Bradley 1972; Page and Burr 1991). Introductions, in combination with other human-induced changes, caused the disappearance of all native fishes from Hiko Spring, Nevada (Courtenay et al. 1985).

Remarks: Concern exists that this aggressive cichlid will compete with native sunfishes for spawning sites (Courtenay and Hensley 1979a). There was early concern that Nevada populations, if they were to become established in Lake Mead, might affect the sport fishery adversely (Deacon et al. 1964). An attempt in December 1963 to eliminate the Nevada population at Rogers Spring failed (Hubbs and Deacon 1964). Barney Creek in Idaho is a tributary of Summitt Creek; the Little Lost River Valley is an isolated, cold, high-altitude drainage whose water sinks into the sand or gravel of the Arco Desert (W. Horton, personal communication; also see Linder 1964).

Voucher specimens: Florida (UF uncatalogued); Idaho (UMMZ 213373; FAU WRC-ID-2).

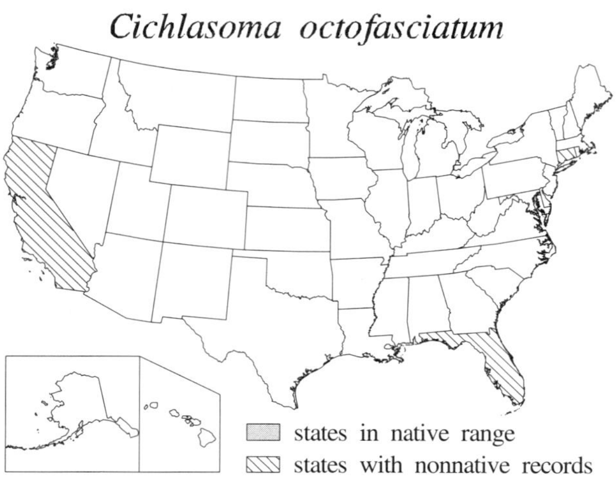

Cichlasoma octofasciatum (Regan 1903)

Common Names: Jack Dempsey (mojarra castarrica).

Identification: Synonyms are *Cichlasoma bifasciatum* and *C. hedricki*. The species is included in identification keys of Greenfield and Thomerson (1997). For distinguishing characteristics and figure see Page and Burr (1991). Color photographs appeared in Konings (1989) and in Conkel (1993). Maximum size: 25 cm.

Native Range: Tropical America. Atlantic Slope drainages in Middle America from Río Paso San Juan, Veracruz, Mexico, south to the Río Ulua basin in Honduras (Greenfield and Thomerson 1997).

Nonindigenous Occurrences: Three specimens were taken from Lafayette Creek, Contra Costa County, **California**, in 1986 (A. J. Cordone, personal communication). Specimens 6–15 cm long have been taken from major drainages in **Connecticut** (Whitworth 1996). This species has been found in several canals, ditches, and quarry ponds in various parts of **Florida** in Dade, Palm Beach, Hillsborough, Levy, and Brevard counties (Courtenay et al. 1974; Hogg 1976a, 1976b; Courtenay and Hensley 1979a; Levine et al. 1979; Dial and Wainright 1983; Gilmore et al. 1983; Jennings 1986; J. D. Williams, personal observation). Courtenay et al. (1974) reported established populations near fish farms west of Lantana, Palm Beach County, but the species was not found during a 1978 survey (Courtenay and Hensley 1979a). An isolated population formerly in a rock pit in southern Levy County was eradicated in 1977 (Levine et al. 1979; Courtenay and Hensley 1979a).

Means of Introduction: This species probably was introduced through a combination of aquarium releases and fish farm escapes. In Florida, introductions into Palm Beach, Hillsborough, and Dade counties reportedly were due to intentional or accidental releases from fish farms (Courtenay et al. 1974; Hogg 1976a, 1976b).

Status: Established in Florida; reported from California and Connecticut.

Impact of Introduction: Unknown.

Remarks: Juvenile fish from the vicinities of Gibsonton, Riverview, and Ruskin in Hillsborough County, identified as *Cichlasoma cyanoguttatum* by Buntz and Chapman (1970), were actually this species (Courtenay et al. 1974). This highly aggressive and combative species possibly competes with native sunfishes for food and spawning sites. The population in Alachua County, the northernmost record for Florida, has persisted for more than 10 years.

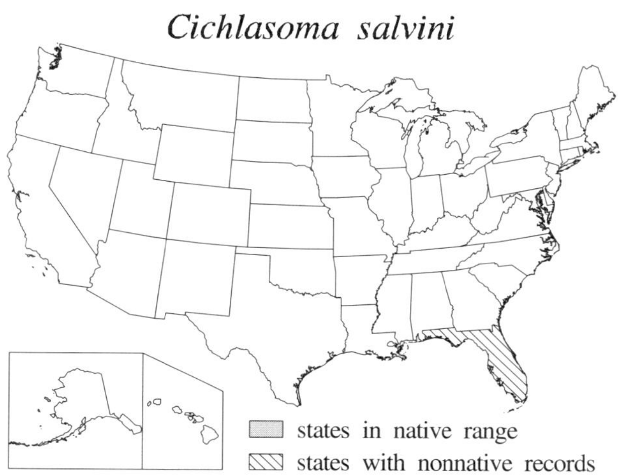

Cichlasoma salvini (Günther 1862)

Common Names: yellowbelly guapote (yellowbelly cichlid, tricolor cichlid, guapote tricolor).

Identification: Distinguishing characteristics were given by Meek (1904). The species is included in keys in Greenfield and Thomerson (1997). A few distinguishing characteristics and color photographs were provided by Conkel (1993). Photographs also appeared in Konings (1989) and Axelrod (1993, referenced as *Herichthys salvini*). Maximum size: 22 cm SL (Conkel 1993).

Native Range: Tropical America. Atlantic Slope drainages of Middle America from Río Papaloapan, Mexico, south to Sulphur River, Guatemala (Conkel 1993; Greenfield and Thomerson 1997).

Nonindigenous Occurrences: A population was established in a rock pit or borrow pit adjacent to the abandoned tourist attraction "Pirate's World" in Dania, southern Broward County, **Florida**, ca. 1980; that population was eradicated by state personnel in 1981 (Courtenay et al. 1984; Courtenay and Stauffer 1990). The borrow pit site later was converted into a parking lot and no longer exists. A second established population was documented in the South New River Canal (C-11 Canal), Broward County, during collecting efforts in 1990 and again in 1993 (Smith-Vaniz, personal communication). Additional specimens were collected from the same canal in 1992, 1993, and 1994 (Shafland 1995b).

Means of Introduction: Probable aquarium release (Courtenay and Stauffer 1990). Shafland (1995b) speculated that the rock pit population eliminated in 1981 could be the source of the South New River Canal population, or it could be the result of a recent illegal introduction.

Status: Established in south Florida.

Impact of Introduction: Unknown.

Remarks: Although information on the existence of the first Florida population was published, the exact location of the borrow pit was mistakenly identified as being in Dade County (e.g., Courtenay and Hensley 1979a, 1980). Shafland (1995b) indicated that the rock pit is located only a few kilometers from the South New River Canal. An underwater photograph of a live *C. salvini* inhabiting the Broward County rock pit was published in Courtenay (1980).

Voucher specimens: Florida (UF 90533, 92188, 93543).

Cichlasoma spilurum

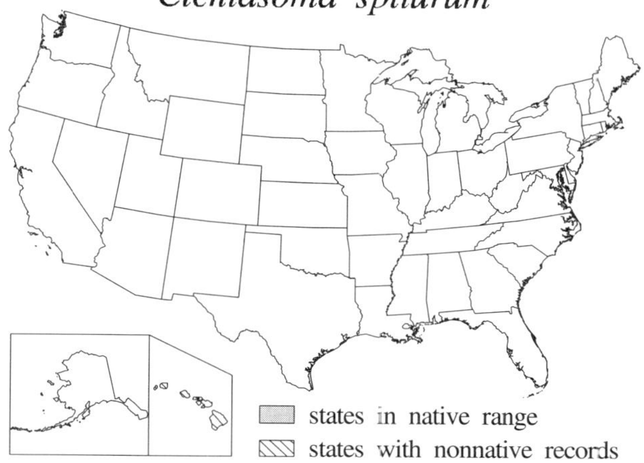

Cichlasoma spilurum (Günther 1862)

Common Names: blue-eye cichlid (Cutter's cichlid).

Identification: Synonyms or other names include *Cichlasoma spilurus* and *C. cutteri*. The species is included in keys of Greenfield and Thomerson (1997). A few distinguishing characteristics were given in Sterba (1973) and in Conkel (1993). Photographs were provided by Konings (1989) and by Conkel (1993). Maximum size: 11 cm SL.

Native Range: Tropical America. Atlantic Slope drainages of Middle America (Conkel 1993); from the Yucatan Peninsula, Mexico, throughout Belize, south to Panama (Greenfield and Thomerson 1997).

Nonindigenous Occurrences: The species is established in reservoir or pond habitats on Oahu, **Hawaii**; it was first released or discovered in 1984 (Devick 1991a, 1991b).

Means of Introduction: Aquarium escape; considered an accidental introduction (Devick 1991a, 1991b).

Status: Established in Hawaii since about 1984 (Devick 1991a, 1991b).

Impact of Introduction: Unknown.

Remarks: There are no known voucher specimens.

Cichlasoma synspilum

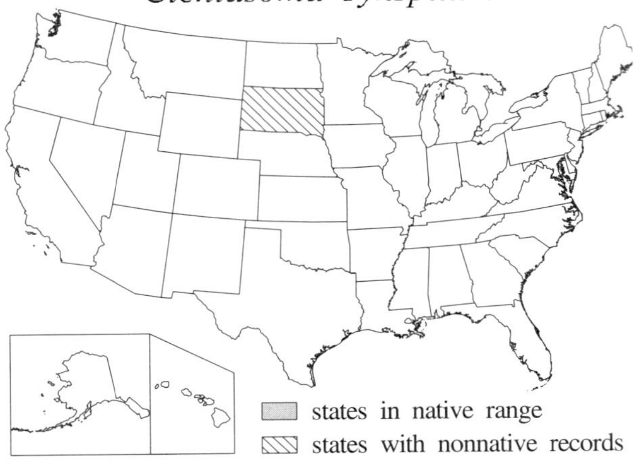

Cichlasoma trimaculatum

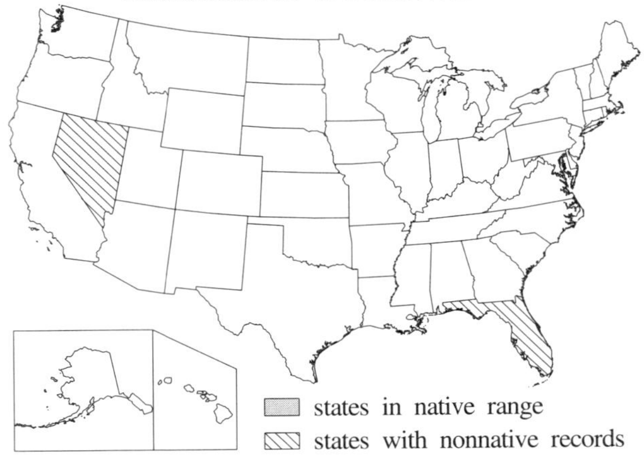

Cichlasoma synspilum Hubbs 1935

Common Names: redhead cichlid (pastel cichlid, quetzal cichlid).

Identification: Synonyms or other names include *Cichlasoma synspilus* and *C. hicklingi*. The species is included in identification keys of Greenfield and Thomerson (1997). Konings (1989) and Conkel (1993) discussed color variants and provided photographs. Photographs also appeared in Axelrod (1993). Maximum size: 30 cm (Conkel 1993).

Native Range: Tropical America. Atlantic Slope of Middle America including southern Mexico (Usumarinata River basin), Guatemala, and Belize (Conkel 1993; Greenfield and Thomerson 1997).

Nonindigenous Occurrences: A single fish (104 mm SL, 58 g) was taken with boat-mounted electrofishing gear from Gateway Lake at Ellsworth Air Force Base, Pennington County, **South Dakota**, on 15 September 1993 (museum specimen).

Means of Introduction: Probable aquarium release.

Status: Reported from South Dakota.

Impact of Introduction: Unknown.

Remarks: Voucher specimens: South Dakota (UF 96525).

Cichlasoma trimaculatum (Günther 1867)

Common Names: threespot cichlid (red eyed cichlid).

Identification: Identifying characteristics were given by Conkel (1993). Color photographs appeared in Loiselle (1980a), Konings (1989), Conkel (1993), and Axelrod (1993). Other names include *Heros trimaculatus* and *Herichthys trimaculatus*. Maximum size: 36.5 cm SL (Conkel 1993).

Native Range: Pacific slope drainages of Middle America from Laguna Coyuca, Mexico to the Río Lempa, El Salvador (Loiselle 1980a; Conkel 1993).

Nonindigenous Occurrences: This species was locally established in a rock pit in Manatee County, **Florida**, during the mid-1970s (Shafland 1976; Courtenay and Stauffer 1990; Smith-Vaniz, personal communication). It was also listed as unsuccessfully introduced into an unspecified site in **Nevada** (Robins et al. 1991b).

Means of Introduction: Probably aquarium release (Courtenay and Stauffer 1990).

Status: The Florida populations was first discovered in the mid-1970s (Shafland 1976); it was eradicated by personnel of the Florida Game and Fresh Water Fish Commission in 1975 (Courtenay and Stauffer 1990). The only published information on the Nevada introductions was a very abbreviated account given by Robins et al. (1991b).

Impact of Introduction: Unknown.

Remarks: According to Loiselle (1980a), the morphology and behavior of this species is almost indistinguishable from *amphilophus* of the *C. labiatum* group. It is common in the aquarium trade (Loiselle 1980a).

Voucher specimens: Florida (UF 92207).

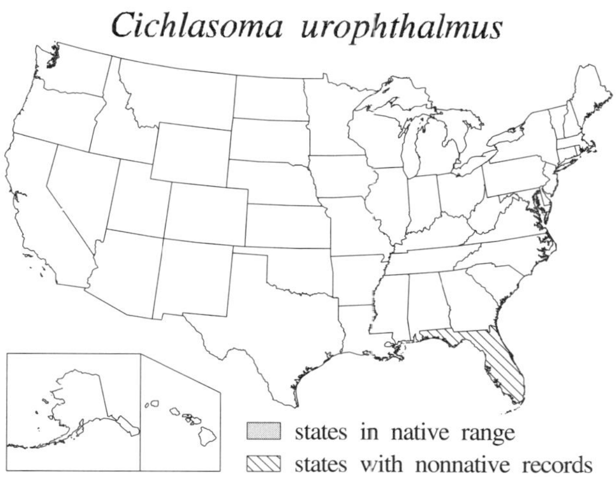

Cichlasoma urophthalmus (Günther 1862)

Common Names: Mayan cichlid (orange tiger, castarrica).

Identification: Other commonly used names or synonyms include *Heros troscheli* and *Herichthys urophthalmus*. For distinguishing characteristics see Page and Burr (1991). The species is included in identification keys of Greenfield and Thomerson (1997). Photographs appeared in Loftus (1987), Konings (1989), Conkel (1993), and Axelrod (1993). Many subspecies have been described (Axelrod 1993). Maximum size: 30 cm.

Native Range: Tropical America. Atlantic Slope drainages from Río Coatzacoalcos basin in Mexico southward into Nicaragua, including Isla Mujeres and Yucatán Peninsula (Conkel 1993; Greenfield and Thomerson 1997).

Nonindigenous Occurrences: This species was first documented in **Florida** when specimens were observed and collected and observed in Everglades National Park in 1983; it is established in several areas in and around the park (Loftus 1987; Lorenz et al. 1997; Smith-Vaniz, personal communication) and Big Cypress National Preserve (Nico, unpublished data). On the east side of Florida it has been recorded from Canal C-111 north to the Little River Canal (C-7 Canal) (Shafland 1995b). A single specimen was taken from a rock pit in Manatee County in October 1975 (Smith-Vaniz, personal communication).

Means of Introduction: The origins and dates of actual Florida introductions are unknown (Loftus 1987). Fish were probably aquarium releases or fish farm escapes.

Status: Established in southern Florida (Loftus 1987; Lorenz et al. 1997); recent survey work in south Florida indicate the species is continuing to expand its range (Nico, unpublished data).

Impact of Introduction: Unknown.

Remarks: Florida populations have persisted in spite of severe cold spells, drought, floods, and salinity fluctuations. Spawning was observed at Snook Creek in South Florida in 1984 and 1985, at salinities of 26 ppt and 10 ppt (Loftus 1987). This voracious predator has the potential to be one of the most damaging introduced cichlids (Courtenay, personal communication).

Voucher specimens: Florida (UMMZ 212535, 212537; UF 42825, 42826, 92163; Everglades National Park Museum 6883, 6888).

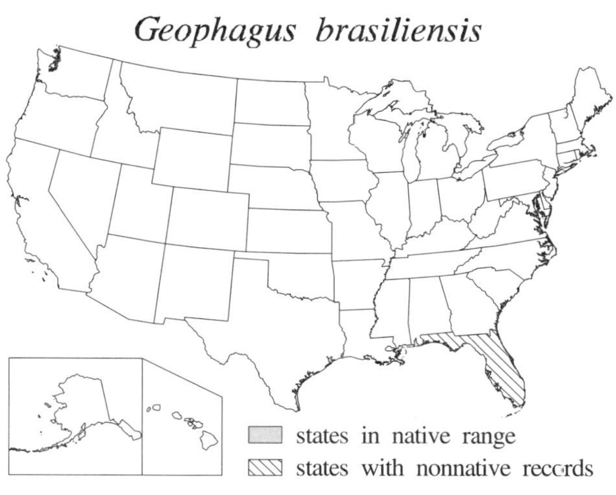

Geophagus brasiliensis (Quoy and Gaimard 1824)

Common Names: pearl eartheater (pearl cichlid).

Identification: A few distinguishing characteristics were provided in Sterba (1973). A key and distinguishing characteristics, in Portuguese, were also presented by Britski et al. (1984). Color photographs appeared in Loiselle (1980b), Axelrod et al. (1985), and Axelrod (1993). Maximum size: 30 cm SL (Loiselle 1980b).

Native Range: Tropical America. Widely distributed in rivers along the Atlantic Coast of Brazil from Bahia State to mouth of Río de la Plata, Argentina (Axelrod 1993), including Rio São Francisco (Britski et al. 1984).

Nonindigenous Occurrences: A single specimen was taken near a fish farm in Lake Worth Drainage District canal L-15 west of Lantana, Palm Beach County, **Florida** (Courtenay et al. 1974; Courtenay and Hensley 1979a).

Means of Introduction: Probable fish farm escape.

Status: Reported from Florida.

Impact of Introduction: Unknown.

Remarks: The first *Geophagus* to enter the aquarium trade, this species has been widely available and bred commercially in Florida (Loiselle 1980b). Captive fish can tolerate temperatures as low as 13°C briefly (Loiselle 1980b). Given the taxonomic confusion associated with the genus *Geophagus*, there is some doubt about correct identification of the Florida specimen (Smith-Vaniz, personal communication).

There are no known voucher specimens.

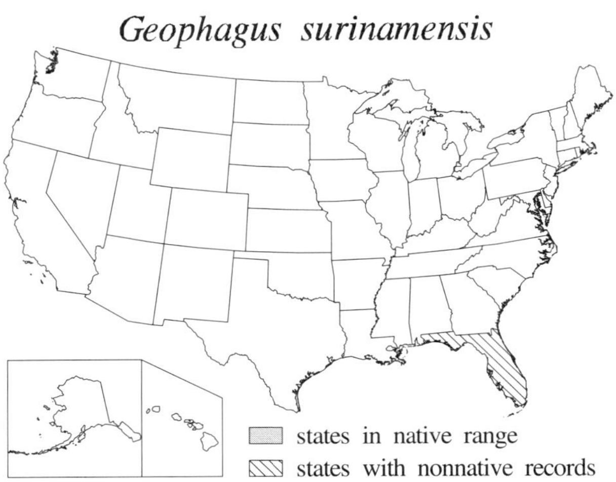

Geophagus surinamensis (Bloch 1791)

Common Names: redstriped eartheater (mother-of-pearl eartheater, opalescent eartheater).

Identification: Along with their recent revision of the genus *Geophagus*, Kullander and Nijssen (1989) redescribed *Geophagus surinamensis*. A few distinguishing characteristics and a figure were also given by Page and Burr (1991). A photograph of *G. surinamensis* holotype appeared in Kullander and Nijssen (1989); additional photographs of this species, or a closely related form, also appeared in Axelrod (1993). Maximum size: 30 cm.

Native Range: Tropical America. Limited to parts of Surinam and French Guiana, northeastern South America (Kullander and Nijssen 1989).

Nonindigenous Occurrences: Two fish were taken from Snapper Creek canal north of its junction with Tamiami Canal in the Miami area of Dade County, **Florida**, in April 1981. A small reproducing population was observed and fry and adults were collected in the same 1-km reach of Snapper Creek during visits in May, and from June to July 1982 (Metzger and Shafland 1984). Additional specimens have been taken from Snapper Creek and one of its tributaries in the late 1980s and early 1990s and from Tamiami Canal (Shafland 1995b; Smith-Vaniz, personal communication)

Means of Introduction: Probable escape or release from fish farm, or aquarium release (Metzger and Shafland 1984).

Status: Recent collections indicate the species is established in Dade County Canals in south Florida (Shafland 1995b; Smith-Vaniz, personal communication).

Impact of Introduction: Unknown.

Remarks: According to Kullander and Nijssen (1989), most past literature on "*Geophagus surinamensis*" actually pertains to other species. Given the past taxonomic confusion associated with the genus *Geophagus*, the Florida form may very well represent another member of the genus and not *G. surinamensis* (Smith-Vaniz, personal communication).

Voucher specimens: Florida (UF 91897, 96391).

Hemichromis elongatus

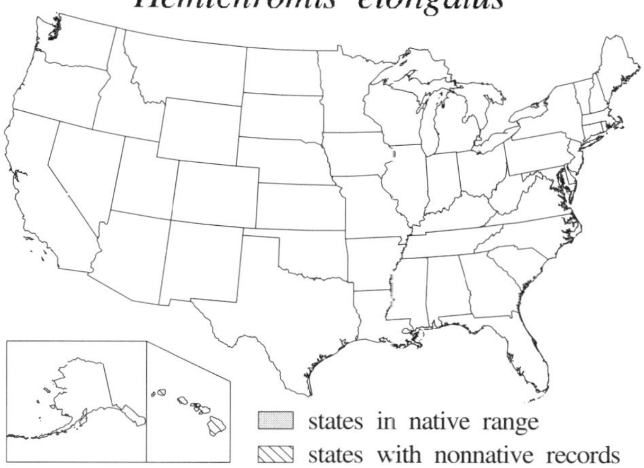

Hemichromis elongatus (Guichenot 1861)

Common Names: banded jewelfish (five-spot cichlid, five-star general).

Identification: Loiselle (1979) revised the genus. Distinguishing characteristics and illustrations were given by Loiselle (1979) and Skelton (1993). Keys were included in Loiselle (1992) and in Skelton (1993). Color illustrations or photographs were provided by Axelrod et al. (1985), Axelrod (1993), Skelton (1993), and Linke and Staeck (1994). Maximum size: 19 cm (Skelton 1993).

Native Range: Western Africa. Widespread in Zaire basin to tropical West Africa, and in Okavango and upper Zambezi systems (Skelton 1993; Linke and Staeck 1994).

Nonindigenous Occurrences: Established in Wahiawa Reservoir, Oahu, **Hawaii**, as of 1994 (A. Tagawa, personal communication).

Means of Introduction: Probable aquarium release.

Status: Established in Hawaii.

Impact of Introduction: Unknown.

Remarks: *Hemichromis elongatus* is found in the aquarium trade. This species has a reputation as an aggressive predator that preys on smaller fishes, shrimps, and insects (Skelton 1993).

There are no known voucher specimens.

Hemichromis letourneauxi

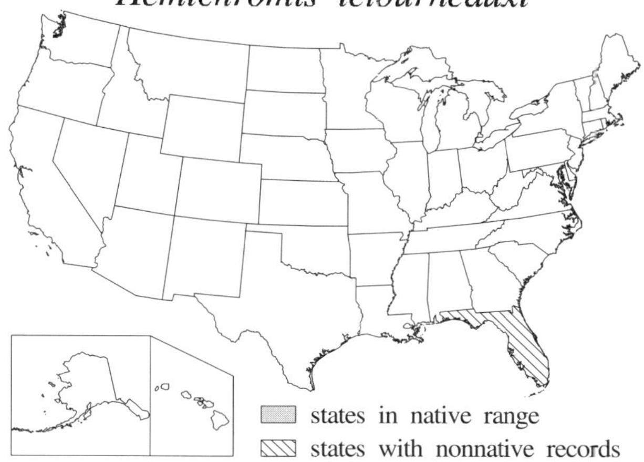

Hemichromis letourneauxi Sauvage 1880

Common Name: African jewelfish.

Identification: Loiselle (1979) revised the genus and provided diagnoses, photographs, and synonyms for *H. letourneauxi*. This species is included in a key provided by Loiselle (1992). Color photograph also provided by Linke and Staeck (1994). Until recently, all published references to introduced populations of this species taken in Florida were incorrectly identified or listed as *Hemichromis bimaculatus* (Smith-Vaniz, personal communication). Maximum size: 12 cm SL (Loiselle 1979).

Native Range: Tropical Africa. Widespread in Central and West Africa (Loiselle 1979, 1992; Linke and Staeck 1994).

Nonindigenous Occurrences: The species was first documented as occurring in South **Florida** in the Hialeah Canal–Miami River Canal system, Miami area, by Rivas (1965). It is now established and abundant in many canals in and around Miami, Dade County, and also in parts of the Everglades (Courtenay et al. 1974; Hogg 1976a, 1976b; Loftus and Kushlan 1987; Nico and Smith-Vaniz, personal communication). The species was taken from canals near fish farms in Palm Beach and Brevard counties in 1970 and 1983 (Smith-Vaniz, personal communication). Specimens also were taken from a drainage ditch on Big Pine Key, Monroe County, on 10 May 1993 (museum specimens). There are unconfirmed reports that the species has occurred and is possibly established in Eureka Springs near Tampa, Hillsborough County (Courtenay et al. 1974).

Means of Introduction: This species was possibly introduced into Florida through aquarium release, perhaps in combination with escapes or releases from fish farms.

Status: Established in Florida. Prior to 1972, found only in Miami Canal and canals on western side of Miami International Airport (Hogg 1976a). Species now abundant and spreading in many Miami canals. Reported and possibly established in other Florida counties.

Impact of Introduction: Unknown.

Remarks: This cichlid possibly competes with native sunfishes for spawning sites. Loftus and Kushlan (1987) reported that this aggressive fish nests near the spawning areas of other introduced cichlids, including the spotted tilapia *Tilapia mariae* and black acara *Cichlasoma bimaculatum*.

Voucher specimens: Florida (UF 34930, 96387, 96408, 96426, many others).

Heros severus

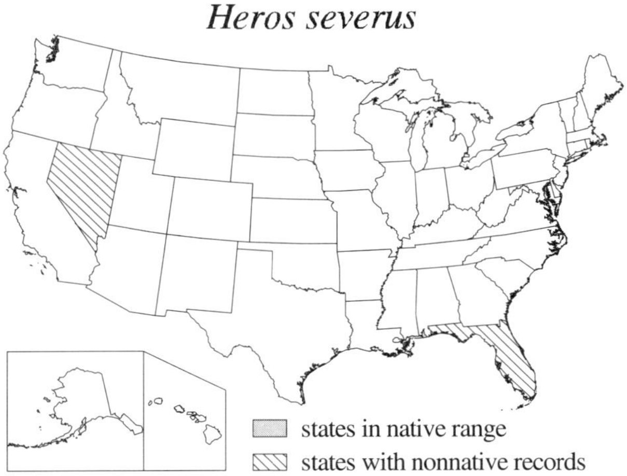

Heros severus Heckel 1840

Common Names: banded cichlid (severum, striped cichlid, sedate cichlid, deacon [juvenile], convict fish).

Identification: Another commonly used name is *Cichlasoma severum*. For distinguishing characteristics and figure see Page and Burr (1991). Color photographs were given in Axelrod (1993). Kullander (1986) reviewed the genus, stating that *Heros* does not appear to be a monotypic genus and that a revision of component species is needed. Maximum size: 20 cm.

Native Range: Tropical America. Genus *Heros* is distributed widely in Atlantic drainages of South America, including much of Amazon and Orinoco basins, Tocantins drainage, and coastal drainages of Guianas. *Heros severus* occurs in upper Rio Negro and upper Rio Orinoco (Kullander 1986).

Nonindigenous Occurrences: This species has been recorded from several sites in Dade County, **Florida** including Tamiami Canal just west of Miami and Snapper Creek (Shafland 1995b; Nico, unpublished data). A single cichlid (tentatively identified as this species) was taken during rotenone treatment of Rogers Spring, a warm spring near Overton and a tributary to Lake Mead, Clark County, **Nevada**, on 8 December 1963 (Courtenay and Deacon 1982; Hubbs and Deacon 1964).

Means of Introduction: Probably aquarium release. Probable introduction by local residents in Nevada (Hubbs and Deacon 1964; Courtenay and Hensley 1979a).

Status: Based on recent collections, this species is apparently established in Florida (Nico, personal communication). Reported from Nevada (Courtenay and Deacon 1982; Deacon and Williams 1984; Courtenay and Stauffer 1990). Contrary to statements that the species is possibly established (e.g., Courtenay and Hensley 1979a; Page and Burr 1991), no additional *Heros severus* have been seen or taken at Rogers Spring since the first capture in 1963 and the species is considered exterminated from Nevada (Courtenay and Deacon 1982, 1983; Deacon and Williams 1984; Courtenay and Stauffer 1990).

Impact of Introduction: Unknown.

Remarks: The statement by Page and Burr (1991) that the species is "apparently established in Rogers Spring" appears erroneous. The rotenone treatment of Rogers Spring in 1963 was undertaken to rid the site of introduced species; until that time it was not known that *H. severus* was even present (Hubbs and Deacon 1964). Concern has been expressed that if *H. severus* survived in Nevada, it, like other introduced species, would compete with local or native fish fauna and sport fish, and that if the cichlid became established in Lake Mead, it potentially would compete with native sunfish for spawning sites (Hubbs and Deacon 1964; Courtenay and Hensley 1979a).

Voucher specimens: Florida (UF uncatalogued).

Labeotropheus sp.

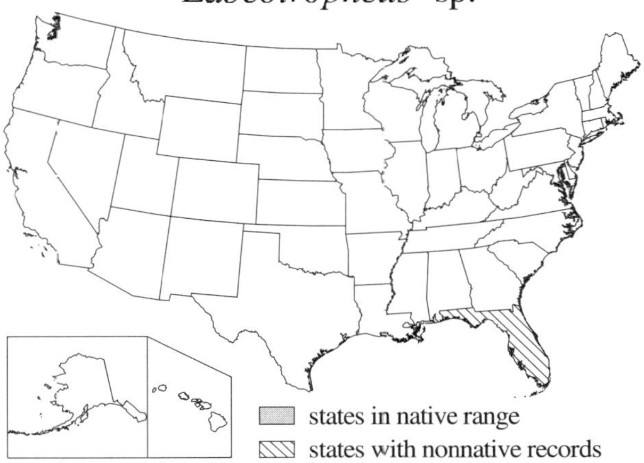

Labeotropheus sp.

Common Names: mbuna (Lake Malawi cichlid).

Identification: The genus is currently recognized as containing two species, blue mbuna *Labeotropheus fuelleborni* Ahl 1927 and scrapermouth mbuna *L. trewavasae* Fryer 1956; both have large, protruding, fleshy snouts and exhibit polychromatism (Ribbink et al. 1983). Ribbink et al. (1983) and Lewis et al. (1986) described the various color morphs and distinguished between the two known species. Color photographs were given in Ribbink et al. (1983), Axelrod et al. (1985), Lewis et al. (1986), and Axelrod (1993). Maximum size: 15 cm (Petrovicky 1988).

Native Range: Tropical Africa. Lake Malawi (Lewis et al. 1986).

Nonindigenous Occurrences: Large numbers of juveniles and adult blue African cichlids were observed by C. R. Robins in large rock pits at or near the MetroZoo in south Miami, Dade County, **Florida**, ca. 1987 (Courtenay, personal communication). Robins initially reported that the species, tentatively identified as *L. trewavasae*, was apparently established; however, no additional reports of this species are known from the site. The MetroZoo record was apparently the basis for the listing of *Lebeotropheus* sp. in later publications as introduced but not established in Florida (e.g., Courtenay and Stauffer 1990; Courtenay et al. 1991).

Means of Introduction: Probable aquarium release.

Status: Reported from Florida.

Impact of Introduction: Unknown.

Remarks: Both known species of *Labeotropheus* are found in the aquarium trade. Axelrod (1993) stated that preserved specimens of *L. trewavasae* and *L. fuelleborni* cannot be differentiated. C. R. Robins (personal communication) recently made clear to us that the identification of these cichlids as *Labeotropheus* should be considered highly tentative. Based on recent fish samples taken by us from moats surrounding some of the exhibits, the fish earlier reported from the Miami MetroZoo rock pits actually may have been one of the other blue cichlids (i.e., *Pseudotropheus* sp.).

There are no known voucher specimens.

Melanochromis auratus

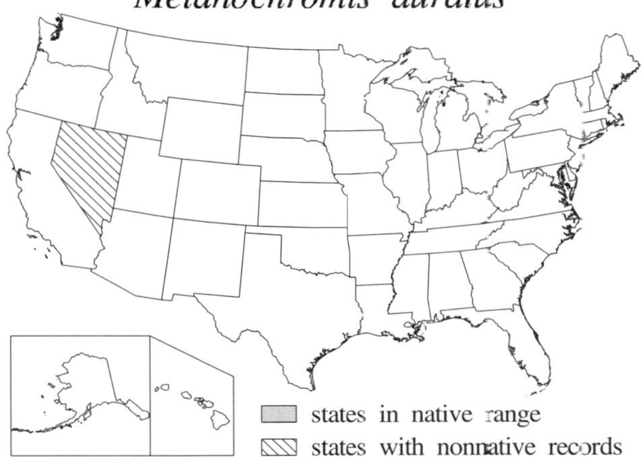

Melanochromis auratus (Boulenger 1897)

Common Names: golden mbuna (golden cichlid, golden pseudotropheus, Nyassa golden cichlid).

Identification: Another commonly used name is *Pseudotropheus auratus* (see discussion by Ribbink et al. 1983). According to Eccles and Trewavas (1989), the genera *Melanochromis* and *Pseudotropheus* are in particular need of revision and critical appraisal. Based on Trewavas (1983b), Bowers and Stauffer (1993) listed the defining characteristics of the genus *Melanochromis*. Distinguishing characteristics of *M. auratus* (mainly color patterns) were provided by Ribbink et al. (1983). Dentition and color patterns associated with this species were described by Fryer and Iles (1972); discussion and photographs of different color patterns were also provided by Lewis et al. (1986). Additional color photographs or illustrations of fish were given by Goldstein (1970), Ribbink et al. (1983), Axelrod et al. (1985), Petrovicky

(1988), and Axelrod (1993). Males are predominantly blue and black; females are gold and black (Ribbink et al. (1983). Maximum size: 11 cm (Petrovicky 1988).

Native Range: Tropical Africa. Lake Malawi (Lewis et al. 1986).

Nonindigenous Occurrences: This species is listed as reported but not established in Rogers Spring near Lake Mead, Clark County, **Nevada** (Courtenay et al. 1984, 1986; also see Deacon and Williams 1984; Courtenay and Stauffer 1990; Courtenay et al. 1991). This species apparently was found first when three specimens were seen (one collected) in Rogers Spring on 1 August 1983 (Courtenay, personal communication).

Means of Introduction: Probable aquarium release.

Status: Reported from Nevada.

Impact of Introduction: Unknown.

Remarks: The species is common in the aquarium trade. Females and breeding or territorial males have very different color patterns (Ribbink et al. 1983; Lewis et al. 1986).

There are no known voucher specimens.

Melanochromis johanni

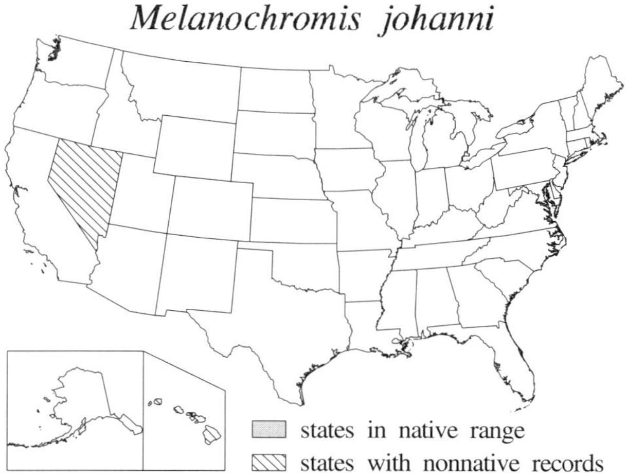

Melanochromis johanni (Eccles 1973)

Common Names: bluegray mbuna (johanni).

Identification: Another commonly used name is *Pseudotropheus johanni*. According to Eccles and Trewavas (1989), *Melanochromis* and the related genus *Pseudotropheus* are in need of revision and critical appraisal. Based on Trewavas (1983b), Bowers and Stauffer (1993) listed the defining characteristics of the genus *Melanochromis*. Detailed color descriptions of *M. johanni* were given by Ribbink et al. (1983). In general, males are black with royal blue bands, and females are yellow-orange or orange, with faint bands (Ribbink et al. 1983). Photographs of live fish appeared in Axelrod et al. (1985), Lewis et al. (1986), and Axelrod (1993); the male was illustrated in Petrovicky (1988). Maximum size: 10 cm (Petrovicky 1988).

Native Range: Tropical Africa. Lake Malawi (Lewis et al. 1986).

Nonindigenous Occurrences: This species was first discovered in Oahu, **Hawaii**, in a stream below a reservoir (Yamamoto, personal communication). The species is listed as introduced but not established in Rogers Spring near Lake Mead, Clark County, **Nevada** (Deacon and Williams 1984; also see Courtenay et al. 1984, 1986; Courtenay and Stauffer 1990; Courtenay et al. 1991). A single fish was observed in Rogers Spring on 1 August 1983, and again (perhaps the same fish) on 15 November 1983 (Courtenay, personal communication).

Means of Introduction: Probable aquarium release.

Status: Established, or possibly established, in Hawaii. Reported from Nevada.

Impact of Introduction: Unknown.

Remarks: There is some uncertainty concerning the reproductive status of this species in Hawaii. Several years ago it was indicated to us that this cichlid had become established in Oahu by 1994 (Yamamoto, personal communication). In the aquarium trade, this species has the reputation of being highly aggressive (Petrovicky 1988).

There are no known voucher specimens.

Oreochromis aureus

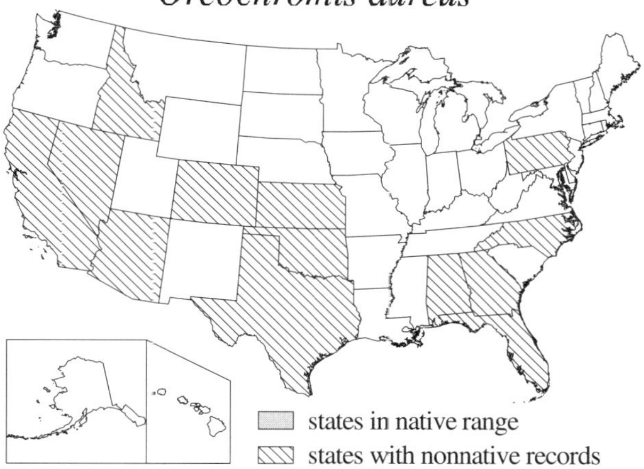

Oreochromis aureus (Steindachner 1864)

Common Names: blue tilapia (Israeli tilapia).

Identification: Commonly used names include *Tilapia aurea* and *Sarotherodon aurea*. Distinguishing characteristics, synonyms, photographs, keys, and discussion of hybrids were provided in Trewavas (1983a); for identification also see Page and Burr (1991), and Skelton (1993). Illustrations and diagnoses of larval and small juveniles of introduced populations were given by McGowan (1988). Color photographs were presented in Axelrod et al. (1985) and Axelrod (1993). Many or most accounts of "*Tilapia nilotica*" in U.S. ponds probably refer to *O. aureus*, likely imported from Israel, before the two species were shown to be distinct (Trewavas 1983a). Maximum size: 51 cm (Lee et al. 1980 et seq.).

Native Range: Tropical and subtropical Africa, and Middle East. Native range includes Senegal, Niger, and many smaller drainages and lakes in Africa and Middle East (Trewavas 1983a; Skelton 1993).

Nonindigenous Occurrences: This species (often identified as *Tilapia nilotica*) was stocked annually by the Alabama Department of Conservation and Auburn University in lakes and farm ponds in **Alabama** during the late 1950s, 1960s, and 1970s (Rogers 1961; Smith-Vaniz 1968; Habel 1975). There are a few records of populations surviving mild winters, such as an account for Crenshaw County Public Lake, a southern Alabama public fishing lake, between 1971 and 1972 (Habel 1975). One recent record is of 25 specimens taken from Saugahatchee Creek in the Tallapoosa drainage, Mobile Basin, near Loachapoka, Lee County, on 2 October 1980 (museum specimens). The species reportedly is reproducing in experimental ponds associated with Auburn University, but there is no evidence of established populations in open waters of the state. It has been established in **Arizona** since about 1975 (Courtenay and Hensley 1979a). This species (and perhaps a hybrid with *O. niloticus*) is established and locally common in various parts of the lower Colorado River in the southwestern part of the state (Grabowski et al. 1984; Courtenay et al. 1984, 1986). Specimens of this species or a possible hybrid were collected from Alamo Reservoir on the Bill Williams River in the Colorado River drainage, Mojave and Yuma counties, ca. 1968 (Grabowski et al. 1984, Courtenay et al. 1986); the likely source of Alamo Lake tilapia was a population stocked in Francis Creek in 1968 that later moved downstream during flood periods (Grabowski et al. 1984). The species apparently is established as far north in the Colorado as Lake Havasu, above Parker Dam (Courtenay et al. 1986). It has been documented as being stocked in Dankworth ponds in Graham County, and in Randolph Park in Tucson, Pima County; many unrecorded stockings, official and unofficial, probably have occurred in various other parts of the state (Grabowski et al. 1984). The species is established in the Gila River north of Yuma (Courtenay et al. 1984, 1986). It was stocked in an irrigation district near Gila Bend in the early 1980s (Courtenay and Hensley 1979a; Courtenay et al. 1986). It is established and locally common in several areas of the lower Colorado River in the southeastern part of **California**, near the Arizona border (Grabowski et al. 1984; Courtenay et al. 1986, 1991; Swift et al. 1993). The species is apparently established as far north in the Colorado as Lake Havasu, above Parker Dam (Courtenay et al. 1986). It also has been reported and taken from the Salton Sea and vicinity (Courtenay et al. 1986, 1991; Swift et al. 1993), although some tilapia taken from the Salton Sea appeared to be hybrids between *O. aureus* and *O. mossambicus* (Swift et al. 1993). Some populations introduced into the lower Colorado River were possibly hybrids between *O. aureus* and *O. niloticus* (Courtenay et al. 1986, 1991). This, or a closely related tilapia, reportedly was raised commercially for food in high-altitude geothermal waters and ponds in the San Luis Valley, part of the Upper Rio Grande River system, near Alamosa, Conejos County, **Colorado** (Courtenay and Hensley 1979a; Courtenay et al. 1984, 1986; Zuckerman and Behnke 1986); it was reported that tilapia escaped in the area and established self-maintaining populations in two earthen ponds in 1977 (Zuckerman and Behnke 1986). This species was listed as not established in a recent account by Courtenay et al. (1991). The first record of this tilapia in **Florida** was that of 3,000 fish stocked in a series of phosphate pits for aquatic plant control experiments at the Pleasant

Grove Research Station in Hillsborough County in August 1961 (Crittenden 1965; Courtenay et al. 1974; Courtenay and Hensley 1979a). The tilapia later spread and reproduced, and subsequent attempts to eradicate it failed (Langford et al. 1978; Hale et al. 1995). The species is now considered the most widespread foreign species in Florida. It has been reported or collected in more than 20 Florida counties, and is established in most of these (Buntz and Manooch 1969; Courtenay et al. 1974, 1984, 1986, 1991; Burgess et al. 1977; Foote 1977; Langford et al. 1978; Courtenay and Hensley 1979a; Kushlan 1986; Loftus and Kushlan 1987; Zale 1987; museum specimens). The northernmost established population in Florida is in Lake Alice in Gainesville, Alachua County, where the fish has been present since about 1969 or perhaps earlier (Burgess et al. 1977). This species also is reproducing in saline waters of Tampa Bay (Lee et al. 1980 et seq.; Courtenay et al. 1986). It was collected from a pond at Musgrove Plantation on St. Simons Island, Glynn County, **Georgia**, during 1980. Although no attempt was made to document reproduction, that population persisted several years but apparently did not survive the severe winter of 1989 (Gennings, personal communication). An unconfirmed report of this tilapia on St. Simons Island also was mentioned by Courtenay and Hensley (1979a) and Courtenay et al. (1984, 1986). Over 35 juveniles were trapped in a Skidaway River tidal creek draining an aquaculture experimental area on Skidaway Island, Chatham County, in July and August 1989 (Hales 1989). Another unconfirmed report indicated that tilapia, possibly this species, had been stocked and presumably were established in golf course ponds at Sea Island, Glynn County (Courtenay and Hensley 1979a; Courtenay et al. 1984, 1986). In reference to the same population, Gennings (personal communication) reported that an unknown species of tilapia, reported from golf course ponds at the Sea Island Golf Club, possibly was present during the late 1970s or mid 1980s, and indicated that the population apparently was extirpated during or before the winter of 1989. As of 1992, state personnel had concluded that the species is no longer established in Georgia (Gennings, personal communication). Specimens of this species recently have been reported as being taken from Lake Seminole, a reservoir on the Florida border in the Apalachicola drainage (Gennings, personal communication); however, all available specimens and photographs of tilapia from that lake have thus far proven to be those of *O. niloticus* (Smith-Vaniz, personal communication). This species has been cultured in **Idaho** in the Hagerman Valley, Twin Falls County, and may have become established following its escape into the Snake River near natural thermal outflows (Courtenay et al. 1987; V. Moore, personal communication). It has been taken in **Kansas** from a farm pond in Hodgeman County in 1967 and from a lake in Pratt County in 1990 (museum specimens). This species is known from the Muddy River system, Clark County, **Nevada** (Scoppettone et al. 1998). It was purposefully introduced into Skyland Lake (now Julian Reservoir), **North Carolina**, a cooling reservoir of the Carolina Power and Light Company located in the French Broad–Tennessee drainage, south of Asheville, Buncombe County, in 1965 (Courtenay and Hensley 1979a; Courtenay et al. 1986). Although some information suggested that it had been replaced by *O. mossambicus* by the late 1970s (Courtenay and Hensley 1979a), recent reports indicated that *O. aureus* has continued to maintain an established population in Julian Reservoir (Menhinick 1991; D. Herlong, personal communication). The species was introduced into Hyco Reservoir in the Roanoke River drainage, Person and Caswell counties, in 1984, where it is established (McGowan 1988; Crutchfield 1995). In a distribution map for this species, Menhinick (1991) indicated this species had been found in the Tennessee River drainage (i.e., Julian Reservoir), the Roanoke River drainage (i.e., Hyco Reservoir), and possibly a lake site in the lower Cape Fear drainage in or near New Hanover County. In his table of fishes introduced into the state, Menhinick (1991) listed this species as having been introduced into the Neuse River drainage but not in the Cape Fear drainage. This tilapia is known from **Oklahoma** in the North Canadian River since 1977, where it was reported as having a confirmed range of 383 km, from Lake Overholser to Lake Eufaula (Pigg 1978). This population has been somewhat unstable, for instance, the species was reported to have died out during cold weather in late 1977 and early 1978 (Courtenay and Hensley 1979a; Lee et al. 1980 et seq.), but specimens were taken there again in 1979 (Courtenay et al. 1986; Courtenay and Williams 1992). Pigg et al. (1992) discovered large numbers in the North Canadian River in 1987, but they have found no additional specimens in the river since. This species has been taken from the Arkansas River in Tulsa (Pigg et al. 1992). It also has been reported from and may have been established in Sooner Lake (Arkansas River drainage), a power plant reservoir about 20 miles north of Stillwater in Noble and Pawnee counties, since the middle or late 1980s (A. V. Zale, personal communication). It was listed as established in Oklahoma by Courtenay et al. (1991). The species became established in **Pennsylvania**, in warmwater effluents of a power plant on the Susquehanna River, after escaping from Pennsylvania Power and Light's Brunner Island Aquaculture Facility sometime after October 1982, possibly in 1984 (Skinner 1984, 1986; Stauffer et al. 1988; Courtenay and Williams 1992). Populations in the vicinity of Brunner Island were eradicated in February 1986, when condenser cooling water was deliberately and temporarily released at lethal, lower temperatures (Skinner 1987; Stauffer et al. 1988; Courtenay and Williams 1992); however, Stauffer et al. (1988) postulated that *O. aureus* may still survive farther downstream based on an earlier report by Skinner (1984) that tilapia had been collected as far downstream as 78 km from the Brunner Island site. This species first appeared in Texas open waters in reservoirs during the 1960s, apparently as a result of fish farm and bait bucket releases (Howells 1992a). Muoneke (1988) reported that its general distribution included all but the northern- and westernmost parts of the state. This species is most common in warmwater reservoirs and has been reported or is established in more than 30 **Texas** counties (Whiteside 1975; Hubbs et al. 1978, 1991; Courtenay and Hensley 1979a; Lee et al. 1980 et seq.; Muoneke 1988; Courtenay et al. 1991; Edwards and Contreras-Balderas 1991; Howells 1991b, 1992a, 1992b). It is established in the Rio Grande, San Antonio, and Guadalupe drainages, and in parts of the Colorado River drainage; this tilapia is most abundant in areas with warmer water temperatures (e.g., in the lower Rio Grande Basin and in power plant reservoirs) (Hubbs et al. 1991). Reservoirs known to contain established populations include Calaveras, Victor Braunig, Fairfield, Tradinghouse Creek, Canyon, Casa Blanca, Nasworthy, Falcon, Colorado City, and Amistad (Muoneke 1988). The species was established in Trinidad Lake, Henderson County, during the late 1960s and early 1970s (Noble and Germany 1986), but has since been extirpated (Hubbs et al. 1978; Noble and Germany 1986). Hybrids with *O. mossambicus* are present in the San Marcos River, and in Canyon and Gibbons Creek reservoirs (Howells 1992b). Listings of this tilapia's distribution in Texas, both before and after 1979, were given by Muoneke (1988).

Means of Introduction: This species has been introduced through a combination of means, including stocking and experimental work by states and private companies (e.g., the electric power industry), and release by individuals seeking to use the species as a sport fish, as forage for warmwater predatory fish, as a food source, and as a means of aquatic plant control. Introductions and spread have resulted by way of escapes or releases from aquaculture facilities and experimental control areas, and from various other holding sites (e.g., zoological parks); through aquarium and bait bucket releases; and by intentional transport by anglers and private individuals (Courtenay and Hensley 1979a; Lee et al. 1980 et seq.; Courtenay et al. 1984, 1986; Muoneke 1988; Courtenay and Williams 1992). The exact reasons for and sources of some introductions are uncertain (e.g., Texas) (Hubbs et al. 1978; Courtenay and Hensley 1979a). Apparently, power companies became interested in using so-called "tropical fishes" for food or sport in heated effluent ponds used to cool effluents from both fossil fuel fired and nuclear generating plants, where temperatures often became too high to support populations of native fishes (Courtenay and Hensley 1979a). Blue tilapia and redbelly tilapia were inadvertently introduced into Hyco Reservoir in North Carolina in 1984 after a small number of fish escaped from a holding cage located in the heated discharge area during an on-site agricultural study (Crutchfield 1995).

Status: Established or possibly established in ten states. Established in parts of Arizona, California, Florida, Nevada, North Carolina, and Texas. Possibly established in Colorado, Idaho, Oklahoma, and Pennsylvania. Reported from Alabama, Georgia, and Kansas. For more than a decade it has been considered the most widespread foreign fish in Florida (Hale et al. 1995).

Impact of Introduction: The blue tilapia is considered a competitor with native species for spawning areas, food, and space (Buntz and Manooch 1969; Noble and Germany 1986; Muoneke 1988; Zale and Gregory 1990). Courtenay and Robins (1973) reported that certain streams where this species is abundant have lost most vegetation and nearly all native fishes. It has invaded the Taylor Slough portion of Everglades National Park where it is considered a major management problem for the National Park Service (Courtenay 1989; Courtenay and Williams 1992). The blue tilapia's local abundance and high densities in certain areas have resulted in marked changes in fish community structure (Muoneke 1988, and citations therein). A dramatic reduction in native fishes in the Warm Springs area of Nevada coincided with invasion of this species (Scoppettone et al. 1998).

Blue tilapia have also been implicated as the cause for unionid mussel declines in two Texas water bodies, Tradinghouse Creek and Fairfield reservoirs (Howells 1995).

Remarks: *Oreochromis aureus* has been used widely in aquaculture and is able to live and reproduce in brackish waters. The origin of the U.S. stocks of *O. aureus*, imported as *Tilapia nilotica*, was Israel (Courtenay and Hensley 1979a). Voucher specimens taken from the lower Colorado river system, Arizona, in 1980 were initially reported as mango tilapia *Tilapia* (= *Sarotherodon*) *galilaea*; but these were later determined by D. Thys van den Audenaerde to be *O. aureus*. Some lower Colorado River populations in California and Arizona may be hybrids with *O. niloticus* (Courtenay et al. 1984, 1986). Although hybridization is common, electrophoretic studies on tilapia sampled from 12 Texas reservoirs indicated that most populations were *O. aureus* without indicating genetic introgression with other tilapia species (Howells 1991b). There is a 1971 record of Alabama fish overwintering in outdoor ponds at Auburn University (Courtenay and Hensley 1979a, Courtenay et al. 1986); however, tilapia introduced into that state typically begin to die each fall when water temperatures reach about 10°C (Smith-Vaniz 1968). This species was stocked in aquaculture ponds in Iowa to test growth potential; although it reproduced there, it did not overwinter (Pelgren and Carlander 1971; Courtenay and Hensley 1979a). In the southwestern United States, the Central Arizona Project canal system is proving to be a major dispersal route for blue tilapia (Courtenay, personal communication). Collection and reported localities were mapped for all or part of the United States, mainly Florida, by Courtenay et al. (1974), Foote (1977), Courtenay and Hensley (1979a), Lee et al. (1980 et seq.), Loftus and Kushlan (1987), and Menhinick (1991). Hale et al. (1995) reviewed the history of blue tilapia in Florida.

Voucher specimens: Alabama (AUM 20907), Florida (UF 23163, 40306, many others), Kansas (KU 13026, 22917), North Carolina (UNCC uncatalogued), Pennsylvania (PSU 2031).

Oreochromis macrochir

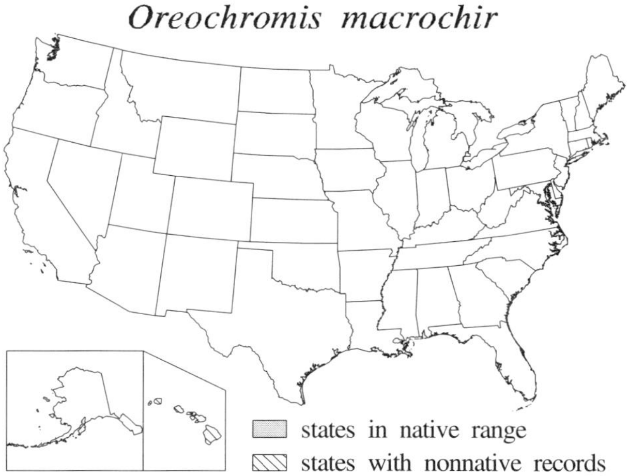

Oreochromis macrochir (Boulenger 1912)

Common Names: longfin tilapia (greenhead bream, greenhead tilapia).

Identification: A commonly used name is *Tilapia macrochir*. Distinguishing characteristics, a key, and illustrations were given by Trewavas (1983a), and by Skelton (1993). Distinguishing characteristics and a figure also were given by Eccles (1992). Color photographs of fish were given by Axelrod (1993). Maximum size: 40 cm (Skelton 1993).

Native Range: Africa. Sections of south central Africa including Cunene, Okavango, upper Zambezi and Kafue rivers (Trewavas 1983a; Skelton 1993).

Nonindigenous Occurrences: The species is established in reservoir or pond habitats on Oahu, Maui, and Hawaii, **Hawaii**; it was first released ca. 1957 or 1958 (Morita 1981; Maciolek 1984; Devick 1991a). This species was reported as common or abundant in Wahiawa Reservoir (= Lake Wilson) on Oahu (Hida and Thomson 1962; Devick 1972).

Means of Introduction: Fish apparently were bred in captivity before being released into the wild, ca. 1958 (Maciolek 1984); this was considered a deliberate introduction (Devick 1991a).

Status: Established in Hawaii.

Impact of Introduction: The overabundance of tilapias in at least one Hawaiian reservoir has possibly had the effect of suppressing production of desirable sport fishes (Devick 1972).

Remarks: *Oreochromis macrochir* has been replaced by *Sarotherodon melanotheron* as the second most abundant tilapia (after *O. mossambicus*?) in many Oahu reservoirs (Devick 1991b). This tilapia species has been widely introduced into Africa and other parts of the world for use in aquaculture (Trewavas 1983a; Axelrod 1993; Skelton 1993). In Hawaii, *O. macrochir* is reported as possibly hybridizing with *O. mossambicus* (Devick 1972).

Oreochromis mossambicus

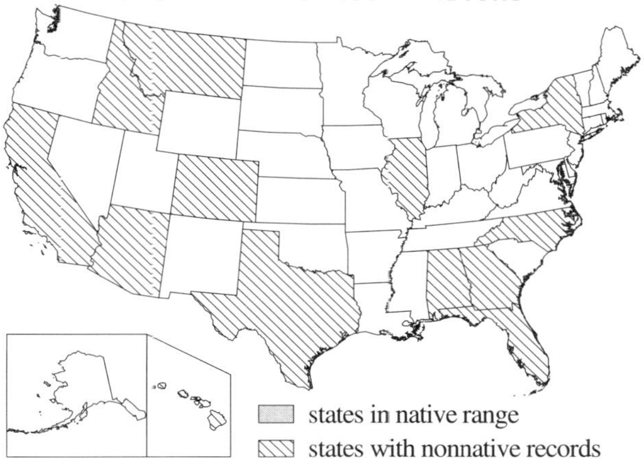

Oreochromis mossambicus (Peters 1852)

Common Names: Mozambique tilapia (Mozambique mouthbrooder, Java tilapia, largemouth kurper).

Identification: Commonly used names include *Tilapia mossambica* and *Sarotherodon mossambicus*. Distinguishing characteristics, synonyms, an illustration, keys, and a discussion of hybrids were provided by Trewavas (1983a); for identification also see Wieland et al. (1982) and Page and Burr (1991). Color photographs appeared in Axelrod et al. (1985). Maximum size: 40 cm SL (Skelton 1993).

Native Range: Tropical and subtropical Africa. Southern Africa from lower Zambezi to Brak River, and Limpopo system (de Moor and Bruton 1988). Occurs in both fresh and brackish waters.

Nonindigenous Occurrences: This species was stocked annually by the Alabama Department of Conservation and Auburn University in lakes and farm ponds in **Alabama** during the late 1950s and 1960s (Rogers 1961; Smith-Vaniz 1968), but such stockings reportedly have not been carried out for some time now (Wieland et al. 1982). The species was introduced into several agricultural drains and mitigation ponds in **Arizona** near Yuma, Yuma County, and has been considered established in the state since the early 1960s (Hoover and St. Amant 1970; Hoover 1971; Minckley 1973; Courtenay et al. 1986). The species also occurs and is established in the Colorado River drainage, mainly the Gila River, from Phoenix to just north of Yuma (Minckley 1973; Lee et al. 1980 et seq.; Courtenay and McCann 1981; Barrett 1983; Grabowski et al. 1984; Courtenay et al. 1986). Populations established at Warm Springs on the San Carlos River, in Gila and Graham counties, and in the Salt River in Tempe, Maricopa County, reportedly were destroyed by floods (Minckley 1973). The first record of this species in **California** was that of a population found in a small pond and its tributary near the Hot Mineral Spa on the east side of the Salton Sea near Niland, Imperial County, on 3 January 1964; more than 5,000 tilapia in the pond were killed with rotenone in an eradication attempt (St. Amant 1966). The species is now broadly established in the southern part of the state and has been in the Santa Ana, San Gabriel, and Los Angeles rivers since 1974 (St. Amant 1966; Hoover and St. Amant 1970; Moyle 1976a; Knaggs 1977; Lee et al. 1980 et seq.; Shapovalov et al. 1981; Courtenay et al. 1984, 1991; Grabowski et al. 1984; Courtenay and Robins 1989; Swift et al. 1993). Two specimens were taken from Lake Success in the San Joaquin Valley, Tulare County, in 1989 (Heyne et al. 1991). The species was found to be locally established in high-altitude, warm-water ponds in **Colorado** in the Upper Rio Grande River system in Conejos County, in 1977. These fish had escaped from a local farm where they were being cultured for food in warm artesian waters at a 7,500-ft elevation (Zuckerman and Behnke 1986). The Mozambique tilapia was first introduced into and became established in **Florida** in Dade County during the 1960s (Courtenay and Stauffer 1990). It is now established or has been reported in five counties including Brevard, Dade, Indian River, Palm Beach, and possibly Hillsborough (Ogilvie 1969; Courtenay et al. 1974, 1984, 1986, 1991; Hogg 1976a, 1976b; Courtenay and Hensley 1979a; Courtenay and McCann 1981; Dial and Wainright 1983; Loftus and Kushlan 1987). Reports of this species from Six-Mile Creek, Hillsborough County (Courtenay et al. 1974), were probably based on misidentifications of *O. aureus* (Courtenay and Hensley 1979a). Individuals of this species were released into a pond in **Georgia** near Athens, but failed to survive (Dahlberg and Scott 1971b). It was indicated that this tilapia may be established in golf course ponds on Sea Island and St. Simons Island, Glynn County (Courtenay and Hensley 1979a; Courtenay et al. 1986); however, *O. mossambicus* was not collected on St. Simons Island during 1980 sampling efforts and as of 1992 the Georgia Department of Natural Resources had concluded that this species was no longer present in state open waters (Gennings, personal communication). This species is established and has large populations in many streams, estuaries, low wetlands, and reservoirs or ponds of **Hawaii**; it is found on all the major islands including Oahu, Kauai, Maui, Hawaii, and Molokai (Brock 1960; Maciolek 1984; Randall 1987; Devick 1991a, 1991b). It also is reportedly established in Pu'uhonua o Honaunau National Historical Park (Courtenay 1989). A reproducing population was found in **Idaho** in Barney Hot Springs and at the upper end of Barney Creek in Custer County, Little Lost River Valley, in September 1985 (Courtenay et al. 1987). The species was listed by Idaho Fish and Game (1990) as being introduced into and confined to geothermal waters of the Snake River below Shoshone Falls. A red hybrid of this species, possibly a cross between *O. mossambicus* and *O. urolepis*, was found in thermally heated sections of the Bruneau River near Bruneau Hot Springs, Owyhee County, after it had escaped from a local aquaculture facility (Courtenay et al. 1987; Courtenay and Stauffer 1990). This species was introduced into ponds in **Illinois**, probably during the early 1960s, but there was no evidence of overwintering (Smith 1965). Four specimens were taken from a high-altitude, thermal spring fed pond of the Columbia River drainage in **Montana** in Bearmouth, Granite County, during 1962–1963 surveys (Brown and Fox 1966), but the pond habitat was later destroyed during highway construction (Brown 1971; Courtenay et al. 1986). In **North Carolina**, the species was introduced into and temporarily established in the Julian Reservoir (formerly Skyland Lake) in French Broad–Tennessee drainage, south of Asheville, Buncombe County, in 1965 (Courtenay and Hensley 1979a), but that population reportedly did not survive beyond the early 1970s (Courtenay et al. 1986). There are unconfirmed reports that this species was introduced into Hyco Reservoir, Roanoke River drainage, in Person and Caswell counties (Menhinick 1991, but see Crutchfield 1995). In **New York**, a single adult fish was seined by state personnel from Hall's Pond in Hall's Pond Park, West Hempstead, Nassau County, in the summer of 1976 (Briggs, personal communication). Probably in reference to the same collection, the species was reported as having been collected in the state by Lee et al. (1980 et seq.). This species has been found in **Texas** in the headwaters of the San Antonio River within the San Antonio Zoo, in Bexar County, since the late 1950s; the first specimen-based record was of a juvenile fish trapped in the San Marcos River in San Marcos, Hays County, in 1959 (Brown 1961). Populations of this species have been identified from several areas along the Balcones fault zone, including tributaries of the San Antonio River within and near the San Antonio Zoo; at the Spring Lake headwaters and in the vicinity of the state fish hatchery of the San Marcos River; and in Canyon Reservoir on the Guadalupe River (Brown 1961; Hubbs et al. 1978, 1991; Muoneke 1988; Howells 1991b, 1992a, 1992b).

Means of Introduction: Similar to *O. aureus*, this species has been introduced for a wide variety of reasons. Most introductions have been the result of intentional stockings for aquatic plant control by state and federal agencies and private companies, but introductions have also come about from stockings for potential use of the species as an insect control agent (e.g., to control mosquitoes and chironomids), as a sport fish, as a bait fish, and as a food or commercial fish, and through aquarium releases; the species also has been introduced through releases or escapes from fish farms, hatcheries, and zoos (Shapovalov et al. 1981; Dial and Wainright 1983; Courtenay et al. 1984, 1986; Grabowski et al. 1984; Courtenay and Stauffer 1990). This species was brought to Hawaii from Singapore in a small shipment of fish in 1951 (Brock 1960; Maciolek 1984; Randall 1987). The Mozambique tilapia was introduced with the expectation that it would be useful for aquatic plant control in irrigation systems, as a food fish, as a sport fish, and as live bait for tuna fishing (Brock 1960); results were only partially successful (Randall 1987). In California, introductions resulted from escapes or releases from fish farms and from intentional stocking by the state (Shapovalov et al. 1981). The Mozambique tilapia's initial introduction into Dade County, Florida, is believed to have been the result of escapes or releases from aquarium fish farms that cultured the species in the 1960s (Courtenay and Stauffer 1990). In some areas of Florida, this species may have been introduced by local anglers to create a commercial fishery (Dial and Wainright 1983), or intentionally stocked by aquarists (Courtenay and Stauffer 1990). In New York, introduction was probably due to aquarium release (Briggs, personal communication). In Texas, this species was introduced as a result of escapes from the San Antonio Zoo in 1956 and also from state and federal hatcheries during the late 1950s and early 1960s (Brown 1961; Courtenay and McCann 1981). Sources and reasons for many of the introductions have been reviewed by Courtenay and McCann (1981), Wieland et al. (1982), and Courtenay and Stauffer (1990).

Status: Established or locally established in seven states including Arizona, California, Colorado, Florida, Hawaii, Idaho, and Texas. Formerly considered locally established but no longer extant in Georgia, Montana, and North Carolina. Reported from Alabama, Illinois, and New York.

Impact of Introduction: In Hawaii, this species is suspected as a threat to native species such as striped mullet *Mugil cephalus* (Randall 1987; Devick 1991b). Tilapia also have been considered a major factor in the decline of the desert pupfish *Cyprinodon macularius* in the Salton Sea area (Courtenay and Robins 1989; Swift et al. 1993). Because of its presence in Dade County, Florida, Courtenay (1989) indicated that the Mozambique tilapia may eventually enter Everglades National Park.

Remarks: Some records of this species apparently are based on incorrect identifications. For instance, recent electrophoretic evidence indicated that populations in the San Marcos River and in Canyon Reservoir were *O. mossambicus* × *O. aureus* hybrids (Howells 1991b, 1992b). With the aid of W. Smith-Vaniz, we examined preserved juveniles catalogued as *O. mossambicus* from the San Marcos River (TCWC 2073.01) and determined them to be *O. aureus* based on their caudal fin patterns and scale and gill raker counts. Some California records of this species may actually be those of *O. urolepis* (= *O. hornorum*) or of hybrids between *O. mossambicus* and *O. urolepis* (Swift et al. 1993). Although considered a hardy species and very tolerant of high salinities (Dial and Wainright 1983), the Mozambique tilapia reportedly does not survive temperatures below about 10°C (Talwar and Jhingran 1992). The occurrence of *O. mossambicus* in the United States was reviewed by Courtenay and McCann (1981), Wieland et al. (1982), and Courtenay et al. (1986). The history of this species introduction into the southwestern United States was reviewed by Hoover (1971), Courtenay and Robins (1989), Shapovalov et al. (1981), and Swift et al. (1993). *Oreochromis mossambicus* has largely replaced redbelly tilapia *Tilapia zillii* in the Salton Sea and possibly other areas in southern California (Swift et al. 1993). Collection sites and reported localities are mapped for the United States (Courtenay and Hensley 1979a; Lee et al. 1980 et seq.), and for the states of Arizona (Minckley 1973; Grabowski et al. 1984) and Florida (Courtenay et al. 1974; Hogg 1976b; Courtenay and Hensley 1979a; Kushlan 1986; Loftus and Kushlan 1987).

Voucher specimens: Alabama (AUM 24075); California (AUM 24653); Florida (UF 34908, 92216), Idaho (UMMZ 213374, FAU WRC-ID-2).

Oreochromis niloticus

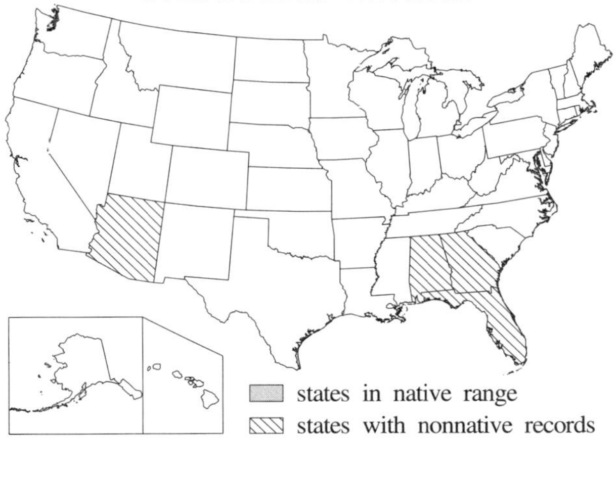

Oreochromis niloticus (Linnaeus 1757)

Common Names: Nile tilapia (Nile mouthbrooder).

Identification: A commonly used name is *Tilapia nilotica*. Distinguishing characteristics, synonyms, an illustration, keys, and a discussion of hybrids in Trewavas (1983a). Distinguishing characteristics and illustration were also given by Eccles (1992). Photographs appeared in Axelrod et al. (1985) and Axelrod (1993). This species closely resembles *Oreochromis aureus*. Before the two species were shown to be distinct, many or most accounts of "*Tilapia nilotica*" in U.S. ponds probably referred to *O. aureus*, likely imported from Israel (Trewavas 1983a). Maximum size: 63 cm (de Moor and Bruton 1988).

Native Range: Tropical and subtropical Africa, Middle East. Widely distributed in Nile and Niger river basins and in lakes Tanganyika, Albert, Edward, and George, as well as in many smaller drainages and lakes in western and eastern Africa; also in Middle East in Yarkon River, Israel (Trewavas 1983a).

Nonindigenous Occurrences: A single fish in cold distress was taken from the Saugahatchee Creek portion of Yates Reservoir, in the Tallapoosa drainage of Mobile Basin, Lee County, **Alabama**, on 12 January 1986 (Hornsby, personal communication; Boschung 1992). Probably in reference to Yates Reservoir, Courtenay and Williams (1992) reported that a reservoir on the Tallapoosa River, where this species has been recorded, receives drainage from the aquaculture ponds of Auburn University. This species reportedly was being reared in fish farms in **Arizona** and apparently was introduced experimentally into the southern part of the state (Minckley 1973; Grabowski et al. 1984); however, the identification of those fish has come into question (Courtenay and Hensley 1979b; Courtenay et al. 1984, 1986). Specific sites of introduction are not mentioned in the literature and there are no recent reports of this species in the state. Anglers have taken this species from Lake Seminole in the Apalachicola drainage, Seminole County, on both sides of the **Florida** and **Georgia** border, since about 1991; it is possibly established in the reservoir (Smith-Vaniz, personal communication).

Means of Introduction: This species was introduced for aquaculture purposes. It was introduced into open waters, likely through escape or release from fish farms.

Status: Possibly established in a large reservoir bordering Florida and Georgia. Reported from Alabama and Arizona.

Impact of Introduction: Unknown. Introduced populations in Lake Victoria, Africa, reportedly displayed intense competition with native species for spawning areas (de Moor and Bruton 1988).

Remarks: A specimen taken from Lake Seminole on the Georgia side of the lake near Saunder's Slough in 1991 was originally reported as *O. aureus* (Gennings, personal communication); however, all available specimens and photographs of tilapia from that lake have thus far proven to be *O. niloticus* (Smith-Vaniz, personal communication). Although *O. niloticus* has been reported from Texas, these reports were based on erroneous identifications of other tilapia species (Hubbs 1982, cited by Muoneke 1988). Reports of this species in Arizona also may be based on a misidentification. Minckley's (1973) figure 122, labeled as "*Tilapia nilotica*," and his description of their young, more closely match *T. mariae* (Courtenay and Hensley 1979b; Courtenay et al. 1984, 1986).

Voucher specimens: Florida (UF many specimens), Georgia (UF, UGAMNH).

Oreochromis urolepis

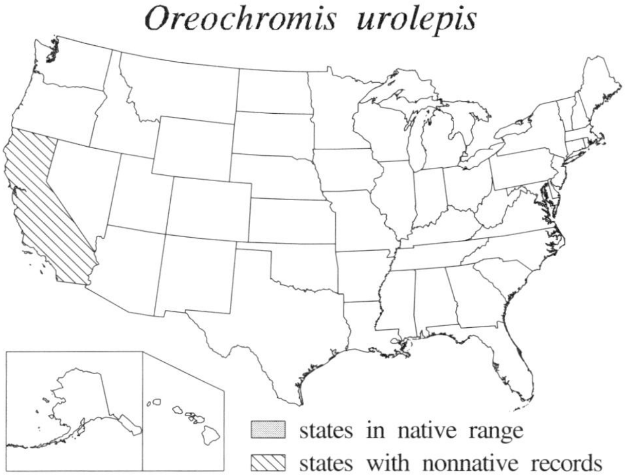

Oreochromis urolepis (Norman 1922)

Common Name: Wami tilapia.

Identification: Other commonly used names or synonyms include *Tilapia hornorum*, *Oreochromis hornorum*, and *Tilapia urolepis*. The Wami tilapia closely resembles *O. mossambicus* and was previously considered a strain (Zanzibar strain) of that species; it was later elevated to species status (i.e., *Tilapia hornorum*). Trewavas (1983a) more recently determined *T. hornorum* to be a subspecies of *Oreochromis urolepis* (i.e., *Oreochromis urolepis hornorum*). Trewavas (1983a) gave distinguishing characteristics, synonyms, illustrations, a key, and a discussion of hybrids. A few distinguishing characteristics were also given by Page and Burr (1991) and by Eccles (1992). A color photograph was given by Axelrod et al. (1993). Maximum size: 30 cm SL (Trewavas 1983a).

Native Range: Tropical Africa. Wami River system, Tanzania (Trewavas 1983a).

Nonindigenous Occurrences: This species was stocked with other tilapia in the Coyote Creek drainage, a tributary of the San Gabriel River in Los Angeles basin, **California**, in 1972; only this species, or a hybrid, persisted and reached high numbers (Legner and Pelsue 1977; Legner et al. 1980). It is established in the Bolsa Chica Flood Control Channel in Huntington Beach, Orange County (Courtenay et al. 1984, 1991). It is also established in the Cerritos Flood Control Channel, Cerritos Lagoon, and in the Coyote Creek–San Gabriel River drainage, Long Beach, Los Angeles County (Knaggs 1977; Courtenay et al. 1984). It is possibly established in Imperial and Riverside counties (Legner and Pelsue 1977; Legner et al. 1980; Courtenay et al. 1984).

Means of Introduction: This species was first brought to California prior to the 1970s and survived in isolated colonies in relatively warm waters of the southern part of the state in the area of the Colorado Desert, Imperial and Riverside counties (Legner and Pelsue 1977; Courtenay et al. 1984), but it is unclear whether these fish were restricted to experimental ponds or whether populations were actually released into open waters. In 1972, *O. urolepis* was stocked into open waters by the University of California, Riverside, and by the Southeast and Orange County Mosquito Abatement districts, to control aquatic plants, mosquitoes, and chironomid midges (Legner and Pelsue 1977; Legner et al. 1980; Courtenay et al. 1984, 1986). These fish were derived from stock obtained from the University of Arizona (Legner et al. 1980).

Status: Established in southern California (Courtenay et al. 1984, 1991; Page and Burr 1991)

Impact of Introduction: Unknown.

Remarks: *Oreochromis urolepis* in the United States represent the subspecies *O. urolepis hornorum* (Courtenay et al. 1991). Because many hybrid *O. urolepis* were introduced into southern California, some doubt remains as to whether the established populations in California open waters represent pure *O. urolepis* strains or hybrids with *O. mossambicus* (Legner et al. 1980; Courtenay et al. 1984, 1991).

There are no known voucher specimens.

Pelvicachromis pulcher

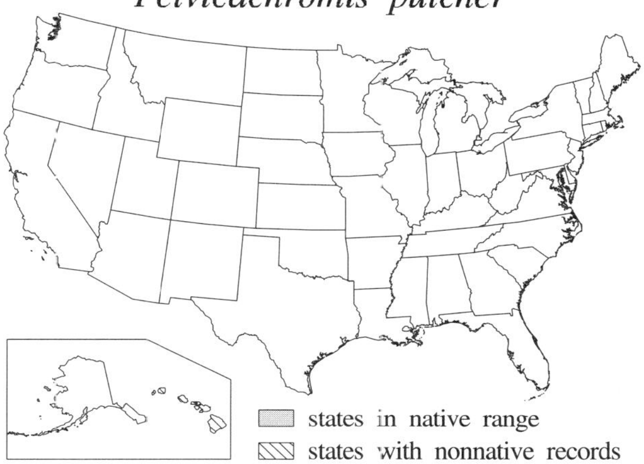

Pelvicachromis pulcher (Boulenger 1901)

Common Names: rainbow krib (kribensis, common krib, pink krib, rainbow cichlid, pelvicachromis).

Identification: Another commonly used name is *Pelmatochromis pulcher*. Many color varieties of *P. pulcher* are common in the aquarium trade (Axelrod 1993). Color photographs were given in Axelrod et al. (1985) and in Axelrod (1993). The genera of pelmatochromine fishes were reviewed by Greenwood (1987). Maximum size: 10 cm (Petrovicky 1988).

Native Range: Tropical Africa. Western Africa; types from Niger delta, Nigeria (Axelrod 1993).

Nonindigenous Occurrences: This species is established in reservoir or pond habitats on Oahu, **Hawaii**; it was first released or discovered in 1984 (Devick 1991a).

Means of Introduction: Aquarium escape (Devick 1991b), accidental introduction (Devick 1991a).

Status: Established in Hawaii.

Impact of Introduction: Unknown.

Remarks: This species is common in the aquarium trade. The many color varieties available are the result of artificial crossbreeding and natural variation (Axelrod 1993).

There are no known voucher specimens.

Pseudotropheus sp.

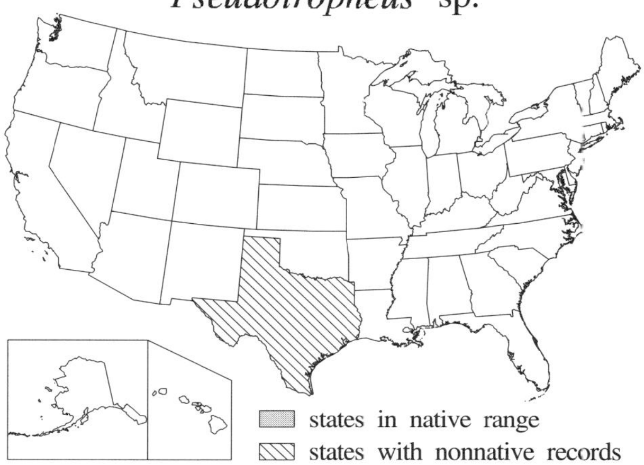

Pseudotropheus sp.

Common Names: African lake cichlid (mbuna cichlid).

Identification: The genus *Pseudotropheus* contains approximately 200 species (Axelrod 1993). Several large species complexes exist and the genus is in great need of systematic revision (Ribbink et al. 1983; Eccles and Trewavas 1989). Ribbink et al. (1983) surveyed Lake Malawi cichlids, providing descriptions of coloration and markings for identification. Many of the different color varieties, including hybrids, are common in the aquarium trade (Axelrod 1993). Color photographs of fish were given in Ribbink et al. (1983), Axelrod et al. (1985), Lewis et al. (1986), and Axelrod (1993). Maximum size: Depending on species, from about 5 to 18 cm (Axelrod 1993).

Native Range: Tropical Africa. Lake Malawi.

Nonindigenous Occurrences: One specimen was taken with a hook and line from the San Marcos River below the Spring Lake dam in Hays County, **Texas**, on 3 September 1991 (Whiteside and Berkhouse 1992).

Means of Introduction: Probable aquarium release (Whiteside and Berkhouse 1992).

Status: Reported from Texas.

Impact of Introduction: Unknown.

Remarks: Many members of the mbuna cichlids are popular in the aquarium trade.

There are no known voucher specimens.

Pseudotropheus zebra

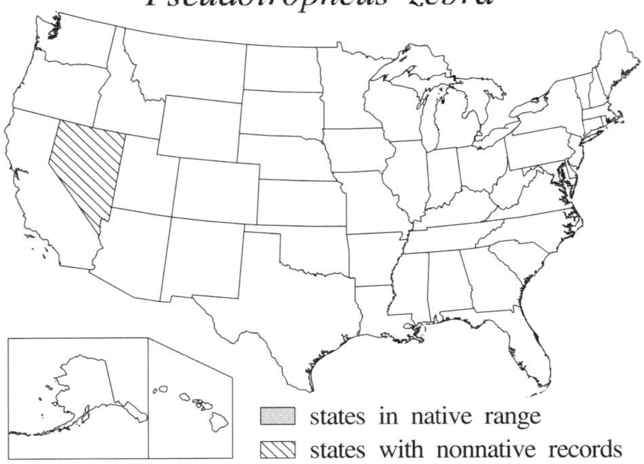

Pseudotropheus zebra (Boulenger 1899)

Common Names: zebra mbuna (blue mbuna, Nyassa blue cichlid).

Identification: According to Eccles and Trewavas (1989), the genus *Pseudotropheus* is in particular need of revision and critical appraisal. *Pseudotropheus zebra* alone represents a complex of about 27 species (Ribbink et al. 1983). Some distinguishing characteristics of the species complex, a description of color patterns, figures, and photographs were given by Ribbink et al. (1983). Color photographs of various species and morphs were given in Ribbink et al. (1983), Axelrod et al. (1985), Lewis et al. (1986), and Axelrod (1993); also see illustrations in Petrovicky (1988). Maximum size: Size varies among members of the complex, with maximum size ranging from about 6 to 14 cm (Axelrod 1993).

Native Range: Tropical Africa. Lake Malawi.

Nonindigenous Occurrences: The first report of this fish in **Nevada** was by a member of the American Cichlid Association, who observed several specimens of the "marmalade" form of *P. zebra* during a visit to Rogers Spring, a tributary of Lake Mead near Overton, Clark County, in July 1981 (Courtenay and Deacon 1982). Later records from Rogers Spring included a single male fish taken on 14 November 1981 (Courtenay, personal communication), and several individuals reportedly found in 1983 (Courtenay, personal communication). These records were apparently the basis for the listing of the species in later publications as introduced but not established in Nevada (e.g., Deacon and Williams 1984; Courtenay et al. 1984, 1986, 1991; Courtenay and Stauffer 1990).

Means of Introduction: Probable aquarium release.

Status: Reported from Nevada.

Impact of Introduction: Unknown.

Remarks: This species or species complex is well known in the aquarium trade. In its native range, the true *P. zebra* is distributed widely in Lake Malawi and shows marked geographical variation in coloration and markings (Ribbink et al. 1983). In contrast to references given in selected published fish lists (e.g., Deacon and Williams 1984), we found no mention of *P. zebra* in Courtenay and Deacon (1983), a review paper on introductions into Rogers Spring, Nevada.

There are no known voucher specimens; however, a photograph is on file to support record.

Pterophyllum scalare

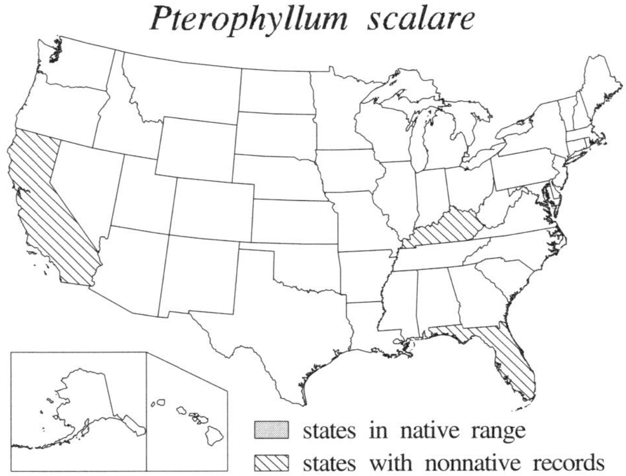

Pterophyllum scalare (Lichtenstein 1823)

Common Names: freshwater angelfish (veil angelfish, black angelfish, longfin angelfish, common angelfish, angel, scalare, leaf fish [many other names used for aquarium-bred color variants]).

Identification: Schultz (1967) reviewed the genus and gave distinguishing characteristics of the species; he considered *Pterophyllum eimekei* a junior synonym. Many color variants are found in the aquarium trade. Photographs of live fish were given in Axelrod et al. (1985), Axelrod (1993), and Bleher (1993). Maximum size: 15 cm (Petrovicky 1988; Axelrod 1993).

Native Range: Tropical America. Guyana, Venezuela, in Amazonas south to Mato Grosso of Brazil and Peru (Schultz 1967; Bleher 1993).

Nonindigenous Occurrences: A single specimen (5 cm long and identified as *Pterophyllum eimekei*) was taken with rotenone from a pond on the eastern side of the Salton Sea near the Hot Mineral Spa in Imperial County, **California**, on 3 January 1964 (St. Amant 1966). There is an unconfirmed report from an unspecified locality in **Florida** (Courtenay et al. 1984, 1991). There are also unconfirmed reports of this species in **Kentucky** (Courtenay et al. 1991).

Means of Introduction: Not known, although Kentucky record likely result of aquarium release.

Status: Reported from three states, California, Florida, and Kentucky.

Impact of Introduction: Unknown.

Remarks: The Florida report of *Pterophyllum* sp. (Courtenay and Hensley 1979a) and that of *Pterophyllum scalare* (Courtenay et al. 1984, 1991) represent the same record (Courtenay, personal communication).

There are no known voucher specimens.

Pterophyllum spp.

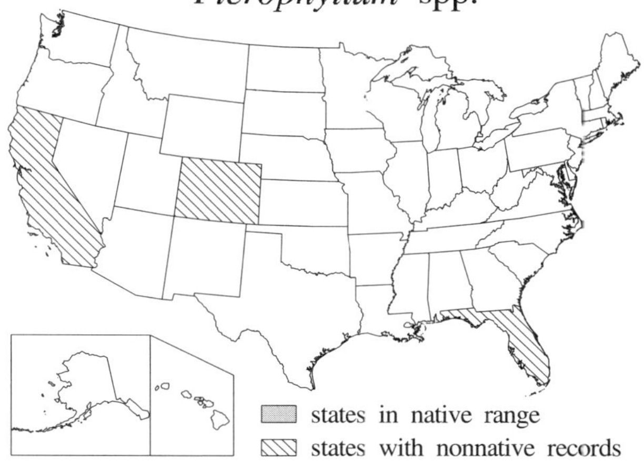

Pterophyllum spp.

Common Names: freshwater angelfish (leaf fish).

Identification: Schultz (1967) reviewed the genus and recognized three species: *Pterophyllum altum*, *P. scalare*, and *P. dumerilii* (= *P. leopoldi*). There are many color variants in the aquarium trade. Photographs of live fish appeared in Axelrod et al. (1985), Axelrod (1993), and Bleher (1993). Maximum size: 18 cm (Petrovicky 1988).

Native Range: Tropical America. Orinoco and Amazon basins, and various coastal drainages of Guianas, South America (Schultz 1967).

Nonindigenous Occurrences: There is a single report from Westminster flood control channel, Orange County, **California**, in 1968 (St. Amant and Hoover 1969; Shapovalov et al. 1981). Specimens were collected in geothermal waters on the Weisbart Swine Farm in the Upper Rio Grande drainage in the San Luis Valley, Conejos County, **Colorado**, on 3 November 1983 (Zuckerman and Behnke 1986; Platania, personal communication). There are unconfirmed and unsubstantiated reports of the species from Lake Worth Drainage District canal L-14, near a fish farm in Palm Beach County, **Florida** (Courtenay and Hensley 1979a), in or before 1979.

Means of Introduction: Probable release or escape from local fish farms (St. Amant and Hoover 1969; Courtenay and Hensley 1979a; Zuckerman and Behnke 1986).

Status: Reported from California, Colorado, and Florida.

Impact of Introduction: Unknown.

Remarks: Florida reports of *Pterophyllum* sp. (Courtenay and Hensley 1979a) and that of *Pterophyllum scalare* (Courtenay et al. 1984, 1991) represent the same record (Courtenay, personal communication).

There are no known voucher specimens.

Sarotherodon melanotheron

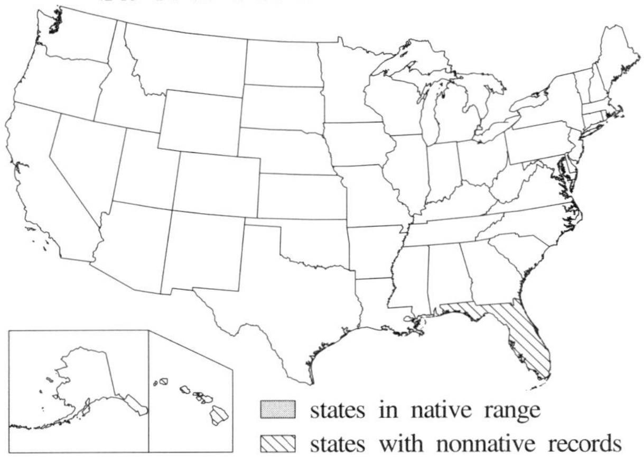

Sarotherodon melanotheron Rüppell 1852

Common Names: blackchin tilapia (blackchin mouthbrooder).

Identification: Names or synonyms for this species used in earlier publications on introduced fish and in aquarium literature include *Tilapia melanotheron*, *Tilapia heudeloti*, and *T. macrocephala*. Distinguishing characteristics and illustrations were given by Trewavas (1983a) and by Page and Burr (1991). Photographs were provided by Axelrod (1993). Maximum size: 26 cm.

Native Range: Tropical Africa. Brackish estuaries and lagoons from Senegal to Zaire (Trewavas 1983a).

Nonindigenous Occurrences: The first documented occurrence of this species in **Florida** was a specimen gillnetted by commercial fishermen in Hillsborough Bay near Tampa, Hillsborough County, in 1959 (Springer and Finucane 1963). Additional records for the western part of the state indicate that this species is established in brackish and freshwaters in eastern Tampa Bay and in adjoining drainages in Hillsborough County, ranging from the Alafia River south to Cockroach Bay. The species has been recorded from the Alafia River from its mouth up to Lithia Springs; from the Hillsborough River, the Palm River, and the Little Manatee River; and from various western drainage and irrigation ditches (Finucane and Rinckey 1967; Buntz and Manooch 1969; Lachner et al. 1970; Courtenay et al. 1974; Courtenay and Hensley 1979a; Courtenay and Kohler 1986; Lee et al. 1980 et seq.; Courtenay and Stauffer 1990; DNR collections; UF museum specimens). There are two records of this species from the west side of Tampa Bay, in Pinellas County: a collection from Lake Maggiore in St. Petersburg, taken in October 1989 (UF museum specimens); and several fish taken from Mud Hole Island, in Weldon Island State Park, in December 1989 (identification based on a photograph of specimens, D. Jennings, personal communication). The species has been established in ditches and saline waters of the Banana and Indian River lagoon system along the eastern side of Florida since or shortly before about 1980 (Dial and Wainright 1983; Gilmore et al. 1983; Courtenay 1989; Snodgrass 1991; Courtenay and Stauffer 1990; Jennings and Williams 1992; Courtenay, personal communication; UF museum specimens). Jennings and Williams (1992) reported that its distribution in eastern Florida had increased significantly since initial introduction, ranging from Vero Beach, Indian River County, north to Whites Point, Brevard County. This species was reported from the Indian River lagoon south of Edgewater, Volusia County, in 1985 (C. Goodrich, personal communication), but there are no known voucher specimens and the report should be considered unconfirmed. The only record of this species in south Florida is that of five specimens taken from the upper Rio Vista Canal, at the Westview Country Club in Dade County, in November 1983 (USGS file record), but the voucher specimens have not been located and this record also should be considered unconfirmed. This species is abundant in brackish-water lagoons in **Hawaii**, including Pearl Harbor and several coastal marine environments (Randall 1987). It has been established and is abundant in Wahiawa Reservoir (= Lake Wilson), a privately owned irrigation reservoir on Oahu, since the mid-1970s; it has been common in other Oahu reservoirs since the late 1970s, and also is found in many lower streams and estuaries and around local coral reefs (Maciolek 1984; Devick 1991a, 1991b).

Means of Introduction: The probable source of Florida fish was Tampa area fish farms, although release of aquarium fish cannot be ruled out (Springer and Finucane 1963). In the eastern part of Florida, this species apparently was introduced or spread by local anglers to create a commercial fishery (Dial and Wainright 1983). Blackchin tilapia were imported into Hawaii in 1962 by the Federal Bureau of Commercial Fisheries (now the

National Marine Fisheries Service); coastal and lagoon introductions were the result of escapes from a fish station on Oahu, where the species was being tested as a baitfish for tuna, in or about 1965 (Randall 1987; Devick 1991b). The fish was introduced accidentally into Wahiawa Reservoir, Hawaii, in the mid-1970s; these reservoir introductions probably resulted from bait bucket releases (Devick 1991a, 1991b).

Status: Established in Florida and Hawaii. Evidence indicates it is spreading rapidly in both fresh and salt water around island of Oahu, Hawaii (Devick 1991b).

Impact of Introduction: In Lithia Springs, Florida, *S. melanotheron* made up about 90 percent of the total fish biomass (Courtenay et al. 1974). An estimated 20,000 tilapia, most of them *S. melanotheron*, were killed in Wahiawa Reservoir, Hawaii, by fungal infections in early 1991; populations of this tilapia appear to be serving as a reservoir for fungal disease and may be responsible for spreading infection to other fish species (Devick 1991a). Devick estimated that the 20,000 dead tilapia represented only about 0.05 percent or less of the total tilapia population in the reservoir.

Remarks: At one time, the species was very popular in the aquarium fish trade (Goldstein 1973). It is very tolerant of high salinities (Dial and Wainright 1983). This species has replaced *Oreochromis macrochir* as the second most abundant tilapia (after *O. mossambicus*?) in many Hawaiian reservoirs (Devick 1991b). Distribution maps for the species have been provided for all or parts of Florida by Courtenay et al. (1974), Lee et al. (1980 et seq.), Dial and Wainright (1983), Jennings (1991), and Jennings and Williams (1992).

Voucher specimens: Florida (USNM 179008, 179009, UF 30876, 84981, 86324, 91156, 91834, 91844, many others).

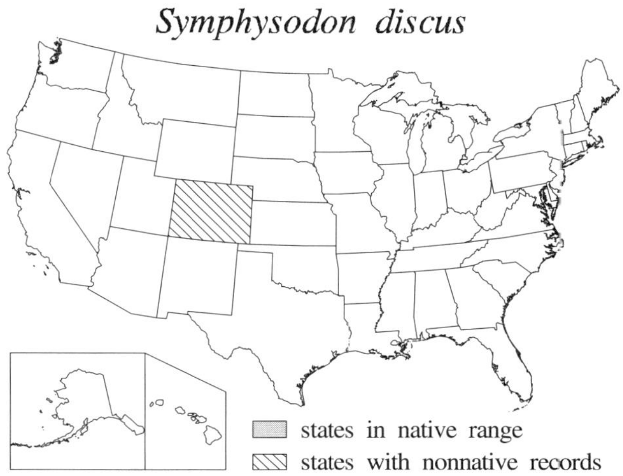

Symphysodon discus Heckel 1840

Common Names: red discus (pompadour fish, Heckel discus).

Identification: For a recent review of the genus, refer to Kullander (1986). The only other species in the genus recognized as being valid is *S. aequifasciatus*. Photographs of live fish were given in Axelrod et al. (1985) and Axelrod (1993). Several subspecies and color variants are found in the aquarium trade (Burgess 1981). Maximum size: 20 cm (Petrovicky 1988).

Native Range: Tropical America. Amazon basin of South America including lower Negro, Trombetas, and Abacaxis rivers (Kullander 1986).

Nonindigenous Occurrences: Specimens were collected in geothermal waters on the Weisbart Swine Farm, in the Upper Rio Grande drainage in the San Luis Valley, Conejos County, **Colorado**, ca. 1983 (Zuckerman and Behnke 1986).

Means of Introduction: Probable escape from local fish farm (Zuckerman and Behnke 1986).

Status: Reported from Colorado.

Impact of Introduction: Unknown.

Remarks: There are no known voucher specimens.

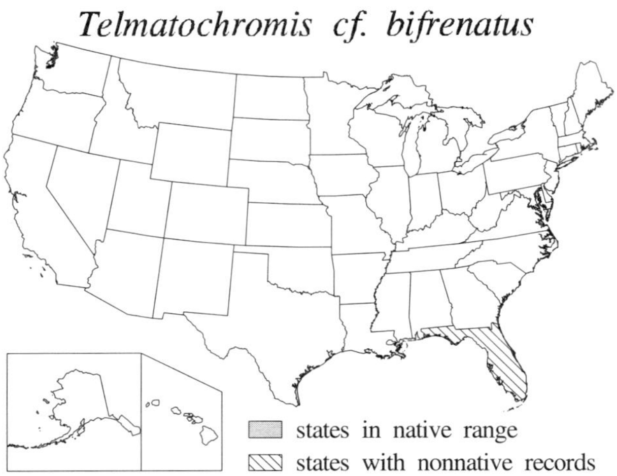

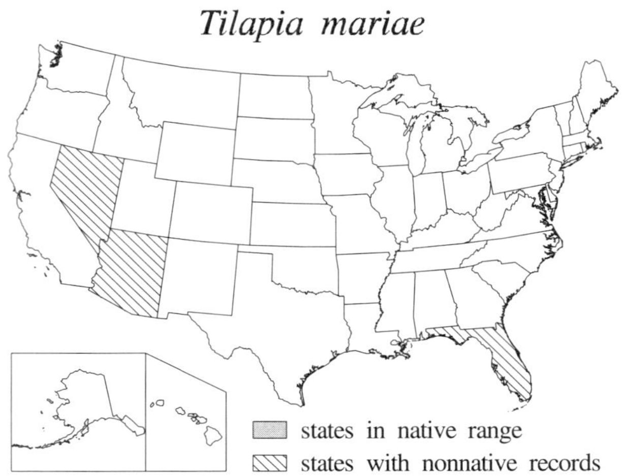

Telmatochromis cf. bifrenatus Myers 1936

Common Name: Lake Tanganyika dwarf cichlid.

Identification: Distinguishing characteristics and keys to Lake Tanganyika cichlid genera and members of the genus *Telmatochromis* were given by Brichard (1978). Color photographs were given in Brichard (1978), Axelrod et al. (1985), and Axelrod (1993). Maximum size: 10 cm (Axelrod 1993).

Native Range: Tropical Africa. Lake Tanganyika (Brichard 1978).

Nonindigenous Occurrences: A single fish was taken from the River Styx, Alachua County, **Florida**, on 6 September 1987 (museum specimen).

Means of Introduction: Probable aquarium release.

Status: Reported from Florida.

Impact of Introduction: Unknown.

Remarks: Voucher specimens: Florida (UF 47323).

Tilapia mariae (Boulenger 1899)

Common Names: spotted tilapia (black mangrove, Niger cichlid).

Identification: Distinguishing characteristics were presented in Taylor et al. (1986) and Page and Burr (1991). Photographs were given by Axelrod et al. (1985), Taylor et al. (1986), and Axelrod (1993). Illustrations were provided by Courtenay and Hensley (1979b), Lee et al. (1980 et seq.), and Page and Burr (1991). Maximum size: 31 cm.

Native Range: Tropical Africa. West Africa from middle Ivory Coast to southwestern Ghana and from southeastern Benin to southwestern Cameroon (Thys van den Audenaerde 1966; Philippart and Ruwet 1982).

Nonindigenous Occurrences: *Tilapia mariae* was possibly introduced for experimental purposes into waters of southern **Arizona** (Courtenay and Hensley 1979b; Lee et al. 1980 et seq.; Courtenay et al. 1984, 1986) and it was considered to be established in the state by Courtenay and Hensley (1979b). The first records in **Florida** were from Snapper Creek Canal in South Miami, Dade County, in April 1974 (Hogg 1974). By the latter half of the 1970s, it had become established in canals throughout most of eastern Dade County (Hogg 1976a, 1976b), in southeastern Broward County, and possibly in southern Collier County (Courtenay and Hensley 1979b; Courtenay et al. 1984, 1986). Previously absent in Everglades National Park (Loftus and Kushlan 1987), it became established in the park and in Big Cypress National Preserve by the late 1980s (Courtenay 1989; Lorenz et al. 1997; museum specimens). This species is now considered established or has been reported from water bodies, mainly canals, lakes, and ponds, in at least eight counties, all in the southern portion of the state; these include Brevard, Broward, Collier, Dade, Indian River, Martin, Monroe, and Palm Beach counties (Hogg 1976b; Courtenay and Hensley 1979a, 1979b; Courtenay et al. 1984, 1986; Clark 1981; Gilmore et al. 1983; Taylor et al. 1986; Loftus and Kushlan 1987; museum specimens; USGS file records). It has been established and reportedly has been abundant in **Nevada** in Rogers Spring, a thermal spring in Lake Mead National Recreation Area above the Overton Arm of Lake Mead, in Clark County, since about 1980 (Courtenay and Deacon 1982, 1983; Deacon and Williams 1984; Courtenay and Stauffer 1990). A single specimen was taken from the Blue Point Spring outlet in Lake Mead National Recreation Area, Clark County, in December 1980, but there was no evidence of reproduction (Courtenay et al. 1986; UF museum specimen).

Means of Introduction: Spotted tilapia was introduced into Florida as a result of escapes or intentional releases from one or several aquarium fish farms in Dade County, probably between 1972 and 1974 (Hogg 1974, 1976a, 1976b; Courtenay and Hensley 1979b; Courtenay and Stauffer 1990). The Nevada introduction was reportedly due to an aquarium release (Courtenay and Deacon 1982, 1983; Courtenay and Stauffer 1990).

Status: Established in Florida; locally established in Nevada; reported from Arizona.

Impact of Introduction: This apparently aggressive species is the dominant fish in many canal systems of southeastern Florida and has the potential to affect other introduced and native fishes (Courtenay and Hensley 1979b). In Nevada, *T. mariae* was reportedly the dominant fish in Rogers Spring and there was concern that this omnivorous cichlid competed with endemic spring fishes for food and also preyed on smaller fishes (Courtenay and Deacon 1982, 1983).

Remarks: Since the early 1970s, *Tilapia mariae* has rapidly dispersed in south Florida, where it has gradually replaced black acara *Cichlasoma bimaculatum* as the most abundant cichlid in area canal systems, possibly through competition for space (Courtenay and Hensley 1979a, 1979b; Kushlan 1986; Loftus and Kushlan 1987). Possession of this species in Florida has been banned since 1974 (Clark 1981). A portion of a population found in a borrow pit in Perrine, Dade County, Florida, included hybrids with redbelly tilapia *Tilapia zillii* (Taylor et al. 1986). Distribution maps for Florida records were given by Hogg (1976b), Courtenay and Hensley (1979b), Lee et al. (1980 et seq.), Clark (1981), Kushlan (1986), and Loftus and Kushlan (1987). The conclusion by some authors that this species was introduced into and is possibly established in Arizona apparently is based entirely on the fact that Minckley (1973) presented a photograph of *Tilapia mariae* (incorrectly identified as "*Tilapia nilotica*") in his book on Arizona fishes. Minckley also described young tilapia that more closely matched *T. mariae* as opposed to *Tilapia nilotica* (= *Oreochromis niloticus*) (see Courtenay and Hensley 1979b; Courtenay et al. 1984, 1986). However, the conclusion that *T. mariae* was ever present in Arizona assumes that the specimens photographed and examined by Minckley actually were collected in Arizona.

Voucher specimens: Florida (UF 30875, 90534, 91946, 92171, 92187, many others; IRCZM 107:5295, 107:5296, 107:5297, 107:5442, 107:5443, 107:5444, 107:05634); Nevada (UF uncatalogued).

Tilapia rendalli

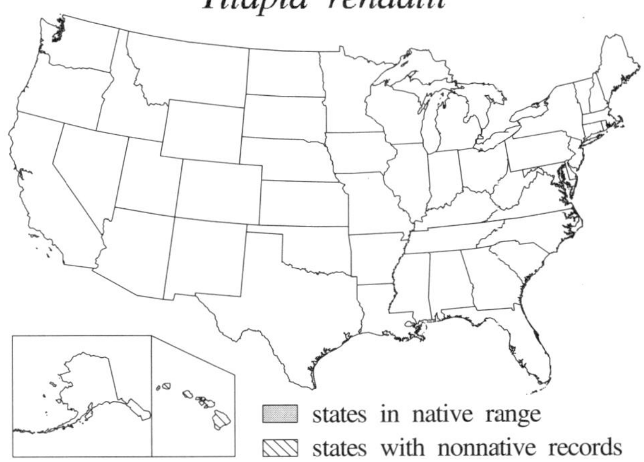

Tilapia rendalli (Boulenger 1896)

Common Names: redbreast tilapia (Congo tilapia, redbreasted bream).

Identification: The commonly used synonym is *Tilapia melanopleura*. Distinguishing characteristics, a key, and a color illustration were given by Skelton (1993). Identifying traits and an illustration were also given by Eccles (1992). Color photographs were given in Axelrod (1993). Maximum size: 40 cm (Skelton 1993).

Native Range: Tropical and subtropical Africa. Cunene, Okavango, Zambezi system, east coastal rivers south to Phongolo, and coastal lakes to Lake Sibaya; also occurs in estuaries in Mozambique and Natal (de Moor and Bruton 1988; Skelton 1993).

Nonindigenous Occurrences: This species has been introduced on the islands of Kauai, Maui, and Oahu, **Hawaii** (Brock 1960; Morita 1981; Maciolek 1984). The species became established in Reservoir No. 44 on Maui during the late 1950s or early 1960s (Hida and Thomson 1962). As of about 1990, it is listed as having an uncertain status and its occurrence is limited to Oahu (Devick 1991a).

Means of Introduction: This species was brought to Hawaii as a shipment of 52 fish from Africa in 1957; these fish were bred in tanks by the state, and resulting offspring were stocked in Wahiawa Reservoir in 1958 and 1959 and on Maui in 1959. According to Devick (1991b), these intentional introductions were undertaken for aquaculture (aku bait) and weed control.

Status: Although this species was formerly considered established and spreading in Hawaii (Maciolek 1984), or at least established (Courtenay et al. 1991), its recent status in the state is considered uncertain (Devick 1991a).

Impact of Introduction: Unknown.

Remarks: This species is referenced as *T. melanopleura* in previous reports concerning introduced species (e.g., Maciolek 1984; Devick 1991a, 1991b). A report of *T. melanopleura* in Alabama by Smith-Vaniz (1968) was based on incorrect identifications of fish later shown to be *T. zillii* (Smith-Vaniz, personal communication). Barrett (1983) tentatively re-identified some California specimens, previously reported to be *T. zillii*, as *T. rendalli*; he based his determination on various morphological characters and color patterns. These fish were taken from near Blythe, Riverside County. *Tilapia rendalli* is a popular angling species in Africa, is important in aquaculture and fisheries, and also is used for weed control in reservoirs (Skelton 1993). The species is tolerant of a broad range of temperatures (11–37°C) and salinity to 19 parts per thousand (Skelton 1993).

There are no known voucher specimens.

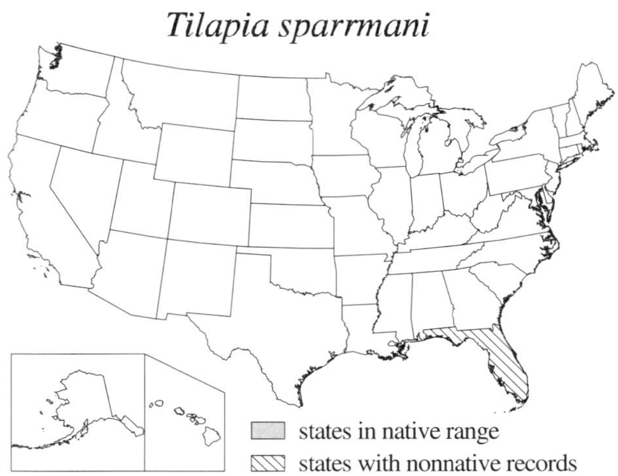

Tilapia sparrmani Smith 1840

Common Names: banded tilapia (banded bream, Sparrman's bream).

Identification: Distinguishing characteristics were given in de Moor and Bruton (1988), Eccles (1992), and Skelton (1993). The species is included in keys of Skelton (1993). It was illustrated in de Moor and Bruton (1988) and in Eccles (1992). For color illustrations or photographs of this fish, see Goldstein (1973), Axelrod (1993), and Skelton (1993). Maximum size: 23 cm SL (Skelton 1993).

Native Range: Tropical and subtropical Africa. Orange River and Natal south coast north to upper reaches of southern Zaire tributaries, Lake Malawi, and Zambezi system (Skelton 1993); distribution map provided in Philippart and Ruwet (1982).

Nonindigenous Occurrences: This species was found in Six Mile Creek above U.S. Highway 92 in **Florida**, and in ditches near Eureka Springs (which connects to Six Mile Creek), Hillsborough County, in 1958 (Burgess 1958; museum specimens).

Means of Introduction: Probable fish farm release or escape (Smith-Vaniz, personal communication).

Status: Reported from Florida but not established (Courtenay and Stauffer 1990; Courtenay and Williams 1992).

Impact of Introduction: Unknown.

Remarks: In response to suggestions that *T. sparrmani* be introduced into California, Pelzman (1972) evaluated its life history and recommended that the species be placed on the prohibited list because of its potential for competition with native sport fishes. In Africa, this species has been introduced outside its native range for use as a forage fish for bass (de Moor and Bruton 1988). It can tolerate low temperatures of about 7°C (de Moor and Bruton 1988) and, based on its distribution in Africa, *T. sparrmani* may be "more restricted by high (above 32°C) than low temperatures" (Skelton 1993).

Voucher specimens: Florida (UF 56197, 56198).

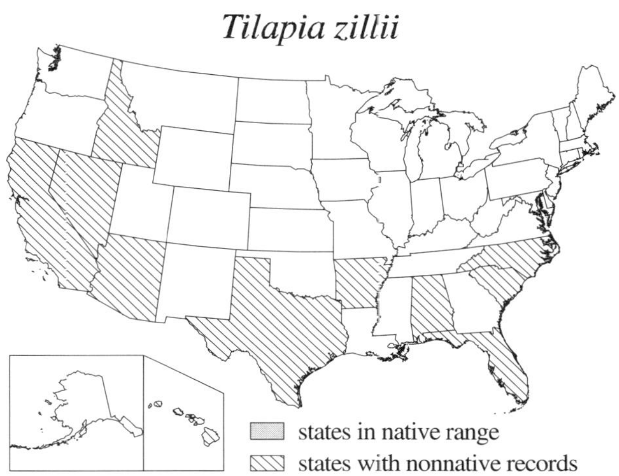

Tilapia zillii (Gervais 1848)

Common Names: redbelly tilapia (Zill's tilapia, striped tilapia, Jordan's St. Peter's fish).

Identification: It is included in identification keys given in Moyle (1976a) and Hubbs et al. (1991). Distinguishing characteristics were given in Taylor et al. (1986), Page and Burr (1991), and Eccles (1992). Photographs appeared in Taylor et al. (1986), and Axelrod (1993). Identification of this species in the United States has been problematic and some reports of this fish in the literature on nonindigenous species may be misidentifications (Lee et al. 1980 et seq.) or reports of hybrids (Courtenay et al. 1984; Taylor et al. 1986; Howells 1991b). For example, in the United States *Tilapia zillii* has occasionally been misidentified as *Tilapia melanopleura* (e.g., Smith-Vaniz 1968), the latter a junior synonym of *Tilapia rendalli*. Maximum size: 30 cm SL (Lee et al. 1980 et seq.).

Native Range: Tropical and subtropical Africa, Near East. West Africa through Chad basin to Nile, Lake Albert, and Lake Turkana into Israel and Jordan Valley (Trewavas 1982).

Nonindigenous Occurrences: This species was annually stocked in state lakes and farm ponds in **Alabama** in past years (Smith-Vaniz 1968 [incorrectly identified as *T. melanopleura*]; Lee et al. 1980 et seq.), but it is not known to be established (Courtenay et al. 1986). It has been collected in **Arkansas** but is not considered established (Carter, personal communication). This species was stocked in **Arizona** by the state government between 1961 and 1980. Sites stocked included Cholla Lake, Apache County, in 1968; Pecks Lake near Cottonwood, Yavapai County, in 1979; Phoenix Zoo, Maricopa County, in 1979; and Sally Ann nos. 1 and 2, Yuma County, in 1980 (Grabowski et al. 1984). This species was introduced into a canal in Mesa, Maricopa County, in early 1973, and the population apparently overwintered (Minckley 1973). It also is established in Papago Park, Scottsdale, and in waters near Phoenix, Maricopa County (Courtenay et al. 1984, 1986). The redbelly tilapia is established in the southern part of **California** (Lee et al. 1980 et seq.). It was first introduced into drainage ditches near Calexico, Imperial County, in May 1972 (Hauser 1975), and into tributaries of the San Gabriel and Santa Ana rivers in 1973 (Knaggs 1977). The species was recorded from irrigation canals in Bard and Imperial valleys, Imperial County, and in the Coachella Valley, Riverside County; it is also recorded from Palo Verde Valley, including the Salton Sea and vicinity, backwaters of the lower Colorado River, and Lake Cahuilla, Riverside County (Moyle 1976a; Courtenay and Hensley 1979a; Schoenherr 1979; Shapovalov et al. 1981; Courtenay and Robins 1989). The species was introduced into 20 ponds, lakes, and creeks in Kern, Santa Clara, Los Angeles, Orange, and Riverside counties, but the populations either did not become established or their status is unknown (Moyle 1976a; Courtenay et al. 1986). Legner and Pelsue (1977) indicated it was no longer extant in Los Angeles County, and Pelzman (1973) reported that fish introduced into Napa County perished during the third winter after introduction. The redbelly tilapia was reported in High Rock Spring, Lassen County, as of January 1983 (Courtenay et al. 1984, 1986). There are recent indications that *T. zillii* has been replaced by *Oreochromis mossambicus* in the Salton Sea and has been partially or completely replaced in many other areas of California (Swift et al. 1993). In **Florida**, this species was stocked in the Bok Tower wildlife pond in Lake Wales, Polk County; it was introduced experimentally at Eustis, Lake County, but that population was exterminated in 1970; this species reportedly was established in a closed ditch on a fish farm near Micco, Brevard County (Courtenay et al. 1974). A population in a 0.2-ha isolated borrow pit (quarry pond) at Wayside Park in Perrine, Dade County, first reported in 1974, was eradicated by the state in December 1975 (Hogg 1976b; Courtenay et al. 1984, 1986; Taylor et al. 1986). In **Hawaii**, this species (identified as both *Tilapia zillii* and *Tilapia melanopleura*) is established in reservoirs on the islands of Hawaii, Maui, Oahu, and Kauai; it first was released in 1957 (Maciolek 1984) or possibly as early as 1955 (Morita 1981). It reportedly is confined to one or a few geothermal waters in south-central **Idaho** in the Snake River region below Shoshone falls (Idaho Fish and Game 1990). There are also unconfirmed reports that this species is cultured as a food fish in some unspecified localities in Idaho (Courtenay et al. 1984, 1986). A reproducing population survived several winters in **Nevada** in a spring-fed golf course pond at Cottonwood Park, Pahrump Valley, Nye County, ca. 1980 (Courtenay and Deacon 1982), but that population is no longer extant (Courtenay et al. 1984, 1986; Deacon and Williams 1984). The species was introduced into various sites in **North Carolina** including Lake Hyco in Person and Caswell counties, in 1984 (McGowan 1988; Crutchfield 1995), and other reservoirs or ponds, also associated with electric power (e.g., Sutton Lake adjacent to the Lumber River), as well as drainage ponds around phosphate mines (e.g., sites near Aurorean addition to Person and Caswell counties, other North Carolina counties involved are Beaufort, Brunswick, New Hanover, and Robeson (Courtenay et al. 1986; Menhinick 1991; J. Crutchfield, personal communication). This species has been cultured for aquatic plant control and for use as a possible food fish in a heated section of the Santee-Cooper Reservoir, **South Carolina** (= Lake Marion?) (Courtenay et al. 1986, 1991). The first report of this species in **Texas** was from spring waters (headwaters of the San Antonio River) within the San Antonio Zoo, Bexar County, in 1978 (Hubbs 1982; Courtenay et al. 1984, 1986; Hubbs et al. 1991). That population was still present as of 1990, but it was not known to be expanding its range and has not been found elsewhere in the state (Howells 1991b, 1992a, 1992b).

Means of Introduction: Introduced in most locations by state agencies, universities, or private companies for control of aquatic plants, to control mosquitoes and chironomid midges, as forage or food fish, and for aquaculture evaluation (Minckley 1973; Legner and Pelsue 1977; Lee et al. 1980 et seq.; Shapovalov et al. 1981; Grabowski et al. 1984; McGowan 1988; Courtenay and Robins 1989; Page and Burr 1991). For example, the species has been stocked into various North Carolina waters for aquatic plant control by Texas Gulf, Inc., and also by Carolina Power and Light Company (Courtenay et al. 1986; J. Crutchfield, personal communication). There have been both authorized as well as illegal releases. Introductions into Dade County, Florida, probably resulted from escapes from nearby fish farms or aquarium releases (Hogg 1976a, 1976b). Redbelly tilapia and blue tilapia were inadvertently introduced into Hyco Reservoir in North Carolina in 1984 after a small number of fish escaped from a holding cage located in the heated discharge area during an on-site agricultural study (Crutchfield 1995).

Agents and reasons for introductions were reviewed by Courtenay et al. (1986).

Status: Established or locally established in southern Arizona and California, Hawaii, North and South Carolina, and Texas. Reported from several other states. Considered eradicated from all sites in Florida (Smith-Vaniz, personal communication) and Nevada (Courtenay et al. 1984, 1986). In California, *Oreochromis mossambicus* has largely replaced *T. zillii* in the Salton Sea (and possibly in coastal southern California) (Swift et al. 1993).

Impact of Introduction: The redbelly tilapia is considered a potential competitor with native fish for food and spawning areas, and is potentially detrimental to California rice crops (Pelzman 1973). Juvenile *T. zillii* were implicated in population declines of desert pupfish *Cyprinodon macularius* inhabiting shallow irrigation canals near the Salton Sea, California (Lee et al. 1980 et seq.; Schoenherr 1985). In Florida, this species was found to be highly aggressive; it is considered a serious threat to native aquatic plants and to fish that rely on plants for cover, foraging, or spawning sites (Courtenay et al. 1974). In Hyco Reservoir, North Carolina, feeding by introduced redbelly tilapia eliminated all aquatic macrophytes from the reservoir within a 2-year period that coincided with declines in populations of several native fishes; tilapia populations continued expanding in the absence of macrophytes because of its ability to switch to alternate food sources (Crutchfield et al. 1992; Crutchfield 1995).

Remarks: Populations introduced into Alabama normally did not survive the winter and required annual restocking (Smith-Vaniz 1968). However, their tolerance to cold temperatures in central California prompted officials to place the species on the prohibited list for portions of the state (Shapovalov et al. 1981). A portion of the population found in a borrow pit in Perrine, Dade County, Florida, included hybrids with *Tilapia mariae* (Taylor et al. 1986). For detailed information on introduction history see Courtenay and Hensley (1979b), Shapovalov et al. (1981), Grabowski et al. (1984), and Courtenay et al. (1986).

Voucher specimens: Florida (UF 39874, 92219), Nevada (UF 91904), North Carolina (UF 92221; UNCC uncatalogued).

Oreochromis, Sarotherodon, Tilapia sp.

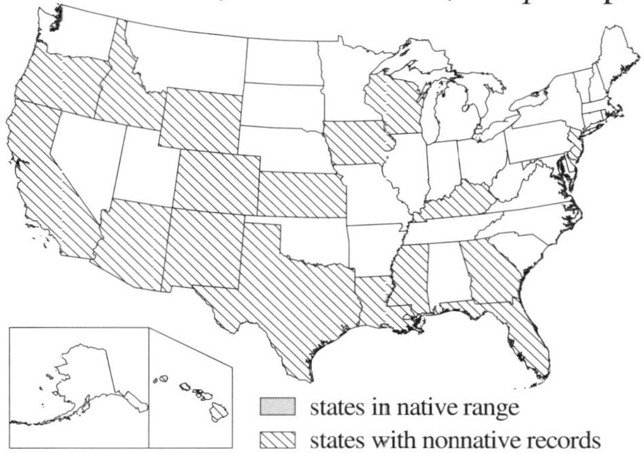

Unidentified Tilapiine Fishes and Hybrid Tilapias

Common Names: tilapias, tilapiine cichlids.

Identification: Tilapiine cichlids are a somewhat poorly defined taxonomic assemblage that includes a number of different genera (possibly eight or more) and well over one hundred described species. Tilapiine fishes introduced into the United States have included members of the genera *Oreochromis*, *Sarotherodon*, and *Tilapia*. Prior to Trewavas (1983), members of all three species were placed in the broad genus *Tilapia*, therefore those listed as *Tilapia* sp. in the older literature (and even in some of the more recent literature) may represent members of any one of these three currently recognized genera. Use of the vernacular name "tilapia" is typically used for members of any of the tilapiine genera. Correct identification of tilapiine fishes to species level has been problematic for several reasons. Of most importance, the literature includes many synonyms;a number of species are difficult to distinguish from one another and have been frequently misidentified or confused; and, hybridization is common among the different tilapiine species. Many tilapia hybrids found in open waters were artificially produced in aquaculture before being released or escaping; however, natural hybridization also has been documented among introduced forms (e.g., Taylor et al. 1986). In addition, not all hybrids are infertile (Barrett 1983). Because of the difficulty associated with tilapia identification, introduced tilapia frequently have not been reported to species level and, unfortunately, voucher specimens often have not been retained. On the other hand, many reported as pure species subsequently were found to represent another species or hybrids. Trewavas (1983) provided information on the morphological and behavioral differences among the three genera and also discussed hybridization in the group. A few selected references that may be of use in identification include Barrett (1983), Howells (1991b, 1992b), Eccles (1992), Axelrod (1993), and Skelton (1993). Maximum size: about 60 cm SL, most species to about 40 cm SL or less.

Native Range: Africa and the Middle East.

Nonindigenous Occurrences: Several unidentified species and hybrids have been reported from **Arizona** (Barrett 1983; Grabowski et al. 1984). Some specific sites mentioned include Cholla Lake, Apache County, 1968; Francis Creek, Yavapai County, 1968; and Redondo Lake, Yuma County, 1984. Based on a literature review, Swift et al. (1993) reported that as many as ten possible hybrids have been tentatively identified in the lower Colorado River area of Arizona and **California**. These included combinations of *Oreochromis mossambicus*, *O. urolepis* (= *O. hornorum*), *O. aureus*, *O. niloticus*, and *Tilapia zillii*. Unidentified tilapia taken from waters (mainly flood control channels) in the Los Angeles area may include *O. mossambicus* as well as hybrids between *O. mossambicus* and *O. urolepis* (Courtenay et al. 1984, 1986, 1991; Swift et al. 1993). Fish that appeared to be hybrids between *O. aureus* and *O. mossambicus* were collected in the Salton Sea in 1992 along with *O. aureus* (Swift et al. 1993). Several tilapiine species were stocked in Coyote Creek drainage, a tributary of the San Gabriel River, Los Angeles basin, in 1972; only *O. urolepis*, or a hybrid, persisted and reached high numbers (Legner and Pelsue 1977; Legner et al. 1980). Because many of the *O. urolepis* introduced into southern California were hybrid with *O. mossambicus* (Legner et al. 1980; Courtenay et al. 1984, 1991), there is still some doubt remaining as to whether or not the established populations in California open waters represent pure *O. urolepis* strains. Unidentified tilapiine forms have been introduced into **Colorado** (Grabowski et al. 1984; Zuckerman and Behnke 1986). Hybrid *Tilapia zillii* × *T. mariae* has been documented in **Florida** (Taylor et al. 1986). There are also unconfirmed reports from Florida of hybrid *Oreochromis* from the state, as well as a number of records of unidentified tilapia. Several unidentified tilapiine cichlids have been reported for **Georgia** (Grabowski et al. 1984; Gennings, personal communication). There is an unconfirmed report that tilapia (possibly *O. aureus*) were stocked and presumably established in golf course ponds at Sea Island, Glynn County, Georgia (Courtenay and Hensley 1979a; Courtenay et al. 1984, 1986). In an apparent reference to the same population, Gennings (personal communication) indicated that an unknown species of tilapia was reported from golf course ponds at the Sea Island Golf Club, possibly present during the late-1970s or mid-1980s, and these were supposedly extirpated during the winter of 1989 or before. Hybrid *O. mossambicus* and *Tilapia macrochir* have been reported for **Hawaii** (Devick 1972). A red hybrid of *O. mossambicus*, possibly with *O. urolepis*, was reported in **Idaho** from thermally heated sections of the Bruneau River near Bruneau Hot Springs, Owyhee County, after it escaped from a local aquaculture facility (Courtenay et al. 1987; Courtenay and Stauffer 1990). There are records of unidentified tilapia from **Iowa** (Carlander 1978; Grabowski et al. 1984). There are scattered records of unidentified tilapias from several states, including **Kansas**, **Kentucky**, **Louisiana**, and **Mississippi** (Grabowski et al. 1984). An unidentified tilapia was also reported from **New Jersey** (R. Soldwedel, personal communication). In **New Mexico**, unidentified tilapia have been introduced on at least two occasions, but the populations did not survive and there are no voucher specimens (Sublette et al. 1990). One or more unidentified tilapiine cichlids also have been recorded from **Oregon** (Grabowski et al. 1984); and from **Wisconsin** (Becker 1983; Grabowski et al. 1984). Electrophoretic evidence suggests that tilapia populations in **Texas** in the San Marcos River and in Canyon Reservoir are hybrid *Oreochromis aureus* × *O. mossambicus* (Howells 1991b, 1992b). There is a single report from **Wyoming** of a tilapia observed in Kelly Warm Springs during the late 1980s (W. Roston, personal communication).

Means of Introduction: Tilapiine cichlids have been widely introduced for avariety of reasons, for instance, because of their potentials as a food fish and also as potential sportfish. They also have been used as agents (often experimental) to control nuisance aquatic plants and insects. Some have been introduced as a result of escapes or releases from fish farms or aquaculture facilities. A few introductions have been the result of aquarium releases. Tilapia have also been transferred from one water to another by anglers. In addition, some hybrids have probably been intentionally or inadvertently introduced during stockings of pure tilapia species (Barrett 1983; Courtenay et al. 1986).

Status: Unidentified, or hybrid, tilapiine cichlids have been recorded from at least 16 states.

Impact of Introduction: See discussions on impacts included in the accounts for the various tilapia species.

Remarks: Barrett (1983) and Grabowski et al. (1984) provided details on the locations of known introductions in Arizona and also gave information on the distribution of unidentified tilapia and tilapia hybrids in the lower Colorado River basin. Dill and Cordone (1997) provided a fairly detailed account of the introduction of tilapias in California. They also briefly remarked on tilapia specimens taken in California from the north end of the Salton Sea that reportedly had been identified by D. F. E. Thys van den Audenaerde as *Tilapia guineensis*; they concluded, without explanation, that the identification was likely in error and omitted any detailed discussion. In any case, the accuracy of many introduced tilapia records are considered suspect because of the difficulties involved in positive identification of tilapiine species and hybrids. Until detailed genetic studies are done to supplement morphological work, we suspect problems associated with the positive identification of tilapiine cichlids will continue. Initial work towards this end with introduced tilapia has been done (e.g., Howells 1991b; Costa-Pierce and Doyle 1996, cited in Dill and Cordone 1997). It also should be noted that many tilapiine species are not naturally sympatric, but because of indiscriminate introductions within Africa, and into other tropical and warm temperature areas throughout the world, they have been brought into contact and many show evidence of introgressive hybridization (e.g., Trewavas 1983).

Embiotocidae - Surfperches

The family Embiotocidae contains thirteen genera with twenty-four species (Nelson 1994). The family is distributed in the northern Pacific including western North America, Japan and Korea (Nelson 1994). Maximum size is 20 cm (Nelson 1994). Surfperches are characterized by a single dorsal fin with 6–11 spines (except in *Hysterocarpus traski* which has 15–19), three anal fin spines, cycloid scales, and a forked caudal fin (Nelson 1994). A unique feature of the family is that they are viviparous, meaning they give birth to live young (Page and Burr 1991; Nelson 1994). These are mostly coastal marine fish but some species get into brackish water and *H. traskii* lives in freshwater (Nelson 1994).

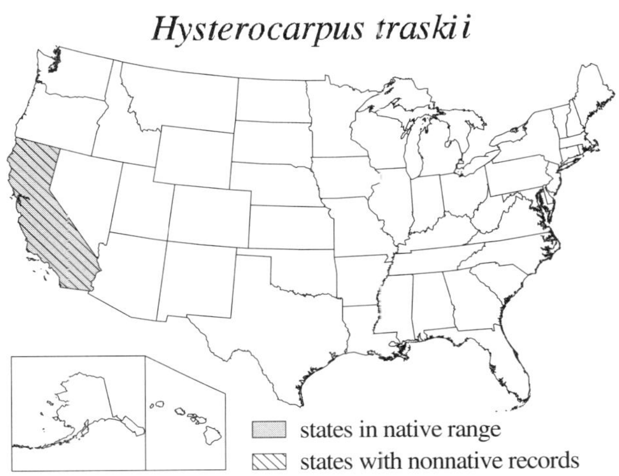

Hysterocarpus traskii Gibbons 1854

Common Name: tule perch.

Identification: Moyle (1976a); McGinnis (1984); Page and Burr (1991). Many recent authors give the name as *Hysterocarpus traski*. Maximum size: 15 cm.

Native Range: Formerly in low-elevation waters of the Sacramento–San Joaquin River drainage, as well as Clear Lake, Coyote Creek, and the Russian, Napa, Pajaro, and Salinas rivers. Now apparently extirpated from the Pajaro, Salinas, and San Joaquin rivers (Moyle 1976a). Lee et al. (1980 et seq.) provided a dot distribution map for the species.

Nonindigenous Occurrences: The tule perch was introduced in the early 1970s into O'Neill Forebay of San Luis Reservoir, and Los Banos Reservoir in Merced County, **California** (Moyle 1976a). McGinnis (1984) reported that it also has been introduced into a number of small lakes and ponds, such as Lake Merced in San Francisco, where the species is doing well.

Means of Introduction: Presumably gained access to the above areas via the California Aqueduct.

Status: Established in parts of California outside its native range (Moyle 1976a).

Impact of Introduction: Unknown.

Remarks: Moyle (1976a) noted that introductions involved what is considered by some to be a subspecies (i.e., *H. t. Triassic*).

Labridae - Wrasses

This family, known as the wrasses, is comprised of approximately 57 genera and 500 species. This is the second largest family of marine fishes. It is also one of the most diversified fish families in shape, color and size. Many species are popular marine aquarium fish (Nelson 1994). Most species have protruding, tucklike teeth; swim by flapping their pectoral fins; and are non-schooling (Robins and Ray 1986). Some have two distinct color phases for breeding males: a less colorful phase associated with group spawning, and a more colorful phase associated with supermales that pair spawn. Sex reversal in known to occur in this family; females can become supermales. Wrasses range in size from less than 30 cm to over 3 meters (Robins and Ray 1986).

Halichoeres semicinctus

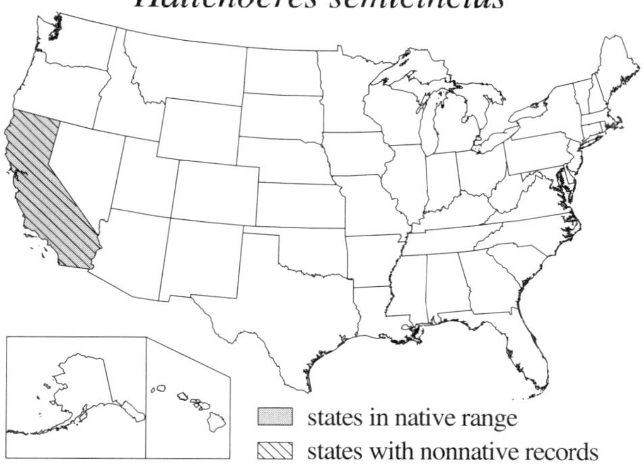

Halichoeres semicinctus (Ayres 1859)

Common Name: rock wrasse (kelp fish).

Identification: Miller and Lea (1972); Thomson et al. (1979); Eschmeyer et al. (1983). Maximum size: 24 cm.

Native Range: Marine; Pt. Conception, California, to Guadalupe Island (off north-central Baja) and Gulf of California (Ramirez-Hernández and Gonzales Pages 1976; Eschmeyer et al. 1983).

Nonindigenous Occurrences: Introduced into Salton Sea, **California**, in 1951 (Walker et al. 1961; Dill and Cordone 1997).

Means of Introduction: Intentionally stocked. Only two fish, obtained from San Felipe, Mexico, were stocked (Walker et al. 1961).

Status: The introduction failed to produce a viable population and the species is extirpated in the Salton Sea.

Impact of Introduction: Unknown.

Remarks: None.

Tautoga onitis

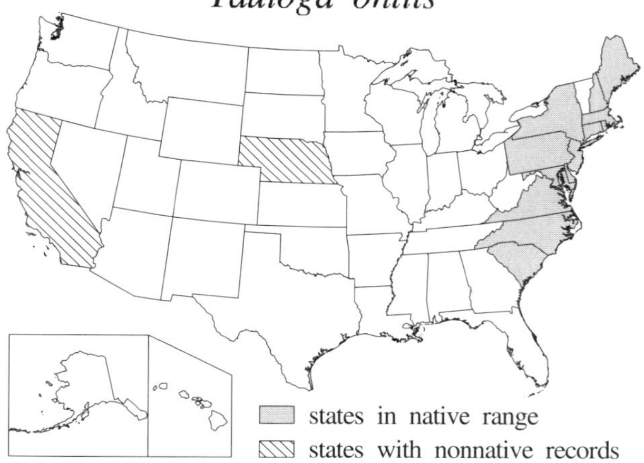

Tautoga onitis (Linnaeus 1758)

Common Name: tautog.

Identification: Manooch (1984); Robins and Ray (1986); Murdy et al. (1997). Maximum size: 91 cm, 10 kg.

Native Range: Marine; tautog are native to the Atlantic coast from Nova Scotia to South Carolina (Robins and Ray 1986).

Nonindigenous Occurrences: Two attempts have been made to stock these fish on the west coast. The first was made in 1873, when Livingston Stone made his famous ill-fated journey across the continent with east coast fishes in an aquarium railroad car. The entire shipment was lost in the Elkhorn River near Omaha, **Nebraska**, when a railroad bridge collapsed (Smith 1896). The tautogs did not survive. The second attempt was a year later. This time he made it to **California**. The tautogs were stocked in San Francisco Bay near Oakland (Smith 1896) but apparently did not take hold. A second attempt was made in 1897 when a few hundred more fish were stocked, again without success (Shebley 1917).

Means of Introduction: Accidental in Nebraska, intentionally stocked in California.

Status: Extirpated in both Nebraska and California.

Impact of Introduction: Unknown.

Remarks: DeKay (1842) wrote of stockings of this species in Charleston Harbor, South Carolina. Reportedly the fish were taken from Rhode Island. Both DeKay (1842) and Goode (1884) stated that this was an attempt to expand the species' range southward, implying they did not already occur there. However, Jordan and Gilbert (1882) reported a specimen "*Hiatula onitis*" which was actually a tautog with a missing anal fin. The specimen was collected from Charleston, South Carolina, and sent to Linnaeus in the 1700s. Linnaeus named the fish *Labrus hiatula* in his work *Systema Naturae* in 1788. Therefore, the species must have been native to the Charleston area and is not considered introduced outside the native range in this location.

Gobiidae - Gobies

The Gobiidae is a very large family comprising about 212 genera and about 1,875 species (Nelson 1994). Gobies are widespread in tropical and subtropical oceans (Berra 1981; Nelson 1994). These fishes are generally small, usually less than about 10 cm in length, with the maximum ca. 50 cm (Nelson 1994). They are bottom-dwelling fishes, mostly carnivorous, and typically inhabit shallow coastal marine or brackish waters. Most gobies of fresh and brackish waters have the pelvic fins fused in the form of a disc. Although the family consists mostly of marine forms, many species enter freshwaters and ascend far up rivers. In many inland freshwaters of oceanic islands, gobies are the most abundant fishes (Allen 1991; Nelson 1994).

Acanthogobius flavimanus

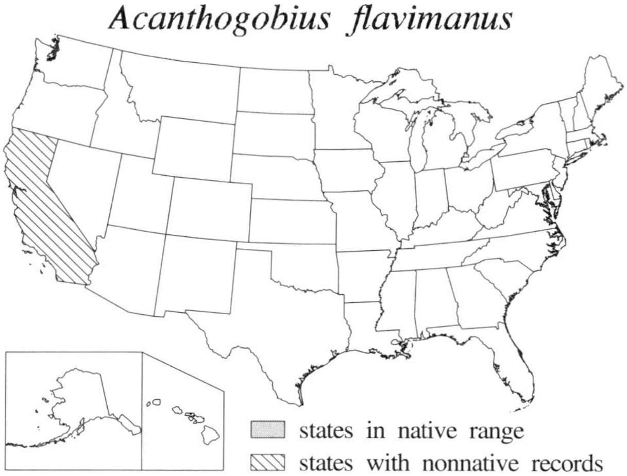

Acanthogobius flavimanus (Temminck and Schlegel 1845)

Common Names: yellowfin goby (mahaze, Japanese river goby).

Identification: Distinguishing characteristics, and illustrations or photographs were provided by Berg (1949); Miller and Lea (1972); Moyle (1976a); Eschmeyer et al. (1983); and Masuda et al. (1984). The species was included in keys of Berg (1949); Miller and Lea (1972); and Moyle (1976a). Maximum size: 30 cm.

Native Range: Fresh, brackish and marine. Japan, Korea, and China (Eschmeyer et al. 1983; Masuda et al. 1984; Meng et al. 1994).

Nonindigenous Occurrences: Introduced to **California**; the first records in that state were based on two specimens found in the Sacramento–San Joaquin Delta region, San Joaquin County, in early 1963. The first of these fish was trawled from the lower San Joaquin River near Venice Island, and the second specimen was taken from the Stockton Deepwater Channel near the Calaveras River (Brittan et al. 1963). The species later was found in surrounding areas including Suisan, San Pablo, and San Francisco bays, the Sacramento Delta, Bolinas Lagoon, Delta-Mendota Canal, and the San Luis Reservoir in Alameda, Contra Costa, Marin, Merced, Napa (possibly), San Francisco, San Mateo, Santa Clara, Solano, and Sonoma counties (Brittan et al. 1970; Moyle 1976a; Courtenay et al. 1986). Specimens also were taken in Elkhorn Slough, Monterey County (Kukowski 1972), and Tomales Bay, Marin County (Miller and Lea 1972). The first records of this species in southern California were from the Los Angeles Harbor area, Los Angeles County, in 1977 (Haaker 1979); subsequently specimens were found in Long Beach Harbor and near the mouth of the Los Angeles River, Los Angeles County; in the San Gabriel River, Upper Newport Bay, and upstream to San Diego Creek, Orange County; and in Ballona Marsh and Mugu Lagoon (Haaker 1979; Allen 1982; Swift et al. 1993). This species was reported, apparently in small numbers, from other coastal areas of southern California including Malibu Lagoon, San Onofre, San Mateo, Las Pulgas, and Santa Margarita lagoons, and Morro Bay (Swift et al. 1993). In 1980, the species was reported as occurring in San Diego (perhaps extending as far south as Baja California Norte, Mexico) (Courtenay et al. 1986). Williams et al. (1998) reported them from San Diego Bay marshes but gave the first date for that area as 1984.

Means of Introduction: Initial and possibly later introductions were probably by way of ballast water carried in transoceanic ships (Brittan et al. 1963). It also is hypothesized that introduced gobies arrived as eggs on fouling organisms, such as oysters, growing on ship hulls (Hubbs and Miller 1965; Eschmeyer et al. 1983). Although first collected in 1963, the yellowfin goby was probably introduced into California in 1959 or 1960, likely about the same time as the chameleon goby (Brittan et al. 1970; Meng et al. 1994). Once established, this species spread in California, probably as a result of its own dispersal abilities, and sometimes with the aid of currents; in addition, dispersal may have resulted from the species' possible use as a baitfish (Brittan et al. 1970; Courtenay and Hensley 1979a).

Status: Established in coastal and inland waters of central and southern California.

Impact of Introduction: In at least one saltwater location, yellowfin gobies were reported to have partially replaced Pacific staghorn sculpins *Leptocottus armatus* (Brittan et al. 1970). There also is concern that the yellowfin goby might outcompete and possibly eliminate freshwater populations of the small and endangered tidewater goby *Eucyclogobius newberryi* (Moyle 1976a). Meng et al. (1994) suggested that environmental disturbances, coupled with the introduction of this and other foreign species, are altering fish communities and hastening declines of native fishes in California. Although Meng et al. (1994) found that the yellowfin goby has an impact on the introduced chameleon goby *Tridentiger trigonocephalus*, recent investigations have shown this species is actually the shimofuri goby *Tridentiger bifasciatus* (not the yellowfin goby) that occurs in Suisun Bay where the study was conducted (Fleming, personal communication). Hence, it is the shimofuri goby the study is documented.

Remarks: The yellowfin goby underwent a population explosion in the San Francisco area in the late 1960s and early 1970s (Brittan et al. 1970). In 1967, a fish kill occurred in the San Luis Reservoir, which receives freshwater from the Sacramento–San Joaquin River Delta. About half of the approximately 10,000 fishes killed in this incident were *A. flavimanus* (Brittan et al. 1970). Apparently another massive die-off occurred in Rodeo Lagoon in 1981 and was thought to be caused by low salinity (<5 ppt) (Wang 1984).

The species first was reported from Suisan Marsh, a portion of the San Francisco Bay estuary, in 1967 (Brittan et al. 1970); by the early 1980s its population in that area had grown and it was reported as the third most abundant fish in 1980–1982 trawl catches by Meng et al. (1994). Recent drought conditions in California have reduced freshwater outflows and may have allowed this goby to gain an advantage over native freshwater and estuarine fishes less able to tolerate high salinity conditions (Herbold et al. 1992; Meng et al. 1994). Meng et al. (1994) presented a figure showing the annual relative abundance of chameleon and yellowfin gobies taken in trawls over the period 1979 to 1992. Brittan et al. (1970), Courtenay and Hensley (1979a), and Lee et al. (1980 et seq.) provided maps showing this species distribution in California.

Voucher specimens: California (CAS 14405, LACM 37711-1, 39764-6, 44205-1).

Gillichthys mirabilis Cooper 1864

Common Name: longjaw mudsucker.

Identification: Moyle (1976a); Eschmeyer et al. (1983). Formerly used synonym is *Gillichthys destrusus*. Maximum size: 21 cm.

Native Range: Marine. Coast of California from Tamales Bay in the north to the Gulf of California in the south (Eschmeyer et al. 1983).

Nonindigenous Occurrences: This species has been reported from Roosevelt Lake on the Salt River in **Arizona** (Minckley 1973), and the Salton Sea of **California** (Barlow 1961; Walker et al. 1961; Moyle 1976a). It is probably also introduced into the lower Colorado River in Arizona–California–Nevada, where it is commonly used as a bait minnow (Miller 1952; Minckley 1973; Minckley, personal communication).

Means of Introduction: The longjaw mudsucker seen in Arizona was probably a bait bucket release. This species is used as bait in central Arizona and in the lower Colorado River (Miller 1952; Minckley 1973). Five hundred fish obtained from San Diego Bay were intentionally stocked into the Salton Sea by the California Department of Fish and Game in 1930 (Barlow 1961). Hubbs (personal communication in Barlow 1961) suggested that some *G. mirabilis* may have been introduced into the Salton Sea by live-bait dealers. However, Barlow (1961) considered existing Salton Sea populations to be derived from the San Diego planting.

Status: Established in California. Reported in Arizona.

Impact of Introduction: Unknown.

Remarks: The longjaw mudsucker is a popular bait fish that can live out of water for six to eight days if kept moist (Eschmeyer et al. 1983), can withstand being hooked, and can tolerate freshwater (Barlow and De Vlaming 1972). Evans and Douglas (1950) reported collections of several specimens, identified by Carl Hubbs as *Gillichthys destrusus*, from the Salton Sea in 1949. Barlow (1961) treated that species as a synonym of *G. mirabilis*.

Gillichthys seta

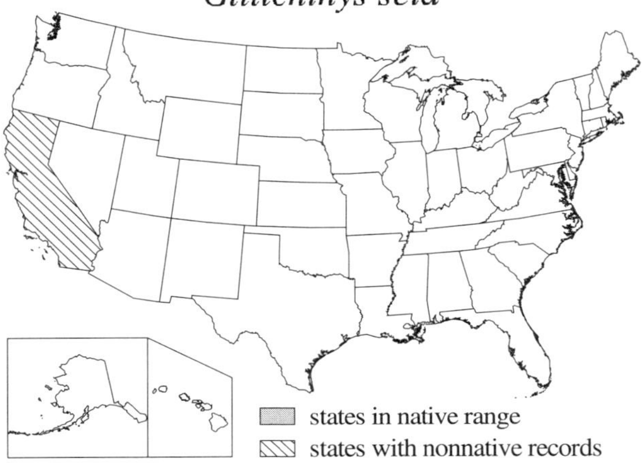

Gillichthys seta (Ginsburg 1938)

Common Name: (mudsucker, shortjaw mudsucker).

Identification: Barlow (1961); Thomson et al. (1979). For many years *G. seta* was confused with *G. mirabilis*, from which it is believed to have evolved (Barlow 1961). Maximum size: 89 mm.

Native Range: Marine; *G. seta* is found only in the northern Gulf of California in the rocky intertidal zone (Barlow 1961; Thomson et al. 1979).

Nonindigenous Occurrences: Introduced into Salton Sea, **California**, in 1951 (Walker et al. 1961; Dill and Cordone 1997).

Means of Introduction: Intentionally stocked as forage. Sixty-three fish were stocked. Stock was obtained from San Felipe, Mexico (Walker et al. 1961).

Status: The introduction failed to produce a viable population and is no longer extant in the Salton Sea.

Impact of Introduction: Unknown.

Remarks: None.

Gobiosoma bosc

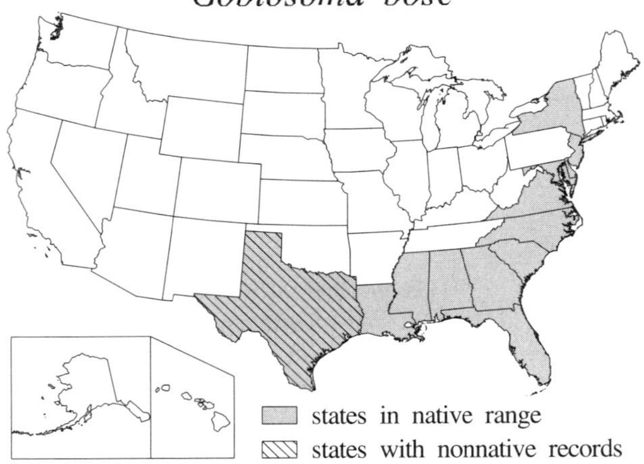

Gobiosoma bosc (Lacepède 1800)

Common Name: naked goby.

Identification: Hoese and Moore (1977); Robins and Ray (1986); Hubbs et al. (1991). Maximum size: 6 cm.

Native Range: Marine. Estuaries and coastal waters from New York to Texas (Robins and Ray 1986).

Nonindigenous Occurrences: The naked goby was introduced into two **Texas** reservoirs, Victor Braunig and Calaveras, both in Bexar County in the San Antonio drainage (Howells, personal communication).

Means of Introduction: Accidentally introduced through stock contamination.

Status: Reported from inland Texas.

Impact of Introduction: Unknown.

Remarks: None.

Neogobius melanostomus

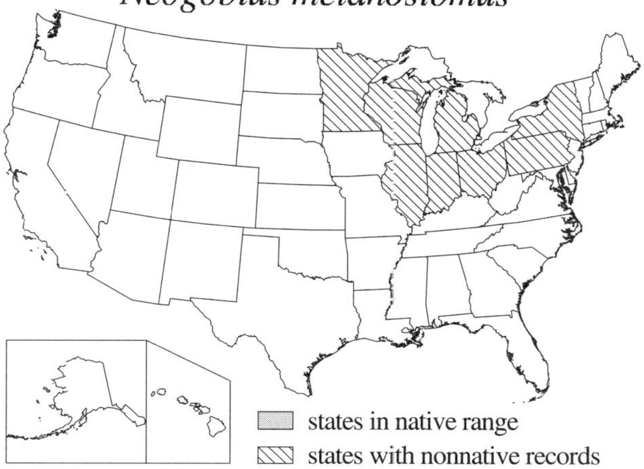

Neogobius melanostomus (Pallas 1814)

Common Name: round goby.

Identification: Distinguishing characteristics have been given by Berg (1949), Miller (1986), Crossman et al. (1992), Jude (1993), and Marsden and Jude (1995). Young round gobies are solid slate gray. Older fish are blotched with black and brown and have a greenish dorsal fin with a black spot. The raised eyes on these fish are also very distinctive (Jude 1993). This goby is very similar to native sculpins but can be distinguished by the fused pectoral fins (sculpins have two separate fins) (Marsden and Jude 1995). Maximum size: 30.5 cm; 17.8 cm maximum seen in United States (Jude 1993).

Native Range: Marine to freshwater. Eurasia including Black Sea, Caspian Sea, and Sea of Azov and tributaries (Miller 1986).

Nonindigenous Occurrences: This species was introduced into the St. Clair River and vicinity on the **Michigan**–Ontario border; several collections were made in 1990 on both the U.S. and the Canadian side (Jude et al. 1992; D. J. Jude and D. Nelson, personal communication). By 1994 the species had spread to the north end of Lake St. Clair at Anchor Bay. It has been taken inland in the Shiawasse and Flint rivers since August 1996 and June 1997, respectively (Jude, personal communication). In 1994, the round goby began appearing in southern Lake Michigan near the Calumet-Chicago area of **Illinois** (Cavender, personal communication). It was first collected in **Indiana** from the Grand Calumet River in 1993 (J. Francis, personal communication). The following year it was taken in Hammond Harbor (Francis and T. Lauer, personal communication); then in the Port of Indiana and East Chicago in 1996 (Francis, personal communication). In 1993, it was collected at Fairport Harbor in Lake Erie, and from the mouth of the Grand River in Lake County, **Ohio** (Knight 1994). In 1994, the species was taken from the lake offshore at depths of 70 feet, and reportedly from Lorain Harbor in Lorain County, Ohio, 60 miles west of Fairport, although there are no vouchers to confirm this location (Cavender, personal communication). The round goby was first reported from **Pennsylvania** in October 1996, in Lake Erie off Walnut Creek, just west of the city of Erie (C. Murray, personal communication). In July 1995, a single individual was collected from **Wisconsin** waters of the St. Louis Bay, Lake Superior, on the Minnesota–Wisconsin state line in a trawl (T. Busiahn, personal communication). In May 1996, a single adult was taken in Duluth Harbor, **Minnesota** (D. Jensen, personal communication). No other specimens have been reported from this area since that time.

There was an unconfirmed report of a round goby in eastern Lake Ontario, **New York**, during the summer of 1995. The first confirmed collection of a round goby in Lake Ontario occurred in July 1998. A single fish was collected at Port Dalhousie at the mouth of the Welland Canal in Ontario, Canada (C. Scobie, personal communication).

Means of Introduction: Introduced into the Great Lakes via freighter ballast. Spread to Lake Superior by freighters operating within the Great Lakes.

Status: This goby is established and apparently undergoing a population explosion in the Great Lakes. The goby was considered extremely abundant in the St. Clair River in 1994. Short trawls made in Lake Erie in October 1994 turned up 200 individuals. Frequent trawling in 1995 collected over 3,000 individuals near Fairport Harbor, Ohio (Knight, personal communication). Densities in Calumet Harbor exceed 20 per square meter (Marsden and Jude 1995). Gravid females and different size classes have been found in Lake Erie (Cavender, personal communication). Only two individuals have been reported from Lake Superior.

Impact of Introduction: The numbers of native fish species have declined in areas where this goby has become abundant (Crossman et al. 1992); sculpins have been particularly affected (Marsden and Jude 1995). This species has been found to prey on darters, other small fish, and lake trout eggs and fry in laboratory experiments. They also may feed on eggs and fry of sculpins, darters and logperch (Marsden and Jude 1995). Adults aggressively defend spawning sites and may occupy prime spawning areas, keeping natives out (Marsden and Jude 1995). Walleye anglers in Detroit report that at times all they can catch are gobies, which eagerly attack bait (Marsden and Jude 1995).

Remarks: The diet of round gobies collected in the United States consists of aquatic insects, zebra mussels, and some native snails. Studies have shown a single goby can eat as many as 78 zebra mussels per day. These gobies are very pugnacious fish that feed voraciously, and, as such, they may prey on the young of other deepwater bottom dwellers such as sculpins, darters, and logperch. Their well-developed lateral line may help them outcompete natives for food in the murky Great Lakes waters. Their aggressive nature may allow individuals to dominate prime spawning sites, making these sites unavailable to natives. There is a long spawning period during which individuals can spawn every 20 days, while they aggressively defend their nests (Jude et al. 1992; Jude 1993). Although Jude (1993) expected introduced round gobies to be restricted to nearshore rocky or weedy habitats, the species has since been captured at depths as great as 21.5 m (Cavender, personal communication). Because the species exhibits two pigmentation morphs, investigations are planned to determine whether more than one introduction of *Neogobius* occurred in the Great Lakes (Cavender, personal communication).

Voucher specimens: Ohio (OSM, UF 98888); Michigan (UMMZ); no voucher specimens exist for Lorain Harbor or Calumet, Illinois (Lake Michigan). Voucher specimens from the Canadian side of the St. Clair River (UMMZ 217682, 218279; ROM 60675); Lake Ontario (Ontario Federation of Anglers and Hunters, Peterborough, Ontario).

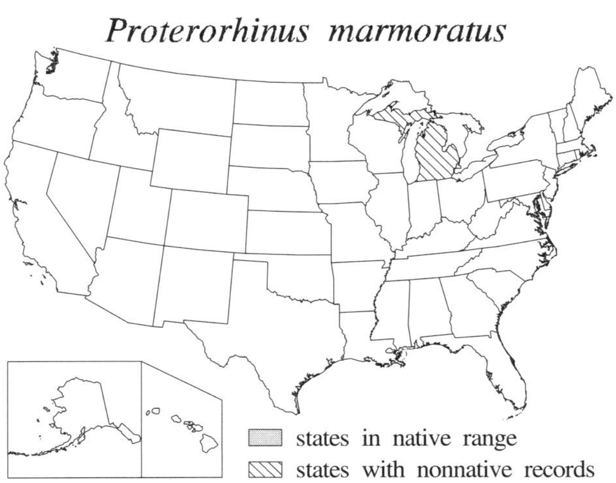

Proterorhinus marmoratus (Pallas 1814)

Common Name: tubenose goby.

Identification: Characteristics were given by Berg (1949) and Miller (1986). This species is distinguished from the round goby by its long anterior nostrils (Miller 1986; Jude 1993). Miller (1986), Crossman et al. (1992), and Jude et al. (1992) provided characteristics to distinguish the round and tubenose gobies. Maximum size: 12.7 cm.

Native Range: Slightly brackish to freshwater. Eurasia, including rivers and estuaries of Black and Caspian seas and Sea of Azov; also in rivers of northern Aegean, and Aral Sea (Miller 1986).

Nonindigenous Occurrences: The species was introduced into the St. Clair River, **Michigan**. It was taken in several 1990 collection samples from the cove next to the Detroit Edison Company's Belle River Power Plant and near the intake structures (Jude et al. 1992; Jude 1993; Jude, personal communication; G. Smith, personal communication). As of 1994, it also had been found at the north end of Lake St. Clair at Anchor Bay (Cavender, personal communication). In July 1997, a single tubenose goby was captured in Lake Erie at Port Glasgow, Canada (ROM 70904) (A. Dextrase, personal communication). Since then, additional specimens have been found in the area (Kingsville Marsh) and the species is believed to be established on the northwestern shore of Lake Erie. These reports indicate the species has spread and may be present but undetected in Ohio waters.

Means of Introduction: Introduced via ballast water.

Status: This species is believed to be established but rare in the St. Clair River, and in Lake St. Clair, Michigan (Jude 1993; Cavender, personal communication). Eggs attached to vegetation brought up during a trawl in 1994 were bought into a laboratory and hatched (Cavender, personal communication).

Impact of Introduction: Unknown.

Remarks: The tubenose goby is considered endangered in its native range. This goby may live as long as five years (Jude 1993).

Tridentiger barbatus

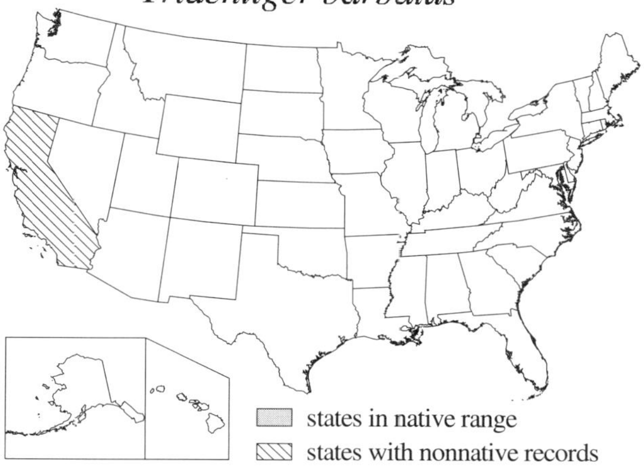

Tridentiger barbatus (Günther 1861)

Common Name: Shokihaze goby.

Identification: Jordan and Snyder (1902); Tomiyama (1936); Masuda et. al. (1984).

Native Range: Marine and brackish waters in China, Japan, and Korea.

Nonindigenous Occurrences: The first report of this species was a single juvenile collected on November 3, 1997, in the San Joaquin River, **California**, near West Island. Three more juveniles were collected the following month in San Pablo Bay. Since that time, approximately a dozen more have been collected in San Francisco Bay from San Pablo Bay to the confluence of the San Joaquin and Scramento rivers at the Delta (K. Fleming, personal communication).

Means of Introduction: Probably introduced via ballast water (Fleming, personal communication).

Status: Possibly established in California. Both young-of-the-year and adults have been collected, suggesting the possibility of reproduction (Fleming, personal communication).

Impact of Introduction: Unknown. However, this species is piscivorous and may prey on native gobies and/or sculpins (Fleming, personal communicaton).

Remarks: These gobies generally breed in shallow water areas. The shallow areas of the Bay have not been sampled. Hence, no definite conclusions can be drawn concerning reproduction in the Bay (Fleming, personal communication).

Voucher specimens: California (UCD; CAS).

Tridentiger bifasciatus

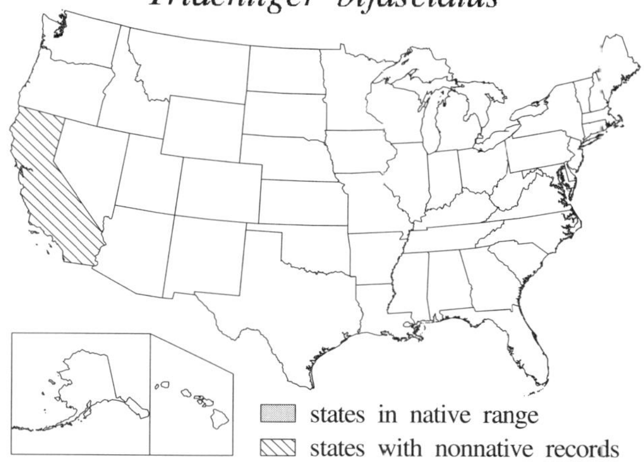

Tridentiger bifasciatus Steindachner 1881

Common Names: shimofuri goby (proposed) (shimofuri shimahaze [Japanese]).

Identification: Akihito and Sakamoto (1989) resurrected this species after over 50 years of synonomy with *T. trigonocephalus*. They provided details on distinguishing characteristics, and also provided color photographs for both species. Distinguishing characteristics were also summarized by Matern and Fleming (1995). According to Matern and Fleming (1995), some published reports of *T. trigonocephalus* in California (e.g., Meng et al. 1994) actually refer to *T. bifasciatus*. Maximum size: 11 cm.

Native Range: Fresh and brackish water. Asia, including Japan, Korea, China, Taiwan, and the former Soviet Union (Akihito and Sakamoto 1989).

Nonindigenous Occurrences: The first records of this species, all in **California**, are based on collections taken from Suisun Marsh in 1985, from John Skinner Delta Fish Protective Facility near Byron, Contra Costa County, in 1987, and from low-salinity regions of the San Francisco Bay estuary. By 1990 the species had dispersed approximately 513 km south to Pyramid Lake in Los Angeles County (October 1990), and then spread to Piru Creek downstream of this lake by June 1992 (as *T. trigonocephalus*) (Swift et al. 1993). It has not been taken from Piru Lake or downstream of the lake, to date (Matern and Fleming 1995).

Means of Introduction: Probably introduced via ballast water circa 1985 (Matern and Fleming 1995).

Status: Based on trawl samples, this species was the most abundant adult fish in Suisun Marsh by 1989 (Meng et al. 1994, as *T. trigonocephalus*; Matern and Fleming 1995), and the third most abundant larval fish (Matern and Fleming 1995). They were the most abundant larvae by 1990 (Matern and Fleming 1995).

Impact of Introduction: A study conducted by Meng et al. (1994) found that yellowfin gobies prey on the eggs and larvae of the introduced chameleon goby *Tridentiger trigonocephalus*. However, their research was conducted in Suisun Bay, an area known to be in the range of the shimofuri goby and not the chameleon goby (Fleming, personal communication). Therefore, it is actually the shimofuri goby that eats the eggs of the chameleon goby. Meng et al. (1994) also suggested this goby may be responsible, at least in part, for the chamelon goby's decline in the San Francisco Bay estuary. However, Fleming (personal communication) disagrees. He believes it is unlikely the shimofuri goby has been in the Bay long enough to have and impact on the chameleon goby's population.

Remarks: The shimofuri goby may compete with the federally endangered tidewater goby *Eucyclogobius newberryi* if the two come in contact. The two species are very similar in habitat and dietary preferences (Matern and Fleming 1995). One potential scenario that would cause the two species to become sympatric is introduction of the shimofuri goby into Lake Cachuma in southern California via the California Aqueduct, and then into the Santa Ynez River (Matern and Fleming 1995). *Tridentiger bifasciatus*, although often sympatric with *T. trigonocephalus* in its native range, apparently prefers freshwater and has never been taken in salt water (Matern and Fleming 1995); whereas, *T. trigoncephalus* is rarely vound in salinities less than 22 ppt. (Fleming, personal communication). Both species are found in brackish water. This species, like the chameleon goby, is capable of rapidly changing color (Matern and Fleming 1995).

Voucher specimens: California (CAS 81588, 81582, 82361).

Tridentiger trigonocephalus

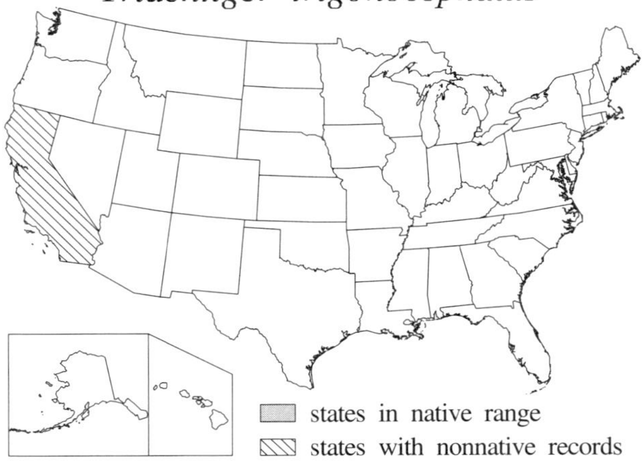

Tridentiger trigonocephalus (Gill 1859)

Common Names: chameleon goby (shimahaze, striped tripletooth goby).

Identification: Eschmeyer et al. (1983). Distinguishing characteristics were also provided by Masuda et al. (1984), Akhito and Sakamoto (1989), and Matern and Fleming (1995). Color photographs appeared in Masuda et al. (1984) and in Akihito and Sakamoto (1989). *Tridentiger bifasciatus*, formerly considered a synonym, was resurrected by Akihito and Sakamoto (1989). Maximum size: 11 cm.

Native Range: Brackish and marine waters of China, Korea, and Japan (Courtenay et al. 1986; Meng et al. 1994).

Nonindigenous Occurrences: The species first was recorded in 1960 when two individuals were observed on a rock jetty in Los Angeles Harbor, **California** (Haaker 1979; Matern and Fleming 1995). In 1962 another specimen was taken from the Redwood City docks in the southern portion of San Francisco Bay (museum specimen; Meng et al. 1994; Matern and Fleming 1995). The fish also occurs in Lake Merritt in Oakland (which is connected to San Francisco Bay) (Courtenay et al. 1986), and in Los Angeles Harbor (Haaker 1979; Eschmeyer et al. 1983).

Means of Introduction: The initial introduction may have been as fertilized eggs on introduced Japanese oysters (Courtenay et al. 1984, 1986, 1991) or from ballast water (Eschmeyer et al. 1983).

Status: Established in California. Recently, chameleon goby populations in San Francisco Bay have plummeted, possibly because of predation by yellowfin gobies (Meng et al. 1994). Because adults spawn 3–4 months after the piscivorous yellowfin goby, their young are vulnerable (Wang 1986).

Impact of Introduction: Unknown.

Remarks: Some reports in the literature referring to *T. trigonocephalus* collections made in freshwater in California (i.e., Meng et al. 1994) should actually refer to *T. bifasciatus* (Matern and Fleming 1995). This goby gets its name from its ability to rapidly change colors from a striped to a barred pattern (Eschmeyer et al. 1983). *Tridentiger trigonocephalus* is more common in marine environments than is *T. bifasciatus*; it is rarely found in salinities less than 22 ppt. (Matern and Fleming 1995; Fleming, personal communication).

Voucher specimens: California (CAS 27011); two other specimens, LACM 45016-1 and LACM 45688-1, should be re-examined in light of the discovery of *T. bifasciatus*.

Scombridae - Mackerels and tunas

The family Scombridae, the mackerels and tunas, contains 15 genera and 49 species (Nelson 1994). They are found in tropical and subtropical seas worldwide (Nelson 1994). Mackerels and tunas are fast-swimming, schooling, usually pelagic fishes. They are characterized by torpedo-shaped bodies, two dorsal fins with finlets behind the second dorsal and anal fins, a slender caudal peduncle, and a lunate caudal fin (Robins and Ray 1986; Nelson 1994). Some of the large tunas can reach 14 ft (4.3m) and 1800 lb. (816 kg) (Eschmeyer et al. 1983). Some species are endothermic (Nelson 1994). Collette and Nauen (1983) provided a synopsis of this family.

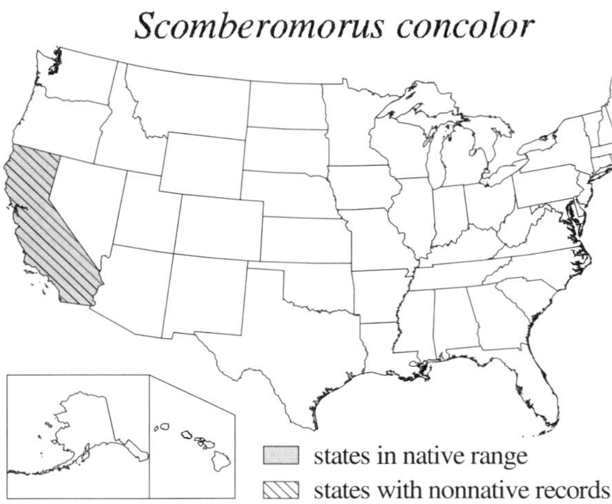

Scomberomorus concolor (Lockington 1879)

Common Name: gulf sierra (Monterey Spanish mackerel).

Identification: Miller and Lea (1972); Collette and Nauen (1983); Eschmeyer et al. (1983); Fischer et al. (1995b). Maximum size: 76 cm.

Native Range: Marine. Monterey Bay to Mexico. Abundant in northern part of Gulf of California; stragglers reach southern California (Eschmeyer et al. 1983). Now confined to the northern Gulf of California (Collette and Nauen 1983).

Nonindigenous Occurrences: Introduced into Salton Sea, **California**, in 1950 (Walker et al. 1961; Dill and Cordone 1997).

Means of Introduction: A single fish was intentionally stocked.

Status: The introduction failed to produce a viable population and the species is extirpated in the Salton Sea.

Impact of Introduction: Unknown.

Remarks: None.

Stromateidae - Butterfishes

The family Stromateidae, also known as butterfishes, contains three genera and about 13 species (Nelson 1994). Eschmeyer et al. (1983) list 12–15 species. Robins and Ray (1986) group them differently and list 67 species, apparently by lumping 5–6 closely-related families. Butterfishes are found in coastal areas of North and South America, western Africa, and southern Asia (Nelson 1994). They are characterized by a very deep, compressed, silvery body with a rounded head; the absence of pelvic fins in adults; a continuous dorsal fin; a forked caudal fin, and 2–6 spines and 30–50 soft rays in the anal fin (Eschmeyer et al. 1983; Nelson 1994). Many species are associated with jellyfishes at some life stage (Robins and Ray 1986).

Peprilus simillimus

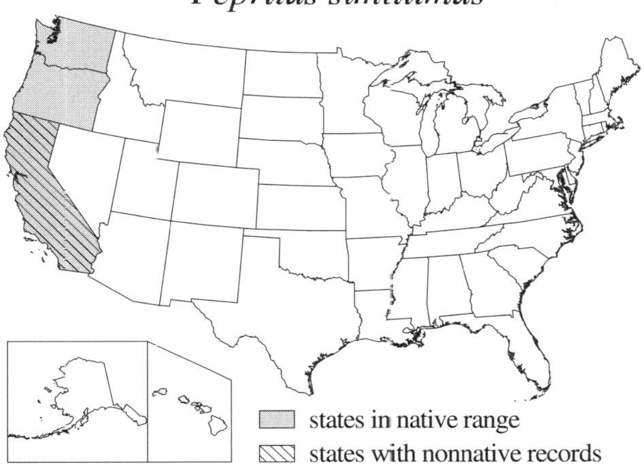

Peprilus simillimus (Ayres 1860)

Common Name: Pacific pompano (Pacific butterfish).

Identification: Miller and Lea (1972); Eschmeyer et al. (1983); Fischer et al. (1995b). Maximum size: 28 cm.

Native Range: Marine; Queen Charlotte Sound, British Columbia, to central Baja and Gulf of California (Miller and Lea 1972; Eschmeyer et al. 1983).

Nonindigenous Occurrences: Introduced into Salton Sea, **California**. See discussion of *Peprilus ovatus* in "Erroneous Records of Introductions."

Means of Introduction: Twelve fish entered the bait tanks through the circulation system from San Diego Bay and were knowingly introduced in 1948 as a result of stock-contamination (Dill and Cordone 1997).

Status: Extirpated in California.

Impact of Introduction: Unknown.

Remarks: None.

Anabantidae - Climbing gouramies

The anabantoid family Anabantidae, whose members are also known as climbing gouramies or climbing perches, consists of three genera, *Ctenopoma* and *Sandelia* in Africa and *Anabas* in Asia (Roberts 1989), with about 30 species (Nelson 1994). Like all anabantoids, these fishes have a labyrinth organ that enables them to survive in oxygen-poor waters. Anabantids are small fish, less than about 25 cm in length, characterized by an elongated, compressed body with well-developed dorsal and anal fins. They inhabit fresh and sometimes brackish waters (Talwar and Jhingran 1992; Nelson 1994).

Anabas testudineus

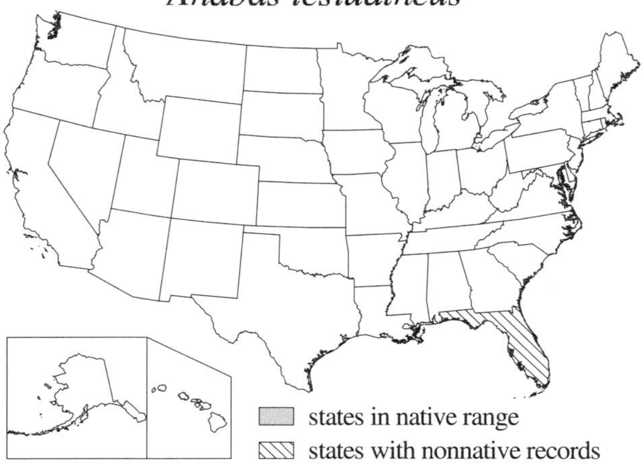

Anabas testudineus (Bloch 1795)

Common Name: climbing perch.

Identification: This species is highly variable in terms of body form and meristics (Roberts 1989). A key, distinguishing characteristics, and figures were given in Talwar and Jhingran (1992); distinguishing characteristics were also provided by Sterba (1973) and Kottelat et al. (1993). See Axelrod et al. (1985) and Kottelat et al. (1993) for color photographs of live fish. Maximum size: 25 cm.

Native Range: Tropical Asia. Native to India, Sri Lanka, Burma, Southeast Asia, southern China, Malaysia, and possibly Philippine Islands (introduced?) (Sterba 1973; Axelrod et al. 1985). Inhabits fresh and brackish waters (Talwar and Jhingran 1992).

Nonindigenous Occurrences: A small population was established near fish farm on Piney Point Road in northwestern Manatee County, **Florida**, in late 1960s or early 1970s; however, it was not found in 1970–1972 surveys. The population is thought to have died out, probably due to cold winter temperatures (Courtenay and Hensley 1979a; Courtenay and Stauffer 1990).

Means of Introduction: Probable escape from local aquarium fish farms (Courtenay and Stauffer 1990).

Status: Extirpated. Formerly established locally in Manatee County, Florida (Courtenay and Hensley 1979a; Courtenay and Stauffer 1990).

Impact of Introduction: Unknown.

Remarks: Roberts (1989) considered *Anabas* to be among the hardiest of fishes.

There are no known voucher specimens.

Ctenopoma nigropannosum

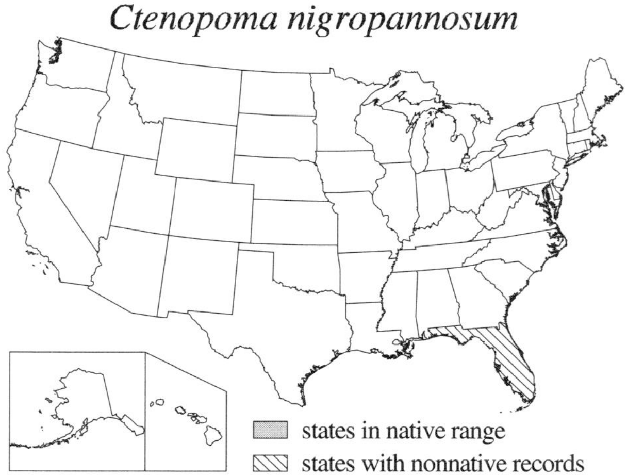

Ctenopoma nigropannosum Reichenow 1875

Common Names: twospot ctenopoma (twospot climbing perch, two-spotted ctenopoma).

Identification: Sterba (1973) gave distinguishing characteristics. See Axelrod et al. (1985) for color photograph of live fish. Genus includes approximately 16 species (Norris et al. 1988). Maximum size: 17 cm.

Native Range: Tropical Africa. Native to Niger River delta and Zaire River basin (Sterba 1973, see Norris et al. 1988)

Nonindigenous Occurrences: A small population was locally established in a canal adjacent to a fish farm in northwestern Manatee County, **Florida**, in the early 1970s (Courtenay et al. 1974); the population survived several years before disappearing (Courtenay and Hensley 1979a; Courtenay et al. 1984; Courtenay and Stauffer 1990).

Means of Introduction: Probable escape from local aquarium fish farm (Courtenay et al. 1974).

Status: Extirpated. In the past, it was locally established in Florida, but no specimens have been collected since the 1970s (Courtenay and Hensley 1979a). Disappearance was possibly the result of cold winter temperatures (Courtenay and Hensley 1979a; Courtenay et al. 1984; Courtenay and Stauffer 1990).

Impact of Introduction: Unknown.

Remarks: The species formerly occupied the same canal, adjacent to the same fish farm, from which *Anabas testudineus* had been reported (Courtenay and Hensley 1979a).

There are no known voucher specimens.

Helostomatidae - Kissing gouramies

The anabantoid family Helostomatidae contains a single species, *Helostoma temminckii*, a highly specialized filter feeder endemic to Southeast Asia (Roberts 1989; Nelson 1994). As part of its feeding apparatus, the fish has unusually fine gill rakers and highly modified soft structures on its gill arches (Roberts 1989). The fish also is armed with horny teeth on the lips for scraping algae off surfaces (Nelson 1994). It is a popular aquarium species, in large part because fish pairs show the unusual behavior of extending their broad-lipped mouths and touching or kissing. It is restricted to freshwater.

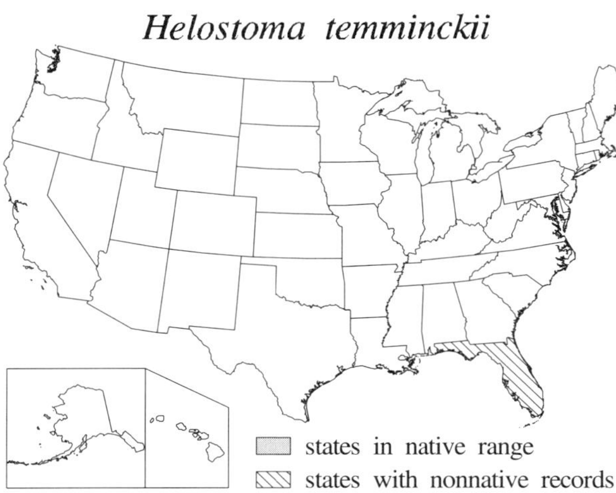

Helostoma temminckii Cuvier 1831

Common Names: kissing gourami (pink kisser, green kisser [varietal names]).

Identification: A key and distinguishing characteristics were given in Smith (1945) and Talwar and Jhingran (1992); distinguishing characteristics were also provided by Sterba (1973); Nelson (1994); and Kottelat et al. (1993). Color photographs appeared in Axelrod et al. (1985) and in Kottelat et al. (1993). Name sometimes given as *Helostoma temmincki*; another incorrect spelling is *Helostoma temminkii* (Roberts 1989). Maximum size: 30 cm.

Native Range: Tropical Asia. Southeast Asia including central Thailand, Malay Peninsula, Sumatra, Borneo, and Java (Berra 1981; Roberts 1989; Talwar and Jhingran 1992).

Nonindigenous Occurrences: The fish was collected in Lake Worth Drainage District Canal L-15, west of Atlantis and Lantana, adjacent to a fish farm in Palm Beach County, **Florida**, in 1969 and 1970 (Courtenay et al. 1974; Courtenay and Hensley 1979a) and in 1971 (museum specimen). It was taken from a canal near a fish farm in Ruskin, Hillsborough County, between 1977 and 1978 (Courtenay and Hensley 1979a).

Means of Introduction: The species was introduced as a result of either an aquarium release or a fish farm escape.

Status: Reported from two sites in Florida.

Impact of Introduction: Unknown.

Remarks: Voucher specimens: Florida (UF 97050, UF 97051).

Belontiidae - Gouramies

The Belontiidae is the largest and most diverse anabantoid family, comprising 12 genera with about 46 species (Nelson 1994). Witte and Schmidt (1992) recently recognized 32 species in the single genus *Betta*, the so-called fighting fishes. Belontiids are distributed throughout the Indian subcontinent and Southeast Asia, including much of China (Berra 1981; Roberts 1989). They are closely related to another anabantoid family, the Osphronemidae (Roberts 1989). All have an aerial-breathing labyrinth organ. Belontiids are small fishes, less than about 24 cm in length, with a body shape that ranges from cylindrical to compressed (Berra 1981). Many species are very popular in the aquarium trade and some of the larger species are used in aquaculture. The family is restricted to freshwater.

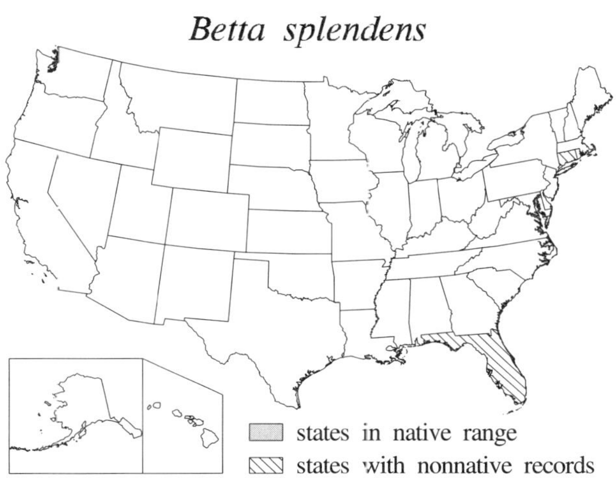

Betta splendens Regan 1909

Common Names: Siamese fightingfish (betta).

Identification: Witte and Schmidt (1992) discussed the genus and provided an identification key to species. They treated *B. splendens* as a valid species, but also recognized a *B. splendens* species complex. Many color variants have been developed by aquarists. See Axelrod et al. (1985) for color photographs of live fish. Maximum size: 6 cm.

Native Range: Tropical Asia. Native to Southeast Asia including the northern Malay Peninsula, central and eastern Thailand, Kampuchea, and southern Vietnam (Witte and Schmidt 1992).

Nonindigenous Occurrences: Specimens ranging from 1–3 cm were taken from the Thames River drainage in **Connecticut** (Whitworth 1996). Specimens were collected from Lake Worth Drainage District canal L-15, adjacent to an aquarium fish farm west of Atlantis in Palm Beach County, **Florida** (Ogilvie 1969; Courtenay et al. 1974), but have not been found at that locality in subsequent years (Courtenay and Hensley 1979a). A population became established in a canal south of Holmberg Road in Parkland, northern Broward County. When first discovered in December 1975, it was a dominant fish; however, the population was killed by extremely cold weather in January 1977 (Courtenay and Hensley 1979a; Courtenay et al. 1984; Courtenay and Stauffer 1990).

Means of Introduction: Florida records are the result of probable escapes from local ornamental fish farms (Courtenay and Stauffer 1990). Connecticut records are probably aquarium releases.

Status: Reported from one site and formerly established locally in another site in Florida (Courtenay et al. 1984). Reported from Connecticut.

Impact of Introduction: Unknown.

Remarks: A supposed report of *Betta splendens* in Hiko Spring, Lincoln County, Nevada, was false (Courtenay, personal communication).

Voucher specimens: Florida (UF 97046).

Colisa fasciata

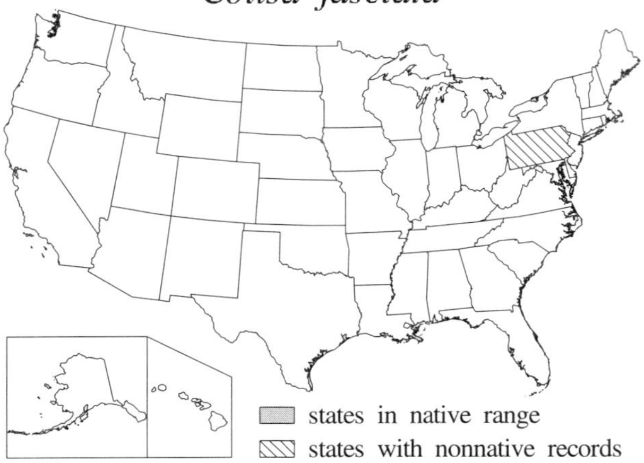

Colisa fasciata (Bloch and Schneider 1801)

Common Names: banded gourami (giant gourami, striped gourami, rainbow gourami).

Identification: A key and distinguishing characteristics were given in Talwar and Jhingran (1992); distinguishing characteristics were also provided by Sterba (1973). A color photograph of live fish appeared in Axelrod et al. (1985). Maximum size: 12 cm.

Native Range: Tropical Asia. Bengal, Burma, Thailand, and the Malay Peninsula (Sterba 1973); northern India, Nepal, and Pakistan, and Bangladesh (Jayaram 1981; Talwar and Jhingran 1992).

Nonindigenous Occurrences: One or possibly a few reports of specimens taken by fisherman from an unspecified locality in **Pennsylvania** in state prior to 1979 (Courtenay and Hensley 1979a; Courtenay et al. 1984, 1991; Courtenay and Stauffer 1990; C. N. Shiffer, personal communication).

Means of Introduction: Probable aquarium release.

Status: Reported from Pennsylvania.

Impact of Introduction: Unknown.

Remarks: No known voucher specimens.

Colisa labiosa

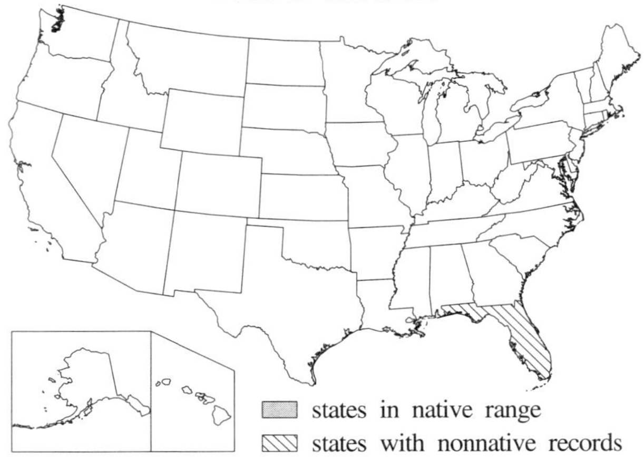

Colisa labiosa (Day 1877)

Common Names: thicklip gourami (thick-lipped gourami).

Identification: Commonly used synonym is *Colisa labiosus*. A key and distinguishing characteristics were given in Talwar and Jhingran (1992); distinguishing characteristics also were provided by Sterba (1973). This species is very similar in appearance to *C. fasciata*. A color photograph of live fish appeared in Axelrod et al. (1985). Maximum size: 8 cm.

Native Range: Tropical Asia. Southern Burma (Sterba 1973; Talwar and Jhingran 1992).

Nonindigenous Occurrences: Collected from a canal adjacent to a fish farm in Ruskin, Hillsborough County, **Florida**, in 1977 (Courtenay and Hensley 1979a; Courtenay et al. 1984; Courtenay and Stauffer 1990; Courtenay et al. 1991).

Means of Introduction: Probable release or escape from fish farm.

Status: Reported from Florida.

Impact of Introduction: Unknown.

Remarks: There are no known voucher specimens.

Colisa lalia

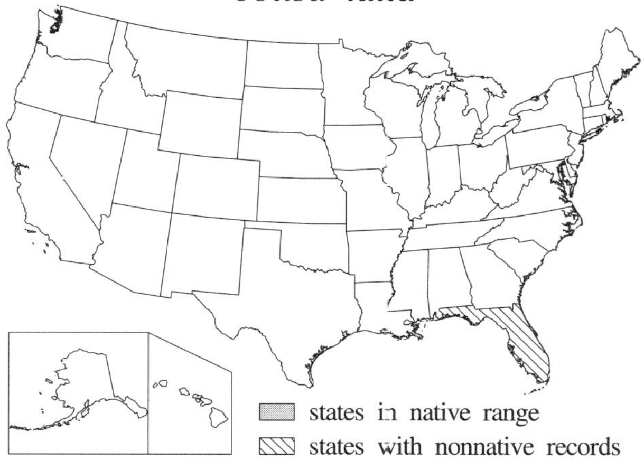

Colisa lalia (Hamilton 1822)

Common Name: dwarf gourami.

Identification: A key and distinguishing characteristics were given in Talwar and Jhingran (1992); distinguishing characteristics were also provided by Sterba (1973). Color photographs of live fish were given in Axelrod et al. (1985). Maximum size: 5 cm.

Native Range: Tropical Asia. India, Pakistan, Bangladesh, and possibly Borneo (Jayaram 1981; Talwar and Jhingran 1992).

Nonindigenous Occurrences: Collected in Lake Worth Drainage District canal L-15, west of Atlantis and Lantana, adjacent to a fish farm in Palm Beach County, **Florida**, in 1969 and 1970 (Ogilvie 1969; Courtenay et al. 1974; Courtenay and Hensley 1979a). Taken from several sites in Hillsborough County including a canal east of Ruskin in 1971 (Courtenay et al. 1974; Courtenay and Hensley 1979a); Bullfrog Creek at U.S. 301, east of Ruskin, on 24 Mar 1971 (museum specimens); a canal adjacent to a fish farm in Ruskin, in 1978 (Courtenay and Hensley 1979a); a drainage ditch west of U.S. 41 in Ruskin, on 26 Oct 1979 (museum specimens); and from a ditch adjacent to the Tampa Bypass Canal in November 1993 (museum specimens).

Means of Introduction: Probable release or escape from fish farms.

Status: Reported from two regions in Florida.

Impact of Introduction: Unknown.

Remarks: Voucher specimens: Florida (UF 97047, 97048, 97049, 98923).

Macropodus opercularis

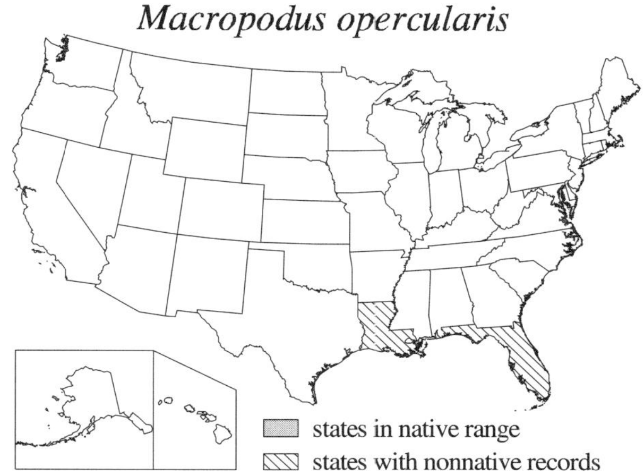

Macropodus opercularis (Linnaeus 1758)

Common Names: paradisefish (taiwan-kingyo, forktail fightingfish).

Identification: Distinguishing characteristics were given by Sterba (1973) and Masuda et al. (1984). Color photographs of live fish appeared in Masuda et al. (1984) and Axelrod et al. (1985). Maximum size: 9 cm.

Native Range: Tropical and temperate(?) Asia. Southern China, Taiwan, Indochina, northern Philippines, and Okinawajima of the Japanese Archipelago (Masuda et al. 1984); also reported from Korea and southern Vietnam (Sterba 1973).

Nonindigenous Occurrences: Unconfirmed report of breeding population in the Everglades, **Florida**, prior to about 1940 (Myers 1940). Based on extensive sampling during the 1970s, established populations, if ever present, no longer exist (Courtenay and Hensley 1979a; Courtenay and Stauffer 1990). Collected in Lake Worth Drainage District canal L-15, west of Atlantis and Lantana, adjacent to a fish farm in Palm Beach County, in 1969 and 1970 (Ogilvie 1969; Courtenay et al. 1974), but species not found in subsequent surveys (Courtenay and Hensley 1979a). Several specimens were taken from a marsh in Orleans Parish, **Louisiana**, in March 1997 (Cashner, personal communication).

Means of Introduction: The Florida records probably result of releases or escapes from fish farms.

Status: There is no recent evidence of a breeding population in the Florida Everglades; more recent reports of fish represent nonestablished populations (Courtenay and Hensley 1979a; Courtenay and Stauffer 1990). Reported from Louisiana.

Impact of Introduction: Unknown.

Remarks: Voucher specimens: Florida (UF 97052); Louisiana (UNO 9488).

Trichogaster leerii

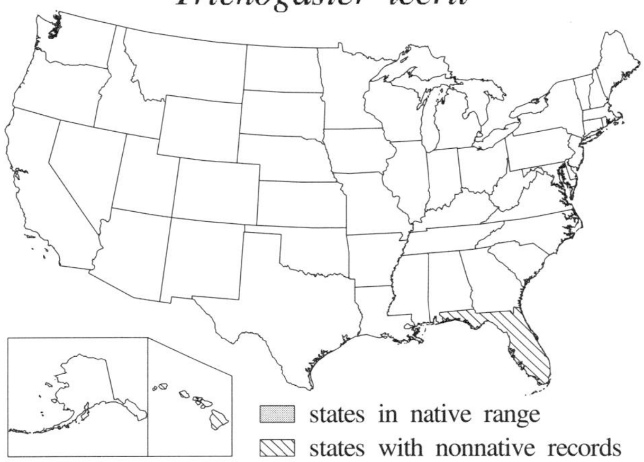

Trichogaster leerii (Bleeker 1852)

Common Names: pearl gourami (leeri, mosaic gourami, lace gourami, diamond gourami).

Identification: For distinguishing characteristics and a key, see Smith (1945); distinguishing characteristics were also provided by Sterba (1973) and Kottelat et al. (1993). Color photographs appeared in Axelrod et al. (1985) and Kottelat et al. (1993). Also appears as *Trichogaster leeri* (Roberts 1989). Maximum size: 12 cm.

Native Range: Tropical Asia. Southeast Asia including Thailand, Malay Peninsula, Sumatra, and Borneo (Roberts 1989).

Nonindigenous Occurrences: Collected in Lake Worth Drainage District canal L-15, west of Atlantis and Lantana, adjacent to a fish farm in Palm Beach County, **Florida**, in 1970 (Courtenay et al. 1974; Courtenay and Hensley 1979a). Released in unspecified area on Oahu, **Hawaii,** in 1940; the introduction failed (Maciolek 1984).

Means of Introduction: In Florida, probable release or escape from fish farm (Courtenay et al. 1974). Hawaii record based on failed introduction of unknown number of fish released on island of Oahu from Asia(?) in 1940 (Maciolek 1984).

Status: Reported from Florida and Hawaii.

Impact of Introduction: Unknown.

Remarks: There are no known voucher specimens.

Trichogaster trichopterus

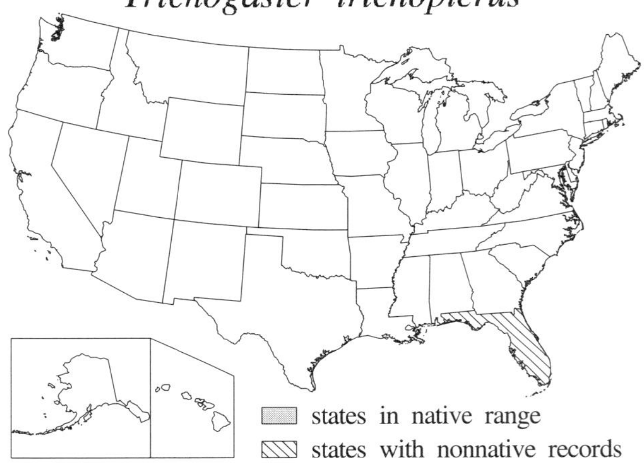

Trichogaster trichopterus (Pallas 1770)

Common Names: threespot gourami (blue gourami, opaline gourami, golden gourami, gold gourami, Cosby gourami).

Identification: Distinguishing characteristics, a key, and illustration were given in Smith (1945); distinguishing characteristics also were given by Sterba (1973) and Kottelat et al. (1993). Color photographs appear in Axelrod et al. (1985) and in Kottelat et al. (1993). Maximum size: 15 cm.

Native Range: Tropical Asia. Natural distribution is Southeast Asia, but exact range is uncertain because of past introductions. Range possibly includes Thailand, Laos, Kampuchea, Vietnam, Malay Peninsula, Sumatra, Borneo, Java, and Madura (Roberts 1989).

Nonindigenous Occurrences: Collected in Lake Worth Drainage District canal L-15, west of Atlantis and Lantana, adjacent to a fish farm in Palm Beach County, **Florida**, in 1970 (museum specimen) and 1971 (Courtenay et al. 1974; Courtenay and Hensley 1979a). Taken from an irrigation ditch along South Canal Drive, east of the Florida Turnpike in Dade County, in 1978 (Courtenay and Hensley 1979a).

Means of Introduction: Probable release or escape from fish farms (Courtenay et al. 1974).

Status: Reported from two counties in Florida.

Impact of Introduction: Unknown.

Remarks: Voucher specimens: Florida (UF 97053, UF 97054, UF 97055).

Trichopsis vittata

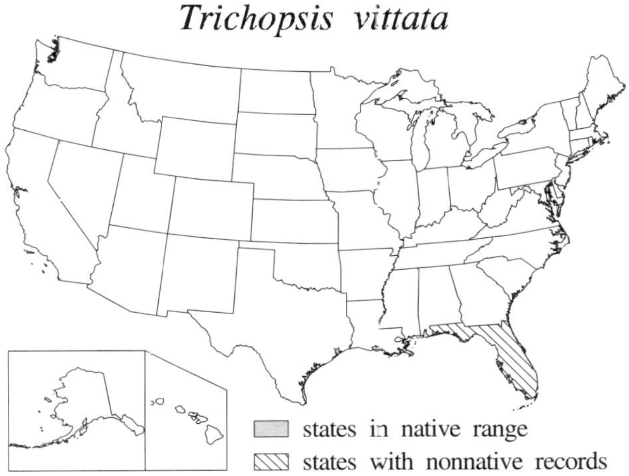

Trichopsis vittata (Cuvier 1831)

Common Names: croaking gourami (talking gourami).

Identification: For distinguishing characteristics and figure, see Smith (1945) and Page and Burr (1991); distinguishing characteristics were also given by Kottelat et al. (1993). Also spelled as *Trichopsis vittatus*. Color photographs were given in Axelrod et al. (1985) and Kottelat et al. (1993). Maximum size: 6 cm.

Native Range: Tropical Asia. Widely distributed in Southeast Asia including Thailand, southern Vietnam, the Malay Peninsula, Sumatra, Java, and Borneo (Sterba 1973; Lee et al. 1980 et seq.; Axelrod et al. 1985).

Nonindigenous Occurrences: A population has been established in a small area (canal system) on the south side of Lake Worth Drainage District canal L-36 in Delray Beach, Palm Beach County, **Florida**, since the late 1970s (Courtenay and Hensley 1979a; Lee et al. 1980 et seq.; Courtenay et al. 1986; Courtenay and Stauffer 1990; Page and Burr 1991; museum specimens).

Means of Introduction: Probable escape from a nearby aquarium fish farm (Courtenay et al. 1986; Courtenay and Stauffer 1990), or possible aquarium release (Courtenay and Hensley 1979a; Lee et al. 1980 et seq.).

Status: Locally established in Palm Beach County, Florida, since at least the late 1970s (Courtenay and Stauffer 1990); the population was first found in 1978 (Lee et al. 1980 et seq.).

Impact of Introduction: Unknown.

Remarks: Florida population has survived several cold winters (Courtenay and Stauffer 1990). Distribution map for Florida was given in Lee et al. (1980 et seq.).

Voucher specimens: Florida (UF 90547, UF 97056).

Osphronemidae - Giant gouramies

The anabantoid family Osphronemidae comprises three species in one genus, all from Southeast Asia (Roberts 1992a; Nelson 1994). The family is closely related to the Belontiidae and there is some question as to whether the two should be treated as separate groups (Roberts 1989). Giant gouramies are medium-sized fishes characterized by compressed bodies, long dorsal and anal fins, and filamentous pelvic rays that are used as feelers (Berra 1981; Roberts 1992a). *Osphronemus goramy*, until recently the only known member of the family, is a hardy, air-breathing species introduced into many tropical areas around the world as a food fish (Roberts 1992a). Giant gouramies are also found in the aquarium trade. They are restricted to freshwater.

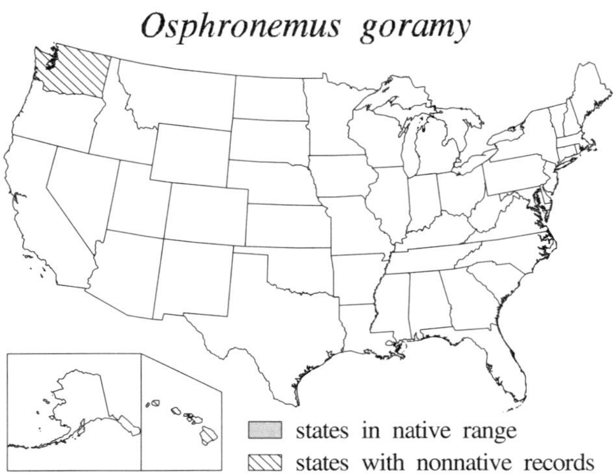

Osphronemus goramy Lacepède 1801

Common Names: gourami (giant gourami, giant gouramy).

Identification: See Roberts (1992a) for recent revision of genus, redescription of *Osphronemus goramy*, and color photographs. Additional color photographs appeared in Axelrod et al. (1985) and Kottelat et al. (1993). Similar species include the recently described *O. septemfasciatus* and *O. laticlavius* (Roberts 1992a). Maximum size: 80 cm SL.

Native Range: Tropical Asia. Natural range probably limited to Thailand, the Malay Peninsula, Sumatra, Java, Borneo, and Indochina (Mekong Basin) (Roberts 1992a; also see Berra 1981).

Nonindigenous Occurrences: Nine small specimens were introduced to **Hawaii**, apparently into pond or lake habitat, in 1950 (Brock 1960); the site was recorded as Oahu, Honolulu County by Maciolek (1984). This species was reportedly taken from an unspecified locality in **Washington**, apparently some time prior to July 1992 (Fletcher, personal communication).

Means of Introduction: According to Brock (1960), the giant gorami introduced to Hawaii came from the Philippines. The reason for the introduction was not given, although it was presumably stocked as another food fish. The Washington fish was a possible aquarium release.

Status: Reported from Hawaii and Washington. Although a few of the fish introduced to Hawaii survived for an extended period, they did not reproduce (Brock 1960).

Impact of Introduction: Unknown.

Remarks: This is a famous food fish in Southeast Asia and important in freshwater aquaculture in tropical Asia (Roberts 1989, 1992a). It has been introduced widely into the Old and New World Tropics, including islands that otherwise have no primary freshwater fish; the introduced stock, in most cases, is probably from Java (Roberts 1992a). Brock (1960) stated that there may have been introductions into Hawaii prior to 1950 which, similar to the one reported, were unsuccessful. During the late 1800s there were plans to introduce this species into California waters, apparently because of it potential as a food fish, but live specimens were never obtained (Dill and Cordone 1997).

There are no known voucher specimens.

Channidae - Snakeheads

The anabantoid family Channidae, whose members are known as snakeheads, contains two genera and about 21 species (Nelson 1994). Three species of *Parachanna* are native to tropical Africa and about 18 species of *Channa* are found in Asia (Roberts 1989; Nelson 1994). Snakeheads are distinguished from other anabantoids by their elongate cylindrical bodies, long dorsal and anal fins (all soft-rayed), large mouths with well-toothed jaws and palates, and simple labyrinth organs (Roberts 1989). Snakeheads are highly predacious and superficially resemble our native bowfin *Amia calva*. They achieve a maximum length of about 1.2 m (Nelson 1994). Their ability to breathe air and survive for long periods out of water makes them popular as a food fish in Asia, but also allows accidental or deliberate introductions (Berra 1981). They are found in freshwater.

Channa micropeltes

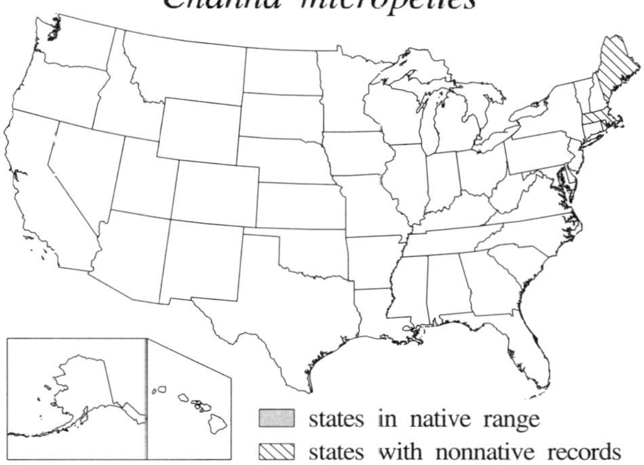

Channa micropeltes (Cuvier 1831)

Other Names: giant snakehead (red snakehead, Malabar snakehead).

Identification: A key and distinguishing characteristics were given in Smith (1945) and Talwar and Jhingran (1992); a few distinguishing characteristics also were provided by Roberts (1989) and Kottelat et al. (1993). Color photographs appeared in Axelrod et al. (1985), Kottelat et al. (1993), and Sakurai et al. (1993). For color photographs of other *Channa* species, see Masuda et al. (1984). Another name is *Ophicephalus micropeltes*. Maximum size: 100 cm and more than 20 kg.

Native Range: Tropical Asia. Southeast Asia including India, Burma?, Thailand, Malay Peninsula, Sumatra, Borneo, Java, Banka, and Billiton (Roberts 1989).

Nonindigenous Occurrences: One fish (272 mm SL) was taken below Springvale Dam on the Mousam River, near Sanford, York County, **Maine**, during the summer of 1976 (USGS/BRD-G database). Two specimens were collected from the Saco River prior to 1979 (Courtenay and Hensley 1979a), but these records require verification. One fish (318 mm SL) was taken by an angler (who claimed to have caught a second specimen), at Pomps Pond (Merrimack drainage) in Andover, Essex County, **Massachusetts**, in August 1990 (Hartel 1992; Cardoza et al. 1993; Hartel et al. 1996). The two reports from **Rhode Island** likely represent duplication of a single record: one fish (209 mm SL) taken from an unspecified pond (Courtenay, personal communication), and one fish taken by an angler at Johnston Pond in Coventry, Kent County, in 1968 (J. A. Stolgitis, personal communication).

Means of Introduction: All probably aquarium releases.

Status: Reported from Maine, Massachusetts, and Rhode Island.

Impact of Introduction: Unknown.

Remarks: Juveniles are sold in the aquarium trade. In its native habitat, this aggressive predator is destructive to other fishes, killing all kinds and sizes in excess of actual needs (Roberts 1989). There is some confusion surrounding the Massachusetts record: the specimen was originally identified as *Ophicephalus marulius* (= *Channa striata*) (Halliwell, personal communication); the same fish was later reidentified as *Channa cf. micropeltes*, but with an error in the collection year [1991] (Hartel 1992), and finally as *Channa* sp. in Cardoza et al. (1993). Identification as *C. micropeltes* has since been verified (Hartel, personal communication).

Voucher specimens: Massachusetts (MCZ 96907); Maine (voucher specimen reported to be in University of Maine collection - UMO); Rhode Island (UMO 343).

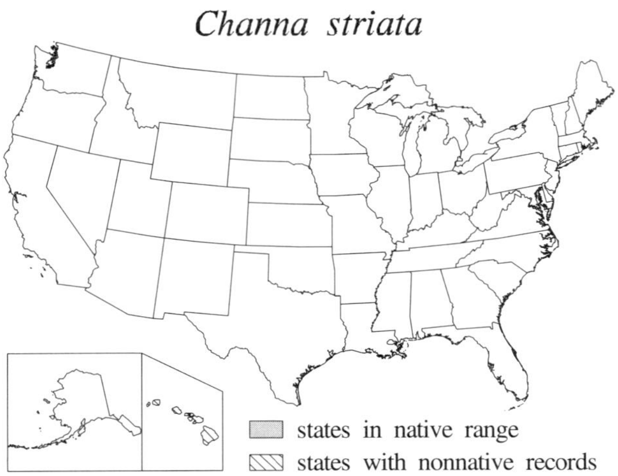

Channa striata (Bloch 1793)

Other Names: chevron snakehead (banded snakehead, striped snakehead, pongee, murel, China-fish).

Identification: A key, distinguishing characteristics, and illustration were given in Talwar and Jhingran (1992); distinguishing characteristics also were given by Smith (1945) and Kottelat et al. (1993). Photographs of live fish appear in Axelrod et al. (1985), Kottelat et al. (1993), and Sakurai et al. (1993). Jordan and Evermann (1905) provided a detailed description of a specimen taken from Hawaii. Another name commonly used in the literature is *Ophicephalus striatus* (e.g., Brock 1960; Devick 1991a, 1991b). Maximum size: 90 cm.

Native Range: Tropical Asia. Range (probably expanded by introductions) includes the East Indies, Malay Peninsula, Thailand, India, Sri Lanka, and the Philippines (Roberts 1989).

Nonindigenous Occurrence: This species has been known from Oahu, **Hawaii** since the 1800s (Jordan and Evermann 1905; Brock 1960; Maciolek 1984; Devick 1991a, 1991b).

Means of Introduction: Cobb (1902) stated that the China-fish or "*Ophicephalus*" was brought to Hawaii from China by the Chinese. However, Jordan and Evermann (1905) suggested that the species had been introduced into Hawaii from Borneo. Devick (1991b) indicated it was a deliberate introduction for use as a food fish (he listed the period of introduction as both 1800 and the 19th century). Jordan and Evermann (1902, 1905) reported that "China-fish" (along with goldfish) were generally sold alive to the Chinese [in Hawaii]. Borneo was mentioned as source by Jordan and Evermann (1902).

Status: The species has been established on Oahu, Hawaii since before 1900 (Brock 1960; Maciolek 1984; Devick 1991a). Cobb (1902) and Jordan and Evermann (1902, 1905) reported that the species was fairly common in irrigation ditches and freshwater ponds in the Honolulu area. Maciolek (1984) noted that it was found in unspecified standing waters (e.g., reservoirs) on Oahu and reported it as actively spreading. Devick (1991a, 1991b) listed the dominant habitat of this species in Oahu to be reservoirs or ponds.

Impact of Introduction: Unknown.

Remarks: Cobb (1902) and Jordan and Evermann (1902) used the name "*Ophiocephalus*" for the snakehead introduced to Hawaii. Jordan and Evermann (1905) recognized it as *Ophicephalus striatus* and reported taking ten specimens from a site in the Honolulu area in 1901. Snakeheads are voracious predators and this species is reputed to attack and eat fish, invertebrates, snakes, frogs, and small mammals (Axelrod et al. 1985). In his book entitled "A history of fishes," Norman (1958, and earlier editions) stated that snakeheads or "*Ophiocephalus*" "are carried alive by the Chinese to San Francisco and to Hawaii, where they are now naturalised and known as China-fishes." This led some to believe that snakeheads had been introduced to California. Dill and Cordone (1997) investigated the report with respect to San Francisco and concluded that, although snakeheads apparently appeared in Chinese markets in that city, it was not known from open waters.

Voucher specimens: Hawaii (USNM 126588). Jordan and Evermann (1905) referenced specimen(s) No. 03585 [probably U.S. Fish Commission or Bureau of Fisheries label].

ORDER PLEURONECTIFORMES

Paralichthyidae - Paralichthyid flounders

The flatfish family Paralichthyidae comprises 16 genera with at least 85 species (Nelson 1994). This family was formerly considered a subfamily of Bothidae. These fishes are distributed widely in tropical and temperate waters of the Atlantic, Indian, and Pacific oceans (Eschmeyer et al. 1983; Nelson 1994). These bottom-dwelling carnivores are asymmetrical, compressed fishes that lack spines in their fins. The pelvic fin bases are short and nearly symmetrical. Both eyes are positioned on the left side of the head in adults (i.e., they are sinistral). Many are important as food or sport fishes (Robins and Ray 1986; Murdy et al. 1997). The family is marine.

Etropus crossotus

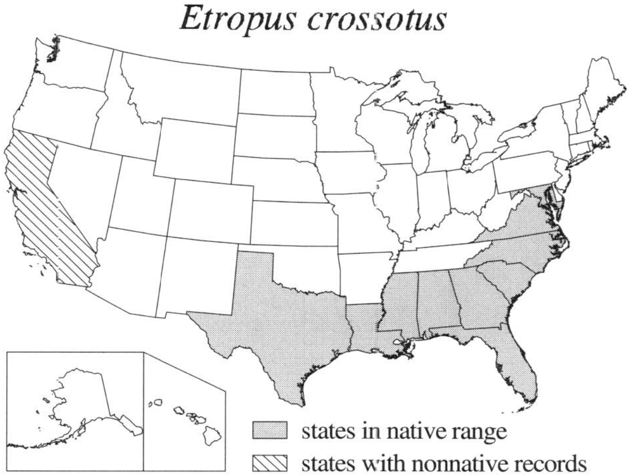

Etropus crossotus Jordan and Gilbert 1882

Common Name: fringed flounder.

Identification: Fischer et al. (1995b). Maximum size: 20 cm.

Native Range: Marine; Gulf of California south to Panama on the Pacific coast (Ramirez-Hernández and Gonzales Pages 1976; Fischer et al. 1995b), and Chesapeake Bay, northern Gulf of Mexico, and the Antilles to Brazil on the Atlantic coast (Ramirez-Hernández and Gonzales Pages 1976; Robins and Ray 1986).

Nonindigenous Occurrences: Introduced into Salton Sea, **California**, in 1951 (Walker et al. 1961; Dill and Cordone 1997).

Means of Introduction: Intentionally stocked with fish from San Felipe, Mexico, on the Gulf of California (Walker et al. 1961).

Status: The introduction failed to produce a viable population and the species is extirpated in the Salton Sea.

Impact of Introduction: Unknown.

Remarks: None.

Paralichthys aestuarius

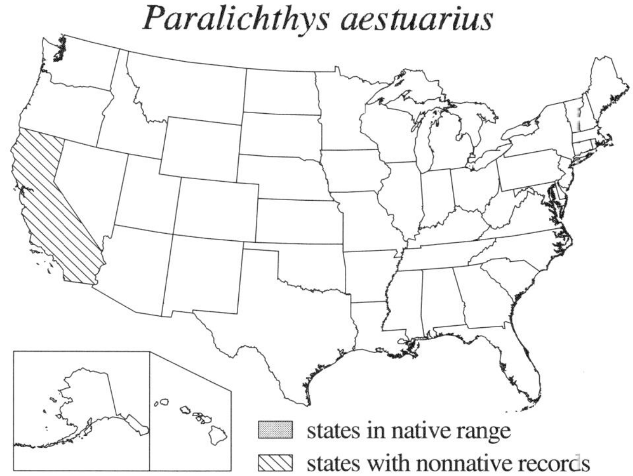

Paralichthys aestuarius (Gilbert and Scofield 1898)

Common Name: (halibut, Cortez flounder, Cortez halibut).

Identification: Fischer et al. (1995b). Maximum size: 45 cm.

Native Range: Marine; Gulf of California and southern Baja on the Pacific side (Fischer et al. 1995b).

Nonindigenous Occurrences: Introduced into Salton Sea, **California**, in 1950 and 1951 (Walker et al. 1961; Dill and Cordone 1997).

Means of Introduction: Intentionally stocked. A single fish was stocked in May 1950. Four more were stocked in March 1951. Stock was obtained from San Felipe, Mexico (Walker et al. 1961; Dill and Cordone 1997).

Status: The introduction failed to produce a viable population and the species is extirpated in the Salton Sea.

Impact of Introduction: Unknown.

Remarks: None.

Paralichthys albigutta

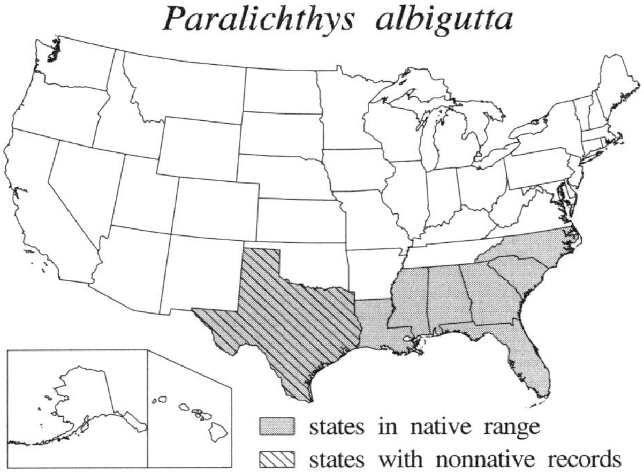

Paralichthys albigutta Jordan and Gilbert 1882

Common Name: gulf flounder.

Identification: Hoese and Moore (1977); Manooch (1984); Robins and Ray (1986). Maximum size: 38 cm.

Native Range: Marine. Western Atlantic from North Carolina to the Gulf of Mexico and Panama, including the Bahamas (Robins and Ray 1986; Boschung 1992).

Nonindigenous Occurrences: A total of 3,500 Gulf flounder were stocked by the Texas Parks and Wildlife Department (TPWD) in Alcoa Lake, Milam County, **Texas**, in 1977 (Luebke 1978).

Means of Introduction: Intentional, authorized stocking for sportfishing (Luebke 1978).

Status: The stocking failed (Luebke 1978).

Impact of Introduction: Unknown.

Remarks: None.

Paralichthys lethostigma

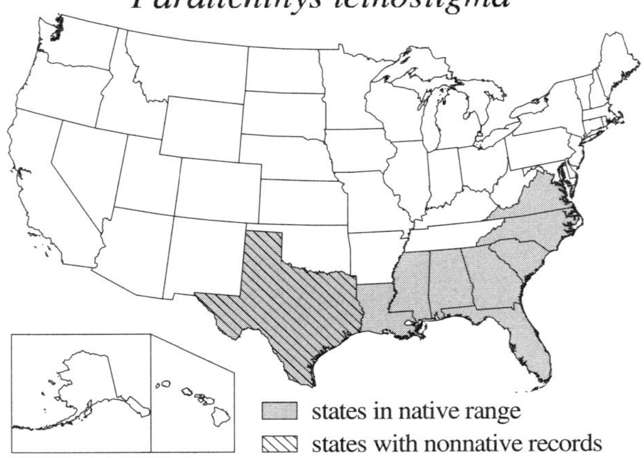

Paralichthys lethostigma Jordan and Gilbert 1884

Common Name: southern flounder.

Identification: Hoese and Moore (1977); Manooch (1984); Robins and Ray (1986). Maximum size: 76 cm.

Native Range: Marine. Atlantic coastal and estuarine waters of the United States from Albemarle Sound, North Carolina, to Jupiter Inlet, Florida, and from Caloosahatchee estuary, Florida, to northern Mexico. South Florida is not included in the range (Lee et al. 1980 et seq.; Boschung 1992). Murdy et al. (1997) reported it as an occasional visitor to lower Chesapeake Bay, Virginia.

Nonindigenous Occurrences: This species was stocked by the TPWD in the middle and late 1970s in five **Texas** reservoirs: Alcoa, W.E. Long, Town Lake, Aboerne City Park Lake, and Casa Blanca (Lasswell et al. 1979; Luebke 1978; Howells and Garrett 1992).

Means of Introduction: Intentional, authorized stocking for sportfishing.

Status: Extirpated from all reservoirs, with the possible exception of Casa Blanca (Luebke 1978).

Impact of Introduction: Unknown.

Remarks: No reproduction has been documented in Texas freshwaters. Limited benefits to anglers resulted in termination of the experimental stocking (Howells and Garrett 1992). Stomach contents of southern flounder taken from Long Lake, Texas, were mostly small sunfishes (Lasswell et al. 1979).

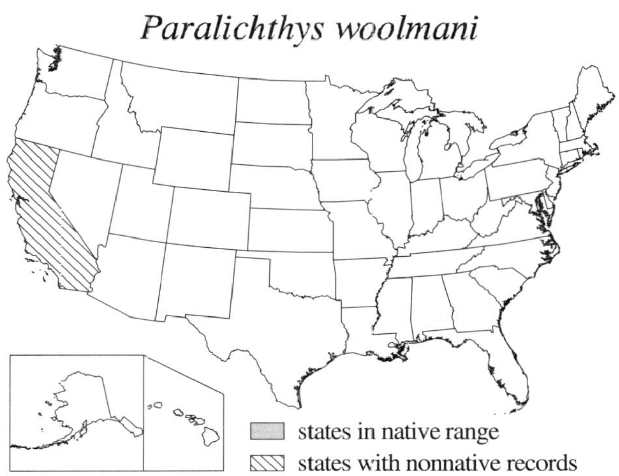

Paralichthys woolmani Jordan and Williams 1897

Common Name: (halibut, speckled flounder).

Identification: Fischer et al. (1995b). Maximum size: 50 cm.

Native Range: Marine; Gulf of California and southern Baja south to Panama and the Galapagos Islands (Ramirez-Hernández and Gonzales Pages 1976; Fischer et al. 1995b).

Nonindigenous Occurrences: Introduced into Salton Sea, **California**, in 1951 (Walker et al. 1961; Dill and Cordone 1997).

Means of Introduction: Intentionally stocked. A single fish obtained from San Felipe, Mexico, was stocked in December 1951 (Walker et al. 1961; Dill and Cordone 1997).

Status: The introduction failed to produce a viable population (not surprisingly) and the species is extirpated in the Salton Sea.

Impact of Introduction: Unknown.

Remarks: None.

Pleuronectidae - Righteye flounders

The flatfish family Pleuronectidae includes 39 genera and about 93 species (Nelson 1994). These fishes are widely distributed in temperate and cold waters of the Arctic, Atlantic, Indian, and Pacific oceans; a few occur in tropical marine waters (Eschmeyer et al. 1983; Robins and Ray 1986; Nelson 1994). Most righteye flounders are under 61 cm long, although the largest species range from 122 to 267 cm (Eschmeyer et al. 1983). Similar to other flatfishes, righteye flounders are asymmetrical, compressed fishes that lack spines in their fins. Both eyes are positioned on the right side of the head (i.e., dextral) in adults. Many of these fishes are commercially important as food sources (Robins and Ray 1986). The family is found primarily in marine waters and occasionally in brackish waters, but rarely in freshwater (Eschmeyer et al. 1983; Nelson 1994).

Hypsopsetta guttulata

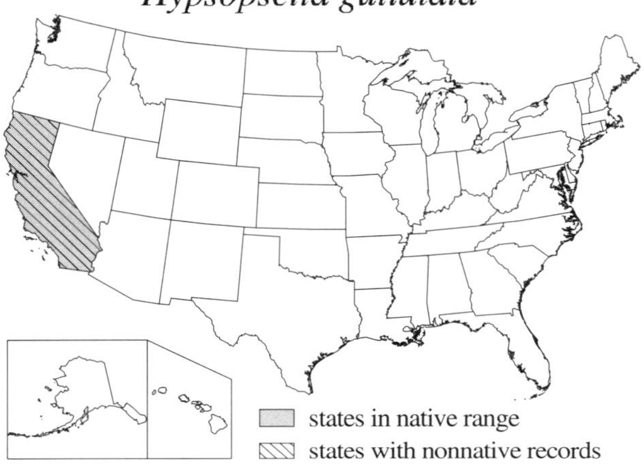

Hypsopsetta guttulata (Girard 1856)

Common Name: diamond turbot.

Identification: Miller and Lea (1972); Eschmeyer et al. (1983); Fischer et al. (1995b). Maximum size: 46 cm.

Native Range: Marine; Cape Mendocino in northern California to southern Baja; isolated population in Gulf of California (Miller and Lea 1972; Ramirez-Hernández and Gonzales Pages 1976; Eschmeyer et al. 1983).

Nonindigenous Occurrences: Introduced into Salton Sea, **California**, in 1951 (Walker et al. 1961; Dill and Cordone 1997).

Means of Introduction: Intentionally stocked.

Status: The introduction failed to produce a viable population and the species is extirpated in the Salton Sea.

Impact of Introduction: Unknown.

Remarks: None.

Platichthys flesus

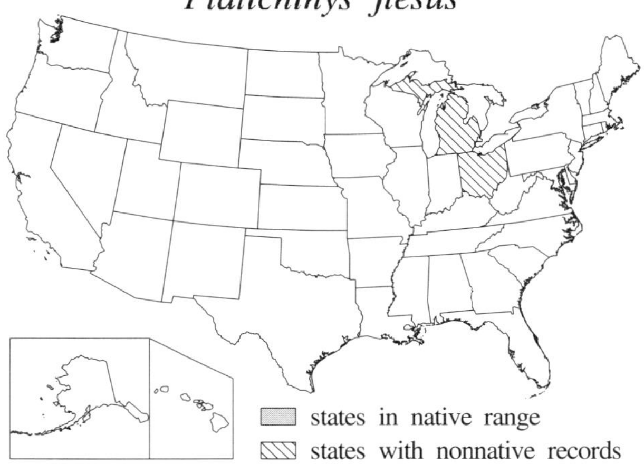

Platichthys flesus (Linnaeus 1758)

Common Name: European flounder.

Identification: Wheeler (1969); Maitland (1977); Wheeler (1978). Maximum size: 50 cm.

Native Range: Marine; can penetrate freshwater. This species is native to the northeast Atlantic Ocean and the Black Sea (Robins et al. 1991b).

Nonindigenous Occurrences: This species was collected at three different localities in the Great Lakes, in areas bordering **Michigan** and **Ohio**. Two fish were taken from Lake Erie; the first was caught off Long Point in 1974, the second taken near Port Dover in 1976 (Emery and Teleki 1978; Crossman 1984). A third specimen was captured in 1981 in Lake Superior (Crossman 1984).

Means of Introduction: Possibly introduced in ballast water from trans-Atlantic vessels (Emery and Teleki 1978; Crossman 1984).

Status: Reported from the Great Lakes near Ohio and Michigan.

Impact of Introduction: Unknown. The stomach contents of one of the specimens taken from Lake Erie included shells of a freshwater clam *Pisidium* sp. (Emery and Teleki 1978).

Remarks: European flounder do not reproduce in freshwater; therefore, they are not expected to become established in the Great Lakes. The three known specimens caught in North America all were taken in Canadian waters (Emery and Teleki 1978: Crossman 1984). Courtenay et al. (1991) included this species in their list of fishes introduced into the continental United States and Canada, but not known to be established. Their list also indicated that European flounder had been collected in Michigan and Ohio as well as to Ontario. It is assumed that Courtenay et al. (1991) included Ohio and Michigan because fish introduced into the Great Lakes do not respect political boundaries.

Voucher specimens: Lake Erie (ROM 30521).

Platichthys stellatus

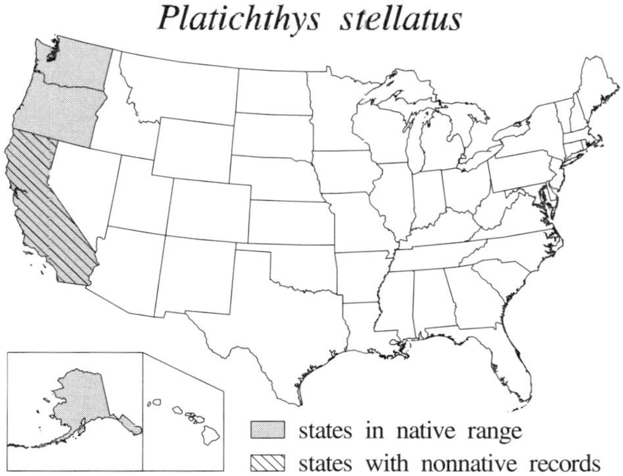

Platichthys stellatus (Pallas 1787)

Common Name: starry flounder.

Identification: Moyle (1976a); Eschmeyer et al. (1983).

Native Range: Marine; can penetrate freshwater. Japan to Bering Sea and Arctic Alaska, and to Santa Barbara in southern California (Eschmeyer et al. 1983).

Nonindigenous Occurrences: Introduced into San Luis Reservoir and O'Neill Forebay, Merced County, **California** (Moyle 1976a).

Means of Introduction: Gained access to these areas via the California Aqueduct (Moyle 1976a).

Status: Collected in California.

Impact of Introduction: Unknown.

Remarks: None.

Erroneous Records of Introductions

The following species are ones with published record of introductions that we consider to be erroneous. These species were not otherwise included in the text. Other records of species' introductions considered erroneous can be found in remarks section of accounts.

Cyprinidae

Luxilus pilsbryi (Fowler 1904)

Common Name: duskystripe shiner.

This shiner was recorded from several sites in the Red River drainage in southeastern **Oklahoma** (Hubbs and Moore 1940; Miller and Robison 1973; reported as *Notropis zonatus pilsbryi* and *Notropis pilsbryi*, respectively). It was suggested that the species may have been introduced to that drainage. However, the recent literature indicates that the classification of this species as introduced to the Red River drainage is unwarranted. That conclusion is the result of a somewhat complex interplay of taxonomic revisions and biogeographic hypotheses. Hubbs and Moore (1940) were the first to suggest the possibility of introduction. These researchers identified specimens from the Red River drainage as a subspecies of the bleeding shiner, *Notropis zonatus pilsbryi*. In addition, Hubbs and Moore noted that the Red River system records were disjunct from other populations of *Notropis zonatus* and so isolated as to suggest the "possibility of an accidental transfer, either of data or of the living fish." Later, the subspecies *N. zonatus pilsbryi* was elevated to species status (i.e., *Notropis pilsbryi*). Still echoing the thoughts of Hubbs and Moore (1940), Miller and Robison (1973) hypothesized that the Red River records of this species (then identified as *Notropis pilsbryi*) were probably the result of bait introductions. Mayden (1988), further revising the taxonomy of the group, concluded that the Red River population to be part of the newly described cardinal shiner *Notropis cardinalis*. In contrast to earlier workers, Mayden (1988) argued that the Red River populations (now thought to be extirpated) were native and part of a relict distribution of *Notropis cardinalis*. The native range for this species (as *Luxilus cardinalis*) given by Page and Burr (1991) is in agreement with Mayden's conclusions. Mayden (1989) placed all three of the species mentioned above (i.e., *pilsbryi*, *zonatus*, and *cardinalis*) in the genus *Luxilus*. Rutherford et al. (1987) noted that a single specimen of *Notropis pilsbryi* (=*Luxilus pilsbryi*) was taken from the Little River system (Red River drainage) in Oklahoma prior to about 1955 and they presumed it to be a stray, or perhaps a released baitfish. We assume the Little River specimen to represent native *Luxilus cardinalis*. Mayden and Matthews (1989) offered hypotheses to explain the historical distribution of *Luxilus cardinalis* in the Red River drainage of Oklahoma.

Notropis fumeus Evermann 1892

Common Name: ribbon shiner.

On the basis of the one known collection at the time, Snelson (1973) suggested that the record of the ribbon shiner from Rough River, **Kentucky**, probably was the result of a bait bucket release. However, Burr et al. (1980) found the species to occur in surrounding areas and concluded that *N. fumeus* is native to the area east of the Coastal Plain in Kentucky.

Phoxinus erythrogaster (Rafinesque 1820)

Common Name: southern redbelly dace.

Caldwell and Collins (1977) reported taking *Phoxinus erythrogaster* from Turkey Creek in the Medicine Lodge River drainage, Pratt County, **Kansas**, in 1976. These authors stated that there was a "reasonable possibility that these specimens are the result of accidental introduction by man." However, Cross and Collins (1995) noted that the southern redbelly dace was much more widely distributed in Kansas prior to European settlement and they apparently considered the record from the Medicine Lodge drainage (and a few other geographically disjunct records) to be part of the species native range.

Catostomidae

Moxostoma carinatum (Cope 1870)

Common Name: river redhorse.

Menhinick (1991) reported the river redhorse was probably introduced into the Yadkin drainage of **North Carolina** in the Pee Dee River near the South Carolina border (Menhinick 1991). Robert Jenkins (personal communication) has determined this species is actually *Moxostoma robustum*, the robust redhorse. Jenkins and Burkhead (1994: 491, 509) discussed the rediscovery of this species. The robust redhorse has been found in the Pee Dee (including the Yadkin), Savannah, and Altamaha (Oconee) drainages of the Carolinas and Georgia (Jenkins and Burkhead 1994).

Cross et al. (1986) reported the river redhorse's status in the Kansas River, Kansas, was uncertain. However, Cross (1967) reported a specimen taken from there in the 1880s. Cross and Collins (1995) reported there have been few collected from the drainage in the past 40 years and noted that the species has declined throughout its range in the Mississippi River Valley. Perhaps this species was more widespread in the past but it is now uncommon and with a fragmented distribution.

Moxostoma duquesnei (Lesueur 1817)

Common Name: black redhorse.

Hocutt et al. (1986) reported the black redhorse as introduced into the Santee drainage. A single juvenile was collected in the Green River, Henderson County, **North Carolina**, in 1980 (R. E. Jenkins, personal communication). At the time of the 1986 publication, that was the only specimen from the drainage. In 1997, Jenkins discovered another population in the Rocky Broad River. Both rivers are in the Broad system of the Santee drainage. As a result, Jenkins now regards this species as native to the Santee. Nevertheless, its presence in the Santee may be a result of bait bucket introduction (Gilbert, personal communication).

Ictaluridae

Noturus funebris Gilbert and Swain 1891

Common Name: black madtom.

The black madtom was reported as introduced into the Apalachicola drainage by Yerger (1977) and a collection was plotted in Uchee Creek in the Chattahoochee drainage, **Alabama**, by Lee et al. (1980 et seq) and Mettee et al. (1996). These reports are based on two misidentified *Noturus leptacanthus* (R. Mayden, personal communication).

Salmonidae

Hucho hucho (Linnaeus 1758)

Common Name: (huchen, Danube salmon).

Although this species, a native of the Danube River, has been introduced into Canada near the U.S. border, there is no confirmed record from U.S. waters. Fry and fingerlings of this species were released into Canadian waters of the St. Lawrence River near Montreal, the Richeleiu River, and the Maskinongé River, Quebec, from 1968 to 1973 (Crossman 1984; Dumont et al. 1988). A small number of hybrids with rainbow trout *Salmo gairdneri* were released in the St. Lawrence River near Montreal, Canada, in 1978 and 1979 (Dumont 1988). According to Dumont et al. (1988), few of these fish were ever recaptured and there are no recent reported capture of huchen or of its hybrid. Thus, it is doubtful they survived.

Hessel (1874) investigated the potential uses of this species and, under the direction of the U.S. Fish Commission, was involved in transporting salmon eggs, identified as '*Salmo hucho*' from Europe to the United States. A few thousand eggs were successfully hatched and the fry later introduced into the Musconetcong, a tributary of the Delaware River in **New Jersey** (Baird 1974b). The date of introduction is not clearly stated, but it apparently took place during the early 1870s, probably 1872 or shortly thereafter. The fish did not survive. Based on those accounts, Radonski et al. (1984) reported that huchen fry were released but failed to survive.

For the present, we have chosen not to include this species because of the somewhat questionable identity of the fish released in the 1800s, and because the release in the 1900s was in Canadian waters with no verified U.S. records. However, our decision may be reversed if it can be shown that the New Jersey fish was *Hucho hucho*.

Oncorhynchus chrysogaster (Needham and Gard 1964)

Common Name: Mexican golden trout.

Introductions of Mexican golden trout reported in **New Mexico** (Sublette et al. 1990) and California (Dill and Cordone 1997) are actually of an undescribed species native to the Rio Truchas (the watershed south of the range of *O. chrysogaster*). The Rio Truchas trout is intermediate in morphology between *O. chrysogaster* and the Rio Hondo rainbow trout of the Rio del Presidio basin, and may be a hybrid. In 1956, more than 700 Rio Truchas trout were brought to a federal and a state hatchery in New Mexico. In 1957, some of these trouth were shipped to a state hatchery in California. All of the trout died and none were ever actually stocked in open waters (Behnke, personal communication). Dill and Cordone (1997) listed this species as *Oncorhynchus mykiss* ssp., but stated that they did not know the fate of the California fish. The researchers presumed the species was never stocked. It was not listed as introduced into California by Shapovalov et al. (1959).

Belonidae

Strongylura kreffti (Günther 1866)

Common Name: long tom.

Courtenay et al. (1991) and Robins et al. (1991b) incorrectly reported the long tom *Strongylura kreffti* as having been introduced into **Hawaii** based on an early tentative identification of *Xenentodon cancila* (Devick, personal communication).

Fundulidae

Fundulus waccamensis Hubbs and Raney 1946

Common Name: Waccamaw killifish.

Bailey (1977) reported the Waccamaw killifish from both Lake Waccamaw and Phelps Lake in **North Carolina**, but did not discuss the possibility of the species being introduced in Phelps Lake. Subsequently, Lee et al. (1980 et seq.) referred to the report by Bailey (1977) and added that "Phelps Lake population may or may not be introduced." Wiley and Mayden (1985:618) believed that the Phelps Lake population was the result of a recent or artificial introduction. Page and Burr (1991) followed suit by saying it was probably introduced to Phelps Lake through use as fishing bait.

The limited distribution of *F. waccamensis* and the wide separation of the two lakes (about 165 air miles) would greatly reduced the likelihood of it being introduced in Phelps Lake. Lee et al. (1980 et seq.) suggested that a comparison of the two known populations (i.e., Lake Waccamaw and Phelps Lake) should show some differentiation if the Phelps Lake population is natural. Shute et al. (1981) did report some morphological differences between the two populations and noted that this differentiation would lessen the possibility that Phelps Lake population of *Fundulus waccamensis* was introduced. Menhinick (1991) did not list the Phelps Lake population as an introduction; he also noted (without providing details) that the Lake Phelps form differs from both *F. waccamensis* and *F. diaphanus*.

Recently, Young et al. (1998) obtained samples thought to be *F. waccamensis* from Lake Phelps and also determined that they differ morphologically from those found in Lake Waccamaw, thus, implying that these fish are not the product of an accidental introduction of *F. waccamensis* into Lake Phelps. The researchers also noted that the large geographical distance between Phelps and Lake Waccamaw and the absence of *F. waccamensis* in intervening bay lakes, suggests that these populations might be remnants of a once widespread species, or, alternatively, might have evolved independently. Examination of allozymes between the two populations revealed no fixed differences (Cashner, personal communication). However, genetic studies with mitochondrial DNA control region sequences supports the hypothesis of an independent origin of *F. waccamensis* in Lake Phelps and Lake Waccamaw (J. Quattro, personal communication).

Percidae

Etheostoma spectabile (Agassiz 1854)

Common Name: orangethroat darter.

Reported as introduced to the Allegheny drainage by Hocutt et al. (1986). No field notes or specimens have been found to document this report and it is considered to be erroneous. Record probably represents a misidentified *E. caeruleum* (P. Ceas, personal communication).

Stromateidae

Peprilus ovatus Horn 1970

Common Name: Gulf butterfish (shining butterfish).

This species was reported as introduced into **California**; however, there is considerable confusion concerning the positive identification of the fish and also some uncertainty about the location of introduction. Hubbs et al. (1979) listed this species as "introduced but not known to occur at present in California" without providing additional information. Dill and Cordone (1997) suggested that the location of the introduction may have been the Salton Sea; however, they noted that Walker et al. (1961) did not include this species in their list of fishes introduced into the Salton Sea. During further investigation of the problem, Dill and Cordone (1997) uncovered 1962 correspondence documenting the introduction of *Palometa simillium* (now *Peprilus simillimus*) the Pacific pompano as a contaminate in a stock of anchoveta *Cetengraulis mysticetus* introduced to the Salton Sea. The letter reported that about 12 pompano, a species native to coastal California, entered the bait tanks through the circulation system with water from San Diego Bay and were knowingly introduced in 1948. However, a 1963 letter of J. A. St. Amant remarked on a fish, found on the shore of the Salton Sea, that was identified by C.L. Hubbs and B.W. Walker as "*Palometa* sp.". This fish was probably found sometime during the period 1961–1963 and, according to Dill and Cordone (1997), would not have been included in Walker et al. (1961). *Peprilus ovatus* is native to the northern Gulf of California (Fischer et al. 1995b). Apparently the *Palometa* sp., mentioned in the St. Amant correspondence, was later identified as *Peprilus ovatus* and subsequently provided the basis for its inclusion in Hubbs et al (1979). Nevertheless, Dill and Cordone (1997) believed that Hubbs wrongly assumed that the fish he and Walker identified had originated in Mexico, thus, Dill and Cordone concluded that inclusion of *Peprilus ovatus* as introduced to California to be open to question. Until contrary evidence comes to light, and based on the information provided by Dill and Cordone (1997), we are treating the true identity of the introduced species as *Peprilus simillimus*.

Literature Cited

Aasen, G. A., D. A. Sweetnam, and L. M. Lynch. 1998. Establishment of the wakasagi, *Hypomesus nipponensis*, in the Sacramento-San Joaquin Estuary. California Fish and Game 84(1):31-35.

Abbott, T. M., K. L. Dickson, and W. A. Potter. 1977. *Notropis cerasinus* (Cope) record from the Appomattox River drainage. Virginia Journal of Science 28:167–168.

Adams, C. C., and T. L. Hankinson. 1928. The ecology and economics of Oneida Lake fish. Bulletin of the New York State College of Forestry at Syracuse Univesity, Volume 1 (Number 4a). Roosevelt Wildlife Annals 1(3/4):239–548.

Akihito and K. Sakamoto. 1989. Reexamination of the status of the striped goby. Japanese Journal of Ichthyology 36(1):100–112.

Allan, J. H. 1977. First report of the tiger trout hybrid, *Salmo trutta* Linneaus × *Salvelinus fontinalis* (Mitchill), in Alberta. Canadian Field-Naturalist 91(1):85–86.

Allen, G. R. 1989. Freshwater fishes of Australia. Tropical Fish Hobbyist Publications, Inc., Neptune City, New Jersey.

Allen, G. R. 1991. Field guide to the freshwater fishes of New Guinea. Christiansen Research Institute Publication 9, Madang, Papua New Guinea.

Allen, G. R., and N. J. Cross. 1982. Rainbowfishes of Australia and Papua New Guinea. Tropical Fish Hobbyist Publications, Inc., Neptune City, New Jersey.

Allen, L. G. 1982. Seasonal abundance, composition, and productivity of the littoral fish assemblage in upper Newport Bay, California. U.S. National Marine Fisheries Service Fishery Bulletin 80:769–790.

Anderson, R. O. 1992. A case for zander: fish for the future? Pages 22–32 *in* The In-Fisherman walleye guide 1992. In-Fisherman, Brainerd, Minnesota.

Anderson, R. S., R. J. Metzger, and P. L. Shafland. 1984. Establishment of the Midas cichlid, *Cichlasoma citrinellum*, in Florida. Florida Scientist 47:263–267.

Anonymous. 1892. Report of distribution of fish and eggs from July 1, 1888, to June 30, 1889. Pages 379–394 *in* Report of the Commissioner for 1888. Part XVI. U.S. Commission of Fish and Fisheries, Washington, D.C.

Anonymous. 1977. Grass carp problems. Sport Fishing Institute Bulletin 288:5.

Anonymous. 1981. Piranha caught in Kentucky. Pet Business 7(11):33, Greensboro, North Carolina.

Anonymous. 1983a. Spiny 'walking catfish' moving upstate, spotted in Vero Beach. Sun-Sentinel (October 19), Ft. Lauderdale, Florida.

Anonymous. 1983b. Iowa fisherman surprised to hook piranha. USA Today (August 4), Arlington, Virginia.

Anonymous. 1987a. Zander program setback. Dakota Country (July), Bismark, North Dakota.

Anonymous. 1987b. Piranha found in San Joaquin River. Monterey Peninsula Herald (July 7):23, Monterey, California.

Anonymous. 1987c. Coastal striped bass history still needed. Texas Parks and Wildlife News 1987(Sept. 30): 14–15.

Anonymous. 1988. Mighty Mississippi is full of secrets. Chester Herald Tribune (Randolph County Herald Tribune) (September 21)., Chester, Illinois.

Anonymous. 1989. The lineup: elusive piranha. Tropical Fish Hobbyist 38(4):105, Neptune City, New Jersey.

Anonymous. 1990. N.H. angler hooks piranha. Newburyport Daily News (August 22), Newburyport, New Hampshire.

Anonymous. 1992a. 1992 stocking records. Texas Parks and Wildlife News, October 2. Austin,Texas.

Anonymous. 1992b. Hooked. Atlanta Journal (November 28):A3, Atlanta, Georgia.

Anonymous. 1992c. Angler lands in hot water over a question of identity. The Philadelphia Inquirer (November 29):A3, Philadelphia, Pennsylvania.

Anonymous. 1993. A new exotic state record. Florida Game and Fish, June 1993:12, Tallahassee.

Anonymous. 1994a. Nature: something's fishy in Yellowstone. Newsweek (August 29).

Anonymous. 1994b. Black carp escape to Missouri waters. The LMRCC (Lower Mississippi River Conservation Committee) Newsletter 1(1) (June):11, Vicksburg, Mississippi.

Anonymous. 1995a. This little fishy is a what? Brownsville States Graphic (April 6):1, Brownsville, Tennessee.

Anonymous. 1995b. Fisherman pulls piranha from river near Easton. Pocono Record (August 1):5, Stroudsburg, Pennsylvania.

Applegate, V. C. 1950. Natural history of the sea lamprey (*Petromyzon marinus*) in Michigan. U.S. Fish and Wildlife Service Special Scientific Report Fisheries 55.

Araujo-Lima, C., and M. Goulding. 1997. So fruitful a fish: conservation biology of the Amazon's tambaqui. Columbia University Press, New York.

Armbruster, J. W. 1997. Phylogenetic relationships of the sucker-mouth armored catfishes (Loricariidae) with particular emphasis on the Ancistrinae, Hypostominae, and Neoplecostominae. Doctoral dissertation. University of Illinois, Champaign-Urbana.

Arthington, A. H., and L. N. Lloyd. 1989. Introduced poeciliids in Australia and New Zealand. Pages 333–348 *in* G. K. Meffe and F. F. Snelson, Jr., editors. Ecology and evolution of livebearing fishes (Poeciliidae). Prentice Hall, Englewood Cliffs, New Jersey.

Ashbaugh, N. A., A. A. Echelle, and A. F. Echelle. 1994. Genic diversity in Red River pupfish *Cyprinodon rubrofluviatilis* (Atheriniformes: Cyprinidontidae) and its implications for conservation genetics of the species. Journal of Fish Biology 45:291–302.

Ashley, K. W., and B. Buff. 1986. Determination of current food habits of flathead catfish in the Cape Fear River. Final Report submitted to the North Carolina Wildlife Resources Commission, Division of Boating and Inland Fisheries, Raleigh.

Associated Press. 1997. Some Sound waters better fishing than usual. The Olympian (July 20), Olympia, Washington.

Atz, J. W. 1962. Effects of hybridzation on pigmentation in fishes of the genus *Xiphophorus*. Zoologica 47:153–181.

Avise, J. C., and M. J. Van den Arvyle. 1984. Genetic analysis of reproduction of hybrid white bass × striped bass in the Savannah River. Transactions of the American Fisheries Society 113:563–570.

Axelrod, H. R. 1993. The most complete colored lexicon of cichlids. Tropical Fish Hobbyist Publications, Inc., Neptune City, New Jersey.

Axelrod, H. R., C. W. Emmens, D. Sculthorpe, W. Vorderwinkler, R. Socolof, and N. Pronek. 1993. Exotic tropical fishes. Tropical Fish Hobbyists Publications, Inc., Jersey City, NJ.

Axelrod, H. R., W. E. Burgess, N. Pronek, and J. G. Walls. 1985. Dr. Axelrod's atlas of freshwater aquarium fishes. Tropical Fish Hobbyist Publications, Inc., Neptune City, New Jersey.

Axon, J. R. 1981. Development of a muskellunge fishery at Cave Run Lake, Kentucky, 1974–1979. North American Journal of Fisheries Management 1:134–143.

Bade, E. 1926. The central European bitterling found in the states. Bulletin of the New York Zoological Society 29:188, 205–206.

Bagarinao, T. 1994. Systematics, distribution, genetics and life history of milkfish, *Chanos chanos*. Environmental Biology of Fishes 39:23–41.

Bailey, J. R. 1950. A new subspecies of the darter *Hololepis barratti* from western North Carolina. Copeia 1950(4):311–316.

Bailey, J. R. 1977. Freshwater fishes. Pages 265–298 *in* J. E. Cooper, S. S. Robinson, and J. B. Funderburg, editors. Endangered and threatened plants and animals of North Carolina. North Carolina State Museum of Natural History, Raleigh.

Bailey, J. R., and J. A. Oliver. 1939. The fishes of the Connecticut watershed. Pages 150–189 *in* H. E. Warfel, editor. Biological survey of the Connecticut watershed. New Hampshire Fish and Game Commission Survey Report 4, Concord.

Bailey, R. M. 1938. A systematic review of the centrarchid fishes, with a discussion of their distribution, variations, and probable interactions. Doctoral dissertation. University of Michigan, Ann Arbor.

Bailey, R. M., and M. O. Allum. 1962. Fishes of South Dakota. Miscellaneous Publications Museum of Zoology, University of Michigan 119:1–131.

Bailey, R. M., and C. Gans. 1998. Two new synbranchid fishes, *Monopterus roseni* from peninsular India and *M. desilvai* from Sri Lanka. Occasional Papers of the Museum of Zoology, University of Michigan 726:1–18.

Bailey, R. M., and six coauthors. 1970. A list of common and scientific names of fishes from the United States and Canada, 3rd edition. American Fisheries Society, Special Publication 6, Bethesda, Maryland.

Bailey, R. M., and G. R. Smith. 1981. Origin and geography of the fish fauna of the Laurentian Great Lakes basin. Canadian Journal of Fisheries and Aquatic Science 38:1539–1561.

Bailey, R. M., and G. R. Smith. 1992. Names of Michigan fishes. Michigan Department of Natural Resources. Pp 1–6.

Bailey, R. M., H. E. Winn, and C. L. Smith. 1954. Fishes from the Escambia River, Alabama and Florida, with ecologic and taxonomic notes. Proceedings of the Academy of Natural Sciences of Philadelphia 106:109–164.

Bailey, W. M. 1975. An evaluation of striped bass introductions in the southeastern United States. Proceedings of the 28th Annual Conference Southeastern Association of Game and Fish Commissioners 28(1974):54–68.

Bain, M. B. 1993. Assessing impacts of introduced aquatic species: grass carp in large systems. Environmental Management 17(2):211–224.

Baird, S. F. 1874a. Report on shad-hatching operations. Part XVIII. U.S. Commission of Fish and Fisheries, Washington, D.C. Pp. 403–418.

Baird, S. F. 1874b. Report of the Commissioner of Fish and Fisheries for 1872–1873. U.S. Commission of Fish and Fisheries, Washington, D.C.

Baird, S. F. 1876. Report of the Commissioner of Fish and Fisheries for 1873–1874. Part III. U.S. Commission of Fish and Fisheries, Washington, D.C. (Not seen; cited in Jones 1963).

Baird, S. F. 1878. Report of the Commissioner of Fish and Fisheries for 1875–1876. Part IV. U.S. Commission of Fish and Fisheries, Washington, D.C.

Baird, S. F. 1879. Report of the Commissioner of Fish and Fisheries for 1876–1877. Part V. U.S. Commission of Fish and Fisheries, Washington, D.C.

Baird, S. F. 1887. Report of the Commissioner for 1885. Part XIII. U.S. Commission of Fish and Fisheries, Washington, D.C.

Baird, S. F. 1889. Report of the Commissioner for 1886. Part XIV. U.S. Commission of Fish and Fisheries, Washington, D.C.

Baker, B. 1993. Flatheads come to Florida. Florida Game and Fish, 1993(June):28–31, 60–61. Tallahassee.

Balon, E. K. 1995. Origin and domestication of the wild carp, *Cyprinus carpio*: from Roman gourmets to the swimming flowers. Aquaculture 129:3–48.

Barber, M. A., W. H. W. Komp, and C. H. King. 1929. Malaria and the malaria danger in certain irrigated regions of southwestern United States. Public Health Reports 44(3):1300–1315.

Barbour, C. D. 1973. The systematics and evolution of the genus *Chirostoma* Swainson (Pisces, Atherinidae). Tulane Studies in Zoology and Botany 18(3):97–141.

Barclay, L. A., Jr., and W. M. Howell. 1973. Ichthyofauna of the Locust Fork of the Black Warrior River system of Alabama. Journal of the Alabama Academy of Science 44(2):120–146.

Barkuloo, J. M. 1967. Florida striped bass. Florida Game and Fresh Water Fish Commission Fishery Bulletin 4.

Barlow, G. W. 1961. Gobies of the genus *Gillichthys* with comments on the sensory canals as a taxonomic tool. Copeia 1961(4):423–437.

Barlow, G. W., and V. L. De Vlaming. 1972. Ovarian cycling in longjaw gobies, *Gillichthys mirabilis*, from the Salton Sea. California Fish and Game 58(1):50–57.

Barlow, G. W., and J. W. Munsey. 1976. The red devil-midas-arrow cichlid species complex in Nicaragua. Pages 359–369 *in* T. B. Thorson, editor. Investigations of the ichthyofauna of Nicaraguan Lakes. University of Nebraska, Lincoln.

Barrett, P. J. 1983. Systematics of fishes of the genus *Tilapia* (Perciformes: Cichlidae) in the lower Colorado River basin. Master's thesis, Arizona State University, Tempe.

Barron, J. L. 1964. Reproduction and apparent over-winter survival of the sucker-mouth armoured catfish, *Plecostomus* sp., in the headwaters of the San Antonio River. Texas Journal of Science 16:449.

Bart, H. L., and five coauthors. 1994. New distribution records of Gulf Slope drainage fishes in the Ocmulgee River system, Georgia. Southeastern Fishes Council Proceedings. 30:4–9.

Barthem, R., and M. Goulding. 1997. The catfish connection: ecology, migration and conservation of Amazon giants. Columbia University Press, New York.

Baugh, T. 1980. *Iotichthys phlegethontis*: the least chub. Part two. Freshwater and Marine Aquarium 3(2):24–26, 81–82. Sierra Madre, California.

Baugh, T. M., J. E. Deacon, and D. Withers. 1985. Conservation efforts with the Hiko White River springfish *Crenichthys baileyi grandis* (Williams and Wilde). Journal of Aquariculture & Aquatic Sciences 4(3):49–53.

Baughman, J. L. 1947. The tench in America. Journal of Wildlife Management 11(3):197–204.

Baughman, J. L. 1950. Random notes on Texas fishes. Texas Journal of Science 2:117–138.

Baxter, G. T., and J. R. Simon. 1970. Wyoming fishes. Wyoming Game and Fish Department Bulletin 4, Cheyenne.

Bay, E. C. 1966. Adaptation studies with the Argentine pearl fish, *Cynolebias bellottii*, for its introduction into California. Copeia 1966(4): 839–846.

Bayless, J. D. 1972. Artificial propagation and hybridization of striped bass, *Morone saxatilis* (Walbaum). South Carolina Wildlife and Marine Resources Department, Charleston.

Bean, B. A., and A. C. Weed. 1911. Recent additions to the fish fauna of the District of Columbia. Proceedings of the Biological Society of Washington 24:171–174.

Bean, T. H. 1882. Movements of young alewives (*Pomolobus* sp.) in Colorado River, Texas. Pp. 69–70 *in* S. F. Baird, editor. Report of the Commissioner of Fish and Fisheries for 1881. Volume I. U.S. Commission of Fish and Fisheries, Washington, D.C.

Bean, T. H. 1884. List of the fishes distributed by the United States Fish Commission. Pages 1039–1044 *in* Report of the Commissioner for 1882, Part X. U.S. Commission of Fish and Fisheries, Washington, D.C.

Bean, T. H. 1892a. Observations upon fishes and fish-culture. Bulletin of the U.S. Fish Commission for 1890, 10:49–61.

Bean, T. H. 1892b. The fishes of Pennsylvania, with descriptions of the species and notes on their common names, distribution, habits, reproduction, rate of growth and mode of capture. Report of the State Commissioners of Fisheries for the years 1889–90–91. Edwin K. Meyers, State Printer, Harrisburg, Pennsylvania.

Bean, T. H. 1896. Report on the propagation and distribution of food-fishes. Pages 20–80 *in* Report of the Commissioner for the year ending June 30, 1894, part XX. U.S. Commission of Fish and Fisheries, Washington, D.C.

Bean, T. H. 1897. Notes upon New York fishes received at the New York Aquarium 1895–1897. Bulletin of the American Museum of Natural History 9:327–375.

Bean, T. H. 1902. The food and game fishes of New York, notes on their common names, distribution, habits and mode of capture. Pages 250–478 *in* Seventh Report of the Forest, Fish and Game Commission. J. B. Lyon Company, Albany, New York.

Bean, T. H. 1903. Catalogue of the fishes of New York. New York State Museum Bulletin 60, Zoology 9. University of the State of New York Bulletin 278.

Becker, G. C. 1983. Fishes of Wisconsin. University of Wisconsin Press, Madison.

Beckman, W. C. 1974. Guide to the fishes of Colorado. University of Colorado Museum, Boulder, Colorado.

Bechler, D. L., and R. C. Harrel. 1994. Notes on the biology and occurrence of *Astyanax mexicanus* (Characidae, Teleostei) in southeast Texas. Texas Journal of Science 46(3):293–294.

Beecher, H. A., and R. F. Fernau. 1983. Fishes of oxbow lakes of Washington. Northwest Science 57(2):125–131.

Behnke, R. J. 1962. Status of the landlocked salmonid fishes of Formosa with a review of *Oncorhynchus masou*. Copeia 1962(2):400–407.

Behnke, R. J. 1992. Native trout of western North America. American Fisheries Society Monograph 6. Bethesda, Maryland.

Behnke, R. J., and R. M. Wetzel. 1960. A preliminary list of the fishes found in the fresh waters of Connecticut. Copeia 1960(2):141–143.

Beland, K. F. , F. L. Roberts, R. L. Saunders. 1981. Evidence of *Salmo salar* × *Salmo trutta* by hybridization in a North American river. Canadian Journal of Fisheries and Aquatic Sciences 38:552–554.

Bell, M. A. 1978. Fishes of the Santa Clara River system, southern California. Natural History Museum of Los Angeles County, Contributions in Science 295:1–20.

Bell, M. A. 1982. Melanism in a high elevation population of *Gasterosteus aculeatus*. Copeia 1982(4):829–835.

Bellrichard, S. J. 1996. Effects of common carp (*Cyprinus carpio*) on submerged macrophytes and water quality in a backwater lake on the upper Mississippi River. Master's thesis, University of Wisconsin-La Crosse. Reprinted by the National Biological Service, Environmental Management Technical Center, LTRMP 96–R008. Onalaska, Wisconsin.

Belshe, J. F. 1961. Observations of an introduced tropical fish (*Belonesox belizanus*) in southern Florida. Master's thesis. University of Miami, Coral Gables, Florida.

Bennett, B. 1988. Bighead carp is latest "exotic Asian invader." Lawrence Journal-World, 24 January 1988. p. 8B.

Benson, N. G., J. R. Greeley, M. I. Huish, and J. H. Kuehn. 1961. Status of management of natural lakes. Transactions of the American Fisheries Society 90:218–224.

Berg, L. S. 1948–1949. Freshwater fishes of the U.S.S.R. and adjacent countries, 4th edition. Three volumes. Translated from Russian, 1962–1965, for the Smithsonian Institution and the National Science Foundation, by Israel Program for Scientific Translations, Jerusalem, Israel.

Berra, T. M. 1981. An atlas of distribution of the freshwater fish families of the world. University of Nebraska Press, Lincoln.

Berry, F. H., and I. Barrett. 1963. Gillraker analysis and speciation in the thread herring genus *Opisthonema*. Inter-American Tropical Tuna Commission Bulletin 7(2):113–190.

Berst, A. H., A. R. Emery, and G. R. Spangler. 1981. Reproductive behavior of hybrid charr (*Salvelinus fontinalis* × *S. namaycush*). Canadian Journal of Fisheries and Aquatic Sciences 38:432–440.

Bestgen, K. R., S. P. Platania, J. E. Brooks, and D. L. Propst. 1989. Dispersal and life history traits of *Notropis girardi* (Cypriniformes: Cyprinidae), introduced into the Pecos River, New Mexico. American Midland Naturalist 122:228–235.

Bickford, W. M. 1914. Notes on the Montana grayling. Transactions of the American Fisheries Society 43:153–155.

Billington, N., and M. J. Maceina. 1997. Genetic and population characteristics of walleyes in the Mobile drainage of Alabama. Transactions of the American Fisheries Society 126:804–814.

Bishop, R. D. 1968. Evaluation of the striped bass (*Roccus saxatilis*) and white bass (*R. chrysops*) hybrids after two years. Proceedings of the 21st Annual Conference Southeastern Association of Game and Fish Commissioners 21(1967):245–254.

Black, G. F. 1980. Status of the desert pupfish, *Cyprinodon macularius* (Baird and Girard) in California. California Department of Fish and Game, Inland Fisheries Endangered Species Program, Special Publication 80–1. Sacramento.

Blatchley, W. S. 1938. The fishes of Indiana: with descriptions, notes on habits and distribution in the state. The Nature Publishing Co., Indianapolis, Indiana.

Bleher, H. 1993. Leaf fish. Aqua Geõgraphia 3 :76–79.

Blinn, D. W., C. Runck, D. A. Clark, and J. N. Rinne. 1993. Effects of rainbow trout predation on Little Colorado spinedace. Transactions of the American Fisheries Society 122:139–143.

Bohn, J. 1988. Toothy fish puts the bite on Burke Lake angler. Washington Post, 4 June 1988. pp. G1, G3–G4.

Boles, G. L. 1976. A range extension for the logperch, *Percina macrolepida*, in California. California Fish and Game 62(2):154.

Boltz, J. M., and J. R. Stauffer, Jr. 1993. Systematics of *Aphredoderus sayanus* (Teleostei: Aphredoderidae). Copeia 1993(1):81–98.

Bond, C. E. 1961. Keys to Oregon freshwater fishes. Oregon State University Agriculture Experimental Station Technical Bulletin 58:1–42.

Bond, C. E. 1973. Keys to Oregon freshwater fishes. Oregon State University Agriculture Experimental Station Technical Bulletin 58:1–42, revised.

Bond, C. E. 1994. Keys to Oregon freshwater fishes. Oregon State University Bookstores, Inc., Corvallis. revised.

Bonn, E. W., C. M. Fetterolf, Jr., and I. B. Byrd. 1955. Panel discussion on white bass. Proceedings of the Eighth Annual Conference Southeastern Association of Game and Fish Commisioners 8(1954):98–99.

Bortone, S. A. 1972. Recent capture of *Phoxinus oreas* (Pisces: Cyprinidae) from the Yadkin-Pee Dee River drainage, North Carolina. Journal of the Elisha Mitchell Scientific Society 88:28–29.

Boschi, E. 1957. Argentine pearl fish *Cynolebias bellottii* Steindachner 1881. Translation from Spanish by W. Voderwinkler. Tropical Fish Hobbyist Publications, Inc., Neptune City, New Jersey.

Boschung, H. T. 1987. Physical factors and the distribution and abundance of fishes in the upper Tombigbee River system of Alabama and Mississippi, with emphasis on the Tennessee-Tombigbee Waterway. Pages 184–192 *in* W. J. Matthews and D. C. Heins, editors. Community and evolutionary ecology of North American stream fishes. University of Oklahoma Press, Norman, OK.

Boschung, H. T. 1992. Catalogue of freshwater and marine fishes of Alabama. Alabama Museum of Natural History Bulletin 14:1–266. Tuscaloosa.

Boschung, H. T., J. D. Williams, D. W. Gotshall, D. K. Caldwell, and M. C. Caldwell. 1983. The Aududon Society field guide to North American fishes, whales and dolphins. Knopf, New York.

Bottroff, L., J. S. St. Amant, and W. Parker. 1969. Addition of *Pylodictis olivaris* to the California fauna. California Fish and Game 55(1):90.

Bouc, K. 1987. The fish book. Nebraskaland Magazine 65(1):1–130. Lincoln, Nebraska.

Bowers, G. M. 1901. Report of the commissioner of fish and fisheries for the year ending June 30, 1900. Part XXVI. U.S. Commission of Fish and Fisheries, Washington, D.C.

Bowers, G. M. 1904. Report of the Commissioner for year ending June 30, 1902. Part XXVIII. U.S. Commission of Fish and Fisheries, Washington, D.C.

Bowers, G. M. 1905. Report of the Commissioner for year ending June 30, 1904. U.S. Commission of Fish and Fisheries, Washington, D.C.

Bowers, G. M. 1906. The propagation and distribution of food fishes in 1905. Bureau of Fisheries Document No. 602, 64 pp. U.S. Commission of Fish and Fisheries, Washington, D.C

Bowers, N. J., and J. R. Stauffer, Jr. 1993. New species of rock-dwelling cichlid (Pisces: Cichlidae) from Lake Malawi, Africa, with comments on *Melanochromis vermivorus* Trewavas. Copeia 1993(3):715–722.

Branson, B. A. 1968. *Notropis venustus*: another introduced species in the overburdened Nevada fish fauna. Copeia 1968(4):870–871.

Branson, B. A., and D. L. Batch. 1971. Stream capture in Kentucky indicated by distributional records of *Fundulus catenatus* and *Etheostoma spectabile*. American Midland Naturalist 86:496–500.

Branson, B. A., and D. L. Batch. 1983. Fishes of the South Fork of the Kentucky River, with notes and records from other parts of the drainage. Proceedings of the Southeastern Fishes Council 4:1–15.

Breder, C. M., Jr. 1933. *Rhodeus amarus* spawning in American mussels. Copeia 1933(3):147–148.

Breukelman, J. 1946. A review of Kansas ichthyology. Transactions of the Kansas Academy of Science 49:51–70.

Brichard, P. 1978. Fishes of Lake Tanganyika. Tropical Fish Hobbyist Publications, Inc., Neptune City, New Jersey.

Britski, H. 1977. Sobre o género *Colossoma* (Pisces-Characidae). Suplemento Ciencia y Cultura, Brasil, 29(7):810.

Britski, H. A., Y. Sato, and A. B. S. Rosa. 1984. Manual de identificação de peixes da região de Tres Marias (com chaves de identificação para os peixes da Bacia do São Francisco). Camara dos Deputados and Companhia de Desenvolvimento do Vale do São Francisco (CODEVASF), Brasília, Brasil.

Brittan, M., A. Albrecht, and J. Hopkirk. 1963. An oriental goby collected in the San Joaquin River delta near Stockton, California. California Fish and Game 49(4):302–304.

Brittan, M. R., and G. D. Grossman. 1979. A pacu (*Colossoma*, family Characidae) caught in the Sacramento River. California Fish and Game 65(3):170–173.

Brittan, M. R., J. D. Hopkirk, J. D. Conners, and M. Martin. 1970. Explosive spread of the oriental goby *Acanthogobius flavimanus* in the San Francisco Bay-Delta region of California. Proceedings of the California Academy of Sciences 38:207–214.

Brock, V. E. 1960. The introduction of aquatic animals into Hawaiian waters. Internationale Revue der Gesamten Hydrobiologie 45(4):463–480.

Brown, B. E. 1962. Occurrence of the walleye, *Stizostedion vitreum*, in Alabama south of the Tennessee Valley. Copeia 1962(2):469–472.

Brown, C. J. D. 1966. Natural hybrids of *Salmo trutta* and *Salvelinus fontinalis*. Copeia 1966(3):600–601.

Brown, C. J. D. 1971. Fishes of Montana. Montana State University, Bozeman.

Brown, C. J. D., and A. C. Fox. 1966. Mosquito fish (*Gambusia affinis*) in a Montana pond. Copeia 1966(3):614–616.

Brown, C. J. D., and R. J. Graham. 1954. Observations on the longnose sucker in Yellowstone Lake. Transactions of the American Fisheries Society 83:38–46.

Brown, D. J., and T. G. Coon. 1991. Grass carp larvae in the lower Missouri River and its tributaries. North American Journal of Fisheries Management 11:62–66.

Brown, L. R., and P. B. Moyle. 1981. The impact of squawfish on salmonid populations: a review. North American Journal of Fisheries Management 1:104–111.

Brown, L. R., and P. B. Moyle. 1991. Changes in habitat and microhabitat partitioning within an assemblage of stream fishes in response to predation by Sacramento squawfish (*Ptychocheilus grandis*). Canadian Journal of Fisheries and Aquatic Sciences 48:849–856.

Brown, W. H. 1953. Introduced fish species in the Guadalupe River Basin. Texas Journal of Science 5:245–251.

Brown, W. H. 1961. First record of the African mouthbreeder *Tilapia mossambica* Peters in Texas. Texas Journal of Science 13:352–354.

Bryant, R. T., B. H. Bauer, M. G. Ryon, and W. C. Starnes. 1979. Distributional notes on fishes from northern Georgia with comments on the status of rare species. Proceedings of the Southeastern Fishes Council 2(4):1–4.

Buchanan, T. M. 1973. Key to the fishes of Arkansas. Arkansas Game and Fish Commission, Little Rock, AR.

Buntz, J., and P. Chapman. 1970. A preliminary report on the increasing establishment of non-native fish in the Tampa Bay area. Report to the Florida Game and Fresh Water Fish Commission.

Buntz, J., and C. S. Manooch. 1969. *Tilapia aurea* (Steindachner), a rapidly spreading exotic in south central Florida. Proceedings of the 22nd Annual Conference Southeastern Association of Game and Fish Commissioners 22(1968):495–501.

Burgess, G. H., C. R. Gilbert, V. Guillory, and D. C. Taphorn. 1977. Distributional notes on some north Florida freshwater fishes. Florida Scientist 40(1):33–41.

Burgess, J. E. 1958. The fishes of Six Mile Creek,

Hillsborough County, Florida, with particular reference to the presence of exotic species. Paper presented at the 12th annual Conference Southeastern Association of Game and Fish Commissioners, Louisville, KY.

Burgess, W. E. 1981. Studies on the family Cichlidae: 10. New information on the species of the genus *Symphysodon* with the description of a new subspecies of *S. discus* Heckel. Tropical Fish Hobbyist 29(7):32–42. Neptune City, New Jersey.

Burgess, W. E. 1989. An atlas of freshwater and marine catfishes: a preliminary survey of the Siluriformes. Tropical Fish Hobbyist Publications, Inc., Neptune City, New Jersey.

Burgess, W. E. 1992. Colored atlas of miniature catfish: every species of *Corydoras*, *Brochis*, and *Aspidoras*. Tropical Fish Hobbyist Publications, Inc., Neptune City, New Jersey.

Burgess, W. E., and H. R. Axelrod. 1974. Pacific marine fishes book 4: fishes of Taiwan and adjacent waters. Tropical Fish Hobbyist Publications, Inc., Neptune City, New Jersey.

Burkhead, N. M., and J. D. Williams. 1991. An intergeneric hybrid of a native minnow, the golden shiner, and an exotic minnow, the rudd. Transactions of the American Fisheries Society 120:781–795.

Burkhead, N. M., and J. D. Williams. 1992. Okaloosa darter, *Etheostoma okaloosae*. Pages 23–30 *in* C. R. Gilbert, editor. Rare and endangered biota of Florida. Volume II: fishes. University of Florida Press, Gainesville.

Burkhead, N. M., R. E. Jenkins, and E. G. Maurakis. 1980. New records, distribution and diagnostic characters of Virginia ictalurid catfishes with an adnexed adipose fin. Brimleyana 4:75–93.

Burns, J. W. 1966a. Sacramento blackfish. Pages 522–525 *in* A. Calhoun, editor. Inland fisheries management. California Department of Fish and Game. Sacramento.

Burns, J. W. 1966b. Threadfin shad. Pages 481–488 *in* A. Calhoun, editor. Inland fisheries management. California Department of Fish and Game. Sacramento.

Burns, J. W. 1966c. Western sucker. Pages 516–517 *in* A. Calhoun, editor. Inland fisheries management. California Department of Fish and Game. Sacramento.

Burns, J. W. 1966d. Sacramento squawfish. Pages 525–527 *in* A. Calhoun, editor. Inland fisheries management. California Department of Fish and Game. Sacramento.

Burns, J. W. 1966e. Tui chub. Pages 528–530 *in* A. Calhoun, editor. Inland fisheries management. California Department of Fish and Game. Sacramento.

Burr, B. M. 1980. A distributional atlas of the fishes of Kentucky. Brimleyana 3:53–84.

Burr, B. M. 1991. The fishes of Illinois: an overview of a dynamic fauna. Proceedings of our living heritage symposium. Illinois Natural History Survey Bulletin 34(4):417–427.

Burr, B. M., and R. C. Cashner. 1983. *Campostoma pauciradii*, a new cyprinid fish from southeastern United States, with a review of related forms. Copeia 1983(1):101–116.

Burr, B. M., R. C. Cashner, and W. L. Pflieger. 1979. *Campostoma oligolepis* and *Notropis ozarcanus* (Pisces: Cyprinidae), two additions to the known fish fauna of the Illinois River, Arkansas and Oklahoma. Southwestern Naturalist 24(2):371–396.

Burr, B. M., and six coauthors. 1996. Nonnative fishes in Illinois waters: What do the records reveal? Transactions of the Illinois State Academy of Science 89(1/2):73–91.

Burr, B. M., and R. L. Mayden. 1980. Dispersal of rainbow smelt, *Osmerus mordax*, into the upper Mississippi River. American Midland Naturalist 104(1):198–201.

Burr, B. M., and L. M. Page. 1986. Zoogeography of fishes of the lower Ohio-upper Mississippi basin. Pages 287–324 *in* C. H. Hocutt and E. O. Wiley, editors. The zoogeography of North American freshwater fishes. Wiley, New York.

Burr, B. M., M. E. Retzer, and R. L. Mayden. 1980. A reasssessment of the distributional status of five Kentucky cyprinids. Transactions of the Kentucky Academy of Science 41(1–2):48–54,

Burr, B. M., and M. L. Warren, Jr. 1986. A distributional atlas of Kentucky fishes. Kentucky Nature Preserves Commission Scientific and Technical Series 4.

Burr, B. M., and M. L. Warren, Jr. 1993. Fishes of the Big Muddy River drainage with emphasis on historical changes. Pages 186–209 *in* L. W.

Hesse, C. B. Stalnaker, N. G. Benson, and J. R. Zuboy, editors. Proceedings of the symposium on restoration planning for the rivers of the Mississippi River ecosystem. Biological Report 19. U.S. Department of the Interior, National Biological Survey, Washington, D.C.

Burr, B. M, M. L. Warren, Jr., and K. S. Cummings. 1988. New distributional records of Illinois fishes with additions to the known fauna. Transactions of the Illinois Academy of Science 81(1&2):163–170.

Busch, W. N., D. H. Davies, and S. J. Nepszy. 1977. Establishment of white perch, *Morone americana*, in Lake Erie. Journal of the Fisheries Research Board of Canada 34:1039–1041.

Busiahn, T. R. 1993. Can the ruffe be contained before it becomes your problem. Fisheries 18(8):22–23.

Buss, K. 1957. The controversial kokanee – a salmon for the lakes of northeastern United States. Pennsylvania Fish Commission Special Purpose Report.

Buss, K., and J. Miller. 1967. Interspecific hybridization of esocids: hatching success, pattern development, and fertility of some F_1 hybrids. Technical Papers of the Bureau of Sport Fisheries and Wildlife 14:1–30. Washington, D.C.

Buss, K., and J. E. Wright. 1958. Appearance and fertility of trout hybrids. Transactions of the American Fisheries Society 87:172–181.

Bussing, W. A. 1987. Peces de las aguas continentales de Costa Rica. Editorial de la Universidad de Costa Rica, San Jose.

Buth, D. G., and C. B. Crabtree. 1985. A comparison of the genetic characteristics of the Zuni sucker (*Catostomus discobolus yarrowi*), the bluehead sucker (*Catostomus discobolus*), and the Rio Grande sucker (*Catostomus plebeius*). Report of the New Mexico Department of Game and Fish, Santa Fe.

Cadwallader, P. L. 1977. Introduction of rudd *Scardinius erythrophthalmus* into New Zealand. Part 1. Review of the ecology of rudd and the implications of its introduction into New Zealand. New Zealand Ministry of Agriculture and Fisheries Fisheries Technical Report 147:1–18.

Cahoon, W. G. 1953. Commercial carp removal at Lake Mattamuskeet, North Carolina. Journal of Wildlife Management 17(3):312–317.

Cahn, A. R. 1925. The European rudd (*Scardinius*) in Wisconsin. Copeia 162:5.

Cahn, A. R. 1927. An ecological study of southern Wisconsin fishes. Illinois Biological Monographs 11(1):1–151.

Caldwell, J., and J. T. Collins. 1977. New records of fishes, amphibians and reptiles. Technical Publication of the State Biological Survey of Kansas 4:63–78.

Call, L. E. 1961. Agricultural research at Kansas State Agricultural College (KSU) before enactment of the Health Act (1887). Kansas Agricultural Experiment Station Bulletin 441:1–43.

Campton, D. E. 1987. Natural hybridization and introgression in fishes: methods of detection and genetic interpretation. Pages 161–192 *in* N. Ryman and F. Utter, editors. Population genetics and fishery management. University of Washington Press, Seattle.

Cardoza, J. E., G. S. Jones, T.W. French, and D. B. Halliwell. 1993. Exotic and translocated vertebrates of Massachusetts, 2nd edition. Fauna of Massachusetts series 6. Massachusetts Division of Fisheries and Wildlife, Publication 17223-110-200-11/93-C.R, Westborough.

Carlander, K. D. 1978. History of the fish population of Lake Laverne, Iowa State University campus. Iowa State Journal of Research 52(4):425–433.

Carp Task Force. 1989. Report to the Louisiana Legislature, March 1989. Three volumes. Louisiana Department of Wildlife and Fisheries, Baton Rouge.

Carter, F. A., and J. K. Beadles. 1983. Range extension of the silver carp, *Hypophthalmichthys molitrix*. Proceedings of the Arkansas Academy of Science 37:80. Fayetteville.

Cashner, R. C., T. M. Berra, and D. G. Cloutman. 1992. Reidentification of William Bartram's Savannah River *Ambloplites*, with early evidence for a Tennessee-Savannah faunal exchange. Proceedings of the Southeastern Fishes Council 26:11–14.

Cashner, R. C., and W. J. Matthews. 1988. Changes in the known Oklahoma fish fauna from 1973 to

1988. Proceedings of the Oklahoma Academy of Science 68:1–7.

Cashner, R. C., and R. D. Suttkus. 1977. *Ambloplites constellatus*, a new species of rock bass from the Ozark upland of Arkansas and Missouri with a review of western rock bass populations. American Midland Naturalist 98(1):147–161.

Cavender, T. M., and C. R. Ciola. 1981. Collection records and distributional atlas of the fishes of the Mukingum River basin, Ohio. Ohio Biological Survey Technical Report 5, Ohio State University Museum of Zoology.

Champ, M. A., J. T. Lock, C. D. Bjork, W. G. Klussmann, and J. D. McCullough, Jr. 1973. Effects of anhydrous ammonia on a central Texas pond, and a review of previous research with ammonia in fisheries management. Transactions of the American Fisheries Society 102(1):73–82.

Chapman, W. M. 1942. Alien fishes in the waters of the Pacific northwest. California Fish and Game 28:9–15.

Cheney, A. N. 1897. Black bass and their distribution in the waters of New York state. Annual report of the Forest, Fish, and Game Commission State of New York 2(1896):176–184.

Chernoff, B., J. V. Conner, and C. F. Bryan. 1981. Systematics of the *Menidia beryllina* complex (Pisces: Atherinidae) from the Gulf of Mexico and its tributaries. Copeia 1981(2):319–336.

Childers, W. F. 1967. Hybridization of four species of sunfishes (Centrarchidae). Bulletin of the Illinois Natural History Survey 29(3):159–214. Champaign.

Childs, M. R., A. A. Echelle, and T. E. Dowling. 1996. Development of the hybrid swarm between Pecos pupfish (Cyprinodontidae: *Cyprinodon pecosensis*) and sheepshead minnow (*Cyprinodon variegatus*): a perspective from allozymes and MtDNA. Evolution 50:2014–2022.

Chilton, E. W., and M. I. Muoneke. 1992. Biology and management of grass carp (*Ctenopharyngodon idella*, Cyprinidae) for vegetation control: a North American perspective. Reviews in Fish Biology and Fisheries 2:283–320.

Christie, W. J. 1972. Lake Ontario: effects of exploitation, introductions, and eutrophication on the salmonid community. Journal of the Fisheries Research Board of Canada 29:913–929.

Christie, W. J. 1974. Changes in the fish species composition of the Great Lakes. Journal of the Fisheries Research Board of Canada 31:827–854.

Christie R. W., and T. A. Curtis. 1983. Establishment of bluefin killifish, *Lucania goodei*, in Cooper River, South Carolina. Georgia Journal of Science 41(3):91–92.

Churchill, E. P., and W. H. Over. 1933. Fishes of South Dakota. South Dakota Department of Game and Fish, Pierre.

Cincotta, D. A., K. D. Bledsoe, and F. Jernejcic. 1990. Addition of *Fundulus catenatus* (Storer) to West Virginia's ichthyofauna. Proceedings of the West Virginia Academy of Science 62(2–4):51–56.

Clark, M. E., and K. A. Rose. 1997. Factors affecting competitive dominance of rainbow trout over brook trout in southern Appalachian streams: implications of an individual-based model. Transactions of the American Fisheries Society 126(1):1–20.

Clark, M. R. 1981. Probable establishment and range extension of the spotted tilapia, *Tilapia mariae* Boulenger (Pisces: Cichlidae) in east central Florida. Florida Scientist 44(3):168–171.

Clarke, R. F., J. Breukelman, and T. F. Andrews. 1958. An annotated list of the vertebrates of Lyon County, Kansas. Transactions of the Kansas Academy of Science 61(2):165–194.

Clay, W. M. 1962. Aquatic-life resources of the Ohio River. Ohio River Valley Water Sanitation Commission, Cincinnati.

Clay, W. M. 1975. The fishes of Kentucky. Kentucky Department of Fish and Wildlife Resources, Frankfort.

Cleary, R. E. 1956. The distribution of the fishes of Iowa. Pages 267–324 *in* J. R. Harlan and E. B. Speaker, editors. Iowa fish and fishing. Iowa Conservation Commission, Des Moines.

Clemmer, G. 1970. Systematics and biology of the *Hybopsis amblops* complex. Doctoral dissertation. Tulane University, New Orleans, LA.

Coates, C. W. 1950. Tropical fishes for a private

aquarium. Liveright Publishing Corporation, New York.

Cobb, J. N. 1902. Commercial fisheries of the Hawaiian Islands. Pages 381–499 *in* Report of the Commissioner for the year ending June 30, 1901. U.S. Commission of Fish and Fisheries, Government Printing Office, Washington, D.C.

Coburn, M. N., and T. M. Cavender. 1992. Interrelationships of North American minnows. Pages 328–373 *in* R. L. Mayden, editor. Systematics, historical ecology, and North American freshwater fishes. Stanford University Press, Stanford, California.

Cochran, B. 1987. Jaws: fish caught in Roanoke River resembles tough cousin but is harmless. Roanoke Times & World-News, 3 October 1987. p. B1.

Cochran, P. A., and P. J. Hesse. 1994. Observations on the white perch (*Morone americana*) early in its invasion of Wisconsin. Transactions of the Wisconsin Academy of Sciences, Arts and Letters 82:23–58.

Cofer, L. M. 1995. Invalidation of the Wichita spotted bass, *Micropterus punctulatus wichitae*, subspecies theory. Copeia 1995(2):487–490.

Cole, L. J. 1905. The German carp in the United States. Pages 523–641 *in* Report of the Bureau of Fisheries for 1904. U.S. Department of Commerce and Labor. Government Printing Office, Washington, D.C

Collette, B. B. 1962. The swamp darters of the subgenus *Hololepis* (Pices, Percidae). Tulane Studies in Zoology 9(4):115–211.

Collette, B. B., and C. E. Nauen. 1983. FAO species catalogue. Volume 2. Scombrids of the world. An annotated and illustrated catalogue of tunas, mackerels, bonitos and related species known to date. FAO (Food and Agriculture Organization of the United Nations) Fisheries Synopsis 125.

Collins, M. R. 1985. Species profiles: life histories and environmental requirements of coastal fishes and invertebrates (South Florida)–Striped mullet. U.S. Fish and Wildlife Service Biological Report 82(11.34). U.S. Army Corps of Engineers, TR EL-82-4.

Colorado Division of Wildlife. 1993. 1993 fish planting schedule. Colorado Division of Wildlife, Denver.

Conkel, D. 1993. Cichlids of North and Central America. Tropical Fish Hobbyist Publications, Inc., Neptune City, New Jersey.

Connecticut State Board of Fisheries and Game. 1959. A fishery survey of the lakes and ponds of Connecticut. Connecticut State Board of Fisheries and Game, Hartford.

Conner, J. V., R. P. Gallagher, and M. F. Chatry. 1980. Larval evidence for natural reproduction of the grass carp (*Ctenopharyngodon idella*) in the lower Mississippi River. Pages 1–19 *in* L. A. Fuiman, editor. Proceedings of the fourth annual larval fish conference. U.S. Fish and Wildlife Service, Biological Services Program, National Power Plant Team. FWS/OBS-80/43. Ann Arbor, Michigan.

Conner, J. V., and R. D. Suttkus. 1986. Zoogeography of freshwater fishes of the western Gulf Slope of North America. Pages 413–456 *in* C. H. Hocutt and E. O. Wiley, editors. The zoogeography of North American freshwater fishes. Wiley, New York.

Contreras, S., and M. A. Escalante. 1984. Distribution and known impacts of exotic fishes in Mexico. Pages 102–130 *in* W. R. Courtenay, Jr. and J. R. Stauffer, editors. Distribution, biology, and management of exotic fishes. Johns Hopkins University Press, Baltimore, Maryland.

Cook, F. A. 1959. Freshwater fishes of Mississippi. Mississippi Game and Fish Commission, Jackson.

Cook, S. F., Jr., and R. L. Moore. 1970. Mississippi silversides, *Menidia audens* (Atherinidae), established in California. Transactions of the American Fisheries Society 99(1):70–73.

Cooper, E. L. 1983. Fishes of Pennsylvania. Pennsylvania State University Press, University Park.

Coots, M. 1956. The yellow perch, *Perca flavescens* (Mitchill), in the Klamath River. California Fish and Game 47(7):219–228.

Coots, M. 1966. Yellow perch. Pages 426–430 *in* A. Calhoun, editor. Inland fisheries management. California Department of Fish and Game. Sacramento.

Cope, E. D. 1870. A partial synopsis of the fishes of the fresh waters of North Carolina. Proceedings of the American Philosophical Society 11:448–495.

Cordone, A. J., and T. C. Frantz. 1966. The Lake Tahoe sport fishery. California Fish and Game 52(4):240–274.

Cordone, A. J., S. J. Nicola, P. H. Baker, and T. C. Frantz. 1971. The kokanee salmon in Lake Tahoe. California Fish and Game 57(1):28–43.

Costa, W. J. E. M. 1990. Classifiçatão e distribuição da família Rivulidae (Cyprinodontiformes, Aplocheiloidei). Revista Brasileira de Biologia (Rio de Janeiro) 550(1):883–889.

Costa, W. J. E. M. 1995. Pearl killifishes: the Cynolebiatinae. Tropical Fish Hobbyist Publications, Inc., Neptune City, New Jersey.

Costa, W. J. E. M. 1996. Phylogenetic and biogeographic analysis of the Neotropical annual fish genus *Simpsonichthys* (Cyprinodontiformes: Rivulidae). Journal of Comparative Biology 1(3/4):129–140.

Costa-Pierce, B. A., and R. W. Doyle. 1996. Species composition of tilapia stocks in southern California. Final report for the California Department of Fish and Game, Inland Fisheries Division contract FG 4054IF-1, Sacramento.

Couch, C. A., J. C. DeVivo, and B. J. Freeman. 1995. What fish live in the streams of metropolitan Atlanta? U.S. Geological Survey, National Water-Quality Assessment Program, Fact Sheet FS-091–95, Atlanta, Georgia.

Countryman, W. D. 1975. Checklist of the recent fishes of Vermont. Norwich University, Northfield, Vermont. Unpublished mimeograph (revised version, October 1975).

Courtenay, W. R., Jr. 1970. Florida's walking catfish. Ward's Natural Science Bulletin 10(69):1, 4, 6.

Courtenay, W. R., Jr. 1978. Additional range expansion in Florida of the introduced walking catfish. Environmental Conservation 5(4):273–275.

Courtenay, W. R., Jr. 1979a. Continued range expansion of the walking catfish. Environmental Conservation 6(1):20.

Courtenay, W. R., Jr. 1979b. Biological impacts on introduced species and management policy in Florida. Pages 237–257 *in* R. Mann, editor. Exotic species in mariculture. MIT Press, Cambridge, MA.

Courtenay, W. R., Jr. 1980. Exotic fish: environmental roulette. Water Spectrum 12(4):10–17.

Courtenay, W. R., Jr. 1985. Florida Atlantic University quarterly reports for 1985 to the U.S. Fish and Wildlife Service, Gainesville, Florida.

Courtenay, W. R., Jr. 1989. Exotic fishes in the National Park system. Pages 237–252 *in* L. K. Thomas, editor. Proceedings of the 1986 conference on science in the national parks, volume 5. Management of exotic species in natural communities. U.S. National Park Service and George Wright Society, Washington, D.C.

Courtenay, W. R., Jr. 1993. Biological pollution through fish introductions. Pages 35–61 *in* B. N. McKnight, editor. Biological pollution: the control and impact of invasive exotic species. Proceedings of a symposium, Indiana University–Purdue University, Indiana Academy of Science, Indianapolis.

Courtenay, W. R., Jr., and J. E. Deacon. 1982. Status of introduced fishes in certain spring systems in southern Nevada. Great Basin Naturalist 42(3):361–366.

Courtenay, W. R., Jr., and J. E. Deacon. 1983. Fish introductions in the American southwest: a case history of Rogers Spring, Nevada. Southwestern Naturalist 28(2):221–224.

Courtenay, W. R., Jr., J. E. Deacon, D. W. Sada, R. C. Allan, and G. L. Vinyard. 1985. Comparative status of fishes along the course of the pluvial White River, Nevada. Southwestern Naturalist 30(4):503–524.

Courtenay, W. R., Jr., and D. A. Hensley. 1979a. Survey of introduced non-native fishes. Phase I Report. Introduced exotic fishes in North America: status 1979. Report submitted to National Fishery Research Laboratory, U.S. Fish and Wildlife Service, Gainesville, Florida.

Courtenay, W. R., Jr., and D. A. Hensley. 1979b. Range expansion in southern Florida of the introduced spotted tilapia, with comments on environmental impress. Environmental Conservation 6(2):149–151.

Courtenay, W. R., Jr., and D. A. Hensley. 1980. Special problems associated with monitoring exotic species. Pages 281–307 *in* C. H. Hocutt and J. R. Stauffer, Jr. Biological Monitoring of Fish. Lexington Books, Lexington, Massachusetts. D.C. Heath and Company.

Courtenay, W. R., Jr., D. A. Hensley, J. N. Taylor, and J. A. McCann. 1984. Distribution of exotic fishes in the continental United States. Pages 41–77 *in* W. R. Courtenay, Jr., and J. R. Stauffer, Jr., editors. Distribution, biology and management of exotic fishes. Johns Hopkins University Press, Baltimore, Maryland.

Courtenay, W. R., Jr., D. A. Hensley, J. N. Taylor, and J. A. McCann. 1986. Distribution of exotic fishes in North America. Pages 675–698 *in* C. H. Hocutt and E. O. Wiley, editors. The zoogeography of North American freshwater fishes. Wiley, New York.

Courtenay, W. R., Jr., D. P. Jennings, and J. D. Williams. 1991. Appendix 2: exotic fishes. Pages 97–107 *in* C. R. Robins, and six coauthors. Common and scientific names of fishes from the United States and Canada, 5th edition. American Fisheries Society, Special Publication 20. Bethesda, Maryland.

Courtenay, W. R., Jr., and C. C. Kohler. 1986. Exotic fishes in North American fisheries management. Pages 401–413 *in* R. H. Stroud, editor. Fish culture in fisheries management. American Fisheries Society, Fish Culture Section and Fisheries Managment Section, Bethesda, Maryland.

Courtenay, W. R., Jr., and J. A. McCann. 1981. Status and impact of exotic fish presently established in U.S. open waters (September 1, 1980; revised April 1981). In-House Report, National Fishery Research Laboratory, U.S. Fish and Wildlife Service, Gainesville, Florida.

Courtenay, W. R., Jr., and G. K. Meffe. 1989. Small fishes in strange places: a review of introduced poeciliids. Pages 319–331 *in* G. K. Meffe, and F. F. Snelson, Jr., editors. Ecology and evolution of livebearing fishes (Poeciliidae). Prentice Hall, Englewood Cliffs, New Jersey.

Courtenay, W. R., Jr., and W. W. Miley, II. 1975. Range expansion and environmental impress of the introduced walking catfish in the United States. Environmental Conservation 2(2):145–148.

Courtenay, W. R., Jr., and C. R. Robins. 1973. Exotic aquatic organisms in Florida with emphasis on fishes: a review and recommendations. Transactions of the American Fisheries Society 102:1–12.

Courtenay, W. R., Jr., and C. R. Robins. 1989. Fish introductions: good management, mismanagement, or no management? CRC Critical Reviews in Aquatic Sciences 1(1):159–172.

Courtenay, W. R., Jr., C. R. Robins, R. M. Bailey, and J. E. Deacon. 1987. Records of exotic fishes from Idaho and Wyoming. Great Basin Naturalist 47(4):523–526.

Courtenay, W. R., Jr., H. F. Sahlman, W. W. Miley, II, and D. J. Herrema. 1974. Exotic fishes in fresh and brackish waters of Florida. Biological Conservation 6(4):292–302.

Courtenay, W. R., Jr., and J. R. Stauffer, Jr., editors. 1984. Distribution, biology and management of exotic fishes. Johns Hopkins University Press, Baltimore, Maryland.

Courtenay, W. R., Jr., and J. R. Stauffer, Jr. 1990. The introduced fish problem and the aquarium fish industry. Journal of the World Aquaculture Society 21(3):145–159.

Courtenay, W. R., Jr., and J. D. Williams. 1992. Dispersal of exotic species from aquaculture sources, with emphasis on freshwater fishes. Pages 49–81 *in* A. Rosenfield and R. Mann, editors. Dispersal of living organisms into aquatic ecosystems. University of Maryland Sea Grant Publication, College Park.

Cowley, D. E., and J. E. Sublette. 1987. Distribution of fishes in the Black River drainage, Eddy County, New Mexico. Southwestern Naturalist 32(2):213–221.

Creaser, C. W. 1926. The establishment of the Atlantic smelt in the upper waters of the Great Lakes. Papers of the Michigan Academy of Science Arts and Letters 5(1925):405–424.

Crittenden, E. 1965. Status of *Tilapia nilotica* Linnaeus in Florida. Proceedings of the 16th Annual Conference Southeastern Association of Game and Fish Commissioners 16(1962):257–262.

Cross, F. B. 1954. Fishes of Cedar Creek and the South Fork of the Cottonwood River, Chase County, Kansas. Transactions of the Kansas Academy of Science 57:303–314.

Cross, F. B. 1967. Handbook of fishes of Kansas. State Biological Survey and University of Kansas Museum of Natural History, Miscellaneous Publication 45, Topeka.

Cross, F. B. 1970. Occurrence of the Arkansas River shiner, *Notropis girardi* Hubbs and Ortenburger, in the Red River system. Southwestern Naturalist 14(3):370.

Cross, F. B. and J. T. Collins. 1995. Fishes in Kansas. University of Kansas Natural History Museum, Lawrence.

Cross, F. B., O. T. Gorman, and S. G. Haslouer. 1983. The Red River shiner, *Notropis bairdi*, in Kansas with notes on depletion of its Arkansas River cognate, *Notropis girardi*. Transactions of the Kansas Academy of Science 86(2–3):93–98.

Cross, F. B., and S. G. Haslouer. 1984. *Pimephales vigilax* (Pisces, Cyprinidae) established inthe Missouri River basin. Transactions of the Kansas Academy of Science 87(3–4):105–107.

Cross, F. B., R. L. Mayden, and J. D. Stewart. 1986. Fishes in the western Mississippi drainage. Pages 363–412 *in* C. H. Hocutt and E. O. Wiley, editors. The zoogeography of North American freshwater fishes. Wiley, New York.

Cross, J. N. 1975. Ecological distribution of the fishes of the Virgin River (Utah, Arizona, Nevada). Western Interstate Commission of Higher Education, Boulder, Colorado.

Cross, J. N. 1976. Status of the native fish fauna of the Moapa River (Clark County, Nevada). Transactions of the American Fisheries Society 105:503–508.

Crossman, E. J. 1978. Taxonomy and distribution of North American esocids. Pages 13–26 *in* R. L. Kendall, editor. Selected coolwater fishes of North America. American Fisheries Society, Special Publication 11, Bethesda, Maryland.

Crossman, E. J. 1984. Introductions of exotic fishes into Canada. Pages 78–101 *in* W. R. Courtenay, Jr. and J. R. Stauffer, Jr., editors. Distribution, biology, and management of exotic fishes. Johns Hopkins University Press, Baltimore, Maryland.

Crossman, E. J., and K. Buss. 1965. Hybridization in the family Esocidae. Journal of the Fisheries Research Board of Canada 22(5):1261–1292.

Crossman, E. J., E. Holm, R. Cholmondeley, and K. Tuininga. 1992. First record for Canada of the rudd, *Scardinius erythrophthalmus*, and notes on the introduced round goby, *Neogobius melanostomus*. Canadian Field-Naturalist 106(2):206–209.

Crossman, E. J., and J. W. Meade. 1977. Artificial hybrids between Amur pike, *Esox reicherti*, and North American esocids. Journal of the Fisheries Research Board of Canada 34:2338–2343.

Crossman, E. J., and D. E. McAllister. 1986. Zoogeography of freshwater fishes of the Hudson Bay drainage, Ungava Bay and the Arctic Archipelago. Pages 53–104 *in* C. H. Hocutt and E. O. Wiley, editors. The zoogeography of North American freshwater fishes. Wiley, New York.

Crowder, L. B. 1984. Character displacement and habitat shift in a native cisco in southeastern Lake Michigan: evidence for competition? Copeia 1984(4):878–883.

Crowder, L. B., and F. P. Binkowski. 1983. Foraging behaviors and the interactions of alewife, *Alosa pseudoharengus*, and bloater, *Coregonus hoyi*. Environmental Biology of Fishes 8:105–113.

Crutchfield, J. U., Jr. 1995. Establishment and expansion of redbelly tilapia and blue tilapia in a power plant cooling reservoir. Pages 452–461 *in* H. L. Schramm, Jr. and R. G. Piper, editors. Uses and effects of cultured fishes in aquatic ecosystems. American Fisheries Society Symposium 15, Bethesda, Maryland.

Crutchfield, J. U., Jr., D. H. Schiller, D. D. Herlong, and M. A. Mallin. 1992. Establishment and impact of redbelly tilapia in a vegetated cooling reservoir. Journal of Aquatic Plant Management 30:28–35.

Dahlberg, M. D., and D. C. Scott. 1971a. The freshwater fishes of Georgia. Bulletin of the Georgia Academy of Science 29:1–64.

Dahlberg, M. D., and D. C. Scott. 1971b. Introductions of freshwater fishes in Georgia. Bulletin of the Georgia Academy of Science 29:245–252.

Danielsen, T. L. 1968. Differential predation on *Culex pipiens* and *Anopheles albimanus* mosquito larvae by two species of fish (*Gambusia affinis* and *Cyprinodon nevadensis*) and the effects of simulated reeds on predation. Doctoral dissertation. University of California, Riverside.

Dawes, J. 1991. Livebearing Fishes: a guide to their aquarium care, biology and classification. Blandford. London.

Day, F. 1958. The fishes of India; being a natural history of the fishes known to inhabit the seas and fresh waters of India, Burma, and Ceylon, volume I. William Dawson and Sons Ltd.,

London.

Deacon, J. E. 1988. The endangered woundfin and water management in the Virgin River, Utah, Arizona, Nevada. Fisheries 13:18–29.

Deacon, J. E., and W. G. Bradley. 1972. Ecological distribution of fishes of Moapa (Muddy) River in Clark County, Nevada. Transactions of the American Fisheries Society 101:408–419.

Deacon, J. E., C. Hubbs, and B. J. Zahuranec. 1964. Some effects of introduced fishes on the native fish fauna of southern Nevada. Copeia 1964(2):384–388.

Deacon, J. E., G. Kobetich, J. D. Williams, and S. Contreras. 1979. Fishes of North America endangered, threatened, or of special concern: 1979. Fisheries 4(2):29–44.

Deacon, J. E., and J. E. Williams. 1984. Annotated list of the fishes of Nevada. Proceedings of the Biological Society of Washington 97(1):103–118.

Dees, L. T. 1961. The mosquitofish, *Gambusia affinis*. Bureau of Commercial Fisheries, U.S. Fish and Wildlife Service Fishery Leaflet 525. Washington, D.C.

DeKay, J. E. 1842. Zoology of New-York, or the New-York fauna. Part IV. Fishes. W. and A. White and J. Visscher, Albany, New York.

DeLong, J. 1997. Walking in 100-year-old footsteps in southern Oregon. (Available from http://www.nanfa.org/reg_spot/klamath/klamath.htm. Page accessed 7 December 1997.)

DeLorme Mapping. 1988. New Hampshire atlas and gazetteer. DeLorme Mapping, Freeport, Maine.

DeLorme Mapping. 1992a. Idaho atlas and gazetteer. DeLorme Mapping, Freeport, Maine.

DeLorme Mapping. 1992b. Wyoming atlas and gazetteer. DeLorme Mapping, Freeport, Maine.

DeLorme Mapping. 1993a. Maryland Delaware atlas and gazetteer. DeLorme Mapping, Freeport, Maine.

DeLorme Mapping. 1993b. Utah atlas and gazetteer. DeLorme Mapping, Freeport, Maine.

DeLorme Mapping. 1996a. Nebraska atlas and gazetteer. DeLorme Mapping, Freeport, Maine.

DeLorme Mapping. 1996b. Nevada atlas and gazetteer. DeLorme Mapping, Freeport, Maine.

DeMarais, B. D., T. E. Dowling, M. E. Douglas, W. L. Minckley, and P. C. Marsh. 1992. Origin of *Gila seminuda* (Teleostei, Cyprinidae) through introgressive hybridization: Implications for evolution and conservation. Proceedings of the National Academy of Science 89(7):2747–2751.

de Moor, I. J., and M. N. Bruton. 1988. Atlas of alien and translocated indigenous aquatic animals in southern Africa. South African National Scientific Programmes Report 144. Foundation for Research Development and Council for Scientific and Industrial Research, Pretoria, South Africa.

Dence, W. A. 1925. Bitter carp (*Rhodeus amarus*) from New York state waters. Copeia 142:33.

Dence, W. A. 1952. Establishment of white perch, *Morone americana*, in central New York. Copeia 1952(3):200–201.

Denoncourt, R. F., C. H. Hocutt, and J. R. Stauffer, Jr. 1975b. Extensions of the known ranges of *Ericymba buccata* Cope and *Etheostoma zonale* (Cope) in the Susquehanna River drainage. Proceedings of the Pennsylvania Academy of Science 49:45–46.

Denoncourt, R. F., C. H. Hocutt, and J. R. Stauffer, Jr. 1975c. Additions to the Pennsylvania ichthyofauna of the Susquehanna River drainage. Proceedings of the Academy of Natural Sciences of Philadelphia 127(9):67–69.

Denoncourt, R. F., T. B. Robbins, and R. Hesser. 1975a. Recent introductions and reintroductions to the Pennsylvania fish fauna of the Susquehanna River drainage above Conowingo Dam. Proceedings of the Pennsylvania Academy of Science 49:57–58.

Dentler, J. L. 1993. Noah's farce: the regulation and control of exotic fish and wildlife. University of Puget Sound Law Review 17:191–242.

Devick, W. S. 1972. Population densities of tilapia in Wahiawa Reservoir. Hawaii Department of Land and Natural Resources, Division of Aquatic Resources, Project F-9-2, Job 1, Study V. Honolulu.

Devick, W. S. 1988. Disturbances and fluctuations in the Wahiawa Reservoir ecosystem. Hawaii Department of Land and Natural Resources, Division of Aquatic Resources, Project F-14-R-12, Job 4, Study I. Honolulu.

Devick, W. S. 1989. Disturbances and fluctuations in the Wahiawa Reservoir ecosystem. Hawaii Department of Land and Natural Resources, Division of Aquatic Resources, Project F-14-R-13, Job 4, Study I. Honolulu.

Devick, W. S. 1991a. Disturbances and fluctuations in the Wahiawa Reservoir ecosystem. Hawaii Department of Land and Natural Resources, Division of Aquatic Resources, Hawaii Department of Land and Natural Resources, Division of Aquatic Resources, Project F-14-R-15, Job 4, Study I. Honolulu.

Devick, W. S. 1991b. Patterns of introductions of aquatic organisms to Hawaiian freshwater habitats. Pages 189–213 *in* new directions in research, management and conservation of Hawaiian freshwater stream ecosystems. Proceedings of the 1990 symposium on freshwater stream biology and fisheries management, Hawaii Department of Land and Natural Resources, Division of Aquatic Resources. Honolulu.

Devick, W. S. 1992. The great piranha hunt. Hawaii Fishing News 17(10)6–7. Honolulu.

Devivo, J. C., and B. J. Freeman. 1995. Impact of introduced *Cyprinella lutrensis* on stream fish assemblages in Georgia. Abstracts of the 56th Annual Meeting of the Association of Southeastern Biologists. Association of Southeastern Biologists Bulletin 42(2):129.

DeWald, L., and M. A. Wilzbach. 1992. Interactions between native brook trout and hatchery brown trout: effects on habitat use, feeding, and growth. Transactions of the American Fisheries Society 121(3):287–296.

Dial R. S., and S. C. Wainright. 1983. New distributional records for non-native fishes in Florida. Florida Scientist 46(1):8–15.

Dill, W. A., and A. J. Cordone. 1997. History and status of introduced fishes in California, 1871–1996. Fish Bulletin 178:1–414. California Department of Fish and Game.

Dimmick, W. W., K. L. Fiorino, and B. M. Burr. 1996. Reevaluation of the *Lythrurus ardens* (Cypriniformes: Cyprinidae) complex with recognition of three evolutionary species. Copeia 1996(4):813–823.

Dodge, J. 1997. Escaped fish open to anglers. The Olympian (July 23), Olympia, Washington.

Dolman, W. B. 1983. Red drum freshwater stocking evaluation. Texas Parks and Wildlife. Federal Aid in Sport Fish Restoration Act, Federal Aid Project F-31-R-9, Objective XLII. Performance Report.

Douglas, N. H. 1974. Freshwater fishes of Louisiana. Claitor's Publishing Division, Baton Rouge, Louisiana.

Douglas, N. H., and J. T. Davis. 1967. A checklist of the freshwater fishes of Louisiana. Louisiana Wildlife and Fisheries Commission. Baton Rouge.

Douglas, N. H., S. G. George, J. J. Hoover, K. J. Killgore, and W. T. Slack. 1996. Records of two Asian carps in the lower Mississippi Basin. Page 127 *in* Abstracts of the 76th Annual Meeting of the American Society of Ichthyologists and Herpetologists, University of New Orleans, New Orleans, Louisiana.

Dowell, V. E., and C. D. Riggs. 1958. Further observation on *Astyanax fasciatus* and *Menidia audens* in Lake Texoma. Proceedings of the Oklahoma Academy of Science 33(1955):52–53.

Dowling, T. E., and B. D. DeMarais. 1993. Evolutionary significance of introgressive hybridization in cyprinid fishes. Nature 362:444–446.

Drewry, G. E., E. A. Delco, Jr., and C. Hubbs. 1958. Occurrence of the Amazon molly, *Mollienesia formosa*, at San Marcos, Texas. Texas Journal of Science 10(4):489–490.

Duffy, J. M., and H. J. Bernard. 1985. Milkfish, *Chanos chanos* (Forsskål, 1775), taken in southern California adds new family (Chanidae) to the California marine fauna. California Fish and Game 71:122–125.

Dumont, P., and six co-authors. 1988. Introduced salmonids: where are they going in Quebec watersheds of the Saint-Laurent River. Fisheries 13(3):9–17.

Durocher, P. P. 1994. Status of the grass carp program in Texas. Pages 16–17 *in* Proceedings of the grass carp symposium. U.S. Army Corps of

Engineers, Waterways Experiment Station, Vicksburg, Mississippi.

Dyche, L. L. 1914. Ponds, pond fish and pond fish culture. Kansas Department of Fish and Game Bulletin 1:1–208.

Dymond, J. R., J. L. Hart, and A. L. Pritchard. 1929. The fishes of the Canadian waters of Lake Ontario. University of Toronto Studies, Biology Series 33:1–35.

Eccles, D. H. 1992. FAO species identification sheets for fishery purposes: field guide to the freshwater fishes of Tanzania. FAO (Food and Agriculture Organization of the United Nations), Rome.

Eccles, D. H., and E. Trewavas. 1989. Malawian cichlid fishes: the classification of some haplochromine genera. Lake Fish Movies, Herten, Germany.

Echelle, A. A., and P. J. Connor. 1989. Rapid, geographically extensive genetic introgression after secondary contact between two pupfish species (*Cyprinodon*, Cyprinodontidae). Evolution 43(4):717–727.

Echelle, A. A., A. F. Echelle, and F. B. Cross. 1977. First records of *Cyprinodon rubrofluviatilis* (Cyprinodontidae) from the Colorado and Arkansas river systems, Texas. Southwestern Naturalist 22(1):142–143.

Echelle, A. A., and A. F. Echelle. 1997. Genetic introgression of endemic taxa by non-natives: a case study with Leon Springs pupfish and sheepshead minnow. Conservation Biology 11:153–161.

Echelle, A. A., C. W. Hoagstrom, A. F. Echelle, and J. E. Brooks. 1997. Expanded occurrence of genetically introgressed pupfish (Cyprinodontidae: *Cyprinodon pecosensis* × *C. variegatus*) in New Mexico. Southwestern Naturalist 42(3):336–339.

Echelle, A. F., and A. A. Echelle. 1994. Assessment of genetic introgression between two pupfish species, *Cyprinodon elegans* and *C. variegatus* (Cyprinodontidae), after more than 20 years of secondary contact. Copeia 1994(3):590–597.

Eddy, S., and J. C. Underhill. 1974. Northern fishes, with special reference to the Upper Mississippi Valley, 3rd edition. University of Minnesota Press, Minneapolis.

Edwards, G. B. 1974. Biology of the striped bass *Morone saxatilis* (Walbaum) in the lower Colorado River (Arizona-California-Nevada). Master's thesis. Arizona State University. Tempe.

Edwards, R. J. 1976. Relative and seasonal abundance of the fish fauna in an urban creek ecosystem. Master's thesis. University of Texas, Austin.

Edwards, R. J. 1979. A report of Guadalupe bass (*Micropterus treculi*) × smallmouth bass (*Micropterus dolomieui*) hybrids from two localities in the Guadalupe River, Texas. Texas Journal of Science 31(3):231–238.

Edwards, R. J., and S. Contreras-Balderas. 1991. Historical changes in the ichthyofauna of the lower Rio Grande (Rio Bravo del Norte), Texas and Mexico. Southwestern Naturalist 36(2):201–212.

Eglin Air Force Base. 1968. Fishing guide for waters in and around Eglin Air Force Base. Conservation and Beautification Committee, Eglin Air Force Base.

Eigenmann, C. H. 1925. A review of the Doradidae, a family of South American nematognathi, or catfishes. Transactions of the American Philosophical Society 22(5):280–365.

Elder, H. S., and B. R. Murphy. 1997. Grass carp (*Ctenopharyngodon idella*) in the Trinity River, Texas. Journal of Freshwater Ecology 12:281–289.

Ellis, M. M. 1974. Fishes of Colorado. University of Colorado Studies, Boulder, Colorado 11(1):1–136.

Elser, A. A., M. W. Gorges, and L. M. Morris. 1980. Distribution of fishes in southeastern Montana. Cooperatively compiled by Montana Department of Fish, Wildlife and Parks and U.S. Department of the Interior, Bureau of Land Management.

Emery, A. R., and G. Teleki. 1978. European flounder (*Platichthys flesus*) captured in Lake Erie, Ontario. Canadian Field-Naturalist 92(1):89–91.

Emery, L. 1981. Range extension of pink salmon (*Oncorhynchus gorbuscha*) into the lower Great Lakes. Fisheries 6(2):7–10.

Emery, L. 1985. Review of fish introduced into the Great Lakes, 1819–1974. Great Lakes Fishery

Commission Technical Report, 45.

Eschmeyer, W. N., editor. 1998. Catalogue of fishes. Three volumes. California Academy of Sciences, San Francisco.

Eschmeyer, W. N., E. S. Herald, and H. Hamann. 1983. A field guide to Pacific Coast fishes of North America. Peterson Field Guide Series. Houghton Mifflin, Boston.

Etnier, D. A., and W. C. Starnes. 1993. The fishes of Tennessee. University of Tennessee Press, Knoxville.

Evans, W. A. 1950. Notes on the occurrence of the bigmouth buffalo in southern California. California Fish and Game 36(3):332–333.

Evans, W. A., and P. A. Douglas. 1950. Notes on fishes recently introduced into southern California. California Fish and Game 36:435–436.

Everhart, W. H. 1950. Fishes of Maine. Maine Department of Inland Fisheries and Wildlife. Augusta.

Everhart, W. H. 1966. Blueback country. Maine Fish and Game 1966 (fall).

Everhart, W. H. 1976. Fishes of Maine. Maine Department of Inland Fisheries and Wildlife. Augusta.

Everhart, W. H., and W. R. Seaman. 1971. Fishes of Colorado. Colorado Game, Fish and Parks Division, Denver.

Everhart, W. H., and C. A. Waters. 1965. Life history of the blueback trout (Arctic char, *Salvelinus* (Linnaeus)) in Maine. Transactions of the American Fisheries Society 94(4):393–397.

Evermann, B. W., and H. W. Clark. 1920. Lake Maxinkuckee, a physical and biological survey. Indiana Department of Conservation. Two volumes. 1:238–451.

Evermann, B. W., and H. W. Clark. 1931. A distributional list of species of freshwater fishes known to occur in California. California Department of Fish and Game Fish Bulletin 35.

Evermann, B. W., and W. C. Kendall. 1895. A list of the species of fishes known from the vicinity of Neosho, Missouri. Bulletin of the United States Fish Commission for 1894, 14(22):469–472.

Evermann, B. W., and C. Rutter. 1895. The fishes of the Colorado Basin. Bulletin of the U.S. Fish Commission for 1894, 14:473–486.

Exotic Species Workgroup. 1994. Pros and cons of stocking triploid grass carp in the Chesapeake Bay Basin: technical considerations. A critical issue forum organized by the Exotic Species Workgroup of the Living Resources Subcommittee, Chesapeake Bay Program, held on January 18, 1994 in Annapolis, Maryland. Unpublished mimeograph recorded by Free State Reporting, Inc. Washington, D.C.

Fago, D. 1992. Distribution and relative abundance of fishes in Wisconsin. VIII. Summary report. Department of Natural Resources, Technical Bulletin, Madison, Wisconsin.

Fago, D. 1993. Skipjack herring, *Alosa chrysochloris*, expanding its range into the Great Lakes. Canadian Field-Naturalist 107:352–353.

Farley, D. G. 1972. A range extension for the logperch. California Fish and Game 58(3):248.

Fausch, K. D. 1988. Tests of competition between native and introduced salmonids in streams; what have we learned? Canadian Journal of Fisheries and Aquatic Sciences 45:2238–2246.

Fausch, K. D., and R. J. White. 1981. Competition between brook trout (*Salvelinus fontinalis*) and brown trout (*Salmo trutta*) for positions in a Michigan stream. Canadian Journal of Fisheries and Aquatic Sciences 38:1220–1227.

Fausch, K. D., and R. J. White. 1986. Competition among juveniles of coho salmon, brook trout, and brown trout in a laboratory stream, and implications for Great Lakes tributaries. Transactions of the American Fisheries Society 115(3):363–381.

Ferguson, T. B. 1876. Report of the Commissioners of Fisheries of Maryland to the General Assembly. 1 January 1876. John F. Wiley, Annapolis, Maryland.

Ferraris, C. J., Jr. 1991. Catfish in the aquarium. Tetra Press, Morris Plains, New Jersey.

Fink, W. L. 1993. Revision of the piranha genus *Pygocentrus* (Teleostei, Characiformes). Copeia 1993(3):665–687.

Fink, W. L., and A. Machado-Allison. 1992. Three new species of piranhas from Brazil and Venezuela (Teleostei: Characiformes). Ichthyological Exploration of Freshwaters 3:55–71.

Finucane, J. H. 1965. Threadfin shad in Tampa Bay, Florida. Quarterly Journal Florida Academy of Science 28(3):267–270.

Finucane, J. H., and G. R. Rinckey. 1967. A study of the African cichlid, *Tilapia heudeloti* Dumeril, in Tampa Bay, Florida. Proceedings of the 18th Annual Conference Southeastern Association of Game and Fish Commissioners 18(1964):259–269.

Fischer, W., and five coauthors. 1995a. Guía FAO para la identificatión de especies para los fines de la pesca. Pacifico centro-criental. Vertebrados - Parte 1, Volume II: 647–1200. Roma.

Fischer, W., F. Krupp, W. Schneider, C. Sommer, K. E. Carpenter, V. H. Niem. 1995b. Guía FAO para la identificatión de especies para los fines de la pesca. Pacifico centro-criental. Vertebrados - Parte 2, Volume III: 1201–1813. Roma.

Fite, K. R. 1973. Feeding overlap between roach and juvenile steelhead in the Eel River. Master's thesis. California State University, Humboldt.

Fletcher, D., and M. Hallock. 1992. Prohibited fish species in Washington as of January 18, 1991: Identification, biology and justification for species restrictions. Washington Department of Fish and Wildlife, Report 92-10. Olympia, Washington.

Florida Game and Fresh Water Fish Commission. 1989. Aquatic plant management utilizing triploid grass carp. 1 July 1988–30 June 1989. Performance report for herbivorous fish project.

Florida Game and Fresh Water Fish Commission. 1994. Proceedings of the grass carp symposium. U.S. Army Corps of Engineers, Waterways Experiment Station, Vicksburg, Mississippi.

Foltz, J. W., and J. P. Kirk. 1994. Aquatic vegetation and water quality in Lake Marion, South Carolina. Pages 93–107 *in* Proceedings of the grass carp symposium. U.S. Army Corps of Engineers, Waterways Experiment Station, Vicksburg, Mississippi.

Foote, K. J. 1977. Annual performance report: blue tilapia investigations. Study I: preliminary status investigations of blue tilapia. (Job I-1 through Job I-7; period July 6, 1976–June 30, 1977). Report to the Florida Game and Fresh Water Fish Commission.

Ford, H. C., and five coauthors. 1892. Report of the Fish Commissioners for the years 1889–90–91. Pennsylvania Report of State Commissioners of Fisheries, Edwin K. Meyers, State Printer, Harrisburg. 19:1–156.

Fossett, J. 1981. Here's one that didn't get away. The Courier-Journal (June 19), Louisville, KY.

Fowler, H. W. 1906. The fishes of New Jersey. Pages 35–477 *in* Annual report of the New Jersey State Museum (1905), part II. MacCrellish and Quigley, State Province, Trenton, New Jersey.

Fowler, H. W. 1907. A supplementary account of the fishes of New Jersey. Pages 251–408 *in* Annual report of the New Jersey State Museum (1906), part III. Academy of Natural Sciences of Philadelphia, Philadelphia.

Fowler, H. W. 1919. A list of fishes of Pennsylvania. Proceedings of the Biological Society of Washington 32:49–74.

Fowler, H. W. 1920. A list of the fishes of New Jersey. Proceedings of the Biological Society of Washington 33:139–170.

Fowler, H. W. 1952. A list of the fishes of New Jersey, with off-shore species. Proceedings of the Academy of Natural Sciences of Philadelphia CIV:89–151.

Frantz, T. C., and A. J. Cordone. 1965. Introductions of the Bonneville cisco (*Prosopium gemmiferum* Snyder) into Lake Tahoe, California and Nevada. California Fish and Game 51(4):270–275.

Frantz, T. C, and A. J. Cordone. 1967. Final introductions of the Bonneville cisco (*Prosopium gemmiferum* Snyder) into Lake Tahoe, California and Nevada. California Fish and Game 53(3):209–210.

Freeze, M., and S. Henderson. 1982. Distribution and status of the bighead carp and silver carp in Arkansas. North American Journal of Fisheries Management 2:197–200.

French, J. R. 1993. How well can fishes prey on zebra mussels in eastern North America. Fisheries 18(6):13–19.

Fries, L. T., and W. D. Harvey. 1989. Natural hybridization of white bass with yellow bass in Texas. Transactions of the American Fisheries Society 118:87–89.

Fryer, G., and T. D. Iles. 1972. The cichlid fishes of the Great Lakes of Africa. Oliver and Boyd,

Edinburgh, UK.

Gach, M. H. 1996. Geographic variation in mitochondrial DNA and biogeography of *Culaea inconstans* (Gasterosteidae). Copeia 1996(3):563–575.

Galat, D. L., and B. Robertson. 1992. Response of endangered *Poeciliopsis occidentalis sonoriensis* in the Rio Yaqui drainage, Arizona, to introduced *Gambusia affinis*. Environmental Biology of fishes 33:249–264.

Ganzhorn, J., J. S. Rohovec, and J. L. Fryer. 1992. Dissemination of microbial pathogens through introductions and transfers of finfish. Pages 175–192 *in* A. Rosenfield and R. Mann, editors. Dispersal of Living Organisms into Aquatic Ecosystems. University of Maryland Sea Grant College Program, College Park, Maryland.

Garrett, G. P. 1982. Status report on peacock bass (*Cichla* sp.) in Texas. Presented at the Annual Proceedings of the Texas Chapter, American Fisheries Society.

George, C. J. 1981. The fishes of the Adirondack Park. New York State Department of Environmental Conservation, Albany, NY.

Gerking, S. D. 1945. Distribution of the fishes of Indiana. Investigations of Indiana Lakes and Streams 3:1–137.

Gerking, S. D. 1955. Key to the fishes of Indiana. Investigations of Indiana Lakes and Streams 4:49–86.

Géry, J. 1977. Characoids of the world. Tropical Fish Hobbyist Publications, Inc., Neptune City, New Jersey.

Gestring, K., and P. Shafland. 1997. Selected life history attributes of the exotic jaguar guapote (*Cichlasoma managuense*) in Florida. Florida Scientist 60(3):137–142.

Gibbs, R. H., Jr. 1963. Cyprinid fishes of the subgenus *Cyprinella* of *Notropis*. The *Notropis whipplei-analostanus-chloristius* complex. Copeia 1963(3):511–528.

Gilbert, C. H., and N. B. Scofield. 1898. Notes of a collection of fishes from the Colorado Basin in Arizona. Proceedings of the U.S. National Museum 20:487–499.

Gilbert, C. R. 1964. The American cyprinid fishes of the subgenus *Luxilus* (genus *Notropis*). Bulletin of the Florida State Museum 8(2):95–194.

Gilbert, C. R. 1980. Zoogeographic factors in relation to biological monitoring of fish. Pages 309–339 *in* C. H. Hocutt, and J. R. Stauffer, Jr. Biological Monitoring of Fish. Lexington Books, Lexington, Massachusetts.

Gilbert, C. R. 1987. Zoogeography of the freshwater fish fauna of southern Georgia and peninsular Florida. Brimleyana 13:25–54.

Gilbert, C. R., editor. 1992. Rare and endangered biota of Florida. Volume II: fishes. University of Florida Press, Gainesville.

Gilbert, C. R. 1993. Geographic distribution of the striped mullet in the Atlantic and eastern Pacific oceans. Florida Scientist 56(4):204–210.

Gilbert, C. R. 1998. Type catalogue of recent and fossil North American freshwater fishes: families Cyprinidae, Catostomidae, Ictaluridae, Centrarchidae and Elassomatidae. Special Publication Florida Museum of Natural History Gainesville. 1:ii + 284 pp.

Gilbert, C. R., and R. M. Bailey. 1962. Synonymy, characters, and distribution of the American cyprinid fish *Notropis shumardi*. Copeia 1962(4):807–819.

Gilmore, R. G., P. A. Hastings, and D. J. Herrema. 1983. Ichthyofaunal additions to the Indian River Lagoon and adjacent waters east-central Florida. Florida Scientist 46:22–30.

Goldschmidt, T., F. Witte, and J. Wanink. 1993. Cascading effects of the introduced Nile perch on the detritivorous/phytoplanktivorous species in the sublittoral areas of Lake Victoria. Conservation Biology 7:686–700.

Goldstein, R. J. 1970. Cichlids. Tropical Fish Hobbyist Publications, Inc., Neptune City, New Jersey.

Goldstein, R. J. 1973. Cichlids of the world. Tropical Fish Hobbyist Publications, Inc., Neptune City, New Jersey.

Gomez, R., and H. L. Lindsay. 1972. Occurrence of Mississippi silversides, *Menidia audens* (Hay), in Keystone Reservoir and the Arkansas River. Proceedings of the Oklahoma Academy of Science 52:16–18.

Goode, G. B. 1884. The fisheries and fishery industries

of the United States. Section I: natural history of useful aquatic animals. Government Printing Office, Washington.

Goodson, L. F. 1966a. Crappie. Pages 312–332 *in* A. Calhoun, editor. Inland fisheries management. California Department of Fish and Game. Sacramento.

Goodson, L. F. 1966b. Redeye bass. Pages 371–373 *in* A. Calhoun, editor. Inland fisheries management. California Department of Fish and Game. Sacramento.

Gordon, B. 1987. Piranhas caught in Fresno County. San Francisco Chronicle, 8 July 1987. p. 31.

Gordon, H., and M. Gordon. 1957. Maintenance of polymorphism by potenetially injurious genes in seven natural populations of the platyfish, *Xiphophorus maculatus*. Journal of Genetics 55(1):1–44.

Gordon, M., and D. E. Rosen. 1951. Genetics of species differences in the morphology of the male genetalia of xiphophorin fishes. Bulletin of the American Museum of Natural History 95:409–464.

Gorges, M. 1994. Resident fish habitat management strategy for Wyoming. U.S. Bureau of Land Management, Wyoming State Office.

Gosline, W. A. 1940. Revision of the neotropical catfishes of the family Callichthyidae. Stanford Ichthyological Bulletin 2(1):1–29.

Gould, W. R. 1981. First records of the rainbow smelt (Osmeridae), sicklefin chub (Cyprinidae) and white bass (Percichthyidae) from Montana. Proceedings of the Montana Academy of Sciences 40:9–10.

Goulding, M. 1981. Fishes of the forest: explorations in Amazonian natural history. University of California Press, Los Angeles.

Grabowski, S. J., S. D. Hiebert, and D. M. Lieberman. 1984. Potential for introduction of three species of nonnative fishes into central Arizona via the Central Arizona Project–A literature review and analysis. U.S. Department of the Interior, Bureau of Reclamation, REC-ERC-84-7. Denver.

Gray, D. L. 1958. Striped bass for Arkansas? Proceedings of the 11th Annual Conference Southeastern Association of Game and Fish Commissioners 11(1957):287–289.

Gray, R. H., and D. D. Dauble. 1977. Checklist and relative abundance of fish species from the Hanford reach of the Columbia River. Northwest Science 51(3):208–215.

Greeley, J. R. 1937. Fishes of the area with annotated list. Pages 45–53 *in* A biological survey of the lower Hudson watershed. Supplemental to twenty-sixth annual report, 1936. New York Conservation Department, Albany.

Greene, C. W. 1935. The distribution of Wisconsin fishes. Wisconsin Conservation Commission.

Greenfield, D. W., and J. E. Thomerson. 1997. Fishes of the continental waters of Belize. University Press of Florida, Gainesville.

Greenwood, P. H. 1987. The genera of pelmatochromine fishes (Teleostei, Cichlidae): a phylogenetic review. Bulletin of the British Museum (Natural History) Zoology 53(3):139–203.

Greger, P. E., and J. E. Deacon. 1988. Food partitioning among fishes of the Virgin River. Copeia 1988(2):314–323.

Guest, W. C. 1983. Blueback herring evaluation. Fort Worth Research Unit, Fort Worth, Texas. Federal Aid in Fisheries Restoration Act, Texas. Project F-31-R-9.

Guest, W. C. 1985. Texas statewide striped bass fishery management program. Fort Worth Research Unit, Fort Worth, Texas. Federal Aid in Fisheries Restoration Act, Texas. Project F-31-R-11.

Guillory, V., and R. D. Gasaway. 1978. Zoogeography of the grass carp in the United States. Transactions of the American Fisheries Society 107:105–112.

Guire, C. R., L. E. Nichols, and R. T. Rachels. 1984. Biological investigations of flathead catfish in the Cape Fear River. Proceedings of the 35th Annual Conference Southeastern Association of Fish and Wildlife Agencies 35(1981):607–621.

Gunning, G. E., and W. M. Lewis. 1956. Recent collections of some less common fishes in southern Illinois. Transactions of the Illinois State Academy of Science 48:23–26.

Haaker, P. 1979. Two asiatic gobiid fishes, *Tridentiger trigonocephalus* and *Acanthogobius flavimanus*, in southern California. Bulletin of the Southern California Academy of Science 78:56–61.

Habel, M. L. 1975. Overwintering of the cichlid, *Tilapia aurea*, produces fourteen tons of harvestable size fish in a south Alabama bass-bluegill public fishing lake. Progressive Fish-Culturist 37:31–32.

Hahn, D. E. 1966. An introduction of *Poecilia mexicana* (Osteichthyes: Poeciliidae) into Colorado. Southwestern Naturalist 11(2):296–312.

Hale, M. M., J. E. Crumpton, and R. J. Schuler, Jr. 1995. From sportfishing bust to commercial fishing boon: a history of the blue tilapia in Florida. Pages 425–430 *in* H. L. Schramm, Jr. and R. G. Piper, editors. Uses and effects of cultured fishes in aquatic ecosystems. American Fisheries Society Sympsium 15, Bethesda, Maryland.

Hales, L. S., Jr. 1989. Occurrence of an introduced African cichlid, the blue tilapia, *Tilapia aurea* (Perciformes: Cichlidae), in a Skidaway River tidal creek. Department of Zoology and Institute of Ecology, University of Georgia, Athens, and Marine Extension Service Aquarium, Georgia Sea Grant College Program, Savanna, GA. Unpublished mimeograph.

Hall, G. E. 1956. Additions to the fish fauna of Oklahoma with a summary of introduced species. Southwestern Naturalist 1(1):16–26.

Haller, W. T. 1994. Probable grass carp scenarios. Pages 236–238 *in* Proceedings of the grass carp symposium. U.S. Army Corps of Engineers, Waterways Experiment Station, Vicksburg, Mississippi.

Hambrick, P. S., C. H. Hocutt, M. T. Masnik, and J. H. Wilson. 1973. Additions to the West Virginia ichthyofauna, with comments on the distribution of other species. Proceedings of the West Virginia Academy of Science 45(1):58–60.

Hardy, J. D. 1978. Development of fishes of the Mid-Atlantic Bight. An atlas of egg, larval and juvenile stages, volume 2: Anguillidae through Syngnathidae. U.S. Fish and Wildlife Services Program FWS-OBS-78/12.

Harlan, J. R., and E. B. Speaker. 1951. Iowa fish and fishing. Iowa Department of Natural Resources, Des Moines, Iowa.

Harlan, J. R., E. B. Speaker, and J. Mayhew. 1987. Iowa fish and fishing. Iowa Department of Natural Resources, Des Moines, Iowa.

Harrison, H. M., and E. B. Speaker. 1954. An annotated list of the fishes in the streams tributary to the Missouri River in Iowa. Proceedings of the Iowa Academy of Science 61:511–523.

Harrison, K. 1991. The taxonomy of East African Nile perch, *Lates* spp. (Perciformes, Centropomidae). Journal of Fish Biology 38:175–186.

Hartel, K. E. 1992. Non-native fishes known from Massachusetts freshwaters. Occasional Reports of the Museum of Comparative Zoology 2:1–9, Cambridge, Massachusetts.

Hartel, K. E., D. B. Halliwell, and A. E. Launer. 1996. An annotated working list of the inland fishes of Massachusetts, University of Massachusetts, Cambridge (Available from http://www.mcz.harvard.edu/fish/ma_fam.htm. Page accessed March 5, 1998.)

Haskell, W. L. 1959. Diet of the Mississippi threadfin shad, *Dorosoma petenense atchafalayae*, in Arizona. Copeia 1959(4):298–302.

Hasse, J. J. 1981. Characters, synonymy and distribution of the middle American cichlid fish *Cichlasoma meeki*. Copeia 1981(1):210–212.

Hatch, M. D. 1985. The native fish fauna of major drainages east of the Continental Divide in New Mexico. Master's thesis. Eastern New Mexico University, Portales.

Hauser, M. 1998. Champlain Canal fish barrier study. Aquatic Nuisance Species Digest 2(3):26–27.

Hauser, W. J. 1975. An unusually fast growth rate for *Tilapia zillii*. California Department of Fish and Game 61(1):54–56.

Havey, K. A. 1973. Effects of a smelt introduction on growth of landlocked salmon at Schoodic Lake, Maine. Transactions of the American Fisheries Society 102:392–397.

Hawkins, L. 1994. Central Florida: area report. Florida Fish and Game Finder Magazine. March, pp. 18, 20–21. (black and white photograph with caption p. 18.)

Hay, O. P. 1894. The lampreys and fishes of Indiana. Report of the State Geologist.

Hayes, M. P., and M. R. Jennings. 1986. Decline of ranid frog species in western North America: are bullfrogs (*Rana catesbeiana*) responsible? Journal of Herpetology:20:490–509.

Haynes, C. M., R. T. Muth, and L. C. Wycoff. 1982. Range extension for the redside shiner, *Richardsonius balteatus* (Richardson), in the upper Colorado River drainage. Southwestern Naturalist 27(2):223.

Haynes, J. L. 1983. Annual reestablishment of mosquitofish in Nebraska. Copeia 1983(1):232–235.

He, X., and J. F. Kitchell. 1990. Direct and indirect effects of predation on a fish community: a whole lake experiment. Transactions of the American Fisheries Society 119:825–835.

Hearn, M. C. 1986. Reproductive viability of sauger-walleye hybrids. Progressive Fish-Culturist 48:149–150.

Helm, W. T. 1958. A "new" fish in Wisconsin. Wisconsin Conservation Bulletin 23(7):1–3.

Hendricks, M. L., J. R. Stauffer, Jr., C. H. Hocutt. 1983. The zoogeography of the fishes of the Youghiogheny River system, Pennsylvania, Maryland and West Virginia. American Midland Naturalist 110(1):145–164.

Hendricks, M. L., J. R. Stauffer, Jr., C. H. Hocutt, and C. R. Gilbert. 1979. A preliminary checklist of the fishes of the Youghiogheny River. Chicago Academy of Sciences, Natural History Miscellanea 203:1–15.

Hendrickson, D. A., and J. E. Brooks. 1991. Transplanting short-lived fishes in North American deserts: review, assessment, and recommendations. Pages 283–298 *in* W. L. Minckley and J. E. Deacon, editors. Battle against extinction: native fish management in the American West. University of Arizona Press, Tucson.

Hendrickson, D. A., W. L. Minckley, R. R. Miller, D. J. Seibert, and P. H. Minckley. 1980. Fishes of the Río Yaqui basin, México and United States. Journal of the Arizona-Nevada Academy of Science 15(3):65–106.

Hendrickson, D. A., and A. Varela-Romero. 1989. Conservation status of desert pupfish, *Cyprinodon macularius*, in Mexico and Arizona. Copeia 1989(2):478–483.

Hensley, D. A. 1976. Collection of postlarval and juvenile *Hoplias malabaricus* (Characoidei: Erythrinidae) in Florida. Florida Scientist 39(4):236–238.

Hensley, D. A., and D. P. Moody. 1975. Occurrence and possible establishment of *Hoplias malabaricus* (Characoidei; Erythrinidae) in Florida. Florida Scientist 38(2):122–128.

Herbold, B. A., A. D. Jassby, and P. B. Moyle. 1992. Status and trends report on aquatic resources in the San Francisco estuary. U.S. Environmental Protection Agency Public Report, San Francisco.

Hergenrader, G. L., and Q. P. Bliss. 1971. The white perch in Nebraska. Transactions of the American Fisheries Society 100:734–738.

Hessel, R. 1874. The salmon of the Danube, or the hucho, (*Salmo hucho*,) and its introduction into American Waters. Report of the Commissioner of Fish and Fisheries for 1872–1873. U.S. Commission of Fish and Fisheries, Washington, D.C. Pages 161-165

Heyne, T., B. Tribbey, M. Brooks, and J. Smith. 1991. First record of Mozambique tilapia in the San Joaquin Valley, California. California Fish and Game 77(1):53–54.

Hickley, P. 1986. Invasion by zander and the management of fish stocks. Philosophical Transactions of the Royal Society of London Biological Sciences 314:571–582.

Hida, T. S., and D. A. Thomson. 1962. Introduction of the threadfin shad to Hawaii. Progressive Fish-Culturist 24:159–163.

Hidalgo, C. 1997. South Florida's peacock bass. CatFish Books, Inc., Tamarac, Florida.

Higgins, E., and R. Lord. 1926. Prelimary report on the marine fisheries of Texas. Appendix IV to the Report of the U.S. Commissioner of Fisheries for 1926. Bureau of Fisheries Document No. 1009. U.S. Department of Commerce, Bureau of Fisheries.

Higuchi, H. 1992. A phylogeny of the South American thorny catfishes (Osteichthyes; Siluriformes, Doradidae). Doctoral dissertation. Harvard University, Cambridge, Massachusetts.

Hildebrand, S. F. 1923. Annotated list of fishes collected in the vicinity of Augusta, Georgia, with a description of a new darter. Bulletin of the U.S. Bureau of Fisheries 39:1–8.

Hindar, K., N. Ryman, and F. Utter. 1991. Genetic effects of cultured fish on natural fish populations. Canadian Journal of Fisheries and Aquatic Sciences 48:945–957.

Hirsch, J. 1998. Nonindigenous fish in inland waters: response plan to new introductions. Minnesota Department of Natural Resources Special Publication 152: 1–21.

Hocutt, C. H., R. E. Jenkins, and J. R. Stauffer, Jr. 1986. Zoogeography of the fishes of the central Appalachians and central Atlantic Coastal Plain. Pages 161–212 *in* C. H. Hocutt and E. O. Wiley, editors. The zoogeography of North American freshwater fishes. Wiley, New York.

Hoese, H. D., and R. H. Moore. 1977. Fishes of the Gulf of Mexico, Texas, Louisiana, and adjacent waters. Texas A&M University Press, College Station.

Hoffman, G. L., and G. Schubert. 1984. Some parasites of exotic fishes. Pages 233–261 *in* W. R. Courtenay, Jr. and J. R. Stauffer, Jr., editors. Distribution, biology, and management of exotic fishes. The Johns Hopkins University Press, Baltimore, Maryland.

Hogg, R. G. 1974. Environmental hazards posed by exotic fish species newly established in Florida. Environmental Conservation 1:176.

Hogg, R. G. 1976a. Ecology of fishes of the family Cichlidae introduced into the fresh waters of Dade County, Florida. Doctoral dissertation. University of Miami, Coral Gables.

Hogg, R. G. 1976b. Established exotic cichlid fishes in Dade County, Florida. Florida Scientist 39(2):97–103.

Holcík, J., and K. Hensel. 1974. A new species of *Gymnocephalus* (Pisces: Percidae) from the Danube, with remarks on the genus. Copeia 1974(2):471–486.

Holden, P. B. 1991. Impacts of Green River poisoning on management of native fishes. Pages 43–55 *in* W. L. Minckley and J. E. Deacon, editors. Battle against extinction: native fish management in the American West. University of Arizona Press, Tucson.

Holden, P. B., and C. B. Stalnaker. 1975a. Distribution of fishes in the Dolores and Yampa River systems of the upper Colorado basin. Southwestern Naturalist 19(4):403–412.

Holden, P. B., and C. B. Stalnaker. 1975b. Distribution and abundance of mainstream fishes of the middle and upper Colorado River basins, 1967–1973. Transactions of the American Fisheries Society 104:217–231.

Holm, E., and J. G. Hamilton. 1988. Range extension for the fourspine stickleback, *Apeltes quadracus*, to Thunder Bay, Lake Superior. Canadian Field-Naturalist 102:653–656.

Holton, G. D. 1990. A field guide to Montana fishes. Montana Department of Fish, Wildlife and Parks, Helena.

Hoover, E. E. 1936. Preliminary biological survey of some New Hampshire lakes. New Hampshire Fish and Game Department, Survey Report 1. Concord.

Hoover, F. G. 1971. Status report on *Tilapia mossambica* (Peters) in southern California. California Department of Fish and Game, Inland Fisheries Administrative Report 716. Unpublished mimeograph.

Hoover, F. G., and J. A. St. Amant. 1970. Establishment of *Tilapia mossambica* (Peters) in Bard Valley, Imperial County, California. California Fish and Game 56(1):70–71.

Hoover, F. G., and J. A. St. Amant. 1983. Results of Mohave chub, *Gila bicolor mohavensis*, relocations in California and Nevada. California Fish and Game 69(1):54–56.

Howard, C. 1998. Wild Atlantic salmon reproducing here. Vancouver Sun (September 23), Vancouver, British Columbia.

Howells, R. G. 1987. Genetic introgression of introduced Florida largemouth bass and coppernose bluegill on native largemouth and bluegill in a new south Texas reservoir. Texas Parks and Wildlife Department, Inland Fisheries Data Series 2, Austin,Texas.

Howells, R. G. 1991a. An introduction of hybrid crappie in a south Texas power plant reservoir. Texas Parks and Wildlife Department, Management Data Series 65, Austin.

Howells, R. G. 1991b. Electrophoretic identification of feral and domestic tilapia in Texas. Texas Parks and Wildlife Department, Management Data Series 62, Austin.

Howells, R. G. 1991c. Identification of orangemouth corvina, spotted seatrout and their hybrids. Texas Parks and Wildlife Department, Management Data Series 57, Austin.

Howells, R. G. 1992a. Annotated list of introduced non-native fishes, mollusks, crustaceans and aquatic plants in Texas waters. Texas Parks and Wildlife Department, Management Data Series 78, Austin.

Howells, R. G. 1992b. Guide to identification of harmful and potentially harmful fishes, shellfishes and aquatic plants prohibited in Texas. Texas Parks and Wildlife Department Special Publication, Austin.

Howells, R. G. 1995. Losing the old shell game: could mussel reproductive failure be linked to tilapia? Info-Mussel Newsletter 3(8):4.

Howells, R. G., R. L. Benefield, and J. M. Mambretti. 1991a. Records of pacus (*Colossoma* spp.) and piranhas (*Serrasalmus* spp.) in Texas. Texas Parks and Wildlife, Management Data Series 70, Austin.

Howells, R. G., and G. P. Garrett. 1992. Status of some exotic sport fishes in Texas waters. Texas Journal of Science 44(3):317–324.

Howells, R. G., R. W. Luebke, B. T. Hysmith, and J. G. Moczygemba. 1991b. Field collections of rudd, *Scardinius erythrophthalmus* (Cyprinidae), in Texas. Southwestern Naturalist 36(2):244–245.

Howells, R. G., and J. A. Prentice. 1991. Performance of Florida largemouth bass from Cuba in Texas waters. Texas Parks and Wildlife Department Management Data Series 59, Austin.

Hoy, J. B., E. E. Kaufmann, and A. G. O'Berg. 1972. A large-scale field test of *Gambusia affinis* and Chlorpyrifos for mosquito control. Mosquito News 32(2):163–171.

Hrabik, T. R., J. J. Magnuson, and A. S. Mclain. 1998. Predicting the effects of rainbow smelt on native fishes in small lakes: evidence from long-term research on two lakes. Canadian Journal of Fisheries and Aquatic Sciences 55:1364–1371.

Hubbs, C. 1954. Corrected distributional records for Texas fresh-water fishes. Texas Journal of Science 6(3):277–291.

Hubbs, C. 1957. Distributional patterns of Texas fresh-water fishes. Southwestern Naturalist 2(2–3):89–104.

Hubbs, C. 1980. Solution to the *C. bovinus* problem: eradication of a pupfish genome. Proceedings of the Desert Fishes Council 10:9–18.

Hubbs, C. 1982. Occurrence of exotic fishes in Texas waters. Texas Memorial Museum, Pearce Sellards Series 36:1–19.

Hubbs, C., and J. E. Deacon. 1964. Additional introductions of tropical fishes into southern Nevada. Southwestern Naturalist 9(4):249–251.

Hubbs, C., R. J. Edwards, and G. P. Garrett. 1991. An annotated checklist of freshwater fishes of Texas, with key to identification of species. Texas Journal of Science, Supplement 43(4):1–56.

Hubbs, C., and A. E. Peden. 1968. Notes on the distribution of blackbass (*Micropterus*) in the San Marcos River, Hays County, Texas. Texas Journal of Science 20(2):193–194.

Hubbs, C., T. and six coauthors. 1978. Survival and abundance of introduced fishes near San Antonio, Texas. Texas Journal of Science 30(4):369–376.

Hubbs, C. L. 1919. The stickleback: a fish eminently fitted by nature as a mosquito destroyer. California Fish and Game 5:21–24.

Hubbs, C. L. 1921. Geographical variation of *Notemigonus crysoleucas* – an American minnow. Transactions of the Illinois State Academy of Science 11 (1918):147–151.

Hubbs, C. L. 1954. Establishment of a forage fish, the red shiner (*Notropis lutrensis*), in the lower Colorado system. California Fish and Game 40(3):287–294.

Hubbs, C. L. 1955. Hybridization between fish species in nature. Systematic Zoology 4(1):1–20.

Hubbs, C. L., and J. D. Black. 1947. Revision of *Ceratichthys*, a genus of American cyprinid fishes. University of Michigan Museum of Zoology, Miscellaneous Publication 66:1–56.

Hubbs, C. L., and K. Bonham. 1951. New cyprinid fishes of the genus *Notropis* from Texas. Texas Journal of Science 5(1):91–110.

Hubbs, C. L., and G. P. Cooper. 1936. Minnows of

Michigan. Cranbrook Institute of Science Bulletin 8:1–95. Bloomfield Hills, Michigan.

Hubbs, C. L., W. I. Follett, and L. J. Dempster. 1979. List of the fishes of California. California Academy Science Occasional Papers 133.

Hubbs, C. L., and K. F. Lagler. 1947. Fishes of the Great Lakes region. Cranbrook Institute of Science. Bulletin 26:1–186. Bloomfield Hills, Michigan.

Hubbs, C. L., and K. F. Lagler. 1949. Fishes of Isle Royale, Lake Superior, Michigan. Reprinted from Papers of the Michigan Academy of Science, Arts, and Letters, volume XXXIII, 1947. Contributions from the Scripps Institute of Oceanography, New Series, volume 420.

Hubbs, C. L., and K. F. Lagler. 1958. Fishes of the Great Lakes region. University of Michigan Press, Ann Arbor.

Hubbs, C. L., and K. F. Lagler. 1964. Fishes of the Great Lakes region. University of Michigan Press, Ann Arbor.

Hubbs, C. L., and R. R. Miller. 1943. Mass hybridization between two genera of cyprinid fishes in the Mohave Desert, California. Papers from the Michigan Academy of Sciences, Arts, and Letters 28:343–378.

Hubbs, C. L., and R. R. Miller. 1948. The zoological evidence: correlation between fish distribution and hydrographic history in the desert basins of western United States. University of Utah Bulletin Biological Series 10(7):17–166.

Hubbs, C. L., and R. R. Miller. 1965. Studies of cyprinodontid fishes. XXII: variation in *Lucania parva*, its establishment in western United States, and description of a new species from an interior basin in Coahuila, Mexico. Miscellaneous Publications of the University of Michigan Museum of Zoology 127:1–104.

Hubbs, C. L., R. R. Miller, and L. C. Hubbs. 1974. Hydrographic history and relict fishes of the north-central Great Basin. Memoirs of the California Academy of Sciences 7:1–259.

Hubbs, C. L., and G. A. Moore. 1940. The subspecies of *Notropis zonatus*, a cyprinid fish of the Ozark upland. Copeia 1940:91–99.

Huber, J. H. 1992. Review of *Rivulus*–ecobiogeography, relationships. Cybium, Societe Francasise d'Ichtyologie, Paris.

Hubert, W. 1994. Exotic fish. Pages 158–174 *in* T. L. Parrish and S. H. Anderson, editors. Exotic species manual. Wyoming Game and Fish Department, Laramie.

Huckins, C. J. F. 1996. Functional relationships between morphology, feeding performance, diet, and competitive ability in molluscivorous sunfish. Doctoral dissertation. Michigan State University, East Lansing.

Huckins, C. J. F. 1997. Functional linkages among morphology, feeding performance, diet, and competitive ability in molluscivorous sunfish. Ecology 78:2401–2414.

Hughes, R. M. 1981. The plains killifish, *Fundulus zebrinus* (Cyprinodontidae), in the Colorado River basin of western North America. Southwestern Naturalist 26(3):321–324.

Humphries, J. M., and R. C. Cashner. 1994. *Notropis suttkusi*, a new cyprinid from the Ouachita uplands of Oklahoma and Arkansas, with comments on the status of Ozarkian populations of *N. rubellus*. Copeia 1994:82–90.

Hurlbert, S. H., J. Zedler, and D. Fairbanks. 1972. Ecosystem alteration by mosquitofish (*Gambusia affinis*) predation. Science 175:639–641.

Ichthyological Associates. 1976. Reports to Delmarva Power, Wilmington, Delaware, of fishes collected during 1973–1975. Ithaca, New York.

Idaho Fish and Game. 1990. Fisheries Management Plan 1991–1995. Idaho Fish and Game. Boise.

Idaho Fish and Game. 1996. Fisheries Management Plan 1996–2000. Idaho Fish and Game. Boise.

Idaho Fish and Game. 1997. Official list of Idaho record fish (as of November 15, 1996). Idaho Fish and Game, Boise. (Available from http://www.state.id.us/fishgame/fishrecs.htm. Page accessed June 26, 1997.)

Illinois Department of Natural Resources. 1997a. North American and Illinois record fish. (Available from http://dnr.state.il.us/ftp/available_documents/records.pdf. Page accessed June 25, 1997.)

Illinois Department of Natural Resources. 1997b. Places to fish. (Available from http://dnr.state.il.us/ftp/available_documents/

places.pdf. Page accessed June 25, 1997.)

Illinois Department of Natural Resources. 1997c. Illinois fish prospects. (Available from http://dnr.state.il.us/ftp/available_documents/prospect.pdf. Page accessed June 25, 1997.)

Indiana Department of Natural Resources. 1997a. Ohio River sport fish. (Available from http://www.dnr.state.in.us/fishwild/fishgide/ohiorivr.htm. Page accessed June 23, 1997.)

Indiana Department of Natural Resources. 1997b. Ohio River sport fish. (Available from http://www.dnr.state.in.us/fishwild/fishgide/general.htm. Page accessed June 23, 1997.)

International Game Fish Association. 1994. World record game fishes. International Game Fish Association, Pompano Beach, Florida.

Internet Waterway, Inc. 1998. Fishing records for Maine. (Available from http://www.iwol.com/iww/national/fishnnet/fortherecords/Maine.html. Page accessed December 8, 1998).

Iowa Department of Natural Resources. 1997. Fishes of Iowa. (Available from http://www.state.ia.us/government/dnr/organiza/fwb/fish/iafish/iafish.htm. Page accessed June 25, 1997.)

Jayaram, K. C. 1981. The freshwater fishes of India: a handbook. Zoological Survey of India, Calcutta.

Jayaram, K. C. 1991. Systematic status of *Danio malabaricus* (Pisces: Cyprinidae). Ichthyological Exploration of Freshwaters 2:109–112.

Jenkins, R. E., and N. M. Burkhead. 1994. Freshwater fishes of Virginia. American Fisheries Society, Bethesda, Maryland.

Jenkins, R. E., E. A. Lachner, and F. J. Schwartz. 1972. Fishes of the central Appalachian drainages: their distribution and dispersal. Virginia Polytechnic Institute and State University Research Division Monograph 4:43–117. (erroneously stamped 1971).

Jenkins, R. E., and T. Zorach. 1970. Zoogeography and characters of the American cyprinid fish *Notropis bifrenatus*. Chesapeake Science 11(3):174–182.

Jenkins, R. M. 1954. The effect of gizzard shad on the fish population of a small Oklahoma lake. Transactions of the American Fisheries Society 85:58–74.

Jennings, D. P. 1986. Characterization of a localized Jack Dempsey, *Cichlasoma octofasciatum*, population in Alachua County, Florida. Florida Scientist 49(4):255–259.

Jennings, D. P. 1988. Bighead carp (*Hypophthalmichthys nobilis*): a biological synopsis. U.S. Fish and Wildlife Service Biological Report 88(29).

Jennings, D. P. 1991. Behavioral aspects of cold tolerance of blackchin tilapia, *Sarotherodon melanotheron* (Pisces, Cichlidae), at different salinities. Environmental Biology of Fishes 31(2):185–195.

Jennings, D. P., and J. D. Williams. 1992. Factors influencing the distribution of blackchin tilapia *Sarotherodon melanotheron* (Osteichthyes: Cichlidae) in the Indian River system, Florida. Northeast Gulf Science 12(2):111–117.

Jennings, M. R., and M. K. Saiki. 1990. Establishment of red shiner, *Notropis lutrensis*, in the San Joaquin Valley, California. California Fish and Game 76(1):46–57.

Jensen, D. A., editor. 1997. International symposium on biology and management of ruffe. Minnesota Sea Grant Publication X48.

Johnson, H. E. 1937. Feeding Montana grayling fry. Progressive Fish-Culturist 30:35–36.

Johnson, J. E., and C. Hubbs. 1989. Status and conservation of poeciliid fishes. Pages 301–331 *in* G. K. Meffe and F. F. Snelson, Jr., editors. Ecology and evolution of livebearing fishes (Poeciliidae). Prentice Hall, Englewood Cliffs, New Jersey.

Johnson, M., and G. C. Becker. 1980. Annotated list of the fishes of Wisconsin. Papers of the Wisconsin Academy of Sciences, Arts and Letters 58:265–300.

Johnson, R. S. 1915. The distribution of fish and fish eggs during the fiscal year 1914. Bulletin of the U.S. Fish Commission for the fiscal year 1914.

Johnson, T. B., and D. O. Evans. 1990. Size-dependent winter mortality of young-of-the-year white perch: climate warming and invasion of the Laurentian Great Lakes. Transactions of the American Fisheries Society 119:301–313.

Johnston, C. E. 1991. Discovery of the threespine stickleback (*Gasterosteus aculeatus*) (Pisces: Gasterosteidae) in the Lake Michigan drainage, Illinois. Transactions of the Illinois State Academy of Science 84(3&4):173–174.

Johnston, C. E., J. S. Ramsey, S. T. Sobaski, and C. K. Swing. 1995. Introduced species of fishes in the Southern Appalachians: consequences for conservation. Journal of the Tennessee Academy of Science 70(304):65–76.

Jones, D. J. 1963. A history of Nebraska's fisheries resources. Nebraska Game, Forestation and Parks Commission. Federal Aid in Fish Restoration Project F-4-R Publication.

Jordan, D. S. 1882. Report on the fishes of Ohio. Report of the Geological Survey of Ohio 4(1):735–1002.

Jordan, D. S., and B. W. Evermann. 1902. Preliminary report on an investigation of the fishes and fisheries of the Hawaiian Islands. Pages 353–380 *in* Report of the Commissioner for the year ending June 30, 1901. U.S. Commission of Fish and Fisheries, Government Printing Office, Washington, D.C.

Jordan, D. S., and B. W. Evermann. 1905. The aquatic resources of the Hawaiian Islands. Part I. The shore fishes. Bulletin of the U.S. Fish Commission for 1903, 23:1–574.

Jordan, D. S., and C. H. Gilbert. 1882. Synopsis of the fishes of North America. Bulletin of the United States National Museum 16. Government Printing Office, Washington, D.C.

Jordan, D. S., and J. O. Snyder. 1902. A review of the gobioid fishes of Japan, with descriptions of twenty-one new species. Proceedings of the United States National Museum 24(1244):33–132.

Jude, D. J. 1993. The alien goby in the Great Lakes Basin. Great Lakes Information Network. (Available from gopher://gopher.great-lakes.net:2200/00/waterairland/exotics/goby/goby.txt. Page accessed October 17, 1994.)

Jude, D. J., R. H. Reider, and G. R. Smith. 1992. Establishment of Gobiidae in the Great Lakes Basin. Canadian Journal of Fisheries and Aquatic Sciences 49:416–421.

Kaeding, L. R., G. D. Boltz, and D. G. Carty. 1996. Lake trout discovered in Yellowstone Lake threaten native cutthroat trout. Fisheries 21(3):16–20.

Kallman, K. D., and Brunetti. 1983. Genetic basis of three mutant color varieties of *Xiphophorus maculatus*: the gray, gold and ghost platyfish. Copeia 1983(1):170–181.

Kanayama, R. K. 1968. Hawaii's aquatic animal introductions. Proceedings of the Annual Conference of Western Association State Game and Fish Commissioners 47:123–131.

Kanazawa, R. H. 1966. The fishes of the genus *Osteoglossum* with a description of a new species from the Rio Negro. Aquarium Journal (April):161–172.

Kansas Department of Wildlife and Parks. 1990. Exotic fish invades Kansas waters. Kansas Department of Wildlife and Parks, Pratt, KS. News Release, Volume 1990:23 (June 6).

Kansas Department of Wildlife and Parks. 1997. 1997 fishing forecast. (Available from http://www.ink.org/public/kdwp/Fishing/Fishforecast/FishFor.html. Page accessed June 26, 1997.)

Karp, C. A., and H. M. Tyus. 1990. Behavioral interactions between young Colorado squawfish and six fish species. Copeia 1990(1):25–34.

Keevin, T. M. 1978. A distributional and faunistic analysis of the small stream fishes of St. Louis and St. Louis County, Missouri. Master's thesis. Southern Illinois University, Edwardsville.

Kempinger, J. J., and W. S. Churchill. 1972. Contribution of native and stocked walleye fingerlings to the anglers' catch, Escanaba Lake, Wisconsin. Transactions of the American Fisheries Society 101:644–649.

Kendall, R. L. 1997. Corrections of and additions to published American Fisheries Society books. Fisheries 22(12):21–22.

Kendall, W. C. 1914a. An annotated catalogue of the fishes of Maine. Proceedings of the Portland Society of Natural History 3:1–198.

Kendall, W. C. 1914b. The Fishes of New England. The salmon family. Part I – the trout or charrs. Memoirs of the Boston Society of Natural History 8(1).

Kendall, W. C. 1917. The pikes: their geographical

distribution habits, culture, and commercial importance. U.S. Commissioners Fisheries Report for 1917. Bureau of Fisheries Document 853. Washington, D.C.

Kennedy, H. 1990. N.H. woman reels in the catch of day: a piranha. The Boston Herald (Boston, Massachusetts), 23 August 1990.

Kennedy, S. E. 1977. Life history of the Leon Springs pupfish, *Cyprinodon bovinus*. Copeia 1977(1):93–103.

Kerby, J. H. 1979. Meristic characteristics of two *Morone* hybrids. Copeia 1979(3):513–518.

Kerby, J. H., V. G. Burrell, Jr., and C. E. Richards. 1971. Occurrence and growth of striped bass × white bass hybrids in the Rappahannock River, Virginia. Transactions of the American Fisheries Society 100:787–790.

Kilby, J. D., E. Crittenden, and L. E. Williams. 1959. Several fishes new to Florida freshwaters. Copeia 1959(1):77–78.

Killgore, K. J., J. J. Hoover, and R. P. Morgan. 1991. Habitat value of aquatic plants for fishes. U.S. Army Corps of Engineers Waterways Experiment Station, Technical Report A-91-5. Vicksburg, Mississippi.

Kimsey, J. B. 1950. Some Lahontan fishes in the Sacramento River drainages, California. California Fish and Game 36(4):438–439.

Kimsey, J. B. 1957. The status of the redeye bass in California. California Fish and Game 43(1):99–100.

Kimsey, J. B., and L. O. Fisk. 1960. Keys to the freshwater and anadromous fishes of California. California Fish and Game 46(4):453–479.

Kimsey, J. B., and L. O. Fisk. 1964. Freshwater nongame fishes of California. California Department of Fish and Game. Sacramento.

Kircheis, F. W. 1975. Draft blueback trout management plan.

Kircheis, F. W. 1976. Reproductive biology and early life history of the sunapee trout of Floods Ponds, Maine. Transactions of the American Fisheries Society 1976(5):615–619.

Kircheis, F. W. 1994. Update on freshwater fish species reproducing in Maine. Maine Naturalist 2(1):25–28.

Kirk, J. P., J. V. Morrow, Jr., and K. J. Killgore. 1994. Grass carp collection, aging, and growth in large water bodies–a status report for 1992–1993. Pages 210–214 *in* Proceedings of the grass carp symposium. U.S. Army Corps of Engineers, Waterways Experiment Station, Vicksburg, Mississippi.

Klindt, R. M. 1990. St. Lawrence River subcommittee report on the distribution of rudd in the St. Lawrence River. New York State Department of Environmental Conservation, Albany.

Knaggs, E. H. 1977. Status of the genus *Tilapia* in California's estuarine and marine waters. Cal-Nevada Wildlife Transactions 1977:60–67.

Knapp, F. T. 1953. Fishes found in the freshwaters of Texas. Rangland Studio and Litho Printing Company, Brunswick, GA.

Knight, C. 1994. The round goby: Lake Erie's newest invader. Ohio Chapter American Fisheries Society Newsletter 21(3):5.

Koehn, R. K. 1965. Development and ecological significance of nuptial tubercles of the red shiner, *Notropis lutrensis*. Copeia 1965(4):462–467.

Konings, A. 1989. Cichlids from Central America. Tropical Fish Hobbyist Publications, Inc., Neptune City, New Jersey.

Koster, W. J. 1957. Guide to the fishes of New Mexico. University of New Mexico Press, Albuquerque.

Kottelat, M. 1985. Freshwater fishes of Kampuchea. Hydrobiologia 121:249–279.

Kottelat, M., A. J. Whitten, S. N. Kartikasari, and S. Wirjoatmodjo. 1993. Freshwater fishes of Western Indonesia and Sulawesi. Periplus Editions, Ltd., Republic of Indonesia.

Kraai, J. E., W. P. Provine, and J. A. Prentice. 1985. Case histories of three walleye stocking techniques with cost-to-benefit considerations. Proceedings of the 37th Annual Conference Southeastern Association of Fish and Wildlife Agencies 37(1983):395–400.

Kristmundsdottir, A. Y., and J. R. Gold. 1996. Systematics of the blacktail shiner (*Cyprinella venusta*) inferred from analysis of mitochondrial DNA. Copeia 1996(4):773–783.

Krueger, C. C., and B. May. 1991. Ecological and genetic effects of salmonid introductions in North America. Canadian Journal of Fisheries and Aquatic Sciences 48(Suppliment 1):66–77.

Krumholz, L. A. 1948. Reproduction in the western mosquito fish, *Gambusia affinis affinis* (Baird and Girard), and its use in mosquito control. Ecological Monographs 18:1–43.

Kuda, D. B., and J. S. Griffith. 1993. Establishment of Shoshone sculpin (*Cottus greenei*) in a spring inhabited by mottled sculpin (*C. bairdi*). Great Basin Naturalist 53(2):190–193.

Kuehne, R. A. 1955. Stream surveys of the Guadalupe and San Antonio rivers. Texas Game and Fish Commission, Inland Fisheries Report Series No. 1.

Kuhne, E. R. 1939. A guide to the fishes of Tennessee and the mid-South. Tennessee Department of Conservation, Nashville.

Kukowski, G. E. 1972. Southern range extension for the yellowfin goby, *Acanthogobius flavimanus* (Temminck and Schlegel). California Fish and Game 58(4):326–327.

Kullander, S. O. 1983. A revision of the South American cichlid genus *Cichlasoma* (Teleostei: Cichlidae). Swedish Museum of Natural History, Stockholm.

Kullander, S. O. 1986. Cichlid fishes of the Amazon River drainage of Peru. Swedish Museum of Natural History, Stockholm.

Kullander, S. O., and K. E. Hartel. 1997. The systematic status of cichlid genera described by Louis Agassiz in 1859: *Amphilophus*, *Baiodon*, *Hypsophrys* and *Parachromis* (Teleostei: Cichlidae). Ichthyological Exploration of Freshwaters 7(3):193–202.

Kullander, S. O., and H. Nijssen. 1989. The cichlids of Surinam, Teleostei: Labroidei. E. J. Brill, New York.

Kushlan, J. A. 1972. The exotic fish (*Aequidens portalegrensis*) in the Big Cypress Swamp. Florida Naturalist 45:29.

Kushlan, J. A. 1986. Exotic fishes of the Everglades: a reconsideration of proven impact. Environmental Conservation 13:67–69.

Kushlan, J. A., and T. E. Lodge. 1974. Ecological and distributional notes on the freshwater fish of southern Florida. Biological Sciences 37(2):110–128.

Kwain, W., and A. H. Lawrie. 1981. Pink salmon in the Great Lakes. Fisheries 6(2):2–6.

Lachner, E. A., and R. E. Jenkins. 1971. Systematics, distribution and evolution of the *Nocomis biguttatus* group (family Cyprinidae: Pisces) with a description of a new species from the Ozark Upland. Smithsonian Contributions to Zoology 91:1–28.

Lachner, E. A., C. R. Robins, and W. R. Courtenay, Jr. 1970. Exotic fishes and other aquatic organisms introduced into North America. Smithsonian Contributions to Zoology 59:1–29.

Ladiges, W., and D. Vogt. 1986. Guida di pesci d'acqua dolce d'Europa. Franco Muzzio Editore, Padova, Italy.

Laird, C. A., and L. M. Page. 1996. Non-native fishes inhabiting the streams and lakes of Illinois. Illinois Natural History Survey Bulletin 35(1):1–51.

Lambou, V. W. 1962. Fishes occurring in Lake Bistineau, Louisiana. Proceedings of the Louisiana Academy of Sciences 25:75–79.

Lampman, B. H. 1946. The coming of the pond fishes. Binfords and Mort, Portland, Oregon.

Lander, A., Jr. 1991. Woman snares piranha in Lake Cumberland. Lexington Herald-Leader (September 29): D14, Lexington, Kentucky.

Langford, F. H., F. J. Ware, and R. D. Gasaway. 1978. Status and harvest of introduced *Tilapia aurea* in Florida lakes. Pages 102–108 *in* R. O. Smitherman, W. L. Shelton, and J. H. Grover, editors. Symposium on the culture of exotic fishes. Fish Culture Section, American Fisheries Society, Bethesda, Maryland.

Lanse, R. I. 1965. The occurrence of warmouth, *Chaenobryttus gulosus* (Cuvier), in the lower Colorado River. California Fish and Game 51(2):123.

La Rivers, I. 1962. Fishes and fisheries of Nevada. Nevada State Fish and Game Commission, Carson City.

La Rivers, I., and T. J. Trelease. 1952. An annotated checklist of the fishes of Nevada. California Fish and Game 38(1):113–123.

Larkin, P. A., and S. B. Smith. 1954. Some effects of introduction of the redside shiner on the Kamloops trout in Paul Lake, British Columbia. Transactions of the American Fisheries Society 83:161–175.

Larsen, A. 1954. First record of the white perch (*Morone americana*) in Lake Erie. Copeia 1954(2):154.

Larsen, L. 1993. Peacock bass explosions! Larsen's Outdoor Publishing, Lakeland, Florida.

Lasswell, J. L., G. Garza, and W. H. Bailey. 1979. Status of marine fish introductions into the freshwaters of Texas. Proceedings of the 31st Annual Conference Southeastern Association of Fish and Wildlife Agencies 31(1977):399–403.

Lawrie, A. H. 1970. The sea lamprey in the Great Lakes. Transactions of the American Fisheries Society 99:766–775.

Leach, G. C. 1921. Distribution of fish and fish eggs during the fiscal year 1919. Appendix I to the Report of the U.S. Commission of Fisheries for 1919. Pp. 1–76.

Leach, G. C. 1923. Propagation and distribution of food fishes, 1922. Appendix XVII to the Report of the U.S. Commission of Fisheries for 1922. Bureau of Fisheries Document No. 941. Pp. 1–100.

Leary, R. F., F. W. Allendorf, and G. K. Sage. 1995. Hybridization and introgression between introduced and native fish. Pages 91–101 *in* H. L. Schramm, Jr., and R. G. Piper, editors. Uses and effects of cultured fishes in aquatic ecosystems. American Fisheries Society Symposium 15, Bethesda, Maryland.

Leberg, P. L., and R. C. Vrijenhoek. 1994. Variation among desert topminnows in their susceptibility to attack by exotic parasites. Conservation Biology 8:419–424.

Lee, D. S. 1976. Aquatic zoogeography of Maryland. Atlantic Naturalist 31(4):147–158.

Lee, D. S., and five coauthors. 1980 et seq. Atlas of North American freshwater fishes. North Carolina State Museum of Natural History, Raleigh. (Cited as a work rather than as individual accounts in the interest of space).

Lee, D. S., A. Norden, and C. R. Gilbert. 1984. Endangered, threatened, and extirpated freshwater fishes of Maryland. Pages 287–328 *in* A. Norden, D. Forester, and G. Fenwick, editors. Threatened and endangered plants and animals of Maryland. Maryland Department of Natural Resources, Annapolis, Maryland.

Lee, D. S., A. Norden, C. R. Gilbert, and R. Franz. 1976. A list of the freshwater fishes of Maryland and Delaware. Chesapeake Science 17(3):205–211.

Lee, D. S., S. P. Platania, C. R. Gilbert, R. Franz, and A. Norden. 1981. A revised list of the freshwater fishes of Maryland and Delaware. Southeastern Fishes Council Proceedings 3(3):1–10.

Lee, M. C. 1991. Angler hooks toothy catch. The Daily News, Jacksonville, NC, 18 October 1991. Pp. B1–2.

Legner, E. F., and R. A. Medved. 1972. Predators investigated for the biological control of mosquitoes and midges at the University of California, Riverside. Proceedings of the Mosquito Control Association 40:109–111.

Legner, E. F., R. A. Medved, and F. Pelsue. 1980. Changes in chironomid breeding patterns in a paved river channel following adaptation of cichlids of the *Tilapia mossambica-hornorum* complex. Annals of the Entomological Society of America 73:293–299.

Legner, E. F., and F. W. Pelsue. 1977. Adaptations of *Tilapia* to *Culex* and chironomid midge ecosystems in south California. Proceedings of the forty-fifth annual conference of the California Mosquito and Vector Control Association, Inc. 45:95–97.

Lemly, A. D. 1985. Suppression of native fish populations by green sunfish in first-order streams of Piedmont North Carolina. Transactions of the American Fisheries Society 114:705–712.

Leventer, H. 1987. The contribution of silver carp *Hypophthalmichthys molitrix* to the biological control of reservoirs. Mikoroth Water Company. Tel Aviv, Israel.

Lever, C. 1996. Naturalized Fishes of the World. Academic Press. San Diego, CA.

Levine, D. S., J. T. Krummrich, and P. L. Shafland. 1979. Renovation of a borrow pit in Levy County, Florida, containing Jack Dempseys (*Cichlasoma octofasciatum*). Non-native Fish Research Laboratory, Florida Game and Fresh

Water Fish Commission, Contribution 21. Boca Raton.

Leviton, A. E., R. H. Gibbs, Jr., E. Heal, and C. E. Dawson. 1985. Standards in herpetology and ichthyology, part I. Standard symbolic codes for institutional resource collections in herpetology and ichthyology. Copeia 1985(3):802–832.

Lewis, D., P. Reinthal, and J. Trendall. 1986. A guide to the fishes of Lake Malawi National Park. World Wildlife Fund, Gland, Switzerland.

Li, H. W., and P. B. Moyle. 1993. Management of introduced fishes. Pages 287–307 *in* C. C. Kohler and W. A. Hubert, editors. Inland fisheries management in North America. American Fisheries Society, Bethesda, Maryland.

Li, H. W., P. B. Moyle, and R. L. Garrett. 1976. Effect of the introduction of the Mississippi silverside (*Menidia audens*) on the growth of black crappie (*Pomoxis nigromaculatus*) and white crappie (*P. annularis*) in Clear Lake, California. Transactions of the American Fisheries Society 105:404–408.

Li, S., and F. Fang. 1990. On the geographical distribution of the four kinds of pond-cultured carps in China. Acta Zoologica Sinica 36(3):244–250.

Linder, A. D. 1963. Idaho's alien fishes. Tebiwa 6(2):12–15.

Linder, A. D. 1964. The guppy, *Lebistes reticulatus* (Peters), from a hot spring in Idaho. Copeia 1964(4):708–709.

Linke, H., and W. Staeck. 1994. African cichlids I: cichlids from West Africa. Tetra Press, Melle, Germany.

Loftus, W. F. 1987. Possible establishment of the Mayan cichlid, *Cichlasoma urophthalmus* (Günther) (Pisces: Cichlidae) in Everglades National Park, Florida. Florida Scientist 50:1–6.

Loftus, W. F., and J. A. Kushlan. 1987. Freshwater fishes of southern Florida. Bulletin of the Florida State Museum Biological Science 31(4):255.

Logan, D. J. 1994. A checklist of the fishes of Benton County, Oregon. American Currents 1994(Summer):16–18.

Logan, D. J., E. L. Bibles, and D. F. Markle. 1996. Recent collections of exotic aquarium fishes in the freshwaters of Oregon and thermal tolerance of Oriental weatherfish and pirapatinga. California Fish and Game 82(2):66–80.

Lohman, J. 1989. Biologists introduce zander into North America. The Forum, Fargo, ND, Saturday, July 22, 1989.

Loiselle, P. V. 1979. A revision of the genus *Hemichromis* Peters 1858 (Teleostei: Cichlidae). Museé Royal de L'Afrique Central Tervuren Belgique Annales Serie in 8 Sciences Zoologiques 228.

Loiselle, P. V. 1980a. The *Cichlasoma labiatum* species complex. Freshwater and Marine Aquarium 3(4)(April):30–35, 67.

Loiselle, P. V. 1980b. South American eartheaters *Geophagus* the genus and its allies. Freshwater and Marine Aquarium 3(6)(June):23–29, 86–87, 89–90.

Loiselle, P. V. 1992. An annotated key to the genus *Hemichromis* Peters 1958. Buntbarsche Bulletin (Journal of the American Cichlid Association) 148:2–19.

Lorenz, J. J., C. C. McIvor, G. V. Powell, and P. C. Frederick. 1997. A drop net and removable walkway used to quantitatively sample fishes over wetland surfaces in the dwarf mangroves of the southern Everglades. Wetlands 17:346–359.

Louisiana Department of Wildlife and Fisheries. 1997. Freshwater recreational fishing. (Available from http://www.wlf.state.la.us/wsdocs/freshrec.htm. Page accessed June 23, 1997.)

LoVullo, T. J., and J. R. Stauffer, Jr. 1993. The retail bait-fish industry in Pennsylvania-source of introduced species. Journal of the Pennsylvania Academy of Science 67(1):13–15.

Loyacano, H. A., Jr. 1975. A list of freshwater fishes of South Carolina. Bulletin of the South Carolina Experimental Station 580:1–8.

Ludlow, M. E., and S. J. Walsh. 1991. Occurrence of a South American armored catfish in the Hillsborough River, Florida. Florida Scientist 54(1):48–50

Ludwig, H. R., Jr., and J. A. Leitch. 1996. Interbasin transfer of aquatic biota via anglers' bait buckets. Fisheries 21(7):14–18.

Luebke, R. W. 1978. Evaluation of a multi-predator introduction. Federal Aid Project F-31-R-4. Texas Parks and Wildlife Department.

Lujan, V. A. 1994. Native nongame fish progress report for region 1: surveys performed July through October, 1994. Nevada Division of Wildlife.

Lynch, J. D. 1988a. Introduction, establishment, and dispersal of western mosquitofish in Nebraska (Actinopterygii: Poeciliidae). Prairie Naturalist 20:203–216.

Lynch, J. D. 1988b. Habitat utilization by an introduced fish, *Gambusia affinis*, in Nebraska (Actinopterygii: Poeciliidae). Transactions of the Nebraska Academy of Sciences 16:63–67.

Lynch, J. D. 1991. A footnote to history: a tale of two fish species in Nebraska. Nebraskaland (July 1991):50–55. Lincoln, Nebraska.

MacCrimmon, H. R., and B. L. Gots. 1979. World distribution of Atlantic salmon, *Salmo salar*. Journal of the Fisheries Research Board of Canada 36:422–457.

MacCrimmon, H. R., and T. L. Marshall. 1968. World distribution of brown trout, *Salmo trutta*. Journal of the Fisheries Research Board of Canada 25:2527–2548.

MacCrimmon, H. R., T. L. Marshall, and B. L. Gots. 1970. World distribution of brown trout, *Salmo trutta*: further observations. Journal of the Fisheries Research Board of Canada 27:811–818.

MacCrimmon, H. R., and W. H. Robbins. 1975. Distribution of black basses *in* North America. Pages 56–66 in R. H. Stroud, and H. Clepper, editors. Black Bass Biology and Management. Sport Fishing Institute, Washington, D.C.

Machado-Allison, A. 1982. Estudio sobre la subfamilia Serrasalminae (Teleostei, Characidae). Prte 1. Estudio comparado de los juveniles de las "cachamas" de Venezuela (generos *Colossoma y Piaractus*). Acta Biologica Venezuelica 11(3):1–101.

Maciolek, J. A. 1984. Exotic fishes in Hawaii and other islands of Oceania. Pages 131–161 *in* W. R. Courtenay, Jr. and J. R. Stauffer, Jr., editors. Distribution, biology, and management of exotic fishes. The Johns Hopkins University Press, Baltimore, Maryland.

Madsen, J. D., G. O. Dick, D. Honnell, J. Shearer, and R. M. Smart. 1994. Ecological assessment of Kirk Pond. U.S. Army Corps of Engineers, Waterways Experiment Station, Aquatic Plant Control Research Program. Miscellaneous Paper A-94-1, Vicksburg, Mississippi.

Magnan, P., and G. J. Fitzgerald. 1982. Resource partitioning between brook trout (*Salvelinus fontinalis* Mitchill) and creek chub (*Semotilus atromaculatus* Mitchill) in selected oligotrophic lakes of southern Quebec. Canadian Journal of Zoology 60:1612–1617.

Magnan, P., and G. J. Fitzgerald. 1984. Mechanisms responsible for the niche shift of brook char, *Salvelinus fontinalis* Mitchill, when living sympatrically with creek chub, *Semotilus atromaculatus* Mitchill. Canadian Journal of Zoology 62:1543–1555.

Maitland, P. S. 1977. The Hamlyn guide to freshwater fishes of Britain and Europe. Hamlyn Publishing Group Limited, New York.

Malick, R. W., Jr. 1978. The mimic shiner, *Notropis volucellus* (Cope), in the Susquehanna River drainage of Pennsylvania. Proceedings of the Pennsylvania Academy of Science 52(12):199–200.

Mandrak, N. E., and E. J. Crossman. 1992. A checklist of Ontario freshwater fishes. Royal Ontario Museum. Toronto, Canada.

Manooch, C. S., III. 1984. Fisherman's guide: fishes of the Southeastern United States. North Carolina State Museum of Natural History, Raleigh.

Mansueti, R. J. 1964. Eggs, larvae, and young of the white perch, *Roccus americanus*, with comments on its ecology in the estuary. Chesapeake Science 5(1–2):3–45.

Marchand, N. 1998. Regional news: settling in. Outdoor Life Magazine 201(3):96.

Marsden, J. E., and D. J. Jude. 1995. Round gobies invade North America. Fact sheet produced by Ohio State University Sea Grant, Columbus.

Marsh, P. C., and J. E. Brooks. 1989. Predation by ictalurid catfishes as a deterrent to re-establishment of hatchery-reared razorback suckers. Southwestern Naturalist 34(2):188–195.

Marsh, P. C., T. A. Burke, B. D. DeMarais, and M. E. Douglas. 1989. First North American record of *Cichlasoma managuense* (Pisces: Cichlidae). Great Basin Naturalist 49(3):387–389.

Marsh, P. C., and M. E. Douglas. 1997. Predation by introduced fishes on endangered humpback chub and other native species in the Little Colorado River, Arizona. Transactions of the American Fisheries Society 126:343–346.

Martin, C. 1995. Fly fishing in northwestern New Mexico: Rio Grande Gorge. (Available from http://www.unm.edu/~datkins/flyfish/areas/nmnw/riog_gorge.html. Page accessed October 31, 1997.)

Massette, B. 1993. Florida's next game fish–the pacu?. Florida Game and Fish, 1993(February):28–31, 60–61. Tallahassee.

Masuda, H., K. Amaoka, C. Araga, T. Uyeno, and T. Yoshino, editors. 1984. The fishes of the Japanese archipelago. Tokai University Press, Tokyo, Japan.

Matern, S. A., and K. J. Fleming. 1995. Invasion of a third Asian goby species, *Tridentiger bifasciatus*, into California. California Fish and Game 81(2):71–76.

Mather, F. 1889. Brown trout in America. Bulletin of the U.S. Fish Commission, 7(1887):21–22.

Mayden, R. L. 1989. Phylogenetic studies of North American minnows, with emphasis on the genus *Cyprinella* (Teleostei: Cypriniformes). University of Kansas Museum of Natural History, Miscellaneous Publication 80, Lawrence.

Mayden, R. L., editor. 1992. Systematics, historical ecology, and North American freshwater fishes. Stanford University Press, Stanford.

Mayden, R. L., B. M. Burr, L. M. Page, and R. R. Miller. 1992. The native freshwater fishes of North America. Pages 827–863 *in* R. L. Mayden, editor. Systematics, historical ecology, and North American freshwater fishes. Stanford University Press, Stanford.

Mayden, R. L., F. B. Cross, and W. T. Gorman. 1987. Distributional history of the rainbow smelt, *Osmerus mordax* (Salmoniformes: Osmeridae), in the Mississippi River Basin. Copeia 1987(4):1051–1054.

Mayden, R. L., and C. R. Gilbert. 1989. *Notropis ludibundus* (Girard) and *Notropis tristis* (Girard), replacement names for *N. stramineus* (Cope) and *N. topeka* (Gilbert) (Teleostei: Cypriniformes). Copeia 1989(4):1084–1089.

Mayden, R. L., and W J. Matthews. 1989. Zoogeographic implications of *Luxilus cardinalis* (Cyprinidae) and *Etheostoma radiosum* (Percidae) in the Washita River system, Oklahoma. Southwestern Naturalist 34(3):415–416.

McAda, C. W. 1983. Colorado squawfish, *Ptychocheilus lucius* (Cyprinidae), with a channel catfish, *Ictalurus punctatus* (Ictaluridae), lodged in its throat. Southwestern Naturalist 28(1):119–120.

McAffee, W. R. 1966a. Rainbow trout. Pages 192–215 *in* A. Calhoun, editor. Inland fisheries management. California Department of Fish and Game. Sacramento.

McAffee, W. R. 1966b. Golden trout. Pages 216–221 *in* A. Calhoun, editor. Inland fisheries management. California Department of Fish and Game. Sacramento.

McAffee, W. R. 1966c. Lahontan cutthroat trout. Pages 225–231 *in* A. Calhoun, editor. Inland fisheries management. California Department of Fish and Game. Sacramento.

McAffee, W. R. 1966d. Landlocked king salmon. Pages 294–295 *in* A. Calhoun, editor. Inland fisheries management. California Department of Fish and Game. Sacramento.

McAffee, W. R. 1966e. Landlocked Atlantic salmon. Pages 297–299 *in* A. Calhoun, editor. Inland fisheries management. California Department of Fish and Game. Sacramento.

McAffee, W. R. 1966f. Mountain whitefish. Pages 299–303 *in* A. Calhoun, editor. Inland fisheries management. California Department of Fish and Game. Sacramento.

McBride, F. T., R. I. Jones, and F. A. Harris. 1982. Growth and food habits of Roanoke bass in the Eno and Tar rivers, North Carolina. Proceedings of the 34th Annual Conference Southeastern Association of Fish and Wildlife Agencies 34(1980):341–348.

McCarraher, D. B., and R. W. Gregory. 1970. Adaptability and status of introductions of Sacramento perch, *Archoplites interruptus*, in North America. Transactions of the American

Fisheries Society 99:700–707.

McCosker, J. E. 1989. Freshwater eels (family Anguillidae) in California: current conditions and future scenarios. California Fish and Game 75(1):4–10.

McCullen, K. 1994. Alien trout puzzle Yellowstone biologists. Rocky Mountain News, 22 September 1994; p. 8A.

McDonald, M. 1887. Report on distribution of fish and eggs by the U.S. Fish Commission for the season of 1885–'86. Bulletin of the U. S. Fish Commission 6(1886):385–394.

McDonald, M. 1893. Report of the Commissioner for 1889 to 1891. Part XVII. U.S. Commission of Fish and Fisheries, Washington, D.C.

McGinnis, S. M. 1984. Freshwater fishes of California. University of California Press, Los Angeles.

McGowan, E. G. 1988. An illustrated guide to larval fishes from three North Carolina piedmont impoundments. Report by Carolina Power and Light Company, Shearon Harris Energy and Environmental Center, New Hill.

McKechnie, R. J. 1966a. Golden shiner. Pages 488–492 *in* A. Calhoun, editor. Inland fisheries management. California Department of Fish and Game. Sacramento.

McKechnie, R. J. 1966b. Spotted bass. Pages 366–370 *in* A. Calhoun, editor. Inland fisheries management. California Department of Fish and Game. Sacramento.

McKechnie, R. J. 1966c. Log perch. Pages 530–531 *in* A. Calhoun, editor. Inland fisheries management. California Department of Fish and Game. Sacramento.

McKechnie, R. J., and R. C. Tharratt. 1966. Green sunfish. Pages 399–401 *in* A. Calhoun, editor. Inland fisheries management. California Department of Fish and Game. Sacramento.

McLean, J., and J. J. Magnuson. 1977. Species interactions in percid communities. Journal of the Fisheries Research Board of Canada 34:1941–1951.

McLean, M. 1993. Ruffe (*Gymnocephalus cernuus*) fact sheet. Minnesota Sea Grant Program, Great Lakes Sea Grant Network, Duluth.

McMahon, T. E., and D. H. Bennett. 1996. Walleye and northern pike: boost or bane to northwest fisheries? Fisheries 21(8):6–13.

McPhail, J. D., and C. C. Lindsey. 1986. Zoogeography of the freshwater fishes of Cascadia (the Columbia system and rivers north to the Stikine). Pages 615–638 *in* C. H. Hocutt and E. O. Wiley, editors. The zoogeography of North American freshwater fishes. Wiley, New York.

Mearns, A. J. 1975. *Poeciliopsis gracilis* (Heckel), a newly introduced poeciliid fish in California. California Fish and Game 61(4):251–253.

Meek, S. E. 1904. The fresh-water fishes of Mexico north of the Isthmus of Tehuantepec. Field Columbian Museum, Zoology Series.

Meffe, G. K. 1983. Attempted chemical renovation of an Arizona spring brook for management of the endangered Sonoran topminnow. North American Journal of Fisheries Management 3:315–21.

Meffe, G. K. 1985. Predation and species replacement in American southeastern fishes: a case study. Southwestern Naturalist 30(2):173–187.

Meffe, G. K., D. A. Hendrickson, and W. L. Minckley. 1983. Factors resulting in decline of the endangered Sonoran topminnow *Poeciliopsis occidentalis* (Atheriniformes: Poeciliidae) in the United States. Biological Conservation 25(1983):135–159.

Meffe, G. K., and F. F. Snelson, Jr., editors. 1989. Ecology and evolution of livebearing fishes (Poeciliidae). Prentice Hall, Englewood Cliffs, New Jersey.

Meng, L., P. B. Moyle, and B. Herbold. 1994. Changes in abundance and distribution of native and introduced fishes of Suisun Marsh. Transactions of the American Fisheries Society 123:498–507.

Menhinick, E. F. 1991. The freshwater fishes of North Carolina. North Carolina Wildlife Resources Commission, Raleigh.

Merrick, J. R., and G. E. Schmida. 1984. Australian freshwater fishes: biology and management. Griffin Press, Netley, South Australia.

Mettee, M. F. 1978. The fishes of the Birmingham-Jefferson County region of Alabama (with ecologic and taxonomic notes). Geological Survey of Alabama Bulletin 115.

Mettee, M. F., E. O'Neil, R. D. Suttkus, and J. M. Pierson. 1987. Fishes of the Lower Tombigbee River System in Alabama and Mississippi. Geological Survey of Alabama Bulletin 107.

Mettee, M. F., P. E. O'Neil, and J. M. Pierson. 1996. Fishes of Alabama and the Mobile Basin. Oxmoor House, Inc. Birmingham, Alabama.

Metzger, R. J., and P. L. Shafland. 1984. Possible establishment of *Geophagus surinamensis* (Cichlidae) in Florida. Florida Scientist 47(3):201–203.

Michigan Department of Natural Resources. 1997. Michigan state record fish. (Available from http://www.dnr.state.mi.us/www/fish/pubs/pdfs/st_record.pdf. Page accessed June 26, 1997.)

Middlemas, K. 1994. Local angler hooks a peculiarity. The News Herald (September 25), Panama City, Florida.

Middleton, K. 1988. Man-eating fish found in river. Athens News Courier (September 1):1–2, Athens, Alabama.

Miley, W. W. 1978. Ecological impact of the pike killifish, *Belonesox belizanus* Kner (Poeciliidae), in southern Florida. Master's thesis. Florida Atlantic University, Boca Raton.

Miller, A. I., and L. G. Beckman. 1996. First record of predation on white sturgeon eggs by sympatric fishes. Transactions of the American Fisheries Society 125:338–340.

Miller, D. J., and R. N. Lea. 1972. Guide to the coastal marine fishes of California. Fish Bulletin of the California Department of Fish and Game 157:1–235. Sacramento.

Miller, D. L. 1983. Rare fishes of the Arkansas River drainage in Colorado. Pages 6–11 *in* Proceedings of the 18th Annual Meeting, Colorado-Wyoming Chapter, American Fisheries Society. Laramie, Wyoming.

Miller, D. L. 1984. Distribution, abundance, and habitat of the Arkansas darter *Etheostoma cragini* (Percidae) in Colorado. Southwestern Naturalist 29(3):496–499.

Miller, D. R. 1955. Two additions to Oklahoma's fish fauna from Red River in Bryan County. Proceedings of the Oklahoma Academy of Science 34(1953):33–34.

Miller, G. L., and S. C. Jorgenson. 1969. Seasonal occurrence and length of frequency distribution of some marine fishes of coastal Georgia. U.S. Fish and Wildlife Service, Data Report 35.

Miller, P. J. 1986. Gobiidae. Pages 1019–1085 *in* P. J. P. Whitehead, M.-L. Bauchot., J.-C. Hureau, J. Nielsen, E. Tortonese, editors. Fishes of the north-eastern Atlantic and the Mediterranean, volume III. United Nations Educational, Scientific and Cultural Organization, Paris, France.

Miller, R. J., and H. W. Robison. 1973. The fishes of Oklahoma. Oklahoma State University Press, Stillwater.

Miller, R. R. 1946. Distributional records for North American fishes, with nomenclatorial notes on the genus *Psenes*. Journal of the Washington Academy of Sciences 36(6):206–212.

Miller, R. R. 1952. Bait fishes of the lower Colorado River, from Lake Mead, Nevada, to Yuma, Arizona, with a key for identification. California Fish and Game 38(1):7–42.

Miller, R. R. 1957. Origin and dispersal of the alewife, *Alosa pseudoharengus*, and the gizzard shad, *Dorosoma cepedianum*, in the Great Lakes. Transactions of the American Fisheries Society 86:97–111.

Miller, R. R. 1966. Geographical distribution of Central American freshwater fishes. Copeia 1966(4):773–802.

Miller, R. R. 1968. Records of some native freshwater fishes transplanted into various waters of California, Baja California, and Nevada. California Fish and Game 54(3):170–179.

Miller, R. R. 1973. Two new fishes, *Gila bicolor snyderi* and *Catostomus fumeiventris*, from the Owens River basin, California. Occassional Papers of the University of Michigan Museum of Zoology, Number 667.

Miller, R. R. 1983. Checklist and key to the mollies of Mexico (Pisces: Poeciliidae, *Poecilia*, subgenus *Mollienesia*) Copeia 1983(3):817–822.

Miller, R. R., and J. R. Alcorn. 1946. The introduced fishes of Nevada, with a history of their introduction. Transactions of the American Fisheries Society 73:173–193.

Miller, R. R., and J. M. Fitzsimons. 1971. *Ameca splendens*, a new genus and species of Goodeid fish from western Mexico, with remarks on the classification of the Goodeidae. Copeia 1971(1):1–13.

Miller, R. R., C. Hubbs, and F. H. Miller. 1991. Ichthyological exploration of the American West: the Hubbs-Miller era, 1915–1950. Pages 19–40 *in* W. L. Minckley and J. E. Deacon, editors. Battle against extinction: native fish management in the American West. University of Arizona Press, Tucson.

Miller R. R., and C. L. Hubbs. 1969. Systematics of *Gasterosteus aculeatus* with particular reference to intergradation and introgression along the Pacific Coast of North America: a commentary on a recent contribution. Copeia 1969(1):52–69.

Miller, R. R., and C. H. Lowe. 1967. Fishes of Arizona. Pages 133–151 *in* C. H. Lowe, editor. The vertebrates of Arizona, part 2. University of Arizona Press, Tucson.

Miller, R. R., and E. P. Pister. 1971. Management of the Owens pupfish, *Cyprinodon radiosus*, in Mono County, California. Transactions of the American Fisheries Society 100:502–509.

Miller, R. R., J. D. Williams, and J. E. Williams. 1989. Extinctions of North American fishes during the past century. Fisheries 14(6):22–38.

Mills, D., and G. Vevers. 1982. The Golden encyclopedia of freshwater tropical aquarium fishes. Golden Press, Racine.

Mills, E. L., J. H. Leach, J. T. Carlton, and C. L. Secor. 1993. Exotic species in the Great Lakes: a history of biotic crisis and anthropogenic introductions. Journal of Great Lakes Research 19(1):1–54.

Mills, E. L., M. D. Scheuerell, J. T. Carlton, and D. L. Strayer. 1997. Biological invasions in the Hudson River basin: an inventory and historical analysis. New York State Museum Circular 57. Albany.

Milner, J. W. 1874a. Report on the propagation of shad (*Alosa sapidissima*) and its introduction into new waters by the U.S. Fish Commissioner in 1873. Part XIX. Report of the Commissioner for Fish and Fisheries for 1872 and 1873. Pp. 419–451.

Milner, J. W. 1874b. Transfer of living fishes. Part XXI, D-2. Report of the Commissioner for Fish and Fisheries for 1872 and 1873. Pp. 524–528.

Minckley, W. L. 1969. Attempted re-establishment of the Gila topminnow within its former range. Copeia 1969(1):193–194.

Minckley, W. L. 1973. Fishes of Arizona. Arizona Fish and Game Department. Phoenix.

Minckley, W. L., and J. E. Deacon. 1968. Southwestern fishes and the enigma of "endangered species." Science 159:1424–1432.

Minckley, W. L., and J. E. Deacon, editors. 1991. Battle against extinction: native fish management in the American West. University of Arizona Press, Tucson.

Minckley, W. L., D. A. Hendrickson, and C. E. Bond. 1986. Geography of western North American freshwater fishes: description and relationships to intracontinental tectonism. Pages 529–614 *in* C. H. Hocutt and E. O. Wiley, editors. The zoogeography of North American freshwater fishes. Wiley, New York.

Minckley, W. L., and L. A. Krumholz. 1960. Natural hybridization between the clupeid genera *Dorosoma* and *Signalosa*, with a report on the distribution of *S. petenensis*. Zoologica 44(4):171–180.

Minckley, W. L., and R. M. McNatt. 1974. Fishes of Hunter's Hole, Yuma County, Arizona. Arizona State University, Tempe.

Minckley, W. L., G. K. Meffe, and D. L. Soltz. 1991. Conservation and management of short-lived fishes: the cyprinodontoids. Pages 247–282 *in* W. L. Minckley and J. E. Deacon, editors. Battle against extinction: native fish management in the American West. University of Arizona Press, Tucson.

Moczygemba, J. H., and D. J. Morris. 1977. Statewide striped bass study for Texas. Texas Parks and Wildlife. Federal Aid Project F-13-R-3.

Moe, M. A. 1964. Survival potential of piranhas in Florida. Quarterly Journal of the Florida Academy of Sciences 27:197–210.

Moffett, J. W. 1943. A preliminary report on the fishery of Lake Mead. Transactions of the Eighth North American Wildlife Conference 1943:179–186.

Moore, G. A. 1968. Fishes. Pages 22–165 *in* W. F. Blair, editor. Vertebrates of the United States,

2nd edition. McGraw-Hill, New York.

Moore, G. A., and C. D. Riggs. 1963. Checklist of known Oklahoma fishes. Pages 41–44 *in* Oklahoma Wildlife Commission. Know your Oklahoma fishes. Department of Wildlife Conservation, Oklahoma City.

Moore, H. H., and R. A. Bream. 1965. Distribution of fishes in U.S. streams tributary to Lake Superior. U.S. Fish and Wildlife Service Special Scientific Report 516.

Moore, R. H., R. A. Garrett, and P. J. Wingate. 1976. Occurrence of the red shiner, *Notropis lutrensis*, in North Carolina: a probable aquarium release. Transactions of the American Fisheries Society 105:220–221.

Morgan, A. R., and A. R. Gerlach. 1950. Striped bass studies on Coos Bay in 1949 and 1950. Report of the Oregon Fish Commission and Oregon Game Commission to the Forty-sixth Legislature.

Morgan, M. D., S. T. Threlkeld, and C. R. Goldman. 1978. Impact of the introduction of kokanee (*Oncorhynchus nerka*) and opossum shrimp (*Mysis relicta*) on a subalpine lake. Journal of the Fisheries Research Board of Canada 35: 1572–1579.

Morita, C. M. 1981. Freshwater fishing in Hawaii. Division of Aquatic Resources, Department of Land and Natural Resources, Honolulu, Hawaii.

Morman, R. H., D. W. Cuddy, and P. C. Rugen. 1980. Factors influencing the distribution of sea lamprey (*Petromyzon marinus*) in the Great Lakes. Canadian Journal of Fisheries and Aquatic Sciences 37:1811–1826.

Morris, J., L. Morris, and L. Witt. 1974. The fishes of Nebraska. Nebraska Game and Parks Commission, Lincoln.

Morrow, J. E. 1980. The freshwater fishes of Alaska. Alaska Northwest Publishing Company, Anchorage.

Morse, S. R. 1905. Fresh and salt water fish found in the waters of New Jersey, part I. Annual Report of the New Jersey State Museum. MacCrellish and Quigley, State Province, Trenton.

Moser, M. L. and S. W. Ross. 1993. Distribution and movements of shortnose sturgeon (*Acipenser brevirostrum*) and other anadromous fishes of the lower Cape Fear River, North Carolina. Final report to the U.S. Army Corps of Engineers, Wilmington District. Wilmington, North Carolina.

Moyle, J. B., and W. D. Clothier. 1959. Effects of management and winter oxygen levels on the fish population of a prairie lake. Transactions of the American Fisheries Society 88:178–185.

Moyle, P. B. 1976a. Inland fishes of California. University of California Press, Berkeley.

Moyle, P. B. 1976b. Fish introduction in California: history and impact on native fishes. Biological Conservation 9:101–118.

Moyle, P. B. 1993. Assessing effects on ecosystem function, structure and resilience. Performance standards workshop, August 18–20, 1993, University of Minnesota.

Moyle, P. B., and R. A. Daniels. 1982. Fishes of the Pit River System, McCloud River System, and Surprise Valley Region. University of California Publications in Zoology 115:1–82.

Moyle, P. B., F. W. Fisher, and H. Li. 1974a. Mississippi silversides and logperch in the Sacramento–San Joaquin River system. California Fish and Game 60(2):145–147.

Moyle, P. B., and N. J. Holzhauser. 1978. Effects of the introduction of Mississippi silverside (*Menidia audens*) and Florida largemouth bass (*Micropterus salmoides floridanus*) on the feeding habits of young-of-year largemouth bass in Clear Lake, California. Transactions of the American Fisheries Society 107:574–582.

Moyle, P. B., and A. Marciochi. 1975. Biology of the Modoc sucker, *Catostomus microps* (Pisces: Catostomidae) in northeastern California. Copeia 1975:556–560.

Moyle, P. B., S. B. Matthews, and N. Bonderson. 1974b. Feeding habits of the Sacramento perch, *Archoplites interruptus*. Transactions of the American Fisheries Society 103(2):399–402.

Moyle, P. B., and R. Nichols. 1973. Ecology of some native and introduced fishes of the Sierra Nevada foothills in central California. Copeia 1973(3):478–490.

Moyle, P. B., and R. Nichols. 1974. Decline of the native fish fauna of the Sierra-Nevada foothills, central California. American Midland

Naturalist 92(1):72–83.

Mulvey, M. 1990. Voracious carp seen in area waterways. Milwaukee Sentinel, 31 May 1990.

Muoneke, M. I. 1988. Tilapia in Texas waters. Texas Parks and Wildlife Department, Inland Fisheries Data Series 9, Austin.

Muoneke, M. I. 1990. Summary of biological information on rudd *Scardinius erythrophthalmus*. Texas Parks and Wildlife Department, Management Data Series, Austin.

Murdy, E. O., R. S. Birdson, and J. A. Musick. 1997. Fishes of Chesapeake Bay. Smithsonian Institution Press, Washington, D.C.

Musick, J. A. 1972. Fishes of Chesapeake Bay and the adjacent coastal plain. Pages 175–212 *in* M. L. Wass, editor. A checklist of the biota of lower Chesapeake Bay. Virginia Institute of Marine Science Special Scientific Report 65.

Muus, B. J., and P. Dahlstrom. 1978. Collins guide to the freshwater fishes of Britain and Europe. Collins, London.

Myers, G. S. 1925. Introduction of the European bitterling (*Rhodeus*) in New York and of the rudd (*Scardinius*) in New Jersey. Copeia 140(1925):20–21.

Myers, G. S. 1940. An American cyprinodont fish, *Jordanella floridae*, reported from Borneo, with notes on the possible widespread introduction of foreign aquarium fishes. Copeia 1940(4):267–268.

Myers, G. S. 1967. *Gambusia*, the fish destroyer. Australian Zoologist 13(2):102.

Naiman, R. J., and E. P. Pister. 1974. Occurrence of the tiger barb, *Barbus tetrazona*, in the Owens Valley, California. California Fish and Game 60(2):100–101.

Neale, G. 1931. Spiny-rayed fresh water game fishes of California inland waters. California Fish and Game 17(1):1–16.

Nelson, J. 1890. Descriptive catalogue of the vertebrates of New Jersey. Geological Survey of New Jersey 1890:489–824.

Nelson, J. S. 1969. Geographic variation in the brook stickleback, *Culaea inconstans*, and notes on nomenclature and distribution. Journal of the Fisheries Research Board of Canada 26:2431–2447.

Nelson, J. S. 1984. Fishes of the world, 2nd edition. Wiley, New York.

Nelson, J. S. 1994. Fishes of the world, 3rd edition. Wiley, New York.

Nelson, J. S., E. J. Crossman, H. Espinosa-Pérez, C. R. Gilbert, R. N. Lea, and J. D. Williams. 1998. Recommended changes in common fish names: pikeminnow to replace squawfish (*Ptychocheilus* spp.). Fisheries (9):37.

Nelson, J. S., and S. D. Gerking. 1968. Annotated key to the fishes of Indiana. Indiana State University, Department of Zoology, Indiana Aquatic Research Unit. Project 342–303–815. Bloomington.

New Jersey Division of Fish, Game and Wildlife. 1997. New Jersey state record freshwater sport fish. (Available from http://www.state.nj.us/dep/fgw/freshrec.htm. Page accessed June 27, 1997.)

Newman, R. M. 1991. The invasion of foreign aquatic plants and animals. Minnesota-Out-Of-Doors 1991(May/June):16–17, 29.

Nichols, J. T. 1943. The freshwater fishes of China. Natural history of central Asia volume IX. The American Museum of Natural History, New York.

Nico, L. G., and P. L. Fuller. 1999. Spatial and temporal patterns of nonindigenous fish introductions in the United States. Fisheries 24(1):16–27.

Nico, L. G., and D. C. Taphorn. 1986. Those bitin' fish from Venezuela. Tropical Fish Hobbyist 34(4): 24–27, 30–34, 36, 40–41, 56–57. Neptune City, New Jersey.

Nico, L. G., S. J. Walsh, and R. H. Robins. 1996. An introduced population of the South American callichthyid catfish *Hoplosternum littorale* in the Indian River Lagoon system, Florida. Florida Scientist 59(3):189–200.

Nico, L. G., and J. D. Williams. 1996. Risk assessment on black carp (Pisces: Cyprinidae). Final Report to the Risk Assessment and Management Committee of the Aquatic Nuisance Species Task Force. U.S. Geological Survey, Biological Resources Division, Gainesville, Florida.

Nijssen, H., and I. J. H. Isbrücker. 1980. A review of the genus *Corydoras* Lacepède, 1803 (Pisces, Siluriformes, Callichthyidae). Bijdragen tot de Dierkunde 50(1):190–220.

Nikolski, G. V. 1956. Fishes of the Amur Basin. Moscow Academy of Science U.S.S.R., Moscow.

Noble, R. L., and R. D. Germany. 1986. Changes in fish populations of Trinidad Lake, Texas, in response to abundance of blue tilapia. Pages 455–461 *in* R. H. Stroud, editor. Fish culture in fisheries management. American Fisheries Society, Fish Culture Section and Fisheries Management Section, Bethesda, Maryland.

Norman, J. R. 1958. A history of fishes. Hill and Wang, New York.

Norris, S. M., R. J. Miller, and M. E. Douglas. 1988. Distribution of *Ctenopoma muriei* and the status of *Ctenopoma ctenotis* (Pisces: Anabantidae). Copeia 1988(2):487–491.

North American Native Fish Association. 1978. Another threat to natives? The Lateral Line (North American Native Fish Association Newsletter), June–July: 1, Semmes, Alabama.

North Dakota Game and Fish Department. 1994. Fishes of the Dakotas. Brochure. North Dakota Game and Fish Department, Bismark.

North Dakota Game and Fish Department. 1997. North Dakota record fish. (Available from http://www.state.nd.us.gnf.fish.html. Page accessed June 26, 1997.)

Nyman, O. L. 1970a. Electrophoretic analysis of hybrids between salmon (*Salmo salar* L.) and trout (*Salmo trutta* L.). Transactions of the American Fisheries Society 99:229–236.

Nyman, O. L. 1970b. Ecological interaction of brown trout (*Salmo trutta*) L., and brook trout (*Salvelinus fontinalis*) (Mitchill), in a stream. Canadian Field-Naturalist 84:343–350.

O'Donnell, D. J. 1935. Annotated list of the fishes of Illinois. Illinois Natural History Survey Bulletin 20(5):473–500.

O'Leary, J., and D. G. Smith. 1987. Occurrence of the first migration of the gizzard shad, (*Dorosoma cepedianum*), in the Connecticut River, Massachusetts. U.S. National Marine Service Fishery Bulletin 85(2):380–383.

O'Malley, H. 1920. The distribution of fish and fish eggs during the fiscal year 1918. Appendix I. Bulletin of the U.S. Fish Commission for 1918, pp. 1–82.

O'Malley, H. 1919. The distribution of fish and fish eggs during the fiscal year 1917. Appendix I. Bulletin of the U.S. Fish Commission for 1917, pp. 1–99.

Ogilvie, V. E. 1969. Illustrated checklist of fishes collected from the L-15 Canal (Lake Worth Drainage District) in Palm Beach County, Florida (collection date November 8, 1969). Unpublished Report for the Florida Game and Fresh Water Fish Commission.

Ogle, D. H., J. H. Selgeby, J. F. Savino, R. M. Newman, and M. G. Henry. 1995. Diet and feeding periodicity of ruffe in the St. Louis River estuary, Lake Superior. Transactions of the American Fisheries Society 124:356–369.

Ogle, D. H., J. H. Selgeby, J. F. Savino, R M. Newman, and M. G. Henry. 1996. Predation on ruffe by native fishes of the St. Louis River estuary, Lake Superior, 1989–1991. North American Journal of Fisheries Management 16:115–123.

Ohio Department of Natural Resources, Division of Wildlife. 1997. Fishing prospects reports 1997. (Available from http://www.dnr.ohio.gov/odnr/wildlife/fishpro/fishpro.html. Page accessed June 23, 1997.)

Oklahoma Wildlife Commission. 1963. Know your Oklahoma fishes. Department of Wildlife Conservation, Oklahoma City, OK.

Ono, R. D., J. D. Williams, and A. Wagner. 1983. Vanishing fishes of North America. Stone Wall Press, Inc., Washington, D.C.

Owen, J. B., D. S. Elsen, and G. W. Russell. 1981. Distribution of fishes in North and South Dakota basins affected by the Garrison Diversion Unit. Fisheries Research Unit, University of North Dakota, Grand Forks.

Page, L. M. 1983. Handbook of darters. Tropical Fish Hobbyist Publications, Inc., Neptune City, New Jersey.

Page, L. M. 1994. Identification of sailfin catfishes introduced to Florida. Florida Scientist 57(4):171–172.

Page, L. M., and B. M. Burr. 1991. A field guide to freshwater fishes of North America north of

Mexico. The Peterson Field Guide Series, volume 42. Houghton Mifflin, Boston.

Page, L. M., and C. A. Laird. 1993. The identification of the nonnative fishes inhabiting Illinois waters. Report prepared by Center for Biodiversity, Illinois Natural History Survey, Champaign, for Illinois Department of Conservation. Center for Biodiversity Technical Report 1993(4). Springfield

Parenti, L. R. 1981. A phylogenetic and biogeographic analysis of cyprinodontiform fishes (Teleostei, Atherinomorpha). Bulletin of the American Museum of Natural History 168(4):335–557.

Parrish, D. L., and F. J. Margraf. 1990. Interactions between white perch (*Morone americana*) and yellow perch (*Perca flavescens*) in Lake Erie as determined from feeding and growth. Canadian Journal of Fisheries and Aquatic Sciences 477:1779–1787.

Parrish, D. L., and F. J. Margraf. 1994. Spatial and temporal patterns of food use by white perch and yellow perch in Lake Erie. Journal of Freshwater Ecology 9(1):29–35.

Parsons, J. W. 1973. History of salmon in the Great Lakes, 1850–1970. U.S. Bureau of Sport Fisheries and Wildlife Technical Paper 68.

Pauley, G. B., and eight coauthors. 1994. An overview of the use and efficacy of triploid grass carp *Ctenopharyngodon idella* as a biological control of aquatic macrophytes in Oregon and Washington state lakes. Pages 46–54 *in* Proceedings of the grass carp symposium. U.S. Army Corps of Engineers, Waterways Experiment Station, Vicksburg, Mississippi.

Pearson, J. G., and F. P. Ward. 1972. A new record of the bowfin, *Amia calva* Linnaeus, in the upper Chesapeake Bay. Chesapeake Science 13(4):323–324.

Pearson, W. D., and L. A. Krumholz. 1984. Distribution and status of Ohio River fishes. Oak Ridge National Laboratory, ORNL/sub/79-7831/1. Oak Ridge, Tennessee.

Pelgren, D. W., and K. D. Carlander. 1971. Growth and reproduction of yearling *Tilapia aurea* in Iowa ponds. Proceedings of the Iowa Academy of Science 78:27–29.

Pelzman, R. J. 1972. Evaluation of introduction of *Tilapia sparrmanii* into California. California Department of Fish and Game, Inland Fisheries Administrative Report 72–3. Unpublished mimeograph.

Pelzman, R. J. 1973. A review of the life history of *Tilapia zilli* with a reassessment of its desirability in California. California Fish and Game Inland Fish Report 74(1):1–9.

Pennington, C. H., G. J. Dahl, and H. L. Schramm, Jr. 1982. Occurrence of the rainbow smelt in the lower Mississippi River. Journal of the Mississippi Academy of Sciences 27:41–42.

Pennsylvania Fish and Boat Commission. 1997. Stocking schedule for warmwater fishes. (Available from http://www.state.pa.us/Fish/ww97ind.htm. Page accessed June 24, 1997.)

Peterson, M. S., and S. T. Ross. 1987. Morphometric and meristic characteristics of a peripheral population of *Enneacanthus*. Proceedings of the Southeastern Fishes Council 17:1–4.

Petrovicky, I. 1988. Aquarium fish of the world. Hamlyn, London.

Pflieger, W. L. 1971. A distributional study of Missouri fishes. University of Kansas Museum of Natural History Publication 20(3):225–570.

Pflieger, W. L. 1975. The fishes of Missouri. Missouri Department of Conservation, Jefferson City.

Pflieger, W. L. 1978. Distribution and status of the grass carp (*Ctenopharyngodon idella*) in Missouri streams. Transactions of the American Fisheries Society 107:113–118.

Pflieger, W. L. 1997. The fishes of Missouri. Missouri Department of Conservation, Jefferson City.

Pflieger, W. L., and O. Fajen. 1970. Natural hybrids, smallmouth bass × spotted bass. Abstracts Fisheries Resource Reports Missouri Department of Conservation Division of Fisheries 9:37.

Philippart, J.-Cl., and J.-Cl. Ruwet. 1982. Ecology and distribution of tilapias. Pages 15–59 *in* R. S. V. Pullin, and R. H. Lowe-McConnell, editors. The biology and culture of tilapias. ICLARM (International Center for Living Aquatic Resources Management), conference proceedings 7, Manila, Philippines.

Phillips, G. L., W. D. Schmid, and J. C. Underhill. 1982. Fishes of the Minnesota region. University of Minnesota Press, Minneapolis.

Phillips, R., and M. Rix. 1985. Freshwater fish of Britain, Ireland and Europe. Pan Books, London.

Pigg, J. 1978. The tilapia *Sarotherodon aurea* (Steindachner) in the North Canadian River in central Oklahoma. Proceedings of the Oklahoma Academy of Science 58:111–112.

Pigg, J., R. Gibbs, and T. Beard. 1996. Observations on two exotic fish species in Oklahoma. Proceedings of the Oklahoma Academy of Science 76:90.

Pigg, J., R. Gibbs, and G. R. Luttrell. 1995. Distribution of the Red River pupfish *Cyprinodon rubrofluviatilis* Fowler, in the South Canadian River in Texas and Oklahoma. Proceedings of the Oklahoma Academy of Science 75:59–60.

Pigg, J., R. Gibbs, and J. Stahl. 1992. Distribution records for three new introduced species of the ichthyofauna of Oklahoma waters. Proceedings of the Oklahoma Academy of Science 72:1–2.

Pigg, J., and T. Pham. 1990. The rudd, *Scardinius erythrophthalmus*, a new fish in Oklahoma waters. Proceedings of the Oklahoma Academy of Science 70:37.

Pigg, J., J. Smith, and M. Ambler. 1997. Additional records of bighead carp, *Hypophthalmichthys nobilis*, in Oklahoma waters. Proceedings of the Oklahoma Academy of Science 77:123.

Pigg, J., J. Stahl, M. Ambler, and J. Smith. 1993. Two potential sources of exotic fish in Oklahoma. Proceedings of the Oklahoma Academy of Science 73:67.

Pimental, R., R. V. Bulkley, and H. M. Tyus. 1985. Choking of Colorado squawfish, *Ptychocheilus lucius* (Cyprinidae), on a channel catfish, *Ictalurus punctatus* (Ictaluridae), as a cause of mortality. Southwestern Naturalist 30(1):154–158.

Pipas, J. C. 1996. Distribution of the redeye bass (*Micropterus coosae*) in Tennessee and its hybridization with smallmouth bass (*Micropterus dolomieu*). Master's thesis. Tennessee Technological University, Cookeville.

Pipas, J. C., and F. J. Bulow. 1998. Hybridization between redeye bass and smallmouth bass in Tennessee streams. Transactions of the American Fisheries Society 127:141–146.

Pister, E. P. 1974. Desert fishes and their habitats. Transactions of the American Fisheries Society 103:531–540.

Platania, S. P. 1991. Fishes of the Rio Chama and upper Rio Grande, New Mexico, with preliminary comments on their longitudinal distribution. Southwestern Naturalist 36(2):186–193.

Plosila, D. S., L. J. Nashett. 1990. First reported occurrence of white perch in Lake Champlain. New York Department of Environmental Conservation, Bureau of Fisheries. Albany.

Polson, J. 1964. New reservoirs new fish species. Kansas Fish and Game 21(3):3–6.

Popov, B. H., and J. B. Low. 1953. Game, fur animal, and fish introductions into Utah. Utah State Department of Fish and Game Publication 4, pp. 1–85.

Poss, S. G., and R. R. Miller. 1983. Taxonomic status of the plains killifish *Fundulus zebrinus*. Copeia 1983(1):55–67.

Power, G. J., and F. Ryckman. 1998. Status of North Dakota's fishes. North Dakota Game and Fish Department, Division Report 27, Jamestown, North Dakota (Available from http://www.npwrc.org/resource/othrdata/fshries/fshries.htm [version 2 April 1998]. Page accessed 4 November 1998.)

Pratt, D. M., W. H. Blust, and J. H. Selgeby. 1992. Ruffe, *Gymnocephalus cernuus*: newly introduced in North America. Canadian Journal of Fisheries and Aquatic Sciences 49:1616–1618.

Prentice, J. A. 1977. Texas statewide walleye stocking evaluation. Texas Parks and Wildlife Department. Federal Aid Project F-31-R-3.

Prentice, J. A. 1985. Texas statewide walleye fishery management program. Texas Parks and Wildlife Department. Federal Aid Project F-31-R-11.

Priegel, G. R. 1963. Dispersal of the shortnose gar, *Lepisosteus platostomus*, into the Great Lakes drainage. Transactions of the American Fisheries Society 92:178.

Pritchard, D. L., O. D. May, Jr., and L. Rider. 1978. Stocking of predators in the predator-stocking-evaluation reservoirs. Proceedings of the 30th Annual Conference Southeastern Association of Fish and Wildlife Agencies 30(1976):108–

113.

Propst, D. L., G. L. Burton, and B. H. Pridgeon. 1987. Fishes of the Rio Grande between Elephant Butte and Caballo reservoirs, New Mexico. Southwestern Naturalist 32(3):408–411.

Propst, D. L., and C. A. Carlson. 1986. The distribution and status of warmwater fishes in the Platte River drainage, Colorado. Southwestern Naturalist 31(2):149–167.

Quadri, S. U. 1974. Taxonomic status of the *Salvelinus alpinus* complex. Journal of the Fisheries Research Board of Canada 31:1355–1361.

Quinn, S. P. 1988. Stomach contents of flathead catfish in the Flint River, Georgia. Proceedings of the 41st Annual Conference Southeastern Association of Fish and Wildlife Agencies 41(1987):85–92.

Raasch, M. S., and V. L. Altemus, Sr. 1991. Delaware's freshwater and brackish water fishes – a popular account. Delaware State College. Dover.

Radcliffe, L., and W. W. Welsh. 1916. A list of the fishes of the Seneca Creek, Montgomery County, Maryland, region. Proceedings of the Biological Society of Washington 29:39–46.

Radonski, G. C., N. S. Prosser, R. G. Martin, and R. H. Stroud. 1984. Exotic fishes and sport fishing. Pages 313–321 *in* W. R. Courtenay, Jr. and J. R. Stauffer, editors. Distribution, biology and management of exotic fishes. Johns Hopkins University Press, Baltimore, Maryland.

Radtke, R.L. 1995. Forensic biological pursuits of exotic fish origins: piranha in Hawaii. Environmental Biology of Fishes 43:393–399.

Raesly, R. L., J. R. Stauffer, Jr., and R. F. Denoncourt. 1990. Hybridization between *Etheostoma zonale* and *Etheostoma olmstedi* (Teleostei: Percidae), following an introduction event. Copeia 1990(2):584–588.

Raibley, P. T., D. Blodgett, and R. E. Sparks. 1995. Evidence of grass carp (*Ctenopharyngodon idella*) reproduction in the Illinois and upper Mississippi rivers. Journal of Freshwater Ecology 10(1):65–74.

Raicu, P., E. Taisescu, and P. Banarescu. 1981. *Carassius carassius* and *C. auratus*, a pair of diploid and tetraploid representative species (Pisces, Cyprinidae). Cytologia 46:233–240.

Raloff, J. 1992. Exotic intruders. Science News 142(4):56–58.

Ramirez-Hernández, E., and A. Gonzales Pages. 1976. Catalogo de peces marinos Mexicanos. Mexico, D. F.: Secretaría de Industria y Comercio, Instituto Nacional de Pesca. Mexico City, Mexico.

Ramsey, J. S. 1965. Zoogeographic studies on the freshwater fish fauna of rivers draining the southern Appalachian region. Doctoral dissertation. Tulane University, New Orleans, Louisiana.

Randall, J. E. 1987. Introductions of marine fishes to the Hawaiian Islands. Bulletin of Marine Science 41(2):490–502.

Raney, E. C. 1938. The distribution of the fishes of the Ohio drainage basin of western Pennsylvania. Doctoral dissertation. Cornell University, Ithaca.

Raquel, P. F. 1992. Record of the alligator gar (*Lepisosteus spatula*) from the Sacramento–San Joaquin Delta. California Fish and Game 78(4):169–171.

Rauchenberger, M. 1989a. Systematics and biogeography of the genus *Gambusia* (Cyprinodontiformes: Poeciliidae). American Museum Novitates 2951:1–74.

Rauchenberger, M. 1989b. Annotated species list of the subfamily Poeciliinae. Pages 359–367 *in* G. K. Meffe and F. F. Snelson, Jr., editors. Ecology and evolution of livebearing fishes (Poeciliidae). Prentice Hall, Englewood Cliffs, New Jersey.

Ravenel, W. C. 1896. Report on the propagation and distribution of food-fishes. Pages 6–72 *in* Report of the Commissioner for the year ending June 30, 1895, Part XXI. U.S. Commission of Fish and Fisheries, Washington, D.C.

Ravenel, W. C. 1898. Report on the propagation and distribution of food-fishes. Pages 11–92 *in* Report of the Commissioner for the year ending June 30, 1896, Part XXII. U.S. Commission of Fish and Fisheries, Washington, D.C.

Ravenel, W. C. 1902. Report on the propagation and distribution of food-fishes. Pages 21–110 *in* Report of the Commissioner for the year ending June 30, 1901, Part XXVII. U.S. Commission of Fish and Fisheries, Washington, D.C.

Ray, J. 1994. Record oscar caught in Okeechobee. Florida Wildlife 48(3):47. Tallahassee.

Redding, J. D. 1884. Character of the carp introduced by Capt. Henry Robinson about 1830. Bulletin of the U.S. Fish Commission 4(1884):266–267.

Rees, D. M. 1934. Notes on mosquito fish in Utah, *Gambusia affinis* (Baird and Girard). Copeia 1934(4):157–159.

Rees, D. M. 1945. Supplemental notes on mosquito fish in Utah, *Gambusia affinis* (Baird and Girard). Copeia 1945(4):236.

Reis, R. E. 1997. Revision of the neotropical catfish genus *Hoplosternum* (Ostariophysi: Siluriformes: Callichthyidae), with the description of two new genera and three new species.

Rhode, F. C., R. G. Arndt, D. G. Lindquist, and J. F. Parnell. 1994. Freshwater fishes of the Carolinas, Virginia, Maryland, and Delaware. University of Carolina Press, Chapel Hill.

Ribbink, A. J., B. A. Marsh, A. C. Marsh, A. C. Ribbink, and B. J. Sharp. 1983. A preliminary survey of the cichlid fishes of rocky habitats in Lake Malawi. South African Journal of Zoology (Zoology Dierkunde) 18(3):149–310.

Ricciardi, A. 1994. Black carp and zebra mussel debate continues. Fisheries 19(9):34.

Richardson, F., and W. C. Carnes. 1964. Survey and classification of the New River and tributaries, North Carolina. North Carolina Wildlife Resources Commission, Federal Aid in Sport Fish Restoration, Project F-14-R, Job I-O. Final Report, Raleigh, NC.

Richardson, M. J., F. G. Whoriskey, and L. H. Roy. 1995. Turbidity generation and biologica impacts of an exotic fish *Carassius auratus*, introduced into shallow seasonally anoxic ponds. Journal of Fish Biology 47:576–585.

Richardson, W. M., J. A. St. Amant, L. J. Bottroff, and W. L. Parker. 1970. Introduction of blue catfish into California. California Fish and Game 56(4):311–312.

Riggs, C. D. 1954. The occurrence of *Astyanax fasciatus mexicanus* in Lake Texoma, Oklahoma. Proceedings of the Oklahoma Academy of Science 33(1952):141.

Riggs, C. D., and G. A. Moore. 1963. A new record of *Moxostoma macrolepidotum pisolabrum*, and a range extension for *Percina shumardi*, in the Red River, Oklahoma and Texas. Copeia 1963(2):451–452.

Rinne, J. N. 1995. The effects of introduced fishes on native fishes: Arizona, southwestern United States. Pages 149–159 *in* D. P. Philipp, J. M. Epifanio, J. E. Marsden, and J. E. Claussen, editors. Protection of aquatic biodiversity. proceedings of the World Fisheries Congress, Theme 3. Science Publishers Inc., Lebanon, New Hampshire.

Rinne, J. N., and W. L. Minckley. 1985. Patterns of variation and distribution in Apache trout (*Salmo apache*) relative to co-occurrence with introduced salmonids. Copeia 1986(2):285–292.

Rivas, L. R. 1965. Florida fresh water fishes and conservation. Quarterly Journal of the Florida Academy of Sciences 28(3):255–258.

Rivas, L. R. 1980. Eight new species of poeciliid fishes of the genus *Limia* from Hispaniola. Northeast Gulf Science 4:28–38.

Rivas, L. R., and W. L. Fink. 1970. A new species of poeciliid fish of the genus *Limia* from the Island of Grand Cayman, B.W.I. Copeia 1970:270–274.

Roberts, T. R. 1989. The freshwater fishes of Western Borneo (Kalimantan Barat, Indonesia). Memoirs of the California Academy of Science 14, San Francisco.

Roberts, T. R. 1992a. Systematic revision of the southeast Asian anabantoid fish genus *Osphronemus*, with descriptions of two new species. Ichthyological Explorations of Freshwaters 2(4):351–360.

Roberts, T. R. 1992b. Systematic revision of the Old World freshwater fish family Notopteridae. Ichthyological Explorations of Freshwaters 2(4):361–383.

Robins, C. R., and six coauthors. 1991a. Common and scientific names of fishes from the United States and Canada, 5th edition. American Fisheries Society, Special Publication 20. Bethesda, Maryland.

Robins, C. R., and six coauthors. 1991b. World fishes important to North Americans exclusive of species from the continental waters of the United States and Canada. American Fisheries

Society, Special Publication 21. Bethesda, Maryland.

Robins, C. R., and E. C. Raney. 1956. Studies of the catostomid fishes of the genus *Moxostoma*, with descriptions of two new species. Cornell University Memoir 343:1–56.

Robins, C. R., and G. C. Ray. 1986. A field guide to Atlantic Coast fishes of North America. The Peterson Guide Series, volume 32. Houghton Mifflin, Boston.

Robinson, J. W. 1995. Missouri commercial fish harvest 1993. Final report, Missouri Department of Conservation, Columbia.

Robison, H. W. 1986. Zoogeographic implications of the Mississippi River basin. Pages 267–286 *in* C. H. Hocutt and E. O. Wiley, editors. The zoogeography of North American freshwater fishes. Wiley, New York.

Robison, H. W., and T. M. Buchanan. 1988. Fishes of Arkansas. University of Arkansas Press, Fayetteville.

Robison, H. W., and T. M. Buchanan. 1993. Changes in the nomenclature and composition of the Arkansas fish fauna from 1988 to 1993. Arkansas Academy of Science Proceedings 47:145–148.

Robison, H. W., and S. Winters. 1978. Occurrence of the slender madtom, *Noturus exilis* Nelson (Osteichthyes: Ictaluridae), in the Little River system, Arkansas. Southwestern Naturalist 23:688–689.

Rogers, W. A. 1961. Second progress report on stocking and harvesting of tilapia and channel catfish in Alabama's state-owned and managed public fishing lakes. Alabama Department of Conservation. Federal Aid in Fish Restoration Project F-10.

Rohde, F. C., R. G. Arndt, D. G. Lindquist, and J. F. Parnell. 1994. Freshwater fishes of the Carolinas, Virginia, Maryland, and Delaware. University of North Carolina Press, Chapel Hill.

Rose, S. 1972. What about the white amur? A superfish or a supercurse? Florida Naturalist 1972:156–157.

Rosen, D. E. 1960. Middle-American Poeciliid fishes of the genus *Xiphophorus*. Bulletin of the Florida State Museum Biological Sciences 5(4):57–242.

Rosen, D. E. 1979. Fishes from the uplands and intermontane basins of Guatemala: revisionary studies and comparative geography. Bulletin of the American Museum of Natural History 162:269–375.

Rosen, D. E., and R. M. Bailey. 1963. The poeciliid fishes (Cyprinodontiformes), their structure, zoogeography, and systematics. Bulletin of the American Museum of Natural History 126:1–176.

Rosen, D. E., and P. H. Greenwood. 1976. A fourth Neotropical species of synbranchid eel and the phylogeny and systematics of synbranchiform fishes. Bulletin of the American Museum of Natural History 157:1–69.

Rosen, P. C., C. R. Schwalbe, D. A. Parizek, Jr., P. A. Holm, and C. H. Lowe. 1995. Introduced aquatic vertebrates in the Chiricahua region: effects on declining native ranid frogs. Pages 251–261 in Biodiversity and management of the Madrean Archipelago: the sky island of the southwestern United States and northwestern Mexico. U.S. Forest Service General Technical Report RM-GTR-264.

Rosen, R. A., and D. C. Hales. 1981. Feeding of the paddlefish, *Polyodon spathula*. Copeia 1981:441–455.

Rosler, H. J. 1994. A pocket-sized "pike." Tropical Fish Hobbyist 42(10):110, 112, 114. Neptune City, New Jersey.

Ross, R. D., and J. E. Carico. 1963. Records and distributional problems of fishes of the North, Middle, and South Forks of the Hoston River, Virginia. Virginia Agriculture Experiment Station Technical Bulletin 161.

Ross, S. T., and W. M. Brenneman. 1991. Distribution of freshwater fishes in Mississippi. manuscript. Mississippi Department of Wildlife, Fisheries and Parks, Jackson.

Rourke, A. W., and R. L. Wallace. 1978. A morphological and electrophoretic comparison of Henry's Lake *Salmo clarki* and *Salmo clarki* × *Salmo gairdneri* hybrids. Comparative Biochemistry Physiology 60(B):447–451.

Rowe, J. W. 1992. The sturgeon chub and the brook silverside in the Platte River of Nebraska. Prairie Naturalist 24(4):281–282.

Rubinshtein, I. 1994. Disagrees with zebra mussel recommendations. Letter to the editor. Fisheries 19(4):40.

Ruffe Task Force. 1992. Ruffe in the Great Lakes: a threat to North American fisheries. Great Lakes Fishery Commission, Ann Arbor, Michigan.

Rule, R. 1885. Southern region. Pages 10–15 *in* J. H. Taggart, editor. Annual report of the Arizona Fish Commission, 1883–1884, to Frederick A. Tritle, Governor of the Territory of Arizona.

Rutherford, D. A., A. A. Echelle, and O. E. Maughan. 1987. Changes in the fauna of the Little River drainage, southeastern Oklahoma, 1948–1955 to 1981-1982: a test of the hypothesis of environmental degradation. Pages 178–183 *in* W. J. Matthews and D. C. Heins, editors. Community and evolutionary ecology of North American stream fishes. University of Oklahoma Press, Norman.

Ryon, M. G., and J. M. Loar. 1988. A checklist of fishes on the Department of Energy Oak Ridge Reservation. Journal of the Tennessee Academy of Science 63(4):97–102.

Sakaizumi, M., K. Moriwaki, and N. Egami. 1983. Allozymic variation and regional differentiation in wild populations of the fish *Oryzias latipes*. Copeia 1983(1):311–318.

Sakurai, A., Y. Sakamoto, and F. Mori. 1993. Aquarium fish of the world: the comprehensive guide to 650 species. Chronicle Books, San Francisco.

Sands, D. D. 1984. Catfishes of the world, volume four, Aspredinidae, Doradidae, and Loricariidae. Dunure Publications, Dunure, Scotland.

Sanford, G. 1995. An illustrated encyclopedia of aquarium fish. Quintet Publishing Limited, London.

Savino, J. F., and C. S. Kolar. 1996. Competition between nonindigenous ruffe and native yellow perch in laboratory studies. Transactions of the American Fisheries Society 125:562–571.

Savitz, J., C. Aiello, and L.G. Bardygula. 1989a. The first record of the white perch (*Morone americana*) in Illinois waters of Lake Michigan. Transactions of the Illinois State Academy of Science 82(1&2):57–58.

Savitz, J., L. G. Bardygula, and L. Scoma. 1989b. The first record of the quillback carpsucker (*Carpiodes cyprinus*) in Illinois waters of Lake Michigan. Transactions of the Illinois State Academy of Science 82(3&4):191–192.

Scarola, J. F. 1973. Freshwater fishes of New Hampshire. New Hampshire Fish and Game Department, Division of Inland and Marine Fisheries.

Schaefer, S. A. 1997. The neotropical cascudinhos: systematics and biogeography of the *Otocinclus* catfishes (Siluriformes: Loricariidae). Proceedings of the Academy of Natural Sciences of Philadelphia 148:1–148.

Schaeffer, J. S., and F. J. Margraf. 1987. Predation on fish eggs by white perch, *Morone americana*, in western Lake Erie. Environmental Biology of Fishes 18(1):77–80.

Scheel, J. J. 1990. Atlas of killifishes of the Old World. Tropical Fish Hobbyist Publications, Inc., Neptune City, New Jersey.

Schmidt, R. E. 1980. New distribution records of freshwater fishes in North Carolina. Journal of the Elisha Mitchell Scientific Society 96(1):36–38.

Schmidt, R. E. 1986. Zoogeography of the Northern Appalachians. Pages 137–160 *in* C. H. Hocutt and E. O. Wiley, editors. The zoogeography of North American freshwater fishes. Wiley, New York.

Schmidt, R. E., and J. M Samaritan. 1984. Fishes of an urban stream: the Bronx River, New York. Northeastern Environmental Science 3(1):3–7.

Schmidt, R. E., J. M. Samaritan, and A. Pappantoniou. 1981. Status of the bitterling, *Rhodeus sericeus,* in southern New York. Copeia 1981(2):481–482.

Schmidt, R. E., and W. R. Whitworth. 1979. Distribution and habitat of the swamp darter (*Etheostoma fusiforme*) in southern New England. American Midland Naturalist 102(2):408–413.

Schneider, C. P., R. W. Owens, R. A. Bergstedt, and R. O'Gorman. 1996. Predation by sea lamprey (*Petromyzon marinus*) on lake trout (*Salvelinus namaycush*) in southern Lake Ontario, 1982–1992. Canadian Journal of Fisheries and Aquatic Sciences 53:1921–1932.

Schoenherr, A. A. 1979. Niche separation within a

population of freshwater fishes in an irrigation drain near the Salton Sea, California. Bulletin of the Southern California Academy of Science 78(1):46–55.

Schoenherr, A. A. 1985. Replacement of *Cyprinodon macularius* by *Tilapia zillii* in an irrigation drain near the Salton Sea. Proceedings of the Desert Fishes Council 13(1981):65–66.

Schoenherr, A. A. 1988. A review of the life history and status of the desert pupfish, *Cyprinodon macularius*. Bulletin of the Southern California Academy of Science 87(3):104–134.

Schultz, E. E. 1960. Establishment and early dispersal of a loach, *Misgurnus anguillicaudatus* (Cantor), in Michigan. Transactions of the American Fisheries Society 89:376–377.

Schultz, L. P. 1929. Check-list of the freshwater fishes of Oregon and Washington. Pub. Fish., University of Washington 2(4):43–50.

Schultz, L. P. 1944. The catfishes of Venezuela, with descriptions of thirty-eight new forms. Proceedings of the United States National Museum 94(3172):173–338.

Schultz, L. P. 1967. Review of South American freshwater angelfishes–genus *Pterophyllum*. Proceedings of the United States National Museum 120:1–10.

Schultz, L. P., and A. C. DeLacy. 1935. Fishes of the American Northwest. First installment. Pan-Pacific Magazine 10(4):365–380.

Schultz R. J., and R. R. Miller. 1971. Species of the *Poecilia sphenops* complex (Pisces: Poeciliidae) in Mexico. Copeia 1971(2):282–290.

Schumacher, R. E., and S. Eddy. 1960. The appearance of pink salmon, *Oncorhynchus gorbuscha* (Walbaum), in Lake Superior. Transactions of the American Fisheries Society 89:371–373.

Schumacher, R. E., and J. G. Hale. 1962. Third generation pink salmon, *Oncorhynchus gorbuscha* (Walbaum), in Lake Superior. Transactions of the American Fisheries Society 91:421–422.

Schwartz, F. 1963. The fresh-water minnows of Maryland. Maryland Conservationist 40(2):19–29.

Schwartz, F. 1964. Several Maryland fishes are close to extinction. Maryland Conservationist 41(3):8–12.

Schwartz, F. J. 1981. World literature to fish hybrids with an analysis by family, species, and hybrid: Supplement 1. NOAA (National Oceanic and Atmospheric Administration) Technical Report NMFS-750, (National Marine Fisheries Service), Seattle, Washington.

Schwartz, F. J., W. T. Hogarth, and M. P. Weinstein. 1981. Marine and freshwater fishes of the Cape Fear Estuary, North Carolina, and their distribution in relation to environmental factors. Brimleyana 7:17–37.

Scoppettone, G. G. 1993. Interactions between native and nonnative fishes of the upper Muddy River, Nevada. Transactions of the American Fisheries Society 122:599–608.

Scoppettone, G. G., P. H. Rissler, M. B. Nielsen, and J. E. Harvey. 1998. The status of *Moapa coriacea* and *Gila seminuda* and status information on other fishes of the Muddy River, Clark County, Nevada. Southwestern Naturalist 43(2):115–122.

Scott, W. B., and W. J. Christie. 1963. The invasion of the lower Great Lakes by white perch, *Roccus americanus* (Gmelin). Journal of the Fisheries Research Board of Canada 20:1189–1195.

Scott, W. B., and E. J. Crossman. 1973. Freshwater Fishes of Canada. Fisheries Research Board of Canada Bulletin 184.

Seal, W. P. 1910. Fishes in their relation to the mosquito problem. Bulletin of the U.S. Bureau of Fisheries 28(1908):831–838.

Seale, A. 1905. Report of Mr. Alvin Seale of the United States Fish Commission, on the introduction of top-minnows to Hawaii from Galveston, Texas. The Hawaiian Forester and Agriculturist 2:364–367.

Seeley, C. M. 1962. A report on the orfe, *Leuciscus idus*. California Department of Fish and Game Inland Fisheries Administrative Report 62–4.

Seeley, C. M., and G. W. McCammon. 1966. Kokanee. Pages 274–295 *in* A. Calhoun, editor. Inland fisheries management. California Department of Fish and Game. Sacramento.

Setzler, E. M., and eight coauthors. 1980. Synopsis of biological data on striped bass, *Morone saxatilis* (Walbaum). NOAA (National

Oceanic and Atmospheric Administration) Technical Report, NMFS (National Marine Fisheries Service) Circular 433.

Sevon, M., and D. Delany. 1987. Development of Blue Link Spring, a refugium for Hiko White River springfish. Proceedings of the Desert Fishes Council 17(1985):168–173.

Shafland, P. L. 1976. The continuing problem of non-native fishes in Florida. Fisheries 1(6):25.

Shafland, P. L. 1995a. Introduction and establishment of a successful butterfly peacock fishery in southeast Florida canals. Pages 443–451 *in* H. L. Schramm and R. G. Piper, editors. Uses and effects of cultured fishes in aquatic ecosystems. American Fisheries Society, Symposium 15.

Shafland, P. L. 1995b. Florida's exotic fishes–1994. Reviews in Fisheries Science 4(2):101–122.

Shafland, P. L., and K. J. Foote. 1979. A reproducing population of *Serrasalmus humeralis* Valenciennes in southern Florida. Florida Scientist 42:206–214.

Shafland, P. L., and W. M. Lewis. 1984. Terminology associated with introduced organisms. Fisheries 9(4):17–18.

Shapovalov, L. 1944. The tench in California. California Fish and Game 30(1):54–57.

Shapovalov, L., W. A. Dill, and A. J. Cordone. 1959. A revised check list of the freshwater and anadromous fishes of California. California Fish and Game 45:159–180.

Shapovalov, L., A. J. Cordone, and W. A. Dill. 1981. A list of freshwater and anadromous fishes of California. California Fish and Game 67(1):4–38.

Shebley, W. H. 1917. History of the introduction of food and game fishes into the waters of California. California Fish and Game 3(1):3–10.

Shields, J. T. 1958a. Experimental control of carp reproduction through water drawdowns in Fort Randall Reservoir, South Dakota. Transactions of the American Fisheries Society 87:23–33.

Shields, J. T. 1958b. Fish management problems of large impoundments on the Missouri River. Transactions of the American Fisheries Society 87:356–362.

Shima, C. S. 1972. Creel census and fishermen checking station operations, Statewide Dingell-Johnson Program. Hawaii Division of Fish and Game, Job Progress Report for Project F-9-2, Job 1 (Study VI).

Shireman, J. V., and C. R. Smith. 1983. Synopsis of biological data on the grass carp *Ctenopharyngodon idella* (Cuvier and Valenciennes, 1844). FAO (Food and Agriculture Organization of the United Nations) Fisheries Synopsis 135. Rome.

Shute, J. R., P. W. Shute, and D. G. Lindquist. 1981. Fishes of the Waccamaw River drainage. Brimleyana 6:1–24.

Sigler, W. F., and R. R. Miller. 1963. Fishes of Utah. Utah Department of Fish and Game, Salt Lake City.

Sigler, W. F., and J. W. Sigler. 1987. Fishes of the Great Basin: A natural history. University of Nevada Press, Reno.

Sigler, W. F., and J. W. Sigler. 1996. Fishes of Utah. University of Utah Press, Salt Lake City.

Sigler, W. F., and G. W. Workman. 1978. The Bonneville cisco of Bear Lake, Utah-Idaho. Research Report 33, Utah Agricultural Experiment Station, Utah State University, Logan.

Silfvergrip, A. M. C. 1996. A systematic revision of the Neotropical catfish genus *Rhamdia* (Teleostei, Pimelodidae). Swedish Museum of Natural History, Stockholm.

Simon, D. C., and D. F. Markle. 1997. Interannual abundance of nonnative fathead minnows (*Pimephales promelas*) in Upper Klamath Lake, Oregon. Great Basin Naturalist 57:142–148.

Simon, J. R. 1946. Wyoming fishes. Wyoming Game and Fish Department Bulletin 4.

Simon, T. P., and J. T. Vondruska. 1991. Larval identification of the ruffe, *Gymnocephalus cernuus* (Linnaeus) (Percidae: Percini), in the St. Louis River estuary, Lake Superior drainage basin, Minnesota. Canadian Journal of Zoology 69:436–441.

Simon, T. P., J. O. Whitaker, Jr., J. S. Castrale, and S. A. Minton. 1992. Checklist of the vertebrates of Indiana. Proceedings of the Indiana Academy of Science 101:95–126.

Simons, L. H. 1987. Status of the endangered Gila topminnow, *Poeciliopsis occidentalis*, in the United States. Arizona Game and Fish Department, Project E-I, Title VI. Phoenix.

Simpson, J., and R. Wallace. 1978. Fishes of Idaho. University of Idaho Press, Moscow.

Skelton, P. H. 1993. A complete guide to the freshwater fishes of southern Africa. Southern Book Publishers, Halfway House, South Africa.

Skinner, J. E. 1962. An historical overview of the fish and wildlife resources of the San Francisco Bay area. California Fish and Game Water Projects Branch Report. 1:225 pp.

Skinner, J. E. 1971. *Anguilla* recorded from California. California Fish and Game 57(1):76–79.

Skinner, J. E. 1972. Ecological studies of the Sacramento–San Joaquin Estuary. California Department of Fish and Game, Delta Fish Wildlife Protection Study Report 8.

Skinner, W. F. 1984. *Oreochromis aureus* (Steindachner; Cichlidae), an exotic fish species, accidentally introduced to the lower Susquehanna River, Pennsylvania. Proceedings of the Pennsylvania Academy of Science 58:99–100.

Skinner, W. F. 1986. Susquehanna River tilapia. Fisheries 11(4):56–57.

Skinner, W. F. 1987. Report on the eradication of tilapia from the vicinity of the Brunner Island Steam Electric Station. Pennsylvania Power and Light Company, Allentown, PA.

Slipp, J. W. 1943. The rock bass, *Ambloplites rupestris*, in Washington state. Copeia 1943(2):132.

Smiley, C. W. 1885a. Notes upon fish and the fisheries. Bulletin of the United States Fish Commission for 1885, 5(7):97–112.

Smiley, C. W. 1885b. Notes upon fish and the fisheries. Bulletin of the United States Fish Commission for 1885, 5(30):465–469.

Smiley, C. W. 1886. Some results of carp culture in the United States. Pages 657–890 *in* Report of the Commissioner of Fish and Fisheries for 1884, Part XII. U.S. Commission of Fish and Fisheries, Washington, D.C.

Smiley, C. W. 1889. Notes upon fish and the fisheries. Bulletin of the United States Fish Commission for 1887, 7(13):33–50.

Smith, B. R., and J. J. Tibbles. 1980. Sea lamprey (*Petromyzon marinus*) in Lakes Huron, Michigan, and Superior: history of invasion and control, 1936–78. Canadian Journal of Fisheries and Aquatic Sciences 37:1780–1801.

Smith, C. L. 1985. The inland fishes of New York state. New York State Department of Environmental Conservation, Albany.

Smith, E. 1905. *Exoglossum* east of the Delaware basin. Science 22(552):110–120.

Smith, G. R. 1966. Distribution and evolution of the North American catostomid fishes of the subgenus *Pantosteus*, genus *Catostomus*. Miscellaneous Publications of the University of Michigan Museum of Zoology 129:1–132.

Smith, G. R. 1992. Phylogeny and biogeography of the Catostomidae, freshwater fishes of North America and Asia. Pages 778–826 *in* R. L. Mayden, editor. Systematics, historical ecology, and North American freshwater fishes. Stanford University Press, Stanford, California.

Smith, H. M. 1896. A review of the history and results of the attempts to acclimatize fish and other water animals in the Pacific states. Bulletin of the U.S. Fish Commission for 1895, 40:379–472.

Smith, H. M. 1945. The fresh-water fishes of Siam, or Thailand. Bulletin of the U.S. National Museum (Smithsonian Institution) 188:1–622.

Smith, H. M., and B. A. Bean. 1899. List of fishes known to inhabit the waters of the District of Columbia and vicinity. Bulletin of the U.S. Fish Commission for 1898, 43:179–487.

Smith, J. J. 1982. Fishes of the Pajaro River system. Pages 83–170 *in* P. B. Moyle, J. J. Smith, R. A. Daniels, T. L. Price, and D. M. Baltz, editors. Distribution and ecology of stream fishes of the Sacramento–San Joaquin drainage system, California. University of California Press, Berkeley.

Smith, J. V. C. 1833. A natural history of the fishes of Massachusetts. A reprint of the 1833 edition. Freshet Press, Inc., Rockville Center, New York.

Smith, M. L, and R. R. Miller. 1986. The evolution of the Rio Grande basin as inferred from its fish fauna. Pages 457–486 *in* C. H. Hocutt and E.

O. Wiley, editors. The zoogeography of North American freshwater fishes. Wiley, New York.

Smith, P. W. 1965. A preliminary annotated list of the lampreys and fishes of Illinois. Illinois Natural History Survey Biological Notes 54.

Smith, P. W. 1979. The fishes of Illinois. University of Illinois Press, Urbana.

Smith, S. H. 1970. Species interactions of the alewife in the Great Lakes. Transactions of the American Fisheries Society 99:754–765.

Smith, S. J. 1995. The orfe: a fish for all seasons. Tropical Fish Hobbyist 64(1):102, 104–105. Neptune City, New Jersey.

Smith-Vaniz, W. F. 1968. Freshwater fishes of Alabama. Auburn University Agricultural Experiment Station, Auburn, Alabama.

Snelson, F. F., Jr. 1968. Systematics of the cyprinid fish *Notropis amoenus*, with comments on the subgenus *Notropis*. Copeia 1968(4):776–802.

Snelson, F. F., Jr. 1971. *Notropis mekistocholas*, a new herbivorous cyprinid fish endemic to the Cape Fear River basin, North Carolina. Copeia 1971(3):449–462.

Snelson, F. F., Jr. 1972. Systematics of the subgenus *Lythrurus*, genus *Notropis* (Pisces: Cyprinidae). Bulletin of the Florida State Museum Biological Sciences 17:1–92.

Snelson, F. F., Jr. 1973. Systematics and distribution of the ribbon shiner, *Notropis fumeus* (Cyprinidae), from the central United States. American Midland Naturalist 89(1):166–191.

Snelson, F. F., Jr. 1990. Redescription, geographic variation, and subspecies of the minnow *Notropis ardens* (Pisces: Cyprinidae). Copeia 1990(4):966–984.

Snelson, F. F., Jr. 1991. Review of "Phylogenetic studies of North American minnows, with emphasis on the genus *Cyprinella* (Teleostei: Cypriniformes)" by R. L. Mayden. Copeia 1991(1):258–274.

Snodgrass, J. W. 1991. Winter kills of *Tilapia melanotheron* in coastal southeast Florida, January 1977. Florida Scientist 41:1–12.

Snyder, J. O. 1935. California fresh water fish. The Aquarium Journal 8(9):146.

Soltz, D. L., and R. J. Naiman. 1978. The natural history of native fishes in the Death Valley system. Natural History Museum of Los Angeles County, Science Series 30:1–76.

South Carolina Department of Natural Resources, Division of Freshwater Fisheries. 1997. Freshwater fish records for South Carolina. (Available from http://www.dnr.state.sc.us/wild/freshrecs/freshrec.html. Page accessed June 25, 1997.)

South Dakota Game, Fish and Parks. 1997. South Dakota state fishing records. (Available from http://www.state.sd.us/state/executive/gfp/fishing/records.htm. Page accessed June 25, 1997.)

Spencer, S. L., W. E. Swingle, and T. Scott, Jr. 1964. Commercial fishing in the Mobile Delta, Alabama, during the period of July 1, 1963 to June 30, 1964. Alabama Department of Conservation 1:17.

Springer, V. G., and J. H. Finucane. 1963. The African cichlid, *Tilapia heudeloti* Dumeril, in the commercial fish catch of Florida. Transactions of the American Fisheries Society 92:317–318.

St. Amant, J. A. 1959. Striped bass introduced into the Colorado River. California Fish and Game 45(4):353.

St. Amant, J. A. 1966. Addition of *Tilapia mossambica* Peters to the California fauna. California Fish and Game 52(1):54–55.

St. Amant, J. A. 1970. Addition of Hart's rivulus, *Rivulus harti* (Boulenger), to the California fauna. California Fish and Game 56(2):138.

St. Amant, J. A., and F. G. Hoover. 1969. Addition of *Misgurnus anguillicaudatus* (Cantor) to the California fauna. California Fish and Game 57(2):330–331.

St. Amant, J. A., and I. Sharp. 1971. Addition of *Xiphophorus variatus* to the California fauna. California Fish and Game 57:128–129.

Starnes, W. C., R. T. Bryant, and G. C. Greer. 1998. Perilous experiment: the Asian rice eel in Georgia. Urban Wildlife (Georgia Wildlife) 7(1):60–70.

Starnes, W. C., and D. A. Etnier. 1986. Drainage evolution and fish biogeography of the Tennessee and Cumberland rivers drainage realm. Pages 325–362 *in* C. H. Hocutt and E.

O. Wiley, editors. The zoogeography of North American freshwater fishes. Wiley, New York.

Starnes, W. C., and L. B. Starnes. 1979. Discovery of the percid genus *Ammocrypta* (Pisces) in the Apalachicola drainage, Florida. Florida Scientist 42(1):61–62.

Stauffer, J. R., Jr., J. M. Boltz, and L. R. White. 1995. The fishes of West Virginia. West Virginia Department of Natural Resources. Academy of Natural Sciences of Philadelphia, Philadelphia.

Stauffer, J. R., Jr., S. E. Boltz, and J. M. Boltz. 1988. Cold shock susceptibility of blue tilapia from the Susquehanna River, Pennsylvania. North American Journal of Fisheries Management 8:329–332.

Stauffer, J. R., Jr., C. H. Hocutt, and D. S. Lee. 1978. The zoogeography of the freshwater fishes of the Potomac River basin. Pages 44–54 *in* K. C. Flynn and W. T. Mason, editors. The freshwater Potomac: aquatic communities and environmental stresses. Interstate Commission on the Potomac River Basin, Rockville, Maryland.

Stauffer, J. R., Jr., C. H. Hocutt, M. T. Masnik, and J. E. Reed, Jr. 1975. The longitudinal distribution of the fishes of the East River, West Virginia-Virginia. Virginia Journal of Science 26(3):121–125.

Stearns, S. C. 1983. A natural experiment in life-history evolution: field data on the introduction of mosquitofish (*Gambusia affinis*) to Hawaii. Evolution 37:607–617.

Stedman, R. M., C. A. Bowen II. 1985. Introduction and spread of the threespine stickleback (*Gasterosteus aculeatus*) in lakes Huron and Michigan. Journal of Great Lakes Research 11(4):508–511.

Stefanovic, D. 1966. Racial and ecological study of the Ohrid salmonids. Translated and published for the U.S. Department of the Interior and the National Science Foundation, Washington, D.C., by the NOLIT Publishing House, Belgrade, Yugoslavia.

Steirer, F. S., Jr. 1992. Historical perspective on exotic species. Pages 1–4 *in* M. R. DeVoe, editor. Introductions and transfers of marine species. South Carolina Sea Grant Consortium. Charleston.

Stepien, C. A., A. K. Dillon, and M. D. Chandler. 1998. Genetic identity, phylogeography, and systematics of ruffe *Gymnocephalus* in the North American Great Lakes and Eurasia. Journal of Great Lakes Research 24(2):361–378.

Sterba, G. 1967. Freshwater fishes of the world. The Pet Library, Ltd., New York.

Sterba, G. 1973. Freshwater fishes of the world. English translation and revision from German. Two volumes. Tropical Fish Hobbyist Publications, Inc., Neptune City, New Jersey.

Stevenson, M. M., and T. M. Buchanan. 1973. An analysis of hybridization between the cyprinodont fishes *Cyprinodon variegatus* and *C. elegans*. Copeia 1973(4):682–692.

Stevenson, M. M., and B. A. Thompson. 1978. Further distribution records for the bigscale logperch, *Percina macrolepida* (Osteichthyes: Percidae) from Oklahoma, Texas and Louisiana with notes on its occurrence in California. Southwestern Naturalist 23(2):303–313.

Stickney, R. R. 1996. Aquaculture in the United States: a historical survey. Wiley, New York.

Stiles, E. W. 1978. Vertebrates of New Jersey. Edmund W. Stiles, Somerset, New Jersey.

Stolte, L. W. 1974. Introduction of coho salmon into the coastal waters of New Hampshire. Progressive Fish-Culturist 36:29–32.

Stolzenburg, W. 1992. Red danger for desert fish. Nature Conservancy Magazine 1992(1):28–29.

Stone, M. D. 1995. Fish stocking programs in Wyoming: a balanced perspective. Pages 47–51 *in* H. L. Schramm, Jr. and R. G. Piper, editors. Uses and effects of cultured fishes in aquatic ecosystems. American Fisheries Society, Symposium 15. Bethesda, Maryland.

Stone, S. 1987. 6-inch piranha found in pit at Indian Lakes. Virginia Pilot and Ledger Star (August 23), Norfolk.

Stroud, R. A. 1976. Ohio piranha. Sport Fishing Institute Bulletin 272:3. Washington, D.C.

Stroud, R. H. 1969. Ohrid trout tests. Sport Fishing Institute Bulletin 210:7–8. Washington, D.C.

Sturgess, J. A. 1976. Taxonomic status of *Percina* in California. California Fish and Game 62(1):79–81.

Sublette, J. E., M. D. Hatch, and M. Sublette. 1990. The fishes of New Mexico. New Mexico Department of Game and Fish, University of New Mexico Press, Albuquerque.

Surber, E. W. 1958. Results of striped bass (*Roccus saxatilis*) introductions into freshwater impoundments. Proceedings of the 11th Annual Conference Southeastern Association of Game and Fish Commissioners 11(1957):273–276.

Suttkus, R. D. 1963. Order Lepisostei. Pages 61–88 *in* Fishes of the Western North Atlantic Part 3. Sears Foundation for Marine Research, Yale University, New Haven, Connecticut.

Suttkus, R. D., and G. H. Clemmer. 1968. *Notropis edwardraneyi*, a new cyprinid fish from the Alabama and Tombigbee river systems and a discussion of related species. Tulane Studies in Zoology and Botany 15:18–39.

Suttkus, R. D., and J. V. Conner. 1980. The rainbow smelt, *Osmerus mordax*, in the lower Mississippi River near St. Francisville, Louisiana. American Midland Naturalist 104(2):394.

Suzuki, R. 1974. Inter-crossings and back-crossings of F_1 hybrids among salmonid fishes. Bulletin of the Freshwater Research Laboratory 24(1):11–31.

Sweeney, Z. T. 1902. Biennial Report of the Commissioner of Fisheries and Game for Indiana. Indianapolis, Indiana.

Swift, C. C., C. R. Gilbert, S. A. Bortone, G. H. Burgess, and R. W. Yerger. 1986. Zoogeography of the freshwater fishes of the southeastern United States: Savannah River to Lake Pontchartrain. Pages 213–266 *in* C. H. Hocutt and E. O. Wiley, editors. The zoogeography of North American freshwater fishes. Wiley, New York.

Swift, C., R. W. Yerger, and P. R. Parrish. 1977. Distribution and natural history of the fresh and brackish water fishes of the Ochlockonee River, Florida and Georgia. Bulletin of Tall Timbers Research Station 20.

Swift, C. C., T. R. Haglund, M. Ruiz, and R. N. Fisher. 1993. The status and distribution of the freshwater fishes of southern California. Bulletin of the Southern California Academy of Science 92(3):101–167.

Taggart, J. H. 1885. Annual Report of the Arizona Fish Commission, 1883–1884, to Frederick A. Tritle, Governor of the Territory of Arizona.

Talwar, P. K., and A. G. Jhingran, editors. 1992. Inland fishes of India and adjacent countries. (Two volumes.)A. A. Balkema, Rotterdam, The Netherlands.

Tanner, V. M. 1936. A study of the fishes of Utah. Utah Academy of Sciences, Arts and Letters 13:155–183.

Taphorn, D. C. 1989. Los peces de la familia Callichthyidae de la cuenca del Río Apure. BioLlania (Guanare, Venezuela) 6:15–50.

Taphorn, D. C. 1992. The characiform fishes of the Apure River drainage, Venezuela. BioLlania (Guanare, Venezuela), special edition 4.

Taphorn, D. C., and A. Barbarino-Duque. 1993. Evaluación de la situación actual de los pavones, (*Cichla* spp.), en el Parque nacional Capanaparo-Cinaruco, Estado Apure, Venezuela. Natura (Caracas) 96:10–25.

Taphorn, D. C., and C. G. Lilyestrom. 1984. Claves para los peces de agua dulce de Venezuela: 1. Las familias de Venezuela. 2. Los generos y las especies de la Cuenca del Lago de Maracaibo. Revista UNELLEZ de Ciencia y Tecnología (Guanare, Venezuela) 2(2):5–30.

Tatum, B. L. 1958. Introduction and success of white bass (*Roccus chrysops*) in North Carolina waters. Proceedings of the 11th Annual Conference Southeastern Association of Game and Fish Commissioners 11(1957):185–192.

Taylor, J. N. 1985. Key to the species of the genus *Colossoma* (Characidae: Serrasalminae) (II:26:85). Unpublished mimeograph.

Taylor, J. N., W. R. Courtenay, Jr., and J. A. McCann. 1984. Known impact of exotic fishes in the continental United States. Pages 322–373 *in* W. R. Courtenay, Jr. and J. R. Stauffer, editors. Distribution, biology, and management of exotic fish. Johns Hopkins University Press, Baltimore, Maryland.

Taylor, J. N., D. B. Snyder, and W. R. Courtenay, Jr. 1986. Hybridization between two introduced, substrate-spawning tilapias (Pisces: Cichlidae) in Florida. Copeia 1986(4):903–909.

Taylor, T. L, P. B. Moyle, and D. G. Price. 1982. Fishes of the Clear Lake basin. Pages 171–224 *in* P. B. Moyle, J. J. Smith, R. A. Daniels, T. L.

Price, and D. M. Baltz, editors. Distribution and ecology of stream fishes of the Sacramento–San Joaquin drainage system, California. University of California Press, Berkeley.

Taylor, W. R. 1969. A revision of the catfish genus *Noturus* Rafinesque with an analysis of higher groups in the Ictaluridae. Bulletin of the U.S. National Museum (Smithsonian Institution) 282:1–315.

Teuscher, D., and C. Luecke. 1996. Competition between kokanees and Utah chub in Flaming Gorge Reservoir, Utah–Wyoming. Transactions of the American Fisheries Society 125:505–511.

Thomas, P. W. 1994. The use of grass carp in Florida: past, present, and future. Program issue and abstracts of the 14th Annual International Symposium of the North American Lake Management Society, Orlando, Florida. Lake and Reservoir Management (Washington, DC) 9(2):119.

Thomerson, J. E. 1969. Variation and relationship of the studfishes, *Fundulus catenatus* and *Fundulus stellifer* (Cyprinodontidae, Pisces). Tulane Studies in Zoology and Botany 16(1):1–21.

Thomson, A. J., and S. McKinnell. 1994. Summary of reported Atlantic salmon (*Salmo salar*) catches and sightings in British Columbia and adjacent waters in 1993. Canadian Manuscript Report of Fisheries and Aquatic Sciences 2246.

Thompson, B. A. 1995. *Percina austroperca*: a new species of logperch (Percidae, subgenus *Percina*) from the Choctawhatchee and Escambia rivers in Alabama and Florida. Occasional Papers of the Museum of Natural Science 69:1–20.

Thomson, D. A., L. T. Findley, and A. N. Kerstitch. 1979. Reef Fishes of the Sea of Cortez. Wiley, New York.

Thorgaard, G. H., and S. K. Allen. 1992. Environmental impacts of inbred, hybrid and polyploid aquatic species. Pages 281–288 *in* A. Rosenfield and R. Mann, editors. Dispersal of living organisms into aquatic ecosystems. University of Maryland Maryland Sea Grant College Program, College Park.

Thys van den Audenaerde, D. F. E. 1966. Les *Tilapia* (Pisces, Cichlidae) du Sud-Cameroun et du Gabon étude systématique. Annales du Mussee Royale de l'Afrique Centreale (Sciencs Zoologiques) 153:1–98.

Titcomb, J. W. 1904. Report on the propagation and distribution of food fishes. Bulletin of the U.S. Fish Commission for the year ending June 30, 1902, pp. 22–110.

Todd, B. 1962. Explosive new fish in Kansas. Kansas Fish and Game 20(1):3–5.

Todd, T. N. 1983. The feasibility of mass-culturing coregonines in the Great Lakes. Research Completion Report (February, 1983). Unpublished mimeograph.

Todd, T. N. 1986a. Occurrence of white bass-white perch hybrids in Lake Erie. Copeia 1986(1):196–199.

Todd, T. N. 1986b. Artificial propagation of coregonines in the management of the Laurentian Great Lakes. Archiv fuer Hydrobiologie Beih./Ergebnisse der Limnologie 22:31–50. Stuttgart, Germany.

Tomelleri, J. R., and M. E. Eberle. 1990. Fishes of the central United States. University Press of Kansas, Lawrence.

Tomiyama, I. 1936. Gobiidae of Japan. Japanese Journal of Zoology 7(1):37-112.

Trautman, M. B. 1957. The fishes of Ohio. Ohio State University Press, Columbus.

Trautman, M. B. 1981. The fishes of Ohio. Ohio State University Press, Columbus.

Trewavas, E. 1982. Tilapias: taxonomy and speciation. Pages 3–13 *in* R. S. V. Pullin, and R. H. Lowe-McConnell, editors. The biology and culture of tilapias. ICLARM (International Center for Living Aquatic Resources Management), conference proceedings 7, Manila, Philippines.

Trewavas, E. 1983a. Tilapine fishes of the genera *Sarotherodon*, *Oreochromis* and *Danakilia*. British Museum (Natural History) Publication No. 878, Cornell University Press, Ithaca, New York.

Trewavas, E. 1983b. Nouvel examen des genres et sous-genres du complexe *Pseudotropheus-Melanochromis* du lac Malawi (Pisces, Perciformes, Cichlidae). Revue Francaise d'Aquariologie 10:97–106.

Trimm, D. L., G. Guillen, C. T. Menn, and G. C. Matlock. 1989. The occurrence of grass carp in Texas water. Texas Journal of Science 41(4):413–417.

Trotter, P. C. 1987. Cutthroat: native trout of the West. Colorado Associated University Press, Boulder.

Truitt, R. V., B. A. Bean, and H. W. Fowler. 1929. The Fishes of Maryland. Maryland Conservation Department, Conservation Bulletin 3, Annapolis.

Tsai, C., and E. C. Raney. 1974. Systematics of the banded darter, *Etheostoma zonale* (Pisces: Percidae). Copeia 1974(1):1–24.

Tucker, J. K., F. A. Cronin, R. A. Hrabik, M. D. Petersen, and D. P. Herzo. 1996. The bighead carp (*Hypophthalmichthys nobilis*) in the Mississippi River. Journal of Freshwater Ecology 11(2):241–243.

Turner, J. M., Jr. 1989. Introgressive hybridization of redeye bass and smallmouth bass in a north-central Tennessee stream. Master's thesis. Tennessee Technological University, Cookeville.

Turner, J. M., F. J. Bulow, and C. J. O'Bara. 1991. Introgressive hybridization of redeye bass and smallmouth bass and its management implications. First International Bass Symposium 1991:143–150.

Turner, J. S., and F. F. Snelson, Jr. 1984. Population structure, reproduction and laboratory behavior of the introduced *Belonesox belizanus* (Poeciliidae) in Florida. Environmental Biology of Fishes 10:89–100.

Tyus, H. M. 1998. Early records of the endangered fish *Gila cypha* Miller from the Yampa River of Colorado with notes on its decline. Copeia 1998(1):190–193.

Tyus, H. M., and five coauthors. 1982. Fishes of the upper Colorado River basin: distribution, abundance, and status. Pages 12–70 *in* W. H. Miller, H. M. Tyus, and C. A. Carlson, editors. Fishes of the upper Colorado River system: present and future, American Fisheries Society Western Division, Bethesda, Maryland.

Uhler, P. R., and O. Lugger. 1876. List of fishes of Maryland. Page 69–176 *in* Report of the Commissioners of Fisheries of Maryland to the General Assembly, January 1, 1876. John F. Wiley, Annapolis.

Underhill, J. C. 1959. Fishes of the Vermillion River, South Dakota. Proceedings of the South Dakota Academy of Science 38:96–102.

Underhill, J. C. 1986. The fish fauna of the Laurentian Great Lakes, the St. Lawrence Lowlands, Newfoundland and Labrador. Pages 105–136 *in* C. H. Hocutt and E. O. Wiley, editors. The zoogeography of North American freshwater fishes. Wiley, New York.

Underhill, J. C. 1989. The distribution of Minnesota fishes and late Pleistocene glaciation. Journal of the Minnesota Academy of Science 55(1):32–37.

Underhill, J. C. , J. T. Hatch, and K. Schmidt. 1997. Fishes of Minnesota: distribution of Minnesota fishes in 8 major drainage basins. (Available from http://www.umn.edu/faculty_staff/hatch/fishes/distribution_table.html. Page accessed July 31, 1998.)

University of Washington Fish Collection. 1996. Main collection. (Available from http://artedi.fish.washington.edu/Main-db20/UW_021588.html Page accessed May 14, 1998.)

U.S. Army Corps of Engineers. 1997. Fishing on the Tenn-Tom Waterway. (Available from http://www.sam.usace.army.mil/sam/op/rec/tenn-tom/fish.htm. Ppage accessed June 26, 1997.)

U.S. Congress, Office of Technology Assessment. 1993. Harmful non-indigenous species in the United States. U.S. Government Printing Office, OTA-F-565, Washington, D.C.

U.S. Fish and Wildlife Service. 1977a. Greenback cutthroat trout recovery plan. U.S. Fish and Wildlife Service.

U.S. Fish and Wildlife Service. 1977b. Recovery plan for unarmoured threespine stickleback, *Gasterosteus aculeatus williamsoni*, an endangered fish. U.S. Fish and Wildlife Service.

U.S. Fish and Wildlife Service. 1979a. Pahrump killifish recovery plan. U.S. Fish and Wildlife Service, Denver, Colorado.

U.S. Fish and Wildlife Service. 1979b. Arizona trout recovery plan. Arizona Trout Recovery Team, U.S. Fish and Wildlife Service, Albuquerque, New Mexico.

U.S. Fish and Wildlife Service. 1980a. Watercress

darter (*Etheostoma nuchale*) recovery plan. U.S. Fish and Wildlife Service.

U.S. Fish and Wildlife Service. 1980b. Devil's Hole pupfish recovery plan. Prepared in cooperation with the Devil's Hole pupfish recovery team. U.S. Fish and Wildlife Service.

U.S. Fish and Wildlife Service. 1982a. Snail darter recovery plan. U.S. Fish and Wildlife Service.

U.S. Fish and Wildlife Service. 1982b. Pecos gambusia (*Gambusia nobilis*) recovery plan. U.S. Fish and Wildlife Service, Albuquerque, New Mexico.

U.S. Fish and Wildlife Service. 1983a. The Moapa dace recovery plan. Prepared in cooperation with the Nevada Department of Wildlife.

U.S. Fish and Wildlife Service. 1983b. Gila and Yaqui topminnow recovery plan. U.S. Fish and Wildlife Service, Albuquerque, New Mexico.

U.S. Fish and Wildlife Service. 1983c. Arizona trout (Apache trout) recovery plan. Arizona Trout Recovery Team, U.S. Fish and Wildlife Service, Albuquerque, New Mexico.

U.S. Fish and Wildlife Service. 1984a. Recovery plan for woundfin, *Plagopterus argentissimus* Cope. U.S. Fish and Wildlife Service, Albuquerque, New Mexico.

U.S. Fish and Wildlife Service. 1984b. Recovery plan for the Mohave tui chub, *Gila bicolor* mohavensis. U.S. Fish and Wildlife Service, Portland, Oregon.

U.S. Fish and Wildlife Service. 1984c. Recovery plan for the Owens pupfish, *Cyprinodon radiosus*. U.S. Fish and Wildlife Service, Portland, Oregon.

U.S. Fish and Wildlife Service. 1985a. Paiute cutthroat trout recovery plan. U.S. Fish and Wildlife Service, Portland, Oregon.

U.S. Fish and Wildlife Service. 1985b. Recovery plan for the Pahranagat roundtail chub, *Gila robusta jordani*. U.S. Fish and Wildlife Service, Portland, Oregon.

U.S. Fish and Wildlife Service. 1990a. Spikedace recovery plan. U.S. Fish and Wildlife Service, Albuquerque, New Mexico.

U.S. Fish and Wildlife Service. 1990b. Colorado squawfish recovery plan. Denver, Colorado.

U.S. Fish and Wildlife Service. 1992. Watercress darter (*Etheostoma nuchale*) recovery plan, revised. U.S. Fish and Wildlife Service, Jackson, Mississippi.

U.S. Fish and Wildlife Service. 1993a. Endangered and threatened wildlife and plants. Federal Register, 50 CFR 17.11 & 17.12. (August 23, 1993). Washington, D.C.

U.S. Fish and Wildlife Service. 1993b. Desert pupfish recovery plan. U.S. Fish and Wildlife Service, Phoenix, Arizona.

U.S. Fish and Wildlife Service. 1993c. Gila trout recovery plan. U.S. Fish and Wildlife Service, Albuquerque, New Mexico.

U.S. Fish and Wildlife Service. 1993d. Big Spring spinedace, *Lepidomeda mollispinis pratensis*, recovery plan. U.S. Fish and Wildlife Service, Portland, Oregon.

U.S. Fish and Wildlife Service. 1993e. Lost River (*Deltistes luxatus*) and shortnose (*Chamistes brevirostris*) sucker recovery plan. U.S. Fish and Wildlife Service, Portland, Oregon.

U.S. Fish and Wildlife Service. 1994a. Endangered and threatened wildlife and plants: animal candidate review for listing as endangered or threatened species. Federal Register, 50 CFR 17.11 & 17.12. (November 15, 1994). 59(219):58982–589028. U.S. Department of the Interior, Washington, D.C.

U.S. Fish and Wildlife Service. 1994b. Yaqui fishes recovery plan. U.S. Fish and Wildlife Service, Albuquerque, New Mexico.

U.S. Fish and Wildlife Service. 1994c. Virgin River fishes recovery plan. U.S. Fish and Wildlife Service, Salt Lake City, Utah.

U.S. Fish and Wildlife Service. 1994d. Lahontan cutthroat trout, *Oncorhynchus clarki henshawi*, recovery plan. U.S. Fish and Wildlife Service, Portland, Oregon.

U.S. Fish and Wildlife Service. 1994e. White River spinedace, *Lepidomeda albivallis*, recovery plan. U.S. Fish and Wildlife Service, Portland, Oregon.

U.S. Fish and Wildlife Service. 1994f. Endangered Species Act Section 7 Biological Opinion: transportation and delivery of Central Arizona Project water to the Gila River basin (Hassayampa, Agua Fria, Salt, Verde, San Pedro, Middle and Upper Gila rivers and associated tributaries) in Arizona and New Mexico. U.S. Fish and Wildlife Service, Ecological Services Office, Phoenix, Arizona.

U.S. Fish and Wildlife Service. 1995. Greenback cutthroat trout recovery plan. U.S. Fish and Wildlife Service, Denver, Colorado.

U.S. Fish and Wildlife Service. 1997. Endangered and threatened species; Review of plant and animal taxa; Proposed Rule. Federal Register, 50 CFR 17. (September 19, 1997). 62(182):49397–49411. Washington, D.C.

Utah Division of Wildlife Resources. 1997. Current angling record fish for the state of Utah. (Available from http://www.nr.state.ut.us/dwr/recfish.htm. Page accessed 26 June 1997.)

Vanicek, C. D., R. H. Kramer, and D. R. Franklin. 1970. Distribution of Green River fishes in Utah and Colorado following closure of Flaming Gorge Dam. Southwestern Naturalist 14(3):297–315.

Van Dine, D. L. 1907. The introduction of top-minnows (natural enemies of the mosquitoes) into the Hawaiian Islands. The Hawaiian Forester and Agriculturist 20:1–10.

Van Oosten, J. 1937. The dispersal of smelt, *Osmerus mordax* (Mitchill), in the Great Lakes region. Transactions of the American Fisheries Society 66:16–171.

Vari, R. P. 1983. Phylogenetic relationships of the families Curimatidae, Prochilodontidae, Anostomidae, and Chilodontidae (Pisces: Characiformes). Smithsonian Contributions to Zoology 378.

Vigg, S., and P. A. Kucera. 1981. Contributions to the life history of Sacramento perch, *Archoplites interruptus*, in Pyramid Lake, Nevada. Great Basin Naturalist 41:278–289.

Villa, J. 1976. Systematic status of the cichlid fishes *Cichlasoma dorsatum*, *C. granadense* and *C. nigritum* Meek. Pages 375–383 *in* T. B. Thorson, editor. Investigations of the ichthyofauna of Nicaraguan Lakes. University of Nebraska, Lincoln.

Vladykov, V. D., and J. R. Greeley. 1963. Order Acipenseroidei. Pages 24–60 *in* Fishes of the western North Atlantic Part 3. Sears Foundation for Marine Research, Yale University, New Haven.

Vladykov, V. D., and E. Kott. 1980. Description and key to metamorphosed specimens and ammocoetes of Petromyzontidae found in the Great Lakes region. Canadian Journal of Fisheries and Aquatic Sciences 37:1616–1625.

von Geldern, C. E., Jr. 1966. The introduction of white bass (*Roccus chrysops*) into California. California Fish and Game 52(4):303.

Vovscko, J. 1997. Salmon cut loose from B.C. farms. National Fisherman 78(2):14.

Wagner, W. C. 1976. Pink salmon in Michigan tributaries of the Great Lakes, 1975. Michigan Department of Natural Resources, Marquette Fisheries Research Station.

Waldrip, L. 1992. Grass carp increasing in Trinity River. Texas Parks and Wildlife News November 20, 1992:5–7.

Wales, J. H. 1962. Introduction of pond smelt from Japan to California. California Fish and Game 48(2):141–142.

Walker, B. W., R. R. Whitney, and G. W. Barlow. 1961. Fishes of the Salton Sea. Pages 77–92 *in* B. W. Walker, editor. The ecology of the Salton Sea, California, in relation to the sport fishery of California. Fish Bulletin of the California Department of Fish and Game 113.

Walker, P. 1993. A list of the endemic and introduced fishes of Colorado – March, 1993. Colorado Division of Wildlife, Aquatic Resources Unit. Unpublished manuscript.

Wallace, R. L., J. S. Griffith, Jr., D. M. Daley, P. J. Connolly, and G. B. Beckham. 1984. Distribution of the Shoshone sculpin (*Cottus greenei*: Cottidae) in the Hagerman Valley of south central Idaho. Great Basin Naturalist 44(2):324–326.

Walters, D. M. 1997. The distribution, status, and ecology of the fishes of the Conasauga River system. Master's Thesis. University of Georgia, Athens.

Walters, L. L., and E. F. Legner. 1980. Impact of desert pupfish, *Cyprinodon macularius*, and *Gambusia affinis affinis* on fauna in pond ecosystems. Hilgardia 48(3):1–18.

Wang, J. C. S. 1984. On the ecological status of the tidewater goby, *Eucyclogobius newberryi* (Girard), in a lagoon and lake of the Golden Gate National Recreation Area, CA. Final Report to the National Park Service. Technical

Report 15. Contribution CPSU/UCD 337/1.

Wang, J. C. S. 1986. Fishes of the Sacramento–San Joaquin Estuary and adjacent waters, California: a guide to early life histories. Interagency Ecological Study Program for the Sacramento–San Joaquin Estuary, Technical Report 9. Sacramento, California.

Ware, F. J. 1971. Some early life history of Florida's inland striped bass, *Morone saxatilis*. Proceedings of the 24th Annual Conference Southeastern Association of Game and Fish Commissioners 24(1970):439–447

Ware, F. J. 1975. Progress with *Morone* hybrids in fresh water. Proceedings of the 28th Annual Conference Southeastern Association of Game and Fish Commissioners 28(1974):48–54.

Warren, M. A., and P. E. Denette. 1992. Addition of the northern studfish, *Fundulus catenatus* (Fundulidae) to the fish fauna of Louisiana. Proceedings of the Louisiana Academy of Sciences 55:31–33.

Warren, M. L., Jr. 1981. New distributional records of eastern Kentucky fishes. Brimleyana 6:129–140.

Warren, M. L., Jr. 1992. Variation of the spotted sunfish, *Lepomis punctatus* complex (Centrarchidae): meristics, morphometrics, pigmentation and species limits. Bulletin of the Alabama Museum of Natural History 12. Tuscaloosa.

Washington Department of Fish and Wildlife. 1997. Washington fishing guide: Sportfish of Washington. (Available from http://www.wa.gov/wdfw/outreach/fishing/wfg97/wfgsport..htm. Page accessed October 31, 1997.)

Waters, T. F. 1983. Replacement of brook trout by brown trout over 15 years in a Minnesota stream: production and abundance. Transactions of the American Fisheries Society 112:137–146.

Wattendorf, R. J., and C. Phillippy. 1996. Administration of a state program. Pages 130–151 *in* J. R. Cassani, editor. Managing aquatic vegetation with grass carp: a guide for water resource managers. American Fisheries Society, Introduced Fish Section, Bethesda, Maryland.

Webb, M. A., H. S. Elder, and R. G. Howells. 1994. Grass carp reproduction in the lower Trinity River, Texas. Pages 29–32 *in* Proceedings of the grass carp symposium. U.S. Army Corps of Engineers, Waterways Experiment Station, Vicksburg, Mississippi.

Weber, C. 1991. Nouveaux taxa dans *Pterygoplichthys* sensu lato (Pisces, Siluriformes, Loricariidae). Revue Suisse de Zoologie 98:637–643.

Weber, C. 1992. Révision du genre *Pterygoplichthys* sensu lato (Pisces, Siluriformes, Loricariidae). Revue Francaise d'Aquariologie 19:1–36.

Webster, D. A. 1942. The life histories of some Connecticut fishes. Pages 122–227 *in* A fishery survey of important Connecticut lakes. State Board of Fisheries and Game. Connecticut Geological and Natural History Survey 63.

Weise, J. G. 1957. The spring cave-fish, *Chologaster papilliferus*, in Illinois. Ecology 38(2):195–204.

Weitzman, S. H., and W. L. Fink. 1987. Neon tetra relationships and phylogenetic systematics. Tropical Fish Hobbyist 1987(October):72–77. Neptune City, New Jersey.

Weitzman, S. H., and W. L. Fink. 1983. Relationships of the neon tetras, a group of South American freshwater fishes (Teleostei, Characidae) with comments on the phylogeny of New World characiforms. Bulletin of the Museum of Comparative Zoology 150(6):339–395.

Welcomme, R. L. 1988. International introductions of inland aquatic species. FAO (Food and Agriculture Organization of the United Nations). Fisheries Technical Paper 294. Rome.

Welsch, R. L. 1996. Something's fishy in the cornbelt: jaws. Natural History 105(8):66–67.

Werner, R. G. 1972. Bluespotted sunfish, *Enneacanthus gloriosus*, in the Lake Ontario drainage, New York. Copeia 1972(4):878–879.

Werner, R. G. 1980. Freshwater fishes of New York state: a field guide. Syracuse University Press, Syracuse, New York.

Wheeler, A. 1969. The fishes of the British Isles and north-west Europe. Michigan State University Press, East Lansing.

Wheeler, A. 1978. Key to the fishes of northern Europe.

Frederick Warne Ltd., London.

Whitehead, J. P. 1985. FAO species catalogue. Volume 7. Clupeoid fishes of the world (suborder Clupeoidei). An annotated and illustrated catalogue of the herrings, sardines, pilchards, sprats, anchovies and wolf-herrings, part 1 – Chirocentridae, Clupeidae and Pristigasteridae. FAO (Food and Agriculture Organization of the United Nations). Fisheries Synopsis 125.

Whitehead, J. P., G. J. Nelson, and T. Wongratana. 1988. FAO species catalogue. Volume 7. Clupeoid fishes of the world (suborder Clupeoidei). An annotated and illustrated catalogue of the herrings, sardines, pilchards, sprats, anchovies and wolf-herrings, part 2 – Engraulidae. FAO (Food and Agriculture Organization of the United Nations). Fisheries Synopsis 125.

Whiteside, B. G. 1975. Additional distribution notes on the Mozambique tilapia. Texas Journal of Science 26(3/4):620.

Whiteside, B. G., and C. Berkhouse. 1992. Some new collection locations for six fish species. Texas Journal of Science 44(4):494.

Whitmore, D. H. 1983. Introgressive hybridization of smallmouth bass (*Micropterus dolomieui*) and Guadalupe bass (*Micropterus treculi*). Copeia 1983(3):672–679.

Whitmore, D. H., and T. R. Hellier. 1988. Natural hybridization between largemouth and smallmouth bass (*Micropterus*). Copeia 1988(2):493–496.

Whitmore, S. 1997. Aquatic nuisance species in region 6 of the Fish and Wildlife Service. U.S. Fish and Wildlife Service, Great Plains Fish and Wildlife Management Assistance Office, Pierre, South Dakota.

Whittier, T. R., and K. E. Hartel. 1997. First records of redear sunfish (*Lepomis microlophus*) in New England. Northeastern Naturalist 4(4):237–240.

Whitworth, W. R., P. L. Berrien, and W. T. Keller. 1968. Freshwater fishes of Connecticut. State Geological and Natural History Survey of Connecticut, Bulletin 101. Hartford.

Whitworth, W. R. 1996. Freshwater fishes of Connecticut. State Geological and Natural History Survey of Connecticut, Bulletin 114. Hartford.

Wick, R. J., T. A. Lytle, and C. M. Haynes. 1981. Colorado squawfish and humpback chub population monitoring. Colorado Division of Wildlife. Federal Aid in Fisheries Restoration Project SE-3-3, Job 1, Progress Report. Denver.

Wieland, W., W. L. Shelton, and J. S. Ramsey. 1982. Biological synopsis of the Mozambique tilapia (*Tilapia mossambica*). Final Report submitted to the National Fisheries Research Laboratory, U.S. Fish and Wildlife Service, Gainesville, Florida.

Wilde, G. R., and A. A. Echelle. 1992. Genetic status of Pecos pupfish populations after establishment of a hybrid swarm involving an introduced congener. Transactions of the American Fisheries Society 121:277–286.

Wilde, G. R., and A. A. Echelle. 1997. Morphological variation in intergrade pupfish populations from the Pecos River, Texas, U.S.A. Journal of Fish Biology 50:523–539.

Wiley, E. O., and R. L. Mayden. 1985. Species and speciation in phylogenetic systematics, with examples from the North American fish fauna. Annals of the Missouri Botanical Garden 72:596–635.

Williams, C. D., and J. E. Williams. 1981. Distribution and status of native fishes of the Railroad Valley system, Nevada. Cal-Nevada Wildlife Transactions 1981:48–51.

Williams, C. D., and J. E. Williams. 1989. Refuge management for the threatened Railroad Valley springfish in Nevada. North American Journal of Fisheries Management 9:465–470.

Williams, G. D., J. S. Desmond, and J. B. Zedler. 1998. Extension of 2 nonindigenous fishes, *Acanthogobius flavimanus* and *Poecilia latipinna*, into San Diego Bay marsh habitats. California Fish and Game 84(1):1–17.

Williams, H. M. 1972. Preliminary fecundity studies of the hybrid (striped bass × white bass) in two South Carolina reservoirs. Proceedings of the 25th Annual Conference Southeastern Association of Game and Fish Commissioners 25(1971):536–542.

Williams, H. M. 1976. Characteristics for distinguishing white bass, striped bass and their hybrid (striped bass × white bass). Proceedings of the 29th Annual Conference Southeastern Association of Game and Fish Commissioners 29(1975):168–172.

Williams, J. D. 1967. Records of three cyprinid fishes from the Tallapoosa River system. Journal of the Alabama Academy of Science 38.

Williams, J. D. 1975. Systematics of the percid fishes of the subgenus, *Ammocrypta*, genus *Ammocrypta*, with descriptions of two new species. Bulletin of the Alabama Museum of Natural History 1.

Williams, J. D., and G. H. Burgess. In prep. The shoal bass, *Micropterus cataractae* (Centrarchidae), a new species from the Apalachicola River basin, Alabama, Florida, and Georgia.

Williams, J. D., and D. P. Jennings. 1991. Computerized data base for exotic fishes: the western United States. California Fish and Game 77(2):86–93.

Williams, J. E. 1991. Preserves and refuges for native western fishes: history and management. Pages 171–189 *in* W. L. Minckley and J. E. Deacon, editors. Battle against extinction: native fish management in the American West. University of Arizona Press, Tucson.

Williams, J. E., D. W. Sada, C. D. Williams, and other members of the Western Division of Endangered Species Committee. 1988. American Fisheries Society guidelines for introductions of threatened and endangered fishes. Fisheries 13(5):5–11.

Williamson, G. R., and O. Tabeta. 1991. Search for *Anguilla* eels on the West Coast of North America and on the Aleutian and Hawaiian Islands. Japanese Journal of Ichthyology 38(3):315–317.

Wiltzius, W. J. 1981. Compendium of introduction date and state and federal annual stocking of various fishes in Colorado, 1972–1978. Colorado Division of Wildlife, Fort Collins, Colorado.

Wiltzius, W. J. 1985. Fish culture and stocking in Colorado, 1872–1978.Colorado Division of Wildlife. Division Report 12.

Winemiller, K. O., D. C. Taphorn, and A. Barbarino-Duque. 1997. Ecology of *Cichla* (Cichlidae) in two blackwater rivers of southern Venezuela. Copeia 1997(4):690–696.

Wing, B. L., C. M. Guthrie, III, and A. J. Gharrett. 1992. Atlantic salmon in marine waters of southeastern Alaska. Transactions of the American Fisheries Society 121:814–818.

Wingate, P. J. 1992. Zander–evaluate carefully before introducing. Page 32 *in* The In-Fisherman Walleye Guide 1992. Brainerd, Minnesota.

Wischnath, L. 1993. Atlas of livebearers of the world. Tropical Fish Hobbyist Publications, Inc., Neptune City, New Jersey.

Wisconsin Department of Natural Resources. 1997. State record fish. (Available from http://www.dnr.state.wi.us/fh/fish/3brecords.htm. Page accessed June 26, 1997.)

Wisconsin Department of Natural Resources. 1998. Wisconsin record fish. (Available from http://www.dnr.state.wi.us/org/water/fhp/fish/3brecords.htm. Page accessed March 24, 1998.)

Witte, K. E., and J. Schmidt. 1992. *Betta brownorum*, a new species of anabantoid (Teleostei: Belontiidae) from northwestern Borneo, with a key to the genus. Ichthyological Exploration of Freshwaters 2(4):305–330.

Wood, R. M., and R. L. Mayden. 1992. Systematics, evolution, and biogeography of *Notropis chlorcephalus* and *N. lutipinnis*. Copeia 1992:68–81.

Woodling, J. 1985. Colorado's little fish: a guide to the minnows and other lesser known fishes in the state of Colorado. Colorado Division of Wildlife, Denver.

Woolcott, W. S. 1962. Infraspecific variation in the white perch, *Roccus americanus* (Gmelin). Chesapeake Science 3(2):94–113.

Wooten, M. C., K. T. Scribner, and M. H. Smith. 1988. Genetic variability and systematics of *Gambusia* in the southeastern United States. Copeia 1988(2):283–289.

Worth, S. G. 1895. Report on the propagaton and distribution of food-fishes. Pages 78–138 *in* Report of the Commissioner for the year ending June 30, 1893, Part XIX. U.S. Commission of Fish and Fisheries, Washington, D.C.

Wright, D. 1992. Zander hot line. American Fisheries Society, Introduced Fish Section Newsletter, 11(3):2. Bethesda, Maryland.

Wright, S. 1995a. A fishy situation. Northwest

Arkansas Times (June 7) 129(350):A1. Fayetteville.

Wright, S. 1995b. Piranha or pacu? it's still a whale of a fish tale. Northwest Arkansas Times (June 7) 129(350):B6. Fayetteville.

Wydoski, R. S., and R. R. Whitney. 1979. Inland fishes of Washington. University of Washington Press, Seattle.

Wyoming Game and Fish, Fish Division. 1997. Wyoming state record fish. (Available from http://gf.state.wy.us/HTML/fish/fishrecs.htm. Page accessed June 25, 1997.)

Yerger, R. W. 1977. Fishes of the Apalachicola River. Florida Marine Research Publications 26:22–33.

Yerger, R. W., and H. A. Beecher. 1975. First records of two percid fishes in Florida freshwaters. Florida Scientist 38(3):142–143.

Young, C. R., J. M. Quattro, and J. M. Grady. 1998. Genetic relationships among disjunct populations of Waccamaw killifish. Program and Abstracts of the 78th Annual Meeting of the American Society of Ichthyologists and Herpetologists, University of Guelph, Guelph, Ontario, Canada.

Yurgens, K. C. 1954. Records of four cyprinid fishes of the genera *Notropis* and *Semotilus* from Central Texas. Copeia 1954(2):155–156.

Zale, A. V. 1987. Periodicity of habitation of a stenothermal spring run in north-central Florida by blue tilapia. North American Journal of Fisheries Management 7:575–579.

Zale, A. V., and R. W. Gregory. 1990. Food selection by early life stages of blue tilapia, *Oreochromis aureus*, in Lake George, Florida: overlap with sympatric shad larvae. Florida Scientist 53(2):123–129.

Zaret, T. M., and R. T. Paine. 1973. Species introduction in a tropical lake. Science 182:449–455.

Zimpfer, S. P., C. F. Bryan, and C. H. Pennington. 1987. Factors associated with the dynamics of grass carp larvae in the lower Mississippi River valley. Pages 102–108 *in* R. D. Hoyt, editor. 10th annual larval fish conference. American Fisheries Society, Symposium 2, Bethesda, Maryland.

Zuckerman, L. D., and R. J. Behnke. 1986. Introduced fishes in the San Luis Valley, Colorado. Pages 435–452 *in* R. H. Stroud, editor. Fish culture in fisheries management. Fish Culture Section and Fisheries Management Section, American Fisheries Society, Bethesda, Maryland.

Personal Communications

Armbruster, J. - Division of Fishes, National Museum of Natural History, Smithsonian Institution, Washington, D.C.

Barnett, B. - Florida Game and Fresh Water Fish Commission, Vero Beach, Florida. 1988.

Baxter, G. - retired biologist, Laramie, Wyoming.

Beam, J. - Chief of Hatcheries, Kansas Department of Wildlife and Parks, Pratt.

Behnke, R. J. - Department of Fishery and Wildlife Biology, Colorado State University, Fort Collins.

Banek, T. J. - Fisheries Management Biologist, Missouri Department of Conservation, Springfield, Missouri.

Beisser, G. - Georgia Game and Fish Commission, Calhoun, Georgia.

Bell, M. - SUNY Stony Brook, Stony Brook, New York.

Bender, T. - Pennsylvania Fish and Boat Commission, State College.

Bennett, C. - Colorado Division of Wildlife.

Bergman, G. T. - Research Technician, U.S. Geological Survey, Biological Resources Division, Florida Caribbean Science Center, Gainesville, Florida.

Bever, C. - Fisheries Biologist, Kansas Department of Wildlife and Parks, Manhatten. 1998.

Bond, C. - Oregon State University, Corvallis.

Bowers, N. - Research Assistant, Portland State University, Portland, Oregon.

Briggs, P. T. - Associate Aquatic Biologist, Finfish and Crustaceans Unit, New York State Department of Environmental Conservation, Stony Brook, (letter dated 19 September 1979 to James McCann)

Britski, H. A. - Museu de Zoologia, Universidade de São Paulo, Brazil.

Bruce, J. - Iowa Department of Natural Resources, Moravia, Iowa. Response to USGS/BRD-G nonindigenous questionaire. 1992.

Bryan, F. - Louisiana State University, Shreveport.

Buell, H. - District Superintendent of Hatcheries, New York Department of Environmental Conservation, Albany.

Burgess, G. - Florida Museum of Natural History, Gainesville.

Burkhead, N. - U.S. Geological Survey, Gainesville.

Burr, B. - Southern Illinois University at Carbondale, Illinois. 1995.

Busiahn, T. - U.S. Fish and Wildlife Service, Ashland, Wisconsin.

Carter, A. - Fisheries Division, Arkansas Game and Fish Comission, Little Rock. Response to USGS/BRD-G nonindigenous questionaire. 1992.

Cashner, R. - University of New Orleans, Louisiana.

Catchins, D. - Alabama Fish and Game, Montgomery.

Cavender, T. - Ohio State University, Museum of Biological Diversity, Columbus.

Ceas, P. - Eastern Kentucky University, Richmond.

Chick, J. - Florida International University, Miami.

Christmas, J. - Maryland Department of Natural Resources, Annapolis.

Churchill, T. - Statewide Reservoir Program Coordinator, Tennessee Wildlife Resources Agency, Nashville.

Cincotta, D. - West Virginia Department of Natural Resources, Elkville.

Clemmer, G. - Nevada Natural Heritage Program, Carson City.

Cordone, A. J. - Department of Fish and Game, Sacramento, California. Response to USGS/BRD-G non-indigenous questionaire. 1992.

Costa, W. J. E. M. - Department of Zoology, Universidade Federal do Rio de Janeiro, Rio de Janeiro, Brazil.

Courtenay, W., Jr. - Florida Atlantic University, Boc, Florida.

Cross, F. B. - Curator of Fishes, Museum of Natural History, University of Kansas, Lawrence.

Cox, K. - Vermont Department of Fish and Wildlife, North Springfield, Vermont.

Crutchfield, J. - Carolina Power and Light Company, Raleigh, North Carolina.

Cummings, K. - Illinois Natural History Survey, Champaigne.

Davis, R. M. - Associate Director, Freshwater Fisheries Division, Maryland Dept. of Nature Resources, Cumberland. Response to USGS/BRD-G nonindigenous questionaire. 1992.

Devick, W. S. - Division of Aquatic Resources, Hawaii Department of Land and Natural Resources, Honolulu.

Dextrase, A. - Ontario Ministry of Natural Resources, Aquatic Ecosystems Branch, Peterborough, Ontario, Canada.

Echelle, A. A. - Oklahoma State University, Stillwater.

Edwards, R. J. - University of Texas/Pan American, Edinburg, Texas.

Erickson, J. - South Dakota Department of Game, Fish and Parks, Rapid City. 1998.

Eschmeyer, W. - California Academy of Sciences, San Francisco.

Etnier, D. - University of Tennessee, Knoxville.

Ferraris, C. - California Academy of Sciences, San Francisco. 1997.

Feinberg, M. N. - Collection manager, American Museum of Natural History, New York. Letter to N. Burkhead dated 15 February 1989.

Fink, W. L. - Museum of Zoology-Division of Fishes, University of Michigan, Ann Arbor.

Fleming, K. - California Department of Fish and Game, Stockton.

Fletcher, D. - Warmwater Fisheries Resource Manager, Washington Department of Wildlife, Olympia, Washington. Response to USGS/BRD-G nonindigenous questionaire and other reports. 1992.

Flickinger, S. - Colorado State University, Ft. Collins, Colorado.

Florida Department of Natural Resources (FL DNR) (currently Florida Department of Environmental Protection), St. Petersburg.

Florida Game and Fresh Water Fish Commission - Tallahassee.

Forrester, J. - U.S. Fish and Wildlife Service, Baton Rouge, Louisiana.

Fox, D. - Florida Game and Freshwater Fish Commission, Okeechobee.

Francis, J. - Indiana Department of Natural Resources, Michigan City.

Fredenberg, W. - U.S. Fish and Wildlife Service, Kalispell, MT.

Freeman, B. - University of Georgia, Athens.

Gennings, R. M. - Georgia Department of Natural Resources, Atlanta. Response to USGS/BRD-G nonindigenous questionaire.

Gerrish, B. - Connecticut Department of Environmental Protection, Hartford.

Geving, S. A. - Natural Resource Specialist, Minnesota Department of Environmental Resources, Duluth. Letter to K. Schmidt dated June 6, 1996.

Gilbert, C. R. - Florida Museum of Natural History, Gainesville.

Gilmore, G. - Harbor Branch Marine Institute, Fort Pierce, Florida.

Good, S. - Vermont Department of Fish and Wildlife, Pittsford. 1998.

Goodrich, C. - Goodrich and Son Seafood, Oakhill, Florida. Conversation with Dawn Jennings 1985.

Goulding, M. - Senior Scientist, Rainforest Alliance, Gainesville, Florida.

Gunderson, J. L. - Sea Grant Extension Program, Duluth, Minnesota. Memo dated 7 July 1994 to Aquatic Nuisance Species Outreach Contacts.

Guttermuth, F.B. - Salt Lake City, Utah.

Halliwell, D. - Massachusetts Division of Fisheries, Westboro, Massachusetts.

Hanten, R. L. - Department of Game, Fish, and Parks, Pierre. Response to USGS/BRD-G nonindigenous questionaire.

Harris, P. M. - Department of Biological Sciences, University of Alabama, Tuscaloosa, Alabama. E-mail message to J. D. Williams on 14 May 1998.

Hartel, K. - Fish Department, Museum of Comparative Zoology, Harvard University, Cambridge, Massachusetts.

Hartley, R. - Massachusetts Department of Fish and Wildlife, Westborough.

Hartman, B. - Kansas Department of Wildlife and Parks, Pratt.

Hatch, M. - New Mexico. Response to USGS/BRD-G nonindigenous questionaire. 1992.

Heard, B. - National Marine Fisheries, Juneau, Alaska.

Heinrich, J. - Nevada Department of Wildlife. Letter dated October 8, 1993 to "Cooperators and Interested Parties".

Hendrickson, D. - Univerisity of Texas, Austin.

Herlong, D. - Carolina Power and Light Company, Apex, North Carolina.

Hess, T. - Vermont Department of Fish and Wildlife, Waterbury.

Hoeft, M. - West Virginia Department of Natural Resources, Pt. Pleasant.

Horak, D. - Colorado Division of Wildlife, Fort Collins. 1995.

Horner, R. - Fish Pathology and Aquaculture Coordinator, State of Illinois, Manito. Response to USGS/BRD-G nonindigenous questionaire. 1992.

Hornsby, J. - Alabama Game and Fish, Montgomery.

Horton, T. - Arkansas Fish and Wildlife Magazine, Little Rock.

Horton, W. - Staff Biologist, Idaho Fish and Game, Boise.

Howell, M. - Samford University, Birmingham, Alabama.

Howells, R. G. - Heart of the Hills Reserach Station, Texas Parks and Wildl. Dept., Ingram, Texas. Response to USGS/BRD-G non-indigenous questionaire. 1992.

Hubbs, C. - University of Texas, Austin, Texas.

Hughes Little, J. - District Fisheries Superintendent, Louisiana.

Icenhour, W. - Local resident of Winter Garden, Florida.
Irwin, E. - U.S. Geological Survey, Alabama Cooperative Fish and Wildlife Research Unit. Auburn.

Jelks, H. - U.S. Geological Survey, Gainesville, Florida.

Jenkins, R. E. - Roanoke College, Roanoke, Virginia.

Jennings, C. - U.S. Geological Survey, Biological Resources, University of Georgia Co-op, Athens.

Jennings, D. - U.S. Fish and Wildlife Service, Gainesville, Florida (formerly).

Jensen, D. - Minnestoa Sea Grant, Duluth.

Jones, R. A. - Fisheries Bureau, Departments of Environmental Protection, Hartford, Connecticut. Response to USGS/BRD-G nonindigenous questionaire. 1992.

Jude, D. J. - University of Michigan and Freshwater Physicians, Inc. Ann Arbor, Michigan.

Keppner, S. - U.S. Fish and Wildlife Service, Amherst, New York.

Kindt, K. - Biological Science Technician, U.S. Fish and Wildlife Service, Ashland, Wisconsin.

Kircheis, F.W. - Fishery Biologist, Dept. of Inland Fisheries and Wildlife, Augusta, Maine. Response to USGS/BRD-G nonindigenous questionaire. 1992.

Knight, C. - Ohio Department of Natural Resources, Fairport Harbor.

Konrad, M. - Iowa Department of Natural Resources, Des Moines.

Kowalski, G. - U.S. Geological Survey, Alabama Cooperative Fish and Wildlife Research Unit. Auburn.

Krummrich, J. - Florida Game and Fresh Water Fish Commission, Lake City. Personal communication to J. P. Clugston, September 1998.

Kullander, S. O. - Department of Vertebrate Zoology, Swedish Museum of Natural History,

Stockholm, Sweden.

Lapin, W. J. - Dept. of Environmental Management, Division of Fish and Wildlife, West Kingson, Rhode Island. Response to USGS/BRD-G nonindigenous questionaire. 1992.

Latta, W.C. - Fisheries Scientist Emeritus, Michigan DNR. Response to USGS/BRD-G nonindigenous questionaire. 1992.

Lauer, T. - Purdue University, West Lafayette, Indiana.

Lea, R. N. - California Department of Fish and Game, Monterey .

Li, H. - Professor, Oregon State University, Corvallis.

Logan, D. - Department of Fisheries and Wildlife, Oregon State University, Corvallis.

Lundberg, J. G. - Department of Ecology and Evolutionary Biology, University of Arizona, Tucson.

Lyons, J. - Wisconsin Department of Natural Resources, Monona, Wisconsin.

MacNeill, D. - New York Sea Grant, State University of New York at Brockport, Brockport.

Marsh, P. C. - Center for Environmental Studies, Arizona State University, Tempe.

Martin, C. - Department of Natural Resources and Environmental Conservation. Dover, Delaware. Response to USGS/BRD-G mailing on non-indigenous species. 1992.

Martin, R. T. - Graduate student, Department of Wildlife and Fisheries, Texas A & M University, College Station.

Mayden, R. - University of Alabama, Tuscaloosa.

McCann, J. - National Biological Service, Gainesville, Florida.

Mettee, S. - Geological Survey of Alabama, Tuscaloosa, Alabama.

Miller, J. B. - Division of Ichthyology, Florida Museum of Natural History, Gainesville. (currently with Division of Recreation and Parks, Florida Park Service, Hobe Sound)

Minckley, W. - Arizona State University, Tempe.

Mitchell, D. - Arkansas Game and Fish Commission.

Moore, V. - Idaho Fish and Game, Boise.

Moreno, D. - Cleveland Metroparks Zoo, Cleveland, Ohio.

Morrison, T. - Louisiana Department of Wildlife and Fisheries, Baton Rouge.

Moser, M. - Center for Marine Science Research, Wilmington, North Carolina.

Mosher, T. - Kansas Department of Wildlife and Parks, Emporia.

Moyle, P. - University of California, Davis.

Murray, C. - Pennsylvania Fish and Boating Commission, Fairview.

Mykre, R. - Army Corps Of Engineers, St. Louis, Missouri.

Nehring, B. - Colorado Division of Wildlife, Montrose.

Nelson, D. - University of Michigan, Ann Arbor.

Nickum, J. - U.S. Fish and Wildlife Service, Fish Technology Center, Bozeman, Montana.

Nico, L. G. - U.S. Geological Survey, Biological Resources Division, Florida Caribbean Science Center, Gainesville, Florida.

Ober, R. - Georgia Department of Natural Resources, Albany.

Owens, R. - National Biological Service, Oswego, New York.

Paruka, F. - USFWS, Panama City, Florida.

Peyton, M. - Central Nebraska Public Power and Irrigation District, Gothenburg.

Pflieger, W. - Missouri Department of Conservation, Columbia.

Pierson, M. - Alabama Power Company, Birmingham.

Pigg, J. - State of Oklahoma Department of Environmental Quality, Oklahoma City.

Platania, S. - University of New Mexico, Albuquerque,.

Pollock, W. Tennessee Wildlife Resources Agency, Nashville, Tennessee. Response to USGS/BRD-G non-indigenous questionaire. 1992.

Quattro, J. M. - Department of Biological Sciences, University of South Carolina, Columbia.

Raasch, M. - Author of Delaware fishes book, Wilmington.

Rauchenberger, M. - Department of Zoology, Fish Section, Natural History Museum, London.

Ray, M. - Texas Parks and Wildlife Department. Austin.

Rendall, J. - Minnesota Department of Natural Resources, St. Paul.

Robins, C. R. - University of Kansas, Lawrence (formerly University of Miami, Miami, Florida).

Robinson, J. W. - Missouri Department of Conservation, Columbia.

Rohde, F. - North Carolina Department of Environment, Health, and Natural Resources, Division of Marine Fisheries, Wilmington.

Ross, S. T. - University of Southern Mississippi, Museum of Ichthyology.

Roston, W. - Underwater photographer, Forsyth, Missouri.

Ruessler, S. - U.S. Geological Survey, Gainesville, Florida.

Saylor, C. - Tennessee Valley Authority.

Scheidegger, K. - Bureau of Fisheries Management, Madison, Wisconsin. Response to USGS/BRD-G non-indigenous questionaire. 1992.

Schiffer, C. N. - Pennsylvania Fish and Boat Commission, Belfonte.

Schmidt, B. - Chief Fisheries Mangement, Division of Wildlife Resources, Salt Lake City, UT. Response to USGS/BRD-G non-indigenous questionaire. 1992.

Schmidt, K. - Native Fish Conservancy, St. Paul, Minneapolis.

Scobie, C. - Ontario Federation of Anglers and Hunters, Peterborough, Ontario, Canada.

Seiler, D. - Atlantic salmon coordinator, Washington Department of Fish and Wildlife, Olympia.

Sheidegger, K. Bureau of Fisheries Managment, Madison, Wisconsin. Response to USGS/BRD-G nonindigenous questionaire. 1992.

Shelton, W. L. Zoology Department, University of Oklahoma, Norman.

Shiffer, C.N. - Herpetology and Endangered Species Coordinator, Pennsylvania Fish Commission, Bellfonte, Pennsylvania. 1992. Response to USGS/BRD-G non-indigenous questionaire and (letter dated 4 Jan 1979 to W. Courtenay).

Silvey, W. - Arizona Game and Fish Department, Phoenix.

Sinnott, T. - New York Department of Environmental Conservation, Albany.

Slade, J. - Fishery Biologist, U.S. Fish and Wildlife Service. Ashland, Wisconsin.

Smith, G. - Museum of Zoology, University of Michigan, Ann Arbor.

Smith-Vaniz, W. - Ichthyologist, U.S. Geological Survey, Biological Resources Division, Florida Caribbean Science Center, Gainesville, Florida.

Snodgrass, J. - Savannah River Ecology Lab, Aiken, South Carolina.

Soldwedel, R. H. - Chief, Bureau of Freshwater Fisheries, Department of Environmental Protection and Energy, Division of Fish, Game and Wildlife, Trenton, New Jersey. Response to USGS/BRD-G non-indigenous questionaire. 1992.

Southwick, R. - District Fisheries Supervisor, Virginia Department of Game and Inland Fisheries. Richmond. Response to USGS/BRD-G non-indigenous questionaire. 1992.

Starnes, W. - North Carolina State Museum of Natural History, Raleigh.

Stefferud, S. - U.S. Fish and Wildlife Service, Arizona Ecological Services Office, Phoenix.

Steinwand, T. North Dakota Game and Fish Department. Bismark.

Stolgitis, J.A. - Rhode Island Division of Fish and Game, (brief response to a letter from W. Courtenay dated 6 September 1979).

Stone, C. - South Dakota Department of Game, Fish and Parks, Chamberlain.

Stone, M. - Wyoming Game and Fish, Cheyenne.

Stroud, R. M. - Fisheries Biologist, South Carolina Wildlife and Marine Resources Department, Rock Hill. Letter to S. J. Walsh dated 25 August 1993.

Tagawa, A. - Division of Aquatic Resources, Hawaii Department of Land and Natural Resources, Honolulu.

Taylor, M. - Curator, Tulane University Museum.

Thomerson, J. E. - Department of Biology, Southern Illinois University at Edwardsville, Edwardsville.

Thomson, A. J. - Pacific Biological Station, Nanaimo, British Columbia.

Tilyou, G. A. - Inland Fish Division, Dept. of Wildlife and Fisheries, Baton Rouge, Louisiana. Response to USGS/BRD-G nonindigenous questionaire. 1992.

Todd, T. - Great Lakes Science Center, U.S. Geological Survey, Biological Resources Division, Ann Arbor, Michigan.

Unkenholz, D. - Chief of Fisheries, South Dakota Department of Game, Fish and Parks, Pierre,. 1998.

USGS/BRD-G - U.S. Geological Survey, Biological Resources Division, Florida Caribbean Science Center, Gainesville, Florida.

Vives, S. - Arizona State University, Tempe.

Walker, P. - Maine Department of Inland Fisheries (formerly); Colorado Division of Wildlife, Brush (currently).

Wakefield, W. - Institute of Marine and Coastal Sciences, Rutgers University, New Brunswick, New Jersey.

Ward, R. - Texas Parks and Wildlife, Palacios.

Wheeler, C. - University of Wyoming, Laramie.

Weitzman, S. H. - Smithsonian Institution, Washington, D.C.

Werner, R. - School of Environmental Science and Forestry, Syracuse University, Syracuse, New York.

White, R. - National Biological Service Co-op Unit, Bozeman, Montana.

Wibur, J. - Bureau of Reclamation, Albuquerque, New Mexico.

Williams, J. D. - U.S. Geological Survey, Biological Resources, Florida Caribbean Science Center, Gainesville, Florida.

Williams, J. E. - Science Advisor, Intermountain Research Station, Bureau of Land Management, Boise, Idaho.

Yamamoto, M. - Biologist, Division of Aquatic Resources, Hawaii Department of Land and Natural Resources, Honolulu.

Zadina, L. - Nebraska Game and Parks, Lincoln.

Zale, A. - Oklahoma State University (formerly), National Biological Service, Bozeman, Montana (currently).

Zuckerman, L. - Aquatic Ecologist, Kansas Department of Wildlife and Parks, Pratt.

Appendix A - List of Species Included in this Report

ORDER PETROMYZONTIFORMES

Petromyzontidae - Lampreys

Petromyzon marinus

ORDER ACIPENSERIFORMES

Acipenseridae - Sturgeons

Acipenser fulvescens

Acipenser transmontanus

Acipenser or *Scaphirhynchus* sp.

Polyodontidae - Paddlefishes

Polyodon spathula

ORDER SEMIONTIFORMES

Lepisosteidae - Gars

Atractosteus spatula

Lepisosteus cf. oculatus

Lepisosteus platostomus

Lepisosteus sp.

ORDER AMIIFORMES

Amiidae - Bowfin

Amia calva

ORDER OSTEOGLOSSIFORMES

Osteoglossidae - Bonytongues

Osteoglossum bicirrhosum

Hiodontidae - Mooneyes

Hiodon alosoides

Notopteridae - Featherfin knifefishes

Chitala ornata

ORDER ELOPIFORMES

Megalopidae - Tarpons

Megalops atlanticus

ORDER ALBULIFORMES

Albulidae - Bonefishes

Albula vulpes

ORDER ANGUILLIFORMES

Anguillidae - Freshwater eels

Anguilla anguilla

Anguilla australis

Anguilla rostrata

Anguilla spp.

ORDER CLUPEIFORMES

Engraulidae - Anchovies

Anchoa mundeoloides

Cetengraulis mysticetus

Engraulis mordax

Clupeidae - Herrings

Alosa aestivalis

Alosa chrysochloris

Alosa pseudoharengus

Alosa sapidissima

Dorosoma cepedianum

Dorosoma petenense

Opisthonema libertate

ORDER GONORYNCHIFORMES

Chanidae - Milkfish

Chanos chanos

ORDER CYPRINIFORMES

Cyprinidae - Minnows

Brachydanio rerio

Campostoma anomalum

Campostoma oligolepis

Carassius auratus

Carassius carassius

Clinostomus elongatus

Clinostomus funduloides

Couesius plumbeus

Ctenopharyngodon idella

Ctenopharyngodon idella × *Hypophthalmichthys nobilis*

Cyprinella analostana
Cyprinella galactura
Cyprinella lutrensis
Cyprinella pyrrhomelas
Cyprinella spiloptera
Cyprinella venusta
Cyprinella whipplei
Cyprinus carpio
Danio malabaricus
Eremichthys acros
Exoglossum laurae
Exoglossum maxillingua
Gila atraria
Gila bicolor
Gila copei
Gila orcutti
Gila pandora
Gila purpurea
Gila seminuda
Hesperoleucus symmetricus
Hybognathus hankinsoni
Hybognathus placitus
Hybognathus regius
Hypophthalmichthys molitrix
Hypophthalmichthys molitrix × *H. nobilis*
Hypophthalmichthys nobilis
Iotichthys phlegethontis
Lavinia exilicauda
Lepidomeda albivallis
Lepidomeda mollispinis
Leuciscus idus
Luxilus albeolus
Luxilus cardinalis
Luxilus cerasinus
Luxilus chrysocephalus
Luxilus coccogenis
Luxilus cornutus
Luxilus zonistius
Lythrurus ardens
Lythrurus atrapiculus
Meda fulgida
Moapa coriacea
Morulius chrysophekadion
Mylopharyngodon piceus
Nocomis biguttatus
Nocomis leptocephalus
Nocomis micropogon
Nocomis raneyi
Notemigonus crysoleucas
Notropis amblops
Notropis amoenus
Notropis ariommus
Notropis atherinoides
Notropis baileyi
Notropis bairdi
Notropis bifrenatus
Notropis blennius
Notropis boops
Notropis buccatus
Notropis buccula
Notropis buchanani
Notropis chiliticus
Notropis dorsalis
Notropis girardi
Notropis harperi
Notropis hudsonius
Notropis hypsilepis
Notropis leuciodus
Notropis longirostris
Notropis ludibundus
Notropis lutipinnis

Notropis oxyrhynchus
Notropis ozarcanus
Notropis potteri
Notropis procne
Notropis rubellus
Notropis rubricroceus
Notropis shumardi
Notropis spectrunculus
Notropis telescopus
Notropis texanus
Notropis volucellus
Notropis xaenocephalus
Orthodon microlepidotus
Phenacobius mirabilis
Phoxinus eos
Phoxinus oreas
Pimephales notatus
Pimephales promelas
Pimephales tenellus
Pimephales vigilax
Plagopterus argentissimus
Platygobio gracilis
Pogonichthys macrolepidotus
Ptychocheilus grandis
Pytchocheilus oregonensis
Ptychocheilus umpaquae
Puntius conchonius
Puntius filamentosus
Puntius gelius
Puntius schwanefeldii
Puntius semifasciolatus
Puntius tetrazona
Relictus solitarius
Rhinichthys atratulus
Rhinichthys cataractae
Rhinichthys chrysogaster
Rhinichthys osculus
Rhodeus sericeus
Richardsonius balteatus
Richardsonius egregius
Scardinius erythrophthalmus
Semotilus atromaculatus
Semotilus corporalis
Tinca tinca

Catostomidae - Suckers

Carpiodes carpio
Carpiodes cyprinus
Carpiodes velifer
Catostomus ardens
Catostomus catostomus
Catostomus commersoni
Catostomus fumeiventris
Catostomus occidentalis
Catostomus platyrhynchus
Catostomus plebeius
Catostomus santaanae
Catostomus sp. (undescribed)
Catostomus tahoensis
Erimyzon sucetta
Hypentelium etowanum
Hypentelium nigricans
Ictiobus bubalus
Ictiobus cyprinellus
Ictiobus niger
Minytrema melanops
Moxostoma carinatum
Moxostoma duquesnei
Moxostoma erythrurum
Moxostoma macrolepidotum
Scartomyzon cervinus
Scartomyzon lachneri
Scartomyzon rupiscartes

Thoburnia rhothoeca

Cobitidae - Loaches

Misgurnus anguillicaudatus

Misgurnus mizolepis

Pangio kuhlii

ORDER CHARACIFORMES

Anostomidae - Headstanders

Leporinus fasciatus

Erythrinidae - Trahiras

Hoplias malabaricus

Characidae - Characins

Aphyocharax anisitsi

Astyanax mexicanus

Colossoma macropomum

Gymnocorymbus ternetzi

Hemigrammus ocellifer

Hyphessobrycon cf. serpae

Metynnis spp.

Moenkhausia sanctaefilomenae

Myleus cf. rubripinnis

Paracheirodon innesi

Piaractus brachypomus

Piaractus mesopotamicus

Pygocentrus nattereri

Serrasalmus rhombeus

Colossoma or *Piaractus* spp.

Pygocentrus or *Serrasalmus* spp.

ORDER SILURIFORMES

Ictaluridae - Bullhead catfishes

Ameiurus brunneus

Ameiurus catus

Ameiurus melas

Ameiurus natalis

Ameiurus nebulosus

Ameiurus platycephalus

Ictalurus furcatus

Ictalurus pricei

Ictalurus punctatus

Noturus exilis

Noturus flavus

Noturus gilberti

Noturus gyrinus

Noturus insignis

Noturus miurus

Pylodictis olivaris

Clariidae - Labyrinth catfishes

Clarias batrachus

Clarias fuscus

Doradidae - Thorny catfishes

Platydoras costatus

Pseudodoras niger

Pterodoras granulosus

Pterodoras sp.

Auchenipteridae - Driftwood catfishes

Parauchenipterus galeatus

Pimelodidae - Long-whiskered catfishes

Perrunichthys perruno

Phractocephalus hemiliopterus

Callichthyidae - Plated catfishes

Callichthys callichthys

Corydoras aeneus

Corydoras sp.

Hoplosternum littorale

Loricariidae - Suckermouth catfishes

Ancistrus spp.

Hypostomus spp.

Otocinclus sp.

Peckoltia sp.

Pterygoplichthys disjunctivus

Pterygoplichthys multiradiatus

Pterygoplichthys pardalis

ORDER ESOCIFORMES

Esocidae - Pikes

Esox americanus

Esox lucius

Esox lucius × *E. masquinongy*

Esox lucius × *E. reicherti*

Esox masquinongy

Esox niger

Esox reicherti

Umbridae - Mudminnows

Dallia pectoralis

Novumbra hubbsi

Umbra limi

ORDER OSMERIFORMES

Osmeridae - Smelts

Hypomesus nipponensis

Osmerus mordax

Plecoglossus altivelis

ORDER SALMONIFORMES

Salmonidae - Trouts

Coregonus albula

Coregonus artedi

Coregonus clupeaformis

Coregonus lavaretus

Oncorhynchus aguabonita

Oncorhynchus clarki

Oncorhynchus clarki × *O. mykiss*

Oncorhynchus gilae

Oncorhynchus gorbuscha

Oncorhynchus keta

Oncorhynchus kisutch

Oncorhynchus masou

Oncorhynchus mykiss

Oncorhynchus nerka

Oncorhynchus tshawytscha

Prosopium abyssicola

Prosopium coulteri

Prosopium gemmifer

Prosopium williamsoni

Salmo letnica

Salmo salar

Salmo salar × *S. trutta*

Salmo trutta

Salmo trutta × *Salvelinus fontinalis*

Salvelinus alpinus

Salvelinus aureolus

Salvelinus fontinalis

Salvelinus fontinalis × *S. namaycush*

Salvelinus malma

Salvelinus namaycush

Thymallus arcticus

ORDER PERCOPSIFORMES

Percopsidae - Trout-perches

Percopsis omiscomacus

Aphredoderidae - Pirate Perch

Aphredoderus sayanus

Amblyopsidae - Cavefishes

Forbesichthys agassizi

ORDER GADIFORMES

Gadidae - Cods

Lota lota

ORDER MUGILIFORMES

Mugilidae - Mullets

Mugil cephalus

Mugil curema

ORDER ATHERINIFORMES

Melanotaeniidae - Rainbowfishes

Melanotaenia nigrans

Atherinidae - Silversides

Atherinops regius

Chirostoma jordani

Labidesthes sicculus
Leurethes sardina
Membras martinica
Menidia beryllina

ORDER BELONIFORMES

Adrianichthyidae - Ricefishes
Oryzias latipes

Belonidae - Needlefishes
Strongylura marina
Xenentodon cancila

ORDER CYPRINODONTIFORMES

Aplocheilidae - Old World rivulins
Aplocheilus lineatus
Nothobranchius guentheri

Rivulidae - New World rivulins
Cynolebias bellottii
Cynolebias nigripinnis
Rivulus hartii
Simpsonichthys whitei

Fundulidae - Topminnows
Fundulus catenatus
Fundulus chrysotus
Fundulus diaphanus
Fundulus grandis
Fundulus heteroclitus
Fundulus lineolatus
Fundulus notatus
Fundulus sciadicus
Fundulus seminolis
Fundulus stellifer
Fundulus waccamensis
Fundulus zebrinus
Lucania goodei
Lucania parva

Poeciliidae - Live bearers
Belonesox belizanus
Gambusia affinis
Gambusia holbrooki
Gambusia geiseri
Gambusia nobilis
Heterandria formosa
Limia vittata
Poecilia formosa
Poecilia latipinna
Poecilia latipinna × *P. velifera*
Poecilia latipunctata
Poecilia mexicana
Poecilia petenensis
Poecilia reticulata
Poecilia sphenops
Poeciliopsis gracilis
Poeciliopsis occidentalis
Xiphophorus hellerii
Xiphophorus hellerii × *X. maculatus*
Xiphophorus hellerii × *X. variatus*
Xiphophorus maculatus
Xiphophorus maculatus × *X. hellerii*
Xiphophorus maculatus × *X. variatus*
Xiphophorus variatus

Goodeidae - Splitfins
Ameca splendens
Crenichthys baileyi
Crenichthys nevadae
Empetrichthys latos

Cyprinodontidae - Pupfishes
Cyprinodon diabolis
Cyprinodon macularius
Cyprinodon nevadensis
Cyprinodon radiosus

Cyprinodon rubrofluviatilis
Cyprinodon salinus
Cyprinodon variegatus
Jordanella floridae

ORDER GASTEROSTEIFORMES

Gasterosteidae - Sticklebacks
Apeltes quadracus
Culaea inconstans
Gasterosteus aculeatus
Pungitius pungitius

ORDER SYNBRANCHIFORMES

Synbranchidae - Swamp eels
Monopterus albus

ORDER SCORPAENIFORMES

Cottidae - Sculpins
Cottus asper
Cottus bairdi
Cottus extensus

ORDER PERCIFORMES

Centropomidae - Snooks
Lates angustifrons
Lates mariae
Lates niloticus

Moronidae - Temperate basses
Morone americana
Morone americana × *M. saxatilis*
Morone chrysops
Morone chrysops × *M. mississippiensis*
Morone chrysops × *M. saxatilis*
Morone mississippiensis
Morone mississippiensis × *M. saxatilis*
Morone saxatilis

Serranidae - Sea basses
Paralabrax muculatofasciatus

Centrarchidae - Sunfishes
Ambloplites cavifrons
Ambloplites constellatus
Ambloplites rupestris
Archoplites interruptus
Centrarchus macropterus
Chaenobryttus gulosus
Enneacanthus gloriosus
Lepomis auritus
Lepomis cyanellus
Lepomis cyanellus × *L. macrochirus*
Lepomis gibbosus
Lepomis humilis
Lepomis macrochirus
Lepomis megalotis
Lepomis microlophus
Lepomis miniatus
Lepomis punctatus
Micropterus coosae
Micropterus dolomieu
Micropterus punctulatus
Micropterus salmoides
Micropterus sp. (undescribed)
Micropterus treculii
Pomoxis annularis
Pomoxis annularis × *P. nigromaculatus*
Pomoxis nigromaculatus

Percidae - Perches
Ammocrypta bifascia
Ammocrypta clara
Etheostoma blennioides
Etheostoma caeruleum
Etheostoma cragini
Etheostoma edwini

Etheostoma exile
Etheostoma fusiforme
Etheostoma nigrum
Etheostoma nuchale
Etheostoma olmstedi
Etheostoma simoterum
Etheostoma zonale
Gymnocephalus cernuus
Perca flavescens
Percina caprodes
Percina macrolepida
Percina maculata
Percina roanoka
Percina shumardi
Percina tanasi
Stizostedion canadense
Stizostedion canadense × *S. vitreum*
Stizostedion lucioperca
Stizostedion vitreum

Nematistiidae - Roosterfish
Nematistius pectoralis

Carangidae - Jacks and Pompanos
Caranx caballus
Trachinotus paitensis

Gerreidae - Mojarras
Eucinostomus argenteus
Eucinostomus gracilis

Haemulidae - Grunts
Anisotremus davidsonii

Sciaenidae - Drums
Aplodinotus grunniens
Bairdiella icistia
Cynoscion nebulosus
Cynoscion nebulosus × *C. xanthulus*
Cynoscion orthonopterus
Cynoscion parvipinnis
Cynoscion xanthulus
Leiostomus xanthurus
Menticirrhus nasus
Menticirrhus undulatus
Micropogonias megalops
Micropogonias undulatus
Pogonias cromis
Pogonias cromis × *Sciaenops ocellatus*
Sciaenops ocellatus
Totoaba macdonaldi

Kyphosidae - Sea chubs
Girella simplicidens

Kuhliidae - Flagtails
Kuhlia rupestris

Cichlidae - Cichlids
Aequidens pulcher
Astronotus ocellatus
Cichla ocellaris
Cichla temensis
Cichlasoma beani
Cichlasoma bimaculatum
Cichlasoma citrinellum
Cichlasoma cyanoguttatum
Cichlasoma managuense
Cichlasoma meeki
Cichlasoma nigrofasciatum
Cichlasoma octofasciatum
Cichlasoma salvini
Cichlasoma spilurum
Cichlasoma synspilum
Cichlasoma urophthalmus
Geophagus brasiliensis
Geophagus surinamensis

Hemichromis elongatus
Hemichromis letourneauxi
Heros severus
Labeotropheus sp.
Melanochromis auratus
Melanochromis johanni
Oreochromis aureus
Oreochromis macrochir
Oreochromis mossambicus
Oreochromis niloticus
Oreochromis urolepis
Pelvicachromis pulcher
Pseudotropheus spp.
Pseudotropheus zebra
Pterophyllum scalare
Pterophyllum spp.
Sarotherodon melanotheron
Symphysodon discus
Telmatochromis cf. bifrenatus
Tilapia mariae
Tilapia rendalli
Tilapia sparrmani
Tilapia zillii

Embiotocidae - Surfperches
Hysterocarpus traskii

Labridae - Wrasses
Halichoeres semicinctus
Tautoga onitis

Gobiidae - Gobies
Acanthogobius flavimanus
Gillichthys mirabilis
Gillichthys seta
Gobiosoma bosc
Neogobius melanostomus
Proterorhinus marmoratus
Tridentiger barbatus
Tridentiger bifasciatus
Tridentiger trigonocephalus

Scombridae - Mackerels and Tunas
Scomberomorus concolor

Stromateidae - Butterfishes
Peprilus ovatus
Peprilus semillimus

Anabantidae - Climbing gouramies
Anabas testudineus
Ctenopoma nigropannosum

Helostomatidae - Kissing gourami
Helostoma temminckii

Belontiidae - Gouramies
Betta splendens
Colisa fasciata
Colisa labiosa
Colisa lalia
Macropodus opercularis
Trichogaster leerii
Trichogaster trichopterus
Trichopsis vittata

Osphronemidae - Giant gourami
Osphronemus goramy

Channidae - Snakeheads
Channa micropeltes
Channa striata

ORDER PLEURONECTIFORMES

Paralichthyidae - Paralichthyid flounders
Etropus crossotus
Paralichthys aestuarius
Paralichthys albigutta
Paralichthys lethostigma
Paralicthys woolmani

Pleuronectidae - Righteye flounders
Hypsopsetta guttulata

Platichthys flesus
Platichthys stellatus

Appendix B - Glossary

alien: see nonindigenous.

aquarium release: a type of intentional introduction whereby a captive or pet fish is released into open waters by an aquarium-fish hobbyist.

aquatic nuisance species: a nonindigenous species that threatens the diversity or abundance of native species; the ecological stability of infested waters; or commercial, agricultural, aquacultural, or recreational activities dependent on such waters.

bait bucket introduction: a type of introduction involving a fish (usually small minnows), used or intended to be used as bait, that is released or escapes into open waters as a result of fishing activities.

contaminant introduction: see stock contaminant.

endemic: a species or any other taxonomic group that is native to a limited geographic area (e.g., a lake, drainage system, biogeographic region, country).

escape: a type of unintentional introduction whereby a fish escapes into open water from captive conditions such as an aquaculture facility, research facility, hatchery, ornamental fish farm, or zoological park.

established: an introduced organism with one or more reproducing or breeding populations in open waters.

exotic: see foreign.

extirpated: no longer present, either as a result of natural causes or because of eradication by humans.

foreign: an organism introduced from a foreign country; a species native to an area outside of, or foreign to, the national geographic area under discussion.

indigenous: occurring or found naturally in a particular area or ecosystem; historically occurring in a geographic range previous to the arrival of the first European settlers; a species that is a member of the native natural community.

intentional stocking: a type of introduction involving a fish intentionally released by humans into open water, usually performed for such purposes as sportfishing, commercial harvest, forage provision, or biological control. Intentional stockings are authorized if carried out by appropriate state or federal agencies. Unauthorized intentional stockings are those performed by individuals or groups without approval from appropriate government regulatory agencies.

introduced: an organism moved by humans (or by human actions) to an ecosystem or region where it was not found historically due to human actions (i.e., an individual, group, or population of organisms that occur in a particular locale because of human actions).

locally established: an introduced organism with one or more naturally reproducing populations but with a very restricted distribution and no evidence of natural range expansion (in general, limited to a relatively confined area, such as a small lake).

native: see indigenous.

naturalized: see established.

nonindigenous: an individual, group, or population of a species that is introduced into an area or ecosystem outside its historic or native geographic range. In this report, the term includes both foreign (i.e., exotic) and transplanted species, and is used synonymously with "alien," "nonnative" and "introduced."

nonnative: see nonindigenous.

open water: in this report, includes all lakes, rivers, streams, and springs; also any water bodies, such as reservoirs, ponds, canals, and drainage ditches, considered to be outside the boundaries or control of captive conditions (e.g., aquaculture facility, research facility, hatchery, ornamental fish farm, or zoological park). An open water may have either a permanent, temporary, or intermittent water connection (e.g., via flooding) with other aquatic systems.

reintroduction: the intentional release by humans of a species into a drainage, or portion of a drainage, in which it was indigenous in historic times but where it subsequently became locally extinct. (Note: Like many terms dealing with a species' geographic range, the term "reintroduction" is scale dependent. This report does not include fish that have been reintroduced into specific waterbodies or watersheds where the species historically occurred; however, the report does include species that have been "reintroduced" into a part of a drainage, even though the fish may have been previously native to another part of the same drainage.

restocking: the deliberate release by humans of a species into an area where it already occurs, usually with the intention of augmenting the existing population.

reported: refers to an introduced species which has been recorded (i.e., collected, stocked, or observed) from open waters but is not as yet known to be established.

stock contaminant: fish unintentionally stocked with, or instead of, another species or taxa (e.g., green sunfish mistakenly mixed with, or misidentified as, bluegill during stocking); fish released into open waters because one or a few individuals were inadvertently mixed in with a larger shipment of fish or fishes intended for stocking.

taxon: group of organisms of any taxonomic rank. Plural is taxa; in this report, it typically refers to a combination of fish species, hybrids, and unidentified fish or fishes.

transplant: an organism moved outside its native geographic range but within a country where it occurs naturally (i.e., one whose native range includes at least a portion of the country where it is found); a species moved by humans, either deliberately or accidentally, from an area where it is native, to another area outside its native distribution but within the same national geographic range. In this report, a transplant is a fish native to the United States that has been introduced into ecosystems within other parts of a state or the United States where the species did not occur prior to European colonization.

unsuccessful introduction: an introduced species that has failed to establish a self-sustaining or reproducing population.

vector: a biological pathway for a disease or parasite (i.e., an organism that transmits pathogens to various hosts).

Appendix C - Foreign Fish Matrix

Nonindigenous foreign fishes (i.e., exotics) introduced to inland waters are listed by state (state abbreviations follow U.S. Postal System - see Table 1 for listing).

Key to symbols:

E - established, one or more breeding populations known to exist (based on most recent available information).

E? - reported from state, possibly established, existence of a breeding population not confirmed.

X - only known established population (or populations) either eradicated or became extirpated.

X? - previously known to be established, most recent information indicates it may be extirpated.

R - reported from one or more localities, no evidence of establishment.

R? - reported, but record or records not fully confirmed (e.g., some doubt remains concerning identification).

asterisk (*) - indicates that we have located (or found published reference to) voucher specimen(s) that support one or more of a state's records (in most cases, the voucher has been deposited and catalogued in a recognized ichthyological collection; however, not all specimens have been examined to verify identification).

Symbols associated with scientific names:

(?) - provisional identification of fish from all or most states where recorded.

(M) - strictly marine species (most marine species introduced to inland waters involved the Salton Sea, California).

For states where a species is listed as established, possibly established, or extirpated, other records may indicate the species has been reported from one or more sites within the same state.

Family	Scientific Name	Common Name	AL	AK	AZ	AR	CA	CO	CT	DE	FL	GA	HI	ID	IL	IN	IA	KS	KY
Adrianichthyidae	*Oryzias latipes*	Japanese medaka					X						R						
Anabantidae	*Anabas testudineus*	climbing perch									X								
Anabantidae	*Ctenopoma nigropanosum*	twospot climbing perch									X								
Anguillidae	*Anguilla anguilla* (?)	European eel					R*												
Anguillidae	*Anguilla australis*	shortfin eel					R*												
Anguillidae	*Anguilla* sp.	unidentified anguillid eel					R*												
Anostomidae	*Leporinus fasciatus*	banded leporinus									R		R						
Aplocheilidae	*Aplocheilus lineatus*	striped panchax											R						
Aplocheilidae	*Nothobranchius guentheri*	redtail notho											R						
Atherinidae	*Atherinops regius* (M)	silverside					R												
Atherinidae	*Chirostoma jordani* (?)	charal																	
Atherinidae	*Leuresthes sardina* (M)	Gulf grunion					R												
Auchenipteridae	*Parauchenipterus galeatus* (?)	driftwood catfish			R														
Belonidae	*Xenentodon cancila*	Asian needlefish											E*						
Belontiidae	*Betta splendens*	Siamese fighting fish							R		X*								
Belontiidae	*Colisa fasciata*	banded gourami																	
Belontiidae	*Colisa labiosa*	thicklipped gourami									R								
Belontiidae	*Colisa lalia*	dwarf gourami									R*								
Belontiidae	*Macropodus opercularis*	paradisefish									R*								
Belontiidae	*Trichogaster leerii*	pearl gourami									R		R						
Belontiidae	*Trichogaster trichopterus*	blue gourami									R*								
Belontiidae	*Trichopsis vittata*	croaking gourami									E*								
Bothidae	*Etropus crossotus* (M)	fringed flounder					R												
Bothidae	*Paralichthys aestuarius* (M)	halibut					R												
Bothidae	*Paralichthys woolmani* (M)	halibut					R												
Callichthyidae	*Callichthys callichthys*	cascarudo									R								
Callichthyidae	*Corydoras aeneus* (?)	green corydoras											E						
Callichthyidae	*Corydoras* sp.	corydoras						R			R								
Callichthyidae	*Hoplosternum littorale*	brown hoplo									E*								
Carangidae	*Caranx caballus* (M)	green jack					R												
Carangidae	*Trachinotus paitensis* (M)	paloma pompano					R												
Centropomidae	*Lates angustifrons*	Tanganyika lates																	
Centropomidae	*Lates mariae*	bigeye lates																	
Centropomidae	*Lates niloticus*	Nile perch																	
Channidae	*Channa micropeltes*	giant snakehead																	
Channidae	*Channa striata*	chevron snakehead											E*						
Characidae	*Aphyocharax anisitsi*	bloodfin tetra									R*								
Characidae	*Colossoma macropomum* (?)	tambaquí					R				R*		R						
Characidae	*Colossoma/Piaractus* sp. (?)	unidentified pacu			R	R					R	R						R	
Characidae	*Gymnocorymbus ternetzi*	black tetra						R			R*								
Characidae	*Hemigrammus ocellifer*	head-and-taillight tetra						R											
Characidae	*Hyphessobrycon serpae*	serpae tetra									R								
Characidae	*Metynnis* sp. (?)	metynnis									R								R?
Characidae	*Moenkhausia sanctaefilomenae*	redeye tetra									R*								
Characidae	*Myleus rubripinnis* (?)	redhook pacu																	
Characidae	*Paracheirodon innesi*	neon tetra						R											
Characidae	*Piaractus brachypomus*	pirapatinga	R*			R	R				R*	R*			R*				R*
Characidae	*Piaractus mesopotamicus* (?)	small-scaled pacu									R?								
Characidae	*Pygocentrus nattereri*	red piranha									R*		R*						
Characidae	*Pygocentrus/Serrasalmus* sp. (?)	unidentified piranha					R		R		R			R	R		R		R
Characidae	*Serrasalmus rhombeus*	redeye piranha									X*								
Cichlidae	*Aequidens pulcher* (?)	blue acara									X?								
Cichlidae	*Astronotus ocellatus*	oscar									E*	R	E						
Cichlidae	*Cichla ocellaris* (?)	peacock cichlid									E*		E						
Cichlidae	*Cichla temensis*	speckled pavon									R								
Cichlidae	*Cichlasoma beani*	green guapote					X												
Cichlidae	*Cichlasoma bimaculatum*	black acara									E*								
Cichlidae	*Cichlasoma citrinellum*	Midas cichlid									E*		E						
Cichlidae	*Cichlasoma labiatum*	red devil											R						
Cichlidae	*Cichlasoma managuense*	jaguar guapote									E*		E*						
Cichlidae	*Cichlasoma meeki*	firemouth cichlid			R						E*		E						
Cichlidae	*Cichlasoma nigrofasciatum*	convict cichlid			E?						X?		E	E*					
Cichlidae	*Cichlasoma octofasciatum*	Jack Dempsey					R		R		E*								
Cichlidae	*Cichlasoma salvini*	yellowbelly cichlid									E*								

LA	ME	MD	MA	MI	MN	MO	MS	MT	NE	NV	NH	NJ	NM	NY	NC	ND	OH	OK	OR	PA	RI	SC	SD	TN	TX	UT	VT	VA	WA	WV	WI	WY
														X																		
										R									R													
																									R?							
																				R												
R*																																
														R?																		
																									R							
																									R							
																									R							
	R		R*																		R*											
			R																						R							
R					R	R								R	R		R		R*			R			R			R	R			
			R?																													
			R*		R	R	R*	R*	R						R				R*						R*			R				
																			R*													
			R*	R*	R*												R	R*		R					R*			R				
						R					R			R				R		R						R			R			
R			R				R*					R								R	R				R			R				
																									X							
																									X?							
																										X?*						
										E															R							R

Family	Scientific Name	Common Name	AL	AK	AZ	AR	CA	CO	CT	DE	FL	GA	HI	ID	IL	IN	IA	KS	KY
Cichlidae	*Cichlasoma spilurum*	blue-eyed cichlid											E						
Cichlidae	*Cichlasoma synspilum*	redhead cichlid																	
Cichlidae	*Cichlasoma trimaculatum*	threespot cichlid									X*								
Cichlidae	*Cichlasoma urophthalmus*	Mayan cichlid									E*								
Cichlidae	*Geophagus brasiliensis* (?)	pearl eartheater									R?								
Cichlidae	*Geophagus surinamensis* (?)	redstriped eartheater									E*								
Cichlidae	*Hemichromis elongatus*	banded jewelfish											E						
Cichlidae	*Hemichromis letourneauxi*	jewelfish									E*								
Cichlidae	*Heros severus*	banded cichlid									E*								
Cichlidae	*Labeotropheus* sp. (?)	scrapermouth cichlid									R?								
Cichlidae	*Melanochromis auratus*	golden mbuna																	
Cichlidae	*Melanochromis johanni*	bluegray mbuna											E?						
Cichlidae	*Oreochromis aureus*	blue tilapia	R		E		E	E?			E*	R		E?				R*	
Cichlidae	*Oreochromis macrochir*	longfin tilapia											E						
Cichlidae	*Oreochromis mossambicus*	Mozambique tilapia	R*		E		E*	E			E*	X	E*	E*	R				
Cichlidae	*Oreochromis niloticus*	Nile tilapia	R		R						E?*	E?*							
Cichlidae	*Oreochromis urolepis*	Wami tilapia					E												
Cichlidae	*Pelvicachromis pulcher*	rainbow krib											E						
Cichlidae	*Pseudotropheus* sp.	African Lake cichlid																	
Cichlidae	*Pseudotropheus zebra*	zebra mbuna																	
Cichlidae	*Pterophyllum scalare*	freshwater angelfish					R				R								R
Cichlidae	*Pterophyllum* sp.	freshwater angelfish					R	R			R								
Cichlidae	*Sarotherodon melanotheron*	blackchin tilapia									E*		E						
Cichlidae	*Symphysodon discus*	red discus						R											
Cichlidae	*Telmatochromis bifrenatus* (?)	Lake Tanganyika dwarf cichlid									R*								
Cichlidae	*Tilapia mariae*	spotted tilapia			R						E*								
Cichlidae	*Tilapia rendalli*	redbreast tilapia											E?						
Cichlidae	*Tilapia sparrmannii*	banded tilapia									R*								
Cichlidae	*Tilapia zillii*	redbelly tilapia	R		E*	R	E				X*		E	R					
Cichlidae	*tilapine cichlids*	unidentified tilapias			R						R	R	R	R			R	R	R
Clariidae	*Clarias batrachus*	walking catfish					R*		R		E*	R							
Clariidae	*Clarias fuscus*	whitespotted clarias											E*						
Cobitidae	*Misgurnus anguillicaudatus*	Oriental weatherfish					E*				E*		E	E*	E*				
Cobitidae	*Misgurnus mizolepis*	Chinese fine-scaled loach																	
Cobitidae	*Pangio kuhlii*	coolie loach									R*								
Cyprinidae	*Brachydanio rerio*	zebra danio					R		R		R*								
Cyprinidae	*Carassius auratus*	goldfish	E?*		E?*	E?*	E*	E?	E	E	R*	E*	E	E	E*	E*	E*	E?*	E
Cyprinidae	*Carassius carassius* (?)	crucian carp																	
Cyprinidae	*Ctenopharyngodon idella*	grass carp	R		R	E	R	R	R	R	R*	R	R	R	E*	R	R	R	E
Cyprinidae	*Cyprinus carpio*	common carp	E*		E*	E*	E*	E*	E	E*	E*	E*	E	E*	E*	E*	E*	E*	E*
Cyprinidae	*Danio malabaricus*	Malabar danio									R*								
Cyprinidae	*Hypophthalmichthys molitrix*	silver carp	R		R	R		R			R*				E*	R		R	R
Cyprinidae	*Hypophthalmichthys nobilis*	bighead carp	R		R	R	R	R			R				E*	R	R	R*	R
Cyprinidae	*Leuciscus idus*	ide							X										
Cyprinidae	*Morulius chrysophekadion*	black sharkminnow									R								
Cyprinidae	*Mylopharyngodon piceus*	black carp																	
Cyprinidae	*Puntius conchonius*	rosy barb									R								
Cyprinidae	*Puntius filamentosus*	blackspot barb											E						
Cyprinidae	*Puntius gelius*	dwarf barb									R								
Cyprinidae	*Puntius schwanenfeldii*	tinfoil barb									R								
Cyprinidae	*Puntius semifasciolatus*	green barb											E						
Cyprinidae	*Puntius tetrazona*	tiger barb					R*				R								
Cyprinidae	*Rhodeus sericeus*	bitterling																	
Cyprinidae	*Scardinius erythrophthalmus*	rudd	R			R		R	R?						R*			X	
Cyprinidae	*Tinca tinca*	tench	R?		R	R?	E	E*	X?	X	R?	R		E*	R	R	R	R	R?
Doradidae	*Oxydoras niger*	ripsaw catfish									R								
Doradidae	*Platydoras costatus*	Raphael catfish									R								
Doradidae	*Pterodoras granulosus*	granulated catfish									R*								
Doradidae	*Pterodoras* sp. (?)	thorny catfish									R?								
Engraulidae	*Anchoa mundeoloides* (M)	Northern Gulf anchovy					R												
Erythrinidae	*Hoplias malabaricus*	trahira									X*								
Esocidae	*Esox reicherti*	Amur pike																	
Gobiidae	*Acanthogobius flavimanus*	yellowfin goby					E*												
Gobiidae	*Gillichthys seta* (M)	mudsucker					R												

LA	ME	MD	MA	MI	MN	MO	MS	MT	NE	NV	NH	NJ	NM	NY	NC	ND	OH	OK	OR	PA	RI	SC	SD	TN	TX	UT	VT	VA	WA	WV	WI	WY
																							R*									
										R																						
										X?																						
										R																						
										R																						
										E					E*			E?		E?*					E*							
								X						R	X										E?*							
																									R							
										R																						
										E*																						
										X*					E*							E			E							
R							R					R	R						R						R						R	R
			R*							R																						
				E*															E*					R*								
																			R*													
													E																			
E?	R	E*	E*	E*	E?	E*	E?*	E?	E*	E*	E*	E?*	E?	E*	E?*	E?	E*	E	E?	E*	E*	E*	E	E*	E*	R	R	E*	E	R*	E	E
																									R?							
E*		R	R	R*	E?	E	E?		R	R	R	R	R	R*	E	R	R	R	R	R		R	R	E	E	R		R	R	R	R	R
E*	R	E*	E*	E*	E*	E*	E*	E	E*	E*	E*	E*	E*	E*	E*	E*	E*	E*	E*	E*	E	E*	E*	E*	E*	E*	E	E*	E*	E*	E*	E*
										R																						
E?*						R																		R								
E?*						E	E?										R*	R						R	R					R		
	E?	X?							R					X*						X				R	R			X				
						R																										
																																R
														E*																		
	E		E*			R			E			X		E*				R		R			E		R		R*	R		R	X	
R		E?*	R?	R		R*			R	R			X*	X?	R		R	R	X*	R		R		R	R	R		X*	E*	R?	R	
																				X?												

Family	Scientific Name	Common Name	AL	AK	AZ	AR	CA	CO	CT	DE	FL	GA	HI	ID	IL	IN	IA	KS	KY
Gobiidae	*Neogobius melanostomus*	round goby													E*	E			
Gobiidae	*Proterorhinus marmoratus*	tubenose goby																	
Gobiidae	*Tridentiger barbatus*	Shokihaze goby					E?*												
Gobiidae	*Tridentiger bifasciatus*	Shimofuri goby					E*												
Gobiidae	*Tridentiger trigonocephalus*	chameleon goby					E*												
Goodeidae	*Ameca splendens*	butterfly splitfin																	
Helostomatidae	*Helostoma temminckii*	kissing gourami									R*								
Kuhlidae	*Kuhlia rupestris*	nato											R						
Kyphosidae	*Girella simplicidens* (M)	opaleye					R												
Loricariidae	*Ancistrus* sp.	suckermouth catfish											E*						
Loricariidae	*Hypostomus* spp.	suckermouth catfish			R			R	R?		X?*		E*						
Loricariidae	*Otocinclus* sp. (?)	suckermouth catfish						R											
Loricariidae	*Peckoltia* sp. (?)	clown knife											R						
Loricariidae	*Pterygoplichthys disjunctivus*	vermiculated sailfin catfish									E*								
Loricariidae	*Pterygoplichthys multiradiatus*	Orinoco sailfin catfish									E*		E						
Loricariidae	*Pterygoplichthys pardalis*	Amazon sailfin catfish																	
Melanotaeniidae	*Melanotaenia nigrans*	black-banded rainbowfish													R				
Notopteridae	*Chitala ornata*	clown knife									E?								
Osmeridae	*Hypomesus nipponensis*	wakasagi					E*												
Osmeridae	*Plecoglossus altivelis*	ayu					R						R						
Osphronemidae	*Osphronemus goramy*	giant gourami											R						
Osteoglossidae	*Osteoglossum bicirrhosum*	arawana					R												
Percidae	*Gymnocephalus cernuus*	ruffe																	
Percidae	*Stizostedion lucioperca*	zander																	
Pimelodidae	*Perrunichthys perruno*	leopard catfish																	
Pimelodidae	*Phractocephalus hemioliopterus*	redtail catfish									R								
Pimelodidae	*Rhamdia quelan*	bagre									R*								
Pleuronectidae	*Platichthys flesus* (M)	European flounder																	
Poeciliidae	*Belonesox belizanus*	pike killifish									E*								
Poeciliidae	*Limia vittata*	Cuban limia											E						
Poeciliidae	*Poecilia latipunctata*	broadspotted molly									R								
Poeciliidae	*Poecilia mexicana* (?)	shortfin molly			R		E*	E					E	E					
Poeciliidae	*Poecilia petenensis*	swordtail molly									R								
Poeciliidae	*Poecilia reticulata*	guppy			E?		R	R	R		E?		E	E					
Poeciliidae	*Poecilia sphenops* (?)	liberty molly					R?				E?		E?						
Poeciliidae	*Poeciliopsis gracilis*	porthole livebearer					E*												
Poeciliidae	*Poeciliopsis* sp.	livebearer							R										
Poeciliidae	*Xiphophorus hellerii*	green swordtail			X		R	E			E*		E*	E?					
Poeciliidae	*Xiphophorus maculatus*	southern platyfish					X*	E?			E*		E						
Poeciliidae	*Xiphophorus variatus*	variable platyfish			X		E?*	R			E*		**E?**						
Rivulidae	*Cynolebias bellottii*	Argentine pearlfish					R												
Rivulidae	*Cynolebias nigripinnis*	blackfin pearlfish					R												
Rivulidae	*Simpsonichthys whitei*	Rio pearlfish					R												
Rivulidae	*Rivulus hartii*	giant rivulus					E*												
Salmonidae	*Coregonus albula*	vendace																	
Salmonidae	*Coregonus lavaretus*	powan																	
Salmonidae	*Hucho hucho* (?)	huchen																	
Salmonidae	*Oncorhynchus masou*	cherry salmon																	
Salmonidae	*Salmo letnica*	Ohrid trout						R											
Salmonidae	*Salmo trutta*	brown trout	R*		E*	E	E*	E*	E	E	R*	E*	R	E	E*	E	E*	R	E?
Sciaenidae	*Bairdiella icistia* (M)	bairdiella					R*												
Sciaenidae	*Cynoscion othonopterus* (M)	scaly corvina					R												
Sciaenidae	*Cynoscion xanthulus* (M)	orangemouth corvina					E*												
Sciaenidae	*Menticirrhus nasus* (M)	corbina					R												
Sciaenidae	*Micropogonias megalops* (M)	croaker					R												
Sciaenidae	*Totoaba macdonaldi* (M)	totuava					R												
Synbranchidae	*Monopterus albus* (?)	swamp eel									E*	E*	E*						

LA	ME	MD	MA	MI	MN	MO	MS	MT	NE	NV	NH	NJ	NM	NY	NC	ND	OH	OK	OR	PA	RI	SC	SD	TN	TX	UT	VT	VA	WA	WV	WI	WY
				E*	E									R?			E*			R											E	
				E*																												
										R																						
R?										E*										R*					E*							
																						R*										
																													R			
										R																						
				E	E*																										E*	
														R		R																
																									R							
				R*													R*															
																									X							
								X		E*															R							
						R		R		E*			E												E*	R					R	E*
								E		E																						
								E		E*								R*							E							E
								R		E?*															R							E
								E																								
	R?																															
	R			R																												
												R?																				
				R																									R*			
					R			R																R								R
	E*	E*	E*	E*	E*	E		E*	E	E	E*	E*	E*	E*	E*	E	R	R	E*	E*	E	E	E*	E?*	R*	E*	E	E*	E*	E*	E*	E*
													R																			
													R												R							

Appendix D - Native Fish Matrix

Native fishes introduced to inland waters outside of their native range (i.e., transplants) are listed by state (state abbreviations follow U.S. Postal System - see Table 1 for listing).

Key to symbols:
N - native to the state
I - introduced to a nonnative area of the state
N/I - native to one area of the state and introduced to a nonnative area in the state
N?/I? - native or introduced status uncertain
N/I? - possibly introduced

Family	Scientific Name	Common Name	AK	AL	AR	AZ	CA	CO	CT	DE	FL	GA	HI	IA	ID	IL	IN	KS
Acipenseridae	*Acipenser fulvescens*	lake sturgeon		N	N							N		N		N	N	N
Acipenseridae	*Acipenser transmontanus*	white sturgeon				I	N/I					I			N/I			
Acipenseridae	*Acipenser or Scaphirhynchus sp.*	sturgeon (eastern species)		N	N					N	N	N		N		N	N	
Albulidae	*Albula vulpes*	bonefish		N			N/I				N							
Amblyopsidae	*Forbesichthys agassizi*	spring cavefish														N/I		
Amiidae	*Amia calva*	bowfin		N	N				I		N	N		N/I		N/I	N	N/I
Anguillidae	*Anguilla rostrata*	American eel		N	N	I	I	I	N	N	N	N		N		N/I	N/I	N
Anguillidae	*Anguilla sp.*	unidentified eel																
Aphredoderidae	*Aphredoderus sayanus*	pirate perch		N	N					N	N	N		N		N	N	
Atherinidae	*Labidesthes sicculus*	brook silverside		N	N						N	N		N		N	N	N
Atherinidae	*Membras martinica*	rough silverside		N						N	N	N						
Atherinidae	*Menidia beryllina*	inland silverside		N	N/I		I		N	N	N	N				N		
Belonidae	*Strongylura marina*	inland silversides		N/I					N	N	N	N						
Bothidae	*Paralichthys albiguttata*	Atlantic needlefish		N							N	N						
Bothidae	*Paralichthys lethostigma*	river carpsucker		N							N	N						
Catostomidae	*Carpiodes carpio*	quillback		N	N			N				N		N		N	N	N
Catostomidae	*Carpiodes cyprinus*	highfin carpsucker		N	N					N	N	N		N		N/I	N	N
Catostomidae	*Carpiodes velifer*	Utah sucker		N	N						N			N		N	N	N
Catostomidae	*Catostomus ardens*	longnose sucker													N			
Catostomidae	*Catostomus catostomus*	white sucker						N/I	I						N	N	N	
Catostomidae	*Catostomus commersoni*	Owens sucker		N	N			N/I	N	N		N/I		N		N	N	N
Catostomidae	*Catostomus fumeiventris*	Sacramento sucker					N/I											
Catostomidae	*Catostomus occidentalis*	dusky mountain sucker					N/I											
Catostomidae	*Catostomus platyrhynchus*	mountain sucker				I	N	N							N			
Catostomidae	*Catostomus plebeius*	Rio Grande sucker				I		N/I										
Catostomidae	*Catostomus santaanae*	Santa Ana sucker					N/I											
Catostomidae	*Catostomus tahoensis*	Tahoe sucker					N/I											
Catostomidae	*Catostomus undescribed sp.*	Little Colorado River sucker				N/I												
Catostomidae	*Chasmistes brevirostris*	shortnose sucker					N											
Catostomidae	*Erimyzon sucetta*	lake chubsucker		N	N						N	N		N		N/I	N	
Catostomidae	*Hypentelium etowanum*	Alabama hog sucker		N								N						
Catostomidae	*Hypentelium nigricans*	northern hog sucker		N	N							N		N		N	N	
Catostomidae	*Ictiobus bubalus*	smallmouth buffalo		N	N	I	I					N		N		N	N	N
Catostomidae	*Ictiobus cyprinellus*	bigmouth buffalo		N/I	N	I	I							N		N	N	N
Catostomidae	*Ictiobus niger*	black buffalo		N	N	I								N		N	N	N
Catostomidae	*Minytrema melanops*	spotted sucker		N	N						N	N		N		N	N	N
Catostomidae	*Moxostoma erythrurum*	golden redhorse		N	N							N		N		N	N	N
Catostomidae	*Moxostoma macrolepidotum*	shorthead redhorse		N	N							N		N		N	N	N
Catostomidae	*Scartomyzon cervinus*	black jumprock																
Catostomidae	*Scartomyzon lachneri*	greater jumprock		N								N/I						
Catostomidae	*Scartomyzon rupiscartes*	striped jumprock										N						
Catostomidae	*Thoburnia rhothoeca*	torrent sucker																
Centrarchidae	*Ambloplites cavifrons*	Roanoke bass																
Centrarchidae	*Ambloplites constellatus*	Ozark bass			N/I													
Centrarchidae	*Ambloplites rupestris*	rock bass		N	I	I	I	I	I	I		N		N	I	N	N/I	I
Centrarchidae	*Archoplites interruptus*	Sacramento perch				I	N/I	I										
Centrarchidae	*Centrarchus macropterus*	flier		N	N						N	N				N	N	
Centrarchidae	*Chaenobryttus gulosus*	warmouth		N	N	I	I	I		I	N	N		N	I	N	N/I	N/I
Centrarchidae	*Enneacanthus gloriosus*	bluespotted sunfish		N						N	N	N						
Centrarchidae	*Lepomis auritus*	redbreast sunfish		N/I	I				N	N	N	N/I						
Centrarchidae	*Lepomis cyanellus*	green sunfish		N/I	N	I	I	N/I	I	I	I	I	I	N	I	N	N	N
Centrarchidae	*Lepomis gibbosus*	pumpkinseed				I	I	I	N	N				N/I	I	N/I	N/I	I
Centrarchidae	*Lepomis humilis*	orangespotted sunfish		N/I	N						I	I		N		N	N/I	N
Centrarchidae	*Lepomis macrochirus*	bluegill		N	N/I	I	I	I	I	I	N	N	I	N	I	N/I	N/I	N/I
Centrarchidae	*Lepomis megalotis*	longear sunfish		N	N						N	N		N		N	N	N
Centrarchidae	*Lepomis microlophus*	redear sunfish		N	N	I	I	I		I	N	N		I		N/I	N/I	I
Centrarchidae	*Lepomis miniatus*	redspotted sunfish		N	N											N	N?/I?	
Centrarchidae	*Micropterus coosae*	redeye bass		N	I		I					N/I						
Centrarchidae	*Micropterus dolomieu*	smallmouth bass		N/I	N/I	I	I	I	I	I		N/I	I	N	I	N/I	N	N/I
Centrarchidae	*Micropterus punctulatus*	spotted bass		N/I	N	I	I	I			N/I	N/I		I		N	N	N/I
Centrarchidae	*Micropterus salmoides*	largemouth bass		N/I	N/I	I	I	I	I	I	N	N/I	I	N/I	I	N/I	N	N/I
Centrarchidae	*Micropterus treculi*	Guadalupe bass																
Centrarchidae	*Micropterus undescribed sp.*	shoal bass		N							N	N/I						
Centrarchidae	*Pomoxis annularis*	white crappie		N/I	N/I	I	I	I	I	I	I	N/I		N	I	N	N/I	N/I
Centrarchidae	*Pomoxis nigromaculatus*	black crappie		N	N/I	I	I	I	I	I	N	N		N	I	N	N/I	N/I
Chanidae	*Chanos chanos*	milkfish					I											
Characidae	*Astyanax mexicanus*	Mexican tetra				I	I											
Cichlidae	*Cichlasoma cyanoguttatum*	Rio Grande cichlid				I					I					I		
Clupeidae	*Alosa aestivalis*	blueback herring							N	N	N	N						
Clupeidae	*Alosa chrysochloris*	skipjack herring		N	N						N	N		N		N	N	N
Clupeidae	*Alosa pseudoharengus*	alewife						I	N	N		I				I	I	
Clupeidae	*Alosa sapidissima*	American shad	I	I	I	I	I	I	N	N	N	N/I		I	I	I	I	I
Clupeidae	*Dorosoma cepedianum*	gizzard shad		N	N			N/I		N	N	N		N		N/I	N	N/I
Clupeidae	*Dorosoma petenense*	threadfin shad		N/I	N/I	I	I	I		I	N	N/I	I			N/I	N	I
Clupeidae	*Opisthonema libertate*	deepbody thread herring					N/I											

KY	LA	MA	ME	MD	MI	MN	MO	MS	MT	NC	ND	NE	NH	NJ	NM	NV	NY	OH	OK	OR	PA	RI	SC	SD	TN	TX	UT	VA	VT	WA	WI	WV	WY
N	N				N	N	N	N				N					N	N			N				N				N		N/I	N	
									N											N/I										N			
N	N	N	N	N	N	N	N	N		N		N	N	N			N	N		I	N	N	N		N			N	N		N	N	
	N							N																		N							
N							N																		N								
N/I	N	I		N/I	N	N	N/I	N		N/I				I			N/I	N	N		N/I		N		N	N		N/I	N		N/I	N/I	
N	N	N	N	N	N	N	N	N		N		N/I	N	N	N	I	N	N/I	N		N/I	N	N	N	N	N	I	N	N		N	N	
																I				I													
N	N			N	N	N	N	N		N				N			N	N	N		N		N		N	N		N			N/I		
N	N				N	N	N	N				I					N	N	N		N		N		N	N		N			N	N/I	
	N			N				N		N				N			N						N			N/I		N					
N	N	N		N			N/I	N		N				N	I		N		N/I		N	N	N		N	N/I		N					
	N	N	N	N				N		N			N	N			N				N	N	N		I	N		N					
	N							N		N													N			N/I							
	N							N		N													N			N/I		N					
N	N					N	N	N	N	N	N	N			N			N/I	N					N	N	N					N	N	N
N	N			N	N	N	N	N		N/I		N		N			N	N			N		N	N	N			N	N		N/I	N	N
N	N					N	N	N		N/I		N						N	N		N		I		N	N					N	N	
									I							N											N/I						N/I
		N	N	N	N	N			N		N	N/I	N				N	N			N			N					N	N	N	N	N/I
N		N	N	N	N	N	N	N	N	N/I	N	N	N	N	N/I		N	N	N		N	N	N	N	N		I	N	N		N	N	N/I
																				N													
									N			N				N/I				N				N			N/I			N			N
															N/I																		
																N/I				N													
																				N/I													
N	N				N		N	N		N		I					N	N			N		N		N	N		N			N	N	
								N																	N/I								
N	N	I		N	N	N	N	N		N				N			N	N	N		N		N		N			N			N	N	
N	N				N	N	N	N	N	I	N	N			N			N	N		N		I	N	N	N					N/I	N	
N	N				N	N	N	N	N	I	N	N						N	N		N			N	N	N		I			N/I	N	
N	N				N	N	N	N				N			N			N	N					N	N	N					N/I	N	
N	N				N	N	N	N										N	N		N		N		N	N					N/I	N	
N				N/I	N	N	N	N		N	N	N					N	N	N		N			N	N			N/I			N	N/I	
N				N	N	N	N	N		N	N						N	N	N		N		N	N	N	N/I		N	N		N	N	
										N																		N/I					
										N/I?													N										
																												N/I				N/I	
										N/I																		N/I					
							N/I																										
N		N/I		N/I	N	N	N/I	N	I	N/I	N	I	N/I	N/I	I		N	N	I	I	N/I		N/I	N/I	N	I	I	N/I	N	I	N	N/I	I
											I	I			I	I				I				I		I	I						
N	N			N/I			N	N		N													N		N	N		N					
N/I	N			I	N	N	N/I	N		N/I					I	I	I	N/I	N	I	N		N		N	N		N/I		I	N/I	N	
				N				N/I		N				N			N/I				N/I		N					N					
I	I	N	N	N						N/I			N	N			N/I		I		N/I	N	N		I	I		N/I	N			I	
N	N	I	I	N/I	N/I	N/I	N	N	I	N/I	I	N		I	N/I	I	N/I	N	N	I	N/I		I	N/I	N	N	I	N/I		I	N	N/I	N/I
N/I		N	N	N/I	N	N/I	N/I		I	N/I	N/I	I	N	N		I	N	N/I		I	N/I	N	N	N/I	I			N/I	N	I	N	N/I	I
N	N				I	N	N	N			N/I	N						N/I	N					N/I	N	N					N	I?	
N	N	I		I	N/I	N/I	N	N	I	N/I	I	N/I	I	I	N/I	I	N/I	N/I	N/I	I	N	I	N/I	N/I	N	N/I	I	N/I	N	I	N/I	N/I	I
N	N	I		I	N	N?/I?	N/I	N		N/I					N/I		N	N	N		N		N		N	N		N/I			N	N/I	
N/I	N				I		N/I	N		N/I		I			I	I		I	N/I	I	I		N/I		N/I	N/I	I	I	I			I	
N/I?	N						N	N											N/I						N	N/I							
I										I													N		N/I	I							
N/I		N/I	I	I	N/I	N/I	N/I	N/I	I	N/I	N/I	I	I	I	I	I	N/I	N	N/I	I	N/I	I	I	N/I	N	I	I	N/I	N/I	I	N/I	N/I	I
N/I	N						N/I	N		N/I		I			I	I		N	N/I				I		N/I	N/I	I	N/I				N/I	
N/I	N	I	I	I	N/I	N/I	N/I	N/I	I	N/I	I	N/I	I	I	I	I	N/I	N/I	N/I	I	N/I	I	N	N/I	N	N/I	I	N/I	I	I	N/I	N/I	I
																										N/I							
N/I	N	I		I	N	N/I	N	N	I	N/I	N/I	N/I		I	I	I	N/I	N/I	N/I	I	N/I		I	N/I	N	N/I	I	N/I	I	I	N/I	N/I	I
N/I	N	I	I	I	N	N	N	N	I	N	N/I	I	I	I	I	I	N/I	N	N/I	I	N/I	I	N	N/I	N	N/I	I	N/I	I	I	N/I	N/I	I
	I														I			I	I							N/I							
	I																									N/I							
		N	N	N						N/I			N	N			N/I				N	N	N			I		N/I	N/I				
N	N					N	N	N				N						N	N		N			N	N	N					N	N/I	
I		N/I	N/I	N	I	I				N		I	N/I	N			N/I	I			N/I	N	N		I			N/I	N/I		I	I	
I	I	N/I	N	N/I	I	I	I	I		N/I		I	N	N/I		I	N/I	I		I	N	N	N			I	I	N/I	N/I	I	I	I	
N/I	N	N		N	N	N	N	N		N	N	N/I		N	N/I		N/I	N/I	N		N/I		N	N	N	N	I	N/I	I		N/I	N	I
N/I	N			I			N/I	N		I		I			I	I		I	N/I	I	I		I		N/I	N	I	I				I	

Family	Scientific Name	Common Name	AK	AL	AR	AZ	CA	CO	CT	DE	FL	GA	HI	IA	ID	IL	IN	KS
Cottidae	*Cottus asper*	prickly sculpin	N				N/I											
Cottidae	*Cottus bairdi*			N		I	I	N/I		N		N		N	N	N	N	
Cottidae	*Cottus extensus*	Bear Lake sculpin													N			
Cottidae	*Cottus greenei*	Shoshone sculpin													N/I			
Cyprinidae	*Campostoma anomalum*	central stoneroller			N			N	I			N		N		N	N	N
Cyprinidae	*Campostoma oligolepis*	largescale stoneroller		N/I	N							N		N		N		
Cyprinidae	*Clinostomus elongatus*	redside dace												N		N	N	
Cyprinidae	*Clinostomus funduloides*	rosyside dace		N						N		N						
Cyprinidae	*Couesius plumbeus*	lake chub							N					N	N	N	N	
Cyprinidae	*Cyprinella analostana*	satinfin shiner								N								
Cyprinidae	*Cyprinella galactura*	whitetail shiner		N	N							N						
Cyprinidae	*Cyprinella lutrensis*	red shiner		I	N	I	I	N/I				I		N		N/I	N	N
Cyprinidae	*Cyprinella pyrrhomelas*	fieryblack shiner																
Cyprinidae	*Cyprinella spiloptera*	spotfin shiner		N	N							N		N/I		N	N	N
Cyprinidae	*Cyprinella venusta*	blacktail shiner		N	N						N	N/I				N		
Cyprinidae	*Cyprinella whipplei*	steelcolor shiner		N	N							N				N	N	
Cyprinidae	*Eremichthys acros*	desert dace																
Cyprinidae	*Exoglossum laurae*	tonguetied minnow																
Cyprinidae	*Exoglossum maxillingua*	cutlips minnow							N/I									
Cyprinidae	*Gila atraria*	Utah chub						I							N/I			
Cyprinidae	*Gila bicolor*	tui chub					N/I								I			
Cyprinidae	*Gila copei*	blue chub													N/I?			
Cyprinidae	*Gila orcutti*	leatherside chub					N/I											
Cyprinidae	*Gila pandora*	arroyo chub						N										
Cyprinidae	*Gila purpurea*	Rio Grande chub				N/I												
Cyprinidae	*Gila robusta*	Virgin River chub				N		N										
Cyprinidae	*Hesperoleucus symmetricus*	California roach					N/I											
Cyprinidae	*Hybognathus hankinsoni*	brassy minnow						N/I						N		N	N	N
Cyprinidae	*Hybognathus placitus*	plains minnow			N			N						N		N		N
Cyprinidae	*Hybognathus regius*	eastern silvery minnow							N	N		N						
Cyprinidae	*Hybopsis amblops*	bigeye chub		N	N				I			N				N	N	
Cyprinidae	*Hybopsis winchelli*	clear chub		N							N/I	N						
Cyprinidae	*Iotichthys phlegethotis*	least chub																
Cyprinidae	*Lavinia exilicauda*	hitch					N/I											
Cyprinidae	*Lepidomeda albivallis*	White River spinedace					I											
Cyprinidae	*Lepidomeda mollispinis*	Virgin spinedace				N												
Cyprinidae	*Luxilus albeolus*	white shiner																
Cyprinidae	*Luxilus cerasinus*	crescent shiner																
Cyprinidae	*Luxilus chrysocephalus*	striped shiner		N/I	N							N				N	N	
Cyprinidae	*Luxilus coccogenis*	warpaint shiner		N								N						
Cyprinidae	*Luxilus cornutus*	common shiner						N	N	N				N		N	N	N
Cyprinidae	*Luxilus zonistius*	bandfin shiner		N/I							N	N/I						
Cyprinidae	*Lythrurus ardens*	rosefin shiner		N/I								N				N	N	
Cyprinidae	*Lythrurus atrapiculus*	blacktip shiner		N/I							N	N/I						
Cyprinidae	*Margariscus margarita*	pearl dace							I									
Cyprinidae	*Meda fulgida*	spikedace				N/I												
Cyprinidae	*Moapa coriacea*	Moapa dace																
Cyprinidae	*Nocomis biguttatus*	hornyhead chub			N			N						N		N	N	N
Cyprinidae	*Nocomis leptocephalus*	bluehead chub		N/I								N						
Cyprinidae	*Nocomis micropogon*	river chub		N								N/I					N	
Cyprinidae	*Nocomis raneyi*	bull chub																
Cyprinidae	*Notemigonus crysoleucas*	golden shiner		N	N/I	I	I	I	N	N	N/I	N		N/I		N	N	N
Cyprinidae	*Notropis amoenus*	comely shiner																
Cyprinidae	*Notropis ariommus*	popeye shiner		N								N					N	
Cyprinidae	*Notropis atherinoides*	emerald shiner		N	N							N		N		N	N	N
Cyprinidae	*Notropis baileyi*	rough shiner		N/I							I	I						
Cyprinidae	*Notropis bairdi*	Red River shiner			N													I
Cyprinidae	*Notropis bifrenatus*	bridle shiner							N	N								
Cyprinidae	*Notropis blennius*	river shiner			N			N						N		N	N	N
Cyprinidae	*Notropis boops*	bigeye shiner		N	N											N	N	N
Cyprinidae	*Notropis buccatus*	silverjaw minnow		N								N/I				N	N	
Cyprinidae	*Notropis buccula*	smalleye shiner																
Cyprinidae	*Notropis buchanani*	ghost shiner		N	N									N		N	N	N
Cyprinidae	*Notropis chiliticus*	redlip shiner																
Cyprinidae	*Notropis chrosomus*	rainbow shiner		N/I								N/I						
Cyprinidae	*Notropis dorsalis*	bigmouth shiner						N/I						N		N		N
Cyprinidae	*Notropis girardi*	Arkansas River shiner			N													N
Cyprinidae	*Notropis harperi*	redeye chub		N							N/I	N						
Cyprinidae	*Notropis hudsonius*	spottail shiner						I	N	N		N/I		N	I	N	N	
Cyprinidae	*Notropis hypsilepis*	highscale shiner		N								N/I						
Cyprinidae	*Notropis leuciodus*	Tennessee shiner		N								N/I						
Cyprinidae	*Notropis longirostris*	longnose minnow		N							N	N/I						
Cyprinidae	*Notropis ludibundus*	sand shiner				I		N/I						N		N	N	N
Cyprinidae	*Notropis lutipinnis*	yellowfin shiner										N/I						
Cyprinidae	*Notropis nubilis*	Ozark minnow			N									N		N	N	N

KY	LA	MA	ME	MD	MI	MN	MO	MS	MT	NC	ND	NE	NH	NJ	NM	NV	NY	OH	OK	OR	PA	RI	SC	SD	TN	TX	UT	VA	VT	WA	WI	WV	WY
																				N										N			
N				N	N	N	N		N	N/I					N	N/I	N	N		N	N		N		N		N	N	N	N	N	N	N
																											N/I						I
N	N			N	N	N	N	N		N/I	N	N			N/I		N/I	N	N		N		N	N	N	N		N/I			N	N	N
N						N	N	N		N?									I						N?						N		
N				N/I	N	N											N	N			N										N	N	
N				N				N		N								N			N		N		N			N/I				N	
		N	N		N	N			N		N	N	N				N				N	N		N					N	N	N		N/I
				N						N				N			N/I				N		N					N				N	
N/I							N	N		N/I													N		N			N/I				N/I	
N	N	I				N	N	N		I		N			N/I	I			N					N	N	N	I						I
										N													N/I										
N				N	N	N	N	N		N	N	I		N			N	N	N		N				N			N/I			N	N	
N	N						N/I	N								I			N						N/I	N							
N	N						N	N										N	N/I						N			N				N	
																N/I																	
										N							N	N			N							N				N/I	
				N						N				N			N				N							N/I				N/I	
									I							I											N/I						N/I
																N/I				N										N			
																											N/I						N/I
															N/I											N							
															N	N/I											N						N
																				I													
					N	N	N		N		N	N					N/I							N	I		I				N		N
							N		N		N	N			N/I				N					N		N	I						N
		N	I	N						N			N	N			N				N		N					N	N			N	
N					N		N			N							N	N	N		N				N			N				N	
																											N/I						
																N/I																	
																N/I											N						
										N/I																		N				N	
										N/I																		N/I					
N	N			N	N		N	N		N							N	N	N		N				N	N		N			N	N/I	
										N/I													N/I		N			N					
		N	N	N	N	N	N				N	N	N	N			N	N			N	N		N			I	N	N		N	N/I	N
N								N		N/I								N/I?							N			N/I				N	
															N																		
																N/I																	
N/I					N	N	N				N	N					N/I	N			N			N			I				N		N
	N							N		N/I													N					N/I					
N				N	N			N		N/I							N	N			N		N/I		N			N				N	
										N/I																		N/I					
N/I	N	N	N	N	N	N/I	N	N	N/I	N/I	N	N/I	N	N	I	I	N/I	N/I	N	I	N	N	N	N/I	N/I	N/I	I	N/I	N		N	N/I	N/I
				N						N/I				N			N/I				N							N				N	
N																		N			N				N							N/I	
N	N	N/I	I	I	N	N	N	N	N	N	N	N/I	N	N			N/I	N	N		N			N	N	N	I	N	N		N	N/I	N/I
								N																									
																			N/I							N							
		N	N	N						I			N	N			N				N	N						N	N				
N	N				N	N	N	N			N	N						N	N/I					N	N						N	N	
N	N						N/I	N										N	N						N								
N	N				N		N	N									N	N			N				N			N/I				N	
																										N/I							
N	N					N	N	N				N						N/I	N		I				N	N					N	N	
										N/I																		N/I					
																									N								
					N	N	N				N	N					N	N			N			N							N/I		N
															N/I				N/I							N							
		N	I	N	N	N	N		N/I	N	N/I		N/I	N			N/I	N			N/I	N	N	N/I			I	N/I	N		N	N/I	I
N										N/I													I		N			N/I					
	N							N																									
N					N	N	N		N		N	N			N		N/I	N	N		N			N	N	N	I	N			N	N	N/I
										N/I													N										
						N	N												N/I														

Family	Scientific Name	Common Name	AK	AL	AR	AZ	CA	CO	CT	DE	FL	GA	HI	IA	ID	IL	IN	KS
Cyprinidae	*Notropis oxyrhynchus*	sharpnose shiner																
Cyprinidae	*Notropis ozarcanus*	Ozark shiner			N/I													
Cyprinidae	*Notropis potteri*	chub shiner			I													
Cyprinidae	*Notropis procne*	swallowtail shiner								N								
Cyprinidae	*Notropis rubellus*	rosyface shiner		N	N				I	N				N		N	N	N
Cyprinidae	*Notropis rubricroceus*	saffron shiner																
Cyprinidae	*Notropis shumardi*	silverband shiner			N									N		N	N	N
Cyprinidae	*Notropis spectrunculus*	mirror shiner										N						
Cyprinidae	*Notropis telescopus*	telescope shiner		N	N							N						
Cyprinidae	*Notropis texanus*	weed shiner		N/I	N						N	N/I		N		N	N	
Cyprinidae	*Notropis volucellus*	mimic shiner		N	N				I			N		N		N	N	N
Cyprinidae	*Notropis xaenocephalus*	Coosa shiner		N								N/I						
Cyprinidae	*Orthodon microlepidotus*	Sacramento blackfish					N/I											
Cyprinidae	*Phenacobius mirabilis*	suckermouth minnow			N			N						N		N	N	N
Cyprinidae	*Phoxinus eos*	northern redbelly dace						N										
Cyprinidae	*Phoxinus oreas*	mountain redbelly dace																
Cyprinidae	*Pimephales notatus*	bluntnose minnow		N	N				I			I		N		N	N	N
Cyprinidae	*Pimephales promelas*	fathead minnow		N/I	N/I	N/I	I	N/I	N/I	N/I	I	N/I		N	N/I	N	N	N
Cyprinidae	*Pimephales tenellus*	slim minnow			N													N/I
Cyprinidae	*Pimephales vigilax*	bullhead minnow		N	N							N		N		N	N	N/I
Cyprinidae	*Plagopterus argentissius*	woundfin				N/I												
Cyprinidae	*Platygobio gracilis*	flathead chub			N			N/I						N		N		N
Cyprinidae	*Pogonichthys macrolepidotus*	splittail					N/I											
Cyprinidae	*Ptychocheilus grandis*	Sacramento squawfish					N/I	N										
Cyprinidae	*Ptychocheilus oregonensis*	northern squawfish													N			
Cyprinidae	*Ptychocheilus umpquae*	Umpqua squawfish																
Cyprinidae	*Relictus solitarius*	relict dace																
Cyprinidae	*Rhinichthys atratulus*	blacknose dace		N					N	N		N		N		N	N	
Cyprinidae	*Rhinichthys cataractae*	longnose dace						N/I	N			N		N	N	N	N	N
Cyprinidae	*Rhinichthys chrysogaster*	longfin dace				N/I												
Cyprinidae	*Rhinichthys osculus*	speckled dace				N	N/I	N							N			
Cyprinidae	*Richardsonius balteatus*	redside shiner				I		I							N			
Cyprinidae	*Richardsonius egregius*	Lahontan redside					N/I											
Cyprinidae	*Semotilus atromaculatus*	creek chub		N	N			N/I	N	N	N	N		N		N	N	N
Cyprinidae	*Semotilus corporalis*	fallfish							N	N								
Cyprinodontidae	*Cyprinodon diabolis*	Devils Hole pupfish					I											
Cyprinodontidae	*Cyprinodon macularius*	desert pupfish				N/I	N/I											
Cyprinodontidae	*Cyprinodon nevadensis*	Amargosa pupfish					N/I											
Cyprinodontidae	*Cyprinodon radiosus*	Owens pupfish					N/I											
Cyprinodontidae	*Cyprinodon rubrofluviatilis*	Red River pupfish																
Cyprinodontidae	*Cyprinodon salinus*	Salt Creek pupfish					N/I											
Cyprinodontidae	*Cyprinodon variegatus*	sheepshead minnow		N					N	N	N	N						
Cyprinodontidae	*Jordanella floridae*	flagfish									N/I							
Embiotocidae	*Hysterocarpus traski*	tule perch					N/I											
Engraulidae	*Cetengraulis mysticetus*	anchoveta					N/I											
Engraulidae	*Engraulis mordax*	northern anchovy					N/I											
Esocidae	*Esox americanus*	grass pickerel		N	N		I	I	N	N	N	N		N		N	N	N
Esocidae	*Esox lucius*	northern pike	N/I		I	I	I	I	N/I			I		N/I	I	N	N	N/I
Esocidae	*Esox masquinongy*	muskellunge		I	I	I	I	I	I	I		N/I		N/I		N/I	N	
Esocidae	*Esox niger*	chain pickerel		N	N/I			I	N/I	N	N/I	N		I		N	I	
Fundulidae	*Fundulus catenatus*	northern studfish		N	N												N	N
Fundulidae	*Fundulus chrysotus*	golden topminnow		N	N						N	N						
Fundulidae	*Fundulus diaphanus*	banded killifish							N	N				N		N	N	
Fundulidae	*Fundulus grandis*	gulf killifish		N							N		I					
Fundulidae	*Fundulus heteroclitus*	mummichog							N	N	N	N						
Fundulidae	*Fundulus lineolatus*	lined topminnow									N/I	N						
Fundulidae	*Fundulus notatus*	blackstripe topminnow		N	N							N		N		N	N	N
Fundulidae	*Fundulus sciadicus*	plains topminnow						I						N				N
Fundulidae	*Fundulus seminolis*	Seminole killifish									N/I							
Fundulidae	*Fundulus stellifer*	southern studfish		N								N/I						
Fundulidae	*Fundulus zebrinus*	plains killifish				I		N/I										N
Fundulidae	*Lucania goodei*	bluefin killifish		N							N	N						
Fundulidae	*Lucania parva*	rainwater killifish		N			I		N	N	N	N						
Gadidae	*Lota lota*	burbot	N						I					N	N	N	N	N
Gasterosteidae	*Apeltes quadracus*	fourspine stickleback							N	N								
Gasterosteidae	*Culaea inconstans*	brook stickleback		I			I	I	I					N		N	N	
Gasterosteidae	*Gasterosteus aculeatus*	threespine stickleback					N/I		N	N						I		
Gasterosteidae	*Pungitius pungitius*	ninespine stickleback						I	N	N						N/I	N	
Gerreidae	*Eucinostomus argenteus*	spotfin mojarra					N/I											
Gerreidae	*Eucinostomus gracilis*	Pacific flagfin mojarra					N/I											
Gobiidae	*Gillichthys mirabilis*	longjaw mudsucker				I	N/I											
Gobiidae	*Gobiosoma bosc*	naked goby		N						N	N	N						
Goodeidae	*Crenichthys baileyi*	Hiko White River spingfish																
Goodeidae	*Crenichthys nevadae*	Railroad Valley springfish																
Goodeidae	*Empetrichthys latos*	Pahrump killifish																

KY	LA	MA	ME	MD	MI	MN	MO	MS	MT	NC	ND	NE	NH	NJ	NM	NV	NY	OH	OK	OR	PA	RI	SC	SD	TN	TX	UT	VA	VT	WA	WI	WV	WY
																										N/I							
							N												I														
	I																		I							N/I							
				N						N/I				N			N				N		N					N/I				N/I	
N				N	N	N	N	N	N	N	N			N			N	N	N		N			N	N			N	N		N	N	
										N/I															N			N/I					
N	N/I						N	N				N							N/I					N	N	N/I							
										N/I													I		N			N					
N							N			N/I															N			N/I				I	
N	N				N	N	N	N																	N/I	N					N		
N	N	I			N	N	N	N		N							N	N	N		N/I				N	N		N	N		N	N/I	
																									N								
																I																	
N	N				I	N	N					N			N/I			N/I	N					N	N	N					N	N	N
		N	N		N	N			N		N	N	N				N	I	I		N/I			N					N		N		
										N/I																		N/I				N	
N	N	I		N	N	N	N	N			N	N		N			N	N/I	N		N			N/I	N		I	N/I	N		N	N/I	
N/I	N/I	N/I	I	N/I	N	N/I	N/I	N/I	N/I	N/I	N	N/I	N/I	N	N/I	N/I	N	N/I	N/I	I	N/I	N		N	N/I	N/I	N/I	N/I	N	I	N/I	N/I	N/I
							N												N														
N	N					N/I	N	N				I			I			N/I	N		N			I	N	N/I	I				N/I	N	
																N											N						
N	N					N/I	N	N	N		N	N			N/I				N					N	N	N							N
															N	N											N						
									N							N				N/I										N			
																				N/I													
																N/I											I						
N		N	N	N	N	N	N			N/I	N	N	N	N			N	N			N	N	N	N	N		I	N	N		N	N	
N/I		N		N	N	N			N	N	N	N	N	N	N		N	N	N	N	N	N	N	N	N	N	N/I	N/I	N	N	N	N	N/I
															N/I																		
															N/I	N/I				N							N			N			N
									N/I							N				N							N/I			N			N/I
																N/I																	
N	N	N	N	N	N	N	N	N	N	N	N	N	N	N	N		N	N	N		N	N	N	N	N	N/I	I	N	N		N	N	N/I
		N	N	N									N	N			N/I				N	N						N/I	N			N	
																N/I																	
																N/I																	
																			N							N/I							
	N	N		N				N		N				I			N				N	N	N			N/I		N					
																							I										
																				N										N			
N	N	N	N	N/I	N		N	N		N		N	N	N			N/I	N	N	I	N/I	N	N		N	N		N	N/I	I	N/I	N	
I		N/I	N/I	I	N	N/I	N/I		N/I	I	N/I	N/I	N/I	N/I	I	I	N	N/I	I	I	N/I	N/I		N/I	I	I	I	I	N	I	N/I	I	I
N/I		I	I	N/I	N	N/I	I			N/I	I	I	N	N/I			N/I	N/I	I		N/I		I	I	N/I	I		N/I	N/I		N/I	N/I	
N/I	N	N/I	N	N/I		I	N	N		N		I	N	N/I			N/I	I	I		N/I	N	N		N	N/I		N/I	N/I			I	
N/I	N						N/I	N											N/I						N			N				I	
N	N						N	N															N		N	N/I							
		N	N	N	N	N				N	N		N	N			N	N/I		I	N/I	N	N	N/I				N	N	I	N	N	
	N							N							I											N/I							
		N	N	N						N			N/I	N			N				I	N	N					N					
										N													N					N					
N	N				N	N	N	N										N/I	N						N	N					N		
							N					N/I							N					N									N
																									N								
							N		I			N			N/I	I			N					N/I		N/I	I						N/I
										I													I										
	N	N		N				N		N				N/I	I	I	N			I		N	N			N/I	I	N					
N			N		N	N	N		N		N	N	N	N/I			N	N/I		N	N			N	N				N	N	N		N
		N	N	N						N			N	N/I			N				N/I	N						N					
I			N/I		N	N			N/I		N	N	N		N/I		N	N	I		N/I			N	I		I		N		N		
		N	N	N	I	I				N			N	N			N	I		N		N						N		N	I		
		N	N		N	N							N	N			N				N	N									N		
																I?																	
	N			N				N		N				N			N						N			N/I		N					
																N/I																	
																N/I																	
																N/I																	

Family	Scientific Name	Common Name	AK	AL	AR	AZ	CA	CO	CT	DE	FL	GA	HI	IA	ID	IL	IN	KS
Haemulidae	*Anisotremus davidsonii*	sargo					N/I											
Hiodontidae	*Hiodon alosoides*	goldeye		N	N									N		N	N	N
Hiodontidae	*Hiodon tergius*	mooneye		N	N							N		N		N	N	
Ictaluridae	*Ameiurus brunneus*	snail bullhead		N							N	N/I						
Ictaluridae	*Ameiurus catus*	white catfish		N/I	I		I		I	N	N/I	N				I	I	
Ictaluridae	*Ameiurus melas*	black bullhead		N	N	I	I	N/I	I		N	N/I		N	I	N	N	N/I
Ictaluridae	*Ameiurus natalis*	northern black bullhead		N	N	I	I	I	I	N	N	N		N	I	N	N	N
Ictaluridae	*Ameiurus nebulosus*	yellow bullhead		N/I	N/I	I	I	I	N	N	N	N	I	N	I	N	N	I
Ictaluridae	*Ameiurus platycephalus*	brown bullhead										N/I						
Ictaluridae	*Ictalurus furcatus*	blue catfish		N/I	N/I	I	I	I			I	I		N	I	N	N	N
Ictaluridae	*Ictalurus pricei*	Yaqui catfish				N/I												
Ictaluridae	*Ictalurus punctatus*	channel catfish		N	N/I	I	I	N/I	I	I	N	N/I	I	N/I	I	N	N	N/I
Ictaluridae	*Noturus exilis*	slender madtom		N	N/I									N		N		N
Ictaluridae	*Noturus flavus*	stonecat		N	N			N				N		N		N	N	N
Ictaluridae	*Noturus gilberti*	orangefin madtom																
Ictaluridae	*Noturus gyrinus*	tadpole madtom		N	N					N	N	N		N	I	N	N	N
Ictaluridae	*Noturus insignis*	margined madtom								N		N						
Ictaluridae	*Noturus miurus*	brindled madtom		N	N											N	N	N
Ictaluridae	*Pylodictis olivaris*	flathead catfish		N	N	I	I	I			I	N/I		N	I	N	N	N/I
Labridae	*Halichoeres semicinctus*	rock wrasse					N/I											
Labridae	*Tautoga onitis*	tautog					I		N	N								
Lepisosteidae	*Atractosteus spatula*	alligator gar		N	N		I				N					N	N	
Lepisosteidae	*Lepisosteus oculatus*	spotted gar		N	N						N	N				N	N	N
Lepisosteidae	*Lepisosteus platostomus*	shortnose gar		N	N									N		N	N	N
Lepisosteidae	*Lepisosteus sp.*	gar		N	N					N	N	N		N		N	N	N
Megalopidae	*Megalops atlanticus*	tarpon		N							N	N						
Moronidae	*Morone americana*	white perch						I	N	N						I	I	
Moronidae	*Morone chrysops*	white bass		N/I	N/I	I	I	I		I	I	I		N		N	N	N/I
Moronidae	*Morone mississippiensis*	yellow bass		N/I	N	I								N/I		N/I	N	I
Moronidae	*Morone saxatilis*	striped bass		N/I	I	I	I	I	N	N	N/I	N/I	I	I		I	I	I
Mugilidae	*Mugil cephalus*	striped mullet		N	N		N/I		N	N	N	N				N		
Mugilidae	*Mugil curema*	white mullet		N			N/I			N	N	N						
Nemastistiidae	*Nematistius pectoralis*	roosterfish					N/I											
Osmeridae	*Hypomesus transpacificus*	delta smelt					N											
Osmeridae	*Osmerus mordax*	rainbow smelt			I			I	N/I			I		I	I	I	I	I
Percidae	*Ammocrypta bifascia*	Florida sand darter		N							N/I							
Percidae	*Ammocrypta clara*	western sand darter			N									N		N	N	
Percidae	*Etheostoma blenniodes*	greenside darter		N	N							N				N	N	N
Percidae	*Etheostoma caeruleum*	rainbow darter		N	N							N		N		N	N	
Percidae	*Etheostoma cragini*	Arkansas darter			N			N/I										N
Percidae	*Etheostoma edwini*	brown darter		N							N/I	N						
Percidae	*Etheostoma exile*	Iowa darter						I						N		N	N	
Percidae	*Etheostoma fusiforme*	swamp darter		N	N				N	N	N	N						
Percidae	*Etheostoma nigrum*	johnny darter		N	N			N/I						N		N	N	N
Percidae	*Etheostoma nuchale*	watercress darter		N/I														
Percidae	*Etheostoma olmstedi*	tessellated darter							N	N	N	N						
Percidae	*Etheostoma simoterum*	snubnose darter		N								N						
Percidae	*Etheostoma zonale*	banded darter		N	N							N/I		N		N	N	N
Percidae	*Perca flavescens*	yellow perch		I	I	I	I	I	N	N	I	I		N/I	I	N/I	N/I	I
Percidae	*Percina caprodes*	logperch		N	N			I				N		N		N	N	N
Percidae	*Percina macrolepida*	bigscale logperch					I											
Percidae	*Percina maculata*	blackside darter		N	N							N		N		N	N	N
Percidae	*Percina roanoka*	Roanoke darter																
Percidae	*Percina shumardi*	river darter		N	N									N		N/I	N	N
Percidae	*Percina tanasi*	snail darter		N								N						
Percidae	*Stizostedion canadense*	sauger		N/I	N/I			I			I	N/I		N	I	N	N	N/I
Percidae	*Stizostedion vitreum*	walleye		N/I	N/I	I	I	I	I	I		N/I		N/I	I	N	N/I	N/I
Percopsidae	*Percopsis omiscomaycus*	trout-perch							I					N		N	N	
Petromyzontidae	*Lampetra appendix*	silver lamprey		N	N				I	N				N		N	N	
Petromyzontidae	*Petromyzon marinus*	sea lamprey							N	N	N	N				I	I	
Pleuronectidae	*Hysopsetta guttulata*	diamond turbot					N/I											
Pleuronectidae	*Platyichthys stellatus*	starry flounder	N				N/I											
Poeciliidae	*Gambusia affinis*	western mosquitofish	I	N/I	N	I	I	I	I	N	N/I	N	I	I	I	N/I	N/I	N/I
Poeciliidae	*Gambusia geiseri*	largespring gambusia																
Poeciliidae	*Gambusia holbrooki*	eastern mosquitofish		N/I						N	N	N						
Poeciliidae	*Gambusia nobilis*	Pecos gambusia																
Poeciliidae	*Heterandria formosa*	least killifish		N							N	N						
Poeciliidae	*Poecilia formosa*	Amazon molly																
Poeciliidae	*Poecilia latipinna*	sailfin molly		N		I	I	I			N	N	I					
Poeciliidae	*Poeciliopsis sp.*	unidentified Poeciliopsis							I									
Polydontidae	*Polydon spathula*	paddlefish		N	N						I	I				N	N	N
Salmonidae	*Coregonus artedi*	cisco														N/I	N	
Salmonidae	*Coregonus clupeaformis*	lake whitefish	N				I	I	I					I	I	N	N/I	
Salmonidae	*Oncorhynchus aguabonita*	golden trout				I	N/I	I							I			
Salmonidae	*Oncorhynchus clarki*	cutthroat trout	N		I	N/I	N/I	N/I	I						N/I			I

KY	LA	MA	ME	MD	MI	MN	MO	MS	MT	NC	ND	NE	NH	NJ	NM	NV	NY	OH	OK	OR	PA	RI	SC	SD	TN	TX	UT	VA	VT	WA	WI	WV	WY
															I																		
N	N					N	N	N	N		N	N/I						N	N		N			N	N	N					N	N	N
N	N			I	N	N	N	N		N		N					N	N	N		N				N	N			N		N		
										N/I													N					N/I					
I		I		N			I	N/I		N/I				N		I	N	I		I	N/I	I	N		I			N		I		N	
N	N	I		I	N	N	N	N	N/I	N/I	N	N/I			N/I	I	N	N	N	I	N			N	N	N/I	I	N/I		I	N/I	N/I	N
N	N	I		N	N	N	N	N	I	N	N	N	I		I	I	N	N	N	I	N		N	N	N	N	I	N/I	N	I	N	N/I	
N/I	N/I	N	N	N	N	N	N/I	N		N	N	I	N/I	N	I	I	N	N/I	N	I	N	N	N	N	N			N/I	N	I	N	N/I	
										N/I													N					N/I					
N	N			I		I	N	N		N/I		N		I	I			N/I	N/I	I	N		I	N	N	N		I				N	
N	N	I		I	N	N/I	N	N	N	N/I	N	N		I	N/I	I	N	N/I	N	I	N/I		N/I	N	N	N/I	I	N/I		I	N/I	N/I	N/I
N						N	N	N											N						N						N		
N				N	N	N	N	N	N	N	N	N					N	N	N		N			N	N			N	N		N	N/I	N
										N																		N/I					
N	N	I		N	N	N	N	N		N	N	N/I	I	N			N	N/I	N	I	N		N	N	N	N/I		N		I	N		
		I		N/I	I					N/I			I	N			N/I				N/I		N		I			N/I				N/I	
N	N				N		N	N									N/I	N	N		N				N							N	
N	N				N	N	N	N		N/I	N	N/I			N/I			N	N	I	N		I	N	N	N		N/I		I	N/I	N	I
		N	N	N						N		I	N	N			N				N	N	N					N					
N	N						N	N										N	N						N	N							
N	N	I			N		N	N										N	N		N				N	N					N		
N	N					N	N	N	N		N	N						N	N					N	N	N					N/I		
N	N			N	N	N	N	N	N	N	N	N		N	N		N	N	N	I	N		N	N	N	N		N	N		N	N	
	N			N				N		N													N			N/I		N					
		N/I	N	N	I	I	I			N		I	N/I	N			N/I	I			N/I	N	N					N/I	N/I		I		
N/I	N/I			I	N	N	N/I	N/I	I	I	I	N/I		I	N/I	I	N	N/I	N/I		N/I		I	N/I	N/I	N/I	I	N/I		I	N/I	N/I	
N/I	N					N	N	N											N						N/I	N					N/I		
I	N/I	N	N	N/I		I	I	N/I		N/I	I	I	N	N/I	I	I	N/I	I	I	I	N/I	N	N/I		I	I	I	N/I		I	I	I	
N	N	N	N	N			N	N		N			N	N			N		N			N	N		N	N		N					
	N	N		N				N		N				N			N				N	N	N			N		N					
																											I						
I	I	N/I	N/I	I	I	I	I		I	I	I	I	N/I	N			N/I	I			N/I	N		I	I			I	N/I		I		
N	N					N	N	N											N						N	N					N/I		
N				N/I	N		N			N							N	N	N		N				N			N/I				N	
N	N				N	N	N	N		N							N/I	N			N				N			N			N	N	
							N												N														
					N	N			N		N	N			I		N	N			N			N							N		N
N	N	N	N	N			N	N		N/I			N	N			N		N			N	N		N	N		N					
N				N	N	N	N	N		N	N	N					N	N	N		N			N	N			N	N		N	N	N
		N		N						N/I			N	N			N				N	N	N					N	N				
N																									N			N/I?				N?/I?	
N				I		N	N			N							N/I	N	N		N/I		I		N			N			N	N	
I		N	N	N/I	N	N/I	N/I	I	I	N/I	N	N/I	N	N/I	I	I	N/I	N/I	I	I	N	N	N/I	N/I	I	I	I	N/I	N	I	N	I	I
N	N			N	N	N	N	N		N	N						N	N	N		N			N	N	N	I	N	N		N	N/I	
	N														N/I				N/I							N							
N	N				N	N	N	N			N	N					N	N	N		N			N	N	N		N			N/I	N	
										N																		N/I				I	
N	N				N	N	N	N			N							N	N						N	N					N/I	N	
																									N/I								
N	N/I				N	N	N	N/I	N	N/I	N	N/I					N	N	N/I		N/I		I	N	N	I		N	N		N/I	N	N
N/I	I	I	I	I	N/I	N/I	N	N/I	N/I	N/I	N	N/I	I	I	I	I	N/I	N/I	I	I	N/I		I	N/I	N/I	I	I	N/I	N/I	I	N/I	N/I	N/I
N		I		N	N	N	N				N	N					N	N			N			N			I	N	N		N	N	
N		N		N	N	N	N			N			N	N			N	N			N				N			N			N	N	
		N	N	N	I	I				N			N	N			N/I	I			N/I	N	N					N	N		I		
																				N										N			
N/I	N	I		N	I	I	N/I	N/I	I	N/I		I		I	N/I	I	I	I	N	I	I		N		N/I	N/I	I	N		I	I		I
																										N/I							
				N				N/I		N				I				I					N		N/I			N/I					I
															N/I											N							
	N							N		N/I				I									N										
																										N/I							
	N							N	I	N					I	I							N			N/I							
N	N					N	N	N	N		N	N					N	N	N		N			N	N	N		N			N	N	
				I	N	N/I			I		I	I					N/I	N			N/I			I	I				N		N		
			N/I		N	N/I			I		I	I	N/I			I	N/I	N		I	N/I			I			I		N	I	N		
									I						I	I				I							I			I			I
				I	I	I			N		I	I		I	N/I	N/I				N/I				I	I		N/I	I	I	N/I	I		N/I

Family	Scientific Name	Common Name	AK	AL	AR	AZ	CA	CO	CT	DE	FL	GA	HI	IA	ID	IL	IN	KS
Salmonidae	*Oncorhynchus gilae*	Gila trout				N												
Salmonidae	*Oncorhynchus gorbuscha*	pink salmon	N				N									I	I	
Salmonidae	*Oncorhynchus keta*	chum salmon	N				N								I			
Salmonidae	*Oncorhynchus kisutch*	coho salmon	N/I			I	N/I	I	I	I				I	N/I	I	I	
Salmonidae	*Oncorhynchus mykiss*	rainbow trout	N/I	I	I	I	N/I	I	I	I	I	I	I	I	N/I	I	I	I
Salmonidae	*Oncorhynchus nerka*	kokanee	N			I	N/I	I	I						N/I			
Salmonidae	*Oncorhynchus tshawytscha*	chinook salmon	N				N	I	I	I			I	I	N/I	I	I	I
Salmonidae	*Prosopium cylindraceum*	round whitefish							N/I							N	N	
Salmonidae	*Prosopium gemmifer*	Bonneville cisco					I	I							N			
Salmonidae	*Prosopium williamsoni*	mountain whitefish					N	I							N			
Salmonidae	*Salmo salar*	Atlantic salmon	I				I	I	N/I					I	I	I	I	I
Salmonidae	*Salvelinus alpinus*	Arctic char	N					I	N/I									
Salmonidae	*Salvelinus aureolus*	blueback trout													I			
Salmonidae	*Salvelinus fontinalis*	brook trout	I	I	I	I	I	I	N/I	I		N/I	I	N	I	N/I	N/I	I
Salmonidae	*Salvelinus malma*	Dolly Varden	N				I	I										
Salmonidae	*Salvelinus namaycush*	lake trout			I		I	I	I	I				I	I	N/I	N/I	I
Salmonidae	*Thymallus arcticus*	Arctic grayling	N/I			I	I	I	I					I	I			
Sciaenidae	*Aplodinotus grunniens*	freshwater drum		N	N			I				N		N		N/I	N	N
Sciaenidae	*Cynoscion nebulosus*	spotted seatrout		N					N	N	N	N						
Sciaenidae	*Cynoscion parvipinnis*	shortfin corvina					N/I											
Sciaenidae	*Leiostomus xanthurus*	spot		N					N	N	N	N						
Sciaenidae	*Menticirrhus undulatus*	California corbina					N/I											
Sciaenidae	*Micropogonias undulatus*	Atlantic croaker		N					N	N	N	N						
Sciaenidae	*Pogonias cromis*	black drum		N					N	N	N	N						
Sciaenidae	*Sciaenops ocellatus*	red drum		N					N	N	N	N						
Scombridae	*Scombreromorus concolor*	gulf sierra					N/I											
Serranidae	*Paralabrax maculatofasciatus*	spotted sand bass					N/I											
Umbridae	*Dallia pectoralis*	Alaska blackfish	N/I															
Umbridae	*Novumbra hubbsi*	Olympic mudminnow																
Umbridae	*Umbra limi*	central mudminnow			N				I					N		N	N	

KY	LA	MA	ME	MD	MI	MN	MO	MS	MT	NC	ND	NE	NH	NJ	NM	NV	NY	OH	OK	OR	PA	RI	SC	SD	TN	TX	UT	VA	VT	WA	WI	WV	WY
															N/I																		
			I	I	I	I											I	I		N	I									N	I		
			I		I											I				N							I			N			
I		I	I	I	I	I			I		I	I	I		I	I	I	I		N/I	I			I	I	I	I	I		N/I	I		I
I	I	I	I	I	I	I	I	I	N/I	I	I	I	I	I	I	N/I	I	I	I	N/I	I		I	I	I	I	I	I	I	N	I	I	I
		I	I		I	I	I		I	I	I	I			I	I	I			N/I	I			I	I		I		I	N/I	I		I
	I	I	I	I	I	I			I		I	I	I	I		N/I	I	I		N	I			I		I	I	I	I	N	I		
		N			N	N							N	N			N												N		N		
																I								I			N/I						I
					I				N							N				N							N			N			N
I		N/I	N/I	I	I	I	I	I	I	I	I	I	N/I	I	I	I	N/I	I		I	I	N/I	I		I	I	I	I	N/I	I	I	I	I
			N		I								N/I	I			I																
		I	N/I										N/I				I												N				
I		N	N	N/I	N/I	N/I	I		I	N/I	I	I	N	N/I	I	I	N	N/I		I	N/I	N	N/I	I	N/I	I	I	N/I	N	I	N	N	I
															I															N			I
I		I	N	I	N/I	N/I			N/I		I	I	N/I	I	I	I	N/I	N		I	N/I			I	I		I	I	N	I	N/I		I
			I		N/I	I	I		N/I			I	I		I	I	I			I	I			I			I	I	I	I	I		I
N	N				N	N	N	N	N	N	N	N					N	N	N		N			N	N	N		N	N		N/I	N	I
	N	N		N				N		N				N	I		N				N	N	N			N/I		N					
	N	N		N				N		N				N			N				N	N	N			N/I		N					
	N	N		N				N		N				N			N				N	N	N			N/I		N					
	N	N	N	N				N		N			N	N	I		N				N	N	N			N/I		N					
	N	N		N				N		N				N	I		N				N	N	N			N/I		N					
																														N/I			
N		I			N	N	N		I								N	N	I		N			N	N	I					N		

Appendix E - Hybrid Fish Matrix

Hybrid fishes introduced into inland waters are listed by state (state abbreviations follow U.S. Postal System - see Table 1 for listing).

Key to symbols:
Species are designated as either N (native to the U.S.) or E (foreign/exotic to the U.S.).
NxN - U.S. native crossed with U.S. native
ExN - foreign (exotic) crossed with U.S. native
I - introduced into the state

Family	Scientific Name	Common Name	N/E		AK	AL	AR	AZ	CA	CO	CT	DE	FL	GA	HI	IA	ID	IL	IN	KS	KY
Centrarchidae	*Lepomis cyanellus x L. macrochirus*	green sunfish x bluegill	NxN							I		I						I	I	I	
Centrarchidae	*Pomoxis annularis x P. nigromaculatus*	crappie hybrid	NxN																		
Cyprinidae	*Ctenopharyngodon idella x Hypophthalmichthys nobilis*	grass carp x silver carp	ExE					I													
Cyprinidae	*Hypophthalmichthys molitrix x H. nobilis*	silver carp x bighead carp	ExE			I															I
Esocidae	*Esox lucius x E. masquinongy*	tiger muskellunge	NxN				I			I		I				I	I	I	I	I	
Esocidae	*Esox lucius x E. reicherti*	northern pike x Amur pike	NxE																		
Moronidae	*Morone americana x M. saxatilis*	white perch x striped bass	NxN									I									I
Moronidae	*Morone chrysops x M. mississippiensis*	white bass x yellow bass	NxN																		
Moronidae	*Morone chrysops x M. saxatilis*	wiper	NxN			I	I		I	I		I	I	I		I		I	I	I	I
Moronidae	*Morone mississippiensis x M. saxatilis*	yellow bass x striped bass	NxN																		
Percidae	*Stizostedion canadense x S. vitreum*	saugeye	NxN				I			I						I		I	I	I	I
Poeciliidae	*Poecilia latipinna x P. velifera*		ExE										I								
Poeciliidae	*Xiphophorus helleri x X. maculatus*		ExE										I								
Poeciliidae	*Xiphophorus helleri x X. variatus*		ExE										I								
Poeciliidae	*Xiphophorus maculatus x X. variatus*		ExE										I								
Salmonidae	*Oncorhynchus clarki x O. mykiss*	cutbow trout	NxN					I	I	I							I				
Salmonidae	*Salmo salar x S. trutta*	Sambrown	NxE								I										
Salmonidae	*Salmo trutta x Salvelinus fontinalis*	tiger trout	ExN						I	I		I						I			
Salmonidae	*Salvelinus fontinalis x S. namaycush*	splake	NxN						I	I							I				
Sciaenidae	*Cynoscion nebulosus x C. xanthulus*		NxE																		
Sciaenidae	*Pogonias cromis x Sciaenops ocellatus*		NxN																		

LA	MA	MD	ME	MI	MN	MO	MS	MT	NC	ND	NE	NH	NJ	NM	NV	NY	OH	OK	OR	PA	RI	SC	SD	TN	TX	UT	VA	VT	WA	WI	WV	WY
						I	I				I							I					I									
																									I							
																									I							
	I	I		I	I	I		I		I	I		I			I				I			I		I	I	I		I	I	I	I
																				I												
		I				I																										
																		I							I							
I						I	I		I		I		I	I	I	I	I	I	I	I		I	I	I	I	I	I				I	
I																																
				I	I			I		I	I						I	I		I			I	I						I	I	
															I																	
								I						I	I				I							I			I			I
				I				I							I	I							I	I		I				I	I	I
		I	I	I	I							I				I							I			I				I		I
																									I							
																									I							

United States Geological Survey
Florida Caribbean Science Center
Nonindigenous Aquatic Species Program
7920 NW 71st Street
Gainesville, Florida 32653
Telephone (352) 378-8181 Fax (352) 378-4956

Nonindigenous Aquatic Species Reporting Form

Common Name ____________________

Genus ____________________ Species ____________________

Subspecies ____________________ Date Collected ____________________

State __________ County __________ Drainage ____________________

Location (please be as specific as possible, e.g. road crossing; distance and direction to nearest town):

____

____

____

Identified by ____________________

Collected by ____________________

Method of collection ____________________

Number collected ________ Age class ____________________ Size ____________________

(Adult, Juvenile, young-of-year) (TL, SL, or FL)

Method of disposal ____________________

(released, discarded, preserved, mounted, deposited in museum)

Specimen storage ____________________

(museum or agency's name and catalogue number)

Comments ____________________

____

____

____

____

Index

Other titles of interest from the American Fisheries Society

Common and Scientific Names of Fishes from the United States and Canada, 5th edition
Paper $34 **$27 AFS members** (Stock #510.16P)
Cloth $43 **$33 AFS members** (Stock #510.16C)

Common and Scientific Names of Aquatic Invertebrates from the United States and Canada: Mollusks, 2nd edition
Paper $59 **$41 AFS members** Includes CD
(See order form for more options)

Common and Scientific Names of Aquatic Invertebrates from the United States and Canada: Cnidaria and Ctenophora
Paper $33 **$26 AFS members** (Stock #510.18P)

Common and Scientific Names of Aquatic Invertebrates from the United States and Canada: Decapod Crustaceans
Paper $25 **$20 AFS members** (Stock #510.13P)

World Fishes Important to North Americans Exclusive of Species from the Continental Waters of the United States and Canada
Paper $39 **$30 AFS members** (Stock #510.17P)
Cloth $47 **$37 AFS members** (Stock #510.17C)

Uses and Effects of Cultured Fishes in Aquatic Ecosystems
Cloth $58 **$36 AFS members** (Stock #540.15C)

Evolution and the Aquatic Ecosystem: Defining Unique Units in Population Conservation
Paper $76 **$46 AFS members** (Stock #540.17P)

Freshwater Fishes of Virginia
Cloth $96 (Stock #550.20C)

Native Trout of Western North America
Paper $38 **$29 AFS members** (Stock #520.06P)

To order any of these books

Fill out the order form on the back of this page and mail to:
American Fisheries Society, Publication Fulfillment
P. O. Box 1020, Sewickley PA 15143

Or FAX it to:
412-741-0609

For information on all AFS books, consult our web site at

http://www.fisheries.org/publications/catalog/catalog_books_products.htm

For information on the Books Program in general, call us at 301-897-8616 (ext. 223)
FAX 301-897-5080

Order Form

Quantity		Stock Number	Member Price	Price	Subtotal
	Report on Nonindigenous Fishes	Paper 510.27P	$34	$49	$
	Common and Scientific Names of Fishes from the United States and Canada, 5th ed	Paper 510.16P Cloth 510.16C	$27 $33	$34 $43	$ $
	Common and Scientific Names of Aquatic Invertebrates from the United States and Canada: Mollusks, 2nd ed	Book & CD 510.26P CD only 699.07	$41 $36	$59 $52	$ $
	Common and Scientific Names of Aquatic Invertebrates from the United States and Canada: Cnidaria and Ctenophora	Paper 510.18P	$26	$33	$
	Common and Scientific Names of Aquatic Invertebrates from the United States and Canada: Decapod Crustaceans	Paper 510.13P	$20	$25	$
	World Fishes Important to North Americans Exclusive of Species from the Continental Waters of the United States and Canada	Paper 510.17P Cloth 510.17C	$30 $37	$39 $47	$ $
	Uses and Effects of Cultured Fishes in Aquatic Ecosystems	Cloth 540.15C	$36	$58	$
	Evolution and the Aquatic Ecosystem: Defining Unique Units in Population Conservation	Paper 540.17P	$46	$76	$
	Freshwater Fishes of Virginia	Cloth 550.20C	$96	$96	$
	Native Trout of Western North America	Paper 520.06P	$29	$38	$

Shipping

Add $4.50 for the first book and $2.00 each additional book shipped inside the U.S.

Add $6.50 for the first book and $2.50 each additional book shipped outside the U.S.

Total $

Discounts: Member prices available to individual AFS members only. A valid member number is required for discount. One copy per member at these prices. With purchase of ten or more AFS books of the same title, member price applies.

☐ Please send me an AFS membership application.

☐ Please send me information on other AFS publications.

Payment: All orders must be prepaid. Payment must be in U.S. currency or equivalent. Make checks payable to **American Fisheries Society**.

☐ Check ☐ MasterCard ☐ Visa Total Amount ____________________

Card # ______________________________ Exp. date ____________________

Signature ____

Ship to:

Name ____

Member # (required for member price) ______________________________

Street address (no PO Box) ______________________________

City ____________________ State/Province __________ Zip __________

Country ______________________ Phone (____) ______________

Mail or Fax Order to: American Fisheries Society, Publication Fulfillment, PO Box 1020, Sewickley, PA 15143

FAX 412/741-0609